Handbook Integral Logistics Management

Paul Schönsleben

Handbook Integral Logistics Management

Operations and Supply Chain Management Within and Across Companies

Sixth Edition

Springer

Prof. Dr. Paul Schönsleben
Professor Emeritus ETH Zürich
Zürich, Switzerland

This 6th English language edition is an extended translation of the 8th German language edition.

ISBN 978-3-662-65624-2 ISBN 978-3-662-65625-9 (eBook)
https://doi.org/10.1007/978-3-662-65625-9

Originally published by CRC Press/Taylor & Francis Group, Boca Raton, 2016

This Springer imprint is published by the registered company Springer-Verlag GmbH, DE, part of Springer Nature.
The registered company address is: Heidelberger Platz 3, 14197 Berlin, Germany

Foreword to the Sixth Edition

With great pleasure I'm able to release this sixth edition of *Integral Logistics Management — Operations and Supply Chain Management Within and Across Companies*, from now on published by Springer. The new sections contain material that kept me busy during the past years in both my research and in the practical application.

First, Integral Logistics Management applies to the classical industry as well as in service industry. A new section shows how in up-to-date product-service systems both tangibles and intangibles *together* make up the offer that meets the demand of the customer. Second, the chapter on location planning contains a new section which allows to systematically elaborate the already introduced *integrated* design of production, distribution, retail, service and transportation networks. Third, the chapter on sustainability in supply chains newly presents some examples of frameworks, standards and indices used by firms today to practically demonstrate their social and environmental performance in their integration with the economic performance. Fourth, The chapter on product families and one-of-a-kind production contains a new section on different types of cooperation between the R&D and Engineering departments in companies with an "engineer-to-order" production environment (i.e. if at least some design or engineering work occurs during delivery lead time, according to customer specification). Fifth, for long-term planning, sometimes the action principles are little or not at all know, e.g. if influence factors of the surrounding systems play a role in an unknown manner. For this, scenario forecasts, based on scenario planning apply. This is discussed in a new section of the chapter of demand planning and demand forecasting.

Furthermore, each section now contains its intended learning outcomes (ILO). Still, working over and tightening of all the other chapters resulted in an only slightly higher number of pages for the sixth edition compared to the previous edition. The interactive elements were newly designed to be run by different end-user devices. You can download them from the website www.intlogman.ethz.ch. In addition, on www.opess.ethz.ch, a case study for some of the chapters is available, both in English and in German.

The material covers most of the key terms in the five CPIM modules contained in the ASCM / APICS *CPIM Exam Content Manual*, as well as in the CSCP program. Also, previously to this sixth English edition of the book, Springer has published the eight edition in German, *Integrales Logistikmanagement — Operations- und Supply Chain Management innerhalb von Unternehmen und unternehmensübergreifend* (ISBN 9783-662-60672-8).

Being now professor emeritus from ETH Zürich, I continue to act as chairman of the board of the A. Vogel group (www.avogel.com). The requirements on a supply chain that processes fresh plants to remedies and food supplements and distributes them globally are continuously increasing and will keep on giving me to a great extent insight in the practical applications and limits of the integral logistics management. Finally yet importantly, I invite readers to send suggestions and comments to me at Paul.Schoensleben@ethz.ch.

Zurich, July 2022 Prof. Dr. Paul Schönsleben

Foreword to the First Edition

Changes in the world outside the company alter the way that we look at problems and priorities in the company itself. This presents new challenges to company logistics and to planning & control of corresponding business processes.

While logistics was once understood as storing and transport, today — in the course of the reorganization of business processes — an integral perspective on company logistics is making headway. Naturally, products must still be stored and transported. But now these processes are seen as disturbing factors that should be reduced as greatly as possible. The current focus lies on that part of the logistics chain that adds value. This chain, from sales logistics to research and design logistics, production and procurement logistics, distribution logistics, service and maintenance logistics, and — a recent development — disposal logistics, now stands as a whole as the subject for discussion. We seek improvements at the level of the comprehensive, coordinated business process. Moreover, more and more networks of companies arise that develop and manufacture products in cooperation. The logistics of these coupled companies must work together closely and rapidly. This also demands integral management of logistics.

These recent tendencies do not only affect the logistics of the flow of goods itself, but rather also its planning & control, or, in other words, *administrative* and *planning logistics*. The term PPC (for production planning & control) has in reality long since been expanded to become planning & control of the entire logistics network.

Changing requirements in the world of practice often call for new theories and methods, particularly if earlier theories seem to have lost their connection to that world. This impression indeed often arises when we look at what is happening in company logistics. Close examination reveals that behind the methods and techniques that are sold on today's market with new and rousing catchwords, there is seldom anything that is really new. It seems reasonable to assume that the attempt to match existing knowledge against the rapidly changing reality and — in the sense of continual improvement — to expand and adapt it has met with failure. Here lies the crux of the challenge to company logistics today.

The methods and techniques implemented in planning & control are, interestingly enough, not dependent on classification of the tasks and competencies in the organization of the company. For example, techniques of capacity planning do not change according to whether control tasks are executed by central operations planning and scheduling or, in a decentralized fashion, by the job shops. The algorithms also remain in principle the same despite being either realized manually or with the aid of software. The algorithms in a comprehensive software package are also the same as those of a locally implemented planning board. In contrast, methods and techniques do indeed change in dependency on the entrepreneurial objectives, which the choice of logistics should support. These objectives relate to key areas such as quality, costs, delivery, or various aspects of flexibility.

The present volume aims to present the differing characteristics, tasks, methods, and techniques of planning & control in company logistics as comprehensively as possible. Development and change in operational management for company performance should become transparent. However, we will not be content with a wide-ranging, general treatment of the subject at the cost of depth and scientific elucidation of the matter at hand. Because of logistics and planning & control take place at the operational level of a company, competency in the details is absolutely necessary. Effective plans at the strategic level should not lead to contradictions and inconsistency at the operational level.

Consultants and the software industry, as well as widespread circles in educational institutions, produce constant pressure for novelty — which should not be confused with innovation. There is no need to allow ourselves to be irritated by such influences, which are often just short-lived trends. As always, after all, broad, detailed, methodological, and operational knowledge continues to lead to competency. It is this competency that makes it possible to classify and relate the various business processes and the tasks people in companies carry out and to continuously adapt this system of relations and categorizations to changing entrepreneurial objectives, market situations, product ranges, and employee qualifications.

Today, IT-supported planning & control enjoys a very high status in small- to medium-sized companies. And this is usually rightly so, for the large amounts of data can often not be handled quickly enough by another means. For this reason, presentation of the methods of planning & control in detail will include references to possible IT support.

The present volume is a textbook for industrial engineers, business managers, engineers and practitioners, and computer scientists as part of their studies. It also aims to serve the further education of professionals in business practice in industry, including the service industries.

The book is a translation of my book *Integrales Logistikmanagement — Planung & Steuerung umfassender Geschäftsprozesse*, published in 1998 by Springer. The first edition has sold out. The second edition will appear simultaneously and with the same content as the English version.

You will find a part of the bibliography referring to German books or papers. This means that I am still looking for English literature on the specific topic. I would be grateful for any additional English sources for the specific topic. In parts, the book reflects the work of my esteemed colleague Prof. Dr. Alfred Büchel, to whom I am greatly obliged. This is the case particularly with regard to the area of his great interest, statistical methods in planning & control. These are treated mainly in Chapter 10 and Sections 11.3, 11.4, and 13.2.

Zurich, January 2000 Prof. Dr. Paul Schönsleben

Acknowledgments (Third to Sixth Editions)

My thanks go first of all to you, my readers, for your numerous suggestions. And then to my colleagues and fellow members of the APICS Curricula and Certification Council: You have enriched my work through your many ideas. Here, special thanks go to Merle Thomas and Roly White. I am grateful to the members of my staff at the BWI Center for Industrial Management at the Swiss Federal Institute of Technology Zurich (ETH) and other colleagues, for their valuable input to the new sections: Ingo Lange (Section 1.1), Matthias Baldinger, Manuel Rippel and Stephan Verhasselt (Section 1.3), Oliver Schneider, Matthias Wandfluh (1.7), Johannes Plehn (2.1.2), Philipp Bremen (2.1.3), Robert Alard (2.2), Arne Ziegenbein (2.4), Gandolf Finke, Sören Günther, Philip Hertz, Nikolai Iliev, Johannes Plehn, Andreas Radke (3.1), Kathatrina Bunse, Josef Oehmen, Johannes Plehn, Alexander Sproedt, and Matthias Vodicka (3.3), Aldo Duchi, Olga Willner (7.4), and Alexander Verbeck (18). And I thank my colleagues Hugo Tschirky (ETH Zurich), Masaru Nakano (Keio University and Toyota), Hans-Peter Wiendahl (University of Hannover), and Markus Baertschi (ETH Zurich) for their continuing support of my work. Roger Cruz, Dipl. Ing., and his team again took on ready-to-print production of these editions. To them also I express my thanks.

Zurich, January 2007, March 2011, October 2015, July 2022 Prof. Dr. Paul Schönsleben

Acknowledgments (First and Second Editions)

For the first and second editions, I wish to thank numerous scientific colleagues here and abroad, for valuable discussions and suggestions. Special thanks are due to many of the active and engaged experts in the APICS community. In particular, I would like to thank

- Barry Firth, CPIM, Melbourne, for his invaluable help in updating the classification of fundamental concepts in logistics management
- Prof. Merle Thomas, CFPIM, Faimont State College, West Virginia, for his ongoing support of my work.
- Prof. Dr. Alfred Büchel and Prof. Markus Bärtschi, my colleagues at the BWI Institute for Industrial Engineering and Management at ETH Zurich.
- Paul Bernard, Rapistan Systems, Grand Rapids, Michigan.
- Prof. Dr. Hans-Peter Wiendahl, University of Hannover, Germany.
- Prof. Dr. Thomas M. Liebling, ETH Lausanne, Switzerland.

For help with the manuscript, particularly for their critical questions, I wish to thank all previous and current scientific associates and postgraduate students of my chair of Logistics and Information Management of the Department of Manufacturing, Industrial Engineering and Management at the Swiss Federal Institute of Technology ETH in Zurich. They make up far too great a number to list individually here. Instead, I am pleased to refer to many of their doctoral theses and further scientific works in the text and bibliography of this book.

And for their untiring help in creating, translating, and correcting the manuscript, I give hearty thanks to Dipl. Ing. Roger Cruz and all the many professionals and assistants that participated in this undertaking.

Zurich, January 2000 and April 2003 Prof. Dr. Paul Schönsleben

Overview of Contents

Introduction 1
1 Logistics, Operations, and Supply Chain Management 7
2 Supply Chain Design: Business Relations and Risks 71
3 Supply Chain Design: Location Planning and Sustainability 127
4 Process Analysis and Concepts for Planning & Control 183
5 The MRP II / ERP Concept: Business Processes and Methods 235
6 The Lean / Just-in-Time Concept and Repetitive Manufacturing 293
7 The Concept for Product Families and One-of-a-Kind Production 335
8 The Concept for the Process Industry 369
9 ERP and SCM Software 397
10 Demand Planning and Demand Forecasting 423
11 Inventory Management and Stochastic Materials Management 467
12 Deterministic Materials Management 513
13 Time Management and Scheduling 549
14 Capacity Management 595
15 Order Release and Control 635
16 Cost Estimation, Job-Order Costing, and Activity-Based Costing 681
17 Representation and System Management of Logistic Objects 715
18 Quality Management — TQM and Six Sigma 759
19 Systems Engineering and Project Management 789
20 Selected Sections of Information Management 815

Index 843

Contents

At a first reading of the book, some sections are optional in the sense that they are not necessary to understand the material that follows. An asterisk (*) indicates these sections.

Introduction **1**
Structure of the Book 2
Notes to the Reader and Additional Teaching Material 6

1 Logistics, Operations, and Supply Chain Management **7**
1.1 Basic Concepts, Issues, and Challenges 7
1.2 Business Objects 23
1.3 Strategies in the Entrepreneurial Context 39
1.4 Performance Measurement 52
1.5 Summary 61
1.6 Keywords 61
1.7 Scenarios and Exercises 62
1.8 References 69

2 Supply Chain Design: Business Relations and Risks **71**
2.1 Ownership and Trade in a Supply Chain 71
2.2 Strategic Procurement 83
2.3 Designing a Partnership Relationship 96
2.4 Supply Chain Risk Management 111
2.5 Summary 119
2.6 Keywords 119
2.7 Scenarios and Exercises 120
2.8 References 124

3 Supply Chain Design: Location Planning and Sustainability **127**
3.1 Design Options for Integrated Production, Distribution, and Service Networks 128
3.2 Location Selection and Location Configuration 152
3.3 Sustainable Supply Chains 162
3.4 Summary 176
3.5 Keywords 177
3.6 Scenarios and Exercises 177
3.7 References 181

4 Process Analysis and Concepts for Planning & Control **183**
4.1 Elements of Business Process Management 185
4.2 Push and Pull in the Design of Business Processes 188
4.3 Important Techniques of Analysis in Business Process Engineering 193
4.4 Characteristic Features Relevant to Planning & Control in Supply Chains 200
4.5 Branches, Production Types, and Concepts for Planning & Control 220
4.6 Summary 229
4.7 Keywords 230
4.8 Scenarios and Exercises 230
4.9 References 234

5 The MRP II / ERP Concept: Business Processes and Methods **235**
5.1 Business Processes and Tasks in Planning & Control 235
5.2 Master Planning — Long-Term Planning 248
5.3 Introduction to Detailed Planning and Execution 262
5.4 Logistics Business Methods in R&D (*) 277
5.5 Summary 283
5.6 Keywords 284

5.7 Scenarios and Exercises ... 285
5.8 References ... 292

6 **The Lean / Just-in-Time Concept and Repetitive Manufacturing ... 293**
6.1 Characterizing Lean / Just-in-Time and Repetitive Manufacturing ... 294
6.2 The Lean / Just-in-Time Concept ... 298
6.3 The Kanban Technique ... 313
6.4 The Cumulative Production Figures Principle ... 320
6.5 Implementing Procedure and Comparison of Techniques ... 322
6.6 Summary ... 327
6.7 Keywords ... 328
6.8 Scenarios and Exercises ... 328
6.9 References ... 334

7 **The Concept for Product Families and One-of-a-Kind Production ... 335**
7.1 Logistics Characteristics of a Product Variety Concept ... 335
7.2 Adaptive Techniques ... 342
7.3 Generative Techniques ... 348
7.4 Generative and Adaptive Techniques for Engineer-to-Order ... 354
7.5 Cooperation between R&D and Engineering in ETO Companies ... 359
7.6 Summary ... 364
7.7 Keywords ... 365
7.8 Scenarios and Exercises ... 365
7.9 References ... 368

8 **The Concept for the Process Industry ... 369**
8.1 Characteristics of the Process Industry ... 370
8.2 Processor-Oriented Master and Order Data Management ... 377
8.3 Processor-Oriented Resource Management ... 382
8.4 Special Features of Long-Term Planning ... 388
8.5 Summary ... 391
8.6 Keywords ... 392
8.7 Scenarios and Exercises ... 393
8.8 References ... 396

9 **ERP and SCM Software ... 397**
9.1 Software in the Area of ERP and SCM: An Introduction ... 398
9.2 Contents of ERP and SCM Software ... 401
9.3 Factors for Successful Implementation of ERP and SCM Software ... 410
9.4 Summary ... 419
9.5 Keywords ... 420
9.6 Scenarios and Exercises ... 420
9.7 References ... 422

10 **Demand Planning and Demand Forecasting ... 423**
10.1 Overview of Demand Planning and Forecasting Techniques ... 424
10.2 Historically Oriented Techniques for Constant Demand ... 431
10.3 Historically Oriented Techniques with Trend-Shaped Behavior (*) ... 438
10.4 Future-Oriented Techniques ... 447
10.5 Using Forecasts in Planning ... 452
10.6 Summary ... 461
10.7 Keywords ... 462
10.8 Scenarios and Exercises ... 463
10.9 References ... 466

11 **Inventory Management and Stochastic Materials Management ... 467**
11.1 Stores and Inventory Management ... 468

11.2 Usage Statistics, Analyses, and Classifications 475
11.3 Order Point Technique and Safety Stock Calculation 480
11.4 Batch or Lot Sizing 495
11.5 Summary 506
11.6 Keywords 507
11.7 Scenarios and Exercises 508
11.8 References 512

12 Deterministic Materials Management 513
12.1 Demand and Available Inventory along the Time Axis 514
12.2 Deterministic Determination of Independent Demand 523
12.3 Deterministic Determination of Dependent Demand 527
12.4 Batch or Lot Sizing 535
12.5 Analyzing the Results of Material Requirements Planning (MRP) 539
12.6 Summary 542
12.7 Keywords 544
12.8 Scenarios and Exercises 544
12.9 Literaturhinweise 548

13 Time Management and Scheduling 549
13.1 Elements of Time Management 550
13.2 Buffers and Queues 557
13.3 Scheduling of Orders and Scheduling Algorithms 567
13.4 Splitting, Overlapping, and Extended Scheduling Algorithms 581
13.5 Summary 587
13.6 Keywords 588
13.7 Scenarios and Exercises 588
13.8 References 594

14 Capacity Management 595
14.1 Fundamentals of Capacity Management 596
14.2 Infinite Loading 600
14.3 Finite Loading 608
14.4 Rough-Cut Capacity Planning 618
14.5 Summary 627
14.6 Keywords 628
14.7 Scenarios and Exercises 628
14.8 References 634

15 Order Release and Control 635
15.1 Order Release 636
15.2 Shop Floor Control — Production Activity Control and the Manufacturing Execution System 651
15.3 Order Monitoring and Shop Floor Data Collection 656
15.4 Distribution Control 662
15.5 Summary 673
15.6 Keywords 674
15.7 Scenarios and Exercises 675
15.8 References 679

16 Cost Estimation, Job-Order Costing, and Activity-Based Costing 681
16.1 Costs, Cost Elements, and Cost Structures 682
16.2 Cost Estimation 690
16.3 Job-Order Costing 694
16.4 Activity-Based Costing 697
16.5 Summary 708
16.6 Keywords 709

16.7 Scenarios and Exercises 709
16.8 References 714

17 Representation and System Management of Logistic Objects 715
17.1 Order Data in Sales, Distribution, Production, and Procurement 716
17.2 The Master Data for Products and Processes 723
17.3 Extensions Arising from the Variant-Oriented Concept 741
17.4 Extensions Arising from the Processor-Oriented Concept 747
17.5 The Management of Product Data and Product Life Cycle Data 749
17.6 Summary 755
17.7 Keywords 756
17.8 Scenarios and Exercises 756
17.9 References 758

18 Quality Management — TQM and Six Sigma 759
18.1 Quality: Concept and Measurement 759
18.2 Quality Management Tasks at the Operations Level 766
18.3 Quality Management Systems 780
18.4 Summary 786
18.5 Keywords 787
18.6 References 788

19 Systems Engineering and Project Management 789
19.1 Systems Engineering 791
19.2 Project Management 799
19.3 Summary 812
19.4 Keywords 813
19.5 Literaturhinweise 814

20 Selected Sections of Information Management 815
20.1 Important Terms in Information Management 815
20.2 Modeling Enterprise Information Systems 817
20.3 Modeling Information Systems in the Data View and Object View 825
20.4 Summary 840
20.5 Keywords 841
20.6 References 842

Index 843

Detailed Contents

At a first reading of the book, some sections are optional in the sense that they are not necessary to understand the material that follows. An asterisk (*) indicates these sections.

Introduction 1

Structure of the Book 2

Notes to the Reader and Additional Teaching Material 6

1 Logistics, Operations, and Supply Chain Management 7

1.1 Basic Concepts, Issues, and Challenges 7

1.1.1 Important Terms of the Working Environment and of Business Life 8
1.1.2 Service, and Service Orientation in the Classical Industry 9
1.1.3 Service Industry, and Product Orientation in the Service Industry 12
1.1.4 The Industrial Product-Service System 13
1.1.5 The Product Life Cycle, and Logistics and Operations Management 15
1.1.6 The Synchronization of Supply and Demand Using Inventories 17
1.1.7 The Supply Chain and the Extended Enterprise, Supply Chain Management and Integral Logistics Management 19
1.1.8 The Role of Planning and Control and the SCOR Model 21

1.2 Business Objects 23

1.2.1 Business-Partner, and Order-Related Business Objects 24
1.2.2 Product-Related Business Objects 28
1.2.3 Process-Related Business Objects 30
1.2.4 Resource-Related Business Objects 33
1.2.5 Rough-Cut Business Objects 36

1.3 Strategies in the Entrepreneurial Context 39

1.3.1 Entrepreneurial Objectives in a Company and in a Supply Chain 40
1.3.2 Resolving Conflicting Entrepreneurial Objectives 41
1.3.3 Customer Order Penetration Point (OPP) and Coordination with Product and Process Design 45
1.3.4 Target Area Flexibility: Investments in Enabling Organizations, Processes, and Basic Technologies 47
1.3.5 Enabling Technologies Toward Personalized Production 50

1.4 Performance Measurement 52

1.4.1 The Basics of the Measurement, Meaning, and Practical Applicability of Logistics Performance Indicators 53
1.4.2 Performance Indicators in the Target Area of Quality 54
1.4.3 Performance Indicators in the Target Area of Costs 54
1.4.4 Performance Indicators in the Target Area of Delivery 56
1.4.5 Performance Indicators in the Target Area of Flexibility 58
1.4.6 Performance Indicators of the Primary Entrepreneurial Objective 60

1.5 Summary 61

1.6 Keywords 61

1.7 Scenarios and Exercises 62

1.7.1 Rough-Cut Business Objects 62
1.7.2 Improvements in Meeting Entrepreneurial Objectives 64
1.7.3 Entrepreneurial Objectives and ROI 64
1.7.4 Assessing the Economic Value Added (EVA) of Supply Chain Initiatives 65

1.8 References ... 69

2 Supply Chain Design: Business Relations and Risks ... 71

2.1 Ownership and Trade in a Supply Chain ... 71
2.1.1 The Make-or-Buy Decision — Transaction Costs as the Basis of Forming Companies ... 72
2.1.2 Global Trading — Value-Content Requirements and Tariff-Orientation in a Supply Chain ... 76
2.1.3 Total Cost of Ownership in a Global Supply Chain ... 79

2.2 Strategic Procurement ... 83
2.2.1 Overview on Strategic Procurement ... 83
2.2.2 Traditional Market-Oriented Relationship Compared with Customer-Supplier Partnership ... 85
2.2.3 Strategic Procurement Portfolios ... 88
2.2.4 Strategic Selection of Suppliers ... 91
2.2.5 Basics of Supplier Relationship Management and E-Procurement Solutions ... 94

2.3 Designing a Partnership Relationship ... 96
2.3.1 Target Area Strategies for Intensive Cooperation ... 96
2.3.2 The Advanced Logistics Partnership (ALP) Model, a Framework for Implementation of Intensive Cooperation in the Supply Chain ... 99
2.3.3 Top Management Level: Building Trust and Establishing Principal Relationships ... 100
2.3.4 Middle Management Level: Working Out Collaborative Processes in the Supply Chain ... 101
2.3.5 Operational Management Level: Collaborative Order Processing — Avoiding the Bullwhip Effect ... 104
2.3.6 Example Practical Application ... 106
2.3.7 The Virtual Enterprise and Other Forms of Coordination among Companies ... 108

2.4 Supply Chain Risk Management ... 111
2.4.1 Identification of Supply Chain Risks ... 113
2.4.2 Assessment of Supply Chain Risks ... 114
2.4.3 Handling Supply Chain Risks ... 117

2.5 Summary ... 119

2.6 Keywords ... 119

2.7 Scenarios and Exercises ... 120
2.7.1 Advanced Logistics Partnership (ALP) ... 120
2.7.2 Evaluate Company Relationships in the Supply Chain ... 121

2.8 References ... 124

3 Supply Chain Design: Location Planning and Sustainability ... 127

3.1 Design Options for Integrated Production, Distribution, and Service Networks ... 128
3.1.1 Design Options for Production Networks ... 129
3.1.2 Design Options for Distribution Networks ... 133
3.1.3 Network Structure for Decentralized Distribution, and Design Options for Retail Networks ... 138
3.1.4 Design Options for Service Networks ... 141
3.1.5 Design Options for Transportation Networks ... 146
3.1.6 Integration of the Portfolios of the Partial Networks ... 150

3.2 Location Selection and Location Configuration ... 152
3.2.1 Location Selection Using Qualitative Methods and Factor Rating ... 153

3.2.2 Location Selection and Location Configuration with Linear Programming ... 161

3.3 Sustainable Supply Chains ... 162
3.3.1 The Changing Concept of Sustainability with Reference to the Triple Bottom Line ... 163
3.3.2 Economic Opportunities for Social Commitment ... 166
3.3.3 Economic Opportunities for Environmental Commitment ... 168
3.3.4 Energy Management Concepts and Measures for Improved Environmental Performance ... 170
3.3.5 The Measurement of the Environmental Performance ... 174
3.3.6 Social and Environmental Dimensions in Industrial Practice ... 175

3.4 Summary ... 176

3.5 Keywords ... 177

3.6 Scenarios and Exercises ... 177
3.6.1 Location Configuration with Linear Programming ... 177

3.7 References ... 181

4 Process Analysis and Concepts for Planning & Control ... 183

4.1 Elements of Business Process Management ... 185
4.1.1 Terms in Business Process Engineering ... 185
4.1.2 Order Management and Graphic Presentation of Logistics Processes ... 185

4.2 Push and Pull in the Design of Business Processes ... 188
4.2.1 Pull Logistics ... 188
4.2.2 Push Logistics ... 190
4.2.3 The Temporal Synchronization between Use and Manufacturing with Inventory Control Processes ... 192

4.3 Important Techniques of Analysis in Business Process Engineering ... 193
4.3.1 Organization-Oriented Process Chart ... 194
4.3.2 Manufacturing and Service Processes in the Company-Internal and Transcorporate Layout ... 196
4.3.3 Detailed Analysis and Time Study of Processes ... 198

4.4 Characteristic Features Relevant to Planning & Control in Supply Chains ... 200
4.4.1 Principle and Validity of Characteristics in Planning & Control ... 200
4.4.2 Six Features in Reference to Customer, and Item or Product or Product Family ... 202
4.4.3 Five Features in Reference to Logistics and Production Resources ... 206
4.4.4 Seven Features in Reference to the Production or Procurement Order ... 211
4.4.5 Important Relationships between Characteristic Features ... 215
4.4.6 Features of Transcorporate Logistics in Supply Chains ... 218

4.5 Branches, Production Types, and Concepts for Planning & Control ... 220
4.5.1 Branches of Industry in Dependency upon Characteristic Features ... 221
4.5.2 Production Types ... 223
4.5.3 Concepts for Planning & Control ... 225
4.5.4 Selecting an Appropriate Branch Model, Production Type, and Concept for Planning & Control ... 227

4.6 Summary ... 229

4.7 Keywords ... 230

4.8 Scenarios and Exercises ... 230
4.8.1 Concepts for Planning & Control within the Company ... 231
4.8.2 Synchronization between Use and Manufacturing with Inventory Control Processes ... 231
4.8.3 Basic Process Analysis and Manufacturing Processes in the Company-Internal Layout ... 231

4.9 References ... 234

5 The MRP II / ERP Concept: Business Processes and Methods ... 235

5.1 Business Processes and Tasks in Planning & Control ... 235
5.1.1 The MRP II Concept and Its Planning Hierarchy ... 235
5.1.2 Part Processes and Tasks in Long-Term and Medium-Term Planning ... 238
5.1.3 Part Processes and Tasks in Short-Term Planning & Control ... 240
5.1.4 Reference Model of Processes and Tasks in Planning & Control ... 242
5.1.5 Beyond MRP II: DRP II, Integrated Resource Management, and the "Theory of Constraints" ... 245

5.2 Master Planning — Long-Term Planning ... 248
5.2.1 Demand Management: Bid and Customer Blanket Order Processing and Demand Forecasting ... 248
5.2.2 Sales and Operations Planning and Resource Requirements Planning ... 250
5.2.3 Master Scheduling and Rough-Cut Capacity Planning ... 255
5.2.4 Supplier Scheduling: Blanket Order Processing, Release, and Coordination ... 260

5.3 Introduction to Detailed Planning and Execution ... 262
5.3.1 Basic Principles of Materials Management Concepts ... 262
5.3.2 Overview of Materials Management Techniques ... 264
5.3.3 Basic Principles of Scheduling and Capacity Management Concepts ... 268
5.3.4 Overview of Scheduling and Capacity Management Techniques ... 271
5.3.5 Available-to-Promise and Capable-to-Promise ... 275

5.4 Logistics Business Methods in R&D (*) ... 277
5.4.1 Integrated Order Processing and Simultaneous Engineering ... 277
5.4.2 Release Control and Engineering Change Control ... 280
5.4.3 Different Views of the Business Object According to Task ... 280

5.5 Summary ... 283

5.6 Keywords ... 284

5.7 Scenarios and Exercises ... 285
5.7.1 Master Scheduling and Product Variants ... 285
5.7.2 Available-to-Promise (ATP) ... 285
5.7.3 Theory of Constraints ... 286
5.7.4 Master Planning Case ... 287

5.8 References ... 292

6 The Lean / Just-in-Time Concept and Repetitive Manufacturing ... 293

6.1 Characterizing Lean / Just-in-Time and Repetitive Manufacturing ... 294
6.1.1 Just-in-Time and Jidoka — Increasing Productivity through Reduction of Overburdening, Unevenness, and Useless Effort, or Waste ... 294
6.1.2 Characteristic Features for Simple and Effective Planning & Control Techniques of Repetitive Manufacturing ... 297

6.2 The Lean / Just-in-Time Concept ... 298
6.2.1 Lead Time Reduction Through Setup Time Reduction and Batch Size Reduction ... 298
6.2.2 Lead Time Reduction Through Manufacturing Segmentation, Cellular Manufacturing and One-Piece Flow ... 301
6.2.3 Further Concepts of Lead Time Reduction ... 304
6.2.4 Line Balancing — Harmonizing the Content of Work ... 306
6.2.5 Just-in-Time Logistics ... 309
6.2.6 Generally Valid Advantages of the Lean / Just-in-Time Concept for Materials Management ... 311

6.2.7 Generally Valid Advantages of the Lean / Just-in-Time Concept for Capacity Management ... 312

6.3 The Kanban Technique ... 313

6.3.1 Kanban: A Technique of Execution and Control of Operations ... 313

6.3.2 Kanban: A Technique of Materials Management ... 316

6.3.3 Kanban: Long- and Medium-Term Planning ... 319

6.4 The Cumulative Production Figures Principle ... 320

6.5 Implementing Procedure and Comparison of Techniques ... 322

6.5.1 Implementing Procedure ... 323

6.5.2 Comparison of Techniques: Kanban versus Order Point Technique (*) ... 324

6.6 Summary ... 327

6.7 Keywords ... 328

6.8 Scenarios and Exercises ... 328

6.8.1 Operation Time versus Operation Cost, or the Effect of Varying Setup Time and Batch Size ... 329

6.8.2 The Effect of Cellular Manufacturing on Lead Time Reduction ... 330

6.8.3 Line Balancing — Harmonizing the Content of Work ... 331

6.8.4 Calculating the Number of Kanban Cards ... 333

6.9 References ... 334

7 The Concept for Product Families and One-of-a-Kind Production ... 335

7.1 Logistics Characteristics of a Product Variety Concept ... 335

7.1.1 High-Variety and Low-Variety Manufacturing ... 336

7.1.2 Different Variant-Oriented Techniques, and the Final Assembly Schedule ... 339

7.2 Adaptive Techniques ... 342

7.2.1 Techniques for Standard Products with Few Variants ... 342

7.2.2 Techniques for Product Families ... 346

7.3 Generative Techniques ... 348

7.3.1 The Combinatorial Aspect and the Problem of Redundant Data ... 348

7.3.2 Variants in Bills of Material and Routing Sheets: Production Rules of a Knowledge-Based System ... 350

7.3.3 The Use of Production Rules in Order Processing ... 351

7.4 Generative and Adaptive Techniques for Engineer-to-Order ... 354

7.4.1 Classical Procedure and Different Archetypes of Engineer-to-Order ... 354

7.4.2 Approach for Basic Engineer-to-Order ... 356

7.4.3 Approach for Repeatable Engineer-to-Order ... 357

7.5 Cooperation between R&D and Engineering in ETO Companies ... 359

7.5.1 Different Means Used for Cooperation between the R&D and the Order-specific Engineering Departments ... 360

7.5.2 The Portfolio of Cooperation Types between R&D and Engineering in ETO Companies ... 361

7.6 Summary ... 364

7.7 Keywords ... 365

7.8 Scenarios and Exercises ... 365

7.8.1 Adaptive Techniques for Product Families ... 366

7.8.2 Generative Techniques — the Use of Production Rules in Order Processing ... 366

7.8.3 Generative Techniques — Setting the Parameters of a Product Family ... 367

7.9 References 368

8 The Concept for the Process Industry 369

8.1 Characteristics of the Process Industry 370
8.1.1 Divergent Product Structures and By-Products 371
8.1.2 High-Volume Line Production, Flow Resources and Inflexible Facilities 374
8.1.3 Large Batch Size, Lot Traceability, and Loops in the Order Structure 376

8.2 Processor-Oriented Master and Order Data Management 377
8.2.1 Processes, Technology, and Resources 377
8.2.2 The Process Train: A Formalized Processor-Oriented Production Structure 379
8.2.3 Lot Control in Inventory Management 381

8.3 Processor-Oriented Resource Management 382
8.3.1 Campaign Planning 382
8.3.2 Processor-Dominated Scheduling versus Material-Dominated Scheduling 385
8.3.3 Consideration of a Nonlinear Usage Quantity and of a Product Structure with Loops 386

8.4 Special Features of Long-Term Planning 388
8.4.1 Determining the Degree of Detail of the Master Production Schedule 388
8.4.2 Pipeline Planning across Several Independent Locations 388

8.5 Summary 391

8.6 Keywords 392

8.7 Scenarios and Exercises 393
8.7.1 Batch Production versus Continuous Production 393
8.7.2 Manufacture of By-Products 394
8.7.3 Production Planning in Process Industries 395

8.8 References 396

9 ERP and SCM Software 397

9.1 Software in the Area of ERP and SCM: An Introduction 398
9.1.1 History and Origin of ERP Software 398
9.1.2 Scope and Range of ERP and SCM Software 400

9.2 Contents of ERP and SCM Software 401
9.2.1 Classical MRP II / ERP Software 401
9.2.2 Software for Customer Order Production or the Variant-Oriented Concept 403
9.2.3 Software for the Process Industry or the Processor-Oriented Concept 404
9.2.4 Software for Transcorporate Planning & Control in a Supply Chain 405
9.2.5 Software for Customer Relationship Management (CRM) 407
9.2.6 Standard or Company-Specific Software? 409

9.3 Factors for Successful Implementation of ERP and SCM Software 410
9.3.1 Possibilities and Limitations of the IT Support of Planning & Control 411
9.3.2 Factors That Influence Individual Acceptance and the Range of Implementation of ERP Software 415

9.4 Summary 419

9.5 Keywords 420

9.6 Scenarios and Exercises 420
9.6.1 Factors That Influence People's Acceptance of ERP Software 420
9.6.2 Standard or Company-Specific Software 421
9.6.3 Software for Transcorporate Planning & Control 421

9.7 References ... 422

10 Demand Planning and Demand Forecasting ... 423

10.1 Overview of Demand Planning and Forecasting Techniques ... 424
10.1.1 The Problem of Demand Planning ... 424
10.1.2 Subdivision of Forecasting Techniques ... 426
10.1.3 Principles of Forecasting Techniques with Extrapolation of Time Series and the Definition of Variables ... 428

10.2 Historically Oriented Techniques for Constant Demand ... 431
10.2.1 Moving Average Forecast ... 432
10.2.2 First-Order Exponential Smoothing Forecast ... 433
10.2.3 Moving Average Forecast versus First-Order Exponential Smoothing Forecast ... 437

10.3 Historically Oriented Techniques with Trend-Shaped Behavior (*) ... 438
10.3.1 Regression Analysis Forecast ... 439
10.3.2 Second-Order Exponential Smoothing Forecast ... 441
10.3.3 Trigg and Leach Adaptive Smoothing Technique ... 443
10.3.4 Seasonality ... 445

10.4 Future-Oriented Techniques ... 447
10.4.1 Trend Extrapolation Forecast ... 447
10.4.2 The Delphi Method ... 449
10.4.3 Scenario Forecasts ... 449

10.5 Using Forecasts in Planning ... 452
10.5.1 Comparison of Techniques and Choice of Suitable Forecasting Technique ... 453
10.5.2 Consumption Distributions and Their Limits, Continuous and Discontinuous Demand ... 454
10.5.3 Demand Forecasting of Variants of a Product Family ... 457
10.5.4 Safety Demand Calculation for Various Planning Periods ... 458
10.5.5 Translation of Forecast into Quasi-Deterministic Demand and Administration of the Production or Purchase Schedule ... 460

10.6 Summary ... 461

10.7 Keywords ... 462

10.8 Scenarios and Exercises ... 463
10.8.1 Choice of Appropriate Forecasting Techniques ... 463
10.8.2 Moving Average Forecasting Technique ... 464
10.8.3 First-Order Exponential Smoothing ... 465
10.8.4 Moving Average Forecast versus First-Order Exponential Smoothing Forecast ... 466

10.9 References ... 466

11 Inventory Management and Stochastic Materials Management ... 467

11.1 Stores and Inventory Management ... 468
11.1.1 Characteristic Features of Stores Management ... 469
11.1.2 Inventory Transactions ... 471
11.1.3 Physical Inventory and Inventory Valuation ... 473

11.2 Usage Statistics, Analyses, and Classifications ... 475
11.2.1 Statistics on Inventory Transactions, Sales, and Bid Activities ... 475
11.2.2 The ABC Classification and the Pareto Chart ... 477
11.2.3 The XYZ Classification and Other Analyses and Statistics ... 479

11.3 Order Point Technique and Safety Stock Calculation ... 480
11.3.1 The Order Point Technique ... 480

11.3.2 Variants of the Order Point Technique ... 482
11.3.3 Safety Stock Calculation with Continuous Demand ... 484
11.3.4 Determining the Service Level and the Relation of Service Level to Fill Rate (*) ... 490

11.4 Batch or Lot Sizing ... 495
11.4.1 Production or Procurement Costs: Batch-Size-Dependent Unit Costs, Setup and Ordering Costs, and Carrying Cost ... 495
11.4.2 Optimum Batch Size and Optimum Length of Order Cycle: The Classic Economic Order Quantity ... 498
11.4.3 Optimum Batch Size and Optimum Length of Order Cycle in Practical Application ... 501
11.4.4 Extensions of the Batch Size Formula (*) ... 503

11.5 Summary ... 506

11.6 Keywords ... 507

11.7 Scenarios and Exercises ... 508
11.7.1 The ABC Classification ... 508
11.7.2 The ABC-XYZ Analysis and its Use in Materials Management ... 509
11.7.3 Safety Stock Variation versus Demand Variation ... 510
11.7.4 Batch Sizing Depending on Stockout Costs (*) ... 511
11.7.5 Effectiveness of the Order Point Technique ... 512

11.8 References ... 512

12 Deterministic Materials Management ... 513

12.1 Demand and Available Inventory along the Time Axis ... 514
12.1.1 Allocated Quantities, and the Projected Available Inventory ... 515
12.1.2 Projected Available Inventory Calculation ... 518
12.1.3 Scheduling and Cumulative Projected Available Inventory Calculation ... 519
12.1.4 Operating Curves for Stock on Hand ... 522

12.2 Deterministic Determination of Independent Demand ... 523
12.2.1 Customer Order and Distribution Requirements Planning (DRP) ... 523
12.2.2 Consuming the Forecast by Actual Demand (*) ... 526

12.3 Deterministic Determination of Dependent Demand ... 527
12.3.1 Characteristics of Discontinuous Dependent Demand ... 527
12.3.2 Material Requirements Planning (MRP) and Planned Orders ... 529
12.3.3 Determining the Timing of Dependent Demand and the Load of a Planned Order ... 533

12.4 Batch or Lot Sizing ... 535
12.4.1 Combining Net Requirements into Batches ... 535
12.4.2 Comparison of the Different Batch-Sizing Policies ... 538

12.5 Analyzing the Results of Material Requirements Planning (MRP) ... 539
12.5.1 Projected Available Inventory and Pegging ... 539
12.5.2 Action Messages ... 541

12.6 Summary ... 542

12.7 Keywords ... 544

12.8 Scenarios and Exercises ... 544
12.8.1 Projected Available Inventory Calculation ... 545
12.8.2 MRP Technique: Determining Net Requirements and Planned Release ... 546
12.8.3 Order Point Technique versus MRP Technique ... 547

12.9 Literaturhinweise ... 548

13 Time Management and Scheduling ... 549

13.1 Elements of Time Management ... 550
13.1.1 The Order of the Operations of a Production Order ... 550
13.1.2 Operation Time and Operation Load ... 552
13.1.3 The Elements of Interoperation Time ... 554
13.1.4 Administrative Time ... 555
13.1.5 Transportation Time ... 555

13.2 Buffers and Queues ... 557
13.2.1 Wait Time, Buffers, and the Funnel Model ... 557
13.2.2 Queues as an Effect of Random Load Fluctuations ... 560
13.2.3 Conclusions for Job Shop Production ... 563
13.2.4 Logistic Operating Curves ... 565

13.3 Scheduling of Orders and Scheduling Algorithms ... 567
13.3.1 The Manufacturing Calendar ... 567
13.3.2 Lead Time Scheduling: Calculating the Manufacturing Lead Time ... 568
13.3.3 Backward Scheduling and Forward Scheduling ... 570
13.3.4 Network Planning ... 573
13.3.5 Central Point Scheduling ... 575
13.3.6 The Lead-Time-Stretching Factor and Probable Scheduling ... 577
13.3.7 Scheduling Process Trains ... 580

13.4 Splitting, Overlapping, and Extended Scheduling Algorithms ... 581
13.4.1 Order or Lot Splitting ... 581
13.4.2 Overlapping ... 582
13.4.3 An Extended Formula for Manufacturing Lead Time (*) ... 584
13.4.4 Extended Scheduling Algorithms (*) ... 586

13.5 Summary ... 587

13.6 Keywords ... 588

13.7 Scenarios and Exercises ... 588
13.7.1 Queues as an Effect of Random Load Fluctuations (1) ... 589
13.7.2 Queues as an Effect of Random Load Fluctuations (2) ... 589
13.7.3 Network Planning ... 590
13.7.4 Backward Scheduling and Forward Scheduling ... 590
13.7.5 The Lead-Time-Stretching Factor and Probable Scheduling ... 593

13.8 References ... 594

14 Capacity Management ... 595

14.1 Fundamentals of Capacity Management ... 596
14.1.1 Capacity, Work Centers, and Capacity Determination ... 596
14.1.2 Overview of Capacity Management Techniques ... 599

14.2 Infinite Loading ... 600
14.2.1 Load Profile Calculation ... 601
14.2.2 Problems Associated with Algorithms for Load Profile Calculation ... 603
14.2.3 Methods of Balancing Capacity and Load ... 605
14.2.4 Order-Wise Infinite Loading ... 607

14.3 Finite Loading ... 608
14.3.1 Operations-Oriented Finite Loading ... 608
14.3.2 Order-Oriented Finite Loading ... 612
14.3.3 Constraint-Oriented Finite Loading ... 617

14.4 Rough-Cut Capacity Planning ... 618
14.4.1 Rough-Cut Network Plans and Load Profiles ... 619
14.4.2 Rough-Cut Infinite Loading ... 622
14.4.3 Rough-Cut Finite Loading ... 624
14.5 Summary ... 627
14.6 Keywords ... 628
14.7 Scenarios and Exercises ... 628
14.7.1 Capacity Determination ... 629
14.7.2 Algorithms for Load Profile Calculation ... 630
14.7.3 Rough-Cut Capacity Planning ... 632
14.8 References ... 634

15 Order Release and Control ... 635

15.1 Order Release ... 636
15.1.1 Order Proposals, Order Release for Procurement and Production, as well as Accompanying Documents and Container Logistics ... 636
15.1.2 Load-Oriented Order Release (Loor) ... 639
15.1.3 Capacity-Oriented Materials Management (Corma) ... 645
15.2 Shop Floor Control — Production Activity Control and the Manufacturing Execution System ... 651
15.2.1 Issuance of Accompanying Documents for Production ... 651
15.2.2 Operations Scheduling, Dispatching, Finite Forward Scheduling, and the (Electronic) Control Board (Leitstand) ... 652
15.2.3 Sequencing Methods ... 655
15.3 Order Monitoring and Shop Floor Data Collection ... 656
15.3.1 Recording Issues of Goods from Stock ... 657
15.3.2 Recording Completed Operations ... 657
15.3.3 Progress Checking, Quality Control, and Report of Order Termination ... 658
15.3.4 Automatic and Rough-Cut Data Collection ... 660
15.4 Distribution Control ... 662
15.4.1 Order Picking ... 663
15.4.2 Packaging and Load Building ... 665
15.4.3 Transportation to Receiver ... 669
15.5 Summary ... 673
15.6 Keywords ... 674
15.7 Scenarios and Exercises ... 675
15.7.1 Load-Oriented Order Release (Loor) ... 675
15.7.2 Capacity-Oriented Materials Management (Corma) ... 676
15.7.3 Finite Forward Scheduling ... 677
15.7.4 Order Picking ... 678
15.8 References ... 679

16 Cost Estimation, Job-Order Costing, and Activity-Based Costing ... 681

16.1 Costs, Cost Elements, and Cost Structures ... 682
16.1.1 Actual, Direct, and Overhead Costs ... 683
16.1.2 Average Costs and Standard Costs ... 684
16.1.3 Variable Costs and Fixed Costs ... 685
16.1.4 Cost Accumulation Breakdown: The Cost Breakdown Structure of a Product ... 686

16.2 Cost Estimation ... 690
16.2.1 An Algorithm for Cost Estimation of Goods Manufactured ... 690
16.2.2 Representation of the Cost Accumulation and Comprehensive Calculation for a Product Line ... 692
16.3 Job-Order Costing ... 694
16.3.1 Actual Quantities and Actual Costs ... 694
16.3.2 Cost Analysis ... 695
16.3.3 The Interface from Order Management to Cost Accounting ... 696
16.4 Activity-Based Costing ... 697
16.4.1 Limits of Traditional Product Costing ... 697
16.4.2 Introducing Activity-Based Costing: Aim, Basic Premise, and Technique ... 699
16.4.3 Typical Processes (Activities) and Process Variables ... 703
16.4.4 Activity-Based Product Cost Estimation ... 703
16.5 Summary ... 708
16.6 Keywords ... 709
16.7 Scenarios and Exercises ... 709
16.7.1 Job-Order Costing ... 709
16.7.2 Activity-Based Costing ... 711
16.7.3 Comparing Job-Order Costing and Activity-Based Costing ... 714
16.8 References ... 714

17 Representation and System Management of Logistic Objects ... 715
17.1 Order Data in Sales, Distribution, Production, and Procurement ... 716
17.1.1 Customer and Supplier Data, and the Corporate Group Structure ... 716
17.1.2 The General Structure of Orders in Sales and Distribution, Production, and Procurement ... 717
17.1.3 The Order and Partial Order Header ... 719
17.1.4 The Order Position ... 720
17.1.5 Inventories and Inventory Transactions ... 722
17.2 The Master Data for Products and Processes ... 723
17.2.1 Product, Product Structure, Components, and Operations ... 723
17.2.2 Item Master ... 725
17.2.3 Bill of Material, Bill-of-Material Position, and Where-Used List ... 728
17.2.4 Work Center Master Data ... 735
17.2.5 The Work Center Hierarchy ... 736
17.2.6 Operation and Routing Sheet ... 737
17.2.7 Production Equipment, Bill of Production Equipment, and Bill of Tools ... 738
17.2.8 Composition of the Basic Master Data Objects ... 740
17.3 Extensions Arising from the Variant-Oriented Concept ... 741
17.3.1 Expert Systems and Knowledge-Based Systems ... 742
17.3.2 Implementation of Production Rules ... 743
17.3.3 A Data Model for Parameterized Representation of a Product Family (*) ... 745
17.4 Extensions Arising from the Processor-Oriented Concept ... 747
17.4.1 Process, Technology, and the Processor-Oriented Production Structure ... 747
17.4.2 Objects for Lot Control ... 748
17.5 The Management of Product Data and Product Life Cycle Data ... 749
17.5.1 Product Life Cycle Management and Engineering Data Management ... 749
17.5.2 The Engineering Database as Part of a Computerized System ... 751
17.5.3 Data and Functional Model for General EDM Tasks ... 752
17.5.4 Object Classes and Functions for Release and Engineering Change Control (*) ... 753

17.6 Summary ... 755
17.7 Keywords ... 756
17.8 Scenarios and Exercises ... 756
17.8.1 Different Forms of Representing Bills of Material ... 756
17.8.2 Where-Used Lists ... 757
17.8.3 Basic Master Data Objects ... 757
17.9 References ... 758

18 Quality Management — TQM and Six Sigma ... 759
18.1 Quality: Concept and Measurement ... 759
18.1.1 Quality of Processes ... 760
18.1.2 Quality of Products ... 761
18.1.3 Quality of Organizations ... 762
18.1.4 Quality and Its Measurability ... 764
18.1.5 Quality Measurement and Six Sigma (6σ) ... 765
18.2 Quality Management Tasks at the Operations Level ... 766
18.2.1 The Deming Cycle (PDCA Cycle) and the Shewhart Cycle ... 767
18.2.2 The Six Sigma Phases ... 768
18.2.3 Quality Planning: PDCA Plan Task — DMAIC Define Phase ... 770
18.2.4 Quality Control: PDCA Check Task, Part 1 — DMAIC Measure Phase and Analyze Phase, Part 1 ... 774
18.2.5 PDCA Do Task — DMAIC Improve Phase ... 775
18.2.6 Quality Assurance: PDCA Check Task, Part 2 — DMAIC Measure Phase and Analyze Phase, Part 2 ... 777
18.2.7 PDCA Act Task — DMAIC Control Phase ... 778
18.2.8 Project Management, Continual Improvement, and Reengineering ... 779
18.3 Quality Management Systems ... 780
18.3.1 Standards and Norms of Quality Management: ISO 9000:2015 ... 781
18.3.2 Models and Awards for Total Quality Management ... 782
18.3.3 Audits and Procedures for Assessing the Quality of Organizations ... 785
18.3.4 Benchmarking ... 785
18.4 Summary ... 786
18.5 Keywords ... 787
18.6 References ... 788

19 Systems Engineering and Project Management ... 789
19.1 Systems Engineering ... 791
19.1.1 Systems Thinking and the Top-Down Approach ... 792
19.1.2 Phases of Life of a System and System Life Cycle ... 793
19.1.3 The Problem Solving Cycle ... 796
19.1.4 Differences between Software Engineering and Classical Systems Engineering ... 797
19.2 Project Management ... 799
19.2.1 Goals and Constraints of a Project ... 800
19.2.2 Project Phase, Project Life Cycle, and Work Breakdown Structure ... 801
19.2.3 Project Scheduling and Effort Planning ... 803
19.2.4 Project Organization ... 806
19.2.5 Project Cost, Benefits, Profitability, and Risk ... 808

19.3 Summary ... 812
19.4 Keywords ... 813
19.5 Literaturhinweise ... 814

20 Selected Sections of Information Management ... 815

20.1 Important Terms in Information Management ... 815
20.2 Modeling Enterprise Information Systems ... 817
20.2.1 Basic Principles of Modeling ... 817
20.2.2 Various Dimensions in the Modeling of Information Systems for Business Processes ... 818
20.2.3 Dimension: Hierarchy Creation ... 820
20.2.4 Dimension: Various Views in Modeling ... 823
20.3 Modeling Information Systems in the Data View and Object View ... 825
20.3.1 Object, Attribute, and Object Class ... 825
20.3.2 View and Primary and Secondary Keys ... 827
20.3.3 Association and (Association) Role ... 829
20.3.4 Single-Valued and Multivalued Roles, and Total and Partial Roles ... 830
20.3.5 Association Types ... 831
20.3.6 Breaking up an n-to-n Association; Association Class ... 833
20.3.7 Different Notations and the Reidentification Key ... 834
20.3.8 Breaking Up a Reflexive n-to-n Association ... 836
20.3.9 Using the Hierarchical Constructs Based on Basic Object Classes: The Company-Wide Generic Object Model ... 839
20.4 Summary ... 840
20.5 Keywords ... 841
20.6 References ... 842

Index ... 843

Introduction

Intended learning outcomes: Describe an enterprise as a company in its economic environment, and namely as a sociotechnical system. Explain three dimensions of business activity. Disclose the structure of the suite of courses. Identify notes to the student.

An *enterprise* is, according to the APICS Dictionary (15th edition, APICS, Chicago, 2016), any undertaking, venture, initiative, or business organization with a defined mission.

In this book, the enterprise is understood as a *firm*, a *company* in its economic environment, and namely as a sociotechnical system. The elements of the system as well as their relationships, both within the system and to surrounding systems, are complex in nature. Various interested parties with different ideas and goals have an impact on the company, which makes company management a complex task. Figure 1 shows three dimensions of business activity. Integral company management means building management systems along these dimensions, which simultaneously interlock.

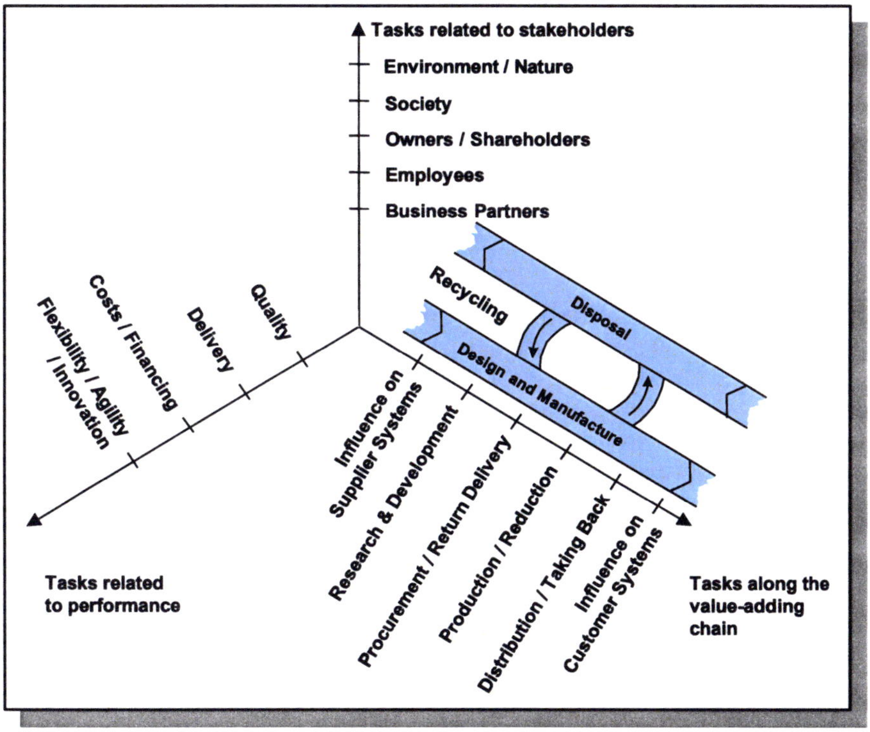

Fig. 1 Three dimensions of business activity.

Today, management systems for tasks *along the value-adding chain* affect customers and especially suppliers, just as customers and suppliers influence the management systems. This close partnership is also necessary from the perspective of the comprehensive product life cycle. Product returns from the customer, disassembly, recycling, and returns to suppliers have to be considered as part of value-adding and paid for accordingly.

Management systems for the *stakeholders* of the enterprise deal with business partners, employees and owners (shareholders). Such individual stakeholders stand opposite the collective stakeholders in the form of society — that is, the macroeconomic system in which the company as a microcosm is embedded. In the figure, the environment (nature) is shown as a personified stakeholder. In practice, of course, the demands of the environment become manifest only through the environmental consciousness of the other stakeholders mentioned.

Priority areas of management systems related to company *performance* focus on the expected quality and delivery as well as required costs and their financing. Flexibility, agility, and innovation are usually potentials that have an indirect impact on business results, via future performance in the other three areas. Tasks related to company performance influence one another mutually and function as tasks that cut across the tasks along the value-adding chain and the tasks related to stakeholders.

Integral logistics management focuses in particular on *expected delivery* — on goals such as customer service ratio, delivery reliability, and short lead times. To achieve the objectives, the corresponding way of thinking has to be successfully anchored in all of the management systems along the entire value-adding chain, also across companies. Integral logistics management monitors value-adding over the entire product life cycle, but also considers the impact on the various stakeholders, especially the business partners.

Integral logistics management means *implementing* ideas, concepts, and methods that have the potential to increase the effectiveness and efficiency of company performance. Here, magic formulas, catchwords, and simplifying theories do not stand much of a chance. The complex reality of day-to-day operation of companies in industry and the service sector demands highly diligent detailed work. In contrast to many a strategic concept in company management, the proof of truth — namely, effectiveness — shows up quickly and measurably. Errors rapidly produce dissatisfied customers and employees, and thus poor business results. This immediacy and measurability do not make it easy to shift the blame to others.

On the other hand, logistics tasks offer a variety of possible solutions. This is an area that calls for human creativity, drive, and perseverance. Methods of planning & control in company logistics, and particularly computer-supported tools, are after all merely supporting aids. Moreover, experience has shown repeatedly that the successful use of methods and tools depends heavily on the people who implement them.

Structure of the Book

The book comprises 4 parts:

- Part A (Chapters 1 through 3): Fundamentals, Strategies, and Design Options in Integral Logistics Management

- Part B (Chapters 4 through 9): Strategic and Tactical Concepts of Planning & Control in Integral Logistics Management
- Part C (Chapters 10 through 17): Methods of Planning & Control in the Operational Routine
- Part D (Chapters 18 through 20): Overview of Adjacent Management Systems in the Enterprise

Parts A examines integral logistics management as embedded in the entrepreneurial activities as well as the strategic design of supply chains.

- Chapter 1 deals with integral logistics management as embedded in the entrepreneurial activities of developing, manufacturing, using, and disposing of goods. The focus is on the business objects, target areas, basic principles, analyses, concepts, systemic and systematic methods, as well as technologies for the management of logistics systems both within and across companies.
- Chapter 2 on supply chain design first considers basic principles for make-or-buy. It discusses models, opportunities and threats for different kinds of partnership between legally independent firms along the supply chain, followed by the management of supply chain risks.
- Chapter 3, also on supply chain design, discusses location planning with its integrated determination of production, distribution, service, and transport networks as well as the sustainability in supply chain management.

Part B introduces the fundamental concepts and tasks of planning & control in logistics, operations, and supply chain management, as well as the corresponding software.

- Chapter 4 starts with methods for business process analysis that are important for the systematic procedure for the design of systems for planning & control in supply chains. It develops a characteristic in planning & control with features that are tailored to the key performance indicators for measuring the degree of achievement of the company objectives in Chapter 1. This characteristic can be different for each product family. Finally, the chapter introduces four concepts for planning & control in supply chains dependent on this characteristic.

- Chapters 5 through 8 introduce the essential business objects and business processes for these four concepts:
    - Chapter 5: the MRP II / ERP concept
    - Chapter 6: the Lean / Just-in-Time concept and repetitive manufacturing
    - Chapter 7: the concept for product families and one-of-a-kind production
    - Chapter 8: the concept for the process industry.

  Chapters 5 through 8 present an overview of the business methods and their relation to the characteristic in planning & control. The respective chapter develop detailed business methods in two simple but important cases: master planning in the MRP II / ERP concept and repetitive manufacturing in the Lean / Just-in-Time concept.

- Finally, Chapter 9 discusses ERP and SCM software for these four concepts as well as success factors for the implementation of this kind of software.

Part C offers a detailed discussion of planning & control methods in the operational routine. The reference model in Figure 2 (introduced in Section 5.1.4) summarizes the presents an overview of the planning processes according to — vertically — temporal range (long, medium, and short term) and — horizontally — all the planning & control tasks. The processes and tasks are shown in the logical temporal sequence.

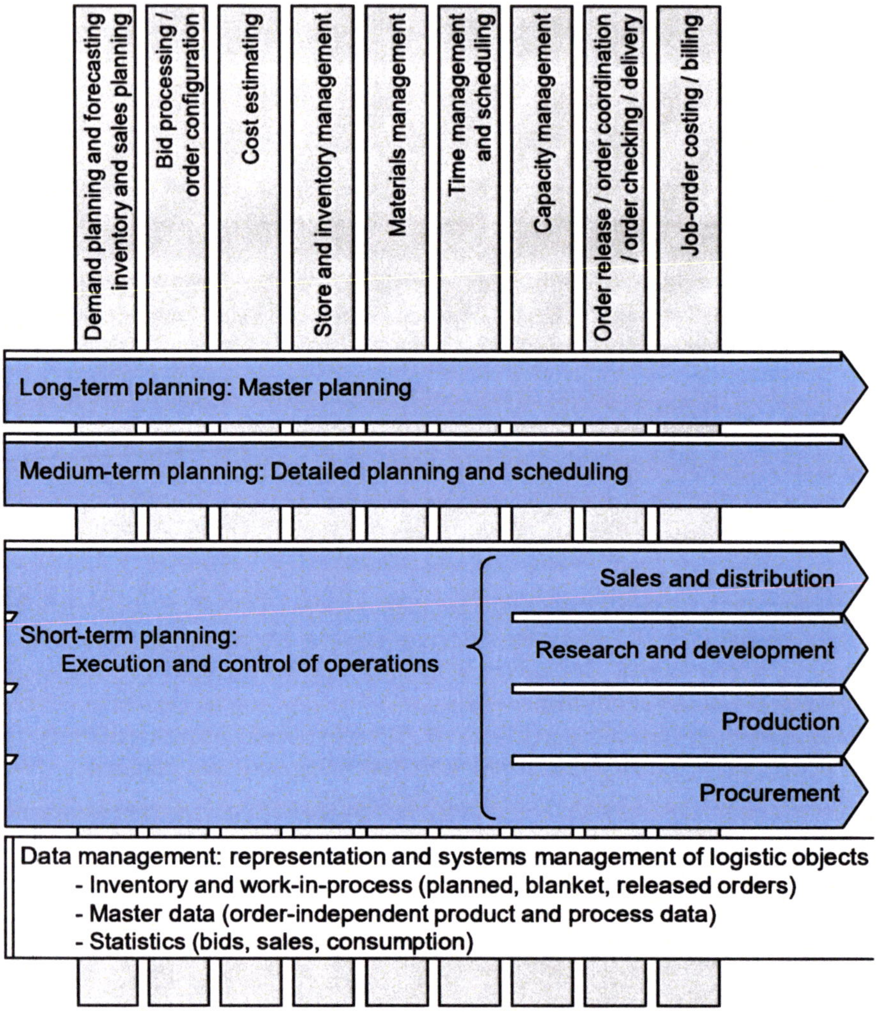

Fig. 2 Reference model of business processes and tasks in planning & control.

- Chapters 10 through 17 examine the individual tasks in succession (exceptions: The discussion of bid processing and customer order configuration starts in Section 5.2.1 and continues in Section 12.1; Chapter 16 discusses cost estimation together with job-order costing):

- Chapter 10: Demand planning and demand forecasting
- Chapter 11: Inventory management and stochastic materials management
- Chapter 12: Deterministic materials management
- Chapter 13: Time management and scheduling
- Chapter 14: Capacity management
- Chapter 15: Order release and control
- Chapter 16: Cost estimation, job-order costing, and activity-based costing
- Chapter 17: Representation and system management of logistic objects

The introduction to each section will refer back to the above reference model and show the task together with the temporal ranges of planning for which the task is particularly pertinent.

- The methods in Chapters 10 through 17 provide a deeper understanding of the concepts in Chapters 5 through 8. They comprise all that is required for designing the logistic systems that is not characterized by frequent order repetition. The detailed discussion of these methods provides the reader with an in-depth methodological foundation for understanding the Kanban and master planning techniques introduced in part B. Many of these techniques have their origins in the MRP II / ERP concepts. However, they also apply to the process industry as well as to product families with many variants, whereby they, of course, are applied to the business objects of those processes.

Part D provides an overview of some adjacent management systems that are closely interconnected with Integral logistics management. Among these are strategic management, technology and product innovation management, the financial and cost accounting system, information, knowledge and know-how management, and system and project management. Part D shows, also and most especially, why and where the linkages exist. In any case, the information provided here is meant to serve as a summary.

- Chapter 18: Integral Logistics Management and Total Quality Management (or Six Sigma) are interconnected. Both management systems focus on fulfillment of concrete customer needs and thus belong to the area of operative implementation in the enterprise. The Japanese approaches give priority to a combination of concepts from both systems. The Toyota Production System for example, combines the lean / just-in-time concept with the jidoka concept, which is a concept for quality management.
- Chapter 19: Systems engineering and project management are also very strongly connected with Integral Logistics Management. First, the associated tasks in their entirety can be understood as management systems. The design, development, and continual improvement of these systems must be approached using the methods of systems engineering and project management. Second, some tasks are unique (one-of-a-kind); for example, in facility location planning, in the project business, or in customer-specific services (production and procurement without order repetition). Third, some techniques are used in both management systems. These are, among

others, scheduling techniques, such as the Critical Path Method (CPM) and the Gantt chart, and methods of financial evaluation of investments, such as the payback or Net Present Value methods.

- *Chapter 20:* In connection with ERP and SCM software systems (see Chapter 9), the connection of information management with Integral logistics management is clear. Information management can provide techniques and methods for realistic modeling of business processes and correct representation of logistic business objects. This makes possible suitable data management, which ensures that the necessary data on objects is available at all times in a detailed and up-to-date form.

Notes to the Reader and Additional Teaching Material

Some notes to the reader:

- Definitions of key concepts and terms usually appear in text boxes, or in special tables or lists, or in footnotes. The terms being defined always appear in *italics*.

- The *definitions of terms* sometimes take the form of an indented bulleted list. This form is useful particularly where one and the same characteristic has varying degrees of expression.

- A gray background highlights important principles, examples, points to remember, prescribed procedures, steps of a technique, or solutions of selected scenarios and exercises. The reader will often find a reference to a figure.

- Some sections of the book are not essential reading for an understanding of the subsequent material. An asterisk (*) identifies these optional sections.

- Also optional in this sense are the additional definitions provided in footnotes. They appear for the sake of completeness or as information for practitioners or for readers coming from related disciplines.

We use the following abbreviations in the text:

- cf. for "compare"
- R&D for "research and development"
- ID for "identification" (for example, item ID)
- IT for "information technology"

For our interactive elements, please refer to: www.intlogman.ethz.ch .

For additional teaching material, please refer to: www.opess.ethz.ch .

Please direct your questions or comments to: Paul.Schoensleben@ethz.ch.

1 Logistics, Operations, and Supply Chain Management

Intended learning outcomes: Describe basic definitions, issues, and challenges. Identify business partners and business objects. Explain strategies in the entrepreneurial context. Disclose how performance is measured.

Logistics, operations, and supply chain management deal with the management of systems that determine the performance of an enterprise or among companies as well as with the corresponding planning & control of daily business operations. These tasks are still managed by human beings who, through intuition and based on experience, find creative solutions. People have unique strategic and operational management abilities, in that they can fill in the blanks accurately and react flexibly to specific situations. However, if processes become more complex, frequent, and rapid, intuition alone does not suffice. Prior experience can also be misleading. In large companies and in transcorporate supply chains, moreover, there are many people involved in the processes, both simultaneously and in sequence. They differ in their experience, knowledge, and intuition. Therefore, logistics, operations, and supply chain management stand in the field of tension of the various stakeholders and contradictory objectives of the company or supply chain.

It is here that the scientific approach of Integral Logistics Management comes into play. After defining basic concepts, issues, and challenges (Section 1.1) and the related business objects (Section 1.2), we examine this field of tension in Section 1.3, paying attention to the various aspects of flexibility, as they represent potentials for future benefits. To measure performance, enterprises or supply chains must select appropriate performance indicators (Section 1.4) that relate to the business objects and objectives. These measures allow firms to evaluate the degree to which objectives are reached and to analyze related causes.

1.1 Basic Concepts, Issues, and Challenges

Intended learning outcomes: Present terms of the working environment and of business life. Explain service orientation in the classical industry, product orientation in the service industry, and the industrial product-service system. Disclose the product life cycle, the synchronization of supply and demand, and the role of inventories. Produce an overview on supply chain management, the role of planning and control as well as the SCOR model.

When looking for solutions of practical problems, people are not generally concerned about definitions. However, clearly defined terms become essential when we seek to gain an understanding of the concepts of Integral Logistics Management. First of all, definitions transmit a picture of the phenomena under study. They also avoid misunderstandings that arise because people and companies make varying usage of technical terms. Finally, definitions are indispensable for structured presentation in a textbook that covers a topic in substantial

P. Schönsleben, *Handbook Integral Logistics Management*,
https://doi.org/10.1007/978-3-662-65625-9_1

detail. However, definitions should not detract from the pleasure of learning new concepts. This section therefore focuses on the issues and challenges connected with the terms defined.

1.1.1 Important Terms of the Working Environment and of Business Life

Intended learning outcomes: Present terms of the working environment, such as work, task, function, order, procedure, process, method, object, business. Produce an overview on terms of business life, such as value-added, business process, business method, business object, goods, item, part, component, material, product, artifact, management, etc.

With designing of organizations, there is (too) frequent confusion among the terms *task*, *function*, and *process* (as well as task orientation, function orientation, and process orientation). By referring to etymological dictionaries, such as [MeWe18], and dictionaries of related words and meanings, we can find out how people *normally* understand the terms. While some branches of science traditionally give terms their own definitions, such definitions are arbitrary. However, a company best takes its orientation from everyday understanding, and uses definitions of terms that will be generally understood.

Figure 1.1.1.1 defines the basic terms of the working environment.

Term	Word origin, definition	Related terms
work	old: travail, toil, drudgery, exertion of strength new: employment, activity that leads to an achievement but also: something produced by mental effort or physical labor	job, task
task	a piece of work that needs to be done regularly	function; order; assignment
function	the particular purpose for which a person or thing is specially fitted or used or for which a thing exists	task; purpose
order	a direction to buy or sell goods	task
procedure	a series of steps followed in a regular order	process
process	a series of actions or operations directed toward a particular result	procedure; course of action
method	a procedure or process for achieving an end also: orderly arrangement	procedure
object	something that may be seen or felt; also: something that may be perceived or examined mentally	thing
business	something to be dealt with; task; concern; new: a commercial or industrial activity or organization	

Fig. 1.1.1.1 Basic terms used in the working environment.

The most important finding here is that the word *work* contains both the character of a process and of content and result. This duality seems to be fundamental. The content of work, that is, its purpose or objective, is often expressed as *task*. The term *function* is clearly related to *task*. *Function* more strongly refers to the result of work, while *task* is more work's content and purpose, whereby each term includes the other. An *order* arises when a task is assigned to someone else.

Procedure and *process* are practically synonymous and stand in duality to the terms *task* and *function*. In most cases, a task or function can be structured as a sequence or as a net of subtasks, or subfunctions, and thus thought of as a process. Turned around, a process is usually seen as various works progressing in a certain sequence. Each of these works may be seen as a task or function, or as a part of such. Of course, there exist tasks and functions that finally are "nuclear" — they cannot be broken down further. In the area of company strategy, but also in R&D, we find tasks that are difficult to break down.

Note that *business* refers to the central term *work*, whereby in today's usage, business means tradable work according to its new definition.

Figure 1.1.1.2 presents additional basic terms that are used in business life. The first four (composed) terms have been defined by the author, as [MeWe18] does not give a definition.

Value added varies in meaning according to the standpoint of either producer or customer. From the traditional perspective, that of the manufacturer, for example, the expense of keeping inventory or work in process is always value-adding. The customer, however, does not normally view such processes as value-adding. With the trend toward customer orientation, it has become increasingly important to take the customer's point of view.

Linked with a *business process* is its order processing. The order fulfilling unit performs not only the value-adding process itself, but also the necessary planning & control of the process. *Business methods* — e.g., methods of order processing — describe how tasks are performed or functions within the company can be achieved. Familiar *business objects* are, for example, customers, employees, products, equipment, and — particularly — orders.

In general, the term *material* is not perceived synonymous to *component*. *Material* generally refers to rather simple initial resources, such as raw material, or information such as documents, evidence, certificates, or similar things, whereas *component as a business object* can also refer to semifinished products.

For the matters covered in this book, the nuances of meaning between the terms *product* and *artifact* are of minor importance. We thus use both terms synonymously.

1.1.2 Service, and Service Orientation in the Classical Industry

Intended learning outcomes: Present terms of the service domain such as service, customer service, service in the originary sense, servitization. Differentiate between a (primary, or core) product, a product in a broad sense, and a product in the most comprehensive sense.

> A *service* is, according to [MeWe18], a performance of official or professional duties, the act, fact, or means of serving.

Term	Definition
value-added (or *value creation*)	(1): a company's own output, including overhead; purchased products or services may complement this (2): value and usefulness of design and production as seen by the customer
business process	process of a company, performed to achieve a potentially tradable outcome that is value added as seen by the customer — internal or external — and that the customer is willing to pay for
business method	an important method in connection with business
business object	an important thing, or a content of thought, in connection with business
good	something that has economic utility or satisfies an economic want
goods	something manufactured or produced (or bought and sold in business)
investment goods	machinery, tools, factories and commodities used in the production of goods (also called *capital goods*)
consumer goods	goods that directly satisfy human needs
item	a separate particular in a list, account, or series
part	a division or portion of a whole
component	one of the parts that make up a whole
material	(1): the elements or substance of which something is composed or made (2, usually plural): items needed for the performance of a task or activity
product	something produced by physical or intellectual effort of *humans*
artifact	something made or modified by *humans* usually for a purpose
management	(1): the act or art of managing: the conducting or supervising of something (2): judicious use of means to accomplish an end

Fig. 1.1.1.2 Additional basic terms used in business life.

The definition of service refers to the beneficiary of the service, generally referred to in the commercial world as the customer. With firms, service often means customer support.

Customer service or customer support is the ability of a company to address the needs, inquiries, and requests from customers ([APIC16]).

A *service in the originary sense* is a process involving a *service object*, that is, an object belonging to the customer that must be brought together with the provider of the service (or vice versa), potentially along with additional customer input.

In many cases, the object, thus the recipient of the service is the customer himself. In other cases, it is technical support and maintenance of machines or plants. Further examples include services in relation to information products, such as correction of software.

Interestingly, [Levi81] points out that "distinguishing between companies according to whether they market services or goods has only limited utility". The author suggests that it is more useful to speak instead of *intangibles* and *tangibles*. He states that "everybody sells intangibles in the marketplace, no matter what is produced in the factory." Remarkably, based on the above mentioned definitions *goods, products,* and *materials* can be either tangible or intangible. On the other hand, the author notes that often "intangible products must be tangibilized. Hotels wrap their drinking glasses in fresh bags or film, put on the toilet seat a sanitized paper band, and neatly shape the end piece of the toilet tissue into a fresh-looking arrowhead". This is particularly important when an otherwise intangible service is part of a more comprehensive service (e.g. a guest's stay in a hotel) and this specific (partial) service can be provided earlier, i.e. before the customer uses the (entire) service.

As [Levi81] states, "tangible products differ in that they can usually, or to some degree, be directly experienced — seen, touched, smelled, or tasted, as well as tested". However, for capital goods in particular, tangibility is not enough. As Figure 1.1.2.1 shows, the decisive factor influencing the purchase is rather the holistic experience.

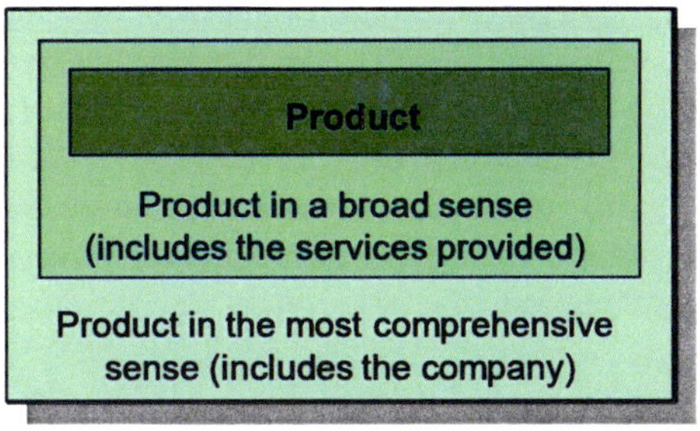

Fig. 1.1.2.1 The product in its holistic experience.

A *product, in a broad sense,* is a (primary, or core) product along with the services provided, where the consumer sees the two as a unit.

For investment goods, additional services increasingly constitute the key sales argument. Here, "services provided" could mean installation instructions or a user manual, training information, or the promise of future (after-sales) services such as maintenance and repairs.

Servitization is an artificial term for the process of systematically designing services that entail considerable additional revenues. Such a design leads to what is called a product-service system. See here Section 1.1.4.

In buyer's markets, individualization of products to customers' requirements and personalized production become more and more important. In particular, an ETO ("engineer-to-order") production environment leads to a service focus and to value co-creation. See here Section 7.4. With a product variety subject to (changing) customer specification (e.g. for the "haute couture" sector of the fashion industry), repeated input from the customer is a key characteristic of the production process. Trying on a half-finished dress is actually a service process in the original sense, i.e. it needs ongoing intensive contact with the beneficiary of the service. It even offers an opportunity for the specification of the product (the dress) to be changed. Then, the quality of such services, or the processes "around the product," can become as much or even more important than the quality of the core product.

A *product, in the most comprehensive sense*, comprises the (primary, or core) product, the services provided, and the company itself, with its image and reputation.

Here, the consumer sees all three as a unit. An example is the concept of *Total Care* in the insurance branch. The aim is to give the customer the idea that the quality of the company, that is of the organization as a whole will provide all-encompassing care. This "all about the customer" process builds trust. Image and reputation of a company are a consequence of stakeholder opinions which, according to Figure 1 in the introduction, include business partners, employees, shareholders, society, environment and nature.

1.1.3 Service Industry, and Product Orientation in the Service Industry

Intended learning outcomes: Differentiate between service industry and classical (or conventional) industry. Produce an overview on industrialization of service.

The *service industry* is, according to [APIC16], in the narrowest sense, an organization that provides *intangibles* (e.g. medical or legal advice). In the broadest sense, all organizations except farming, mining, and manufacturing. It includes retail trade, wholesale trade, transportation and utilities, finance, insurance and real estate; construction; professional, personal, and social services; and local, state, and federal governments and also provides intangible *goods* such as information.

Using this definition, examples of a *classical (or conventional) industry* would include organizations like farming, mining, and manufacturing. Companies working in this sector mainly produce *tangible goods*, or *tangibles*.

Beside the need for service orientation in the classical industry, there also is a need for product orientation in the service industry. According to the hospitality sector example quoted above from [Levi81], "tangibilization of an intangible should ideally be done as a matter of routine on a systematic basis." Such hotels have "*industrialized* the delivery (of their promise of service)."

Industrialization of service means, like in classical industries, standardization and automation of its performance. Section 7.3 describes an example from the insurance industry. The use of a product configurator allows (intangible) elementary insurance services (insurance specialists call them elementary *products*) to be modularly assembled to form a variety of combined products, and these combined products to be put together to form contracts. Catering is another example: there is a lot of similarity between standardized recipes in the catering sector and recipes used in food or chemical-pharmaceutical production.

Industrialization also means some *standardized* service components can be developed and prepared in advance. This applies for hotel and catering services every bit as much as for spare parts. Also, the costs for these components that form part of a more comprehensive service can be calculated in advance. In this context, it becomes clear why an *industrialized* service, although intangible, is often perceived as a product (and is referred to as one), for example in the above-mentioned hospitality or insurance sector. Industrialization of services offers efficiency gains, without any loss of effectiveness. This is an area where the service sector is learning from classical production used for tangible goods. Then, the performance of the (part) service can be perceived as a "production" of intangibles (sometimes, called service production), and the result can be perceived as an (intangible) product or commodity.

The service industry can also provide entire services that are similar to the supply of tangible products. Simple spare parts delivery is often perceived as a service, but is actually no different to producing standard products, where these are kept in stock to ensure fast delivery. And although delivery of a passport is considered to be a service, these days it is actually no different (even in the degree of industrialization) to the supply of a make-to-order product, for which the beneficiary has to enter their personal data (including a facial photograph). Thus, the customer focus will be on acquiring a product rather than receiving a service.

This means that although the so-called *IHIP characteristics* of services (Intangibility; Heterogeneity, i.e. uniqueness of service processes; Inseparability (or simultaneity) of provision and consumption; and Perishability (e.g. exclusion from inventory)) are popular for practical applications in service-oriented companies, their suitability is limited. According to [Hert13], IHIPs are "not defining characteristics, but simply symptoms."

1.1.4 The Industrial Product-Service System

Intended learning outcomes: Present the industrial product-service system. Explain the product-oriented, use-oriented, and result-oriented service, and their degree of intangibility.

According to [MeRo10], an *industrial product-service system IPSS* (or IPS^2) is characterized by the integrated and mutually determined planning, development, provision and use of product and service shares including its immanent software components in business-to-business applications and represents a knowledge-intensive socio-technical system.

An IPSS is thus focused on capital goods such as machinery and equipment which are produced and sold by classical industry, and which are used over a long time period. In that model, customers are businesses, rather than individuals. In addition, although the service recipient is a tangible (core or primary) product, this definition shows that in an IPSS there are no "add-on" services to this product. To create value for the customer and to sell well, the original equipment manufacturer (OEM) network and their suppliers must design the services with the customer (beneficiary) right from the start, as shown in Figure 1.1.4.1. That is the thinking behind the term *value co-creation* (see also [KaNi18]).

Industrial product-service systems have become more important particularly in classical industries, since quality and costs of the core product from many OEM have become almost indistinguishable from the customer's perspective. Offering something different in terms of additional services can be the distinguishing feature that makes what a company is offering stand out, especially in competitive markets.

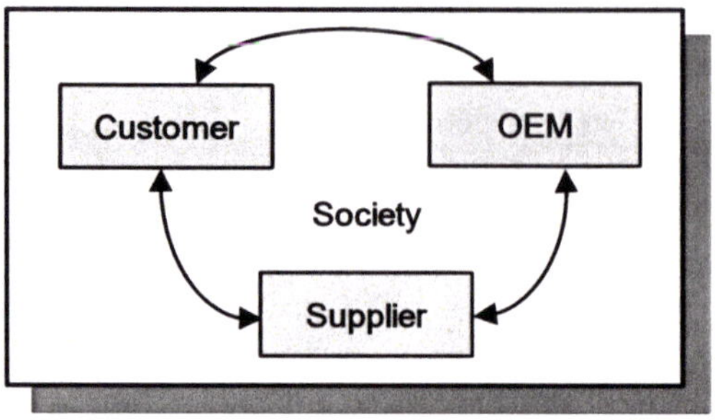

Fig. 1.1.4.1 Industrial product-service system stakeholders (cf. [MeRo10],)

Product-service systems can be categorized into product-oriented, use-oriented, and result-oriented product-service systems. Across these three categories, the degree of integration in the customer's process is increasing. Figure 1.1.4.2 assigns possible services in an industrial product-service system for packaging machines to these categories.

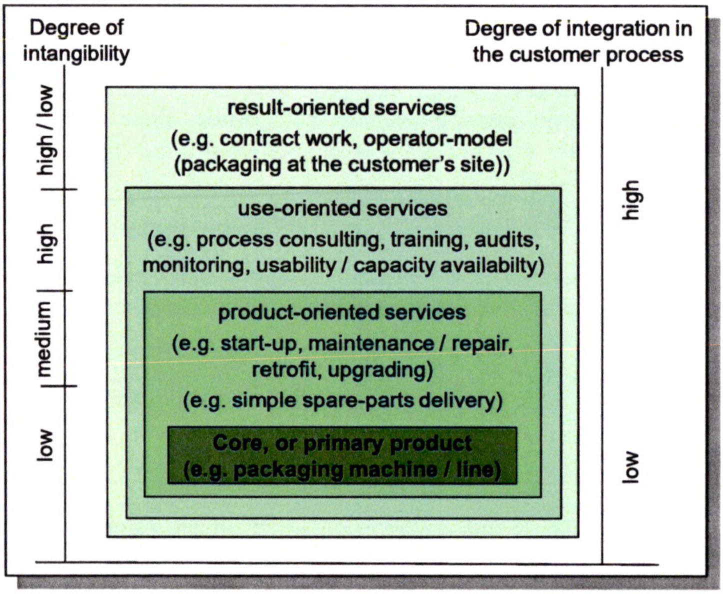

Fig. 1.1.4.2 Categories of possible industrial services (cf. [Lang09])

Some product-oriented services are of a material nature. As an example, there is no real distinction between the simply supply of spare parts without any additional services and supplying and charging for some form of material goods. Other product-oriented services tend to be of an intangible nature and are usually charged for in time units, whereby (material) consumable materials or components can also be charged for. Examples of this would include commissioning the machine, maintenance and repair, retrofit, or upgrading.

A use-oriented service is generally intangible. Examples include process consulting, training, audits, monitoring, and usability of the machine. Such processes are normally charged for in time units or on the basis of a performance indicator that is measured as a percentage.

The degree of intangibility for a result-oriented service can be high or low, depending on the agreed result. As an example, for cleaning services, the result could be an agreed upon level of cleanliness, which is an intangible result. Photocopier manufacturers supply the machine, paper and consumables. As a result, they can determine the number of copies produced. Whether the result counts as a tangible or intangible will depend on whether attention is focused on the photocopy that is produced, or whether the significant fact is simply that a copy has been taken. For industrial services, the focus for *contract work* (an *external* manufacturer executes just one or more operations for making a product) or the *operator model* (the manufacturer not only supplies but also operates the machine under contract to the customer) is on the items that are produced, which the customer assesses the quality of and pays for accordingly. Those are tangibles. In the case of the packing machines, the manufacturer of the packing machines as well produces packaging.

Conclusion: The oft-formulated distinction between "product = tangible" and "service = intangible" is becoming ever more blurred. What the customer actually perceives as value-added depends on a number of tangible and intangible elements that work together as a system and offer the customer the desired benefits. The elements and the system as a whole are perceived to be a product or a service, depending on the customer's focus (cf. [Schö19]).

1.1.5 The Product Life Cycle, and Logistics and Operations Management

Intended learning outcomes: Produce an overview on the product life cycle. Differentiate between terms such as logistics, operations, logistic management, operations management, and value-added management.

Products are made by converting goods. The use or utilization of products leads to their consumption or usage.

Consumption of goods (by the *consumer*) means, according to [MeWe18] the act of consuming or using up, and also, according to [Long15], the amount of goods that are used (up).

Goods that are used up must be disposed of properly. There is thus a life cycle to products.

Put simply, the *product life cycle* consists of three stages: *design and manufacturing*, *use (and ultimately consumption)*, and *disposal*, which can be connected with *recycling*.[1]

Figure 1.1.5.1 shows the product life cycle. Design, manufacturing, service, and disposal are seen as value-adding processes,[2] symbolized by an arrow pointing in the direction of value-adding. Use is itself a process; however, it is a value-consuming one.

The life cycle of *material products* generally begins with nature and leads from *design and manufacturing* to the consumer. A consumed product must then be disposed of, possibly

---

1 This definition follows the definition of the term *life cycle* given by *Merriam-Webster*, i.e. "a series of stages through which something passes during its lifetime." See Section 4.4.4 for another possible definition, and Section 17.5.1 for the definition of the term "product life cycle management (PLM)."

2 Also, disposal is a value-adding process. After use (or being used up), a product even has a negative value as soon as disposal involves costs, such as fees for trash disposal.

connected with recycling of components. In the most general case, the life cycle ends once again with nature, in that the materials are returned to the earth.

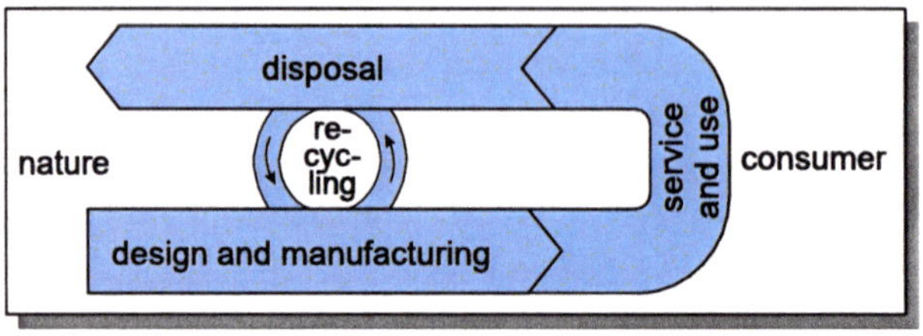

Fig. 1.1.5.1 The product life cycle.

The life cycle of *nonmaterial products* begins with an issue about which something is declared. This issue, in a broad sense, can also be seen as ultimately connected to things in nature, whether to objects or at least to human thinking about objects. Disposal ends with the information being erased or deleted. In the broadest sense, then, it is also returned to nature.

Logistics is involved with products over their entire life cycle:

Logistics is the organization, planning, and realization of the forward and reverse flow and storage of goods, data, and control[3] along the entire product life cycle.

Logistics management is the efficient and effective management of logistics activities to meet customers' requirements.

The definition of operations management is very similar to logistics management.

Operations, according to [RuTa17], is a function or a system that transforms input to output of greater value.

Operations management, according to [APIC16], is the planning, scheduling, and control of the activities that transform input into finished goods and services.

Operations Management also denotes concepts from design engineering to industrial engineering, management information systems, quality management, production management, accounting, and other functions as they affect the operation. According to [RuTa17], it denotes the design and operation of productive systems — systems for getting work done.

It also makes sense to view the other functions found all along the company's value chain, namely, *procurement*, *production*, and *sales*, from the management perspective. In the literature, the *functional* terms are usually defined clearly and distinctly. In contrast, for *management* terms — like procurement *management*, production *management*, and sales *management* — there often are no formal definitions. In practical usage, however, these terms are increasingly similar to the definitions given above for logistics or operations *management*. This is not surprising, for it is impossible to conduct successful operations management if it is applied to only a part of the value chain. For this reason, we assume in the following that there are no significant differences among all these management terms.

---

3 See Section 4.1.2 for definitions of flow of goods, data, and control.

Value-added management can thus be used as a generalized term for all the types of management mentioned above.

Figure 1.1.5.2 shows a graphical representation of how the terms fit the company's world.

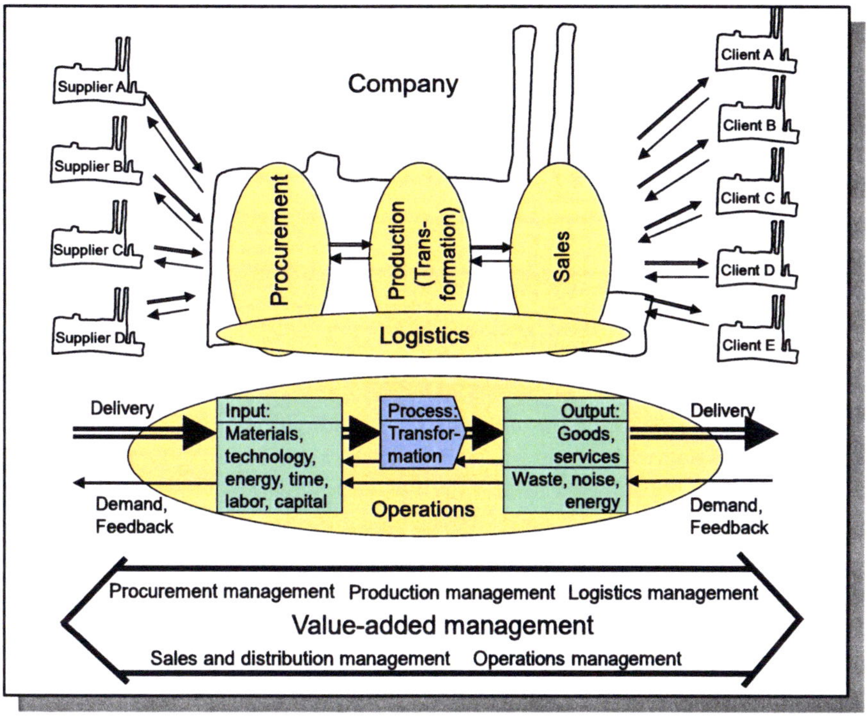

Fig. 1.1.5.2 Assignment of terms to value-added management.

1.1.6 The Synchronization of Supply and Demand Using Inventories

Intended learning outcomes: Describe supply, demand, lead time, and customer tolerance time. Explain the problem of temporal synchronization between supply and demand as well as the role of various kinds of inventories in solving this problem.

A fundamental problem in logistics management is *temporal synchronization between supply and demand*. Here are some basic definitions, according to [APIC16]:

Supply is the quantity of goods available for use.

Demand is the need for a particular product or component. The demand could come from any number of sources, e.g., customer order or forecast, an interplant requirement, or a request from a branch warehouse for a spare part, or for manufacturing another product.

Actual demand is composed of customer orders, and often allocations of components to production or distribution.

Demand forecasting is an estimation of future demand. *Demand prognosis* is used here synonymously.

Lead time is a span of time required to perform a process (or a series of operations). In a logistics context, it is the time between the recognition of the need for an order and the receipt of goods.

Customer tolerance time, or *demand lead time*, is the time span the customer will (or can) tolerate from order release to the delivery of the product or the fulfillment of the service.

Delivery lead time is the total time required to receive, fill, and deliver an order, from the receipt of a customer order to the delivery of the product or the fulfillment of the service. *Delivery cycle*, *delivery time*, or *time to delivery* are used synonymously.

The *delivery policy* is the company's objective for the time to deliver the product after the receipt of a customer's order.

In a market-oriented economy, the consumer expresses a need as demand for a product. A manufacturer then attempts to fulfill the demand. In principle, design and manufacturing are thus controlled by demand: They should begin only when the need has been validly formulated. In the world of practice, this ideal orientation of the producer toward the consumer is often not possible. On the one hand the delivery lead time may be longer than the customer tolerance time. Obvious examples are medications, groceries, or tools. On the other hand, in nature, many basic materials are ready at a point in time that does not coincide with the timing of the consumer's need. Obvious examples are foodstuffs and energy. In many cases, however, the relationship between supply and demand determines whether customers can enforce their required delivery lead times.

Storage of goods over time plays an important role in solving this problem, allowing temporal synchronization between supply and demand.

Storage is the retention of goods (i.e., parts or products) for future use or shipment (cf. [APIC16]). *Warehouse*, *store*, or, more precisely, *goods store* are possible terms for the infrastructure for the storage of goods.

Inventory includes all physical items in any form that can be found in the company:

- On-hand balance is the inventory of stored items, for example, items used to support production (raw materials), customer service (end products or spare parts), and supporting activities (MRO items).
- In-process inventory, or work-in-process (WIP) means goods in various stages of completion throughout the plant.
- In-transit inventory, or transportation inventory means goods moving between two locations and owned by the company in accordance with the agreed-upon incoterms, which are terms used in international commercial transactions.

Inventory at sufficiently high levels in the value-adding process may allow the company to meet the customer tolerance time. But there are also disadvantages. Inventory ties up capital and requires space. Because of limited *shelf life* (that is, the length of time an item may be held in inventory before it becomes unusable), goods may perish, become obsolete, damaged, or destroyed. Keeping an inventory only makes sense where stored goods will be turned over rapidly enough. In order to minimize these disadvantages, inventory must therefore be positioned at the right levels during design and manufacturing (and, analogously, disposal).

This means that goods to be stored should ideally involve none of the disadvantages mentioned above. In Figure 1.1.6.1, there are two stores within design and manufacturing.

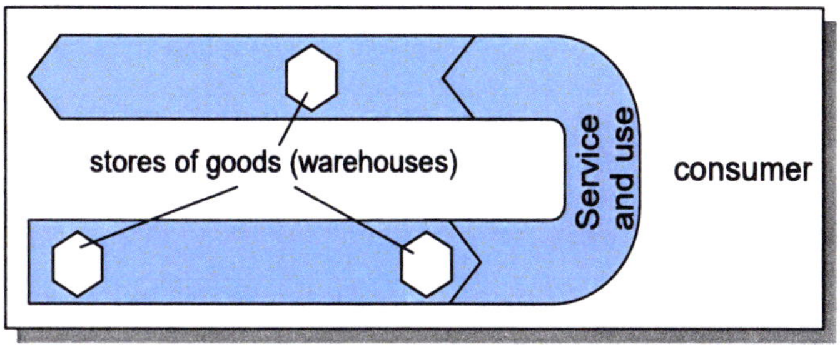

Fig. 1.1.6.1 Storage of goods within logistics.

A goods store decouples the processes upstream and downstream from this point, and therefore demand from supply. The following definitions reflect this point of view:

Decoupling is the process of creating independence between use and supply of material. *Decoupling inventory* is the amount of inventory kept at a decoupling point ([APIC16]).

A *decoupling point* is a location along the value-added process where inventory is placed to create independence between processes or entities ([APIC16]).

Decoupling points constitute a degree of freedom in logistics and operations management. Their selection is a strategic decision that determines delivery lead times and the *inventory investment* — that is, the dollars that are in all levels of inventory ([APIC16]).

1.1.7 The Supply Chain and the Extended Enterprise, Supply Chain Management and Integral Logistics Management

Intended learning outcomes: Describe the reasons for an extended enterprise. Differentiate between a logistics network, a production network, a procurement network, a distribution network, and a service network. Describe the concept of the supply chain. Produce an overview on supply chain management and on integral logistics management.

For products of a certain complexity, design and manufacturing are often distributed among several companies or among different organizational units within a company. From the perspective of the individual manufacturer, the reasons for this are, for example:

- *Quality*: Some technologies or processes may not be mastered successfully enough (problem of *effectiveness*, that is, achieving the given or expected quality standard).
- *Costs*: Certain technologies or processes cannot be implemented economically (problem of *efficiency*, that is, the actual output compared with the output expected, with regard to the use of means).
- *Delivery*: Some processes are not rapid enough, or they are unstable over time.
- *Flexibility*: Customer demand may show rapid variation; the company's own competencies or capacity cannot be adapted quickly enough.

An *extended enterprise*, i.e. a network of the sublogistics of a number of legally independent companies that participate in design and manufacture, is formed as a result.

A chain or a tree structure leading to a product are simple forms of such a network.

A *logistics network* is the joining of the logistics of several organizational units, that is, companies or parts of companies, to form comprehensive logistics. *Production network*, or *production system*, and *procurement network* can be used as synonyms of logistics network.

The *end user* of a logistics network is the consumer.

Figure 1.1.7.1 shows an example where three organizational units form a logistics network.

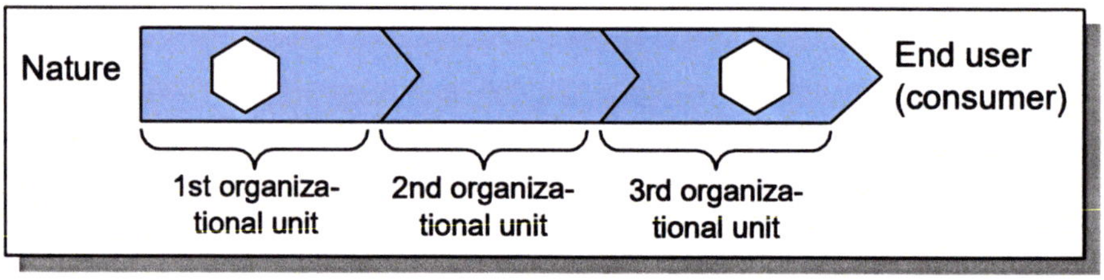

Fig. 1.1.7.1 Three organizational units in a logistics chain.

The logistics chain between two stores is crucial. The logistics of the second organizational unit in Figure 1.1.7.1 must not be viewed in isolation. The logistics of the first organizational unit and the logistics of the third organizational unit will have a direct influence on the logistics of the second organizational unit, since there is no buffer between the two stores.

From the perspective of the end user, distribution and service networks also belong to value-added, for only with delivery and possibly service is the customer order fulfilled.

A *distribution network*, or *distribution system*, is a group of interrelated facilities — manufacturing and one or more levels of warehousing — linking the production, storage, and consumption activities for spare parts and finished goods inventory ([APIC16]).

A *service network is a group of interrelated facilities for performing all services in connection with tangibles or intangibles.*

This leads to the following general terms used for all the types of networks mentioned above:

A *supply chain* is the global network used to deliver products and services from raw materials to end customers through an engineered flow of information, physical goods, and cash ([APIC16]). A comprehensive definition of supply chain also includes the networks for disposal and recycling.

The *extended enterprise* is the notion that supply chain partners form a larger entity. These partners form the *supply chain community* (cf. [APIC16]).

Figure 1.1.7.2 shows how, e.g. with investment goods, supply chains can be connected.

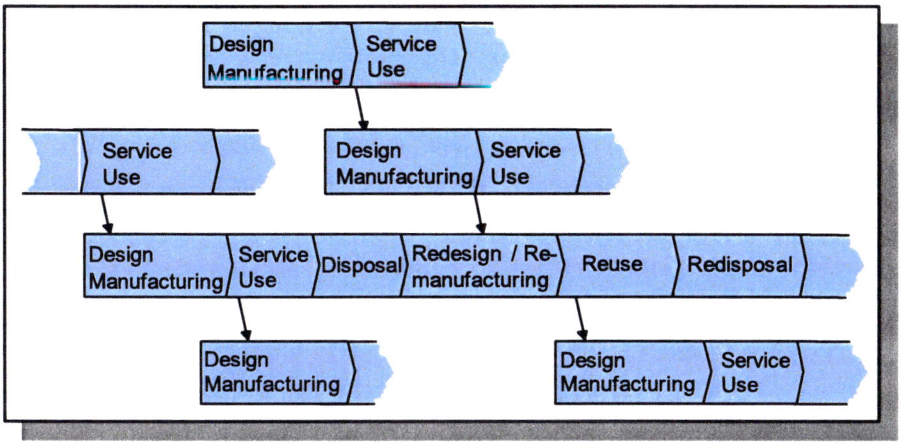

Fig. 1.1.7.2 Connected supply chains for the design and manufacturing of investment goods.

One dimension is the *multilevel* nature. The user is in turn part of another supply chain. That network may produce other investment goods, and so on. For example, with a machine tool, products may be manufactured that are used in the manufacture of other machines.

Another dimension is the product life cycle. A close look shows that *reverse logistics*, that is, a supply chain dedicated to the reverse flow of the product, such as through returns, disassembly, and recycling, can lead to a further life cycle — through *redesign* and *remanufacturing* to *reuse* — as another product, if need be.

Supply chain management (SCM) is the design, planning, execution, control, and monitoring of supply chain activities with the objective of creating net value, building a competitive infrastructure, leveraging worldwide logistics, synchronizing supply with demand, and measuring performance globally ([APIC16]).

Integral logistics management is the management of the comprehensive supply chain, that is, along the entire product life cycle, within and across companies.

1.1.8 The Role of Planning and Control and the SCOR Model

Intended learning outcomes: Produce an overview on supply chain planning and planning & control system. Differentiate between production planning and control (PPC) and a PPC system. Present the Supply Chain Operations Reference (SCOR) model. Describe levels 1 and 2 of the actual SCOR model.

Planning (in an enterprise) is the process of setting goals for the organization and choosing various ways to use the organization's resources to achieve the goals ([APIC16]).

Supply chain planning is the determination of a set of policies and procedures that govern the operation of a supply chain. It includes the determination of marketing channels, promotions, respective quantities and timing, inventory and replenishment policies, and production policies. Planning establishes the parameters within which the supply chain will operate ([APIC16]).

Supply chain planning aims at having the right item in the right quantity at the right time at the right place for the right price in the right condition for the right customer — every time. This task must include the entire supply chain. Once the weighing of the entrepreneurial objectives is done (Section 1.3.1), supply chain planning entails a number of principles, methods, and procedures in order to accomplish the following subtasks:

- Evaluate the various possibilities of distribution, production, and procurement that may be utilized to achieve set objectives.
- Create a program in suitable detail, i.e., determine salable products, their quantities, and deadlines. Revise the plans periodically in response to changing constraints.
- Elaborate and realize distribution, production, and procurement plans derived from the program, in suitable detail and in consideration of objectives and constraints.

Planning decisions thus concern logistics issues, such as: When, how, and in what quantities will goods be procured, produced, or distributed? Will inventory be inserted between storehouse, factories, and the supply chain community? Which services will be performed, when, where and how? What personnel and what assets will be used? When will delivery to customers and subsidiaries take place?

The suitable *information logistics* for this task can be put into a system for *planning & control.* Here, the term *"control"* should not be interpreted in a technical sense as complete mastery of a controlled process. In companies, the term indicates regulation or even just coordination. Because of the established use of the term (for example, "production planning and control"), however, we maintain it here. In the manufacturing stage within the product life cycle, such a system is frequently called *PPC*, or *production planning and control. MPC (manufacturing planning and control)* is another classic abbreviation. See [VoBe18].

The term *PPC* sometimes leads to misunderstandings, because the term *PPC system* is used to refer to both the logistics task and the computer software supporting the task. These two meanings are often mixed deliberately. Upon the background of misplaced optimism with regard to PPC software, demagogues — when the use of PPC software fails — tend to declare that the entire scientific body of knowledge on planning & control is "useless." They overlook the fact that the liability for understanding methods and their practical application always falls upon the people in the company. Chapter 9 examines these issues in more detail.

Figure 1.1.8.1 shows the planning of the comprehensive supply chain as an ongoing synchronization of supply with demand in the comprehensive supply chain. The organizational units involved can be independent companies, or profit or cost centers within a company.

This task is based on the internal chain of "source," "make," and "deliver" in each of the organizational units involved. All demand and capabilities of fulfilling them are carried by the network as a whole and reconciled jointly. Based on this idea, the Supply Chain Council (SCC), founded in 1996, merged with ASCM (the Association for SCM) in 2014, published the Supply Chain Operations Reference model (SCOR®) (see also www.supply-chain.org).

SCOR, the Supply Chain Operations Reference model, is an aid to standardization of process chains within and across companies.

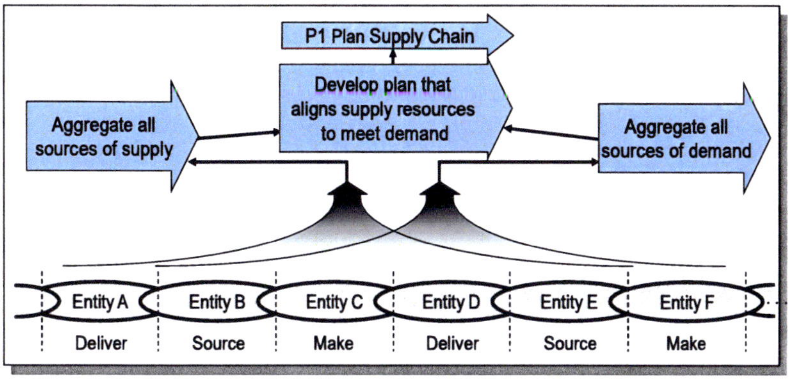

Fig. 1.1.8.1 Ongoing synchronization of supply with demand in the comprehensive supply chain.

Figure 1.1.8.2 shows level 1 of the actual SCOR model.

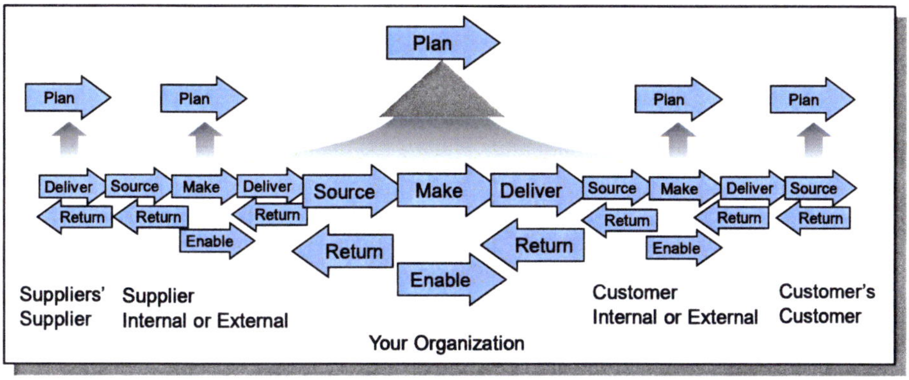

Fig. 1.1.8.2 The SCOR model, version 12.0, level 1.

The aim of SCOR is to foster a common understanding of processes in the various companies participating in a supply chain. Figure 1.1.8.3 shows the six process categories and 32 reference processes defined by level 2 of the actual SCOR model.

1.2 Business Objects

Intended learning outcomes: Present business-partner and order-related business objects in detail. Explain product-related, process-related, and resource-related business objects. Produce an overview on rough-cut business objects.

As Section 1.1 did, also this Section contains numerous definitions. The business objects presented here will be familiar to some readers because of their own professional experience. They may want to skim through these sections and go directly to the sections in 1.3 and 1.4 that deal with the challenges the enterprise faces, keeping in mind that they can always return here for definitions of terms that require some explanation. The important business objects will be described below in principle, that is, in their perception as total objects.

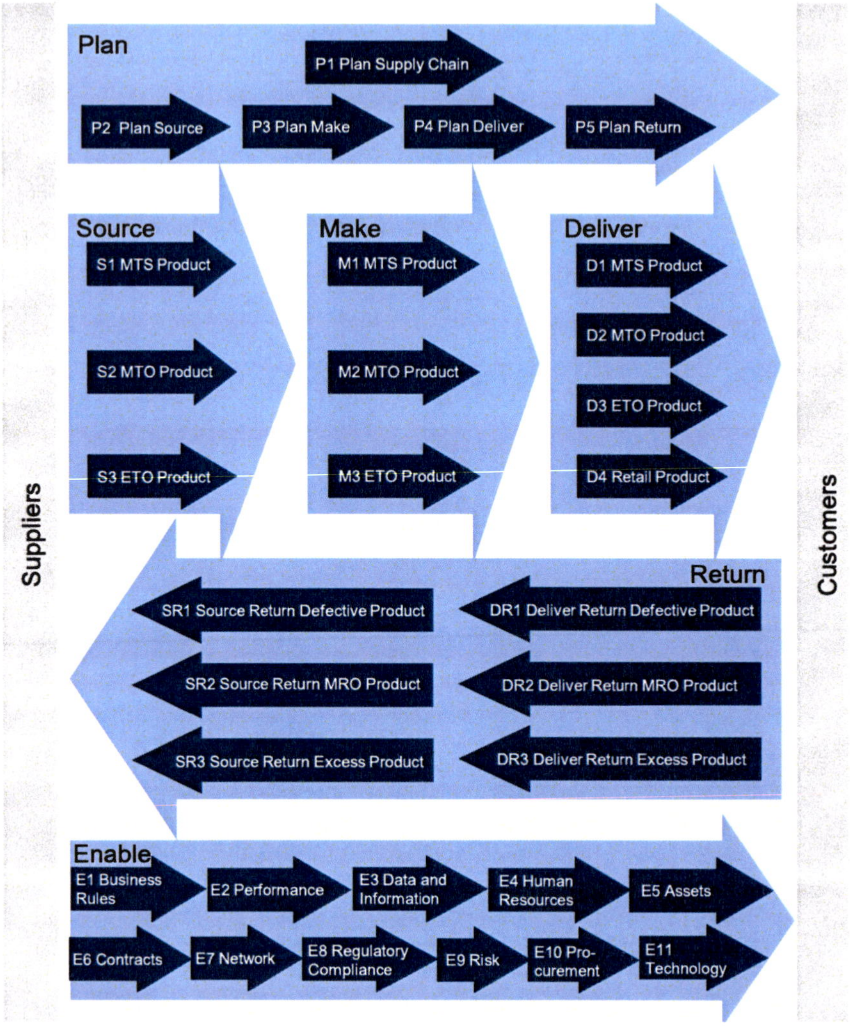

Fig. 1.1.8.3 The 6 process categories and 32 reference processes of SCOR version 12.0, Level 2, toolkit.

1.2.1 Business-Partner, and Order-Related Business Objects

Intended learning outcomes: Present in detail the order as a business object. Produce an overview on terms such as business partner, due date, order promising, order confirmation, order status. Differentiate between a simple sales order and a complex sales order. Differentiate between order positions for items (to be) delivered and labor (to be) performed.

The order serves as an instrument both in the legal sense and with regard to process organization, within and across companies. Basic business objects for the definition of an order are:

A *business partner* of a company is a general term for a *customer*, recipient or beneficiary of a good or a service, or a *supplier*, selling or providing a good or service.

A *date* is a fixed point in time at which an event occurs. It is normally expressed as day and time of day ([MeWe18]).

A *due date* is a date on which something is scheduled, i.e., expected in the prescribed, normal, or logical course of events ([MeWe18]).

A *time period* is a period on the time axis. The *start date* is the beginning, and the *end date* is the end of the time period. In a logistics environment, it is mostly a completion date.

An order as a business object is rather complex. It contains all the information required for planning & control of the flow of goods.

An *order as a business object* consists at minimum of a *business partner* (in addition to the company itself) and a *date* and sets binding obligations with regard to:

- Who the business partners are: the customer and the supplier
- When the order is issued, or what the *order validity date* is
- Within what time period the order is fulfilled (*order start date* and *order completion date* or *order end date* — in general the *order due date*).

Depending on the purpose, the order, with its constituent *order items* or *order position*s, also sets binding obligations with regard to

- The items (ID, quantity, due date) that must be manufactured or procured.
- The components (ID, quantity, due date) that must be ready for use or building in.
- The work, labor, tasks or services that must be performed and in what sequence; this also includes transport, inspection, and other similar tasks.
- Whether and how order tasks are linked to other orders.

These definitions hold for all kinds of orders, both in industry and the service sector.

The *kind of order* classifies an order according to its business partners.

- A *customer order* or a *sales order* is an order from an *external customer* (i.e., a customer that is *not* part of the company) to the company.
- A *procurement or purchase order* is an order from the company to an *external* supplier.
- A *production order* or a *manufacturing order* or a *job order* or a *shop order* is an *internal order*, or order from an *internal customer* (i.e., a customer that is part of the company) to manufacture a good.
- An *overhead order* or a *work order* is a company internal order, e.g., for R&D, for items to be manufactured (such as tools) or for services that concern the infrastructure of the company (such as equipment maintenance and repair).

An order becomes legally binding by order promising and confirmation.

Order promising is the process of making a delivery commitment, i.e., answering the question: When can you ship how much ([APIC16])?

An *order confirmation* is the result of order promising.

An order runs through several phases.

Order status is a phase in the carrying out of the order. We distinguish among four phases:
1. Planning or bid status
2. Order confirmation status
3. Order execution status
4. Billing status (calculation or invoice)

In the first status (planning or bid), the order data represent projections. In the second and third status, they are projections (budgets or cost estimations) that will be replaced gradually with real data. In the fourth status (billing), we find the effective data associated with a concrete order, tapped through some kind of recording of shop floor data.

Figure 1.2.1.1 shows an example of a simple sales order, an order form used by an Internet company. This order is a typical example of a sales order or also simple purchase order in all areas of business.

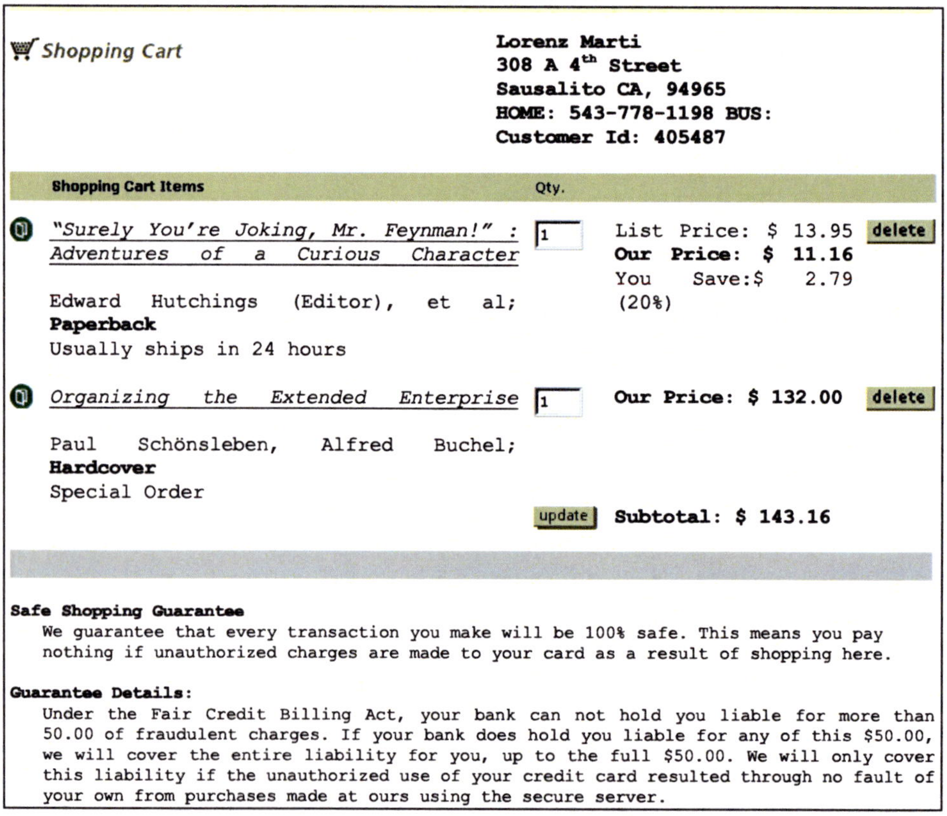

Shopping Cart

Lorenz Marti
308 A 4th Street
Sausalito CA, 94965
HOME: 543-778-1198 BUS:
Customer Id: 405487

Shopping Cart Items	Qty.		
"Surely You're Joking, Mr. Feynman!" : Adventures of a Curious Character Edward Hutchings (Editor), et al; **Paperback** Usually ships in 24 hours	1	List Price: $ 13.95 **Our Price: $ 11.16** You Save:$ 2.79 (20%)	delete
Organizing the Extended Enterprise Paul Schönsleben, Alfred Buchel; **Hardcover** Special Order	1	**Our Price: $ 132.00**	delete
	update	**Subtotal: $ 143.16**	

Safe Shopping Guarantee
We guarantee that every transaction you make will be 100% safe. This means you pay nothing if unauthorized charges are made to your card as a result of shopping here.

Guarantee Details:
Under the Fair Credit Billing Act, your bank can not hold you liable for more than 50.00 of fraudulent charges. If your bank does hold you liable for any of this $50.00, we will cover the entire liability for you, up to the full $50.00. We will only cover this liability if the unauthorized use of your credit card resulted through no fault of your own from purchases made at ours using the secure server.

Fig. 1.2.1.1 Simple sales order form used by an Internet company: status "order."

- The upper portion, the heading, contains customer data and supplier, that is, company data. Order validity date in this case is understood implicitly as the date the order is received by the company.
- The main body of the sales order represents its positions and lists the items to be delivered, that is, their identification and quantities.
- Finally, the footer contains the delivery address.

Here, the delivery due date is understood to be "as soon as possible." Thus, with little data a practical order comes about. Because the company usually has the items in stock, this sales order serves as distribution control to the customer. The invoice is usually produced — following delivery — within the same structure. Billing information, in most cases, will correspond to order information. Deviations might occur due to delayed deliveries or backorders.

Figure 1.2.1.2 shows a more complicated example from the service industry, an invoice for auto repair. This invoice is the result of an order that was placed previously within the same structure: usually in verbal, sometimes in written form.

JOHN DOE
PO BOX 9999
W BARNSTABLE, MA
HOME: 508-362-9999
BUS:

INVOICE

PAGE 1

SEARS AUTO SALES INC.
JEEP / EAGLE
499 Rt. 6A
E. Sandwich, MA 02537
(508) 888-0175 * 775-7972

SERVICE ADVISOR: 112 PATRICIA STARBARD

COLOR	YEAR	MAKE / MODEL	VIN	LICENSE	MILEAGE IN / OUT	TAG
MAROON	1990	JEEP CHEROKEE	1J4FJ58L0LL205432	639BBW	138000 / 138000	

DEL. DATE	PROD. DATE	WARR. EXP.	PROMISED	PO NO.	RATE	PAYMENT	INV.DATE
01JAN1990			16:30 10MAY99		60.00	CASH	10MAY1999

R.O. OPENED	READY	OPTIONS:
07:54 10MAY99	16:20 10MAY99	STK: / 1) CHEROK

LINE OPCODE TECH TYPE HOURS LIST NET TOTAL

A CHECK FOR NOISE IN LOWER PART OF ENGINE - USUALLY WHEN WARM - RATTLE -
00 REPLACED ALTERNATOR
110 CP 1.00 60.00 60.00
1 JR775126 REMAN-ALTERNTR- 175.00 175.00 175.00
PARTS: 175.00 LABOR: 60.00 OTHER: 0.00 TOTAL LINE
A: 235.00

****

B TAILGATE HATCH HAS A WATER LEAK - ROOF HINGE HAS MOVED SO THERE IS A
GAP WHEN CLOSE DOOR - MOSTLY ON RIGHT SIDE
00 RESEALED LIFTGATE WINDOW
110 CP 1.00 60.00 60.00
PARTS: 0.00 LABOR: 60.00 OTHER: 0.00 TOTAL LINE
B: 60.00

****

CUSTOMER PAY SHOP SUPPLIES FOR REPAIR ORDER 3.95
THANK YOU FROM SEARS AUTO SALES INC.
FOR AN APPOINTMENT CALL 508-888-0175, FAX # 508-888-8841

ON BEHALF OF SERVICING DEALER, I HEREBY CERTIFY THAT THE INFORMATION CONTAINED HEREON IS ACCURATE UNLESS OTHERWISE SHOWN. SERVICES DESCRIBED WERE PERFORMED AT NO CHARGE TO OWNER. THERE WAS NO INDICATION FROM THE APPEARANCE OF THE VEHICLE OR OTHERWISE, THAT ANY PART REPAIRED OR REPLACED UNDER THIS CLAIM HAD BEEN CONNECTED IN ANY WAY WITH ANY ACCIDENT, NEGLIGENCE OR MISUSE. RECORDS SUPPORTING THIS CLAIM ARE AVAILABLE FOR (1) YEAR FROM THE DATE OF PAYMENT NOTIFICATION AT THE SERVICING DEALER FOR INSPECTION BY MANUFACTURER'S REPRESENTATIVE.

(SIGNED) DEALER, GENERAL MANAGER, AUTHORIZED PERSON
(DATE)

STATEMENT OF DISCLAIMER
The factory warranty constitutes all of the warranties with respect to the sale of this item/items. The Seller hereby expressly disclaims all warranties either express or implied, including any implied warranty of merchantability or fitness for a particular purpose. Seller neither assumes nor authorizes any other person to assume for it any liability in connection with the sale of this item/items.

CUSTOMER SIGNATURE

DESCRIPTION	TOTALS
LABOR AMOUNT	120.00
PARTS AMOUNT	175.00
GAS, OIL, LUBE	0.00
SUBLET AMOUNT	0.00
MISC. CHARGES	3.95
TOTAL CHARGES	298.95
LESS INSURANCE	0.00
SALES TAX	8.95
PLEASE PAY THIS AMOUNT	307.90

Fig. 1.2.1.2 Example of a complex sales order at an auto garage: status "billing."

- The heading contains company and customer data, complemented by the delivery date and the characteristic object related to the service (the car). As this is an invoice, the billing date is also given.
- The main body includes order positions for work, or labor performed. The (spare) parts list lists the items used to complete the labor. These items may be listed as in-stock shop supplies or items ordered specially for the job. Quantity and price relate to definite defined units, such as pieces and hourly labor rates. Comments on the invoice aid communication between customer and service provider.
- The footer of the invoice contains specific billing information, such as the total amount, broken down into the various charges, conditions of sales, and sales tax. Bids and order confirmations, or, in other words, the first and second statuses that preceded the billing status, would contain similar data.

1.2.2 Product-Related Business Objects

Intended learning outcomes: Produce an overview on business objects such as item, specializations thereof (such as part, component, assembly, raw material, spare part, MRO item), and item family. Explain the product structure and the bill of material. Identify the quantity per and the low-level code.

According to [APIC16], some terms get a specialized definition (cf. Section 1.1.1):

- An *item as a business object* is any unique manufactured or purchased part, material, intermediate, subassembly, or product. A part as a business object is generally an item that is used as a component and is not an assembly, subassembly, blend, intermediate, etc.
- A *component as a business object* is the raw material, part, or subassembly that goes into a higher assembly, compound, or other item.

From the company's perspective, the collective term *item* includes the following specializations, also called *item types*:

- An *end product* or *end item* or *finished good* is a completed item that generally does not serve as a component of another product.
- An *intermediate product* or a *semifinished good* is stored or awaits final operations in the production process. It can be used in the assembly of a higher-level product and is thus also a component.
- An *assembly* is an intermediate product and is composed of at least two components.
- A *single part* (often simply called *part*) is either produced in-house (*in-house part*) or purchased (*purchased part*) and is used in an end item. An in-house part is produced from only one component.
- A *raw material* is, for the company, a purchased item or an original material that is converted via the manufacturing process.
- A *service part* or a *spare part* is a component that can be used without modification to replace a part or an assembly.

- A *MRO item* is an item for maintenance, repair, and operating supplies. It supports activities in the company and is, in general, not used as a component for products.

As the majority of the basic descriptions (or attributes) of all these types of items are the same, such as iID, description, inventory, costs, and price, they are often grouped together in the generalized object *item*. Figure 1.2.2.1 shows the item types as specializations of *item*.

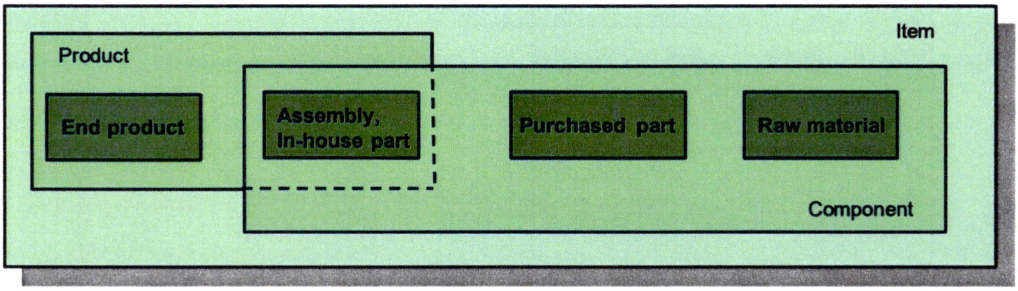

Fig. 1.2.2.1 The business object *item* as a generalization of various goods objects.

An *item family* is a group of items having similar features (such as form, material) or a similar function.

For example, different kinds of screws may be grouped together as an item family and viewed as a (composed) business object, "family of screws."

Product structure is the structured list of components to be used in order to manufacture a product, understood as a whole-part hierarchy.

A *structure level*, or simply *level*, is assigned to every part or assembly in a product structure. It signifies the relative level in which that part or assembly is used within the product structure ([APIC16]).

The (structure) level stands in inverse relation to the relative depth of the components in the product structure. End products generally have the level 0. The direct components of an end product have the level 1. A component in an assembly has a level code one unit higher than the assembly.

A *design structure level* is a structure level from the point of view of product design.

Bill of material and *nomenclature* are other terms for a *convergent* product structure (in contrast to *divergent* product structure, where we usually speak of *recipes*; see also the definition of these concepts in Section 4.4.2).

The *quantity required* or *quantity per* or *usage quantity* is the number of components per unit of measure of the next higher level product into which the component is built. The *cumulative quantity per* of each component in the end product is thus the product of quantities required along the product structure.

The example in Figure 1.2.2.2 shows a bill of material, that is, a convergent product structure with two (structure) levels.

Item 107421 is the end product composed of the two assemblies 208921 and 218743. Each assembly, in turn, has two components. The quantity required is given in parentheses. As an example of cumulative quantity per, in 107421 there are 2 · 3 = 6 components 390716.

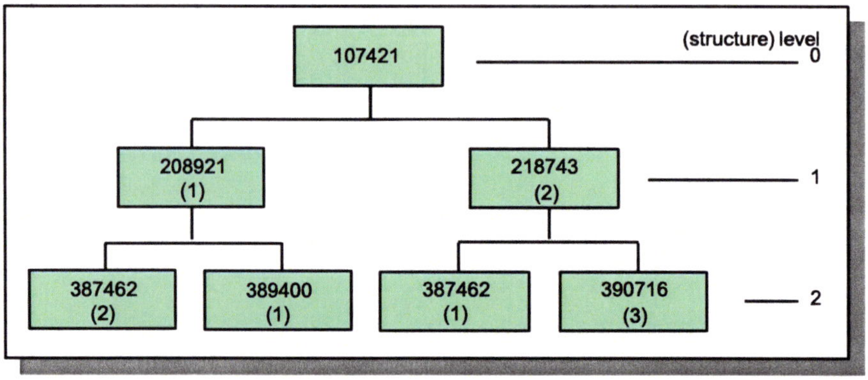

Fig. 1.2.2.2 A product structure (bill of material) with two (structure) levels.

The *low-level code* is a number that identifies the lowest level in any bill of material at which a particular component appears ([APIC16]).

The low-level code is generally calculated by a computer program.

A *product family*, or *product line*, is a group of products having similar features (such as form or material) or similar functions, similar product structure with a high percentage of the same components or components from the same family, and a high percentage of the same processes in the process plan. Cf. [APIC16].

A *variant*, or a *product variant*, is a specific product in a product family.

An *option* is a choice — often mandatory and from a limited selection — that must be made by the customer or company when customizing the end product. Cf. [APIC16].

Commonality is a condition where a given component is used in multiple products. Cf. [APIC16].

A product family is designed as such as early as the product design phase. Throughout its life cycle, it will be expanded where desired. The product structure of each variant is different, but according to its definition, it is based on a high degree of commonality.

1.2.3 Process-Related Business Objects

Intended learning outcomes: Present in detail business objects such as operation, setup and run time. Explain the routing sheet, the critical path, interoperation time and the production lead time. Identify a product module. Describe the production structure and the cumulative lead time. Disclose the process plan and the lead-time offset.

An understanding of the composition of the lead time is fundamental — particularly in terms of *short* lead times. The most detailed business object to examine is the operation. Factors affecting this building block of a business process have a strong influence on lead time.

An *operation* in logistics is a step in a process that is required for the design and manufacturing of a product. Another term used is *routing-sheet position* or *basic manufacturing step*. Examples of operations are "cut," "stamp," or "bend" in industrial areas, or "serve," "maintain," "advise," or "repair" in service industries.

Setup, or *changeover*, is the work required to change or prepare the production infrastructure (machines, tools, and other resources) for the next order.

Operation time is the time required to complete an operation. In the simplest case, operation time is the sum of *setup time*, or *setup lead time*, that is, the time required for setup, and run time for the actual work on the order.

Run time is, in the simplest case, the product of the size of the *lot*, or *batch*, that is, the number of the units of measure produced together, and the *run time per unit*, that is the total treatment time for one unit of the batch.

When the run times are planned as a series after setup time, the simplest formula for operation time is as shown in Figure 1.2.3.1.

Operation time = (setup time) + (lot size) · (run time per unit)

Fig. 1.2.3.1 The simplest formula for operation time.

Operation time can refer to either planned or real manufacturing processes.

Standard time, or *standard hours*, is the length of time that should be required to set up and run an operation. It assumes average efficiency of people and production infrastructure and is also frequently used as a basis for planning and incentive pay systems as well as a basis for allocating overhead costs.

Actual time is the actual length of time for the execution of an operation in a particular order. It is often used as a basis for job-order costing.

The *routing sheet* of a product, also called *route sheet*, *operation sheet*, or sinply *routing* is a complex object; it is a list of the operations required to manufacture a particular item from its components. It includes information on the work centers involved (see the definitions in Section 1.2.4 and also [APIC16]).

The *critical path* is the set of activities, or operations, that defines the (planned) duration of the network of operations. These activities usually have very little slack time, close or equal to zero.

The *production lead time*, or *manufacturing lead time*, is the total time to manufacture an item, exclusive of lower-level purchasing lead time.

Production lead time is measured along the critical path. It is made up of the three following categories of time:

- Operation time, defined as above
- *Interoperation time*, which can occur either before or after an operation and may be wait time, that is, the time a job remains at a work center before or after execution of the operation, or transportation time (move time, or transit time)
- *Administration time*, the time required to release and complete an order

Lead time projected on the basis of these three categories is a probable value only, because it is based on time averages, particularly for interoperation time. Wait times depend on the current situation in production and its physical organization. In typical job shop production (see Section 4.4.3), interoperation time and administration time make up more than 80% of lead time and are thus its main determinants.

A sequence of operations is the simplest and most important order of the operations. A more complex order of the operations makes up a network or repeatedly executed sequences of operations (see Section 14.1.1).

The *production structure* of a product is the combination of its product structure and the routing sheet for the product itself and for its assemblies and its single parts.

Through combining routing sheets with product structure in production structure, we gain a useful rationale for integration into a structure level, and thus for differentiating an intermediate product from a subsequent, higher structure level.

A *production structure level* is determined by the arguments shown in Figure 1.2.3.2.

- The last operation results in a ***product module***, that is, a semifinished good that can be built into various further products as components.
- The last operation results in a semifinished good that is to be stored.
- The operations are required for a particular process technology.
- The last operation results in an intermediate state that is seen as an object or ***entity***, that is, as a self-contained thing or object.

Fig. 1.2.3.2 Useful rationale for combining operations in a product structure level and thus for differentiating an intermediate product.

Within a production structure level, there is no storage. Components needed for this production structure level are drawn from storage or from the immediately preceding production structure level.

The *purchasing lead time* is the total time required to obtain a purchased item. Included here are order preparation and release time; the *supplier lead time* (that is, the amount of time that normally elapses between the time an order is received by a supplier and the time the order is shipped); transportation time; and receiving, inspection, and putting into storage (*put away time*) ([APIC16]).

The *cumulative lead time*, or *critical path lead time*, is the longest planned length of time to accomplish the value-adding activity in question, with respect to the time to deliver to the customer, the lead time for all production structure levels, as well as the purchase lead times.

Depending on the context, *lead time* denotes either the cumulative lead time, the lead time required for one production structure level, or the purchasing lead time.

The *process plan* of a product is the total production structure on the time axis.

The process plan is a very complex business object that shows the cumulative lead time to manufacture a product. Figure 1.2.3.3 serves as an example for a product P.

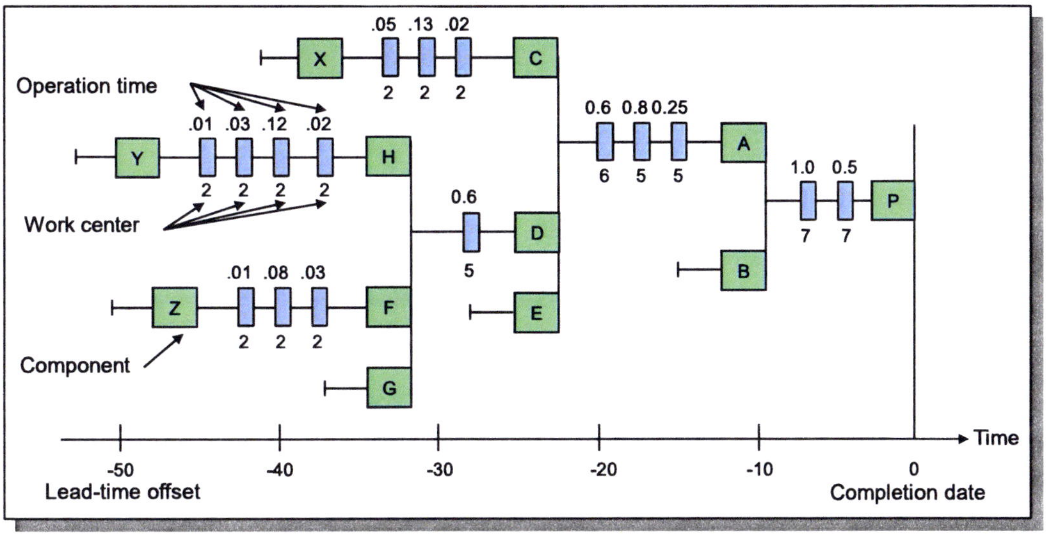

Fig. 1.2.3.3 Process plan for a product P (detailed structure).

The process plan corresponds, as does product structure, to the way that the workers view customer order processing (their *scheme*, or natural conception of the process).

Lead-time offset is the period of time of preponing a resource requirement (component or capacity) relative to the completion date of a product, based on the lead time for that product.

For each component, we can calculate the lead-time offset. To do this, the proportion of lead time must be calculated along the corresponding branch of the process structure. Throughout the total working process time, this time period is dependent on batch size.

1.2.4 Resource-Related Business Objects

Intended learning outcomes: Produce an overview on terms such as employees, facilities, workstations, production equipment, work center. Present terms such work-center capacity, load, work-center load. Differentiate between standard load and actual load. Disclose capacity utilization. Explain the load profile of a work center. Identify work center efficiency and rated capacity. Differentiate between rated capacity and theoretical capacity.

Employees, or *workers*, in an enterprise are all those people involved directly and indirectly in a company's performance.

The term *facilities* refers to the physical production plant, distribution and service centers, offices, R&D laboratories, and related equipment ([APIC16]).

Facility location is the physical location (for example, a region or a city) where the facilities come to be. In the following, we use the abbreviated term, location, synonymously.

The *production infrastructure* comprises all available *production facilities*, that is, factories and their workstations, as well as other production equipment.

A *workstation* is an assigned location where a worker performs the job; it can be a machine or a workbench ([APIC16]).

The *production equipment* includes machines, appliances, *devices* (such as jigs, fixtures), and tools.

Employees and the production infrastructure of an enterprise make up work centers:

A *work center*, or *load center* (or a *machine center*, if consisting mainly of machines), is an organizational unit of production within the chosen organization of the production infrastructure (see Section 4.4.3). It comprises the totality of employees and production infrastructure required to complete a quantity of work considered to be one unit for the purposes of higher-level planning & control. Internal planning & control of a work center is not necessary or takes place autonomously under consideration of the higher-level orders. Cf. [APIC16].

The *capacity* of a work center is its potential to produce output. This potential is always related to a time period. The unit of measure is called the *capacity unit*, and it is mostly a unit of time (hours of work).[4]

Theoretical capacity is the maximum output capacity, determined by the number of shifts, the number of workers or machines, and the theoretically available capacity per shift. Theoretical capacity can vary from week to week due to *foreseen*, overlapping changes, such as vacation time, additional shifts, overtime, or preventive maintenance requirements.

The *capacity profile* of a work center represents its capacity over time. See Figure 1.2.4.1.

Economic use of capacity by workload is fundamental in Integral Logistics Management.

Load is the amount of work planned for or released to a facility, work center, or operation for a specific span of time, measured in capacity units.

To calculate load, we must first — once again, as in Section 1.2.3 — take a closer look at the detailed object *operation*.

---

[4] There are other possible measures that could be used as the unit. See also Section 16.4 on activity-based costing.

Operation load is the work content of the operation, measured in the capacity unit of the work center carrying out the operation.

The terms *setup load*, *run load*, and *run load per unit* are defined in the following analogously to setup time, run time, etc., in Figure 1.2.3.1, with "work content" in the place of "time." The formula for operation load is analogous to the formula for operation time in Figure 1.2.3.1.

Load can refer to either planned or real manufacturing processes.

Standard load is the given, probable content of work.

Actual load is the actual content of work, the use of capacity by the content of work.

Standard load of an operation and actual load of an operation are defined in a similar way. The following definitions are again related to the work center.

Work center load is the sum of the load of all operations for orders processed by the work center.

(Capacity) utilization is a measure of how intensively a resource is being used to produce a good or service. Traditionally, it is the ratio of its actual load to its theoretical capacity.

The *load profile*, or *load projection*, of a work center is a display of work center load and capacity over a given span of time. See Figure 1.2.4.1.

Figure 1.2.4.1 shows a typical picture of a load profile, under the assumption of continuous or rectangular distribution within a time period. Similar to product structure and the process plan, the load profile represents a scheme, or natural conception, from the perspective of the people responsible for the processing of the production order.

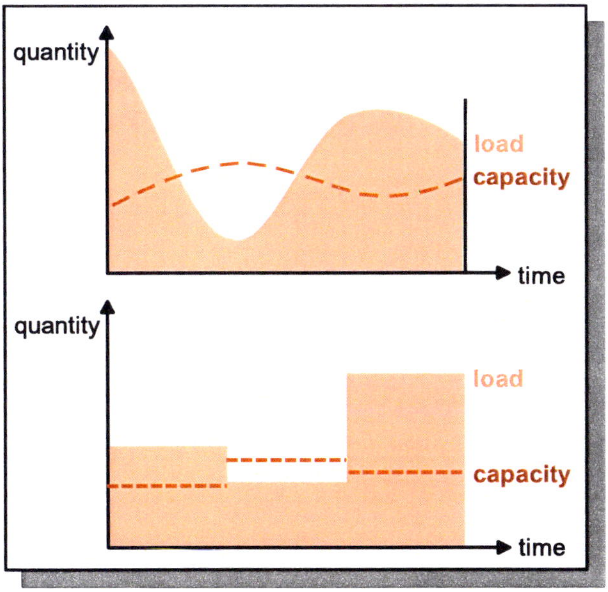

Fig. 1.2.4.1 The load profile of a work center (continuous and rectangular distribution).

It is also a common practice to set the capacity profile at 100%, that is, to make it the horizontal value and to express load as percentages thereof.

The production lead time ignores the actual capacity utilization of the work center, although utilization can strongly influence queue times. Mostly, for long-term planning, the lead time calculation according to Section 1.2.3, which is based on the average duration of operations and interoperation times, is sufficiently accurate. The shorter the planning term, the more important it is to consider capacity utilization when calculating lead time.

For a detailed analysis of the load profile, it is necessary to take a closer look at *capacity*.

Work center efficiency, or the *efficiency rate* of a work center, is a percentage, namely, "standard load divided by actual load" or — equivalently — "actual units produced divided by standard units to produce" (cf. [APIC16]). This is calculated as the average of all operations performed by a work center.

Rated capacity, or *calculated capacity*, is the expected output capability of a work center, that is, theoretical capacity times capacity utilization times work center efficiency.

An example of theoretical capacity and rated capacity, along with detailed explanations of the terms, is shown in Figure 14.1.1.1. The above definitions, however, provide a basis for understanding important aspects for planning & control:

Standard load to be scheduled should always refer to *rated capacity*. To compare capacity with *standard load*, the capacity profile should always show *theoretical capacity multiplied by efficiency*.

1.2.5 Rough-Cut Business Objects

Intended learning outcomes: Identify reasons and principles for defining rough-cut business objects. Disclose the rough-cut product structure and the rough-cut process plan. Explain a way to derive a rough-cut process plan from a detailed process plan. Identify the bill of resource and the product load profile. Describe a way to establish a product load profile from a rough-cut process plan.

To estimate the requirements for goods and capacity *quickly*, planning cannot reach the detail of the precise number of screws or the minutest task. Sometimes only partial data are needed, because

- Only relatively few purchased items, such as raw materials or semifinished goods, are expensive or difficult to procure (have very long procurement lead times).
- For a great percentage of work centers, load is not critical, because for technical reasons, overcapacity is the rule (for example, replacement machines or special machines that are not available with low capacity).
- Various processes are very short and do not affect the total load of a work center.

Furthermore, it can suffice to use item families or product families as the business object rather than individual items or products. In analogous fashion, the following rough-cut business objects may be defined:

Rough-cut product structure is the structured makeup of the product from its components, whereby both product and components may be an item or product family. For convergent product structure (see Section 4.4.2), the term *rough-cut bill of material* is also used.

A *rough-cut work center* is composed of the total of work centers that do not have to be further differentiated by rough-cut planning & control.

A *rough-cut operation* is composed of the total of operations, not further differentiated by rough-cut planning & control.

A *rough-cut routing sheet* for a product (family) is the overall chain of operations, not broken down further by rough-cut planning & control.

The *rough-cut production structure* of a product (family) is the combination of its rough-cut product structure and the rough-cut routing sheets of the product or product family itself, as well as associated (rough-cut) assemblies and single parts.

The *rough-cut process plan* of a product is the rough-cut production structure plotted on the time axis.

One way to derive a rough-cut process plan from a detailed process plan involves three steps:

1. Determine an item's item family. Determine the item families to be included for the rough-cut product structure.
2. Determine the work centers or rough-cut work centers to be included and assign the work centers to the rough-cut work center. Determine a time length for operation time under which a (rough-cut) operation can be omitted in a rough-cut structure. Instead, determine a percentage for the reduction of capacity that will be caused by these short operation times, and use this percentage to take these into account.
3. Determine rough-cut product structure (rough-cut bill of material) and the rough-cut routing sheet for each product or product family, often by contraction of several structure levels into one.

Example: The following list contains measures that allow, departing from the detailed process plan in Figure 1.2.3.3, the formulation of the rough-cut process plan. The numbered steps refer to steps 1 and 2. Figure 1.2.5.1 shows the result.

1a. Purchased components X, Y, and Z form a single item family Y'.

1b. Component E is not included in the rough-cut structure.

1c. Components G and B form the single item family B'.

2a. Work center 6 is not included in the rough-cut structure.

2b. Work centers 5 and 7 join to form a single rough-cut work center 7'.

2c. All operations having an operation time of less than 0.1 hours are not included in the rough-cut structure.

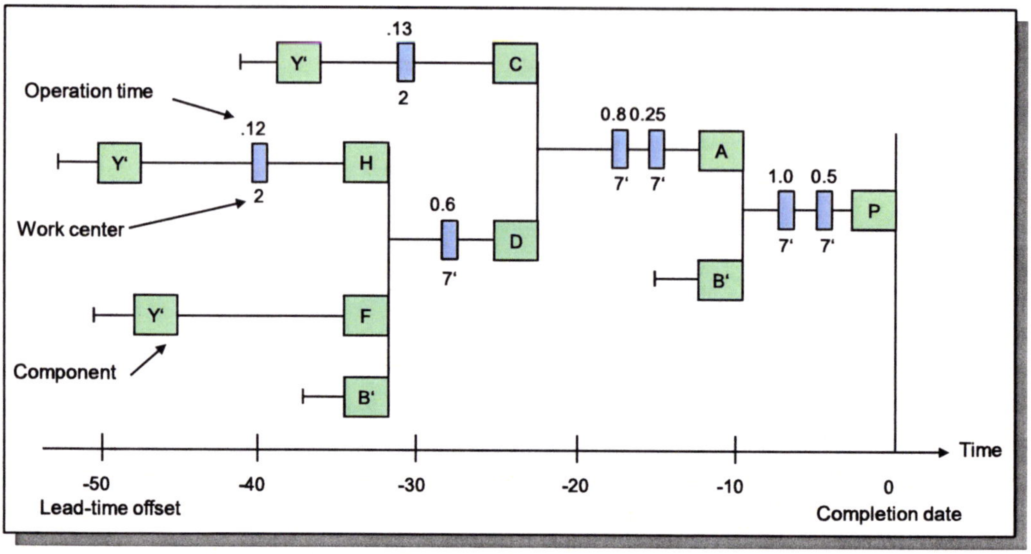

Fig. 1.2.5.1 Rough-cut process plan for product P.

The rough-cut process plan must, of course, include (rough-cut) interoperation times, which are no longer apparent once individual (rough-cut) operations have been excluded. Otherwise, lead-time calculation will be wrong. Every (rough-cut) component gains the correct lead-time offset in relation to the completion date of the product. Setup time and setup load are divided by a norm batch size and added to run time and run load. The lead-time offset then refers to that batch size.

A *bill of resources* is a listing of the required key resources (components and capacities) needed to manufacture one unit of a selected product or product family.

A *product load profile* is a bill of resources where the resource requirements are defined by a lead-time offset.

In general, this is a matter of one-level rough-cut bills of material and/or routing sheets.

Example: Figure 1.2.5.2 shows two variants of a product load profile for the example in Figure 1.2.5.1. Notice the contraction to one structural level. To do this, lead-time offset must be calculated for each operation. The second variant additionally joins together all positions that load the same rough-cut resource within 10 units of time. This further reduces the complexity of the rough-cut business object.

In some cases, rough-cut business objects can be derived from detailed business objects in a direct fashion. *Manual*, synchronous modification of rough-cut and detailed business objects is difficult and expensive. Therefore, rough-cut business objects are often kept general enough that they will not be affected by changes in the detailed business object.

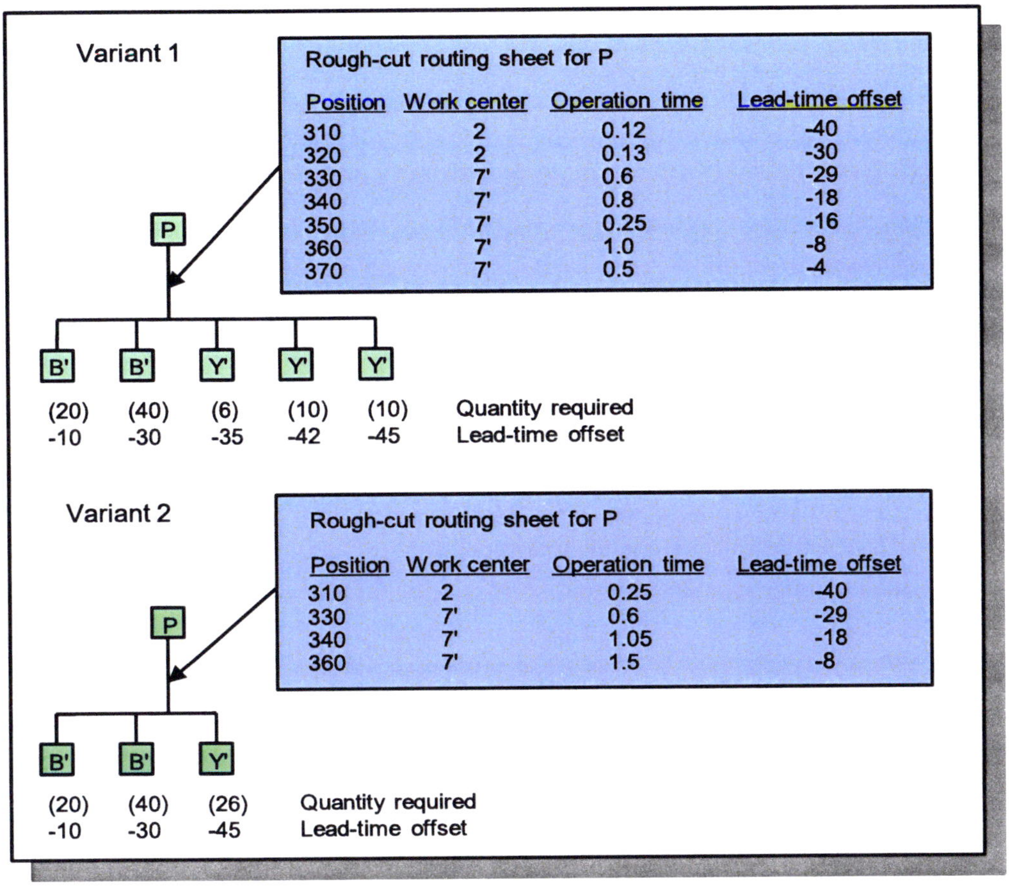

Fig. 1.2.5.2 The product load profile: a one-level rough-cut bill of material and one-level rough-cut routing sheet.

1.3 Strategies in the Entrepreneurial Context

Intended learning outcomes: Differentiate between various entrepreneurial objectives in a company and in a supply chain. Explain resolving conflicting entrepreneurial objectives. Describe the customer order penetration point (OPP) and the coordination with product and process design. Produce an overview on the target area flexibility: investments in enabling organizations, processes, basic technologies, and technologies toward personalized production.

Logistics, operations, and supply chain management can best be understood as management systems for performance, according to Fig. 1 in the introduction. Particular attention will be paid to delivery, that is, high fill rate, delivery reliability, and short lead times. No other management system within and across companies focuses to this extent on delivery. In addition, it is also concerned with how the various stakeholders of the company, in this case especially the business partners, experience the way in which delivery is handled.

1.3.1 Entrepreneurial Objectives in a Company and in a Supply Chain

Intended learning outcomes: Produce an overview on company performance, or supply chain performance. Identify entrepreneurial objectives affected by logistics, operations, and supply chain management in the target areas of quality, costs, delivery, and flexibility (QCDF). Describe target areas in supply chain performance across companies.

Company performance, or *supply chain performance*, comprises the achievement of *entrepreneurial objectives* in the target areas of quality, costs, delivery, and flexibility.[5]

In part, logistics, operations, and supply chain management have a significant influence on entrepreneurial objectives in all the four areas. Figure 1.3.1.1 identifies, both main and partial entrepreneurial objectives in the four areas.

- Target area quality:
  - Main objective: to meet high demands for product quality
  - Main objective: to meet high demands for process quality
  - Main objective: to meet high demands for organization quality
  - Partial objective: high transparency of product, process, and organization
- Target area costs:
  - Main objective: low net working capital (on-hand balance, low in-process or in-transit inventory)
  - Main objective: high capacity utilization
  - Main objective: low cost rates for administration
  - Partial objective: complete and detailed bases for calculation and accounting
- Target area delivery:
  - Main objective: high fill rate (high customer service ratio or short delivery lead time)
  - Main objective: high delivery reliability rate
  - Main objective: short lead times in the flow of goods
  - Partial objective: short lead times in the data and control flow
- Target area flexibility:
  - Main objective: high degree of flexibility to enter as a partner in supply chains
  - Main objective: high degree of flexibility in achieving customer benefit, e.g., by product and process innovation (that is, by innovative power)
  - Main objective: high degree of flexibility in the use of resources

Fig. 1.3.1.1 Entrepreneurial objectives affected by logistics, operations, and supply chain management.

[5] Performance includes more than *productivity* (that is, the actual output of production compared with the actual input of resources), measured by quality and costs, but less than *competitiveness*, which in addition includes the economic environment.

Objectives in the target area flexibility aim to enable performance excellence in the other three target areas. They are called *enablers* or *enabler objectives*; they lay the foundation for achieving *results* or *results objectives* in the target areas of quality and delivery (that is, effectiveness) as well as costs (that is, efficiency) in later business periods. In the ongoing business period, they represent investments and thus cost money and tie up capacities; for this reason, the number of measures to achieve flexibility objectives is limited.

Following [Hieb02], supply chain performance across companies includes, in addition, the achievement of entrepreneurial objectives in the three target areas of supply chain collaboration, coordination, and changeability. Figure 1.3.1.2 lists the objectives within each target area that can all be assigned to the target area of flexibility, as follows:

- Target area supply chain collaboration:
    - Main objective: to achieve a high degree of strategic alignment in the supply chain
    - Main objective: to achieve highly integrated business processes, both in planning and in execution
- Target area supply chain coordination:
    - Main objective: to achieve seamless goods, data, and control flow among the supply chain partners
    - Main objective: to achieve a high degree of information transparency
- Target area supply chain changeability:
    - Main objective: to achieve a high degree of flexibility in (re-)configuration of supply chains for customer responsiveness

Fig. 1.3.1.2 Target areas in supply chain performance *across* companies (according to [Hieb02]).

1.3.2 Resolving Conflicting Entrepreneurial Objectives

Intended learning outcomes: Produce an overview on the supply chain strategy and the business plan. Explain opportunity, opportunity cost, and the potential for conflicting QCDF objectives. Present terms such as return on net assets (RONA), net income, profit after tax, net working capital, and primary entrepreneurial objective. Disclose the objective of a supply chain initiative (SCI).

The relative weighting of target areas and the individual objectives of companies or supply chains are determined by the strategy.

A *company's strategy*, or a *supply chain strategy*, identifies how the company or the supply chain will function in its environment with regard to product line, fill rate, make-or-buy decisions, distribution and supplier channels, people and partnership, organizational development, and financial objectives ([APIC16]).

Strategic planning is the corresponding planning process.

Strategic planning develops the *strategic plan*, the *business plan* for the whole company. The plan reflects top management's view of

- The market and other companies in the market

- Product and service positioning[6] in the market segment
- Competitive advantages and product differentiation[7]
- Order qualifiers and order winners[8]
- The type of production and procurement

The surrounding systems that influence the company's perspective include economic considerations (such as the relationship between supply and demand), probable customer behavior (whether the products will be seen as investment goods or consumer goods, for example), competition, available suppliers, the costs of short- and long-term financing, and expected economic and political trends.

The actual quantitative weighting of these areas and objectives represents a challenge to the company. Objectives are not readily comparable. One method of comparison is to translate objectives outside of the area of costs into monetary values.

Opportuneness is the suitability of an action in a particular situation. *Opportunity cost* is defined by [APIC16] as the return on capital that could have resulted had the capital been used for some purpose other than its present use.

Opportunity costs arise when for some reason customer demand cannot be fulfilled. In this case, the invested capital is used for something other than the gain that would have been made through meeting customer demand. Such costs result if entrepreneurial objectives with regard to concrete demand have not been weighted appropriately.

As an example of translating non-cost objectives into monetary values in order to determine the opportunity cost, let us take the objective of high fill rate. What does it cost to be unable to deliver? There can be loss of:

1. The non-deliverable order item.
2. The complete order, even though other items can be delivered.
3. The customer, even if other orders can be filled.
4. All customers, due to the company's resulting poor reputation.

This example shows how difficult it is to determine opportunity cost. Translating other non-cost objectives into monetary values is just as complex. Thus, the weighting of objectives is

---

6 *Product* or *service positioning* is the marketing effort involved in placing a product or a service in a market for a particular niche or function ([APIC16]).

7 A *competitive advantage* is an edge, i.e., a process, patent, management philosophy, or system that enables the company to have a larger market share or profit than it would have without that advantage. *Product differentiation* is a strategy of making a product different — the best or unique — from the competition with regard to at least one feature or goal of a target area (see [APIC16]).

8 *Order qualifiers* are those competitive characteristics that a firm must exhibit to be a viable competitor in the marketplace. This means that the firm has to be within a certain range in all four target areas, even when it may be leading in a specific target area. *Order winners* are those competitive characteristics that cause a firm's customer to choose that firm's goods and services. They can be considered as competitive advantages and focus on one, rarely two, of the four target areas. *Order losers* are capabilities in which poor performance can cause loss of business (see [APIC16]).

unquestionably an entrepreneurial matter that must be conducted within the framework of the normative and strategic orientation of the enterprise.

In contrast to objectives in the areas of costs and delivery, logistics, operations, and supply chain management have only a limited influence on the achievement of objectives with regard to quality and flexibility.

- Target area quality: Clearly, product and process design as well as the choice of production infrastructure, employees, and the supply chain community are the main determinants of the quality of products, processes, and the organization.
- Target area flexibility: Flexibility to enter as a partner in supply chains is first of all a question of the culture of an enterprise. The potential for flexibility in achieving customer value develops through product and process design and the choice of production infrastructure. Flexibility in the use of resources is determined initially by the qualifications of personnel and by the choice of product infrastructure.

Section 1.7.1 contains a scenario on possible improvements in achieving entrepreneurial objectives in the different areas.

Some possible strategies, shown as example profiles in Figure 1.3.2.1, illustrate that the four target areas result in a potential for conflicts among objectives. There are even conflicts within the area of costs itself: As we will show later, reduction of inventory with a simultaneous increase of capacity utilization can result in goal conflict.

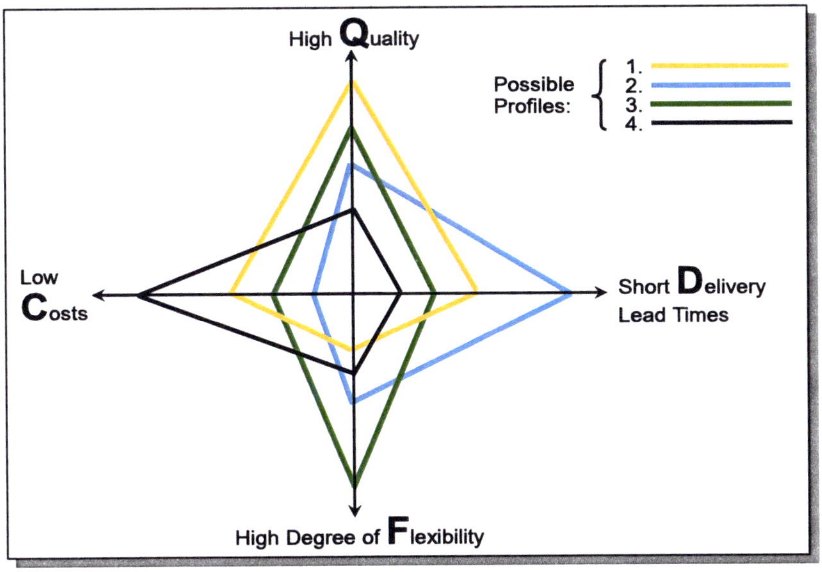

Fig. 1.3.2.1 Potential for conflicting entrepreneurial objectives.

1. High quality of product or process tends to result in high costs and long lead times. There is also a tendency toward repeatable processes, and thus to a low degree of flexibility.

2. The shorter the delivery lead times, the higher the costs: To achieve short delivery lead times, stock or overcapacity is a must. Short lead times can result in cost of poor quality and reductions in flexibility (for example, reduced product variety).
3. A high degree of flexibility in achieving customer value, through product variety, for example, leads either to long delivery lead times (as little inventory can be stocked) or to high costs due to unusable inventory of product variants.
4. Low costs, due to high capacity utilization and simultaneous avoidance of stock, result in long delivery lead times, cost of poor quality, and reductions in flexibility in the range of goods.

Determining opportunity cost implicitly determines the relationship between entrepreneurial objectives in the four areas in Figure 1.3.1.1 and the primary entrepreneurial objective.

The *primary entrepreneurial objective*, the achievement of which logistics, operations, and supply chain management are intended to support, is maximization of return on net assets (RONA).

Return on net assets (RONA) is defined as (net income) / (fixed assets + net working capital). Note: only fixed *operating* assets should be included in the calculation. *Operating assets* are the resources owned by a company for productive purposes. [9]

Net income is the profit that a firm has after subtracting costs and expenses from the total revenue. *Profit after tax* is also called *bottom line.*[10]

Net working capital is defined as current assets minus current liabilities, that is, the capital that is locked up in the "short-term" operating business. It is the fraction of the operating assets that can be converted into cash within the normal business cycle, yet is secured by long-term liabilities or equity.[11]

A particular goal in any of the four areas does not always support the primary entrepreneurial objective. This is true not only for daily decisions but also for supply chain initiatives.

A *supply chain initiative* (SCI) is a project or an investment aiming to improve supply chain performance.

For example, if investments to reduce lead time do not result in increased demand or a larger market share, then return decreases rather than increases (see Section 1.7.2). Section 1.7.3 presents a method for assessment of the economic value (added) of SCIs.

---

9 For definitions of basic terms in finance (not provided here) such as *balance sheet*, *income statement*, *profit*, *profit margin*, *gross margin*, *cash flow*, *current* and *fixed assets*, *current* and *fixed liabilities*, and *equity*, see, for example, [APIC16].

10 For companies with no debt and thus no interest expense, *net operational profit after tax* (NOPAT) is equal to profit after tax.

11 Current liabilities are for the most part accounts payable. Strictly speaking, current assets should not include advance payments or any liquid assets that are not necessary for the operating business.

1.3.3 Customer Order Penetration Point (OPP) and Coordination with Product and Process Design

Intended learning outcomes: Explain the determination of the (customer) order penetration point, or customer order decoupling point, and the stocking level. Describe the modular product concept, the concepts of customization and late customization as well as the postponement approach.

Resolving the problem of the conflicting objectives of "high fill rate" versus "low costs for inventory on stock and in process" is one of the main tasks of logistics, operations, and supply chain management. This problem is equivalent to the basic problem of temporal synchronization between supply and demand described in Section 1.1.6.

The *(customer) order penetration point (OPP)* is a key variable in a logistics configuration; it is the point in time at which a product becomes earmarked for a particular customer. Downstream from this point, the system is driven by customer orders; upstream processes are driven by forecasts and plans ([APIC16]).

Figure 1.3.3.1 shows the situation applied to the process plan of Figure 1.2.3.3, reduced to the product structure in the time axis. The issue here is the relation of customer tolerance time, which is a result of the market situation, to cumulative lead time.

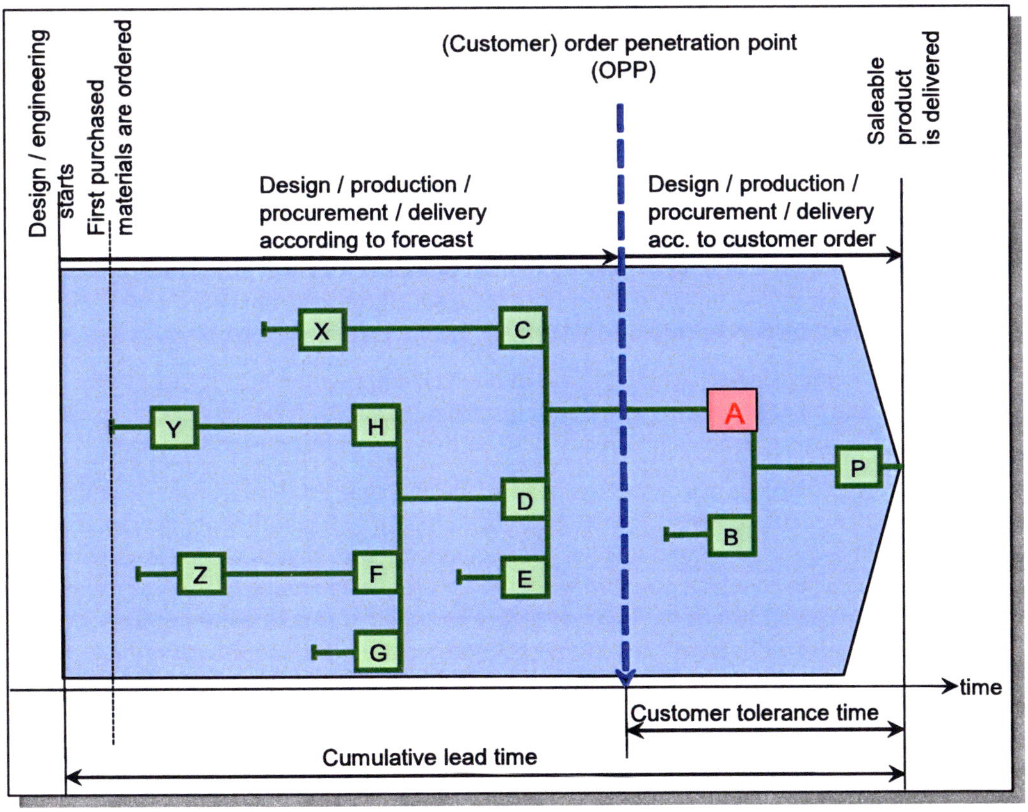

Fig. 1.3.3.1 The (customer) order penetration point.

If the customer tolerance time is at least as long as the cumulative lead time, the product can be engineered, procured, produced, or delivered when actual demand in the form of a customer order is placed. Otherwise, all goods (such as semifinished goods, single parts, raw materials, and information) from which the end product cannot be manufactured and delivered within the customer tolerance time must be ordered on the basis of forecast. If the customer tolerance time is zero, even the end product must be ordered and, if need be, stocked before demand is known. The OPP corresponds to the level in the product structure where a buffer of inventory should be located.

The *stocking level* defines that level in the bill of material above which a product can be engineered, procured, produced, or delivered within the customer tolerance time. For goods below and at the stocking level, no exact demand is known. Demand forecast is required.

Therefore, the OPP entails decoupling points, each with its decoupling inventory. In general, decoupling points in the product structure are items downstream from or at the OPP with at least one direct component upstream from the OPP. In the example in Figure 1.3.3.1, item A (shaded in red) is the only decoupling point. It is downstream from the OPP, whereas its direct components C, D, and E are upstream from the OPP. The necessary inventory at the stocking level (in the example, of item A) is determined by estimation of the opportunity costs according to the required fill rate.

In general, the *customer order decoupling point (CODP)* is used synonymously to the OPP.

However, as each decoupling point is linked with inventory (see the definition in Section 1.1.6), it is identical to the stocking level. An equalization of the CODP with the OPP only applies if the stocking level is located exactly at the OPP.

Figure 1.3.3.1 already points in the direction of SCIs. Many SCIs require close coordination with product and process design. Here some examples:

- Reduction of cumulative lead time at all levels. In this way, the stocking level can be set lower and thus carrying cost reduced. Short lead times are a topic in lean-/just-in-time concepts (lean/JIT), which are covered in Chapter 6.
- A *modular product concept* is based on standardizing the components and operations and on commonality and building product families. Product variants are decided on the basis of a concept already defined in marketing and product design.
- *Customization* is a concept in product design that produces products tailored to customer requirements. In late customization, product modules are characterized by commonality up to high level of product structure, so that the many variants can be produced as much as possible within the customer tolerance time. This reduces the supply chain risks of too high inventory and shelf warmers upstream from or at the (customer) OPP. Here, the number of significantly different process variants should be kept small. On customization, see the variant-oriented concept in Chapter 7.
- *Postponement* is an approach in product design. It shifts product differentiation closer to the customer by postponing identity changes, such as assembly (*finish-to-order*) or packaging (*package to order*), to the last possible supply chain position ([APIC16]). Postponement makes sense, if, for example, longer transport distances

from the producer to the customer can be resolved through lower-value semifinished items. Efficient postponement can also support the ability for late customization. See [SwLe03]. For postponement in pharmaceutical supply chains, see [Verh12].

1.3.4 Target Area Flexibility: Investments in Enabling Organizations, Processes, and Basic Technologies

Intended learning outcomes: Present flexibility, agility and different aspects thereof. Differentiate between short, medium and long-term potentials developed by agile companies. Describe various enabling information technologies allowing worldwide running order coordination.

Flexibility is the capability to adapt to new, different, or changing requirements [MeWe18].

This general term is variously defined and used in the literature on logistics, operations, and supply chain management. In any case, however, it refers to making appropriate investments to proactively build up certain potentials for better fulfillment of objectives in the target areas of quality, costs, and delivery in future and thus to achieve future benefits. Figure 1.3.1.1 shows three specific kinds of flexibility that have proved to be important in this target area.

[APIC16] distinguishes six categories of flexibility: mix flexibility, design changeover flexibility, modification flexibility, volume flexibility, rerouting flexibility, and material flexibility. [Gots06] provides a comprehensive list of kinds of flexibility. [WiMa07] introduces *changeability* as a general concept; other concepts, including *transformability* and flexibility concepts, fall under the general concept as specializations. In Figure 1.3.4.1 specializations of changeability are assigned to the various levels of the product and productions systems within a factory and its surrounding field.

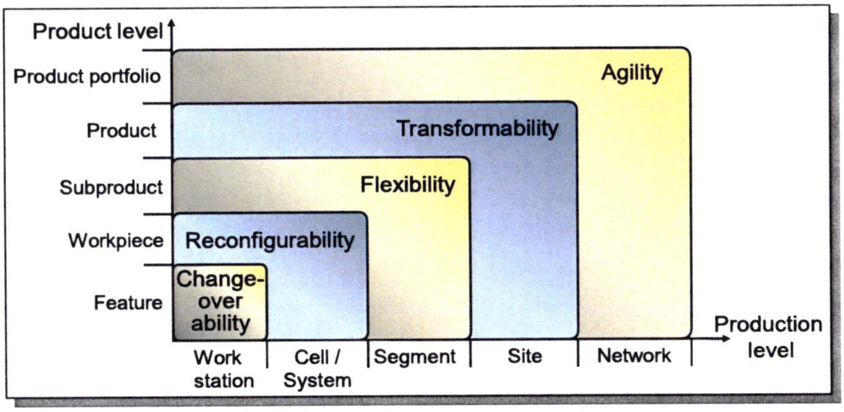

Fig. 1.3.4.1 Classes of changeability within a factory and in its surrounding field (according to [WiMa07]).

Aspects of flexibility that describe the company, network, or supply chain as a whole are adjectives like lean, agile, adaptable, changeable, resilient (see [Shef07]). The examples that are given for these terms are all quite similar and are now discussed using the term agility.

Agility is defined in [MeWel8] as ability to move; dexterity. *Agile manufacturing* refers to the building up of potentials or scope (or "play") in the right place at the right time and in the right amount. *Agile companies* are companies that apply the principles of agile manufacturing to all areas in their organizations.

Increasingly, the buyer's market demands personalized production even of consumer goods. The product is ever more frequently defined in direct interaction with the customer. At the same time, clearly identifiable market segments disappear. Brand names serve increasingly to reflect the personality of the customer instead of providing a function as they used to do.

Agile competitors [PrGo97] understand how to remain competitive by means of *proactive* amassing of information, knowledge, know-how, and competency.

Agile companies develop potentials and play or slack in the supply chain that the end customer might not pay at the moment. In the long term, though, these potentials allow opportunities to be seized or supply chain risks (see Section 2.4) to be reduced. Examples:

- Short term: Formation of parallel management of orders and running order coordination in the supply chain linked by pull logistics (see Section 4.2.1); establishment of overlapping activities in part-processes to coordinate push logistics (see Section 4.2.2). Each of these makes possible rapid business processes of high quality.
- Medium to long term: Building a staff capable of qualitatively flexible work assignments by means of training qualifications and coordination in groups; setting up a production infrastructure that can be implemented flexibly. Both measures lead to flexibility in the use of resources.
- Medium term: Building overcapacity or quantitatively flexible capacity and/or inventories, allowing a response to unplanned demand or shifts in demand with short delivery lead times. In the production of capital goods, capacity measures that reduce ordering deadlines take precedence.
- Medium term: Development of competency in proactive service. The maker gathers "life data" on the product. By evaluating this data, he is able to recognize changes in customer demands. The maker can proactively offer an upgrade or new product, before the customer is aware of need. In this way, the maker sells to the customer a solution rather than a single product. The customer feels cared for (total care).
- Long term: Development of know-how and methods to develop and manufacture products in manifold variants. This knowledge allows flexibility in achieving customer benefit. The maker is able to give positive answers to customer requests at crucial moments, which increases bid proposal and order success rate.
- *Long term*: Developing know-how for supply chain changeability, e.g., know-how in reconfiguring the supply chain. This knowledge also comprises the flexibility to enter as a partner in supply chains. According to the product, departments take on a new structure and cooperate with other organizations. Thus, for example, a company can become a partner of a virtual enterprise at the right moment.

Automation with broad implementation of enabling information technologies allowing worldwide running order coordination supports agility. Here are some examples:

- *Protocols*, that is, a set of rules for defining formats for data transfer: In the manufacture of cars or airplanes, the partners represent their business objects in a standardized form. They developed special standards of EDI,[12] among others, IGES (later STEP, for engineering purposes), and EDIFACT[13] with its variations (Odette in the automotive industry, for example) for order processing. There are also various Internet protocols, such as VoIP (Voice over Internet Protocol) or file transfer protocols (FTPs) and XML. The electronic signature also belongs to this domain.
- *E-commerce* to conduct business via electronic transfer of data and documents: e-mail, internet, company-internal intranets, transcorporate extranets, and electronic communities, that is, communities of people who communicate only electronically.
- *Business-to-business commerce (B2B) / business-to-consumer sales (B2C): business via the Internet between companies or with consumers. For example,* in jeans or shoe stores, customer measurements are taken and transmitted directly to production. Later, the customer receives the finished, made-to-order product. Or, customers may configure specific insurance policies via the Internet. They type in the desired parameters, which they may vary according to need. See here also [MaSc04].
- *Automatic identification and data capture (AIDC) is a set of technologies that collect data about* objects and send these data to a computer without human intervention. Examples are the RFID technique, bar code scanners, badges. See Section 15.3.4.
- *Tracking and tracing*: Transport companies supply the customer with information on the location of their deliveries via the internet and the World Wide Web. This is enabled through self-identification of goods by means of attached transponders, e.g. an RFID sensor.
- *ERP and SCM software* are used to describe software for information systems within and across companies. For details, see Section 9.2. In the example of proactive service mentioned above, automobile manufacturers have access to the product and service data bank of their customers via service centers.
- *Supply chain event management (SCEM)*: Suitable SCEM software allows the users of an application to mark the occurrence of events in the supply chain and to create event alerts for notification or action in other applications. In this way, business processes such as procurement, production, and delivery can be monitored. SCEM can then also be used for business intelligence applications, for example when unexpected events occur. Another utilization is the direct transmission of order data to control processes (such as in machines).

---

12 *Electronic data interchange (EDI)* is a term for the transmission of trading documents, such as purchase orders, shipment notices, and invoices, via telecommunications.

13 *EDIFACT* (electronic data interchange for administration, commerce, and transport) is a set of United Nations rules for EDI. The data are standardized in such a way that all companies involved in the order exchange can process them in the same format.

1.3.5 Enabling Technologies Toward Personalized Production

Intended learning outcomes: Produce an overview on terms such as cyber-physical system, Industry 4.0, smart sensor, internet of things (IoT), big data. Disclose the potentials of additive manufacturing (3D printing) and personalized medication.

The ideas outlined below show how, in the future, information technologies (IT) will play a substantially more significant role in the production of physical goods than they do today.

In a *cyber-physical system*, IT devices that control physical objects (e.g., mechanical and electronic objects) work together over a communication network.

In industry, the trend is increasingly towards a complete network that covers all of the relevant machines, both within a company and across companies, and regardless of the machine's manufacturer. The digital components will allow automated production to adapt increasingly quickly to changing requirements. In the USA, a first step in this direction was taken in 2014 when the *Industrial Internet Consortium* (IIC) was founded. Many large companies work together in this organization to draw up common standards.

The term *Industry 4.0* was coined by Acatech (German Academy of Science and Engineering) in 2011. It postulates a fourth industrial revolution, following on from the previous revolutions of mechanization, electrification, and computerization.

This digital revolution should replace many production technologies that have been used until now, and should do so in a disruptive manner, as happened within a very short time when digital photography came in. At the same time, individualization of products to customers' requirements will become more widespread, without significantly increasing costs. The building blocks that will enable this to happen include intelligent sensors, the Internet of Things, big data, and additive manufacturing. These are described below.

In practice, overall development tends to be a continuous progression. But companies that are built around specific analogue technologies, such as Kodak or the Swiss firm Gretag in the analogue photography sector, are exposed to substantial risk. The same would also apply, for example, to companies that currently offer product lines for analogue telephony, which will be replaced by VoIP relatively quickly. Another sector is offset printing, which is constantly losing ground to digital printing, where many of the steps relating to setting the copy (i.e., the printing plate) are no longer necessary. That significantly shortens the production process, and it is possible to print something different on each sheet in a short space of time (the keyword being "personalized printing").

Apart from simply measuring things (as a conventional sensor would), a *Smart Sensor* can also process the measured data and make the results available in the required form.

The "intelligence" is provided by a microprocessor, a result of microsystem and nanosystem technologies. Here again, the driving force is the need for personalized production. Examples include accelerometers, motion sensors, and magnetic field sensors for functional movement therapy. The use of sensor technology helps improve accuracy in the observation of the patients' movements. The data collection results can be processed by the sensor, and immediately converted to movement objectives that are tailored for the patient and which can be displayed on a mirror, for example.

The *Internet of Things* is a network of tangible or intangible goods or objects ("things") that are connected to each other and that can exchange data. An integrated computer identifies each "Thing" and can communicate via the Internet infrastructure.

As a sensor, or with the help of a sensor, the "Thing" can capture useful data, which it can then independently send on to other objects, either humans or machines. One example would be Internet-based building management systems (e.g., light, temperature, and humidity) that are adapted to each resident. The postulate is that the Internet of Things also allows large volumes of data to be gathered from remote locations, which in turn enables big data.

Big data is a broad term for data sets that are so large or complex that traditional data processing applications are unable to capture, store, and process them.

The term is therefore relative to technologies that are currently used, and like several other terms in this article, it appears to be more of a postulate than a reality at the moment. It is often applied to "advanced" methods of evaluating data with the aim of improving decision making, which in turn reduces risk or increases efficiency. The idea is that the large volumes of data will allow statistical analysis to identify previously unknown correlations and trends, both in physical systems (e.g., meteorology, environmental research) and in socio-technical systems (e.g., businesses, government). Data protection is an important issue here.

Additive manufacturing is nowadays more widely known as *3D printing* (although, in doing so, there is actually no printing involved). It is a process that offers the possibility of creating three-dimensional objects. A 3D model created by CAD software can be used to produce an item by building up successive layers of a material (plastic or metal).

One obvious benefit of this process lies in the efficiency and speed with which the first item of a production batch can be produced: The slow, expensive mould-making process is no longer needed. That means this process is particularly well-suited to making prototypes. Conventional (e.g., abrasive) methods may still be more cost-effective for mass production, even assuming that they are not a necessity for quality reasons. A second benefit is that lot size one (batch size one) can be produced cheaply. Totally different 3D shapes can even be produced in a container, i.e., in a single production batch. This makes the process especially attractive for spare parts. Again, the process has to provide sufficiently high quality. Identifying the optimum arrangement needs an algorithm, in much the same way as a cutting optimizer is needed for 2D cutting, such as when using a laser cutting machine to trim sheet metal. For further information, see www.additively.com. The use of 3D printing to make toys in the private sector underlines 3D printing's potential for personalized production.

Personalized medication is a patient-focused approach that incorporates both medication and the dispensing process.

This is an important concept for health care of the future. The processes rely heavily on the use of various information technologies. In the case of solid forms (e.g., tablets), this can involve automatically dividing the blister packs up into individual doses and packing the individual doses according the patient's prescription for delivery to the patient at the appropriate time, with integrated tracking and tracing. For liquids, it involves the automatic production, tailored to the patient, of liquid medicines (e.g., for cancer treatment), again organized by date and time for delivery to the patient and, if necessary, with an appropriate *cold chain*, i.e., an end-to-end cooling system for transport from the producer to the user. Last but not least: 3D printing could in the future be used to produce biomedical fibre, which could enable dispensing devices for medication to be tailored very specifically to the patient.

1.4 Performance Measurement

Intended learning outcomes: Present the basics of the measurement, meaning, and practical applicability of logistics performance indicators. Describe performance indicators in the target areas of quality, costs, delivery, and flexibility. Produce an overview on performance indicators of the primary entrepreneurial objective.

A *performance indicator* or *performance criterion* is the specific characteristic to be measured for estimating performance.

A *performance measurement system* collects, measures, and compares a measure to a standard for a specific performance indicator.

A *performance measure*, or *performance measurement*, is the actual value measured for the indicator ([APIC16]).

Appropriate performance indicators are meant to show the degree to which entrepreneurial objectives (see Figure 1.3.1.1) are fulfilled or not fulfilled.

Logistics performance indicators analyze the effect of logistics on entrepreneurial objectives in the four target areas of quality, costs, delivery, and flexibility.

Descriptions of logistics performance indicators can be found in [OdLa93] or [Foga09], Chapter 5, and in the following. Whenever possible, a logistics performance indicator will give direct indication of fulfillment of one of the individual objectives within a target area. A performance indicator relates to a logistics object and thus becomes an attribute of that object — and sometimes it becomes a logistics object in its own right.

Global measures measure the overall performance of a company or supply chain (such as cash flow, throughput, utilization, inventory).

Local measures relate to a single resource or process and usually have a small influence on global measures (i.e., volume discount on an item, lead time for stock entries, utilization of a storage location).

In the following, we introduce a balanced set of *global* measures *from a logistics perspective*. This balance is one of the requirements of the *balanced scorecard*, an approach that pointed out the one-sidedness of performance indicators in the financial sector, which (too) often only refer to the company's *primary* entrepreneurial objective of return (see [KaNo92] and

Section 1.3.2). A systematical derivation of the balanced set of performance indicators from the company's strategy can be found in [Schn07]. Together with indicators from other areas of the firm, such as finance, marketing, and R&D, the logistics indicators form a complete set of measurements of performance, and thus a basis for improving company performance.

1.4.1 The Basics of the Measurement, Meaning, and Practical Applicability of Logistics Performance Indicators

Intended learning outcomes: Identify central problems in terms of the meaning and applicability of performance indicators in practice.

In actual practice, the *measuring of logistics performance* varies in difficulty and usually requires that certain aspects be counted. With the exception of local measures, it is generally not possible to assess these aspects without expending a lot of time and energy. In addition, integrating and compressing the local measures into global measures, covering several levels, for example, can be very problematic.

The following sums up central problems in terms of the meaning and practical applicability of performance indicators in the form of practical methods. The problems are typical of any quality measurement system and, in part, costing systems as well.

- General performance indicators: Simple, measurable performance indicators are often so general and qualitative in meaning that no practical steps can be derived from them without making additional, nonquantitative, and implicit assumptions. An example of such a performance indicator is customer satisfaction.
- Lack of comprehensive measurement methods: Simple, applicable performance indicators often cannot be measured directly. They require various, at times complicated or inexact measurements that are combined with nonmeasured, implicit methods to yield the desired indicator. A good example is flexibility potential.
- Distortion of the processes: Each measurement affects the process being measured. The disturbance can be so great that the process would behave differently under nonmeasurement conditions.
- Meaning of the performance indicators: The absolute value of a performance indicator has little meaning as such. Only repeated comparison of measurements of the same performance indicator over time can make the performance indicator an instrument of continual process improvement (CPI).
- Comparability of performance indicators: Benchmarking with other companies, in the same supply chain, has meaning only if the competitor has used the same bases of measurement. In practice, it is common to find that companies use different *reference objects*, the objects to which certain performance indicators refer. An example is *fill rate* or *customer service ratio* (see Section 1.4.4). Fill rate can refer to either order positions or items; its measurement can be based on quantity units or value units. Before making comparisons, therefore, it is essential to know how another enterprise defines the performance indicator.

It makes sense to weigh the value of the potential application of the measurement against the time and effort required by the measurement. In practice, a few simply measured performance indicators have proven worthwhile. Employees must then apply the measurement using a multitude of means that cannot be directly derived from the measurement.

1.4.2 Performance Indicators in the Target Area of Quality

Intended learning outcomes: Differentiate between the indicators scrap factor and yield factor. Present the indicator complaint rate.

Performance of logistics, operations, and supply chain management can have an impact, although a rather low impact, on indicators in the target area of quality; for example, on scrap factors and complaint rates of all kinds. The causes can be many and difficult to pinpoint. They may even be insufficient in quality of information.

Complaint rate and scrap factor in Figures 1.4.2.1 and 1.4.2.2 are related. The source of a complaint may turn out to be a cause that, if discovered sooner, would have entailed scrap. Scrap can lead to supplier complaints. The yield factor is complementary to the scrap factor. Hence, for a given reference object, the scrap factor plus the yield factor is equal to 1.

Indicator	Scrap factor (or yield factor)
Definition	Number of rejected (or accepted) facts divided by number of facts
Why measure?	A high scrap factor indicates insufficient quality and leads to opportunity cost
Reference object	(a) process, (b) components, (c) part logistics (e.g., production)
Fact to measure	For (a): item demand or order position For (b) and (c): order position or order

Fig. 1.4.2.1 The indicators *scrap factor* and *yield factor*.

Indicator	Complaint rate
Definition	Number of rejected facts divided by number of facts
Why measure?	A high complaint rate indicates insufficient quality and leads to opportunity cost
Reference object	(a) item, (b) business partner, (c) part logistics (such as sales)
Fact to measure	For (a): item demand or order position For (b) and (c): order position or order

Fig. 1.4.2.2 The indicator *complaint rate.*

1.4.3 Performance Indicators in the Target Area of Costs

Intended learning outcomes: Present the performance indicators stock-inventory turnover, work-in-process-inventory turnover, work center efficiency, capacity utilization, and administration cost rate.

The influence of logistics, operations, and supply chain management in the target area of costs is significant. The performance indicators in Figure 1.4.3.1, Figure 1.4.3.2, Figure

1.4.3.3, and Figure 1.4.3.4 are direct measures of the target objectives involved. For terms, definitions, and arguments, see Sections 1.2.1, 1.2.3, and 1.2.4.

Indicator	Stock-inventory turnover
Definition	Annual cost of inventory issues divided by average inventory
Why measure?	Carrying cost increases as average inventory increases or stock-inventory turnover decreases
Reference object	(a) item and item group, (b) time period
Fact to measure	Annual inventory issues and average inventory (e.g., based on standard cost)

Fig. 1.4.3.1 The performance indicator *stock-inventory turnover*.

Indicator	Work-in-process-inventory turnover
Definition	Sales divided by average work in process
Why measure?	Production infrastructure costs increase for high-level work in process and low work-in-process-inventory turnover
Reference object	(a) work center, (b) time period, (c) combination of the two
Fact to measure	Sales and work in process (e.g., based on cost price)

Fig. 1.4.3.2 The performance indicator *work-in-process-inventory turnover*.

Indicator	Work center efficiency
Definition	Standard load divided by actual load = actual units produced divided by standard units to produce
Why measure?	High work-center efficiency leads to lower costs through better use of investment costs
Reference object	(a) work center, (b) time period, (c) combination of the two
Fact to measure	Production orders load (planned / actual, for setup and run)

Fig. 1.4.3.3 The performance indicator *work center efficiency*.

Indicator	Capacity utilization
Definition	Actual load divided by theoretical capacity (= standard load divided by efficiency rate divided by theoretical capacity)
Why measure?	High capacity utilization tends to result in lower costs through better use of investment costs
Reference object	(a) work center, (b) time period, (c) combination of the two
Fact to measure	Load by production orders (planned / actual, setup and run), work center capacity

Fig. 1.4.3.4 The performance indicator *capacity utilization*.

Further performance indicators relate to administration costs for purchase, sales, work preparation, and so on. They are all of the type in Figure 1.4.3.5.

Indicator	Administration cost rate (such as purchasing)
Definition	Costs of administration divided by turnover
Why measure?	Administration costs should be kept as low as possible
Reference object	(a) organizational unit, (b) time period
Fact to measure	Sales of the organizational unit, costs of the organizational unit for administration

Fig. 1.4.3.5 The performance indicator *administration cost rate.*

1.4.4 Performance Indicators in the Target Area of Delivery

Intended learning outcomes: Present performance indicator fill rate, also called customer service ratio, or OTIF (on-time and in-full). Identify the performance indicators delivery reliability rate, batch size or lot size, capacity utilization, value-added rate of lead time, variance in work content response time and order confirmation time.

Logistics, operations, and supply chain management have a direct effect on the target area of delivery. The performance indicators in Figure 1.4.4.1 and Figure 1.4.4.2 are direct measures of objectives.

Indicator	Fill rate or customer service ratio
Definition	Number of facts delivered in full on desired date divided by number of facts ordered
Why measure?	Poor fill rate entails opportunity cost and, depending on contract, penalty costs
Reference object	(a) item, (b) business partner, (c) part logistics (e.g., sales)
Fact to measure	For (a): item demand or order position. For (b) and (c): order position or order

Fig. 1.4.4.1 The performance indicator *fill rate, customer service ratio*, or *OTIF* (*on-time and in-full*)

Indicator	Delivery reliability rate
Definition	Number of facts delivered in full on confirmed delivery date divided by number of facts confirmed
Why measure?	Poor delivery reliability rate entails opportunity cost and, depending on contract, penalty costs
Reference object	(a) item, (b) business partner, (c) part logistics (e.g., sales)
Fact to measure	For (a): item demand or order position. For (b) and (c): order position or order

Fig. 1.4.4.2 The performance indicator *delivery reliability rate.*

The performance indicators in Figure 1.4.4.3 through Figure 1.4.4.8 are connected with lead time. For terms, definitions, and arguments, see Sections 1.2.1, 1.2.3, and 1.2.4.

Indicator	Lot size (batch size)
Definition	Average order quantity
Why measure?	Large batch size may result in longer lead time
Reference object	(a) process, (b) product
Fact to measure	Order quantity of the order position

Fig. 1.4.4.3 The performance indicator *batch size* or *lot size.*

Indicator	Capacity utilization
Definition	Actual load divided by theoretical capacity (= standard load divided by efficiency rate divided by theoretical capacity)
Why measure?	High loading of the work center may result in longer queue time
Reference object	(a) work center, (b) time period, (c) combination of the two
Fact to measure	Load by production orders (planned and actual, for setup and run), work center capacity

Fig. 1.4.4.4 The performance indicator *capacity utilization.*

Indicator	Value-added rate of lead time
Definition	Value-added part of lead time divided by lead time
Why measure?	Non-value-added parts of lead time should be reduced
Reference object	(a) process and product, (b) business partner, (c) part logistics (e.g., production)
Fact to measure	Value-added (e.g., operation time) and non-value-added parts (e.g., interoperation times, administration time) of lead time

Fig. 1.4.4.5 The performance indicator *value-added rate of lead time.*

Indicator	Variance in work content
Definition	Standard deviation of operation times
Why measure?	A high degree of variance in work content may result in longer queue time
Reference object	(a) work center, (b) time period, (c) product, (d) order
Fact to measure	Actual operation time for a reference object or combination of reference objects

Fig. 1.4.4.6 The performance indicator *variance in work content.*

And, finally, there are two performance indicators in Figure 1.4.4.7 and Figure 1.4.4.8 for data and control flow.

Indicator	Response time
Definition	Time from order entry up to temporary order confirmation divided by total lead time
Why measure?	Long response time results in long lead time, but also in opportunity cost
Reference object	(a) order, (b) business partner, (c) part logistics (e.g., sales)
Fact to measure	Time from order entry up to temporary order confirmation

Fig. 1.4.4.7 The performance indicator *response time.*

Indicator	Order confirmation time
Definition	Time from temporary order confirmation up to order confirmation divided by total lead time
Why measure?	Long order confirmation time results in long lead time
Reference object	(a) order, (b) business partner, (c) part logistics (e.g., sales)
Fact to measure	Time from temporary order confirmation up to order confirmation

Fig. 1.4.4.8 The performance indicator *order confirmation time.*

Additional performance indicators may reflect the time required for product design or maintenance time of the production infrastructure.

The SCOR model contains the following performance indicators in the area of delivery. The first concerns the goal of high fill rate, the second the goal of short lead times.

- *Perfect order fulfillment*: The percentage of orders meeting delivery performance with complete and accurate documentation and no delivery damage. This includes all items and quantities on-(customer's)-time, and documentation.
- *Order fulfillment cycle time*: The average actual cycle time consistently achieved to fulfill customer orders. For each individual order, this cycle time starts from the order receipt and ends with customer acceptance of the order.

1.4.5 Performance Indicators in the Target Area of Flexibility

Intended learning outcomes: Present the performance indicators bid proposal success rate, order success rate, breadth of qualifications, and temporal flexibility. Describe the SCOR indicators upside supply chain flexibility, upside supply chain adaptability, downside supply chain adaptability, and overall value at risk (VaR).

Examples of performance indicators in the target area of flexibility are the success rates in Figure 1.4.5.1 and Figure 1.4.5.2:

Indicator	Bid proposal success rate
Definition	Number of bid positions proposed divided by nbr. of customer requests for quotations
Why measure?	A high bid proposal success rate demonstrates high flexibility to create customer value
Reference object	(a) item, (b) business partner, (c) part logistics (e.g., sales)
Fact to measure	For (a): items in bid position, or bid positions; For (b) and (c): bid position, or bid The same for requests for quotations

Fig. 1.4.5.1 The performance indicator *bid proposal success rate.*

Indicator	Order success rate
Definition	Number of order positions divided by number of bid positions
Why measure?	A high order success rate is a measure of high flexibility in achieving customer value
Reference object	(a) item, (b) business partner, (c) part logistics (e.g., sales)
Fact to measure	For (a): item demand or order position; For (b) and (c): order position or order. The same for bid positions and bids

Fig. 1.4.5.2 The performance indicator *order success rate.*

The performance indicators in Figure 1.4.5.3 and Figure 1.4.5.4 rather measure the past. In order to determine potentials, additional assumptions are required.

Indicator	Breadth of qualifications
Definition	Number of different operations that can be executed by an employee or a production infrastructure
Why measure?	Broad qualifications raise the potential for flexibility in the use of resources
Reference object	Workers and production infrastructure, or organizational units
Fact to measure	Different operations executed by the reference object or combination of objects

Fig. 1.4.5.3 The performance indicator *breadth of qualifications.*

Indicator	Temporal flexibility
Definition	Short-term possible percentage of deviation from an employee's or a production infrastructure's average capacity
Why measure?	Temporal flexibility raises the potential for flexibility in resource use
Reference object	Workers and production infrastructure, or organizational units
Fact to measure	Actual load in time periods of a reference object or combination of objects

Fig. 1.4.5.4 The performance indicator *temporal flexibility.*

The SCOR model contains the following performance indicators in the area of flexibility:

- *Upside supply chain flexibility*: The number of days required to achieve an unplanned sustainable 20% increase in quantities delivered. The new operating level needs to be achieved without a significant increase of unit costs.

- *Upside supply chain adaptability*: The maximum sustainable percentage increase in quantity delivered that can be achieved in 30 days. The new operating level needs to be achieved without a significant increase of unit costs.
- *Downside supply chain adaptability*: The reduction in quantities ordered sustainable at 30 days prior to delivery with no inventory or cost penalties.
- *Overall value at risk (VaR)*: The sum of the probability of risk events times the monetary impact of the events for all the supply chain functions.

As performance indicators of the flexibility to enter as a partner in supply chains, the following are possible (see [HuMe97], p. 100):

- Reduction of the company's part in value-adding in the various supply chains.
- The number of partners in a supply chain community and the turnover achieved by the supply community.

Additional performance indicators measure what has been achieved by the enabler objectives that were introduced in Figure 1.3.1.2. But as those are qualitative goals, it is generally not possible to calculate the degree achieved. Usually, a value is determined ranging from "insufficient" to "perfect." See here [Hieb02].

1.4.6 Performance Indicators of the Primary Entrepreneurial Objective

Intended learning outcomes: Present the performance indicators cash-to-cash cycle time and return on net assets (RONA).

We noted above (see Section 1.3.2) that the *primary* entrepreneurial objective is the return on net assets (RONA). To measure performance, the indicators in Figure 1.4.6.1 and Figure 1.4.6.2 are used:

Indicator	Cash-to-cash cycle time
Definition	Duration (measured in days) from the time capital is tied up in investment to the time the organization receives cash as a return on that investment
Why measure?	To know how effectively assets are managed to improve cash flow
Reference object	(a) item, (b) order
Fact to measure	(On-hand plus in-process plus in-transit inventory days) + days of accounts receivable outstanding minus days of accounts payable outstanding

Fig. 1.4.6.1 The performance indicator *cash-to-cash cycle time.*

Indicator	Return on net assets (RONA)
Definition	A measure of financial performance that takes the use of assets into account
Why measure?	To know how the primary entrepreneurial objective is met
Reference object	(a) order, (b) period of time
Fact to measure	(Net income) / (fixed assets + net working capital)

Fig. 1.4.6.2 The performance indicator *return on net assets (RONA).*

1.5 Summary

The chapter defines basic terminology, including product, components, and the product life cycle. A supply chain is the global network used to perform services and deliver products, from raw materials to consumers, to disposal and recycling. Supply chain management (SCM) is the design, planning, execution, control, and monitoring of supply chain activities with the objective of creating net value, building a competitive infrastructure, leveraging global supply chains, synchronizing supply with demand, and measuring performance globally. SCM is applied along the entire product life cycle, within and across companies.

Business objects correspond to the way the persons involved naturally envision them. Some objects are relatively simple, such as worker, business partner, date, time period, item, product, and product family. In addition there exist several quite complicated objects. To these belongs the order itself. Further complex business objects include product structure, process plan, and load profile. We also introduced the similarly complex objects of operation, production infrastructure, work center, capacity, and load. Through reducing the degree of detail, we can derive rough-cut business objects, to process certain tasks more simply.

Logistics, operations, and supply chain management highly affect entrepreneurial objectives, such as fill rate, delivery reliability rate, and costs. There are basic conflicts among entrepreneurial objectives. Opportunity cost should be estimated and compared with the primary entrepreneurial objective, return on net assets (RONA). Objectives in the target area of flexibility and agility are important enablers for effective supply chains, as they lay the foundation for future benefits. Agile companies are able to remain competitive by means of the *proactive* development of knowledge and competence. Timely implementation of appropriate temporal, local, and quantitative potentials as well as enabling technologies toward personalized production will be competitively decisive.

Appropriate performance indicators are connected to entrepreneurial objectives and business objects. Such indicators measure the degree to which objectives in the four areas of quality, costs, delivery, and flexibility are met. The best-known indicators are inventory turnover, utilization, fill rate, and delivery reliability rate. All of these indicators can refer to various business objects or combinations of these objects.

1.6 Keywords

3D printing, 51
actual demand, 17
additive manufacturing, 51
agile company, 48
assembly, 28
automatic identification and data capture, 49
big data, 51
bill of material, 29
business plan, 41
business-to-business, 49
business-to-consumer, 49
capacity utilization, 55
cash-to-cash cycle time, 60
commonality, 30
company performance, 40
component, 10
  as a business object, 28
consumption, 15
customer tolerance time, 18
customization, 46
cyber-physical system, 50
decoupling, 19
delivery lead time, 18
delivery reliability rate, 56
demand, 17
demand forecasting, 18

due date, 25
e-commerce, 49
end user, 20
file transfer protocol, 49
fill rate, 56
forecast, 18
Industrie 4.0, 50
Internet of things, 51
in-transit inventory, 18
inventory, 18
inventory turnover, 55
item, 10
as a business object, 28
late customization, 46
lead time, 18
load profile, 35
logistics management, 16
logistics performance indicator, 52
material, 10
modular product concept, 46
net working capital, 44
operations management, 16
opportunity cost, 42
order, 8, 25
order losers, 42
order penetration point, 45
order qualifiers, 42
order winners, 42
OTIF (on-time and in-full), 56
part, 10
as a business object, 28
performance measurement, 52
personalized medication, 51
planning & control, 22
postponement, 46
process, 8
product, 10
product family, 30
product life cycle, 15
product module, 32
product structure, 29
production structure, 32
return on net assets, 44
reverse logistics, 21
rough-cut bill of material, 36
SCOR model, 23
service, 9
in the originary sense, 10
smart sensor, 50
stocking level, 46
strategic planning, 41
supply, 17
supply chain, 20
supply chain management, 21
supply chain performance, 40
tracking and tracing, 49
value-added, 10
variant, 30
work in process (WIP), 18

1.7 Scenarios and Exercises

Intended learning outcomes: Derive rough-cut business objects from detailed business objects. Describe improvements in meeting entrepreneurial objectives. Differentiate between entrepreneurial objectives and the ROI. Assessing the Economic Value Added (EVA) of Supply Chain Initiatives.

1.7.1 Rough-Cut Business Objects

Intended learning outcomes: Explain how to determine the process plan, the rough-cut process plan, and a possible load profile for an example product P.

Determine the process plan, the rough-cut process plan, and a possible load profile for the following product P. If not specified differently, the operations for each (intermediate) product are the same as in Figure 1.2.3.3. The lead time at every level and for purchasing adds up to 10 time units.

- P is produced from one unit of components A and B.
- A is produced from one unit of component C.
- B is produced from one unit of components X and Y, by the same operations as for producing C.
- C is produced from the components X and Z.
- X, Y, and Z are purchased components.

Apply the technique presented in Section 1.2.5, using the same rules as shown in the example there, but assuming that components C and B form the single item family B'.

Solutions:

a) Process plan:

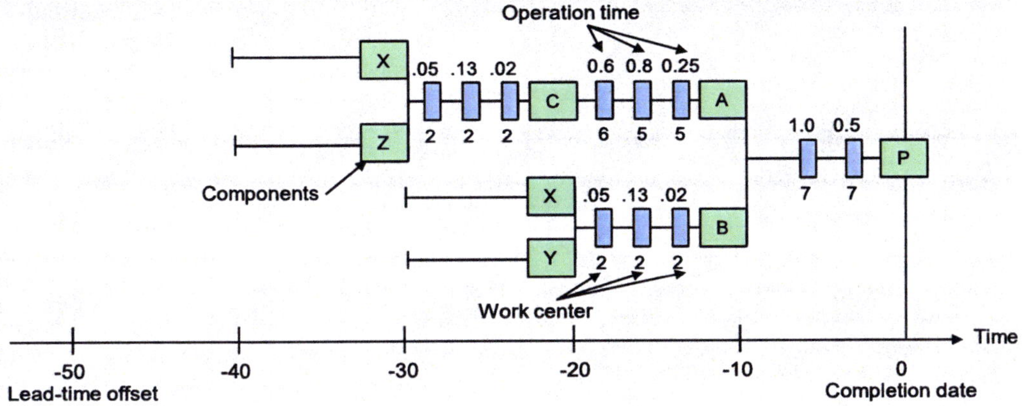

b) Rough-cut process plan:

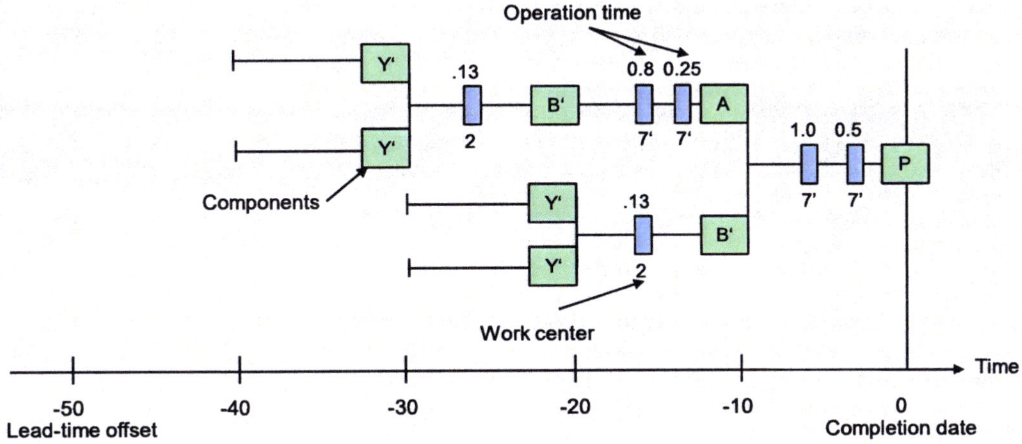

c) Possible product load profile:

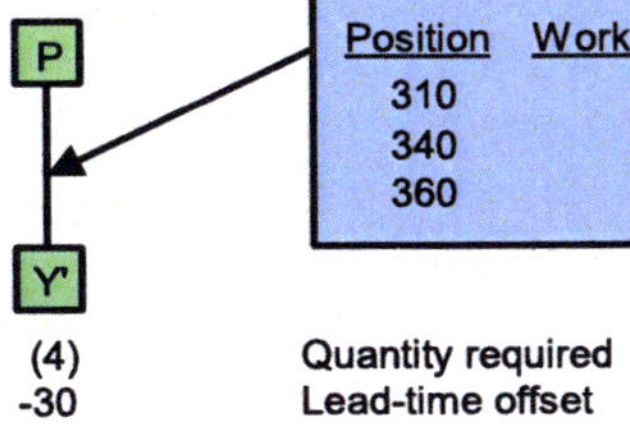

Position	Work center	Operation time	Lead-time offset
310	2	0.26	-26
340	7'	1.05	-18
360	7'	1.5	-8

1.7.2 Improvements in Meeting Entrepreneurial Objectives

Intended learning outcomes: Identify several problems with regard to the entrepreneurial objectives for an example company that manufactures a single product. Disclose and discuss possible measures in each of the four target areas (quality, costs, delivery, and flexibility).

Review the entrepreneurial objectives in the target areas quality, costs, delivery, and flexibility in Section 1.3.1. Your company manufactures a single product from easily obtainable components in four operations with a lot 5. You identify the following problems:

- Your product does not meet the demands for product quality; returns of delivered products are frequent.
- When demand is high, you regularly run into delivery difficulties. In addition to the problem of insufficient quality — which results in frequent rework — delivery difficulties are being caused mainly by poor coordination of the manufacturing departments among themselves and with the sales department. Moreover, production at the first work center is too slow, and in-house transport cannot keep up the pace. In other areas, there tend to be too many employees, particularly in sales and distribution and quality assurance.
- You think that there is a strong fluctuation of demand per period. However, you do not have the figures to back this up. You also do not know whether you can predict future demand reliably from the sales figures of past periods.

Discuss possible measures in each of the four target areas. For each measure, consider the amount of required investments. Decide the order in which the measures will be realized.

1.7.3 Entrepreneurial Objectives and ROI

Intended learning outcomes: Explain, using examples, how to determine exactly whether the increase in revenue will be cancelled out by the increased costs for improvement measures in the four target areas (quality, costs, delivery, and flexibility).

The following exercise was developed in communication with Prof. Dr. Peter Mertens (University of Nuremberg-Erlangen) to whom we express many thanks.

When we looked at opportunity cost in Section 1.3.2, we mentioned that a particular objective in the four target areas (quality, costs, delivery, and flexibility) does not always support the *primary* entrepreneurial objective, which a firm can seek to fulfill through maximum "return on investment" (ROI). For example, if investments to reduce lead time do not result in increased demand or a larger market share, then ROI decreases rather than increases.

How can this be shown more exactly, correlating the objective *short lead time* to factors in ROI? ROI can be expressed as follows:

ROI	= earnings / (investment or assets)
	= (revenue minus costs) / (current assets + fixed assets).

A possible solution is based on the following line of thinking: Reduction of lead time can have the following consequences:

- It can increase the number of customer orders and thus revenue.
- It requires the elimination of bottlenecks. This can have the following consequences:
  - It generally requires investment, which increases fixed assets and therefore capital costs.
  - It can reduce inventories of work in order, which reduces current assets and therefore capital costs.

In this case, it is important to determine exactly whether the increase in revenue will be cancelled out by the increased costs (taking into account the increase and decrease in capital costs according to the line of thinking above). Since total assets appear in the denominator of the division, ROI decreases even when total assets increase with constant earnings.

Now, use similar arguments to try to elaborate the correlation of the following performance indicators in Section 1.4 (each corresponding to a different objective of the target areas in Section 1.3.1) to the factors in ROI:

- Scrap factor (objective: meet high demands for product quality)
- Inventory turnover (objective: low physical inventory)
- Capacity utilization (objective: high capacity utilization)
- Fill rate (objective: high fill rate)
- Delivery reliability rate (objective: high delivery reliability rate)

1.7.4 Assessing the Economic Value Added (EVA) of Supply Chain Initiatives

Intended learning outcomes: Describe selected supply chain events and their relationship to current assets. Explain an example representation of the value of a supply chain initiative (SCI) in the form of the Economic value added (EVA).

Supply chain managers often have problems in communicating quantitative benefits of their management decisions, which go further than reporting service level improvements and cost reductions, to the boardroom. On the other hand, financial managers have problems assessing the real contribution to enterprise value of *supply chain initiatives* (SCIs). Many assumptions have to be made when, for instance, calculating the economic value added (EVA) of such projects. As a result, investment decisions about SCIs carry a certain level of risk.

> *Economic value added (EVA)* is a metric for representing enterprise value. EVA is positive, i.e., value is generated, when an investment activity leads to higher NOPAT than the weighted average costs of capital (WACC) invested in the assets required for generating that income.

In other words, value is only generated when the investment is expected to provide more profit than the stockholders would get by alternative investments on the market. The equation is therefore:

$$EVA = NOPAT - WACC \times \textit{value of fixed and current assets}$$

Hence, the challenge is to provide transparency on the benefits and risks of the various supply chain structures and SCIs used for improving the performance of the supply chain — generally the reduction of inventory and reduction of lead time — in terms of the financial variables like EVA.

The *supply chain value contribution (SCVC)* method as described by [Schn10] is a tool for providing transparency about the cause and effect of SCIs on supply chain performance and the utilization of assets, with a focus on the working capital.

The integrated view of the SCVC method links logistics' and financial managers' perspectives on the supply chain. This provides the required common language for assessing the contribution of SCIs to enterprise value. Hence, investment decisions can be made on a more profound basis and related risks mitigated. The key element of the SCVC method is the use of the SCOR model and identified relationships between specific supply chain events and the different elements of net working capital. Figure 1.7.4.1 shows an example visualization of these relationships in a make-to-stock production environment.

The upper part of the figure lists identified supply chain events with a direct relationship to the elements of working capital, displayed in the middle part, and relevant metrics used by logistics and financial management shown in the lower part. The horizontal axis represents the flow of time, which is followed by the flow of material, information, and cash. The vertical axis is relevant for the middle part and represents the valuation of material and the amount of the unpaid received or sent invoices and of the cash in- and outflows.

All material that is inventory, either in transit, raw material, in process, or finished goods, appears on the balance sheet, valued at actual total costs. This has two important consequences. First, where costs are reduced, the valuation is decreased by the same amount. Second, considering that the balance sheet is a snapshot of the company's asset and capital situation at a given point in time, there is material in the whole supply chain, which appears as inventory in the corresponding stage. The shorter the supply chain, the less material adds up in the inventory account. In Figure 1.7.4.1, the heights of the inventory rectangles represent the actual total costs of the material until the particular stage. The widths represent the duration of time (i.e., cycle time) the material remains (on average) within the particular stage of the supply chain at a given point in time. The area of the rectangles is the value of the particular element of inventory appearing on the balance sheet. By reducing costs, the height of the rectangles is decreased. By reducing cycle times, the width of the rectangles is reduced, resulting in a smaller area and therefore a lower valuation of the particular element of inventory. Therefore, both cost and cycle time reductions reduce the amount of capital lockup within a company.

A similar logic can be applied to the accounts payable and receivable, respectively. The widths of these boxes represent payment terms, or, as a performance measure, the days payable/receivable outstanding. The heights represent the amount appearing on the corresponding inbound or outbound invoices. Cash inflows and outflows do not have a "cycle time," but they have a value, and influence the amount of cash the company holds in the balance sheet. Figure 1.7.4.1 does not show the cash pool. Because production processes are here considered a "black box," the figure shows this element not as a rectangle but as a trapezoid. This represents the value-adding activities taking place in this phase, with costs incurred appearing in accounts payable. However, in a more detailed analysis, these aspects could be included.

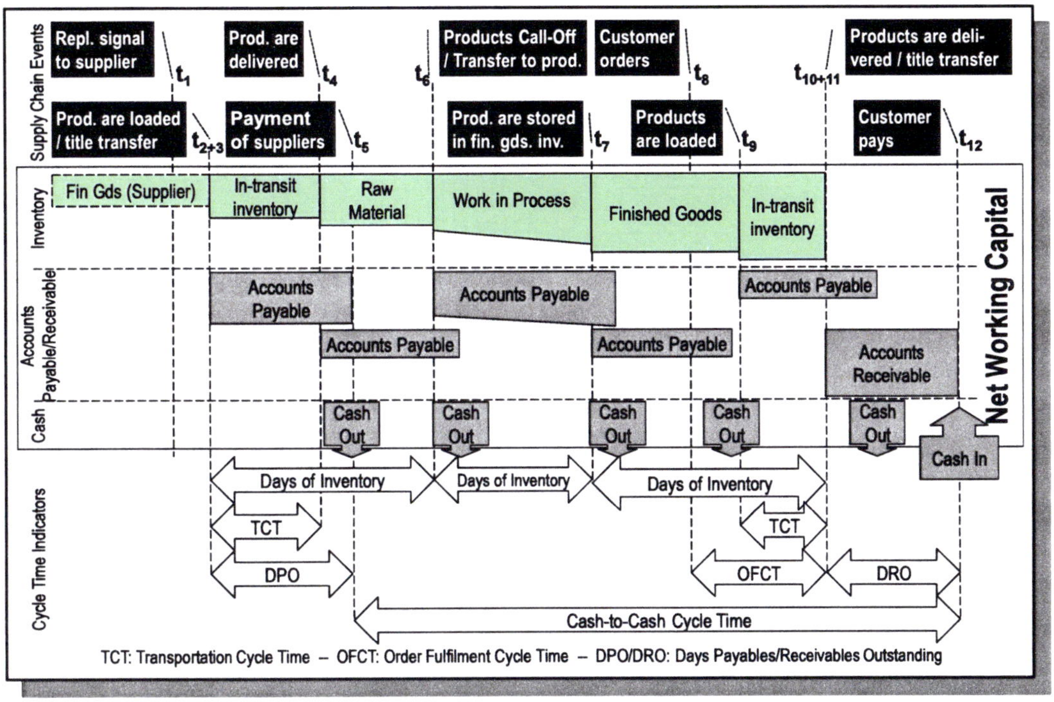

Fig. 1.7.4.1 Selected supply chain events and their relationship to current assets.

Any supply chain can be displayed accordingly, providing an integrated view of operational supply chain performance and the resulting utilization of working capital. Using this visualization to display a change in supply chain performance, resulting from the implementation of an SCI, makes it possible to calculate the contribution of the initiative to enterprise value, based on EVA. This is done by entering the data of the corresponding analysis into the structure as shown in Figure 1.7.4.2.

The data in the figure are based on an example scenario fully described in [Schn10]. In short, the scenario describes an SCI in which a logistics service provider takes over supply and inventory-related management activities. From a logistics perspective, this increased the reliability of shipping and storing processes, with shorter lead times at lower costs per product. Because of lower return rates and higher product availability, the number of sold products and therefore turnover increased by \$22,980. The higher volume results in higher total SC cost of \$4,589 and higher taxes of \$5,517. From a financial perspective, the NOPAT is disproportionately higher (\$12,874).

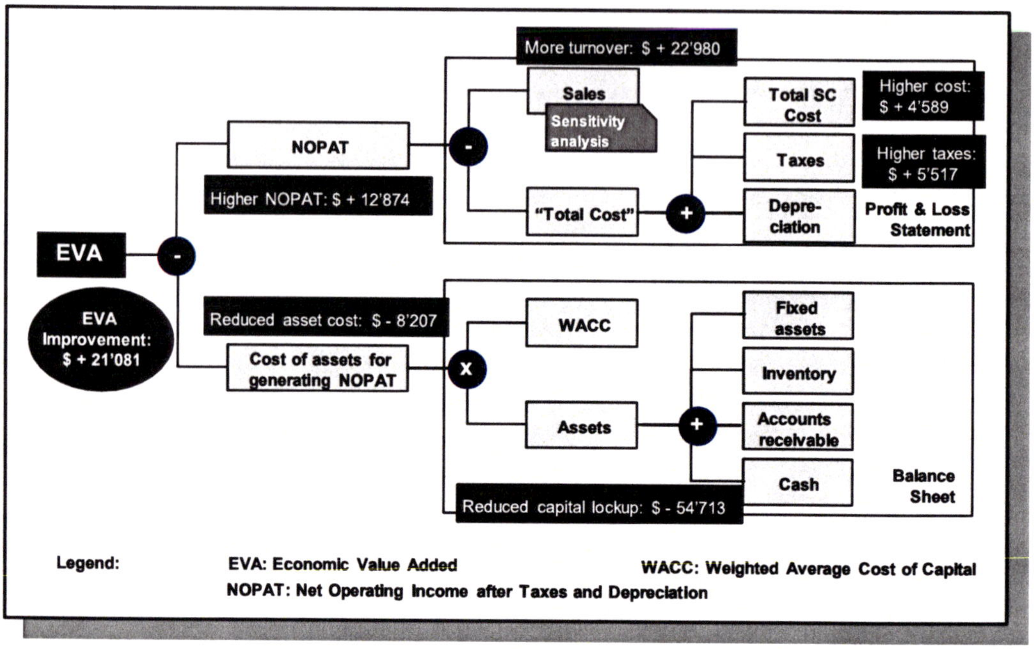

Fig. 1.7.4.2 Example representation of an SCI value in the form of the EVA.

In addition to the perspective of the profit and loss statement, EVA integrates the changes on the balance sheet. The logistics performance improvements described above affect working capital in two dimensions. Shorter lead times reduce the cash-to-cash cycle time, representing the time capital is locked up as material in the supply chain. In addition, because costs per product could be reduced at several stages of the supply chain, the valuation of the material in the different inventory accounts is reduced, too. These relations are visible in Figure 1.7.4.1. Both effects result in reduced capital lockup of $54,713. After being multiplied with the WACC of the company of 15%, this value and the NOPAT effect make up the total EVA contribution of the SCI of $21,081.

Consider the following scenario: A central distribution center (CDC) located in Switzerland wants to evaluate whether it would be beneficial to change the mode of transport to the regional distribution center (DC) located in the south of Norway. Currently, transportation is by truck in order to achieve short transportation cycle times (3 days). Transportation by ship would take 7 days but is cheaper. The title of inventory is transferred as soon as the products arrive at the DC. The relevant average inventory value at the CDC is $300,000 in the finished goods warehouse, plus average $25,000 in-transit inventory with transportation by truck. The average in-transit inventory would double when changing the transportation to ships. At the same time, the annual transportation cost would decrease from $20,000 to $15,000, with payment terms toward any carrier of 60 days. The WACC of the firm is 8%.

What is the effect of the change of the mode of transport on NOPAT and EVA after one year? Would you advise changing the transportation mode? Include a sensitivity analysis in your reasoning, as the values of the initial variables can vary in practice. Hint: As the SCVC method only calculates the change of the EVA contribution from a baseline to a changed scenario, you need to consider only values that differ between the scenarios. Solution:

- NOPAT: +$5,000 (same sales – $5,000 less transportation cost)
- Average value of accounts payable: from $3,333 ($20,000 / 12 months * 2 months payment terms) to $2,500
- Capital lockup: +$25,833 ($25,000 higher average in-transit inventory plus $833 lower accounts payable)
- EVA change: $5,000 - $25,833 * 8% (WACC) = +$2,933

Approaches for sensitivity analysis:

1) Sales could drop because of longer order fulfillment cycle time
2) Higher/lower WACC

1.8 References

APIC16 Pittman, P. et al., APICS Dictionary, 15th Edition, APICS, Chicago, IL, 2016

Foga09 Fogarty, D.W., "Production Inventory Management," 3rd Edition, South Western Publishing, 2009

Gots06 Gottschalk, L., "Flexibilitätsprofile: Analyse und Konfiguration von Strategien zur Kapazitätsanpassung in der Produktion," Diss. ETH Zurich, 2006

Hert13 Hertz, P., Industrial field service network planning. Diss. ETH Zürich, 2013

Hieb02 Hieber, R., "Supply Chain Management — A Collaborative Performance Measurement Approach," vdf-Verlag, Zurich, 2002

HuMe97 McHugh, P., Merli, G., Wheeler, W.A., "Beyond Business Process Reengineering — Towards the Holonic Enterprise," John Wiley & Sons, 1997

KaNi18 Kaihara T., Nishino, N. et al., Value creation in production: Reconsideration from interdisciplinary approaches. CIRP Annals, 2018; 67(2), p. 791-813

KaNo92 Kaplan, R., Norton, D., "The Balanced Scorecard — Measures That Drive Performance," *Harvard Business Review*, Jan./Feb. 1992, pp. 71–79

Lang09 Lange, I., "Leistungsmessung industrieller Dienstleistungen," Diss ETH Zürich, 2009

Levi81 Levitt, T., "Marketing intangible products and product intangibles," Cornell Hotel and Restaurant Administration Quarterly, 1981, 22.2: 37-44

Long15 Longman, "Dictionary of Contemporary English," 6th Edition, Pearson Education, Harlow, England, 2015

MaSc04 Manecke, N., Schönsleben, P., "Cost and Benefit of Internet-Based Support of Business Processes," *International Journal of Production Economics*, 87, 2004, pp. 213-229

MeRo10 Meier, H., Roy, R., Seliger, G., "Industrial product-service systems — IPS 2," CIRP Annals-Manufacturing Technology; 2010, 59.2: 607-627

MeWe18 Merriam-Webster, "Merriam Webster's Collegiate Thesaurus," 2nd Revised Edition, Merriam-Webster, Inc., Springfield, MA, 2018

OdLa93 Oden, H.W., Langenwalter, G.A., Lucier, R.A., "Handbook of Material & Capacity Requirements Planning," McGraw-Hill, New York, 1993

PrGo97 Preiss, K., Goldman, S.L., Nagel, R.N., "Cooperate to Compete — Building Agile Business Relationships," Van Nostrand-Rheinhold, New York, 1997

RuTa17 Russell, R.S., Taylor, III, B.W., "Operations and Supply Chain Management," 9th Edition, John Wiley & Sons, 2017

Schn07 Schnetzler, M.J., Sennheiser, A., Schönsleben, P., "A Decomposition-Based Approach for the Development of a Supply Chain Strategy," *International Journal of Production Economics*, 105(1) 2007, pp. 21–42

Schn10 Schneider, O., "Adding Enterprise Value — Mitigating Investment Decision Risks by Assessing the Economic Value of Supply Chain Initiatives," vdf-Verlag, Zurich, 2010

Schö19 Schönsleben, P., "Tangible Services and Intangible Products in Industrial Product Service Systems," Procedia CIRP 83, 2019, pp. 28-31

Shef07 Sheffi, Y., "The Resilient Enterprise: Overcoming Vulnerability for Competitive Advantage," The MIT Press, Cambridge MA, 2007

SwLe03 Swaminathan, J.M., Lee, H., "Design for Postponement," in Kok, A.G., Graves, S.C., "Handbooks in Operations Research & Management Science," Vol. 11, Elsevier, 2003

Verh12 Verhasselt, S., "Postponement Strategies in Pharmaceutical Supply Chains", 23rd POMS Conference, Chicago, 20.-23. April 2012

VoBe18 Jacobs, R., Berry, W., Whybark, D.C., Vollmann, T.E., "Manufacturing Planning and Control for Supply Chain Management — The CPIM Reference," 2nd Edition, McGraw-Hill, New York, 2018

WiMa07 Wiendahl, H.-P., ElMaraghy, H.A., et al., "Changeable Manufacturing — Classification, Design, and Operation," Annals of the CIRP, Vol. 56/2, 2007, pp. 783–796

2 Supply Chain Design: Business Relations and Risks

Intended learning outcomes: Explain concepts such as the make-or-buy decision, tariff-orientation and total cost of ownership in a global supply chain. Differentiate strategic design options for the relationships with and the selection of suppliers. Disclose strategies and a framework for implementation of intensive cooperation in the supply chain. Describe identification, assessment and handling of supply chain risks.

For the design and manufacturing of products of a certain degree of complexity, the value added is distributed among different companies or organizational units of a company, which become parts of the supply chain. Only in this way are the necessary competencies brought together for high-quality, rapid, and profitable value adding.

Supply chain design is the determination of how to structure a supply chain. Design decisions include the selection of partners, the location and capacity of warehouse and production facilities, the products, the modes of transportation, and supporting information systems (cf. [APIC16]).

Section 2.1 introduces various ways of thinking and initial tools for supply chain design. The topic is why and how companies form, change in their boundaries to the outside and in their inner structure, and how global trade supply chains should be handled with regard to tariffs. All of these factors must be included in an assessment of the total cost of ownership.

Section 2.2 deals with strategic procurement and shows criteria and design possibilities for the relationships with and the selection of suppliers.

Section 2.3 discusses the design of intensive cooperation with suppliers, which is required when the focus is not on a firm in competition with its suppliers but instead on the entire supply chain in competition with another supply chain for the favor of the *end* customer.

Finally, Section 2.4 turns to risk management of supply chains.

2.1 Ownership and Trade in a Supply Chain

Intended learning outcomes: Present the concept of the make-or-buy decision. Explain value content requirements, tariff-orientation and total cost of ownership in a global supply chain.

Figure 2.1.0.1 shows a possible strategic process for fundamental designing of the supply chain. Here we will not discuss the macro-environment analysis (see also Section 10.4.3), the development of product and market strategy, competitive analysis and the subsequent phase of product design in R&D. For the requirements analysis, a demand forecast for each product and market (see Chapter 10) as well as a rough-cut planning of the type and amount of the necessary resources to meet the demand are worked out (section 5.2).

P. Schönsleben, *Handbook Integral Logistics Management*,
https://doi.org/10.1007/978-3-662-65625-9_2

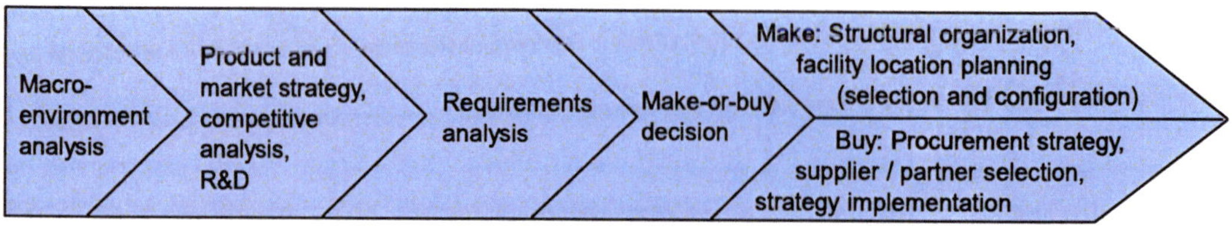

Fig. 2.1.0.1 Strategic process of designing the supply chain.

Supply chain design is concerned with what organizational units will be responsible for meeting the demand, and what factors in effective and efficient value adding must already at this point be taken into consideration in the case of both the make and the buy process.

2.1.1 The Make-or-Buy Decision — Transaction Costs as the Basis of Forming Companies

Intended learning outcomes: Produce an overview on outsourcing and transaction costs. Disclose some factors that lead to a buy, or to a make decision. Describe counterdeals and protectionism. Differentiate between various forms of company-internal organization.

A *make-or-buy decision* is the choice between outsourcing and insourcing.

In *outsourcing*, parts of the value-added chain are turned over to other companies. *Insourcing* refers to the formation or expansion of companies by means of taking parts of the value-added chain into the company.

In what cases will an organization disband or become re-dimensioned by means of outsourcing? Keeping in mind the objectives of the enterprise outlined in Section 1.3.1, a company will choose outsourcing whenever a product or product part can be produced on the whole in better quality and more cheaply, rapidly, reliably, and flexibly by a third party than when produced in-house. If the contrary is the case, the decision is made to form or expand a company through insourcing.

In the following, we will assume that the same quality product may be procured on the market as could be produced by the company itself. The crucial factor in forming a a firm or a company under this condition is transaction costs, according to Nobel Prize winner Ronald H. Coase ([Coas93]; this fundamental work was actually written in 1937). For the *transaction-costs approach*, see also [Pico82].

The *transaction process* is the transmission of goods from *vendor* (that is, a seller of a good in the marketplace) to *buyer*. *Transaction costs*, or *market transaction costs* of goods, are the costs of the organization as a production factor. These include all costs of the transaction process that are not set in price by the market.

Transaction costs thus arise when price does not reflect all the necessary information on goods; for example, due to inability, opportunism, uncertainty, or market distortions. Transaction costs are thus the cost of information and include the following types of costs:

- *Search and initiation costs*: These are, for example, the costs of locating and obtaining information on potential business partners and the conditions involved.
- *Negotiation costs* include the actual costs of negotiation and decision making, legal counsel, and fees.
- *Control costs* include expenditures necessary to coordinate orders so as to maintain quality, quantity, costs, and delivery dates as well as eventual costs to adapt to changes in orders. In addition, there are costs to ensure other contractual agreements, in particular patent protection, licensing and security agreements, and so on.

Transaction costs are comparable to friction loss in the relations in a supply chain, influenced by factors of the "specificity" and "risk" (uncertainty) type (see here also [Port04a] and [Port04b]). For each factor the following list contains examples that lead to a *buy decision*, or outsourcing, as well as examples that speak well for a *make decision*, or insourcing.

The *specificity of product and processes or location* makes a first factor:

- Outsourcing: Product and process are not specific. For design and manufacturing, a number of bidders are on the market. Those companies already have specialists and specific infrastructures at their disposal. Moreover, transport is not a problem.
- Insourcing: Transaction costs rise, whenever specific investitures in production infrastructure and the qualification of employees are required, or the supplier needs to be in a proximate location. On the other hand, product specificity creates better product differentiation and thus the building up of a trade name and market share.

Complexity of product and processes, and *time-to-product* make a second factor:

- Outsourcing: The projects are too complex or too extensive to be realized on time by the company with the capabilities and capacities of its personnel. Small enterprises often face this problem, for customer-order-specific tasks in the value-added process.
- Insourcing: Order coordination and control become more costly and more difficult. The danger of opportunistic behavior on the part of the supplier increases.

Core competencies, greater degree of innovation in product and process, and *time to market* (time to product innovation) make a third factor:

- Outsourcing: Superior quality products require technology that is more difficult to command. Within increasingly shorter time periods, new technologies have to produce goods that succeed on the market. Procuring some of the company's competencies from a third party does not cause any critical problems, even if know-how that exists may be lost. In addition, the company desires access to the know-how of another organization.
- Insourcing: A lead in know-how through the development of core competencies and the achievement of innovation secure the survival of the company. A great store of know-how results in short lead times and flexibility. Continuing to give work to third parties involves too great a risk and high control costs.

Capital requirements and cost breakdown structure make a forth factor:

- Outsourcing: The company cannot afford the cash requirements for amassing and maintaining company know-how. Specialists do not fit into the payroll or the company culture. The firm cannot fully utilize their specific abilities. The same holds for the infrastructure.
- Insourcing: The company's favorable size and structure permit the advantages of in-house design and manufacturing. The cash requirements are affordable.

Lack of trust and *lack of stability* make a fifth factor:

- Outsourcing: The company is dependent on too few or even individual persons. It cannot build up a culture of sufficient capacity in the respective area. Remedial action, such as cooperating with several like-minded companies, is not possible.
- Insourcing: Changes of partners within a supply chain lead to an increase in transaction costs. Are the relationships to the crucial individuals stable? Is quality maintained at a certain level? Does the supplier retain customer focus and user orientation? Do the supplier's prices reflect a *learning curve* (that is, the supplier's rate of improvement due to the frequently repeated transaction) and decrease?

In addition, outsourcing can be forced in the following contexts:

- *Counterdeals:* With their various subsidiary companies, corporations can be both potential customers and suppliers of a manufacturer. If a company wishes to gain them as customers, it may have to agree to a counterdeal stipulating that one of their subsidiaries will supply certain components, even if it could produce these itself.
- *Protectionism:* Certain markets elude the laws of a free economy. Political decisions can force manufacturers — in order to gain market access — to form "joint ventures" with companies in other countries. This type of cooperation then involves parts of the supply chain that manufacturers could actually process themselves.

Thorough evaluation of all factors thus helps the firm to determine the *optimum depth of added value* of the company.

- *Vertical integration* is the degree to which a firm has decided to directly produce multiple value-adding stages from raw material to the sale of the product to the end user ([APIC16]).
- *Backward /forward integration* is the process of buying or owning elements of the production cycle and the channel of distribution, backward toward raw material suppliers / forward toward the final consumer (cf. [APIC16]).

Clearly, there are also friction losses within a company.

Internal transaction costs are all costs related to the processing of company-internal transactions among the organizational units involved. These are all costs that would not arise if one single individual could do the processing. Internal transaction costs arise from a lack of mutual information due to, e.g., inability, opportunism, uncertainty, or diverging interests.

Internal transaction costs are thus the price of information. They include types of costs that are similar to market transaction costs. These are the costs of shaping an organization, the ongoing coordination of workers, planning and control costs, and flexibility costs, as well as costs of lead times. [1]

The right form of organization can be decisive in supporting company-internal cooperation, especially when the facility locations are at a great distance from one another or handle different product families or services. Various forms of organization are possible:

- *Profit center* within a decentralized or product-focused organization: In its pure form, a profit center plans and acts just like an independent subcompany. It carries comprehensive responsibility and also has the authority to accept or reject orders from other organizational units of the company.
- *Cost center* within a centralized or process-focused organization: In its pure form, a cost center receives clearly formulated orders: due date, type, and quality of products. The often complex and capital-intensive processes are triggered by order management of a central department. The cost center then has the task of fulfilling quality and quantity requirements, and due dates; it does not, however, carry responsibility for these, as it does not have its own resources (personnel and production infrastructure) nor does it manage resources procured from the outside (for example, information, semifinished goods, and raw materials).
- A *semiautonomous organizational unit* is linked to a company-wide strategy. For example, it must accept orders or must procure certain components from other organizational units of the company or from a central department. However, it is autonomous with regard to order processing and fulfillment. It negotiates due dates, type, and quantity of goods with customers and suppliers and realizes order processing under its own direction. At the level of company strategies, framework conditions are set within which the unit has some degrees of freedom. In dependency upon framework conditions, semi-autonomous organizational units may act as profit centers or cost centers. For this reason, they are seldom stable in the long term.

Internal transaction costs are particularly dependent on the persons involved. Here the human factor has to be closely attended to. The reader is referred to [Ulic11]. Two common mistakes result when all-too-human characteristics find expression:

- *"Kingdoms" within departments or foreman areas:* Decentralized organizational units take on authority without perceiving the conjugate, necessary responsibility. For example, they might set order due dates autonomously without taking the superordinate interests of the total supply chain into consideration. "Kingdoms" such as these also arise when decentralized organizational units are evaluated on the basis of isolated objectives, such as capacity utilization.
- *Centralistic kingdoms:* Central management delegates responsibility to decentralized organizational units without giving them the necessary authority. Holding companies, for example, may turn over cost and profit responsibility to

[1] In addition to internal transaction costs, there are *"agency costs,"* i.e., costs of coordinating the interests of the owner with those of decision makers in management.

subsidiaries and affiliated companies while maintaining the right to choose these companies' suppliers and customers.

2.1.2 Global Trading — Value-Content Requirements and Tariff-Orientation in a Supply Chain

Intended learning outcomes: Produce an overview on terms such as tariff, free trade agreement (FTA), and free trade area. Explain the concepts of rules of origin (RoO), value content, and tariff heading. Describe various scenarios for a truck manufacturer within the EU that wants to sell trucks in the NAFTA area.

Globalization trends in recent decades have exposed companies to international competition and high cost pressure, but they also make savings possible through global sourcing. From the point of view of a local economy, this is not always favorable, since shifting value adding elsewhere also entails loss of jobs, which can also result in negative impacts on gross domestic product. That is why there are now increasing attempts at the political level to tie multinational companies to the local economy, but without inhibiting trade flows between economies. Possible tools for resolving this conflict of aims are free trade agreements and free trade areas in combination with Rules of Origin (see below).

A *tariff* is, according to [APIC16], an official schedule of taxes and fees imposed by a country on imports or exports.

As a result of a *free trade agreement (FTA)*, a *free trade area (FTA)* is a form of preferential trading area where there is fully free trade among members, but each country is free to levy different external tariffs against nonmember countries.

FTAs are used to stimulate trade among member countries and therefore support the local companies and economies. Two of the most prominent FTAs are the European Union (EU) and the North American Free Trade Agreement (NAFTA — now replaced by the USMCA agreement). See also Figure 2.1.2.1 and [LiLi07].

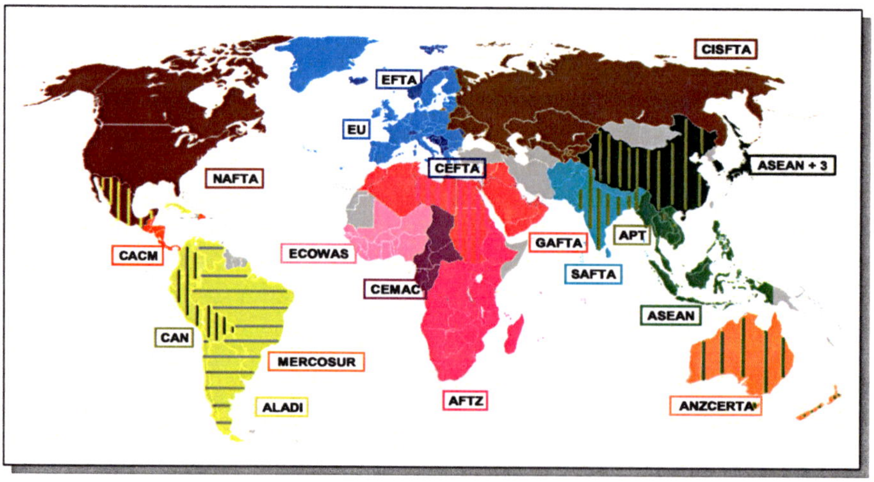

Fig. 2.1.2.1 Some of the most important FTA.

When there is no common external tariff among the member countries of an FTA, companies can enter the FTA at the lowest tariff entry point. Goods can then be shipped to the final destination within the FTA without additional tariff costs. This results in tariff revenue transfer effects, since this trade deflection transfers tariff revenue to the country with the lowest external tariff within the FTA. To reduce trade deflection, Rules of Origin are used.

Rules of Origin (RoO) define the extent of a product's local content needed to be regarded as local and therefore the prerequisites for trading goods tariff-free within the FTA. Additionally, RoO are used to induce companies from nonmember countries to shift value-adding activities toward the FTA to be able to benefit from the preferential tariff treatment.

There are two ways to treat a good as "local." First, it can be "wholly obtained or produced," meaning the good has been wholly grown, harvested, or extracted from the soil of the member country or has been manufactured from any of these products. This type of RoO is not further examined here. Second, the good has gone through a "substantial local transformation." This criterion is important for global SCM. It has three aspects that can be used in a combination or on a stand-alone basis (see also [EsSu05]):

- *Value content* (VC_{FTA}): Value content defines, for a specific FTA, the minimum percentage of local value added that is required for the product to be considered local.
- *Change in tariff heading* (CTH): This criterion is fulfilled as soon as the imported good has been processed in a way that alters the tariff code in the harmonized system (HS).
- *Technical requirements* (TECH): This criterion defines a list with processes and/or inputs that are prohibited or that need to be realized within the FTA.

What is the implication of FTAs and RoO for global SCM? An *opportunity* occurs when the companies meet the FTA's requirements and are allowed to trade tariff-free within the economic area. In most cases, this requires an amount of value adding (VC_{FTA}) to take place within the domestic area. The following example illustrates the issues: A truck manufacturing firm either produces trucks P (tariff code HS8701.20) completely in the EU or produces semifinished products S of P in the EU and assembles them to finished trucks in the NAFTA. In both cases, the company ships its finished goods to its customers in the NAFTA member countries, the United States (USA), Mexico, and Canada. If the value content of the manufacturer's products, namely, $VC_{P,FTA}$ or semifinished products, namely, $VC_{S,FTA}$ is less than the VC_{FTA}, the company faces the tariff situation shown in Figure 2.1.2.2.

		FTA	VC_{FTA}	To: Canada	To: Mexico	To: USA
From	EU	no FTA	n.a.	6.1 %	n.a.	4.0 %
From	EU	EC - Mexico	50 %	n.a.	30.0 %	n.a.
From	Canada	NAFTA	62.5 %	n.a.	30.0 %	4.0 %
From	Mexico	NAFTA	62.5 %	6.1 %	n.a.	4.0 %
From	USA	NAFTA	62.5 %	6.1 %	30.0 %	n.a.

Fig. 2.1.2.2 Tariffs of code HS8701.20 for shipments from one country to another, if there is no FTA in place or $VC_{P,FTA} < VC_{FTA}$ or $VC_{S,FTA} < VC_{FTA}$ (n.a. = "not applicable").

The tariffs, which can differ depending on type of products, may have a significant impact on the *cost optimal* supply chain design. Figure 2.1.2.3 shows a case-by-case analysis of the VC_{FTA} threshold fulfillment in a simplified but still realistic supply chain structure for the truck manufacturer.

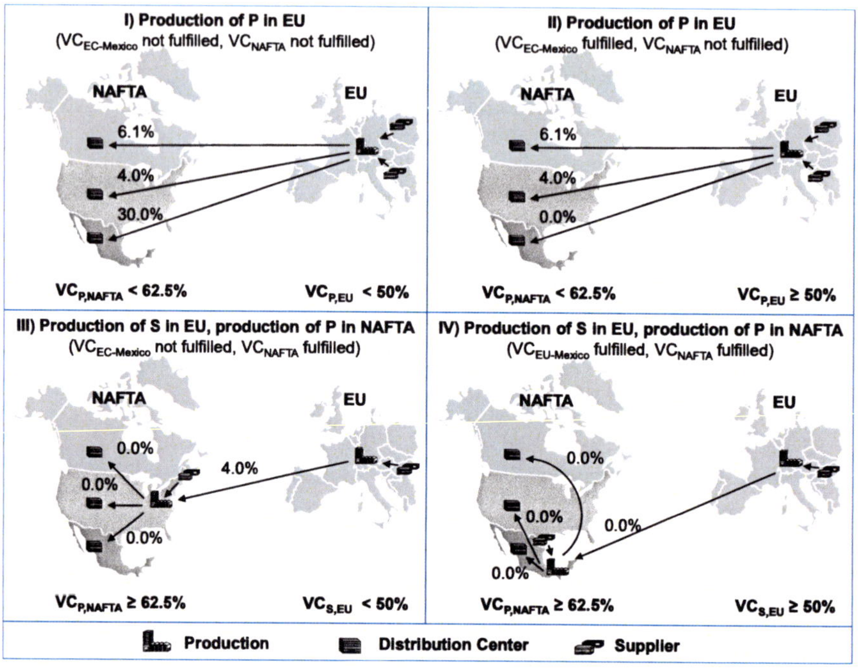

Fig. 2.1.2.3 Different supply chain design options regarding value content (VC_{FTA}) fulfillment.

The truck manufacturer can either produce P wholly in the EU (scenarios I, II) or meet the VC_{NAFTA} threshold by shifting production of P to the NAFTA regions (scenarios III, IV) and reducing value adding in the EU to production of semifinished products. To stress the influence of tariff costs, this example disregards sourcing, production, and transportation costs and the sales quantity of the destination markets.

- In scenario I, the manufacturer produces all trucks P in the EU and then ships the finished products to the destination markets Canada, USA, and Mexico. Because of the sourcing from non-EU members and no value adding in the NAFTA, it cannot meet either the $VC_{EC\text{-}Mexico}$ threshold (50%) or the VC_{NAFTA} requirements (62.5%). From a tariff perspective, the best solution would be direct shipment of P to, and payment of the external tariff of each country.

- In scenario II, the manufacturer is able to fulfill the $VC_{EC\text{-}Mexico}$ (50%) by executing 50% or more of its value adding of P in the EU. It can therefore exploit the tariff-free shipment to Mexico.

- Scenario III is possible if the manufacturer decides to produce only semifinished goods S in the EU and finished products P in the NAFTA (principle of "completely knocked down" or "semi knocked down"). This makes sense as soon as the value content of trucks P ($VC_{P,NAFTA}$) becomes greater or equal to VC_{NAFTA} (62.5%). As for the semifinished goods S, in this scenario $VC_{S,EU}$ is smaller than $VC_{EC\text{-}Mexico}$. Therefore, it is optimal to enter the NAFTA with the semifinished goods S through the USA, which in this case is the lowest tariff entry point, assemble them to P in the USA, and distribute the finished goods P to the USA, Canada, and Mexico.

- In scenario IV, the manufacturer allocates its value adding activities for semifinished goods S in a way that $VC_{S,EU} \geq VC_{EC\text{-}Mexico}$, and for finished product P in a way that $VC_{P,NAFTA} \geq VC_{NAFTA}$. From a tariff perspective, the assembly of P in Mexico would be the optimum.

There are considerable *threats* in a tariff-oriented supply chain. This is because the basis for the decision-making, that is, the value content of a product, the calculation method of VC_{FTA}, and even TECH can alter rapidly. In addition, the degree of CTH can increase (here see [PlFi10]). Once a company has made long-term investments in facility locations and sourcing decisions based on an optimum tariffs scenario, it might suddenly be locked in a local supply chain (scenarios III, IV). As a consequence, it might no longer be able to compete with global competitors (scenarios I, II) due to relatively high sourcing costs. The following three scenarios have been picked out of a multitude of possibilities, in which a changing environment causes problems with VC_{FTA} fulfillment.

- *Scenario "supplier shortfall":* When a company has decided on local sourcing to match VC_{FTA} requirements, a shortfall by a local supplier can have a crucial impact on tariff costs. If there are no alternative local suppliers, the company will be forced to switch to a global sourcing strategy. This would decrease the amount of originating materials and thus decrease the $VC_{P,FTA}$ of the product.
- *Scenario "increasing capacity utilization":* A problem from the perspective of the VC of the product emerges as soon as increased capacity utilization (a much desired effect) reduces the total cost per product. This could happen because of, for example, manufacturing overheads being allocated to a greater amount of final products. As a decrease in total cost, this could cause a reduction of $VC_{P,FTA}$.
- *Scenario "rise of the value content (VC_{FTA}) requested by the government":* This kind of increase could mean that RoO requirements cannot be met. The relevance of this factor is shown by the trend of the VC_{FTA} in the NAFTA region for light vehicles set out in NAFTA Article 403. Whereas VC_{NAFTA} was at 50% on Jan. 1, 1994, it increased to 56% on Jan. 1, 1998, and to 62.5% on Jan. 1, 2002 [Cana19].

The example shows that the potential benefit of a tariff-oriented supply chain structure is significant. Nevertheless, companies that seek to exploit the advantages need to be aware of the underlying risks. These risks need to be evaluated and tested before decision investments are made.

2.1.3 Total Cost of Ownership in a Global Supply Chain

Intended learning outcomes: Explain the elements that make up the total cost of ownership (TCO). Disclose a method for analysis of the total cost of ownership.

Total cost of ownership (TCO) is a concept for analysis of acquisition costs; it is used for both make-or-buy decisions and for deciding between potential suppliers.

Total cost of ownership (TCO) of the supply delivery system is the sum of all the costs associated with every activity of the supply stream.

The main insight that TCO offers to the supply chain manager is the understanding that the acquisition cost can be a small portion of the total cost of ownership (cf. [APIC16]). The crux of the concept is that decisions should be made not only based on the purchase price of a procured good but also based on consideration of all costs associated with acquiring the good ([Ellr93]). In addition to the purchase price of a good, the TCO is made up of a number of different cost elements, as shown in Figure 2.1.3.1. For clarity, the costs are shown subdivided in four categories:

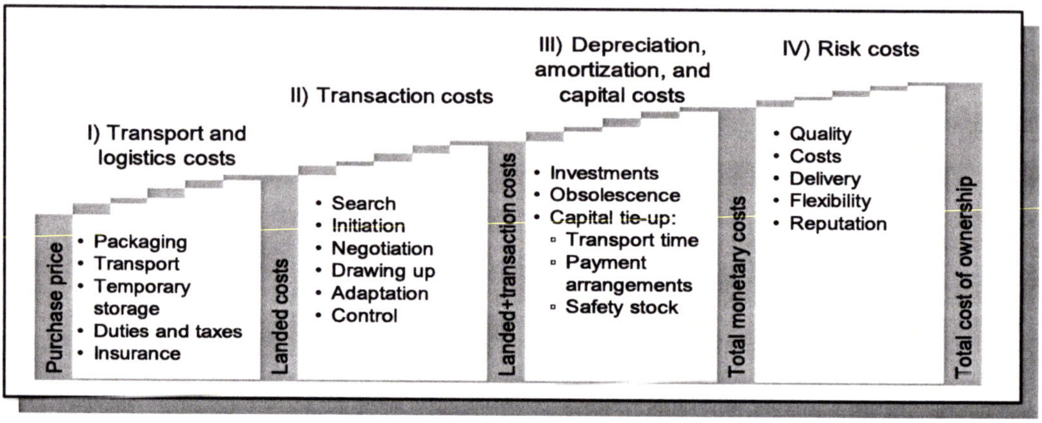

Fig. 2.1.3.1 Elements that make up the total cost of ownership.

- *Transport and logistics costs (category I)* include the cost elements packaging, transport, temporary storage, duties and taxes, and insurance.
- *Landed costs* generally denotes the sum of the purchase price and the transport and logistics costs.
- *Transaction costs* (category II) comprise company-internal expenditures for organization of the buyer-supplier relationship. They include costs for the processes of searching, initiation, negotiation, drawing up contracts, adaptation, and control.
- *Depreciation, amortization*, and *capital costs* (category III) comprise, for one, the cost elements investments and obsolescence and, for another, the costs for tied-up capital owing to transport times, payment arrangements, and safety stocks.
- *Total monetary costs* is the sum of the landed costs, the transaction costs, and the depreciation and capital costs.
- *Risk costs (category IV)* comprise risks concerning the company's objectives in the target areas quality, costs, delivery, flexibility, and reputation.

Figure 2.1.3.2 shows the relative importance of the cost elements based on a survey of 178 Swiss companies conducted in 2010, mainly in the machine, electrical, and metal industries.

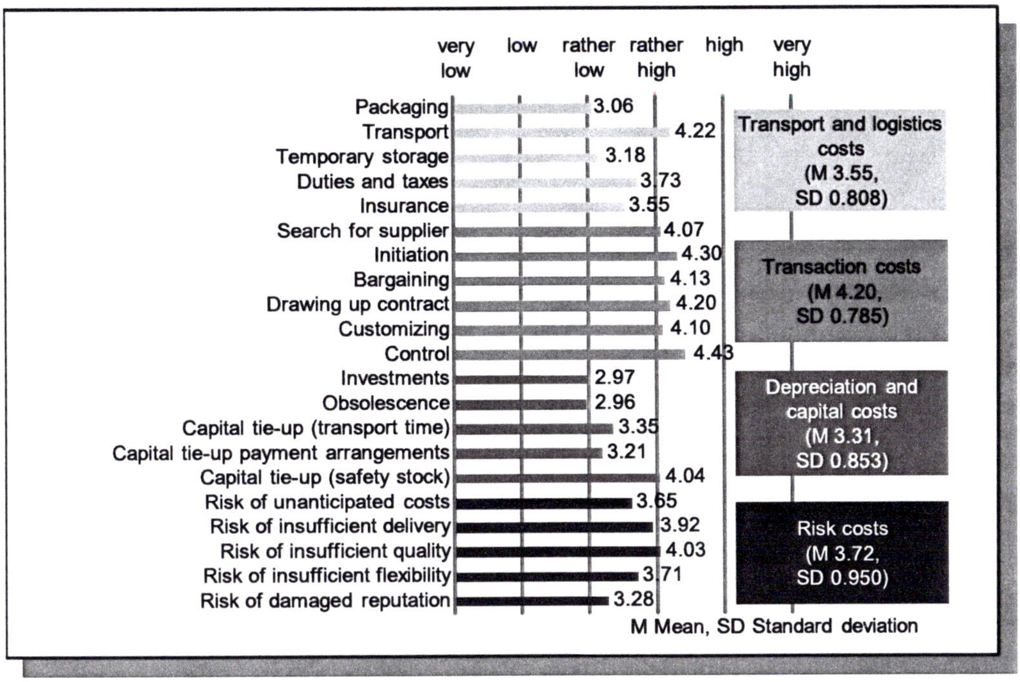

Fig. 2.1.3.2 Importance of the cost elements.

The subdivision and examination of the costs in the four categories show that the companies participating in the survey rated all transaction costs and the cost elements transport, capital tied up in safety stock, and risk of insufficient quality as high. Quantitative assessments result in total additional costs for the four categories mentioned in Figure 2.1.3.1 on average of 24.6% of the purchase price of the good when procuring from low-wage countries ([Brem10]).

Figure 2.1.3.3 shows the method developed by [Brem10] for company-specific analysis of TCO.

Here are examples of calculation functions.

1.) Cost element "freight costs":

- Cost category I, transport and logistics costs
- Belongs to variable costs, calculated per piece
- p1: amount, p2: exchange rate
- Calculation function: $f_1 = p_1 * p_2$

2.) Cost element "traveling expenses":

- Cost category II, transaction costs
- Belongs to fixed costs, calculated per PQ; that is, the estimated project quantity of goods to produce/procure

- p_1: number of persons; p_2: costs for residence; p_3: flight expenses
- Calculation function: $f_2 = p_1 * (p_2 + p_3) / PQ$

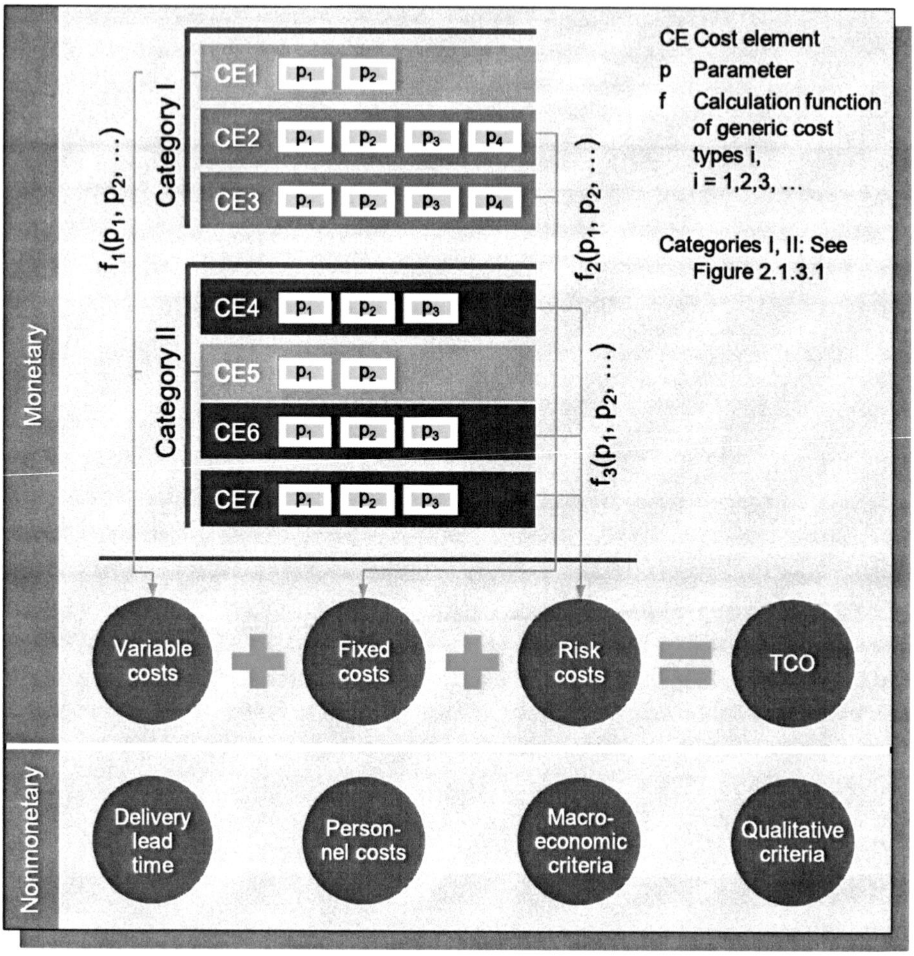

Fig. 2.1.3.3 Method for analysis of TCO.

In the model, generic cost types comprise those cost elements that have the same number and types of parameters and calculation functions. The calculation function determines the assignment to one of the monetary assessment dimensions (variable costs, fixed costs, and risk costs). Using a simple procedure, specific cost elements are derived from the generic cost elements and put together in an overall individual TCO model. In addition to the monetary view, the method is extended to also include non-monetary assessment criteria in the assessment dimensions delivery lead time, personnel costs, macroeconomic criteria, and qualitative criteria. The combination of monetary and nonmonetary assessment criteria produces an extensive and transparent database, based on which complex procurement decisions can be made. The method thoroughly incorporates the special demands of global procurement with regard to longer delivery lead times, dynamic factors, and supply chain risks as well as their implications for the profitability of procurement projects.

2.2 Strategic Procurement

Intended learning outcomes: Differentiate between traditional market-oriented relationship and customer-supplier partnership. Explain strategic procurement portfolios and selection of suppliers. Present supplier relationship management and e-procurement solutions.

In the mid-1970s, there was a shift in many areas of the economy — caused by the law of supply and demand — from sellers' markets to buyers' markets. This had big consequences for procurement management, or supply management, in particular for collaboration strategies. In this section, we compare procurement via the traditional marketplace with customer-supplier partnership. We then introduce strategic procurement portfolios and methods for strategic selection of suppliers. This is followed by a look at supplier relationship management, and in particular, e-procurement solutions.

2.2.1 Overview on Strategic Procurement

Intended learning outcomes: Disclose the supplier structure follows the product structure. Differentiate between direct material, indirect material, commodities, and various demand patterns. Describe various traditional procurement strategies.

Procurement strategies in supply chains differ from those in simple trade relations in that the manufacturer attempts to realize a *supplier structure that follows the product structure* (see Figure 1.2.2.2). *From the viewpoint of the end product manufacturer, the suppliers are arranged accordingly in tiers. Figure 2.2.1.1 shows an example of this for end product* A.

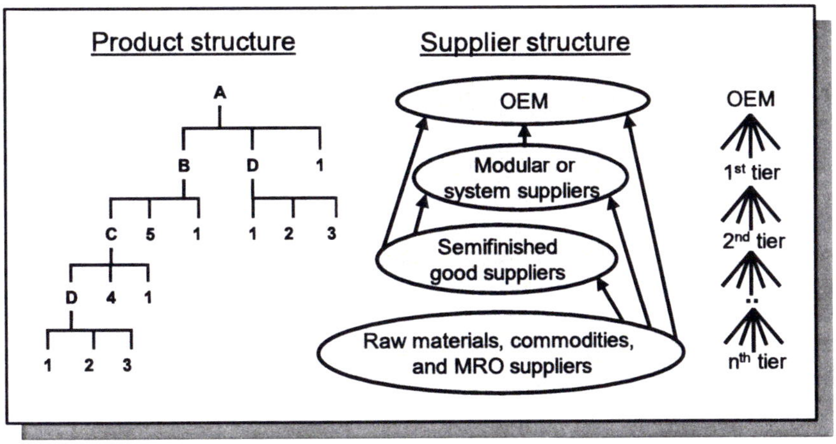

Fig. 2.2.1.1 Procurement strategy in a supply chain: the supplier structure follows the product structure.

- The *original equipment manufacturer* (OEM) produces the finished product, or end product (in the example, product A), and sells it to the consumer.
- *First-tier suppliers*, or *modular suppliers* or *system suppliers*, are responsible for several structure levels that together yield an assembly ready for final assembly, and supply this directly to the assembly line of the original equipment manufacturer. In

automobile manufacturing, for instance, this can be complete car door modules or instrument panels. In the example in Figure 2.2.1.1, assembly B is this kind of module or system.[2]

- *Second-tier suppliers*, or *semifinished good suppliers*, supply simple semifinished goods, both to the OEM and also to the system suppliers. In the example in Figure 2.2.1.1, assembly D is a semifinished good.
- *Raw materials, standard parts, and maintenance, replenishment, and operating (MRO) suppliers* supply these items both to OEM and also to the system and second tier suppliers. In the example in Figure 2.2.1.1, these are items 1, 2, 3, 4, and 5.

Procurement strategies in the supply chain are generally different for each of these levels of suppliers. The companies involved in such a channel of distribution[3] form customer-supplier links. With the exception of the end user, each customer is also a supplier within the supply chain.

The procurement strategies also differ dependent on the logistics characteristics of the goods that flow from supplier to customer.

- *Direct material* is material procured that becomes a part of the product or is required for the execution of an order. Typically, direct material includes components, documents, receipts, proofs, or similar materials but also, in the broadest sense, external operations. As a rule of thumb, direct material is material that would not be procured if there were no production or selling.[4]
- *Indirect material* is all material that is not direct material. More specifically, indirect material includes all material that must be procured to maintain company performance and efficiency, e.g., MRO materials.
- *Commodities* are entirely standard and can come from a sufficient variety of sources. Custom semifinished goods are manufactured to spec. Printed circuit boards and injection-molded plastic pieces are good examples.
- *Demand pattern: Items have continuous demand* if demand is approximately the same in every observation period. Items have *discontinuous demand* if many observation periods with no or very little demand are interrupted by few periods with large, for example, 10-times-higher demand, without recognizable regularity.

The following is a list of some of the frequently mentioned traditional procurement strategies. Combinations of these can also appear.

---

2 Module suppliers supply components that for the most part have been developed by the buyer; system suppliers do a great of deal development work of their own.

3 A *channel of distribution* is a series of firms or individuals that participate in the flow of goods and services from the raw material supplier and producer to the final user or customer ([APIC16]). A *distribution channel* is the route that products take along the channel of distribution (cf. [APIC16]).

4 What is said in the following about direct material also holds — particularly in the case of services — for capacity that is to be procured externally and required for producing a product or carrying out a contract.

Multiple sourcing or *multisourcing* refers to the search for the greatest number of sources of a service. This strategy reduces the risk of too great a dependency on another company. This is a common strategy in traditional market-oriented relations.

Single sourcing refers to the search for one single source of a certain service, such as a *single-source supplier*, i.e., a unique supplier per item or family of items. This strategy lowers transaction costs and speeds up order processing; it becomes imperative if short lead times are important.

Dual sourcing plans for two suppliers for an item or an item family. This strategy can reduce the risk of disruption in production due to delivery difficulties on the part of the supplier, while retaining most of the advantages of single sourcing.

Sole sourcing refers to the situation where the supply of a product is available from only one supplier. Usually barriers such as patents preclude other suppliers from offering the product.

With modular or system suppliers, we also speak of *modular sourcing* or *system sourcing*. Here the buyer has the advantage of not having to assemble a great number of parts and components from various suppliers that have no contact with each other. The task of technical coordination and procurement of the required components is taken over by the modular or system supplier.

Global sourcing refers to the search for the best source worldwide of a particular service. This sourcing strategy may be necessary with products and processes involving high technology.

Local sourcing is the search for local sources of a certain service. Intensive cooperation entailing personal meetings or large transports may require this strategy.

2.2.2 Traditional Market-Oriented Relationship Compared with Customer-Supplier Partnership

Intended learning outcomes: Present target area strategies for the traditional market-oriented relationship as well as for the customer-supplier partnership. Disclose possible supply chain risks entailed.

The traditional *market-oriented relationship* is determined by the law of supply and demand. Suppliers are chosen on the basis of low prices. Cost reductions are achieved as suppliers play off one another.

Market-oriented relationships show low intensity in terms of entrepreneurial cooperation. In principle, the duration of the relationship is indefinite, but in fact it is calculated to be short term: the supplier network is flexible, and any relationship may be replaced with another. Related to the entrepreneurial objectives in Figure 1.3.1.1, strategies arise between the producer as buyer and the producer's suppliers as shown in Figure 2.2.2.1.

In sum, price and quality arguments, or productivity in the narrow sense, determine supply and demand. Where friction loss is too high, the customer tends to use insourcing. This is also why there has been a trend in the past toward large and even multinational corporations.

- Target area quality:
  - The supplier is responsible for meeting the customer's quality specification.
  - The customer is responsible for the acceptance and must check the meeting of the specification.
- Target area costs:
  - The customer chooses a supplier, where quality is sufficient, primarily according to the lowest prices, following the law of supply and demand.
- Target area delivery:
  - The customer awards a contract stating desired product, quantity, and delivery due date.
  - Safety stock is necessary in order to avoid the problems caused by delivery delays.
- Target area flexibility:
  - The customer aims for multiple sourcing. This helps offset demand fluctuations and offers better protection against dependencies on individual suppliers.
  - If transaction costs become too high, a make decision is made.
- Entrepreneurial cooperation in the supply chain:
  - Starting from raw materials and standardized parts, it is the customer who develops all products and processes in the supply chain.
  - The customer delegates the manufacturing of semifinished goods or part of the manufacturing process to suppliers. The customer controls the quality, particularly of first deliveries.

Fig. 2.2.2.1 Target area strategies for the traditional market-oriented relationship.

The market-oriented relationship tends to have the following supply chain risks, which on the whole have to be smaller than the risk of over-dependence on one supplier:

- Relatively high costs for the ordering process, due to the great expense of frequent information gathering and contract negotiations. The selection of the supplier must be made within a short time and based on fewer criteria.
- For custom material, changing the supplier can result in significant adaptation costs on the buyer side; for example, the cost of changing production or logistics processes. For this reason, companies should purchase only commodity material via the marketplace whenever possible.
- In a buyer's market, the suppliers can absorb the pressure to lower prices by drastically reducing costs on their side. They minimize costs via lower quality, long delivery times, and low delivery reliability. This can impact the buyer's service level. Therefore the buyer, even though dominant, cannot push prices down too low.

The *customer-supplier partnership*, or simply *customer partnership*, is the strategic and long-term reduction of the number of suppliers to achieve fast and easy operational order processing. The choice of a supplier is made in view of total cost of ownership.

The term stands for an approach to supply and demand that functions not only according to price and quality, because delivery unreliability on the part of suppliers results in opportunity cost for the manufacturer, if it is then unable to supply its own customers. Related to the entrepreneurial objectives, strategies arise between the producer as buyer and the producer's suppliers as shown in Figure 2.2.2.2.

- Target area quality.
  - The supplier achieves a minimum level of quality (according to its own quality evaluation or external certification). Defects and flaws are corrected immediately.
  - To control the quality of the supplier, the customer has access to its production facilities. Both parties mutually improve quality in a supply chain.
- Target area costs:
  - Through single sourcing, greater business volume and thus lower-cost prices are achieved.
  - (Long-term) blanket orders allow intermediate stores to be reduced.
  - The choice of a supplier is made according to total cost of ownership; that is, in consideration of opportunity cost.
- Target area delivery:
  - (Long-term) blanket orders reduce total lead time (supplier and customer).
  - There is now direct delivery on demand to the production facilities of the buyer.
- Target area flexibility:
  - In a buyer-dominated relationship, the buyers' market secures the robustness of the relationship: transaction costs are small, and it is relatively easy to secure a replacement supplier (buy decision).
  - The result of a sole sourcing situation can be a supplier-dominated relationship. In this case in particular, a stable and long-term relationship is important for the buyer.
- Entrepreneurial cooperation in the supply chain:
  - Demands on products and processes to be delivered are mutually defined.
  - The supplier is consulted about each (further) development.

Fig. 2.2.2.2 Target area strategies for the customer-supplier partnership.

The customer-supplier partnership, in short-term order processing, leads to the elimination or reduction of friction loss caused by order negotiations or incoming inspection. With this, many of the advantages of company-internal production for fast lead time can be retained. This type of cooperation with suppliers demands extensive preparations. For this reason, long-term relationships of this kind cannot be established and maintained with a large number of partners. They have to be laid out for the long term but show, however, rather low intensity in terms of entrepreneurial cooperation. Thus, they can be checked again and again with regard to their validity.

Quality goals are achieved through certification of the suppliers; cost goals are achieved through closing blanket order contracts across entire item families or material groups.

A *certified supplier* is a status awarded to a supplier that constantly achieves a minimum level of quality as well as other objectives in other target areas, such as cost or delivery (see also [APIC16]).

An *(item) family contract* is a purchasing order grouping a family of items or a material group together to obtain pricing advantages and a continuous supply of material ([APIC16]).

The customer-supplier partnership tends to have the following supply chain risks, which must be smaller than the advantages mentioned:

- Dependence on one supplier can prove to be too strong (delivery failures, lack of flexibility when demand fluctuates, changes in company ownership on the supplier side). If there is no sole sourcing situation, a switch to dual sourcing may be possible, which can lead to higher unit costs.
- The long-term nature of the relationship and the costs incurred for changing suppliers can lead to a lack of adjustment to pricing developments on the market. After a sufficiently long period, for this reason, continuance of the relationship must be examined and, if necessary, new terms must be negotiated.
- A buyer-dominated relationship can transition into a supplier-dominated relationship unexpectedly, that is, the buyer's market can become a supplier's market. This is, for example, the case with system suppliers if they take over technological leadership, but it can also occur in raw material procurement due to bad natural phenomena or due to speculative manipulation. The relationships must then be renegotiated.

2.2.3 Strategic Procurement Portfolios

Intended learning outcomes: Explain the supplier portfolio describing the degree of mutual dependence between buyer and supplier. Present procurement strategies for material groups in dependency on their logistics characteristics.

In procurement, the material portfolio and the supplier portfolio are established tools for classing strategies. See here, for example, [Alar02], [Kral83], [Bens99]. The attempt is to use simple visual mapping to distinguish between objects and suppliers that are "important" for procurement or have risks and objects and suppliers that are "less important" or associated with fewer procurement risks. The idea is to use this as a basis to derive procurement strategies and recommendations to organizations and IT.

Today, companies seldom procure single items from suppliers and instead procure entire item families or even more comprehensive material groups, or planning groups. This is the case not only for high-cost items *or high-turnover items* (A items — see Figure 11.2.2.1) but also particularly for low-cost items or low-turnover items (C items).

With this, the material portfolio shifts more and more to a material *group* portfolio. On this level, however, only a few groups of indirect material, such as office materials, are "less important." As companies generally work with a few or only one supplier per material group, *all* material groups are in addition associated with certain procurement risks. In such cases, the traditional material portfolio does not yield much information. If that is so, then the *supplier portfolio* becomes more useful. Figure 2.2.3.1 shows a possible supplier portfolio.

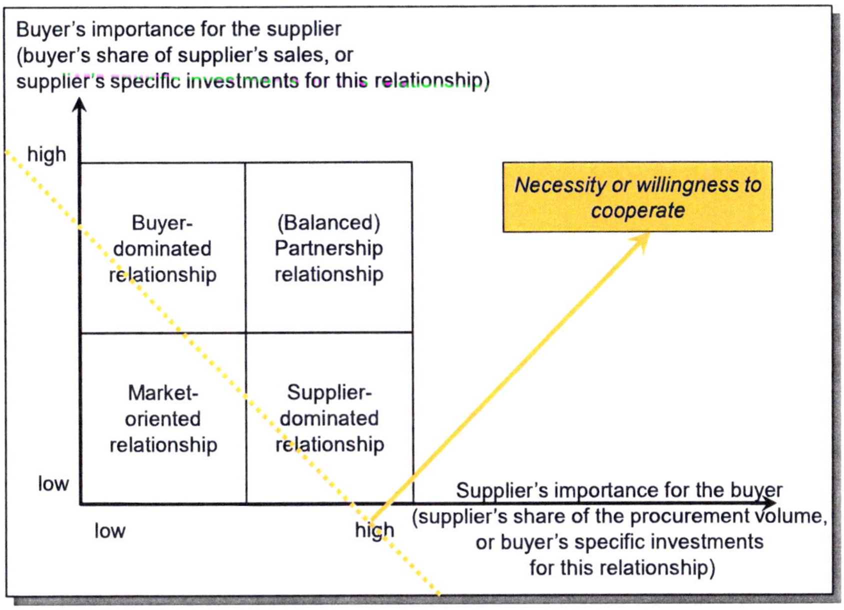

Fig. 2.2.3.1 Supplier portfolio.

This supplier portfolio describes the degree of mutual dependence between buyer and supplier. As companies become aware of their dependence, they generally recognize that a certain degree of cooperation is necessary and are then mostly willing to cooperate.

- In *market-oriented relationships*, that is, via the traditional marketplace, goods or capacities are procured that are unimportant in the eyes of both the buyer and the supplier. Either side, buyer or supplier, can easily change business partners.
- In *buyer-dominated relationships*, goods or capacities are procured that are major in the eyes of the supplier but not for the buyer. Or, the supplier must make one-sided investments related to only the one buyer. These can be, for example, buyer-specific devices, IT platforms, consignment inventories, or know-how on customer-specific processes and business practices. Thus, this type of relationship should be designed for the long term. This means that there must be a customer-supplier partnership to a certain degree but not necessarily a balanced one. Here see Section 2.2.2.
- In *supplier-dominated relationships*, goods or capacities are procured, with the one-sidedness of the investments on the opposite side: the buyer's side.
- In *(balanced) partnership relationships*, goods or capacities are procured that are seen as important in the eyes of both buyer and supplier. Both parties make considerable and mutually demonstrable investments in this relationship to the one business partner. This type of relationship should in every case be designed for the long term, mostly with intensive cooperation. Here see Section 2.3.

Figure 2.2.3.2 shows that — in dependency on the logistics characteristics of the material group — between one and the same pair of business partners a different positioning in the supplier portfolio and related procurement strategies can occur.

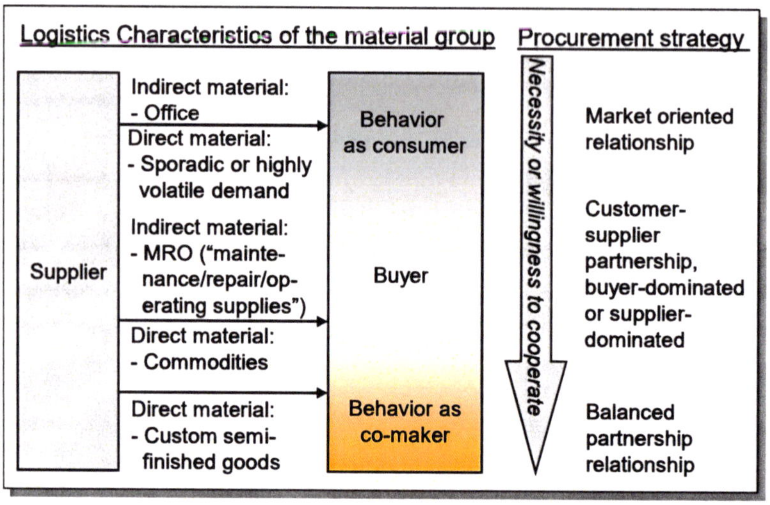

Fig. 2.2.3.2 Procurement strategies for material groups in dependency on their logistics characteristics.

- The procurement of indirect material in the case of office supplies is usually handled via a market-oriented relationship. The behavior of the buyer is like that of any consumer: The buyer follows the classical law of supply and demand. The buyer will try to achieve a position of strength by bundling the requirements. If material of this type shows a sporadic pattern of demand, it is necessary to find especially inexpensive and rapid solutions for their operative procurement, called buy-side solutions (see Section 2.2.5).
- The same can be said of the case with MRO items. However, here the buyer depends on high-quality and on-time deliveries to maintain performance levels towards the buyer's own customers. For material groups of this type, companies will strive to design longer-term relationships; depending on the conditions, a buyer-dominated or supplier-dominated relationship.
- For procurement of direct material, what was said about MRO items largely holds for commodities. Experience has shown that suppliers are quite willing to enter into buyer-dominated relationships. For example, suppliers of material groups of C items (e.g. screws, nuts) invest in efficient delivery logistics such as Kanban or vendor-managed inventory (VMI) due to the large delivery volumes for an extensive range of products. Also, for the business of the buyer, long-term relationships are certainly important also in the case of a supplier-dominated relationship. A supplier-dominated relationship can occur, for example, in the case of use of patents or proprietary technologies of the supplier or with large suppliers of electronic components.
- For *customer-specific semifinished goods,* joint product innovation can also stand in the foreground. One example is assembly modules from system suppliers. Here it is best for the business partners to choose the (balanced) partnership relationship.
- For direct material with highly volatile or lumpy demand, the buyer — in some circumstances also where there is a well-established buyer-dominated or even

partnership relationship — may have to procure a part of the demand, sometimes the entire demand, via a market-oriented relationship, often with the aid of one or more brokers. Here the buyer is often not in a strong position.

Treating material groups instead of individual items is also called *material group management*. Assigning the procurement items to material groups, evaluating suppliers in supplier portfolios, and choosing the procurement strategy are prone to errors. To *reduce* these *risks*, these tasks should generally be performed by a team made up of people from R&D, logistics, production, strategic and operative procurements, and quality assurance.

2.2.4 Strategic Selection of Suppliers

Intended learning outcomes: Describe possible supplier-evaluation criteria for each target area. Explain the score and the gap method for supplier evaluation, using an example with two suppliers. Disclose possible supply chain risks entailed using these methods.

After establishing the procurement strategy, the task is to find appropriate suppliers. The process involved is called procurement market research.

Procurement market research is the systematic gathering of information for an object to be procured. It entails supply market analysis, identification of potential suppliers, and request for quotations.

Procurement market research leads to transaction costs. These are search and initiation costs, as mentioned in Section 2.1.1. They are incurred:

- either ongoing, for example, costs for updating or enlarging the existing market information,
- or due to a special occurrence such as a shortage of raw materials, bankruptcy of a previous supplier, the introduction of a new product, new regulations, or cost-savings measures.

Next, there are supplier evaluation and supplier selection. The supplier's quotations have to be analyzed and compared. Then it is time to conduct negotiations — and if necessary renegotiations — and to make decisions. The costs incurred here belong to negotiation costs.

Supplier evaluation is a method for evaluating the performance of a supplier generally and relative to the required object.

In practice, there are various methods for evaluating suppliers. See here also [WaJo04] or [DoBu03]. One possibility is to measure the performance of a supplier in the same way that the company's own performance is measured, that is, by assessing performance in achieving targets in the areas of quality, costs, delivery, and flexibility (here see Figure 1.3.1.1). As a category in its own right, general working cooperation with the potential supplier is evaluated. Figure 2.2.4.1 shows possible evaluation criteria for each target area.

- Target area quality:
  - Existing quality infrastructure, guidelines, recording of data, quality certifications and awards, qualifications of personnel, CPI programs.
  - Product and process: patents and licenses, mastery of product and process technologies, system supplier, results of examination of first samples.
  - Organization: service (sales and technical support, handling of complaints, fair dealing).
  - Triple bottom line: ethical standards (according to the supplier code of conduct (SCoC) and environment (standards, environmental performance evaluation, emissions certificate, location and transport, packing).
- Target area costs:
  - Unit costs, additional delivery costs, terms of acceptance, delivery, and payment. Ongoing efforts to reduce costs.
  - Stability of exchange rate, inflation in the procurement country.
- Target area delivery:
  - Fill rate, delivery reliability, short lead times in flow of goods, data, and control, terms of transport, and delivery.
  - Ability to handle logistics concepts (such as blanket order processing and blanket releases, VMI, Just-in-Time, Just-in-Sequence).
- Target area flexibility:
  - Customer benefit: product innovation capability, integration of external know-how, capability for custom production, and joint product innovation.
  - Use of resources: technology and production infrastructure, way of dealing with fluctuating order quantities.
- Business cooperation in the supply chain:
  - Importance of the buyer to the supplier and vice versa, location proximity.
  - Information on development of the company (e.g., business development, financial stability, product range, organization, international support).
  - Reduction of total costs (for example, passing on cost savings to customers, joint value analyses, transparency on pricing changes).

Fig. 2.2.4.1 Evaluation criteria for supplier evaluation.

Supplier selection includes preselection of the supplier (or possibly a number of suppliers) and possibly postselection negotiations.

The best suppliers are generally *preselected* using a factor rating. First, the criteria are weighted. The potential suppliers are then rated as to the degree to which they fulfill the individual criteria. This fulfillment rating can be an absolute value (*scoring method*) or a relative value compared to a maximum possible value (*gap method*). Triple bottom-line arguments (see Section 3.3) are usually evaluated separately, as framework conditions.

In many cases, postselection negotiations are conducted with the best-rated suppliers to establish definitive prices or other conditions. This can change the degree of fulfillment of some of the criteria. Finally, the supplier with the highest total score (or possibly, several suppliers with the highest scores) is awarded the contract.

Procurement market research, supplier evaluation, and supplier selection can be supported today in many cases by information technology. Here the transaction costs can be decisively lowered, especially in the case of global sourcing (on this, see the next section below).

Figure 2.2.4.2 shows as an example the score and gap method with two potential suppliers and a limited number of criteria.

Criterion	Buyer's weighting 1=low, 2=medium, 4=high	Degree of fulfillment 1=low, 2=medium, 4=high	
		supplier A	supplier B
Product technology	4	2	4
Process technology	2	4	2
Acquisition and additional costs	2	2	1
Delivery and payment terms	1	2	4
Service level and delivery reliability	4	2	4
Short lead times in flow of goods	2	4	2
Capability for custom production	1	4	2
Capability for joint R&D	2	2	4
Financial stability	4	4	2
Importance of buyer rather low	2	4	1
International support	2	4	2
Total score (number of points)	Max. = 104 (= 26·4)	78	70
100% minus gap	Max. = 100%	75%	67%

Fig. 2.2.4.2 Supplier evaluation: score and gap method with two suppliers.

Evaluation according to criteria as shown in Figure 2.2.4.1 is not a guarantee but rather an aid. There will still be risks entailed in the selection of suppliers. Some of the possible risks are the following:

- The factor rating does not yield an unambiguous decision. Sensitivity analysis can provide clarification here. On this, see also Figure 3.2.1.12 and the explanation in the text.
- The evaluation may be based upon the wrong criteria. Teamwork can help here (second-set-of-eyes principle). It also makes sense to evaluate fewer criteria if the importance of the supplier is low for the buyer than if the importance of the supplier is high (keyword: one-factor comparison versus multifactor comparison).
- Conditions can change at the supplier company, particularly when key persons leave the company.
- If a supplier-dominated relationship is foreseeable or if this is a case of sole sourcing, supplier selection will already involve high costs. Here, personal relationships will be important from the start.

2.2.5 Basics of Supplier Relationship Management and E-Procurement Solutions

Intended learning outcomes: Produce an overview on several categories of e-procurement solutions. Describe various classifications of an electronic marketplace according to the institutional provider, the degree of "openness", and the range.

Supplier relationship management (SRM) is, according to [APIC16], a comprehensive approach to managing an enterprise's interactions with the organizations that supply the goods and services the enterprise uses.

The goal of SRM is increased efficiency of the processes between the company and its suppliers. Technologies, guidelines, and methods that support the procurement processes are used. At the center are IT platforms and *SRM software* for automation of processes, from requests for quotations obtaining blanket orders, up to paying and evaluating the performance of the supplier. A further aim is to exchange information on products and processes to be supplied as early as possible. Also falling under the heading of SRM software are e-procurement solutions.

E-procurement refers to electronic, particularly Internet-based, procurement solutions.

E-procurement solutions can be grouped in several categories according to the institutional provider of the application, as shown in Figure 2.2.5.1 (see also [AlHi01], [BeHa00]).

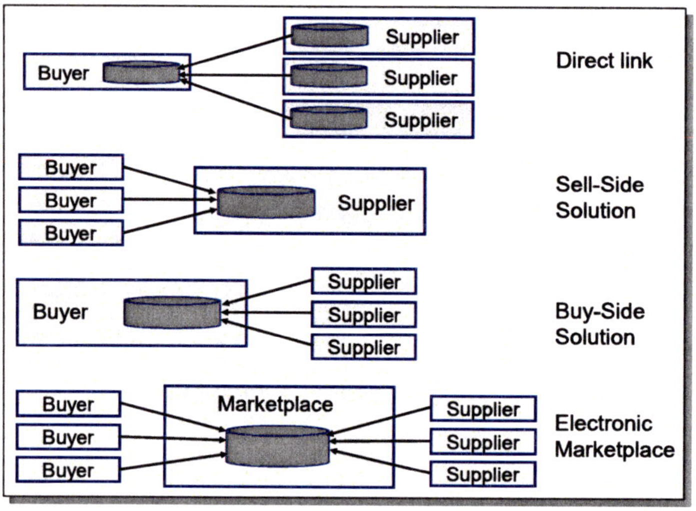

Fig. 2.2.5.1 Categories of e-procurement solutions (the cylinder stands for electronic procurement solutions of the trading platform).

- A direct link connects existing electronic procurement solutions of buyers and suppliers. With the introduction of MRP II / ERP software, modern IT-based systems were already redesigning communication between business partners. EDI solutions, which used standards like EDIFACT, were developed to improve the exchange of data and information among strategic partners in the supply chain. Today, Internet-based solutions have become firmly established because of the use of *XML* (Extensible Markup Language) technologies.

- With a *sell-side solution*, or shop system, the supplier provides access to a catalog of products and ordering procedures on the World Wide Web. This is also typical of B2C applications in the area of consumer goods (for example, amazon.com and dell.com). However, for industrial purchasers, these applications are of limited value, as they do not offer uniform accessibility to the offerings of various sellers. For their decision-making process, each time, the purchasers have to take the trouble to navigate through a number of vendor Web sites.
- For a *buy-side solution*, a standard software, e.g., Ariba, is installed on the buyer's side. The purchasing department uses the software to set up a uniform catalog of products from a number of suppliers. The user within the buyer's company can then select products directly from this catalog and, via an interface to ERP software, place orders. Internal company procedures, such as obtaining approvals from the cost center, are also processed by the system. The systems thus simplify internal company processes and prevent individual orders from being placed with suppliers not in the preferred vendor pool (maverick buying). This reduces transaction costs, but actual purchasing costs remain essentially the same, excepting discounts that can be obtained by concentrating on a few suppliers. Furthermore, the building and updating of company-internal catalogs can require major work efforts, and the required IT environment is comparatively demanding. Therefore, buy-side solutions are more practicable for medium-to large-sized enterprises than for small-sized firms.
- An *electronic marketplace* brings together a comparatively large group of participants and provides a high degree of transparency in real time to all who are taking part; in that sense, an electronic marketplace enables to come a step closer to optimal market conditions. See also [Gull02].

The types of electronic marketplaces are currently being differentiated according to the institutional provider:

- A *dependent marketplace* is financed and managed by a single company or a group of companies. It will therefore tend to be a buy-side or sell-side solution.
- A *neutral, or independent, marketplace* is provided by an independent third party, meaning a neutral party, which can also aggregate and edit the data. In addition, it can add additional services to the marketplace.
- A *consortium marketplace* is built by a consortium and can take on any of the forms described above.

Marketplaces also have differing degrees of "openness":

- A *public marketplace* is open to any company and accessible without proprietary software. A valid e-mail address is often the only thing that is required.
- A *private marketplace*, or a *private trading exchange (PTX)*, is not open to all companies. Participation in a private marketplace or a PTX hinges on certain conditions, such as a membership in a certain trade association. In other cases,

certain companies (such as partners in a supply chain) will exchange data like forecasts or cooperate in some other form (for instance, in the areas of product design, project planning, and project processing).

In the area of investment goods, a third distinguishing feature of electronic marketplaces is the range:

- A *horizontal marketplace* cuts across industries to offer products and services to support general operations and maintenance in many sectors. As a rule, these marketplaces are channels for the buying and selling of indirect materials, such as MRO items or office supplies. Two examples are alibaba.com and wlw.com.
- A *vertical marketplace* is sector specific. Companies in the same sector come together to conduct business, for communication purposes, or to call up industry-specific information. Examples of vertical marketplaces are chemnet.com (chemical industry), efresh.com (food industry), VWGroupSupply.com, supplyon.com (automotive industry), and MFG.com (manufacturing industry).

2.3 Designing a Partnership Relationship

Intended learning outcomes: Present target area strategies for intensive cooperation. Explain the Advanced Logistics Partnership (ALP) model with concepts such as building trust and working out collaborative processes in the supply chain. Describe the virtual enterprise.

According to Figure 2.2.3.1, a partnership relationship is based on weighty and mutually demonstrable investments by customer or supplier — or in the case of a balanced partnership, by both sides — in the relationship with this one business partner. This can be the case, for example, with products and services that must be customized for the user, or also where joint product innovation stands in the foreground — for example, with system suppliers.

This section begins with a discussion of the target area strategies for intensive cooperation in a partnership relationship. A framework is then proposed that brings together the various tasks, methods, and techniques in designing a partnership relationship at all levels of the companies involved. Section 2.3.7, finally, looks at virtual organizations.

2.3.1 Target Area Strategies for Intensive Cooperation

Intended learning outcomes: Produce an overview on the social competency of a company. Present target area strategies for an intensive cooperation in the partnership relationship, describe tasks and investment areas, and disclose possible supply chain risks entailed.

In any case, a partnership relationship must be designed as long term. In contrast to relationships that are buyer-dominated or supplier-dominated, in a *balanced* partnership relationship, the intensity of the cooperation can be significantly greater. For successful partnerships, an enabler objective within the target area of flexibility stands in the foreground. We propose calling this the social competency of a company. This could make sense because of the definition of a company as a *socio*technical system (see the introduction to Chapter 19).

> The *social competency of a company* comprises the flexibility to enter as a partner in supply chains and to link others in a supply chain.

This demands a high degree of social competency, particularly of the leading partner in the supply chain. For many companies, acquiring social competency requires some changes in behavior. Similar to the way that individuals develop social competency for a balanced partnership, a company must develop, first, the ability to play a part in cooperation with others, and, second, the ability to engage others as partners in a trustworthy way, that is, without using coercion.

Related to the entrepreneurial objectives, strategies arise between the producer as buyer and the producer's suppliers as shown in Figure 2.3.1.1. They are complementary to the strategies shown in Figure 2.2.2.2.

- Target area quality:
  - Each partner *feels responsible* for the satisfaction of the end user.
  - Quality requirements are developed and improved mutually.
- Target area costs:
  - All advantages of the customer-supplier partnership are maintained. This leads generally to lower transaction costs.
  - Sharing of methods and know-how among partners reduces costs.
  - Each partner is active in its area of core competence. This yields the best possible return from the resources implemented (including time).
  - Modular sourcing or system sourcing results in fewer ordering processes: instead of many items, only a few modules or systems have to be procured.
- Target area delivery:
  - The same logistics are necessary for all partners (same operational procedures, documents, and so on).
  - Planning & control systems are linked (for example, via EDI).
  - The choice of partners depends with chief importance on speed and; that is, the partner's contribution to short lead times.
  - Local sourcing increases speed and reduces unproductiveness due to misunderstandings.
- Target area flexibility:
  - All partners give impetus toward product design.
  - Customer and supplier invest considerably in the partnership relationship. In a buyer's market, a change in suppliers is possible, but it is connected with the associated costs.
- Entrepreneurial cooperation in the supply chain:
  - All partners are involved in product and process design as well as in planning & control from the start.
  - Friction losses during procurement are eliminated or reduced. In principle, the advantages of a profit-center organization are carried over to independent companies.

Fig. 2.3.1.1 Target area strategies for an intensive cooperation in the partnership relationship.

In buyer's markets, the demand for short product innovation times (time to market) has come to the fore. Cross-company product and process development with partners can be advantageous. When product design becomes more costly, entrepreneurial risk may in this way be more widely distributed. Reducing the time for R&D and production demands more intensive cooperation with partners — and this at all levels of the supply chain (see [Fish97]). This means that partners have insight into the participating companies. One absolute prerequisite is the long-term formation of trust.

Figure 2.3.1.2 groups tasks in which both supplier and buyer invest in three areas, namely, supply chain structure, supply chain organization, and the required information technology.

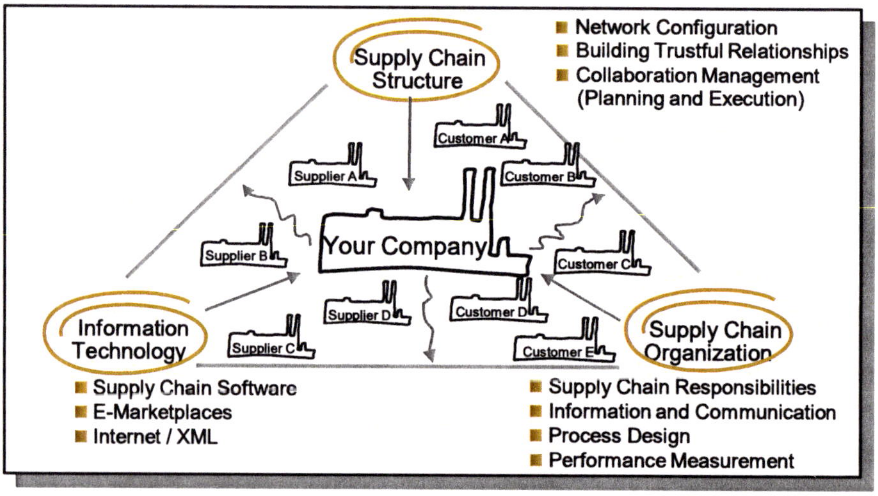

Fig. 2.3.1.2 Tasks and investment areas for intensive cooperation in the partnership relationship.

To support the requirements, specific SCM software has been developed. See also Section 9.2.4. Good communication paths are necessary, both technical (phone, fax, ISDN, EDI, transponders [e.g., RFID sensors]) and personal (regular meetings at all hierarchical levels).

Intensive cooperation in the partnership relationship tends to entail the following supply chain risks — in addition to those of the customer-supplier partnership (see Section 2.2.2), which must be in total smaller than the advantages mentioned.

- Abuse of the knowledge gained from cooperation with partners to enter into business relationships with their competitors.
- Investment by partners that — due to too brief cooperation periods — is not profitable.
- Dependence on a system or modular supplier, because of the very close link, can prove to be excessively strong, but dual sourcing cannot be considered.
- Local sourcing can result in higher prices, suboptimal product quality, and a lack of quantitative capacity flexibility.

2.3.2 The Advanced Logistics Partnership (ALP) Model, a Framework for Implementation of Intensive Cooperation in the Supply Chain

Intended learning outcomes: Produce an overview on the ALP model: a framework for implementation of an intensive cooperation in the supply chain.

A distinguishing feature of a partnership relationship is its long-term nature. The stability and, for a balanced relationship, also the intensive cooperation are guaranteed only if each partner perceives the situation as "win–win." Achieving a win–win situation is the guiding principle in implementing a partnership. The *Advanced Logistic Partnership (ALP) model*[5] puts this basic principle into concrete terms. The ALP model is a framework that describes three management levels of interactions among suppliers and customers:

- Top management level: building trust and establishing principal legal relationships
- Middle management level: working out collaborative processes on the supply chain
- Operational management level: order processing

ALP distinguishes among three phases in the relationship between suppliers and customers:

- Intention phase: choice of potential partners
- Definition phase: search for possible solutions, decision making
- Execution phase: operations and continual improvements

Figure 2.3.2.1 shows the nine fields that result from this structuring. Marked in the fields is the basic sequence of implementation of a partnership among companies.

The top management level supplies the requirements for the middle level, which in turn sets requirements for the operational level. As cooperation on all levels is the key condition for the partnership, it is important to involve all participants early on. Only in this way will the consensus and team spirit, essential to transcorporate cooperation, develop within an organization. With this, the operational and middle management levels also influence the top level, as indicated in the figure by means of the thin arrow. The bigger letters in the fields along the axis from the top left to the bottom right indicate that the main work is performed for the relevant activities. The top right and bottom left fields enclose activities mostly overlooked in practice.

In recent discussion on supply chain management, attention was shifted to the four fields at the bottom right of Figure 2.3.2.1 (highlighted by dark shading). Through an integral perspective and a focus on all business processes in the supply chain, a company aims to coordinate its own planning and execution with that of suppliers and customers to achieve the optimum in the entire supply chain. All the tasks are oriented toward the darkest field, at bottom right, of the nine fields in Figure 2.3.2.1: cooperative order processing in the network. For that is where the value adding takes place that is of interest to the end user. In general, SCM software manages only the tasks in this ninth field. Adequate and efficient

---

5 The ALP model was developed at the BWI Center for Industrial Management at ETH Zurich in cooperation with several firms. See [AlFr95].

implementation of IT support is a necessary, but by itself not sufficient, prerequisite for the success of all other components of supply chain management.

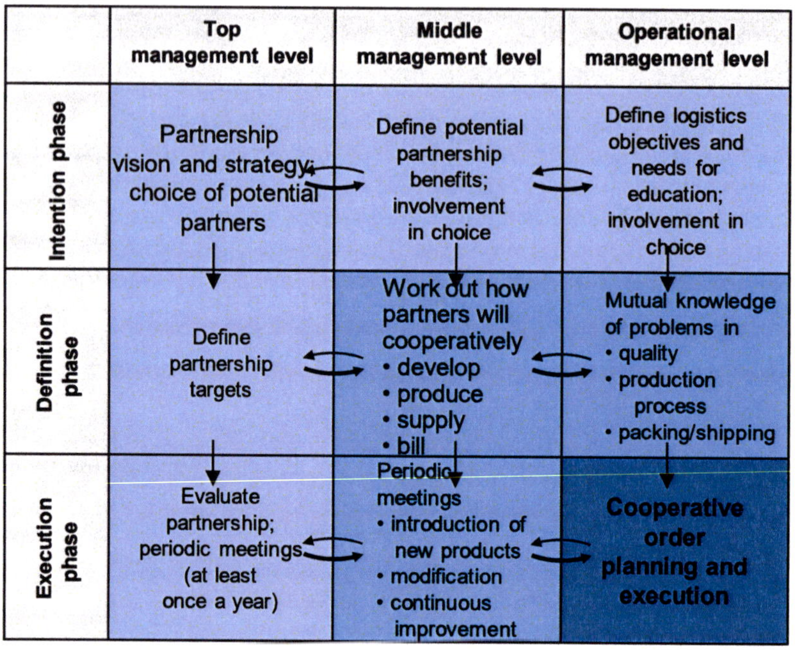

Fig. 2.3.2.1 The ALP model: a framework for implementation of an intensive cooperation in the supply chain.

2.3.3 Top Management Level: Building Trust and Establishing Principal Relationships

Intended learning outcomes: Present trust-building measures in partnership relationships. Disclose reasons for placing emphasis on local networks (local sourcing), and not exploiting strengths in your company's negotiating position.

In the selection of potential partners, the fundamental consideration is whether a partner can fulfill the required goals. Partners must formulate these objectives in a sufficiently clear manner, in order to master deviations in results from the contractual agreements. However, to cooperate long-term and intensively in a supply chain, our research found the *trust-building measures* in Figure 2.3.3.1 to have proven significance.

Even this incomplete list shows that a great deal of social competency is demanded of each partner in the supply chain. But *exploitation* of strengths in a company's negotiating position is in accordance with the traditional buyer mentality. Many supplier relationships that go under the term partnership probably do not deserve that term. In many cases, the positioning of the supplier in the supplier portfolio reveals this fact, which can be definitely correct with regard to the current business activity. See also [Hand95], [WaEg10]. However, if the importance of supplier and buyer are mutually high, only the *balanced* partnership relationship will result in competitiveness of the entire supply chain. People working in procurement learn to become supply chain managers. And employees in sales, production, and logistics can also acquire the required knowledge from procurement and the other areas and successfully become supply chain managers.

Create the required conditions in your own company first.

- The necessary mentality for a mutual win–win situation
- Openness to suggestions from internal and external participants
- Orientation toward procedures and value-adding tasks
- Delegation, teamwork, and the like

Where possible, place emphasis on local networks (local sourcing).

- Local proximity affects not only logistics favorably (speed, transport, and carrying cost), but also has a particularly favorable effect on relationships among the participants.
- The persons participating speak the same language and possibly also see each other outside the business relationship. Such informal contacts are often crucial to the success of a network.
- If there are no "world-class suppliers" in the region, and none can be brought to locate in the area, it is sometimes advantageous to help a local company to become one. It is then called a "world-class local supplier."

Do not exploit strengths in your company's negotiating position.

- Present all intentions openly (no hidden agendas).
- Formulate the objectives of the cooperative venture clearly for all. These objectives may include, for example, achieving a leadership position in a certain market segment or reaching a certain sales volume of an item group.
- The primary competition is competition of the entire supply chain against other supply chains for the favor of the user. Competition between buyer and supplier within the supply chain is of secondary importance. The optimum depth of added value of a partner in a supply chain is not necessarily optimal for the *total* supply chain.
- It is advisable to distribute gains from a cost reduction or increase in earnings equally, because it is the partnership that is the primary factor in success and not the individual contribution of a partner. A balanced win–win situation for all companies involved is prerequisite to long-term or intensive cooperation.

Fig. 2.3.3.1 Trust-building measures in partnership relationships.

2.3.4 Middle Management Level: Working Out Collaborative Processes in the Supply Chain

Intended learning outcomes: Present collaborative processes in the supply chain. Produce an overview on collaborative processes in participative design/engineering. Describe contract issues for a partnership relationship.

At the middle management level, the task is to work out collaborative processes that fulfill the required objectives outlined in Figure 2.3.1.1.

A *collaborative process* is a process in which supply chain trading partners collaborate.

Figure 2.3.4.1 shows processes in the supply chain in the company (value-adding entity) view. To optimize the *entire* supply chain, the processes have to be worked out jointly with the supply chain partners: that is, the customer chain — the customer and the customer's customers, and the supplier chain — the supplier and the supplier's suppliers.

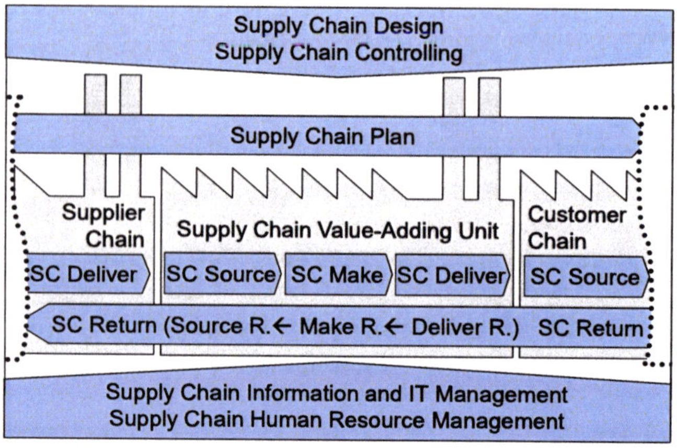

Fig. 2.3.4.1 Cooperative processes in the supply chain.

- Supply chain design comprises the selection of partners in the network and location planning. Defining the controlling processes in the supply chain serves evaluation of the degree of fulfillment of the postulated value. These processes can work out performance indicators of the kind introduced in Section 1.4. Both the design and controlling of the supply chain are processes that determine strategically the subsequent planning and operations processes all along the chain.
- SC Plan means processes for comprehensive planning of demand and the resources in the network, in particular long-term planning. Also belonging here are processes for cooperatively billing. SC Source, SC Make, SC Deliver, and SC Return describe the specific planning and execution tasks in the value-adding entities. Also a part of these tasks is the influence and impact on the adjoining areas of the supply chain, on the side of the network of suppliers and the network of customers. Also compare here the Supply Chain Operations Reference model, SCOR (see Section 1.1.8).
- Processes integrated across the entire network in the area of IT support are a further key to successful supply chains. In addition, education and training of employees at all levels is fundamental, and includes both specialist competencies in the field and social competency. Both of these categories of processes are support processes that determine what the customer views as the value-adding planning and operations processes along the entire supply chain.

Figure 2.3.4.2 takes the example of concurrent, or participative design / engineering, and shows how the increased demands for speed result in challenges for cooperative processes.

Early supplier (or customer) involvement means involving suppliers (or customers) early in product design, drawing on their expertise, insights, and knowledge to generate better designs in less time and designs that are easier to manufacture with high quality ([APIC16]).

Suppliers in particular have to possess know-how of the logistics processes in temporally coordinated design, and production and delivery of components. The transparency of planning and control systems is crucial. All necessary information must be freely exchangeable among the partners. See also the Figure 2.3.1.1.

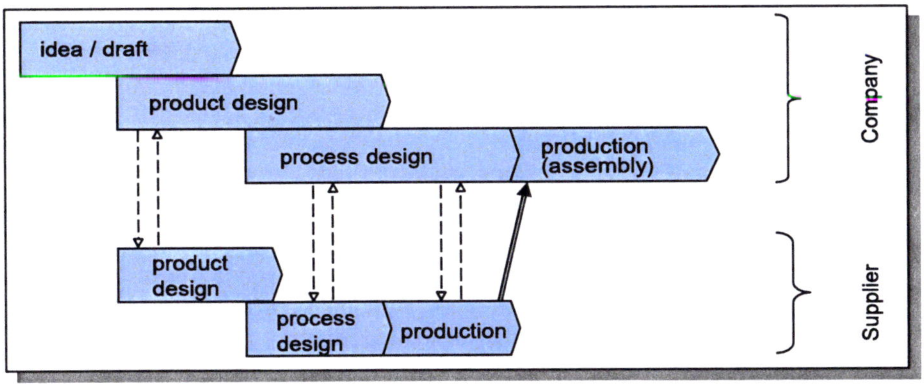

Fig. 2.3.4.2 Collaborative processes in participative design/engineering.

An example of concurrent engineering is the Boeing Company in Seattle, Washington. For some time now, Boeing has worked with partners in the Pacific area, in particular in Japan. These companies manufacture the greater part of the airplane bodies. The cooperation was undertaken with an explicit view to the Asian market. Potential customers are airline companies that belong for the most part to national governments. For decision makers, it is crucial that a part of the value-added chain take place in their own countries. Initial cooperation experience gained with the B747 was then applied to the successful and cost-effective manufacture of the B777. This airplane design was conceived from the start according to the principle of concurrent engineering.

With a view to long-term win–win collaboration, the processes as well as all further agreements should be documented in writing. Here, contracts should be drawn up that address the points outlined in Figure 2.3.4.3:

- *Fundamentals:* Duration, procedure upon liquidation, confidentiality and secrecy, point of arbitration.
- *Quality:* Specification of products and processes,[6] quality management, and measures to handle deviation.
- *Costs:* Distribution of investments in facilities and communication systems.
- *Delivery:* Delivery procedures (normal and rush), batch size and packing, responsibility and cost distribution for warehousing.
- *Flexibility:* Performance indicators and improvement objectives (quality, costs and delivery).
- *Entrepreneurial cooperation:* Project management of new products and production technologies, copyrights and rights of ownership, liability and guarantees.

Fig. 2.3.4.3 Contract issues for a partnership relationship.

For the selection of suppliers, the steps and evaluation criteria in Figure 2.2.4.1 are again useful. Supplier selection often takes place in steps:

---

6 A *specification* is a clear, complete, and accurate statement of the technical requirements of a material, product, or service, and of the procedure to determine whether the requirements are met ([APIC16]).

- A first call for tenders can define the detailed product requirement specifications as well as the rules for detailed cooperation. (Here see Section 2.3.4). The result serves as the basis for the contractual details in the further requests for quotations.
- The next request can be a request for design and production of a prototype, or the first conducting of a service. Especially for service products or for software production, this can allow for testing the feasibility as well as the validity of the rules and agreements.
- In the case of repetitive services or for large batch production of tangible goods, there may be one more selection process. Here the supplier that was responsible for the prototype will not automatically be selected again.

2.3.5 Operational Management Level: Collaborative Order Processing — Avoiding the Bullwhip Effect

Intended learning outcomes: Present collaborative planning, forecasting, and replenishment. Describe vendor-managed inventory and continuous replenishment planning. Identify planning & control tasks for a partnership relationship. Explain the bullwhip effect.

To fulfill the objectives of the collaboration, not only must planning & control systems be linked, but close contact among the participants is also key.

Collaborative planning, forecasting, and replenishment (CPFR) is a process whereby supply chain partners can jointly plan key supply chain activities from production and delivery of final products to end customers. Collaboration comprises business planning, forecasting, and all operations required to replenish raw materials and finished goods (cf. [APIC16]).

A high degree of responsibility and powers of authorization for well-trained transcorporate teams are typical of functional supply chains. Such teams have a mutual understanding of problems with regard to quality, production processes, and delivery, and they strive toward continual improvement of order processing, following the idea of a learning organization.[7] Further measures include techniques of transcorporate data accessing and data revising:

- *Vendor-managed inventory (VMI)*, or *supplier-managed inventory (SMI)*: The supplier has access to the customer's inventory data and is responsible for managing the inventory level required by the customer. This includes in-time inventory replenishment as well as removal of damaged or outdated goods. The vendor obtains a receipt for the restocked inventory and invoices accordingly. Cf. [APIC16].
- *Continuous replenishment (CRP)*: The supplier is notified daily of actual sales or warehouse shipments and commits to replenishing these sales without stockouts and without receiving replenishment orders. Cf. [APIC16].

---

[7] A *learning organization* is a group of people who have woven a continuous, enhanced capacity to learn into the corporate culture ([APIC16]). Each of the individuals of the group is engaged in problem identification and solution generation.

An implementation of such procedures entails a lowering of associated costs and an improvement in speed and stock-inventory turnover.

The planning & control system for customer order processing is composed of the tasks shown in Figure 2.3.5.1. Systemic aspects and the systematic of planning & control within a supply chain will be examined in later sections. Here, we present the terms without further definition or commentary.

Long- and medium-term planning:

- System of blanket orders either for design and manufacturing of products or "only" for capacities to be reserved, such as production infrastructure and staff members

  Procedures:
- Rolling planning and supplier scheduling
- Continual fine-tuning along the short term

Short-term planning and control:

- System of short-range blanket orders for products and capacities reserved by medium-term and long-term planning, now for concrete products or processes

  Procedures:
- Rapid data and control flow
- Blanket releases and delivery schedules to deliver directly on the production site
- In extreme cases, also delivery on the basis of unplanned demand

Fig. 2.3.5.1 Planning & control tasks for a partnership relationship.

It is very important in supply chain management to implement countermeasures to prevent the bullwhip effect (also called the Forrester effect).

The *bullwhip effect* is an extreme change in the supply position upstream generated by a small change or no change in customer demand. Inventory can shift quickly from being highly backordered to being excess.

Observations show that the variation of inventory and order quantities increases up the supply chain from customers to the various tiers of suppliers. In addition, the longer the lead times of goods, data, and control flow, the stronger the bullwhip effect. See [Forr58], [LePa97], and [SiKa19]. Figure 2.3.5.2 shows this effect.

A famous example, analyzed and published by Procter & Gamble, is demand for Pampers® disposable diapers. The bullwhip effect is caused mainly by information processing obstacles in the supply chain; the obstacles are information time lag and distortion (by the actual orders). A countermeasure is adapting manufacturing lead times (see [SöLö03]), based on rapid information exchange on consumption, or demand, by point-of-sale scanning.

Point of sale (POS) is the relief of inventory and computation of sales data at the time and place of sale, generally through the use of bar coding or magnetic media and equipment ([APIC16]).

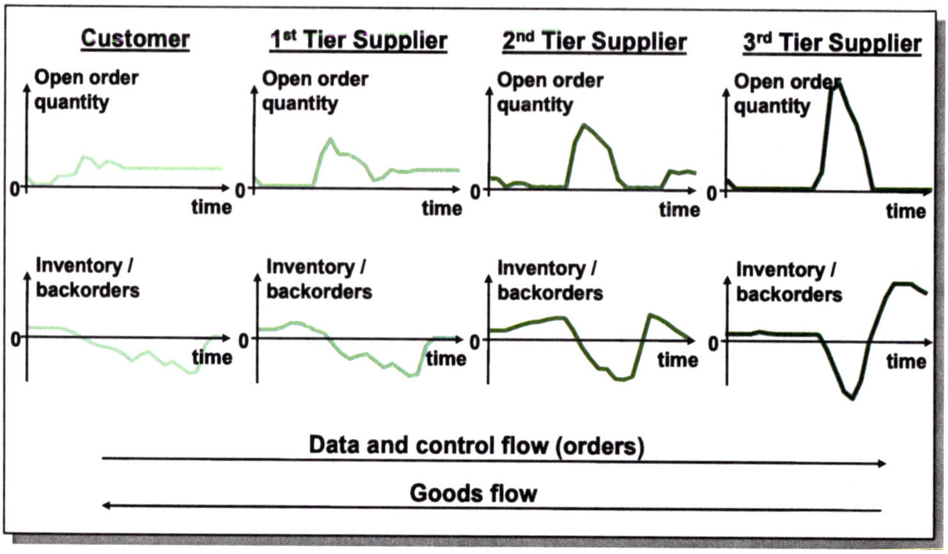

Fig. 2.3.5.2 Open order quantities and inventories/backorders in a supply chain: the bullwhip effect (or Forrester effect).

In distribution control, the term *quick response program (QRP)* stands for an information system that links retail sales along with the production and shipping schedules back through the distribution chain. At the point of sale, it employs electronic scanning and data transfer. It may use direct shipment from a factory to a retailer ([APIC16]).

This type of information system can transmit information on demand from end user back to the highest tier in the supply chain. All partners in the network can rapidly adapt their capacities to current demand and thus avoid large fluctuations in inventory. Experience has shown that this type of information is exchanged only in networks characterized by complete trust.

2.3.6 Example Practical Application

Intended learning outcomes: Present the ALP phases in the case of a high-tech Swiss machine tool manufacturer with a world market presence who wanted to introduce an intensive partnership relationship with suppliers of important assemblies.

Agie-Charmilles SA, a high-tech Swiss machine tool manufacturer with a world market presence (www.agie-charmilles.com), wanted to introduce an intensive partnership relationship with suppliers of important assemblies. Its objective was to reduce the number of partners while improving quality, keeping costs the same, receiving reliable delivery, and gaining more flexibility in terms of quantity and delivery date. Even more important to the company, however, was the creation of conditions that would allow it to focus on its core competencies in developing and assembling its products.

The various suppliers differed in terms of degree of independence and depth of value added. For example, the circuit board manufacturers were all pure subcontractors for performing single operations: The machine tool manufacturer provided not only design and engineering of the circuit boards, but also the production materials required. The manufacturers of metal casing for the encasement of the benches for workpiece processing, while they procured their own materials, did not do their own development. At the foreground stood local suppliers, in most cases small firms with 50 or so employees and individual departments of medium-sized companies. The following outlines the relevant phases of the project.

Intention phase:

The firm's management met for several rounds of discussion with the management of each supplier. Some of the meetings included various employees from affected offices and shopfloors. Emphasis was placed on the win–win principle. A strategic gain for the supplier was greater competitive advantage achieved through taking on additional competencies. Clearly, each supplier was free not to participate. However, it had to reckon with the possibility that it would lose its client to a competitor willing to cooperate.

The circuit board manufacturer, in addition to building its own purchasing department, was to achieve delivery quality of virtually 100% while meeting delivery quantity and delivery timing demands. Successive steps toward reaching these objectives were planned out. The machine tool manufacturer promised full assistance in transferring know-how in these areas.

For the metal casing manufacturer, the objective was to build up an R&D department having "time to market" priorities that matched those of the machine tool manufacturer. Prerequisites with regard to quality, cost, and delivery were defined more precisely.

Officials met four times a year to examine strategies and objectives. Management of the firms met once a year in order to monitor progress. A serious difficulty arose when the production manager of the machine tool manufacturer, who had lent strong ideological support to the project, left his firm. Serious doubts about the continuity of the project made themselves felt among the suppliers. Things only calmed down once a successor to the production manager was chosen — who was known to support the chosen policy. It quickly became apparent that such demanding forms of cooperation do not generally just continue to run at the operational level. Repeated confirmation by responsible officials at the participating companies is essential.

Definition phase:

During this phase, products and processes are developed and introduced. This is the level where it becomes clear whether the trust-building measures were just talk or were instituted solidly. Let's look at subsequent steps by taking as an example one particular circuit board manufacturer and one metal casing manufacturer.

The *metal casing manufacturer* insisted on a minimum sales quantity, set in advance for a period of several years, to have some measure of security in the face of the large investment in CAD. The machine tool manufacturer was not prepared to agree, as this did not accord with its own view of the meaning of a balanced partnership. He argued that it was also incurring an associated risk: namely, potential abuse of the knowledge on the part of the supplier, gained from cooperation with the machine tool manufacturer, used to enter into business relationships with the competitors of the latter. Finally, the attempt at close cooperation had to be abandoned. The supplier had reckoned with this result. This was not a problem, because its volume of business with the machine tool manufacturer made up only 4% of its turnover, and its main business was booming. The machine tool manufacturer soon found other metal casing manufacturers with which it realized its partnership satisfactorily.

The *circuit board manufacturer* saw the requirement to build up its own purchasing department as an opportunity to acquire know-how in qualified office work. Even though, or perhaps because, 80% of its turnover fell to the machine tool manufacturer, it became convinced by the idea that new know-how could in the future be used with other clients as well. Finally, the machine tool manufacturer made up only 20% of its turnover, proving the success of the strategy for the supplier. Yet the required investment in employees who were only indirectly productive was not without risk. Throughout the entire design phase, officials of the two companies paid each other visits to better understand the processes of shared production, procurement, delivery, calculation and associated problems. This led the supplier to initiate a complete redesign of its procedures and the layout of its production infrastructure. But the machine tool manufacturer also had to modify some of its procedures.

Execution phase:

For planning and execution of the machine tool manufacturer's orders to the circuit board manufacturer, they chose a *supplier scheduling* system, that is, a system of long-, middle-, and short-term blanket orders as well as blanket releases with quantities and time periods. This was a method previously unfamiliar to the supplier. But it soon recognized that only improved planning on both sides would allow adherence to the drastically reduced delivery lead times that were now demanded. And only in this way could the supplier, for its part, procure the necessary electronic components from its own supplier in time. A system like this, with continuous planning of ever-more-precise blanket orders and blanket releases, demanded significant investments in rapid and efficient communication techniques of both the company and its supplier.

In the example, the machine manufacturer orders the *exact* required quantity only for the next month, by placing a short-range blanket order. The exact points in time for individual blanket releases during the next month result in this case from a Kanban control principle. In the course of the monthly period, requirements arise unpredictably, so that if the company has not given precise dates for probable delivery, the supplier will have to ready the entire quantity of the short-range blanket order at the start of the month.

2.3.7 The Virtual Enterprise and Other Forms of Coordination among Companies

Intended learning outcomes: Produce an overview on the virtual enterprise and underlying long-term network of potential partners. Present target area strategies for a virtual enterprise and disclose possible supply chain risks entailed.

Are there any possible forms of temporally restricted and yet intensive cooperation, such as for nonrepetitive production or a service that solves a customer's specific problem? A virtual enterprise is a potential answer. The reader is referred to [DaMa93] and [GoNa94].

The adjective *virtual* means, according to [MeWe18], "possessed of certain physical virtues." In reference to the business world, this means that a company functions as such, even though it is not a company in a legal sense.

The concept of virtuality aims to fulfill individual needs of a customer defined on short notice. To this aim several partners, with some of their departments, join together. Toward the customer, they stand as a single company, but later they will separate again. These same companies may then join with other companies to form a new virtual enterprise.

A *virtual enterprise* is a short-term form of cooperation among legally independent partners of various types in a network of long-term duration of potential business partners. Partners cooperate on the basis of mutual values and act toward the third party as a single organization. Each partner is active within the area of its core competence. The choice of a partner depends on its innovative power and its flexibility to enter as a partner into supply chains.

The strength of virtual enterprises lies in their ability to form quickly. In practice, partners must already be familiar with one another. Figure 2.3.7.1 illustrates this concept.

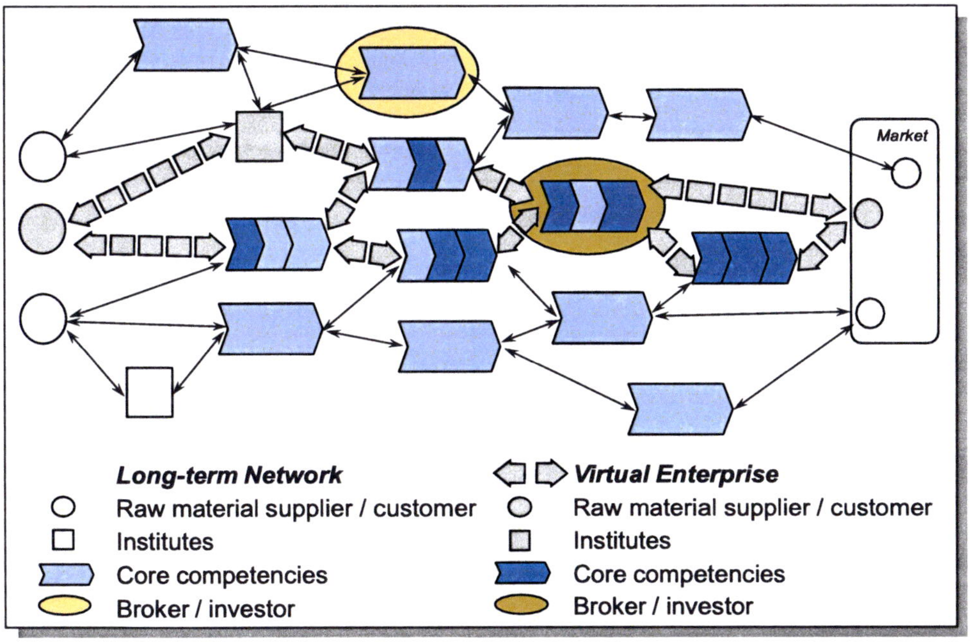

Fig. 2.3.7.1 The virtual enterprise and underlying long-term network of potential partners (from [Brue98]).

The potential partners in the virtual enterprise strive toward a community of interests in the form of a *long-term* network (see thin arrows in Figure 2.3.7.1) that gives each partner competitive advantages. Any obstructions to cooperation must be removed during this phase, so that the individual network participants can develop a relationship of trust. This requires the establishment of good communication channels, both technical and personal. For development cooperation, it makes sense in some cases to stipulate contractual terms.

Figure 2.3.7.2 shows strategies related to the entrepreneurial objectives that arise between the partners of a virtual enterprise. They complement the strategies shown in Figure 2.3.1.1.

- **Target area quality:**
  - Each partner is also co-entrepreneur, i.e., the partner shares the entrepreneurial risks within the entire supply chain. Thus, the partner carries *full responsibility* for end-user satisfaction.
  - Action guidelines, structures, and processes of the virtual enterprise are developed mutually, as is the basic network of potential partners.
- **Target area costs:**
  - All advantages of the intensive cooperation in a partnership relationship are retained. This leads to lowest costs.
- **Target area delivery:**
  - The supply chain for a specific order is formed rapidly.
  - The same operational procedures, documents, etc., are prerequisites.
  - Identical information systems allow maximal exchange of information during mutual product design and production.
- **Target area flexibility:**
  - Criteria for the choice of a partner are (1) its flexibility to enter as a partner in supply chains; (2) its *innovative power*, that is, its flexibility in achieving customer benefit by product and process innovation; and (3) the extent of shared value orientations.
- **Entrepreneurial cooperation in the supply chain:**
  - All potential partners form a long-term network. Friction losses that arise from procurement negotiations are eliminated or reduced. One partner has the role of a broker that puts together the virtual enterprise according to a concrete demand.
  - All partners supply product and process design, and planning & control, from the start. They share mutual involvement and responsibility for success or failure.

Fig. 2.3.7.2 Target area strategies for a virtual enterprise.

Of all entrepreneurial objectives, a company's flexibility is particularly important here. To form a virtual enterprise rapidly, the company boundaries of the potential partners must already be open. Again, as a prerequisite, trust must develop long-term. As a general principle, competition within the network is often ruled out.

A broker is required for the rapid formation of networks. In the case of nonrepetitive production, the broker often also serves as a center for order processing; that is, for planning & control. If lead time must be very short, the planning autonomy of the participating companies must be curtailed. The decisive factor here is the degree of flexibility of the cost center to contribute to the entrepreneurial objectives of the virtual enterprise.

A virtual enterprise tends to have the following supply chain risks — in addition to the risks of intensive cooperation in the partnership relationship (see Section 2.3.1), which must be in total smaller than the advantages mentioned.

- A lack of competition with regard to potential partners in the network means that certain orders cannot be taken on.

- Legal problems (loss / gain distribution, copyrights, rights of ownership) can arise.
- The volume of business is too small to justify the long-term expense involved.

To reduce the risk of a lack of business volume, each of the partners must attempt to anticipate the customers' needs. This demand on agile companies requires them to study the actual use of products to develop proactive proposals for the implementation of new products that have not even occurred to the customer. See Section 1.3.3.

There are many other forms of cooperation as well. For some of these, specific terms have been coined. The following outline places some of these in relation to the strategies and action plans presented, in particular, in relation to the virtual enterprise. See also [MeFa95].

- A *consortium* has a horizontal effect, as the member companies work on partial lots of a total order, but do not — as in supply chains — supply one another. Examples of consortia are found in the building and construction industry. Banks may form a consortium to issue securities. *Supplier partnerships*, where several supplier organizations act as one, can also be a consortium.
- The *strategic alliance* focuses on particular business areas and thus on similar competencies. Also, it is formed as an addition to a company's actual core business, while the virtual enterprise is related directly to a company's core competence.
- A *corporate group* is characterized by dominating the companies of the group via contracts. Such contracts are not necessary in a virtual enterprise. However, companies of a group can certainly take on the role of partners within a virtual enterprise.
- *Joint ventures* involve re-formations and financial participation. These are not necessary in a virtual enterprise.
- *Keiretsu* is a form of cooperation in Japan in which companies remain legally and economically largely independent, even though they are woven together in various ways. The difference between keiretsu and the virtual enterprise is that, in the Japanese variant, membership is permanent.
- *Virtual service organizations* apply the principle of the virtual organization to the structural installation of large international firms manufacturing machines and plant facilities to manage industrial services. Here see [Hart04].

2.4 Supply Chain Risk Management

Intended learning outcomes: Present supply chain risk management. Explain identification and assessment of supply chain risks. Disclose how supply chain risks can be handled.

For some years now, the trend in supply chain management has been fragmentation of the added value. The individual steps are placed globally, that is, placed where the needed skills are the best value for the money. This course of action is connected with an increased number of transports. A second trend has been to apply lean production and just-in-time concepts, in connection with a sustained reduction in the number of suppliers, inventory, and capacity. These two trends result in increased vulnerability of the flow of goods and information.

Owing to the continuing buyer's market, especially for investment products, again for some years now there has been increasing fragmentation of demand for numerous product variants and new products, frequently associated with declining customer loyalty. There are also dangers connected with the increasing scarcity of critical raw materials and with unpredictable open markets, natural catastrophes, and terrorism.

These causes lead to an increase in unexpected disruptions to supply chains, such as loss of suppliers, larger fluctuations in demand, or unsuitable planning. Because at the same time demands and expectations for smooth functioning of supply chains is rising in the wealthy nations, these disruptions can result in great financial losses for companies.

Supply chain risks are potential disturbances and disruptions within the supply chain that are caused by process-inherent or external sources and that negatively impact the objectives of the supply chain.

Supply chain risk management (SCRM) is a methodological procedure for handling supply chain risks.

[Cran03] provides a simple approach to SCRM, and [NoJa04] provides processes and tools for SCRM. Figure 2.4.0.1 shows the elements of a simple procedure suitable for SME that comprises both aspects. This and the following figures were taken from [Zieg07].

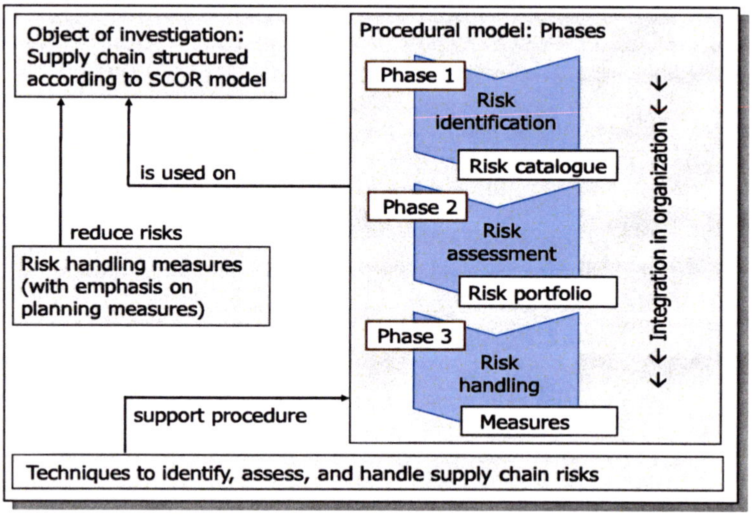

Fig. 2.4.0.1 Supply chain risk management: elements of the methodology.

The crucial element of the methodology is the procedural model, with the three phases risk identification, risk assessment, and risk handling. The following three sections present each of these phases with their techniques and results. The procedural model also includes measures for successful integration of SCRM in the organization of the company.

The causes of supply chain risks can be structured according to the SCOR model, at the network, main processes, and subprocesses levels. Effects are negative deviations from company goals in the areas of quality, delivery reliability, flexibility, operational costs, and investments in current assets and fixed assets. This quickly finds expression in company performance measures such as Economic Value Added (EVA) or Earnings Before Interest and Taxes (EBIT). See here also [Zieg07], [JuPe03], or [WaBo08].

2.4.1 Identification of Supply Chain Risks

Intended learning outcomes: Disclose the process for identifying supply chain risks. Describe the supply chain portfolio, suitable for risk management. Present the depiction of the supply chain using the SCOR model. Explain the supply chain risk catalog.

Figure 2.4.1.1 shows the process of identification of supply chain risks.

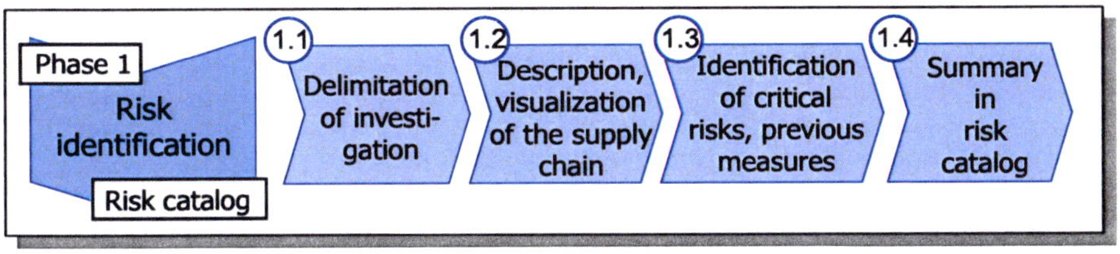

Fig. 2.4.1.1 Process for identifying supply chain risks.

Experience has shown that only a few supply chains can be investigated. The point of delimiting the investigation is so that efforts are not made in the wrong spot. Delimitation is done, for instance, with the aid of a supply chain portfolio.

A *supply chain portfolio* is a description of all of a company's supply chains using criteria such as finished product, customers, and suppliers.

Figure 2.4.1.2 shows a supply chain portfolio that is suitable for SCRM. The two axes are supply chain *vulnerability* and *strategic significance*.

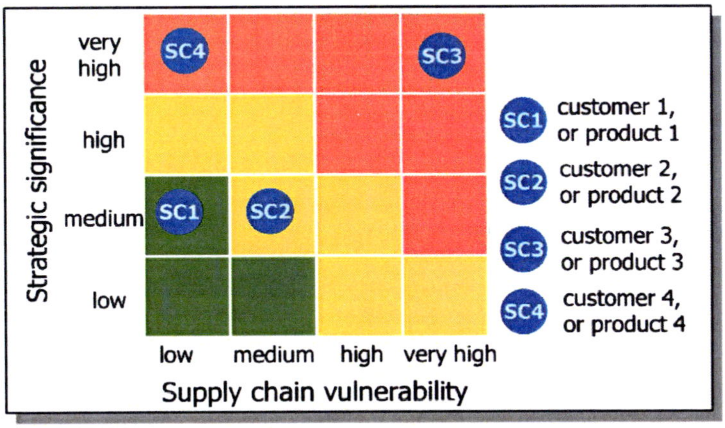

Fig. 2.4.1.2 Supply chain portfolio, suitable for risk management.

In view of risk management, supply chain SC1 in Figure 2.4.1.2 will not require attention. Supply chain SC3 will require a lot of attention, as it is assessed to be both very significant and very vulnerable.

The selected supply chains are then depicted using the SCOR model (for a definition, see Section 1.1.8). This creates transparency, as Figure 2.4.1.3 shows in principle.

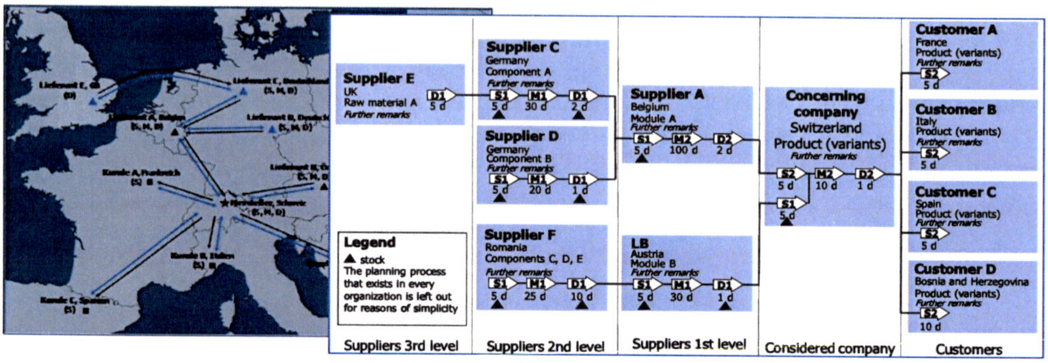

Fig. 2.4.1.3 Depiction of the supply chain using the SCOR model.

The purpose of the step *identification of critical risks and previous measures* is to draw up a catalog of supply chain risks.

A *supply chain risk catalog* is a drawing up of risks relevant to selected supply chains, for example, structured as in Figure 2.4.1.4.

ID	Risk	Risk causes	Risk effects	Previous measures
K1	Customers – inaccurate planning	Unexpected fluctu-ations in the demand for product variants	Inaccurate sales data -> Poor supply chain planning -> …	Increased safety stocks of end products
L1	Supplier C – make – failure	Lack of production machines; equipment is irreparably damaged	Supplier C absent for several months -> … -> Supply chain loses market share	none

Fig. 2.4.1.4 Supply chain risk catalog.

For putting together the risk catalog, it is advisable to use risk checklists, structured according to SCOR processes and the target areas of the company goals. With systematic use of the lists, risks are less likely to be overlooked.

2.4.2 Assessment of Supply Chain Risks

Intended learning outcomes: Disclose the process for assessing supply chain risks. Explain FMEA and FTA techniques for assessing the probability of occurrence qualitatively or quantitatively. Describe FMEA and business interruption value techniques for assessing the business impact qualitatively or quantitatively. Present the supply chain risk portfolio.

Figure 2.4.2.1 shows the process of assessment of supply chain risks.

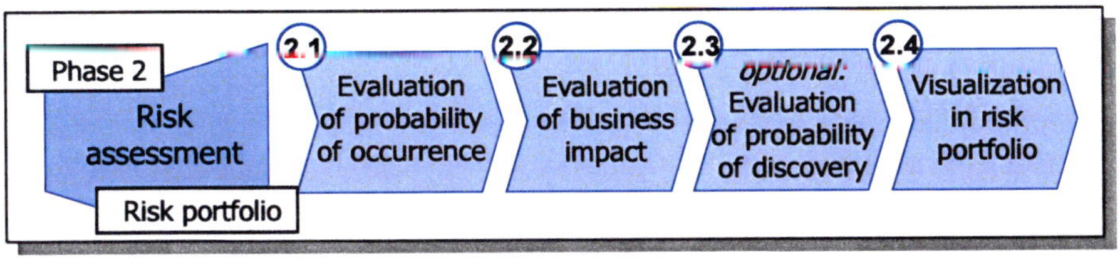

Fig. 2.4.2.1 Process for assessing supply chain risks.

The tools mentioned in the following come mainly from the area of product design and production management. They were adapted for use in supply chain management.

Probability of occurrence can be assessed qualitatively or quantitatively. For qualitative assessment, a suitable tool is *failure mode and effects analysis (FMEA)*, as shown in the example in Figure 2.4.2.2. It is based directly on the risk catalog.

Probability of occurrence scale	
1-2	Very seldom (every few years)
3-4	Seldom (yearly)
5-6	Sometimes (twice a year)
7-8	Frequently (monthly)
9-10	Very frequently (weekly)

Fig. 2.4.2.2 Supply chain FMEA for qualitative assessment of the probability of occurrence of supply chain risks.

For quantitative assessment, fault tree analysis (FTA) can be used, as shown in the example in Figure 2.4.2.3. FTA can be costly, but it yields an in-depth understanding of causes and reasons. For further techniques, also see Section 18.2.4.

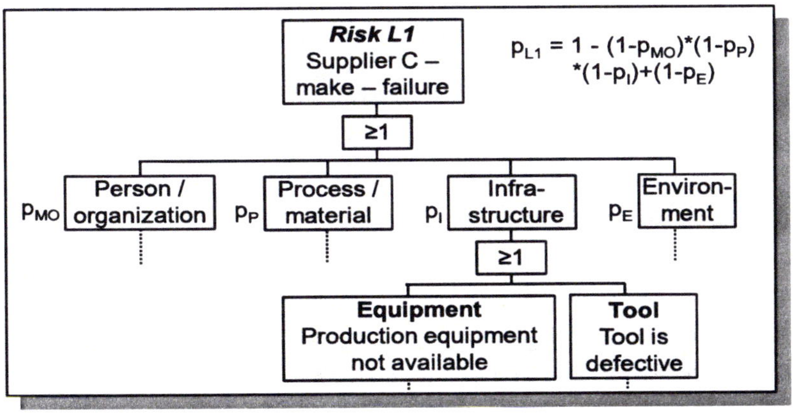

Fig. 2.4.2.3 Failure tree analysis (FTA) for quantitative assessment of the probability of occurrence of supply chain risks.

Business impact can also be evaluated qualitatively or quantitatively. For qualitative assessment, FMEA is again suitable, as shown in the example in Figure 2.4.2.4. Supported by *event tree analysis (ETA)*, it is efficient and delivers useful results. For quantitative assessment, the *business interruption value* can be computed, the main components of which can be seen in the example in Figure 2.4.2.5. Experience has shown that the value depends strongly on one or two parameters, the values of which — similar to the case with qualitative methods — can be determined only imprecisely in practice.

Business impact scale	
1-2	Minor effects on costs/ investments
3-4	Major effects on costs/ investments
5-6	Loss of profit margin
7-8	Loss of orders and customers
9-10	Serious loss of image

Fig. 2.4.2.4 Supply chain FMEA for qualitative assessment of the business impact of supply chain risks.

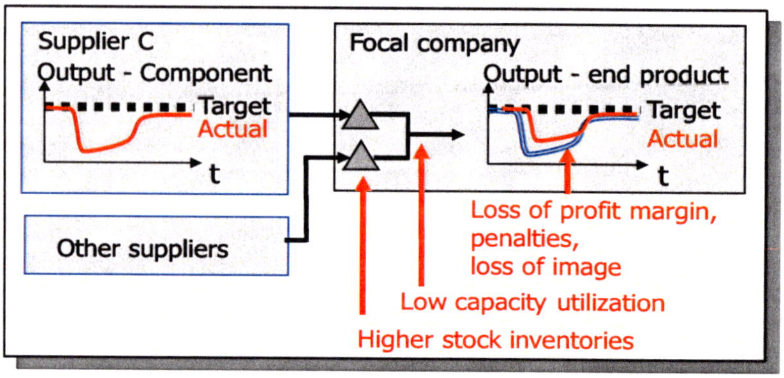

Fig. 2.4.2.5 Business interruption value for quantitative assessment of the business impact of supply chain risks.

Probability of discovery is not evaluated as frequently. Here see [Zieg07].

In practice only relatively few risks can be examined, but these must be examined as in-depth as is necessary. The purpose of the risk assessment is to not expend money and effort in the wrong spot. Condensing the analysis results yields a supply chain risk portfolio.

The *supply chain risk portfolio* shows the supply chain risks on the two axes *probability of occurrence* and *business impact*, as shown in Figure 2.4.2.6.

In the example in Figure 2.4.2.6, risks K1 and L1 will require a lot of attention. Risk K1 has a high probability of occurrence and medium business impact. Risk L1 has a lower probability of occurrence, but it has very high business impact.

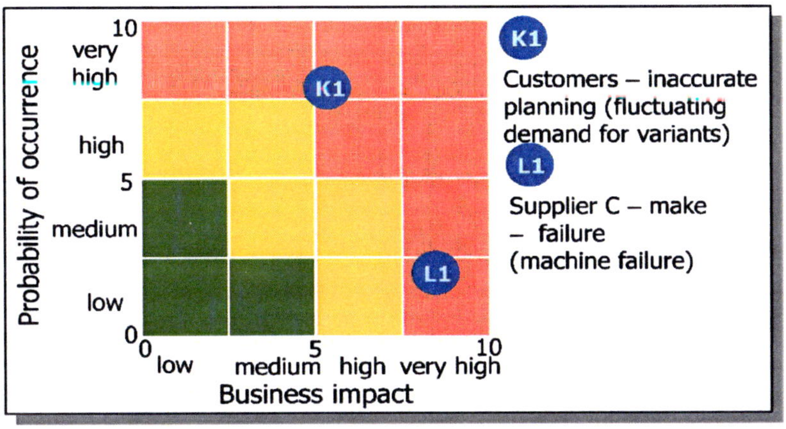

Fig. 2.4.2.6 Supply chain risk portfolio.

2.4.3 Handling Supply Chain Risks

Intended learning outcomes: Disclose the process for handling supply chain risks. Describe possible search fields and measures for an example supply chain risk. Disclose how to evaluate the possibilities for action.

Figure 2.4.3.1 shows the process of handling supply chain risks.

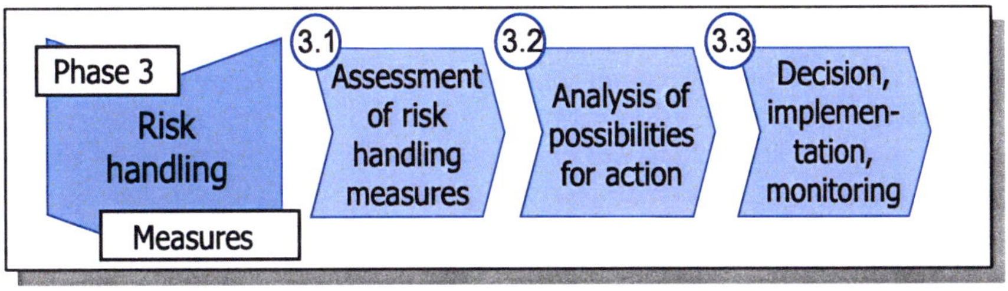

Fig. 2.4.3.1 Process for handling supply chain risks.

Possible measures for handling the risk are dependent on where the risk is positioned in the supply chain risk portfolio.

- A high probability of occurrence requires cause-related measures to reduce the risk, such as more robust planning and more flexible design of the supply chain, including postponement and cooperative processes within the company and with suppliers.
- A high business impact requires effects-related measures to reduce or shift the risk, such as emergency plans (keyword *business continuity and resilience planning* (BCRP), a concept that shows the threats for the company as well as measures to prevent these and to restore operating performance), flexibility through supply chain event management (SCEM) or through *risk pooling* by manufacturers and retailers by pooling together inventory of product families with high demand variety, or redundancy (that is, through alternative suppliers, or higher capacity, or inventory).

Practical experience has shown that the concepts mentioned above are only useful when they are given content in concrete cases. Figure 2.4.3.2 therefore shows possible search fields and measures in the case of supply chain risk K1 from Figure 2.4.2.6, namely, inaccurate planning owing to fluctuating demand for variants.

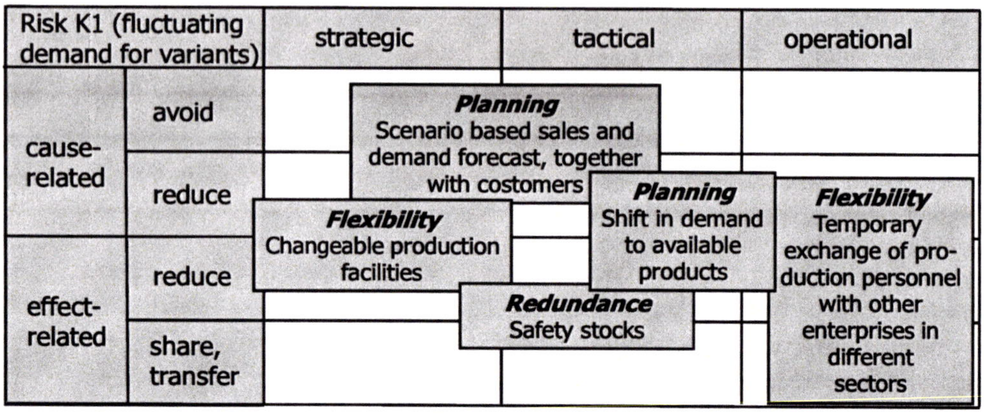

Fig. 2.4.3.2 Search fields and measures for an example supply chain risk.

As the probability of occurrence of K1 is very high, cause-related measures stand in the foreground. In the concrete case, these are planning measures.[8] As K1 also has medium business impact, effect-related measures also come into play, such as flexibility and redundancy measures.

The analysis of possibilities for action includes a cost-benefit analysis. This can be done through a reassessment of all risks using the supply chain FMEA or also with a utility analysis or capital budgeting. Practical experience shows here that planning measures are generally less costly than measures to increase flexibility, which in turn are less costly than measures to increase redundancy. On the other hand, redundancy measures can often be put into effect quickly.

Decision, implementation, and monitoring comprise a listing of the selected measures with an implementation plan for each. Both the position of the risks in the risk catalogue and the selection of the supply chains to be examined must then be updated on a rolling basis.

---

[8] A scenario in sales and demand forecasting comprises a particular group of variants and is estimated with a particular probability. A weighted amalgamation of all scenarios allows more robust planning of the variants than before.

2.5 Summary

Estimation of transaction costs leads to a make-or buy decision. A buy decision results in outsourcing: Parts of the supply chain are turned over to other companies. A make decision results in insourcing; here there are various possibilities of forming organizational entities within the company. With global supply chains, the assessment of the tax situation and the total cost of ownership also inform the make-or-buy decision.

For business relationships with suppliers, concepts such as multiple sourcing, single sourcing, and so on describe only one aspect. The objects to be procured are grouped into material groups. The supplier portfolio distinguishes market-oriented, buyer-dominated, supplier-dominated, and (balanced) partnership relationships. Based on their logistics characteristics, the appropriate supplier relationships are determined for the material groups. As possible business relationships, the traditional marketplace, the customer-supplier partnership, the intensive cooperation in a partnership relationship, and the virtual enterprise are introduced. In all cases, the target area strategies and the supply chain risks that develop between the producer as buyer and the producer's suppliers are discussed.

Supplier selection is then performed via procurement market research, supplier evaluation, and the actual selection of suppliers. All of the steps of search and initiation and negotiation, in particular, can today be supported by e-procurement; for example, by sell-side or buy-side solutions as well as horizontal and vertical marketplaces. In partnership relationships, in particular, there can be several requests for bids: a first one for product requirement specifications and the rules of cooperation, a second one for prototyping, and a third one for repetitive procurement.

The proposed Advanced Logistics Partnership (ALP) Model is a basic framework for the implementation of intensive cooperation in the supply chain. The required tasks, methods, and techniques at all levels of the companies involved are introduced, with examples of practical application.

Supply chain risk management is a methodological procedure for handling supply chain risks. The crucial element of the methodology is the procedural model with the three phases risk identification, risk assessment, and risk handling.

2.6 Keywords

advanced logistic partnership (ALP), 99
backward integration, 74
bullwhip effect, 105
cost center, 75
CPFR (collaborative planning, forecasting, and replenishment), 104
CRP (continuous replenishment planning), 104
customer-supplier partnership, 86
direct material, 84
electronic marketplace, 95
forward integration, 74
free trade area (FTA), 76
global sourcing, 85
horizontal marketplace, 96
indirect material, 84
insourcing, 72
landed costs, 80
local sourcing, 85
make-or-buy decision, 72
multiple sourcing, 85

outsourcing, 72
point of sale (POS), 106
private trading exchange (PTX), 95
process-focused organization, 75
product-focused organization, 75
profit center, 75
quick response program (QRP), 106
rules of origin (RoO), 77
single sourcing, 85
social competency of a company, 97
sole sourcing, 85
supplier relationship management (SRM), 94
supply chain design, 71
supply chain risk, 112
supply chain risk managment, 112
tariff, 76
total cost of ownership, 80
transaction costs, 72
vendor-managed inventory (VMI), 104
vertical marketplace, 96
virtual enterprise, 109

2.7 Scenarios and Exercises

Intended learning outcomes: Explain various aspects of the advanced logistics partnership (ALP) model. Evaluate different company relationships in the supply chain.

2.7.1 Advanced Logistics Partnership (ALP)

Intended learning outcomes: Identify examples of local sourcing with world-class local suppliers. Disclose and analyze the situation of three companies making up a supply chain for processing a particular kind of timber, which grows in a particular region.

a. Figure 2.3.3.1 presented arguments for the emphasis on local networks (local sourcing with world-class local suppliers) that is a feature of the ALP model. Do you know of any companies (including some in the service industry) that follow this principle? Do some Internet research and find out whether these companies address the issue of local sourcing on their Web sites.

b. A supply chain processes a particular kind of timber with special qualities, which grows in a particular region. The following companies make up the supply chain: (1) a lumber mill with various forest owners as potential suppliers, (2) a wood planing mill, and (3) a company that provides surface treatments and finishes and handles distribution. For the wood planing mill, how would you take into consideration and cope with the following risks involved in forming this supply chain:

   b1. There is a risk that the lumber mill could be bought out by a paper factory that requires the entire production for its own use. (*Hint*: Compare this situation with Sections 2.2.2 and 2.3.1.)

   b2. Storms could cause widespread destruction of the forests, resulting in a sharp rise in the price of this type of wood on the free market. (*Hint*: Compare this situation with the argumentation on “trust-building measures in partnership relationships” presented in Figure 2.3.3.1.)

2.7.2 Evaluate Company Relationships in the Supply Chain

Intended learning outcomes: Describe a supply chain with extensions in the wood industry. Explain and perform a classification of company relationships in this supply chain.

You are commissioned to conduct an analysis of a supply chain in the wood and furniture industry. The IGEA Company is a furniture company known mainly for its successful cash-and-carry furniture retail business. Faced with enormous cost pressures, IGEA management has decided to explore the possibility of forming a supply chain. Internal company improvement measures simply do not promise more than marginal cost savings, and prices paid to suppliers cannot be lowered any further without risking losing some suppliers, which would mean that IGEA could no longer offer some of its products.

IGEA managers have read a study that you published on cost savings achieved through transcorporate supply chain management. They believe that the savings they could achieve would give them an edge over their main competitor, the INFERNIO Company. IGEA will therefore head the supply chain project, taking on the role of integrator. Because of its dominant position on the market, IGEA succeeds in convincing its main suppliers and some of the affiliated subsuppliers to join them in taking this transcorporate step.

Figure 2.7.2.1 shows the interrelationships among the companies concerned. The companies highlighted in gray will be integrated into the new supply chain described below. As of now, five companies have agreed to form the supply chain:

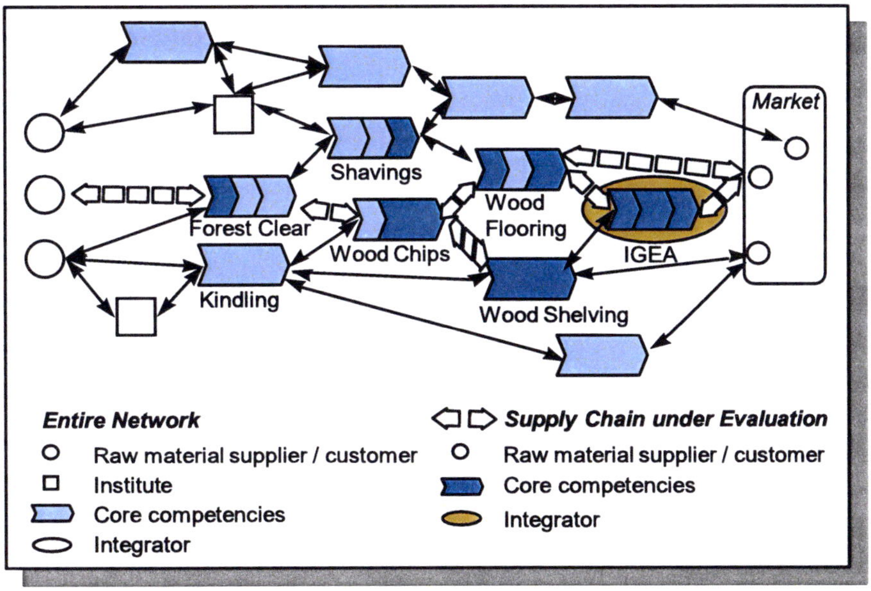

Fig. 2.7.2.1 A supply chain in the wood industry (cf. Fig. 2.3.7.1).

- Forest Clear Co.
- Wood Chips Co.
- Wood Flooring Co.

- Wood Shelving Co.
- IGEA Co.

For the following analyses and considerations, however, it is important not to lose sight of other existing company relationships, since it might make sense to include additional companies as partners in the cooperative project or to sever some of the existing company relationships (for example, Kindling Co., Shavings Co., and other possible companies).

You will need the following details of some of the company relationships to conduct your analysis and identify potential improvements:

- Business relationship between Forest Clear Co. and Wood Chips Co.: Forest Clear, based in Finland, is known for its bold dealings with its customer, the Wood Chips Co. Delivery agreements are very short term, which necessitates frequent, tough negotiations. Still, the excellent quality of the Forest Clear material forces Wood Chips to continue doing business with them. However, delivery delays are becoming more and more frequent, to the point that this is now affecting Wood Chips's own fill rate. The chief buyer at Wood Chips has invested many hours in meetings with the wood supplier in an attempt to improve the situation, but Forest Clear is resistant to showing its cards. The Forest Manager does not encourage visits, and the company will not reveal their long-term product and capacity planning. Although Forest Clear had been asked repeatedly to develop a concept for eliminating the problems, they have produced no proposals.
- Business relationship between Wood Chips Co. and Wood Flooring Co.: The relationship between Wood Chips and Wood Flooring is very tense. The delivery reliability of Wood Chips, as subsupplier of high-quality boards, is seriously deficient, which is having an extremely negative effect on Wood Flooring's own service level. For this reason, Wood Flooring is often forced to procure products from another subsupplier, Shavings Co., which entails considerable additional costs and effort. Another factor is the tense relationship between the chief buyer at Wood Flooring and management at Wood Chips. Because of the very large volume of material purchased, Wood Flooring has not been able to find another, equivalent supplier. In addition, because it procures such vast amounts of material, Wood Flooring has a strong enough position in the market that it can often dictate prices. And naturally, over the years, it has frequently exploited this advantage. Blanket contracts with a 5-year duration thus contain a 2.5% discount annually, based on forecasted productivity increases and a learning curve on the part of the supplier. This is another reason why Wood Chips does not want to work with Wood Flooring.
- Business relationship between Wood Chips Co. and Wood Shelving Co.: Wood Shelving and Wood Chips enjoy a very friendly and constructive business partnership. Wood Shelving is one of Wood Chips' most important customers, and Wood Chips is willing to respond promptly and without complications to any special requests. The business relationship has advanced to the point where monthly product management meetings at Wood Chips are attended by a purchaser from Wood Shelving, who reports on forecasts and trends in the sales market. For delivery, 1- to 2-year contracts are concluded. There are problems, however, with operational order processing. Orders are made by fax and by mail, but also by telephone, which

results in a lot of redundant data, and no one is sure what the correct figures are. The business relationship is supported by the geographical proximity of the two companies (within 20 miles of each other).

- Business relationship between Wood Shelving Co. and IGEA Co.: IGEA is known for its readiness to invest very heavily in new technologies. For instance, IGEA has already set up an EDI system with its main suppliers. As soon as a certain number of products are rung up at the cash registers or withdrawn from stock, automatic orders are placed with suppliers. The order quantity is then subtracted from the agreed-upon blanket order purchasing quantity. In selecting its suppliers, IGEA also has strict criteria: suppliers have to satisfy IGEA's environmental concept, but they also have to meet high quality standards. Wood Shelving Co. has been able to meet these initial demands, but it is experiencing big difficulties in fulfilling the quantity demanded and adapting to the strong fluctuations in the demand. The consequences for Wood Shelving are serious earnings losses, which have led to overtime and special shifts as well as enormous quantities of inventory. The two companies have engaged in heated discussions and mutual recriminations. Because of the unpredictable fluctuations, they have mandated a task force to examine the problem. Despite the frequent bottlenecks, IGEA wants to continue doing business with Wood Shelving. The product quality is high, and the company shows positive cooperation when it comes to new projects.
- Business relationship between Wood Flooring Co. and IGEA Co.: Wood Flooring and IGEA also have a mutual information exchange program. Because demand does not fluctuate and sales processing of these higher-quality products are stable, the exchange of forecast information and planning is optimal. Advertising campaigns are planned cooperatively, and the two companies split the necessary costs as well as the additional earnings. However, as the product assortment of IGEA has a low demand for such high-quality products, the companies cooperate mainly for short-term products or particular partnerships of convenience. For this reason, Wood Flooring is also very active in the international market, and, due to its flexibility, it is highly esteemed as a business partner.
- *Other company relationships, which are not being considered in the start phase of the new supply chain project (shown in white):* Kindling Co. and Shavings Co. have only recently entered into IGEA's supply chain conglomerate. They partially supply to Wood Chips and Wood Flooring, but there are efforts underway to have them supply directly to Wood Shelving Co. IGEA has initiated this and wants to further expand its role as an integrator in the entire network.

Your task: Position the five business relationships listed above and enter your results into the portfolio shown in Figure 2.7.2.2.

Evaluate the individual companies' potential development opportunities and development strategies within this supply chain. Indicate the trend (using an arrow) that best describes the future directions of each company. Write a one-page explanation of the positions and the corresponding trends of the customer-supplier business relationships. Include possible future business relationships with Wood Chips and Wood Flooring.

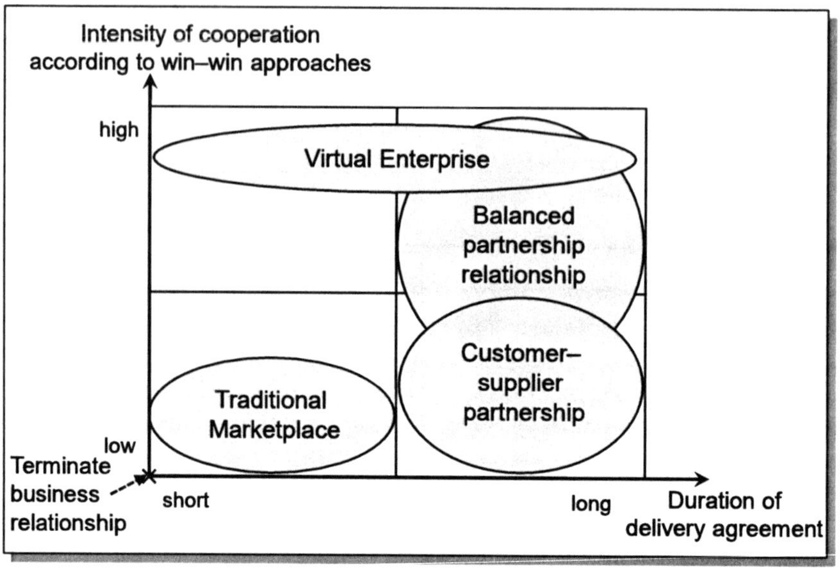

Fig. 2.7.2.2 Classification of company relationships in the supply chain.

2.8 References

Alar02 Alard, R., "Internetbasiertes Beschaffungsmanagement direkter Güter — Konzept zur Gestaltung der Beschaffung durch Nutzung internetbasierter Technologien," Dissertation ETH Zurich, No. 14772, 2002

AlFr95 Alberti, G., Frigo-Mosca, F., "Advanced Logistic Partnership: An Agile Concept for Equitable Relationships between Buyers and Suppliers," Proceedings of the 1995 World Symposium of Integrated Manufacturing, APICS, Auckland, 1995, pp. 31–35

AlHi01 Alard, R., Hieber, R., "Electronic Procurement Solutions for Direct Materials," International Conference on Industrial Logistics, July 9–12, 2001, Okinawa, Japan

APIC16 Pittman, P. et al., APICS Dictionary, 15th Edition, APICS, Chicago, IL, 2016

BeHa00 Berrymann, K., Harrington, L.F., Layton-Rodin, D., Rerolle, V., "Electronic Commerce: Three Emerging Strategies," *The McKinsey Quarterly*, 2000, No. 3 Strategy, pp. 129–136

Bens99 Bensaou, M., "Portfolios of Buyer-Supplier Relationships," *MIT Sloan Management Review*, No. 40, pp. 36–44, Cambridge, 1999

Brem10 Bremen, P., "Total Cost of Ownership — Kostenanalyse bei der globalen Beschaffung direkter Güter in produzierenden Unternehmen," Diss. ETH Zurich, 2010

Brue98 Bruetsch, D. et al., "Building Up a Virtual Organization," in Schönsleben, P., Büchel, A., Eds., "Organizing the Extended Enterprise," Chapman & Hall, London, 1998

Cana19 Canadian Border Service Agency, "Trilateral Customs Guide through NAFTA," http://publications.gc.ca/collections/collection_2017/asfc-cbsa/PS38-83-2015-eng.pdf, accessed August 2019

Coas93 Coase, R.H., "The Nature of the Firm," in Williamson, O., Winter, S., "The Nature of the Firm — Origins, Evolution, and Development," Oxford University Press, London, 1993, p. 18

Cran03 Cranfield University School of Management, "Understanding Supply Chain Risk: A Self-Assessment Workbook," Department for Transport, Cranfield, UK, 2003

DaMa93 Davidow, W.H., Malone, M.S., "The Virtual Corporation," HarperCollins, New York, 1993

DoBu03 Dobler, D.W., Burt, D.N., Starling, S., "World Class Supply Management," 7th Edition, McGraw-Hill, New York, 2003

Ellr93 Ellram, L. M., "A Framework for Total Cost of Ownership," *The International Journal of Logistics Management*, No. 4, S. 49–60, 1993

EsSu05 Estevadeordal A., Suominen K., "Multilateralism and Regionalism. The New Interface — Chapter V — Rules of Origin," United Nations Conference on Trade and Development, 2005, pp. 50–78

Fish97 Fisher, M.L., "What Is the Right Supply Chain for Your Product?" *Harvard Business Review*, March–April 1997, pp. 105–116

Forr58 Forrester, J.W., "Industrial Dynamics: A Major Breakthrough for Decision Makers," *Harvard Business Review*, 36(4) 1958, pp. 37–66

GoNa94 Goldman, S.L., Nagel, R.N., Preiss, K., "Agile Competitors and Virtual Organizations, Strategies for Enriching the Customer," John Wiley & Sons, New York, 1994

Gull02 Gulledge, T.R., "B2B eMarketplaces and Small- and Medium-Sized Enterprises," *Computers in Industry*, 49 (2002), pp. 47–58

Hand95 Handy, Ch., "Trust and the Virtual Organization — How Do You Manage People You Don't See?" *Harvard Business Review*, 5/6 1995, pp. 40–50

Hart04 Hartel, I., "Virtuelle Servicekooperationen — Management von Dienstleistungen in der Investitionsgüterindustrie," BWI-Reihe Forschungsberichte für die Unternehmenspraxis, vdf Hochschulverlag, Zurich, 2004

JuPe03 Jüttner, U., Peck, H., Christopher, M., "Supply Chain Risk Management — Outlining an Agenda for Future Research," *International Journal of Logistics: Research and Applications*, 6(4) 2003, pp. 197–210

Kral83 Kraljic, P., "Purchasing must become supply management," *Harvard Business Review*, 61 1983, pp. 109–117

LePa97 Lee, H.L., Padmanabhan, V., Whang, S., "Information Distortion in a Supply Chain: The Bullwhip Effect," *Management Science*, 43(4) 1997, pp. 546–558

LiLi07 Li, Y., Lim, A., Rodrigues, B., "Global Sourcing Using Local Content Tariff Rules," *IIE Transactions*, 39(5) 2007, pp. 425–437

MeFa95 Mertens, P., Faisst, W., "Virtuelle Unternehmen, eine Organisationsstruktur der Zukunft," Festschrift Universität Erlangen, Deutschland, 1995

MeWe18 Merriam-Webster, "Merriam Webster's Collegiate Thesaurus," 2nd Revised Edition, Merriam-Webster, Inc., Springfield, MA, 2018

NoJa04 Norrman, A., Jansson, U., "Ericsson's Proactive Supply Chain Risk Management Approach after a Serious Sub-supplier Accident," *International Journal of Physical Distribution and Logistics Management*, 34(5) 2004, pp. 434-456

Pico82 Picot, A., "Transaktionskostenansatz in der Organisationstheorie: Stand der Diskussion und Aussagewert," *Die Betriebswirtschaft*, 42. Jahrgang 1982, pp. 267–284

PlFi10 Plehn, J. F., Finke, G. R., Sproedt, A., "Tariff Oriented Supply Chain Management: Implications of Trade Agreements on Strategic Decisions," Proceedings of the 21. POMS Conference, May 7–10, 2010

Port04a Porter, M., "Competitive Strategy: Techniques for Analyzing Industries and Competitors," Simon & Schuster Verlag, 2004

Port04b Porter, M., "Competitive Advantage," Simon & Schuster Verlag, 2004

SöLö03 Schönsleben, P., Lödding, H., Nienhaus, J., "Verstärkung des Bullwhip-Effekts durch konstante Plan-Durchlaufzeiten — Wie Lieferketten mit einer Bestandsregelung Nachfrageschwankungen in den Griff bekommen," PPS Management (2003) 1, pp. 41–45

SiKa19 Simchi-Levi, D., Kaminsky, P., Simchi-Levi, E., "Designing and Managing the Supply Chain. Concepts, Strategies, and Case Studies," 4th Edition, Irwin McGraw-Hill, 2019

Ulic11 Ulich, E., "Arbeitspsychologie," 7th Edition, vdf Verlag, Zurich, 2011

WaBo08 Wagner, S., Bode, C., "An Empirical Examination of Supply Chain Performance along Several Risk Dimensions," *Journal of Business Logistics*, 29(1) 2008, pp. 307-325

WaEg10 Wagner, S., Eggert, A., Lindemann, E., "Creating and Appropriating Value in Collaborative Relationships," *Journal of Business Resarch*, 63(8) 2010, pp. 840–848

WaJo04 Wagner, S., Johnson, J., "Configuring and Managing Strategic Supplier Portfolios," *Industrial Marketing Management*, 33(8) Nov. 2004, pp. 717–730

Zieg07 Ziegenbein, A., "Supply Chain Risiken — Identifikation, Bewertung und Steuerung," vdf Hochschulverlag an der ETH Zurich, 2007Advanced Logistics Partnership (ALP)

3 Supply Chain Design: Location Planning and Sustainability

Intended learning outcomes: Produce design options for global production, distribution, service, and transportation networks. Describe location selection and location configuration for a production network using qualitative and quantitative methods. Explain the concept of sustainability with reference to the triple bottom line. Disclose economic opportunities for social and environmental commitment as well as energy management concepts.

Location planning, or *facility location planning*, is the planning of locations for company facilities.

Location planning is a strategic task, and it is closely associated with a "make" decision. The first steps in location planning sometimes make "make-or-buy" decisions possible at all (see Section 2.1.1). Location decisions also have to be reviewed periodically. Globalization in particular is leading companies to revise their location strategies. For manufacturers of investment goods, the reasons for this are, among others:

- Globalization of the targeted market segment requires the local presence of production and distribution facilities, due to official regulations, for example, or because the customer demands it.
- Entry into new market segments: The creation of new production facilities or a distribution center, i.e. a warehouse with (limited) inventory of end products and/or service items, is necessary.
- Cost pressures due to the market and a focus on core competencies and core businesses: Due to these, individual steps in value added are moved to locations with specific know-how or lower costs.
- Increasing importance of the time factor in development, manufacturing, and service in order to achieve short delivery times in distant markets. One solution can be decentralized adaptation of products and services by completing them locally.
- Current location's disadvantages (personnel, finance, legal, aids for exports, tax, patent system, customer basis, mentality, unions).

These reasons change continuously due to the changing global environment. However, location planning is a long-term task. Generally, mistakes cannot be rectified quickly and are very costly. Figure 3.0.0.1 shows the dynamics of the problem as revealed by a survey of medium-sized companies of the mechanical or electrical industry (M&E) in Germany from 1999 to 2012 (see [Kink12]). The survey investigated the companies' reasons for both offshoring and backshoring production activities (mostly from Eastern Europe or Asia). The statistics demonstrate the importance of systematic location planning.

Section 3.1 deals with the strategic decisions in company-internal value adding. We look mainly at location planning in production, distribution, service, and transport networks.

P. Schönsleben, *Handbook Integral Logistics Management*,
https://doi.org/10.1007/978-3-662-65625-9_3

Section 3.2 introduces possible arrangements, location factors, and criteria for location selection, thereafter location configuration, that is, the assignment of products and services to a location. The solution methods are of both a qualitative and a quantitative kind.

And finally, Section 3.3 looks at the current challenge of design-sustainable supply chains with a view to the "triple bottom line."

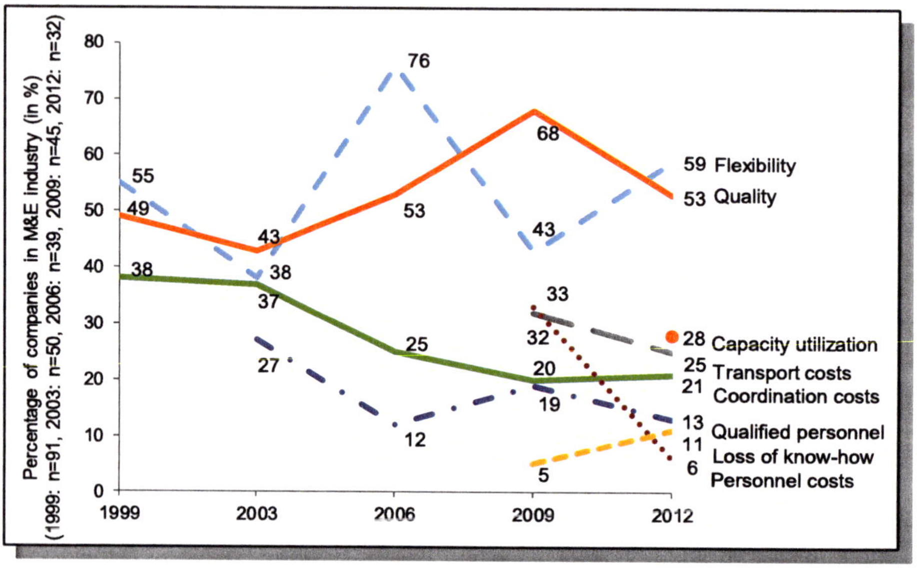

Fig. 3.0.0.1 Reasons for moving facilities back (data according to [Kink12] and previous publications of Fraunhofer ISI, Karlsruhe).

3.1 Design Options for Integrated Production, Distribution, and Service Networks

Intended learning outcomes: Explain design options for global production networks, distribution networks, service networks, and transportation networks. Describe the network structure for decentralized distribution, and design options for retail networks. Disclose the integration of the portfolios.

At the strategic level, the value-added network of the vendor has to fulfill the requirements of the consumer. This is true for any kind of product or service, be it for consumer goods such as furniture and cars or for investments goods such as machine tool. The Figure 3.1.0.1 shows the typical phases of the product life cycle at the consumer's site and the classical, natural breakdown of the value-added of the vendor in the complementing processes.

After design, a product is produced by the vendor or his suppliers. A number of transport processes link the different value-added steps during this phase of the product life cycle. Subsequently, the product is either transported directly to the consumer or to a distribution network that finally delivers the product to the consumer via a number of echelons and warehouses. Decentralized distribution can include a retail network, owned or mandated to 3rd party providers. During the use phase, a service network maintains the product, then called "service object". For this, owned or mandated transport networks pick up and later bring the service object at / to the customer or a collection point. If the product is used up, an adequate network will take back the product from the consumer. As a service network usually can also take back products, there is no need for a separate network for this task.

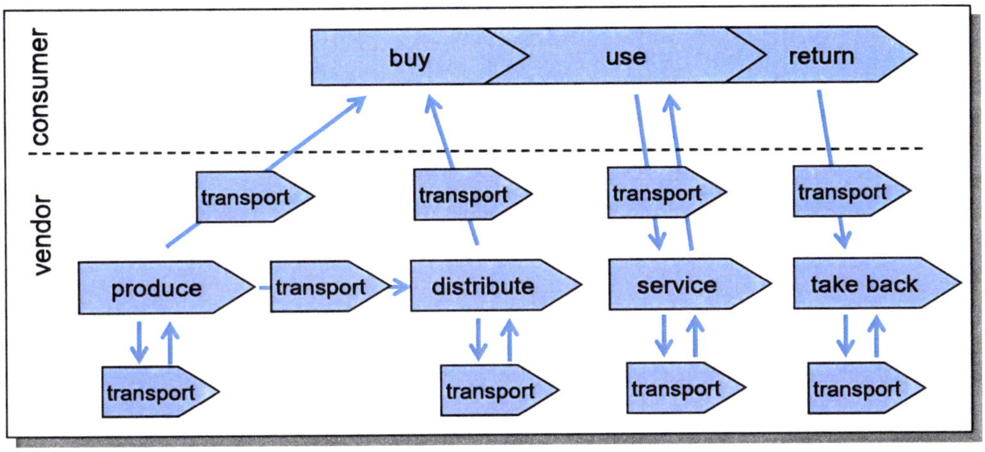

Figure 3.1.0.1 Breakdown of the value-added of the vendor in the complementing processes in function of the phases of the product life cycle at the consumer's site.

The adequate concept may differ for each product line or product family. In making this decision, attention must also be paid to — from the very beginning — tax aspects (see Section 2.1.2). The analysis must then be integrative: a combination of designs of the production, distribution, service and transport networks must both suit the product and the targeted customer segment, and fit well together. Beginning with Section 3.1.2, the presentation follows to a large extent the one in [ScRa15].

3.1.1 Design Options for Production Networks

Intended learning outcomes: Differentiate between centralized production and decentralized production. Present features such as demand volatility, supply chain vulnerability, economies of scale, demand for consistent process quality, customer proximity, market specificity of products, value density. Explain design options for global production networks. Describe some company cases.

In a first approach it is possible to distinguish two fundamental types of production networks.

> In *centralized production,* a product is manufactured at only one location or through a chain of single stations, one station per operation at one location. In *decentralized production,* a product or certain operations of a product are manufactured at several locations.

Figure 3.1.1.1 shows a simple example.

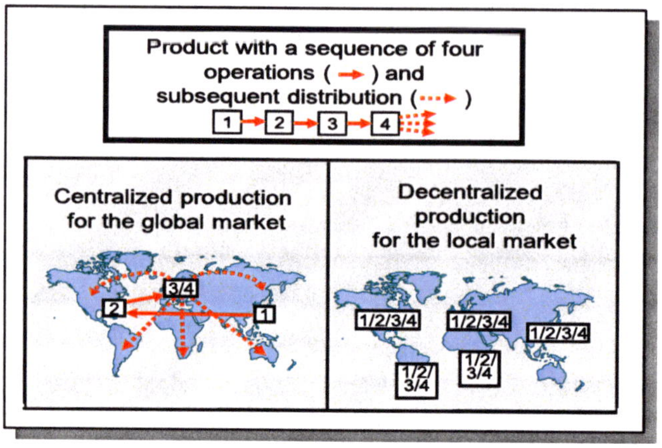

Fig. 3.1.1.1 Centralized versus decentralized production: an example.

Taking a product with four operations (or four production levels) and subsequent distribution, Figure 3.1.1.1 shows centralized production (obviously for the global market) and decentralized production (more for the local or regional market). For the decision as to centralized / decentralized, there are *features for designing production networks*, including:

- *Demand volatility*: Items have continuous demand if demand is approximately the same in every observation period. Items have discontinuous, or highly volatile demand if many periods with no or very little demand are interrupted by few periods with large demand; for example, ten times higher, without recognizable regularity.
- *Supply chain vulnerability*: Unplanned events can disrupt a supply chain. These disruptions can arise from either the supply chain community or the macro-economic environment.
- *Economies of scale*, that is, an effect whereby larger production volumes reduce unit cost by distributing fixed costs over a larger quantity; and economies of scope, that is, when different products can be produced in a changeable factory at lower costs than when each product is produced in its own factory.
- Demand for *consistent process quality*: Can customer needs be satisfied despite differing process quality?

An interesting observation is that these four features are highly correlated: Centralized production is an advantage for high economies of scale or scope and for a high demand for consistent process quality. Decentralized production is an advantage in the case of high demand volatility and in the case of high supply chain vulnerability.

Further important features for designing production networks are:

- *Customer proximity*: To sell a product it can be necessary to locate the value-adding processes close to the customers.
- *Market specificity of products*: Adapting products to the market is needful for functional requirements, such as voltage, electrical connections, packaging, and documentation. But it also applies for the appearance of products in the broadest sense.

- *Customer tolerance time*, as defined in Section 1.1.6.
- *Value density*, that is, product value — or item costs — per cubic meter or kilogram: Transport costs are of greater consequence if value density is low than if it is high.

The above four features are also highly correlated: If customer tolerance time or value density is high, there will be a tendency to centralize production. If high customer proximity or high market specificity is necessary, decentralizing production will be advantageous.

However, the two groups of features unfortunately often stand in opposition to one another. There are examples of this:

- Appliances (specialized machining equipment, but adaptation due to voltage, connections, packaging, documentation): high necessity of economies of scale (in favor of centralized production), however also high market specificity of product (in favor of decentralized production)
- Bakery products with a brand promise regarding quality: high demand for consistent process quality (in favor of centralized production), low value density (in favor of decentralized production)
- High value components with variants (e.g., electronic chips, engines, pumps, inject-ions): high value density (in favor of centralized production), but also high demand volatility and high supply chain vulnerability (in favor of decentralized production)
- Important raw material (such as steel), perishable foodstuffs: low market specificity of product (in favor of centralized production), also, however, high demand volatility and high supply chain vulnerability (in favor of decentralized production)

Here, a company must make a strategic decision, which sometimes differs for each product family. The portfolio in Figure 3.1.1.2 is based on an idea in [AbNu08]. It shows, in addition to the two classical designs (the two sectors in the one-dimensional space in Figure 3.1.1.1, namely centralized or decentralized production), two possible mixed designs. The four possible designs lie in four sectors in a two-dimensional space, spanned by the dimensions that correspond to the two (conflicting) groups of features.

The sector P1 describes the centralized production for the global market. This option is advantageous where economies of scale or economies of scope are strong and, in addition, when there are advantages to having well-established partnerships for the added value of the various production levels. In this way, there is a greater possibility to maintain consistent process quality, which is important mainly for validation of production processes (keyword: GMP, Good Manufacturing Practices). Here it is not essential whether added value occurs within a company or across companies. Distribution takes place from the location that manufactures the last production level, or last operation. Required for this is, in any case, high value density as well as high customer tolerance time and low vulnerability of the (only) supply chain. The products tend to be standard products. Some examples here are electronic components, liquid crystal displays (LCD), consumer electronics, chemicals and pharma-ceuticals, fine chemicals, giant aircraft, standard machines, and standard facilities.

The opposite sector P4 describes the decentralized production for the local market. This option is advantageous when high proximity to customers is required, when products must be modified for the local market, and when customer tolerance time and value density are very low. What is needed is a supply chain that is not strongly dependent on economies of scale or economies of scope. Qualitative differences, however, will result. Some examples here are household appliances, building materials (gravel, cement), and products connected with services, e.g., clothing that must be adapted continually to the object during the manufacturing process.

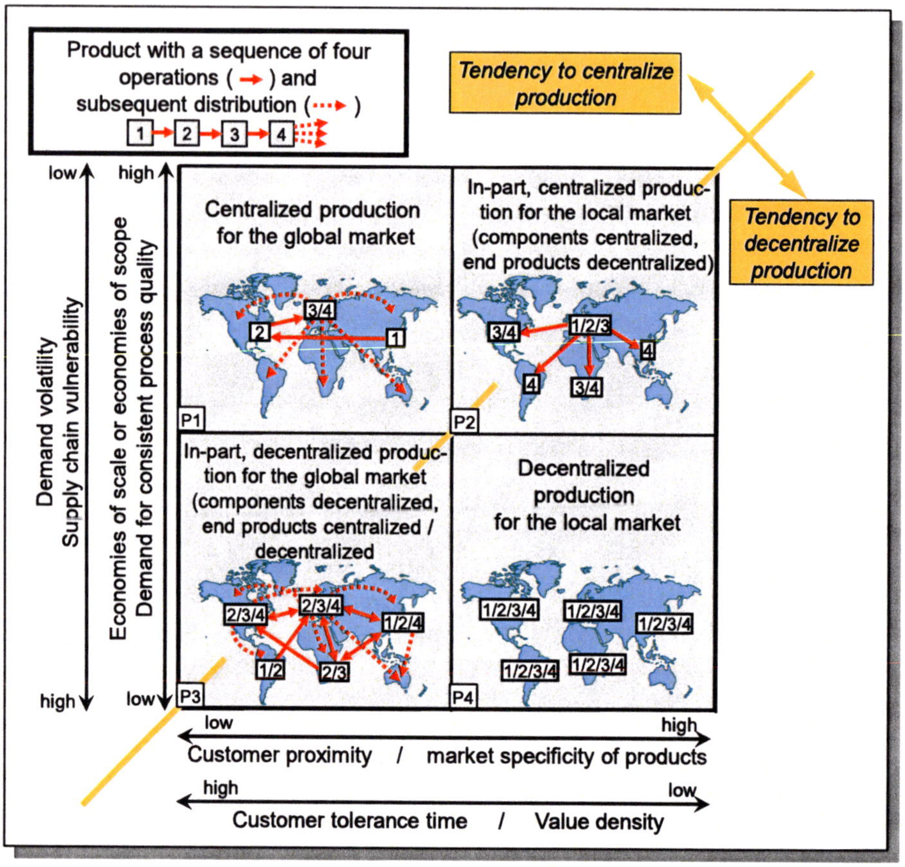

Fig. 3.1.1.2 Features of and design options for production networks.

The intermediate sector P2 describes the in-part centralized production for the local market. If semifinished items are produced centrally, and if the last value added steps are performed at decentralized locations, important economies of scale or scope can be exploited while at once having proximity to market. Examples here are strategies for local end production for all consumer goods, such as "late customization" or "postponement" (see Section 1.3.3).

The intermediate sector P3 describes the in-part decentralized production for the global market. If the same components and/or end products are manufactured at different locations, and if at various production levels they can be moved to different locations and distributed globally, this brings advantages in the case of volatile demand as well as for a supply chain that is vulnerable to disruptions, in that the capacities in the network are utilized more evenly or can even substitute for one another. This makes sense, however, only for standard products with high value density and sufficient customer tolerance with regard to delivery times, such as, for example, for components or end products in the automotive industry, perishable foodstuffs, or important raw materials (such as steel).

There are, of course, mixed forms of production networks that lie between these four main designs. This is particularly the case when the characteristics are not significantly pronounced on the abscissa or ordinate of Figure 3.1.1.2.

Company Cases: When features for designing production networks change, it is appropriate to consider changing the production networks. However, the financial investments required often quickly set limits to changeability.

For example, the production costs of cement are on the rise today, due to rising costs of both energy and CO_2 emissions (see the discussion of the triple bottom line in Section 3.3). As a consequence, value density increases, so that centralized production becomes more and more an option (i.e., option P3 instead of the traditional option P4). But that requires new cement works and additional logistics infrastructure for supply of raw materials and distribution of the cement. As an example, Holcim, a Swiss-based cement manufacturer opened its new production plant in Ste. Genevieve, Missouri, in 2009, with its own port and loading facilities on the Mississippi. Production cost and thus value density increased, as the new plant is aimed to reduce CO_2 emissions significantly. At the same time, the waterway network allows transportation to ten of the twenty largest cities in the USA at lower cost than before. Thus, a more centralized production concept became possible.

Increased demand volatility makes it necessary to produce two engine variants at each of two locations (option P3) instead of producing only one of the variants at each (option P1). Although this entails considerable investments for equipment, the result is much better use of the capacities. As an example, Daimler, a Germany based car manufacturer, produces its four and six cylinder engines in the USA as well as in Germany. The benefit of the flexibility to cope with volatile demand is greater than the cost for double toolsets and facilities as well as for transportation of some of the finished engines between the USA and Germany.

Increased necessity for economies of scale, due to massive competition, forced Hilti, a Liechtenstein-based manufacturer, to centralize the production of drilling machines, despite its "the construction site is the point-of-sales"-driven sales strategy. Today, each drilling machine type is produced at exactly one site (option P1). Here, each site holds specific technology competences. Different fastening consumables, however, continue to be produced in different factories, close to the local markets (option P4). Still, semi-finished items that need expensive or important technologies are produced centrally (option P2).

3.1.2 Design Options for Distribution Networks

Intended learning outcomes: Differentiate between centralized distribution and decentralized distribution. Present features such as demand variety, need for efficient returns, and degree of customer involvement in picking up. Explain design options for global distribution networks. Describe some company cases.

In a first approach and in analogy to production networks, it is possible to distinguish two fundamental types of distribution networks for delivery of a customer's order.

In *centralized distribution* a customer order for a given product is fulfilled directly from the production plant or from one or a few of the manufacturer's central warehouses. In *decentralized distribution*, the given product is stored in several decentralized warehouses or with (often independent) distributors, from where the customer order is fulfilled.

The advantages of centralized distribution, from a central warehouse or the manufacturing plant to the customer, are obvious: a bigger selection of products (which, in part, can also be made per customer order, that is, "assemble-to-order," "make-to-order," or even "engineer-to-order"), a greater availability, and lower total costs for inventories, plants, and handling.

The advantages of decentralized distribution are also easy to see: shorter delivery lead times and a more efficient possibility for product returns (e.g., at the retailer's site). Besides this, transport costs are somewhat lower: first, the decentralized warehouse, which is located nearer to the customer, can be served at lower cost, e.g., using large transport units and (full) truckloads ((F)TL); second, a customer order comprising multiple articles can be bundled into a single delivery at the decentralized warehouse. Finally, order tracking is only needed between the decentralized warehouse and the customer; if the order is raised directly in the store, tracking may even prove unnecessary. Combined with the necessary exchange of information between the manufacturer and the decentralized warehouses, the total cost for information systems is generally lower than the costs of real-time order tracking between the manufacturer and customer.

The following *features for designing distribution networks* have proved to be important:

- *Demand volatility*, as defined in Section 3.1.1
- *Demand variety*: High demand variety means that customers demand many different products. For these products the demand volatility is mostly high as well.
- *Value density*, as defined in Section 3.1.1
- *Customer tolerance time*, as defined in Section 1.1.6. In global distribution, delivery lead time also includes the time for customs procedures, which can disadvantage centralized distribution.

These features are highly correlated: In general, centralized distribution is advantageous for products with high value density, high demand variety and volatility, and for high customer tolerance time. If the values are the opposite, decentralized delivery will be advantageous.

Two further features for the design of distribution networks, which are not correlated with the above features, however, are:

- *Need for efficient returns* via the same network: Is it important that the customer be able to return goods efficiently through the same distribution network and that the network be able to handle these returns efficiently (keyword: reverse logistics)?
- *Degree of customer involvement in picking up*: To what extent are customers willing and able to picking up the product themselves?

$For cases where the need for an efficient product-returns process using the same distribution network is key, and for customers both willing and able to pick up the product themselves, various designs of decentralized distribution are advantageous.

As in the case of production networks, the two groups of features often stand in opposition to one another. There are examples of this:

- The distribution of many rather cheap items, like heavy or bulky items (e.g., beverages), fresh produce (e.g., flowers), express delivery items (e.g., medicaments) or fast moving items to the point of use in companies (e.g., C items like screws, nuts, bolts, etc.): low value density (in favor of decentralized distribution), however, rather low degree of customer involvement in picking up (in favor of centralized distribution)
- The distribution of vehicles or on-line orders: high demand variety or demand volatility (in favor of centralized distribution), however, some degree of customer involvement in picking up, as long as the pickup site is close enough (in favor of decentralized distribution)

Again a firm must make a strategic decision, which at times differs for each product family. In addition, if a company delivers to different customer segments, it will at the same time have to use different routes or channels for distribution. The channels or distribution centers need not necessarily be owned by the manufacturer. An additional required channel usually entails extra costs. In addition, existing channels may change over time. For example, postal services may expand the offering of their mail rooms to become local shops, while shopping points may also offer a reduced set of services traditionally provided by a postal service.

Based on an idea in [Chop03], the portfolio in Figure 3.1.2.1 shows, in addition to the two classical designs (centralized or decentralized distribution), two possible mixed designs. The four possible designs lie in four sectors in a two-dimensional space, spanned by the dimensions that correspond to the two (conflicting) groups of features.

Sector D1 describes centralized storage near the producer or — in the case of make-to-order — delivery directly from production, with direct shipping to the customer or his unloading point. This design option is advantageous for products with high value density and high demand volatility. With this characteristic, customers are mostly willing to tolerate some time to delivery due to the generally long transport routes. This design makes possible a large selection of products, high fill rate and relatively low costs for inventory, installation, and handling. However, transport costs are rather high, as are costs for possible returns, and there are high costs for information systems for transmitting orders from the point of sale and for order tracking *during* shipments. This is the classic design for distribution of investment goods (such as machines), as well as for *drop shipping* (i.e., direct delivery from the manufacturer to the customer of the order entering party, e.g., a wholesaler, or a spare-parts or online retailer. Here, the order entering party can avoid its carrying cost but must factor in the cost of integrating its information system with that of the manufacturer and is also unable to monitor the quality of the delivered product).

The opposite sector D4 describes decentralized storage at a retailer with customer pickup. This design option is suitable if customers are able and willing to pick up the desired products, as well as products from different manufacturers. It generally offers a great amount of flexibility in terms of time for this. Beforehand, customers must also do the order pick-up themselves. This design option is transparent and requires rather simple information systems for tracking orders and delivery, and it also allows returns of products or packaging material. However, as stressed in the figure, it requires an adequate retail network. Here see Section 3.1.3, particularly Figure 3.1.3.2 and the respective examples.

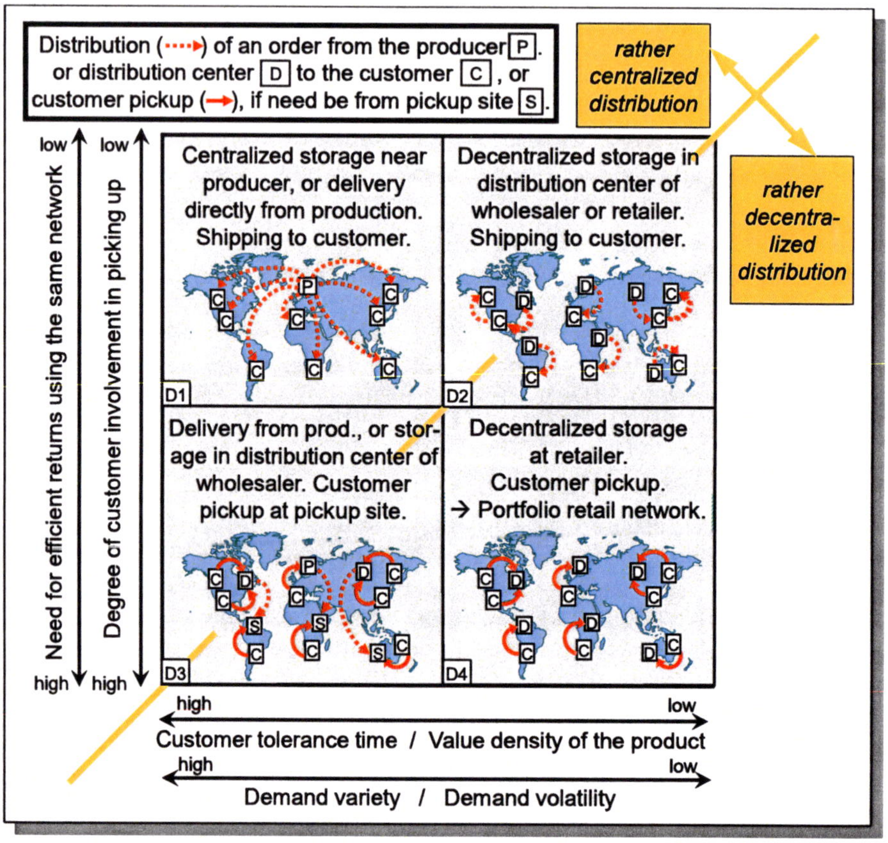

Fig. 3.1.2.1 Features of and design options for distribution networks.

The intermediate sector D2 describes decentralized storage in the distribution center of a wholesaler or retailer with shipping to the customer or his unloading point. This is the most convenient design option for customers. But it requires rather low demand volatility as well as the customer's presence at the unloading point. Otherwise, it entails high transport costs, also owing to irregular delivery tours. The problem of *optimum routing and scheduling* often occurs anyway, especially with respect to efficiency for *last mile delivery* or *same day delivery*. Using lockboxes for unloading, similar to post office boxes, the customers' presence may not be necessary in every case. Storage by the wholesaler copes with smaller stock levels than a corresponding retailer network, but generally does not permit same day delivery. Where the unloading point is set up to handle returns, these can also be processed by this system (e.g., the return of empty bottles on a milk run). However, normally returns must take place via a different network (e.g., via the postal service's network). This solution is suitable for the delivery of heavy articles such as beverages, of fresh produce such as flowers, of express-delivery items such as medicines, or of *fast moving items*, e.g., the distribution of ranges of C item goods (screws, nuts, bolts, etc.) to the point of use in firms. In the latter case, stocks may be managed by the customer (vendor-owned inventories, VOI).

The intermediate sector D3 describes delivery directly from production, or centralized storage in the distribution center of the wholesaler, with shipping to the pickup site. This design option can be selected if customers are willing and able to pick up the goods and thus profit from considerably lower transport costs. Examples include the shipping of vehicles or online orders (click and collect). But this places higher demands on the accompanying information systems than in the case of shipping to the customer. If the pickup site entails high costs, this solution will, in addition, tend not to be cost effective. For this reason, pickup sites should be able to be combined with existing distribution centers for other products or service centers (e.g., a car showroom or a supermarket chain such as Coop or 7eleven). In this case, they are also suitable for product returns or return of packaging material. Storage in the distribution center of the wholesaler reduces delivery lead times. But it also either reduces product selection and availability or increases inventory costs. The costs anyhow increase for installation and handling owing to the costs of the distribution center. Distribution centers, including those located directly in the factory, may also act as a pickup site. An example of this is the pickup of cars from the plant.

Company Cases: When characteristic features change, it is appropriate to consider changing the distribution network design. In the example of Holcim mentioned in Section 3.1.1, not only a more centralized production network design but also a more centralized distribution network design became possible. Still, decentralized storage for basic demand of common products at various so-called "terminals" is part of Holcim's distribution concept in the US Midwest (i.e., Sector D4 or D2 in Figure 3.1.2.1), right down to the Gulf of Mexico. However, for products with volatile demand, Holcim rather uses concepts with more *centralized* storage (Sector D3, where the "terminals" serve as pickup sites, or even Sector D1).

At Hilti, decentralized storage in the distribution center of the wholesaler or retailer (i.e. sector D2) and subsequent delivery to the production site is executed in order to offer short delivery lead times to the customer ("last mile"). Although the inventories and respective current assets are high, the availability of the products is more important in order to satisfy the customers' demand as fast as possible.

Further Correlations That Should Be Considered for an Integrated Determination of the Design Options: The four design options cannot be selected without giving consideration to the design of the production network. *For production destined for the global market* (sectors P1 and P3 in Figure 3.1.1.2), all four design options for the *global* distribution network come into question. In the case of decentralized storage, there is a need for a distribution network structure, possibly with multiple structure levels or echelons (see Section 3.1.3). *For production destined for the local market* (sectors P2 and P4 in Figure 3.1.1.2), only the design options in the sectors D2 and D4 in Figure 3.1.2.1 come into question from a *global perspective*, i.e., decentralized storage. From a *local perspective*, it is naturally possible to view storage in close proximity to a (local) manufacturer as "central." In such a case, all four design options can come into question, albeit *only for the local* distribution network.

3.1.3 Network Structure for Decentralized Distribution, and Design Options for Retail Networks

Intended learning outcomes: Disclose the distribution network structure and describe decision variables in its design. Present features such as available time for shopping, and simultaneously, capacity of an available means of transport of the customer, as well as the required geographical catchment area. For decentralized distribution, explain the portfolio for designing retail networks.

If a decentralized distribution network design is chosen (Sector D4 in Figure 3.1.2.1), the next thing to be done is to define its structure.

The *distribution network structure* defines the planned channels of distribution of goods. Figure 3.1.3.1 shows an example.

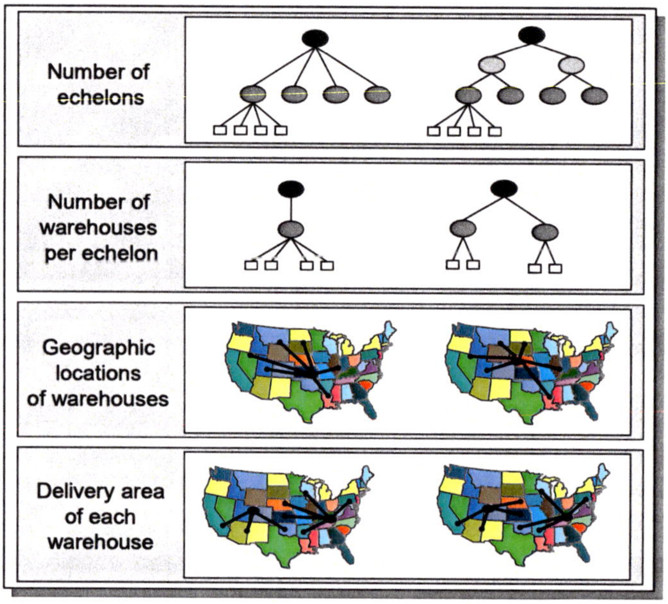

Fig. 3.1.3.1 Decentralized distribution-decision variables in the design of the distribution network structure. (Following [Stic04].)

The distribution network thus comprises, first of all, the number of *structure levels* or *echelons*, e.g., a *multi-echelon* structure with four levels: 1. central warehouse, 2. regional distribution center, 3. wholesaler or distributor, and 4. retailer. Secondly, it comprises the number of warehouses per echelon, thirdly the geographic location of each warehouse, and fourthly the delivery area of each warehouse. The result is a geographically ramified distribution network. At every echelon, by a process called *break-bulk*, (full) truckloads of homogeneous items can be divided into smaller, more appropriate quantities for use ([APIC16]). For transparency of on-hand balances and transportation inventories throughout the entire network, an information system is required. Today, distances are determined more and more automatically, using distance tables or geographic information systems (GIS), for which in many countries commercial software is available.

The lower the customer's tolerance in terms of lead time, the greater the number of decentralized warehouses and the smaller the delivery area for each warehouse. Points of sale (POS) must be located at convenient locations close to customers. Therefore, the crucial point in the selection of new locations of a retail network is the prediction of the number of potential customers.

With respect to the design of *retail networks,* in the first approach it is possible to distinguish *POS with a smaller volume of goods available on-site*, i.e., at the POS, *from those with a larger volume.* "Volume" may relate to the number of different items, and/or the quantity per item. For this approach, important *features for designing retail networks* are the following:

- *Available time for shopping*, and simultaneously, *capacity of an available means of transport of the customer*: For private consumers (B2C), a car has a high capacity. On foot or by bicycle, the capacity of transport is, in contrast, low. If time is limited, or the car is unavailable at the appropriate time, then the purchase option is restricted to a local outlet and limited size and weight. For commercial purchasers (B2B) — depending on the transaction — a lorry offers high capacity. A small car can then only be used to purchase items of limited size and limited weight.
- *Demand variety*: as defined in 3.1.2.
- The *required geographical catchment area* for the product range on offer: This characteristic assesses the size of the catchment area in which a "sufficient" number of customers are based, for whom the offered product range represents a good fit in terms of product quality and price. This assessment is carried out in consideration of purchasing power, time available, and the choice of means of transport. "Sufficient" means that the frequency of purchases multiplied by the average value of each sale corresponds to a minimum sale value per time unit that is required in order to make the operation of the POS a profitable venture.

In dependency of these three features, the portfolio in Figure 3.1.3.2 shows the design options for retail networks with rather smaller or larger volume of goods available on site.

Sector R1 describes the situation with normally no point of sale, since the required geographical catchment area is too large to make it profitable to maintain a POS. Thus, the intended design option of the distribution concept (i.e., Sector D4 in Figure 3.1.2.1) is not realistic and must therefore be abandoned. This, in turn, can entail modifications of the distribution network structure. The customer, if not ceded to a competitor, must place an order that is fulfilled directly from production or delivered from a distribution center, and shipped to a pickup site (e.g., general delivery or poste restante). See the sectors above or to the left in the portfolio in Figure 3.1.2.1. This is the case for most commercial (B2B) purchases.

The opposite sector R4 describes the shopping mall of large stores. On these expensive sales floors, an extended product range can be offered. This increases the number of customers, so that this higher-value product range moves sufficiently quickly and the volatility of demand remains low. This is the design of the big supermarkets for private consumers, for example, or of cash-and-carry wholesale, for commercial purchases. For the shopping experience, competition by different shops is desired. If such centers have a thematic range of products (e.g., clothes, or furnishings), they try to host as many competitive shops as possible. As automobiles are necessary anyway due to the amount or size of goods to be transported, these shopping malls are outside residential neighborhoods at convenient locations easily reached by car.

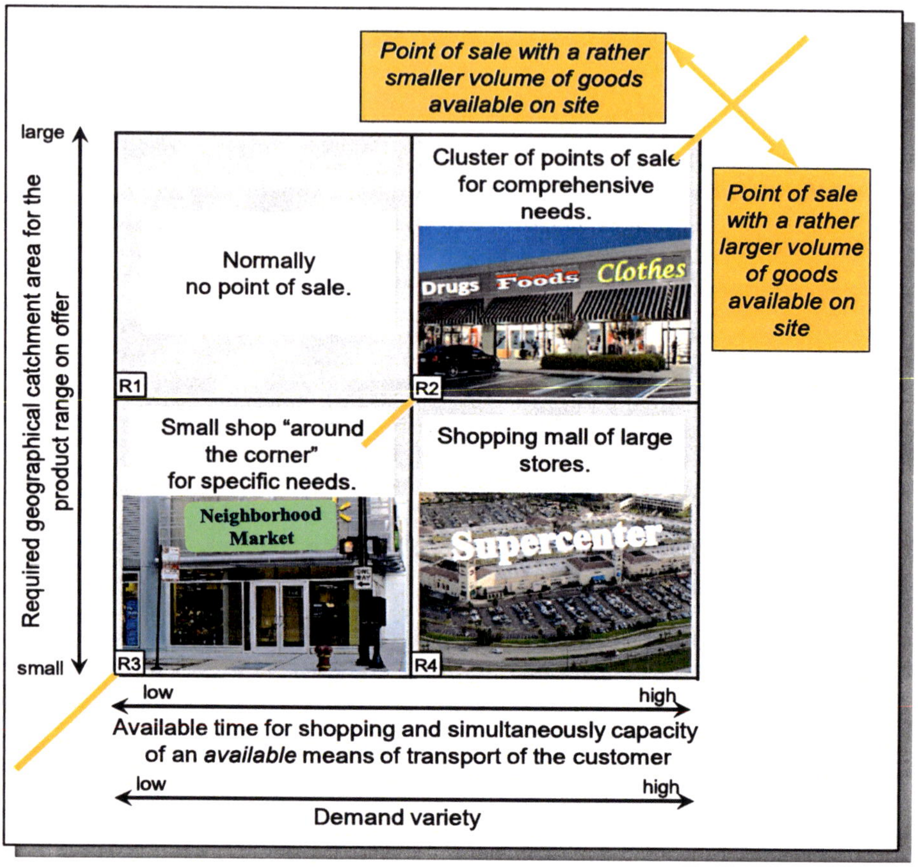

Fig. 3.1.3.2 Decentralized distribution: portfolio for designing retail networks.

The intermediate sector R2 describes the cluster of points of sale or small stores for comprehensive needs, found in low-population-density areas. It can only be accessed by most potential customers using a car. In this case multiple specialist points of sale exist largely with no overlap of offerings. When considered together, these form an offering that meets the comprehensive needs of the largest possible number of customers who are able to access this cluster of points of sale.

The intermediate sector R3 describes the small shop "around the corner". As compared to larger stores, this solution offers customers more comfort, but comfort that must be paid for with higher transport costs and often higher installation and handling costs. The product range comprises basic items for specific needs. Among private consumers, these often comprise purchases of food or items for a specific customer group. For commercial purchases they include frequently used spare parts for renowned vehicle models, for example. In the case of food, this may relate to basic daily needs. These are covered by retailers whose shops can be accessed on foot or by bicycle. For this design option, a minimum number of potential customers with the corresponding purchasing power living or working close by the shop is required. This design option can be chosen in some areas of cities or for locations with a high frequency of visits by specific groups of people (e.g., in schools or sports facilities), for example. Stock costs can be kept at a low level by means of an efficient, normally IT-based replenishment system. An example of this is used by chemists or pharmacies, which hold only one unit of certain medicines in stock, the movement of which is then communicated directly to the distribution center by sensors. A fast logistics system ensures that replenishment can be guaranteed within a few hours.

Company Cases: In the example mentioned in Section 3.1.1 and 3.1.2, customers of Holcim operate quite nearby to its "terminals" (Sector R3 in Figure 3.1.3.2). In the cement industry it is advantageous to cede customers that are far away from such "terminals" (Sector R1) to competing manufacturers.

In some industries, such as food retail, clothing or furniture, big retail chains like Walmart or Swiss-based Migros have points of sale of different sizes and that carry a different range of products. Especially in large conurbations, they use *both* design options R4 *and* R3 across one single area, each for a different size of store as well as product range. Such retail chains are aware that many customers have a choice of modes of transport with different ranges or capacity, and can also make a choice depending on available time and personal sentiment.

The aforementioned Hilti company owns its entire distribution network structure. Its market organizations are the wholesalers that own one or several warehouses. De facto, the sales representatives on site act as retailers. They are in close contact with potential customers and deliver directly at the construction sites (sector D2 in Figure 3.1.2.1). So there is no need for "Hilti stores" or the partnership with a third-party retail chain. Still, Hilti's distribution network structure is changing. Actually, a VMI (vendor-managed inventory) concept should result in a more efficient inventory management at the different echelons.

3.1.4 Design Options for Service Networks

Intended learning outcomes: Differentiate between centralized service and decentralized service. Present features such as the mobility cost ratio of the service, the degree of customer involvement in bringing and picking up the service object, as well as the need for repeated transfer of the service object. Explain design options for global service networks of services in direct contact with the object. Describe some company cases.

A *service in the originary sense* can also be described as a process with direct contact with the *service object* (see the definitions in Section 1.1.2). Due to technological development and industrialization, the following two kind of (sub)processes have developed for which contact with the object in the same location is no longer necessary to the same extent.

- *Process with indirect contact with the object* ranks services that take reservations or orders, such as travel agencies, car rental, or mail order companies; also services that deliver information and thus support the actual products or services, both before and after sales, e.g., call centers or hot lines. As the delivery costs of information do not differ greatly for different locations, the locations can in principle be anywhere in the world where the production costs — for the required quality — are minimal.

- *Process with no object contact* often ranks sub-processes of the service as a whole that bear similarities to classical production and that — for example, due to efficiencies (economies of scale or economies of scope) or difficulty — must be carried out at a centralized location. Some examples of these processes are the "back offices" of banks (for example, in the mortgage or loan business), insurance companies (for example, for policies that cover special risks), or credit card billing companies. Delivery costs do not play a role as soon as goods can be transmitted digitally; these centers can in principle be located anywhere in the world, as long as quality is assured. Other processes that also belong here are the mere delivery of spare parts, or activities in centralized picking locations, e.g., for catering businesses, or processes along the distribution network structure for tangible goods; in these cases there are delivery costs in addition to production costs, so that the facilities cannot be sited at just any locations. With this, under certain conditions multi-level service networks will form, in which the individual locations are linked together.

For networks of *services in the originary sense*, it is possible to distinguish, in a first approach and in analogy to production and distribution networks, two fundamental types:

In *centralized service* a specific service is provided directly at one or a few central service centers. In *decentralized service*, the service is provided at or from several service centers, located, as close to customers as possible.

The advantages and disadvantages of centralized service versus decentralized service arise in a similar manner to the advantages and disadvantages of centralized distribution versus decentralized distribution. It is simply necessary to assume that the object will be transferred to the service provider, potentially also at the place at which the object is located; this must be the first sub-process of the service and is critical to the effectiveness of service delivery. The important *features for designing service networks* are in principle the same as those for the distribution networks. However, the meaning of the following features changes:

- The value density of the product becomes the *mobility cost ratio of the service*, i.e., the mobility costs for the service provider (to bring people, equipment, and materials such as spare parts to the object), in comparison with the mobility costs of the object. The latter include costs of transporting the object (generally dependent on size and weight) and for preparing the object for transportation at its base location. In the event of a complete overhaul or retrofitting of a machine or piece of equipment, preparations also include the dismounting and subsequent re-mounting. When the object is actually a person, the cost is measured in terms of the subjective value placed on loss of comfort, for example, due to a stay away from home in a hospital.
- The degree of customer involvement in picking up becomes the *degree of customer involvement in bringing and picking up*: To what extent are customers willing and able to bring and pick up the object?

- The need for efficient returns becomes the *need for repeated transfer of the service object*. Some objects must be treated repeatedly by the same service provider, e.g., vehicles at a garage or patients by a general practitioner.

As with production or distribution networks, the two groups of features often stand in opposition to one another. There are examples of this:

- The classical maintenance and repair or operator models on site, insurance services, simple home care, medical services provided by general practitioners in the home, home tutoring: low mobility cost ratio of the service (in favor of decentralized service), however, a rather low degree of customer involvement in bringing and picking up (in favor of centralized service).
- Major repairs to tools and equipment, the operation of traditional schools with collective transportation of schoolchildren, group trips: high mobility cost ratio of the service (in favor of centralized service), however, a high degree of customer involvement in bringing and picking up, as long as the pickup site is close enough (in favor of decentralized service).

Again, a company must make a strategic decision, which sometimes differs for each product family. The portfolio in Figure 3.1.4.1 shows, in addition to the two classical designs (centralized or decentralized service), two possible mixed designs. The four possible designs lie in four sectors in a two-dimensional space, spanned by the dimensions that correspond to the two (conflicting) groups of features.

The sector S1 describes the option of a centralized service at the manufacturer's or specific service provider's location, with the object being picked up, and later (that is, after the service provision), brought from or to the location by the service provider. It is advantageous where the mobility cost ratio and demand volatility are high. In such cases, customers are usually also prepared to tolerate generally longer transport routes and longer lead times until delivery or execution of the service. This option permits a wide selection of services and a high fill rate as well as relatively low facility and handling costs. Nevertheless, these are countered by high transport costs, often associated with complex preparation and special modes of transport. In addition, there is a rather higher level of complexity for information systems, in relation to both the transmission of orders from the service point of sale and to order tracking during the service provision. The location of the object before and after the service need not necessarily be the same, and need not necessarily be that of the ordering party (e.g., dealer in second-hand machinery). However, in such cases, even greater complexity must be taken into account for the information systems described above. This is the classical design option for comprehensive refurbishment and modernization of capital goods (normally by the manufacturer, e.g., machinery, aircraft, and vehicles), for contract work, and for major operations at specialized hospitals in the health sector.

The opposite sector S4 describes the decentralized service in the service center. The object is brought and later (that is, after the service provision) picked up by the customer. This design option is suitable where the customer is prepared and able to bring the object to the service center and pick it up again. First, the customer may or must schedule the execution of the service — potentially involving several visits. This design option is transparent, requires much simpler information systems for ordering and order tracking of the service, and allows repeated transfer of the same service object. Examples here are simple repairs to items of everyday use, such as vehicles, shoes, devices, and simple services delivered to people, such as in the hairdresser's, at the bank, at the doctor's surgery, or at a kindergarten.

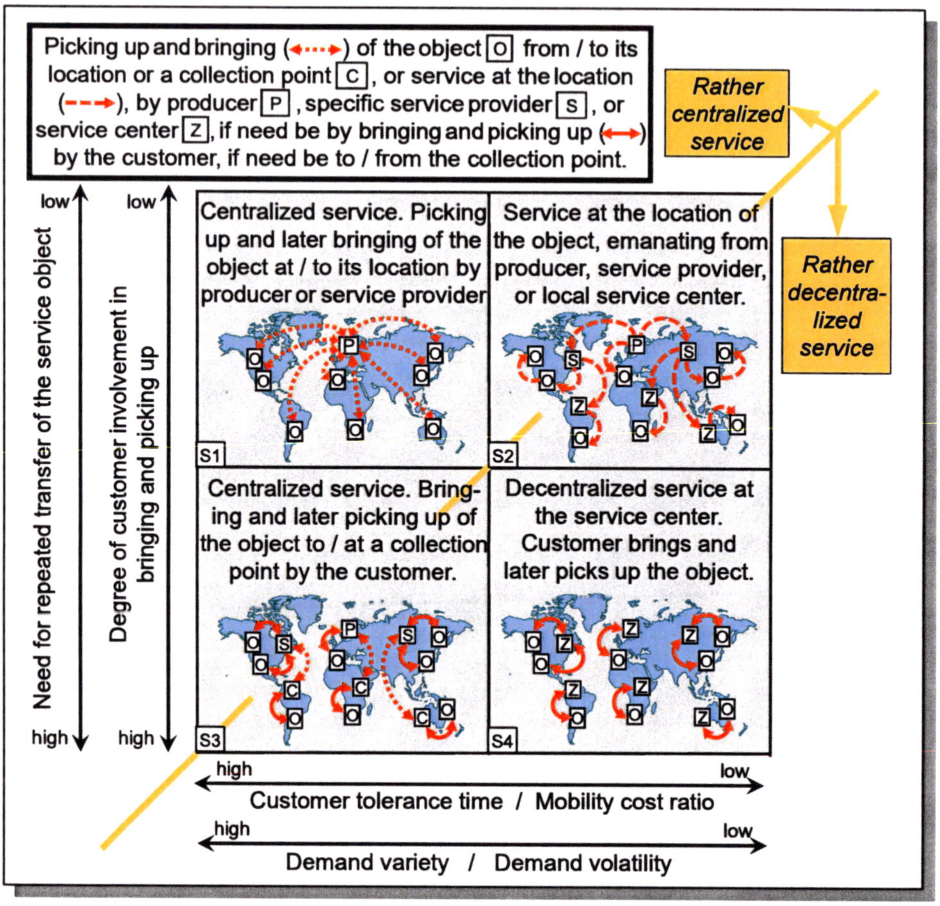

Fig. 3.1.4.1 Features of and design options for service networks for services in direct contact with the object.

The intermediate sector S2 describes the service provided by the manufacturer / specific service provider or from a local service center, with provision of the service at the location of the object: this is the easiest option for the customer. However, it requires rather low demand volatility and accessibility to the object at the agreed time. If this is not the case, the consequences will be high stand-by costs and transport costs that are even higher than their currently high levels owing to futile journeys. Often, the challenge of optimum routing and scheduling also presents itself. The service is provided from a local center. For rarely executed or difficult services it may be better if a specialist service provider or even the manufacturer is deployed. Examples here are the classical maintenance and repair or operator models on site, as well as insurance services, simple home care, medical services provided by general practitioners in the home, and home tutoring.

The intermediate sector S3 describes the centralized service at the manufacturer's location or servicing in a major service center. The customer brings to and later (that is, after the service provision) picks up the object from a collection point. This design option can be selected if the customer is prepared and able to bring and pick up the object and can thereby benefit from significantly lower transport costs. Examples here are major repairs to tools and equipment or the operation of traditional school (collective transportation of schoolchildren) or group trips. The requirement in terms of accompanying information systems for this is even higher than for option S1. If the collection point incurs significant costs, this solution becomes economically unviable. Thus, collection points should be combinable with existing service centers or product distribution centers (e.g., in-store). Such collection points permit repeated transfer of the same service object. Collection points in relation to schools or tourism can be combined with a stop on the public transport network, for example. Provision of the service in service centers instead of the producer reduces the transport times. In turn, either the selection of services and their availability is reduced, or stand-by costs rise. The costs for plant and handling rise further anyway, due to the costs of the service center. This solution is suitable if the objects accumulate in specific regions. Larger service centers and the specific service provider may also act as collection centers. An example of this would be accident and emergency departments in a hospital.

Similarly to the network structure for decentralized distribution and the design options for retail networks, decentralized service concepts (sectors S2, S3, S4) require a suitable "multi-echelon" structure and a network of service providers, service centers, and collection points. The degree of similarity to the shapes of retail networks shown in Figure 3.1.3.2 is high.

Company cases: In the prior example, Hilti owns local service and repair centers as part of the different sales organizations. The customer tolerance time is very low, as is the degree of customer involvement in bringing and picking up. Due to the direct delivery concept, the sales representatives are close to the customer. In case of a defect, e.g., of a drilling machine, Hilti's fleet management quickly delivers replacement, taking back the defective equipment at the same time. Thus S2 is the preferred design option.

For the equipment they previously sold, the aforementioned big retail chains like Walmart or Migros offer collection points right at their larger points of sale. Sometimes, there is also an on-site service shop (option S4). More commonly, they use the transportation network that delivers products via the different echelons of their distribution network structure for transporting the defective part to a larger service center or to the manufacturer (option S3).

Further Correlations That Should Be Considered for an Integrated Determination of the Design Options: Through skillful redesign of a complex service, parts of the service can possibly take on a more decentralized character. For example, extensive revision of a machine at the manufacturer's site can be carried out more as a sequence of simplified service variants at the operator's site, without missing the desired goal of the revision. Prior to the performing of these simpler services, the necessary repair parts can be delivered via the distribution network. Furthermore, the degree of decentralization of services is, in general, at least as high as the degree of decentralization of the distribution. Actually, it would not make sense to the customer, why she or he should accept a longer way for maintenance and repair than for delivery. So a point of sale can often be used as a collection point, sometimes even as a local service center.

For a service that is related to a previously manufactured product (e.g., the classical maintenance and repair of installed appliances), then these four design options cannot be selected without giving consideration to the design of the production network. *For production destined for the global market* (sectors P1 and P3 in Figure 3.1.1.2), all four design options for the *global* service network come into question. *For production destined for the local market* (sectors P2 and P4 in Figure 3.1.1.2) only the design options in the sectors S2 and S4 in Figure 3.1.4.1 come into question from a *global perspective*. From a *local perspective*, it is naturally possible to view service by the (local) manufacturer as "central." In such a case, all four design options can come into question, albeit *only for the local* service network.

3.1.5 Design Options for Transportation Networks

Intended learning outcomes: Differentiate between direct transport and indirect transport. Present features such as size or weight of the delivery, possibility of using an existing transport network, and need for merged transport. Explain design options for global transportation networks. Describe some company cases.

Production and distribution networks for physical products must be designed in consideration of the possibilities for transporting the goods. For each *means of transport*, e.g. lorry, railway wagon, ship or aircraft, the infrastructure must be available in terms of the suitable *mode of transport*, i.e., road, rail, water, or air; this means a network of transport channels with the necessary interchanges, i.e., loading yards, railway stations, harbors, or airports.

Independent carriers may be used in the transportation network. A *third-party logistics (3PL) provider* offers product delivery services. It may provide added supply chain expertise *([APIC16])*. Such logistics services comprise classical services such as transport, reloading, and storage, but also secondary packaging, the insertion of an information sheet, simpler assembly or repair work, and the acceptance of returned products.[1]

In a first approach, it is possible to distinguish two fundamental types of transport networks.

In the case of *direct transport,* the transport between two sites takes place without changing the *primary means of transport*, i.e., the means of transport into which the load unit was directly loaded. In cases where a lorry drives independently onto a train (known as a *"rolling motorway"*) or a ferry in the form of a secondary means of transport, this still counts as direct transport, i.e., there is no change in the mode of transport.

In the case of *indirect transport,* the transport between two sites uses more than one *primary means of transport*. Therefore it is possible to exploit the individual strengths of different means of transport for the individual transport segments, and thereby increase their utilization levels. At the same time, however, the cost and time for reloading the load units from one means of transport to another (possibly also involving a change in mode of

[1] *First-party logistics (1PL) providers* are internal departments of producing businesses or local transporters. A *second-party logistics (2PL) provider* takes over functions such as transport, reloading, and storage as an external provider for producing businesses. A *fourth-party logistics (4PL) provider* is a 3PL-provider that does not have its own infrastructure, but has competence for integration of 3PL tasks throughout the entire supply chain.

transport) at *transshipment centers* — i.e., distribution centers, the purpose of which is solely to reload goods — must be factored in.

The following *features for designing transportation networks* have proved to be important:

- *Size or weight of the delivery* (in kilograms or cubic meters): How do the suitable means of transport match up to this?
- *Possibility of using an existing transport network*: Can the delivery specify a means of transport that is already carrying deliveries between the point of dispatch and the recipient, and that is not yet at full capacity? This calculation could include means of transport that have a known timetable

Observation of activity in the field shows that these two characteristics correlate highly with each other: Large dimensions/high weight or a high possibility of using an existing transport network tend mainly to be served best by direct transport. In the opposite case, indirect transport is more advantageous. Both apply irrespective of the value density. For example, foam packaging and gravel are better transported directly, owing to their volume and weight, respectively. Diamonds, on the other hand, are better suited to indirect transport owing to their low volume and weight, by plane if greater distances are involved.

Two further characteristics for the design of transport networks that correlate with each other, but not with the previous pair, are:

- *Need for merged transport:* To what extent will delivery be made together with products or service objects from another manufacturer or service provider? In the case of returns, to what extent must several products or parts of them be sent back to a number of manufacturers at the same time? To what extent must several service objects or parts of them be transferred to multiple service providers at the same time?
- *Customer tolerance time*, as defined in Section 1.1.6.

If the need for merged transport is on the high side, or if there is greater customer tolerance time, different forms of indirect transport are advantageous.

As with production, distribution, or service networks, the two groups of features often stand in opposition to one another. Examples are

- The transportation of very high-value goods (e.g., money, precious gems, or precious metals), or express transportation (e.g., for spare parts): low size or weight of delivery (in favor of decentralized service), however, low customer tolerance time (in favor of centralized service)
- Regular deliveries to points of sale by a large wholesaler, on-line orders that are delivered to pickup sites, or regular transport of groups of people to events at specific locations: high size or weight of delivery (in favor of centralized service), however, high need for merged transport (in favor of decentralized service)

Again, a company must make a strategic decision, which sometimes differs for each product family. The portfolio in Figure 3.1.5.1 shows, in addition to the two classical designs (direct or indirect transport), two possible mixed designs for the transport between two locations L1 and L2. The four possible designs lie in four sectors in a two-dimensional space, spanned by the dimensions that correspond to the two (conflicting) groups of features.

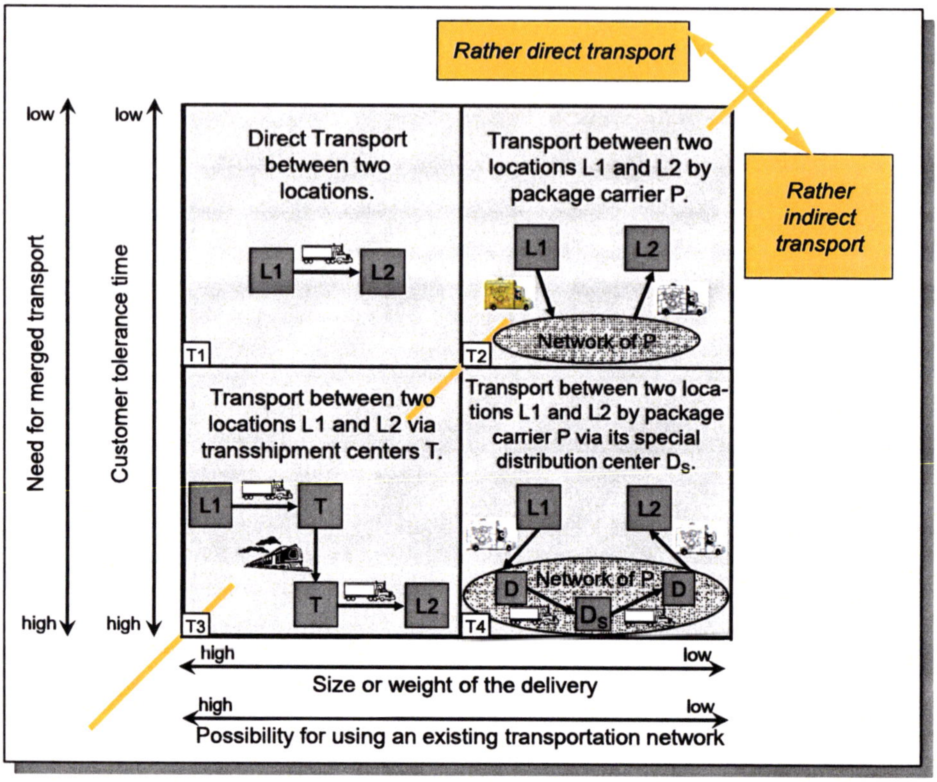

Fig. 3.1.5.1 Features of and design options for transportation networks.

The sector T1 describes the design option direct transport between two locations. It is beneficial if a *(full) truckload lot* — i.e., a single delivery of minimum size or minimum weight that is sufficient for the rate for a full load by the selected means of transport — is to be transported between two locations (e.g., the manufacturer's warehouse and a pickup site). This can result in lower transportation costs. The means of transport can be provided from the dispatcher's own fleet, or by *(full-)truckload (TL or FTL) carriers* — i.e., transportation companies that bill for the full utilization of the means of transport.

Last mile delivery represents a challenge. In this case the distributor supplies products itself to a range of customers on a route, instead of using a package delivery company. As the delivery vehicles used on at least part of the route are equivalent to "LTL carriers" (less than truckload), the transportation costs are correspondingly higher. Providing the distance between the distributor's depot and the customer is short and the *routing and scheduling* are effective, this is countered by a very short delivery lead time that generally cannot be matched by a package carrier company. Examples here are delivery of daily provisions and the supply of medicines to pharmacies. When the service is provided at the location of the object (e.g., maintenance and repair of installed appliances), the service provider must resolve a similar problem in terms of optimum *route planning*.

The opposite sector T4 describes the transportation between two locations by a package carrier via a special distribution center: Third-party logistics (3PL) providers in particular may possess distribution centers that offer a service infrastructure that extends beyond traditional transportation services; for example, for *in-transit merge*, or merging products from several manufacturers. Instead of receiving multiple deliveries, customers need only receive one shipment, which reduces their costs for transportation, handling goods in and putting the orders together themselves. However, this is offset by the longer duration owing to the special distribution center, where additional costs are also incurred for the work involved in combining the products. An example of this is the delivery of computers that are merged from components from a range of manufacturers (e.g., processor from brand x and screen from brand y).

The intermediate sector T2 describes the transportation between two sites by a package carrier. This option is selected if the size or weight of the delivery is too small or low to justify ordering transport specifically for this delivery. The package carrier helps in collecting additional shippers and therefore spreading the burden of costs for the full transport over a broader customer base. Such package carriers can also access their own network of transshipment centers, about which the customer need have no knowledge. Depending on the type of goods and specific transportation requirements, there may also be specialized carriers such as those that carry very high-value goods (e.g., money, precious gems, or precious metals), or those that specialize in express transportation where the customers' tolerance time is short (e.g., for spare parts).

The intermediate sector T3 describes the transportation between two locations via transshipment centers. This option may be selected if an order comprises several products from different manufacturers and must be delivered without interim storage, or if an order is to be supplied by a given transportation that is approaching the destination location anyway. Cross-docking principles, which are designed for fast transfer through a transshipment center, are therefore vital from the customer's viewpoint. This design possibility leads to better utilization of LTL carriers along the whole route. However, it is even possible to use TL carriers. The pros are offset by a normally longer lead time owing to the diversion via the transshipment center and complex planning for load building and cross-docking operations. Examples here include regular deliveries to individual points of sale by a large wholesaler, or online orders that are delivered to pickup sites, or regular transport of groups of people to events at specific locations.

The four design options for the transport network in Figure 3.1.5.1 can, in principle, be selected for each of the four design options of production networks in Figure 3.1.1.2 to transport between two operations. The same holds for distribution networks in Figure 3.1.2.1 to deliver from a warehouse or directly from the production line to a customer or the customer's unloading point or a pickup site. The same holds both for product returns and when the transport is not carried out to the end-customer, i.e., the consumer, but "only" to the depot at the next echelon of the distribution network structure, e.g., from the manufacturer to the wholesaler's distribution center, or from a central distribution center to a retailer's warehouse — see also Section 3.1.3. And the same also holds for service networks in Figure 3.1.4.1 for the transfer of the service object or its parts to multiple manufacturers or service providers, or for delivery of the object or parts back to the location after provision of the service, as well as for the transport of the service provider and its infrastructure closer to the object.

Company Cases: In the prior example, Hilti prefers option T3 for transport between the plant and the warehouses in the local market. This is due to the possibility for using an existing transport network, as the weight of the delivery is high. If the delivery volume of one production plant for one market is high enough, and accordingly the need for merged transport of products from different production plants is low, the direct transport between production plant and local market (option T1) also takes place. Direct transport by Hilti's well-known red station wagons is used for transport from the local warehouse to the construction site (option T1).

De facto, the available transport network design options validate the selected design of the distribution and production network. Should the transport network from the last production or storage location to the customer fail to enable an efficient design for which the delivery lead time can be kept shorter than or equal to the customer's tolerance time, a competing solution could clearly better serve the customer's needs. If the provider still wishes to serve the customer, there is a need to redesign the distribution network, and possibly even the production network. Holcim, for example, deliberately cedes a customer to competitors as soon as it has no chance to meet the customer's tolerance time with its actual production, distribution, and transport network. However, as the example of the plant in Ste. Genevieve (Missouri) shows, long-term investments in transportation infrastructure can change this situation. With the Panama Canal extension, Holcim considers whether or not to distribute cement to the West Coast of the USA as well from this plant. Likewise, with the availability of the new Gotthard railroad base tunnel, Holcim considers to distributing cement to Northern Italy from its plants in northern Switzerland instead of producing locally, as soon as an existing quarry will be exploited, and instead of looking for a new quarry in northern Italy.

Not every activitiy along the supply chain can be clearly identified as part of production, distribution, or service. Certain activities, such as secondary packaging, adding information leaflets, or affixing a local sticker to electrical appliances, can be carried out both in the factory and in a suitable distribution center. Services such as cleaning or battery charging can be handled by a pickup site. Other services, such as preproduction of catering products, can be performed at a production site with immediate distribution. Thus, a distribution network or also a service network can potentially develop into a production network. As for many other companies, this is also true in the case of Holcim, where the local "terminals" are not only elements of the distribution or service network structure, but can also be used to assemble specific or higher-level finished products (e.g., concrete) for local customers.

3.1.6 Integration of the Portfolios of the Partial Networks

Intended learning outcomes: Describe the interrelation between and integration of the production, transport, distribution and retail network.

An adequate design of the value-added network is decisive for customer satisfaction, given a sufficient quality of the product or service. As Figure 3.1.6.1 shows how the production, distribution, retail (if needed), and various transport networks have to fit together in order to achieve the goal of a satisfied customer.

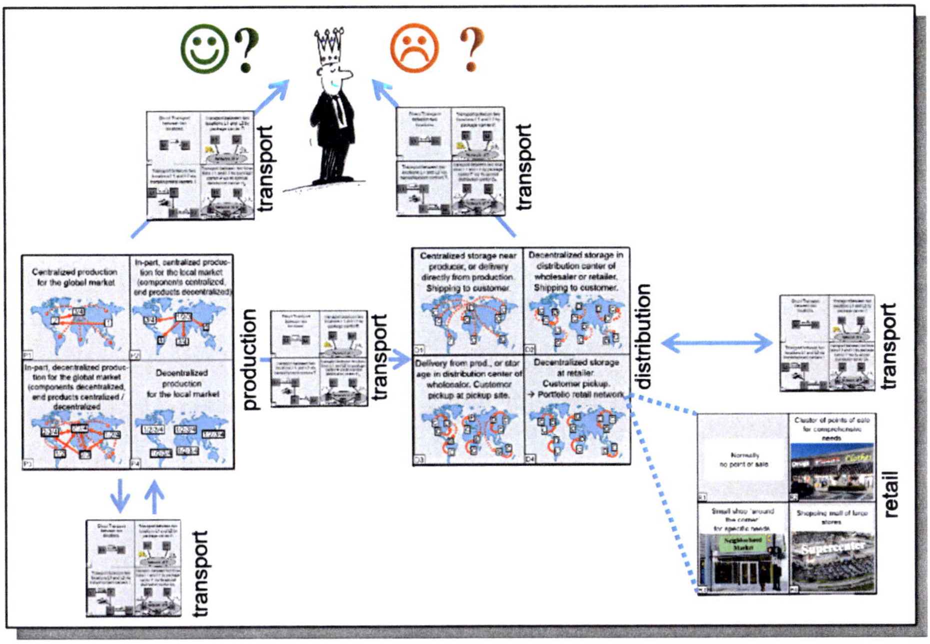

Figure 3.1.6.1 Interrelation between and integration of the production, transport, distribution and retail network

There are dependencies between the design options of the different portfolios that should be considered for their integration. Here are some examples:

As the customer tolerance time is a characteristic for the design of both transport networks (see Figure 3.1.5.1) and production, distribution, and service networks (see Figure 3.1.1.2, Figure 3.1.2.1, and Figure 3.1.4.1), there are naturally close combinations when it comes to integrating the networks. This is the case when the customer tolerance time is low in both portfolios, or high in both portfolios. In the case of the distribution network (Figure 3.1.2.1), this means:

- Decentralized distribution is the preferred combination for direct transport, with the aim of reducing delivery lead time to a minimum; and
- Centralized distribution is the preferred combination with indirect transport — i.e., via transshipment centers — since a short delivery lead time is not the priority and it is preferable to ensure that the means of transport is operating at better capacity utilization levels across the route to achieve lower transportation costs.

However, the other combinations are nevertheless possible:

- If the combination of centralized distribution and direct transport is advantageous, demand is highly varied and/or volatile. The reduced storage costs obtained from centralized warehousing thereby outweigh the disadvantage of a lengthier delivery lead time. If value density of the product is high, it is possible at any rate to select a means of transport that is fast enough:

- If the combination of decentralized distribution with indirect transport is advantageous, the ability to reduce transportation costs or increase simplicity of a merged transport (the aforementioned "in-transit merge" where the customer receives just one complete delivery) outweigh the disadvantages of a longer delivery lead time from the customer's perspective.

3.2 Location Selection and Location Configuration

Intended learning outcomes: Differentiate between location selection and location configuration. Explain location selection using qualitative methods and factor rating. Describe location selection and location configuration with linear programming.

Figure 3.2.0.2 shows the difference between location selection and location configuration.

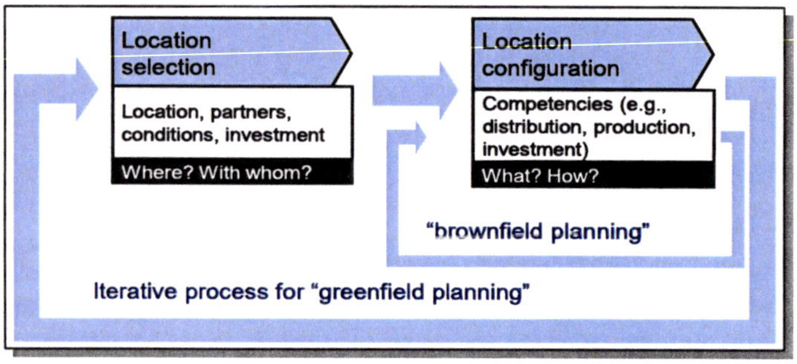

Fig. 3.2.0.2 Location selection and location configuration.

Location selection is selection of new locations, also in the case of moving facilities from existing locations to new locations.

Location selection is a part of what is called "*greenfield planning,*" that is, planning with new locations and therefore the fewest possible given framework conditions. Location selection involves investments in basic infrastructures as well as the building of possible company partnerships (for example, joint ventures). Official regulations often play a large role here.

Location configuration is assignment of products to an existing location. The assignment is made for each new product, or service, and can be reviewed periodically. A certain product or service can be assigned to several locations, not only in distribution but also in production.

Location configuration forms part of "*brownfield" planning*, i.e., planning with predefined locations and possibly other boundary conditions. Location configuration is associated with further investment, this time in personnel, machinery and the establishment of supplier relationships. Conditions imposed by government also play a large role here.

Thus, the starting points and the objectives of location *selection* and location *configuration* are not the same. As a consequence, the methods differ.

- Location selection is often handled using catalogues of criteria, such as morphological schemes, followed by a mostly rather qualitative evaluation.
- In location configuration, in addition, suitable methods of *mathematical programming* are used, such as linear and nonlinear programming or heuristic methods.

3.2.1 Location Selection Using Qualitative Methods and Factor Rating

Intended learning outcomes: Disclose a model for systematic reduction of possible locations / partners. Produce an overview on steps in location selection and evaluation of a concrete case, namely a joint venture partner in China. Present factors for location selection used in this case. Explain the results of factor rating and the sensitivity analysis in this case.

This is primarily concerned with the location selection of a *production network*. Instead of listing all possible criteria for location selection, in the next section below we will discuss a specific case, namely, the evaluation of a joint venture in China by a European company that constructs plants.

It is best to conceive of location selection as a project, with the associated tasks of project initiation, project management, and project realization. Figure 3.2.1.1 shows the steps in the concrete case mentioned above. Noticeable in this specific case is the long time period for location selection. Almost two years were required for the evaluation.

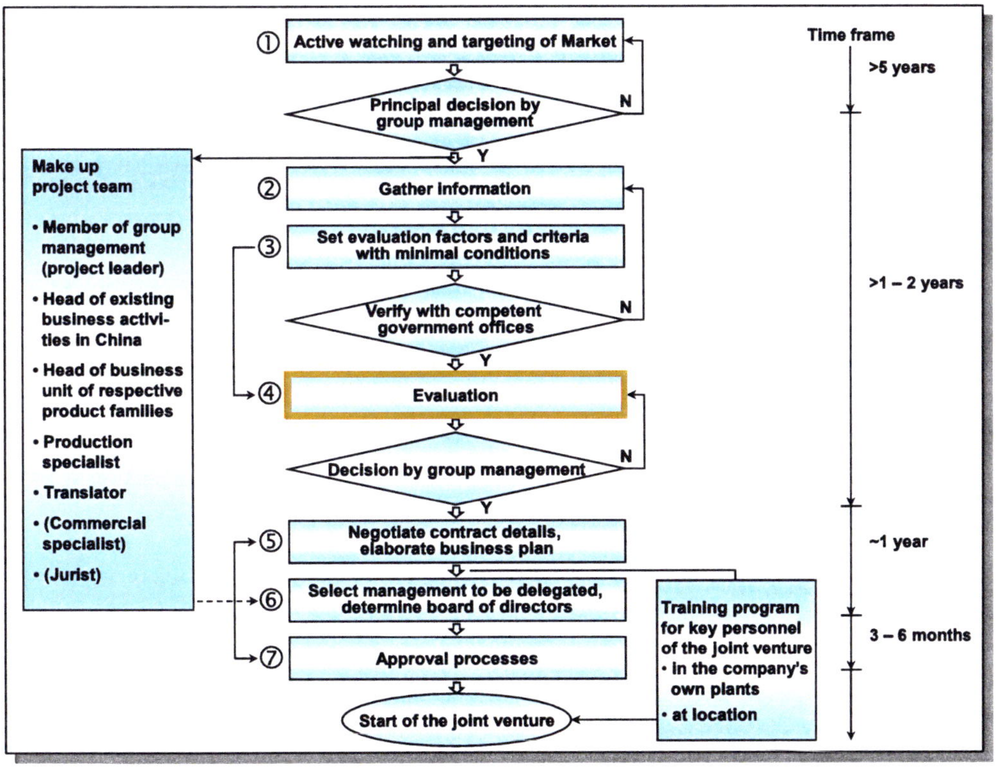

Fig. 3.2.1.1 Steps in location selection and evaluation of a joint venture partner in China.

In practice, there are many *location factors* critical to success, and each factor has criteria. For a comprehensive view, a complete set of factors with individual criteria is required. Evaluating the degree to which the criteria are fulfilled is then dependent on the strategy chosen in the specific case. Figure 3.2.1.2 shows the location factors examined in this case.

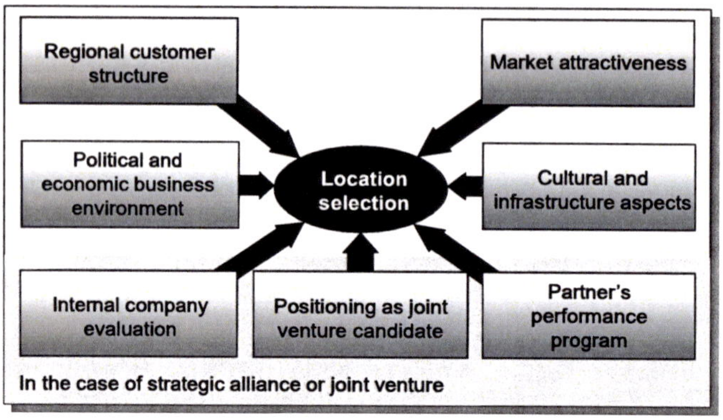

Fig. 3.2.1.2 Factors for location selection.

In addition to four factors that relate to the actual location, three factors are shown that relate to the contemplated joint venture partner. The individual criteria in these three factors can implicitly also characterize — although not only — the location of the partner.

To proceed effectively and also efficiently, whereas as many locations as possible are considered, they are reduced as rapidly as possible to a few candidates through examination of an appropriate sequence of location factors. One model for this is shown in Figure 3.2.1.3.

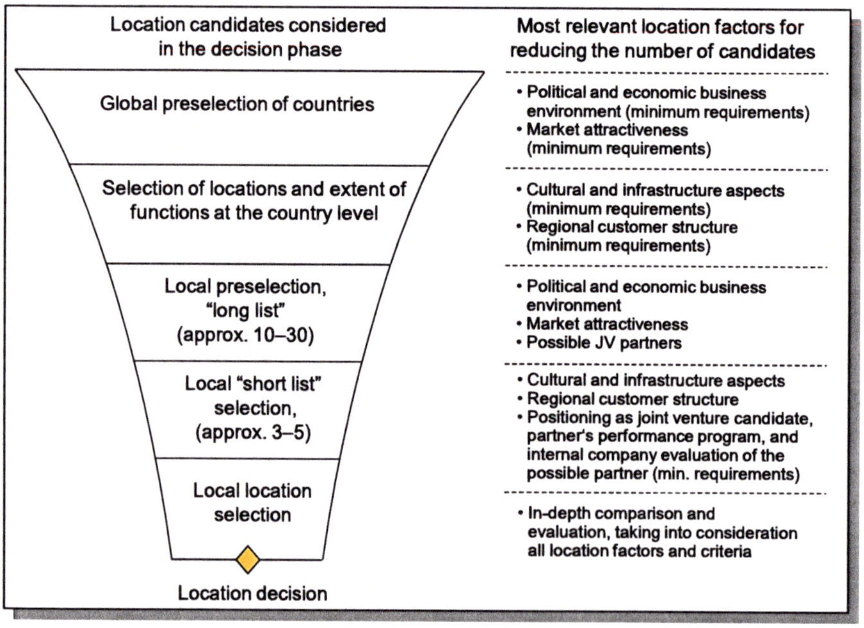

Fig. 3.2.1.3 Systematic reduction of possible locations / partners.

This funnel was based on an idea in [AbNu08]. The location factors are considered in an order that allows systematic reduction of locations. The final step in Figure 3.2.1.3 is discussed in the following. The figures show the criteria for each location factor that a European plant manufacturer rated in the concrete case of evaluating joint venture partners in China. The criteria were rated in the order shown in Figure 3.2.1.2. Each criterion was rated. Of course, the range of values for the factors chosen here are just examples; in other cases, a different range of values might be used.

Figure 3.2.1.4 shows the criteria of the factor "political and economic business environment."

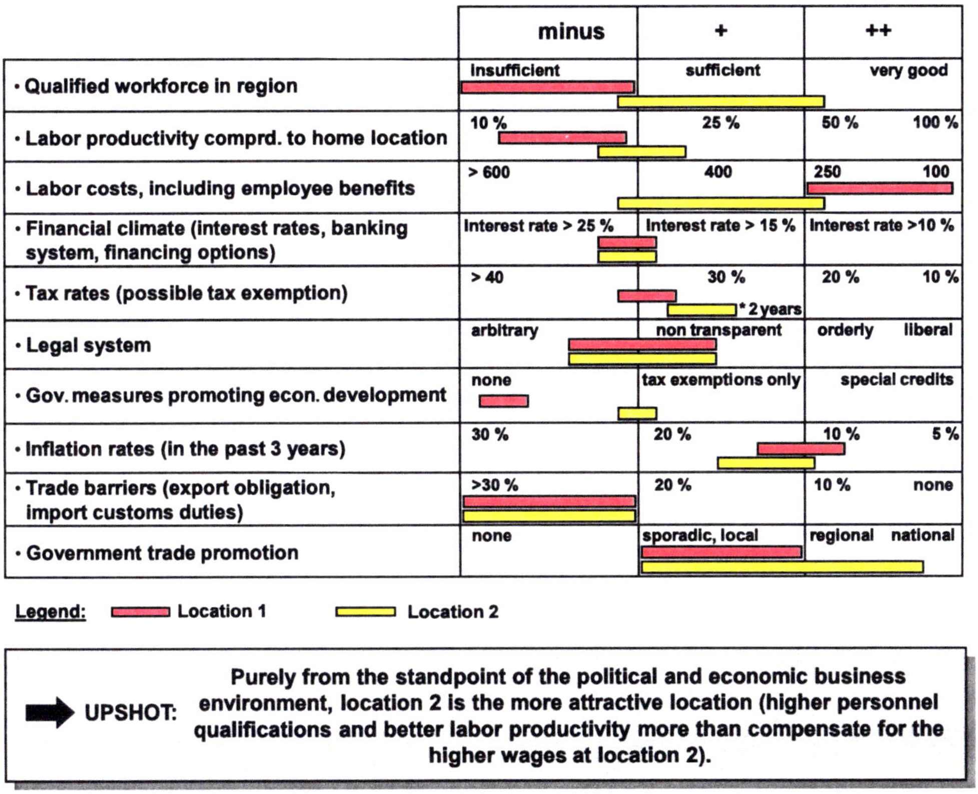

Fig. 3.2.1.4 Evaluation of a joint-venture candidate in China: Criteria of the location factor "political and economic business environment."

As a further criterion for the location factor "political and economic business environment," political stability (unrest, corruption, and strikes) could be evaluated, for example.

Figure 3.2.1.5 shows the criteria of the location factor "cultural and infrastructure aspects."

Other criteria under "cultural and infrastructure aspects" could also be work ethic, availability and skills of workers, and the telecommunications infrastructure or water availability.

Figure 3.2.1.6 shows the criteria of the location factor "regional customer structure."

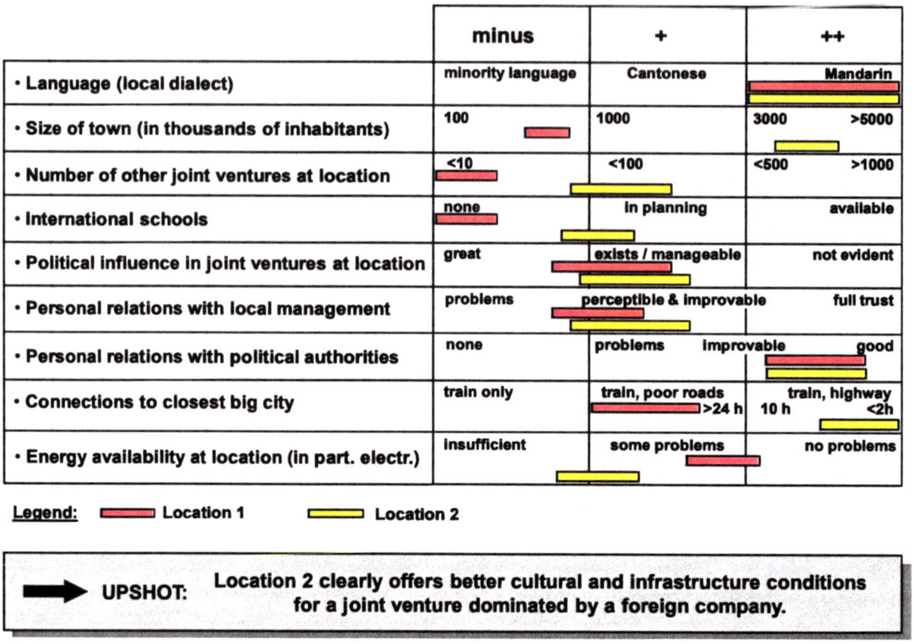

Fig. 3.2.1.5 Evaluation of a joint-venture candidate in China: Criteria of the location factor "cultural and infrastructure aspects."

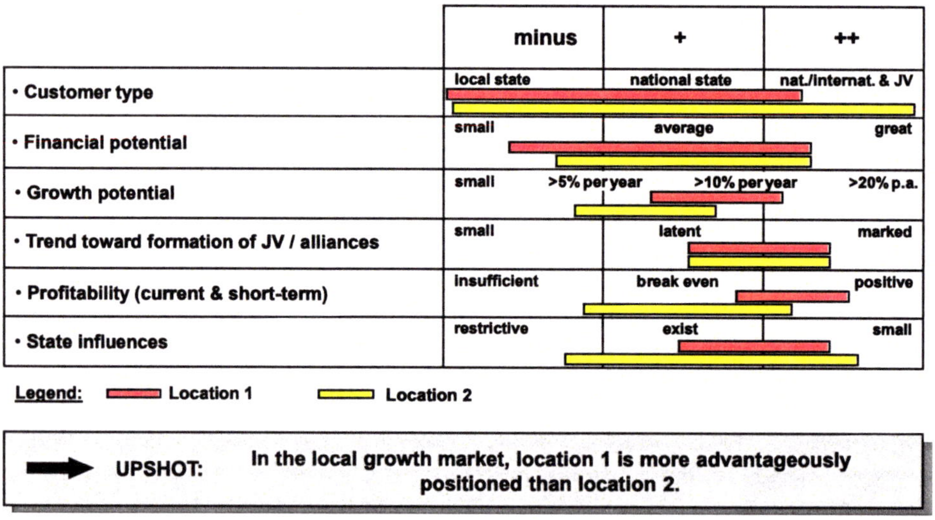

Fig. 3.2.1.6 Evaluation of a joint-venture candidate in China: Criteria of the location factor "regional customer structure."

Further criteria under "regional customer structure" can be the proportion of customers in the region that already are being supplied by the home base, the market power of customers, purchasing power of customers, customer and buying behavior, and the specific product and delivery time requirements of the customers.

Figure 3.2.1.7 shows the criteria of the factor "medium-term attractiveness of the market."

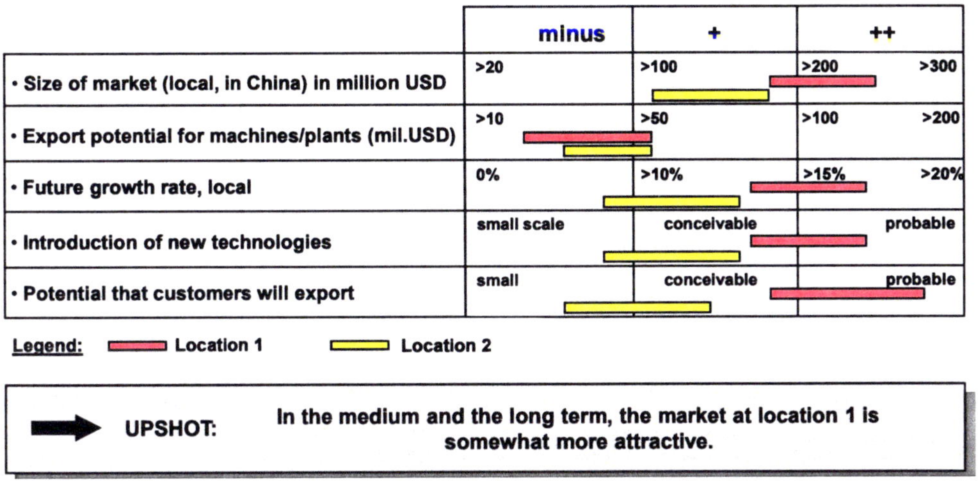

Fig. 3.2.1.7 Evaluation of a joint-venture candidate in China: Criteria of the location factor "medium-term attractiveness of the market."

Further criteria under "medium-term attractiveness of the market" can also be examined: the expected market position, the origins of the competitors (possibly from home), the market segments, the company's own potential for exporting products (transport, customs duties, and so on), and possible substitution products by competitors.

Figure 3.2.1.8 shows the criteria of the location factor "internal company evaluation of a joint venture candidate."

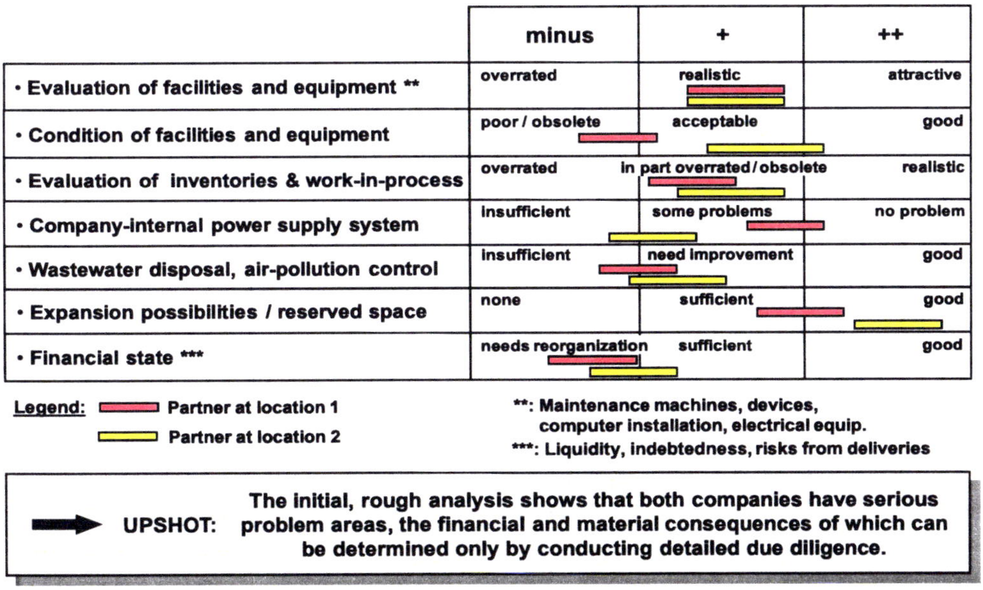

Fig. 3.2.1.8 Evaluation of a joint-venture candidate in China: Criteria of the location factor "internal company evaluation of a joint-venture candidate."

Figure 3.2.1.9 shows criteria of the factor "general positioning as joint venture candidate."

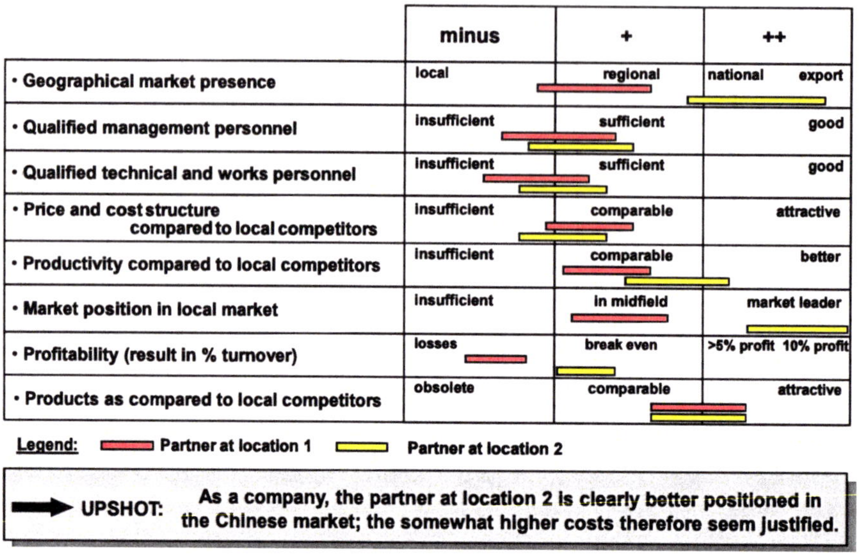

Fig. 3.2.1.9 Evaluation of a joint-venture candidate in China: Criteria of the location factor "general positioning as joint venture candidate."

Further criteria of the location factor "general positioning as joint venture candidate" that can be rated are: regional presence (production / distribution / sales / service), innovation behavior, and strategic orientation.

Figure 3.2.1.10 shows the criteria of the location factor "performance program of the potential joint venture partner."

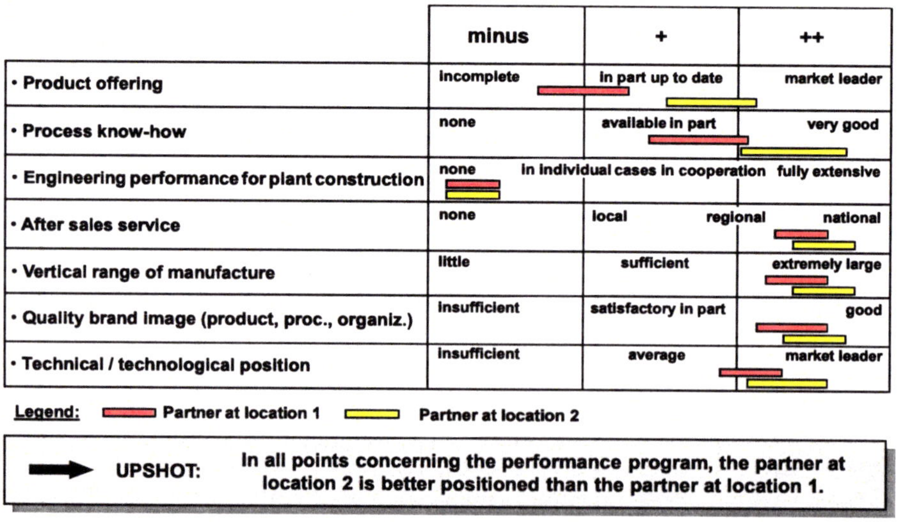

Fig. 3.2.1.10 Evaluation of a joint-venture candidate in China: Criteria of the location factor "performance program of the potential joint venture partner."

Further criteria that can be examined under the location "performance program of the potential joint venture partner" are also the specific process know-how in individual areas of sales and distribution, R&D, production, and installation.

To evaluate alternative locations, a cost-benefit analysis is often performed. A simple, rather qualitative tool for this is factor-rating, for example.

Factor-rating [DaHe05, p. 382], is a decision method for evaluating several possible solutions to a problem that can be characterized using factors or features.

Figure 3.2.1.11 shows the results of a rough-cut factor rating in the form of a graphical summary of the criteria that were applied in the concrete case.

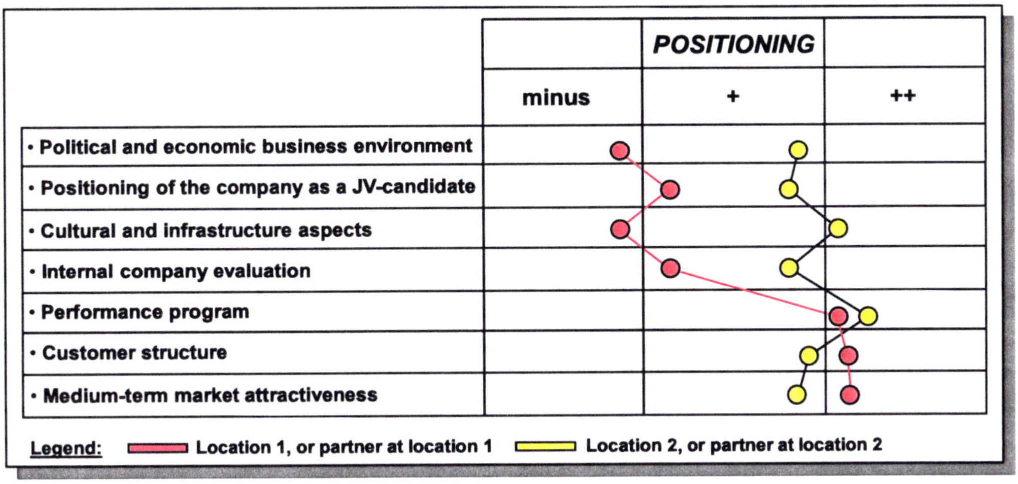

Fig. 3.2.1.11 Results of factor rating.

The overall rating can be determined qualitatively using the graphical representation in Figure 3.2.1.11 as an aid. First, starting from the rating of the individual criteria in Figures 3.2.1.4 to 3.2.1.10, simple graphical averaging, or interpolation, allows the determining of the positions in Figure 3.2.1.11. As the great majority of the values for location 2 are higher than the values for location 1 — also in both cases where better evaluations resulted for location 1, but were only insignificantly better than location 2 — the decision will be made that location 2 is the better location; in the case at hand, that was indeed the company's decision.

Generally, however, it is necessary in the last step to quantify the ratings and to assign weights to both the individual criteria within a location factor and the location factors in relation to one another (reflecting their relative importance to the company) to be able to compare the alternative locations. Instead of estimating the ratings qualitatively (from minus to plus to double-plus), we can determine the degree to which a criterion is fulfilled, expressed, for example, as a percentage of maximum fulfillment of the criterion. Suppose

- n is the number of locations,
- m_i is the number of criteria per location i, $1 \leq i \leq n$,

- $F_{i,j}$ is the degree of fulfillment of the criterion (i,j) of location factor i, $1 \le j \le m_i$, $1 \le i \le n$,
- $W_{i,j}$ is the weight assigned to the criterion (i,j), $1 \le j \le m_i$, $1 \le i \le n$,
- Wi is the weight assigned to the location factor i, $1 \le i \le n$.

Figure 3.2.1.12 shows the formula for calculating the overall benefit B for each location and thus the ranking of the locations. For greater certainty in ranking the alternative locations, in addition to determining the overall benefit, a sensitivity analysis is also necessary.

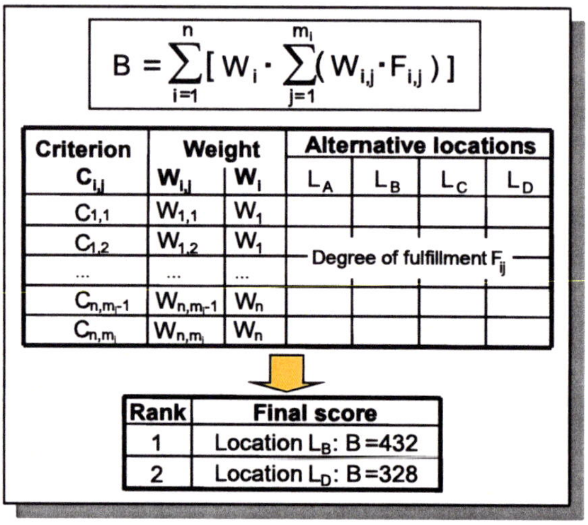

Fig. 3.2.1.12 Factor rating with degrees of fulfillment and weightings.

A *sensitivity analysis* determines how much an expected outcome or result will change in response to a variation of the input variables.

For an identification of locations that are close in the overall score, it is important to vary both the degree of fulfillment and the weightings.

The quantification of ratings can also be set by considering costs and investments associated with the locations. This can, for example, concern the criteria under the location factor "internal company evaluation" (see Figure 3.2.1.8). In this case, to discount the benefits and costs over time, commonly used methods of investment analysis can be applied, such as the net present value technique, NPV (see Figure 19.2.5.3).

Likewise, for location selection for both distribution and service networks, location factors with catalogues of criteria are used.

- For distribution or service processes in direct contact with the customer or the object, the main factors are criteria such as access routes (pedestrians, cars, public transportation), population density, size of families, and annual family income. See the discussion in relation to Figure 3.1.3.2.

- For distribution or service processes in indirect contact with the customer or the object, there are often additional factors, such as the availability of lower-cost temporary personnel that can be flexibly (time) employed. As an example, airlines locate call centers in Cape Town or Dublin, due to the plentiful supply of well-trained, multilingual exchange students present in these places who can provide a low priced, qualified workforce for these part-time jobs.
- For distribution or service processes with no customer or object contact, in contrast, the location factors and criteria are in principle the same as those described above for production networks.

Assessment of the various options for the location selection can be carried out using qualitative methods such as those above for production networks. Sometimes quantitative methods are also used, such as linear programming.

3.2.2 Location Selection and Location Configuration with Linear Programming

Intended learning outcomes: Produce an overview on linear programming (LP). For the design option of "in part decentralized production for the global market", explain the use of mixed-integer linear programming (MILP).

Probably the most challenging configuration of a production network as shown in Figure 3.1.1.2 is *in part decentralized production for the global market.* In this case especially the task of location configuration is difficult to solve, but it is also not insignificant in the other cases. It is the task of determining a global production plan: What products and — in the face of limits of capacity — how much of what product will be manufactured for what markets at what level at what locations? A similar question can also arise for *decentralized distribution* or *decentralized service*: What customers will be served by what distribution and service locations?

Very many influencing variables soon lead to a complex problem. Decision making can be supported — often with simplified model assumptions — by linear programming (LP).

In *linear programming (LP)* the task is to solve a problem that can be expressed as in Figure 3.2.2.1.

1. Objective function: *OF* = *max!*
2. Decision variables $x_i, \; 1 \le i \le n$
3. Restrictions/conditions as constraints for x_i
   Sum notation:

$$OF = \sum_{i=1}^{n} c_i x_i = \text{max!}$$

$$\sum_{i=1}^{n} A_{ij} x_i \le b_j \qquad ,1 \le j \le m$$

$$x_i \ge 0 \qquad ,1 \le i \le n$$

Fig. 3.2.2.1 Problem formulation in linear programming: Maximize the objective function OF and solve for x, subject to the constraints.

If the number of decision variables is two, the problem can be solved using a simple graphical method. With a greater number of variables, the use of an algorithm is recommended, such as, for example, the simplex algorithm. The complexity of the problem increases with increasing values of the number of variables (n) and the number of constraints (m). Computation time does not increase polynomially with n and m: the simplex algorithm is what is called an "NP-hard" algorithm. With high values of n and m, the procurement of data is also a problem.

It is mostly larger companies that use this quantitative method in practice.

- The automobile manufacturer Daimler uses software called "network analyzer" to determine, among other things, what products should be manufactured at what locations. The software uses mixed-integer linear programming (MILP). The constraints are, for example, market guidelines that must be met and the limits of capacity that must be considered. Entering into the objective function are, among other things, the production costs at the locations and the transport costs from the locations to the markets.
- In a similar way, the cement company Holcim uses MILP for location selection of their works. A special application was, for example, the case where a cement works had to be closed because of exhausted raw materials. The issue was whether the other works could deliver the required quantities. In the optimization, the increased transport costs stood vis-à-vis the fixed costs to set up a new works and the different production costs in the other works.

For some years now, MS Excel has offered a Solver tool that can be used to solve an LP problem with (in the current release) 200 variables.

A word of caution: A basic problem is that new roads and concentration centers are always being built. Moreover, important customers can move away, or the political and economic business environment can change. Then, any one selected location can prove to be suboptimal. If the high building and equipment costs have not yet been written off, the facility cannot simply be changed to fit the new data, or in other words, it cannot simply be moved to a new location or re-equipped. *In the long-term view*, therefore, simple, robust methods do not always have to be a priori at a disadvantage compared to complicated optimization algorithms (for example, nonlinear programming or heuristics as well).

3.3 Sustainable Supply Chains

Intended learning outcomes: Explain the concept of sustainability with reference to the triple bottom line. Disclose economic opportunities for social commitment and for environmental commitment. Describe energy management concepts and measures for improved environmental performance. Produce an overview on the measurement of the environmental performance. Present social and environmental dimensions in industrial practice.

In the second half of the twentieth century, business decisions made by the strongly growing, changing, and developing industries did not take environmental aspects into great consideration. Legislation and the needs and requirements of direct stakeholders were taken into account, but the interest of the public and society carried less weight if it did not affect the competitiveness or profitability of the business. Supply chains grew with target areas such as costs, quality, delivery, and later flexibility as the main drivers. This development had a major impact on the carbon footprint distribution of today's industry sectors. Figure 3.3.0.1 shows the economic sectors and their recent specific carbon footprint. Manufacturing and transport are responsible for more than half of the global CO_2 emissions.

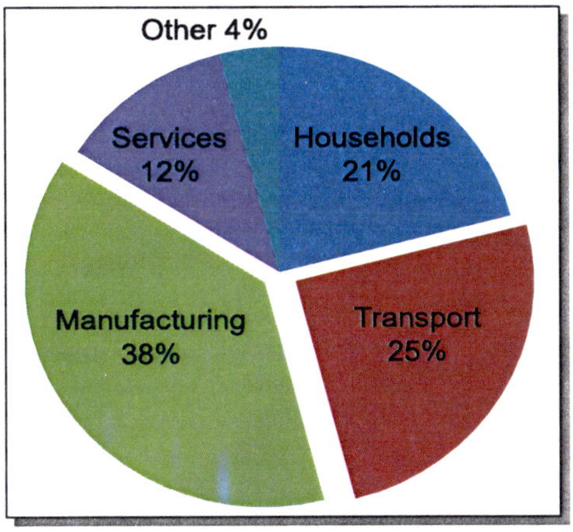

Fig. 3.3.0.1 Shares of global CO2 emissions in 2005 by sector (total direct and indirect CO2 emissions: 21 Gt CO2). (Source [IEA08]).

As supply chains consist of networks of actors and value-adding activities, supply chain management represents an important perspective to support sustainable development.

Sustainability and *sustainable development* can be defined as "meeting the needs of the present without compromising the ability of future generations to meet their own needs" (see [UN87], also known as the Brundtland Report).

Section 3.3.1 discusses the historical development of the concept of sustainability as seen from the viewpoint of industry. Sections 3.3.2 and 3.3.3 present the changing business conditions and the emerging economic drivers that foster social and environmental commitment. Section 3.3.4 gives examples of possible improvements in industry.

3.3.1 The Changing Concept of Sustainability with Reference to the Triple Bottom Line

Intended learning outcomes: Produce an overview on the concept of the triple bottom line. Present the paradigm change that correlates to the evolution of sustainability aspects and their interaction.

Triple bottom line (TBL) refers to measuring success of today's business activities not only in economic terms but also according to environmental and societal criteria, with the aim of "full costing" (see, for example, [GRI02], [WiLe09]).

The annual performance of an organization in the economic, environmental, and social areas may be reported to the public and its impact on its surroundings evaluated. Figure 3.3.1.1 illustrates the concept with the three areas, also called the "three pillars of sustainability."

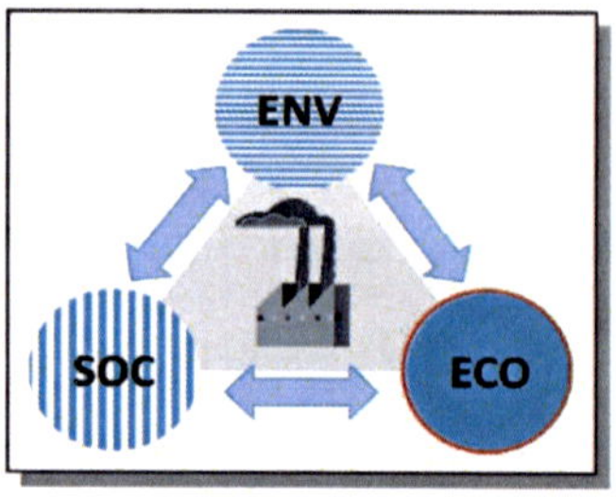

Fig. 3.3.1.1 The concept of the triple bottom line is based on the three pillars of sustainability — namely, economy, society, and environment, which interact with companies (see [ScVo10]).

From the point of view of industry, the three pillars may be defined as follows:

- *Environmental aspects (ENV)* refer to nature as a closed system, with its limited resources and regenerative capacities (e.g., for greenhouse gases) as a base for any business activity.
- *Social aspects (SOC)* refer to society represented by governmental organizations (GOs), nongovernmental organizations (NGOs), individuals, employees, and customers.
- *Economic aspects (ECO)* refer to events and developments relevant to manufacturing industries' competitiveness and their strategic and operational business practices.

The three aspects influence the development of each interdependently. Industry was affected in different ways, depending on the specific importance of each aspect as it changed over time. Every business activity, company, industry, supply chain, and economy is dependent in some way on the availability of (often limited) resources. Companies' competitiveness is affected by these conditions and thus by the three pillars of sustainable development. Performance indicators such as quality, costs, delivery reliability, and flexibility undoubtedly remain relevant, but their importance has to be evaluated as the conditions, under which businesses operate, aim toward sustainable development.

From the perspective of industry, the change in the manufacturing paradigm correlates to the evolution of sustainability aspects and their interaction. Figure 3.3.1.2, taken from [ScVo10], shows this evolution on a time line of selected events and developments. In the figure, a larger bubble depicts the specific relevance of a sustainability aspect. The larger arrows depict increasing impact and the specific direction of impact.

In the past 50 years, various events influenced industrial business conditions. The events can be categorized in the three aspects of sustainability, but because of the interdependencies, the classification is rather soft. Some of these events may be categorized in one of the other aspects as well.

In the 1960s, environmental issues peaked, with noticeable impacts on society. In some regions, environmental movements emerged. For example, the Clean Air Act in the UK regulated smoke from furnaces, as the coal smoke from households and industries affected everyday life negatively. The oil crisis in the 1970s created — as artificial scarcity — worldwide attention to the existing oil dependency and triggered energy efficiency improvements in industry. The toxicity of wastes and the chemicals used in production of goods and food was being recognized. Directives were developed that impacted the business conditions of companies, such as the Dangerous Substances directive (on classification, packaging, and labeling of dangerous substances) and the Waste Framework Directive (on disposal, and prevention of harmful effects through collection, transport, storage, and tipping of waste). The growing public interest led to the foundation of, for example, the US Environmental Protection Agency. The "Limits to Growth" commissioned by the Club of Rome pointed out the fundamental issue of the contrast between the growing need for resources and the scarcity of those resources.

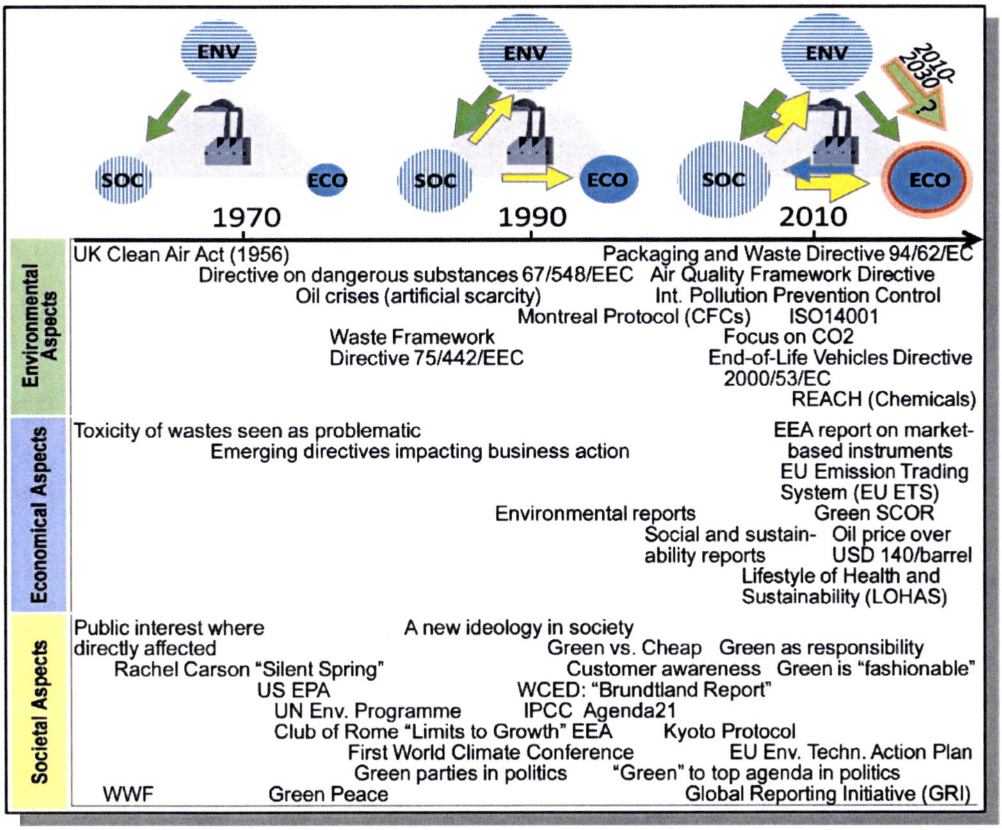

Fig. 3.3.1.2 Companies impacted by the three pillars: The paradigm change correlates to the evolution of sustainability aspects and their interaction [ScVo10].

In the 1980s "green parties" achieved breakthroughs in politics, and environmental protection moved up on the political agendas. The interaction between GOs, NGOs, stakeholders in society, and industrial companies increased, which was partly due to the media attention to the topic. As a result, emerging regulations and pressures further impacted business conditions. The implemented Montreal Protocol (phasing out ozone depleting substances, like CFCs [chlorofluorocarbons]) became an example of a very successful international treaty for which some industries were required by law to implement changes in production facilities and products. Environmental reports found their way into the annual reporting of companies, representing the basis for the sustainability reports later to come. In the 1990s, the European Environment Agency, as a counterpart to the US Environmental Protection Agency, was founded. The Global Reporting Initiative formed in 1997 developed sustainability standards for organizations. In that same year, the well-known Kyoto Protocol was set up, under which participating countries commit to reduce greenhouse gas emissions by a specified level. The Kyoto Protocol offers countries various mechanisms to reach the targets, such as the emissions trading (the EU Emission Trading System ETS commenced operation in 2005) and the Clean Development Mechanism.

The environmental management standard ISO 14000, complementing quality management standard ISO 9000, was published in 2004. Economic changes, such as the oil price (e.g., the peak in 2008 exceeded U.S. $140/barrel) on the supply side and the Lifestyles of Health and Sustainability (LOHAS) demographic on the demand side, represent important economic drivers. From a regulatory perspective, increasing the ecological awareness and the facilitation of eco-innovation remain important. Based on the recent developments, it seems that companies are required to react when environmental pressure increases. The pressure comes from the public (GOs, NGOs, customers) as well as from the environment (as certain resource-depleting activities, e.g., conventional oil production, will not be sustainable in the medium and long term), and thus creates economic impact.

For further reading, see [Pack60], [Cars02], [Mead77], [UN87], and Web sites such as www.eea.europa.eu, and www.epa.gov.

3.3.2 Economic Opportunities for Social Commitment

Intended learning outcomes: Disclose the term "double bottom line". Produce an overview on ethical standards, or code of conduct (CoC). Differentiate between groups of company-internal ethical standards and groups of company-external ethical standards. Present the supplier code of conduct (SCoC) and the certificate of compliance.

A more limited approach as compared to the triple bottom line is called the double bottom line. In the area of the environment, the two terms overlap somewhat. But the point of view is not always the same.

> The term *double bottom line* is about what business or supply chains contribute toward positive social effects.

Some companies see the double bottom line as a marketing approach, in that it gives donations and gifts that foster the development of a community that lives in the environment of the company or is affected by the emissions of the company. This way of proceeding can be useful for the company, which through this purchased reputation can then enlarge or retain its customer base. If the object is to divert attention away from poor treatment of employees or a minority of local residents, this cannot be viewed as anything different than a salesman who uses bribes to push through his goals.

This understanding can be seen in any case as a side effect of business activity. For the customers should actually be won over by the core activity of the company, its products and services. Social commitment in the *primary* business activity — that is, in the manufacturing, sale, and disposal of products or services, means rather adhering to ethical standards.

Ethical standards are a set of guidelines, or a *code of conduct (CoC)*, for impeccable conduct by professionals.

These standards can be summarized in groups. Figure 3.3.2.1 shows company-*internal* ethical standards, and Figure 3.3.2.2 shows company-*external* ethical standards. They were taken from [OeNa10].

However, it is not enough for companies to draft and implement ethical guidelines in their own company. The ethical standards must also be maintained upstream from the company in the supply chain. Particularly, in low-wage countries, this is not a matter of course, as examples in recent years have shown, such as the raw materials from areas of civil war in the Congo or in West Africa, the coal mine disaster in Dongfeng, or toy manufacturer Mattel's problems with toxic materials.

Labor	Health and Safety
▪Forced labor ▪Child labor ▪Juvenile labor ▪Discrimination ▪Harassment, inhumane treatment ▪Respect and dignity ▪Freedom of association ▪Working hours, rest periods, and breaks ▪Minimum wages and benefits ▪Overtime compensation ▪Recorded terms of employment ▪Employee privacy	▪Emergency (preparedness and response) ▪Occupational injury and illness ▪Machine safeguarding / workplace safety ▪Implement safety management system ▪Sanitary infrastructure (incl. potable water) ▪Food preparation and storage facilities ▪Industrial hygiene / safe and healthy work environment ▪Product safety ▪Comply with local health and safety laws

Fig. 3.3.2.1 Groups of company-internal ethical standards.

Environment	Business Partner and Community
▪Wastewater and solid waste ▪Air emissions ▪Noise ▪Safeguarding the health / safety of the public ▪Chemical and hazardous materials ▪Environmental management system ▪Comply with existing legislation and regulation ▪Environmental permits and reporting ▪Product content restrictions ▪Minimize waste / resources used ▪Maximize recycling ▪Environmentally friendly processes ▪Policy (responsibility, stated) ▪Policy promotion / employee education	▪Fair business / competitiion ▪Protection of intellectual property ▪Corruption, extortion, embezzlement ▪Gifts and hospitality (receive) ▪Gifts and hospitality (give) ▪Whistle-blowers ▪Conflict of interest ▪Disclosure of information ▪Accordance with local / international laws ▪Human rights ▪Terrorism ▪Community engagement

Fig. 3.3.2.2 Groups of company-external ethical standards.

A *supplier code of conduct (SCoC)* outlines expectations regarding ethical standards for direct suppliers.

A *certificate of compliance* is a supplier's certification that the supplies or services in question meet the specified requirements ([APIC16]).

In addition to the four groups of standards above, the SCoC contains a fifth group, which describes the aspects in meeting the expectations of the buyer company. An important aspect that can be easily forgotten is that the SCoC also applies to the supplier's suppliers, that is, further upstream in the supply chain.

3.3.3 Economic Opportunities for Environmental Commitment

Intended learning outcomes: Produce an overview on energy-intensive industries. Disclose examples of using alternative fuels and raw materials in order to decrease the carbon footprint and the amount of fossil fuels required in the cement industry. Differentiate between opportunities and threats favoring proactive and reactive environmental involvement.

While the regulatory landscape is changing and forcing firms to do *environmentally responsible business*, research and practice is showing that measures can be taken that improve environmental and economic performance at the same time. However, identifying viable improvement opportunities to increase energy efficiency remains a challenge in daily business. For one, energy is still (in 2010) relatively low priced. In the conventional manufacturing industry, energy cost can make up 2 to 3% of operating costs. For another, for investment decisions, the opportunities and risks that may be caused by regulations, prices, and markets are difficult to estimate: The core competencies and priorities of most companies are not in the field of energy saving (and buying know-how from the outside is also connected with costs). Clearly, for energy-intensive industries, such as chemicals and petrochemicals, iron and steel, cement, and pulp and paper, the situation is different than for the conventional manufacturing industries.

Energy-intensive industries (EIIs) are industries where energy costs make up a significant part of the operating costs (possibly up to 60%) and thus represent a major competitive factor.

As the fuels are regularly mostly fossil fuels, EIIs emit a considerable amount of CO_2, which makes them vulnerable to carbon footprint regulation. EIIs made significant improvements in the past, especially the chemical and petrochemical industry. The following example from the cement industry may be taken for illustration: The cement industry requires a considerable amount of energy for the clinkering process (the chemical process transforming limestone into clinker, a basic element of cement). Fuel makes up to 30 to 40% of the total operating costs. At the same time, the chemical reaction produces CO_2 as a by-product (worldwide, the cement industry is responsible for more than 5% of the man-made CO_2 emissions). Figure 3.3.3.1 shows actions that were taken to reduce both costs and CO_2.

By using by-products (or waste) from other industries, it becomes possible to both reduce the amount of fossil fuel and the amount of clinker required for the production of cement.

An approach like this is called co-processing, and it is an important way to approach the challenges in the cement industry. However, a general criticism may be that using wastes in incineration (as fuel) can lead to toxic emissions and promote more production of waste. Life cycle considerations and pollution prevention need to be taken into account before deciding whether measures are suitable and, because of the complex economy and business activities, a challenging undertaking.

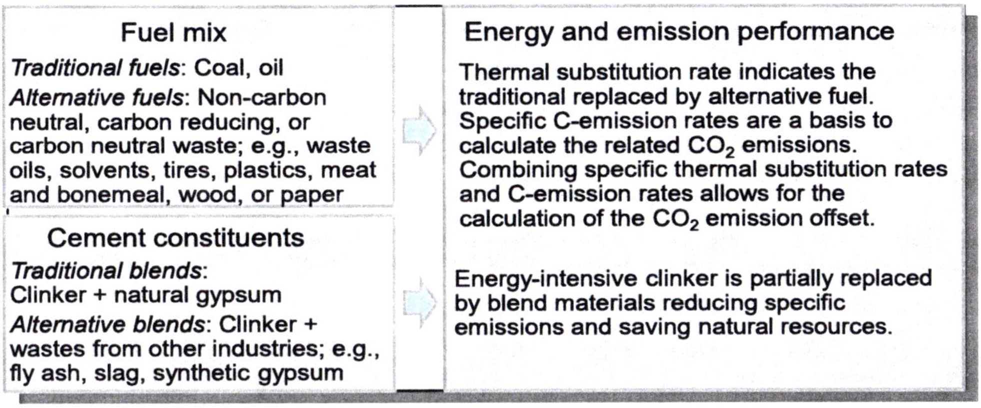

Fig. 3.3.3.1 Example of using alternative fuels and raw materials in order to decrease the carbon footprint and the amount of fossil fuels required in the cement industry [ScVo10].

The situation for companies can be presented with two general options (with many intermediate levels in real application). As Figure 3.3.3.2 shows, companies may decide to be proactively environmentally committed (light gray fields) or take a reactive position (dark gray fields). For both options, uncertainties lead to opportunities as well as threats.

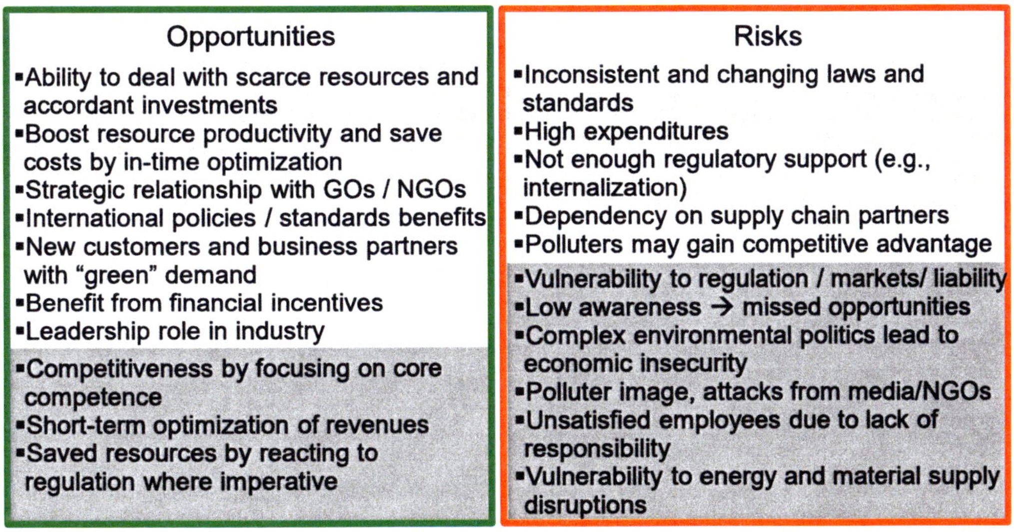

Fig. 3.3.3.2 Selection of opportunities and threats favoring proactive rather than reactive environmental involvement. Adapted from [ScVo10].

When being environmentally proactive, the human factor is very important for creating opportunities. Internally, employees' awareness can be raised, and the capability to deal with challenges and changing environmental conditions can be increased. Further, strategic relationships with stakeholders from GOs, NGOs, and new customers can be developed. Also, productivity can be increased and international standards fulfilled, resulting in financial savings. Thus, being active increases planning security and lessens dependency on volatile prices.

Taking the reactive role may be advantageous in the short and medium term. Focusing on core competencies strengthens competitiveness, as investments in environmental changes may be postponed to a later point in time when technology is "adult" and reliable ("no experiments"). Company resources are used only where it is imperative to satisfy regulatory measures, which allows conservative budgeting.

Proactive commitment entails several risks for which competing polluters may gain competitive advantage (in the medium term). For example, a company may invest in a technology that reduces a certain kind of pollution. When regulation does not implement liability costs for this pollution, it results in a disadvantage for the company. Fulfilling standards may differ greatly from region to region, which hinders knowledge transfer. When engaging in (and marketing) "green" practices and products, the noncompliance of supply chain partners has a stronger impact and they may be more difficult to replace ("captive buyer situation"). The identified risks are rather of a company external nature. Perhaps a reason is that integrated approaches go hand in hand with higher interdependencies between, for example, companies, regulatory bodies, and supply chain partners.

There are various risks when taking the reactive environmental engagement position. A company may become vulnerable to regulation, market development, and liability costs as unanticipated developments emerge. Vulnerability to price shocks and supply disruptions is higher if no countermeasures have been taken (e.g., increased efficiency or alternative feedstock). Lower level of employees' awareness leads to potentially missing out on cost-saving opportunities, and internal satisfaction may suffer due to a lack of environmental responsibility. With regard to a company's visibility, a polluter image and negative reporting from the media and NGOs may lead to disadvantages in the long term.

Although this analysis considered mostly industries in the developed countries, many of the risks and opportunities may be applicable in other regions as well. Environmental regulations change and become more important as wealth increases. The relevance of these economic drivers may apply to virtually all industry sectors (with EIIs being especially affected). Therefore, searching for new approaches and solutions is an essential part of working towards long-term competitiveness. Also, earlier approaches need to be re-evaluated. Although they may not have been accepted in the past, they can become viable under the changed economic conditions. For further reading, see [Sriv07].

3.3.4 Energy Management Concepts and Measures for Improved Environmental Performance

Intended learning outcomes: Describe energy management in production systems. Differentiate between energy-aware manufacturing processes and integrating energy efficiency in production information systems. Produce an overview on major aims of industrial symbiosis. Present measures such as enhanced utilization of wastes, the recovery of medium and low temperature waste heat, and the framework for alternative fuels and resources.

Considering the two sections above, there are various opportunities for industries to improve their performance with the aim of sustainability. As a basis, *energy management* has to take place at different company levels.

According to [Pato01], *energy management* may apply to resources as well as to the supply, conversion, and utilization of energy. Essentially, it involves monitoring, measuring, recording, analyzing, critically examining, controlling, and redirecting energy and material flows through systems, so that the least power is expended to achieve worthwhile aims.

Energy management is an enabling and supporting activity for energy efficiency. Its integration into production management may allow implementation of further improvement measures in production systems (see Figure 3.3.4.1). In energy management, these activities aid detection of viable improvement areas in manufacturing (see [BuVo11]).

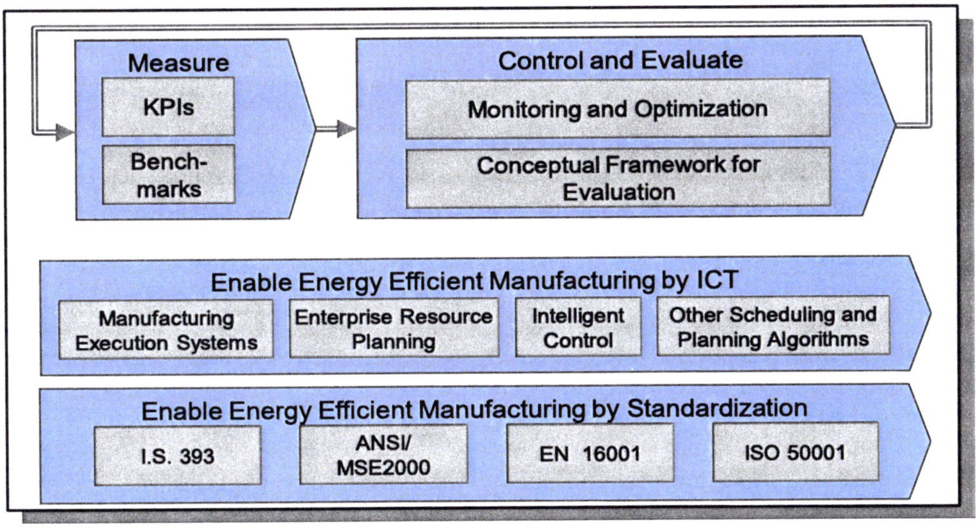

Fig. 3.3.4.1 Energy management in production systems [BuVo11].

Firstly, *energy-aware manufacturing processes:* An effective energy control system has to be developed, using information from in-process and performance measurement. This control system needs to focus on concepts that facilitate the evaluation, control, and improvement of energy efficiency in manufacturing processes.

- Appropriate and standardized energy efficiency metrics on machine, process, and plant level are needed.
- New sensor and in-process measurement technology should be integrated in existing monitoring and control mechanisms to feed decision support tools for production management.

- Benchmarks for production performance with regard to machine/equipment energy efficiency and energy profiles are required. Standardized energy efficiency KPIs are the basis for effective benchmarking across plants and companies.

Secondly, *integrating energy efficiency in production information systems:* A framework that manages and optimizes energy efficiency with respect to production planning and control needs to be developed and implemented in enterprise control and information systems, such as enterprise resource planning (ERP), manufacturing execution systems (MES), and distributed control systems (DCS).

- Information and communication technologies (ICT) tools and standardization can be significant enablers for supporting the measurement, control, and improvement of energy efficiency in manufacturing processes, as software can support visualization and simulation of energy efficiency.
- Energy performance evaluation in real-time facilitates more effective business decisions based on accurate and timely information. Energy efficiency-adapted MES and ERP systems and simulations can deliver appropriate information.

Once viable improvement areas are identified, there may be barriers to implementation. To name a few: decisions based on payback periods instead of interest rate calculations, unrealistically high implicit discount rates, difficult-to-measure components of energy investments (such as transaction or monitoring costs), and limited capital or a low priority given to energy efficiency by the management. The human factor can be a barrier, as bounded rationality, principal-agent problems, and moral hazards represent obstacles to energy efficiency improvement measures (see [BuVo11]).

The following presents exemplary approaches for improvements that can be classified under the term industrial symbiosis.

Industrial symbiosis is defined as an approach for companies, where a by-product (or waste) of one company serves as feedstock to another company. See, for example, [ChLi08].

Figure 3.3.4.2 shows the major aim of industrial symbiosis, namely, the implementation of cycle flows that reduce material and energy waste. Similarly, Figure 3.3.3.1 showed an example of an industrial symbiosis kind of concept that can reduce the amount of waste and simultaneously achieve cost savings. To become a viable option for a broader spectrum of companies, the alternative materials need to fulfill certain criteria and be less costly than virgin raw materials. Besides processing issues, several risks from Figure 3.3.3.2 need to be taken into account.

The following examples are measures that may be taken in the field of industrial symbiosis (based on [ScVo10]).

Firstly, the *enhanced utilization of wastes:* Industry is increasingly interested in access to by-products, which were previously considered “wastes.” Pretreatment, transport, storage, and an efficient use of alternative fuels in existing processes enables higher substitution rates of scarce resources and fossil fuels to be obtained.

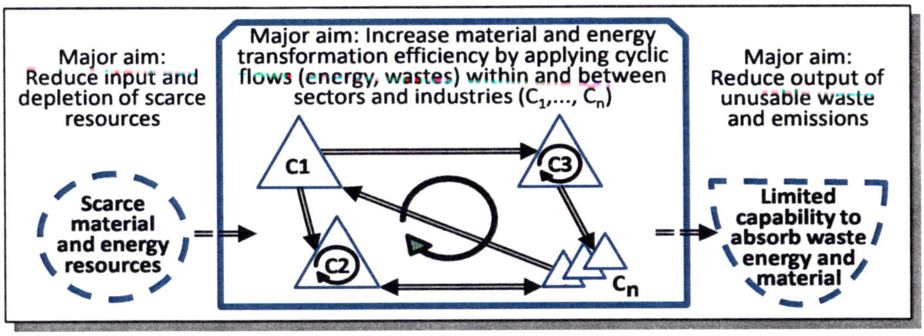

Fig. 3.3.4.2 Major aims of industrial symbiosis. Adapted from [KoMa04].

- Production processes need to cope with alternative feedstocks related to product quality, energy efficiency, and emissions.
- Mapping and integration of the possible flows of materials and energy is required, while efficiently identifying the best possible reuse (in both economic and environmental matters).
- Complexity in the market needs to be reduced on different levels to detect sources and sinks of by-products.

Secondly, the *recovery of medium and low temperature waste heat, i.e., heat around and below 150 °C:* The respective amount of heat is significant. In contrast to current approaches, the analysis needs to take place at various production plants from different sectors.

- A suitable method needs to be developed for plant, industry, and cross-industry analysis to detect heat recovery potentials.
- Collaboration potentials need to be explored and promising partnerships between heat sources and sinks identified to apply advanced technologies for heat recovery, transport, and exchange and benefit from synergies

Thirdly, the *framework for alternative fuels and resources:* This approach is reminiscent of the eco-industrial parks, in which nearby located plants share and use their by-products, energy, information, and capacities in order to increase overall efficiency and productivity. Planning an industrial park of this kind seldom resulted in real ecological and economic benefits. This approach therefore aims at supporting existing industries in efficiently sharing and distributing information and by-products.

- Research should address the collaboration of alternative fuel and resource (AFR) suppliers and users on a cross-sectoral basis to learn about the amounts and suitability of by-products.
- Integrated process chains across industries should be formed in networks of industrial partners to increase the AFR availability. The risk of dependencies requires attention.
- Awareness needs to be fostered so that available materials find their way into a suitable reuse as a standardized commodity, in spite of the fact that today's waste market is rather localized.

3.3.5 The Measurement of the Environmental Performance

Intended learning outcomes: Produce an overview on ecoefficiency. Describe an indicator system for the costs, quality, delivery, and environmental impact performance dimensions.

Ecoefficiency compares economic performance and environmental performance.

The comparison can be expressed as a quotient, or as a (e.g., linear) utility function. Section 1.4 considered the indicators for measuring economic performance. [Pleh13] shows a number of indicators relating to the measurement of environmental performance. For the purposes of the comparison, both groups of performance indicators must be measured using suitable units, or derived from measured intermediate statistics, then weighted and — finally — converted to a scalar value by means of a formula. From the range of available environmental indicators, [Pleh13] has selected the following top-level performance indicators:

- HH — Human Health, measuring unit “Disability Adjusted Life Years,” DALY.
- ED — Ecosystem Diversity, measuring unit “Potentially Disappeared Fraction of Species,” PDF
- RA — Resource Availability (e.g., energy, water), measuring unit $
- UBP06 — “Umweltbewertungspunkte” (a method of the Swiss Government)
- CO_2e — CO_2-Äquivalent, measuring unit tons

Fig. 3.3.5.1 shows these five indicators as an integrated model.

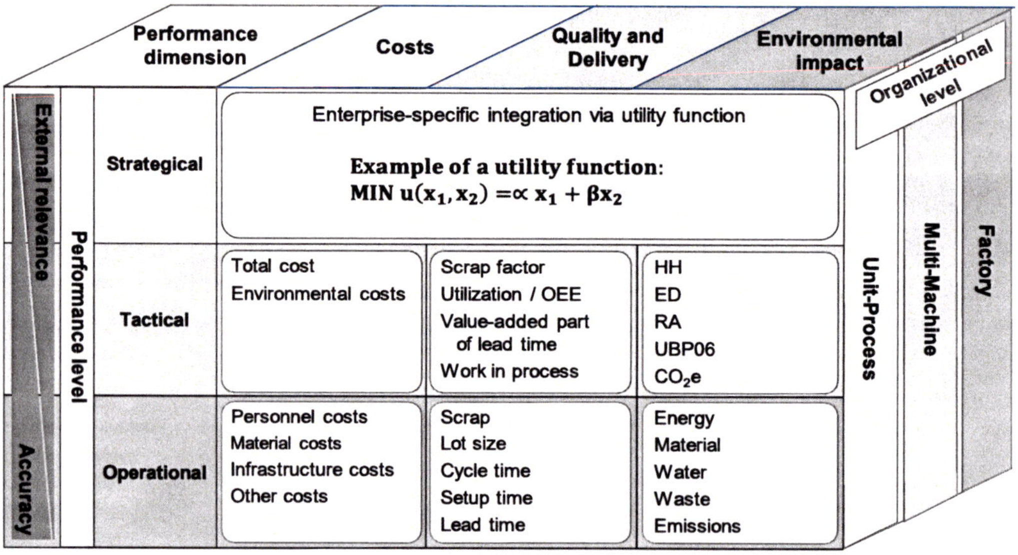

Fig. 3.3.5.1 Indicator system for the costs, quality, delivery, and environmental impact performance dimensions (adapted from [Pleh13]).

The economic indicators are grouped using the target areas for business performance from Figure 1.3.1.1. The actual measurement of the indicators is carried out at the operational level. A suitable formula is then applied to convert the measured results into indicators at a tactical level, which are then used for the ecoefficiency comparison. The comparison itself, shown in the diagram at strategic level, is expressed by a utility function, where x1 represents the environmental indicator and x2 the economic indicator.

The ecoefficiency can be measured for each work process or manufacturing process. It takes into account both the machine and the material that is used. On the one hand, the ecoefficiency for a group of machines or for a whole factory is of interest. On the other hand, it is also interesting to include all the work processes and components (i.e., to add them together) that go into a product, and to compare it with an alternative manufacturing process that would lead to a functionally equivalent product. The benefit of an environmentally advantageous material can be more than cancelled out by less efficient work processes, for example — and vice versa.

3.3.6 Social and Environmental Dimensions in Industrial Practice

Intended learning outcomes: Produce an overview on Corporate Social Responsibility (CSR). Present the integrated profit and loss statement (IPL) of Holcim Global.

The awareness from various stakeholders for Triple-Bottom-Line (TBL) thinking and acting becomes evident when looking at the increasing effort of multinational corporations related to Corporate Social Responsibility (CSR). Sustainability reporting can be misused to "greenwash" the image of a company. To prevent this, reporting reqires transparency, completeness, relevance, and auditability. See here also [SuRi16].

An example, which is currently applied in industrial practice, for the above mentioned "full costing" method is the integrated profit and loss statement (IPL) of Holcim Ltd. [Holc14], a global leader in the building materials industry (cement, concrete, aggregates, and asphalt). Holcim has been included in the Dow Jones Sustainability Index for more than 10 consecutive years. The goal of this index is to track financial performance of best-in-class companies worldwide assessing economic, environmental and social criteria with a focus on long-term shareholder value. The IPL of Holcim was introduced in 2014 for the first time in an effort to quantitatively measure the TBL. Figure 3.3.6.1 shows the IPL as a waterfall chart, which represents the different indicators that have an impact on the TBL.

In [Holc14], Holcim indicate that the "Triple bottom line can be used to assess opportunities beyond compliance", whereby the corporate mentions "Compliance with governance, social and environmental requirements and standards". Basically, the IPL emphasizes the company's objective in measuring its sustainability aspirations in a consistent way and to measure its progress over time. Furthermore, it can already be used "to identify where 1 US dollar invested would bring the highest societal return".

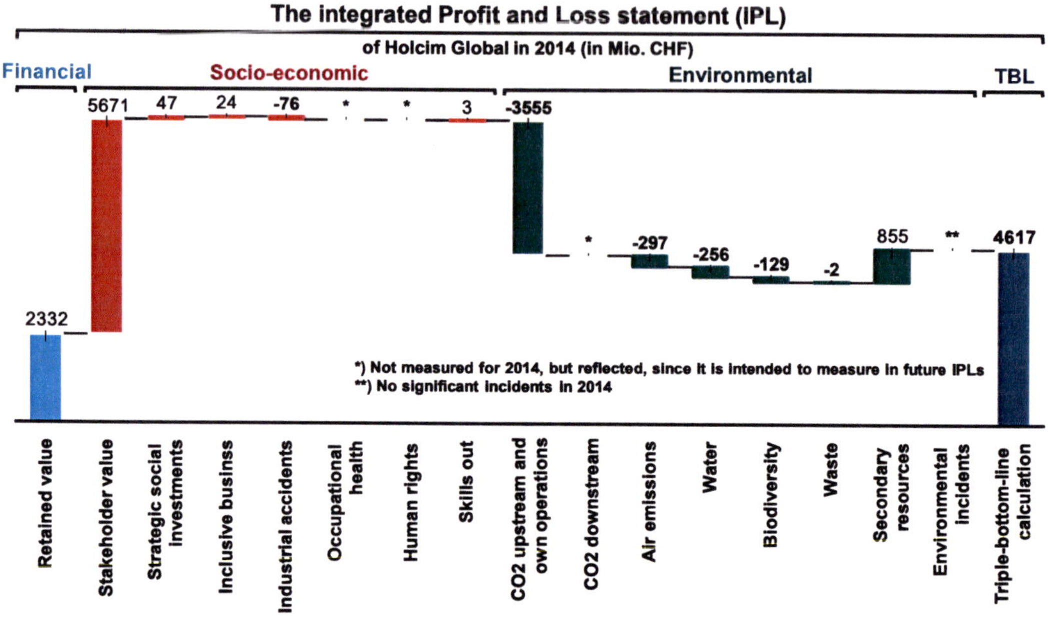

Fig. 1.3.6.1. The integrated profit and loss statement (IPL) of Holcim Global [Holc14].

3.4 Summary

A make decision requires subsequent planning of facility locations. Features for the design of rather centralized or decentralized production networks are value density, customer proximity, customer tolerance time, markets requiring specific products, volatility of demand, the supply chain's vulnerability to disruptions, and economies of scale and scope. Features for the design of distribution and service networks are volume, volatility, and variety of demand, customer tolerance time, value density (in the case of service networks, the value density of the service), and the customer pickup tolerance (in the case of service networks, the customer bring and pickup tolerance). For the selection of new locations, the chapter introduces seven possible location factors (three of these for the selection of joint venture partners) with 5 to 10 criteria per factor and a procedure for systematic reduction of possible locations. Locations for distribution and service networks are selected according to the degree of customer contact. If only indirect contact, or even no contact at all, with customers is necessary (as is the case, for example, with "back offices"), the location criteria are basically the same as those for production locations. For selection of new locations, factor rating is frequently used. For location configuration, that is, for assigning products or services to an existing location, linear programming can be used.

The triple-bottom-line concept is based on the three pillars of sustainability — namely, economy, society, and environment, which interact with companies. Energy-intensive companies especially are increasingly concerned with improving their energy efficiency. This leads to energy management proper. As a solution for this, industrial symbiosis is discussed.

3.5 Keywords

centralized distribution, 134
centralized production, 129
centralized service, 142
certificate of compliance, 167
decentralized distribution, 134
decentralized production, 129
decentralized service, 142
direct transport, 146
distribution center, 127
distribution network structure, 138
double bottom line, 166
ecoefficiency, 173
ethical standard, 166
factor rating, 159
feature for designing
  distribution networks, 134
  production networks, 130
  retail networks, 139
  service networks, 142
  transportation networks, 146
indirect transport, 146
linear programming (LP), 161
location configuration, 152
location factor, 153
location planning, 127
location selection, 152
supplier code of conduct (SCoC), 167
sustainability, 163
triple bottom line, 163
value density, 131

3.6 Scenarios and Exercises

Intended learning outcomes: Elaborate an example of a location configuration with linear programming.

3.6.1 Location Configuration with Linear Programming

Intended learning outcomes: Disclose and solve the planning problem at the Ironer Company, which produces two products for two markets, given a maximum total capacity in a year.

The Ironer Company, a manufacturer of ironing machines, has its facilities at one single location. The Ironer Company markets two different products in two regions. Once a year, the company performs rough-cut capacity planning based on sales forecasts. In addition, it must answer the following important question for marketing: With the given capacity situation, what quantity of what product should be offered in which market in order to maximize the contribution margin? While demand for New Product P1 is increasing sharply in Market M2, sales of Predecessor Product P2 are declining as the market becomes saturated (decline stage). Here the assumed market demand reflects the maximum saleable number of pieces. The contribution margins of the two markets differ, in part considerably, due to the differing cost and price structures. Figure 3.6.1.1 shows the details:

Ironer Company requires 4 hours to manufacture Product P1 and 2.4 hours to manufacture Product P2. The total yearly capacity is 15,000 hours. Please answer the following questions:

1. What quantities of P1 and P2 should be put on the two markets in order to maximize the contribution margin?
2. A consulting firm is proposing, by introducing lean-/just-in-time concepts (lean/JIT), to increase the contribution margin by 5% and lower the capacity required for P1 by 60 minutes and the capacity required for P2 by 24 minutes. What should the

maximum cost of introducing JIT concepts be? And how will this improve the company situation?

3. In addition, the marketing department decides to increase market penetration of P1 and, to maximize profits, to intensify the decline of P2. To do this, sales of P2 in Market M1 must rise to 4,000, while for Market M2 complete product withdrawal is planned. What are the advantages and disadvantages of this strategy?

Input data	Product P1	Product P2
Contribution margin Market M1	80	70
Contribution margin Market M2	70	40
Maximum demand Market M1	1000	3000
Maximum demand Market M2	5000	2000
Capacity required, in hours	4.0	2.4
Total capacity	15000	

Fig. 3.6.1.1 Input data for the planning problem at the Ironer Company.

Proceed as follows:

A) Define the decision variables. Possible solution:

$X\_P_i\_M_j$, $1 \leq j \leq 2$, $1 \leq i \leq 2$ stands for the number of products P1 that will be delivered to Market M_j

B) Formulate the target function. Possibility: contribution margin = max!

$$= (DB\_P1\_M1 \cdot X\_P1\_M1) + (DB\_P1\_M2 \cdot X\_P1\_M2) + (DB\_P2\_M1 \cdot X\_P2\_M1) + (DB\_P2\_M2 \cdot X\_P2\_M2)$$

Figure 3.6.1.2 shows how you can perform these first steps utilizing MS Excel Solver, Microsoft Excel's tool for solving linear optimization.

1. Open an **.xls spreadsheet**
2. Choose a spreadsheet cell to hold the value of each **decision variable** in your model. These are called changing cells. You can define up to 200 changing cells.
3. Create a spreadsheet formula in a cell that calculates the **target function** in your model. This is called the **target cell**. The target function is directly related to the changing cells.

Input data	Product P1	Product P2
Contribution margin Market M1	80 €	70 €
Contribution margin Market M2	70 €	40 €
Maximum demand Market M1	1000	3000
Maximum demand Market M2	5000	2000
Capacity required, in hours	4.0	2.4
Total capacity	15000	

=B11*B2+B3*C11+C2*D11+C3*E11

Decision variables	x_P1_M1	x_P1_M2	x_P2_M1	x_P2_M2		Target function of contribution margin
Results	1000	950	3000	0	=	356'500 €

Fig. 3.6.1.2 Solver tool in MS Excel, part 1.

Formulate all side conditions:

a) Demand: Maximum coverage of market demands

* X_P1_M1 ≤ maximum demand for Product P1 in Market M1=1000
* X_P1_M2 ≤ maximum demand for Product P1 in Market M2=5000
* X_P2_M1 ≤ maximum demand for Product P2 in Market M1=3000
* X_P2_M2 ≤ maximum demand for Product P2 in Market M2=2000

b) Capacity: Restricted total capacity

* $X\_P1\_M1 \cdot \text{capacity required}_{P1} + X\_P1\_M2 \cdot \text{capacity required}_{P1} + X\_P2\_M1 \cdot \text{capacity required}_{P2} + X\_P2\_M2 \cdot \text{capacity required}_{P2} \leq \text{total capacity}$

c) Variable non-negativity

* X_P1_M1 ≥ 0; X_P1_M2 ≥ 0; X_P2_M1 ≥ 0; X_P2_M2 ≥ 0

Figure 3.6.1.3 shows how MS Excel Solver handles the formulation of side conditions. The constraint operators (e.g. ≤, ≥) were entered only as text, for purposes of clarity to the reader.

4. Create a formula in a cell to **calculate side condition** 1, the restriction you place on the changing cells. In the cell next to it, input the **max./min. constraint value of side condition** 1.
5. Repeat step 4 for each side condition, including the **non-negativity requirements** for decision variables.

Input data	Product P1	Product P2
Contribution margin Market M1	80 €	70 €
Contribution margin Market M2	70 €	40 €
Maximum demand Market M1	1000	3000
Maximum demand Market M2	5000	2000
Capacity required, in hours	4.0	2.4
Total capacity	15000	

Decision variables	x_P1_M1	x_P1_M2	x_P2_M1	x_P2_M2		Target function of contribution margin
Results	1000	950	3000	0	=	356'500 €

Side conditions	Computed values		Constraint values
Demand for product P1 in market M1	1000	<=	1000
Demand for product P1 in market M2	950	<=	5000
Demand for product P2 in market M1	3000	<=	3000
Demand for product P2 in market M2	0	<=	2000
Restricted total capacity	15000	<=	15000
x_P1_M1	1000	>=	0
x_P1_M2	950	>=	0
x_P2_M1	3000	>=	0
x_P2_M2	0	>=	0

Fig. 3.6.1.3 Solver tool in MS Excel, part 2.

Figure 3.6.1.4 shows how you must actually enter the decision variables, target function, and side conditions using the Solver tool in MS Excel. Solutions:

6. Run Solver by clicking **Solver** on the **Tools** menu. (To install Solver, click Add-Ins on the Tools menu, and then select the Solver Add-in check box. Click OK, and Excel will install the Solver.)
7. In the **Solver Parameters dialog box**, input the target cell, changing cells (decision variables) and side conditions that apply to your optimization model. Choose Max to maximize the target cell.

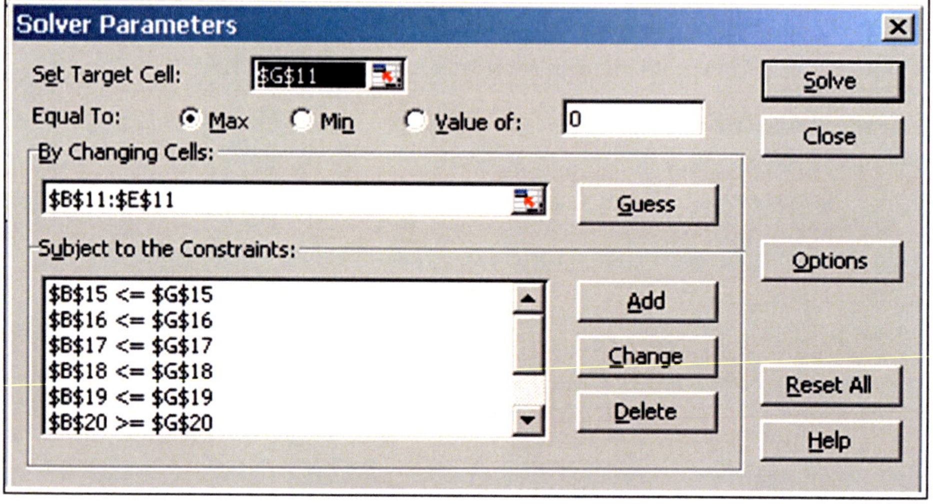

Fig. 3.6.1.4 Solver tool in MS Excel, part 3.

Task 1):

- Click Solve (see Figure 3.6.1.4) to display the results. If you have entered everything correctly, the maximum contribution margin achievable should be 356.500€, as it is in Figure 3.6.1.2.

Task 2), the introduction of JIT:

- Start out from the basic case in Task 1, copying and pasting it into a new Excel spreadsheet. Now change the values accordingly.
- Through the higher contribution margins and lower capacity required, the total contribution margin and service level for P1 can be increased.
- If everything is set correctly, you can determine the cost ceiling for introducing the JIT concept as the contribution margin difference: 451.500€ - 356.500€ = 94.900€.

Task 3), the additional marketing measure:

- Start out from the basic case in Task 2) (JIT), copying and pasting it into a new Excel spreadsheet. Now change the values accordingly.
- By intensifying the decline of Product P2, the total contribution margin can be increased even more, to 476.000€. The service level for P1, however, drops in comparison, since the available capacities produce Product P2 for Market M1. For this reason, it is necessary to consider to what extent the increase in the number of pieces of P1 will take place and the extent to which P1 will at best replace product P2.

3.7 References

AbNu08 Abele, E., Näher, U., Strube, G., Sykes, R. (Hrsg.), "Global Production: A Handbook for Strategy and Implementation," Springer, Berlin, 2008

APIC16 Pittman, P. et al., APICS Dictionary, 15th Edition, APICS, Chicago, IL, 2016

BuVo11 Bunse, K, Vodicka, M., Schönsleben, P., Brühlhart, M., Ernst, F., "Integrating Energy Efficiency Performance in Production Management — Gap Analysis between Industrial Needs and Scientific Literature," Journal of Cleaner Production, 19 (6), 667-679, 2011

Chop03 Chopra, S., "Designing the distribution network in a supply chain," Transportation Research: Part E: Vol. 39, No. 2, pp.123–140, 2003

Cars02 Carson, R., "Silent Spring," 40th Edition, Houghton Mifflin, Boston, 2002

ChLi08 Chertow, M., Lifset, R., "Industrial Symbiosis," 2008, https://editors.eol.org/eoearth/wiki/Industrial_symbiosis, accessed August 2019

DaHe05 Davis, M., Heineke, J., "Operations Management — Integrating Manufacturing and Services," 5th Edition, McGraw-Hill, 2005

GRI02 Global Reporting Initiative, "Sustainability Reporting Guidelines," Boston, 2002, www.globalreporting.org

IEA08 IEA, "Worldwide Trends in Energy Use and Efficiency: Key Insights from IEA Indicator Analysis," Paris, 2008, https://webstore.iea.org/worldwide-trends-in-energy-use-and-efficiency, accessed August 2019

Holc14 Holcim. Integrated Profit & Loss Statement 2014, Available online. www.holcim.com/uploads/CORP/Holcim_IPL_Final2.pdf. accessed 5 Nov. 2015

Kink12 Kinkel, S., "Trends in production relocation and backshoring activities," International Journal of Operations & Production Management, Vol. 32, Issue 6, pp. 696–720, 2012

KoMa04 Korhonen, J., von Malmborg, F., Strachan, P.A., Ehrenfeld, J.R., "Management and Policy Aspects of Industrial Ecology: An Emerging Research Agenda," Business Strategy and the Environment, 13(5) 2004, pp. 289–305

Mead77 Meadows, D.H., "The Limits to Growth: A Report for the Club of Rome's Project on the Predicament of Mankind," 2nd Ed.., Universe B., N.Y., 1977

OeNa10 Oehmen, J., DeNardo, M., Schönsleben, P., Boutellier, R., "Supplier Code of Conduct — State-of-the-Art and Customisation in the Electronics Industry," Production Planning & Control, 21(7), pp. 664–679, 2010

Pack60 Packard, V., "The Waste Makers," Van Rees Press, New York, 1960

Pato01 Paton, B., "Efficiency Gains Within Firms under Voluntary Environmental Initiatives," Journal of Cleaner Production, 9(2), pp. 167–178, 2001

Pleh13 Plehn, J., "Ein Leistungsmesssystem zur integrierten Bewertung der Ökoeffizienz von Produktionsunternehmen — Reduktion der Unsicherheit bei der Kennzahlenauswahl und –interpretation," Dissertation ETH Zürich, Nr. 21225, Zürich, 2013

ScRa15 Schönsleben, P., Radke, A., Plehn, J., Finke, G., Hertz, P., "Toward the integrated determination of a strategic production network design, distribution network design, service network design, and transport network design for manufacturers of physical products," ETH e-collection, 2015, http://dx.doi.org/10.3929/ethz-a-010423875

ScVo10 Schönsleben, P., Vodicka, M., Bunse, K., Ernst, F. O., "The Changing Concept of Sustainability and Economic Opportunities for Energy-intensive Industries," CIRP Annals — Manufacturing Technology, 59(1) 2010, pp. 477–480

Sriv07 Srivastava, S. K., "Green Supply-Chain Management: A State-of-the-Art Literature Review," Int. Journal of Management Reviews, 9(1) 2007, p. 53–80

Stic04 Stich, V., "Industrielle Logistik," 8th Ed., Wissenschaftsverlag Mainz, 2004

SuRi16 Sutherland, J, Richter, J.S., et.al., "The Role of Manufacturing in Affecting the Social Dimension of Sustainability," CIRP Annals, 65/2, pp. 1-24, 2016

UN87 UN Report of the World Commission on Environment and Development, "Our Common Future," Annex to General Assembly Doc. A/42/427, 1987

WiLe09 Wiedmann, T., Lenzen, M., "Unravelling the Impacts of Supply Chains — A New Triple-Bottom-Line Accounting Approach and Software Tool," Environmental Management Accounting for Cleaner Production, edited by Schaltegger, S., Bennett, M., Burritt, R., Jasch, C., Springer Netherlands, 2009

4 Process Analysis and Concepts for Planning & Control

Intended learning outcomes: Explain important techniques in business process engineering. Describe characteristic features relevant to planning & control in supply chains. Disclose fundamental concepts in logistics and operations management.

Chapter 1 of this book showed how, and to what extent, logistics, operations, and supply chain management as well as planning & control of daily processes can contribute to the fulfilling of entrepreneurial objectives. Appropriate performance indicators are connected to individual entrepreneurial objectives (see Section 1.4). These measures allow a company to evaluate the degree to which objectives are reached. The present chapter presents a systematic procedure for the continuative analysis of business processes and the design of systems for planning & control in supply chains. Figure 4.0.0.1 diagrams an overview.

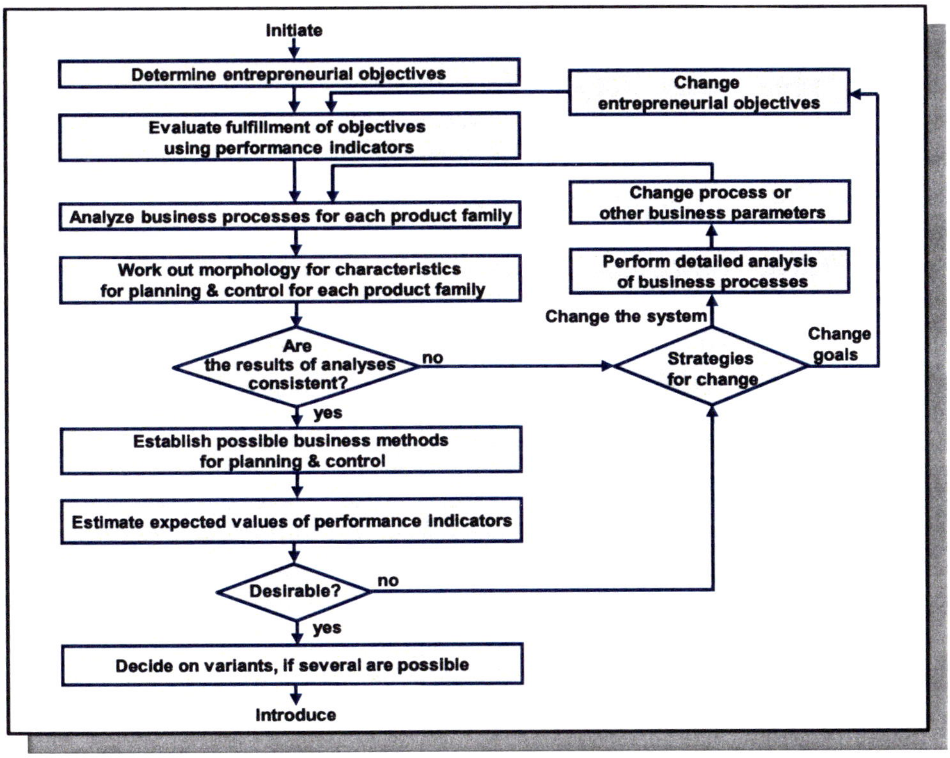

Fig. 4.0.0.1 Procedure for analysis of business processes and design of systems for planning & control in supply chains.

With the initiation of the procedure, company management must set the objectives of the enterprise. The basis of any improvement to the success and efficiency of an enterprise is analysis of the current situation. The following will introduce some methods of analysis that are also (and particularly) relevant to logistics, spotlighting different aspects of the situation.

P. Schönsleben, *Handbook Integral Logistics Management*,
https://doi.org/10.1007/978-3-662-65625-9_4

- Logistics, operations, and supply chain management have as their object business processes within and across companies. The first sections below will focus on the basics of process management.
- A next step is business process analyses. The processes, as well as further analyses, are usually worked out for each product family. Section 4.3 outlines possible methods of analysis of processes and procedures. For now, we will begin with less detailed representations, such as organization-oriented process charts.
- Section 4.4 outlines characteristic features that are relevant to planning & control in supply chains. Different product families, and sometimes even different products, and upstream and downstream from decoupling points, will have varying values for these features. The features are related to entrepreneurial objectives and thus must consider these objectives. Examination of the features allows us to discover inconsistencies with business processes as revealed by process analysis.
- Section 4.5 presents fundamental concepts in their dependency on the characteristic features. Here, a company has to position itself within a selection of different production types, and concepts for planning & control.

The results of the various analyses are checked with regard to consistency, both among themselves and in relation to entrepreneurial objectives and desired results. If no general consistency is found, either the system or the company's objectives must be changed.

- Change of the system requires detailed process analyses, using, e.g., production-infrastructure layouts or process plans. Section 4.3 deals with such techniques. Then, in a design step, process or design parameters (such as infrastructure, employee qualifications, product variety concept, relationship to business partners) are changed.
- Changing the objectives can be necessary, if current options involve too many inconsistencies (see Section 1.3.1).
- In either case, many steps of analysis may have to be repeated.

If the results are consistent, a design step can work out possible business processes and methods of planning & control. Chapters 5 through 8 discuss the principal methods and their dependency upon analysis results. Chapters 10 through 15 outline more detailed methods.

Once concrete methods of planning & control have been worked out, a company can attempt to estimate the expected values of performance indicators. These estimates will be checked against desired values. If the result is negative, then once again either the system or the objectives of the enterprise must be changed (see above). If the results are positive, the company may have a choice of various possibilities. For example, variants emerge as the result of a possible different view of the product family. In other cases, differing methods of planning & control may be implemented.

The approach proposed here does not depend on any particular type of project organization. However, logistics systems function satisfactorily only when the people who use them (want to) understand them well. For this reason, it is advantageous for the people who use the system to be actively involved in the design process.

4.1 Elements of Business Process Management

Intended learning outcomes: Produce an overview on terms in business process engineering. Describe order management and graphical representation of logistics processes.

Effective and efficient business processes are a key factor with regard to a company's performance. See the discussion in [Dave93] and [HaCh06], for example. The following Sections 4.1 and 4.2 examine the elements and design of business processes.

4.1.1 Terms in Business Process Engineering

Intended learning outcomes: Present concepts in business process engineering and management. Differentiate between core competencies and core processes of a company.

Figure 4.1.1.1 shows terms used in business process engineering. Cf. Section 1.1.1.

Term	Word origin, definition	Related terms
state, status	mode or condition of being	quality; the way things stand
event	something that happens; archaic: outcome	occurrence
core competency	significant or crucial ability, capability, or skill	
core process	a process for which a company has competitive competencies	
logistics system	comprises logistics tasks, functions and methods, processes, states, flow, and its trigger event. Has its order and process management .	

Fig. 4.1.1.1 Concepts in business process engineering and management

Looking at the pair of terms *state* and *event*, we see that each task or subprocess describes an *action state* within the whole process, in which the goods being processed (material or information) exist. Between two tasks or subprocesses, there is a transition. If processing does not continue immediately, the transition ends in a *waiting state*. An example would be a buffer or an in-box in an office. The *event* is then a special process through which a person or a sensor registers the waiting state and then triggers the next process or task.

It is generally easier to identify the *core competencies* of a company than to derive *core processes* from them. A core competency may consist in a function that occurs in various business processes that themselves do not have to constitute core processes. Also, other functions of the business processes do not have to be core competencies. Indeed, it is not always easy to distinguish between important and less important business processes.

A *logistics system* is like an independent supplier, responsible for fulfilling the order itself.

4.1.2 Order Management and Graphic Presentation of Logistics Processes

Intended learning outcomes: Describe MEDILS (Method for Description of Integrated Logistics Systems), its symbols and connections of the symbols.

The order is the main instrument of logistics, and its processing is the control flow of logistics, both within and among companies. The form of the contract is unimportant: it may be a detailed written contract, or a simple card in a pull system (a Kanban).

Order processing can be compared to a freight train. The train moves along a certain route. As it goes, goods or information are added to the train. Stopping at stations, it signals to other trains to start out and supply goods or information. Before ending its journey, our freight train also delivers goods and information to trains traveling farther on. An observer could sit in the locomotive and observe these happenings. *MEDILS* (Method for Description of Integrated Logistics Systems) was designed from this observation point. MEDILS aids to better understand processes, showing flows, tasks, waiting states, storages, and so on. Figure 4.1.2.1 introduces the symbols used in MEDILS.

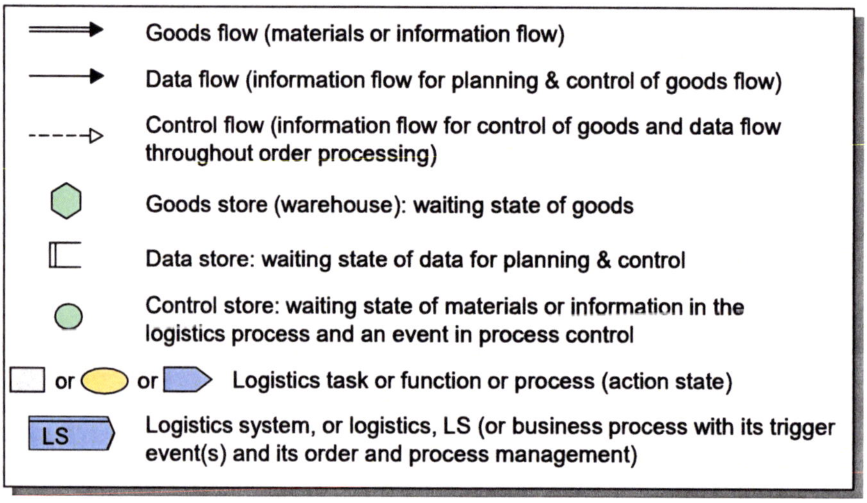

Fig. 4.1.2.1 MEDILS symbols.

- A double arrow represents the *flow of goods*. In the industrial sector, goods are usually tangible, but they can be information that go with the product from the start, such as drawings or specifications. In the service sector, goods are often intangible. In banks or insurance firms, for example, goods are often composed of information.
- A single arrow denotes the *flow of data* for planning & control. This is the flow of information required for administrative and planning purposes. Data describe the characteristics of goods in an appropriate way. Every goods flow is a self-description and thus is also data flow, although it is not drawn separately as such.
- A broken arrow represents the *control flow*. This is made up of information that deals with control of the flow of goods and data. Every goods flow and every data flow are also control flow, although they are not drawn separately as such.
- A hexagon stands for a *goods store*. Depending on the kind of goods, this may be a warehouse, information store, and so on. An object in this store stands for certain goods and thus represents a waiting state in the flow of goods. In principle, it may stay in this state for an indefinite length of time in the store.

- A rectangle with a double line on the left represents a *data store*. An object in this store stands for a certain quantity of data (for example, an order), and it is a waiting state in the flow of data. It may remain in store in this state for an indefinite period of time. The object can be described in more detail by the symbol.
- A circle stands for a *process store*, a kind of intermediate store in the logistics process. We can think of a process store in the flow of data or intangible goods (information) as a mailbox. An object is the envelope addressed with control information, while the data are found inside the envelope. A process store in the flow of tangible goods can be seen as a buffer or transit camp. An object is a crate inscribed with control information, while the goods are found inside the crate.
- A process store stores tasks waiting in line to be processed. The impetus for processing an object is given by an event: a sensor, such as the human eye, registers a state and finds an envelope in the mailbox. Thus, the event is an implicit part of process storage.
- A rectangle represents a *logistics task* that may be described in detail within the rectangle. If the effect of a task is the important aspect, the rectangle stands for a function. If procedure according to plan is the focus, the rectangle stands for a method. If the focus is the way of implementation, the well-known value-adding arrow, which stands for a process, is used instead of the rectangle. A task or process can be "nuclear" or comprise subtasks or subprocesses, which are connected via flows.
- The rectangle in the shape of an arrow represents a *logistics system (LS)* in the direction of the temporal axis. The logistics system includes logistics tasks, states, flows, and sublogistics. It has its own order and process management, which is indicated by the doubled top line. As compared to the simple value-adding arrow, a logistics system includes not only the process itself, but also the process store containing the trigger event(s), that is, the impetus to start the process.

Logistics systems are represented in graphic form by using and connecting the symbols. Figure 4.1.2.2 shows the connections used conventionally in MEDILS.

- Goods or data along with control information or control information alone flow from storage into a task or function, or process. Execution of the task, function, or process transforms the goods or data, and they are then moved to new storage points. Multiple flows to a task must be coordinated at the start of the task. Depending upon the context, related flows may be combined in the sense of "and" connections. Flows that need to be separated in the sense of an "or" or "exclusive or" connection are handled separately. Flows leading out from the task are handled analogously.

- Goods, data, or control flow originate in a task outside the logistics system LS into a process store in the LS or from a task within the LS to a process store outside the LS. We can think of this as follows: goods or information in the order processing "train" are transferred to a transport "train" and delivered to another logistics systems "train." This takes place, for example, when production turns over a completed customer order to distribution.

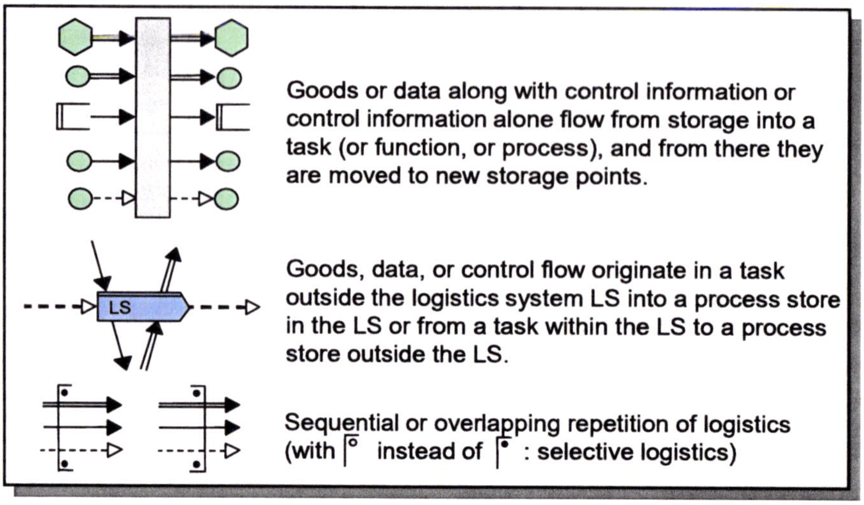

Fig. 4.1.2.2 MEDILS: connecting the symbols.

- Special brackets stand for sequential or overlapping repetition of (sub)logistics, for as many times as demanded by the situation (even zero times). The flows leading into the brackets must be of the same type as those leading out of the brackets. The contents within the brackets can also be executed selectively, i.e., under conditions.

4.2 Push and Pull in the Design of Business Processes

Intended learning outcomes: Differentiate between pull and push logistics. Describe the temporal synchronization between use and manufacturing with inventory control processes.

4.2.1 Pull Logistics

Intended learning outcomes: Produce an overview on the subprocesses of a business process in the company, from order acquisition to fulfillment. Describe the interface between subprocesses according to the model "customer-supplier relationship with an internal order".

Process transitions arise when several people or groups work independently of one another within a business process. In general, for a value-adding process of any complexity, a business process must be divided into a number of subprocesses. Crucial are the states of goods between subprocesses and particularly the event (see above) that detects the state, or momentary standstill. Two subprocesses must be connected by an interface. This guarantees that the they cannot be torn apart in time, but will take place one right after the other.

Figure 4.2.1.1 emphasizes the fact that the customer's logistics are ongoing throughout this whole period. The customer keeps track of order fulfillment, as the goods ordered are needed for fulfilling the customer's own tasks (in design and manufacturing) or for use.

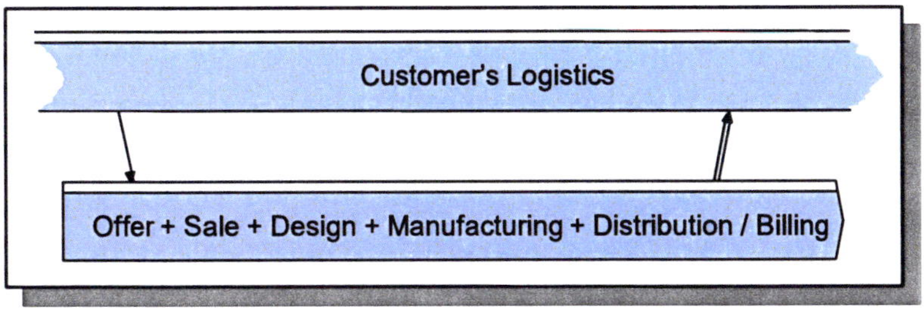

Fig. 4.2.1.1 Business process in the enterprise from order acquisition to fulfillment.

How should the business process be organized into subprocesses, as soon as more people are needed for order acquisition and fulfillment than can be incorporated into one single group? Experience has shown that each transition from subprocess to subprocess is critical. This is the reason why the design of the interfaces is so important.

Figure 4.2.1.2 shows a common solution to the problem. This example has been taken from a midsized company in the metals industry. Transitions are defined by the way that an order arises or is formulated between the persons or groups of persons involved.

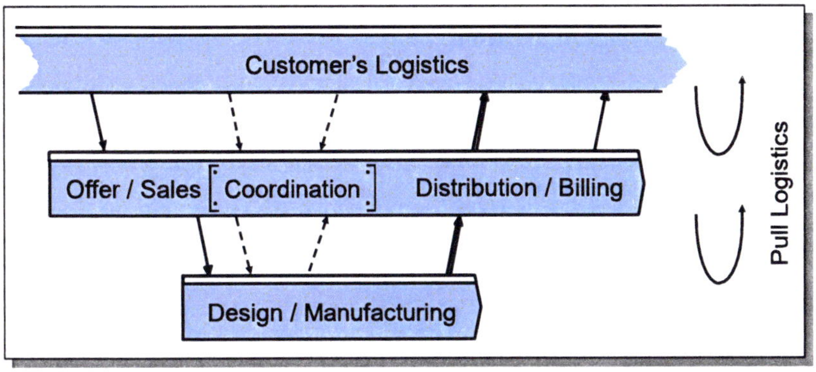

Fig. 4.2.1.2 Interface between subprocesses: "customer-supplier relationship with an internal order" model and pull logistics.

Design and manufacturing is viewed here as its own business process, as the sales department has issued an *internal order* to the design/manufacturing departments. Here, sales is the internal customer, and design/ manufacturing the internal supplier. Sales, however, remains responsible to the customer for order fulfillment during the entire design and manufacturing period. Through continuing coordination, or in other words, the exchange of control information, the order is eventually fulfilled. The "customer," whether internal or external, places an order and "pulls" the logistics in such a way that the logistics produce the goods ordered for delivery. The customer remains an active monitor, at least potentially, throughout the entire delivery lead time.

This results in *cascades*, that is, a number of process levels in the process model. *Pull logistics* is the name for this system: Value-adding takes place only on customer demand or to replace a use of items. Its characteristic is that several parallel order processes arise. This means that several order managing persons concern themselves with thc value-adding process simultaneously. With regard to delivery reliability rate, each customer, through coordination with the supplier, "pulls" the order on up through the process levels.

This kind of logistics ensures that nothing is "forgotten." Parallel order management in multiple levels is in itself, of course, not value-adding. From a lean production perspective in a narrow sense, it may even be wasteful. However, from an agility perspective, this slack is necessary if logistics are to be effective in this model. The interface in the cascade model is formed mainly through the formulation of the order. Customer and supplier must reach an agreement. These negotiations represent slack, and thus unnecessary expenditure, but they do result in an overall effective business process.

4.2.2 Push Logistics

Intended learning outcomes: Describe the interface between subprocesses according to the model "simple sequence". Explain the interface between subprocesses according to the model "partner relationship with overlapping subprocesses for handing over the order".

An alternative solution to the design of the business process in Figure 4.2.1.1 is a type of logistics that is shown in Figure 4.2.2.1 — a simple sequence of subprocesses.

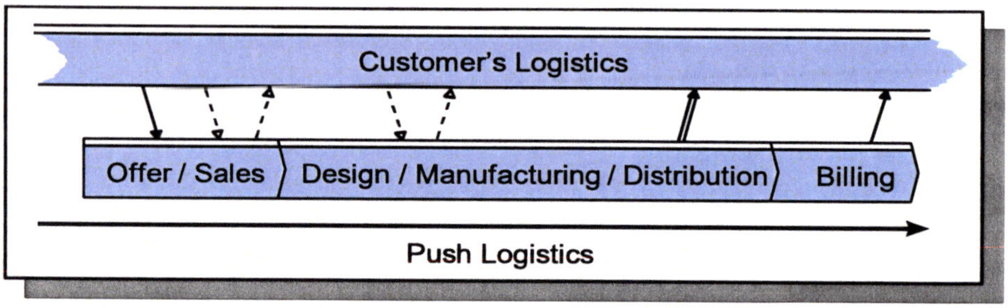

Fig. 4.2.2.1 Interface between subprocesses: the "simple sequence" model.

This "simple sequence" model is common and effective, as long as order management does not change and remains in the hands of the same person. This person is the supplier responsible for all subprocesses; he or she manages the executing organizational units in a central fashion, one after the other. This is the model of push logistics.

With *push logistics*, you push the order based on a given schedule planned in advance in the direction of the added value, without need of customer influence or a definite customer order.

If *decentralized* control by the executing organizational units themselves is desired, the "simple sequence" model can hardly be utilized. First, there are no indications of how states between the subprocesses might be registered so that the next subprocess will be initiated. Between subprocesses, order management must be somehow shifted from one processing facility to the next. Responsibility then lies in the hands of the organizational unit that executes that next subprocess. Second, the external customer in our example must first deal with sales and later with design and manufacturing units. But how will the customer know when these transitions occur? Misunderstandings become inevitable. For these reasons, the "simple sequence" model — although "lean" — is bound to fail. Figure 4.2.2.2 shows that only careful designing of the transitions between subprocesses, that is, the interfaces, can make uninterrupted order fulfillment processing possible using push logistics.

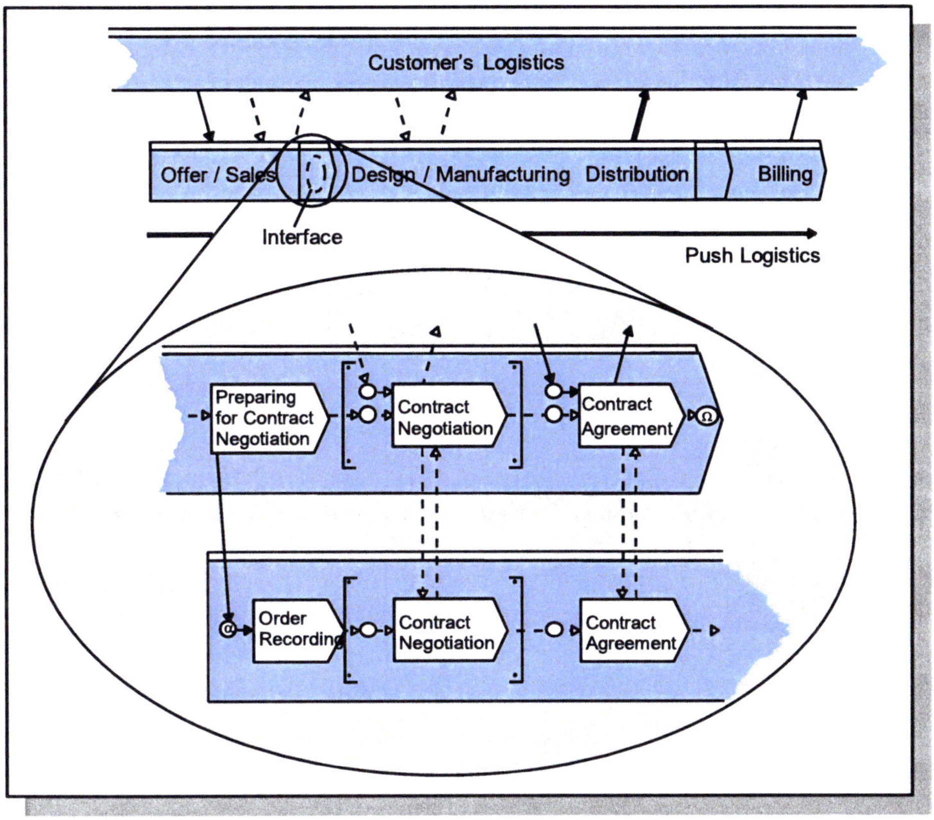

Fig. 4.2.2.2 Interface between subprocesses: "partner relationship with overlapping subprocesses for handing over the order" model.

The practical example in Figure 4.2.2.2 is taken from a consulting firm. In the company's past, vendors had made agreements with customers that the executing units could not fulfill. This, of course, had a negative effect on customer satisfaction. The firm recognized that during contract negotiations, and at the conclusion of the agreement itself, at least one person should take part that will actually perform the services. This type of organization ensures that nothing will be sold that cannot be produced. Conversely, the executing unit commits itself at the right point in time in direct contact with the customer.

With push logistics, it is crucial that the two part processes overlap, that is, that the next part process begins parallel to the end of the preceding part process. This link is established by having people in the organizational unit handling the first part process conduct their last task in coordination with representatives of the organizational unit that will begin the second part process. This second group takes over process management — the responsibility as supplier with regard to quality, cost, delivery, and flexibility. At the same time, the party placing the order knows its "new" business partner, and order fulfillment can be coordinated.

In this model, the organizational units of subprocesses do not stand in a customer-supplier relationship, but rather stand in a partnership. The overlap of the subprocesses is the necessary slack. It is true that more persons than actually necessary perform certain subtasks. But it is this very redundancy that ensures a smooth takeover of the order by one organizational unit from the other. The two subprocesses become sewn together, and this is what makes for an overall effective business process.

It is not necessary to play off the two models in Figures 4.2.2.1 ("customer-supplier relationship with an internal order") and 4.2.2.2 ("partner relationship with overlapping subprocesses for handing over the order") against each other. Both the multiple process levels model with its pull logistics and the flat model with its push logistics have their justifications. For fast, uninterrupted pull-through of complex value-adding processes, enough slack, or non-value-adding activity, must anyway and always be built in at process transition points.

The more employees are capable of handling "longer" processes, the faster and cheaper the processes become. The reason is that there is less necessity for slack times and redundant work in order to join subprocesses in smooth transition. Of course, qualifying employees to do this and coordinating them in the group entails costs. From this, we can derive a guideline for the design of process organization. Division into short subprocesses may be necessary in order to achieve certain quality demands. As soon as several people show competency in the handling of a number of related subprocesses, it is correct — with a view to reducing transition points — to make a long process out of the short subprocesses and to organize these persons into a group (see also [Ulic11]).

4.2.3 The Temporal Synchronization between Use and Manufacturing with Inventory Control Processes

Intended learning outcomes: Differentiate between different inventory-control processes for temporal synchronization between use and supply. Describe the example of pull logistics for order processing with end-product inventory replenishment after consumption.

Section 1.1.6 introduced temporal synchronization between supply and demand as a fundamental problem in logistics. Warehouses serve the storage of goods when manufacturing or procurement is too slow or too early. Figure 4.2.3.1 shows the MEDILS notation for logistics with stocking. Depending on the point of view, or the type of order (see Figure 4.4.4.1), certain cases of inventory control processes result as shown.

1. Manufacturing / procurement takes place only upon actual customer demand. Storage is necessary only if order receipt is too early.
2. Manufacturing / procurement is released based on forecast, before there is a definite customer and without need to replace items taken for use. These products are then held in inventory until required by a user. They can then be delivered immediately.
3. Demand is filled immediately from inventory. The items taken for use will then be replaced by *stock replenishment* afterward. The replacing items remain in inventory for an indefinite length of time.

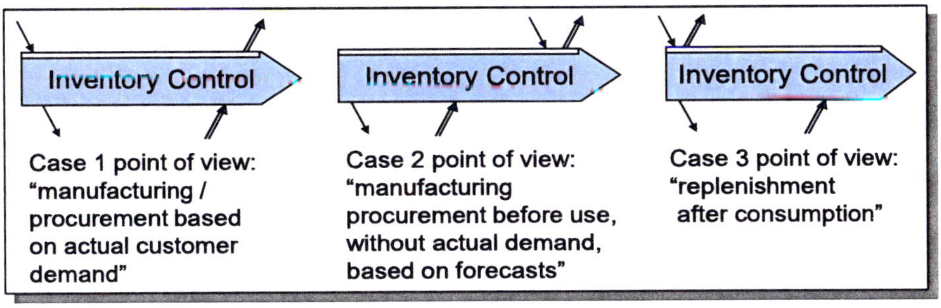

Fig. 4.2.3.1 Different inventory control processes for temporal synchronization between use and manufacturing / procurement.

Cases 1 and 3 can be considered to be pull logistics. Case 2 is mostly solved by push logistics, as long as there is no definite customer. However, for each case it is clear that carrying inventory only makes sense if goods in stock will be used within a reasonably short period of time. Figure 4.2.3.2 shows the example from the above section once again, this time incorporating inventory control for end products following the Case 3 point of view, that is, "inventory replenishment after consumption."

Note that "design" is missing here, because this case concerns the selling of already manufactured and stored products.

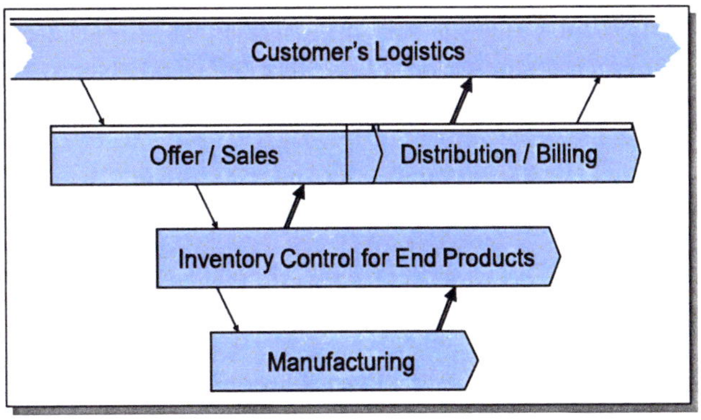

Fig. 4.2.3.2 Pull logistics with inventory: order processing with end product inventory.

4.3 Important Techniques of Analysis in Business Process Engineering

Intended learning outcomes: Explain the organization-oriented process chart as well as the detailed analysis and time study of processes. Disclose manufacturing and service processes in the company-internal and transcorporate layout.

Business process engineering is the discipline of design and improvement of business processes.

The term "engineering" underlines that this is an engineering science approach. This shows in the emphasis on the constructive aspect and on methods, models, and techniques.

An analysis of the processes and their representation with a process map form the basis for all necessary changes both within and across companies.

A *process map* is a diagram of the flow of a production or service process through the production system. Standardized symbols designate the different aspects of the process.

The process analysis requires examination of processes and procedures with regard to their success (effectiveness) and their efficiency. Like any systems analysis, analysis of processes gives us a picture of ancillary constraints and yields initial suggestions for improvement.

Various techniques of process analysis yield different ways of viewing logistics contents. In addition, each technique has its own character with regard to the way data are collected (for example, interviewing experts and participants, observations throughout the course of order processing). These factors can influence the results. Redundant findings using various methods are desirable, for they ensure the soundness of the conclusions.

In the following, we will introduce simple and often-used techniques. They can be used for the description of every kind of output (service, product, or product family) in appropriate detail. Whenever possible, the findings should be complemented as early as this stage with information on 1.) lead times of the processes, 2.) frequency and periodicity of the processes, and 3.) states that launch the processes or part processes.

4.3.1 Organization-Oriented Process Chart

Intended learning outcomes: Describe the organization-oriented process chart for the case of both pull logistics and push logistics.

The *organization-oriented process chart* shows a process with its part processes, tasks, or functions (1) through the course of time (horizontal axis), and (2) in its embeddedness in the structural organization (vertical axis).

There are various ways to draw an organization-oriented process chart. Generally, the diagram corresponds to usual practice in a company. Here is a version based on MEDILS (see Section 4.1.2), incorporating the constructs defined in Section 4.2 into the chart.

For pull logistics (Section 4.2.1), the cascading can be used again unaltered, for vertical cascades necessarily lead to the transition to another organizational unit. Figure 4.3.1.1 shows the example used in Figure 4.2.1.2 in an organization-oriented process chart.

For the push logistics in Section 4.2.2, it makes sense to put the part processes on the vertical as soon as the organizational unit changes. A vertical connection produces the connection in the model of the "simple sequence." The model "partnership relationship with overlapping part processes," on the other hand, shows two vertical connections. Figure 4.3.1.2 shows the example used in Figure 4.2.2.1 in an organization-oriented process chart.

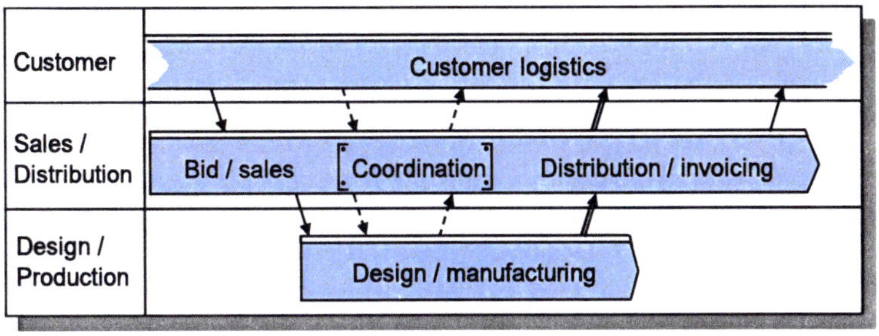

Fig. 4.3.1.1 Pull logistics: organization-oriented process chart.

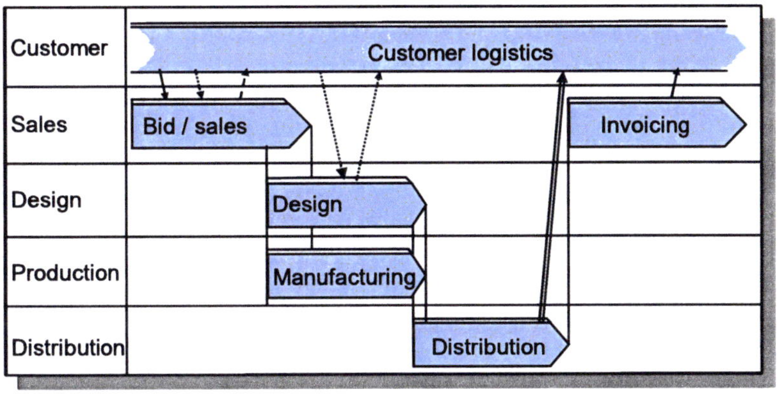

Fig. 4.3.1.2 Push logistics: organization-oriented process chart.

The transition from sales to design/production is represented as an overlapping part process, and the transition to invoicing is shown as a simple sequence. The chart shows parallel part processes for various organizational units involved in design and production.

With the help of persons that are involved, you can use the organization-oriented process chart to analyze and chart the formal flow. Through interviews or brainstorming sessions with the peo involved, you can identify and chart processes, tasks, or functions, each with their incoming and outgoing flows and their origins and destinations. The findings of the various interviews can be placed in proper succession and integrated into a single diagram. Employees generally make quick sense of the charts, for they can identify themselves within the structural organization. Iteratively, the results can now be verified and improved. Employees can determine whether the part processes are indeed executed as diagrammed in the chart and whether goods, data, and control flow have been charted correctly.

One disadvantage is that the organization-oriented process chart may not correspond to reality if it was constructed on the basis of interviews and the know-how of the engineer doing the analysis. That is why additional on-site analyses are necessary. In addition, complex order processes are reflected in complex diagrams that include many organizational units or the same organizational unit involved repeatedly in the process.

For a historical example of the organization-oriented process chart, see [Grul28] on the "division of labor in the company" (note Figures 156/157).

4.3.2 Manufacturing and Service Processes in the Company-Internal and Transcorporate Layout

Intended learning outcomes: Differentiate between a manufacturing and a service process. Describe the company-internal layout. Explain the collaborative service blueprint.

A *manufacturing process* is the series of operations performed on material to convert it from the raw material or a semifinished state to a state of further completion ([APIC16]).

A *service process* is the series of operations performed for customer service or customer support.

A *layout* shows the "geography" of resources involved in the manufacturing or service process — both company-internal and transcorporate.

One layout may show, for example, the "geography" of a supply chain, while another may depict company-internal "geography." The actual course of an order is then drawn into the layout. From this, it is easy to see intuitively the limits of the production infrastructure and to spot areas for improvement. After changing the layout, the new process is then charted. The "new" can now be compared to the "old." Transcorporate layouts are usually diagrammed as maps showing the various sites. The flow of an order is drawn in with arrows that connect the sites. Figure 4.3.2.1 shows a company-internal layout.

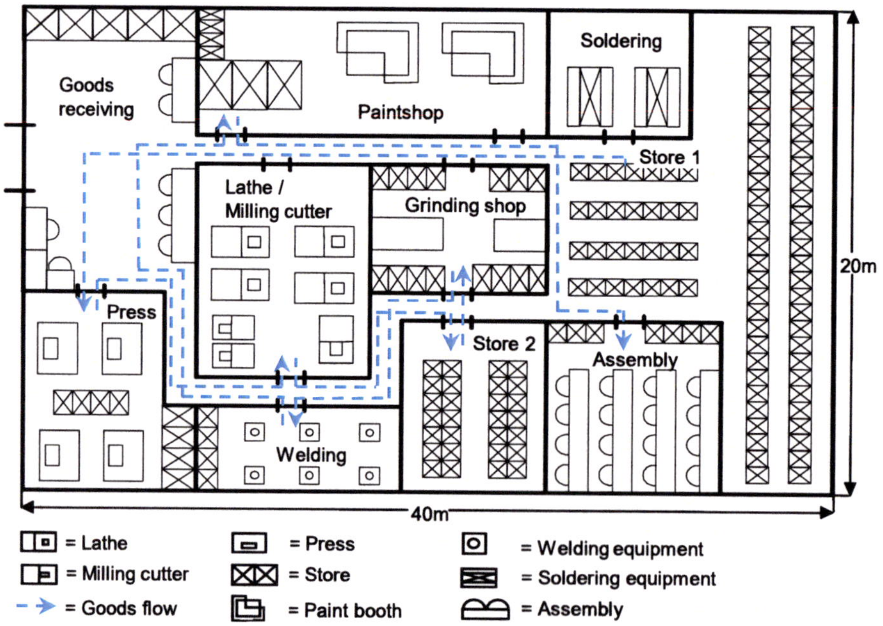

Fig. 4.3.2.1 Company-internal layout with an example process.

Service blueprinting is a technique used for analyzing service processes. The service blueprint usually makes a distinction between customer actions and service provider actions that are visible to the customer and service provider actions that are backstage, or not visible to the customer. Lines of interaction show the form of customer involvement and points of customer contact. Beyond that, using special symbols, the tool can identify frequently occurring fail points and important decisions in the service process (cf. [APIC16]).

Figure 4.3.2.2 shows a "collaborative (service) blueprint" following [Hart04]. This type of service blueprint is especially useful when service delivery is distributed across several companies or cooperation partners, which is a development that is frequently seen today especially in service in the area of machinery and plant manufacturing.

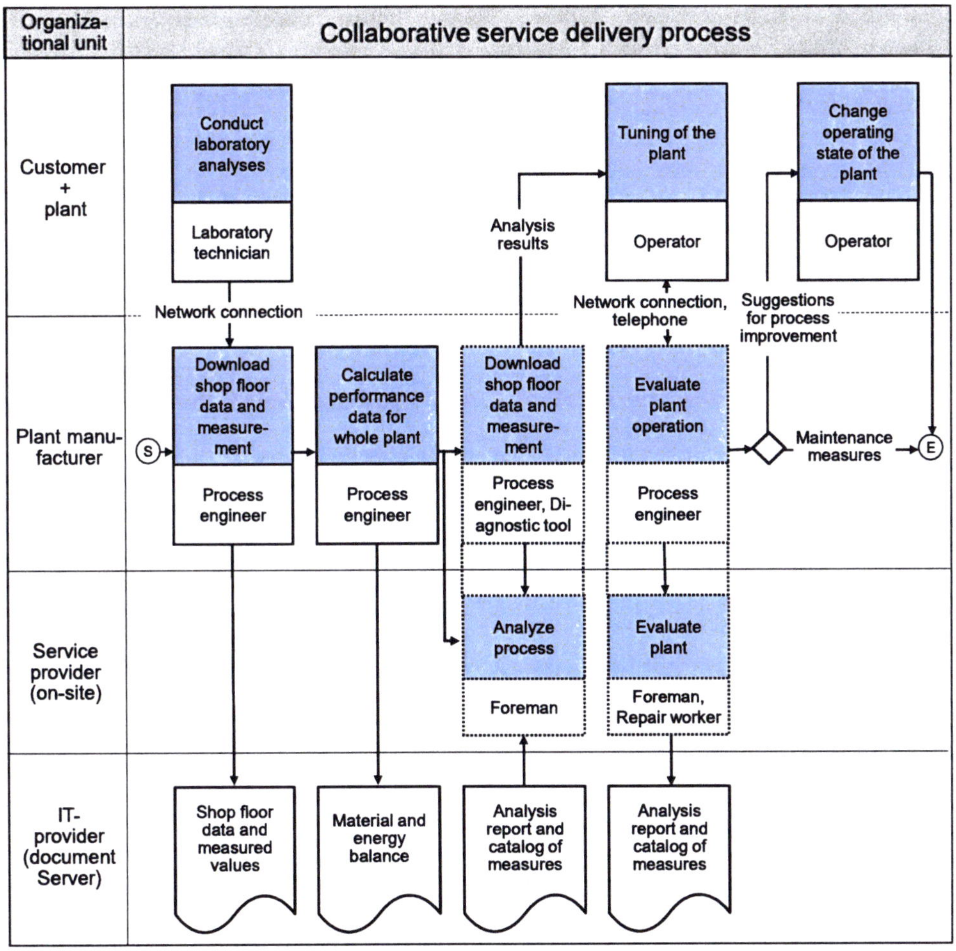

Fig. 4.3.2.2 Example of a service process represented graphically as a collaborative service blueprint following [Hart04].

The horizontal axis shows the course of service provision over time. The part-processes are assigned to the companies involved, the organizational units, on the vertical axis. The arrows between individual actions indicate the information flows. This creates transparency about the division of tasks between the companies involved in the service processes and their resource needs.

4.3.3 Detailed Analysis and Time Study of Processes

Intended learning outcomes: Explain basic process analysis and value-stream mapping.

A *time study* is a representation of the exact temporal sequence of the operations of a process.

Time study can be required in order to identify improvement or optimization potentials. Time study is one of the typical tasks of an industrial engineer. Actual stopwatch timing is often used to establish standard hours, or standard times. These are also often needed for capacity planning (also see here Chapter 13).

A suitable technique of detailed process analysis documents the results of a time study. The same technique can be used for the documentation of the process improvement.

A basic process analysis is a detailed analysis of the process plan on-site that, operation by operation, explains the exact percentages of the total lead time.

Figure 4.3.3.1 shows an example of a basic process analysis. Its form has been chosen similar to [Shin89], and it thus belongs to the Toyota Production System.

(Design of the part)				
	Process ID	451	Batch size	20
	Part name	Transmission	Part ID.	ABC-123
	Material	AC-2		
	Inspector	Smith	Inspect. date	2020.06.15

Quantity	Distance	Time	Symbol	Process (place)	Operator	Machine	Type of storage	Operating conditions, developments, etc.
		2 days	▼	Store 1				
60	40 m		⬇		Transporter			
		3 h	▼	Pressing			Pallet on ground	
		20 s	●	Pressing	Operator	Press		20% parts defective
		20 min	✱	Pressing			Pallet on ground	
	25 m		⬇		Transporter			
		3 h	●	Milling		Milling cutter		

✱ = batch-size-dependent wait time ⬇ = transport
● = process ▼ = wait time ■ = control

Fig. 4.3.3.1 Example of a basic process analysis.

For purposes of illustration, we show only the most important columns. By this tool, the information from the more general tools can also be verified on-site. In a practical sense, this means that you must physically follow the course of the data flow and flow of goods of an order. At the same time, the people processing the order can give information on the flow. By gathering and comparing all this information, you also gain insight into the degree to which employees have mastered the process.

The "basic process analysis" technique was further developed, mainly in the USA, to what is known today as "value stream mapping."

The *value stream* encompasses the processes of creating, producing, and delivering a good or service to the market. It may be controlled by a single business or a network of several businesses ([APIC16]).

Value stream mapping is a paper and pencil tool that helps you to see and understand the flow of material and information as a product or service through all of the value-adding and non-value-adding process steps.

Figure 4.3.3.2 shows an example of a value stream map of a mortgage deal with normal priority together with the value-added rate of the lead time of this process.

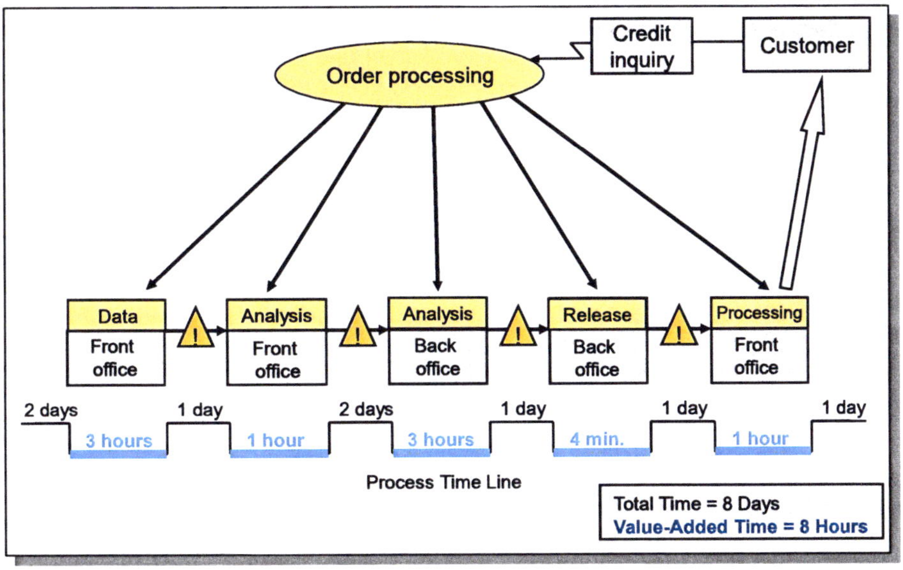

Fig. 4.3.3.2 Value-stream mapping.

- The value-adding activities are the procurement of all relevant data, analysis in the front office, and analysis in the back office required for larger business, followed by release and, finally, processing in the front office (that is, preparation of all documents and forms).
- The non-value-adding activities are shown together with their duration on the upper part of the typical *process time line*. They will be given special attention — as

indicated by the triangular symbols. As with basic process analysis, the objective is to identify *avoidable* non-value-adding activities and thus to identify "waste."

4.4 Characteristic Features Relevant to Planning & Control in Supply Chains

Intended learning outcomes: Produce an overview on principle and validity of characteristics in planning & control. Explain six features in reference to customer, and item or product or product family, five features in reference to logistics and production resources, as well as seven features in reference to the production or procurement order. Describe important relationships between characteristic features of transcorporate logistics in supply chains.

4.4.1 Principle and Validity of Characteristics in Planning & Control

Intended learning outcomes: Describe the characteristic in planning & control in a supply chain. Explain the use of the results of the analysis.

Any enterprise striving to reach its objectives cannot dispense with individual logistics. This alone is not satisfactory, of course, as it is safe to assume that there are principles common to whole branches of business. With the help of a morphological scheme, the weighting of entrepreneurial objectives can be translated into appropriate logistics.

A *characteristic in planning & control* in a supply chain is the sum of all values, that is, one value per *feature in the morphological scheme*. It relates to a product or product family.

Each product or product family, both upstream and downstream from decoupling points, can have a different characteristic for planning & control. This type of schema can be found in [Hack89] or [LuEv01]. Our discussion will include similar features and values. But we will also consider some important changes and additions in reference to transcorporate cooperation, nonrepetitive production, and the process industry. The 18 features are divided into three groups, namely:

- Features pertaining to the user and the product or product family
- Features pertaining to logistics and production resources
- Features pertaining to the production or procurement order

The following describes each feature and its values and defines the terms. While the features are independent of one another, individual values can certainly relate to other values. For example, the value of a feature can result in a particular value of another feature or preclude that value. But there are cases where there are no such dependent relationships. The totality of all features and values therefore shows redundancies. This situation is actually desirable, for it allows, at least to some extent, testing for plausibility.

Logistics analysis works out a characteristic for planning & control for each product or product family. For each company in a supply chain, a company-internal analysis is carried out.

The determination of the value of a feature is often the result of estimation, probability, or even an intuitive grasp of the situation. Such decisions are a matter for top management. Operations management must insist that top management make the decisions here. To do this, it will need the help of operations management to foresee the repercussions of the decision for one or the other value of a feature. Obviously, it is advantageous to have persons in top management who are experienced in operations.

The results of the analysis can be used as follows:

1. A comparison of results in the supply chain reveals potential problems for efficient logistics.

* *Within the company:* If features for the product families are too different, differing business methods of planning & control will be used. The coexistence of differing methods causes problems and diminishes the efficiency of logistics.
* *Transcorporate:* As described in Sections 2.2.3 and 2.2.4, the same logistics and information systems should be implemented in a supply chain wherever possible. With this, the characteristic for planning & control should be the same all along the network. If not, inefficiency will result.

2. Once the characteristic has been determined, it will indicate the appropriate business methods and techniques for planning & control.

The following sections will derive business methods for planning & control from the features. They all have advantages and disadvantages as well as limits to their implementation. They cannot be employed for all types of business processes. They may be incompatible with the business processes determined through process analysis, making it necessary to change the business processes or to alter entrepreneurial objectives. This kind of feedback also shows whether enterprise objectives and actual business processes cohere.

If planning & control in a company follows an outdated philosophy, this is often because enterprise objectives have not been reviewed and given new weightings. Had the company used changed characteristics for planning & control, it would have been in a position to institute new business methods for planning & control in a timely fashion.

3. The features making up a characteristic have an influence on logistics performance indicators.

Various characteristics can result in varying values of performance indicators. To compare performance indicators among companies effectively, the features making up the characteristics must be taken into account. As for a systematic comparison procedure see [FIR97b].

4.4.2 Six Features in Reference to Customer, and Item or Product or Product Family

Intended learning outcomes: Present features referring to the user and the product or product family. Differentiate between a convergent product structure and a divergent product structure. Differentiate between continuous demand and discontinuous demand. Explain the product variety concept. Disclose the T analysis within the VAT analysis and its relation to the product variety concept. Describe unit cost and transportability of an item.

Figure 4.4.2.1 shows the first group of features.

Features referring to customer and item, product or product family						
Feature	→	**Values**				
Depth of product structure	→	many structure levels	some structure levels	one-level production		
Orientation of product structure	→	▲convergent	combination ▲upper / ▼lower structure levels	▼divergent		
Frequency of customer demand	→	unique	discontinuous (lumpy, sporadic)	regular	continuous (steady)	
Product variety concept	→	according to (changing) customer specification	product family with many variants	product family	standard product with options	individual or standard product
(Item) unit cost	→	low	high			
Transporta-bility	→	not transportable	transportable	portable	digitally transmittable	

Fig. 4.4.2.1 Important features and possible values referring to the user and the product or product family.

The *depth of product structure* is defined as the number of structure levels within the total supply chain for the product, whether company-internal or transcorporate.

Product structure and structure level are defined in Section 1.2.2. The depth of product structure is dependent on the product. A deep product structure is usually also "wide": in each structure level, many components are put together. Such complex products usually entail complex planning & control. The depth of product structure is thus also a measure of the complexity of planning & control in the supply chain (see also [Albe95]). This complexity influences planning & control in each of the companies involved in the supply chain. See the feature *depth of product structure in the company* in Section 4.4.3.

The *orientation of product structure* indicates whether in *one single* production process *a certain* product is manufactured from *various* components (symbol ▲, convergent product structure), or whether in *one single* production process *various* products are made out of *a certain* component (symbol ▼ , divergent product structure).

- *Convergent product structure* is often used as a synonym for discrete manufacturing, that is, the production of distinct items such as machines or appliances. It is also called assembly orientation. The triangle pointing up symbolizes a tree (or arborescent) structure, as the product structure, such as that in Figure 1.2.2.2.
- *Divergent product structure* is often used as a synonym for by-products arising in continuous production (see Section 4.4.3). In chemical or oil production, which are typical examples from the process industry, processing of the basic material yields — in one single process — several active substances as well as waste or by-products. In the food industry, there are by-products that, through recycling, can be used as basic materials in another production process (such as scrap chocolate). The triangle pointing down symbolizes an upside-down, arborescent structure as the product structure. Note that a divergent product structure should not be confused with the multiple use of a component in different products.
- "▲ on ▼": This is a product with divergent product structure at lower structure levels, and convergent product structure at higher structure levels. The (lower) chemical level of pharmaceutical products, for example, has a divergent product structure, while the (higher) pharmaceutical level has a convergent structure. Other examples are products made from sheet metals. Many semifinished goods arise simultaneously from the sheet metal through pressing or laser cutting, and they are then used for various end products.

By the way: Determining the values of this feature corresponds exactly to a part of the VAT analysis (the "VA part"):

VAT analysis determines the general flow of parts and products from raw materials to finished products. A V structure corresponds to the divergent product structure (the letter V has the same shape as the symbol ▼). An A structure corresponds to the convergent product structure (the letter A has the same shape as the symbol ▲). A T structure consists of numerous similar finished products assembled from common assemblies, subassemblies, and parts. See the feature *product variety concept* below.

- A note on "▼ on ▲": This sometimes symbolizes an end product having many variants and therefore addresses the T structure mentioned above. In the lower structure levels, semifinished items are put together as modules. In assembly, many variants of end products are built from the semifinished goods or subassemblies. This is the case with automobiles. But because final assembly is clearly based on an assembly-oriented, convergent product structure, it should not be represented by the upside-down triangle. It is *not* the case that several products will arise from a particular semi-processed item. Although the symbol may be used quite commonly for this case, it is used incorrectly. A separate feature for describing the variant structure is the product variety concept. See below.

Frequency of customer demand means the number of times within defined observation time periods that the entirety of the (internal or external) customers demand a product or product family. Demand is

- *Unique*, if it occurs only once within an observation period

- *Discontinuous*, *lumpy*, or *highly volatile*, if many observation periods with no or very little demand are interrupted by few periods with large, for example, times higher demand, without recognizable regularity
- *Regular*, if it can be calculated for every observation period according to a certain formula
- *Continuous* or *steady*, if the demand is about the same in every observation period (e.g., daily)

This feature determines the options for repetitive frequency of the corresponding production and procurement orders. This in turn will determine the basic business methods and procedures for planning & control.

If longer observation periods are chosen, the frequency of customer demand can change, tending toward continuous demand. However, shifts and dips in demand within the observation period in this case will be unknown. For its purposes, planning & control can assume that the total demand could occur at the start of the observation period.

The *product variety concept* determines the strategy for developing the product and offering it to the customer. Where applicable, there may also be a product variety concept for semifinished goods.

The product variety concept allows the producer to respond to customer requests to varying degrees of *variant orientation*:

- An *individual* or *standard product* is offered to the customer "in isolation," that is, with no reference to other products in the range. These are "off the rack" products, or "standard menus." These products have their own complete product structure.
- *Standard product with options*: Here, the number of variants is small. A variant can be an additional feature of one and the same basic product. Each variant has its own product structure along with that of the standard product. Many examples are found in the machine industry.
- *Product family*: Compare here the definition in Section 4.1.2. In gastronomy, this value of the product variety concept is comparable to combining various appetizers, main dishes, and desserts to form an individual menu. Example industrial products are appliances and tools.
- *Product family with many variants*: The potential number of various products that can be produced in a product family can lie in the thousands or even in the millions. Production starts with raw materials or various components, but with an identical process. Variability of the process is achieved by CNC machines or by the workers themselves. Representation of the product structure requires a generic structure to overcome data redundancy problems and to reduce the administrative efforts for defining orders and maintaining the product structure. Product families with many variants are comparable to prêt-à-porter in the fashion industry. Some examples are automobiles, elevators, appliances, and machines with variable specifications, complex furniture, or *insurance* contracts.

- *Product according to (changing) customer specification:* In contrast to the product family, here at least some design work occurs during delivery lead time, according to customer specification. Usually, the product will be similar to a "mother product," meaning a product that has been delivered before. The product structure and the process plan will be derived and adapted from the "mother version." This value of the product variety concept is comparable to *haute couture*, whereby a creation is made to order for the individual customer. Examples can be found in the manufacturing of facilities (plants), such as the building of exteriors or refineries.

  A subcategory of this value is the *degree of change in customer orders*, where product and process structures change *after* the start of production.

Elaborating the values of the feature *product variety concept* can be considered to be a more detailed analysis of the T structure within VAT analysis.

T analysis describes the product variety. Qualitatively, the length of the crossbeam of the T stands for the number of product variants.

Figure 4.4.2.2 shows the idea of T analysis.

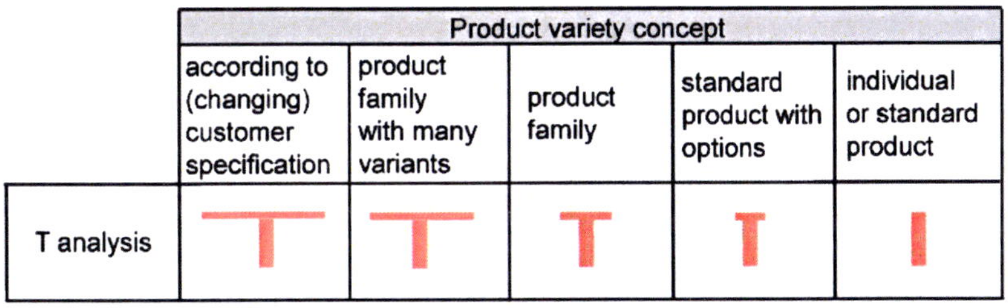

Fig. 4.4.2.2 T analysis within the VAT analysis and its relation to the product variety concept.

The product variety concept relates to other features. This will be discussed in the next section. As a rule, the complexity of planning & control increases with the number of different products produced. It is not, however, dependent on the number of variants, but rather on the number of product families having differing characteristics. Based on the definition of a product family, it is clear that all of its members can be described by one and the same characteristic. However, planning & control becomes more complicated with an increasing degree of product variety and, of course, with the degree of change in customer orders.

An item's *unit cost* is defined as the total cost for producing or purchasing one unit of measure of the item, e.g., one part, one gallon, one pound. It includes labor, material, and overhead cost.

- A *high-cost item* is an item with a relatively high unit cost compared with the unit cost of a *low-cost item.*

  For many important decisions in logistics and operations management, a very rough classification in low- and high-cost items is sufficient. However, an ABC classification considering sales and projected volume would allow a finer distinction. See Section 11.2.2.

The *transportability* of an item is actually a statement on the size and weight per unit of measurement. If the item is a service, transportability refers to the object on which the service is carried out.

- A *nontransportable item* is an item with a size or weight that permits no transport. These are items or objects, for example, of a size greater than 50 m3 or a weight greater than 200 metric tons. An example here is manufacture or maintenance of large plants.
- A *transportable item* is an item with a size or weight that permits transport using technical aids, such as helicopters, heavy goods vehicles, airplanes, or several people working together.
- A *portable item* is an item with a size or weight that permits transport (over a longer period of time) by means of the strength of one person. These are items or objects, for example, of a size smaller than 0.01 m3 or weighing less than 15 kg per unit. Letters sent out by courier are an example.
- A *digitally transmittable item* is an item that may be transmitted, or transported, using a digital communication protocol. Such an intangible good has a size or weight of zero.

The division of unit costs by the size or weight of the object leads to the concept of *value density*, that is, product value per kilogram or cubic meter. Digitally transmittable items have a value density approaching infinity. Value density plays a central role in the design of supply chains. See here also Section 3.1.1 and [Senn04]. It is a challenge, for example, if low-cost services are to be provided for nontransportable objects. This is the case with services for plants that have been installed worldwide. In designing the services, the proportion of services that can be transmitted digitally is particularly important.

4.4.3 Five Features in Reference to Logistics and Production Resources

Intended learning outcomes: Present features in reference to logistics and production resources. Explain the production environment: make-to-stock, assemble-to-order, make-to-order, and engineer-to-order. Describe fixed-position manufacturing, job shop production, as well as single-item-oriented, high-volume and continuous line production. Differentiate between flexible capability of capacity and (quantitatively) flexible capacity.

Figure 4.4.3.1 shows the second group of features.

Production environment or *manufacturing environment* refers to whether a company, plant, product, or service is organized to fulfill orders downstream from a specific (customer) order penetration point (OPP). The organization involves methods and techniques of planning & control of development, procurement, production, and delivery.

This feature is naturally closely connected with the (customer) order penetration point (OPP) *and the* stocking level (see Fig. 1.3.3.1):

- *Make-to-stock* is a store at the level of the end product. Delivery takes place from the end products store according to customer order.

An *order picking store* or a *commission stock* represent a status between actual stocking and use. Here all items or products are brought together that will be used for a certain production or sales order. They are stocked until final use in production or in the form of delivery to the customer. See Section 15.4.1.

Features referring to logistics and production resources						
Feature	→	**Values**				
Production environment	→	engineer-to-order	make-to-order	assemble-to-order (from single parts)	assemble-to-order (from assemblies)	make-to-stock
Depth of product structure in the company	→	many structure levels	few structure levels	one-level production	trade (including external production)	
Facility layout	→	fixed-position layout for site, project, or island production	process layout for job shop production	product layout for single-item-oriented line production	product layout for high-volume line production	product layout for continuous production
Flexible capability of capacity	→	applicable for many processes	applicable for few processes	applicable for only one process		
(Quantitatively) flexible capacity	→	not flexible in terms of time	hardly flexible in terms of time	flexible in terms of time		

Fig. 4.4.3.1 Important features and their possible values in reference to logistics and production resources.

- *Assemble-to-order*, or *finish-to-order*, is stocking at the level of assemblies or single parts. Upon receipt of a customer's order, a customized product is assembled using key components from the *assemblies store* or from the *single parts store* (that is, from the *in-house parts store* or *purchased parts store*).

  Package-to-order is a production environment in which a good can be packaged during the customer tolerance time. The item itself is the same for all customers. However, (only) packaging determines the end product.

- *Make-to-order* involves stocking at the level of raw materials or direct purchasing of material from suppliers after receipt of a customer's order. The final product is produced to meet the special needs of the customer using materials from the *raw materials store* or acquired through customer procurement orders. In both cases, the starting point is completed design and manufacturing process design. Thus, we can speak of stocking at the level of product and process design.

  Consigned stocks, or *consignment inventory*, or *vendor-owned inventory (VOI)* are inventories that legally still belong to the supplier, but have already been physically moved to the company. (A *consignment* is the process leading to consigned stock.)

- *Engineer-to-order* involves no stocking at all, at least for parts of a customer order. These must be developed or engineered prior to procurement and production.

The *depth of product structure in the company* is defined as the number of structure levels within the company.

This feature describes the degree to which the company's logistics resources must work toward the inside and toward the outside of the company. In regard to the supply chain within a company, the following is possible:

- In a pure *trading company* the number of structure levels, and thus the depth of product structure, is zero. Note: A company is still a trading company if it administrates a supply chain but contracts the production processes to third parties. Actually, though, the underlying basis is a one-level process plan with all external operations.
- Pure *assembling companies* or producers of single parts generally have at least one-level production, with mainly outside suppliers.
- A *supplier* may produce preassemblies or single parts or perform individual operations (such as surface treatments). Here, again, one-level production is the general rule. Suppliers are forced, however, to depend on producers further along the supply chain. Sometimes they function as system suppliers.
- The greater the number of structure levels the company itself "makes," the fewer components it will purchase from outsidc suppliers, and the greater the depth of product structure in the company.

This feature goes hand in hand with the feature *depth of product structure within the total supply chain* (Section 4.4.2). The less depth of product structure in a company as compared to that in the entire supply chain, the more strongly the company is bound to the transcorporate supply chain. In other words, with less depth of product structure, the greater the necessity for transcorporate cooperation. Experience has shown that deep product structure of the entire supply chain is also "wide," in the sense that many components enter into each structure level. This extends the range of procurement tasks.

With great depth of production structure, a company may attempt to reduce the complexity of the network by turning over structure levels to third parties (buy decision). This reduces complexity within the company itself, but complexity is not reduced within the total supply chain. Each company should contribute toward mastering the total complexity. Outsourcing must result in lower transaction costs (see also Section 2.1.1). The general rule is that outsourcing replaces long push logistics with pull logistics, through augmenting the number of independent partners and thus the number of process levels in the process model. In consequence, more persons become involved in planning & control. As they stand closer to their part of the entire process, the quality of planning & control can increase.

The *facility layout* describes the physical organization of the production infrastructure (the spatial arrangement and grouping of production equipment in work centers), the degree of division of labor among workers, and the course that orders take through the work centers.

- *Fixed-position layout* for fixed-position manufacturing, (fixed-)site production, project manufacturing, or island production: Here one work center carries out all operations to produce a product. All persons involved work here. All the production equipment is found at this work center or supplied to it. From the outside, the sum of all operations has the appearance of one gross operation. Workers exercise extensive autonomous control at the construction site. Typical examples include plant and facility construction, shipbuilding, large aircraft, very specific car production, automobile repair service, service at tables in a restaurant, and operations in a hospital. Examples for island production include the pilot tests[1] and specific product families, in particular with group technology.[2]
- *Process layout*, also called job shop layout or functional layout, for job shop production, or simply job shop: Similar production equipment is grouped together spatially at one work center. Only one operation is carried out at the work center, usually by one person (division of labor). The product moves from shop to shop according to the process plan, which lists all individual operations to be carried out. Certain persons are responsible for control. Typical examples include the production of appliances, electrical devices and electronics, furniture, pharmaceuticals, radiology and specific analysis in a hospital, and traditional forms of education.[3]
- *Product layout* for single-item-oriented line production: Here, the product moves through all work centers, which are ordered along the process to produce the product. Depending on the product, individual work centers or operations may be omitted. Generally, the line processes several variants of a product family in rather small batches, or a large variety of variants in lot size one, often with high value-added for each unit. The quantity produced by the line is determined by the actual demand. The fewer the number of variants produced, the more that production control can be based on production rates.[4] Setup times between batches, if required, are very short. All the required production equipment is found along the line. Ideally, workers are capable of executing neighboring operations in the process, whereby they move along the line. To the outside, the sum of all these operations looks like one rough-cut operation.[5] In group production, the group autonomously exercises control to a large degree within the group. Sometimes, the offices for planning & control as well as those for product and process design can be found close to the line, too. Typical examples include the assembly of automobiles, catamarans,

---

1 A *pilot test* is the production of a quantity to verify the manufacturability, customer acceptance, or other requirements before implementation of the ongoing production ([APIC16]). This book uses *production of prototypes* as an equivalent term.

2 *Group technology (GT)* identifies product families via a high percentage of the same processes in the process plan and establishes their efficient production. Group technology facilitates cellular manufacturing.

3 *Intermittent production* is a term used by many people as a synonym for job shop production.

4 A *production rate* is the rate of production expressed in simple quantities for a period of time, e.g., a day, a week, or a month. *Rate-based scheduling* is scheduling and controling based on production rates.

5 *Process flow production* describes the case where queue time is virtually eliminated by integrating the movement of the product into the actual operation of the resource performing the work (cf. [APIC16]).

motors and axles, machines, personal computers, and aircraft (the Boeing 717-200, for example). Other examples are a modern cafeteria line or office administration.[6]

- *Product layout for high-volume line production*: Here we find the same arrangement as in single-item-oriented line production. However, the operations are generally more detailed. Whole sequences of operations are carried out in direct succession. At times, the course of the process is rhythmic, following a strict time schedule. The work centers form a chain or a network with fixed, specifically designed facilities, sometimes linked by conveyors or pipes. Generally, the production line produces only a few different products, whenever possible in large batches of discrete units or nondiscrete items (for example, liquids). That is, the line produces with long runs, but the material flow is discontinuous. Setup times between batches are typically very high, because of cleaning or major adjustments of the production equipment, for example. The facility is built in order to obtain very low unit costs. Typical examples include the production of food, general chemicals, and transportation.
- *Product layout for continuous production* or *continuous flow production* is an extreme form of line production, namely, a production system without lots or batches, i.e. where material flow is continuous during the production process (cf. [APIC16]). The process is halted only if required by the transportation infrastructure or if resources are unavailable. The production line generally processes a commodity such as sugar, petroleum, and other fluids, powders, and basic materials.

The latter three kinds of facility layout have a common spatial arrangement:

A *line* is a specific physical space for the manufacture of a product that in a flow shop layout is represented by a straight line. In actuality, this may be a series of pieces of equipment connected by piping or conveyor systems ([APIC16]).

The work centers are arranged along the process, that is, according to the sequence of operations required to produce a product or a product family. A line in the manufacturing environment is often called *assembly line* (particularly in the case of single-item-oriented line production) or *production line* (particularly in the case of high-volume line production). In practice, a line can take any form or configuration, such as straight, U-shaped, or L-shaped (see Section 6.2.2).[7]

From the term *line*, used to describe this particular spatial arrangement, stems the term *line production*. For high-volume line production or continuous production, the terms *flow shop* or *flow manufacturing* are sometimes used synonymously.

The facility layout can be dependent on the structure level. For example, facility layout may differ for assembling and parts production. In addition, a subcategory here is the *degree of structuring of the process plan*. This degree of structuring tells us the number of operations divided up in the process plan for one structure level. Site production and single-item-

---

6 *Mixed-model production* is a term similar to single-item-oriented line production. It stands for a factory producing close to the same mix of different products that will be sold that day (see [Foga09]).

7 A *dedicated line* is a production line permanently configured to run well-defined parts, one piece at a time, from station to station ([APIC16]). It thus is a simple kind of single-item-oriented line production.

oriented line production generally have a low degree of structuring, as the operations defined are considerably less detailed.

The *flexible capability of capacity* determines whether capacity can be implemented for various or for particular processes only.

A producer's flexible capability of capacity is made up of the flexible capability of its employees and of its production infrastructure. This is the feature that sets a company's possible range with regard to the target area of flexibility. If employees have broad qualifications and the production infrastructure can be widely implemented, there will be great flexibility in the use of resources. This is also the necessary prerequisite for a wide product range and thus for flexibility in achieving customer benefit.

In practical application, this feature can be broken down further into sub-categories, if the different types of capacity show differences in flexible capability. The main differentiation is between the *flexible (capability of the) workforce* and the *flexible capability of the production infrastructure*. The flexible workforce deserves special attention (*job enlargement* is also often used). First of all, it can normally be achieved to a far greater degree than flexible capability of the production infrastructure. Second, in contrast to the production infrastructure, employees do not simply represent a production factor, for they are themselves stakeholders.

The feature *(Quantitatively) flexible capacity* describes its temporal flexibility.

Temporal flexibility of capacity along the time axis is a significant factor in the target areas of delivery and cost. As follows, it even becomes a crucial feature when choosing planning & control methods, particularly in capacity.

If different types of capacity show varying quantitative flexibility, it will be necessary to differentiate subcategories. The main differentiation is between the *Quantitatively flexible workforce* and the *(Quantitatively) flexible production infrastructure*.

People have far greater possibilities to achieve quantitative flexibility than machines. (Quantitatively) flexible machines can only be reached by means of maintaining overcapacity. People, however, are to a certain degree able to adapt their efforts to the current load.

Moreover, if capacity has a flexible capability that transcends the "home" work center (that is, employees can be deployed for processes outside the "home" work center), flexibility along the time axis is increased. For example, if workers can be moved from one work center to another, this is the same as flexibility in deployment of the employees at both work centers. Depending on load in the areas, the employees can be deployed flexibly.

4.4.4 Seven Features in Reference to the Production or Procurement Order

Intended learning outcomes: Present features in reference to production or procurement order. Differentiate between order release according to demand, prediction, and consumption. Differentiate between production or procurement without, with infrequent, and with frequent order repetition. Identify flexibility of the order due date and type of long-term order. Explain lot size, lot traceability, and loops in the order structure.

Figure 4.4.4.1 shows the third group of features.

<table>
<tr><th colspan="14">Features referring to production or procurement order</th></tr>
<tr><th>Feature</th><th>→</th><th colspan="12">Values</th></tr>
<tr><td>Reason for order release (type of order)</td><td>→</td><td colspan="4">demand / (customer production or procurement order)</td><td colspan="4">prediction / (forecast order)</td><td colspan="4">consumption / (stock replenishment order)</td></tr>
<tr><td>Frequency of order repetition</td><td>→</td><td colspan="4">production / procurement without order repetition</td><td colspan="4">production / procurement with infrequent order repetition</td><td colspan="4">production / procurement with frequent order repetition</td></tr>
<tr><td>Flexibility of order due date</td><td>→</td><td colspan="4">no flexibility (fixed delivery date)</td><td colspan="4">not very flexible</td><td colspan="4">flexible</td></tr>
<tr><td>Type of long-term orders</td><td>→</td><td colspan="4">none</td><td colspan="4">blanket order: capacity</td><td colspan="4">blanket order: goods</td></tr>
<tr><td>(Order) lot or batch size</td><td>→</td><td colspan="3">One (single item production / procurement)</td><td colspan="3">Single item or small batch (production / procurement)</td><td colspan="3">large batch (production / procurement)</td><td colspan="3">(production / procurement) without lots or batches</td></tr>
<tr><td>Lot traceability</td><td>→</td><td colspan="4">not required</td><td colspan="4">lot / batch / charge</td><td colspan="4">position in lot</td></tr>
<tr><td>Loops in the order structure</td><td>→</td><td colspan="6">Product structure without loops, and directed network of operations</td><td colspan="6">Product structure with loops, or undirected network of operations</td></tr>
</table>

Fig. 4.4.4.1 Important features and possible values in reference to production or procurement order.

The *reason for order release* is the origin of the demand. The *type of order* indicates the origin of demand that resulted in the order.

- *Order release according to demand* and *customer production order* or *customer procurement order:* A customer has placed an order. It may be a classic (single) order, for a car, for example, or it may be a blanket order, such as for electronic components. In the latter case, customer production orders can follow at different points in time, released according to the delivery agreements. This is also called *demand-controlled materials management*, using pull logistics.
- *Order release according to prediction* and *forecast order:* Future demand has been estimated, such as demand for a machine tool. Customer orders for the machine tool have not yet been received. To meet forecasted demand, a production or procurement order is released. This is also called *forecast-controlled materials management* using push logistics.

Order release according to consumption, and *(stock) replenishment order:* A customer places an order for a product in stock, for example, in the retail trade. In response to the demand, stock must be reordered. Actually, this is a response to forecasting future need in the quantity that is reordered. This is also called *consumption-controlled materials management* using pull logistics.Compare these definitions to Figure 4.2.3.1. By the way:

The trigger for the release of orders can be different for end products, semifinished goods, and raw materials. It is dependent on the (customer) order penetration point (OPP).

The *frequency of order repetition* tells us how often within a certain time period a production or procurement order for the same product will be made. The time period chosen should be sufficiently long.

- *Production without order repetition* or *procurement without order repetition* means that an order for the same physical product will practically never be placed again.
- *Production with infrequent order repetition* or *procurement with infrequent order repetition* means that, with a certain probability, an order for the same physical product will be placed again.
- *Production with frequent order repetition* or *procurement with frequent order repetition* means that orders for the same physical product will be very frequent.

Note: The adjective *physical* is used here to underline that this feature refers to the product level, and not to the product family level. Therefore, if an order produces a physically different product of the same family compared to another order, this is *not* considered to be production with order repetition.

The *flexibility of the order due date* indicates whether customers (internal or external) are flexible when stipulating the delivery due date.

The flexibility of the order due date is of great importance to methods of planning & control, particularly with regard to the target area of delivery. With regard to the target area of cost, it is connected to the (quantitatively) flexible capacity of workforce and production infrastructure, on-hand balance, and in-process inventory.

The feature *type of long-term order* describes the manner in which long-term planning is done in the supply chain.

A *blanket order*, for example, is a long-term agreement for a great number of deliveries.

A *minimum blanket order quantity* is — for a blanket order — a long-term minimum volume of business for a particular period of time.

Long-term orders are in the best interests of both parties. The customer profits from lower pricing and a higher fill rate from the supplier. The supplier in turn can depend on a minimum blanket order quantity and gains the advantage of increased planning capability.

We distinguish the following values, which correspond generally to the values of the features *frequency of customer demand* and *product variety concept* in Figure 4.4.2.1:

- *Blanket orders for goods* are long-term binding commitments in the supply chain for products and their components. Assured sales are necessary and are guaranteed by continuous customer demand. If the minimum blanket order quantity is zero, then demand is only a forecast. If the forecast is relatively reliable, production planning for partners in the supply chain will be better than without the forecast. For example, if customer demand is discontinuous, forecasts will be used for long-term planning.

- *Blanket orders for capacity* are long-term binding agreements on reserving capacity. This may be in reference to a product family, for example, for which at least regular customer demand is guaranteed and which is produced in the main according to the same production process. The products are ordered short term, and they are to be produced using the reserved capacity within the delivery lead time. Again, if the minimal order quantity is zero, the same applies as described above.
- "None" means that, in the supply chain, neither blanket orders nor forecasts are made. This is appropriate when actual customer demand is nonrepetitive.

Lot size or *batch size* is the *order quantity* of an ordered item (and vice versa).

- *Lot size one*, or *batch size one*, or *single-item production*, or *single-item procurement*, means that only one unit of the product is produced or procured for an order.
- *Small batch production* or *small batch procurement* indicates that for an order only a few units of the product will be produced or procured.
- *Large batch production* or *large batch procurement* means that a high quantity of units of the product will be produced or procured for one order.
- *Production without lots* or *procurement without lots* means that no specific quantity is linked with the order. Rather, after order opening, production/procurement continues until an explicit order stop is given. *Lotless production* or *lotless procurement* are possible synonyms.

Note: There is no correlation between the values in nearby columns of the feature *batch size* and of the feature *frequency of order repetition*. For example, single-item production with frequent order repetition is quite common (for example, in machine tool production). Conversely, there can be production (without order repetition) of exactly *one batch for the entire product life cycle*[8] (such as when an active substance in the chemical industry, for cost reasons, is produced only once in the product life cycle, or in the case of special components that are very difficult to procure).

Lot traceability is information on the production and procurement of a product, in particular about the components used in the product.

Lot traceability is often required by law or can be important with regard to liability and problems associated with recalling a product. It generally asks for records about every production or procurement lot, batch, or charge:

- A *charge*, according to [APIC16], is the initial loading of ingredients or raw materials into a processor, such as a reactor, to begin the manufacturing process. It has become a synonym for a number or quantity of goods produced or procured together that, for the purposes of the lot traceability, are identical.

[8] Here, a second definition of the term *product life cycle* is used: the market stages a new product goes through from the beginning to the end, i.e., introduction, growth, maturity, saturation, and decline ([APIC16]). See Section 1.1.5 for the first definition.

- *Position in lot* refers to the successive numbering of the individual items in a lot.

The lot traceability requirement makes planning & control well more complicated. Nevertheless, lot traceability plays a notably important role in the process industry. See Chapter 8.

Loops in the order structure is a situation in resource planning, where business objects have to be considered an indefinite number of times.

- A *product structure with loops* means a situation where a product is its own component — either directly or via intermediate products. It plays an important role again in the process industry, where production yields important quantities of by-products that are reused, such as scrap chocolate or energy.
- An *undirected network of operations* means a process plan where sequences of operations may be repeated. It is found in the precision industry, where individual operations are repeated until the required degree of quality is reached. In addition, it plays an important role in the process industry, where a mixing operation may be repeated as often as is necessary to ensure the desired level of homogeneity.
- A *product structure without loops*, as well as a *directed network of operations*, are free of the above-mentioned effects.

Planning of loops in the order structure is relatively complicated. See Sections 8.1.3, 8.3.3, and 13.4.4.

4.4.5 Important Relationships between Characteristic Features

Intended learning outcomes: Identify links among the features facility layout, orientation of product structure, and batch size. Disclose links among the features product variety concept, production environment, and frequency of order repetition. Explain why frequency of customer demand and frequency of order repetition do not necessarily need to correspond.

In some cases, there is a relationship among characteristic features, which can even be a positive correlation. For example, the feature *facility layout* is — according to Figure 4.4.5.1 — closely related to other features.

The figure shows that, in a *first approximation*, the different values of the features in the same columns appear together. For example:

- Site production, job shop production, and single-item-oriented line production have a tendency to appear together with a convergent product structure and production or procurement of single items or small batches.
- High-volume line production and continuous production tend to appear together with a combination of convergent product structure on upper levels and divergent product structure on lower levels, or a fully divergent product structure, and with large-batch production or procurement, or production or procurement without lots.

Both observations also hold in the reverse direction. This means that in all the following figures in Section 4.5, we can replace the feature *facility layout* with one of the two features *orientation of product structure* and *(order) batch size.*

Features referring to logistics and production resources						
Feature	→	**Values**				
Facility layout	→	fixed-position layout for site, project, or island production	process layout for job shop production	product layout for single-item-oriented line production	product layout for high-volume line production	product layout for continuous production
Features referring to user and product or product family						
Feature	→	**Values**				
Orientation of product structure	→	▲ convergent			combination ▲upper/▼lower structure levels	▼ divergent
Features referring to production or procurement order						
Feature	→	**Values**				
(Order) lot or batch size	→	one (single item production / procurement)	single item or small batch (production / procurement)		large batch (production / procurement)	(production / procurement) without lots

Fig. 4.4.5.1 Links among facility layout, orientation of product structure, and (order) batch size. The horizontal distribution of the values was effected to indicate the correlation of the features.

A further observation is that the product variety concept is — according to Figure 4.4.5.2 — closely related to other features:

Features referring to user and product or product family						
Feature	→	**Values**				
Product variety concept	→	according to (changing) customer specification	product family with many variants	product family	standard product with options	individual or standard product
Features referring to logistics and production resources						
Feature	→	**Values**				
Production environment	→	engineer-to-order	make-to-order	assemble-to-order (from single parts)	assemble-to-order (from assemblies)	make-to-stock
Features referring to production or procurement order						
Feature	→	**Values**				
Frequency of order repetition	→	production / procurement without order repetition		production / procurement with infrequent order repetition		production / procurement with frequent order repetition

Fig. 4.4.5.2 Links among the features product variety concept, production environment, and frequency of order repetition.

The figure shows that, in a *first approximation*, the different values of features in the same columns appear together. For example:

- *Product variety concept versus production environment*: A product variety concept according to customer specification (such as the manufacturing of plant facilities) means that part of the customer order has to run through design prior to procurement or production. This is the exact meaning of engineer-to-order. Product families with many variants are generally produced using raw materials (make-to-order). The variants in a product family concept with a restricted number of variants are normally produced during assembly (assemble-to-order). Standard products are stocked at the level of end products (make-to-stock).
- *Product variety concept versus frequency of order repetition*: Production / procurement without order repetition is generally typical for a product variety concept according to customer specification or for product families with multiple variants. Production / procurement with infrequent order repetition is found with product families. Production / procurement with frequent order repetition is the rule with individual or standard products and with a small number of variants.

Thus, in Section 4.5, *product variety concept* can be replaced with either of the two features *production environment* or *frequency of production or procurement order repetition.*

It is also interesting to compare the feature *frequency of customer demand* in Figure 4.4.2.1 with the feature *frequency of order repetition* as shown in Figure 4.4.5.3. It is noteworthy that the values of the features in the same columns do not necessarily have to correspond.

Features referring to user and product or product family					
Feature	→	**Values**			
Frequency of customer demand	→	Unique	discontinuous (lumpy, sporadic)	regular	continuous (steady)
Features referring to production or procurement order					
Feature	→	**Values**			
Frequency of order repetition	→	production / procurement without order repetition	production / procurement with infrequent order repetition	production / procurement with frequent order repetition	

Fig. 4.4.5.3 The features *frequency of customer demand* and *frequency of order repetition* do not necessarily need to correspond.

Indeed, procurement and production can be decoupled from demand via storage:

- Storage can provide a buffer for discontinuous demand, so that there can be more frequent production. For example, a product can be manufactured throughout the year that will be in demand mainly at a yearly holiday time. Through this, capacities can be utilized more evenly. On the negative side, carrying costs are incurred.
- On the other hand, if demand is continuous, delivery can also be made from storage, and usage can be replenished through less frequent orders in large batches. This course of action is sometimes unavoidable, due to both technical constraints (if, for

example, such as in the process industry, certain production facilities allow production in specific batch sizes only) and economic reasons (if, for example, the ordering of a small quantity makes no sense, because transport costs — in production setup costs — are too high in relation to the unit costs of the small quantity).

However, the more storage should be avoided, the more there has to be a connection between values of the features in the same columns: Unique demand occurs together with production or procurement without order repetition, discontinuous demand together with infrequent order repetition, and continuous demand with frequent order repetition.

Similarly, the choice of the planning & control concept (see Section 4.5.3) as well as materials management techniques (see Section 5.3.2) must first be made on the basis of the frequency of customer demand. If a number of techniques are possible, the choice is determined by the selected frequency of production or procurement order repetition.

4.4.6 Features of Transcorporate Logistics in Supply Chains

Intended learning outcomes: Present important features, possible values, and increasing complexity of supply chain collaboration, of supply chain coordination, and of the configuration of the supply chain.

Cooperation among all participants is the key prerequisite for effective operation of the supply chain (see Sections 2.2 and 2.3). For this reason, the characteristic features of supply chains include various aspects of cooperation. A morphological scheme proposed in [Hieb02] encompasses three groups of features that are closely linked to the Advanced Logistics Partnership (ALP) model (see Section 2.3).

Figure 4.4.6.1 presents *features referring to supply chain collaboration.*

Features referring to supply chain collaboration

Feature	→	Values			
Alignment of network strategy and interests	→	common network strategy	common network interests	divergence of network interests	
Orientation of business relations	→	cooperation-oriented	opportunistic	competition-oriented	
Mutual need in the network	→	high; sole sourcing	single sourcing	multiple sourcing	low, highly substitutable
Mutual trust and openness	→	high	low		
Business culture of network partners	→	homogeneous / similar	comparable in size, structure, or volume of sales	heterogeneous / highly different	
Balance of power	→	high dependency / hierarchical	equal / heterarchical		

Increasing complexity of supply chain collaboration.

Fig. 4.4.6.1 Features and possible values referring to supply chain collaboration. The horizontal distribution of the values in the morphological scheme indicates their relation to the increasing degree according to the given criterion.

These features describe the degree and kind of partnership among the participants on a high level as well as the fundamental commitment of the companies to pursue a common "network strategy."

The columns to the left contain values that indicate that the companies have already expended efforts toward strategic collaboration or that there is an inherent alignment from the start. The columns at the right contain values that indicate increasing complexity of the common operation of value-added processes.

Figure 4.4.6.2 presents *features referring to supply chain coordination* that describe the type of the daily operations in shared transcorporate processes and methods.

<table>
<tr><th colspan="9">Features referring to supply chain coordination</th></tr>
<tr><th>Feature</th><th>→</th><th colspan="7">Values</th></tr>
<tr><td>Intensity of information sharing</td><td>→</td><td>limited to needs of order execution</td><td colspan="2">forecast exchange</td><td>order tracking and tracing</td><td colspan="2">sharing of inventory / capacity levels</td><td>as required for planning & control processes</td></tr>
<tr><td>Linkage of logistics processes</td><td>→</td><td>none, mere order execution</td><td colspan="2">integrated ex-ecution, (e.g., consigned inventory)</td><td>vendor-managed inventory</td><td colspan="2">collaborative planning</td><td>integrated planning and execution</td></tr>
<tr><td>Autonomy of planning</td><td>→</td><td colspan="2">heterarchical, locally in-dependent, autonomous</td><td colspan="3">local, with central guidelines</td><td colspan="2">hierarchical, led by strategic center</td></tr>
<tr><td>Variability of consumption</td><td>→</td><td>rather stable consumption</td><td colspan="2">variability in time</td><td>variability in amount</td><td colspan="3">high variability in time and amount</td></tr>
<tr><td>Type of long-term orders</td><td>→</td><td>None</td><td colspan="3">blanket order: capacity</td><td colspan="3">blanket order: goods</td></tr>
<tr><td>Degree of communica-tion among multiple tiers and channels</td><td>→</td><td>single contact for the transaction</td><td colspan="2">regular network meetings (e.g., supplier days)</td><td>central coordination (e.g., supply chain manager)</td><td colspan="3">multiple contacts among levels and channels</td></tr>
<tr><td>Use of information technology (IT)</td><td>→</td><td colspan="2">IT use only to support internal business processes</td><td colspan="3">IT use to support network coordination mechanisms (e.g., EDI)</td><td colspan="2">IT use to support execution and planning mechanisms; SCM-software</td></tr>
</table>

Increasing complexity of supply chain coordination.

Fig. 4.4.6.2 Features and possible values referring to supply chain coordination.

Figure 4.4.6.3 presents *features referring to the configuration of the supply chain.* They describe the modeling of the existing business relationships among the network entities and the setup, meaning the physical structure as well as temporal and legal business relationships. The values of these features determine supply chain changeability to a great extent.

Just as in Figures 4.4.2.1, 4.4.3.1, and 4.4.4.1, the features are — as a whole — independent of one another. However, individual values can certainly relate to values of other features.

Features referring to the configuration of the supply chain					
Feature	→	**Values**			
Multitier network (depth of network)	→	2 value-adding tiers	3–5 value-adding tiers	>5 value-adding tiers	
Multichannel network (breadth of network)	→	1–2 logistics channel(s)	3–5 logistics channels	>5 logistics channels	
Linkage among the partners	→	simple relationship, segmentation	complex relationship, ramifications		
Geographical spread of network	→	local	regional	national	global
Time horizon of business relationship	→	short-term, less than 1 year	mid-term, 1–3 years	long–term, >3 years	
Legal business relation (financial autonomy)	→	independent business partners	alliances, joint ventures	group / combine	

Increasing complexity of the configuration of the supply chain.

Fig. 4.4.6.3 Features, and possible values referring to the configuration of the supply chain.

[Hieb02] defines all of these features in detail. Some of the definitions are readily understood in a common sense, but others have a very specific meaning. However, what is important is that all partners seeking jointly to start a supply chain initiative (SCI) examine the morphological scheme — including the exact definition of each feature. The scheme must be discussed, completed, and agreed on. This can culminate in common performance metrics for the entire network. It can be the first step toward a common understanding of the network and deeper knowledge of the interactions among its members.

Often, a supply chain is already in place when morphological schemes are applied. In that case, the scheme proposed above can support achievement of network objectives. It can also be a very helpful tool when replacing a partner in the supply chain.

4.5 Branches, Production Types, and Concepts for Planning & Control

Intended learning outcomes: Describe branches of industry in dependency upon characteristic features. Explain production types and concepts for planning & control. Disclose selecting an appropriate branch model, production type, and concept for planning & control.

Strikingly, standardization of branches of industry, production types or concepts for planning & control mainly depend on two characteristic features discussed in Section 4.4. Figure 4.5.0.1 spans a two-dimensional space with the coordinates *facility layout* from Figure 4.4.3.1 (features related to logistics and production resources), and *product variety concept* from Figure 4.4.2.1 (features related to user and product or product family).

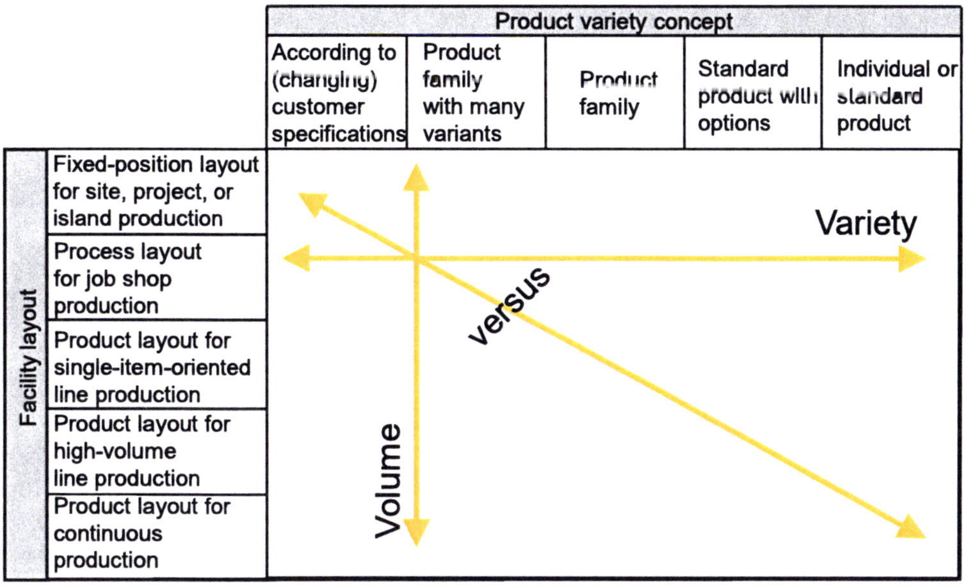

Fig. 4.5.0.1 Standardization of branches in industry, production types, and concepts for planning & control: Positioning products or product families within the space spanned by volume versus variety.

This two-dimensional space allows, for a specific product or product family, to position the volume of produced items ("things" or "stuff," as some would say) in demand against the need for product variety: "volume versus variety".

4.5.1 Branches of Industry in Dependency upon Characteristic Features

Intended learning outcomes: Explain different branches of industry in dependency upon the two features facility layout and product variety concept.

> A *branch* of industry is the sector or segment of business a company engages in.

Definitions of various branches of industry and areas of business can be found in governmental statistics on economics and industry, for example. Typical industrial branches include the chemical industry, plastics industry, electronics and electrical industries, aircraft and automobile industries, engineering and metal industries, watchmaking industry, paper industry, and textile industry. Typical branches in service-providing businesses include banking, insurance, consulting, computer software, trust companies and private management, and care agencies (for people and things).

An obvious approach is to seek branch-dependent concepts.

> A *branch model* of planning & control groups together with concepts appropriate to specific branches, including suitable types of business processes and business methods.

The branch of industry or service is indeed related to many of the characteristic features of planning & control. The corresponding business methods, however, are usually too general to be ideally suited to a particular branch. For this reason, it has been useful to go beyond those concepts and develop branch models.

Figure 4.5.1.1 shows different branches in dependency upon upon the same characteristic features as in Figure 4.5.0.1, that is, *facility layout* and *product variety concept*.

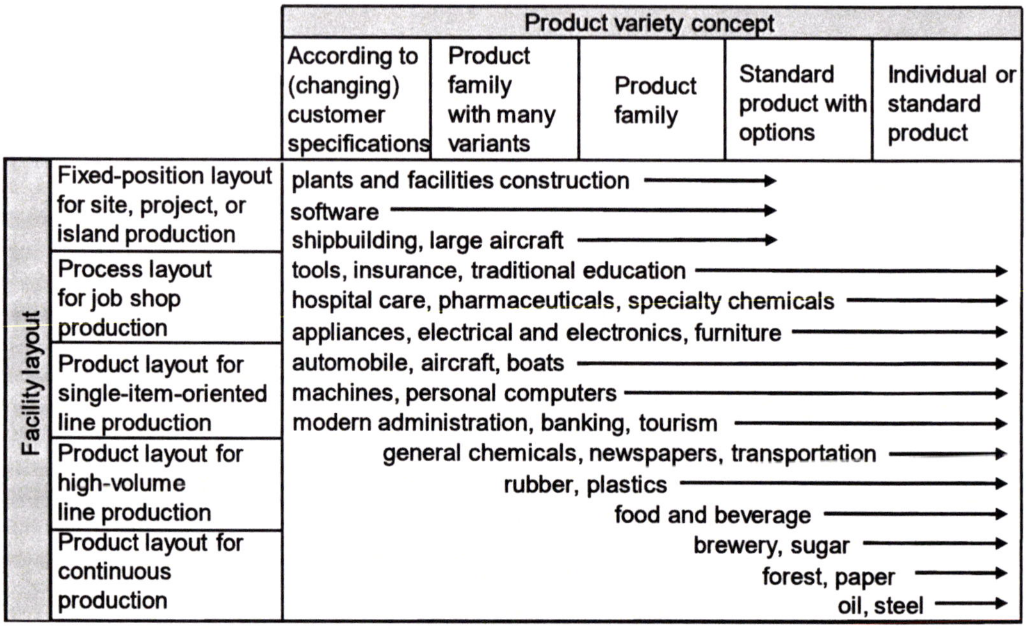

Fig. 4.5.1.1 Different branches in dependency upon the two features *facility layout* and *product variety concept*.

The figure shows that:

- In a first approximation, branches can be readily positioned according to the feature facility layout. This indicates that there is a clear relation here.
- A number of branches, particularly those in the process industry (branches producing "stuff" rather than "things," as some would say), can be distinguished along the values of the feature *product variety concept* relatively clearly. In nearly all branches, however, we find product variety concepts of "according to (changing) customer specification" all the way to "individual or standard product" — with some exceptions. Exceptions are the production of plants and facilities, shipbuilding, large aircraft, and software: there are no examples positioned in the top right-hand corner of the matrix. Other exceptions are the production of rubber, plastics, food and beverage, brewery, sugar, forest, paper, oil, steel: there are no examples positioned in the bottom left-hand corner of the matrix. The relation here, therefore, is less clear than the relation to the feature facility layout.

The feature *product variety concept* is therefore largely independent of the feature *facility layout* (as well as of volume, understood as batch size).[9] This important observation leads to the matrix. The following are some examples that support this observation:

- Take a company producing standard machines. This is done with frequent order repetition, but in single units, either according to arrival of a customer order, or in advance (because it is a standard machine, inventory risk is small: the machine will be sold sooner or later). This is the "nearly top" right-hand corner in the matrix: job shop production or single-item-oriented line production.
- Another company produces standard screws. Again, this is done with frequent repetition, but each time in large batches. This is the bottom right-hand corner in the matrix (as production without lots is also very possible): high-volume line production or continuous production.
- A company in the chemical branch produces a large batch of a specific active substance only once for the whole life cycle of the product, because of the high setup and order administration costs. This is the nearly bottom left-hand corner in the matrix: high-volume line production.
- Still another company produces a plant as a single unit and only once, according to customer specification. This is the top left-hand corner in the matrix: site or project production.

4.5.2 Production Types

Intended learning outcomes: Explain the different production types and the relation of some of them to various kinds of facility layouts.

A *production type* encompasses a set of manufacturing technologies and methodologies, having specific importance with regard to logistics management and planning & control.

In the world of practice, the understanding of the different values of the feature *facility layout* introduced in Figure 4.4.3.1, namely:

- Fixed-position layout for site, project, or island production
- Process layout for job shop production
- Product layout for single-item-oriented line production
- Product layout for high-volume line production
- Product layout for continuous production

is not limited to the physical organization of the production infrastructure or the process design. Beyond this, from a systems capability viewpoint, these values are often also seen as production types.

---

9 A dependency exists, as mentioned, in the top right-hand corner and in the bottom left-hand corner of the matrix — that is, in the blank areas in the figure.

However, a number of new terms have come into use in recent years, each standing for a specific process technology and methodology.

- *Batch production*, or *batch processing*, is production or procurement of a generally wide variety of standard products or variants of a product family manufactured in batches either to order or to stock (see [Foga09],). Because of batching, precise timing and sizing of component lots are essential.
- *Mass production* is high-quantity production characterized by specialization of equipment and labor ([APIC16]).
- *Repetitive manufacturing* is "the repeated production of the same discrete products or families of products. Repetitive methodology minimizes setups, inventory, and manufacturing lead times by using production lines, assembly lines, or cells. Work orders are no longer necessary; production scheduling and control are based on production rates (*flow control*). Products may be standard or assembled from modules. Repetitive is *not* a function of speed or volume" ([APIC16]).
- *One-of-a-kind production* is the production or procurement of an engineer-to-order, in rare cases, make-to-order product, generally according to customer specification, often derived from an earlier customer order.
- *Mass customization* is a production or procurement principle that emphasizes customized products that do not cost more than mass-produced products. According to [APIC16], it is "the creation of a high volume product with large variety so that a customer may specify an exact model out of a large volume of possible end items while manufacturing cost is low due to the large volume." Having some characteristics of repetitive manufacturing with regard to the facility layout, mass customization could be seen as "high volume repetitive manufacturing with high variety" [PtSc03]. In this context, "high volume" means either "high number of orders" or "high work content," but *not* (!) "large batch." It is repetitive manufacturing on the family level, but *not* on the product level: each product (unit) produced is, while belonging to the same family, generally physically different.

  Therefore, techniques of repetitive manufacturing can be used for those aspects of planning & control that refer to the product family as a whole, but they can*not* (!) be used for those aspects of planning & control that refer to a specific product variant. In particular, a specific work order is required for each product produced. The work order includes the configuration of the customer-ordered variant out of a product family with a very large variety of products, with its specific components and operations, if need be with omissions or insertions of positions. Furthermore, long lead times may entail the increasing use of project management techniques rather than rate-based scheduling techniques.

It is not possible to line up these additional production types according to a single feature. In fact, in a systems capability perspective, many of them overlap, just as do some of the different facility layouts already mentioned. Fortunately, however, as Figure 4.5.2.1 demonstrates, all of these additional production types can be shown in dependency upon the same characteristic features as in Figure 4.5.1.1; that is, *facility layout* and *product variety concept*.

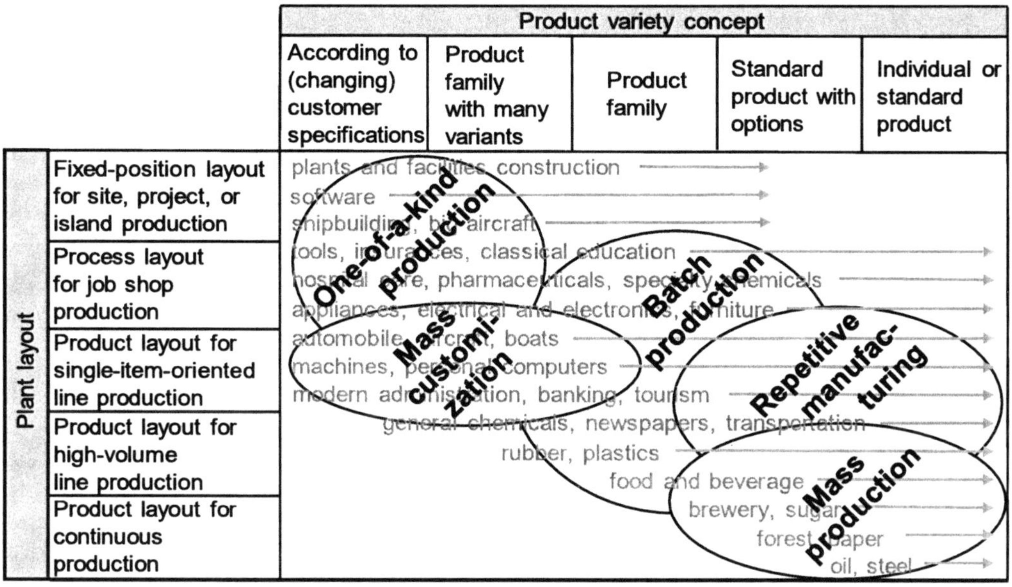

Fig. 4.5.2.1 The different kinds of facility layouts seen — from a systems capabilities viewpoint — as production types together with other production types.

4.5.3 Concepts for Planning & Control

Intended learning outcomes: Explain different concepts of planning & control in dependency upon the features facility layout and product variety concept.

A *concept for planning & control* is made up of particular types of business processes and business methods for order planning and fulfillment.

Recent decades saw the development of different concepts of planning & control in supply chains. Each was developed in a particular area and so represents to a certain degree a model for a branch of industry. Some of the concepts arose in powerful industries, such as the automobile or machine industry. The concepts were systemized and given brand names.

- *The MRP II (manufacturing resource planning) concept*[10] originated in North America in the late 1960s. See [Wigh95] and [VoBe18]. MRP II was developed in branches of industry having clearly convergent product structures, such as for the construction of big machines and in the automobile and aircraft industries. Three temporal ranges of planning & control (short, medium, and long range) were basic to the MRP II concept that quite early on went beyond matters of production. Further development of the concept led to the *ERP (enterprise resources planning) concept* in order to include all areas of a company. See Chapter 5.

[10] Important note: The MRP II concept should not be confused with the MRP method of material requirements planning. See Section 5.3.2 and Chapter 12.

- The Japanese *just-in-time concept*, today also known as lean/JIT (*lean production*), aimed at improving the flow of goods. Marketed in the late 1970s as a contrasting alternative to the MRP II concept, the lean/JIT concept has also turned out to be generally valid and fundamental to planning & control in ERP when *delivery* becomes a targeted company priority. The Kanban technique, often linked with lean/JIT, however, is applicable — along with other simple techniques for repetitive manufacturing — only to standard products or product families with very few variants. The lean/JIT concept and all these techniques form an important *extension* to the MRP II concept and its techniques. See Chapter 6.

The details of resource management developed by the MRP II/ERP concept remain fundamentally valid in the extended concepts below. These extensions differ from the MRP II / ERP concept mainly in the modeling of logistics business objects and, accordingly, in order configuration, order processing, and order coordination in all temporal ranges of planning.

- The *variant-oriented concept* originated particularly in Europe in the late 1970s. It was developed in connection with the product variety concept of *product family*, with *one-of-a-kind production* and *production without order repetition*. It is a necessary *extension* of previous concepts. See Chapter 7. Depending on the product variety concept, different characteristics of planning & control arise often and typically together. See here also Figure 4.4.5.2.

- In the late 1980s, the *processor-oriented concept* was developed in North America for process industries. This concept *extended* MRP II, but it has not yet found complete systematization. It comes with the term *process flow scheduling*. Besides techniques for continuous production and campaign concepts (to handle high setup costs), the processor-oriented concept considers *divergent product structures*, a phenomenon that was not covered adequately by earlier concepts. See Chapter 8.

Figure 4.5.3.1 summarizes the different concepts. It is interesting to see that they, again, can be shown in dependency upon the two characteristic features of planning & control in supply chains that were already showcased in Figure 4.5.1.1, that is, *facility layout* and *product variety concept*. The colored areas indicate the areas of application of the underlying basic MRP II concept (black) and the extended concepts mentioned above (blue, green, gold).

A rough-cut comparison of Figure 4.5.3.1 with Figure 4.5.1.1 shows that the different concepts for planning & control — in a *first approximation* — can be applied to the production types in the following way:

- The MRP II / ERP concepts are well suited to batch production for all facility layouts with the exception of continuous production.

- The lean/JIT concept applies to nearly all production types. It is a prerequisite for mass customization and repetitive manufacturing. However, the Kanban technique and other simple techniques for repetitive manufacturing often linked with Lean/JIT are applicable only to standard products or product families with few variants.

- The variant-oriented concept applies to batch production and all facility layouts designed for single items or small batches. It is a prerequisite for one-of-a-kind production and mass customization.

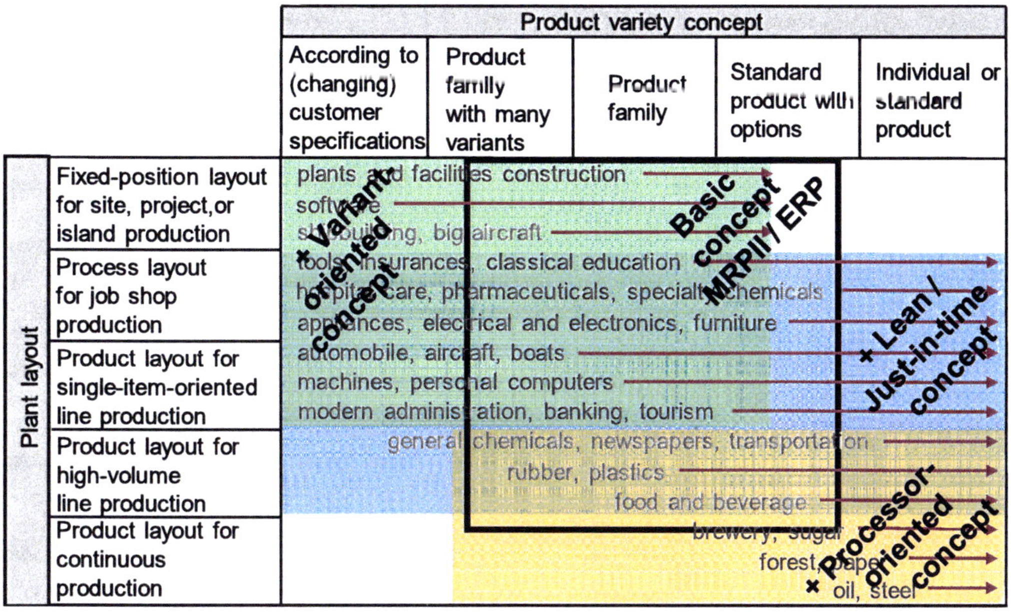

Fig. 4.5.3.1 Different concepts of planning & control in dependency upon the features *facility layout* and *product variety concept*.

- The processor-oriented concept applies to continuous or (discontinuous) high-volume line production, in particular to mass production.

By the way: The process categories in the SCOR model are differentiated according to the *production environment* (refer to Figure 1.1.8.3 and to the definitions in Section 4.4.3). According to Figure 4.4.5.2, there is a close correlation between the two features *production environment* and *product variety concept*. Therefore, the same characteristic feature that now allows differentiation among the various concepts of planning & control already differentiates process categories of the SCOR model.

4.5.4 Selecting an Appropriate Branch Model, Production Type, and Concept for Planning & Control

Intended learning outcomes: Identify the concept of branch model in logistics and operations management. Explain why it is not possible to identify simply a branch model with a specific production type and a specific concept of planning & control.

As the figures in the previous sections show, the branch of industry or service is indeed related to many of the characteristic features of planning & control. *Facility layout* and the *product variety concept* thus prove to be the most important features with regard to the pragmatic development of concepts for planning & control.

A *branch model* in logistics and operations management encompasses concepts appropriate to specific branches, including suitable types of business processes and business methods.

Do such branch models really exist? Let us take as an example the company ABB Turbo Systems (www.abb.com/turbocharging) near Zurich, Switzerland. ABB produces turbochargers for ship motors, each unit according to customer order. A turbocharger is *de facto* a machine with high value-added. ABB produces many production structure levels in-house. What we find is that the application of a unique production type or a unique concept for planning & control would lead to problems in many domains of the enterprise operations:

The main business is the sale of customized machines with multiple variants. The appropriate production types are one-of-a-kind production and — from a systems capability viewpoint — single-item-oriented line production. Thus, the variant-oriented concept has to be applied for planning & control.

Many components and semifinished goods are variant independent and can be produced for a large span of the value-adding chain independent of any customer order; that is, make-to-stock, with frequent order repetition. The appropriate production type is batch production, or — from a systems capability viewpoint — job shop production. The appropriate concept for planning & control can be a simple pull principle (reorder after consumption), which is listed in the figure above under the Lean/JIT concept.

The service parts business, finally, is considered to be just as important as the main business, and this with reason. There, characteristic features are important, such, as for example, backtracking down the history of the machine configuration to the one used for the original production order. The availability of service parts stands in the foreground. The service parts are often just one production structure level above the components and semifinished goods for the main business. But, in contrast to those, the consumption of service parts is lumpy. Thus, the simple pull principle cannot be applied for planning & control. MRP, or the time-phased order point technique of the MRP II concept, based on appropriate forecasting techniques, can be used here. Again, job shop and small batch production is an appropriate production type.

This example clearly illustrates that it is not possible to simply identify a branch model with a specific production type and a specific concept of planning & control. Generally, several production types and concepts for planning & control have to be implemented in parallel in a given company. Vice versa, a specific production type or concept of planning & control is generally valid in different branches. This is one of the reasons why researchers and professionals emphasize the standardization of these production types and concepts of planning & control rather than encourage branch models. Of course, for a given branch, it can be useful to adapt some of the terminology to the common usage in that branch, as well as to further develop the general planning & control techniques with a view to the specific needs and terminology of that branch.

Chapter 9 will present a similar discussion with regard to MRP II and ERP software. At present, there seems to be no simple software available that covers all kinds of production types or concepts for planning & control. Moreover, simple reorder for the components after consumption can be controlled by the Kanban technique (see Section 6.3), which in the eyes of many professionals requires no software at all. As is the case for the underlying production types and concepts for planning & control, a specific MRP II / ERP software package — such as SAP R/3 — can generally be used by different branches. Again, branch packages are available — for example, for furniture production — where specific techniques are implemented in a "branch-customized" way, using branch-customary terminology and graphical user interfaces that represent familiar business objects in the branch.

4.6 Summary

The fundamental elements of process management are based on the terms *work*, *task*, *function*, and *process*. The event is a special process that determines the states of goods. A logistics system encompasses both a process and the order and process management connected with it. Together with business processes, logistics systems form the focus of the logistics perspective. Whether linked or integrated, business processes result in the characteristic pattern. Push logistics are distinguished from pull logistics. Temporal synchronization between manufacturer and user is realized by means of inventory control processes. In the network of customer orders, production orders and procurement orders, inventory, and lead times are the classic design elements of logistics.

Instruments of logistics analysis make up business process analyses in differing degrees of detail. The organization-oriented process chart is an old method that corresponds closely to the way people naturally view the processes. Layouts of the production infrastructure are useful aids to visualizing restrictions and new possibilities. The detailed analysis of the process plan — the basic process analysis — finally allows more precise mapping of the facts and thus helps to qualify the natural view of the processes held by the persons involved.

Logistics analysis works out a characteristic for the planning & control of each product or product family. For each company within a supply chain, a company-internal analysis must be carried out, and final comparisons will reveal areas for potential improvement. Comparison of the findings within and across companies shows potential hindrances to effective logistics, both within the company and in the supply chain as a whole.

By establishing the characteristic features in planning & control, we have already gained indications for appropriate business methods. The characteristic features can also be seen as influences on logistics performance indicators. The chapter discusses six features referring to the user and the product or product family, five features in reference to logistics and production resources, and seven features of the production or procurement order.

Using the three morphological schemes describing features of transcorporate logistics in supply chains, we can obtain an overview of the current state and the specific type of the supply chain and gain some insights into the appropriateness of transcorporate methods and concepts.

Fundamental concepts in logistics and operations management can be distinguished within a matrix of two dimensions: the product variety concept and the facility layout. A first example showed the branches in dependency upon these two characteristics, in a first approximation. The second example, as an extension to the production types already defined by the facility layout, positions additional production types — mass production, repetitive manufacturing, batch production, mass customization, and one-of-a-kind production — using the additional dimension of the product variety concept. The third example positioned four different concepts for planning & control within the matrix. Each comprises particular types of business processes and business methods for order planning and fulfillment: the basic MRP II / ERP concept, and — as extensions — the lean/JIT concept, the variant-oriented concept, and the processor-oriented concept.

4.7 Keywords

assemble-to-order, 207
assembly line, 210
basic process analysis, 198
batch production, 224
batch size, 214
blanket order, 213
branch model, 221
business process engineering, 193
consigned stocks, 207
continuous demand, 203
continuous production, 210
core competency, 185
core process, 185
customer production order, 212
dedicated line, 210
depth of product structure, 202
discontinuous demand, 203
discrete manufacturing, 202
divergent product structure, 202
engineer-to-order, 207
facility layout, 208
fixed-position layout, 208
fixed-position manufacturing, 208
flexible capability of capacity, 210
flexible capability of the production infrastructure, 211
flexible capacity, 211
flexible workforce, 211
flow manufacturing, 210
forecast order, 212
high-volume line production, 209
island production, 208
job shop production, 209
large batch production, 214
layout, 196
line production, 210
logistics system, 185
lot size one, 214
lot traceability, 214
make-to-order, 207
make-to-stock, 206
mass customization, 224
mass production, 224
one-of-a-kind production, 224
order quantity, 214
process layout, 209
process map, 194
process store, 187
product layout, 209
product life cycle, 214
product variety concept, 204
production environment, 206
production line, 210
production without lots, 214
project manufacturing, 208
pull logistics, 189
push logistics, 190
reason for order release, 212
regular demand, 203
repetitive manufacturing, 224
replenishment order, 212
single-item production, 214
single-item-oriented line production, 209
site production, 208
small batch production, 214
standard product, 204
state, 185
stock replenishment order, 212
unique demand, 203
unit cost, 205
value stream mapping, 199
vendor-owned inventory, 207

4.8 Scenarios and Exercises

Intended learning outcomes: Differentiate between various concepts for planning & control. Using process charts for synchronization between use and manufacturing. Elaborate a basic process analysis as well as manufacturing processes in the company-internal layout.

4.8.1 Concepts for Planning & Control within the Company

Intended learning outcomes: Identify different companies and their products where different concepts would be adequate for planning & control. Disclose the branch of industry the company is in and the production type(s) the company does implement.

a. Figure 4.5.3.1 showed different concepts for planning & control in dependency upon the features *facility layout* and *product variety concept*. Using the Internet, try to find three different companies together with their products or product families, where (1) the lean / just-in-time concept, (2) the variant-oriented concept, and (3) the processor-oriented concept would be adequate for planning & control. As you browse the companies' Web sites, try to base your reasoning on *facility layout* and *product variety concept*.

b. For the three companies that you found in (a), what branch of industry, as shown in Figure 4.5.1.1, is the company in? What production type(s), as shown in Figure 4.5.2.1, does the company implement for these products or product families? In reference to the discussion in Section 4.5.4, try to decide whether these companies implement in parallel several production types and concepts for planning & control.

Present your findings for group discussion.

4.8.2 Synchronization between Use and Manufacturing with Inventory Control Processes

Intended learning outcomes: Explain decoupling of procurement / production from demand.

Figure 4.2.3.1 discussed stocking with different inventory control processes for temporal synchronization between supply and demand. Using that kind of process chart, represent decoupling of procurement or production from demand for the two examples discussed with Figure 4.4.5.3, namely:

1. Manufacturing throughout the year to meet demand occurring mainly at a holiday time
2. Manufacturing in large batches where demand is continuous; delivery can be made from storage

4.8.3 Basic Process Analysis and Manufacturing Processes in the Company-Internal Layout

Intended learning outcomes: Describe the basic process analysis using the example of an aluminum frame. Present the company-internal layout using this example process.

Figure 4.8.3.1 shows the company-internal layout at Pedal Works Company, a bicycle manufacturer.

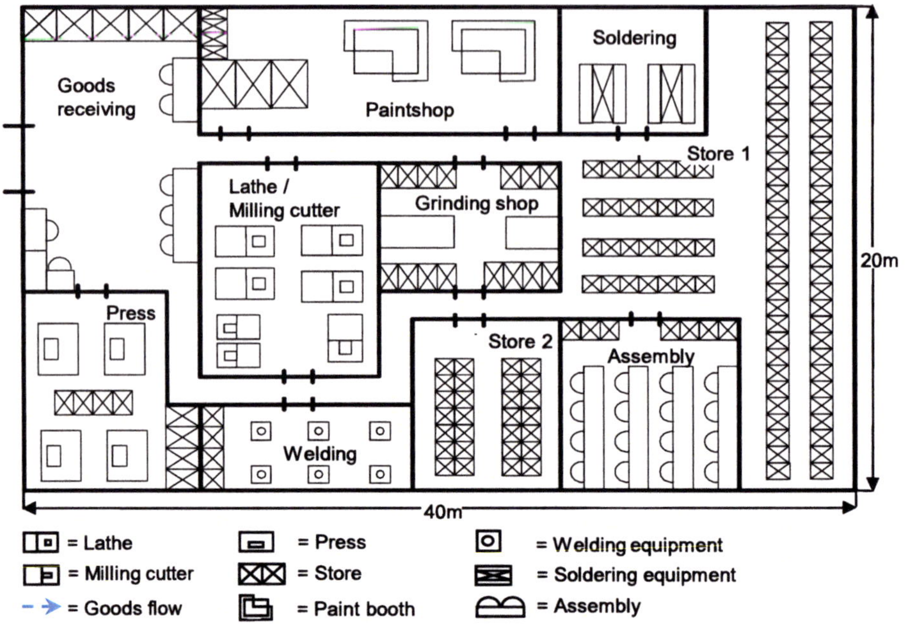

Fig. 4.8.3.1 Company-internal layout with an example process.

Based on the basic process analysis shown in Figure 4.8.3.2, pencil in the paths an aluminum bicycle frame will take through the layout.

Solution:

Compare your results to the result in Figure 4.3.2.1.

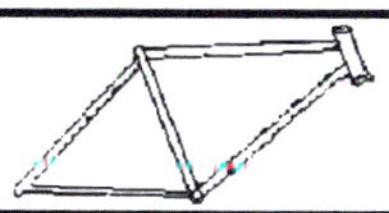

Process ID	451	Batch size	20
Part name	Frame	Part ID	ABC-123
Material	AC-2		
Inspector	Smith	Inspect. date	2020.06.15

Quantity	Distance	Time	Symbol	Process (place)	Operator	Machine	Type of storage	Operating conditions, developments, etc.
		5 days	▼	Store 1				
60 fittings	40 m		⬇		Transporter			
		3 h	▼	Pressing			Pallet on ground	No 100% control; 20% parts defective
		20 s	●	Pressing	Operator	Press		Tool change takes 40 minutes
		20 min	✸	Pressing			Pallet	
	25 m		⬇		Transporter			
		3 h	▼	Milling				
		20 min	●	Milling	Operator	Milling cutter		
		20 min	✸	Milling				
	20 m		⬇		Transporter			
		16 h	▼	Store 2				
		2 min	●	Store 2				
		20 min	✸	Store 2				
20 frames	5 m		⬇		Transporter			
		20 s	●	Grinding	Operator			
		20 min	✸	Grinding			Template	
	20 m		⬇		Transporter			
		2 h	▼	Welding				
		10 min	●	Welding	Welder	Welding tool		
		3.5 h	✸	Welding	Welder			
		2 min	■	Welding				Random sampling inspection of welding joints
	30 m		⬇		Transporter			
		3 min	●	Washing				
		1 min	■					Control if frames clean
		20 min	✸	Washing				
		3 h	▼	Paintshop	Painter			
		30 min	●	Paintshop	Painter	Oven		
		6 h	✸					
	30 m		⬇		Transporter			
		3 h	▼	Assembly				
		10 min	●	Assembly				

✸ = batch-size-dependent wait time ⬇ = transport

● = process ▼ = wait time ■ = control

Fig. 4.8.3.2 Basic process analysis of an aluminum frame

4.9 References

Albe95 Alberti, G., "Erarbeitung von optimalen Produktionsinfrastrukturen und Planungs- und Steuerungssystemen," Diss. ETH Zurich, No. 11841, 1996

APIC16 Pittman, P. et al., APICS Dictionary, 15th Edition, APICS, Chicago, IL, 2016

Dave93 Davenport, T., "Process Innovation — Reengineering Work Through Information Technology," Harvard Business School Press, Cambridge, MA, 1993

FIR97b Forschungsinstitut für Rationalisierung FIR, „Logistik-Richtwerte für den zwischenbetrieblichen Vergleich", Eigenverlag, Aachen, 1. Auflage, 1997

Foga09 Fogarty, D.W., "Production Inventory Management," 3rd Edition, South Western Publishing, 2009

Grul28 Grull, W., "Die Organisation von Fabrikbetrieben," Gloeckners Handelsbücherei BD.11/12, 3. Auflage, Gloeckner, Leipzig, 1928

HaCh06 Hammer, M., Champy, J., "Reengineering the Corporation — A Manifesto for Business Revolution," Harvard Paperbacks, Cambridge, MA, 2006

Hack89 Hackstein, R., "Produktionsplanung und -steuerung (PPS) — Ein Handbuch für die Praxis," 2. Auflage, Düsseldorf, 1989

Hart04 Hartel, I., "Virtuelle Servicekooperationen — Management von Dienstleistungen in der Investitionsgüterindustrie," BWI-Reihe Forschungsberichte für die Unternehmenspraxis, vdf Hochschulverlag, Zurich, 2004

Hieb02 Hieber, R., "Supply Chain Management — A Collaborative Performance Measurement Approach," vdf Hochschulverlag, Zurich, 2002

LuEv01 Luczak, H., Eversheim, W., "Produktionsplanung und -steuerung — Grundlagen, Anwendungen und Konzepte," 2nd Edition, Springer, Berlin, 2001

PtSc03 Ptak, C.A., Schragenheim, E., "ERP: Tools, Techniques, and Applications for Integrating the Supply Chains," 2nd Edition, St. Lucie Press, Boca Raton, FL, 2003

Senn04 Sennheiser, A., "Determinant Based Selection of Benchmarking Partners and Logistics Performance Indicators," Diss. ETH Zurich, No. 15650, 2004

Shin89 Shingo, S., "A Study of the Toyota Production System from an Industrial Engineering Viewpoint," Revised Edition, Productivity Press, Cambridge, MA, 1989

Ulic11 Ulich, E., "Arbeitspsychologie," 7th Edition, vdf Hochschulverlag, Zurich, 2011

VoBe18 Jacobs, R., Berry, W., Whybark, D.C., Vollmann, T.E., "Manufacturing Planning and Control for Supply Chain Management — The CPIM Reference," 2nd Edition, McGraw-Hill, New York, 2018

Wigh95 Wight, O.W., "MRP II: Unlocking America's Productivity Potential," Revised Edition, Oliver Wight, New York, 1995

5 The MRP II / ERP Concept: Business Processes and Methods

Intended learning outcomes: Disclose business processes and tasks in planning & control. Explain master planning or long-term planning, Describe detailed planning and execution.

Section 4.5 introduced different concepts of planning & control for logistics systems. This chapter takes a closer look at the first and fundamental concept, the MRP II / ERP concept.

Enterprise resources planning (ERP) is a framework for organizing, defining, and standardizing the business processes necessary to effectively plan and control an organization so the organization can use its internal knowledge to seek external advantage ([APIC16]).

Beyond the MRP II concept and other concepts of planning & control for logistics systems, the ERP concept deals with financial management, controlling, and human resource management. In this book, however, these themes are only considered marginally.

Section 5.1 presents definitions of the different tasks within these processes and then derives a reference model for business processes and tasks in planning & control. The chapters in Part C (Chapters 10 through 17) of the book follow the structure of this model.

Section 5.2 shows business objects and business methods in the business process of long-term planning. Long-term planning is a fundamental requirement of long-term business relations in the supply chain. In most cases, long-term planning is rough-cut planning.

Section 5.3 presents an overview of business methods for medium and short-term planning & control in the areas of distribution, production, and sales. These business methods will be examined in more depth in later chapters.

Section 5.4 treats business methods for planning & control in R&D.

5.1 Business Processes and Tasks in Planning & Control

Intended learning outcomes: Describe the MRP II concept and its planning hierarchy. Explain the processes and tasks in long-term, medium-term, and short-term planning & control. Present the reference model of processes and tasks in planning & control. Produce an overview on DRP II, integrated resource management, and the theory of constraints.

5.1.1 The MRP II Concept and Its Planning Hierarchy

Intended learning outcomes: Explain the business processes in logistics and operations management of an enterprise, structured according to temporal range. Describe the different degrees of detail in planning. Disclose the aim of data management.

P. Schönsleben, *Handbook Integral Logistics Management*,
https://doi.org/10.1007/978-3-662-65625-9_5

The *MRP II concept (manufacturing resource planning)*[1] encompasses a set of processes, methods, and techniques for effective planning of all resources of a manufacturing company (cf. [APIC16]).

Often, design and manufacturing must be planned long before there is customer demand. Therefore, a typical feature of the MRP II concept is the three-level planning according to temporal range as shown in Figure 5.1.1.1.

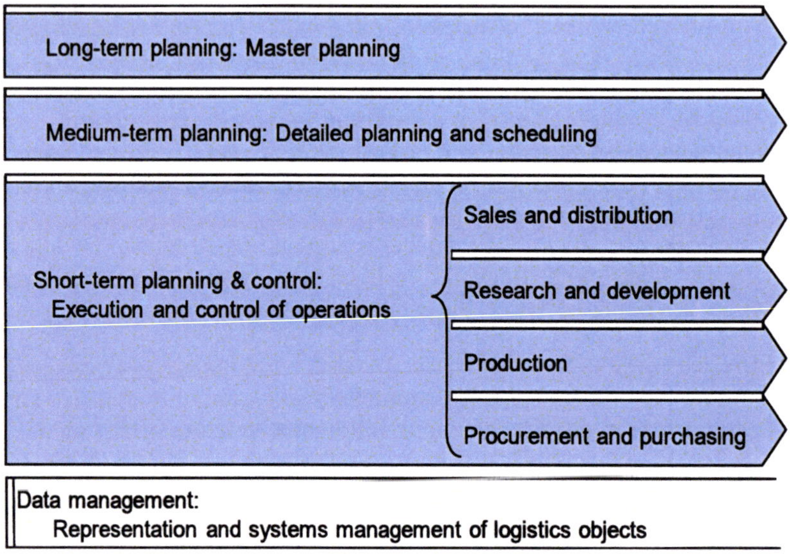

Fig. 5.1.1.1 Business processes in logistics and operations management of an enterprise, structured according to temporal range, with data management.

The aim of *long-term planning* is, firstly, to forecast the total demand for products and processes that will be placed on the enterprise or on the supply chain from the outside, and secondly to derive quantities and gain the resources — persons, production infrastructure, or deliveries from third parties — necessary to fulfill demand.

- *Master planning* is another term used for long-term planning. Both terms emphasize that this type of planning sets the cornerstones for logistics. These cornerstones determine the marginal conditions and limitations of shorter-term planning.

The aim of *medium-term planning* is to forecast demand more precisely along the time axis. Demand must correspond to the resources probably available at certain times. Thereby, sourcing agreements that were reached during long-term planning might have to be precision-tuned or modified.

- *Detailed planning and scheduling* is another name for medium-term planning. It emphasizes the more detailed level of medium-term planning. It often involves only certain areas of production — assembly or parts production, in industry, for example

[1] Important note: The MRP II concept should not be confused with the MRP method of material requirements planning. See Section 5.3.1 and Chapter 12.

— and of procurement. But it may also involve the areas of product and process design — particularly for customer order production.

Short-term planning and control concerns the actual servicing of orders. Within this time horizon also fall capital-intensive investiture in bought goods and value-added from the consumer's view.

- *Execution and control of operations* is another name for short-term planning & control. During execution, the controlled system does yield feedback to the persons controlling the system. Thus, control takes the form of coordination, which is performed by all persons involved. With a view to the company as a sociotechnical system, however, more apt terms are "coordination" or "regulation."

Long-term and medium-term planning are reviewed cyclically or periodically, to adjust planning to the changing estimates of demand. The planning process should be appropriate for each domain. Particularly short-term planning must take account of the actual flow of goods (e.g., in sales and distribution, R&D, production [shop floor], or purchasing). A natural solution for the three temporal ranges of planning is to distribute the associated tasks among different persons. This ensures that the various perspectives are taken into account in the planning. This can contribute to quality and feasibility of the planning.

The different temporal ranges in planning are not equally important in all supply chains. Although the tasks are basically the same, they vary in content. Thus business processes will also vary. In addition, the degree of detail in planning is not the same as the temporal ranges.

Rough-cut planning refers to rough-cut business objects. *Detailed objects planning* refers to detailed business objects.

Rough-cut planning of goods aids rapid determination of the procurement situation for critical item families. Rough-cut planning is indispensable where there are numerous orders to plan. It allows quick calculation of different variants in long-term planning.

In general, the level of detail of planning increases with decreasing temporal range. While rough-cut planning is usually conducted in long-term planning, short-term planning refers to detailed objects. This is not always the case, however. At least some short-term planning can be conducted in a rough-cut manner. In sales, for example, checking the load on rough-cut work centers and the availability of item families of raw materials allows quick decisions on whether to accept customer order production. Conversely, long-term planning in process industry often refers to detailed objects.[2]

Rough-cut and detailed business objects are also objects of data management. See Section 1.2, in particular Section 1.2.5, and Chapter 17.

Data management ensures that the necessary data on objects is available at all times in a detailed and up-to-date form.

---

[2] The term *fine planning* has been avoided. In practice, it is applied to both short-term and detailed objects planning and has led to confusion and misunderstandings.

5.1.2 Part Processes and Tasks in Long-Term and Medium-Term Planning

Intended learning outcomes: Present long-term planning, also called master planning. Disclose medium-term planning & control, also called detailed planning and scheduling. Produce an overview on sales and operations planning, resource requirements planning, requests for quotations, blanket order processing, and order proposal.

Figure 5.1.2.1 shows the sequence and tasks in *long-term planning* in MEDILS form (for an explanation of MEDILS symbols, see Section 4.1.2).

Definitions of the tasks in Figure 5.1.2.1 follow here. For the methods and techniques used for long-term or master planning, see Section 5.2.

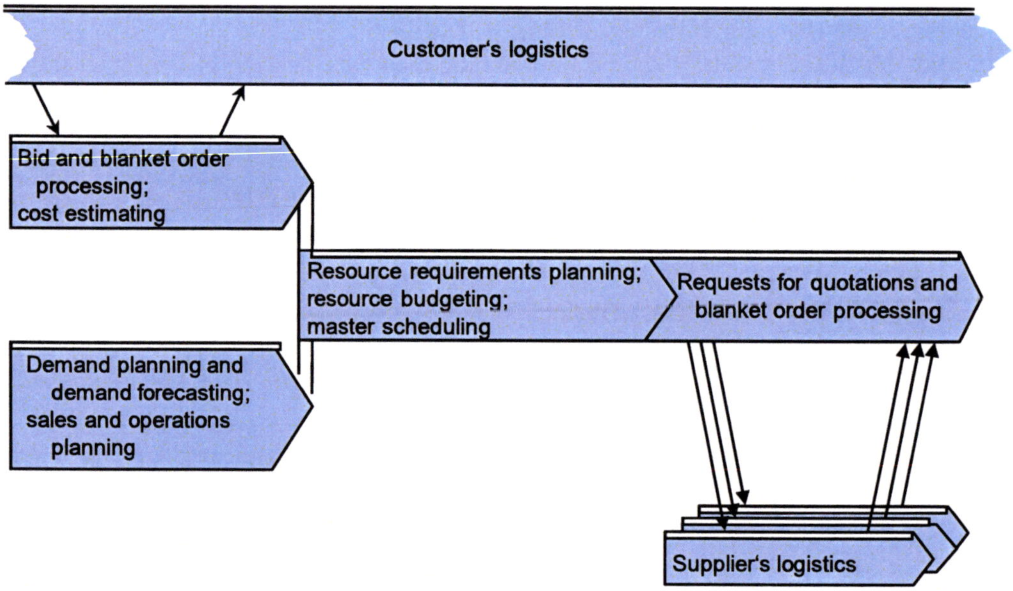

Fig. 5.1.2.1 Long-term planning: master planning.

Bid processing handles a customer request for quotations and determines delivery (labor or product or product family, quantity, and due date). For details, see Section 5.2.1.

A *customer blanket order* determines the scope of delivery. It can then be described by rough-cut business objects, or through product families or rough-cut work centers. In that case, the order due date is only as a time period. For details, see Section 5.2.1.

Demand forecasting, as defined in Section 1.1.6, estimates future demand. *Demand planning* combines forecasting techniques and judgement to estimate demand for products and services along the supply chain (cf. [APIC16]). For details, see Section 5.2.1 and Chapter 10.

Sales and operations planning brings together all the plans for the business (marketing, development, sales, manufacturing / production, sourcing, and financial) in one integrated set of plans. It is performed at least once per month and is reviewed by management at an *aggregate* (product family) level ([APIC16]). For details, see Section 5.2.2.

Resource requirements planning (RRP), or *resource planning*, calculates the requirements for components and capacity (persons and infrastructure), based on the production plan, generally divided up along the time axis, and through analytical explosion of product structures (also called *explosion of bill of material*) and routing sheets. RRP is *gross requirements planning*; inventory and open orders are *not* taken into consideration. For details, see Section 5.2.2.

The output of RRP includes in particular a *procurement plan for components and materials.*

Resource budgeting calculates the procurement (or materials) and capacity budget (direct costs and overheads), and the budget for other overheads. For details, see Section 5.2.2.

Thus, master planning yields the quantities of the resources to be used in the long-term planning horizon and calculates financial implications.

The *planning horizon* is the future time period included in planning.

The planning horizon for master scheduling must be at least as long as the cumulative lead time to manufacture all units in the master schedule (MS). This lead time encompasses production, procurement of all components, and customer-specific design.

Master scheduling is establishing a plan to produce *specific* products or provide *specific* services within a *specific* time period.

See Section 5.2.3. Master scheduling outputs the master production schedule (MPS), a disaggregated version of a production plan, expressed in specific products, configurations, quantities, and dates. The MPS serves as input for rough-cut capacity planning (RCCP).

Requests for quotations and blanket order processing turn over the procurement plan for salable products, components, and materials as well as the requirements for external capacities to suppliers in the supply chain. This task includes supplier selection, call for bids, and the processing of the supplier blanket orders.

For details, see Section 5.2.4. In data management, each blanket order is a business object, an *order* (see Section 1.2.1). If the minimum blanket order quantity on the blanket order is zero, the blanket order is a prediction only.

Figure 5.1.2.2 shows the process and tasks of *medium-term planning* in MEDILS form. The part processes and tasks in medium-term planning are similar to those in long-term planning. Precision-tuning accomplishes more exact determination of bids and blanket orders as well as the schedules (particularly the *production schedule* and the *purchase schedule*, that is, the plan that authorizes the factory to manufacture — or the purchasing department to purchase — certain quantities of specific items within a specific time (cf. [APIC16]).

Detailed resource requirements planning calculates detailed material and components requirements and detailed capacity requirements (persons and infrastructure), divided up along the time axis, usually based on the master production schedule MPS (the disaggregated version of the production plan), divided up along the time axis, and through analytical explosion of detailed product structures and routing sheets. This is *net requirements planning*; inventory and open orders are taken into consideration. Order proposals for R&D, production, and procurement for covering these requirements, are worked out.

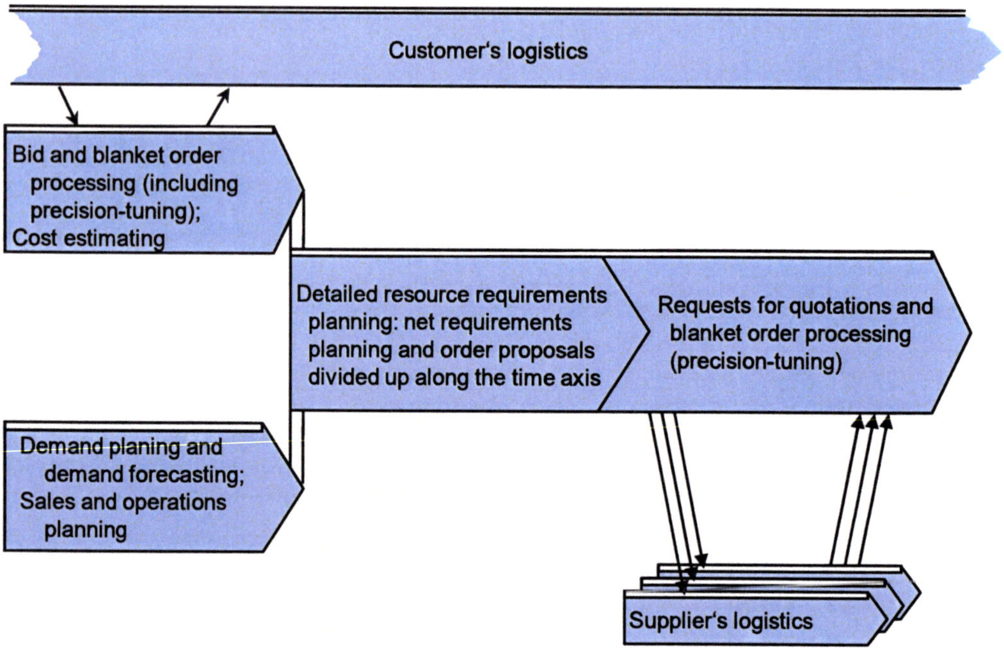

Fig. 5.1.2.2 Medium-term planning & control: detailed planning and scheduling.

An *order proposal*, or *planned order*, sets the goods to be produced or procured, the order quantity, the latest (acceptable) completion date, and — often an implicit given — the earliest (acceptable) start date.

On the basis of the order proposals, blanket order planning can be defined more precisely.

5.1.3 Part Processes and Tasks in Short-Term Planning & Control

Intended learning outcomes: Present short-term planning & control, also called execution and control of operations. Produce an overview on order release, order coordination, order monitoring and order checking, delivery, job-order costing, and billing.

Figure 5.1.3.1 shows, in MEDILS form, part processes and tasks in *short-term planning and control*, or *execution and control of operations*.

The figure shows only one production structure level in the supply chain. The order originates from a sales, production, or procurement department of an internal or external customer. The production structure level itself places orders to suppliers, either components warehouses or lower production structure levels. The second part process can be repeated. For a production order, for example, first all components will be procured. Then all operations can be executed. Orders can be released either separately for each part process or all together. Order coordination can also be repeated (broken arrows in Figure 5.1.3.1).

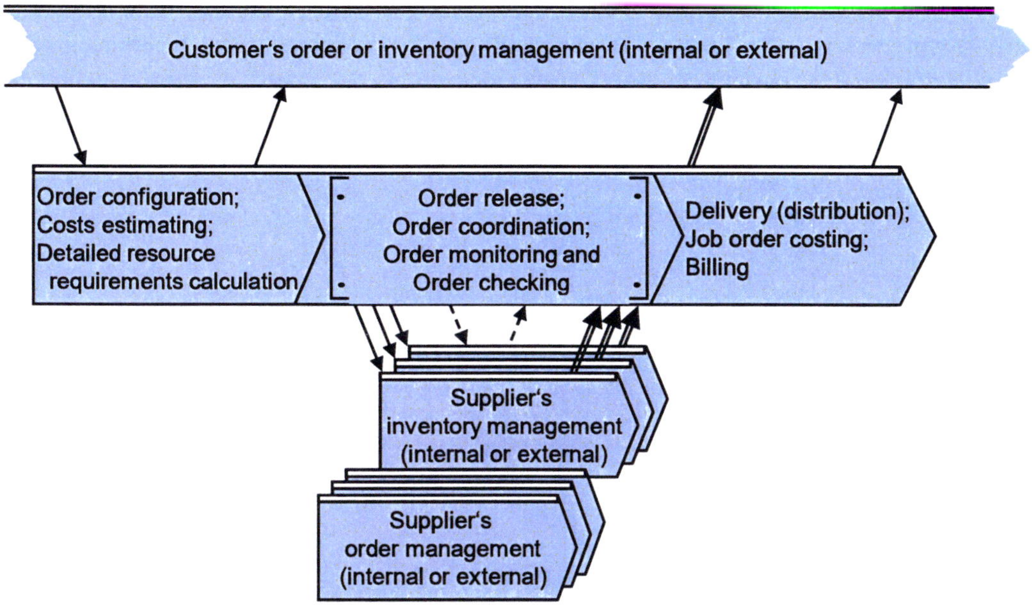

Fig. 5.1.3.1 Short-term planning & control: execution and control of operations.

Order configuration handles an order proposal from medium-term planning or an order from an external or internal customer. It determines delivery (work, or product, quantity, and due date).

Order configuration compares the order to any existing bid or blanket order. In the case of R&D orders, order configuration consists in *planning the volume of the release*. This is part of engineering change control (ECC). See Section 5.4.

Detailed resource requirements calculation calculates, firstly, for an unplanned order, through analytical explosion of product structures and routing sheets (see Sections 1.2.2 and 1.2.3), the detailed requirements of products, material, components and capacity (persons and infrastructure), divided up along the time axis. Secondly, it tests — for a planned or unplanned order — the *availability* of resources; that is, whether the requirements for resources are covered; where necessary, order proposals for covering requirements are worked out.

If resources are not available at the required times, lead time must be increased. Here, techniques such as available-to-promise (ATP) or capable-to-promise (CTP) can be of use. For details refer to Section 5.3.5.

Order release is the decision to execute order proposals or orders originating from higher-level logistics. It produces all administrative documents required for order confirmation, order execution (for example, in production), or for communication with suppliers. Necessary transportation means will also be secured.

A *released order* is a production or procurement order with ongoing production or procurement (in contrast to a planned order).

Order coordination coordinates the order with other connected orders. For example, a customer order may require a development order and several levels of production and procurement orders. These make up further short-term processes of the type shown in Figure 5.1.3.1, arranged in a multilevel cascade. Figure 4.2.1.2 shows a simple example. Normally, there are several levels and, at each level, several parallel part processes to coordinate.

Order monitoring and *order checking*: *Progress checking* monitors execution of all work according to plan in terms of quantity and delivery reliability. (If deviations from the plan are too great, this may lead to recalculation of the rest of the process plan.) *Quality control* means checking the quality of all incoming goods from production and procurement. Quality control has become an extensive process that is based upon specific quality control sheets.

In data management, all types of orders each are business objects, of the *order* class (see Section 1.2.1).

For *delivery*, or *distribution*, the products are issued from stock (*order picking*) and prepared for shipment; the required transportation means and accompanying documents are made available; and delivery is executed.

Job-order costing evaluates data captured by shop floor data collection (that is, mainly resource use).

Billing transmits the results of cost accounting to the customer (e.g. in the form of an invoice) and, where required, adjusts data management's projected values for the business objects.

5.1.4 Reference Model of Processes and Tasks in Planning & Control

Intended learning outcomes: Explain the manufacturing planning & control processes and priority planning within the temporal ranges in the MRP II concept. Describe the reference model of business processes and tasks in planning & control. Differentiate between materials management, time management and scheduling, and capacity management.

Figure 5.1.4.1 summarizes the concepts presented in the previous sections, showing the relation between the planning processes within the temporal ranges. This type of representation is common in teaching materials explaining the MRP II/ERP concept. [3]

Figure 5.1.4.2 presents an overview of the planning processes according to — vertically — temporal range (long, medium, and short term) and — horizontally — all the planning & control tasks. The processes and tasks are shown in the logical temporal sequence that derives from Figures 5.1.2.1, 5.1.2.2, and 5.1.3.1.

Representing the processes in this condensed form allows us to conceive of planning & control tasks as *cross-sectional tasks*. In principle, these tasks appear in the same forms in

[3] *Priority planning* is the function of determining what material is needed and when (cf. [APIC16]).

all time ranges and in all kinds of orders, although there are some deviations. However, a particular task may not arise in every time frame or in every business process. Also, during execution of a particular task, not every logistics object is required in data management. The reference model characterizes the elements of a planning & control system as well as various options for forming part systems — either along the business processes or along equal tasks.

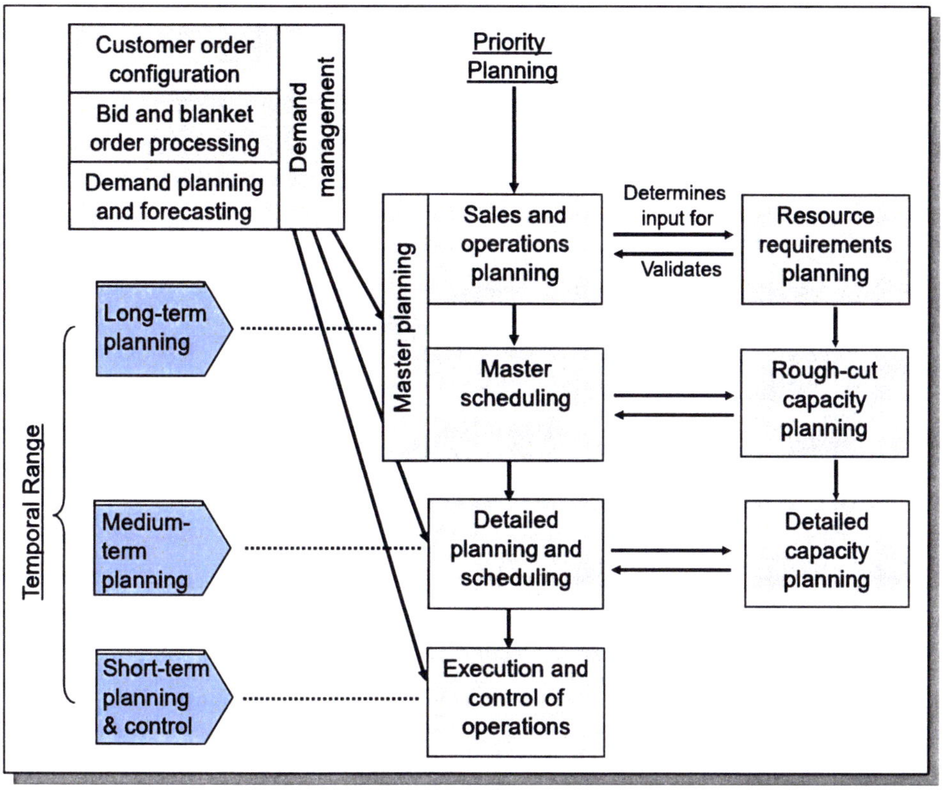

Fig. 5.1.4.1 Manufacturing planning & control processes within the temporal ranges in the MRP II concept.

Demand forecast, stock planning and sales planning, bid processing and order configuration, and cost estimation accord with the definitions in Sections 5.1.2 and. 5.1.3. Furthermore, the model breaks down resource requirements planning into three classical planning & control tasks.

1. *Materials management* ensures that the goods required by demand are provided cost effectively and according to schedule (such as end products, semifinished goods, single parts, raw materials, and information). [4]

[4] Some authors, e.g. [ArCh16], use the term *materials management* in a larger sense, equivalent to the overall concept of Integral logistics management or operations management. However, this book uses the term in the restricted sense, related only to goods and materials.

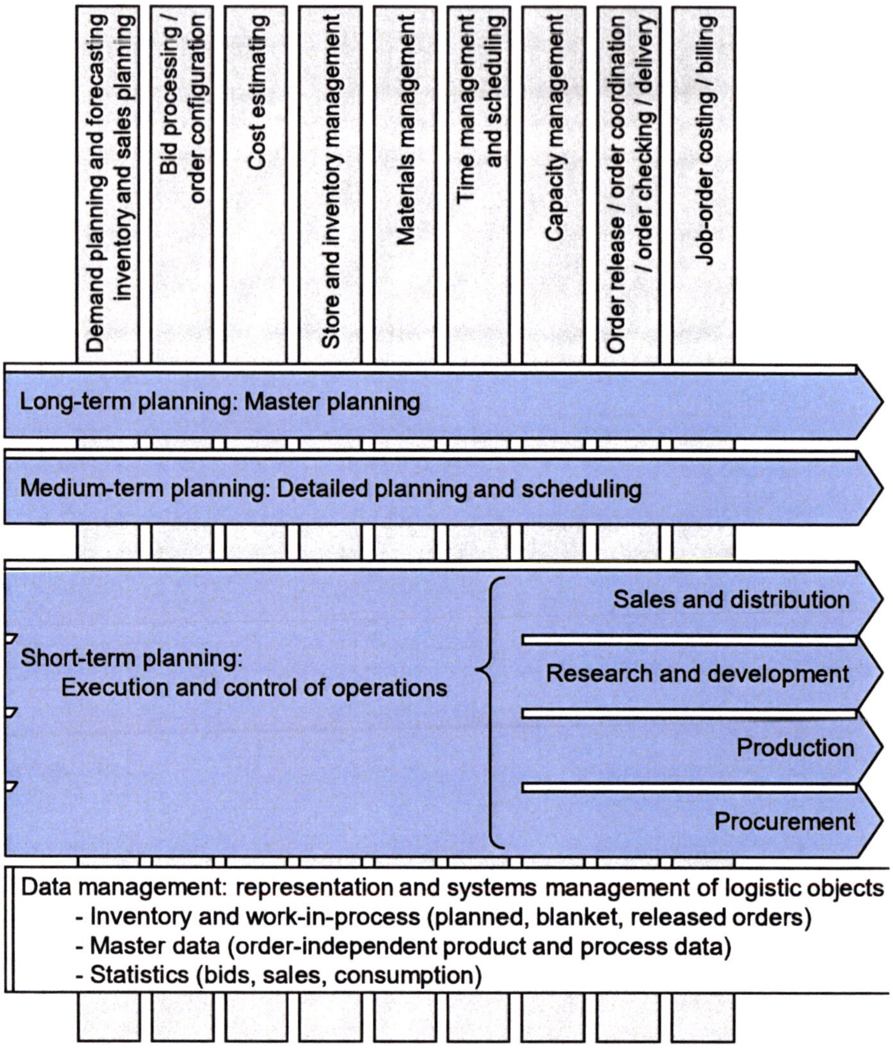

Fig. 5.1.4.2 Reference model of business processes and tasks in planning & control.

2. *Time management and scheduling* and 3. *capacity management* ensure cost-effective and timely provision of the capacities needed to cover the load on persons and the production infrastructure as determined by orders.

The distinction among these planning & control tasks is based on the fact that goods can usually be stocked for a certain period of time (with the exception of continuous production; see Chapter 8), whereas this is generally not the case for time and capacity (see the end of Section 5.3.3). As a natural consequence, the business methods for these resources differ.

The term *management* keynotes the fact that our view has broadened from pure requirements or costs considerations to the more comprehensive task of improving company performance (see Section 1.3.1). *Goods management* would actually be a more appropriate term than materials management, for this task also handles finished products.[5] But as *materials management* is the term commonly used, we retain it here.

Stores management comprises tasks involved in storing goods as well as in handling inventory transactions, such as delivering stocked goods to user sites or receiving goods from suppliers.

Inventory management is the branch of business management concerned with planning and controlling inventories ([APIC16]). It includes all tasks involved in the inventory control process within the supply chain.[6]

Inventory control includes the activities and techniques of maintaining the desired levels of items, such as those shown in Figure 4.2.3.2.

Figure 5.1.4.2 does not show explicitly the process of store and inventory management itself; instead, it defines the *task* in these processes, or *store and inventory management*. In data management, all inventories of stored items or work-in-process items build a business object. Depending on the degree of detail of inventory management, the business object type assigned may be *item* (see Section 1.2.2) or *order* (see Section 1.2.1). Figure 5.1.4.2 also introduces two further categories of objects in data management.

The term *master data* comprises the data of all order-independent business objects, as described in Section 1.2.

The term *statistics* refers to appropriately combined data on consumption, as well as data on bid and sales activity.

The values of performance indicators in Section 4.2 can be derived from certain statistical data. See here Section 11.2. For an in-depth description of master data, see Chapter 17.

Order release, order coordination, order checking, and delivery correspond to the definitions in Section 5.1.3. As we examine business processes and methods in more detail in the following, we will refer again and again to the reference model on planning & control shown in Figure 5.1.4.2. Moreover, the reference model serves as the starting point for a detailed look at the individual planning & control tasks in Chapters 10 through 17.

5.1.5 Beyond MRP II: DRP II, Integrated Resource Management, and the "Theory of Constraints"

Intended learning outcomes: Produce an overview on distribution resource planning (DRP II). Disclose the concept of (integrated) resource management. Explain bottleneck capacity as well as the five focusing steps in the theory of constraints (TOC) approach.

---

5 In accounting, *materials* relates to purchased and production materials rather than to semifinished and end products. See also the distinction between the terms *materials* and *components* in Section 1.1.1.

6 Some authors, for example, [Bern99], use the term *inventory management* in a larger sense, equivalent to the overall concept of Integral logistics management or operations management. However, in this book, the term is used in the restricted sense, related only to goods and materials.

Distribution planning is part of ERP, but goes beyond MRP II. It covers the planning activities associated with site and location planning, transportation, warehousing, inventory levels, materials handling, order administration, industrial packaging, data processing, and communications networks to support distribution ([APIC16], sequence of tasks modified).

Distribution planning determines the distribution network structure, often a multi-echelon structure (see Section 3.1.3). Inventory management in this chain can be handled in principle in the same way that it is for the chain from raw material to final product, via the various structure levels. A central task of distribution planning is resource management in the distribution system, in particular inventory management.

Distribution inventory is inventory, usually spare parts and finished goods, located in the distribution system (e.g., in warehouses and in-transit between warehouses and the consumer [APIC16]). The terms *pipeline inventory*, or *pipeline stock*, are used simultaneously.[7]

Distribution resource planning (DRP II) is distribution planning of the key resources contained in a distribution system: warehouse space, workforce, money, trucks, freight cars, etc. ([APIC16]).

The term *DRP II* developed as an extension of *DRP* (*distribution requirements planning*; see Section 12.2.1), which stands for a method of *management of distribution inventory*. The term *DRP II* was coined in analogy to the term *MRP II*, an extension of MRP. The techniques of management of distribution inventory do *not* differ essentially from those for production and procurement inventory. Thus, they will not be treated in a separate section. However, distribution control is examined in Section 15.4. There you will find a description of important distribution planning tasks and results, such as transport planning and scheduling.

Resource management is, according to [APIC16], the effective identification, planning, scheduling, execution, and control of *all* organizational resources to produce a good or service. *Integrated resource management* is used as a synonymous term.

Today, the ordered sequence of the three classical tasks as shown in Figure 5.1.4.2 — materials management, time management and scheduling, and capacity management — is used mainly for teaching purposes only. This sequence came about because the — original — MRP II concept is about *material-dominated scheduling*. In Figure 5.1.4.1 these are the tasks of the column *priority planning*. In the classical MRP II concept, there is no routing sheet. There exists only the attribute lead-time offset, which is assigned to each item. This perspective also made concessions to the very limited processing capacity of computers of the day, when the materials management planning process of large firms (the so-called MRP run) often took an entire weekend. It took that much time again to then complete the planning process for scheduling and capacity management (the so-called CRP run). This meant that it had to be possible to perform this process separately from materials management.

When capacities are important, however, all tasks must be performed in an integrated way as resource management, in parallel fashion, in dependency upon each other. This is the case

7 In contrast to these terms, *in-transit inventory*, or *transportation inventory*, is limited to inventory that is moving between locations.

with more recent concepts, including the lean/JIT, the variant- and the processor-oriented concept, as well as advanced planning and scheduling (APS) software. Moreover, limitations on computer capacity no longer exist. There is also another impetus for the (integrated) resource management approach, namely, the more in-depth consideration of throughput and bottlenecks and — finally and more comprehensively — the theory of constraints:

Throughput is the rate at which the (production) system generates a desired output. It is expressed for a given time period (cf. [APIC16]).

A *bottleneck*, or a *bottleneck capacity*, is a work center where the required capacity is greater than the available capacity. Cf. [APIC16].

A high capacity utilization tends to result in lower costs (see here Fig. 1.4.3.4). However, well-utilized capacity can also represent a bottleneck. Whenever capacity is not available to work, it directly reduces the throughput of the company and thus its output, its performance. Therefore, effective *bottleneck management* (and also the TOC approach) proposes:

- Utilization of the bottlenecked work center during breaks and with the greatest possible overtime. In addition, buffer stores, both downstream and upstream of the work center, should buffer the bottlenecked work center. On the one hand, this allows maximum utilization, because the bottlenecked work center does not have to wait for delayed delivery of materials. On the other hand, if downtime occurs in the bottlenecked work center, this will not directly affect the fill rate. In addition, through some increased administrative effort, various customer orders for the same item can be produced together at the bottleneck, which increases batch size, so that machine setup time and thus load are reduced.
- Production at non-bottlenecked work centers only when there are actual customer orders. Work centers should not make to stock — thus keeping work in process low. The reason for this is that too-early order releases do not improve capacity utilization; as a result, the work center simply does not work at a later time. In addition, goods will pile up that are not immediately required, implying carrying cost.

The *theory of constraints (TOC)* is an approach to integrated resource management that addresses the problem of bottlenecks in a logistics system, or — more generally — the factors that limit or constrain the throughput in the system.

The TOC was developed in the 1980s and early 1990s in North America by E. M. Goldratt ([GoCo14]). The basic premise of a theory views the planning problem in logistics and operations management as a solution space limited by constraints.

A *constraint* is any element or factor that prevents a system from achieving a higher level of performance with respect to its objective ([APIC16]).

Constraints can take the form of limited capacity, a customer requirement such as quantity or due date, or the availability of a material, for example. They can also be managerial.

The concept of a solution space limited by constraints originated in operations research, which also supplies algorithms for solutions. However, the difficulty often does not lie in the algorithms, but rather in the solution space itself, which may not allow for reasonable solutions. Here, TOC attempts to exploit and expand the solution space, successively and in targeted fashion, according to the *five focusing steps* shown in Figure 5.1.5.1. A production control technique that goes with TOC is called *drum-buffer-rope*.[8] See Section 14.3.3.

1. Identify the most serious constraint — that is, the constraint that is unduly constraining the solution space. This can be a bottleneck, for example.
2. Exploit the constraint: For example, a bottleneck capacity should be utilized during breaks by rotating crews so that the capacity is never idle.
3. Subordinate everything to the constraint: For example, good utilization of other than bottleneck capacities is secondary.
4. Elevate the constraint: Make capacity available, for example.
5. Return to step one — that is, to the next iteration.

Fig. 5.1.5.1 The five focusing steps in the theory of constraints (TOC) approach.

5.2 Master Planning — Long-Term Planning

Intended learning outcomes: Describe demand management, sales and operations planning as well as resource requirements planning. Explain master scheduling and rough-cut capacity planning. Disclose blanket order processing, release, and coordination.

This section highlights in depth the long-term business process in planning & control, that is long-term planning, or master planning. One the one hand, long-term planning is important due to its temporal range. On the other hand, presenting the long-term business process here makes it possible to explain how the planning & control tasks in Figure 5.1.4.2 act together without having to resort to an overly complex method of presenting the material. This section contains detailed information on the different tasks presented in Figure 5.1.2.1.

5.2.1 Demand Management: Bid and Customer Blanket Order Processing and Demand Forecasting

Intended learning outcomes: Describe demand management, customer bid, order success probability and customer blanket order. Present some aspects of demand forecasting.

Demand management is, according to [APIC16], the function of recognizing all demands for goods and services to support the marketplace.

According to Figure 5.1.4.1, this task comprises — among others — the following part task and processes of long, middle, and short-term planning (see Section 5.1):

[8] Drum–buffer–rope represents a synchronized production control approach. *Synchronized production* is a manufacturing management philosophy that includes a consistent set of principles and techniques supporting the global objective of the system (cf. [APIC16]).

- Bid and blanket order processing
- Demand planning and forecasting
- Order entry and order configuration

A *customer order* is a deterministic independent demand. Quantity, due date, and other facts are completely known. *(Customer) order service*, according to [APIC16], encompasses order receiving, entry,[9] configuration, and confirmation of orders from customers, distribution centers, and interplant operations.[10]

Order service is responsible for responding to customer inquiries during delivery lead time as well as for interacting with master scheduling regarding the availability of products. One important factor when scheduling customer demand is the organization's distribution network structure. See Section 3.1.3.

What precedes the status of order confirmation of a customer order are — in the case of investment goods — various bid statuses.

A *customer bid* is a *quotation*, a statement of price, terms of sale, and description of goods or services given to a customer in response to a customer request for quotations.

The bid statuses are of differing duration, during which requirements are defined more and more precisely. In this case, the requirements are not absolutely definitive, but they will guide the planning of production and procurement. For customer order production (often single-item production), there is a certain probability that a bid will lead to an order as it is already defined at this point. The simplest technique of including bids in planning is to multiply the requirements by the probability of their success.

Order success probability devalues the demand defined by the customer bid. Only demand reduced in this way is included as independent demand for resource requirements planning.

Continuous adaptation of order success probability to real conditions, e.g., by continuous measuring of the order success rate, with decreasing temporal range of planning, is crucial to this simple technique. In addition, bids must be confirmed, or removed, early enough that definitive orders can be scheduled even if bottlenecks occur in procurement. For this, an *expiration date* must be assigned to the order, from which time onward the confirmed delivery date may be postponed or the order termed inactive. This function can be automated in an IT-supported system.

If bottlenecks occur in procurement or production, it is difficult to set a reliable delivery date for a bid that is to be planned. If many other bids have been planned, a completion date that has been calculated by placing the new bid in this limited resource situation is only a probable completion date. This date needs to be complemented by a latest (maximum) completion date, calculated on the assumption that all bids, or at least the majority of them, will be realized. To do this, the portion of demand not reserved for each bid on the basis of

---

9 *Order entry* is the translation of the customer order into terms used by the manufacturer or distributor.

10 *Interplant orders* are orders received by another plant or division within the same organization.

order success probability is totaled up and used in the resource requirements management of capacity. The lead time for required but not available components yields the "maximum" completion date for the new bid. While this method, described here only in its rudiments, involves a great deal of complex calculating in detailed planning, it is often an appropriate technique for rough-cut planning with acceptable levels of calculation.

A customer bid often concerns and results in a *customer blanket order*. Here, the delivery quantity is often set by a long-term minimum and maximum blanket order quantity for a particular period of time.

If the minimum blanket order quantity is zero, it is merely a forecast.

- Uncertain quantities in a blanket order can be handled in a way similar to bids, that is, through continuous precision-tuning of their success probability with decreasing temporal range. In short-term planning, a certain quantity is ordered through a short-range blanket order for a defined period of time, but exactly when and in what breakdown the blanket releases will be made is left open.
- For uncertain dates, some additional information is usually available. This information will express, for example, the quantities that will be called for in the future, together with an estimate of the deviation factor in percent. These values allow partial demand to be distributed along the time axis. Here again it is important to continue to adapt the breakdown of the demand to reality or at least to the customer's increasingly precise requests. For more on blanket orders, see Section 5.2.4.

Demand forecasting is, as in Section 1.1.6, the process that estimates the future demand.

Demand forecasting is a necessary process as soon as items upstream from the (customer) order penetration point (OPP) must be procured or produced (see Section 4.4.3). The need for forecasting varies throughout the course of time and depending on the industry, market, and product. Examples of buyers' markets with a great need for forecasting include trade in consumer goods or the provision of components needed for a service or for investment goods. Before a customer places a definitive order, for example, single parts of a machine or "frameworks" containing data descriptions and programs for a software product must have already been produced or procured.

There are simple techniques of forecasting, including those based on judgment and intuition, but there are also some very complicated techniques. Chapter 10 presents many of them.

5.2.2 Sales and Operations Planning and Resource Requirements Planning

Intended learning outcomes: Explain sales and operations planning as an iterative master planning process. Disclose an example of iterative planning by comparing three production plans, with zero, two or four changes in production rhythm per year.

The sales and operations planning process (see definition in Section 5.1.2) is an approach for the tactical planning of a company or a supply chain.

Tactical planning is the process of developing a set of *tactical plans*, e.g., the sales plan, the inventory plan, or the production plan (see [APIC16]).

A *sales plan* is a time-phased statement of expected customer orders anticipated to be received (incoming sales, not outgoing shipments) for each major product family or item.

A sales plan is more than a forecast. It represents sales and marketing management's commitment to achieve this level of customer orders and can be dependent on forecast. It is expressed in units or in gross income, on an aggregate level.

A *production plan* is the agreed-upon plan that comes from the overall level of manufacturing output planned to be produced ([APIC16]). *Production planning* is the process of developing the production plan.

The production plan is usually stated as a monthly rate for each product family. Various units of measure can be used to express the plan: pieces, tonnage, standard hours, number of workers, and so on.

Similarly, a *procurement plan for salable products* is the agreed-upon plan for product families or products to be purchased, that are intended to be sold directly, that is, without being used by the company itself or built as components into products.

Generally, a sales plan does not reflect a steady demand. However, the capacities (workers and production infrastructure) tend to be available at a steady rate. Therefore, if the demand pattern cannot be changed — by offering complementary products or price incentives or simply changing the due dates, for example — there are in principle two possible manufacturing strategies[11] (or a combination of them) to manage supply:

- (Quantitatively) flexible capacity in order to match the demand fluctuations.
- Store products to meet peak demand, while continuing production at a steady rate.

The first option entails so-called costs of changing production rhythm, or production rate change costs. These may include the costs of overtime and undertime, more facilities and equipment, part-time personnel, hiring and releasing employees, subcontracting, or agreements to use infrastructure cooperatively. See the discussion in Section 14.2.3.

The second option incurs — as already discussed in Section 1.1.6 — carrying costs, in particular costs of financing or capital costs, storage infrastructure costs, and depreciation risk. For details, see Section 11.4.1.

An *inventory policy* is a statement of a company's objectives and approach to the management of inventories ([APIC16]).

An inventory policy expresses, e.g., the extent to which either one or both of the above options will be followed. It can include a decision to reduce or increase inventory in general.

[11] A *manufacturing strategy* is a long-term decision on the definition and use of manufacturing resources.

An *inventory plan* determines the desired levels of stored items, mostly end products, according to the company's inventory policy.

The production plan can thus be obtained from the sales plan via the desired inventory plan. Or turned around, a desired production plan implies a corresponding inventory plan. By changing the inventory policy iteratively, a different production plan as well as the corresponding inventory plan (or vice versa) can be obtained.

Once the production plan is established, the process of *resource requirements planning* (RRP) begins. Resource requirements are calculated for each product family in the production plan through simple explosion of product structures (bills of material) for components requirements (dependent demand) and routing sheets for capacity requirements. To do this, the process uses bills of resources or product load profiles (see Figure 1.2.5.2).

If gross requirement for each purchased item calculated in this way is weighted by purchase price, the result is a good approximation that can serve as the procurement budget. Other resource requirements can be estimated analogously. For the planning horizon covered by the production plan, there now result:

- Components requirements, procurement plan for components and materials, and the corresponding procurement or materials budget
- Capacity requirements and the capacity budget (direct and overhead costs)
- Budget for overhead costs (overhead budget)

In the case of rough-cut planning, sales and operations planning produces an *aggregate plan* based mainly on aggregated information (rough-cut business objects such as product families, rough-cut product structures, *aggregate forecast and demand* [that is, forecast and demand on product families]) rather than on detailed product information.

It is in the case of rough-cut planning in particular that long-term planning lends itself well to the simulation and the what-if analysis of several variants of the production plan.[12] For this, company management (or a team caring about supply chain coordination) comes together for a half-day meeting, for example, in order to simulate the various possible patterns of demand and to examine their repercussions with regard to the realization of production and procurement in the supply chain. As some components or operations have not been considered, the budgets can by multiplied by historical figures to obtain expected budgets. In a similar process, sensitivity analysis can take into consideration the effect of demand variation and thus control the whole process with regard to feasibility. Management will then choose and release one of the variants calculated in the above manner and initiate the necessary measures to fulfill the production plan in a timely fashion. For capacity, blanket orders can be given to external production, and orders can be made for the purchase

[12] A *simulation* is a model-based reproduction of various conditions that are likely to occur in the actual performance of a system. A *what-if analysis* is the evaluation of the consequences of alternative strategies, e.g., of changes of forecasts, inventory levels, or production plans.

of machinery and buildings or the acquisition of personnel. To procure goods or capacity, blanket orders can be placed with suppliers, or existing supply agreements can be modified.

Figure 5.2.2.1 shows a typical algorithm used within sales and operations planning to determine the production plan and the procurement plan for salable products. It accords with the concept of integrated resource management: All resources are planned simultaneously.

1. *Sales plan:* Determine forecast or demand pattern.
2. *Production plan, procurement plan for salable products, and inventory plan:* Set inventory policy with regard to change of production rhythm and inventory level. Determine the inventory levels and calculate the corresponding production plan (analogically, the procurement plan for salable products) or vice versa.
3. *Resource requirements planning and budgeting:* Calculate the procurement budget for components and materials, the capacity budget, and overhead costs budget. Take into account macro costs due to change of production rhythm and inventory.
4. Compare budget figures with actual possible realization and, if necessary, begin again with steps 1, 2, and 3 for each desired variation.

Fig. 5.2.2.1 Iterative master planning: integrated resource management.

As mentioned above, this technique usually handles rough-cut business objects of the type discussed in Section 1.2.5, so that various iterations can be calculated rapidly. Resource requirements planning of this kind (rolling planning) must be repeated regularly (for example, monthly), and must include the whole planning horizon.

Figures 5.2.2.2 through 5.2.2.4 illustrate iterative planning of this kind. Using forecasted sales figures, the objective is to produce an optimal production plan. To estimate the consequences of different manufacturing strategies for total production, different variants are calculated. Thus, only steps 2 and 3 of the steps shown in Figure 5.2.2.1 are iterated.

Month	Sales		Production		Inventory
	monthly	cumulative	monthly	cumulative	at end of month
December					200
January	500	500	1000	1000	700
February	600	1100	1000	2000	1100
March	600	1700	1000	3000	1500
April	800	2500	1000	4000	1700
May	900	3400	1000	5000	1800
June	1000	4400	1000	6000	1800
July	600	5000	1000	7000	2200
August	400	5400	1000	8000	2800
September	600	6000	1000	9000	3200
October	600	6600	1000	10000	3600
November	1800	8400	1000	11000	2800
December	3000	11400	1000	12000	800

Fig. 5.2.2.2 Plan 1: production plan at a constant level.

Month	Sales		Production		Inventory
	monthly	cumulative	monthly	cumulative	at end of month
December					200
January	500	500	600	600	300
February	600	1100	600	1200	300
March	600	1700	600	1800	300
April	800	2500	900	2700	400
May	900	3400	900	3600	400
June	1000	4400	900	4500	300
July	600	5000	600	5100	300
August	400	5400	600	5700	500
September	600	6000	600	6300	500
October	600	6600	1900	8200	1800
November	1800	8400	1900	10100	1900
December	3000	11400	1900	12000	800

Fig. 5.2.2.3 Plan 2: production plan with four changes in production rhythm per year.

Month	Sales		Production		Inventory
	monthly	cumulative	monthly	cumulative	at end of month
December					200
January	500	500	800	800	500
February	600	1100	800	1600	700
March	600	1700	800	2400	900
April	800	2500	800	3200	900
May	900	3400	800	4000	800
June	1000	4400	800	4800	600
July	600	5000	1200	6000	1200
August	400	5400	1200	7200	2000
September	600	6000	1200	8400	2600
October	600	6600	1200	9600	3200
November	1800	8400	1200	10800	2600
December	3000	11400	1200	12000	800

Fig. 5.2.2.4 Plan 3: production plan with two changes in production rhythm per year.

Many products, such as toys or chocolate, have a seasonal demand pattern like the one shown in the example. Should planners choose regular production, which will create inventory, or should production be a function of the demand, which will incur the costs of changing production rhythm? These costs go beyond micro costs, such as machine setup costs. Macro costs will be incurred, such as the costs of making changes to personnel or machinery. In the example, planners should calculate the following three production plans:

1. Maintain the production rhythm throughout the whole year.
2. Change production rhythm frequently — in this case, four times a year.
3. Attempt to find an optimal compromise between plans 1 and 2.

The planners can now compare the three variants with respect to budget, assuming the following cost rates:

- Number of hours required to manufacture one unit: 100Cost per hour: $100
- Carrying cost: 20% of inventory value
- Cost of changing production rhythm: $800,000 (at least once a year, according to the new sales plan).

Figure 5.2.2.5 shows that the third solution results in the lowest total costs.

Plan	Average inventory (in hours)	Average inventory (in 1000s of $)	Carrying cost (in 1000s of $)	# of production rhythm changes	Cost of change	Total costs
1	200000	20000	4000	1	800	4800
2	65000	6500	1300	4	3200	4500
3	14000	14000	2800	2	1600	4400

Fig. 5.2.2.5 Comparison of the three production plans.

5.2.3 Master Scheduling and Rough-Cut Capacity Planning

Intended learning outcomes: Present the master production schedule as a disaggregated version of the production plan. Explain the tasks for establishing a master production schedule. Describe the process of rough-cut capacity planning.

Sales and operations planning works mainly with product families, that is, at an aggregate level of information. However, there will be a need for more specific information for individual products.

The corresponding planning process at the level of the individual product is called *master scheduling.*[13]

The most important output of master scheduling is the master production schedule.

A *master production schedule (MPS)* is the disaggregated version of a production plan, expressed in specific products, configurations, quantities, and dates.

Figure 5.2.3.1 shows an example of an MPS as derived from a production plan (shown here only for the first four months of a year).

As the figure shows, the MPS is not only more detailed for individual products rather than product families, but it also yields much more detail for the time period for which the quantities are aggregated. It is thus a link between the production plan, which is relatively close to the sales plan, and the products the manufacturing department will actually build. The MPS is the input to all planning actions in the shorter term.

---

[13] *Scheduling* is the act of creating a schedule, such as a master, shipping, production, or purchasing schedule (cf. [APIC16]). The *master schedule (MS)* is the result of master scheduling.

Product family \ Month	Jan.	Feb.	March	April
...				
P	100	100	150	120
...				

Product \ Week	1	2	3	4	Total
P_1	25	25			50
P_2			25	5	30
P_3				20	20
Total	25	25	25	25	100

Fig. 5.2.3.1 The MPS as a disaggregated version of the production plan (an example of a product family P with three different products, P_1, P_2, P_3).

The *planning time fence* corresponds to the point in time denoted in the planning horizon of the master scheduling process that marks a boundary inside of which changes to the schedule may adversely affect customer deliveries, component schedules, capacity plans, and cost ([APIC16]).[14]

Planned orders outside the planning time fence can be changed automatically by the planning logic of a software. Inside the time fence, the *master scheduler*, that is the person charged with the responsibility of managing the master schedule for select items, must deal with changes manually.

Establishing a master production schedule entails a number of tasks:

First task: *Selection of the master schedule items*, that is, the items managed by the master scheduler and not by the computer. Taking the example in Figure 5.2.3.1, if the difference between the products of the family P is due to three different variants of a assembly (namely, V_1, V_2, and V_3) and if the delivery lead time allows assembling to customer order, then the best choice for the (customer) order penetration point (OPP) is the assembly level. The final products P_1, P_2, and P_3 are then produced to customer order, according to the final assembly schedule (FAS) (see Section 7.1.5). If the usage quantity is 2 for each variant, then Figure 5.2.3.2 shows the MPS corresponding to the production plan.

[14] In general, a *time fence* can be understood as a policy or guideline established to limit changes in operation procedures. In contrast to this, the term *hedge* is used in logistics and operations mangement similar to safety stock, in order to protect against an uncertain event such as a strike or price increase. It is planned beyond some time fence such that, if the hedge is not needed, it can be rolled forward before major resources must be committed to produce the hedge and put it in inventory (cf. [APIC16]).

Assembly \ Week	1	2	3	4	Total	%
V_1	50	50			100	50
V_2			50	10	60	30
V_3				40	40	20
Total	50	50	50	50	200	100

Fig. 5.2.3.2 The MPS on the level of subassemblies V_1, V_2, and V_3.

Second task: *Break down the production plan quantity for a product family into quantity for each product of the family*. We often do not know the exact percentage for splitting the total product family demand into individual product or variant demands. To cover this uncertainty, we increase the percentage of each variant. This percentage is called the *option percentage* (see here Section 10.5.3). This procedure results in overplanning, which yields protection in the form of safety demand. Figure 5.2.3.3 shows example overplanning in the MPS, assuming an uncertainty of 20%.

Assembly \ Week	1	2	3	4	Total	%
V_1	60	60			120	50
V_2			60	12	72	30
V_3				48	48	20
Total	60	60	60	60	240	100

Fig. 5.2.3.3 The MPS for the first four weeks on the level of subassemblies V_1, V_2, V_3, including overplanning due to variant uncertainty.

This safety demand is in effect safety stock, or reserved stock, for the entire planning horizon to be covered. For details, see Section 10.5.4. The safety demand has to be planned at the beginning of the planning horizon. If the forecast indicates a large demand in one of the subsequent periods, the additional safety demand can be planned for that planning period. Figure 5.2.3.4 shows the first overplanning for January. An additional overplanning takes place for March, but only for the part that is not already overplanned in January.

For the rest of the planning period, the safety stock in the system corresponds to the safety demand for the maximal monthly demand. Because of the general uncertainty in the system, it is sometimes easier to plan the whole quantity at the start of the planning period. A coordinated final assembly schedule (FAS, see Section 7.1.5) maintains the service level at 100%, meaning that consumption of the subassemblies stays within the limits of the safety stock. For more details, the reader may refer to Section 7.2, where it is also explained that this kind of master scheduling is valid only as long as the number of variants to be planned in the MPS is significantly lower than the total demand quantity for the product family. Otherwise, a (customer) order penetration point (OPP) more upstream must be chosen.

Product family \ Month	Jan.	Feb.	March	April
...				
P	100	100	150	120
...				

Assembly \ Month	Jan.	Feb.	March	April
V_1	100+20	100	150+10	120
V_2	60+12	60	90+6	72
V_3	40+8	40	60+4	48
Total	200+40	200	300+20	240

Fig. 5.2.3.4 The MPS on the level of subassemblies V_1, V_2, and V_3, including safety demand (due to variant uncertainty) during the planning horizon.

Third task: Verify the feasibility of the MPS by rough-cut capacity planning.

Rough-cut capacity planning (RCCP) is the process of converting the master production schedule into *required capacity*, that is, capacity of (key) resources to produce the desired output in the particular periods. Comparison to available or demonstrated capacity (with regard to feasibility) is usually done for each key resource ([APIC16]).

As the planning is more detailed, RCCP yields more precise information on the work centers and the capacities to be used than does resource requirements planning (RRP). It therefore allows more precise control of the feasibility of the production plan. Figure 5.2.3.5 shows the average (∅) load of the MPS in comparison to the weekly average (∅) capacity of a work center called WC-A.

Assembly \ Week	1	2	3	4	Load per unit	∅ Load / capacity
V_1	60	60			0.75	
V_2			60	12	0.6	
V_3				48	0.5	
Load (in h) (= capacity required)	45	45	36	31.2		39.3
Capacity (in h)	40	40	40	40		40
Over-(+)/under-(–)capacity (in h)	-5	-5	+4	+8.8		+0.7

Fig. 5.2.3.5 RCCP on the level of subassemblies V_1, V_2, and V_3: load and capacity on work center WC-A.

For balancing load with capacity, the following strategies are possible:

- The chase production method maintains a stable inventory level that corresponds to load. To do this, (quantitatively) flexible capacity — as is the case in Figure 5.2.3.5 — is a must.
- The *level production method* maintains a *level schedule* (a master production schedule that generates a load that is spread out more evenly over the time period) corresponding to capacity. This can go as a far as requiring *linearity*, or the production of a constant quantity (or the consumption of a constant quantity of resources) in every period (such as daily). Figure 5.2.3.6 shows a possible solution.

Assembly \ Week	1	2	3	4	Load per unit	Ø Load / capacity
V_1	54	54	12		0.75	
V_2			50	22	0.6	
V_3				48	0.5	
Load (in h) (= capacity required)	40.5	40.5	39	37.2		39.3
Capacity (in h)	40	40	40	40		40
+= Over / - = under (in h)	-0.5	-0.5	+1	+2.8		+0.7

Fig. 5.2.3.6 RCCP on the level of subassemblies V_1, V_2, and V_3: load and capacity on work center WC-A, load leveled.

- *Hybrid production method*: Companies can combine chase and level production methods.
- It is a question of an *overstated master production schedule*. The quantities are greater than the ability to produce, given current capacity and material availability (cf. [APIC16]). The MPS has to be modified.

Figure 5.2.3.6 shows that load leveling is a time-consuming procedure even for just *one* work center. Finite loading algorithms, often developed within operations research (such as linear programming), have to be used. In the face of the degree of uncertainty of the (mostly forecast-based) production plan as well as of the demand breakdown from the family level to the level of individual products, it is often not worth putting too much effort into more detailed calculation. If there is (as in our example) a 20% uncertainty in the distribution of the demand of the family among the single products or subassemblies, a deviation of 10% of the average capacity (as in Figure 5.2.3.5) is probably precise enough. Investing great efforts in detailed calculation will often be useless at this level of planning. In contrast, the importance of investing in (quantitatively) flexible capacities increases with a growing degree of variability of the product concept.

In more complicated cases, the MPS must divide the production plan into individual production or procurement lots. Then, just as in medium-term planning, net requirements planning over the time axis, rather than gross requirements planning, is needed. An example of this is long-term planning that aims explicitly to achieve high-capacity utilization, particularly in the process industry. In that case, RCCP (rough-cut capacity planning) seems to be a good solution:

- Quick calculation of alternative order quantities or subdivisions in part orders with shifted completion dates is possible.
- The number of planning variables is small, and sometimes the whole plan can be displayed on a large monitor. This provides excellent support to the human ability to make situation-appropriate decisions intuitively even when the data are incomplete and imprecise. These intuitive decisions take into account a multitude of nonquantifiable factors and implicit knowledge. This is a very important aspect of future-oriented forecasting techniques. Knowledge about the development of a forecast can influence our evaluation of planning results, particularly interpretations of capacity overload and underload.

See Section 14.4 for a detailed description of rough-cut capacity planning techniques.

5.2.4 Supplier Scheduling: Blanket Order Processing, Release, and Coordination

Intended learning outcomes: Produce an overview on supplier scheduling. Explain the systematics of blanket orders and blanket releases with quantities and time periods.

One objective of resource requirements planning in long-term, or master, planning, is to prepare the channels for later procurement. In the case of goods, the challenge is to determine what suppliers can fulfill the company's requirements in terms of quantity, quality, delivery, and delivery reliability. It is in this phase that the purchasing budget should also be set.

Experience in recent years — particularly in connection with the demand for faster delivery at lower procurement costs — has shown that for efficient logistics, a company must work together more closely with its suppliers.

Supplier scheduling is a purchasing approach using blanket agreements, discussed below when viewing the company as a customer (it has a corresponding significance to the company in its role as a supplier in a supply chain).

The supplier has to have some knowledge of the company's master planning so that its own master planning can allow fast delivery. This exchange of information is a matter of trust, and it cannot be practiced with all or even very many suppliers (see Section 2.2.2).

Gross requirement calculated by resource requirements planning is, after all, a forecast that can be placed with suppliers as blanket orders. A *blanket order* is, in nonbinding cases, a "letter of intent." A minimum blanket order quantity for a planned time period, together with a maximum quantity, increases the binding nature of the agreement and thus also raises planning security.

In medium-term planning, blanket purchase orders are defined ever more precisely, step by step. In agreement with the supplier, a company sets procurement quantities per period in medium-term planning (such as for three months hence, for two months hence, for the next month) with a decreasing range of deviation. From a certain point in time onward, the part of the blanket order planned for "next month" becomes a short-range blanket order.

A *short-range blanket order* is only for a set quantity. A company gradually sets due dates for parts of the order by means of an fitting technique of execution and control of operations.

A *blanket release* is the authorization to ship and/or produce against a (short range) blanket agreement or contract ([APIC16]). It sets the maximum quantity, e.g., per week or per day.

A *delivery schedule* is the required or agreed time or rate of actual delivery of goods. A systems supplier, for example, may be requested by the company to deliver to the assembly line of an automobile manufacturer or machine builder in synchrony with production.

Figure 5.2.4.1 shows an example system of blanket orders and blanket releases. In this case, the two overlap.

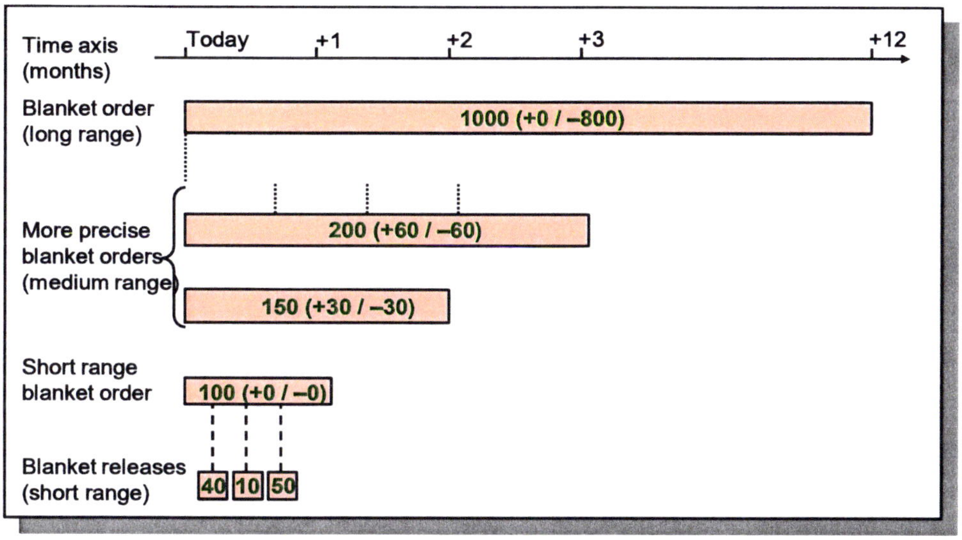

Fig. 5.2.4.1 Systematics of blanket orders and blanket releases with quantities and time periods (example).

The idea is that both the long-range blanket order and the medium-range, more precise blanket orders will be brought up to date on a rolling basis. In the example, the rolling cycle is one month. Blanket orders are given a plus or minus deviation. Each month's continuation of the blanket order must not contradict earlier agreements as to the acceptable range of deviation.

In this example, the company orders the exact required quantity for the next month, or, in other words, it places a short-range blanket order. The delivery schedule during the next month will be determined by a control principle such as a Kanban. In the course of the monthly period, unpredictable requirements arise, so that if a company has not given precise dates for probable delivery, the supplier will have to ready the entire quantity of the short-range blanket order at the start of the month. Additional quantification of a short-range blanket order could also set maximum requirements for blanket releases in that month.

A system like this, of continuous, ever more precise blanket orders and blanket releases, demands investiture in logistics, and planning & control, between a company and its suppliers. Therefore, the system is economically feasible only with a certain number of suppliers. Rapid and efficient communication techniques for information exchange and for updating the planning data are not only an advantage but also often a requirement of coordination. In some cases, a supplier may even have access to the company's database, while the company may check the status of the supplier's planning and implementation of procurement orders. See also Section 2.3.5.

5.3 Introduction to Detailed Planning and Execution

Intended learning outcomes: Disclose basic principles and produce an overview of materials management, scheduling and capacity management concepts and techniques. Differentiate between available-to-promise and capable-to-promise.

This section gives a brief overview of logistics business methods and techniques used for detailed planning and scheduling, as well as for execution and control of operations in the areas of distribution, production, and procurement. We will show the basic considerations that lead to various methods of solving the tasks presented in the reference model in Figure 5.1.4.2. The methods themselves will be the subject of more detailed, later chapters.

5.3.1 Basic Principles of Materials Management Concepts

Intended learning outcomes: Present the objectives of materials management. Differentiate between deterministic and stochastic materials management. Differentiate between independent demand and dependent demand. Produce an overview on quasi-deterministic materials management, fill rate, cumulative fill rate, stockout, backorder.

Materials management must provide the goods required by demand both cost effectively and according to schedule. The objectives of materials management are similar for supply chains in industry and in the service sector. The objectives are (see also Section 1.2.2):

- Avoidance of disruptions in delivery or production due to shortages
- Lowest possible costs for the administration of production and goods purchased externally
- Lowest possible carrying cost caused by goods procured too soon or even unnecessarily

The more exact our knowledge of inventory in stock and of open orders and due dates, the better the problem can be solved. It is even more important, however, to have exact information on demand. There are two possible ways to classify demand: with respect to accuracy or with respect to its relationship to other demand.

Classification of demand according to accuracy is defined as follows:

Deterministic demand is demand downstream from the (customer) order penetration point.
Stochastic demand is demand upstream from the (customer) order penetration point (OPP).

Classification of demand according to accuracy is thus dependent on the OPP, or, in other words, on the relationship between the customer tolerance time and (cumulative) lead time, as shown in Figure 1.3.3.1. Accordingly, the following sections will discuss two classes of methods and techniques in materials management.

Deterministic materials management utilizes a number of deterministic methods and techniques. In principle, these methods and techniques take demand as their starting point to calculate the necessary resources requirements on the basis of current conditions.

Stochastic materials management involves a number of stochastic methods and stochastic techniques. The methods and techniques utilize demand forecasts and buffer forecasting errors by building safety stock into the resource requirements.

Classification of demand according to its relationship is defined as follows:

Independent demand is the demand that is unrelated to the demand for other items.

Dependent demand is demand that is directly related to or derived from the demand for other items (cf. [APIC16]).

Company-external demand, or (customer) demand for end products or service parts, is independent demand, as is also a company's own internal demand for office supplies or — partly — indirect materials. The demand for assemblies, semifinished goods, components, raw materials, and — in part — auxiliary materials are examples of dependent demand.

There is an important subclass of stochastic materials management:

Quasi-deterministic materials management utilizes stochastic methods to determine independent demand. However, it utilizes deterministic methods and techniques to determine dependent demand, e.g., the bill-of-material explosion.

For stochastic demand, the practice is to avoid quasi-deterministic materials management whenever possible and to employ pure stochastic materials management. Here the fill rate plays a decisive role.

The *fill rate* used here is that percentage of demand that can be satisfied through available inventory or by the current production schedule.

This is the definition used as in Figure 1.4.4.1, whereas item demand is measured.

A *stockout* is a lack of materials, components, or finished goods that are needed ([APIC16]).

A *backorder* is an unfilled customer order or commitment, an immediate (or past due) demand against an item whose inventory is insufficient to satisfy the demand ([APIC16]).

The *stockout quantity* or backorder quantity is the extent of demand, that is, the quantity that cannot be covered during a stockout condition.

The *stockout percentage* or backorder percentage is the complementary percentage remaining when the fill rate is subtracted from 100 %.

The *cumulative fill rate* is the probability that several different components will be available simultaneously on demand.

If the fill rate for a component is not very close to 100%, then the probability that several items of a product will be available from inventory simultaneously will be very low. For example, if we need to have 10 components from inventory for an assembly, and the fill rate is 95%, the cumulative fill rate is only 60% ($\approx 0.95^{10}$), which usually will not suffice. Figure 5.3.1.1 illustrates this phenomenon.

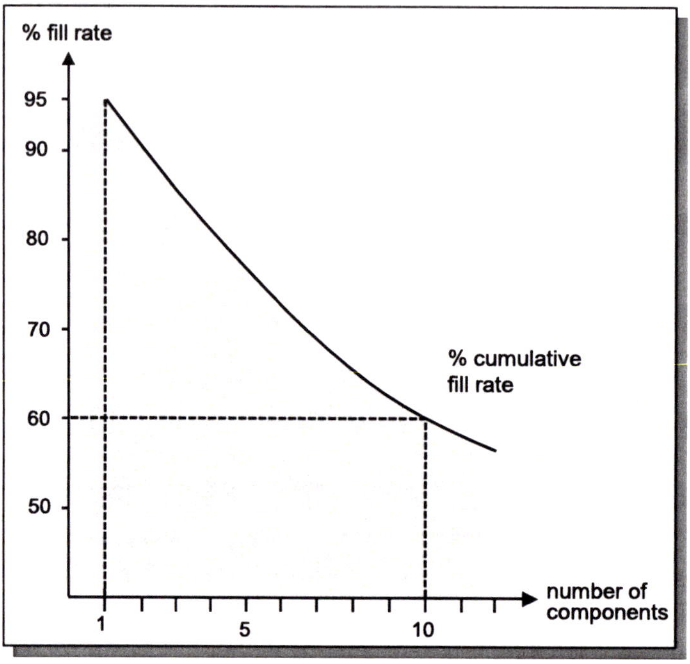

Fig. 5.3.1.1 Cumulative fill rate with components required simultaneously.

Complex products such as machines are very often made up of a large number of components. In this case, avoiding planning errors means ensuring a high fill rate for each component. Materials management, both in techniques and in form, is very dependent on the characteristic features of planning & control.

5.3.2 Overview of Materials Management Techniques

Intended learning outcomes: Disclose the basic classification of detailed planning techniques in materials management. Explain the additional classification for unique demand or demand for high-cost items with a discontinuous demand pattern. Produce an overview on various materials-management techniques.

Figures 5.3.2.1 and 5.3.2.2 distinguish among the common techniques of detailed planning techniques in materials management. First, Figure 5.3.2.1 classifies planning techniques according to the characteristic features frequency of customer demand and unit cost of items (as defined in Figure 4.4.2.1).

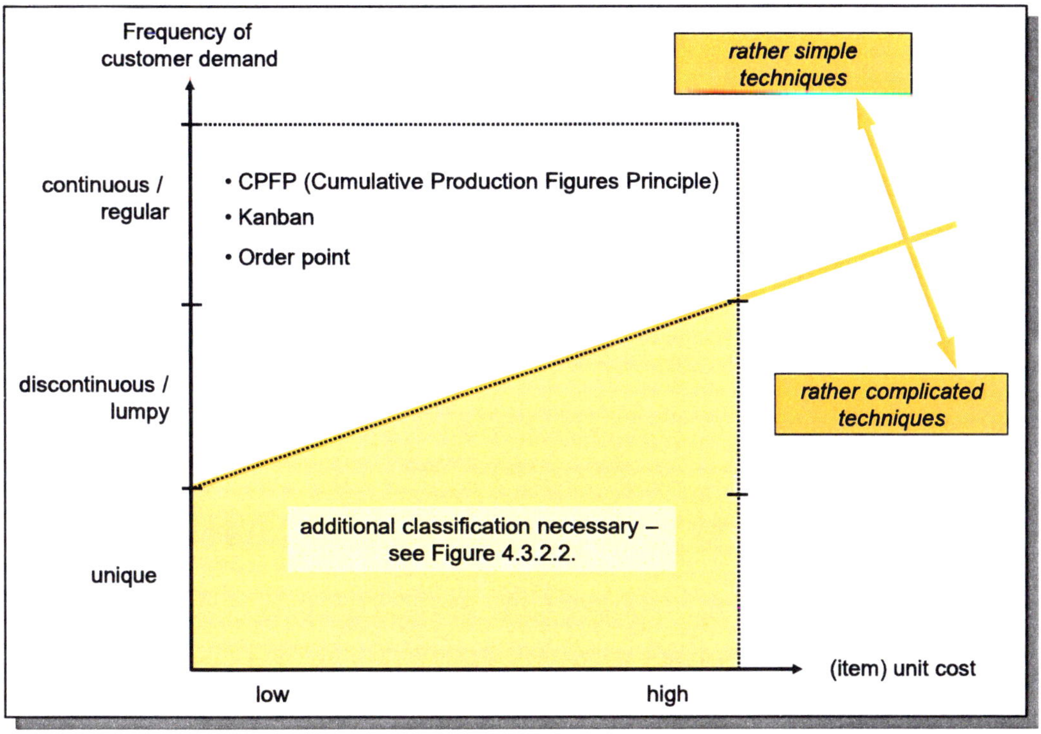

Fig. 5.3.2.1 Classification of detailed planning techniques in materials management.

Demand for low-cost items (with the exception of unique demand) or *demand for high-cost items with a continuous or regular demand pattern* is determined using stochastic techniques.

- In general, forecasting techniques determine future demand analytically or intuitively. From this perspective, demand forecasting is a technique for determining stochastic independent demand and is thus part of stochastic materials management. Once demand has been forecasted, different stochastic planning techniques exist, all being relatively simple. They are described at a first glance below.
- Dependent demand is calculated as if it were independent demand — that is, ignoring its possible derivation from independent demand.
- For low-cost items, a very high service level has priority. This holds true especially in the event where the item appears on the bill of material with many components (see the case mentioned previously in Fig. 5.3.1.1). Low stock inventory is, due to the low carrying cost involved, of secondary importance.
- For high-cost items, short lead times in the flow of goods, meaning rapid value-adding and administrative processes, take priority, requiring simple data and control flow. Inventory is possible: the continuous or regular demand pattern guarantees a

future demand (for end products: a customer order)[15] within a short time. However, because of the high unit cost, the inventories should be low, which generally requires small batch sizes.

For all other kinds of demand, that is for *unique demand* or demand for *high-cost items with a lumpy demand pattern,* Figure 5.3.2.2 shows an additional classification of planning techniques, this time according to the accuracy of the demand and its relationship with other demand (see the definitions above).

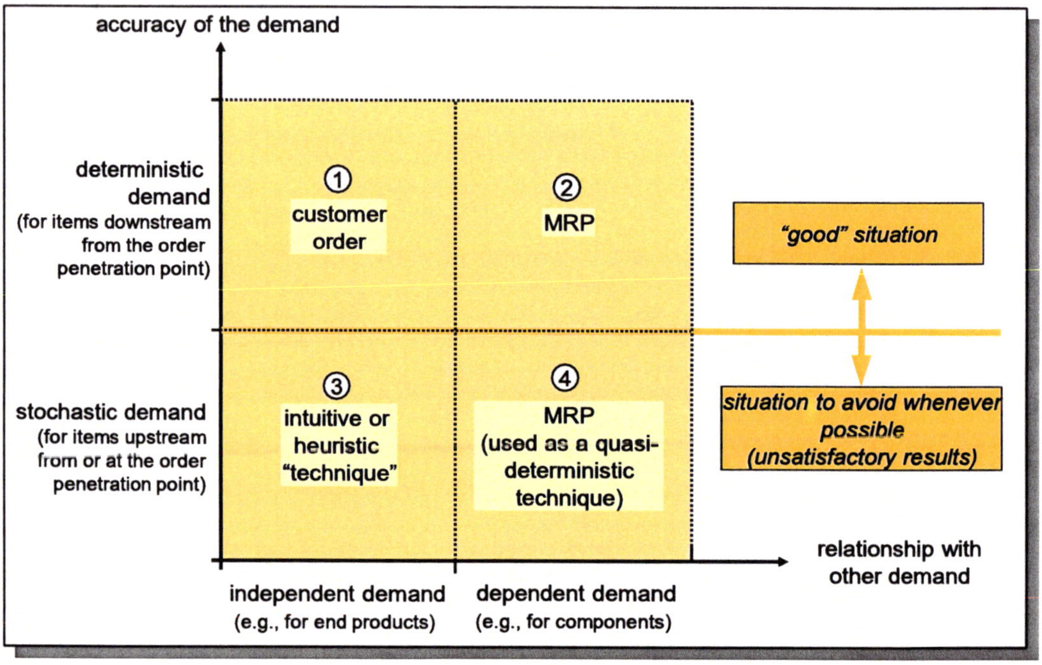

Fig. 5.3.2.2 Additional classification of detailed planning techniques in materials management for unique demand or demand for high-cost items with a lumpy demand pattern.

1. *Deterministic independent demand* can be met according to the actual demand, that is, according to the customer order.

- From this perspective, customer order processing and customer blanket order processing are *techniques for determining deterministic independent demand* and thus in a sense also belong to deterministic materials management.

2. *Deterministic dependent demand* can be calculated from higher-level independent demand.

---

15 Downstream from the order penetration point (OPP), the customer order can also show specific features that lead to a custom-made product (keywords mass customization and late customization). In general, the demand for this custom-made product is then sporadic, while the demand for the underlying product family may be continuous.

- The algorithm called MRP uses "explodes" of the bill of material, that is, the product structure, into its components.
- This type of demand calculation is a relatively complicated procedure. However, because of the priority of both high delivery reliability rate and low or even no inventory, it is appropriate.

Thus, while sometimes being rather complicated, planning techniques for items downstream from the order penetration point with unique demand or — for high-cost items — with a lumpy demand pattern present no great difficulty. However, planning of such items upstream from the OPP generally leads to unsatisfactory results:

3. *Stochastic independent demand* is determined more or less intuitively.

- As demand is lumpy, forecasting techniques tend to be inaccurate, and therefore ask for a lot of additional intuition. The materials management "technique" is often a manual procedure performed by the scheduler using a very personal heuristic. It is often a risky technique that should be avoided whenever possible.

4. *Stochastic dependent demand* is derived by quasi-deterministic techniques.

- Here, independent demand is determined using demand forecast techniques. Calculation of dependent demand is then based on independent demand by means of explosion of the bill of material. This is called *quasi-deterministic explosion of the bill of material.*
- As this demand pattern requires forecasting, there is a risk of a low service level or high carrying cost due to capital costs or risk of depreciation as a consequence of technical obsolescence or expiration due to perishability. As a consequence, any materials management technique handling this case will generally yield unsatisfactory results. Therefore, it should be avoided whenever possible. However, for many businesses, being in that situation is a fact of life.
- It is interesting to consider that, because of the dependent nature of the demand, the value-adding processes are under the control of the company. A thorough analysis of these processes can lead to appropriate modifications that entail more items downstream from the OPP, or a more continuous demand pattern — both situations being desirable. See Chapter 5 on the just-in-time / lean concept.

Forecasting techniques will be discussed in Chapter 9. The planning techniques mentioned are explained in different chapters. At a first glance, they are described in brief as follows.

- *Kanban* is a simple technique for stochastic materials management, but it requires invested capital. Small buffer storages kept close to the user operation will contain a maximum number of standard containers or bins holding a fixed number of items. The Kanban card is a means to identify the contents of the container and to release the order. The order batch size will be one or more empty containers, which are either sent directly by work center employees to the supplier or collected by one of the supplier's employees. The supplier executes the implied stock replenishment order and delivers it directly to the buffer. The Kanban feedback loop is then closed.

One of the tasks of long- and medium-term planning is to determine the type and number of Kanban cards for each feedback loop. See Section 6.3.

- The *cumulative production figures principle* (CPFP) is another simple technique. In the manufacturing process of a certain product, the technique in essence counts the number of intermediate products at particular count points. It compares this amount to the planned flow of goods, through putting the two cumulative production figure curves, or whole cumulative production figure diagrams — the projected diagram and the actual diagram — one on top of the other. The object is to bring the actual diagram closer to the projected diagram, which can be accomplished by speeding up or slowing down the manufacturing process. See Section 6.4.

- The (stochastic) *order point technique* compares goods on hand — plus open orders and, sometimes, minus allocated quantities (reservations) — with a certain level called the *order point*. If the quantity calculated in this manner is no greater than the order point, the system generates orders to replenish stock. These replenishment orders can then be released. The order point is normally calculated as average usage (a forecast!) during the replenishment lead time plus safety stock, or reserved stock, to compensate for forecast errors. The "optimum" order quantity or batch size, called the economic order quantity (EOQ), can be determined through comparing ordering and setup costs to carrying cost. See here Chapter 11.

- The (deterministic) *MRP-technique (material requirements planning)*[16] calculates, starting from higher-level independent demand, dependent demand by exploding the bill of material. The individual dependent demands are grouped together according to certain batch sizing policies and planned for timely production or procurement. In the deterministic case, the safety stock of components can be very small; inventory is kept to a minimum. In the quasi-deterministic case, safety demand at the level of the independent demand determines the safety stock of components. Deterministic materials management produces order proposals and the information required to control the processing of those orders. See here Chapter 12.

Section 6.5.2 discusses further a possible strategy for choosing one of these techniques and gives tips for implementing procedures.

5.3.3 Basic Principles of Scheduling and Capacity Management Concepts

Intended learning outcomes: Present the objective of and challenges linked with scheduling and capacity management. Disclose to which extent capacity can be stored.

The type of business or company makes no difference when it comes to time management and scheduling and capacity management. Industrial and service companies alike face essentially the same challenges:

- How can individual order processing tasks be synchronized in time?

---

16 It is important not to confuse the MRP technique with the MRP II concept (manufacturing resource planning).

- What capacities must be available to realize master planning?
- Where and when must special shifts and overtime (or short-time work or part-time work) be put in place? What jobs, or whole orders, must be turned over to subcontractors (due to overload) or taken over from them (due to underload)?
- Where can the rhythm of production be brought into balance? Can short-time work in one area be compensated for by overtime in another?
- When and where can capacity or orders be shifted? For example, what shifts can be made from one shop, production line, office group, team, and so on to another?
- Can lead times and the number of orders in process be reduced?

The objectives of the tasks of *time management and scheduling* and *capacity management* are similar to the objectives of *materials management* (see Section 5.3.1):

- High service level, short delivery times, high delivery reliability rate, and, at the same time, flexibility to adapt to customer requests
- Low tied capital: minimal inventory of work in process; optimization of wait times
- Efficient use of available capacity through good utilization at a constant level; prediction of bottlenecks
- Flexibility and adaptability of capacity to changing conditions
- Minimal fixed costs in production administration and in production itself

Finding solutions for these issues requires consideration of large bodies of order data. Often, IT-supported processing of data is required. The planning problem becomes complicated because some of the above objectives, such as the first and the third, contradict each other.

Figure 5.3.3.1 shows the result of *not* planning capacity. If capacity is inadequate (here, too low), a vicious circle of actions results. To gain an understanding of how this can arise, begin with "increased number of orders in the factory" at the bottom right of the figure.

1 If the number of customer orders increases, the number of orders released to production also increases, thus increasing the load on capacity.
2 If the number of orders exceeds capacity, queues will form behind the work centers.
3 In consequence, orders must wait and their actual lead times lengthen. Orders cannot be met at their due date, that is, not within the customer tolerance time.
4 Standard lead times, particularly the interoperation times, are prolonged to gain more realistic planning.
5 As a consequence, orders are released earlier, which in turn causes additional load in the form of released orders. The "game" begins all over again at point 1.

In this example, increasing the capacity could be a way to break out of the vicious circle.

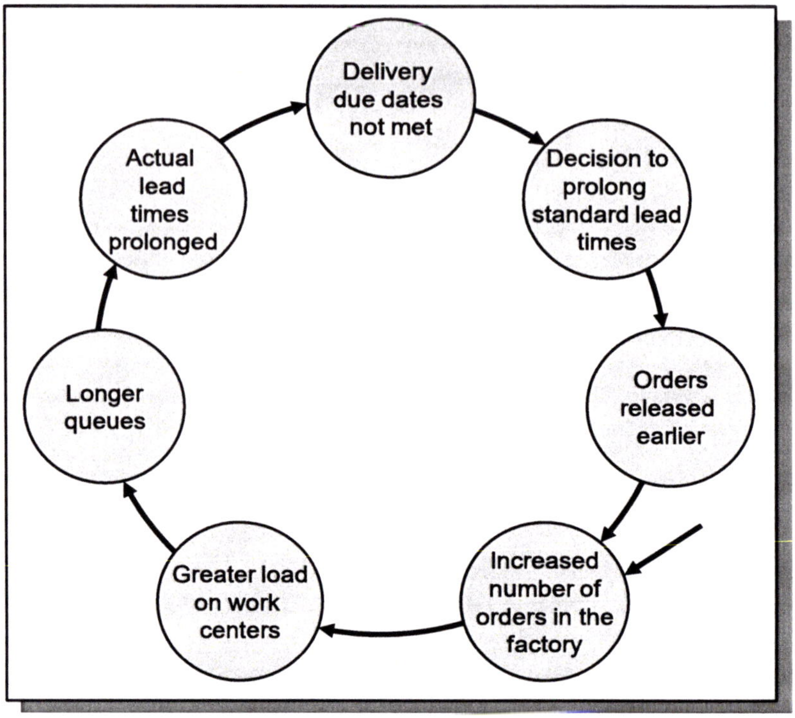

Fig. 5.3.3.1 A vicious circle caused when capacity bottlenecks prolong the planned production lead time. (From [IBM75]).

The *overall objective* of scheduling and capacity management is to *balance load* arising through orders with *capacity available* to process those orders. Figure 5.3.3.2 shows a chance-produced situation through the course of time (above) and, in contrast, an idealized conception of the possible result of planning (below).

The problem to be resolved is basically the same in any of the temporal ranges of planning & control. However, the measures taken for capacity planning — such as procuring additional capacity — are very different in master planning and detailed planning and scheduling.

- In long-term planning, the company can procure additional production means, such as production facilities or persons. In addition, it can make comprehensive arrangements to subcontract to the outside. Or, if capacity must be reduced, this can all be accomplished in reverse.
- In medium-term planning, on the other hand, a company will attempt to gain at least some measure of elasticity of capacity through scheduling overtime or arranging rush subcontracts to the outside. Medium-term planning, however, cannot correct major errors in long-term planning. These planning errors result in late deliveries.

Capacity is a potential factor. Can capacity be stored? A firm may think that this can be accomplished by producing ahead, thus creating inventory. However, inventory cannot be reconverted into capacity. Therefore, the firm has to be very sure to produce ahead only items that will be used within a reasonably short time frame. There are capacity management techniques that use this strategy, such as Corma. In other cases, however, producing ahead in order to "store capacity" may simply be a manifestation of a "just in case" mentality. As a result, the wrong items will be produced, and eventually the capacity is lost.

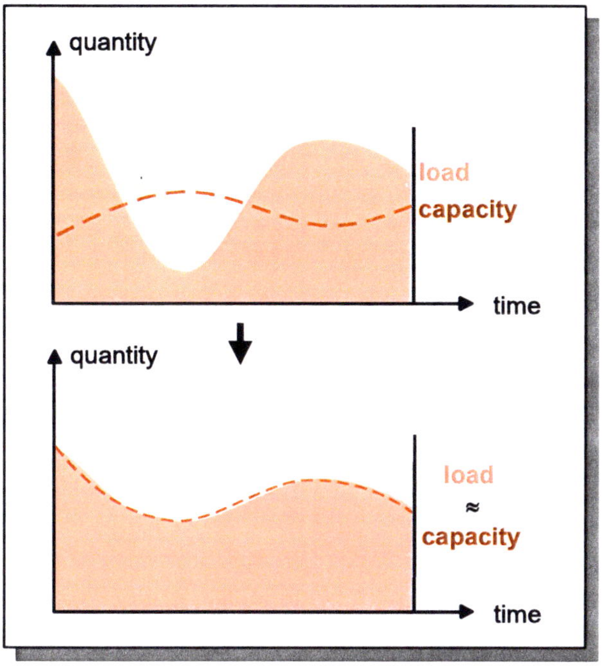

Fig. 5.3.3.2 Objective of time management and scheduling and of capacity management: balancing load with capacity available.

Somewhat "storable" is capacity in the form of personnel — if employees' presence along the time axis is somewhat flexible. For instance, say that an employee has to work only five hours instead of the usual eight on a specific day. If she or he is willing to go home but to work the three hours on another day where there is overload, you could say that three hours of capacity were stored. While this strategy is quite common, it is very limited with regard to the total capacity. Moreover, a company normally has to pay the worker for her or his (quantitatively) flexible capacity.

Generally, capacity cannot be stored effectively. Because this is so, planning must address two dimensions simultaneously; capacity (quantity axis in Figure 5.3.3.2) and dates (time axis) must be planned *together*.

5.3.4 Overview of Scheduling and Capacity Management Techniques

Intended learning outcomes: Differentiate between infinite and finite loading. Explain the classification of techniques for capacity management dependency on flexibility of capacity and of order due date. Produce an overview on various capacity-management techniques.

Depending on the main objectives of the firm (see Section 1.2.2), the values for some of the characteristic features of planning & control as in Figures 4.4.3.1 and 4.4.4.1 will differ.

- If a company puts the focus on flexibility in the utilization of resources, then *flexible capability of capacity* (workforce and production infrastructure) is vital.
- If high capacity utilization is required, there will be no (quantitatively) flexible capacity. This is particularly the case for the production infrastructure.
- If high service level and delivery reliability rate are required, there will be no *flexibility of the order due date* of the production or procurement order.

Flexible capability in capacity, meaning that capacity can also be applied for processes outside a particular work center, can increase its quantitative flexibility, or temporal flexibility regarding assignments. For example, if employees can be moved from one work center to another, this is the same as if each work center was (quantitatively) flexible.

There are various techniques for scheduling and capacity management. The techniques can be grouped into two classes based on the two planning dimensions shown in Figure 5.3.3.2: infinite and finite loading.

- *Infinite loading* means calculating the work center loads by time period, at first without regard to capacity. The primary objective of infinite loading is to meet dates as scheduled, with greatest possible control of fluctuations in capacity requirements. Therefore, infinite loading is most useful when meeting due dates must take priority *over* high capacity utilization, such as is the case in customer order production in a job shop production environment. The planning techniques are rather simple.
- *Finite loading* considers capacity from the start and does not permit overloads. To prevent overloads, the planner changes start dates or completion dates. The primary objective of finite loading is good use of the capacity available through the course of time, with the greatest possible avoidance of delays in order processing. Therefore, finite loading is most useful if limited capacity is the major planning problem, such as in the process industry in a continuous production environment. Often, this condition is given in very short-term planning, in execution and control. The planning techniques are rather complicated.

In addition to these *two classes of techniques*, Figure 5.3.4.1 groups techniques for scheduling and capacity management into *nine sectors* in dependency upon (quantitatively) flexible capacity and flexibility of the order due date. The techniques can be compared with respect to their overall capacity planning flexibility.

Overall capacity planning flexibility is defined as the "sum" of the quantitative flexibility of capacity along the time axis and the flexibility of the order due date.

- Note that there is no technique in the three sectors at top right: Here, the overall capacity planning flexibility is high enough to accept and execute any order at any time. This case is very advantageous with regard to capacity planning, but it is usually too expensive due to overcapacity.
- Note the numerous techniques in the three sectors from top left to bottom right. Here, there is *sufficient* overall capacity planning flexibility to allow a computer algorithm to plan all the orders without intervention by the planner. After completion, the

computer program presents unusual situations to the planner as selectively as possible in the form of lists or tables. The planner will intervene to execute appropriate planning measures — perhaps daily or weekly.

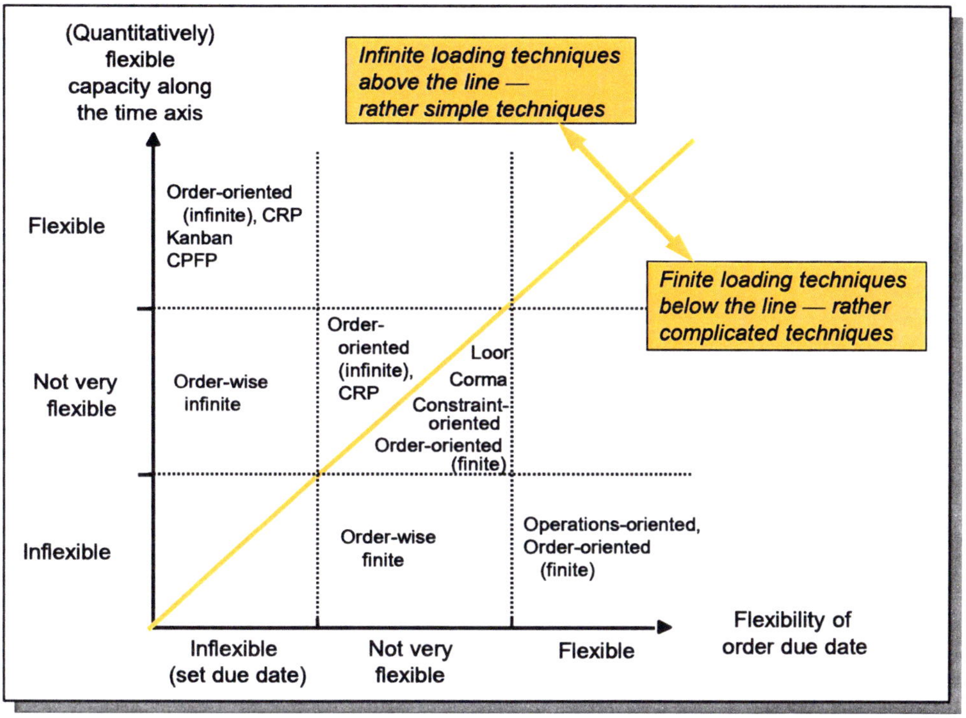

Fig. 5.3.4.1 Classes of techniques for capacity management in dependency upon flexibility of capacity and flexibility of order due date. The abbreviation "CPFP" stands for cumulative production figures principle (see text).

- Note that there are few techniques in the two sectors above and to the right of the bottom left sector. Here, there is no flexibility on one axis and only low flexibility on the other. Thus, there is *little* overall capacity planning flexibility. Planning takes place "order for order" (order-wise). Each new order must be integrated individually into the already planned orders. The planner may, in extreme cases, have to intervene following each operation and change set values for planning (completion date or capacity). Already planned orders may have to be replanned. This procedure is usually very time consuming and is therefore efficient only for orders with considerable added value.
- Finally, note that there is no technique in the sector at bottom left. Here, there is no flexibility of capacity or due date. As a consequence, there can be none of the required balancing, and the planning problem cannot be resolved.

The following describes infinite loading techniques. Infinite loading is frequently the best capacity planning method. In many companies, it is possible to modify labor capacities within one day by more than 50%.

- *Order-oriented infinite loading* aims to achieve a high delivery reliability rate, or to meet the due date for production or procurement orders. In favor, overcapacity is often maintained intentionally. After scheduling (backward or forward, for example) all the orders, each scheduled operation represents a load at the specified work center and in the time period containing its start date. The sum of all these loads is compared to the available capacity for each time period. This yields load profiles showing the overcapacity or undercapacity for each work center and time period. The subsequent planning then attempts to balance capacity against load. This technique for infinite loading is also called *capacity requirements planning (CRP)*, particularly in connection with software for capacity management. Some variations of CRP also exist. See Section 14.2.
- *Kanban* and the *cumulative production figures principle* (CPFP) were introduced above in Section 5.3.2. These two simple materials management techniques serve at the same time as simple capacity management techniques. Execution control by the Kanban technique is a form of infinite loading. It assumes a very high level of flexibility of capacity in the immediate term. See Sections 6.3 and 6.4.
- *Order-wise infinite loading* (order for order, individually): For firms handling small numbers of high-value-adding orders, such as for the production of special-purpose machines, planning takes place after loading each new order, or even after each new operation. As soon as an overload is detected, all work centers are checked, and load and capacity are adjusted until a feasible schedule is obtained. See Section 14.2.

The following describes finite loading techniques:

- *Operations-oriented finite loading* aims to minimize the average delay of the production orders. The individual operations are planned time period by time period on the basis of orders, starting from the start date determined by lead-time scheduling. This means establishing meaningful rules of priority for the sequence in which operations are scheduled (sequencing rules), with the aim of achieving maximum throughput. The queues waiting upstream of the work centers are monitored and adjusted. This type of planning provides a *process simulation* for the coming days and weeks, that is, an actual working program for the shop floor, according to the planning horizon. See Section 14.3.1.
- *Order-oriented finite loading* ensures that as many orders as possible are executed on time with a low level of work in process. Orders are scheduled in their entirety, one after the other, in the time periods. Priority rules should enable as many orders as possible to be scheduled. Those orders that cannot be scheduled for completion on time by a computerized algorithm are highlighted for attention by the planner, who may decide to change order completion dates. See Section 14.3.2.
- As bottlenecks control the throughput of a production system, *constraint-oriented finite loading* plans orders around bottleneck capacities. It follows a theory of constraints (TOC) approach. An application of this is *drum-buffer-rope*. Work centers feeding bottlenecks are scheduled at the rate the bottleneck can process. A time buffer inventory should be established before the bottleneck. A space buffer should be established after the bottleneck. Work centers fed by the bottleneck have their throughput controlled by the bottleneck. See Section 14.3.3.

- *Load-oriented order release (Loor)* has high load as its *primary objective*. Equally important are its *secondary objectives* of low levels of work-in-process, short lead times in the flow of goods, and delivery reliability. The aim of Loor is to adapt the load to the capacity that is actually available. Thanks to a heuristic, the matching of load to capacity can be limited to one time period. See Section 15.1.2.
- *Capacity-oriented materials management* (Corma) plays off work-in-process against limited capacity and lead time for customer production orders. Corma makes intelligent use of capacity that is generally fully utilized, but available short term, by releasing stock replenishment orders earlier than needed. Thus, Corma follows the natural logic of production management as it is implemented in practice in many medium-sized companies that view stock replenishment orders as "filler" loadings. However, the benefit of improved utilization of capacity demands a price, as work-in-process increases. See Section 15.1.3.
- *Order-wise finite loading* (order for order, individually): This can be considered to be similar to order-wise infinite loading, with more flexibility in time axis.

All of these techniques can be used independently of company-organizational implementation of planning & control. Thus, they can be found in software packages of many kinds (ERP software or electronic control boards [Leitstand], simulation software, and so on). In one and the same enterprise, it is quite possible that the company will use different techniques for short-term planning and long-term planning.

5.3.5 Available-to-Promise and Capable-to-Promise

Intended learning outcomes: Explain available-to-promise and the determination of ATP quantities. Produce an overview on multilevel available-to-promise and *capable-to-promise.*

For the short-term planning of customer orders (here see Figure 5.1.3.1) the detailed resource requirements calculation must answer the question of whether a given quantity of a product will be available at a given time. The way to do this is by using predictive simulation with the available-to-promise or capable-to-promise procedures, which use some of the detailed planning techniques discussed in Section 5.3. This way is even necessary when customer orders are not covered by long-term or medium-term demand forecasts. The available times will not be known until all the resource requirements are covered, possibly through new planned orders, which then leads to confirmation and to approval of the customer order.

(Order) backlog is all the customer orders received but not yet shipped. Sometimes referred to as *open customer orders*. Cf. [APIC16].

Available-to-promise (ATP) is the uncommitted — that is not yet assigned to an open customer order — portion of a company's inventory and planned production (cf. [APIC16]).

The ATP quantity is maintained in the master schedule to support customer-order promising. It is normally calculated for each event or each period in which an MPS receipt is scheduled ([APIC16]). However, it is cumulative ATP that is of practical importance. Figure 5.3.5.1 illustrates the definition and calculation of discrete ATP and cumulative ATP.

Product PR

Physical inventory	= 12
Safety stock	= 0
Batch size	= 30
Lead time	= 3 periods

Feature \ Period	0	1	2	3	4	5
Master production schedule			30		30	
Allocated to customer order		5	3	25	20	10
Projected available inventory	12	7	34	9	19	9
Cumulative ATP	7	7	9	9	9	9
ATP per period	7		2			

Fig. 5.3.5.1 Determination of ATP quantities.

We will begin formal calculation of ATP quantities with some definitions:

For i = 1, 2, …, let

$ATP_i \equiv$ ATP of period i.

$ATP\_C_i \equiv$ cumulative ATP of period i.

$MPS_i \equiv$ MPS quantity of the period i.

$QA_i \equiv$ quantity allocated to customer orders in period i.

Now, let ATP_C_0 and ATP_0 be equal to the physical inventory. According to the definition above, the following algorithm, done subsequently for i = 1, 2, …, yields the ATP quantities.

$ATP\_C_i = ATP\_C_{i-1} + MPS_i - QA_i$.
j = i
While $ATP\_C_j < ATP\_C_{j-1}$ and j>0, revise the ATP quantities as follows:
  $ATP\_C_{j-1} = ATP\_C_j$
  $ATP_j = 0$.
  j = j–1
end (while).
If j > 0, then $ATP_j = ATP\_C_j - ATP\_C_{j-1}$.
If j = 0, then $ATP_0 = ATP\_C_0$.

In our example, for the product PR, seven units are available-to-promise from stock. Two additional units become available-to-promise in period 2.

Determining ATP quantities supports decision making regarding whether and for which due dates an order can be accepted. For make-to-stock-products, order promising is a direct consequence of comparing the order quantity with the ATP quantities. For more detailed information on availability and calculating projected available inventory, see Section 12.1.

A small exercise: Taking the example in Figure 5.3.5.1, determine whether 8 units can be promised for period 1. Also, how would you promise delivery of an urgent order of 10 units to an impatient customer waiting on the phone for your answer?

For make-to-order or assemble-to-order products, the processes are more complicated than in the calculation shown above, and cannot be shown in a simple overview. Two options are discussed in depth below, and references are given for more detailed information.

For components with insufficient inventory, the *multilevel available-to-promise (MLATP) check* uses an explosion of product structures by means of the MRP technique according to Sections 5.3.2 and 12.3. Thereby, the lead-time offset of the dependent demand is a lot-size independent production lead time (variant 1 in Figure 12.3.3.1).

The *capable-to-promise (CTP)* technique develops the classic MRP technique for checking availability by adding in not just stock levels but also capacity and other resources, possibly even for suppliers. For determining the timing of dependent demand, routing sheets are used, and the production lead times can be selected according to the lot size (variant 2 in Figure 12.3.3.1). For availability testing of the capacities, the classical order-oriented finite loading according to Section 5.3.4 and 14.3.2 or finite forward scheduling according to Section 15.2.2 can be applied.

It is evident that *MLATP* gives quicker results, but CTP gives more accurate results.

5.4 Logistics Business Methods in R&D (*)

Intended learning outcomes: Produce an overview on integrated order processing and simultaneous engineering. Describe release control and engineering change control. Differentiate between various views of the business object according to task.

In R&D, the individual processes do repeat themselves, but always with new products. This section will present some important concepts common to planning & control in this area. The concepts will be treated in Section 17.5 with regard to computer-supported processes.

5.4.1 Integrated Order Processing and Simultaneous Engineering

Intended learning outcomes: Differentiate between time-to-market and time-to-product. Differentiate between order processing via serial processing and with overlapping phases. Describe some aspects to consider for integrated order processing with overlapping individual phases. Produce an overview on simultaneous or concurrent engineering.

Time-to-market is the total lead time through R&D for new products: the time required for *product innovation*; that is, from product idea to introduction of the product to the market.

Short lead time through R&D is seen today as a strategy toward success. Because significant product ideas will be made ripe for the market by competitors with only slight delays, a few months' difference in the time required for R&D can be crucial to the success of a new product. An extra issue is the requirement of global R&D project organization ([BoGa08]).

Time-to-product is the total time required to receive, fill, and deliver an order for an existing (that is, entirely developed) product to a customer, timed from the moment that the customer places the order until the moment the customer receives the product.

This definition corresponds to the concept of delivery lead time. More and more, customers demand shorter delivery lead times, not only for products with a solid place in the market that are repeated, but also for custom orders, such as single-item or one-of-a-kind production orders. In many cases, such orders involve some design. Figure 5.4.1.1 shows, for example, the departments such an order must pass through during processing.

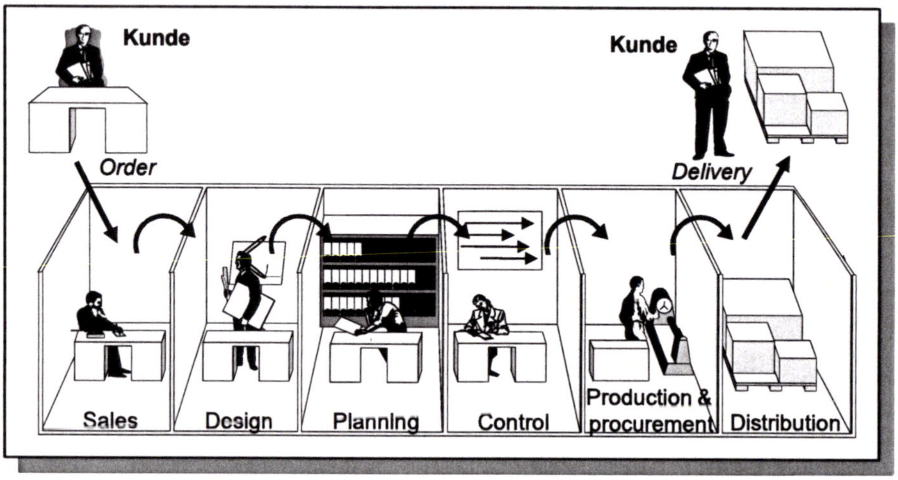

Fig. 5.4.1.1 Order processing of customer orders with specific R&D, production, and procurement (also see [Schö95a]).

If the delivery lead time required by the customer allows enough time, most companies tend toward serial processing of the various R&D, production, and procurement orders required by the customer order. Individual departments are informed about the order only when it is passed along by the upstream department. The information available is limited to the original order data and the specifications followed to date, as well as any documents on previous orders that may exist in the department. Similar observations can be made in R&D activities during the time-to-market. Figure 5.4.1.2 shows how this way of proceeding must change if the customer tolerance time does not allow enough time for serial processing.

Simultaneous engineering, *concurrent engineering*, or *participative design/engineering* refer to an overlapping of the phases of R&D and, in addition, an overlapping with earlier phases of procurement and production.

For the necessary integrated order processing with overlapping individual phases, some four aspects must be considered:

- The social and organizational aspects demand appropriate structural and process organization for rapid business processes, based on cooperative learning and acting. The walls between the departments shown in Figure 5.4.1.1 must come down. All persons involved in the customer order, whether in sales, development, or production, must be grouped "around the product." This means that the organization must be business-process oriented.

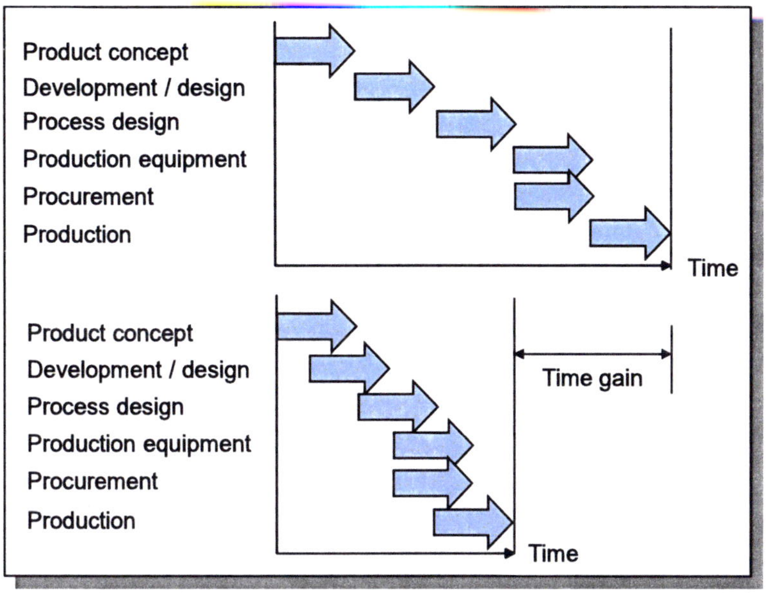

Fig. 5.4.1.2 Order processing via serial processing and with overlapping phases

- The *conceptual-logical aspect* requires that the *content* of information systems must be linked in a way that allows the exchange of data or even allows for commonly shared data management. Integration means that a unit must process data that another unit will require. The design engineer, for instance, must include data on the blueprints that allow identification for the bill of material. And, conversely, data must be kept on the item that is of relevance to design management. See further discussion in Section 5.4.3.
- The *technical-physical aspect* demands that the various hardware and software components be linked, that is an integration of the systems so that there will be a common, or at least commonly accessible, database. See here Section 17.5.

Design for the supply chain means enhancement of a firm's product design in consideration of the issues that will arise in the supply chain, from raw materials to the final stage of the product life cycle ([APIC16]).

Such demands are actually not new. In many small- and midsized companies, work has always been done in this way. This has been the case particularly where there is a large proportion of "one-of-a-kind" production orders, such as in plant and facilities construction or in structural and civil engineering. Companies specializing in these areas have been leaders in the integration of organization and in the integration of their IT-supported information systems as well. See Section 1.4.2 in [Schö01].

5.4.2 Release Control and Engineering Change Control

Intended learning outcomes: Describe step-wise release and engineering change control (ECC). Present procedures in engineering for new product design or a new product release.

Release and engineering change control (ECC) is an organizational concept for the process of the design and manufacture of a new product or of a new release of an existing product.

Release control and engineering change control (ECC) coordinate the production or modification of all blueprints, bills of material, routing sheets, and all other common documentation on a product and its manufacture. The procedure is project oriented and *step-wise releases* new developments, or changes to existing products, to production. Project management of this kind includes the following tasks:

- *Coordination of development and design:* Planning the volume of the release, labeling of all items; stopping the use of these items for planning & control; request for change or new concept of products; quality control; design release of individual items; design release of all items belonging to the volume of the release.
- *Procedures for production release*: Transfer of bills of material and routing sheets; release of all items belonging to the volume of the release.

Step-wise release is particularly important to provide for the principle of simultaneous engineering (see Figure 5.4.1.2). For this reason, we often distinguish between:

- *Rough-cut release for the production of a new development project or a new release:* The data released base upon preliminary drawings. They are provisional and pertain only to the most important products and rough-cut bills of material and routing sheets. They include the most important components that allow activation of the procurement and production process at lower design-structure levels. Depending on work progress, several rough-cut releases are conceivable.
- *Detailed production release with detailed documents:* Project management of the new release ensures that all required documents, such as blueprints, bills of material, routing sheets, and numerical control programs, are available in detailed form. Project management then releases individual items, or all items, to detailed production.

This kind of step-wise release corresponds to common practices in planning & control, which works with various temporal ranges of planning and rough-cut or detailed structures.

Figure 5.4.2.1 presents the different tasks and phases that must be handled by the (systems) engineering (see Section 19.1) for a new product or a new release of a product.

Preliminary study and main study can result in provisional releases, while detailed study leads to final release.

5.4.3 Different Views of the Business Object According to Task

Intended learning outcomes: Disclose examples of different views of a business object. Present business objects and attributes in the areas of design, release control and engineering change control, and planning & control.

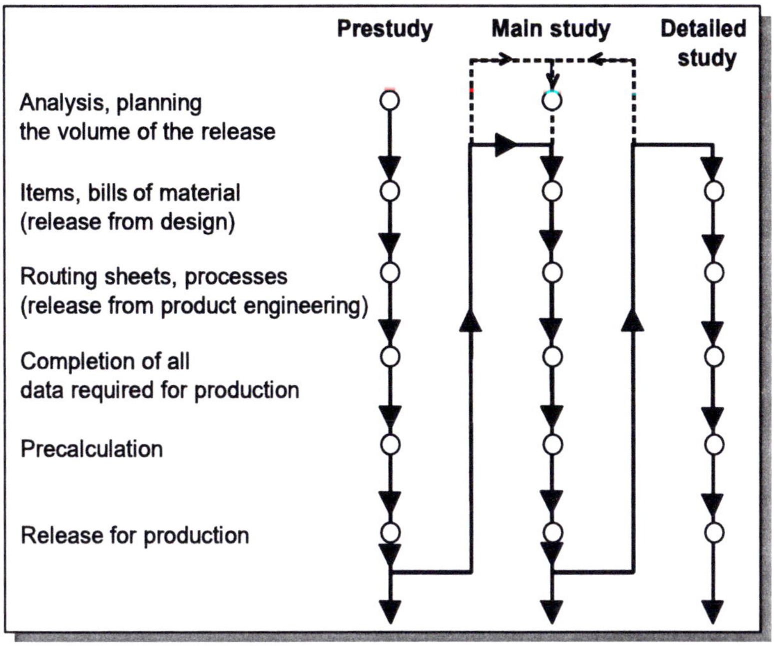

Fig. 5.4.2.1 Procedures in engineering for new product design or a new product release.

The people involved in a business process generally have different viewpoints with regard to the business objects the process handles. Their particular viewpoints depend on the specific tasks their departments must perform. This becomes very apparent whenever persons are moved from their departments to new forms of organization based on a business-process orientation. Problems in mutual understanding arise immediately, and they can only be overcome by means of appropriate training and qualifications combined with a heavy dose of goodwill. It is important that such problems are resolved by the time that a common database is created for purposes of integration of IT-supported tools. The business objects described by the data are, after all, often the same, such as end products, components, production facilities, and so on. However, individual viewpoints in terms of use and task result in only partial descriptions of these objects.

For example, the design department will describe a particular, clearly identified item in terms of its geometry, while the manufacturing process design department — in connection with IT-supported production machines — will describe the same item in terms of numerical control techniques. Figure 5.4.3.1 shows another example, the object "operation."

- The item or article viewpoint shows the state and extract of the product to be manufactured according to the operation.
- The material planning aspect gives the order of operations as well as a description of the operation.
- The production facilities view shows the tools or facilities to be used.
- The capacity viewpoint describes the workstation as a whole at which the operation will be executed.

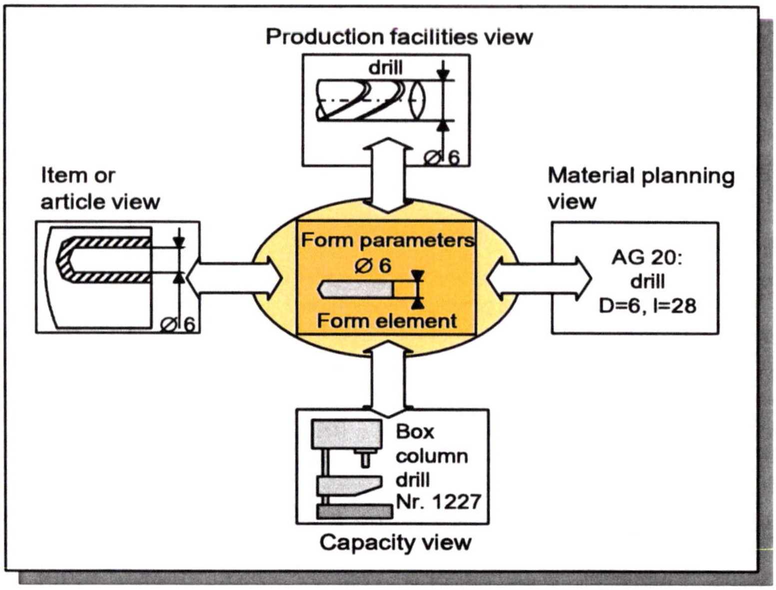

Fig. 5.4.3.1 Examples of different views of a business object (see [Schö95a]).

Figure 5.4.3.2 illustrates the above with objects from design, release control and engineering change control, and planning & control. In many cases, the business objects are identical. Only the points of view differ.

To integrate business processes within the company, these viewpoints must become linked. All departments require access to data from the other areas. For example:

- For the sake of cost and flexibility, the design engineer should preferably select, for his or her design components that are already being used in the current product family as semifinished goods, single parts, or raw materials. To do this, the design engineer needs to have a classification system for items that already exist in the planning & control database. Here see Section 17.5.3.
- Bills of material drawn up by the design department should be automatically entered into the planning & control database in all phases as discussed in Section 5.4.2.
- Conversely, when production orders are released, planning & control may request blueprints from design to add them to the work documents. With parametrically described items, all necessary parameter values on the customer order are passed along to the design department, so that it may create new blueprints according to the parameters for a specific order.

In spite of these differing viewpoints, can the same business objects be represented applicably and comprehensively for all the departments? While there is usually no great difficulty in agreeing on the definition of the objects, this is not the case for attributes of the objects. The same content of information may be represented from the one viewpoint with two attributes, but from another aspect, the information may be represented by three or four attributes. Redundant listings of attributes are generally not a reasonable solution, because this leads to consistency problems when modifying the data. Only a common definition reached by everyone involved in the business process can remedy the matter. In addition, the definition must be supported by appropriate IT systems. Here see Section 17.5

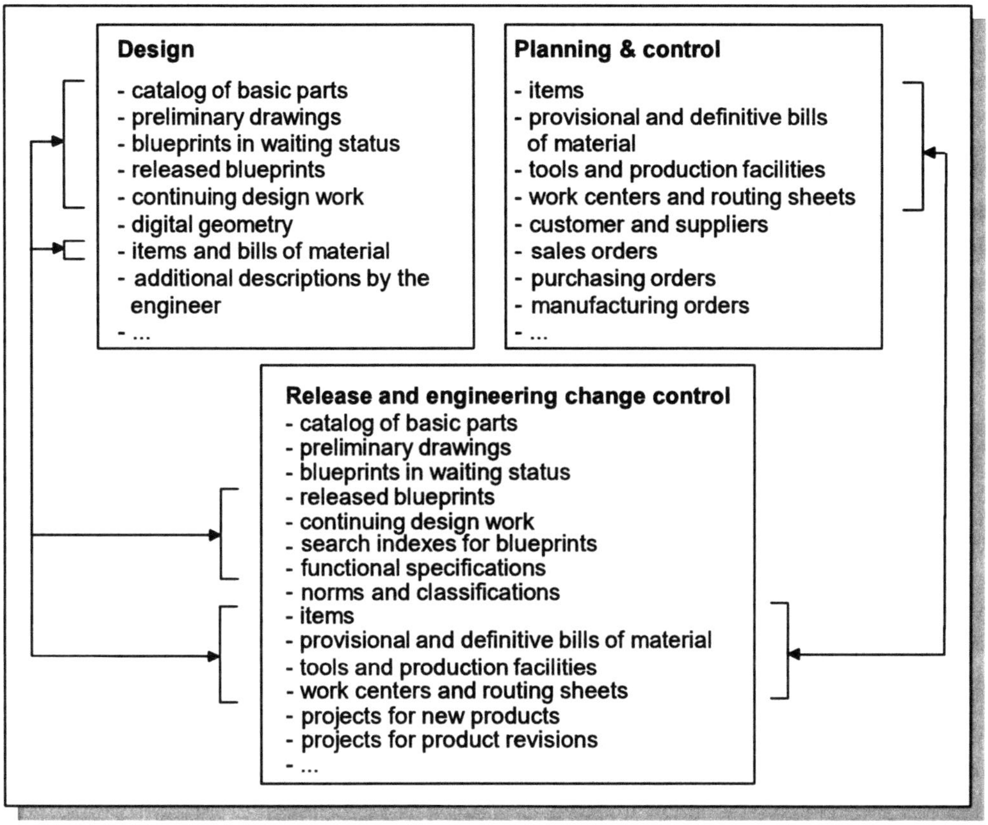

Fig. 5.4.3.2 Business objects and attributes in the areas of design, release control and engineering change control, and planning & control.

5.5 Summary

Business processes of planning & control in the MRP II / ERP concept can be classified as long, medium, or short term. There is a distinction between rough-cut and detailed planning. Tasks involved in the business processes are demand forecasting, bid processing, and order configuration; resource management; order release, order coordination, and order checking, as well as delivery and billing. The processes and tasks are shown in a reference model.

A first subtask of master planning is sales and operations planning and resource requirements planning. In the case of rough-cut planning, sales and operations planning produces an aggregate plan, which is a plan based on aggregated information (such as rough-cut business objects like product families, rough-cut product structures, or gross requirement) rather than on detailed product information. This allows quick calculation of different possible variants of the production plan. Another subtask of master planning is master scheduling and rough-cut capacity planning. The master production schedule (MPS) is the disaggregated version of a production plan, expressed in specific products, quantities, and dates. The appropriate level for scheduling (end products or assemblies) has to be chosen.

Customer blanket orders and blanket orders to suppliers are important instruments of planning & control in supply chains. These agreements set intervals for delivery dates and order quantities. In their most nonbinding form, they are purely forecasts. The intervals will be made more precise with decreasing temporal range. In the short term, precise short-range blanket orders replace blanket orders. Their quantities are set, and delivery dates will be set and confirmed by blanket releases as this becomes possible.

Business methods for detailed planning and scheduling as well as for execution and control of operations include tasks in materials management, scheduling, and capacity management. In materials management, techniques are classed as deterministic or stochastic. Scheduling and capacity management should be integrated, because capacity can generally not be stored. In dependency upon the (quantitatively) flexible capacity as well as the flexibility of the order due date, techniques can be classed as infinite and finite loading. Individual techniques, however, handle either quantity (capacity) or time (dates). Techniques like available-to-promise (ATP) and capable-to-promise (CTP) allow an answer to the question of whether a specific quantity of a product is available at a specific date.

Business methods of planning & control in the area of R&D comprise the integration of the various tasks all along the business process — particularly overlapping execution (simultaneous engineering) — both during the time to market and time to product. The different viewpoints of all those involved in the business object make integration difficult.

5.6 Keywords

aggregate plan, 252
available-to-promise (ATP), 275
backorder, 263
bid, 249
bill-of-material explosion, 239
blanket order processing, 239
bottleneck (capacity), 247
capable-to-promise (CTP), 277
capacity management, 244
constraint, 247
cumulative fill rate, 264
customer blanket order, 238
demand management, 248
dependent demand, 263
  deterministic, 266
  stochastic, 267
detailed planning and scheduling, 236
distribution planning, 246
distribution resource planning, 246
ERP (enterprise resources planning), 235
execution and control of operations, 237
gross requirements planning, 239
independent demand, 263
  deterministic, 266
  stochastic, 267
inventory management, 245
inventory plan, 252
manufacturing resource planning (MRP II), 236
manufacturing strategy, 251
master data, 245
master planning, 236
master production schedule (MPS), 255
master scheduling, 239
materials management, 243
  deterministic, 263
  quasi-deterministic, 263
  stochastic, 263
medium-term planning, 236

multilevel available-to-promise (MLATP), 277
net requirements planning, 240
order checking, 242
order configuration, 241
order coordination, 242
order entry, 249
order monitoring, 242
overall capacity planning flexibility, 272
pipeline inventory, 246
planned order, 240
planning horizon, 239
product family forecast, 252
production plan, 251
production planning, 251
production schedule, 239
purchase schedule, 239
quotation, 249
released order, 242
resource management, 246
resource requirements planning, 239
rough-cut planning, 237
sales and operations planning, 239
sales plan, 251
scheduling, 244
short-term planning and control, 237
stockout, 263
stockout percentage, 264
stockout quantity, 263
stores management, 245
tactical planning, 251
theory of constraints, 247
throughput, 247
time fence, 256
time to market, 277

5.7 Scenarios and Exercises

Intended learning outcomes: Disclose master scheduling for product variants. Calculate the quantity available-to-promise (ATP). Explain an example of the theory of constraints. Elaborate the master planning case.

5.7.1 Master Scheduling and Product Variants

Intended learning outcomes: Determine the degree of overplanning of the number of variants in in the master production schedule (MPS).

Your company produces scissors for left- and right-handed customers. While both models have the same blades, the handles differ. Blade and handle are assembled after you have received customer orders. You can assume that approximately 12% of your customers are left-handed. If you produce 100 blades, how many handles for each type of scissors should you produce?

Solution:

Since the actual option percentage is not known in advance, overplanning in the master production schedule (MPS) is necessary to cover the uncertainty. A safety demand of 25% would result in 12 * 1.25 = 15 handles for left-handled scissors and 88 * 1.25 = 110 handles for right-handed scissors to be produced. Because only 100 blades are produced, it makes no sense to have more than 100 handles of either type. Thus, a good decision would be to produce 15 handles for left and 100 handles for right-handed scissors.

5.7.2 Available-to-Promise (ATP)

Intended learning outcomes: Calculate the quantity available-to-promise, whereupon the master production schedule and a list of orders that have already been promised are given.

Sales employees in your company would like to know whether their customers' orders for can openers can be fulfilled. In long-term planning for the next half-year, you have put up the master production schedule provided below. Furthermore, your sales department has given you a list of customers' orders that have already been promised. At the beginning of the year, you have 800 can openers in stock.

Master Production Schedule:

January	February	March	April	May	June
600	600	600	600	450	450

Promised orders: 1200 pieces on February 14, 1400 pieces on April 5, 450 pieces on June 9.

a. How many can openers can your sales employees promise to customers in the next six months? (Assume that the amount planned to be produced in the master production schedule is available at the beginning of each month.)

b. Is the master production schedule feasible?

c. On January 7, a customer asks for 600 can openers to be delivered instantly. How do you react?

Solution:

a.

Feature \ Month		January	February	March	April	Mai	June
Master production schedule		600	600	600	600	450	450
Allocated to customer orders			1200		1400		450
Inventory available	800	1400	800	1400	600	1050	1050
ATP	600					450	
Cumulative ATP	600	600	600	600	600	1050	1050

b. Yes, because in each period cumulative ATP is greater than zero.

c. Though the amount the customer asks for is generally available, fulfilling this order would mean that the company would not be able to accept any further orders for four months, from January to April. Your decision will depend on how likely it is that this would result in the loss of long-term customers.

5.7.3 Theory of Constraints

Intended learning outcomes: Explain an example of the theory of constraints, whereupon you produce two products, which use the machine capacity of three machines with a certain load. Identify and speed up the bottleneck.

You produce two products A and B, which use the machine capacity of your production according to the following table:

Product \ Machine	I	II	III
A		1.5 hours	2.0 hours
B	1.6 hours	1.0 hours	

a. If per working day (eight hours), you start producing three products A and five products B, what will happen? What will the buffer in front of machine II look like after one week (five working days)? What measures do you suggest to take if you cannot invest any money?

b. A consulting firm offers to speed up your machines, so that the time it takes to machine any product is reduced by a quarter of an hour. To which machine would you apply this measure first, to which next? (Your only objective is to increase the amount of production.)

Solution:

a. The capacity of machine II is not sufficient: (3 * 1.5 hours) + (5 * 1.0 hours) = 9.5 hours. Therefore, the buffer in front of machine II will fill with the speed of 1.5 hours of workload per day, which is equivalent to five products A per week. To reduce work-in-progress, the company should decide to release fewer production orders, e.g., for two products A and five products B, per working day only.

b. The bottleneck is machine II, so it would be desirable to increase its speed. After implementing the consulting firm's measures, the work on products A and B takes (3 * 1.25 hours) + (5 * 0.75 hours) = 7.5 hours. Machine I with a workload of 5 * 1.6 hours = 8 hours will become the new bottleneck.

5.7.4 Master Planning Case

Intended learning outcomes: Work out and explain various variants of the production plan and inventory plan as well as the resulting procurement plan, on the basis of a long-term sales plan of a company in the wood industry.

On the basis of a long-term sales plan of a company in the wood industry, your task — with regard to resource management — will be to work out various variants of the production plan and inventory plan as well as the resulting procurement plan.

The case: The Planing Co. manufactures wood paneling in many different variants. Variants occur, of course, in the dimensions, but also in the profiled edges and the wood finishes. The company offers panels in both natural wood and in painted finishes. The Planing Co. has only one timber supplier, Forest Clear Co. in Finland.

As manager of the Planing Co., you are faced with the task of producing a master schedule for one year in preparation for a management meeting tomorrow morning. You are expected to provide information on capacity load and, in addition, on the quantities of raw material to be procured from your timber supplier.

Your job is to do the planning only for the four most important final products in Planing Co.'s varied product assortment. These four products are shown in Figure 5.7.4.1 below and fall into two product segments: painted finish panels (panel "tradition") or natural wood panels (bio panel).

Product segment	End product	width	length	height
Panel "tradition"	Top finish (profile 4)	97 mm	5 m	20 mm
Panel "tradition"	Top resin (profile 9)	97 mm	5 m	13 mm
Bio panel	Nordic spruce (profile 4)	97 mm	5 m	20 mm
Bio panel	Nordic spruce (profile 9)	97 mm	5 m	13 mm

Fig. 5.7.4.1 Final products requiring master planning.

These panels, already precut to size, are planed down to specific profiled panels at a number of processing centers. As Figure 5.7.4.2 shows, during the planing process there is a material loss of 3 mm to the width and of 2 mm to the height of a precut panel.

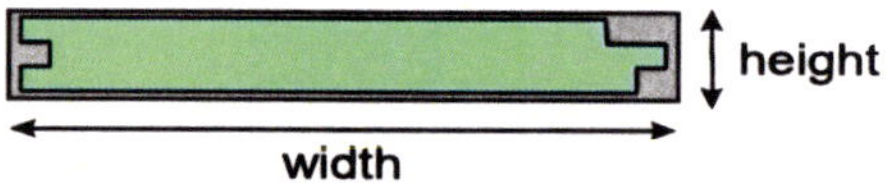

Fig. 5.7.4.2 Profiled edge of a finished panel.

The Planing Co. has machines to plane down the precut panels to specific profiled panels for a total of 2.7 million square meters of precut panels per year. The capacity unit, which comprises several machines, is given as square meters of material to be planed. You can assume that the same amount of material is processed every month.

a. *Production and inventory plan:* You will base your master planning on available data in the cumulative sales plan for the next 12 months (see Figure 5.7.4.3).

Product family	**End product**	**Sales plan, June – Nov. (m²)**					
		June	**July**	**Aug.**	**Sept.**	**Oct.**	**Nov.**
Panel "tradition"	**Top finish (profile 4)**	62,085	65,269	46,166	76,413	85,964	63,677
Panel "tradition"	**Top resin (profile 9)**	59,943	63,017	44,573	73,776	82,998	61,480
Bio panel	**Nordic spruce (profile 4)**	48,969	51,480	36,413	60,269	67,803	50,224
Bio panel	**Nordic spruce (profile 9)**	70,392	74,002	52,343	86,637	97,466	72,197

Product family	**End product**	**Sales plan, Dec. – May (m²)**					
		Dec.	**Jan.**	**Feb.**	**Mar.**	**Apr.**	**May**
Panel "tradition"	**Top finish (profile 4)**	41,390	42,982	52,534	58,901	50,942	63,677
Panel "tradition"	**Top resin (profile 9)**	39,962	41,499	50,721	56,869	49,184	61,480
Bio panel	**Nordic spruce (profile 4)**	32,646	33,901	41,435	46,457	40,179	50,224
Bio panel	**Nordic spruce (profile 9)**	46,928	48,733	59,563	66,783	57,758	72,197

Fig. 5.7.4.3 Sales plan for the next 12 months.

Taking into account the loss of material during the planing process, calculate the load profile according to Figure 1.2.4.1 and enter it into Figure 5.7.4.4. Discuss the result: Is there sufficient capacity?

Based on the load profile, create for the four products the following three variants of the production plan and enter them into Figure 5.7.4.4:

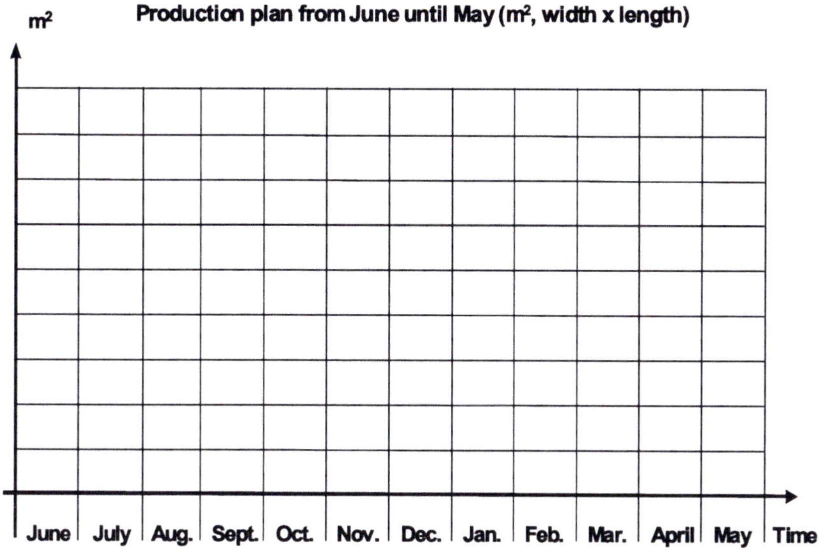

Fig. 5.7.4.4 Production plan for the next 12 months.

1. Each month the quantity produced is exactly the planned load that results from the planned demand. As a result, no inventory stock is produced, but costs are engendered for (quantitatively) flexible capacity (see the definition in Section 4.4.3).
2. Each month the quantity produced is the average load. Fluctuations in demand have to be covered by inventory. To ensure delivery reliability, initial inventory stocks of 180,000 m^2 must be carried (for the sake of simplicity, assume that there is appropriate inventory for all four final products). However, no costs arise for (quantitatively) flexible capacity.
3. Half of the capacity is adapted to the load. This means that each month, the quantity produced is one-half the difference between planned load (that results from the planned demand) and the average load. To ensure delivery reliability, initial inventory stocks of 90,000 m^2 must be held. Again, costs are engendered for (quantitatively) flexible capacity, but the costs are lower than in variant 1, above.

Conduct a qualitative comparison of the total costs of the three solutions above, by comparing the following two aspects. On the one hand, the inventory carrying cost:

- Unit cost: $2 per m2
- Annual carrying cost rate: 30%

On the other hand, the costs for flexibility of capacity:

- Labor cost: \$1 per m^2
- Flexibility percentage required =
  (maximum monthly load – average load) / average load
- Flexibility costs = flexibility percentage * labor cost per year

b. *Procurement plan:* The management at Forest Clear Co. has asked you to give them a rough estimate of the quantity of raw material that Planing Co. will order from them in the next 12 months. As upper management at Planing Co. has just recently decided to build a partnership relationship with this timber supplier, they expect you to respond to Forest Clear by tomorrow at the latest. Your answer will depend on which of the three variants of the production plan you decide is the best.

The raw material — the timber — is the same for all four final products. It is procured and calculated in units of cubic meters. However, as Forest Clear supplies boards of 100-mm width, 50-mm height, and 5-m length only, Planing Co. has to cut the boards to precut panels (see Figure 5.7.4.1) before the precut panels can be planed. Because of the dimensions of the final products, two to three precut panels can be obtained from each raw board (see Figure 5.7.4.5). The raw material must be available in the same month as the final products.

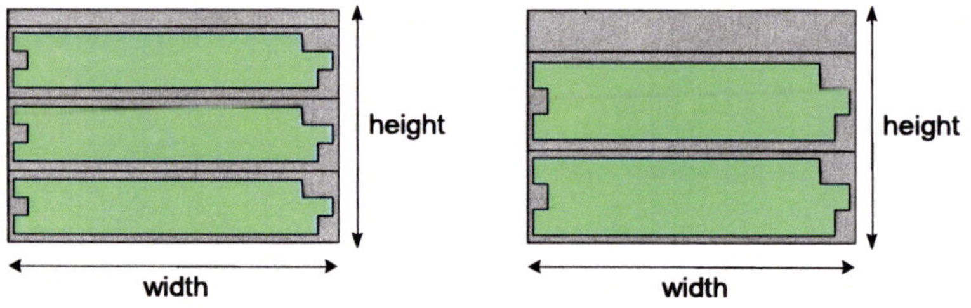

Fig. 5.7.4.5 Possible ways to cut the raw boards into panels.

Create a formula for calculating the raw material requirements for a given production plan. *Hint:* Derive the quantity of raw material in cubic meters (the wood boards) in dependency upon the specific final product, which is given in units of square meters. Company management is only interested in the total raw material requirements per month in Figure 5.7.4.6 (the raw material requirement per product is important only to establish the subtotals).

Raw material requirement for product	Procurement plan June – Nov. (m^2)					
	June	July	Aug.	Sept.	Oct.	Nov.
Top finish (profile 4)						
Top resin (profile 9)						
Nordic spruce (profile 4)						
Nordic spruce (profile 9)						
Total raw material requirement						

Raw material requirement for product	Procurement plan, Dec. – May (m^2)					
	Dec.	Jan.	Feb.	Mar.	Apr.	May
Top finish (profile 4)						
Top finish (profile 9)						
Nordic spruce (profile 4)						
Nordic spruce (profile 9)						
Total raw material requirement						

Fig. 5.7.4.6 Procurement plan: raw material requirements.

Solution:

a. The average load per month is about 237,000 m^2, slightly exceeding the available capacity of 225,000 m^2. Therefore, overtime of about 5% will be necessary to fulfill the demand (about 2,844,000 m^2 per year).

- Variant 1 results in flexibility costs of about $1,300,000. The maximum load is in October (about 345,000 m^2); its production requires a flexibility percentage of (345,000 – 237,000)/237,000 = 46%.
- Variant 2 of the production plan (production of 237,000 m^2 each month) results in a carrying cost of about $80,000. Carrying cost is calculated on the basis of the inventories at the beginning of each of the 12 months in the inventory plan.
- Variant 3 results in a carrying cost of about $40,000 and flexibility costs of about $650,000. Maximum production is in October (about 291,000 m^2), which requires a flexibility percentage of (291,000 – 237,000)/237,000 = 23%.

You can view the animated solution on the Internet at URL:

https://intlogman.ethz.ch/#/chapter05/master-planning/show

For all calculations, click on the "calculate" icon.

- Variants between the two extremes of Variant 1 and Variant 2 — as well as the variants themselves — can be produced by entering a value for alpha between 1 and 0 in the formula Av + alpha * ($Load_i$ – Av), where Av is the average load. $Load_i$ is the planned load that results from the planned demand.

- To calculate the costs of each variant, the parameters for carrying cost and flexibility cost can be changed.

b. For Variant 2 of the production plan, production per month is one-twelfth of the total annual demand. This results in raw material requirements of about 4900 m^3 per month.

A mouse click on the icon "go to procurement plan" takes you to calculation of the procurement plan for the chosen variant; once there, click on "calculate." The upper section shows the production plan for all variants; the lower section shows the raw material requirements. Run the mouse over the product identification numbers in the left-most column to see whether two to three precut panels can be cut out of a raw board.

To create another variant of the production plan, you can click again on the icon "return to production plan". The raw material requirements can be calculated also for that plan.

5.8 References

APIC16 Pittman, P. et al., APICS Dictionary, 15th Edition, APICS, Chicago, IL, 2016

ArCh16 Arnold, T., Chapman, S., Gatewood, A., Clive, L., "Introduction to Materials Management," 8th Edition, Pearson, Harlow, England, 2016

Bern99 Bernard, P., "Integrated Inventory Management," John Wiley & Sons, New York, 1999

BoGa08 Boutellier, R., Gassmann, O., von Zedtwitz, M., "Managing Global Innovation," 3rd Edition, Springer, Berlin, 2007

GoCo14 Goldratt, E., Cox, J., "The Goal: A Process of Ongoing Improvement," 30th Anniversary Edition, North River Press, Norwich, CT, 2014

IBM75 IBM, "Executive Perspective of Manufacturing Control Systems," Brochures IBM G360-0400-12, 1975

Schö95a Schönsleben, P. (Hrsg.), "Die Prozesskette 'Engineering,' Beiträge zum Stand von Organisation und Informatik in produktionsvorgelagerten Bereichen Schweizerischer Unternehmen 1995," vdf Hochschulverlag, Zurich, 1995

Schö01 Schönsleben, P., "Integrales Informationsmanagement: Informationssysteme für Geschäftsprozesse — Management, Modellierung, Lebenszyklus und Technologie," 2nd Edition, Springer, Berlin, 2001

6 The Lean / Just-in-Time Concept and Repetitive Manufacturing

Intended learning outcomes: Produce an overview on lean / just-in-time and repetitive manufacturing. Explain the lean / just-in-time concept. Describe the Kanban technique. Identify the cumulative production figures principle. Disclose an implementing procedure and a comparison of techniques.

In the 1970s, the seller's market changed to a buyer's market in many branches of the capital goods market. As a consequence, the weighting of entrepreneurial objectives (see Section 1.3.1) changed from stressing best possible capacity utilization to a focus on short delivery lead times. At the same time, however, companies had to avoid physical inventory. Inventory proved to be increasingly risky, because technological advances turned goods into nonsellers often overnight. Thus, *short lead time* became a strategy toward success in entrepreneurial competition.

To handle all of these aspects, concepts were developed — mainly in Japan — and grouped together under the term just-in-time, or JIT (pronounced as one word). More recently, the contents of JIT were re-launched under the new catchphrase "lean." Lean/JIT aims towards the fastest possible flow of goods while reducing overburdening, unevenness, and useless effort, or waste.

The lean / just-in-time concept has advantages for all other concepts and all characteristics of planning & control in Section 4.5.3. For this reason, we will give the methods associated with the just-in-time philosophy preferential treatment. The well-known and simple production and purchase control technique in this connection is the *Kanban technique*. Kanban, however, takes care of only short-term planning & control and can be used only in production or procurement with frequent order repetition — that is, in the manufacture of standard products, if need be with a few variants.

The Lean/JIT concept is not only an aid but is indeed a prerequisite for efficient use of all simple planning & control techniques in logistics. Figure 6.0.0.1 shows some of the characteristic features of planning & control, taken from Section 4.4. The values of the features as arranged from left to right correspond to an increasing degree of the suitability for simple techniques of planning & control in supply chains. On the table showing the most important features, the characteristic value is marked with a black background.

Simple techniques of planning & control are therefore particularly well-suited for manufacture of standard products, if need be with a few variants, and thus for *repetitive manufacturing*. As Kanban and the cumulative production figures principle are probably the most easily understood control techniques, they are discussed here in Part B. Moreover, the two techniques are techniques for short-term planning of materials, schedules, and capacity, whereby capacity must adapt to load. For long-term planning in these cases, methods appropriate to the MRP II concept (see Section 5.2) are used. If medium-term planning is necessary at all, methods will correspond to simple techniques of long-term planning.

P. Schönsleben, *Handbook Integral Logistics Management*,
https://doi.org/10.1007/978-3-662-65625-9_6

Feature	→	Values				
Features referring to user and product or product family						
Frequency of customer demand	→	unique		discontinuous (lumpy, sporadic)	regular	continuous (steady)
Product variety concept	→	according to (changing) customer specification	product family with many variants	product family	standard product with options	individual or standard product
Features referring to logistics and production resources						
Production environment	→	engineer-to-order	make-to-order	assemble-to-order (from single parts)	assemble-to-order (from assemblies)	make-to-stock
(quantitatively) flexible capacity	→	not flexible in terms of time		hardly flexible in terms of time		flexible in terms of time
Features referring to production or procurement order						
Reason for order release (type of order)	→	demand / [customer production (or procurement) order]		prediction / (forecast order)		use (stock replenishment order)
Frequency of order repetition	→	production / procurement without order repetition		production / procurement with infrequent order repetition		production / procurement with frequent order repetition

Increasing degree of suitability for simple techniques of planning & control

Fig. 6.0.0.1 Degree of suitability for the simple techniques of planning & control.[1]

6.1 Characterizing Lean / Just-in-Time and Repetitive Manufacturing

Intended learning outcomes: Explain just-in-time and jidoka: Increasing productivity through reduction of overburdening, unevenness, and waste. Describe characteristic features for simple and effective planning & control techniques of repetitive manufacturing.

6.1.1 Just-in-Time and Jidoka — Increasing Productivity through Reduction of Overburdening, Unevenness, and Useless Effort, or Waste

Intended learning outcomes: Identify the basic concepts of the Toyota Production System. Produce an overview on lean production and a lean enterprise. Differentiate between the conventional view of inventory and the Japanese view.

---

[1] The horizontal distribution of the values in the morphological scheme indicates their relation to the increasing degree according to the given criterion.

The origin of the just-in-time concept is in the Toyota Production System (see [Toyo98]).

The *Toyota Production System* (TPS) is a framework of concepts and methods for increasing productivity and quality.

The minimization of the so-called *3M*, "muri", "mura", and "muda" is a basis of TPS.

Overstraining or *excessive stress* (Japanese *"muri"*) refers to an unreasonable *overburdening* of human beings (physically or mentally) or machines.

With human beings, "muri" can entail exhaustion, injuries, unplanned absence, diseases, and even burnout. With machines, "muri" can entail interruptions and decreased availability.

Variation (Japanese *"mura"*) describes *unevenness* in the production system.

"Mura" can result, for example, from discontinuous demand, but also from changing product mix, differing times required for individual operations, or badly organized workplaces. "Mura" can spread out to the whole supply chain, which can entail, among other things, the bullwhip effect (here see Section 2.3.5). Leveling of the production along the entire supply chain (Japanese "heijunka") as an important tool for reducing "mura" requires a reduction of the lead time. For this, the following sections will deal with reduction of inventories, mixed-model and mixed production, as well as lot size reduction.

Waste, or *useless effort* (Japanese *"muda"*) *is seen as* all activities in development and manufacturing within the entire supply chain, extending to and including the consumer, that are non-value-adding from the customer's point of view.[2]

Ohno's seven wastes are overproduction, waiting, transportation, unnecessary inventory, unnecessary motion, making defects, inappropriate processing (e.g., physical work not suited to human beings or overprocessing that will not be paid by the customer). See [Ohno88].

The 3Ms interact mutually. Reducing "muda" without simultaneously reducing "mura" can result in "muri." For example, reducing inventories and simultaneously satisfying heavily discontinuous customer demand will overburden the production system all too often. This will, among other things, decrease quality, and thus entail "muda." Thus, reduction of "mura" is a prerequisite of a durable reduction of "muda." As "muri" entails, in general, "mura" and "muda," avoiding "muri" has priority. An example is the famous cord, by which employees of Toyota's assembly line can stop the line, not only because of defects ("muda") but also if they cannot follow the takt time, for instance because of being overburdened. The short-term stop of the line will then not be considered as "mura" or "muda." However, the reason for the overburdening as well as a feasible solution must be found quickly.

TPS encompass a set of methods, techniques, and tools to increase quality and speed without increasing "muri" or decreasing productivity. Jidoka and just-in-time are the two pillars of the TPS. Both pillars encompass methods, techniques, and tools for reducing all 3Ms.

---

[2] *Value-added from the customer's view* is the benefit view. This is different from value-adding from the view of company accounting, or costs view. See Sections 4.1.2 and 16.1.4.

The *jidoka concept* comprises approaches and techniques for immediately halting production when abnormal conditions occur. Jidoka means "automation (Jap. 自動化) with manlike sensors," which is rendered by the artificial term *"autonomation"* (Jap. 自働化).[3]

Jidoka therefore is aimed at the elimination of production of defective products by building quality into the production process. This way of thinking stems from 1902, when Sakichi Toyoda, who later founded the Toyota Motor Corporation, patented a device to stop a weaving loom as soon as a thread broke. This prevented the weaving of defective fabrics and allowed operators to fix the problem itself, thread breakage, as soon as possible. Section 18.2.5 shows some of the techniques of the Jidoka concept.

The *just-in-time (JIT) concept* encompasses a certain set of approaches, methods, and techniques for planned elimination of all waste. The primary elements are to have only the required inventory when needed; to improve quality to zero defects; to reduce lead times by reducing setup times, queue times, and lot sizes; and to incrementally revise the operations themselves. Cf. [APIC16].

The just-in-time concept thus increases the potential for short delivery lead times for all types of production and for many service lines of business.[4]

The terms *stockless production* or *zero inventories* as synonyms for just-in-time are misleading and thus not used in this work. After all, the Kanban technique does require inventory in buffers at all production structure levels. The misunderstanding resulting from this misrepresentation is probably also responsible for the fact that JIT was frequently understood and applied incorrectly and that, finally, the new catchword "lean" took the place of the term JIT.

Lean production emphasizes the minimization of the amount of all required resources (including time) for the various activities of the company (cf. [APIC16]). It involves identifying waste (see definition above) and eliminating them.

A *lean enterprise* applies the principles of lean production to all areas within the organization.

Since the time it was introduced, the philosophy of lean production [WoJo07] has often been taken to extremes. It served as a convenient justification for firing and not replacing staff members. Some people postulated polemically that the contradictory objectives of the company, as outlined in Section 1.3.1, could be resolved. They did not consider at all explicitly the target area of flexibility, for its objectives are usually long term in nature. The customer does not readily recognize that the building of such competencies is value-adding. This view led to company anorexia and resulting paralysis. It proved to be wrong at the very latest when companies became no longer capable of achieving innovations.

---

3 The difference is just the symbol 人 in the second "character," which means "human."

4 The term JIT has been somewhat unjustly, in its exclusiveness, assigned to the Kanban technique. Even a deterministic technique such as MRP aims to — and without inventory — procure/produce what is required at the moment, just in time.

The JIT concept originally introduced by the Japanese corresponds most extensively with the current concept of “lean,” which once again gives the target area of flexibility the consideration that it deserves. See here also Section 1.3.3. In any case, the aim is still to reduce useless effort, or waste. As to reducing inventory, Figure 6.1.1.1 shows the change in view of inventory that took place between 1970 and 1990.

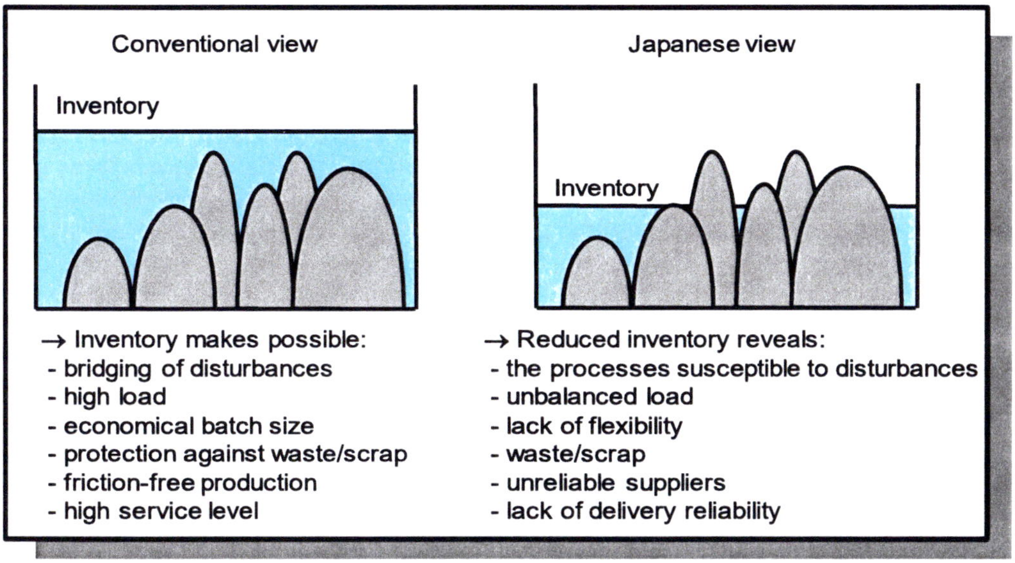

Fig. 6.1.1.1 Alternative views of inventory.

High inventory acts as a high water level (light background) in a lake that has shallows and shoals (dark background). If the water level falls, the obstacles will be felt and must either be removed or avoided through a change in course. Reducing inventory exposes problems that must be corrected by means of appropriate concepts. Japan gained this insight early on (see also [Suza12]).

6.1.2 Characteristic Features for Simple and Effective Planning & Control Techniques of Repetitive Manufacturing

Intended learning outcomes: Produce an overview on how the lean / just-in-time concept tends towards more simple techniques in place of more complicated techniques of materials and capacity management.

Figure 5.3.2.1 pointed out the reasons for simple or rather complicated techniques in materials management. Simple planning & control techniques require, as shown in the figure, *low-cost items* or at least *continuous frequency of customer demand*. In the case of dependent demand for expensive components, more continuous demand can be achieved through, for example, reducing lot sizes but also through a product concept with fewer variants or even standard components. More simple techniques can then be implemented in place of more complicated techniques of materials management. To do this, some important methods were developed within the JIT concept that reduce "mura." They lead to production or procurement with frequent order repetition — that is, to *repetitive manufacturing*.

The repetition of the same processes creates a potential for automation in administration. Continuous frequency of customer demand allows production or procurement with order release according to consumption, or a simple (stock) replenishment.

Figure 5.3.2.2 pointed out the reasons for a "good" situation or a situation that should be avoided whenever possible in materials management. Following that figure, the situation becomes "better" the further upward that the order penetration point (OPP) can be set, that is, the more there is *deterministic demand*. As the assumption must be that the customer tolerance time does not lengthen, the cumulative lead time must be reduced. Some of the best-known JIT/Lean methods to reduce "muri," "mura," and "muda" are precisely those that increase the potential for short lead times.

Figure 5.3.4.1 pointed out the reasons for simple or rather complicated techniques in capacity management. As the Figure shows, simple planning & control techniques require flexibility of capacities along the time axis. Accordingly, the JIT concept contains important methods and guiding principles to reduce "muri," in order to achieve (quantitatively) flexible capacity and thereby, in turn, reduce "mura" and finally "muda."

6.2 The Lean / Just-in-Time Concept

Intended learning outcomes: Explain lead time reduction through reduction of setup time and batch size as well as further concepts. Describe line balancing through harmonizing the content of work. Disclose just-in-time logistics. Present generally valid advantages of the lean / just-in-time concept for materials management and for capacity management.

The following discusses the most important of the methods and techniques of the lean / just-in-time concept. See here also [Wild01].

6.2.1 Lead Time Reduction Through Setup Time Reduction and Batch Size Reduction

Intended learning outcomes: Identify a simple formula for operation time. Produce an overview on setup-friendly production facilities. Present cyclic production planning and leveling of the production ("heijunka"). Describe harmonizing the product range through reduction of variants and a modular product concept. Explain single-minute exchange of dies (SMED).

Most simply reckoned, lead time is the sum of *operation times* and *interoperation times* plus *administration time*. In job shop production, operation time determines in part queue time at a work center, which makes up a significant portion of interoperation time. Reducing operation time, therefore, has both a direct and indirect effect. The simplest definition of operation time can be expressed as the formula in Figure 6.2.1.1. This definition appeared in Figure 1.2.3.1, but here the figure shows commonly used abbreviations that will be useful later on.

(Operation time) = (setup time) + [(lot size) (run time per unit)]
or OT = ST + (LOTSIZE RT)

Fig. 6.2.1.1 The simplest formula for operation time. [5]

The simplest way to reduce operation time is through reduction of batch or lot size. A company can even aim at batch sizes that fulfill only the demand of a day or a few days. Then, the same order is repeated at short intervals, which leads to processes that can be better automated.[6] Smaller batch size, however, does result in more setup (if producing the same overall quantity) and thus greater capacity utilization. In case of high utilization, this increases lead time (here see Section 10.2.3). Increased setup also causes higher costs. Conversely, a significant reduction in setup time allows — with keeping utilization constant — to reduce lot size, thus operation time and finally also lead time. The following shows how setup-time reduction can be achieved.

1. *Setup-friendly production facilities:*

The construction of specific devices (such as gauges or dies) for setup sometimes allows drastic reduction in setup time even where there are existing specialized machines. Another possibility is to use the machines by means of programmable systems such as computer numerical control (CNC) machines, industrial robots, or flexible manufacturing systems (FMS).

2. *Cyclic planning:*

Cyclic planning attempts to sequence the products to be manufactured by a machine in such a way as to keep total setup time at a minimum.[7]

Cyclic planning is an example of *sequencing*, the planning of optimum sequences. Cyclic planning yields a basic cycle, as Figure 6.2.1.2 shows.

In a cyclic manner, batches of parts A, B, E, D, and C are manufactured. It is simple to introduce variations in order quantities; additional batches are planned for a part at the same point that has been planned for that part in the basic cycle. Varying the quantity according to current requirements could also result in a cycle of A, E, E, D, and again A.

Reducing the setup time allows for reducing the lot size. Therefore, instead of producing a big lot of each product (e.g., 1000 A, 4000 E), several cycles of smaller lots can be produced (e.g., 4 * 250 A, 4 * 1000 E). This leads to the principle of leveling of the production.

---

5 For definitions of these terms, see Sections 1.2.3 and 13.1. For detailed explanations of the following relationships, see Sections 13.2.2 and 11.3.

6 From this idea stems the concept of *one less at a time*, that is, the process of gradually reducing the lot size to expose, prioritize, and eliminate waste (cf. [APIC16]).

7 Well qualified for it are *co-products*, that is, products that are usually manufactured together or sequentially because of product or process similarities ([APIC16]).

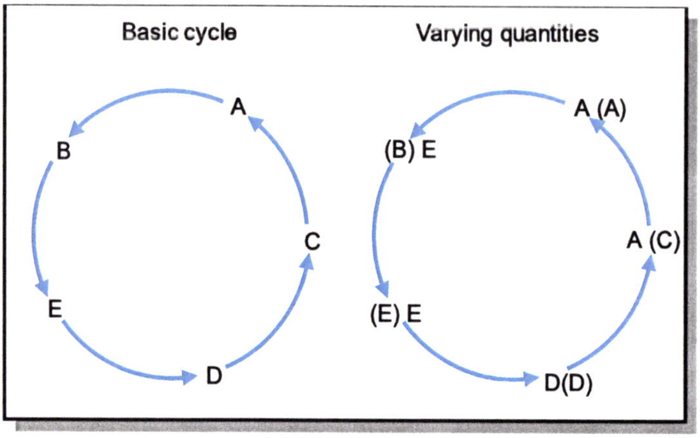

Fig. 6.2.1.2 Cyclic production planning.

Leveling of the production (Japanese *"heijunka"*) is an approach to level highly discontinuous production orders throughout the supply chain to match the planned rate of more continuous customer demand. It is an important tool for reducing "mura."

"Ideally," a product should be produced on the day it will be shipped.

3. *Harmonizing the product range through a modular product concept:*

Harmonizing the product range is reducing the number of different components and process variants required to manufacture a range of products, at times involving the reduction of the product range itself.

Harmonizing the product range thus means *reduction of variants*. The cost advantage is a reduction in overhead (see Section 16.4). Moreover, it simplifies logistics, because it leads to a more balanced flow of goods. A reduction in product variants results, namely, in goods production in sequences of similar operations. With identical goods, this reduction will even result in production with frequent order repetition. Each of these allows successive orders to be processed without major change in equipment, such as machines, for example. *Setup times* in the system decrease. In addition, because of fewer different processes, setup tasks become easier, because they repeat themselves and can be better automated. Also, frequent order repetition entails a more continuous demand of components, and thus a reduction of "mura," as well as more simple techniques of materials management (see Figure 5.3.2.1).

Conversely, a modular product concept (here see Section 1.3.3) allows offering larger product families without increasing the number of components and operations. By standardization of interfaces between the (families of) components and the product family, variants of one component family can be combined with variants of another component family on a bigger scale.

4. *Reducing idle time of production facilities:*

The term *single-minute exchange of dies (SMED)* refers to methods aimed at reducing idle time of production facilities, according to Figure 6.2.1.3.

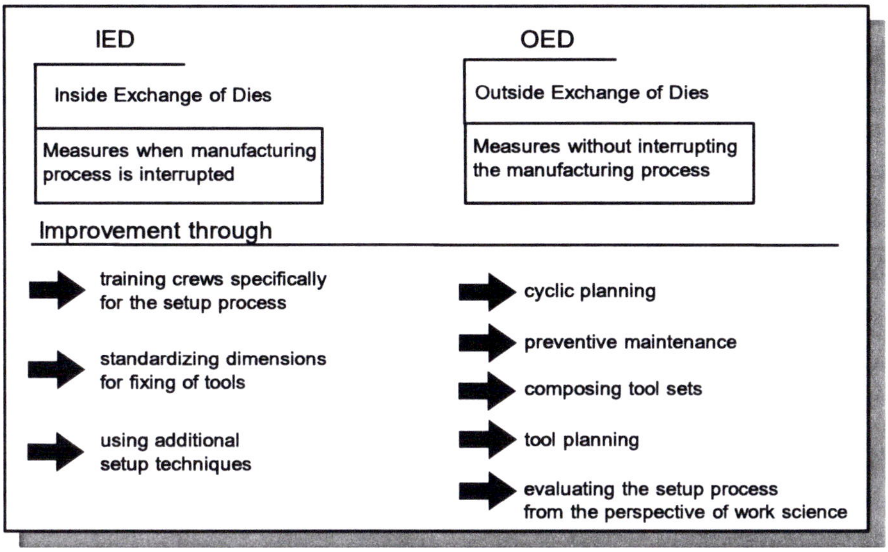

Fig. 6.2.1.3 Concepts of reducing setup time. (Source: [Wild89]).

These methods were developed primarily in Japanese industry (see [Shin85] or [Shin89]). In principle, there are two kinds of setup operations:

- *Internal setup (time)* or *inside exchange of dies (IED)* takes place when the workstation is stopped or shut down.
- *External setup (time)* or *outside exchange of dies (OED)* takes place while the workstation is still working on another order.

SMED is composed of the entire setup process, including insertion and removal of special setup devices, or dies. SMED reduces idle time of the system by means of shifting portions of IED to OED. This method is comparable to a pit stop during a formula-one race. SMED encompasses measures for reducing all 3Ms.

6.2.2 Lead Time Reduction Through Manufacturing Segmentation, Cellular Manufacturing and One-Piece Flow

Intended learning outcomes: Produce an overview on production or manufacturing segmentation. Explain cellular manufacturing, one-piece flow, and the formula for lead-time calculation with cellular manufacturing.

Production segmentation, or *manufacturing segmentation*, is the formation of organizational units according to product families instead of job shop production.

Elimination of organizational boundaries interrupting flow allows autonomous responsibility for products to arise (similar to line production). Figure 6.2.2.1 shows:

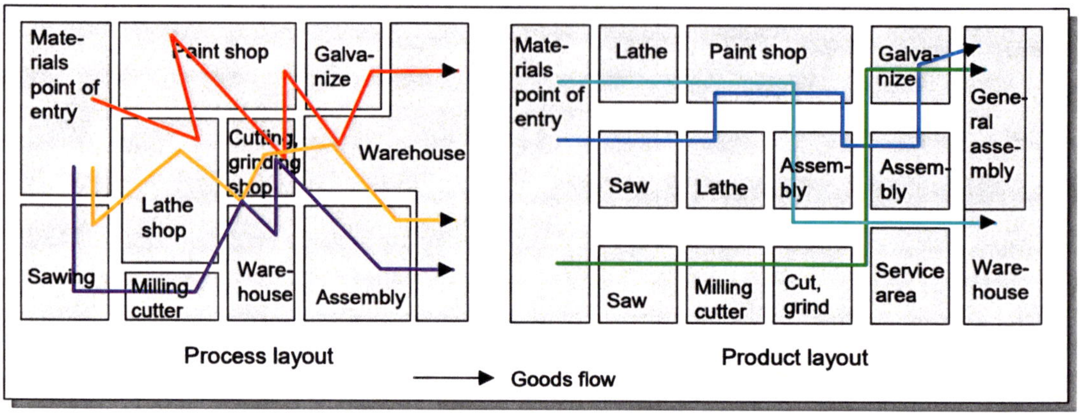

Fig. 6.2.2.1 Production or manufacturing segmentation. (Example taken from [Wild89]).

- In the upper section, an example of a process layout: Operations of a similar nature or function are grouped together, based on process specialty (for example, saw, lathe, mill).
- In the lower section, an example of a product layout: For each product (here with the exception of painting and galvanizing) there is a separate production line, or manufacturing group, but no longer any central job shop for each task.

There are cost factors that restrict the splitting of certain areas (such as galvanizing, painting, tempering), but an appropriate total layout and capacity reserves will ensure rapid throughput. Small- and medium-sized companies are often faced with the problem of special treatments for which they must rely on external refining and finishing companies. Because of the recent weight placed on setup time, however, ever more new facilities for such areas are being offered, such as paint shops that set up lacquer colors in a matter of minutes.

Applying production or manufacturing segmentation consistently leads to focused factories.

A *focused factory* is a plant established to focus on a limited set of products or product families, technologies, and markets, precisely defined by the company's competitive strategy and economics (see [APIC16]).

A further application of manufacturing segmentation leads to cellular manufacturing.

A *cell* is, according to [APIC16], a manufacturing or service unit consisting of a number of workstations, and the materials transport mechanisms and storage buffers that interconnect them.

A *work cell* is, according to [APIC16], a physical arrangement where dissimilar machines are grouped together into a production unit to produce a family of parts having similar routings.

The process of cellular manufacturing is closely linked with work cells.

In *cellular manufacturing*, or *near-to-line production*, workstations required for successive operations are placed one after the other in succession, usually in an *L-shaped line* or *U-shaped line* configuration. Preferably, the individual units of a batch go through the cell according to the one-piece-flow concept.

One-piece flow is a concept that processes items directly from one step to the next, one unit at the time, that is, without having to wait for the other units of the batch between any two operations.

Figure 6.2.2.2 illustrates this concept, showing the change from job shop production to cellular manufacturing.

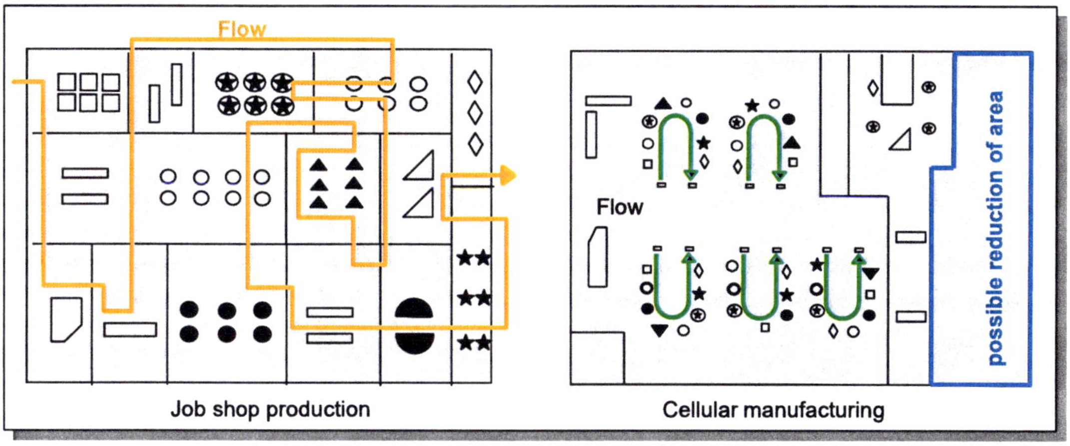

Fig. 6.2.2.2 Changeover to cellular manufacturing.

As cellular manufacturing may require multiple machines, it is not unusual to find older machines, retrieved from the "cellar" so to speak, in these lines. While this is specialized machinery that has a dedicated capacity,[8] it is inexpensive enough, for generally it has already been depreciated.

Efforts to identify business processes and reorganize them (business process reengineering) can also lead to the distributing of machines in lines that correspond to the new business processes. Cellular manufacturing is, moreover, significantly easier to control than job shop-type production. And, in many cases, less area is required for the machines.

Cellular manufacturing and one-piece flow can achieve a lasting reduction of lead time, and thereby also of work-in-process and many other forms of "muda." On the one hand,

[8] A *dedicated capacity* is a work center that is designated to produce a single item or a limited number of similar items. A *dedicated equipment* is equipment whose use is restricted to specific operations on a limited set of components (cf. [APIC16]).

interoperation times can be reduced to zero. On the other hand, cellular manufacturing allows overlapping operations (Section 13.4.2), as shown in the following.

Using the definition in Figure 6.2.1.1, the lead time of an order — assuming a *sequence of operations* and omitting interoperation times and administration times — is the sum of all n operation times, as shown in Figure 6.2.2.3 (for details, see Section 13.3.2).

$$\text{LTI} = \sum_{1 \le i \le n} \text{OT}[i] = \sum_{1 \le i \le n} \{\text{ST}(i) + \text{LOTSIZE} \cdot \text{RT}[i]\}$$

Fig. 6.2.2.3 Formula for lead time with a sequence of operations.

With cellular manufacturing, the estimate in Figure 6.2.2.4 is calculated:

$$\max_{1 \le i \le n} \{\text{ST}(i) + \text{LOTSIZE} \cdot \text{RT}[i]\} \le \text{LTI} \le \max_{1 \le i \le n} \{\text{ST}[i] + \text{LOTSIZE} \cdot \text{RT}[i]\} + \sum_{1 \le i \le n^*} \{\text{ST}[i] + \text{RT}[i]\}$$

*but without the longest operation

Fig. 6.2.2.4 Formula for lead time with cellular manufacturing.

To understand this formula intuitively, consider the following: The longest operation, the so-called *cell driver*, provides the minimum lead time. The other operations overlap. Lead time then increases at most by setup and *one* run time per unit of all other operations. In concrete cases, lead time will fall at some point between the minimum and the maximum.

6.2.3 Further Concepts of Lead Time Reduction

Intended learning outcomes: Disclose the effect of standardizing the production infrastructure and flexible capacities. Describe structuring assembly processes and complete processing. Identify point-of-use inventory and point-of-use delivery.

The following approaches also require adaptation of the production infrastructure. The first approach reduce wait times, the second approach reduces operation time, the third reduces lead time for several operations, and the fourth reduces transport time.

1. *Standardizing the production infrastructure, (quantitatively) flexible capacity, and increasing the flexible capability of capacities:*

Close-to-maximum capacity utilization results in a strong increase in wait time.[9] Overcapacity brings load variation under control and allows short lead times. If capacity is costly, however, overcapacity must be carefully reviewed.[10] First, the following measures should be examined:

---

[9] Section 13.2.2 explains this important phenomenon in detail.

[10] As in the TPS the yield factor *is almost 100%, O*verall equipment effectiveness (OEE), see Section 14.1.1) is not a focus of TPS. Rather, if customer orders fail to appear, there should be *no* production.

- Can we standardize the machinery, tools, and devices — either through greater versatility or by means of standardizing operations? This would allow broader implementation of personnel, which would result in fewer workstations and simpler planning. Airlines, for example, strive toward identical cockpits in their fleets of planes.
- Can the flexible capability of the workforce be increased through training and broader qualifications? If so, employees can be implemented in a more balanced fashion along the time axis, because if there is underload at their own work centers, they can be moved to overloaded work centers.
- Can we increase the availability of production facilities, particularly tools? The employees at a work center can also be trained to do their own repairs and maintenance jobs, as the necessity arises.

2. *Structuring assembly processes:*

In the assembly process, staggered supply of components reduces lead times, as shown in Figure 6.2.3.1. This is a well-known measure, especially in connection with customer order production.

The inbound deliveries in Figure 6.2.3.1 may be preassemblies or assemblies. Preassembly made parallel to assembly reduces the number of storage levels. If quality control is integrated into assembly, lead time can be reduced even further.

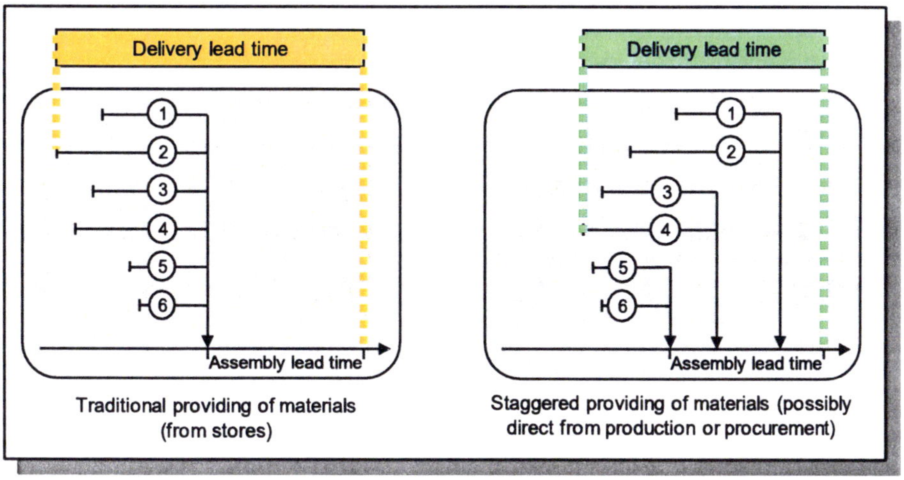

Fig. 6.2.3.1 Assembly-oriented providing of components.

3. *Complete processing:*

Complete processing is the execution of several different operations at a stretch — if possible, all the way up to completion of the product.

The newer tool machines often allow complete processing. With computer numerical control (CNC, DNC), they are versatile in implementation. Moreover, they are more independent in terms of cost as well as output and quality of employee performance.

There are fewer stations to run through with complete processing, so that there are no inter-operation times. Reduced lead times should result. But for this to have a true advantage over the segmentation in approaches 1 and 2, the complete processing duration must be significantly shorter than the sum of operation times with a sequence of machines. Otherwise, the result would be simply that several shorter wait times would be replaced with one single wait time. This time would be just as long as the sum of the shorter times, however.

For complex workpieces, a firm could investigate the possibilities of automation of production with flexible manufacturing systems (FMS) and automation of transport and handling. Modern technological machines are designed to reduce setup time and achieve greater variant flexibility. Automated processes also reduce the problems of 24-hour shift work.

4. *Organizing supply and buffer storage to support the flow of goods:*

The *point of use* is in the focus of delivery and storage.

- *Point-of-use inventory*: Buffer storage is placed directly at the spot where the components will be used (inbound stockpoint). Each container of components has its own specified physical location. On the assembly line, for example, it will stand at the location where the components will be installed.
- *Point-of-use delivery*: Fast connections are set up between suppliers and users. Components are delivered right to the buffer storage at the user workstation. The workstation can transmit its needs via electronic mail.

6.2.4 Line Balancing — Harmonizing the Content of Work

Intended learning outcomes: Disclose how tasks of the same duration at each production structure level result in a rhythmic flow of goods. Explain why the various operations at a workstation (for all products) as well as the various operations for a single product should be of about the same duration. Present measures for changing lead time of operations.

Line balancing balances the assignment of the tasks to workstations in a manner that minimizes the number of workstations and the total amount of idle time at all stations for a given output level (cf. [APIC16]).

Line balancing is particularly important for *line manufacturing*, that is, repetitive manufacturing performed by specialized equipment in a fixed sequence (i.e., an assembly line). Line balancing is an important tool for reducing "mura" and can be realized by harmonizing the content of work.

Harmonizing the content of work means to design the following so that they require the same length of time: (1) the various production structure levels, and (2) the times required for individual operations within a production structure level.

This concept can — by the way — also be very useful in a job shop production environment.

With regard to (1), production structure levels must be designed or redefined in such a way that lead times at the individual levels are either identical or multiples of each other.

Harmonization thus demands close cooperation between design and product engineering (simultaneous engineering). Product and process must be designed together from the start. Figure 6.2.4.1 illustrates this principle at the levels of assembly, preassembly, and parts production. The lead time for parts production is half as long as that for the levels of preassembly and assembly. In the example, the batch size at the part production structure level comprises half the usage quantity for a batch in preassembly or assembly.

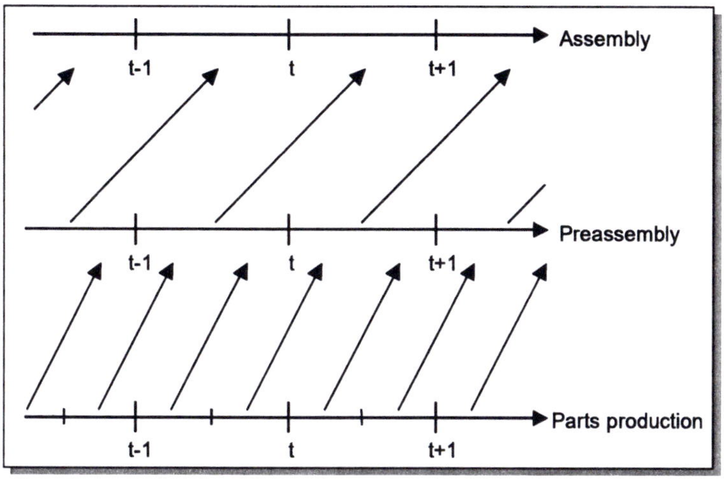

Fig. 6.2.4.1 Harmonizing the content of work: tasks of the same duration at each production structure level result in the rhythmic flow of goods.

With regard to (2), the following should be of the same approximate duration: the various operations at a workstation for all the products, and all the operations for a single product. Figure 6.2.4.2 illustrates this principle.

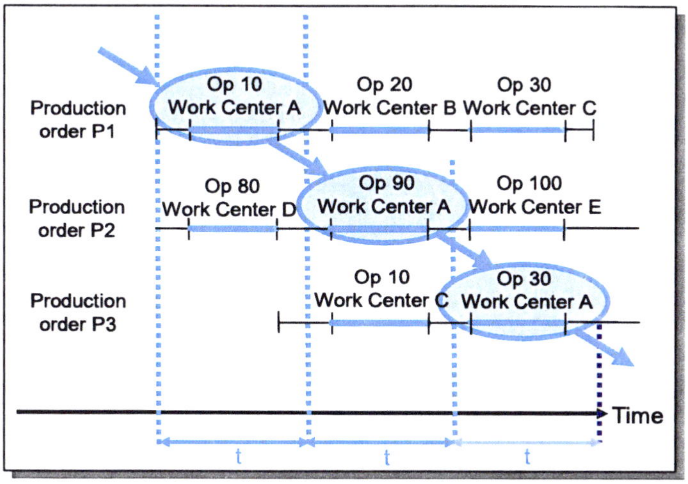

Fig. 6.2.4.2 Harmonizing the content of work: The various operations at a workstation (for all the products) as well as the various operations for a single product should be of the same approximate duration.

There will be little variation of the operation time, and this results in turn in a reduction of lead time. Queue time, except for its dependency on capacity utilization and average operation time, is, namely, a function of the variation coefficients of operation times.[11] In job shop production, queue times at the workstations to a large extent determine inter-operation times, which themselves have a significant effect on lead time.[12]

Such harmonization of the content of work within a production structure level and throughout all levels of production results in a rhythmic flow of goods. Batch size reduction alone cannot achieve this. Workstations and the content of the individual operations must be newly defined. This is a very difficult task that can only be surmounted when product engineering cooperates with development and design. New technologies may be used for certain operations in order to change lead time at the very location where harmonization is required.

To complete the task, product engineering in cooperation with design must repeat the following two steps until sufficient results are achieved:

1. Determine the *duration of one unit of harmonized content of work*, that is, the operation time of the harmonized content of work including necessary inter-operation times before and after the (internal or external) operation. To start, experienced personnel in product engineering determine this time unit empirically. For further iterations, the new time unit will result from correction of a previously unsatisfactory result. The shorter the harmonized content of work is, the more flexibly processes can be put together.
2. Perform measures to change lead times of operations, chosen from the various possible measures in Figure 6.2.4.3.

- Combine operations through automation, thus reducing total lead time of previous operations. Or, analogously, split an operation, thus lengthening the lead time of the operation.
- Change the process by changing the production technique (sticking instead of screwing, a different surface hardening technique, etc.)
- Reduce setup times in order to reduce batch sizes. If capacity is not being utilized, batch size can also be reduced directly. The advantage of harmonizing, however, must supersede the resulting increase in setup costs.
- Purchase different components that allow for a different process, where, in a targeted way, operations will be longer or shorter. The components will be either more or less expensive.
- Purchase semifinished goods in order to avoid operations that do not allow harmonization. Possibly the supplier can perform such an operation better within its own order and production infrastructure.
- Assign operations to subcontractors or take over operations from subcontractors, if this results in a better unit of harmonized content of work. Change subcontracting concepts and subcontract to subcontractors that are better suited with regard to better lead time.

Fig. 6.2.4.3 Measures for changing lead time of operations.

[11] For a detailed explanation of this important phenomenon, see Section 13.2.2.

[12] These phenomena are explained in Sections 13.2.1, 13.1.3, and 13.1.1.

Because of measures connected with suppliers, harmonizing work contents leads to closer cooperation with other companies. Because control must no longer observe priority rules, the advantage resulting from all these measures to harmonize the content of work is very simple management of queues. Such detailed and comprehensive measures to harmonize the content of work are comparable to the design of a railway timetable of departures at regular intervals: Investiture in new lines takes place as a function of postulated rhythms in the regular interval departure plan. As a result, processes in the railway net can be automated, and there is maximum throughput through the net.

The strategic considerations underlying these measures are long term in nature and can be put into practice only if a company's financial policies are in agreement with them. Investiture in capital assets will not accord well with savings. Whenever possible reduction of delivery lead time is projected, the response of the customer to the improvement becomes the important factor. Estimating possible effects of this kind is a matter for decision making at the company level. It is for this reason that traditional profitability calculations generally fail here. However, the calculations are often not necessary at all. A company attempting to stay in competition will be forced to make the investment.

6.2.5 Just-in-Time Logistics

Intended learning outcomes: Produce an overview on measures for motivation, qualification, and empowerment of employees as well as employee involvement (EI) and quality circles. Describe concepts such as genchi genbutsu, kaizen, poka-yokero, Andon, 5S.

Just-in-time logistics comprises the measures to reduce the 3Ms and the lead time discussed in Sections 6.2.1 to 6.2.4. Beyond these, comprehensive concepts and measures will be required in the following areas:

Motivation, qualification, and empowerment of employees: In a just-in-time environment, operators' jobs no longer include only direct productive labor, but also planning & control tasks. As a consequence, their jobs are enriched (this is best described by the term *job enrichment*), but the importance of training and motivation increases. In Japan, a complicated system of bonuses, public commendation, promotions, and so on supports personnel motivation. The result of this type of personnel management, rarely seen in Europe or in North America, is devotion of employees to their duties and to their companies. In the framework of JIT logistics anywhere, a Japanese way of thinking appears. This Japanese approach is summarized in brief in Figure 6.2.5.1.

This kind of motivation leads ultimately to a quantitatively flexible workforce through the course of time. This allows some control of fluctuations in a logistics system set up for continuous demand. There are cases where 25% of overload can be handled by "normal" overtime by employees, 25% by "special" overtime, and 50% by scheduling employees' hours according to need.

Quality assurance is performing actions to ensure the quality of the goods:

- Quality at the source: As buffers at user sites are minimal and the order quantities correspond exactly to the demand, no faulty products may leave the producer.

The group takes priority (the individual "disappears" within the group).

- A "sense of the whole" makes conflict among different areas much less frequent than, for example, in Europe or in North America. At Toyota, for example, university graduates in all fields undergo a 2-year training program through all areas of production.
- *Employee involvement (EI)* — such as in quality circles — promotes acceptance of innovations and expands the quality concept to total quality management (TQM).
- Cultivating a problem-solving orientation, based on reality.
  - "gembutsu": the real thing.
  - "gemba": the place where the truth can be found (e.g., at the customer's site, at the shop floor)
  - "genchi genbutsu": go and see for yourself
- Continual improvement involving everyone (*kaizen*, see Section 18.2.8) is a major element. This may be supported by a corresponding system for improvement suggestions.
- Waste or non-value added is eliminated, and this forms the basis for increased profit.
- Shortages and defects become visible (preferably by means of sensors), so that they can be eliminated.
  - In the case of defects, production stops.
  - Continual process improvement eliminates the causes of defects.
- Simple, "foolproof" techniques (poka-yoke) are preferred; visual control systems (andon) are more effective than numbers and reports. For details, see Section 18.2.5.
- Order and cleanliness improve the morale of the operators. White work uniforms are worn on the shop floor. The *five Ss* (see also [APIC16]):
  - "seiri" (sort): separate needed items from unneeded ones and remove the latter.
  - "seiton" (simplify): neatly arrange items for use.
  - "seiso" (scrub): clean up the work area.
  - "seiketsu" (standardize): do first three S daily.
  - "shitsuke" (sustain): always follow the first four S.
- "Even small details are important."

Fig. 6.2.5.1 Japanese approach.

- *Quality circles* of employees build quality consciousness and achieve the desired level of self-control of quality. They evaluate the measures set to ensure quality and the objectives achieved. Employees are thus encouraged to identify with their tasks and the quality of items that they produce and thus develop a feeling of responsibility for the products they manufacture.

- *Integrated procurement and supply chain management:* These are measures to reduce purchasing lead time. Suppliers are included in planning, sometimes as early as the development phase (see Section 2.3). The flow of information to suppliers includes long-term components, such as blanket orders (see Section 5.2.4), and short-term components for blanket release (see Section 6.3). To be able to issue blanket orders, the user must have reliable long-term planning for the components and work to be purchased. Suppliers are not only selected on the basis of low prices, but also according to the criteria of delivery reliability, quality, and short delivery lead times. There is an advantage to having local suppliers (distance, strikes, etc.).

6.2.6 Generally Valid Advantages of the Lean / Just-in-Time Concept for Materials Management

Intended learning outcomes: Describe the effect of forecast errors through the combining of requirements in batches across many production structure levels. Explain the effect of longer and shorter lead time on the (customer) order penetration point.

The lean / just-in-time concept discussed in Sections 6.2.1 to 6.2.5 can also aid the MRP technique (see Section 12.3), which often does not achieve satisfactory results in the quasi-deterministic case. Lean/JIT corresponds exactly to the above two demands. Thus, production and procurement costs decrease.

1. *Reduction of batch or lot size* results in combining fewer requirements in production or procurement batches at all levels. This is particularly important for lower production structure levels, where forecast errors will affect orders for components that will be required for end products far in the future. Figure 6.2.6.1 shows the positive effect that results if a batch sizing policy of *lot-for-lot* — every requirement is translated into exactly one order — can be achieved (see also Section 12.4.1).

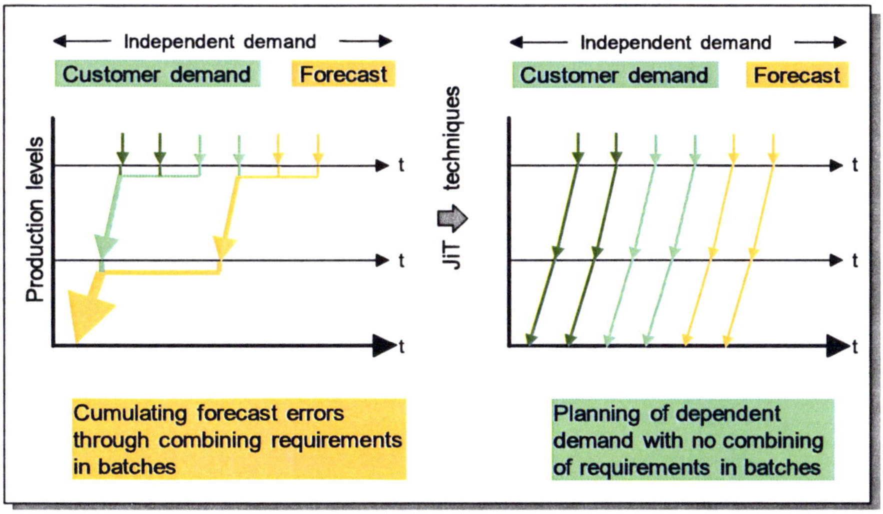

Fig. 6.2.6.1 Effect of forecast errors through the combining of requirements in batches across many production structure levels.

On the one hand, demand at lower production structure levels becomes more continuous, which with any stochastic technique results in smaller safety stocks. In the quasi-deterministic case, it is sometimes even possible to change over to purely stochastic techniques. On the other hand, the probability of production or procurement errors due to forecast errors decreases, because time buckets are reduced and orders are released only for requirements forecasted for the near future.

With product families with many variants (and thus nonrepetitive production to customer order), the prerequisite is lot size one (or batch size one). Here, companies have always been faced with the problem of how to reduce setup time. Lean/JIT is thus also advantageous for deterministic materials management.

2. *Lead time reduction* allows a (customer) order penetration point lower in the product structure. Figure 6.2.6.2 shows this positive effect.

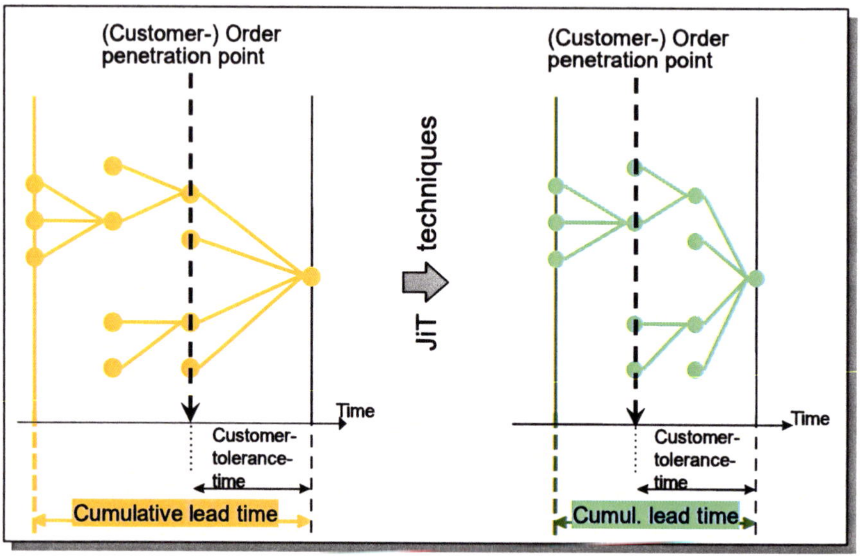

Fig. 6.2.6.2 (Customer) order penetration point with longer and shorter lead time.

The customer tolerance time now corresponds to a greater portion of the — now shortened — cumulative lead time. Thus, a larger part of value-adding processes lie within a deterministic area. Forecast errors affect a smaller part of the value-adding chain. Because forecasts pertain to the nearer future, forecasted demand is smaller.

Through increased production within the required delivery lead time, certain orders can now be produced — thanks to lead time reduction — for which there can be no stockpiling for economic reasons. This is the case with nonrepetitive production. In this way, additional sales can be realized. This is a further example of the advantages of the lean / just-in-time concept for deterministic materials management.

6.2.7 Generally Valid Advantages of the Lean / Just-in-Time Concept for Capacity Management

Intended learning outcomes: Explain how the lean /JIT concept reduces queue time. Describe how the lean /JIT concept allows for simpler control techniques.

The setup time reduction in Section 6.2.1 can achieve sufficiently short setup times even without cyclic planning. In this case, it is no longer necessary to reduce setup times by forming groups of orders for which a queue and thus some buffer inventory is required. Because priority rules for the waiting orders before the workstations are less necessary or fall away altogether, control in job shops becomes less complex.

(Quantitatively) flexible capacity was discussed in the first part of Section 6.2.3 as the lean /JIT concept. What is meant is the practice of scheduling extra capacity or flexible capacity that can continually adapt to load. This measure has the following positive consequences:

- It reduces queue time. This is crucial if the focus is on delivery reliability, given that queue time is one of the least predictable factors in job shop production. If queue time varies little or is very short, this improves planning, particularly for several production structure levels. Decreased size of production areas, thanks to smaller inventory in queues to the workstations, represents a further advantage.
- It allows for simpler control techniques, such as the Kanban technique. But any technique at all of capacity planning will function better. In addition, when an IT-supported control technique (e.g., with ERP or SCM software) is introduced, there will quickly be external costs of at least $100,000 and total costs of at least three times that amount. Therefore, increasing capacity can prove to be a viable alternative, particularly when it can be implemented more rapidly than a computer control technique. In a medium-sized Swiss electronics manufacturing firm, for example, the purchase of two additional coiling machines resolved a bottleneck in capacity. In this way, the firm was not only able to avoid investing in an expensive control technique, but also chose a measure that could be put into effect immediately.

6.3 The Kanban Technique

Intended learning outcomes: Explain Kanban as a technique of execution and control of operations as well as of materials management. Disclose long-term planning for Kanban.

Kanban is a production control technique that is fitting with the JIT concept. The Toyota Company began to develop Kanban in the 1960s. It became well known in connection with the TPS (here see [Ohno88]). Orders to withdraw required parts from suppliers and feeding operations are released directly by the work centers. Kanban represents the control portion of a planning system that is often not mentioned in the literature. The Kanban technique presupposes demands to be as continuous as possible along the entire value-adding chain. In other words, this should be production or procurement with frequent order repetition.

6.3.1 Kanban: A Technique of Execution and Control of Operations

Intended learning outcomes: Describe the Kanban card. Explain the Kanban feedback loop. Present Kanban rules of order release and control of the feedback control system.

Kanban (Japanese for card, or visible record) is a reusable signal card that passes back and forth between two stations. It is thus a kind of traveling card.

Buffers are kept at the user operation. These stores will contain, for example, a maximum number of standard containers or bins (A) holding a fixed number of items (k). The order batch size will be a set of containers (A). The *Kanban card* is a means to identify the contents of the container and to release the order. The card will look alike the one in Figure 6.3.1.1.

STOCK LOCATION	: ***5E215***	SUPPLIER (OPERATION):
ITEM ID.	: ***366'421'937***	
DESCRIPTION	: ***gear***	*lathe*
MODEL TYPE	: ***Z 20***	USER (OPERATION):
CONTAINER CAPACITY: ***20***	CONTAINER TYPE: ***B***	CARD NUMBER: ***4 of 8***
		grind

Fig. 6.3.1.1 Example Kanban card. (Taken from: [Wild89]).

The term Kanban, meaning signboard, is formed from the characters for "to look at closely" and for "wooden board," as shown in Figure 6.3.1.2. Kanban was the word used for decorated shop signs that came into use in merchant towns in the late 1600s in Japan.

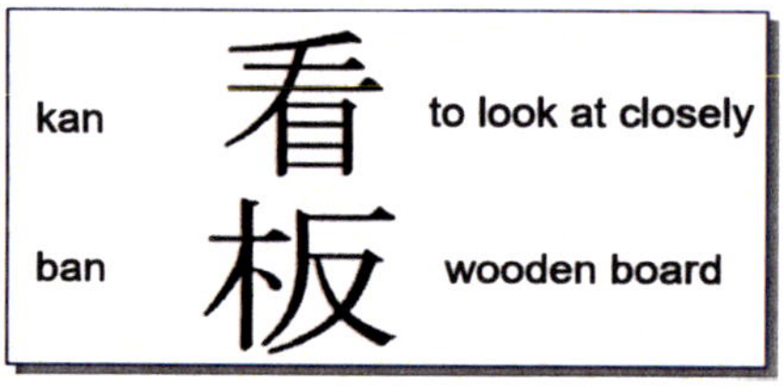

Fig. 6.3.1.2 The word Kanban (explanation by Tschirky; see footnote).[13]

The *Kanban technique* is defined as a Kanban feedback loop and rules for Kanban use.

Figure 6.3.1.3 defines a *Kanban feedback loop*, between parts production and preassembly as well as a *two-card Kanban system*:

1. If in preassembly the contents of the container have been used up, an employee goes with the container and the *use* (or *move*) *card* to the buffer and takes a full container of the required items. He removes the *production card* attached to the full container and places it in a mailbox. He attaches the move card from the empty container to the full container. The empty container remains in the buffer, while the full container with its move card goes to assembly.
2. An employee in parts production routinely goes to the buffer and collects the production cards and empty containers. The Kanbans collected are the orders for manufacturing the corresponding number of items (an order may also comprise several containers). The release of the order is registered by passing through a shop

---

[13] In a personal communication, Prof. Hugo Tschirky (Swiss Federal Institute of Technology ETH Zurich) kindly explained the origin of the Japanese word *Kanban*. The character "kan" is made up of the symbols hand and eye and is derived from the pictograph of a man holding his hand to the brow to shade the eyes in order to better look at something. "Ban," meaning wooden board, contains the symbols for tree, wood, and wall (a wooden board supported against a wall).

floor data collection device, such as a bar code scanner. There is no due date on the Kanban, for each order is to be filled immediately.

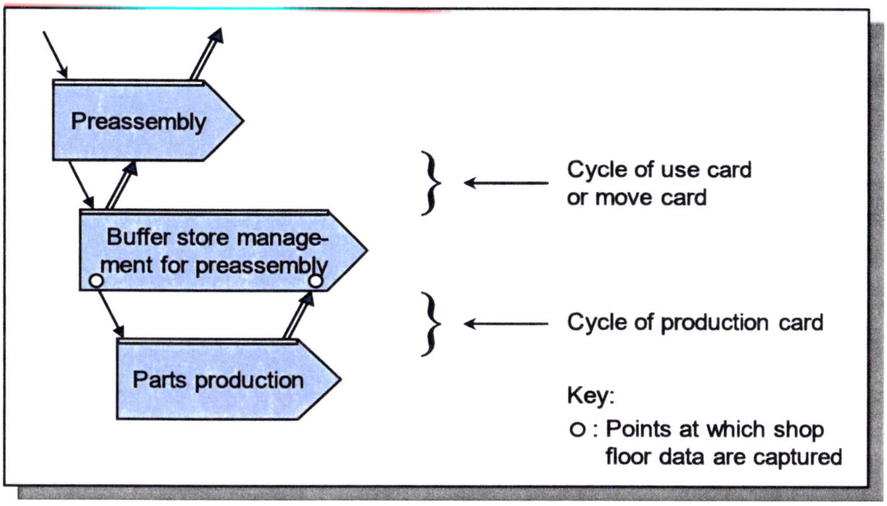

Fig. 6.3.1.3 Basic principle of the Kanban technique: the Kanban feedback loop.

3. Once the items have been manufactured, the production card is attached to the full container and the container is moved to the buffer. Again, passing through a shop floor data collection device serves to register entry of the order.

The buffer usually stands at the *inbound stockpoint* of the user, that is, a defined location next to the place of use on a production floor. The buffer only seldom stands at the *outbound stockpoint* of the manufacturer, that is, a defined location next to the place of manufacturing on a production floor.

As a variant, there may also be a *one-card Kanban system*, where the Kanban remains fixed to the container. Another variant transmits the Kanban of the empty container by fax or by an automatic scanner via telecommunications. This avoids transport time for the return of the empty container in the case of spatially remote sites. Here, the essence of the traveling card is lost: As it becomes "copied" in each cycle, it does not remain the identical physical card. This creates the danger of duplicate orders.

The *Kanban rules* or *rules for Kanban use* are defined in Figure 6.3.1.4 as a process strategy.

The Kanban rules ensure, in their pure application, that no reserves will form and that orders are processed *immediately*. The order is registered immediately as an event, and it sets off the production process.[14] This means, however, that adequate capacity must be available and that it can adapt flexibly to load.

---

[14] Kanban represents a synchronized production control approach. *Synchronized production* is a manufacturing management philosophy that includes a coherent set of principles and techniques supporting the global objective of the system (cf. [APIC16]).

The Kanban technique can be applied across numerous production structure levels or operations. This results in chains of Kanban feedback loops. A comprehensive system includes external suppliers, so that close cooperation with producers is required. For purchased parts, the order to the supplier is a move card. Here again, the Kanban card can be registered and transmitted by means of bar codes.

The user operation may never

- Order more than the required quantity.
- Order at a point in time earlier than required.

The supplying or producing operation may never

- Produce more than what has been ordered.
- Go into production before an order is received.
- Not produce, or produce late, what has been ordered.
- Deliver scrap or insufficient quality.

The planning operation (usually organized as a planning center) takes care of

- Medium- and long-term balancing of load and capacity.
- Keeping a suitable number of Kanban cards in the feedback control system (the smallest number possible) by means of adding and removing cards.

Fig. 6.3.1.4 Kanban rules of order release and control of the feedback control system.

The Kanban card operates as *pull signal*: it entails an order release according to consumption and a (stock) replenishment order. In [Ohno88], Ohno described the Kanban technique as the *supermarket principle*, as the user serves himself with the things he needs; once the shelves are emptied, they are refilled. Because of standard locations for the containers and their contents, goods provision does not involve a lot of effort. This allows small production batches.

6.3.2 Kanban: A Technique of Materials Management

Intended learning outcomes: Present in detail the basic data for calculating the number of Kanban cards. Identify the number of Kanban cards in the system. Explain Formula to calculate the number of Kanban cards.

As each container must be accompanied by a Kanban, the number of Kanban cards in the feedback loop determines the amount of work in process. We can distinguish

- containers in issue by the user operation
- containers in buffer at the user operation
- containers in transport
- containers being filled by work at the feeding operation
- containers queued at feeding operation[15]
- containers that represent safety stock

---

[15] According to Kanban rules, this queue should be of length zero. Orders received are filled on the same day. See the Kanban rules in Figure 6.3.1.4.

To calculate the optimum number of Kanban cards in a feedback control system, the data are first defined in Figure 6.3.2.1.

–	A:	number of Kanban cards
–	k:	number of parts (units) per container
–	LT:	manufacturing or procurement lead time
–	ULT	usage during lead time
–	TP:	length of the statistical period
–	UP:	usage during the statistical period (= expected value of demand)
–	SF:	percent needed as safety factor (for fluctuation of demand and delivery delays)
–	w:	number of containers per transport batch (=1, if possible)
–	SUMRT:	sum of the run times per unit
–	SUMST:	sum of setup times (independent of batch size)
–	SUMINT:	sum of interoperation times plus administrative time

Fig. 6.3.2.1 Basic data for calculating the number of Kanban cards.

How many Kanbans must flow in a feedback control system to guarantee the availability of components? Figure 6.3.2.2 examines the role of all Kanbans that lie "in front of" an emptied container and illustrates the situation formulated below:

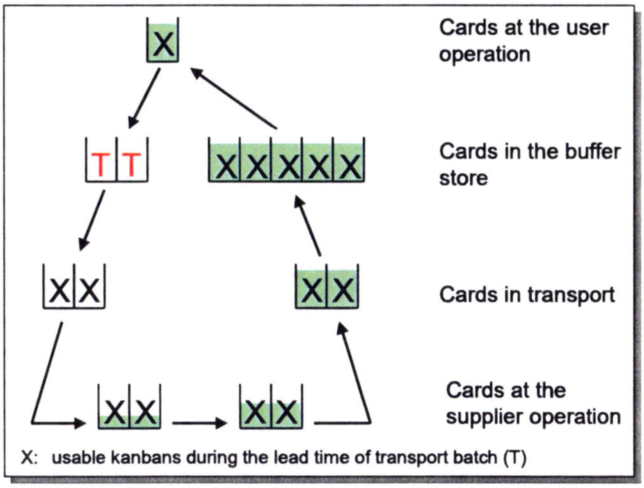

Fig. 6.3.2.2 Number of Kanban cards in the system.

At the moment an order is signaled (i.e., a transport batch of empty containers), the quantity in the buffer and in process — that is, number of Kanban cards multiplied by the contents of a container — must correspond to *the expected usage during lead time.* To the number of cards calculated in this way, the number of containers of the transport batch itself is added.[16]

The value of A can thus be calculated using the formula in Figure 6.3.2.3. Here, w*k is the transport batch size, which is also the batch size of procurement or production. Batch size can therefore be larger than the quantity that fills one container.[17] The value of w is at least 1, and for high-cost items, it should not — if possible — be any larger than 1.

$$\begin{aligned} A \cdot k &= ULT \cdot (1+SF) + w \cdot k \\ &= \frac{UP \cdot LT}{TP} \cdot (1+SF) + w \cdot k \\ &= \frac{UP \cdot (w \cdot k \cdot SUMRT + SUMST + SUMINT)}{TP} \cdot (1+SF) + w \cdot k \end{aligned}$$

Fig. 6.3.2.3 Formula to calculate the number of Kanban cards.

Notice the similarity of this formula to a further technique of order release according to consumption: the order point technique (Figure 11.3.1.3). The way that the Kanban feedback loop functions, the Kanban rules, and now also the formula to calculate the number of Kanban cards, all indicate a technique of stochastic materials management.

To avoid large safety stocks due to demand fluctuations, such techniques must be as continuous as possible at all production structure levels.[18] The Kanban technique allows no large safety stock. For that reason alone, buffers are set up right on the shop floor and have to be kept to small dimensions. The number of Kanban cards can also not be changed frequently because of the great administrative effort involved. Moreover, Kanban rules allow no degrees of freedom for delivery delays. This results in the following:

The Kanban technique guarantees availability only if there is the most continuous possible demand; that is, with limited fluctuations in all Kanban feedback loops. The same holds for customer demand. Thus, this is production or procurement with frequent order repetition and small batch sizes.

The most interesting products when it comes to improving logistics techniques are, of course, those that add high value-added. Such products are often A items in an ABC classification. The ABC classification can be complemented by an XYZ classification, which yields a measure of the continuity of demand (see Section 11.2.3). X items are those items having the greatest continuity, and Z-items are those in lumpy demand. Kanban items are therefore typical A and X items.

---

16 If, in addition, the order rhythm of the Kanban order does not consider the exact point in time, but only time periods, then the period length must be added to the lead time (administrative wait time).

17 Measures to reduce batch size were described in Section 6.2.

18 See also a detailed discussion in Section 11.3.3.

6.3.3 Kanban: Long- and Medium-Term Planning

Intended learning outcomes: Describe the role of a long-term plan for resources according to an MRP II concept. Identify the lean / JIT principles that must be implemented as prerequisites for a successful implementation of the Kanban technique.

The last rule for Kanban use in Figure 6.3.1.4 indicated that Kanban requires some long- and medium-term planning tasks. This planning is independent of the Kanban feedback loop. In detail, planning must fulfill the following tasks:

- Devise a long-term plan (and, if required, a medium-term plan) for resources according to an MRP II concept (manufacturing resource planning): Firstly, determine the master plan (independent demand) based on forecast (ad hoc or using techniques described in Chapter 10) or based, occasionally, on customer demand (see Sections 5.2.1 and 12.2.1). Secondly, calculate gross requirement to determine required resources in the form of purchased goods and capacity (Section 5.2.2). Thirdly, develop long-term contracts with suppliers (blanket orders; see Section 5.2.4); if necessary, fine-tune release quantities in medium-term planning.
- Determine the type and number of Kanban cards for each feedback loop (see Section 6.3.2). Analyses of deviations will reveal those feedback loops that require reexamination of the number of Kanban cards so that overstock in buffers and interruptions in the loop can be corrected. This is done by means of the targeted addition or removal of Kanbans.
- Control actual load through the Kanban systems by, for example registering Kanbans dispatched from the buffer (order releases), and registering incoming Kanbans in the buffer (incoming material in the buffer).

As mentioned in Section 6.2.5, Kanban techniques cannot be simply grafted onto an existing organization of production (such as job shop production). JIT principles, listed below in brief, must be implemented first:

- Clear layout of the organization; that is, the workstations and machines required to make the product are located close together and in the sequence that corresponds to the flow of goods (see Section 6.2.2).
- Small batch sizes, connected with a drastic reduction in setup times (see Section 6.2.1).
- Adherence to exact quantities. The scrap factor aims toward a "zero defects" program, with workstations personally responsible for quality control of components they produce (see Section 6.2.5 on quality assurance).
- Preventive maintenance forestalls machine downtime. It should increasingly eliminate the need for repairs that traditionally take place only once the machine breaks down (endangering delivery). Interdisciplinary troubleshooting teams provide help here (see Section 6.2.1, approach 4).
- *Adherence to short delivery lead times*. This demands adequate capacity and operator flexibility (see Section 6.2.3, approach 1).

6.4 The Cumulative Production Figures Principle

Intended learning outcomes: Explain the cumulative production figures diagram and the cumulative production figures principle.

The cumulative production figures principle originated, like the Kanban technique, within the automobile industry. It aids control of a supply chain with regard to deliveries by system suppliers and to coordination among different manufacturing companies. It is a simple technique that combines long-term resource management with short-term materials management and scheduling. In essence, in the manufacturing process of a certain product, it counts the number of intermediate products or states in the flow at certain *count points* and compares this amount to the planned flow of goods. Depending on the result, the work system can be sped up or slowed down.

For the manufacture of different products, and different variants of products, a particular quantity of cumulative production figures is required. The cumulative production figures principle is best suited to a product concept of standard products or standard products with (few) options and to batch production. The most important prerequisite is the same as that for the Kanban technique: continuous demand along the value-added chain, or production and procurement with frequent order repetition.

The following discussion of the cumulative production figures principle is based mainly on [Wien19].

A *cumulative production figure* (abbreviated below as CPF) is the cumulative recording of the movement of goods over time.

Figure 6.4.0.1 shows an example of cumulative production figures along a sample manufacturing process. The process has been divided into the part processes, also called *control blocks*, which in this case are parts production and assembly.

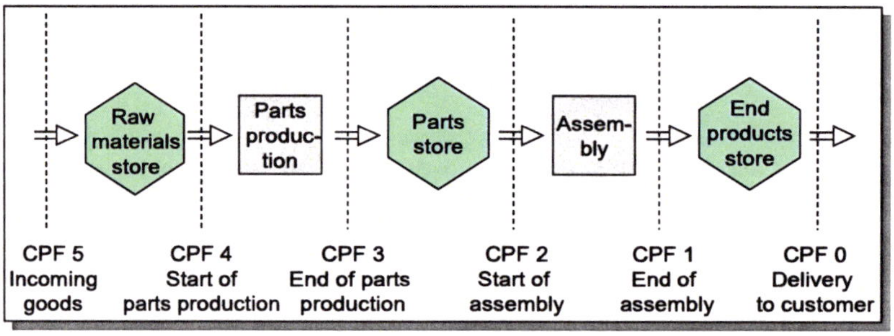

Fig. 6.4.0.1 The definition of cumulative production figures along the manufacturing process.

At the start and end of each part process, a cumulative production figure is defined: the entry cumulative production figure and the issue cumulative production figure. This is based on the assumption that there is always a process store, or buffer, between two part processes. For processing time within a part process, or control block, planning uses average lead time. In planning, this is also called the *control block time offset*.

The *cumulative production figures curve* is a graph of the measurement of a cumulative production figure along the dimensions of amount and time.

A *cumulative production figures diagram* is a summary of cumulative production figures curves throughout the manufacturing process for a particular product.

Each product or product variant has its own pair of cumulative production figures diagrams:

- The *target cumulative production figures diagram* describes the planning based on demand forecast or blanket order and the subsequent resource requirements planning on the time axis. Batch size need not be taken into account, so that between two points in time a cumulative production figure will take a linear course. The difference in amount corresponds to gross requirement during the time period defined by the two count points. The rest follows long- and medium-term planning in the MRP II concept.
- The *actual cumulative production figures diagram* describes the measurement of the actual manufacturing process. The diagram shows the actual, current progress in production, lead times, and inventory in work-in-process and in buffers. Jumps in the lines are caused by batch sizes.

Figure 6.4.0.2 shows an example of a possible target and actual cumulative production figures diagram for the manufacturing process in Figure 6.4.0.1.

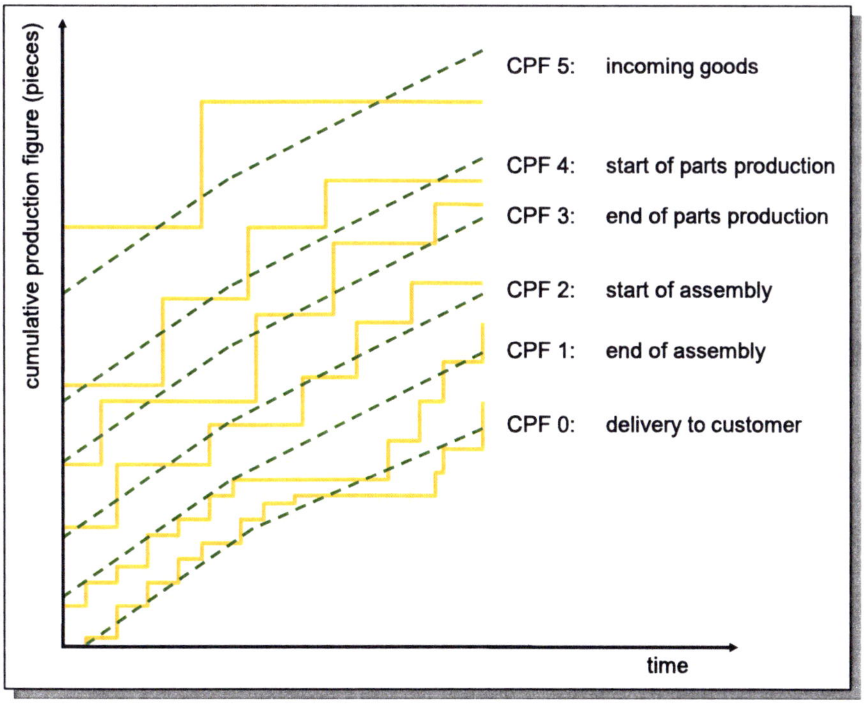

Fig. 6.4.0.2 Cumulative production figures curves and target (dotted) and actual cumulative production figures diagram. (Example is based on [Wien19]).

The *cumulative production figures principle* (CPFP) is the planning & control of the manufacture of a product by means of comparing the target cumulative production figures diagram to the actual cumulative production figures diagram.

Through putting the two cumulative production figures curves, or whole cumulative production figures diagrams, one on top of the other, it is possible to bring the actual diagram closer to the target diagram through speeding up or breaking the manufacturing process. However, the following must be kept in mind:

- The diagrams give no information on the actual operation times and the current load on the work system: Incoming goods to a part process do not necessarily start the process immediately. In addition, there may be several different products being manufactured in the system. Therefore, the cumulative production figures principle cannot provide the basis for capacity management.
- Thus, for capacity management, the accuracy of the lead times, particularly inter-operation times, is an absolute prerequisite.
- Count points must be placed in a way that guarantees accurate counts. A good point in time is at quality control: Here, both the amount of scrap and yield or good quantity are registered. The actual cumulative production figures can now be corrected accordingly at the already measured points, or appropriately marked special demand orders can be released.

In practice, it becomes clear that to keep to the target diagram, sufficient capacity reserves, or capacity that can be implemented flexibly in time, must be available. This is even more the case because capacity management in short-term planning (that is, control) is not possible with the cumulative production figures principle. Only continuous demand along the entire value-added chain will ensure that these reserves will not have to be tapped often.

6.5 Implementing Procedure and Comparison of Techniques

Intended learning outcomes: Present procedures in implementing effective logistics. Differentiate between Kanban and order point technique through a comparison of the techniques.

We have introduced two techniques of materials management: the Kanban technique and the cumulative production figures principle. Two further important techniques, the order point technique and the MRP technique, will be discussed in detail later in Sections 11.3 and 12.3. There are also techniques in connection with capacity management, such as Corma, that influence materials management (see Section 15.1).

Important principles behind these techniques, however, have already been introduced (for example, stochastic versus deterministic techniques). Now we will compare these very principles and, at the same time, compare the particular techniques themselves. We advise the reader to return to these comparisons again after reading the detailed sections mentioned above.

6.5.1 Implementing Procedure

Intended learning outcomes: Identify Features of various techniques of materials management. Describe Procedures in implementing effective logistics: lean/JIT.

Figure 6.5.1.1 lists as features some important prerequisites and effects that can be used as criteria when choosing one or the other of the techniques of materials management discussed above. The cumulative production figures principle does not appear as a separate technique. Its effects are approximately analogous to the Kanban technique with continuous demand.

Technique / **Feature**	Order point	Kanban	MRP
Prerequisites: • Frequency of demand: • Batch size: • Production levels linked: • A system must be set up for ... :	 regular/continuous — no (decoupled) .... control of inventory and goods on order	 continuous small (if high-cost item) via Kanban chains ..... planning a harmonized goods flow	 — — via bills of material ... planning of orders at all structure levels
Measures to undertake if demand fluctuates:	increase safety stock	adapt capacity to load; increase safety factor	change safety requirements; frequent net change or recalculation
Risk incurred if demand is much lower than forecast:	inventory at all levels in the magnitude of the batch	inventory in all Kanban feedback loops in the magnitude of the batch	inventory at all levels below or at the stocking level
Risk incurred if demand is much greater than forecast:	medium stockout risk at all levels	medium stockout risk at all levels	high stockout risk below or at the stocking level
Realization: • Conception (organizational / technical): • IT aids: • Execution/control:	 simple simple simple	 complicated to difficult simple very simple	 simple complicated complicated

Fig. 6.5.1.1 Features of various techniques of materials management.

Figures 6.5.1.2 and 6.5.1.3 show strategy and a way of proceeding when implementing effective logistics. This is based on the JIT concepts in Section 6.2 and the comparison of techniques in Section 5.3.2. The considerations shown in the figures hold for the entire value-adding chain, independently of whether the chain is within a single company or is a transcorporate chain.

The lean / just-in-time concept should be implemented first and independently of the technique to be chosen for materials management. The points raised in Figure 6.5.1.3 then serve to distinguish among the individual techniques of materials management. [Will15] develops practical methods and tools for assessing the progress of implementation of „lean thinking" in manufacturing.

Implementation procedures for effective logistics (part 1): lean / just-in-time concept

1. *Introduce measures to raise the level of quality.* Processes must be so precise that no scrap will be delivered to the site executing the order.
2. *Examine the number and frequency of processes, particularly layouts.* Set up segmented or cellular manufacturing and implement logistics that reduce administration and transport times.
3. *Reduce batch-size-independent production or procurement costs, in particular setup time.* The latter must be very carefully checked at capacity-critical workstations. Implement modern setup technology.
4. *Consider the implementation of CNC machines, industrial robots, and flexible manufacturing systems (FMS).* These allow several operations to be combined into one (complete processing). The opposite may also prove advantageous, particularly in connection with segmenting; consider the use of several simple machines in various segments instead of a single-operation machine that transcends the segments.
5. *Achieve realization of rhythmic and harmonious production.* Production structure levels should be designed in such a way that lead times for the various levels are identical or multiples of one another.
6. *Determine batch size.* As small as possible. (Using the Kanban technique, the batches should cover one day or just a few days.)

Steps 3, 4, 5, and 6 do not have to be performed in strict order.

Fig. 6.5.1.2 Procedures in implementing effective logistics: lean/JIT.

Implementation procedures for effective logistics (part 2): materials management techniques

7. For low-cost items or items with continuous or regular demand:
   a. Install the cumulative production figures principle if a number of successive levels were designed at point 5, through which large batches of relatively few products are manufactured according to forecast or blanket orders and for which capacity can be adapted to actual load despite mild fluctuations.
   b. Set up a chain of Kanban feedback loops if a number of successive levels were arranged at point 5, which can — if, at the same time, demand is sufficiently regular — all be controlled according to use and for which capacity can be adapted to actual load.
   c. Otherwise, use the order point technique, together with the various techniques of scheduling and capacity management.
8. For items where demand is unique or for high-cost items with no regular demand — even after having implemented points 1 to 6:
   a. If the article can be produced downstream from the (customer) order penetration point (OPP) for a number of levels, planning & control should be deterministic, and as based on the customer order using a configuration of the customer production order over various levels with the MRP technique (material requirements planning).
   b. If the item lies at or upstream from the (customer) order penetration point (OPP) and demand is independent, a procedure that makes intuitive sense should be followed.
   c. Otherwise, the MRP technique can (must) be used in the quasi-deterministic case. The calculation should be updated daily or as often as several times per day ("online").

Fig. 6.5.1.3 Procedures in implementing effective logistics: techniques of materials management.

6.5.2 Comparison of Techniques: Kanban versus Order Point Technique (*)

Intended learning outcomes: Explain the development of buffer inventories on several production structure levels when production is rhythmic. Differentiate between order point and Kanban technique when defining production structure levels and (buffer) storage.

The implementation of the JIT concept also entails advantages to the order point technique (see Section 11.3). Indeed, short setup times result in smaller batch sizes, shorter lead times, and thus a lower order point. Smaller batch sizes lead to more frequent repetition of the same orders (which will increasingly overlap). Defining work contents of approximately the same length per production structure level improves the flow of goods.

Figure 6.5.2.1 shows the physical inventories on several production structure levels.

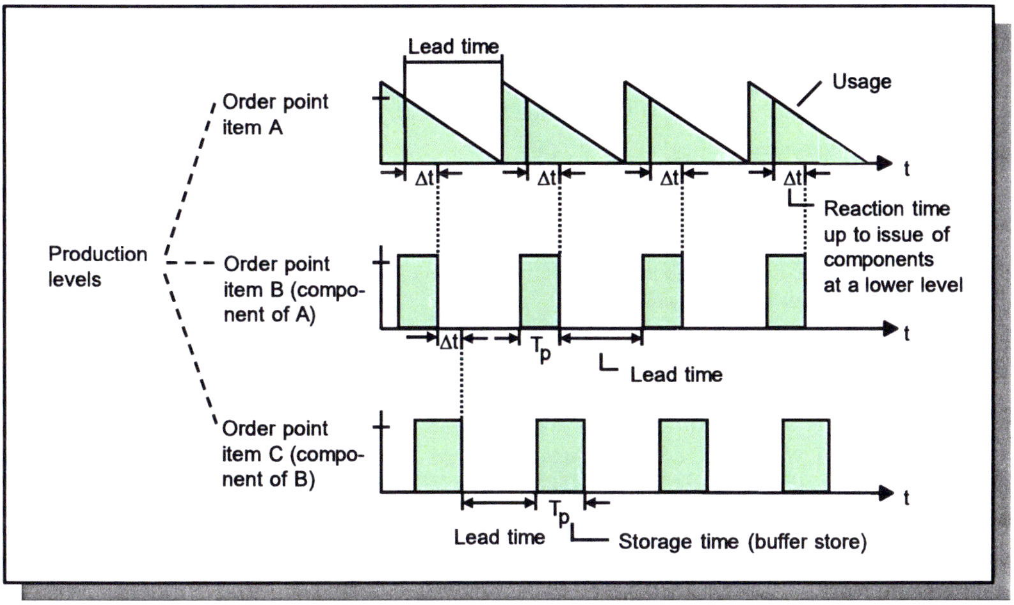

Fig. 6.5.2.1 Development of the buffers when production is rhythmic.

The symbol Δt stands for the necessary reaction time between reaching the order point (or, in Kanban, registering that a container is empty) and withdrawing components from the next lower production structure level. With the lean / just-in-time concept, Δt is as small as possible, due to direct communication between supplier and user operation. T_P is the wait time of the item in the buffer or intermediate store. With just-in-time production, the buffer is located directly at the workstation, or user operation. T_P is thus time in storage.

Order release according to consumption is common to both techniques. Storage time functions as a time buffer. If usage is smaller than forecasted over a longer period, the production or procurement cycle will be triggered less often. In the Kanban technique, fewer and fewer containers will be sent back and forth. But inventory in the buffer increases. From this, the same effect results as with the order point technique. In the reverse case, if usage is greater than predicted over a longer period, safety stock in the buffer would ensure delivery capability. Thus, the percentage of stock for safety stock in the formula in Figure 6.3.2.3 and the number of Kanban cards must then be increased.

So much for the common effects of both techniques. Now let's look at the *differences*. One feedback loop in the Kanban technique will usually encompass only a few operations. There is a buffer between each loop. Consequently, a production structure level controlled by the order point system is divided into a number of Kanban feedback loops, ideally of the same length, as illustrated in Figure 6.5.2.2.

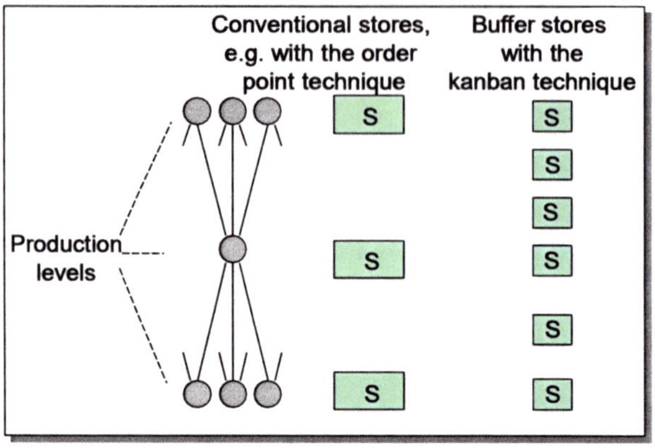

Fig. 6.5.2.2 Definition of production structure levels and (buffer) storage: order point technique versus Kanban technique.

This results in the following advantages of the Kanban technique:

- Inventory tends to be shifted to lower production structure levels, which is important for items with great added value. These are usually high-cost items, too (A items).
- Lead time through a Kanban feedback loop is reduced for two reasons. First, a Kanban feedback loop includes only a few operations. Second, there are no administrative expenses, as the buffers are located directly at the user operations.
- It is a *visual control system*, that is, stockkeeping takes place "at a glance," and there is no paperwork or need for any other organizational unit to intervene.
- The process can be automated, because approximately the same quantities are produced again and again in short, sequential periods of time.
- Batch size in each feedback loop is small, because there are fewer operations and less setup time to consider.

It would be possible to design production structure levels controlled by the order point technique whose value-adding would equal that of the Kanban feedback loops. However, because of the large number of small orders being processed all at once, the amount of administrative effort required to control production remotely (away from the flow of goods) — through consulting computer lists, for example — would be prohibitive. The sensor that registers the Kanban order (an event) to be released (state) is the simplest and most rapid sensor imaginable: the human eye.

The large number of intermediate stores with the Kanban technique can, of course, be seen as a disadvantage, as in the extreme case a buffer must be set up for each operation. However, the large number of stores is only a problem if an external agent or expensive measuring devices (such as tallying by hand) must be used to perform inventory control. Figure 6.3.1.3

has already suggested that automatic data collection is a good way to register the Kanban process, including both open Kanban orders and inventory in the buffers.

There are further important differences between the order point technique and the Kanban technique with regard to flexibility and to assigning requisition control tasks:

- With Kanban, control is decentralized. The shop floor workers take over requisitioning activities, which encourages their autonomy. But one of the rules for Kanban use demands that capacity be adapted to load, which is infinite capacity planning. The due date for all Kanban orders is "now" — which actually restricts autonomy.
- The order point technique can be implemented with either centralized or decentralized control. The more infinite capacity planning is possible, the more requisitioning with the order point technique can be turned over directly to the workers.[19] If the interoperation times are short, or if capacity limits must be considered, possible resulting interruptions in the order cycle could affect the entire value-adding chain. If the value-adding chain is made up of many organizational units, central control (with central order release) may be more flexible, but it also involves greater effort.

6.6 Summary

Lean/JIT aims to reduce the 3Ms ("muri," "mura," "muda") and the lead time and, at the same time, to minimize stored and in-process inventory. The most significant measure to reduce lead time is setup time reduction. A significant reduction in setup time allows — with keeping utilization constant — reduction of lot size, thus operation time, and, finally, lead time as well. Concepts for setup time reduction can be grouped under the term *SMED* (single-minute exchange of die), but also under the terms setup-friendly production facilities, cyclic planning, or modularization.

Further concepts for lead time reduction are production or manufacturing segmentation, cellular manufacturing, complete processing, and structuring of the assembly process. Harmonizing the product range and work contents helps, in addition, to achieve repetitive processes and a balanced flow of goods in production. This increases the degree of automation and reduces wait times. Additional Lean/JIT concepts ensure high-quality as well as rapid administrative connections between feeder and user operations, for example, combined with blanket order processing. And, there is also the availability of resources: overcapacity of machines and tools as well as flexible personnel. Further elements of the Japanese way of thinking include group thinking, elimination of waste, *kaizen*, *poka-yokero*, order, and cleanliness. Lean/JIT concepts improve the quality of all techniques of resource management. Shorter lead times allow the (customer) order penetration point (OPP) to be set lower in the product structure, thus increasing the potential for use of deterministic methods in requirements planning. Through smaller batch sizes, or even "make to order,"

---

[19] There is no strict line dividing this from the Kanban technique if stocks below order point are no longer checked according to lists, but rather are registered "visually" right at the buffer.

inexact forecasts of requirements upstream from or at the (customer) order penetration point (OPP) lead to fewer long-term materials planning errors.

Simple techniques of planning & control are possible with production or procurement with frequent order repetition. The best-known technique is Kanban. Between each user operation and supplier operation, a certain number of Kanbans pass back and forth. Each Kanban refers to a container that stands at the user operation in a clearly marked location. It is managed visually and sent back to the supplier operation as soon as it is empty. It is important to follow the Kanban rules of use. The number of Kanban cards must be determined on the basis of medium- or long-term planning according to the MRP II concept. The cumulative production figures principle represents another simple technique of materials and scheduling management. Work progress is recorded at set count points.

Special considerations are necessary where different techniques coexist. Kanban is similar to the order point system, with the difference that control, in particular order release, is always decentralized. This allows for significantly more orders than with centralized control. The Kanban cycles generally are composed of fewer operations at a product level than the order point system, which tends to result in less (buffer) inventory.

6.7 Keywords

cell, 302
cellular manufacturing, 303
complete processing, 305
cumulative production figures, 320
dedicated capacity, 303
dedicated equipment, 303
external setup (time), 301
five Ss, 310
flexible manufacturing systems (FMS), 324
harmonizing
the content of work, 306
the product range, 300
heijunka, 300
inside exchange of dies (IED), 301
internal setup (time), 301
jidoka concept, 296
job enrichment, 309
just-in-time concept, 296
just-in-time logistics, 309
Kanban, 313
Kanban feedback loop, 314
Kanban rules, 315
Kanban technique, 314
line balancing, 306
move card, 314
muda (syn. waste), 295
mura (syn. variation), 295
muri (syn. overstraining), 295
Ohno's seven wastes, 295
one-piece flow, 303
outside exchange of dies (OED), 301
preventive maintenance, 319
quality at the source, 309
quality circle, 310
single-minute exchange of dies (SMED), 301
stockless production, 296
work cell, 302

6.8 Scenarios and Exercises

Intended learning outcomes: Show the effect of setup time and batch size on operation time and cost. Calculate the effect of cellular manufacturing on lead-time reduction. Perform line balancing through harmonizing the content of work. Determine the number of Kanban cards.

6.8.1 Operation Time versus Operation Cost, or the Effect of Varying Setup Time and Batch Size

Intended learning outcomes: Explain the need to find a balance between short lead time and low cost, for any operation. These two factors are determined by setup time and batch size.

This exercise will help to illustrate the need to find a balance between (1) short lead time, and (2) low cost, for any operation. These two factors are determined by setup time and batch size. You will find the effect of setup time and batch size on

a. The operation time, which is a measure of the lead time of the order.

b. The *operation time per unit* (that is, operation time divided by batch size), which is a measure of the cost of the operation and therefore of the cost of the production or procurement order.

Solve the following tasks:

(0) First, suppose a setup time of 200, a run time per unit of 100, and a batch size of 4. Calculate the operation time and the operation time per unit.

(1) If batch size is increased to 20, what are the effects on operation time and operation time per unit? In your opinion, what effects are positive or negative?

(2) Suppose that because of the hard work of the process engineers (e.g., by applying SMED measures), setup time could be reduced to 100. What is the effect of this, if the batch size is maintained at 20?

(3) To what extent can the batch size be reduced after the reduction of setup time to 100, so that the operation time does not exceed the original operation time of 600? What will the operation time per unit be?

(4) To what extent can the batch size be reduced after the reduction of setup time to 100, so that the operation time per unit does not exceed the original time per unit of 150? What will the operation time be?

Solution:

(0) Operation time: 600; operation time per unit: 150.

(1) Positive: operation time per unit clearly reduced to 110.
negative: operation time very much extended to 2200.

(2) Positive: operation time per unit slightly reduced to 105.
negative: operation time only very slightly reduced to 2100.

(3) Batch size = 5; operation time per unit = 120.

(4) Batch size = 2; operation time = 300.

You can view the animated solution on the Internet at URL:

https://intlogman.ethz.ch/#/chapter06/operation-time/show

Try out different values for setup time, run time per unit, and batch size.

6.8.2 The Effect of Cellular Manufacturing on Lead Time Reduction

Intended learning outcomes: Present the routing sheet for production of shafts. Explore the effect of cellular manufacturing on lead-time reduction.

Figure 6.8.2.1 shows a possible routing sheet for production of shafts. The batch size is 10.

Operation	Setup time	Run time per unit
Millcut	0.02	0.02
Lathe	0.6	0.06
Millcut nut	1.6	0.6
Pregrinding	1.2	0.12
Final grinding	1.2	0.16

Fig. 6.8.2.1 Routing sheet for production of shafts.

a. Calculate the lead time in traditional job shop production. *Hint*: For job shop production, lead time has to be calculated assuming a sequence of operations. Therefore, you can use the formula in Figure 6.2.2.3.

b. Calculate the maximum lead time for the case of cellular manufacturing, that is, using the formula in Figure 6.2.2.4. (*Hint*: First determine the cell driver).

c. For the given routing sheet shown in Figure 6.8.2.1, and for cellular production, find a temporal order of operations that yields minimum lead time.

d. For the given routing sheet shown in Figure 6.8.2.1, and for cellular production, find a temporal order of operations that yields minimal load (or minimum allocated time for the operation, that is, operation time plus wait time between the units of the batch) at the workstations.

Solution:

a. 14.22.

b. 10.98 (the cell driver is the operation "millcut nut" with an operation time of 7.60; setup time plus run time of all the other operations is 3.38).

c. Minimum total lead time is 7.88. The setup and the first unit of the batch of operations "millcut" and "lathe" can be fully executed during the setup of the cell driver, as well as the setup of the operations "pregrinding" and "final grinding". Each unit of the batch can be run directly after its run on the cell driver operation. Thus, the run times for one unit for "pre-grinding" and "final grinding" have to be added to the cell driver operation time, or 0.12 + 0.16 + 7.6, making 7.88.

d. Lead time with minimum load is 8.24. Again, the setup and the first unit of the batch of operations "millcut" and "lathe" can be fully executed during the setup of the cell driver. For "pregrinding," to be ready to execute the last unit of the batch just after the completion of the cell driver operation, the 9 units of "pregrinding" must have been just completed at point 7.6 in the time axis. Thus, the latest start date of "pregrinding" must be 7.6 – 9 * 0.12 – 1.2 = 5.32. For “final grinding,” the first unit of the batch can be executed directly after the first unit of “pregrinding” has been executed, that is, at 5.32 + 1.2 + 0.12 = 6.64. This implies that the latest start date of “final grinding” is at 6.64 – 1.2 = 5.44, and its completion date is at 5.44 + 1.2 + 10 * 0.16 = 8.24.

You can view the animated solution on the Internet at URL:

https://intlogman.ethz.ch/#/chapter06/cell-driver/show

By modifying setup and run times of the operations, change the cell driver. Try to find a combination where the variant “minimum total lead time” tends toward the “maximum lead time” value of the lead time formula for cell manufacturing.

6.8.3 Line Balancing — Harmonizing the Content of Work

Intended learning outcomes: Present the routing sheets for parts production out of sheet metal of three products. Explain possible ways for line balancing through harmonizing the content of work.

Figure 6.8.3.1 shows a possible routing sheet for parts production out of sheet metal. Three different products are produced: items 1, 2, and 3. All have a similar routing sheet. For the different operations, the number in the table is the operation time, and the number in parentheses is the setup time.

In accordance with the discussion in Section 6.2.4, assume a duration of one unit of harmonized content of work of 12 time units. The task is to perform measures to change lead times of operations, chosen from the various possible measures to line balance or harmonize the content of work listed according to Figure 6.2.4.3.

a. Suppose that the first two operations can be combined into one (why is this a feasible assumption?). Item 3 seems — at first glance — to fit quite well into three units of harmonized content of work. Thus, according to the first one of the measures listed in Figure 6.2.4.3, try to change lot sizes of items 1 and 2 (use the empty columns in Figure 6.8.3.1), to obtain for each of them a total operation time on the order of 36 units of time.

b. Is it possible, in practice, to combine the last two operations into one, fitting them into one unit of harmonized content of work?

c. For item 1, the third and fourth operations do not fit into a unit of harmonized content of work, despite significant changes to the batch size. What other possible measures listed in Figure 6.2.4.3 could be implemented?

d. After implementing all these measures, are there still problems?

Product ID	1			2		3	
Lot size / Process	400	?	?	50	?	10	?
Cut *work center A*	10 (2)			5 (1)		6 (1)	
Press *work center B*	6 (2)			15 (1)		6 (1)	
Bend *work center C*	2 (2)			20 (2)		12 (2)	
Treat surface *work center D*	18 (10)			—		9 (7)	
Test *work center E*	2 (2)			9 (5)		—	
Preassemble *work center F*	16 (0)			3 (1)		—	
Σ operation times	54 (18)			52 (10)		33 (11)	
(Σ setup times) / (Σ operation times)	1/3			1/5 (ca.)		1/3	

Fig. 6.8.3.1 Harmonizing the content of work: routing sheets for three products.

Possible solution:

a. There are machines that perform both operations in one step (e.g., laser-cutting machines). Changing the lot size for product 1 to 200 results in a total operation time of 36, with 18 units of time for setup. Furthermore, the length of the combination of the two first operations is now 10 (or even less, due to complete processing), and this fits well into one unit of harmonized content of work. Changing the lot size for item 1 to 100 would result in a total operation time of 27, with 18 units of time for setup. Thus, a batch size of 200 is the better choice. In addition, changing the lot size for product 2 to 25 results in a total operation time of 31, with 10 units of time for setup. Again, the combination of the two first operations fits well into one unit of harmonized content of work, its total length being 11.

b. Yes. Testing and preassembly can be done at the same physical work center. Furthermore, with a lot size of 200 for item 1, the combination of the two last operations would fit well into one unit of harmonized content of work.

c. Considering the very small run time per unit, the bending operation seems to be very simple (also, the second operation, pressing, appears to be rudimentary, as compared to the process for item 2). Thus, it might be possible to purchase sheet metal that is already profiled (bent). Another solution would be to combine this short process into the same unit of harmonized content of work together with cutting and pressing, using a dedicated (simple, but cheap) machine that could be installed not at work center C, but close to work center B.

   Surface treatment is most likely a subcontracted process. This is probably the reason behind the long setup time, which may actually reflect transportation time rather than setup time at the supplier's site. If so, why not look for a faster transportation vehicle or for a subcontractor in greater geographical proximity to the factory?

d. Yes. For product 1, setup time is now 50% of the operation time. If setup time is not reduced significantly with the measure in point c, then additional measures must be found to reduce the setup time (e.g., by implementing SMED techniques).

6.8.4 Calculating the Number of Kanban Cards

Intended learning outcomes: Present data on three products for calculation of the number of Kanban cards. Calculate the required number of Kanban cards before and after the process improvement.

An automotive company has implemented a JIT program using Kanbans to signal the movement and production of product. The average inventory levels have been reduced to where they are roughly proportional to the number of Kanbans in use. Figure 6.8.4.1 shows the data for three of the products.

Item ID.	Lead time	Length of the statistical period	Usage during statistical period	Number of parts (units) per container	Safety factor (%)	Number of containers per transport batch
1	36	20	600	200	0	1
2	36	20	100	25	0	1
3	36	20	50	10	0	1

Fig. 6.8.4.1 Data on three products for calculation of the number of Kanban cards.

a. The process engineers have been hard at work improving the manufacturing process. They have initiated a new project to reduce lead time from 36 days to 21 days. What would the percentage change in average inventory be for each item?

b. Calculate the number of Kanban cards using other data values. Try to answer the following questions:

- What is the minimum number of Kanban cards required in any case?
- How do the safety factor and the number of containers per transport batch influence the number of Kanban cards required?

Solution for a:

Using the formula in Figure 6.3.2.3, calculate the required number of Kanban cards before and after the process improvement. As inventory is proportional to number of Kanbans, the inventory reduction corresponds to the reduction of the number of Kanban cards.

- Item 1: before, 7; after, 5 → Inventory reduction: 29%
- Item 2: before, 9; after, 6 → Inventory reduction: 33%
- Item 3: before, 10; after, 7 → Inventory reduction: 30%

6.9 References

APIC16 Pittman, P. et al., APICS Dictionary, 15th Edition, APICS, Chicago, IL, 2016

Ohno88 Ohno, T., "Toyota Production System: Beyond Large-Scale Production," Productivity Press, Cambridge, MA, 1988

Shin85 Shingo, S., "A Revolution in Manufacturing: The SMED System," Productivity Press, Cambridge, MA, 1985

Shin89 Shingo, S., "A Study of the Toyota Production System from an Industrial Engineering Point," Productivity Press, Cambridge, MA, 1989

Suza12 Suzaki, K., "The New Manufacturing Challenge," Reprint, Free Press, New York, 2012

Toyo98 "The Toyota Production System," Toyota Motor Corporation, Public Affairs Divison, Toyota City, Japan, 1998

Wien19 Wiendahl, H.P., "Betriebsorganisation für Ingenieure," 9th Edition, Hanser, München, 2019

Wild89 Wildemann, H., "Flexible Werkstattsteuerung durch Integration von Kanban-Prinzipien," CW-Publikationen, München, 1989

Wild01 Wildemann, H., "Das Just-In-Time Konzept: Produktion und Zulieferung auf Abruf," TCW Transfer-Centrum für, München, 5. Auflage, 2001

Will15 Wille, T., "Konzeption eines Bezugssystems zur Bewertung des Einführungs-fortschritts von „Lean Thinking" in produzierenden Unternehmen," Diss ETH Zürich, 2015

WoJo07 Womack, J.P., Jones, D.T., Roos, D., "The Machine That Changed the World," Revised Edition, Simon & Schuster, UK, 2007

7 The Concept for Product Families and One-of-a-Kind Production

Intended learning outcomes: Present logistics characteristics of a product variety concept. Explain adaptive and generative techniques. Describe their use for engineer-to-order. Differentiate various ways of cooperation between R&D and Engineering in ETO Companies.

In buyers' markets, customers ask producers to meet their specific requirements with regard to product composition and quality. Customers do not want to have to adapt their own processes to standard products. Instead, they demand adaptation of the product to their own specific requirements. This has given rise to a tendency toward product *families* and one-of-a-kind production, which requires appropriate product and process concepts as well as logistics concepts. The traditional MRP II concept does not suffice.

The *variant-oriented concept* does not aim toward reduction of the number of product variants, but instead aims toward mastering a variety of variants.

For many, particularly medium-sized companies, being *market driven*, that is, fulfilling customer specification, through flexibly offering product families with many variants, is the main market strategy. Globally known examples are turbochargers of ABB Turbo Systems (Baden) or elevators of Schindler (Lucerne-Ebikon), both Switzerland-based companies. Service industries show similar tendencies. In the insurance sector, e.g., in addition to mass business that focuses on low costs, "custom" insurance policies are offered with flexible terms and customer design ease. Mass customization is the corresponding production type that emphasizes custom products that do not cost more than mass-produced products.

The Lean/JIT concept as presented in Chapter 6 is also useful in the case of production with infrequent or without order repetition: With short lead times, the (customer) order penetration point (OPP) can be set as far as possible upstream in the supply chain. This reduces the need of forecasting. The Kanban technique, however, cannot be used: it requires production with frequent order repetition. Often, customer order-specific drawings must be completed as early as the bidding phase. In actual production, the problem to be faced is how to set up the machines rapidly for a new variant. Moreover, variant-specific work documents must also be produced. Here, the variant-oriented concept is required (also compare Figure 4.5.3.1). It is also called *product family orientation*, *variant orientation*, *variant production*, and *customer order production*. It affects virtually all planning & control tasks (see Figure 5.1.4.2). What stands at the fore is discrete manufacturing, or *convergent product structure*.

7.1 Logistics Characteristics of a Product Variety Concept

Intended learning outcomes: Differentiate between high-variety and low-variety manufacturing. Describe different variant-oriented techniques, and the final assembly schedule.

P. Schönsleben, *Handbook Integral Logistics Management*,
https://doi.org/10.1007/978-3-662-65625-9_7

Flexibility to fulfill customer demands varies in degree. In the fashion industry, for example, there are "off-the-rack" products, *prêt-à-porter* (ready-to-wear) products, and *haute couture*, or creations made for individual customers. In gastronomy, there are standard dinner menus, *à la carte* concepts, and even customer-specific menu creations. Other industries, including service industries distinguish similar levels of adapting to customer demand using their own terminology.

7.1.1 High-Variety and Low-Variety Manufacturing

Intended learning outcomes: Identify values of characteristic features for high-variety manufacturing. Explain long- and medium-term planning for manufacturing according to customer specification or of product families with many variants. Disclose values of characteristic features for low-variety manufacturing.

In a first attempt, we distinguish high-variety manufacturing from low-variety manufacturing. For each characteristic, the main values of certain features of planning & control, taken from Section 4.4, are shown in black. Frequent values are shown in gray. Figure 7.1.1.1 shows the characteristics of *high-variety manufacturing*, in the extreme case *manufacturing according to (pure) customer specification*.

Features referring to user and product or product family						
Feature	→	**Values**				
Product variety concept	→	according to (changing) customer specification	product family with many variants	product family	standard product with options	individual or standard product
Features referring to logistics and production resources						
Feature	→	**Values**				
Production environment	→	engineer-to-order	make-to-order	assemble-to-order (from single parts)	assemble-to-order (from assemblies)	make-to-stock
Features referring to production or procurement order						
Feature	→	**Values**				
Reason for order release (type of order)	→	demand / (customer production (or procurement) order)		prediction / (forecast order)		use / (stock replenishment order)
Frequency of order repetition	→	production / procurement without order repetition		production / procurement with infrequent order repetition	production / procurement with frequent order repetition	
Type of long-term orders	→	none	blanket order: capacity		blanket order: goods	

Fig. 7.1.1.1 Values of characteristic features for high-variety manufacturing.

Customer tolerance time is usually long enough to manufacture all production structure levels according to customer order. Thus, products are manufactured in almost the entire supply chain according to demand, with no stockkeeping. Exceptions are general-use raw materials and purchased parts. Here, inventory is replenished as it is consumed.

In long- and medium-term planning, we can revise the generalized presentation in Figure 5.1.2.1 to make it more specialized. Not all tasks are equally important or pronounced. Figure 7.1.1.2 shows this revised form.

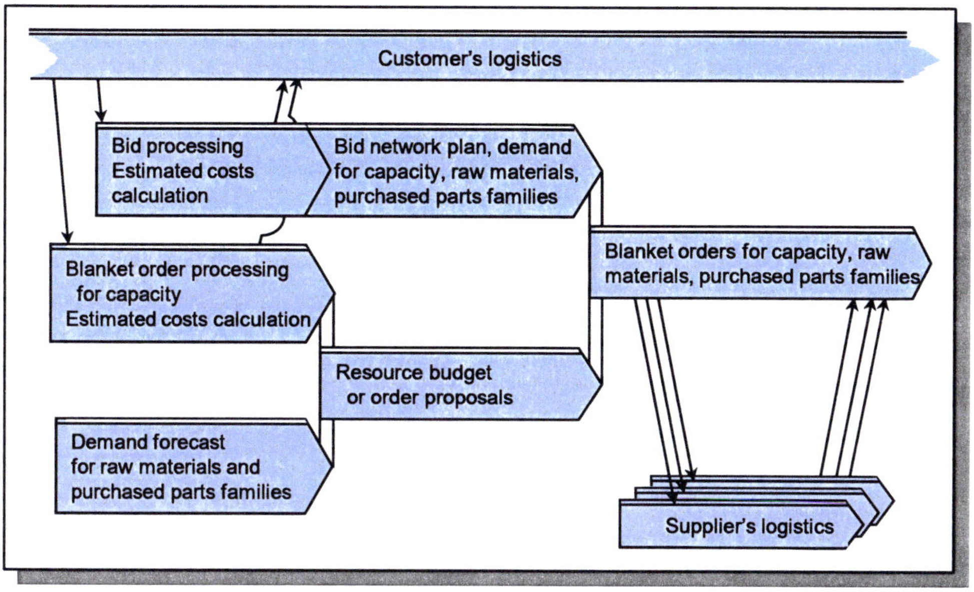

Fig. 7.1.1.2 Long- and medium-term planning for manufacturing according to customer specification or of product families with many variants.

- In long-term planning, a master plan does not make sense. At best, forecasts can be made for raw materials or purchased parts families.
- Forecasts for capacity are necessary, however. Blanket orders for capacity at all production structure levels of the supply chain result in improved planning power.
- An order must first be translated into a process plan for the planning of capacity, raw materials, or purchased parts families. This is usually a process network plan, such as is commonly used in project management.
- Medium-term planning is at most a fine-tuning of the long-term planning network plans for the orders.

In short-term planning & control, the tasks in Figure 5.1.3.1 are — for high-variety manufacturing — rather complex.

- For order configuration according to customer specification, first the process network plan must be refined (see, for example, Section 14.4). Raw materials or purchased parts must be made available.
- Detailed resource requirements calculation should happen quickly. Rule-based product and process configurators are used here. Generally, a multilevel order must be calculated with all its production documents, often including the drawing.

- Order coordination is required for all part orders over the supply chain, that is, over all orders for components or processes at a lower level and over several production structure levels. Generally, there is no flexibility with regard to start and completion date for part orders. Any small disturbances on the user side or in the production infrastructure of a partner have rapid repercussions within the entire supply chain.

Figure 7.1.1.3 shows the characteristics of *low-variety manufacturing*. Customer tolerance time allows usually to produce some of the highest structure levels according to customer order. Ideally, variants arise only at these structure levels, e.g., assembly. The characteristic of low-variety manufacturing is somewhere between the two characteristics of standard product manufacturing (Figures 5.1.2.1, 5.1.2.2, and 5.1.3.1) and high-variety manufacturing.

Features referring to user and product or product family						
Feature	→	**Values**				
Product variety concept	→	according to (changing) customer specification	product family with many variants	product family	standard product with options	individual or standard product
Features referring to logistics and production resources						
Feature	→	**Values**				
Production environment	→	engineer-to-order	make-to-order	assemble-to-order (from single parts)	assemble-to-order (from assemblies)	make-to-stock
Features referring to production or procurement order						
Feature	→	**Values**				
Reason for order release (type of order)	→	demand / (customer production (or procurement) order		prediction / (forecast order)		use (stock replenishment order)
Frequency of order repetition	→	production / procurement without order repetition		production / procurement with infrequent order repetition	production / procurement with frequent order repetition	
Type of long-term orders	→	none	blanket order: capacity		blanket order: goods	

Fig. 7.1.1.3 Values of the characteristic features for low-variety manufacturing.

- Products downstream from the (customer) order penetration point (OPP) are preferably manufactured according to the batch size of the customer order, sometimes with small batch production to stock. Here, there is production with infrequent order repetition, as there are only a limited number of variants.
- Upstream from or at the (customer) order penetration point (OPP), products are manufactured and stored prior to customer demand. Here, order processing has the character of standard products manufacturing. If variants are produced downstream from the order penetration point, production upstream from the order penetration point is production with frequent order repetition. Otherwise, it is production with infrequent order repetition.

- Reason for order release and type of long-term orders: For goods downstream from the order penetration point, we find the character of high-variety manufacturing. Upstream from or at the order penetration point, order processing has the character of standard products manufacturing. Forecast and blanket orders refer here to product families.

7.1.2 Different Variant-Oriented Techniques, and the Final Assembly Schedule

Intended learning outcomes: Differentiate between adaptive and generative variant-oriented techniques. Disclose typical sets of characteristics and production types that arise frequently with the four product variety concepts. Describe how the MPS concerns the highest structure level still having a small number of different items. Identify FAS / MPS / OPP patterns in dependency on the product variety concept and their relation to the patterns of the T analysis.

Variant-oriented techniques are techniques for the planning & control of a product variety concept with low or high variety.

The subsequent Sections will present different variant-oriented techniques. They can best be grouped in *two classes*.

Adaptive techniques entail two steps. The first step determines a suitable "parent version" from the existing variants. Secondly, this parent version is adapted, or specified in detail, according to the actual requirements.

Adaptive techniques are expensive in terms of administrative cost and effort. For use of these techniques to be economically feasible, the value added must be high. The techniques are implemented in the product variety concepts *standard product with options* and *product family*. See Section 7.2.

Generative techniques are variant-oriented techniques that configure the process plan for each product variant during order processing from a number of possible components and operations. Generative techniques use rules that already exist in an information system.

With generative techniques, order administration is quick and inexpensive, so that the product variety concept *product families with many variants*, even when value added is often low, can be handled efficiently in terms of operations. See Section 7.3. *Products according to (changing) customer specification* require additive and generative techniques to a different extent. Here see Section 7.4.

Adaptive and generative techniques are closely associated with the product variety concepts. For further details, also see [Schi01]. Figure 7.1.2.1 summarizes four sets of characteristics that are typically and commonly found together with a particular product variety concept.

Each set of characteristics has a production type and values for the features frequency of order repetition and production environment (which, according to Figure 4.5.2.1 and Figure 4.4.5.2, are closely associated with the product variety concept) as well as order batch size.

Product variety concept / Variant oriented technique	...and typically associated characteristics and production types
Standard product with (few) options / Adaptive techniques	Repetitive manufacturing or batch production Production with frequent order repetition Make-to-stock or assemble-to-order (from assemblies) Small batch production possible
Product family / Adaptive techniques	Repetitive manufacturing, seldom batch production Production with infrequent order repetition Assemble-to-order (from single parts or subassemblies) Mostly single-item production to customer order
Product family with many variants / Generative techniques	Mass customization Tendency toward production without order repetition Make-to-order Mostly single-item production to customer order
Product according to (changing) customer specification / Generative and adaptive techniques	One-of-a-kind production, often together with mass customization Production without order repetition Engineer-to-order or make-to-order Mostly single-item production to customer order

Fig. 7.1.2.1 Typical sets of characteristics and production types that arise frequently with the four product variety concepts.

A *final assembly schedule (FAS)* is a schedule of end items to finish the product for specific customers' orders in a make-to-order or assemble-to-order environment. It is also referred to as the *finishing schedule* because it may involve operations other than the final assembly. Also, it may not involve assembly (e.g. final mixing, cutting, packaging). Cf. [APIC16].[1]

The type of FAS depends on the selection of items to be part of the master production schedule (MPS; see Section 5.2.3) and the production environment, as follows:

- Make-to-stock: The MPS comprises end products. In effect, the FAS is the same as the MPS.
- Assemble-to-order, or package-to-order: The MPS comprises (sub-) assemblies. The FAS assembles the end product (a variant of a product family) according to customer order specification.

Make-to-order: The MPS includes raw materials or components. The FAS fabricates the parts or subassemblies and assembles the end product according to customer order specifications.

In general, the MPS tends to concern the highest structure level still having a small number of different items. If this level corresponds to the (customer) order penetration point (OPP), only a minimum number of different items have to be stocked, and ideally — that is thanks to standardization — each item has a high degree of commonality. This corresponds to the

1 *Finishing lead time* is the time allowed for completing the good based on the FAS.

concept of late customization (see Section 1.3.3) — a desired effect. Figure 7.1.2.2 shows this situation together with the corresponding FAS and MPS levels.

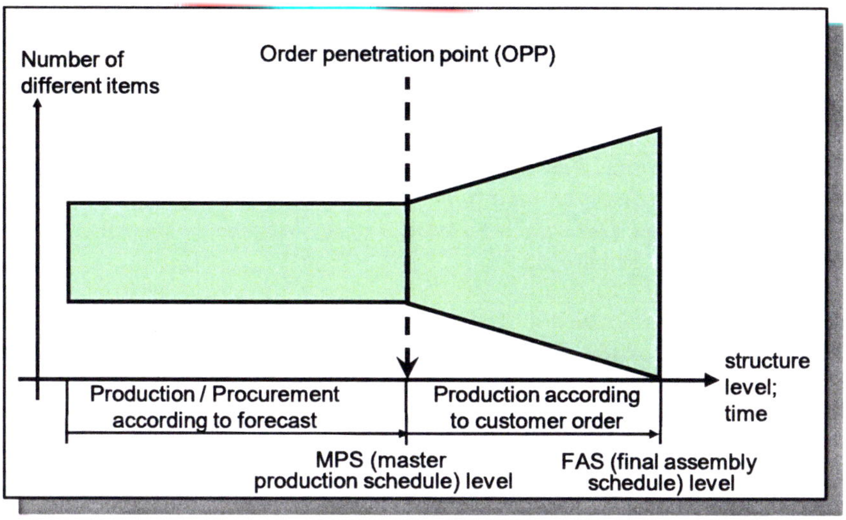

Fig. 7.1.2.2 The MPS concerns the highest structure level still having a small number of different items.

Figure 7.1.2.3 shows typical different patterns of MPS / FAS level and order penetration point in dependency on the product variety concept, or the four different classes of variant-oriented techniques: These patterns correspond to the different pattern of the T analysis within the VAT analysis.

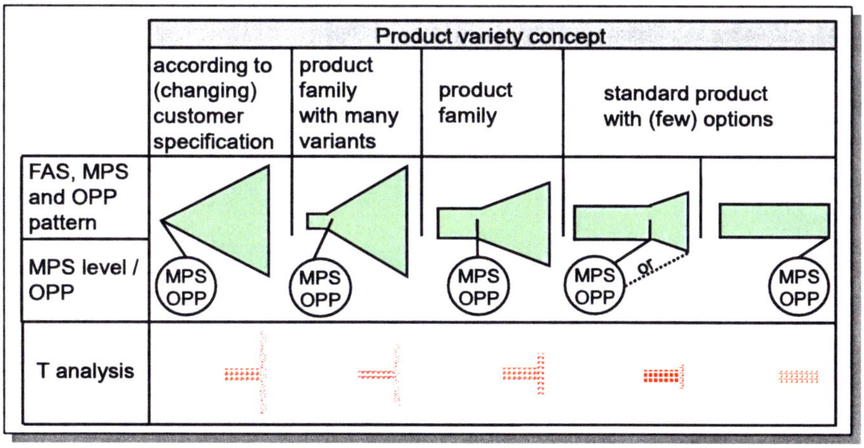

Fig. 7.1.2.3 FAS/MPS/OPP patterns in dependency on the product variety concept and their relation to the patterns of the T analysis. The FAS level is at the right of each pattern.

In the case of products according to (changing) customer specification, an engineer-to-order production type may mean that no MPS can be established. The planning activities then address capacities (personnel hours) rather than parts or material (cf. Figure 7.1.2.2).

7.2 Adaptive Techniques

Intended learning outcomes: Explain techniques for standard products with few variants as well as techniques for product families.

7.2.1 Techniques for Standard Products with Few Variants

Intended learning outcomes: Present the conventional variant structure for a few, stockable variants. Explain the production plan and its corresponding MPS at the end product level and at the assembly level. Describe the revision of the MPS according to actual splitting of family demand as given by the FAS.

A *variant bill of material* is the bill of material for a product family containing the necessary specifications indicating how the bill of material for a variant of the product family is derived. A *variant routing sheet* is defined analogously.[2]

Standard products with few variants are produced repetitively and possibly stored. Here, conventional representations of product structure using bill of material and routing sheets can be used. Figure 7.2.1.1 shows that a variant in stock corresponds to a different *item*. Variant-specific components are grouped in their own variant assembly, called V_1, V_2,..., while the general components form their own assembly G. Variants in stock (P_1, P_2,...) contain as components the general assembly G and the corresponding variant-specific assembly V_1, V_2,....

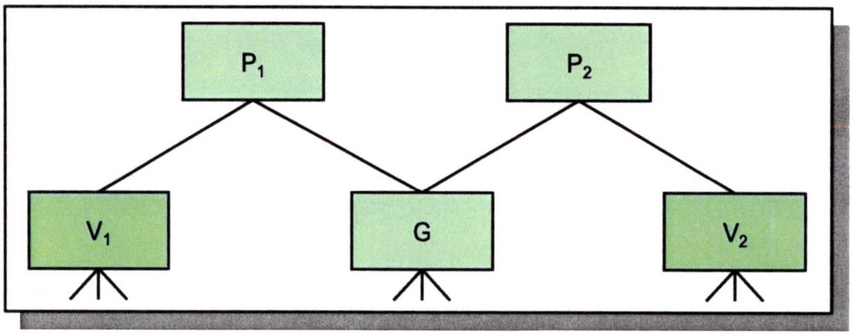

Fig. 7.2.1.1 Conventional variant structure for a few, stockable variants.

The (independent) demand for the product family, weighted by the *option percentage*, results in the independent demand for variants P_1, P_2,... The option percentage, like independent demand, is a stochastic variable (see Section 10.5.3). Because of a necessary safety calculation (see Section 10.5.4), the sum of the independent demand for the variants is greater than the independent demand for the product, or product family. To put it another way, the sum of the option percentages, under consideration of a safety factor, is greater than 1.

2 These definitions are deliberately more comprehensive than is usually the case. They are not restricted to methods for simple variant problems. The aim is to also encompass the methods for product families with many variants or of products according to customer specification, using the same terminology.

In the case of assemble-to-order (ATO), deriving dependent demand for the general assembly G yields an amount that is too large. This is corrected by entering negative independent demand for general assembly G. This negative number equals the sum of the safety demand for the variants P_1, P_2,… minus the safety demand for the product family.

This technique is easy to apply to a range of several dozen variants, which can be found, for example, in the manufacture of large machinery. For planning aspects, it may use different kinds of particular bills of material:

- Both the general assembly G and the variant assemblies V_1, V_2,… can be phantom assemblies, which are transient (nonstocked) subassemblies.

> A *phantom bill of material* represents an item that is physically built, but rarely stocked before being used in the next step or level of manufacturing (cf. [APIC16]).[3]

- A position of a variant-specific assembly can also (or partly) represent the subtraction of a position of the general assembly. This can be achieved through a negative quantity per in the variant-specific assembly, for example.

> A *plus/minus bill of material* is a variant bill of material with added and subtracted positions. A *plus/minus routing sheet* is defined analogously.

- Both the general assembly G and the variant assemblies V_1, V_2,… can be — and in particular the "parents" of a plus/minus bill of material are — pseudo items.

> A *pseudo bill of material* is an artificial grouping of items that facilitates planning ([APIC16]).

- Phantom and pseudo bills of material facilitate the use of common parts bills of material.

> A *common parts bill of material* groups common components of a product or product family into one bill of material, structured to a pseudo parent number (cf. [APIC16]).
> A *modular bill of material* is arranged in product modules or options. It is useful in an assemble-to-order environment, i.e., for automobile manufacturers (cf. [APIC16]).

> A *variant master schedule* is a master (production) schedule for products with few variants or product families.[4]

There are two possibilities for the level of the variant master schedule. Figure 7.2.1.2 shows an example MPS at the end product level, supposing a quantity per of 1 for the general assembly G and an equal option percentage in the demand — with a deviation of 20% — of the two variants at the product family P level. For teaching purposes, the example does not take into consideration safety demand for the product family P.

---

3 Linked with the concept of phantom bill of material is the *blowthrough technique*. See Section 12.4.1.
4 The term *mixed-model master schedule* can be used synonymously.

Product family \ Month	Jan.	Feb.	March	April
..				
P	100	100	150	120
..				

Product \ Month	Jan.	Feb.	March	April
P_1	50+10	50	75+5	60
P_2	50+10	50	75+5	60
Total	100+20	100	150+10	120
(Assembly G, in case of ATO)	(-20)		(-10)	

Fig. 7.2.1.2 The production plan and its corresponding MPS at the end product level (example of a product family P with two different products, P_1 and P_2).

Note the negative demand on the level of the general assembly G, as discussed above. As for distribution of the deviation in the two periods of January and March, the reader can refer to Figure 5.2.3.4.

The associated final assembly schedule (FAS) modifies the MPS according to the actual customer orders. If in January the actual demand is 60 units of P_1 and 40 units of P_2, then the MPS for February must be revised to replenish first the excess use of P_1 in January (20 units). Figure 7.2.1.3 shows this situation, extended for several months.

FAS \ Month	Jan.	Feb.	March	April
P	100	100	150	120
Actual P_1	60	45	60	
Actual P_2	40	55	90	

Product \ Month	Jan.	Feb.	March	April
P_1	50+10	60	45+25+5	60-15
P_2	50+10	40	55+25+5	90-15
Total	100+20	100	150+10	120
(Assembly G, in case of ATO)	(-20)		(-10)	

Fig. 7.2.1.3 Revision of the MPS according to actual splitting of family demand as given by the FAS.

Figure 7.2.1.4 shows the second possibility for the level of the master production schedule (MPS): the MPS at the assembly level. We suppose a quantity per of 2 and, again, an equal option percentage — with a deviation of 20% — for each variant-specific assembly V_1 or V_2. Again, the example does not consider safety demand for product family P. In this case, there is no need to deal with the (tricky) negative demand of general assembly G.

Product family \ Month	Jan.	Feb.	March	April
P	100	100	150	120
..				

Assembly \ Month	Jan.	Feb.	March	April
G	100	100	150	120
V_1	100+20	100	150+10	120
V_2	100+20	100	150+10	120
Total V_1 + V_2	200+40	200	300+20	240

Fig. 7.2.1.4 The production plan and its corresponding MPS at the assembly level (example of a product family P with two variants, V_1 and V_2).

The revision of the MPS according to actual splitting of family demand given by the FAS would result in a table similar to the one in Figure 7.2.1.3.

A *planning bill of material* is an artificial grouping of items that facilitates master scheduling and material planning (cf. [APIC16]).

A planning bill of material can facilitate the management of a variant master schedule. A planning bill of material may include historical option percentages of a product family as the quantity per.

A *production forecast* is a projected level of customer demand for key features (variants and accessories).[5]

A production forecast is calculated by using the planning bill of material.

A *two-level master schedule* uses a planning bill of material to master schedule an end product or product family, along with selected key features (variants and accessories).

[5] Disaggregating a product family forecast into production forecast (or in individual item forecasts) is also called the *pyramid forecasting technique*.

A *product configuration catalog* is a listing of all upper-level configurations contained in an end-item product family. It is used to provide a transition linkage between the end-item level and a two-level master schedule (cf. [APIC16]).

7.2.2 Techniques for Product Families

Intended learning outcomes: Present the super bill of material with option percentages x_1, $x_2, \ldots, x_n$. Describe the production plan and its corresponding MPS at the assembly level, using the example of a product family P with a number of variants in the order of the total demand quantity for the product family.

Generally, a product family can have hundreds of variants. In this case, a super bill of material is an appropriate planning structure.

A *super bill of material* is a planning bill of material for product family P, divided in one common parts and several modular bills of material. The common parts bill of material G, together with one of the modular bills of material $V_1, V_2, \ldots, V_n$, forms one possible product variant. The quantity per (x_i) of each modular bill of material (V_i) is then multiplied by the expected value of the option percentage corresponding to the variant, plus safety demand for the deviation of the option percentage (as was also necessary in the case with few variants).

Figure 7.2.2.1 illustrates the example in the definition above.

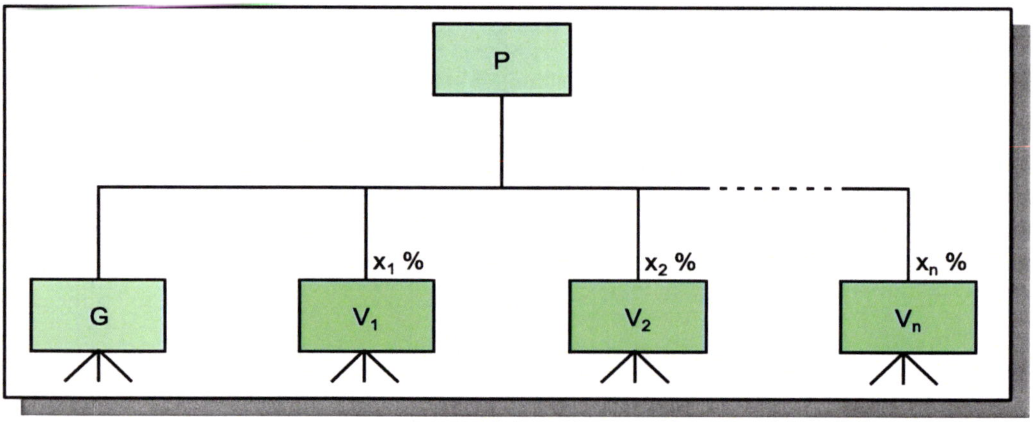

Fig. 7.2.2.1 Super bill of material with option percentages $x_1, x_2, \ldots, x_n$.

The (independent) demand for the product family is the forecast for the entire product family plus eventual safety demand (see Section 10.5.4). In general, the sum of all demand on variant assemblies is — even with a quantity per of 1 — by far greater than the demand for the product family.

A structure like this is also called *one-dimensional variant structure* (variable bill of material and variable routing sheet), because the variants are simply counted *de facto*. $V_1, V_2, \ldots, V_n$ may lie in the form of a plus/minus bill of material.

In contrast to the case with few variants in Section 7.2.1, requirements planning now yields *dependent* demands. In order configuration, a variant number must be added to the product family, so that the correct product variant can be selected and put into a production order.

The number of variants per product family that can be managed practicably with this technique is as high as several hundred. For larger numbers of variants, it becomes very difficult to determine the correct variant. Administrative search efforts become unwieldy, and there is the danger that one and the same variant will be stored as master data more than once. Moreover, many of the bill-of-material positions and routing-sheet positions saved under the variant assemblies are redundant; they exist in the various variants in multiple fashion. In most cases, there is a multiplicative explosion of the quantity of the positions in the bill of material and routing sheet; the same components and operations appear — often except for one — in almost every variant. This redundancy causes serious problems for engineering change control (ECC).

Figure 7.2.2.2 shows an example of the variant master schedule at the assembly level. For this case, let the quantity per for each variant be only 1. In addition, let the number of variants be 100, and let the demand quantity of the whole family P be 100, too. Again, we suppose an equal option percentage — with a deviation of 20% — of the variants of the demand at the product family P level. Again, for teaching purposes, the example does not take into consideration safety demand for product family P.

Month / Product family	Jan.	Feb.	March	April
P	100	100	150	120
..				

Month / Assembly	Jan.	Feb.	March	April
G	100	100	150	120
V_1	1+1	1	2	1
V_2	1+1	1	2	1
...				
V_{100}	1+1	1	2	1
Total $V_1 + V_2 + \ldots + V_{100}$	100+100	100	200	100

Fig. 7.2.2.2 The production plan and its corresponding MPS at the assembly level (example of a product family P with a number of variants in the order of the total demand quantity for the product family).

The revision of the MPS according to actual splitting of family demand given by the FAS would result in a table similar to the one in Figure 7.2.1.3, but it is more complicated to calculate.

Furthermore, the example reveals that, if the number of variants becomes as high as the total demand quantity for the product family, the option percentages become small. In addition, their deviation from the mean becomes so large that no forecast for the variant assemblies with economically feasible consequences is possible. For each variant, demand tends to be lumpy. For this reason, it will be necessary to apply one of the deterministic techniques that are described in the following.

7.3 Generative Techniques

Intended learning outcomes: Disclose the combinatorial aspect and the problem of redundant data. Present variants in bills of material and routing sheets as production rules of a knowledge-based system. Explain the use of production rules in order processing.

Generative techniques prove to be appropriate for *production with many variants*, that is, where there may well be millions of possible variants, but where the entire range of variants can be determined *from the start*, through a combination of possible values of relatively few parameters. Although each product variant results in a qualitatively different product, all stem from the same product family (see definition in Section 1.2.2). The production process for all product variants is principally the same. According to Figure *4.4.5.2,* product families with many variants are strongly related to *the* "make-to-order" *p*roduction environment. According to Figure 4.5.2.1, mass customization is the main production type. See also Figure 7.1.2.1.

7.3.1 The Combinatorial Aspect and the Problem of Redundant Data

Intended learning outcomes: Present setting the parameters of the fire damper. Disclose the number of possible combinations with n parameters as well as an example for number of identical bill-of-material positions.

Let us examine the problem using the example of a fire damper built into a ventilation duct, as shown in Figure 7.3.1.1. In the case of fire, the damper automatically stops the ventilation that would promote the spread of the fire.

Because ventilation ducts must fit the building, fire damper manufacturers must be able to offer a damper for every conceivable cross-section of "width * height * depth." To reach flexibility in achieving customer benefit, the damper is manufactured only to customer's order. Only certain semifinished goods, such as side pieces, strips, and drive kits, are premanufactured in small-sized production according to frequently requested variants. There are approximately 30 to 50 bill-of-material positions. Product features such as damper type, height, width, depth, and type of connection profiles are called *parameters.*

The *product solution space* is the combined variety of the customization domains to form a desired commercial product variety.

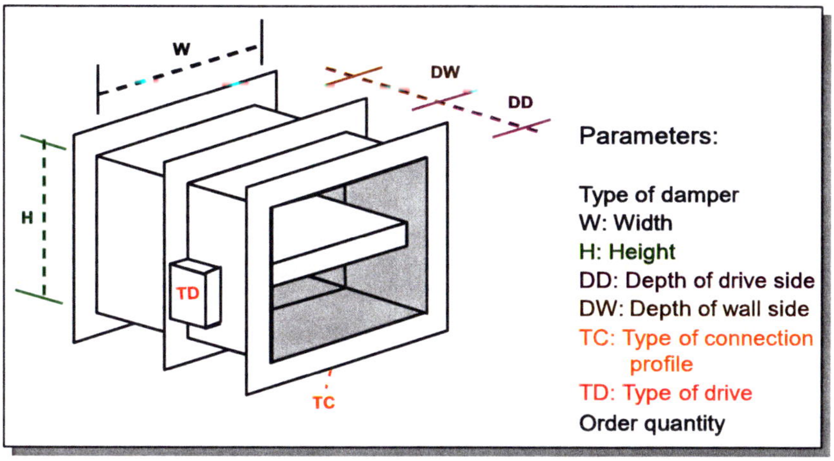

Fig. 7.3.1.1 Setting the parameters of the fire damper.

Therefore, the customer can specify any combination of parameter values of the value domains of the defined solution space. A group of parameters like this determines the type and quantity of required components (such as sheet metal, strip, connecting, and drive materials) as well as the operations with regard to production facilities and setup and run time (or load), and even with regard to description (for example, the number of fastening holes and the distance between holes in the connection profiles). Actually, four types of dampers are offered with widths from 15 to 250 cm and heights from 15 to 80 cm. With measurement increments of 5 cm, there would be 2688 variants (4 * 48* 14). A free combination of parameter values results in a theoretical number of tens of thousands of variants. The number of different dumpers in the practical world reaches thousands.

Let p(i) be the parameter of i (for example, type, width, height, depth, options, accessories) and $|p(i)| \geq 1$ be the number of possible values of the parameter p(i). Figure 7.3.1.2 shows the formula for the number of theoretically possible combinations. Of these, each has a bill of material and a routing sheet and differs — as a whole — from all others. A certain component can, however, be used to build many of these combination possibilities.

$$\prod_{1 \leq i \leq n} |p(i)| \equiv |p(1)| \cdot |p(2)| \cdot \ldots \cdot |p(n)|$$

Fig. 7.3.1.2 Number of possible combinations with n parameters.

For the fire damper, let p(1) and p(2) be the parameters width and depth. As a semifinished good independent of the other parameters, sheet metal pieces are cut to a width of 800 mm and a depth of 240 mm. This item is used as a component in all bills of material for dampers having a width of 800 mm and a depth of 240 mm. The number of these bills of material is calculated according to the formula in Figure 7.3.1.3.

All bills of material and routing sheets for the product family are similar. Their being nearly identical is typical of this type of production. If you were to keep a bill of material and a routing sheet for every single possible combination of parameter values, the greater part of the stored data would be redundant.

$$\prod_{3\leq i\leq n} |p(i)| \equiv |p(3)| \cdot |p(4)| \cdot \ldots \cdot |p(n)|$$

Fig. 7.3.1.3 Example for number of identical bill-of-material positions.

Classical aids to product configuration with the business objects item, bill of material, routing sheet, and work center (see Section 1.2, or as detailed objects in Section 17.2) do not allow the definition and storing of parameters and dependencies.

In such traditional systems, it would be possible to derive the variant from a "parent version" using adaptive techniques according to Section 7.2. However, with very many positions on the bill of material and many operations, this would place a heavy burden on qualified employees. This is not feasible for products with low value added. If, however, bills of material and routing sheets were created from the start in all their possible combinations, e.g., as one-dimensional variant structures (see Section 7.2.2), the multiplicative explosion of quantities of the positions on bills of material and routing sheets to be saved as data would make relocating efforts enormous and unfeasible. Engineering change control (ECC) for these thousands of bills of material and routing sheets would be highly problematic.

7.3.2 Variants in Bills of Material and Routing Sheets: Production Rules of a Knowledge-Based System

Intended learning outcomes: Differentiate between design rules and process rules. Explain the concept of design rules or process rules.

The key to a solution is to extend the business objects by adding a suitable representation of the knowledge about when certain components are built into a variant of a product family and when certain operations become part of the routing sheet. This is accomplished by implementing knowledge-based information systems. For a detailed description of these tools, see Section 17.3.1 and [Schö88b]. For the sake of simplicity, let us explain these systems using our introductory example of the fire damper.

From the perspective of product design, a product family is a single product. For example, there is one single set of drawings for the entire product family. There is one single corresponding bill of material, and it contains all possible components (such as raw materials and semifinished goods) just once; in similar fashion, the single routing sheet contains all possible operations listed just once. By inserting tables or informal remarks, the documents will indicate that certain components or operations will occur only under certain *conditions*. This characteristic is expressed in design rules or process rules.

A *design rule* is a position of the bill of material that is conditional as specified by an *if-clause,* which is a logical expression that varies in the parameters of a product family.

A *process rule* is a position on the routing sheet that is defined analogously.

Following these definitions, the rules in the fire damper example may be structured like those in Figure 7.3.2.1.

Design rule:

"Component X (such as a sheet metal semifinished good) in quantity per of 1 is used in the product,
- If the product width is 800 mm and depth is 240 mm."

Design rule:

"Component Z (for example, for an option) in quantity per of 1 is used in the product,
- If type = 2 and option = x, or type = 1 and option = y."

Process rule:

"Operation 030 (for example, a cutting operation) is carried out
- With description d(1), time t(1), at work center wc(1),
  - If type = 2 and order quantity ≥ 100 and width ≥ 400,

or
- with description d(2), time t(2), at work center wc(2),
  - If type = 2 and order quantity < 100,
  - or type = 2 and width < 400,

or
- with description d(3), time t(3), at work center wc(3) in all other case."

Fig. 7.3.2.1 Design or process rules.

A position of the bill of material or the routing sheet thus becomes a production rule in the actual sense, that is, of a product to be manufactured. These rules are applied to facts, such as the data on item, facilities, and work center in the production database or on parameter values in a query (for example, for a current customer order for a specific product of the product family).

Product designers and process planners in the company function as experts. When they put their rules on paper for variant bills of material and variant routing sheets, they use — unconsciously — expressions that are very similar to production rules. It is evident that these are experts expressing expert knowledge, for no two product designers will deliver precisely the same design for a particular product. In the same way, two process planners will seldom produce exactly the same routing sheet.

The users of the system are those persons who release, control, and produce the orders. The exact realization of production rules in an information system is treated in Section 17.3.2.

7.3.3 The Use of Production Rules in Order Processing

Intended learning outcomes: Present an excerpt from the parameterized bill of material for the fire damper. Identify data storage complexity for the fire damper example. Disclose the use of generative techniques in connection with CAD and CAM as well as in the service industries.

Figure 7.3.3.1 shows an excerpt from the product structure of the fire damper in Figure 7.3.1.1. This part of the bill of material lists some attributes and if-clauses important to an understanding of production rules.

Posi-tion	Variant	Quantity per / unit	Component identification	Component description
130	01 condition	2 PC if type = 1 and width ≥ 150	295191	Distance pipe FD1 D8/10/40
130	05 condition	2 PC if type = 2	295205	Damper axle FD2 D14/18/18
140	01 condition	2 PC if type = 2	295477	Sealing plate FD2 60/6/64
150	01 condition	1 PC if type = 1 and height < 150	296589	Angular console H100 FD1
150	03 condition	1 PC if height ≥ 150	295108	Angular console general FD 1/ 2
150	05 condition	2 PC if type = 2 and width > 1300	295108	Angular console general FD 1/ 2
155	01 condition	1 PC if type = 1 and height < 150	494798	Pivot form B galvanized
160	01 condition	1 PC if type = 1 and drive = "left"	295167	Bearing fixture left FD1
160	03 condition	1 PC if type = 2 and drive = "left" and width < 1300 or type = 2 and width ≥ 1300	295183	Bearing fixture left FD2
160	07 condition	1 PC if type = 1 and drive = "right"	295175	Bearing fixture right FD1
160	09 condition	1 PC if type = 2 and drive = "right" and width < 1300	295191	Bearing fixture right FD2

Fig. 7.3.3.1 Excerpt from the parameterized bill of material for the fire damper.

For the query, the facts — the product identifiers, order quantity, and all parameter values — have been added. Through comparison of these facts with the rules stored for the product family, program logic determines for each position the first variant of the bill of material or routing sheet for which evaluation of the rule results in the value "true."

Try the following exercise: In Figure 7.3.3.1, what variants are selected, given the following parameter values: Type = 1, drive = left, width = 400, height = 120?

Solution: Position/variant: 130/01, 150/01, 155/01, 160/01. Also compare the exercise in Section 7.6.2.

Storing parameterized positions on the bill of material and routing sheet in the form of production rules has key advantages over conventional positions. Each potential position is, in *one* comprehensive, *maximal bill of material* or in *one* comprehensive, *maximal routing sheet*, listed exactly *once*, but it is listed together with the condition under which it will appear in a concrete order. This means that there is no longer the stored data redundancy found in the classical case without parameterizing. In terms of the combinatorial aspect, rather than having a storage problem growing multiplicatively, we now have just additive increases. For a detailed comparison of data storage complexity, see [Schö88a], p. 51 ff.

Figure 7.3.3.2 shows actual, rounded comparative numbers for the data storage necessary for the fire damper in our example.

Version	Number of item identifications	Number of positions (bill of material or routing sheet), or number of production rules
Classical	200 + 1000*	10,000**
Parameterized	200 + 1	400

* The number of 1000 is more or less the number of combinations produced during the observation period. The theoretically possible number is >15,000, when increments of only 5 cm are considered. The number of 200 is more or less the number of semifinished goods in stock.

** Theoretically >30,000. Through intelligent choosing of a phantom bill of material, or intermediate products with no operations, this can be reduced to 10,000.

Fig. 7.3.3.2 Comparing data storage complexity for the fire damper example.

With minimal data storage problems, any number of orders with all possible combinations of parameter values can be transposed into production orders in a simple manner. One only needs to enter the values of the parameters. All these orders contain the correct components and operations, each with correctly calculated attribute values. Moreover, all possible combinations have been defined previously and automatically. Engineering change control (ECC) is also simple. If, for example, a new component is introduced, with a typical bill-of-material mutation, the component identification is added as a position to the (unique) bill of material. If it is a variant, its use dependent on parameters will be given an if-clause. Qualified employees familiar with the design and production process perform all of these tasks.

There is an advantage to the use of knowledge-based product configurators when PPC software is used in connection with CAD and CAM. With CAD, only one unique drawing is produced for all variants, but as above, it is parameterized. Within CAM, there is also only one unique, parameterized program controlling the machines. With this knowledge-based representation, PPC also now keeps only one unique bill of material and routing sheet for all variants. If there is a suitable, parameter-based CAD program package, a parameterized bill of material with a drawing can be exported from CAD to the PPC software. More important, however, is the reverse direction with an order. The parameter values of the production order can be exported from the order to CAD in the bid phase (or at the latest at order release). CAD then produces an order-specific drawing. In practice, this option is used in bids for products in the construction industry, for example. In analogous fashion, linking an order to CAM means that the same set of parameter values can serve as input to a CNC program.

And, finally, the generative technique is used successfully in the service industries, such as in the insurance branch and in banking. A family of insurance products can be seen as a product with many variants. Here, again, we find a clear case of nonrepetitive production. The setting up of a policy, or order processing, is at the same time the production of the product. The parameters are the features of the insured object as well as the types of coverage to be provided (e.g., sum insured, excess, type of compensation). The production rules of the configurator assign the elementary products to possible contracts. Concrete entry of a set of parameters ultimately yields a concrete insurance policy and includes all calculations, particularly the premium. Here see [SöLe96]. Those readers interested in an application in banking or in uncertainty may wish to refer to [Schw96].

7.4 Generative and Adaptive Techniques for Engineer-to-Order

Intended learning outcomes: Differentiate between the classical procedure and different archetypes of engineer-to-order. Describe the approach for basic and for repeatable engineer-to-order.

Although *mass customization* products are nearly all physically different, many companies that consider *mass customization* to be one of their core competencies view such products as "standard" products. This is because they can all be produced using a make-to-order process. For such companies, "non-standard" or "customized" are terms that refer to products that cannot be described with the configurator, and thus need an engineer-to-order environment. According to Figure *4.4.5.2, the* engineer-to-order production environment is strongly related to products according to (changing) customer specification. According to Figure 4.5.2.1, nonrepetitive or "one-of-a-kind" production, often together with mass customization, are the corresponding production types. The customers' requirements here, in terms of speed and cost, are not as stringent as for "standard" (mass customized) products, but in many branches they are getting closer to mass customization requirements. Also cf. Figure 7.1.2.1.

7.4.1 Classical Procedure and Different Archetypes of Engineer-to-Order

Intended learning outcomes: Explain different archetypes in engineer-to-order: basic, repeatable, complex, non-competitive ETO.

In the plant manufacturing industry, many areas of a plant facility are customer specific and produced in nonrepetitive production. With an intelligent product concept, however, it is usually possible to determine similarities of the plant facility to previously produced plants. For instance, during processing of an order, the vendor recalls previous "similar" problems. Derivation can thus often be performed on the basis of a previous customer order as a "parent version," using adaptive techniques according to Section 7.2, at worst position by position. Such orders generally require a high degree of order-specific engineering. They have to be developed and built to exactly match the customer's specifications. Other examples of this classic situation would be customer-specific modernization of an aircraft, an oil platform, or a nuclear power plant. If there are very many bill-of-material and routing-sheet positions, this will entail a high load on qualified employees. In addition, the lead time is long. Either is only justified for high-value-added products.

In the plant manufacturing industry, there have been attempts to restrict order specific engineering to a minimum and to use generative techniques for the larger part of the customer order (also see Section 7.3). This has worked well in the exterior construction business; for example, where certain elements of building exteriors are selected from a preset range of variants. Combining the elements for the whole face of a building, however, may well require order specific engineering. Except in the area of project management, or general procedures, this minimum level of ETO offers little or no potential for standardization and automation. The consequence is a low degree of industrialization. As well as this classic case of engineer-to-order, there are also situations where a higher degree of industrialization is necessary (and also possible) for reasons of cost or throughput time. The Portfolio in Figure 7.4.1.1 shows different archetypes of engineer-to-order (ETO).

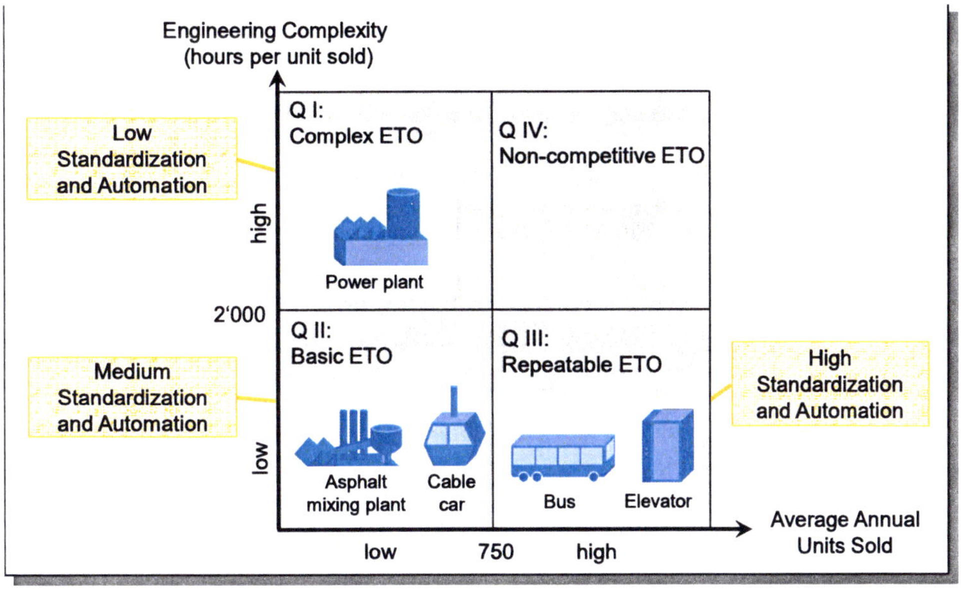

Figure 7.4.1.1 Different archetypes in engineer-to-order (cf. [WiPo16]).

The "Annual units sold" dimension shows how many units of a product family are sold on average per year. A unit is generally the lot that is ordered as one order unit and therefore manufactured in this lot size. Often, lot size one (batch size one) is essential; then, the units sold are the pieces sold. The threshold of 750 units/year is based on the assumption that making at least two units/day offers a sufficiently high degree of repetition for cost-effective automation.

The "Engineering Complexity" dimension measures the number of necessary engineering hours per unit sold. In practice, order-specific engineering is one of the biggest drivers of complexity in the ETO environment. Consequently, the possible benefit of an automation depends on it. The boundary between the quadrants has been set at 2000 engineering hours/unit, as that corresponds to one person year.

The classic "engineer-to-order" situation described above is called *"Complex ETO"* in Figure 7.4.1.1. Only a small degree of industrialization is possible, since the repeat rate is very low at both the product and the process levels. *"Non-competitive ETO"* relates to firms that deploy over 750 full-time positions for customer-specific engineering (≥ 750 units/year times ≥ 2000 h per unit). For firms of this size, it was previously possible to combine them as a conglomerate of subsidiaries, with each company supplying less than 750 order items per year. The subsidiaries can therefore be included in the *"Complex ETO."* We suspect that it is not profitable over the long term for a company to position itself in this quadrant, unless it has a monopoly and the company is therefore not exposed to competition. The two others, *"Basic ETO"* and *"Repeatable ETO,"* are covered in the next two sections.

7.4.2 Approach for Basic Engineer-to-Order

Intended learning outcomes: Explain the template for bill of material and routing sheet used to work out similar variants.

The lower the value added to the order, the lower the effort that should be expended on customer-specific adaptations. Hence the term *"Basic"* engineer-to-order for these archetypes, where engineering complexity needs to be kept to a minimum. Cable cars and asphalt mixers are good examples of this. On the one hand, cable car operators expect a cable car design that is specific to their ski area. On the other hand, the equipment as a whole regularly has to be adapted to the topographical conditions. Typical reasons for customer-specific adaptations to asphalt mixers are regional environmental standards, or harsh climatic conditions. Examples where even less engineer-to-order can be carried out for each order would include fastening tools or air diffusion grilles. But customers often order an entire lot of items that are specifically adapted for a particular construction project.

In order to reduce the engineering complexity, a medium degree of standardization and automation is recommended here. However, the low order repeat rate is a limitation. Standardization is primarily based on commonality of components and a modular product concept. Building on that, a product family is often developed for the production environment on a make-to-order basis and a mass customization production type. If the product and process configurator cannot completely specify the customer order, the provisional configuration result is adapted using adaptive techniques, where the bill-of-material positions and operations can be added, amended, or deleted.

A limited degree of automation is determined by the definition of product families with an unfinished product structure, and the development can then be finalized according to customer requirements. Figure 7.4.2.1 illustrates the technique.

The unfinished product structure looks like a *template* for a product structure. It contains materials that a company (e.g., a sheet metal working firm) typically uses as a starting point for a product (e.g., various aluminium or steel sheets), and a sequence of loosely described operations in which the firm has recognized expertise (e.g., cutting, bending, assembly) or that they get done by external suppliers (e.g., surface treatment). Components or operations may also be entered as dummy positions. The customer's order parameters are used by the configurator as much as possible to decide the positions that have to be part of the order. The configurator stops wherever in the template a "?," i.e., the symbol for an information gap, is encountered and asks for entry of the attribute value specific to the current customer order, e.g., the quantity per for the components C_1, C_2,…, C_n. The result, although intermediate, is often already useful for initial cost calculations (for example, where +/- 10% accuracy is sufficient) and for logistical control during order execution. The intermediate result is then updated as necessary using adaptive techniques, according to the data from the customer, by adding, changing, and removing bill of material and routing positions.

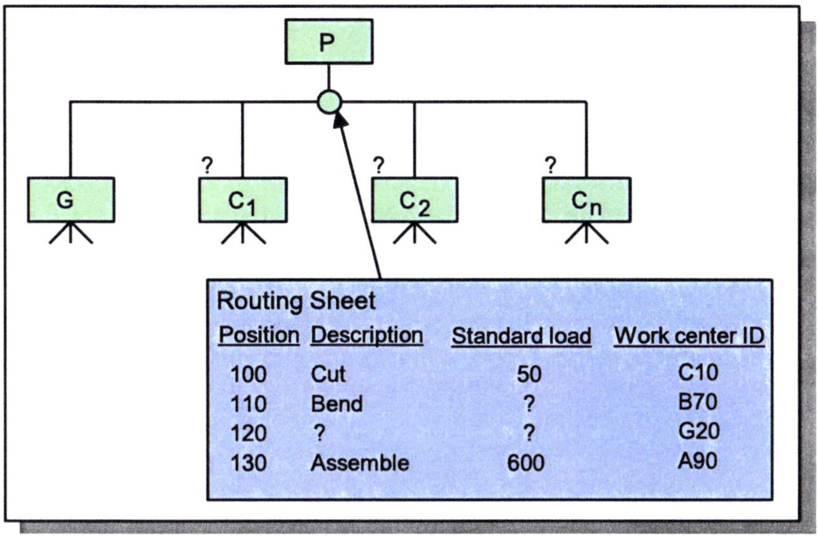

Fig. 7.4.2.1 Template for bill of material and routing sheet used to work out similar variants.

This technique allows a certain degree of automation, as the same unfinished product structure can be used for many customer orders. In practice, it has been observed that automation in "Basic" engineer-to-order is primarily achieved through the use of product configurators in the tendering process.

7.4.3 Approach for Repeatable Engineer-to-Order

Intended learning outcomes: Describe a typical engineer-to-order business process and its permanent enabling process. Explain role of the permanent engineer-to-order enabling process.

This archetype would include elevators or buses, for example. As with the "Basic" engineer-to-order, a product family is often developed for the production environment on a make-to-order basis and a mass customization production type. If necessary, the result of the configuration process is adapted according to the customer order. This approach is naturally only suitable for product families that only need partial customer-specific adaptation.

Since these are often "engineer-to-order" jobs, the level of competition is high. On the other hand, there is often repetition, at least at process level. This fact can and must be used for a high degree of standardization and automation. For example, in the segment of skyscraper elevators, the elevator in the top floor must fit with the architect's personal concept. The "engineer-to-order" process requirements are thus similar to those of the fashion sector. In addition, selling the "non-standard" (i.e., "engineer-to-order") elevator on the top floor is often the order winner for the many "standard" (i.e., "make-to-order") elevators in the whole building. Therefore, the specific engineering of the top floor elevator is not adequately remunerated. Thus, the engineer-to-order process must be fast and efficient.

At the fundamental level, the use and integration of product configurators in the ERP and CAD software systems can help automate the tendering and ordering process from end to end. Expertise and experience with suitable techniques and tools (e.g., product configurators) are used in the individual sell, engineer, and make processes. Even as early as the sales process, configuration allows for an initial cost calculation and for a virtual product, which allows the customer to experience something as close as possible to the physical product.

Fast and efficient engineer-to-order also imposes high organizational requirements. That is a key reason why customer-specific adaptations are normally not handled by product design, but as a separate process by a department that specializes in carrying out customer-specific adaptations. Figure 7.4.3.1 shows a typical engineer-to-order business process in a company.

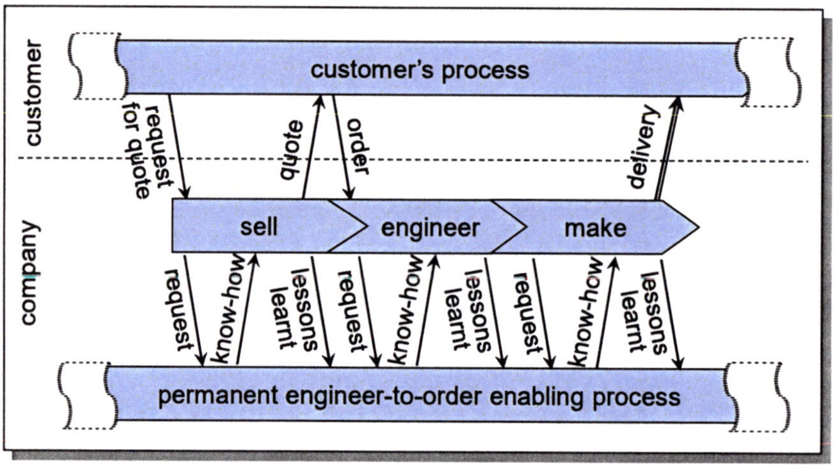

Fig. 7.4.3.1 A typical engineer-to-order business process and its permanent enabling process (cf. [Schö12]).

The sales phase (including quoting) is followed by receipt of the customer's order, followed in turn by design, production, and delivery. In practice, fast and efficient engineer-to-order calls for the consistent, long-term use of an enabling process. Figure 7.4.3.1 calls this process a *permanent engineer-to-order enabling process.* Queries from the ongoing engineer-to-order business process are answered through a form of know-how transfer. If additional know-how is gained during order execution, it is fed back to the enabling process in the form of lessons learnt. At the organization level, this means expertise and experience of dealing with the customer's engineer-to-order requirements. This involves the business models between the company and external customers and suppliers, but it also means the "business model" around the internal customer-supplier relationships between sales, engineering, and production, at all levels. At process level, this means expertise and experience of managing how the company works with external customers during the product specification and manufacturing phases, but also how internal customers work with suppliers. Another point is know-how of user interaction with the product in a virtual status, i.e., before and during the physical manufacturing.

Whilst in the classic mass-customization culture it has been possible over the years to concentrate the mass-customization expertise across fewer people and to focus on the design process (and, to an extent, on the sales process), a quick and efficient engineer-to-order system needs this expertise shared amongst more people, and it must in turn be extended to the workshop. So we can talk here of a distinct engineer-to-order culture.

For companies that consider mass-customization products as their "standard" products, it is notable that an engineer-to-order ("non-standard") customer order (all too) often results in a new parameter, which controls the new customer-specific components. The overhead of introducing the new parameter is only covered if the same customer orders their special variant several times. To solve this problem, we observe that "non-standard" products will use the principles of commonality and of modular product concept that have led to increased productivity in the case of mass customization. Here, these principles can and must also be applied for parameterized component families, as soon as components of product families are families in their own right. However, their parameters may have a different name or semantic structure than that of the product family, perhaps because the component family already existed beforehand for another product family. Transforming the parameter values from the product family efficiently to those of the component family increases the commonality of the component family. This often helps in avoiding new parameters or components. One more thing: The increased commonality of the component family also encourages product innovation.

Therefore, an important role of the permanent engineer-to-order enabling process is to carefully determine the parameterization for product families, particularly for component families. This task includes (1) determining and maintaining the component families and their set of parameters, (2) increasing re-usability of parameterized component families, thereby ensuring their commonality, and (3) encouraging colleagues to use existing parameters or to suggest sensible enhancements (especially for the value ranges). This task is somewhat more difficult than the comparable task of ensuring that single components used for newly constructed products can be re-used. The leaders of the teams that carry out these centrally organized tasks are very experienced product or process developers. They understand the reasons for the existing para meterization, and can follow the thinking of their colleagues when developing. They also have excellent social skills, to encourage colleagues to support their standardization efforts.

7.5 Cooperation between R&D and Engineering in ETO Companies

Intended learning outcomes: Describe different means used for cooperation between the R&D and the order-specific engineering departments. Present the portfolio of cooperation types between R&D and engineering in ETO companies.

Mass customization is widely used in sectors such as construction, machinery and equipment, automotive and fashion. The "standard" product solution space is ultimately determined by the R&D department, as a part of the product design (PD) process, by determining the parameterization. The Engineering department (generally separate from the

R&D department) sets up the order specific engineering (OSE) process, which covers all the design activities aimed at a specific customer order. This process should extend the product solution space quickly and efficiently so that it can cater for non-standard requirements.

A lack of cooperation between the R&D and Engineering departments in companies is one possible cause for delays and cost overruns. Engineer-to-order (ETO) firms have to find ways to quickly adapt and integrate engineering, manufacturing and supplier capacities. In practice, however, there are significant differences in cooperation between the R&D and Engineering departments. [SöWe17] addresses these differences in firms in the construction and machinery industries. Section 7.5.1 summarizes the observations in the case of two companies — details about the various means (practices and tools) used can be found in [SöWe17]. Section 7.5.2 identifies factors that influence the cooperation and thus the choice of such means. From this, it develops a portfolio with four sectors of fundamental types of cooperation in ETO firms. Companies in the same sector can benefit from sharing their methods and tools.

7.5.1 Different Means Used for Cooperation between the R&D and the Order-specific Engineering Departments

Intended learning outcomes: Present in detail design and specification of a product in an ETO environment as five phases of potential cooperation between R&D and Engineering along the value chain. Explain the ETO businesses of industrial steam turbines and high-rise elevators.

It is practical to classify the means utilized, among other criteria, by five consecutive phases of potential cooperation between R&D and Engineering along the value chain, as shown along the green line in Figure 7.5.1.1. Thereby, the typical market-driven PD differs from *customer-driven PD* (i.e. when a customer order triggers conventional PD activities). Also, customer-driven PD and (pre- and post-award) OSE differ in their primary scope: the design of a standard component and the customization of a component, respectively.

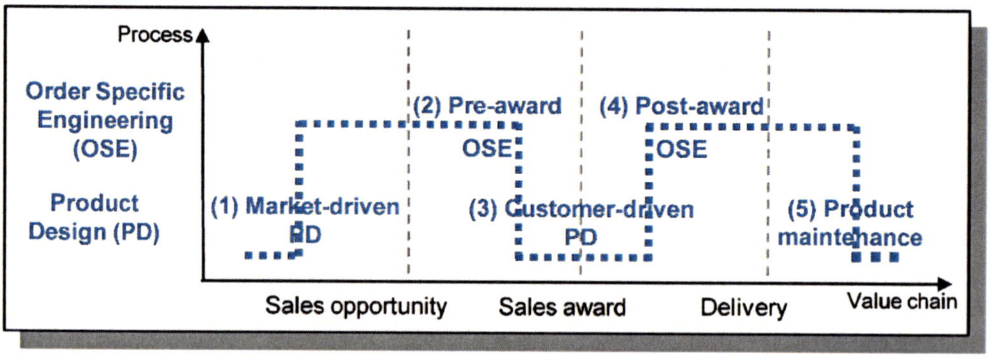

Figure 7.5.1.1 Design and specification of a product in an ETO environment (green line): Five phases of potential cooperation between R&D and Engineering along the value chain

[SöWe17] describes several *formalized* practices and tools used in two companies. These means allow the companies both to fulfil actual customer requirements via OSE and, regularly, to integrate the newly developed product specifications into the standard solution space for future reuse. In both companies, there are also *informal* means of communication. These are just as important. However, their investigation is not within the scope of this Section.

The Industrial Steam Turbines business unit of General Electric (GE) has about 90 R&D and OSE engineers. Engineering is active during 70% of the delivery lead time. Almost all of the 10-20 orders per year follow specific customers' requirements such as output capacity, pressure and temperature levels for the turbine. Each engineering project is divided into subtasks according to knowledge domains and to specific components of the final product. The assembly of the different modules also requires order-specific interfaces. Such complexity results in, on average, more than 6,000 (sic!) man hours per project. GE often benefits from quick retrieval of product-related information out of ad-hoc repositories, and developing search engines aimed at navigating through the company's product solution space and relevant design instructions. R&D and Engineering cooperate using several formalized organizational practices and IT tools. R&D and Engineering jointly develop and maintain these tools and processes. Overall, there is a *high degree of cooperation* between the two departments.

The high-rise elevator business unit is a relatively small entity within Schindler Europe and Asia-Pacific. The use of a sophisticated product configurator allows a high degree of standardization, and therefore offers high efficiency of the order acquisition and fulfilment process. This IT tool was developed in view of the substantial number of orders sold per year. It enables identification of commonalities in the customer's requests, and in turn expansion of the company's standard product solution space. For many orders, the classical make-to-order approach is adequate. If customers' requirements entail a non-standard solution, the Engineering team attempts to generate new designs and associated technical solutions. The majority of orders require less than 100 hours of engineering work. Very few tools are used for cooperation between R&D and Engineering. Most of the practices and IT tools are used within the high-rise elevator business unit, often only within Engineering. Overall, and in particular during the OSE process, there is only a *low degree of cooperation* between R&D and Engineering. Also, phase 3 in Fig. 7.5.1.1 is not used by Schindler.

Furthermore, the majority of the means used by GE are organizational practices and short-term oriented, whereas for Schindler they are IT tools and long-term oriented.

7.5.2 The Portfolio of Cooperation Types between R&D and Engineering in ETO Companies

Intended learning outcomes: Identify influencing factors of cooperation between R&D and Engineering. Disclose two fundamental cooperation types in ETO companies. Explain the portfolio of cooperation types between R&D and Engineering in ETO companies.

In terms of cooperation between R&D and Engineering, the following influencing factors were noted from the observations in companies and from the literature. Thereby, the influence of the factors are not of equal weight in each company.

1. *Sales volume*, or the *number of orders* (for lot size one (batch size one), this is equivalent to the number of units). According to [WiPo16], two or more orders per day counts as a high sales volume, and a dozen orders per year as a low sales volume.
2. *Level of integration of product design and guidelines*: A high level means that knowledge about the method of generation of customer-specific components using OSE has been integrated into the PD process, and therefore into the product design guidelines and the product configurator.
3. *Degree of product customization*: From the company's perspective, a high degree means that it involves developing new components, rather than just adapting existing ones.
4. *Level of component engineering complexity*: A high level means that the OSE contains a high measure of time and effort to adapting components, or developing new ones.
5. *Level of architecture engineering complexity*: A high level means that the combination and interfaces of (main) components in the product have to be newly developed and produced each time, as is often the case in e.g. plant engineering.
6. *Accuracy of estimating the duration of OSE*. A high accuracy also involves a high level of component modularization and commonality, making it easy to estimate time and effort needed to select them and position them in the product (whether time and effort needed are low or high).

Being more precise about "high" or "low" for all six influencing factors (e.g. using mathematical formulae) depends on the product and on the competitive situation, and is usually difficult to implement in practice. However, even using just two simple values, i.e. "high" and "low", the examples of GE and Schindler show clear differences in two fundamental types of cooperation between R&D and Engineering, as shown in Figure 7.5.2.1.

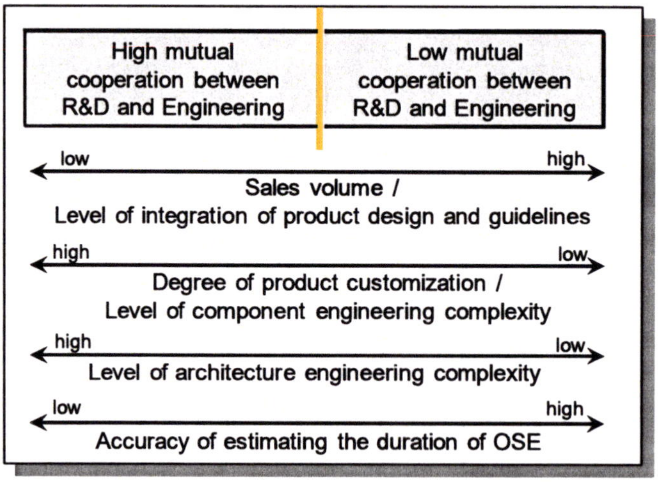

Figure 7.5.2.1 Two fundamental cooperation types in ETO companies

GE Industrial Steam Turbines belongs to the high cooperation type, whilst Schindler high rise elevators is of the low type. [SöWe17] present further companies that fit this division.

[SöWe17] also contains examples of ETO companies with two other types of cooperation, which fall between the two types in Figure 7.5.2.1. This situation entails a two-dimensional classification where influencing factors 1.) to 4.) above represent one dimension, and factors 5.) and 6.) represent the other. Figure 7.5.2.2 shows the resulting portfolio.

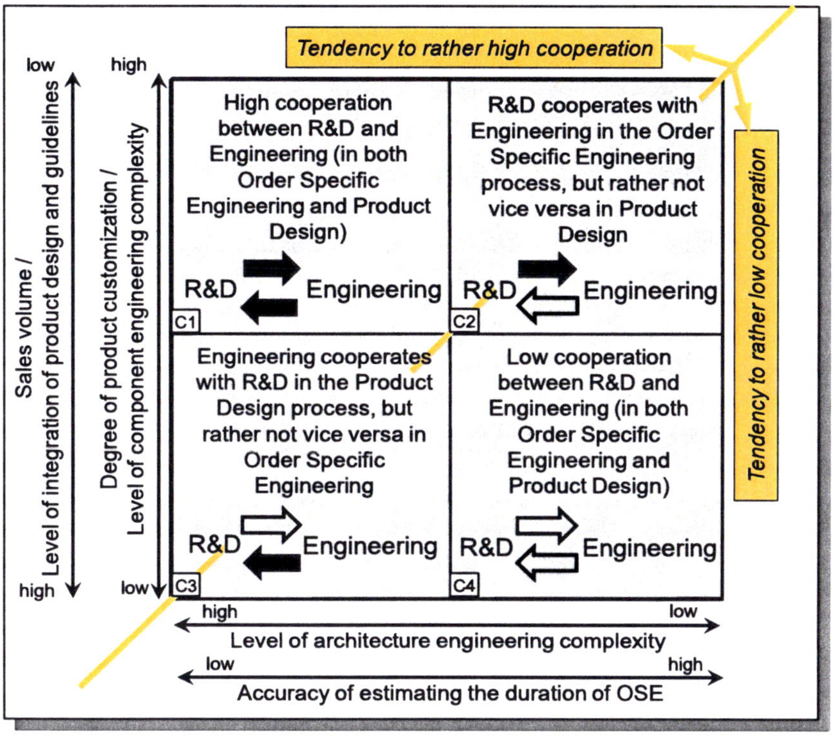

Figure 7.5.2.2 The portfolio of cooperation types between R&D and Engineering in ETO companies. (A filled arrow shows high cooperation in the respective direction, an empty arrow shows low cooperation).

Sector C1 describes the high, and Sector C4 the low mutual cooperation between R&D and Engineering in both the PD and OSE processes, as discussed with Figure 7.5.2.1.

The cooperation type in sector C2 represents a situation where R&D regularly cooperates with Engineering in OSE by supporting the development of new product specifications, while, vice versa, Engineering cooperates rather less with R&D in PD.

The cooperation type in sector C3 represents a situation where Engineering contributes to defining the new variants or components by updating the product family in PD, whilst, vice versa, R&D is largely not included in OSE activities.

Workshops showed that comparing practices and tools becomes fully useful when it involves two firms that use the same cooperation type according to the classification in Figure 7.5.2.2. But even then, the practices and tools must be adapted to suit the specific situation of the relevant company and product family.

7.6 Summary

Variant-oriented techniques are required when the market demands flexibility in meeting customer specifications. Today, this is frequently the case for the investment goods market. Some of the techniques also support production without order repetition, in particular the production types *mass customization* and *one-of-a-kind-production*. There are adaptive and generative techniques. Adaptive techniques determine a “parent version,” from which the bill of material and the routing sheet for the actual customer production order are derived. Subsequently, positions are added, modified, or deleted. Generative techniques use rules that already exist in an information system and that select the product variant, starting with data of the customer order, out of a set of possible components and operations.

The master production schedule (MPS) is best established at the level of the (customer) order penetration point (OPP). Downstream from this point, a final assembly schedule (FAS) is a possible tool to make the end items according to specific customers’ orders.

For low-variety manufacturing, there are in the simplest instance standard products with only a few options (in the dozens). This results in a tendency toward a rather high stocking level. For the demand of each variant a percentage of the total demand, called option percentage, is predicted. Because this is also a stochastic variable, the standard deviation of the demand for a variant is greater than that of the demand for the product family. The sum of independent demands for the variants is thus greater than the independent demand for the product family. In the more difficult case, the number of manufactured products is still much greater than the number of variants, which, however, can lie in the hundreds. This case can be handled in a manner similar to the case above. However, data redundancy in the representation of products and processes increases, and this also raises the efforts required to search and maintain master data and order data.

For high-variety manufacturing, that is, for products to customer specification or for product families with many variants, the number of variants increases to the magnitude of the demand. The use of stochastic methods would lead to high safety demand in variants and thus high inventory. Because in the best case there remains only potential repetitive production, we must move from stochastic to deterministic methods. Through almost the entire supply chain, the products are manufactured according to demand, with no stock-keeping. Inventory in raw materials and in purchased parts is replenished after use.

A product family with many variants is the typical case with mass customization. Here, the order can be produced directly, because all possible variants of the product family have already been included in product and process design. There can be millions of variants, that is, production with many variants. Each variant results in a different product. However, in characteristic areas, all product variants and also the production process are the same. Such product families are based on a concept in which the manifold variants are generated through combination of possible values of relatively few parameters. In principle, there is only one (maximal) bill of material and only one (maximal) routing sheet. To select positions for an order and to check compatibility of parameter values, knowledge-based techniques are used. Production rules then contain an if-clause, which is a logical expression that varies in the parameters.

Products according to (changing) customer specification are closely related to *the* engineer-to-order *production environment.* There are various different archetypes. In the classic case, known as Complex ETO, adaptive techniques are used. Basic ETO and Repetitive ETO are based on generative techniques (and therefore on *mass customization as a production type),* with adaptive techniques applied afterwards. With Basic ETO, a limited degree of automation can be achieved by defining product families with an unfinished product structure that looks like a template. The result of the configuration is often already useful for initial cost calculations and for logistical control during order execution. In the case of Repetitive ETO a fast and efficient engineer-to-order is a prerequisite. Here, a permanent enabling process is required. Additional know-how that is gained during order execution is fed back to the enabling process. The parameterization for product families, particularly for component families, must be determined carefully, thereby ensuring their commonality.

There are significant differences in cooperation between the R&D and Engineering departments in companies. Factors that influence the cooperation entail the development of a portfolio with four sectors of fundamental types of cooperation in ETO firms.

7.7 Keywords

adaptive technique, 339
common parts bill of material, 343
customer order production, 335
design rule, 350
engineer-to-order enabling process, 358
final assembly schedule (FAS), 340
generative technique, 339
high-variety manufacturing, 336
low-variety manufacturing, 338
modular bill of material, 343
phantom bill of material, 343
planning bill of material, 345
plus/minus bill of material, 343
process rule, 350
product configuration catalog, 346
product family orientation, 335
product solution space, 348
super bill of material, 346
unfinished product structure, 356
variant bill of material, 342
variant master schedule, 343
variant orientation, 335
variant production, 335
variant-oriented concept, 335

7.8 Scenarios and Exercises

Intended learning outcomes: Apply adaptive techniques for product families. Disclose the use of production rules in order processing. Elaborate the setting the parameters of a product family.

7.8.1 Adaptive Techniques for Product Families

Intended learning outcomes: Disclose a better understanding of the adaptive technique, by calculating the number of variants in the master production schedule.

Figure 7.2.2.2 showed an example of the variant master schedule. The example revealed that, in practice, this technique would not be applied for that case, because the number of variants turns out to be too high. However, the present exercise is aimed to aid better understanding of the technique, and it is thus useful for all cases where the number of variants is significantly smaller than the total demand quantity for the product family.

a. Suppose that the demand of the product family P for January was 200 instead of 100. Again, suppose an equal option percentage — with a deviation of 20% — of the variants of the demand at the product family P level. What would have been the total number of variants $V_1 + V_2 + \ldots + V_{100}$ in the master production schedule for January?

b. For the month of March, where the demand of the product family was 150, can you explain why two units have to be considered in the MPS for each variant?

c. For April, where the demand of the product family was 120, can you explain why only one unit has been considered in the MPS for each variant?

Solution:

a. 300. If a deviation of 20% has to be considered for each variant, an (equal) option percentage would result in 1.2%. A total demand of 200 units results in 2.4 units per variant. Because no fraction of a unit can be ordered, this value has to be rounded up to the next integer value, which is 1. Therefore, for each variant, 3 units will be in the MPS for January, or 300 in total.

b. Again, a deviation of 20% has to be considered for each variant. Again, this results in an (equal) option percentage of 1.2%. A total demand of 150 units results in 1.8 units per variant. This value is rounded up to the next integer, or 2 units.

c. Again, a deviation of 20% has to be considered for each variant. Again, this results in an (equal) option percentage of 1.2%. A total demand of 120 units results in 1.44 units per variant. As the units were rounded up by 0.8 in January and 0.2 in March, 0.44 units in April are covered in any case. Therefore, it is sufficient to have only 1 unit in the MPS for April.

7.8.2 Generative Techniques — the Use of Production Rules in Order Processing

Intended learning outcomes: Disclose a better understanding of the generative technique, selecting the positions/variants of an order by evaluating production rules.

Look at the excerpt from the parameterized bill of material for the fire damper in Figure 7.3.3.1. What are the positions/variants selected in Figure 7.3.3.1 with the following parameter values?

Type = 2, drive = right, width = 1000, height = 200

Solution:
Position/variant: 130/05, 140/01, 150/03, 160/09

7.8.3 Generative Techniques — Setting the Parameters of a Product Family

Intended learning outcomes: Explain how to determine possible parameters of a product family by examining five product variants of this family.

Figure 7.8.3.1 shows a product family (umbrellas) with some of the possible individual products.

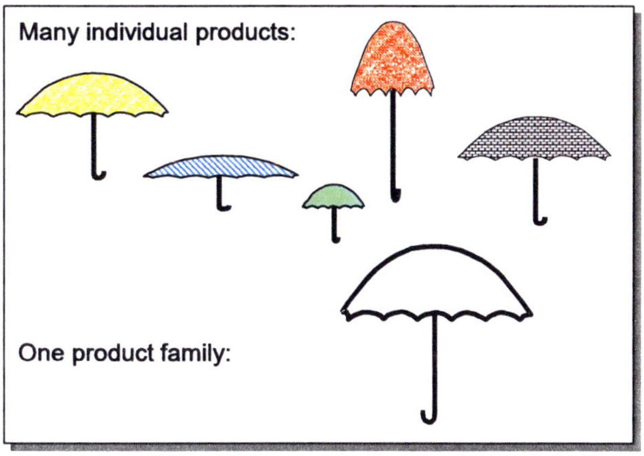

Fig. 7.8.3.1 A product family and five product variants of this family.

Questions:

a. What are the parameters that generate the product family, if they should generate the five variants at the least?

b. What are possible ranges of values for these parameters?

c. How many physically different umbrellas can be generated within that product family?

d. Are there incompatibilities, that is, ranges of values that a parameter can assume, that are partly dependent on other parameters?

Problem-solving hints:

a. There are at least 6 parameters. The diameter of the umbrella is one parameter, for example.

b. For "continuous" parameters (e.g., diameter), assume reasonable increments (e.g., 10 cm), as well as a reasonable minimum (e.g., 60 cm) and maximum (e.g., 150 cm). For parameters representing a set of discrete values (e.g., pattern), assume a reasonable number of different values (e.g., 30).

c. Combine each value of a parameter with each value of another parameter (cf. Figure 7.3.1.2). Your result depends on the number of parameters you detected in question a., as well as the ranges of values you determined in question b. Thus, your answer will be different from your colleagues' results.

d. For example, if the diameter of the umbrella is greater than 120 cm, then the handle of the umbrella must be longer than 100 cm.

7.9 References

APIC16 Pittman, P. et al., APICS Dictionary, 15th Edition, APICS, Chicago, IL, 2016

Schi01 Schierholt, K., "Process Configuration — Mastering Knowledge-Intensive Planning Tasks," vdf Hochschul-verlag, Zurich, 2001

Schö88a Schönsleben, P., "Flexibilität in der computergestützten Produktionsplanung und -steuerung," 2nd Edition, AIT-Verlag, D-Hallbergmoos, 1988

Schö88b Schönsleben, P., "Expertensysteme als Hilfsmittel der variantenreichen Produktkonfiguration," in Informatik, Forschung und Entwicklung, Springer, Berlin, 1988

Schö12 Schönsleben, P., "Methods and tools that support a fast and efficient design-to-order process for parameterized product families," *Annals of the CIRP*, 61/1:179-182, 2012

Schw96 Schwarze, S., "Configuration of Multiple Variant Products," BWI-Reihe Forschungsberichte für die Unternehmenspraxis, vdf Hochschulverlag, Zurich, 1996

SöLe96 Schönsleben, P., Leuzinger, R., "Innovative Gestaltung von Versicherungs-produkten: Flexible Industriekonzepte in der Assekuranz," Gabler, Wiesbaden, 1996

SöWe17 Schönsleben, P., Weber, S. et al., "Different types of cooperation between the R&D and Engineering departments in companies with a design-to-order production environment," CIRP Annals 66 (1), p. 405-408, 2017

WiPo16 Willner, O., Powell, D., Gerschberger, M., Schönsleben, P., "Exploring the Archetypes of Engineer-to-Order: An Empirical Analysis," International Journal of Operations and Production Management, 36 (3), p. 242-264, 2016

8 The Concept for the Process Industry

Intended learning outcomes: Produce characteristics of the process industry. Disclose processor-oriented master and order data management. Explain in detail processor-oriented resource management. Describe special features of long-term planning.

The MRP II concept, which has been supported by ERP software for nearly 40 years, originated in in the fields of mechanical and apparatus engineering and in the automobile and aircraft industries. With MRP II and ERP software, *a de facto* standard emerged, consisting of common terminology, use of similar representations of the logistics objects, and principally similar planning & control methods. However, for repetitive manufacturing as well as for one-of-a-kind production, the MRP II concept has already required extensions in terminology, logistics objects, and methods. The same is taking place for process industries.

Process industries or *basic producer industries* are manufacturers that produce products by process manufacturing.

Process manufacturing is production that adds value by mixing, separating, forming, and/or chemical reactions. Process manufacturing may be done in either batch production or in continuous production (cf. [APIC16]).

Process industries comprises manufacturers of chemical products, paper, food, mineral oil, rubber, steel, and so on. In these industries, it became increasingly clear that the terminology, logistics objects, and fundamental methods of the MRP II concept could not always be applied without adaptation. Many aspects of process manufacturing are simply not comparable to the production of aircraft, cars, or machines ([Hofm92]). It is interesting, however, that no uniform standard had been accepted within the process industry ([Kask95]) and that efforts toward standardization have been made only in the past 10 to 20 years. Clearly, there is a need for more scientific research in this area.

A *processor* in the process industries is the processing unit, or production infrastructure, that is, the production equipment (machines, appliances, devices) and the capacities.

The *processor-oriented concept* aims toward mastering pronounced high-volume line or continuous production and specialized, expensive production equipment (or processors) with a focus on maximizing processor capacity utilization.

Process manufacturers make significant investments in specialized, often single-purpose production equipment. This makes utilization of capacity the key criteria for planning & control purposes, capacity taking precedence over materials, components, and the fastest possible flow of goods.

Figure 8.0.0.1 shows some of the characteristic features for planning & control in supply chains, taken from Section 4.4. The typical values of the feature of greatest importance for this concept are highlighted in black. The further to the *right* that these values appear in the table, the better candidate the industry is for the use of the processor-oriented concept.

P. Schönsleben, *Handbook Integral Logistics Management*,
https://doi.org/10.1007/978-3-662-65625-9_8

After identifying the characteristics of the process industry, the next step is to derive appropriate processor-oriented techniques for planning & control.

Features referring to user and product or product family						
Feature	→	**Values**				
Orientation of product structure	→	▲ convergent			combination ▲upper/▼lower structure levels	▼ divergent
Features referring to logistics and production resources						
Feature	→	**Values**				
Production environment	→	engineer-to-order	make-to-order	assemble-to-order (from single parts)	assemble-to-order (from assemblies)	make-to-stock
Facility layout	→	fixed-position layout for site, project, or island production	process layout for job shop production	product layout for single-item-oriented line production	product layout for high-volume line production	product layout for continuous production
Flexible capability of capacity	→	applicable for many processes	applicable for few processes	applicable for only one process		
Features referring to production or procurement order						
Feature	→	**Values**				
Reason for order release (type of order)	→	demand / (customer production (or procurement) order)	prediction / (forecast order)	use (stock replenish-ment order)		
(Order) lot or batch size	→	"1" (single item production / procurement)	single item or small batch (production/procurement)	large batch (production / procurement)	(production / procurement) without lots	
Lot traceability	→	not required	lot/batch / charge	position in lot		
Loops in the order structure	→	Product structure without loops, and directed network of operations	Product structure with loops, or undirected network of operations			

Increasing suitability for the processor-oriented concept.

Fig. 8.0.0.1 Degree of suitability for the processor-oriented concept. (The horizontal distribution of the values in the morphological scheme indicates their relation to the increasing degree according to the given criterion).

8.1 Characteristics of the Process Industry

Intended learning outcomes: Explain divergent product structures and by-products. Describe high-volume line production, flow resources and inflexible facilities. Produce an overview on large batches, lot traceability, and loops in the order structure.

8.1.1 Divergent Product Structures and By-Products

Intended learning outcomes: Differentiate between primary product, by-product, and waste product. Explain the manufacture of by-products in chemical production. Describe the manufacture of by-products in sheet metal working. Identify the production of collets from a steel cylinder as well as the "saucepan and lid" problem linked with temporary assembly.

One of the characterizing features of the processor-oriented concept is *divergent product structure*. This type of structure is an *upside-down arborescent structure* with by-products.

A *primary product* is the product that the production process is designed to manufacture. A *by-product* is a material of value produced as a residual of or incidental to the process producing the primary product. A *waste product* is seen as a by-product without any value.

Manufacture of by-products is the simultaneous creation — that is, in the same manufacturing step — of further products in addition to the primary product.

The process often starts with a single commodity (raw material or intermediate product), although sometimes several commodities are processed together. The resulting products can be either intermediate products or end products. In some cases, a number of by-products (frequently steam or power) arise in addition to the primary product(s). By-products do not go directly into other products, but they can be recovered, utilized, and recycled in subsequent production processes. In contrast to by-products, which can reenter into the production process either directly or after appropriate treatments, waste products must be disposed of. Waste treatment and disposal engender additional costs.

The first example in Figure 8.1.1.1 stems from the chemical process industry.

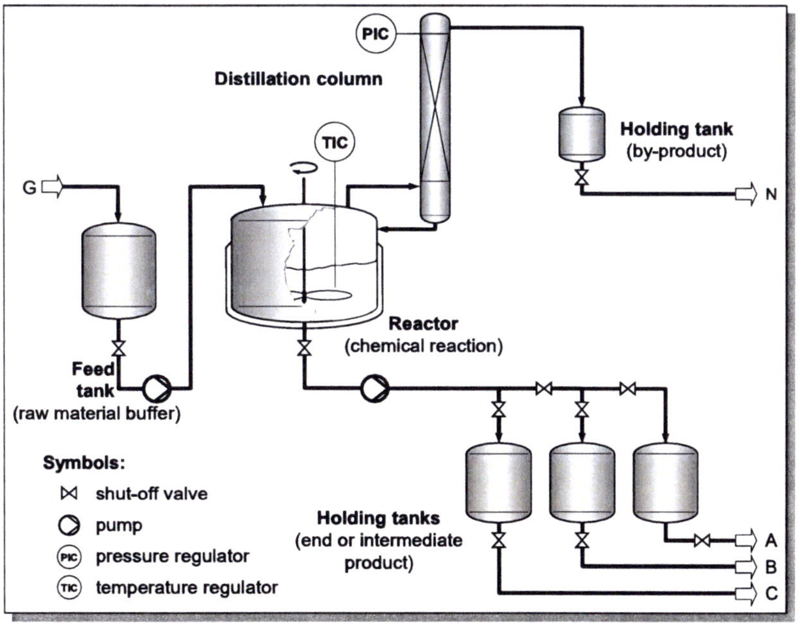

Fig. 8.1.1.1 Chemical production process: reactor with distillation column.

Here, the production of by-products is the result of physical and chemical reactions, or occurs through the changeable operating states of the production equipment. The processor can produce three grades (A, B, and C) of a certain fluid product. Basic material G moves from a feed tank (buffer) to the reactor. The chemical reaction produces the desired material, but also by-product N, which is separated out through the aid of a distillation column, by supplying heat and generating vapor. N exits the distillation column and the production unit.

A change of product from one grade to another without shutting down the reactor involves resetting temperature and pressure. Transitional materials are obtained as a result of these changes. These materials are of a lesser quality, and later they will have to be mixed with a sufficient quantity of high-grade materials, which will be produced once operations reach a stable state. This means that a large quantity of each grade must be produced before the next change of product. Figure 8.1.1.2 shows the flow of goods using MEDILS notation (see Section 4.1.2).

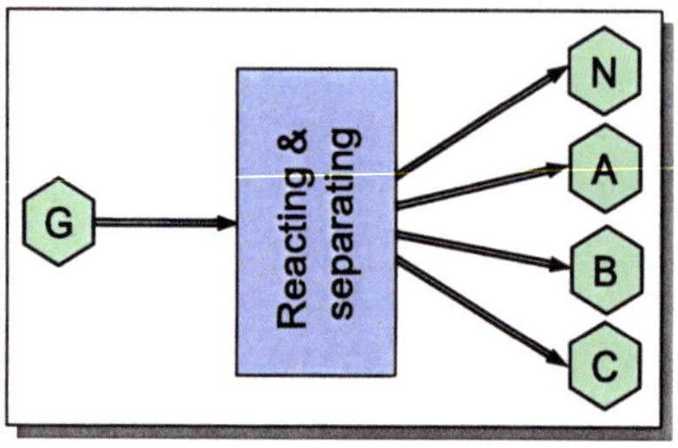

Fig. 8.1.1.2 The manufacture of by-products in chemical production.

The second example is taken from sheet-metal working. Here, washers are stamped from a strip of metal. In this case, beyond the technical process itself, by-product production makes economic sense: it allows the fullest possible utilization of the raw material. Figure 8.1.1.3 shows a section of the metal strip after a typical stamping operation.

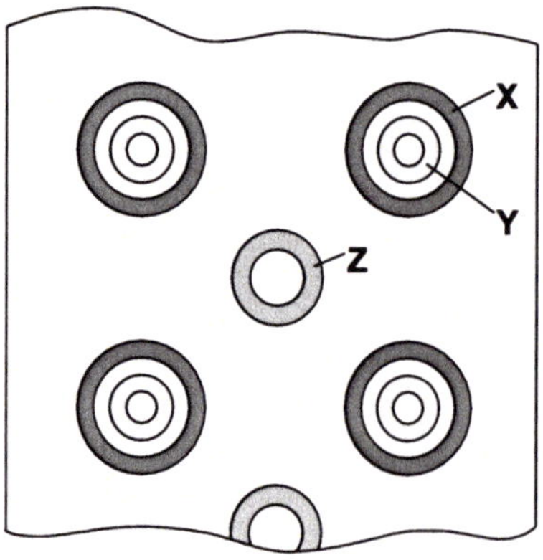

Fig. 8.1.1.3 Washers stamped from a strip of sheet metal by a stamping press.

To utilize more of the strip when producing washer X, a small washer Y is stamped inside each large washer. In addition, the press stamps other washers, of a size determined by the honeycomb principle, between the larger washers. As a result, 5 parts are obtained from each pass of the stamping machine: 2 each of part X and part Y and 1 of part Z. This can be expressed as the goods flow shown in Figure 8.1.1.4.

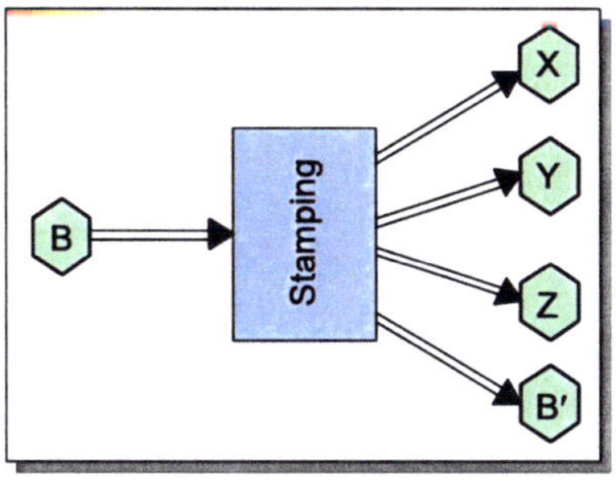

Fig. 8.1.1.4 The manufacture of by-products in the sheet-metal working industry.

The waste product obtained is the stamped sheet metal strip B′. There is an interesting parallel here to our first example: This stamping procedure makes sense only if the washers are separated out according to size. In the first example, it was necessary to separate the primary products (A, B, and C) from by-product (N).

The third example shows the production of split steel collets, which are used for tool holding and disengaging. Figure 8.1.1.5 shows a typical production process that yields a number of different sizes of collets. Here, reasons of economy dictate the production of by-products.

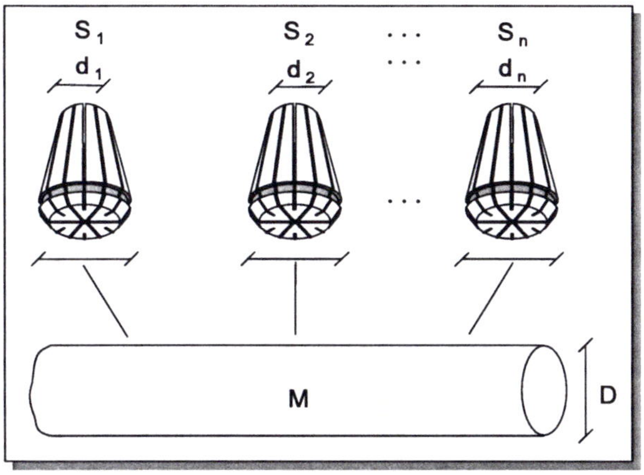

Fig. 8.1.1.5 Production of collets from a steel cylinder.

Collets S1, S2,..., Sn, each of different diameter d1, d2,..., dn, can be produced from a round bar M of diameter D. Here, again, the decision to produce by-products is based on economy. Once production has been set up, collets of various diameters can be produced with negligibly short setup times. Since various collet diameters are produced together, the possible batch size is relatively large. This minimizes the share of setup for each collet. At the same time, only a few collets of each size are produced, which keeps down the carrying cost for each size and for production as a whole. Figure 8.1.1.6 shows the flow of goods for collet production.

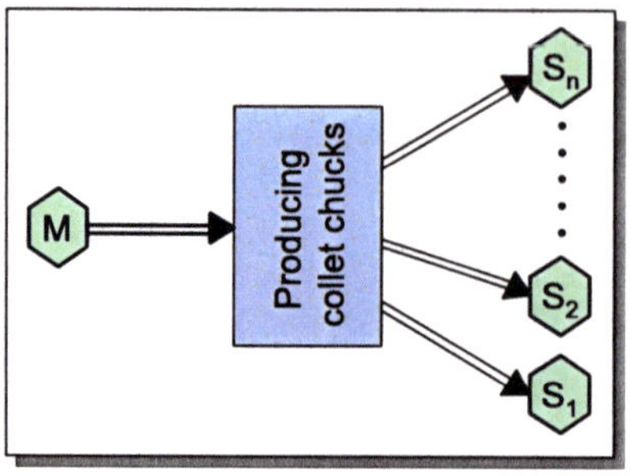

Fig. 8.1.1.6 Production of collets from a steel cylinder.

The fourth and last example is *temporary assembly*, taken from the manufacture of precision machines. Here, components at low production structure levels may have to be put together for mutual adjustment, disassembled again, and sent on for further processing. At the latest at final assembly, the fitted components are rejoined. This is the typical "saucepan and lid" problem, as formally shown in Figure 8.1.1.7.

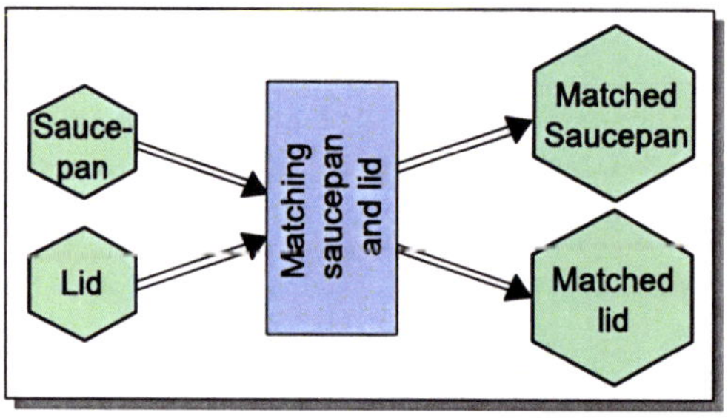

Fig. 8.1.1.7 Temporary assembly: the "saucepan and lid" problem.

The saucepan and the lid have to be produced at the same time since they have to be matched to each other. However, they may then pass through other, quite different orders before they are finally assembled.

There are thus a number of reasons for producing by-products in the process industries. In many cases, the reason lies in the nature of the chemical, biological, or physical processes in the various stages of processing. However, there may be economic factors that demand appropriate processing techniques.

8.1.2 High-Volume Line Production, Flow Resources and Inflexible Facilities

Intended learning outcomes: Describe flow resources within a process stage. Explain the need for flexible capability of the production infrastructure in chemical production.

The following values of characteristic features indicate processor-oriented methods as the appropriate business methods for planning & control:

Production environment: In the process industry, *end products stores*, and thus make-to-stock are widespread and important. Chemical, pharmaceutical, or grocery products are, ultimately, stocked at the shelf in retail shops. Upstream added value stages are also kept in stock, where efficient.

Facility layout: Here, we find high-volume line production, and — in particular — continuous production. Production processes in process industries (producing chemicals, paint, oil, and so on) usually have to carry out an entire sequence of operations (a process stage; see the definition below), that is, one operation after another in a continuous fashion.

A flow resource F is an intermediate product that should not or cannot be stored during the process stage and therefore flows through the process continuously.

An intermediate product becomes a flow resource mainly because of its physical nature or condition. An example is the active substances produced in the chemical industry. As a data element in the product structure, a flow resource is at the same level as the component materials for the subsequent operation or (basic) manufacturing step, and it facilitates modeling and monitoring of the balance of material inputs and outputs of individual manufacturing steps. Figure 8.1.2.1 shows a product Z produced from starting material G.

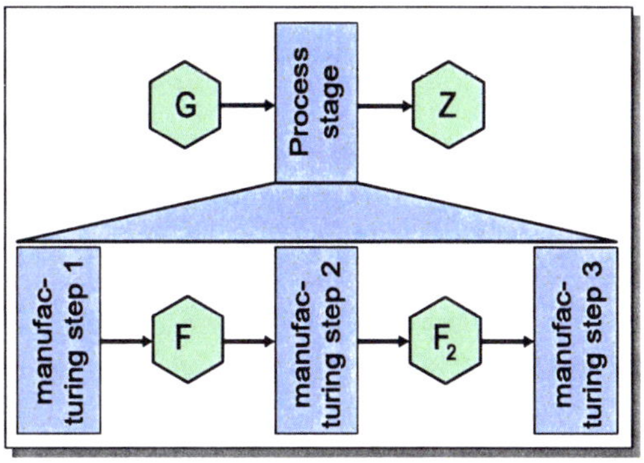

Fig. 8.1.2.1 Flow resources within a process stage.

The intermediate states F_1 and F_2 "flow," meaning that they are not, or cannot, be stored in containers or tanks. Thus, they cannot be in buffers at these work centers.[1] This also means that storable work in process cannot build up at these work centers. This reduces the degree of freedom for capacity planning (that leeway is utilized in the conventional MRP II concept; see the comments on queues in Section 13.2).

Flexible capability of the production infrastructure: Single-purpose facilities were common in chemical production for a long time. For very large-scale mass production, there are sound economic reasons for this type of structure. However, to adapt capacity to load more flexibly and particularly to facilitate change of product on the same production resources, multi-purpose facilities composed of modules became more frequent. Nevertheless, they are far away from achieving the flexibility of mechanical production. Inflexible facilities still exist, not least because of conditions imposed by government regulations. See also [Hübe96], p.

[1] Actually, many technical plants for continuous production have buffers at work centers. Their main purpose is to ensure process stability rather than to maintain degrees of freedom for planning & control.

23 ff. Food and drug production are subject to strict quality control by bodies such as the FDA (U.S. Food and Drug Administration). Production of foodstuffs and drugs must follow a set of guidelines known as Good Manufacturing Practices, or GMP (also known as Quality System Regulation). Under GMP, manufacturing practices are inspected and approved at each plant, which means that it is not possible to simply switch production between facilities in response to temporary capacity shortages or mechanical faults, for example. The production process would also have to be validated at the alternative facility.

8.1.3 Large Batch Size, Lot Traceability, and Loops in the Order Structure

Intended learning outcomes: Explain the need for lot traceability. Present reasons for loops in the product structure.

Customer tolerance time for chemical and pharmaceutical as well in for food products is minimal, and production time is often very long. Thus, the *reason for order release* is mostly a *forecast*. Long lead times make any planning system extremely susceptible to demand fluctuations. which can make incorrect predictions expensive. On the other hand, if there is a continuous usage on a production structure level, the prediction can be related directly to this level and does not have to be derived from the predictions for higher production structure levels. In this case, the reason for release is *consumption*, leading to a *stock replenishment.*

Large batch size, or large lot size of an order: Some processes require large quantities to be produced to obtain the desired quality. Preparation and setup times (such as for cleaning reactors) are generally very long in the process industry, and, strictly speaking, the process start-up should be included in the setup time. Furthermore, the quantities required by the market are sometimes extremely large, as is the case in the food production industry, for example. Here the products are essentially mass produced.

Lot traceability is required by the governing regulations, but also due to product liability and problems associated with recalling a product. Control of lots, batches, or charges or even *positions in lots* serve this purpose. For further information on lot control, see Section 8.2.3. Lot control is also practiced for the following reasons:

- Active substances have a limited shelf life. If a batch results in various units, such as different drums of fluids, they must be labeled for identification (numbered individually in ascending order, for instance). For further processing, this procured or produced material must be identified by means of this relative position.
- To ensure uniform quality within a batch. This is frequently the case in the chemical and pharmaceutical industries and sometimes in the metal- or steel-working branches. It is particularly useful if the product characteristics change from one pass through the process to the next, or if products are produced by mixing or merging different materials, and the starting materials do not affect the characteristics of the end product in a linear manner. One example of this is the mixing of fuels, where the addition of high-octane materials does not have a linear effect on the increase in the octane level.

Another feature of the process industry is a *product structure with loops.* The chemical reactor example shown in Figure 8.1.1.1 might use catalyst K to influence the reaction rate.

Catalyst K does not get used up, and it becomes available again as soon as the reaction has ended. This creates the goods flow shown in Figure 8.1.3.1. As the output and input store for catalyst K are the same, a loop results.[2]

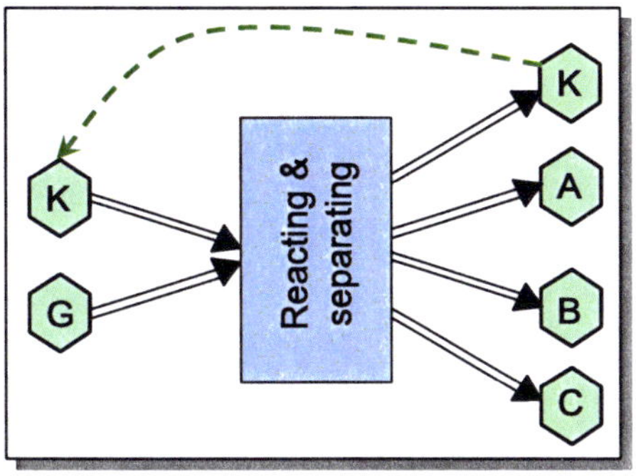

Fig. 8.1.3.1 Product structure with loops.

Another example of a product structure with loops is when co-products are treated, and the recycled material reenters the processor as a starting material. In another case, amounts of product that have already been mixed in a mixing process stage can be returned to the process as often as is necessary to ensure the desired level of homogeneity. An *undirected network of operations* results. This is typical in the production of paints or pharmaceutical products.

8.2 Processor-Oriented Master and Order Data Management

Intended learning outcomes: Produce an overview on processes, technology, and resources. Present the process train: a processor-oriented production structure. Disclose lot control in inventory management.

In Section 1.2, the product structure, production structure, and process plan were each "attached" to a product. This conventional, assembly-oriented arrangement it is not suitable for the process industry. The process industry requires extended business objects that essentially reflect an order structure with various possible products. This section introduces some new business objects and extensions to objects already discussed. Detailed modeling of these business objects is discussed in Section 17.4.

8.2.1 Processes, Technology, and Resources

Intended learning outcomes: Present processor-oriented master data and give examples of resources.

---

[2] *Catalysts* are used in chemical reactors for increasing production, improving the reaction conditions, and emphasizing a desired product among several possibilities.

In the process industry, product design simultaneously means the design of proccsses. There is no clear separation between these two steps, as is the case for mechanical production, for example. Product design is based entirely on the knowledge of the technologies that can be used in production processes. In mechanical production, there are technologies and machines for cutting, milling, electroerosion, and other operations, but the technologies involved in the process industry utilize biological, chemical, or physical reactions.

The object *technology* describes process-independent properties and conditions, that is, all the knowledge contained in a given technology.

The object *process*, on the other hand, describes the possible input, the effect of the process, and the resulting output independently of a given technology.

See also Section 17.4.1. A process may be implemented using different technologies, and, conversely, a technology may be used in various processes.

The object *process with technology* describes the technique that can be implemented during the actual production process.

It is this business object in logistics that ultimately appears in the production structure as a basic manufacturing step.

Resources are all the things that are identified, utilized, and produced in a value-adding process. The term is used in a generalized way here, that is, to represent products, materials, capacities (including personnel), facilities, energy, and so on.

One peculiarity of the process industry is that all resources are regarded as being of equal priority. Thus, materials are no more important than capacity or production equipment. This is reflected in the fact that a production structure is expressed solely in terms of resources, and all the possible types of resources are described in greater detail by appropriate specialization. Figure 8.2.1.1 shows the business object *resource* as a generalization of the business object *item* in Figure 1.2.2.1 and the business objects discussed in Section 1.2.4.

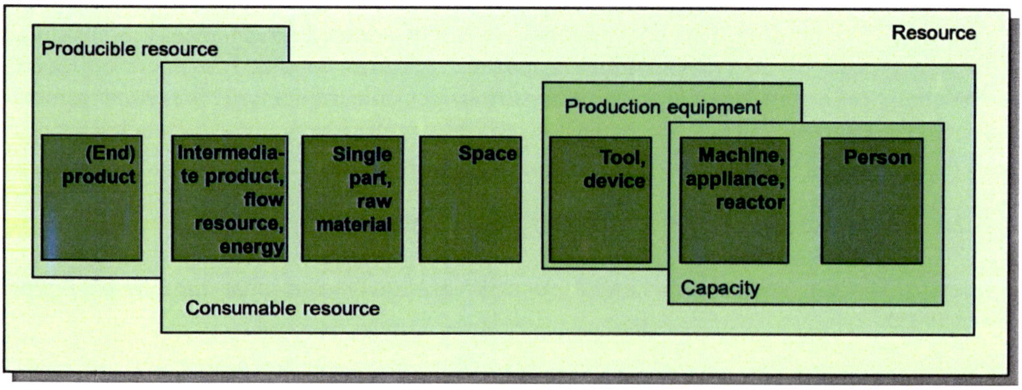

Fig. 8.2.1.1 Processor-oriented master data: examples of resources.

The business object *item*, or *assembly*, or *product*, or *component* is a specialization of the object *resource*, or *intermediate product*, or *producible resource*, or *consumable resource*. A further resource is *energy*, such as electricity, steam, and so on. These resources can also be described as items. They are often produced as by-products. The business object *capacity* also is a specialization of the object *resource*. Capacity can mean employees or automated equipment, such as machines and reactors. The latter resources are grouped with tools, devices, and the like, under the term *production equipment*. They describe the investment in physical plant that is required for the manufacturing process.

8.2.2 The Process Train: A Formalized Processor-Oriented Production Structure

Intended learning outcomes: Describe the production structure in chocolate production. Explain the various manufacturing steps and stages in chocolate production. Explain the formalized process train with stages and basic manufacturing steps. Identify the recipe.

In the process industries, the conventional production structure consisting of bills of material and routing sheets (see Sections 1.2.2 and 1.2.3) has been replaced today, as mentioned above. Close examination of the new structure in current use reveals it to be a more generalized form of the conventional bill of material and routing sheet concept. See also [TaBo00], p. 178 ff., [Loos95]; and [Sche95b].

Figure 8.2.2.1 shows, as an example, a typical production structure in chocolate production.

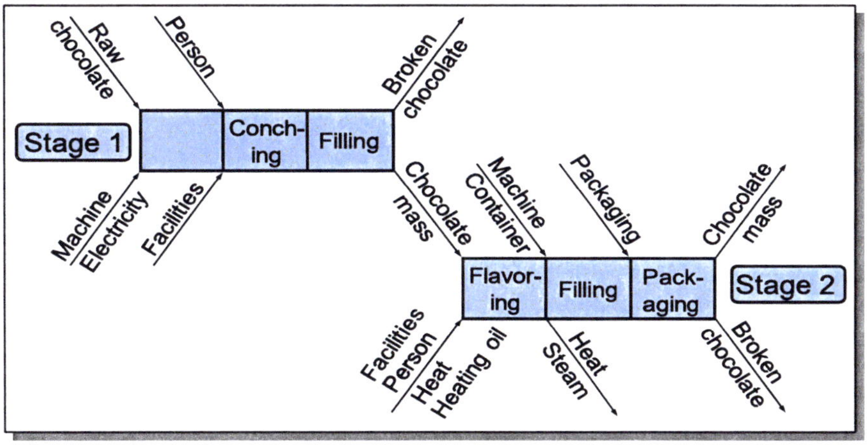

Fig. 8.2.2.1 A process train, here in chocolate production.

The first stage of processing consists of rolling the raw material between rollers, conching,[3] and filling. The resources consumed during rolling are the cocoa mass, the machines

---

[3] *Conching* is a process of rolling and kneading chocolate that gives it the smooth and rich quality that eating chocolate is known for today. The name comes from the shell-like shape of the rollers used. Typically, the process takes the best part of 24 hours. It is judged to be complete when the required reduction in size of the sugar crystals has been achieved — this is what makes the chocolate "smooth."

required, and power. This stage results in an intermediate product, in this case a chocolate mass that is subsequently used for further processing. The by-product is broken chocolate.

The second stage consists of the processes of producing the flavored mass, filling, and packaging. The primary product is the packed, semifinished, flavored product (again a chocolate mass). By-products such as broken chocolate and energy (heat, steam) are also produced. In addition to the material used, the consumed resources include capacity and equipment.

Figure 8.2.2.2 represents the process train concept in a formalized way. This structure is the basic concept behind the data management of both master data objects and order objects in the process industries.

A *process train* is a representation of the flow of materials through a process industry manufacturing system that shows equipment and inventories (cf. [APIC16]).

The term *process unit* stands for the (production) equipment that performs a basic manufacturing step, or operation, such as mixing or packaging.

Resources such as incoming and outgoing items, capacity, and production equipment are allocated to the basic manufacturing steps.

A *process stage* is a combination of (generally successive) process units.

In Figure 8.2.2.2, several (generally successive) stages are combined into process trains. Inventories in intermediate stores decouple the scheduling of sequential stages within a process train. However, if there is an intermediate product between two successive manufacturing steps of a stage, it is “only” a flow resource, which cannot nor should not be stored.

Processor-oriented production structure and *production model* are other terms used for process train.

Recipe or *formula* is the term commonly used to describe the content of a processor-oriented production structure.[4]

A *processor-oriented order structure* is a processor-oriented production structure associated with a specific (production) order, in which quantities and dates are specified.

The process train thus defined can be regarded as an extension of the production structure underlying the process plan shown in Figure 1.2.3.3, but without showing the individual time periods that make up the lead time along the time axis.

As is every production structure, a process train may be the object of cost estimation. The associated processor-oriented order structure will be the object of job-order costing. One feature of such a calculation is that the costs incurred are distributed among the various resources produced, that is, primary and by-products. In the simplest case, this involves allocating a predetermined percentage to each resource produced by the production structure.

---

4 For work on standardizing the terminology used, see also [Namu14], AK 2.3.

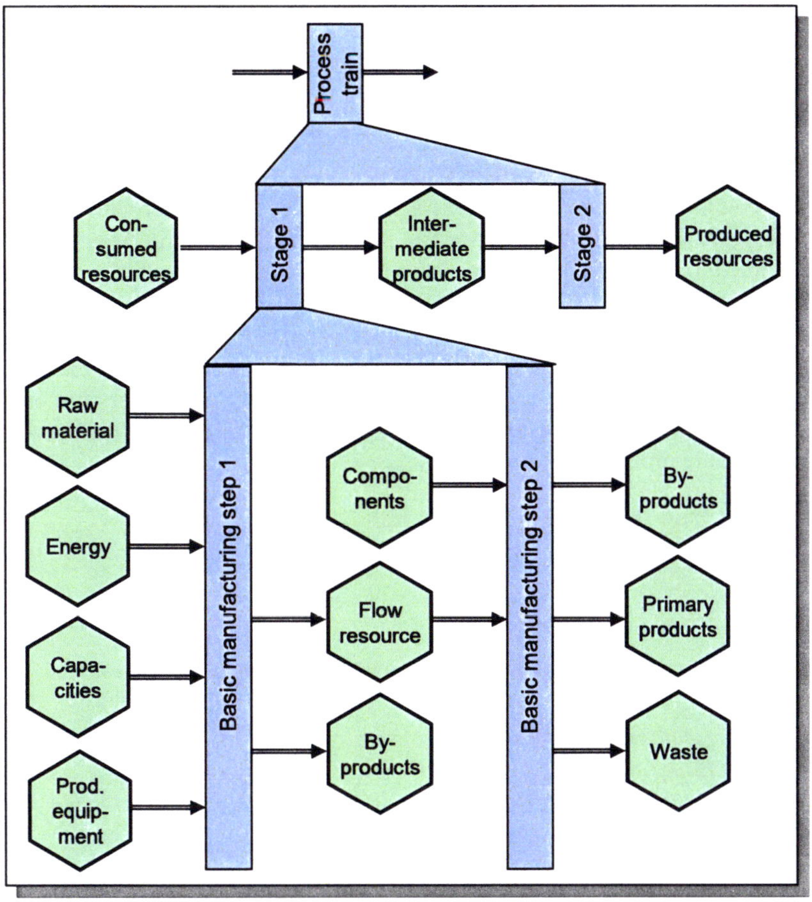

Fig. 8.2.2.2 Process train (formalized) with stages and basic manufacturing steps.

8.2.3 Lot Control in Inventory Management

Intended learning outcomes: Describe the steps of lot control in inventory management. Identify the batch ID.

As mentioned in Section 8.1.3, many process industries require a lot traceability for the ingredients used in a product to satisfy the governing regulations. This requirement is most frequently met by assigning an identification number to every lot, batch, or charge that is produced or procured. The batch thus becomes an object in the company. In the production of by-products, products that are produced at the same time using the same resources may be given the same identification.

Lot control establishes production batch identification for each resource taking the following steps:

1 Each batch is given a *lot number* or *batch identification*, or batch ID, at the time that it is produced. The batch ID is also recorded as a "completed resource transaction" and entered as a receipt into stock. Apart from the batch ID, the attributes of this

object include resource identification, quantity moved, order ID, position of the process in the order structure, and transaction date.

2 The physical inventory of a particular resource consists of the batches described in step 1 minus any quantities already issued from these batches in accordance with step 3.

3 The batch identification for an issue from stock is determined by allocating the issue to a physical inventory as per step 2. The batch ID (determined originally in step 1) assigned to this stock also becomes the batch ID for the issue from stock. The issue from stock is also a "completed resource transaction." The attributes are then the same as those described under step 1. If the quantity issued originates from different receipts into stock, then the same number of issues from stock must be recorded, each with the associated batch ID and the corresponding quantity issued from stock. However, in many cases, the rule "same batch" is required. This prevents an issue being made up from different batches.

See Section 17.4.2 for the objects used for administering batches.

8.3 Processor-Oriented Resource Management

Intended learning outcomes: Explain campaign planning. Differentiate between processor-dominated Scheduling and material-dominated scheduling. Describe a nonlinear usage quantity and a product structure with loops.

8.3.1 Campaign Planning

Intended learning outcomes: Explain campaign planning using an example of a process chain in chemical production. Identify campaign cycles for the example and a minimum campaign of one day's production. Describe how the process stages could be synchronized.

Section 8.1.3 described large lots as a consequence of setup costs due to stopping, cleaning, and restarting processes. The changeover processes for transporting flow resources are of lesser significance. In processor-oriented resource management, the objects concerning capacity management and production control are not equivalent to materials management objects.

- For control, the planning unit is the machine or facility, such as the reactor, which thus also becomes the actual planning object. The technically feasible batch size is calculated by the quantity of goods that should ideally be processed by this facility. The batch thus produced is also used for accounting, stockkeeping, and archiving information for the subsequent lot traceability, for instance.
- From the materials management viewpoint, the emphasis is placed on demand. For technical reasons, a production lot can only be a multiple of a production batch. "Optimum" batch sizes, whether calculated using stochastic or deterministic methods (see Sections 11.3 and 12.3), often have to be rounded up considerably due

to the high setup costs and the required utilization of capacity. Such hidden formation of batch sizes results increasingly in block demand for, and thus a decidedly quasi-deterministic form of, materials management.

A *campaign* is an integer multiple of production batches of a certain item, the batches being produced one after another.

A *campaign cycle* is a sequence of campaigns during which all the important products are produced up to a certain capacity and in the quantity required by demand.

The sequence of campaigns is used to reduce setup costs. As soon as the optimum batch size from the materials management viewpoint consists of several batches, it is then combined to form a campaign. Under certain circumstances, it is then advisable to produce a batch of a different product immediately afterward, if this will avoid the need for a cleaning process, for example. The formation of campaigns in this way is a characteristic feature of processor-oriented resource management. This means that the entire campaign must be considered, rather than just the individual batches, when scheduling capacity. A campaign can, of course, be split back into its constituent batches if necessary.

Campaign planning aims to create optimum campaign cycles.

Campaign planning is one type of *sequencing,* or the combination of optimum sequences. Optimization can target various areas: production costs, manufacturing time, or product quality. Figure 8.3.1.1 shows the example introduced in Figure 8.1.1.2, with the addition of a packaging process. The example is taken from [TaBo00], p. 18 ff.

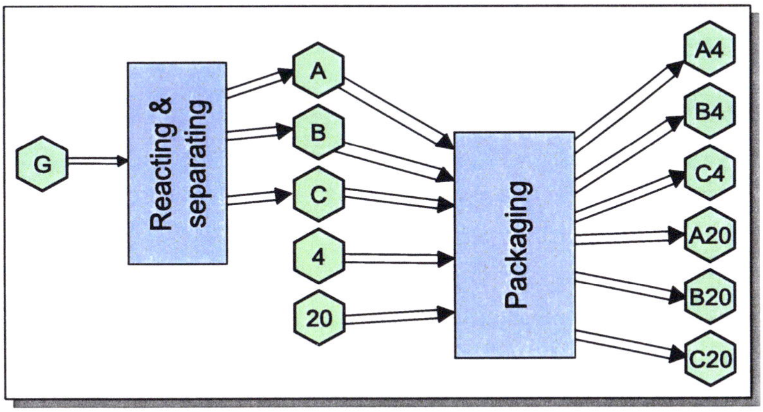

Fig. 8.3.1.1 Example of a process chain in chemical production (see Figure 8.1.1.2).

The grades A, B, and C produced at a plant are packed into two drum sizes (4 liters and 20 liters) in the subsequent packaging process. The demand is for the 6 end products (3 grades times 2 packaging sizes). To simplify the example, the minimum batch is assumed to be one day's production. The demand for an end product is specified in relation to the overall demand: A4, 30%; B4, 20%; C4, 10%; A20, 20%; B20, 10%; and C20, 10%. Bill-of-material explosion results in the proportionate demand for the intermediate products obtained from the reactor: 50% A, 30% B, and 20% C.

Assuming that the long time required to set up the packaging process arises when the packaging size is changed and that the reactor setup costs can be minimized by the sequence A, B, and C, as well as the specification of a minimum campaign of one day's production, the campaign cycles shown in Figure 8.3.1.2 result.

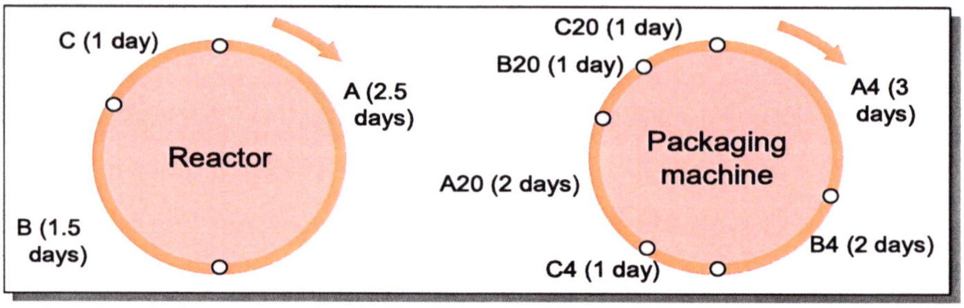

Fig. 8.3.1.2 Campaign cycles for the example in Figure 8.3.1.1 (see Figure 6.2.1.2) and a minimum campaign of one day's production.

The reactor rhythm is determined by the minimum proportion of the demand, 20% for C. The campaign cycle thus lasts 5 days. The packaging rhythm is determined by the minimum proportion of the demand of 10% for C4 or C20. This campaign cycle thus lasts 10 days.

The ideas behind processor-oriented resource management thus correspond in some respects to those of the lean / just-in-time concept (see Section 6.2), in which the optimum sequence of operations is important with a view to reduce the setup times (see also Figure 6.2.1.3). The reduced setup times should result in small lots and, therefore, continuous demand. Only then will it be possible to totally separate the processes that make up the various production structure levels, which will allow the use of the Kanban technique in the process industry.

If continuous demand cannot be achieved, then quasi-deterministic techniques will still be required. In this case, the response to a net demand will be to schedule at least one campaign for production, rather than just a batch. A batch also results in by-products. Both of these contradict the simple pull logistics of the Kanban technique, since production is determined by the technical process and savings in terms of setup time, rather than in response to consumption. The dominating factor is capacity management.

The conventional MRP II / ERP concept of resource management does not incorporate the processor-oriented techniques, such as manufacture of by-products and campaigns, making them less suitable for the process industry. Campaign planning enables demand to be synchronized in terms of quantities with the goods to be produced at all production structure levels, particularly with respect to end products. Where synchronization is not possible, buffers must be kept to absorb any shortfall. The aim of campaign planning is thus to minimize the inventories that have to be kept in the intermediate stores by synchronizing the various (process) stages as accurately as possible. Figure 8.3.1.3 shows how the two (process) stages (or production structure levels) could be synchronized for the above example.

The diagram shows the start and end of the overall campaign for each product, but not the individual batches. It can be used to calculate the resulting stock curves for the end and intermediate product stores for given quantities. The inventory curves are of the type discussed in detail in Chapter 11 for determining the available stock. They are used as the basis for troubleshooting, particularly for determining the buffers that will be needed.

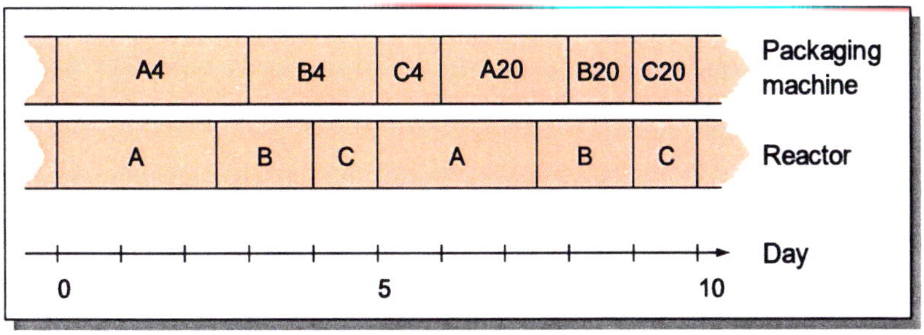

Fig. 8.3.1.3 Campaign planning: how the (process) stages could be synchronized.

The campaign planning technique described here is modified finite capacity planning (see also Section 14.3) that requires continuous intervention by the scheduler. The planning diagrams are similar to the Gantt charts or planning boards used in finite capacity planning (see the illustrations in Sections 14.3 and 15.2.2). The only difference is that they include — as well as individual batches — entire campaigns or even campaign cycles.

8.3.2 Processor-Dominated Scheduling versus Material-Dominated Scheduling

Intended learning outcomes: Differentiate between processor-dominated scheduling and material-dominated scheduling. Produce an overview on the use of the MRP II / ERP concept or of the lean / just-in-time concept compared to the processor-oriented concept.

Processor-dominated scheduling (PDS) is a technique that schedules equipment or capacity (processor) before materials. This technique facilitates scheduling equipment in economic run lengths and the use of low-cost campaign cycles ([APIC16]).

See also [TaBo00], p. 30 ff. The campaign principle outlined in Section 8.3.1 is an example of processor-dominated scheduling. Indeed, capacity management has priority over materials management for scheduling. Finite loading is used as the scheduling principle. Materials are planned according to the results of finite loading.

Processor-dominated scheduling is characteristic of the processor-oriented concept. It is typically used to schedule manufacturing steps within a process stage. However, the process industry does not use it in every situation.

Material-dominated scheduling (MDS) is a technique that schedules materials before processors (equipment or capacity). This technique facilitates the efficient use of materials ([APIC16]).

Material-dominated scheduling can be used to schedule each stage within a process train. Typically, the MRP II / ERP concept as well as the lean / just-in-time concept use material-dominated scheduling logic. In the process industry, they have their significance as well.

The problem in the process industry is to identify the point at which the processor-oriented concept replaces the other concepts. Figure 8.3.2.1 provides a simplified rule of thumb. This line of reasoning is similar to that followed in [TaBo00]. In addition, see also Figure 4.5.3.1.

The MRP II / ERP concept or the lean / just-in-time concept may be used if
- Materials are expensive related to manufacturing costs.
- There is overcapacity.
- Setup times and costs tend to be negligible.
- There is job shop production rather than line or flow shop.

The processor-oriented concept may be used if
- Capacity is expensive related to costs of goods manufactured.
- There are capacity bottlenecks.
- The one-off costs for each lot produced are relatively high.

Fig. 8.3.2.1 Use of the MRP II / ERP concept or of the lean / just-in-time concept compared to the processor-oriented concept.

8.3.3 Consideration of a Nonlinear Usage Quantity and of a Product Structure with Loops

Intended learning outcomes: Present the quantity of a manufactured product P as a nonlinear function of the usage quantity of a resource R. Identify possible solutions of issues entailed by a nonlinear usage quantity.

In the process industry, the quantity per, or usage quantity, corresponds to the selective use of starting materials to produce intermediate, end, or by-products.

The *operation/process yield* is the relationship of usable output from a process, process stage, or operation to the input quantity (cf. [APIC16]).

Operation/process yield can often be expressed by a ratio, usually as a percentage. However, chemical and biological processes are subject to conditions that cannot always be predicted accurately (for example, external influences like the weather). In addition, the technologies and production processes used, as well as variations in the quality of the raw materials, have an effect on the consumption of resources that is not quantifiable in every respect. For example, excessive use may be made of certain materials in the startup phase of a process or in the course of the process — namely, as the produced quantity increases. In such cases, the usage quantity ceases to be a linear function of the quantity produced.

A *nonlinear usage quantity* is an operation/process yield that cannot be expressed by a linear function of the quantity produced.

Just as with the usage quantity, the duration of the process is no longer proportional to the quantity produced. Thus, the effective consumption could change, as shown in Figure 8.3.3.1. See also [Hofm95], p. 74 ff.

In some cases, the nonlinear function for calculating the nonlinear usage quantities may be known in advance. The problem is then solved using a parameterized formula, rather than as a constant value for the *usage quantity* or *quantity per* attribute. In the event of a transition from the production structure to an order structure, the formula is evaluated using the parameter values associated with the order (e.g., the batch size), and the appropriate demand for the resource is thus determined. This procedure is exactly the same as described in Section 7.3 for one-of-a-kind production of products with many variants. There, formulas are linked with various attributes, and not just with the usage quantity and the operation load.

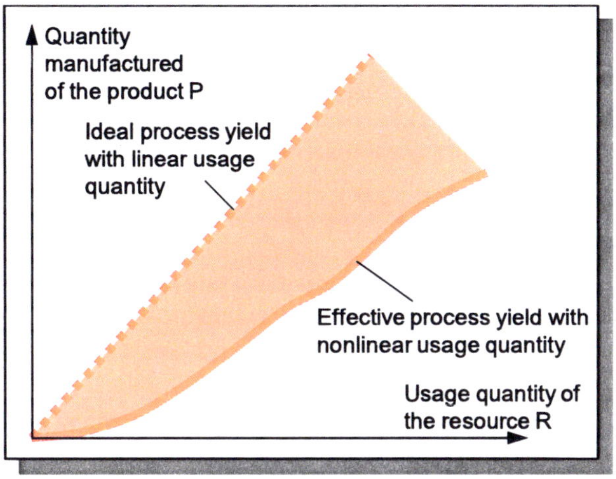

Fig. 8.3.3.1 Quantity of a manufactured product P as a nonlinear function of the usage quantity of a resource R.

Most products with a *product structure with loops* are those that can be returned to the production process. These may be by-products (such as broken chocolate or energy in the form of steam or heat) or processing aids (catalysts, for example) that can be used for further production. It thus follows that the by-products or waste products are not subject to external demand, and their use can therefore be optimized internally. There are, however, certain quantity or time-related marginal conditions concerning usability (spoilage, deterioration) or storability and shelf life. Most of the software packages based on the conventional MRP technique, as described in Section 12.3, do not allow loops. This is because the technique deals with the individual items in the order of their low-level code. In a production structure with loops, the low-level code would be regarded as "infinite."

One possible solution to this problem is to identify such items (by-products or waste products) and then to omit them from the structure-level code calculation or to allocate to them a maximum structure-level code. The MRP technique should then be used to schedule such by-products or waste products only at the end. At this time, all the demand is already known, as are also all the planned receipts in response to planned orders. Any net requirements for such by-products or waste products would then have to be produced or procured. Consequently, an additional production structure without further by-products should be allocated to each of these products. This is then converted into an order structure.

8.4 Special Features of Long-Term Planning

Intended learning outcomes: Disclose the determination of the degree of detail of the master production schedule. Describe pipeline planning across several independent locations.

8.4.1 Determining the Degree of Detail of the Master Production Schedule

Intended learning outcomes: Explain the need for using detailed production structures for long-term or master planning in chemical production.

Companies that process basic materials (basic producers) manufacture a number of different end products from relatively few raw materials. The number of end products is small, however, when compared to the number of products that are manufactured by assembly-oriented production companies. For example, in the chemical industry, part of the pharmaceutical division (formerly Ciba-Geigy) of the Novartis Group produces “only” about 150 active substances, and they are produced from just a few raw materials. Some of these active substances have a cumulative lead time of up to two years. There are also large safety stocks in intermediate stores. The number of different work centers that have to be scheduled corresponds roughly to the number of products and intermediate products. The number can be counted in the hundreds, but not in the thousands, if we consider all the process chains.

Experience shows that with such quantities there are no meaningful rough-cut business objects. Long-term or master planning is therefore carried out using detailed production structures (see Section 5.1.1). This is unavoidable, since resource requirements planning cannot be carried out using gross figures. Even at the long-term planning stage, the campaigns must be offset against the available capacity because, as mentioned above, a campaign cannot simply be interrupted or partly outsourced for economic reasons. This cannot happen at all for batches. In addition, with flow resources, successive processes cannot be interrupted.

Here, demand forecasts are absolutely essential, for this type of production involves stocking at the end product level. The demand for raw materials must also be derived quasi-deterministically, because the components are required block-wise, and there are limits on their use in other products. More and more companies in the process industry are faced with having to review their logistics costs, so that their stocks and lead times must also be reduced. The loss of buffers resulting from this reduction makes any interruptions very visible, particularly if demand fluctuates greatly. Deterministic resource management models can then result in a shortfall of resources. The robustness needed due to changing demand and rescheduling asks for increased flexibility — capacity flexibility in particular. Increasingly, capacity has to be adapted to demand. As a result, finite loading in the sense of comprehensive advance planning is no longer possible. It must be replaced by a greater ability to respond to changing situations. Greater emphasis is thus placed on the interaction between all the people involved in production and in planning and executing the process.

8.4.2 Pipeline Planning across Several Independent Locations

Intended learning outcomes: Describe a typical production pipeline showing its production locations. Explain the master production scheduling process and the planning group for several locations that operate independently of one another.

Globalization of the markets has meant that firms produce at different locations around the world. There are different reasons for this: For example, trade barriers may force companies to establish facilities in countries with important markets (see Section 2.1.1). The buying up of foreign companies is an increasingly frequent phenomenon. Validation requirements of the FDA encourage the centralization of certain production processes at a single location.

All these conditions result, however, in major disadvantages for efficient logistics: Intermediate products and active substances have to be moved from one location to another and from one country to another. Figure 8.4.2.1 shows a practical example of a production structure called, in technical jargon, a *production pipeline*. See also [HüTr98].

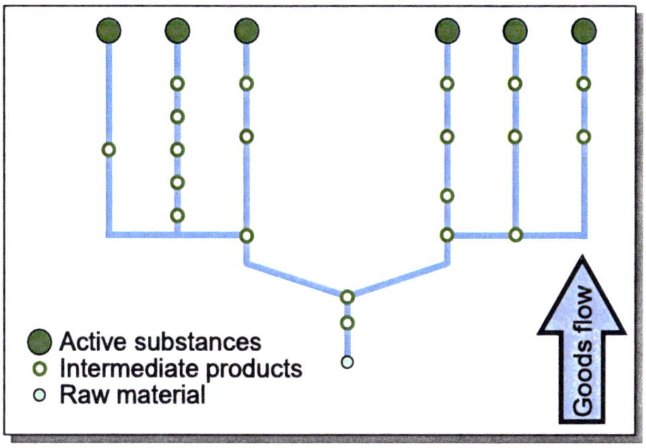

Fig. 8.4.2.1 A typical production pipeline.

The different process stages in this pipeline involve different volumes and process units. Some stages produce large volumes in dedicated single-purpose facilities. Others result in small volumes and take place in multipurpose facilities. Figure 8.4.2.2 shows the same pipeline with the various production locations highlighted in different shades of gray.

This distributed production system can be regarded as a customer-supplier relationship among the individual production locations. In the example, the pipeline even links production sites in different countries. Each of these locations has its own planning process for its logistics systems, which makes it more difficult to schedule the entire pipeline efficiently, since each location aims to optimize different aspects when creating its long-term plan. Products that simply pass through the location (in the pipeline) are not taken into account in this optimization, with the result that, for pipeline products, large stocks build up in the intermediate stores, and long lead times are required.

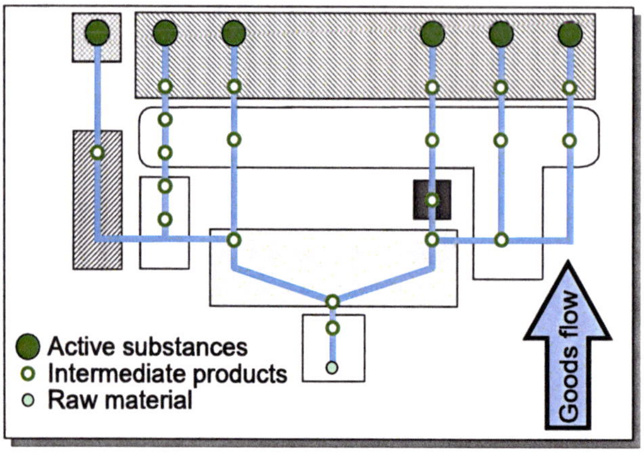

Fig. 8.4.2.2 A typical production pipeline showing its production locations.

This structure is not comparable with a company and its departments, because, in this case, the "departments" are independent companies or profit centers within a group of companies. Here, an intensive cooperation is necessary, particularly the following:

- A partnership relationship between the people involved. This applies to the schedulers at the company that manufactures the active substances (the pipeline products) as a whole, and also to schedulers at the companies involved in producing the product. There is no point in any of the parties overplaying their negotiating position, since the entire pipeline is under the control of people. Mutual respect and consideration do not simply foster good relationships among all parties involved; they also increase people's willingness to attempt to understand specific problems. See also Section 2.3.3.[5]
- Information systems must be networked, to exchange forecasts and other planning data. The results of the central coordinating pipeline planning process must be fed back to the companies involved in the pipeline. See Section 2.3.5.

Figure 8.4.2.3 shows the process for master planning.

The central planning office sends the result of master planning, that is, the master production schedule (MPS) for the entire pipeline, to the individual companies involved, where it will be adjusted to suit local scheduling needs. The result of this process is then returned to the central pipeline planning department, and so on. The planning process is organized on a rolling basis, and the planning horizon may be as long as one or two years hence. Figure 8.4.2.4 shows suggested scheduling groups.

---

[5] In the world of practice, production of the desired pipeline products frequently goes hand-in-hand with the production of by-products and waste products. The economic efficiency of the main process and thus its feasibility often depend upon efficient distribution of the by-products. It follows that in addition to the companies directly involved in producing the pipeline products, planning must certainly also take into account the buyers of the by-products and waste products.

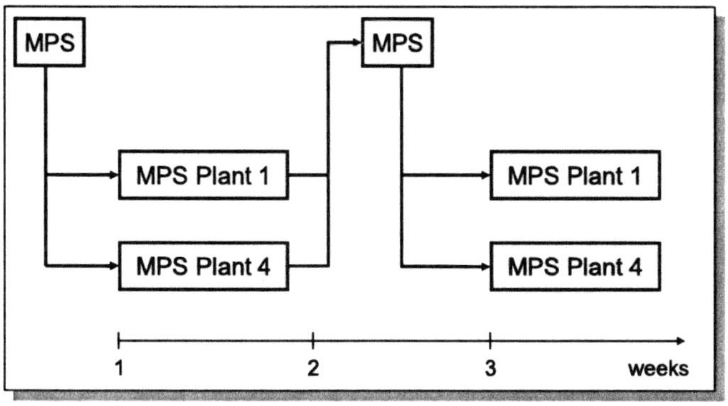

Fig. 8.4.2.3 Master production scheduling process for several locations that operate independently of one another.

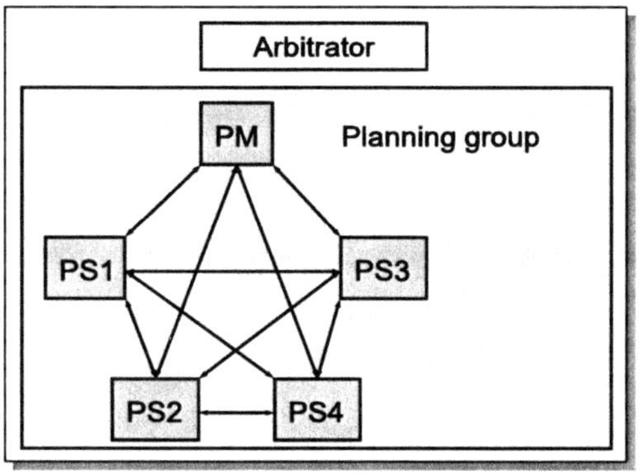

Fig. 8.4.2.4 Planning group for several production locations operating independently of one another.

The scheduling group comprises a (central) pipeline manager (PM) and representatives from the scheduling groups of all the plants involved (plant schedulers, PS). It is important to ensure that all the schedulers constantly exchange information with one another. It can also be useful to have an independent arbitrator. The presence of an arbitrator is a typical indicator of the weakness of every model of this kind, whenever the pipeline or network develops no self-understood culture of cooperation.

8.5 Summary

The ERP / MRP II and lean / just-in-time concepts that are now standard practice in mechanical and apparatus engineering companies and in the automobile and aircraft industries are unable to fully handle the special requirements of the process industry. Manufacture of by-products, high-volume line or continuous production, large batch production, production without lots, mass production, and production structures with cycles are just some of the typical characteristics of the process industry.

Conventional concepts for master data and inventory management must therefore be extended. Materials and capacities become resources of equal value. Processor orientation is the dominating factor. Process trains are therefore defined. A process train comprises several process stages that, in turn, are broken down into several basic manufacturing steps, or operations. A manufacturing step is linked to the resources required, particularly to the equipment. Lot control is indispensable to meet the product traceability requirements imposed by government bodies (especially the FDA).

Manufacture of by-products is not simply the consequence of certain chemical or physical aspects of the process (for instance, the simultaneous production of two substances during a chemical process by means of parallel or overlying chemical reactions during a production process). By-products can also be produced in a targeted fashion for reasons of economy. For example, a single production process may manufacture different products from sheet steel or steel bars to save setup costs.

The decisive factors for planning added value are often the actual production process and the required capacity, rather than the materials used. Such processes typically require a few, but significant, active substances, which are often kept in stock in large quantities. The value of the basic raw materials is often tiny compared to the overall production costs, which essentially means that adding value can only be economically viable if production facilities are utilized efficiently. Processor-dominated scheduling, the campaign principle in particular, respects this situation. The considerable setup costs that are often associated with production facilities for large batch production give rise to campaign cycles. Loss of materials caused by start-up and shutdown processes, shifting operating conditions in the production plant, or variations in the quality of raw materials result in nonlinear functions for the quantity of resources required in relation to the quantities of the product produced. The scheduling of production structures with cycles presents yet another challenge.

Long-term planning generally involves detailed data structures on account of the relatively small number of products to be scheduled and to incorporate high-volume line or continuous production and campaigns. One particular feature is known as pipeline planning, or scheduling across different locations that operate independently of one another. This type of planning environment is a frequent occurrence due to the cost of capacity and the regulation of the markets associated with the process industry.

8.6 Keywords

basic producer, 369
batch identification, 381
by-product, 371
campaign, 383
campaign cycle, 383
campaign planning, 383
catalyst, 377
energy (as a resource), 378
flow resource, 375
formula (syn. recipe), 380
lot control, 381
manufacture of by-products, 371
material-dominated scheduling (MDS), 385
nonlinear usage quantity, 386
operation/process yield, 386
primary product, 371
process industry, 369
process manufacturing, 369

process stage, 380
process train, 380
process unit, 380
processor-dominated scheduling (PDS), 385
processor-oriented concept, 369
processor-oriented production structure, 380
recipe, 380
resource, 378
waste product, 371

8.7 Scenarios and Exercises

Intended learning outcomes: Differentiate between batch production and continuous production. Calculate an example of manufacture of by-products. Elaborate an example of production planning in process industries.

8.7.1 Batch Production versus Continuous Production

Intended learning outcomes: Differentiate between batch production and continuous production. Explain the criteria that are important for the decision for one or the other of these concepts.

As a producer of fine chemicals, you plan to introduce a new type of solvent to the market. It is suitable for use in the production of adhesives for the automobile industry. The corporate marketing department estimates that 5000 to 10,000 tons of the product can be sold per year. The product design with laboratory tests has been completed. But the industrial production concept for the product remains to be determined. While most of the production processes are actually done on the batch principle (discontinuous or batch production), your engineers now suppose continuous production for this product.

a. What are the differences between these two concepts? What criteria are important for the decision for one or the other of these concepts?

b. What is your suggestion regarding the new solvent? Explain the reasons for your decision.

Solution:

a.

Continuous production	Discontinuous production (batch)
Production facility (apparatus, reactor, ...) allows steady flow through by feed material and product.	Production time intervals — filling, process (e.g., chemical reaction), discharge.
Products (flow resources) are not stored under normal conditions.	Products are often stored between process steps.
Hardly flexible regarding production volume and other products.	Facilities and equipment are relatively flexible (e.g., in multipurpose plants).
Start-up and shutdown processes cause product loss.	Proof of origin for single batches is procurable.

In selecting the appropriate production principle, the following points have to be considered:

- Production volume and regularity of demand
- Need for flexibility
- Requirements in terms of proof of origin and quality control
- Technological conditions and safety requirements

b. In the case of the solvent, the preferred principle could be continuous flow production. The production volume is of adequate size for small facilities for continuous production. Furthermore, it can be assumed that the consumption of the new product will run relatively regularly. At least a proof of origin is not necessary.

8.7.2 Manufacture of By-Products

Intended learning outcomes: Estimate how much money can be invested in a distillation device, if management sets a limit of 2 years maximum for payback on this kind of investment.

In the production of 300 kg per hour of an active substance for the manufacturing of photographic paper, 20 tons of sewage water accrue per day. The sewage flow is contaminated with an organic dissolver, which is needed for the production of the active substance. The purchase price of the dissolver is $1.30 per kg. The current production process has about 6000 operating hours per year and runs on the principle of continuous production. The sewage water needs to be disposed of as waste product. Because of dissolver contamination of approximately 5% (mass percent), extra costs of $5.50 per m^3 are caused in comparison to wastewater without organic impurities.

On the basis of thermodynamic calculations and laboratory tests, it was estimated that it would be possible to separate almost all of the dissolver by adding a simple distillation column as a further process step. For the distillation, 80 kg heating steam (cost: $20 per ton) is needed per m^3 of sewage water. The regained dissolver can be reintroduced into the production process without any additional effort.

The plant engineer now attempts to estimate how much money can be invested in the distillation device, if management sets a limit of 2 years maximum for payback on this kind of investment. Can you help?

Solution:

- 6000 operating hours equals 250 days (continuous production!)
- 20 * 250 = 5000 tons of sewage water accrued per year
- Loss of dissolver: 250 t/a, → savings from recovery: $325,000/a
- Savings from lower cost for wastewater treatment: $27,500/a
- Additional cost for heating steam: $8000/a
- Total savings: $344,500 per year
- Payback time: max. 2 years → about $689,000 available for investment

8.7.3 Production Planning in Process Industries

Intended learning outcomes: Determine the needed quantity of feed product and the required number of batches per stage for the production of the desired quantity of the active substance.

For the production in a three-step batch process of 500 tons of an active substance for use in pharmaceutical products, chemical reactors of different sizes come into operation. Figure 8.7.3.1 describes the production sequences with batch size and yield in each process step. Please note that the figure does not show a mass balance or a bill of material.

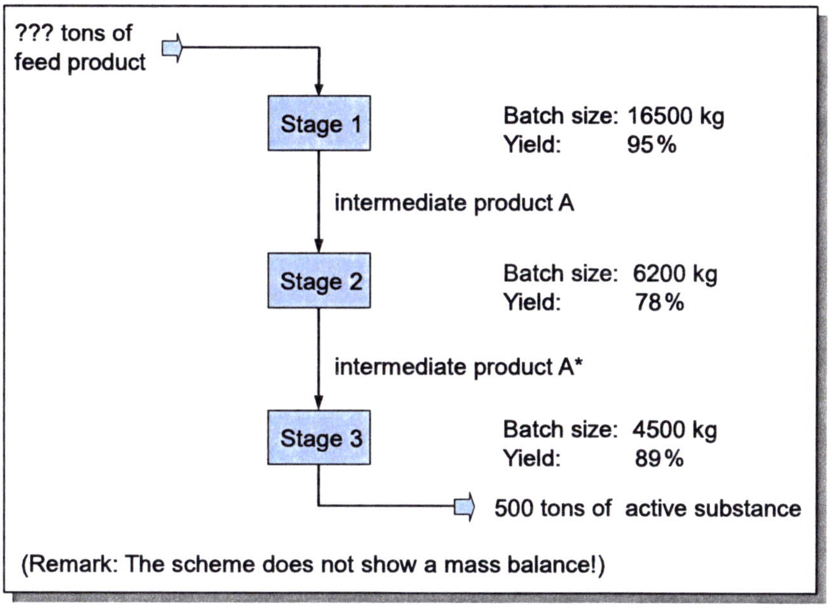

Fig. 8.7.3.1 Batch size and yield in each process step for an active substance.

Determine the needed quantity of feed product and the required number of batches per stage for the production of the desired quantity of the active substance. Please be aware that only complete batches can be produced.

Solution:

- Production quantity of active substance: 500 t
- Stage 3:
  - Yield: 89% → demand for A*: 562 t
  - Batch size: 4.5 t → number of batches: 124.9 → 125
  - → Actual demand for A*: 562.5 t
- Stage 2:
  - Yield: 78% → demand for A: 722 t
  - Batch size: 6.2 t → number of batches: 116.5 → 117

- → Actual demand for A: 725.4 t
- Stage 1:
  - Yield: 95 % → demand for feed product: 764 t
  - Batch size: 16.5 t → number of batches: 46.3 → 47
  - → Actual demand for feed product: 775.5 t

8.8 References

APIC16 Pittman, P. et al., APICS Dictionary, 15th Edition, APICS, Chicago, IL, 2016

Hofm92 Hofmann, M., "PPS — nichts für die chemische Industrie?" in Management Zeitschrift Bd. 61, No. 1, Zurich, 1992

Hofm95 Hofmann, M., "Konzeption eines Prozessinformations- und Managementsystems," Gabler Edition, Wissenschaft, Wiesbaden, 1995

Hübe96 Hübel, S., "Unterstützung zeitkritischer Dokumentationsprozesse in der Pharmaindustrie," BWI-Reihe Forschungsberichte für die Unternehmenspraxis, vdf Hochschulverlag, Zurich, 1996

HüTr98 Hübel, S., Treichler, J., "Organizational Concepts for Production Planning and Resource Allocation in a Multi-national Pharmaceutical Enterprise," in Brandt, D., Cernetic, J., Eds., "Automated Systems Based on Human Skills," IFAC/Pergamon Press, Oxford, 1998

Kask95 McKaskill, T., "Process Planning — In Search of a Standard," *Proceedings of the APICS World Symposium*, Auckland, Australasian Production and Inventory Control Society, 1995

Loos95 Loos, P., "Information Management for Integrated Systems in Process Industries," in Brand, D., Martin, T., Eds., "Automated Systems Based on Human Skills," IFAC/Pergamon Press, Oxford, 1995

Namu14 NAMUR-Recommendation, Standardization Committee for Measuring and Control Engineering in the Chemical Industry, 2014

Sche95b Scherer, E., "Approaches to Complexity and Uncertainty of Scheduling in Process Industries," in Brand, D., Martin, T., Eds., "Automated Systems Based on Human Skills," IFAC/Pergamon Press, Oxford, 1995

TaBo00 Taylor, S.G., Bolander, St.F., "Process Flow Scheduling Principles," APICS — The Association for Operations Management, Chicago, IL, 2000

9 ERP and SCM Software

Intended learning outcomes: Describe software used for logistics purposes. Explain contents of logistics software packages. Disclose factors for successful implementation of logistics software.

In the world of systems at operational level, a system for planning & control is known as an *information system* because it contains information in a structured form concerning future, current, and past events associated with the provision of goods.

One entrepreneurial objective of logistics, operations, and supply chain management is to ensure short lead times (see Section 1.3.1). In doing so, a partial objective is to ensure short lead times in the data and control flow. Small companies can effectively and economically practice information logistics without having to resort to information technology (IT). Today, however, in companies above a certain size, the use of a computer is nearly self-evident. The *information system* is then called *IT-supported.*

Chapter 20 covers basics of *IT-supported information systems. Here is an important insight:*

An information system cannot be *IT-supported* unless all the information to be contained in the system is available in a clear and quantifiable form, that is, unless (1) the system elements or objects can be represented on the hardware, and (2) the information flow can be expressed by software algorithms that handle the objects, i.e., the information flow is "programmable".

The subject of this chapter is the specific IT-support of information logistics of systems at operational level for planning & control.

PPC software, *ERP software*, and *SCM software*, or *APS software*, are terms frequently used to describe software for IT-support of information logistics in the area of logistics, operations, and supply chain management.

In practice, people (whether intentionally or unintentionally) often draw no distinction between the actual planning & control system and the software for the *IT-support of this task.* In recent years, this has led to unnecessary misunderstandings, and even to arguments and decisions based on prejudice. This chapter focuses on the main possibilities and limitations of the *IT-support* of tasks and processes associated with planning & control. We shall first consider the historical development of corresponding software and the extent to which it is used at present. This will be followed by a discussion of this kind of software and an attempt at classification. The chapter ends with some important notes concerning the implementation of this kind of software.

P. Schönsleben, *Handbook Integral Logistics Management*,
https://doi.org/10.1007/978-3-662-65625-9_9

9.1 Software in the Area of ERP and SCM: An Introduction

Intended learning outcomes: Produce an overview on history and origin of ERP software. Disclose scope and range of ERP and SCM software.

9.1.1 History and Origin of ERP Software

Intended learning outcomes: Differentiate between things that have changed and things that have not changed since the introduction of data processing. Describe the use of the Hollerith system at operational level.

Software as a technology first gained acceptance in the late 1950s, as soon as data could be stored on suitable media, rather than having to "plug in" the computer programs every time they were used. During this period, the computer world was dominated by a single company — IBM (International Business Machines).

IBM was founded by an American named Hollerith who had introduced a system to classify the data obtained from the American census in the second decade of the 20th century. This was based on light and electrical circuits (and subsequently electronic circuits). It was also the origin of the term "electronic data processing." The medium used to store the data was the famous punched card. The way in which information was encoded on a punched card was a clever invention, as were the machines for punching and then reading the punched card. Essentially, every character (or "byte"), whether it was a letter, number, or special character, was assigned a unique sequence of six holes. The two states, "hole" or "no hole," thus formed the smallest unit of information — a binary number (0 or 1) known as a "bit." This sequence of 6 bits could be combined in $2^6 = 64$ different ways, enabling 64 different characters to be represented (which had to include the control characters for processing). Figure 9.1.1.1 shows an extract translated from *Hollerith-Mitteilungen* (*Hollerith News*), 1913 ([IBM83], translated by the author). The reference list shows that the system had also quickly gained wide acceptance in Europe because of its ability to perform logistics tasks within a company. The "Hollerith variations" show two possible applications. The second example also directly highlights implicitly an important problem associated with data processing, that is, data protection.

Right from the start, Hollerith's idea was intended to process large quantities of data quickly and accurately. This saved an incredible amount of time, resulting in greater productivity and ultimately a new industrial revolution. The basic idea was perfected over the following decades. For example, the character code was extended from 6 to 7 or 8 bits, that is, 256 possible combinations per byte (ASCII[1] or EBCDIC[2] code) so that lowercase and special characters could be included. The hole in the card was gradually replaced by a two-digit state on a magnetic disk or tape, which resulted in the development of suitable searching and reading devices.

---

1 Abbreviation for "American Standard Code for Information Interchange," 7-bit code.
2 Abbreviation for "Extended Binary Coded Decimal Interchange Code," IBM's 8-bit code.

Here is a list of just some of the major companies that use the Hollerith system to **organize their workshops**:

Allgemeine Elektrizitätsgesellschaft, Kabelwerk Oberspree, Berlin
Siemens & Halske A.-G., Berlin, Askanischer Platz 3
Farbwerke vorm. Meister Lucius & Brüning, Höchst/Main
Accumulatoren-Fabrik Akt schaft, Hagen/Westf.
Waldes & Ko., Prag-Wrs
Brown Boveri & Cie. A.-Baden/Switzerland
Gebr. Sulzer, Winterthur/
Aktienbolaget Seperator, Sweden
Bell Telephone Manufactu Antwerp
Deutsche Gasglühlicht-Ak schaft (Auergesellschaf
Städt. Elektrizitätswerke,
Kaiserliche Werft, Kiel
Witkowitzer Bergbau- und hütten-Gewerkschaft, V
Central Foundry Company
Crucible Steel Company
Miehle Ptg. Press and Mg
Scully Steel & Iron Co.
American Can Co.
American Fork and Hoe
American Iron & Steel Co
American Radiator Co.
American Sheet and Tin
American Steel Foundries
Bridgeport Brass Co.
Bullard Machine Tool Co.
Carnegie Steel Co.
De Laval Separator Co.
Illinois Steel Co.

Jones & Laughlin Steel Comp.
Link Belt Co.
Lodge & Shipley Mach. Tool Co.
McCaskey Register Co.
Marshall Wells Hdw. Co.

Hollerith variations.

Counting individual and specific cards. For the internal waterways statistics kept by the Imperial Statistical Office in Berlin, a card is punched for each consignment of freight and only the first consignment from a certain ship's cargo is punched with the load-bearing capacity of the ship. To determine the number of ships, the return cable of the load-bearing capacity counter is connected to a card counter so that the card counter only moves on one if a load-bearing capacity is actually added. It does not move on for cards in which no load-bearing capacity is added. This means that the cards do not have to be sorted, which would otherwise be necessary in order to separate those cards on which the load-bearing capacity is punched, i.e., which represent ships, from the others that only represent consignments.

Separation of abnormal cases. At the Statistical Bureau in Copenhagen, under the management of its Director, Mr Koefoed, an extremely sensible precautionary measure was taken by the head of the Hollerith Department, Mr Elberling. This has eliminated the need to sort 4.7 million cards. Around 100,000 people in Denmark were classified as abnormal because they fell into one of three different categories — with respect to affliction, religion and military situation. Using the normal sorting method, all the cards would have to be sorted three times in order to separate out the three abnormalities. Since Denmark has a population of around 2½ million, this would have meant sending around 7½ million cards through the sorting machine.

A special sorting machine brush holder was therefore produced. Instead of just one brush, this had three brushes, arranged so that they touched the three columns of the abnormal categories. Since the sorting machine then always sorted by the hole that closed the current circuit first, the consequence of this arrangement was that, even though the 2½ million cards were only sorted once, those with no abnormalities, i.e., which were not punched in the three rows, fell into the "R" hole, while the others were sorted into one or the other compartment. After that, the 100,000 abnormal cards had to be sorted three times because they were all muddled from the first sorting process. As a result of producing this device, it was only necessary to sort 2.8 million cards, rather than 7.5 million.

Fig. 9.1.1.1 Early software: use of the Hollerith system at operational level.

Since the introduction of data processing in the early years of the twentieth century, the *quantity and the speed of the processed data* have increased dramatically. However, the *logical principle used to display and process information* and the *conditions for IT support of an information system* (see the introduction of this chapter) have not changed at all.

These facts are important if we are to understand the possibilities and limitations of data processing. This ingenious idea combined with Hollerith's business sense enabled IBM to hold the monopoly of the commercial use of this technology for many years. The early ERP software also originated from IBM. COPICS (Communication-Oriented Production Information and Control System) was the most well-known standard software package of the 1960s, and for a long time was the standard for further developments in this field. See also [IBM81]. This software was designed, in particular, to meet the needs of the major industries of that period — mechanical engineering and automobile construction.

9.1.2 Scope and Range of ERP and SCM Software

Intended learning outcomes: Disclose issues and challenges of ERP software and SCM software throughout the last decades.

ERP software was originally designed for modeling products and production processes, administering orders, and preparing accounts. These tasks were soon supplemented with planning functions for resource management (goods and capacity).

Between 1960 and 1980, many companies developed their own company-specific software that was precisely tailored to their needs. Data were then transferred from forms to punched cards and processed in batches at computer centers. The range of such software was generally just a few years. A new software generation came in the late 1970s, with the development of the computer monitor (character format with 24 rows * 80 characters). Relational databases that could handle large volumes of data first appeared at roughly the same time. These provided users with direct access to the data and to the programs that process this data (online, or interactive techniques). The software from this period is sometimes still in use today, enhanced by the graphical user interface (GUI) introduced in the late 1980s.

ERP standard software was only seen in large companies up until the mid-1970s. The most common applications could be found in companies with logistics characterized by convergent product structures, batch production, and production with order repetition and with high utilization of capacity as their entrepreneurial objective. The first generation of standard software, such as the COPICS package, was also developed for this user profile. Since then, standard software gradually improved to also meet at least partially the needs of small- and medium-sized companies and their flexible forms of organization. Since the mid-1990s, software for SCM has also been available.

Current software challenges in the areas of ERP and SCM are due, for one thing, to technological developments. Mobile end user devices such as iPhone and iPad increase the demand for mobile applications, making it possible to use various ERP or SCM software functionalities independent of location. "Cloud computing" means that the data and programs are no longer stored on an individual computer. Large main storages allow "in-memory computing," which means that the data no longer have to be read by a hard disk storage unit and instead are kept in the main storage.

In the organizational area, software is currently challenged by the globalization of firms. The software has to be able to be linked up globally, which is not easy owing to different languages, some of them not alphabetic, and local differences in the qualifications of personnel. In addition, there continues to be a demand for increased customization of functionality. *Service-oriented architecture (SOA)* has been developed to meet this demand. SOA is the concept of encapsulating functionalities in loosely coupled services, which can then be flexibly combined for new application functionalities.

ERP and SCM software is still surprisingly long-lasting. The typical life span is 10 years or more, and 20 years is not unusual.[3] What is the reason for this longevity? A change of hardware or software is both expensive and full of risks: for integrated packages, in particular, it would affect every operational order processing system — and thus a large number of users and a wide range of IT-supported processes. Any mistakes would immediately have a detrimental effect on the company's ability to add value and thus to do business. In addition, the ERP and SCM software developers who enter the market take too little notice of the requirements of the users, with the result that packages created using the latest technology have inadequate data models and functionality.

9.2 Contents of ERP and SCM Software

Intended learning outcomes: Describe classical MRP II / ERP software. Present software for customer order production, for the process industry, for transcorporate planning & control in a supply chain, and for Customer Relationship Management (CRM). Differentiate between standard and company-specific software.

Every ERP or SCM software package has developed in a slightly different way. Some were designed for specific branches of industry, products, or production characteristics. The developers also learned their craft in a certain type of company environment, which shows in the features of the software.

9.2.1 Classical MRP II / ERP Software

Intended learning outcomes: Present SAP R/3 as a typical example of a classical, generally applicable ERP software package.

Classical software was developed in mechanical engineering and automobile construction companies, with discrete manufacturing, batch production, and production with order repetition as characteristic, and with high utilization of capacity as the entrepreneurial objective. Extension of this functionality resulted in what we now know as *MRP II software* or *ERP software*. This type of software essentially supports the concept described in Chapter 5.

The first package in this category was COPICS from IBM. Other examples are Manufacturing from Oracle, J.D. Edwards, Infor XA (in the past MAPICS from IBM) and. The actual market leader is SAP, with its R/3, mySAP™ software and the follow-up products. Software houses like SAP offer a comprehensive and integrated package that supports all the business processes within a company. Figure 9.2.1.1 contains an overview of the R/3 structure.

---

3 These software systems in use for a long time are sometimes called *legacy systems*. They often do not interface well with new applications.

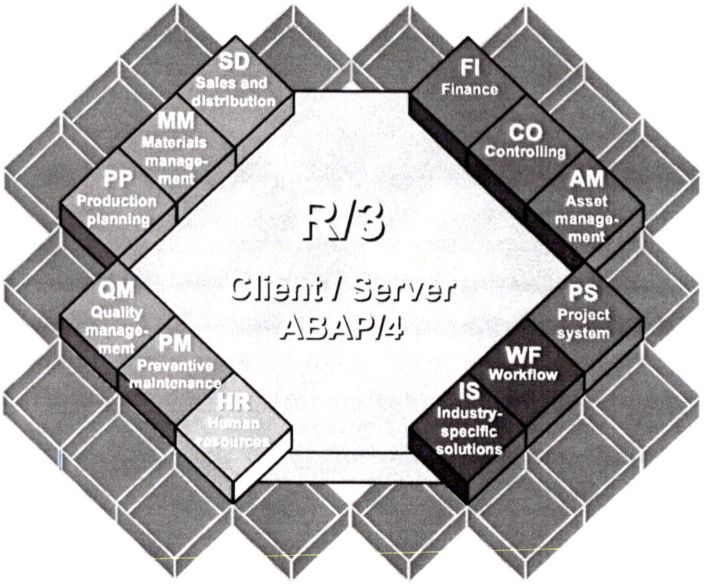

Fig. 9.2.1.1 SAP R/3 as a typical example of a classical, generally applicable ERP software package.

The abbreviations that designate the modules, which are oriented toward specific functions within a company, consist of two letters: SD for sales and distribution, MM for procurement and stochastic materials management, and PP for deterministic materials management, scheduling, and capacity management. The modules contain submodules for the three temporal ranges (long, medium, and short term) and for the individual tasks. The functional separation between MM and PP emphasizes the distribution of users between trade and production. It also betrays the fact that R/3 started out as an MRP II package.

SAP developed R/3 with a view to covering and integrating every function within a company. The finance and accounting functions have always been the driving force behind the development of ERP software, since detailed job-order costing requires efficient administration of all types of orders within the company. This simple aspect of corporate policy explains why the emphasis always has to be placed on certain areas when developing ERP software. The decision will ultimately depend on whether the finance function can be integrated, rather than on the quality of support provided for planning & control.

SAP R/3 can be customized to take account of different values for the features relating to planning & control described in Section 4.4. R/3 specialists configure the software by setting a large number of parameters. It is not enough just to have a knowledge of logistics, planning & control, and the actual company, which means that R/3 is rather suitable for medium-sized and large companies. Since the software was developed from the MRP II concept, the limitations of usability indicated in Figure 4.5.3.1 also apply for this kind of ERP software.

The Lean/JIT concept and all the techniques for production with frequent order repetition are oriented toward the needs of manual organizations. In the best-case scenario, such organizations can manage without software altogether. ERP software can be introduced when the volume of data becomes too large, e.g., a package on a PC with a simplified master data management system. This will enable the number of Kanban cards to be calculated, for example. It could also be an ERP software package extended to include this type of function.

In contrast, the variant-oriented and processor-oriented concepts require adequate software, as discussed in the rest of this chapter. Together with the software for the MRP II concept, such concepts also provide fundamental typologies for ERP software for planning & control.

9.2.2 Software for Customer Order Production or the Variant-Oriented Concept

Intended learning outcomes: Describe typical software modules for customer order or variant-specific production. Identify specific software packages.

Software for customer order production and the variant-oriented concept, that is, for products according to (changing) customer specification or for product families with many variants, has been specially designed for and developed in conjunction with make-to-order producers. Often, bills of material are customer-specific or order-specific. Such companies need the variant-oriented concept for single-item production or nonrepetitive or "one-of-a-kind" production. The different techniques identified in Chapter 7 all place different requirements on the software and, in the most extreme situation, could even lead to different subtypes of ERP software for the variant-oriented concept. Equally, a package may only be suitable for one of the techniques within the variant-oriented concept.

Software for customer order production or the variant-oriented concept was mainly developed in Europe, particularly for small- and medium-sized companies (SME). This software includes PSIpenta from PSI, ProConcept from ProConcept SA, and, in the past, MAS90 from IBM, IPPS from NCR, and many niche products. Packages that are particularly suitable for product families with a wide range of variants include Infor LN (in the past Baan) as well as Expert/400 (which was developed by the author). There are also a number of industry-specific products, e.g., for window and furniture production. For bid processing of engineer-to-order products, the Leegoo Builder Software of EAS GmbH is well known.

Figure 9.2.2.1 shows, by way of example, the PSIpenta software module for product families with a wide range of variants. It also provides an overview of the level of detail below that illustrated in Figure 9.2.1.1.

Some of the modules, such as "Customer order archive," "Create order package," and "Network planning module," suggest that the software is particularly suitable for customer order production. Within the order structure, the product that is ordered or offered may be greatly modified for a particular customer. One particular characteristic is the processing of "exotic" items that are only needed for a specific order and for which it can be said with certainty that there will be no order repetition. In this case, there is no need to store master data for the item or to allocate an item ID.

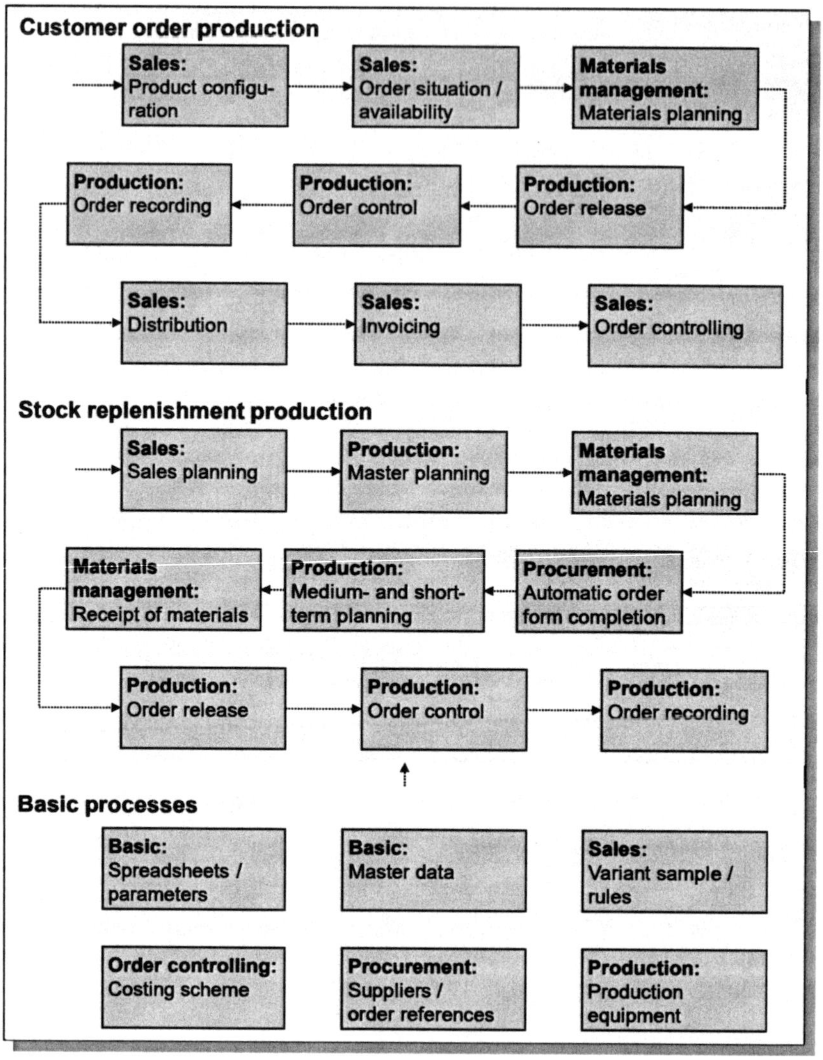

Fig. 9.2.2.1 Typical software for customer order or variant-specific production: the PSIpenta modules.

9.2.3 Software for the Process Industry or the Processor-Oriented Concept

Intended learning outcomes: Describe some typical modules of software for the process industry. Identify specific software packages.

The processor-oriented concept for the process or basic producer industries requires appropriate ERP software, that is, in which the emphasis is placed on mixing ratios and recipes, rather than on bills of material.

Software for the processor-oriented concept largely originates from the chemical, pharmaceuticals, and food industries in the United States or Germany. It includes software such as

Blending from Infor, Infor LX (in the past Bpics), CIMPRO from Palomino Computer Solutions, Ross ERP from Apten, MFG-PRO from QAD and, in the past, Protean (once Prism from Marcam).

Figure 9.2.3.1 shows typical modules that make up such software by way of example. The make up highlights the emphasis placed on resources and on the production model (processor-oriented production structures as described in Chapter 8).

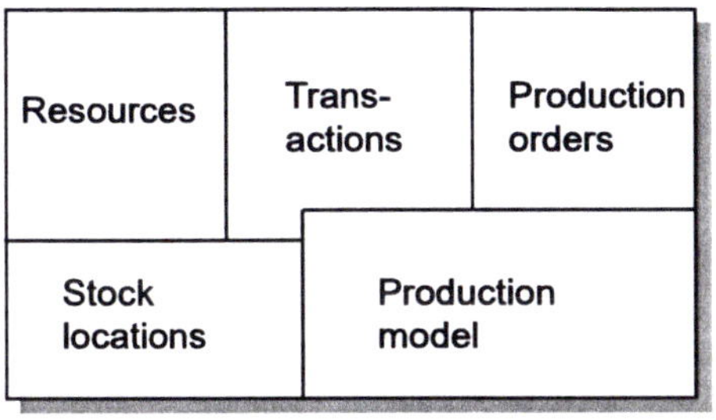

Fig. 9.2.3.1 Software for the process industry: some typical modules.

Problems specific to the process industry that are covered include:

- Different lots of a bought-in product have different characteristics and must thus be handled differently (e.g., production of tomato products: addition of sugar according to the sugar content of the tomatoes, use of different grades for different products).
- The process industry often uses by-products, recycled products, or waste products. The traditional representation of product structures in the form of bills of material is not suitable for such cases.
- Planning & control do not just apply to materials. They are of equal importance for capacity and production equipment (e.g., molds for manufacturing chocolate bars).

Electronic control boards ("Leitstand") software packages are used for IT support of master production scheduling. These packages take account of the limited capacity typical of such industries and, by changing these limitations, allow reliable and appropriate production schedules to be created (constraint-based techniques).

9.2.4 Software for Transcorporate Planning & Control in a Supply Chain

Intended learning outcomes: Describe the concept and some of the tasks performed by SCM software. Identify specific software packages.

The term *SCM software* or *advanced planning and scheduling (APS) software* is used to describe software that supports transcorporate planning & control.

SCM software has been available for several years. Developments are moving in three different directions:

1. Electronic control boards ("Leitstand") software supplemented with modules for logistics and production networks. Software such as JDA solutions places (by the takeover of Manugistics) particular emphasis on distribution networks; that is, the

distribution of end products produced by different companies via various sales channels (e.g., national companies).

2. Conventional MRP II software or ERP software supplemented with company-specific or bought-in modules. These include APO (advanced planner and optimizer) from SAP or the equivalent products from PeopleSoft (by the takeover of Red Pepper). The "problem solver" software kernels from ILOG are often integrated for scheduling tasks. These modules work using constraint propagation techniques.

3. Niche software specially designed for transcorporate planning & control.

Figure 9.2.4.1 illustrates the concept and some of the tasks of SCM software.

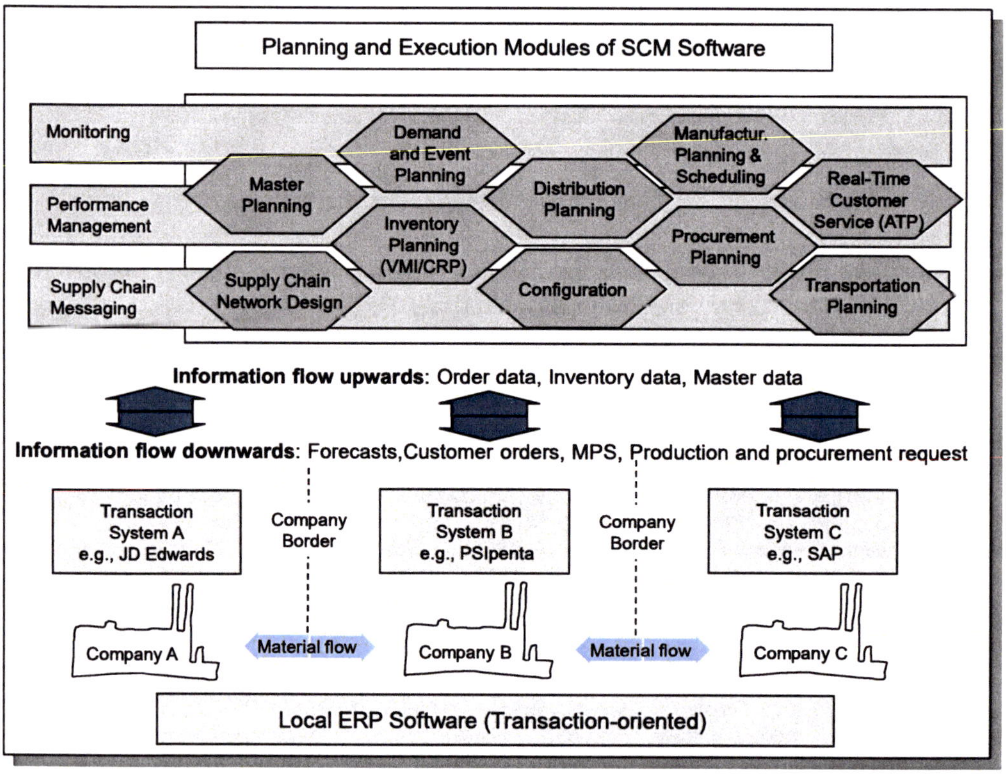

Fig. 9.2.4.1 Concept and some of the tasks performed by SCM software.

The master and order data are still administered by the local planning & control software of the individual companies involved in the logistics and production network. The data are periodically downloaded by the SCM software. The network planning then takes place and the results are returned to the local software.

The actual planning functions of SCM software are similar to those of traditional ERP and electronic control boards ("Leitstand") software, supplemented with new modules that meet the typical needs of networks:

- Supply chain network design to describe the logistics and production network
- (Network) inventory planning for tasks like replenishment of the customer's stocks by the supplier (VMI, vendor-managed inventory; CRP, continuous replenishment planning). To be able to do this, the supplier must have access to the customer's inventory and order data (and the data of any customers downstream in the network).
- Real-time customer service to be able to assess the fill rate of open orders with suppliers in advance. To be able to do this, the customer must have access to the supplier's inventory and order data (and the data of any suppliers upstream in the network).

These concepts are still at the field trial stage, but the sales network software is likely to be implemented first. This is not surprising since the organizational concepts for sales networks are older than those for joint R&D and production. In this context, the author of [Nien04] presents an approach for designing SCM-Software that also considers aspects like robustness, tangibility, and efficiency.

9.2.5 Software for Customer Relationship Management (CRM)

Intended learning outcomes: Produce an overview on the representation of the objects and their interrelationships of CRM software.

Customer relationship management (CRM) is the collection and analysis of information designed for marketing, sales, and service decision support (as contrasted to ERP information) to understand and support existing and potential customer needs (cf. [APIC16]).

CRM software is software that supports CRM.

CRM software development began in the 1980s, when companies first used computer-aided selling (CAS) software for rationalization of their distribution. The shift from this focus on rationalization to an emphasis on quality improvement of customer relationships required an extension to include marketing and services and led to today's generation of software (e.g., Siebel from Oracle).

CRM software provides functionality in two areas:

- The functions of operational CRM facilitate the business processes behind interactions with customers at the point of contact ("front office"). Tasks arising from the interaction processes and the required information are delivered to appropriate employees ("back office") for processing, interfaces are provided for further applications (word processing, e-mail client), and customer contacts are documented.
- Analytical CRM solutions, on the other hand, analyze the data created on the operational side of CRM, particularly for purposes of customer analysis and segmentation of the customer base (for instance, identifying potential failures of customer retention) or to exploit cross- and up-selling potentials.

CRM software supports all staff interactions with the customer in a number of ways (see Figure 9.2.5.1):

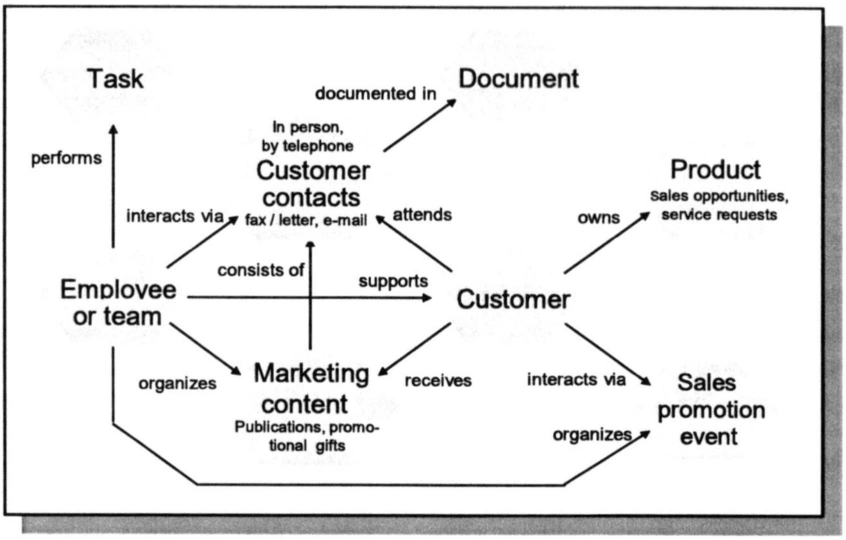

Fig. 9.2.5.1 CRM software representation of the objects and their interrelationships.

- CRM software provides a representation of all interactions between staff and customer. Staff members are always informed about who is responsible for supporting a customer and whom they should inform about contacts with the customer.
- CRM software supports staff in organizing, executing, and documenting customer contacts. These may be contacts with an individual customer (in person or by telephone, e-mail, or fax/letter), the sending of marketing content addressed to several customers, or a sales promotion event to which many customers are invited.
- CRM also provides functions for product-related interactions, such as customer service inquiries or sales opportunities. The system captures the sales opportunities and the corresponding order success probability, allowing the company to forecast expected sales.

The data required by CRM software for the most part already exist in the company, but they are located within various applications:

- Product-related data (such as customer orders) in ERP software or legacy systems
- Customer addresses in personal information management (PIM) software — sometimes at decentralized workplaces; the PIM file may also document some of the interactions (appointments, e-mail)
- Customer-related documents (such as bids, invoices, invitations) produced by word processing and sometimes administered by document management systems

When implementing CRM software, the challenge is to integrate all of the data and the available interfaces. It is important to evaluate whether the systems work together in a coherent and consistent way. Today, many of the complete CRM software systems are being replaced by applications that make use of a company's existing PIM applications as the basis or by applications that are components of enterprise software packages. *Portals*, that is multiservice Web sites, can be used for this task.

9.2.6 Standard or Company-Specific Software?

Intended learning outcomes: Explain various reasons why some companies still need company-specific software. Disclose aspects that should be taken into account when choosing between standard and company-specific software.

Standard software is software designed to meet the needs of different companies. It is developed and sold by a specialist software house. *Company-specific software* is created for a specific company and thus precisely meets the needs of that company. It is either developed within the company or the work is commissioned from a software house.

Many companies had their own company-specific software by the end of the 1980s, since the standard packages, which were solely MRP II oriented, did not meet their needs. In time, more logistics packages with most of the required functionality became available on the market. It was also recognized that the cost of maintaining company-specific software is extremely high. As a result, there has been a massive trend toward the use of standard software over recent years, which has contributed to the success of SAP R/2 and R/3. Nevertheless, some companies still need company-specific software for various reasons:

1. *Unsuitable processes:* When standard software is implemented companies often find, particularly with respect to order processing, that they have to cut down processes forming part of their core processes and not just their antiquated legacy procedures. If core processes have to be adapted to conform to the "standard," then the company is likely to lose its competitive edge. Then, the software must be examined to determine just how modular it is, that is, whether the data model and process model have interfaces that will allow a company-specific program to be integrated in place of the unsuitable module supplied with the standard software. That is, only some modules would then be company specific, rather than the entire package. Such changes are expensive, and often time consuming and difficult.

2. *Inadequate functionality:* Certain object classes or attributes may be missing from or inappropriately defined in the data model. This means that additional classes or attributes must be added or existing ones changed to modify the function model to suit the desired functionality. Today, this type of change can usually be carried out by simply generating the code from a definition language.

4. *The user interface cannot be integrated into the company's processes and way of working:* For example, a variant generator of a well-known ERP-software was awkward to use and required IT-oriented thinking. In one case, careful reprogramming of the user interface provided the design engineers with a simple interface that works well in their language. They are now able to make use of the needed IT support as part of their job description. However, the need for a simpler process must be offset against the increased cost of adapting the user interface. Such changes are often not difficult to implement, but are "merely" time consuming and thus expensive.

There are two other aspects that should be taken into account when choosing between standard and company-specific software:

- Risk of error: The number of man years invested in the production of company-specific software will be less than that required to produce standard software of the same scope. It is also likely that the former will contain more bugs than the latter. On the other hand, standard software is not always completely stable; new software releases often have to be installed in quick succession, even though most of the changes are not relevant to a particular company. This is usually regarded as an unnecessary expense. Poor standard software can contain more bugs than good company-specific programs.
- *Continuity:* Here, again, it is not possible to give generally applicable advice. The pros and cons must be considered in each case. Although the teams involved in developing company-specific software are generally smaller, they also tend to be more committed to their program. Experience shows, however, that practically none of the companies that have produced ERP software packages have managed to issue a second generation of their successful package without going into liquidation or being taken over by another company. Both situations have direct consequences on the continuity of standard software packages.

To summarize, a standard software package can rarely be implemented without adaptation if the entire logistics task is taken into consideration. A commercial decision must be taken to set the priorities: will the benefits of greater user friendliness, greater transparency, and faster lead times for the data and control flow outweigh the longer implementation time and higher costs?

New basic technologies offer great potential for the development of both company-specific and standard software. The benefits of *standard PC software*, such as word processors, spreadsheets, project planning software, etc., can already be used to implement much of the functionality provided by ERP software. See also [MöMe96], for example. The internet, the Java programming language, and a standard for a company's objects (such as CORBA) enable software modules from various sources to be linked to one another.

9.3 Factors for Successful Implementation of ERP and SCM Software

Intended learning outcomes: Explain possibilities and limitations of the IT support of planning & control. Disclose factors that influence individual acceptance and the range of implementation of ERP software.

For many years, contradictory opinions have been expressed about the effectiveness and the efficiency of ERP or SCM software. This contradiction can be illustrated by two extreme and opposite views:

- "There is no satisfactory ERP or SCM software package."

- "Every ERP or SCM software package is good."

If we examine these two statements in greater detail, we discover some interesting and somewhat surprising results that show that the contradictory views are due to different starting positions. The first statement concerns the limitations of any ERP or SCM software package, whereas the second relates to the essential success factors.

The following illustrations relate to both company-specific and standard software, although some of the comments concerning the choice of logistics package will of course apply only to the standard software.

9.3.1 Possibilities and Limitations of the IT Support of Planning & Control

Intended learning outcomes: Explain the influence of ERP or SCM software on the extent to which corporate objectives are fulfilled. Disclose why objectives aimed at improving the company's performance can only partly be affected by the ERP or SCM software. Identify situations where ERP or SCM software can be used to good effect.

"There is no satisfactory ERP or SCM software package." Within companies, this type of view is generally expressed in departments involved in the strategic or overall management of the company, rather than operational management. The problem is often that such people have the wrong expectations of what ERP or SCM software can and cannot do.

These unrealistic expectations may be explained by the abbreviation *PPC*, which stands for Production Planning & Control, and by the term *PPC system*. These are used to describe both the actual task of planning & control and the software used to support this task. The same is true for the abbreviations, or the terms

- *ERP* (Enterprise Resource Planning), or *ERP system*,
- *SCM* (Supply Chain Management), or *SCM system*,
- *APS* (Advanced Planning and Scheduling, or *APS system*.

An opinion about one cannot be applied to the other. The mistake is still made, however, often unintentionally but sometimes intentionally, as well (for both positive and negative purposes). The term *software* is therefore used below in association with IT support.

The acronyms PPC, SCM, or APS can nevertheless be misleading when used in association with the software. This misunderstanding may even be encouraged by the software vendors, but unfortunately, it leaves a large area open to attack by anyone looking for an argument.

- The first letters in PPC and in SCM have extended meanings. PPC software packages no longer relate solely to production or supply, but rather as ERP software to the entire logistics chain from sales, production, and procurement, through distribution and maintenance. In addition, new requirements have arisen in association with return and recycling. It is also no longer possible to equate PPC software with MRP II packages since it incorporates the just-in-time, the variant-oriented, and the processor-oriented concept and with varying levels of quality, just like the MRP II concept. Similarly, SCM software is as useful for demand chain planning.

- The letter "P" in PPC or APS for "planning": Neither a PPC software nor an ERP software nor an SCM software nor an APS software does planning in the strict sense of the word. It simply supports the planning function, for example, by showing the availability of components and capacity along the time axis. Then comes the planning, e.g., action to change stocks, capacity, or order dates. Every attempt to hand this planning step over to the computer, e.g., through the use of simulation software, has ultimately failed, because the software is unable to cope with the day-to-day problems of decision making, either because the relevant parameters were not all known or because they could not be reliably shown along the time axis.
- The letter "C" in PPC for "control" or "S" in APS for "scheduling": Neither PPC nor ERP nor SCM nor APS software controls or schedules anything in the strict sense of the word. In the best-case scenario, it merely provides a snapshot of the current status of order processing in the various domains in the company and recommends options for control or regulation. The actual control or scheduling task still has to be carried out by people. Production and procurement in the manufacturing and service industries cannot be compared to the control of a machine or production system, since the equation inevitably includes people whose behavior finally cannot be predicted or simulated. On the other hand, although the inclusion of people as a production factor appears to be a disadvantage, it is also an advantage: No automated control system will ever be able to match the capabilities and potential of a human in control or scheduling, however flexible and autonomous it might be.

So what are the consequences with respect to the influence of ERP or SCM software on a company's ability to fulfill the entrepreneurial objectives? Figure 9.3.1.1 lists the four target areas discussed in Figure 1.3.1.1 and shows, for each primary and secondary objective, the extent to which ERP or SCM software can help to fulfill the objective.

If we consider the extent to which software influences the various objectives, we see that the objectives aimed at improving the company's performance can only partly be affected by the ERP or SCM software.

- *Quality:* The advantage of using ERP or SCM software is that a company has to explicitly store its products and services, and the processes by which they are created, in the form of master data or, more precisely, in the form of bills of material, routing sheets, or master data on technology and the network. In this way, products, processes, and organization are made transparent and easy to understand for all employees. However, this is only an aid to description and thus has only a minor influence over quality. The quality of products, processes, and the organization is more substantially improved by design, development of processes, and through the choice of production infrastructure, employees, and partners in the supply chain.

- *Costs:* Reduction of inventory in store and in process and increasing the utilization of capacity lead to conflicting objectives. ERP or SCM software cannot resolve these conflicts, but it makes the processes faster, more comprehensive, and transparent to more people. As indicated above, decisions concerning scheduling and materials planning and the actual control cannot be left to the software, so the increased transparency must be converted into better decisions by the people involved. The software thus has only an indirect influence.

Possible strategic objectives	Influence*
Target area quality	
To improve the transparency of product, process, and organization	++
To improve product quality	+
To improve process quality	+
To improve organization quality	+
Target area costs	
To improve input for calculation and accounting	++
To reduce cost rates for administration	++
To reduce physical inventory and work in process	+
To increase capacity utilization	+
Target area delivery	
To reduce lead times in the data and control flow	++
To reduce lead times in the goods flow	+
To increase delivery reliability rate	+
To improve fill rates or customer service ratios, or the potential for short delivery lead times	+
Target area flexibility	
To increase flexibility to enter as a partner in supply chains	+
To increase flexibility in achieving customer benefit	+
To increase flexibility in the use of resources	+

* The influence of ERP or SCM software over the strategic objective:
++: high / direct
+: some / indirect / potential

Fig. 9.3.1.1 Influence of ERP or SCM software on the extent to which corporate objectives are fulfilled.

ERP and SCM software require a complete and accurate management of master and order data. The software thus has a direct influence on improving the input for costing and accounting. The software supports the automation of the processes. It thus has a direct influence on reducing the cost rates for administration. However, stocks and utilization are also subject to macroeconomic influences, such as the employment market and the competitiveness of an entire national economy.

- *Delivery:* Information on orders in progress or stocks can be quickly called up by anyone involved in the process. Software thus directly reduces lead times within the data and control flow. Experience shows, however, that this does not necessarily affect lead times within the goods flow. This can be illustrated by an example in which it took just a few seconds to identify the physical location of a delayed order within the factory. The check demonstrated that the information was correct and reliable, but the goods had been left there because the operator was unavailable. This meant that the promised delivery date could not be met.

  Shorter overall lead times and increased delivery reliability therefore require a firm foundation within the company's internal organization. Simply holding the data on the computer is not enough to improve fill rates or customer service ratios — action must be taken in practice, as well. Thus, the software has only an indirect influence on the target area of delivery as well.

- *Flexibility:* As a first aspect of flexibility, today's software allows product families with a wide range of variants to be managed efficiently. In fact, this is essential to be able to respond flexibly to customers' requirements. However, the potential for flexibility is determined more by the way processes and the production infrastructure are designed and planned. ERP or SCM software are a less important factor.

  The same applies to the other aspect of flexibility — the utilization of resources. ERP or SCM software quickly provides comprehensive information on the needs and options arising from a given situation. It will rarely be able to make the decision to move resources without human input, however. It is worth repeating that the ability to use people flexibly and the capacity of machines to be used flexibly will essentially depend on the qualifications of those people and on the way in which the production infrastructure was planned.

If we consider these points together with Figure 9.3.1.1, we draw the following conclusion:

ERP or SCM software provides IT *support* for planning & control the way in which a company provides its services. However, an ERP or SCM software package is used first and foremost — and, in most cases, successfully — for representing products and their production and procurement processes (make-or-buy) and to administer orders, and thus for administration and preparing accounting.

ERP or SCM software ultimately links people together by the way it uses information. If we assume that sufficient numbers of people have been adequately trained and are given enough time, then they could manually do everything that the software can do.

ERP or SCM software can be used to good effect in situations where human skills and capabilities are insufficient, typically because of:

1. The increasing complexity of products and the product mix
2. Increased volumes of data and frequency of orders (or processes)
3. Greater requirements placed on the speed of process administration

To summarize, ERP or SCM software will always be able to do exactly what Hollerith intended data processing to do right from the start; that is, fast and accurate processing of large quantities of data. It is thus not a replacement for the task at operational level. It is merely used to automate this task. It would be wrong to expect any more of it.

Each ERP or SCM software package has roughly the same influence over whether entrepreneurial objectives are achieved. This means that if a certain package does not fulfill the objectives that a company has set itself, then these objectives will not be achieved by using a different package. If the software is then investigated as the cause of failure, people will be all too ready to say that, "There is no satisfactory software package." They will view this as a welcome opportunity to pass the buck outside the company.

If the processes at operational level have to be reorganized, it is therefore advisable to divide the procedure into two steps, each with its own break-even analysis. This procedure requires attention to be paid to the training of people who will carry out the task within the company.

- The first step is to change the organization and the processes. Can the existing organization and processes actually be carried over to the new? What will this cost? IT support should intentionally be left out of the equation because, as mentioned above, all the tasks of software can, at least theoretically, be carried out by people. The break-even analysis for this first step must then consider the cost of the training required to cope with all aspects of the new organization. Consideration should also be given to how the entrepreneurial objectives (e.g., to reduce lead times in the goods flow) can actually be achieved by the changed processes.
- Only the second step, and thus the second break-even analysis, considers the precise value of IT support with ERP or SCM software. Here, again, there will be costs associated with training employees in the correct use of the hardware and software. On the other hand, in this case it will also be possible to reduce staffing numbers since the flow of information will no longer be processed manually.

This type of procedure can disprove the view that there is no suitable ERP or SCM software (which is sometimes used as a convenient excuse). The problems really arise because the people involved have insufficient knowledge of the processes and the associated tools.

9.3.2 Factors That Influence Individual Acceptance and the Range of Implementation of ERP Software

Intended learning outcomes: Describe factors that influence individual acceptance of ERP software. Disclose factors that influence the range of implementation of ERP software. Differentiate between important factors and less important factors.

It is not easy to quantify the success of implementing an ERP software package. Figure 9.3.1.1 has already shown that success should not be measured against explicitly worded corporate objectives, since these are influenced by the logistics used, the product design process, and factors outside the company's control, rather than by the software. One study [Mart93] adopted "PPC acceptance" and "Range of PPC implementation" as its measured variables. Here, PPC means PPC software and, more broadly, ERP software. Consequently, it is better to speak of the acceptance and range of implementation of ERP software below. Many of the factors can as well be transferred to SCM software. The study was carried out in 100 firms. 900 people were surveyed, particularly those who regularly work with the software. Analysis of the questionnaires revealed extremely high acceptance of ERP software at the individual level: The people questioned felt that the package more or less met their expectations. Figure 9.3.2.1 shows the factors that influence individual acceptance.

Under *personal features*, education, vocational training, experience, and position within the company had no significant influence over the individual acceptance of ERP software, whereas it was affected by general data processing knowledge and experience and the support of colleagues.

Factors that influence individual acceptance	Influence*
Personal features	
School education	+
Vocational training	+
Number of years in the job	
Position within the company	+
General data processing knowledge	+++
Data processing experience	++
Support from colleagues	++
Support for employees during implementation	
Training: duration	++
Training: breadth of training	++
Training: satisfaction	++
Information concerning the reasons for implementation	+++
Participation: range	++
Participation: opportunity to put forward suggestions	++
Participation: desire for opportunity to put forward suggestions	+
Extent to which data had to be revised	+
Cooperation between departments	++
Planning and organization	++
Time available out of daily work	++
Support from senior management	+
Internal contact	
User's opinion of the ERP software	
General suitability for own work	+++
System availability	+
Relevance of information on screen	+
Relevance of information in lists	
Scope for action: in determining time	+++
Scope for action: in determining processes	++
Scope for action: changes	+++
User friendliness: help functions	+
User friendliness: error messages	++
User friendliness: familiarization period	+
User friendliness: error correction	+

* Extent of influence over individual acceptance

+++: High
++: Significant
+: Insignificant
(blank): Minimal or no influence

Fig. 9.3.2.1 Factors that influence individual acceptance of ERP software. (From [Mart93]).

Of the factors that influenced the *support for employees during implementation*, the duration and breadth of training, satisfaction with the training, and the opportunity for participation all had significant influence over acceptance, which rose steadily as the number of days of training increased. No "saturation point" was identified, even with a high number of training days ([Mart93], p. 102). It also appears that certain deficits in the software can be overcome with the aid of training.

The most important factors appeared to be *information on the reasons for implementing ERP software*, combined with cooperation between departments, planning and organization, and the time available out of normal daily work. The extent to which the data had to be revised and, unexpectedly, supported from senior management appeared to be much less important.

For the *user's opinion of the ERP software*, the most important factor was whether the individual agreed that the adopted software was generally suitable for his or her own work. Work psychology concepts expressed by the scope for action also played a central role. This means that users are given the freedom to decide the order in which they perform their tasks and the sequence of activities within each task, even after implementation. On the other hand, the layout of screens and lists and, with the exception of error messages, other components associated with user friendliness (help functions, familiarization period, error correction) appeared to be less important.To summarize, the reasons for implementation, good training, freedom of choice in work, and suitability for an employee's own work are all important factors in the acceptance of an ERP software package.

The range of implementation of the ERP software was then identified with reference to the factors of "time since implementation started," "number of functions implemented," and "degree of distribution." For the first factor, the sobering result from the questionnaire was an average time of 4.3 years, even though all the companies questioned were either in the process of implementation or had just completed this phase. The number of functions implemented was derived by counting the number of modules, such as Sales, Stockkeeping, and so forth. Thirteen such functions were implemented on average. The degree of distribution was calculated by dividing the number of people working with the ERP software by the total number of people working in the operational departments. The range of implementation was derived from the combination of the three values. Figure 9.3.2.2 shows a selection of the factors that might influence the range of implementation.

The *company features* (total number of employees, influence from the group level, company type, and branch of industry) had just as little influence over the range of implementation as the data processing equipment used (hardware, operating system, or cost of software). The selected ERP software also had no influence over the range of implementation, although it did appear to matter whether it was the first implementation of such software or a replacement for an existing package. This result is particularly interesting in view of the opinion that, "Every ERP software package is good."

Of the *project features*, the importance of "ownership" of the project was key. The most successful projects were those in which responsibility was held solely by the Organization and Data Processing Department, rather than by a specialist department or two or more departments. This is one of the most unexpected results of the survey. It can be explained by the fact that, in an SME environment (small- or medium-sized enterprise), responsibility for the ERP software probably lies with employees in the Organization and Data Processing Department, rather than the specialist departments.

The number of levels of the management hierarchy that receives training is also very important. Training must be received by at least the top level (board) and the bottom level (group leader). It is also important to adopt a professional procedure for evaluating standard software (visiting reference customers, vendor tests using the company's own data, analysis of the current situation, list of requirements) and clear project management (appointing employees for the project, establishing a control committee and project team). In contrast, the number of project teams, team members and represented departments, and the project leaders and the amount of time they are able to devote to the project are less important.The average acceptance of the ERP software, which is derived from the individual acceptance scores, also has a significant influence over the range of implementation.

Factor of influence (brief description)	Influence*
Company features	
Number of employees	
Influences from the group	
Features of the type of company, branch of industry	
Data processing equipment	
Hardware, operating system	
Cost of software	
Initial situation	++
ERP software	
Project features	
Project leaders and time they are able to devote to the project	+
Reason for implementation (e.g., replacement, guideline, improvements)	
Control committee	++
Project team	++
Number of project teams	
Number of team members	
Number of departments represented	
Regular project team meetings	++
Project "owner" (specialist dept. / mixed / Organization-Data Processing)	++++
External consultants and number of consultants	
Reference customers visited	++
Vendor tests using company's own data	++
Current situation analyzed	++
Weak points documented	+
List of requirements drawn up	+++
Employees appointed to project	++
Number of trained hierarchical levels	+++
Board trained	+
Departmental manager trained	+++
Section manager trained	++
Project leader trained	++
Group leader trained	
Average acceptance of the ERP software in the company	++

* Extent of influence over the range of implementation

++++: Very high
+++: High
++: Significant
+: Insignificant
(blank): Minimal or no influence

Fig. 9.3.2.2 Factors that influence the range of implementation of ERP software. (From [Mart93]).

To summarize, the survey shows that, for the acceptance and range of implementation of ERP software, the characteristics of the software are important with regard to two points. First, individuals must believe that it is suitable for their own work and that they will retain freedom of choice in their work. Equally important is the support provided during implementation, the employee training, and the quality of the project management in general. If these requirements are fulfilled, it is possible to gain acceptance for and implement any of a number of ERP software products, which ultimately leads to the view that "Every ERP software package is good." This opinion is normally expressed by those who work with the ERP software every day and is not necessarily applicable to people who only use it sporadically.

9.4 Summary

With the benefit of Hollerith's ingenious idea and business sense, IBM long held the monopoly over the commercial use of data processing technology. The early ERP software also originated from IBM. Today, the most widely used ERP software packages are based on the MRP II concept and, increasingly, incorporate just-in-time and both the variant- and the processor-oriented concept. Although the trend is toward standard software, company-specific packages are still very important.

If we consider the question of the quality of ERP or SCM software, it is easy to draw different conclusions, depending on the initial viewpoint. Many misunderstandings are caused by the terms *PPC system*, which is used for both the task of logistics and for the software used to support it. It is therefore important to understand the background behind the arguments used to defend the positions that are adopted — which can lead to totally contradictory views.

Thus, many businesspeople will find it difficult to understand how an ERP or SCM software package, which is an expensive tool, is unable to influence their most important corporate objectives. This is still largely attributable to the vendors who promise too much in this respect because they know what the boss wants to hear. Thus, it is important not to raise any false expectations concerning the possibilities of ERP or SCM software. Its strengths are that it can be used to represent products and production and procurement processes (make-or-buy) and to administer orders, and thus for administration and preparing accounting. By recording and processing the data (and compressing it statistically), ERP or SCM software thus provides the information needed to make decisions concerning planning & control.

Acceptance and the range of implementation of ERP software depend on the way in which the software is implemented, the support given to employees during implementation, and the training they receive. Employees must feel that the actual software is suitable for their own work and that they will retain freedom of choice in their work. People who work with ERP software every day must not assume, however, that it is sufficient just to master the IT aspects. They must also have a thorough understanding of the company's processes and continuously adapt them to the needs of the market and individual products.

9.5 Keywords

acceptance of ERP software, 415
APS software, 405
company-specific software, 409
ERP software, 401
legacy system, 401
MRP II software, 401
portal, 408
PPC software, 397
range of implementation of ERP software, 415
SCM software, 405
service-oriented architecture (SOA), 400
software for customer order production, 403
software for the processor-oriented concept, 404
software for the variant-oriented concept, 403
standard software, 409

9.6 Scenarios and Exercises

Intended learning outcomes: Review factors that influence people's acceptance of ERP software. Differentiate between standard and company-specific software. Describe basic issues of software for transcorporate planning & control.

9.6.1 Factors That Influence People's Acceptance of ERP Software

Intended learning outcomes: Describe for each area the factors that have the greatest effect on a person's acceptance of ERP software.

Look again at Figure 9.3.2.1 and recall the three main areas that have an influence on a person's acceptance of ERP software. Please describe for each area the factors that have the greatest effect on acceptance. Does your personal experience match the findings presented in Figure 9.3.2.1? Please discuss this with your colleagues.

Solution:

a. *Personal features:* In this area, general knowledge of data processing and experience and support from colleagues have the highest impact on acceptance.

b. *Support for employees during implementation:* In this area, the most important factor is being informed of the reasons for implementing the ERP software. Other factors that influence employee acceptance are duration and breadth of training and satisfaction with training. Also important are cooperation between departments, planning and organization, and time to learn the software aside from normal daily work activities during the implementation phase.

c. *User's opinion of the ERP software:* From the user's point of view, the most important factor is whether the user feels that the adopted software is generally suitable for his or her own work. Another factor with high impact is the "scope for action," meaning that the software gives users the freedom to decide the order in which they perform their tasks and the sequence of activities within each task.

9.6.2 Standard or Company-Specific Software

Intended learning outcomes: Explain why a lot of company-specific software is still being produced in the field of ERP software.

Today, ERP standard packages like mySAP™ or J.D. Edwards are used to support planning & control activities by IT. However, a lot of company-specific software is still being produced in this field. Can you explain why?

Hint: Ask people in companies using company-specific software about this issue and compare their responses to the arguments given in Sections 9.1.2 and 9.2.6.

9.6.3 Software for Transcorporate Planning & Control

Intended learning outcomes: Evaluate the claim of some SCM software salespeople that SCM software at last solves the problems that ERP could not handle. Identify the real reasons for the success of SCM software implementations.

Figure 9.2.4.1 illustrated the SCM software concept and some of the tasks it performs. In this exercise, you will examine this concept further and look at some success factors.

How do you evaluate the claim of some SCM software salespeople that SCM software at last solves the problems that ERP could not handle, such as:

a. Taking into account capacity constraints when creating production schedules. (*Hint*: Compare especially the planning principles of the processor-oriented and the variant-oriented concept.)

b. Finding the correct solution. (*Hint*: Look very carefully at the structure of Figure 9.2.4.1)

c. Finding best solutions rapidly (real-time planning).

Finally, consider the more general question that is raised in Section 9.3.1, which discussed possibilities and limitations of the IT support of planning & control:

d. What are the real reasons for the success of SCM software implementations?

Solutions:

a. When proclaiming the advantages of modern SCM software, salespeople often contrast SCM with older, outdated versions of ERP software. Ask a salesperson if he or she is familiar with any software for internal enterprise planning & control besides MRP II. Many software packages for the variant-oriented concept (for example, also project management software) and particularly for the processor-oriented concept, subsumed under ERP software, do indeed take capacity constraints into account.

b. Figure 9.2.4.1 shows that SCM software must get the planning data from a company's ERP system. This means that the same errors in master and order data are generated in enterprise planning with SCM that were generated using ERP software. Ask the software salesperson about the consequences of erroneous data on lead time in the master data of ERP software for the quality of planning through SCM software. After all, the following principle will hold: "garbage in, garbage out." Claims that SCM software eliminates the need for ERP software are true only in theory or in very specific cases. Ask the salesperson for examples that correspond closely with your own company's situation.

c. Rapid planning through the use of SCM software is generally only the case for variants of a plan that has already been calculated. Ask the SCM software salesperson how long it takes to transfer greatly changed master or order data from ERP to SCM software. Ask for a reference from a company similar to your own in order to learn about their experience with the data transfer.

d. As in the case of ERP software, the decisive factors in success with SCM software lie in the company culture and the organization of supply chain collaboration. For implementation, therefore, the task is to find appropriate measures for all of the nine fields in the framework of Figure 2.3.2.1, and not for the ninth field alone.

9.7 References

APIC16 Pittman, P. et al., APICS Dictionary, 15th Edition, APICS, Chicago, IL, 2016

IBM81 IBM, "Communications Oriented Production Information and Control System (COPICS)," Brochures Vol. 1–8, IBM G320-1974-0 to G320-1981-0, 1981

IBM83 IBM, copied from Hollerith-Mitteilungen, No. 3 (June 1913), copied and translated from IBM-Nachrichten 33, 1983

Mart93 Martin, R., "Einflussfaktoren auf Akzeptanz und Einführungsumfang von Produktionsplanung und -steuerung (PPS)," Verlag Peter Lang, 1993

MöMe96 Möhle, S., Weigelt, M., Braun, M., Mertens, P., "Kann man ein einfaches PPS-System mit Microsoft-Bausteinen entwickeln?" 12(5) 1996, pp. 47–52

Nien04 Nienhaus, J., "Modeling, Analysis, and Improvement of Supply Chains — a Structured Approach," Dissertation ETH Zurich, No. 15809, 2004

10 Demand Planning and Demand Forecasting

Intended learning outcomes: Produce an overview of forecasting techniques. Explain history-oriented techniques for constant demand as well as with trend-shaped behavior. Describe future-oriented techniques. Disclose how to use forecasts in planning.

Whenever cumulative lead time exceeds customer tolerance time, production or procurement must take place on the basis of a demand forecast. The dark background in Figure 10.0.0.1 shows this task and the planning processes that require forecasting. The figure refers to the model for business processes and tasks of planning & control in Figure 5.1.4.2.

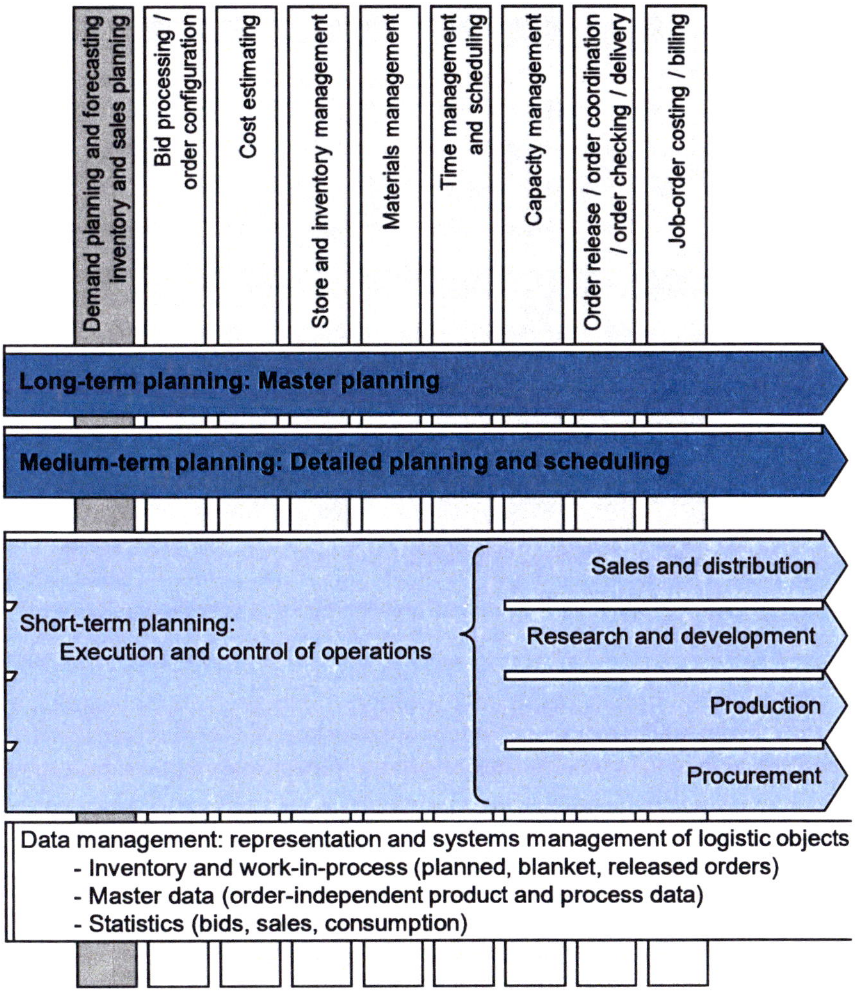

Fig. 10.0.0.1 The darker background shows the tasks discussed in this chapter.

Sections 5.3.1 and 5.3.2 provide a good basis for the material in this chapter. We recommend that you reread these Sections before studying Chapters 10 to 12.

P. Schönsleben, *Handbook Integral Logistics Management*,
https://doi.org/10.1007/978-3-662-65625-9_10

The need for forecasting varies over time, depending on market and product. Examples of markets with a great need for forecasting include trade in consumer goods or provision of the components needed for a service or for investment goods. Before receiving any definite customer orders, the company must produce or procure, for example, machine parts in advance.

The following sections classify forecasting techniques and describe the procedures in general. They describe and compare individual techniques in detail. They define the consumption distribution as an overlay of the distribution of consumption events and of the quantity consumed per event. This will allow us to derive safety demand and the limits of determining stochastic demand. We will also take a look on how to manage the transition from forecast values to independent demand. The material is both qualitative and quantitative in nature. It demands not only intuitive knowledge, but also a grasp of elementary statistical methods.

10.1 Overview of Demand Planning and Forecasting Techniques

Intended learning outcomes: Produce an overview on the problem of demand planning. Present the subdivision of forecasting techniques. Disclose principles of forecasting techniques with extrapolation of time series and the definition of variables.

10.1.1 The Problem of Demand Planning

Intended learning outcomes: Disclose the difference between human and IT-supported forecasting with regard to a forecast's precision. Present a demand planning procedure.

Demand forecasting is, according to Section 1.1.6, estimating the future demand.

A *forecast error* is the difference between actual demand and demand forecast. It can be stated as an absolute value or as a percentage.

A *forecasting technique* is a systematic procedure for forecasting demand according to a particular model.

A certain degree of uncertainty and therefore forecast errors characterize every forecast, regardless of whether people or IT-supported techniques do the forecasting. The latter are a complement to human intuition and creativity. When planning the demand, we should make appropriate use of both according to the situation.

If there are only a few items and only a limited amount of information that can be stated explicitly, human forecasting tends to be more precise. This is because human intelligence can process fragmentary information as well as knowledge derived by analogy, thus taking many further factors necessary for forecasting into account. This can be important, e.g., in rough-cut planning, where we need only forecast few demands for item families.

When there are many items, or when we can use explicit information on demand, an IT-supported forecasting technique generally provides more precise forecasting. This is due to the capacity of computers to process large quantities of data accurately.

- Trends, such as seasonality, can be calculated from consumption statistics. The length of the time frame to be observed makes this a difficult task for human beings.
- People tend to weigh unusual events too heavily. In this case, an IT-supported forecasting technique is more neutral in its "reactions."
- People tend to focus overly on the recent past. If a forecast proves too high for the current period, they tend to forecast a demand that is too low for the next period, even though this is not justified from the medium-term perspective.

Demand planning is always based upon certain fundamental assumptions and constraints. Parameters are used to keep their selection as general or flexible as possible. If the demand situation changes, demand planning should reexamine the choice of both parameters and technique and change them if necessary. Figure 10.1.1.1 shows a possible procedure for choosing the forecasting technique and its parameters:

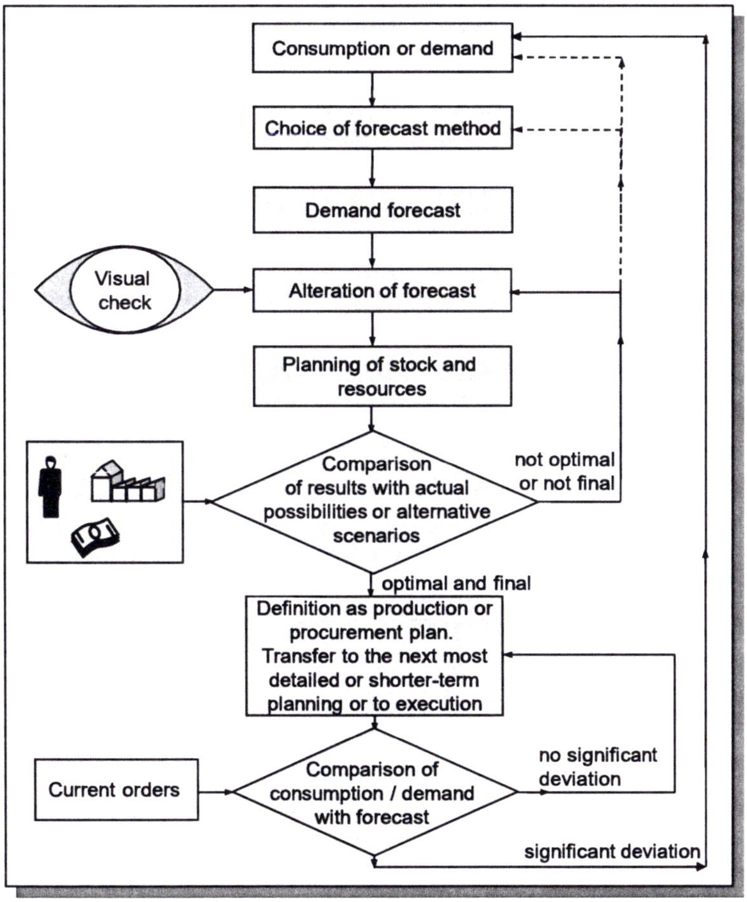

Fig. 10.1.1.1 A possible demand planning procedure.

- Choose a demand forecasting technique based on existing consumption or on partially known demand figures.

- Produce a forecast for future demand by applying the technique.
- When possible, make a visual check of the forecast and, if necessary, correct forecast values that vary too widely from intuitive assumptions. This check allows input of human knowledge of the behavior of the market into automated forecasting techniques.
- Break the demand forecast down into the needed resources — goods and capacity — according to temporal range and level of detail. This allows planners to estimate the consequences of implementing a forecast and to work out better variations if necessary.
- Adopt the optimal variant of the forecast as the production plan or procurement plan. These plans represent the independent demand; they are subsequently provided either to the next most detailed or shorter-term planning or to execution.
- At certain intervals in time, perform an analysis to see whether the course of demand or consumption agrees with the forecast. If the deviation analysis reveals too great a difference, repeat the cycle.

10.1.2 Subdivision of Forecasting Techniques

Intended learning outcomes: Explain possible breakdowns of forecasting techniques.

Figure 10.1.2.1 shows one possible subdivision of forecasting techniques.

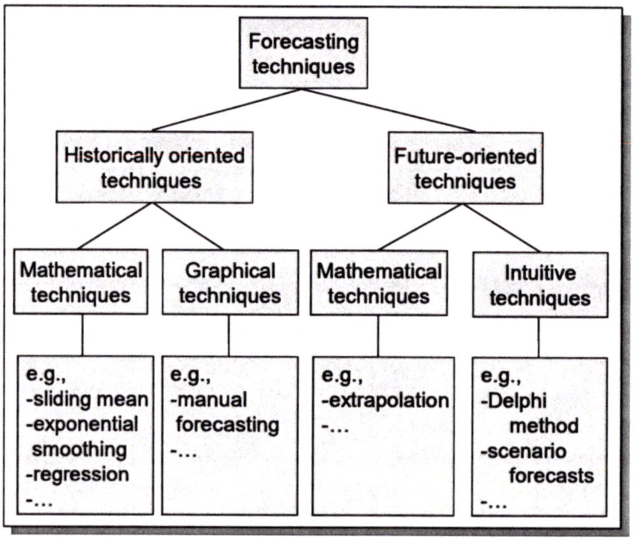

Fig. 10.1.2.1 Breakdown of forecasting techniques.

- *Historically oriented forecasting techniques* predict future demand based on historical data, for example, on consumption statistics. If a forecast can be made only for an item family or a rough-cut item, then the predicted quantity must subsequently be applied to the detailed items with the use of an allocation key. Historically oriented forecasting techniques can be further subdivided into:

- *Mathematical forecasting techniques*, predominant among which is the extrapolation of a time series. Future demand is calculated by extrapolating a series of demands in the past. Such procedures are used widely.
- *Graphical forecasting techniques*, where a time series is represented graphically; a mean course and width of deviation are judged by "eyeballing" and are projected into the future based on past experience.

- *Future-oriented forecasting techniques* take information already at hand about future demand into account, such as bids, firm orders, orders in the concluding phases, or surveys of consumer behavior. Such techniques are further subdivided into:
  - *Mathematical forecasting techniques* attempt to conceive the future behavior of target customers in mathematical formulas. An example is extrapolation. Beginning with confirmed orders, future order volume is calculated empirically.
  - *Intuitive forecasting techniques* attempt to estimate the future behavior of target customers in an intuitive way. These techniques are particularly useful when new or significantly enhanced products are introduced to the market, or if surrounding systems significantly change. Surveys, juries of executive opinion, or estimations are simple intuitive techniques. Relevant information can be provided by the sales department, the sellers, or market research institutes that use surveys to assess customer behavior, or by customers themselves (direct contact). Rather technical methods are expert systems, neural networks, decision support systems (DSS), or other statistics and operations research techniques . As typically intuitive techniques, the Delphi method and the Scenario-based forecasts will be presented.

A *combination* of these techniques is also thinkable. For example, forecasts produced using a mathematical technique may be "eyeballed" for accuracy using a graphical representation.

Another possible subdivision of forecasting techniques is the following (see [APIC16]):

- *Qualitative forecasting techniques* based on intuitive expert opinion and judgment (e.g., manual forecast, Delphi method)
- *Quantitative forecasting techniques* using historical demand data to project future demand; these techniques are further subdivided as follows:
  - *Intrinsic forecasting techniques* are based on internal factors, such as an average of past sales, and are useful for individual product sales.
  - *Extrinsic forecasting techniques* are based on a correlated *leading indicator* (a business activity index that indicates future trends), such as estimating sales of disposable diapers based on birth rates or estimating furniture sales based on housing starts ([APIC16]). Extrinsic forecasts tend to be more useful for large aggregations, such as total company sales.

10.1.3 Principles of Forecasting Techniques with Extrapolation of Time Series and the Definition of Variables

Intended learning outcomes: Present an example of a time series. Explain possible demand models. Produce an overview on statistical methods to determine mean and dispersion. Identify definitions of variables, each calculated at the end of a statistical period.

Particularly for forecasting based on historical data, statistical techniques are used that are based on a series of observations along the time axis (see [BoJe15], [IBM73], or [WhMa97]). The following values are fundamental to the determination of stochastic requirements:

A *time series* is the result of measurement of particular quantifiable variables at set observation intervals equal in length.

The *statistical period* or *observation interval* is a time unit, namely, the period of time between two measurements of the time series (e.g., 1 week, 1 month, 1 quarter).

The *forecast interval* is the time unit for which a forecast is prepared ([APIC16]). This time unit best corresponds to the statistical period.

The *forecast horizon* is the period of time into the future for which a forecast is prepared ([APIC16]). It is generally a whole number multiple of the statistical period.

As an example, Figure 10.1.3.1 shows the frequency distribution[1] of the observed variable "customer order receipts" during the most recent statistical period as a histogram.[2]

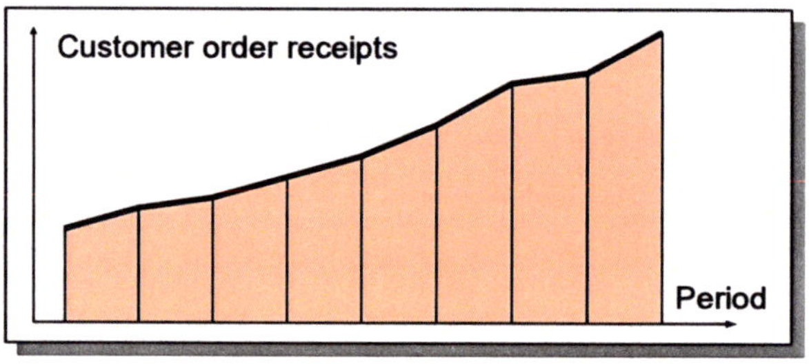

Fig. 10.1.3.1 Example of a time series.

A *demand model* attempts to represent demand by drawing the curve that shows the least scattering of the measured values.

Curve fitting is the process performed to obtain that curve, by means of a straight line, polynomial, or another curve.

---

[1] A *frequency distribution* indicates the frequency with which data fall into each of any number of subdivisions of the variable (cf. [APIC16]).

[2] A *histogram* is a graph of contiguous vertical bars representing a frequency distribution. The subdivisons of the variable are marked on the x axis, and the number of items in each subdivision is indicated on the y axis (cf. [APIC16]).

We assume that the scattering (dispersion) of values is random and, most often, distributed normally. This presupposes that while demand values do indeed have a fluctuating pattern, it is possible to make fairly good approximations. Figure 10.1.3.2 presents some common cases of demand models.

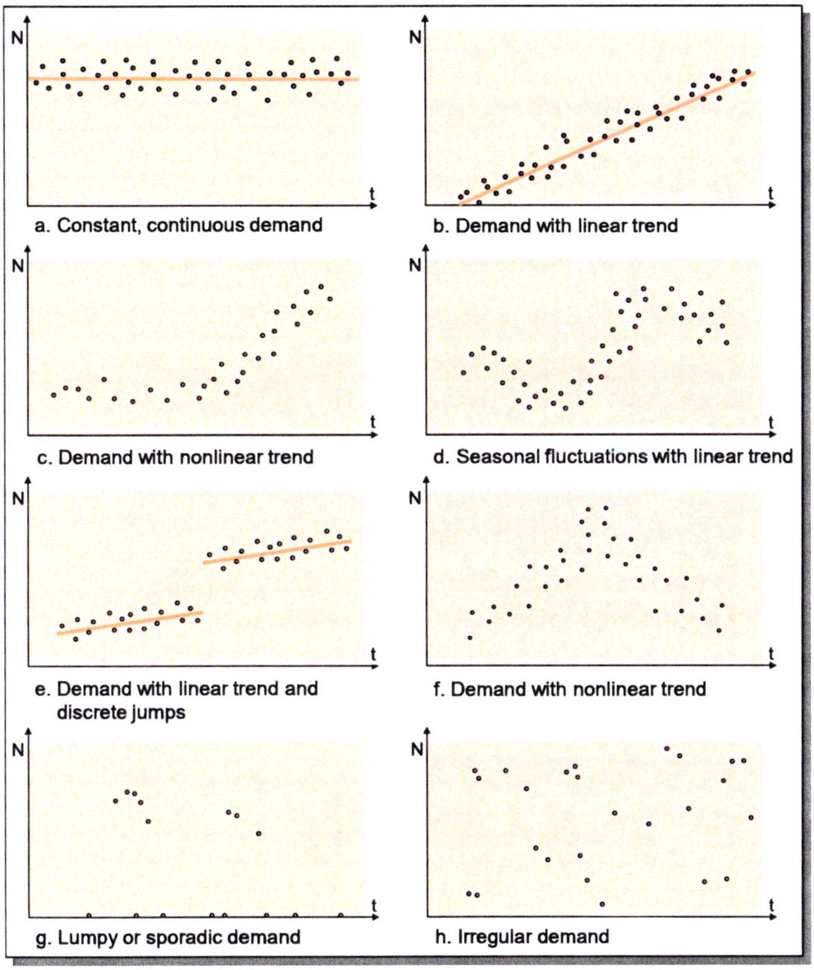

Fig. 10.1.3.2 Possible and common demand models.

Matching a particular demand model to a particular time series leads to a forecasting technique. The forecasting technique is thus based on a concept of the course of demand. This concept forms the basis for the perception of regularity or a regular demand. The model is

- Either an *econometric model*, mostly defined by a set of equations, formulating the interrelation of collected data and variables of the model of the course of the demand as a mathematical regularity,
- Or an *intuitive model* as an expression of the perception of an intuitive regularity.

It is quite possible that for a single time series several models will overlap.

(Statistical) decomposition or *time series analysis* is a breakdown of time series data into various components by analysis; for example, into:

- (Long-term) trend component
- Seasonal component
- Nonseasonal, but (medium-term) cyclical component
- Marketing component (advertising, price changes, etc.)
- Random component (nonquantifiable phenomena), e.g., due to *noise*, that is random variation or a random difference between the observed data and the "real" event.

Mathematical statistics offers various methods for determining the mean, deviation, expected value, and dispersion (scattering)[3] of measured values for a time series. Its ability to reproduce the demand for a demand model accurately depends upon the situation. Figure 10.1.3.3 shows a morphology of possible statistical features and the statistical methods that they characterize.

Degree of freedom		Expressions	
Calculation of dispersion	➡	Extrapolation from the past	Determination of forecast error
Measure of dispersion	➡	Mean square deviation	Absolute deviation
Weighting of historic values	➡	Equally weighted	Exponentially declining

Fig. 10.1.3.3 Statistical methods to determine mean and dispersion.

1. *Calculation of dispersion*. Two basic methods are used:
   - *Extrapolation*, or estimation by calculation of deviations of individual values in the previous statistical periods from the mean, postulated by the demand model.
   - Direct, that is, retrospective *determination of the forecast error* as the difference between actual demand and projected demand according to the demand model.
2. *Measure of dispersion*. There are two standards here:
   - *Mean square deviation*: σ (*sigma*) (i.e., standard deviation)
   - *Mean absolute deviation (MAD)*
3. *Weighting of values*. Most commonly encountered are:
   - *Equal weighting* of all measured values
   - *Exponential weighting* of measured values in the direction of the past

---

[3] An (arithmetic) *mean* is the arithmetic average of a group of values. The *deviation* is the difference between a value and the mean, or between a forecast value and the actual value. An *expected value* is the average value that would be observed in taking an action an infinite number of times. *Dispersion* is the scattering of observations of a frequency distribution around its average (cf. [APIC16]).

In most cases, we only measure *satisfied demand* for all models. This equates consumption with demand. The basic problem with this measurement is that real demand is not taken into account. The customer order receipts mentioned in Figure 10.1.3.1 may have been higher, for example, if a better demand model had resulted in better availability. Strictly speaking, the customer orders that could not be filled should have been measured as well. The problem with this, however, is that the customer orders may be filled at a later time period. At that time, there may be other orders that will then be unfilled, etc. Determining the exact amount of demand in the past by employing a "what would have happened if" method rapidly proves itself redundant; later demand on the time axis is most likely dependent on satisfied demand in the preceding periods on the time axis.

The following sections use the variables defined in Figure 10.1.3.4. The nomenclature was chosen in such a way that the index always shows the point at the end of the statistical period in which a value is calculated. The period to which the value refers is shown in parentheses.

M_t	=	Mean value, calculated at the end of period t
$P_t(t+k)$	=	Prediction (or forecast value) for period t+k, calculated at the end of period t
$\sigma_t(t+k)$	=	Forecast error for period t+k, calculated at the end of period t
N_i	=	Demand in period i, calculated at the end of period i
t	=	Current period or period just ended
n	=	Constant number of periods (the smaller the n is chosen to be, the more quickly the forecast will react to demand fluctuations)
k	=	Distance of a future period from the period just ended

Fig. 10.1.3.4 Definitions of variables, each calculated at the end of a statistical period.

10.2 Historically Oriented Techniques for Constant Demand

Intended learning outcomes: Describe the moving average forecast. Explain the first-order exponential smoothing forecast. Differentiate between the moving average forecast and the first-order exponential smoothing forecast.

In a *forecasting model for constant demand*, planners obtain the forecast value for a future period using a mean from past consumption.

Figure 10.2.0.1 shows the forecast curve resulting from two techniques discussed in the following. The actual events — "damped" or "smoothed"[4] — are projected into the future.

[4] *Smoothing* means the process of averaging data, by a mathematical method, for example.

However, smoothing always lags one statistical period behind, since it is a historically oriented forecast.

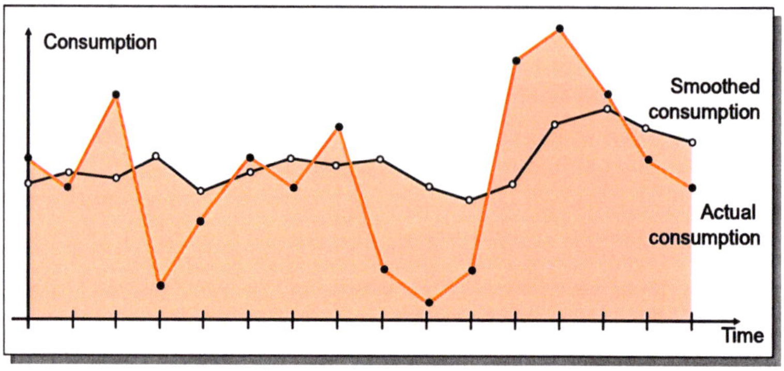

Fig. 10.2.0.1 Smoothing of consumption.

Despite the assumption of constant demand, we should always reckon that demand changes over the course of time. To take this into account, the mean is recalculated at the end of every statistical period, although the characteristic parameter of the mean calculation, that is, the number of the periods in the past included in the calculation or the smoothing constant, is usually kept constant.

10.2.1 Moving Average Forecast

Intended learning outcomes: Explain mean and standard deviation in the moving average forecasting technique. Disclose the average age of the observed values. Present an example of determining the forecast value using moving average.

The *moving average forecasting technique* considers the individual values of a time series as *samples* from the universe, or parent population, of a *sample distribution* with constant parameters and performs periodic recalculations according to the principle of the moving average.

Moving average is the arithmetic average of a certain number (n) of the most recent observations. As each new observation is added, the oldest observation is dropped ([APIC16]).

The technique uses the classic repertoire of mathematical statistics, that is, the mean of a sample and, as a measure of dispersion, the standard deviation.

Figure 10.2.1.1 shows the calculation of *mean* and *standard deviation* in the moving average forecasting technique. The variables are set according to the definitions in Figure 10.1.3.4. The formulas are independent of k; that is, we interpret the determined parameters as the expected value and dispersion of forecast demands. These remain valid for any periods of time in the future.

$$P_t(t+k) = M_t = \frac{1}{n}\sum N_{t-i}$$

$$\sigma_t(t+k) = \sqrt{\frac{1}{n-1}\sum (N_{t-i} - M_t)^2}$$

$$\text{where } 0 \leq i \leq n-1, 1 \leq k \leq \infty$$

Fig. 10.2.1.1 Mean and standard deviation in the *moving average forecasting* technique.

The average age of the values included in the calculation is shown in Figure 10.2.1.2. Thereby, the age of N_{t-i} is i for $0 \leq i \leq n-1$.

$$\bar{n} = \frac{1}{n}(0+1+...+(n-1)) = \frac{n-1}{2}$$

Fig. 10.2.1.2 Average age of the observed values.

The larger the value chosen for n, the more exact the mean becomes, but because the moving average reacts more slowly to alterations in demand, so does the forecast; n should be set so that a rapid adaptation to systematic changes is possible, without causing a significant reaction to a purely random variation in demand. See also Section 10.2.3. Figure 10.2.1.3 shows an example of moving average calculation that includes nine periods in the past.

The calculation formulas and results are valid independent of the underlying consumption distribution, although a particular distribution is assumed for implementation. Forecast calculations often assume a normal distribution as probability distribution. We discuss this assumption in Section 10.5.2.

A *probability distribution* is a table of numbers, or a mathematical expression, that indicates the frequency with which each event out of a totality of events occurs. The mathematical *probability* is a number between 0 and 1 that expresses this frequency as a fraction of all occurring events.

The statement in the last column of Figure 10.2.1.3, that the demand value N_t has a 95.4% probability within the confidence interval "forecast value (= mean) ± 2 * forecast error (= standard deviation)," is only valid in a normal distribution.

10.2.2 First-Order Exponential Smoothing Forecast

Intended learning outcomes: Identify the weighted mean as well as exponential demand weighting. Explain first-order exponential smoothing: mean, and MAD. Disclose the average age of the observed values. Describe how the smoothing constant α determines the weighting of the past. Present an example of first-order exponential smoothing.

Period t	Forecast value $P_{t-1}(t)$	Actual demand N_t	$\sum_i (N_{t-1-i} - M_{t-1})^2$, $0 \le i \le n-1$	Forecast error $\sigma_{t-1}(t)$	Confidence interval 95.44% $I_{t-1}(t)$
1		104			
2		72			
3		110			
4		108			
5	Ø=91	70	$\sum$=3036		
6		86			
7		85			
8		66			
9		118			
10	91	115	3036	19.48	52 – 130
11	92	85	3430	20.71	50 – 134
12	94	105	3055	19.54	55 – 133
13	93	90	2913	19.08	55 – 131
14	91	75	2665	18.25	54 – 128
15	92	130	2477	17.60	57 – 127
16	97	—	3700	21.51	54 – 140

Sample calculation:

$$P_{15}(16) = \frac{85+66+118+115+85+105+90+75+130}{9} = \frac{869}{9} = 96{,}6 \approx 97$$

$$\sigma_{15}(16) = \sqrt{\frac{(85-97)^2 + (66-97)^2 + \ldots\ldots + (130-97)^2}{8}}$$

$$= \sqrt{\frac{3700}{8}} = 21{,}51$$

$$I_{15}(16) = P_{15}(16) \pm 2\sigma_{15}(16) = \langle \frac{54}{140}$$

Fig. 10.2.1.3 Example: determining the forecast value using moving average (n = 9).

If we wish to adapt the forecasting technique to actual demand, the demand values for the last periods must be weighted more heavily, according to the principle of the weighted moving average. The formula in Figure 10.2.2.1 takes this weighting into account; the variables were chosen according to the definitions in Figure 10.1.3.4 and include an indefinite number of periods. $G_{t\text{-}i}$ always expresses the weighting of demand in the period (t–i). [5]

$$M_t = \frac{\sum G_t \cdot N_{t-i}}{\sum G_t} \qquad 0 \le i \le \infty$$

Fig. 10.2.2.1 Weighted mean.

[5] *Weighted moving average* is an averaging technique in which the data are given values according to their importance (cf. [APIC16]).

In the *first-order exponential smoothing forecast technique*, or *single (exponential) smoothing*, the weights are in an exponentially declining relationship and adhere to the definitions in Figure 10.2.2.2.

$$G_y = \alpha \cdot (1-\alpha)^y$$

where

y = age of the period, $0 \le y \le \infty$ (whole number)

G_y = weight of the period demand with age y

α = smoothing factor, $0 < \alpha < 1$

$$\sum_y G_y = \frac{\alpha}{1-(1-\alpha)} = 1, \quad 0 \le y \le \infty$$

Fig. 10.2.2.2 Exponential demand weighting.

Figure 10.2.2.3 shows the calculation of *Mean smoothed consumption* as measure of mean, and *Mean absolute deviation (MAD)* as measure of dispersion. See also the definitions of indexes and variables in Figure 10.1.3.4.

$$M_t = P_t(t+k) = \alpha(1-\alpha)^0 \cdot N_t + \underbrace{\alpha(1-\alpha)^1 \cdot N_{t-1} + \alpha(1-\alpha)^2 \cdot N_{t-2} + \ldots}_{(1-\alpha)\cdot M_{t-1}}$$

$$= \alpha \cdot N_t + (1-\alpha) \cdot M_{t-1}$$

$$MAD_t(t+k) = \alpha(1-\alpha)^0 |N_t - M_{t-1}| + \underbrace{\alpha(1-\alpha)^1 |N_{t-1} - M_{t-2}| + \alpha(1-\alpha)^2 |N_{t-2} - M_{t-3}| + \ldots}_{(1-\alpha)\cdot MAD_{t-1}(t)}$$

$$= \alpha \cdot |N_t - M_{t-1}| + (1-\alpha) \cdot MAD_{t-1}(t)$$

$$\sigma_t(t+k) \approx 1.25 \cdot MAD_t(t+k) \quad \text{where } 1 \le k \le \infty$$

Fig. 10.2.2.3 First-order exponential smoothing: mean, MAD, and standard deviation.

Since the weighting G_y follows a geometric series, the recursive calculation indicated in the formulas is self-evident. These formulas allow us to perform the same calculation as in moving average using only the past values for mean and MAD and the demand value for the current period instead of many demand values. With a normal distribution, standard deviation and mean absolute deviation (MAD) stand in the same relationship as that given in Figure 10.2.2.3.

The recursion to M_{t-1} results by factoring out (1-α) of the part of the formula that is emphasized by the horizontally cambered bracket. Factual equality between σ and MAD*1.25 requires n>30 or α < 6.5%. Figure 10.2.2.4 shows the average age of the observed values. The age of N_{t-i} is i for $0 \le i \le n-1$.

$$
\begin{aligned}
\bar{n} &= 0 \cdot \alpha(1-\alpha)^0 + 1 \cdot \alpha(1-\alpha)^1 + 2 \cdot \alpha(1-\alpha)^2 + \ldots \\
&= \sum y \cdot \alpha \cdot (1-\alpha)^y, 0 \le y \le \infty \\
&= \frac{(1-\alpha)}{\alpha}
\end{aligned}
$$

Fig. 10.2.2.4 Average age of the observed values.

The choice of *smoothing constant α* or *alpha factor* determines the weighting of current and past demand according to the formula in Figure 10.2.2.3.

Figure 10.2.2.5 shows the effect of α = 0.1, a value often chosen for well-established products, and α = 0.5 for products at the beginning or the end of their life cycles.

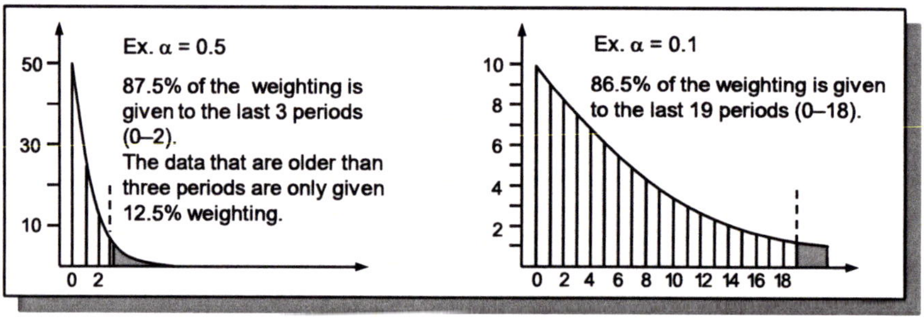

Fig. 10.2.2.5 The smoothing constant α determines the weighting of the past.

Figure 10.2.2.6 shows the behavior of the forecast curve with various values of the smoothing constant α. A high smoothing constant results in a rapid but also nervous reaction to changes in demand behavior. See also Sections 10.2.3 and 10.5.1.

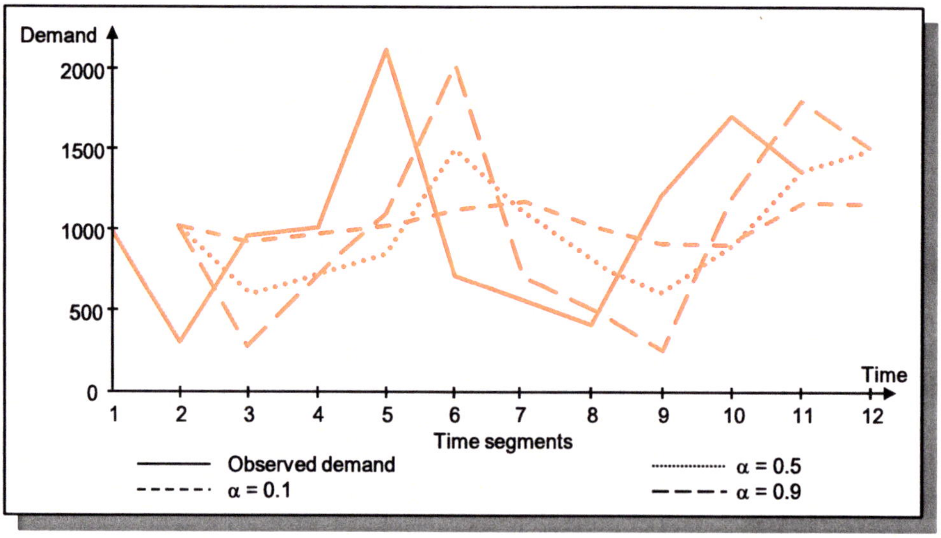

Fig. 10.2.2.6 Forecasts with various values of the smoothing constant α.

Using exponential smoothing techniques, we can determine the uncertainty of a forecast by extrapolating the forecast error. To do this, we calculate the mean absolute deviation (MAD). Figure 10.2.2.7 is an example of exponential smoothing with smoothing constant $\alpha = 0.2$. It was chosen in a way similar to the example of moving average calculation in Figure 10.2.1.4.

Period	Forecast value $P_{t-1}(t)$	Actual demand N_t	Deviation $N_t - P_{t-1}(t)$	Forecast error $MAD_{t-1}(t)$	Confidence interval 95.44% $I_{t-1}(t)$
⋮					
10	91	115	24	17	48 – 134
11	96	85	-11	18	51 – 141
12	94	105	11	17	51 – 137
13	96	90	-6	16	56 – 136
14	95	75	-20	14	60 – 130
15	91	130	39	15	53 – 129
16	99	70	-29	20	49 – 149
17	93	100	7	22	38 – 148
18	94	95	1	19	46 – 142
19	94	120	26	15	56 – 132
20	99	—	—	17	56 – 142

Sample calculation:

$$P_{14}(15) = P_{13}(14) + 0.2 \cdot (N_{14} - P_{13}(14)) = 95 + 0.2 \cdot (-20) = 91$$

$$MAD_{14}(15) = MAD_{13}(14) + 0.2 \cdot (|N_{14} - P_{13}(14)| - MAD_{13}(14)) = 14 + 0.2 \cdot 6 = 15$$

Fig. 10.2.2.7 First-order exponential smoothing with smoothing constant $\alpha = 0.2$.

10.2.3 Moving Average Forecast versus First-Order Exponential Smoothing Forecast

Intended learning outcomes: Disclose formulas for the relationship between α and n. Present the relationship between α and n. Present an example of linear regression.

The results of *moving average* and *first-order exponential smoothing* are comparable, to the extent that the mean age of the observed values corresponds mutually. Figure 10.2.3.1 shows the relationship between the number of observed values and the smoothing constant α.

Figure 10.2.3.2 shows the same relationship between α and n, using a tabular comparison of individual values.

$$\frac{1-\alpha}{\alpha}=\bar{n}=\frac{n-1}{2}$$
$$\alpha=\frac{2}{n+1}$$
$$n=\frac{2-\alpha}{\alpha}$$

Fig. 10.2.3.1 Formulas for the relationship between α and n.

Number of periods n	Smoothing constant α	Reactivity	Adaptation to systematic changes
3 4 5 6 9 12 19 39	0.50 0.40 0.33 0.29 0.20 0.15 0.10 0.05	rapid, nervous reaction ↕ Leveling reaction	rapid ↕ slow

Fig. 10.2.3.2 Relationship between α and n in tabular form.

10.3 Historically Oriented Techniques with Trend-Shaped Behavior (*)

Intended learning outcomes: Explain the regression analysis forecast and the second-order exponential smoothing forecast. Describe the Trigg and Leach adaptive smoothing technique. Produce an overview on seasonality.

Forecast values produced by techniques for a constant demand do not reflect actual demand in cases where the demand follows a trend. For this reason, a number of trend forecasting techniques have been developed.

A *trend forecasting model* takes into account stable trends in demand. [6]

In Figure 10.3.0.1, all demand values fluctuate within the confidence limit around the calculated mean. Nevertheless, there is a systematic error (δ_v) in extrapolation of the mean. Regression analysis shows a rising demand trend. We can avoid the systematic error by extrapolating the regression lines.

To detect a trend in advance, we could, for example, tighten the control limits, (+/– 1 * standard deviation). As soon as the limits have been exceeded a particular number of times, a correction is made.

---

6 A *trend* is a general upward or downward movement of a variable over time.

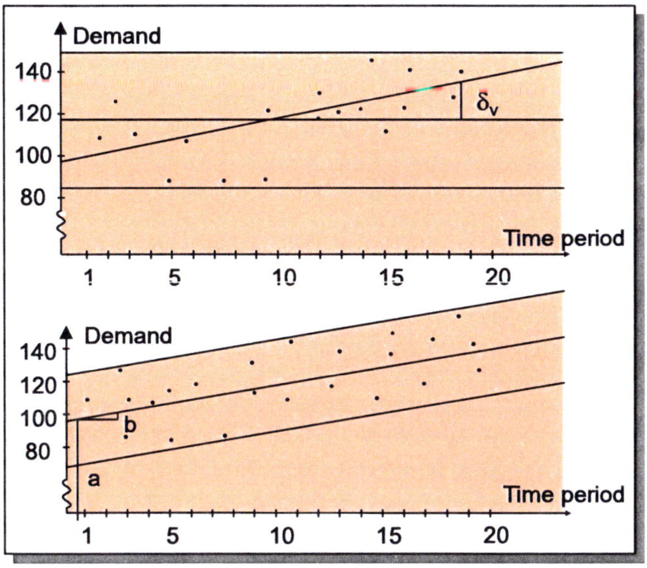

Fig. 10.3.0.1 Demand with linear trend: comparison of extrapolation of the mean with that of regression.

10.3.1 Regression Analysis Forecast

Intended learning outcomes: Explain mean, standard deviation, and forecast error in linear regression.

Regression analysis, or *linear regression*, is often described as trend analysis. It is based on the assumption that demand values appear as a function of time, such as a linear function.

This means that a number of points represented on the *x–y* plane can be approximated by a line. Figure 10.3.0.1 shows demand as a function of time period. Given a *y*-axis value of a and a slope of b, we can determine the mean line (regression line) sought between the two pairs of values. Figure 10.3.1.1 provides the formulas for determining this, along with the values a and b. To perform the calculation, we need to know the values for at least n periods preceding time t. See also the definitions of indexes and variables in Figure 10.1.3.4. The derivation of the formulas is taken from [Gahs71], p. 67 ff.

Because of uncertainty in the determination of a and b, the forecast error is larger than the standard deviation, as shown in Figure 10.3.1.1. The term 1/n in the formula for forecast error represents the uncertainty in determining a, while the other term represents slope b. The influence of the slope b increases with increased forecast distance k. In this situation, therefore, we determine the forecast error by extrapolation of the deviations of individual values from the past value of the regression curve. Figure 10.3.1.2 shows a sample calculation of linear regression with n = 14.

$$
\begin{aligned}
P_t(t+k) &= a_t + b_t \cdot (n+k) \\
a_t &= \frac{1}{n}\sum_i N_{t-i} - b_t \frac{n+1}{2} \\
b_t &= \frac{12 \cdot \sum_i ((n-i) \cdot N_{t-i}) - 6(n+1)\sum_i N_{t-i}}{n(n^2-1)} \\
s_t &= \sqrt{\frac{1}{n-2}\sum_i (N_{t-i} - P_t(t-i))^2} \\
\sigma_t(t+k) &= s_t \cdot \sqrt{1 + \frac{1}{n} + \frac{3 \cdot (n+2k-1)^2}{n(n^2-1)}}
\end{aligned}
$$

where $0 \le i \le n-1$ and $1 \le k \le \infty$

Fig. 10.3.1.1 Mean, standard deviation, and forecast error in linear regression.

Period i	N_t	a_{t-1}	b_{t-1}	$P_{t-1}(t)$	s_{t-1}	$\sigma_{t-1}(t)$
1	110					
2	120					
3	100					
4	85					
5	100					
6	120					
7	90					
8	130					
9	120					
10	90					
11	140					
12	120					
13	135					
14	125					
15	150	98.1319	2.011	128.2969	24.58	32.292
16	130	89.945	3.4835	142.1975	35.1365	46.3345
17	110	90.6588	3.4835	142.911	34.7262	45.7935
18	140	96.3165	2.7363	137.363	29.8761	39.3977
19	130	104.1208	2.3077	138.7363	26.7943	35.3336
20	150	109.7249	1.8462	137.4179	24.4796	32.2813

Sample calculation: estimated values for period 19 (in period 18)

1st step: $\sum N_{t-i} = 1700$, $\sum((n-i) \cdot N_{t-i}) = 13275$

2nd step: $b_{18} = \frac{12 \cdot 13275 - 6 \cdot 15 \cdot 1700}{14(14^2-1)} = 2.3077$

3rd step: $a_{18} = \frac{1}{14} \cdot 1700 - 2.3077 \cdot \frac{14+1}{2} = 104.1208$

4th step: $P_{18}(19) = 104.1208 + 2.3077 \cdot (14+1) = 138.7363$

Fig. 10.3.1.2 Linear regression: sample calculation with n = 14.

10.3.2 Second-Order Exponential Smoothing Forecast

Intended learning outcomes: Disclose the determination of trend lines in second-order exponential smoothing. Explain the formulas for calculation of the trend line and forecast error in second-order exponential smoothing. Present an example of determination of forecast value using second-order exponential smoothing.

Second-order exponential smoothing forecast technique extends first-order exponential smoothing to create a technique capable of capturing linear trend.

Second-order exponential smoothing starts out from:

1. The mean, calculated using first-order smoothing
2. The mean of this first-order means, calculated according to the same recursion formula

These two means are the estimated values for two points on the trend line. Figure 10.3.2.1 shows an overview of this technique, which is elaborated in the following discussion. The exact derivations can be found in [Gahs71], p. 60 ff., and in [Lewa80], p. 66 ff.

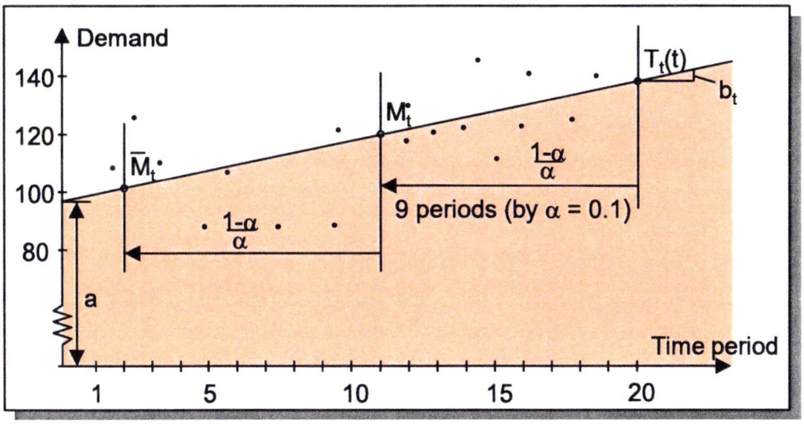

Fig. 10.3.2.1 Determination of trend lines in second-order exponential smoothing.

Figure 10.3.2.2 shows the formulas necessary for calculating the trend line; this gives us the second-order forecast value for subsequent periods as well as the corresponding forecast error. See also the definitions in Figure 10.1.3.4.

The following numbered explanations correspond to those presented in Figure 10.3.2.2:

1 The previous formula to determine first-order mean.

2 The new formula to determine the second-order mean, as the mean of the first-order means. The second-order mean lies at the same distance from the first-order mean as does the latter from the current period.

$$1:\quad M_t = \alpha \cdot N_t + (1-\alpha) \cdot M_{t-1} \qquad \text{mean age}: \frac{1-\alpha}{\alpha} \text{ periods before t}$$

$$2:\quad \overline{M_t} = \alpha \cdot M_t + (1-\alpha) \cdot \overline{M_{t-1}} \qquad \text{mean age}: 2 \cdot \frac{1-\alpha}{\alpha} \text{ periods before t}$$

$$3:\quad b_t = \frac{M_t - \overline{M_t}}{1-\alpha / \alpha} = \frac{\alpha}{1-\alpha} \cdot \left(M_t - \overline{M_t}\right)$$

$$4:\quad T_t = \overline{M_t} + 2 \cdot \left(M_t - \overline{M_t}\right) = 2M_t - \overline{M_t}$$

$$5:\quad P_t(t+k) = 2 \cdot M_t - \overline{M_t} + b_t \cdot k, \qquad 1 \le k \le \infty$$

$$6:\quad MAD_t(t+1) = \alpha \cdot \left|N_t - P_{t-1}(t)\right| + (1-\alpha) \cdot MAD_{t-1}(t)$$

$$7:\quad a, b, T_t = a + b \cdot t \qquad \text{(calculate from linear regression)}$$

$$M_t = T_t - b \cdot \frac{1-\alpha}{\alpha}$$

$$\overline{M_t} = T_t - 2 \cdot b \cdot \frac{1-\alpha}{\alpha}$$

Fig. 10.3.2.2 Trend line and forecast error in second-order exponential smoothing.

3 Slope of the trend line to time t, when two means are given.

4 Starting value T_t for the forecast at time t.

5 Forecast for subsequent periods.

6 Forecast error for the next period t + 1. Because a linear trend entails that the forecast error is dependent on k, the same formula does not automatically hold for period t + k, although it is often used.

7 The determination of the starting value that can be calculated, for example, by means of regression analysis.

Figure 10.3.2.3 provides an example of the determination of the forecast value using second-order exponential smoothing for the smoothing constant $\alpha = 0.2$. We calculated the same demand value as the one in linear regression for the first 14 periods in order to obtain the same starting values.

Period t	Actual demand N_t	First-order mean M_t	Second-order mean $\overline{M}_t$	Slope of trend line b_t	Trend line value $T_t(t)$	Second-order forecast value $P_{t+1}(t)$
1	110					
2	120					
3	100					
4	85					
5	100					
6	120					
7	90					
8	130					
9	120					
10	90					
11	140					
12	120					
13	135					
14	125	118.3	110.2			
15	150	124.6	113.1	2.9	136.1	128.3
16	130	125.7	115.6	2.5	135.8	139.0
17	110	122.6	117.0	1.3	128.2	138.3
18	140	126.0	118.8	1.8	133.2	129.5
19	130	126.8	120.4	1.6	133.2	135.0
20	150	131.4	122.6	2.2	140.2	134.8
21						142.4

Calculation of the beginning value for period 14 using regression analysis:

$$b = \frac{12*12345 - 6*15*1535}{2730} = 2.01$$

$$a = \frac{1}{14} * 1585 - 2.01 * \frac{15}{2} = 98.14$$

$$T_{14} = 98.14 + 2.01 * 14 = 126.3$$

$$M_{14} = 126.3 - 2.01 * \frac{1 - 0.2}{0.2} = 118.3$$

$$\overline{M}_{14} = 126 - 2 * 2.01 * \frac{1 - 0.2}{0.2} = 110.2$$

Sample calculation:

1st step:	M_{18}	= 122.6 + 0.2 * (140 – 122.6)	= 126.0
2nd step:	$\overline{M}_{18}$	= 117.0 + 0.2 * (126.0 – 117.09)	= 118.8
3rd step:	b_{18}	=	= 1.8
4th step:	T_{18}	= 2* 126.0 – 118.8	= 133.2
5th step:	$P_{18}(19)$	= 133.2 + 1.8	= 135.0

Fig. 10.3.2.3 Determination of forecast value using second-order exponential smoothing ($\alpha = 0.2$).

10.3.3 Trigg and Leach Adaptive Smoothing Technique

Intended learning outcomes: Identify forecast errors and their exponential weighting (mean deviation). Explain the tracking signal following Trigg and Leach. Describe the determination of the smoothing constant in first-order exponential smoothing.

Adaptive smoothing is a form of exponential smoothing in which the smoothing constant is automatically adjusted as a function of forecast error measurement.

A good forecasting technique is not biased:

A *(forecast) bias* is a consistent deviation of the actual demand from the forecast in one direction, either high or low.

If forecast values exceed the control limits of, for example, +/– the standard deviation from the mean several consecutive times, we must alter either the parameters or the model. Trigg and Leach ([TrLe67]) suggest the following method for continuous adjustment of the exponential smoothing parameter:

The *smoothing constant* γ or *gamma factor* smoothes forecast errors exponentially according to the formula in Figure 10.3.3.1.

$$MD_t(t) = \gamma \cdot (N_t - P_{t-1}(t)) + (1-\gamma) \cdot MD_{t-1}(t-1) \qquad 0 \le \gamma \le 1$$

Fig. 10.3.3.1 Forecast errors and exponential weighting (mean deviation).

A mean calculated in this way is also referred to as *mean deviation*.

The formula in Figure 10.3.3.2 defines the *tracking signal* and its standard deviation.

$$TS_t = \frac{MD_t}{MAD_t}$$

$$\sigma(TS_t) = \frac{\sigma(MD_t)}{MAD_t} = 1.25 \cdot \sqrt{\frac{\gamma}{2-\gamma}}$$

Fig. 10.3.3.2 Tracking signal following Trigg and Leach.

Lewandowski shows the nontrivial result of the standard deviation ([Lewa80], p. 128 ff.). According to that source, the deviation signal is a nondimensional, randomly distributed variable with a mean of 0 and the standard deviation described above. Because of the manner of its calculation, the absolute value of the deviation signal is always ≤ 1.

Trigg and Leach also developed forecasting techniques that use the deviation signal to adjust the smoothing constant α automatically. Particularly when the mean of the process to be measured changes, a large deviation signal results. In that case, we should choose a relatively large smoothing constant α, so that the mean adjusts rapidly.

In first-order exponential smoothing, it is reasonable to choose a smoothing constant that is equal to the absolute value of the deviation signal, as in Figure 10.3.3.3. The result is a forecast formula with the variable smoothing constant α_t. The factor γ used to smooth forecast errors remains constant and is kept relatively small, between 0.05 and 0.1 for example. This forecasting technique is not only adaptive but also simple from a technical calculation standpoint.

$$\alpha_t = |TS_t|$$

Fig. 10.3.3.3 Determination of the smoothing constant in first-order exponential smoothing.

10.3.4 Seasonality

Intended learning outcomes: Identify the seasonal index S_f. Explain forecasting that considers seasonality. Differentiate between additive seasonality and multiplicative seasonality.

Seasonal fluctuations in the demand for specific items are brought about by factors such as weather, holidays, and vacation periods. Restaurants and theaters experience weekly and even daily "seasonal" variations.

The best way to forecast and take seasonality into account is to compare the pattern of demand over multiple years.

We speak of *seasonality* or a *seasonal demand pattern* when the following conditions hold:

1. Growth in demand occurs in the same time frame for every seasonal cycle.
2. Seasonal fluctuations are measurably larger than the random demand fluctuations.
3. A cause that explains demand fluctuations can be found.

Seasonality does not always have a yearly pattern. In the retail trade, particularly in the grocery industry, there is a commonly observed effect at the end of each month when people receive their monthly salary payments.

Figure 10.3.4.1 shows the definition of the *seasonal index*, which is necessary to accommodate seasonal effects.[7]

$$\begin{aligned} &SC = \text{Length of the seasonal cycle} \\ &S_f = \text{Seasonal index}, 0 \le f \le (SZ-1), f = (t+k)_{\text{mod}\,SC} \end{aligned}$$

Fig. 10.3.4.1 Seasonal index S_f.

The term *base series* stands for the succession of the f seasonal indices. Their average value will be 1.0.

Figure 10.3.4.2 shows the two basic models that superimpose the base series upon the trend in demand (that is, without respecting seasonality) for an item in question. *Additive seasonality* refers to an influence independent of the level of sales, whereas *multiplicative seasonality* refers to an influence that increases with the mean of sales.

$$\begin{aligned} &\text{additive:} && P_t(t+k) = M_t + S_f \\ &\text{multiplicative:} && P_t(t+k) = M_t \cdot S_f \end{aligned}$$

Fig. 10.3.4.2 Forecasting that takes seasonality into account.

Figures 10.3.4.3 and 10.3.4.4 provide qualitative examples of demand adjusted for additive and multiplicative seasonality, respectively.

7 The operation "mod z" upon a number x calculates the remainder when x is divided by z.

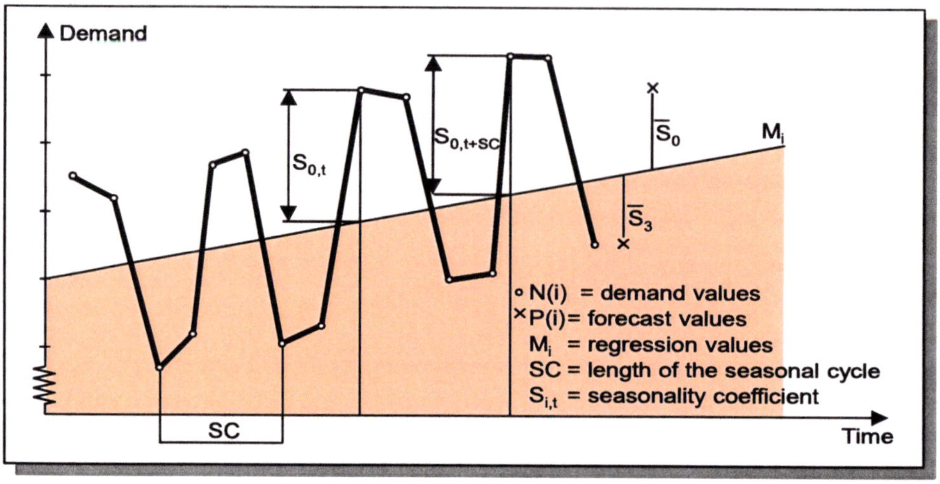

Fig. 10.3.4.3 "Additive seasonality" formulation.

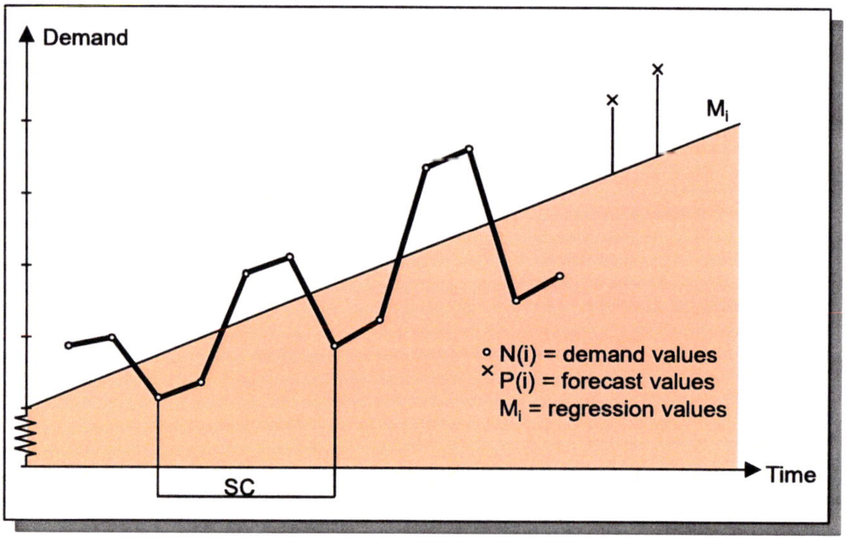

Fig. 10.3.4.4 "Multiplicative seasonality" formulation.

Various techniques that account for seasonal influences can be found in the literature. As an example see [GaKe89]. The following is an example of a simplified procedure:

1. Calculate the seasonal mean.
2. Calculate the trend line from the seasonal means.
3. Determine the base series or the succession of seasonal indices as the average deviation of demand from the trend lines for mutually corresponding periods.
4. Calculate the forecast value from the trend lines and the seasonality coefficient for the corresponding periods in the seasonal cycle.

10.4 Future-Oriented Techniques

Intended learning outcomes: Explain the trend extrapolation forecast and the Delphi method. Describe scenario forecasts.

During the various phases of the product life cycle, different forecasting techniques are used.

Life-cycle analysis is a quantitative forecasting technique based on applying past patterns of demand data covering introduction, growth, maturity, saturation, and decline of similar products to a new product family ([APIC16]).

For the phases of introduction and decline, particularly future-oriented forecasting techniques are used, both quantitative and qualitative. Here, the Delphi method is a well known technique. In its principle, it is a rather qualitative technique.

An additional area for the use of future-oriented forecasting techniques is the perception of a possibly disruptive change in the macro environment of the company. As an example, scenaario-based forecasts can be used.

10.4.1 Trend Extrapolation Forecast

Intended learning outcomes: Identify base demand $B_0(k)$ for period $k>0$ known at time $t=0$. Explain the calculation of the quotient "actual demand in period $t+k$" divided by "base demand known for period $t+k$ at the end of period t", $k>1$. Describe smoothing of quotient means for extrapolation leading to extrapolated forecast values for forecast distance k.

A trend extrapolation forecast attempts to estimate a variable in the future based on the same variable as known at a specific point in time.

In materials management, it may happen that the demand known at a particular point in time t encompasses only a portion of the demand needed for the coming period. Figure 10.4.1.1 provides an example.

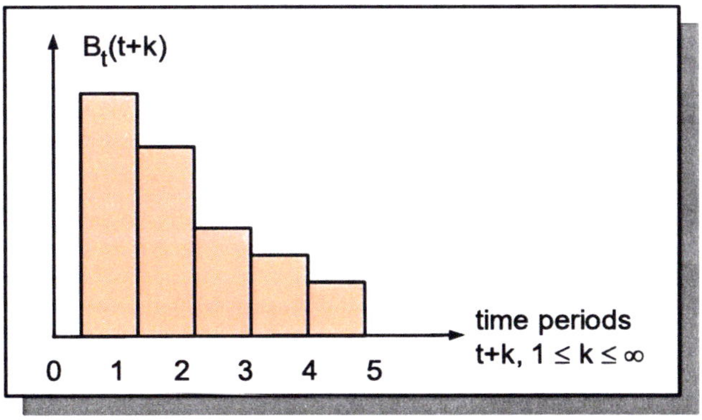

Fig. 10.4.1.1 (Base) Demand $B_0(k)$ for period $k>0$ known at time $t=0$..

Extrapolation calculates the total anticipated demand from the demand already known for a product or product family. It compares the *base demand* $B_t(t+k)$, $1 \leq k \leq \infty$, known at time t, to the demand N_{t+k} observed after the closing of a delivery period t+k. This is shown in Figure 10.4.1.2. The variables for the calculation are chosen either as defined in Figure 10.1.3.4 or in a similar fashion. k stands for the *forecast distance*.

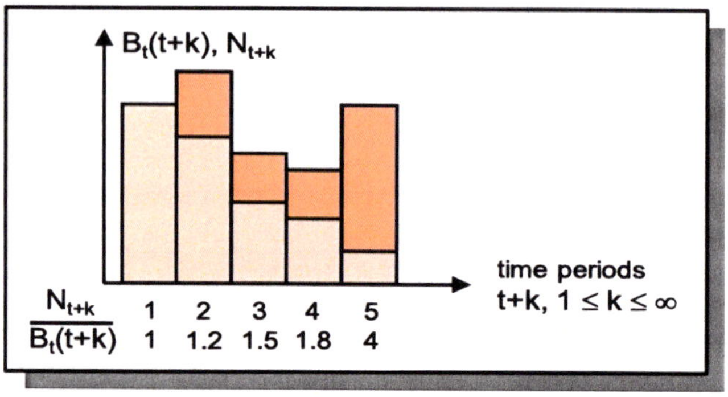

Fig. 10.4.1.2 Actual demand N_{t+k}, divided by base demand $B_t(t+k)$.

This quotient, $\lambda_t(k) = N_{t+k} / B_t(t+k)$, $1 \leq k \leq t$ is called *the extrapolation constant*. The dilemma of this definition? Not until the end of period t+k can we determine the *actual* value of the extrapolation constant, namely $\lambda_{t+k}(0) = N_t / B_{t-k}(t)$, that we had to estimate k periods ago, that is, as $\lambda_t(k)$. Hence, the idea is to smooth the quotients over several periods using exponential smoothing. From now on, let $\lambda_t(k)$, $1 \leq k \leq t$, be the mean after period t for forecast distance k. The previous mean is used to calculate the new mean using exponential smoothing with smoothing constant α according to the formula in Figure 10.4.1.3.

$$\lambda_t(k) = \alpha \left[\frac{N_t}{B_{t-k}(t)} \right] + (1-\alpha)\,\lambda_{t-1}(k), \quad \text{where } 1 \leq k \leq t,\ \ 1 \leq t \leq \infty$$

Fig. 10.4.1.3 Smoothing of quotient means for extrapolation.

The extrapolation constant is defined for every forecast distance and can be used to extrapolate total demand, at the moment not completely known, from the base demand. Figure 10.4.1.4 gives the forecast value $P_t(t+k)$ for the forecast distance k at the end of period t.

$$P_t(t+k) = B_t(t+k)\,\lambda_t(k) \quad \text{where } 1 \leq k \leq t,\ \ 1 \leq t \leq \infty$$

Fig. 10.4.1.4 Extrapolated forecast values for forecast distance k.

The technique described here assumes that the customers' basic order behavior does not change on the time axis or that it does so very slowly. This means that from a change in customer orders on hand, we can infer a proportional change in total demand. Since this assumption is often invalid in the average case, the technique will yield useful results only when used in combination with other forecasting techniques, such as intuitive ones.

The planner can use this same technique to forecast seasonal components. In the grocery industry, for example, the retailer must give orders to the producers early enough to ensure that shipments arrive on time. Assuming that the retailers' order behavior does not change significantly from year to year, the producer can derive standardized quotients from sales over multiple years; the probable total demand for the season in a future year can be extrapolated from the demand already known at a specific point in time.

10.4.2 The Delphi Method

Intended learning outcomes: Explain the Delphi forecasting method as a series of successive surveys that increase consensus.

In the *Delphi method forecast* (the name refers to the oracle at Delphi in antiquity), "expert opinion" is gathered through several structured anonymous rounds of written interviews.

The method generally proceeds in various iterations. Figure 10.4.2.1 shows the desired progression during the successive rounds of questioning.

The mean of the answers shifts in a specific direction. At the same time, when the dispersion of the answers narrows, there is an increase in the consensus about the direction taken. To arrive at this result, a single iteration should include the following steps:

- The questionnaire is meaningfully constructed or altered. The questionnaires are distributed and completed once again.
- The answers are statistically evaluated by determining mean and dispersion. The results of the evaluation are sent to the experts.
- All the experts are asked to defend their views against extreme arguments. Those who, thereupon, change their opinion must provide justifications. The "extreme" respondents must either support their theses with arguments or abandon them.

As for all surveys, the problem for the Deplphi method also lies in formulating the right questions, quantifying the answers, and identifying extreme responses.

The experts are chosen from various areas of an organization, including sales and marketing. They are selected for their competence in the field and their broad vision, not for their hierarchical position within the company. The composition of the group should remain anonymous so that the experts cannot identify and be influenced by the responses of other individuals.

10.4.3 Scenario Forecasts

Intended learning outcomes: Describe scenario forecasts and scenario planning along a time axis. Present the procedure for scenario planning.

Scenario forecasts (or *Scenario-based forecasts*) are plans for how an organization will respond to anticipated future situations ([APIC16]).

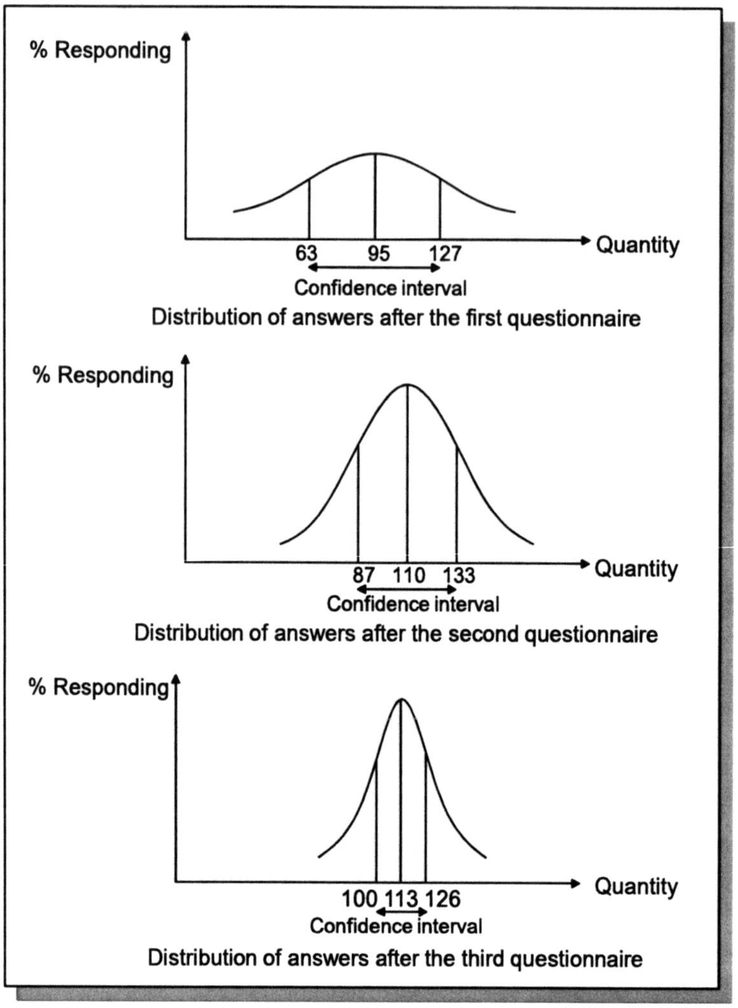

Fig. 10.4.2.1 Delphi forecasting method: increasing consensus.

Scenario planning is a planning process that identifies critical events before they occur and uses this knowledge to determine effective alternatives ([APIC16]).

A *scenario driver* is a key factor or key parameter in determining how the future environment that the organization works in will look.

Scenario planning and the scenario forecasts that the planning leads to can be used as tools for use when dealing with situations where the *long-term* mechanisms of action are either unknown or not fully understood. This applies in particular where various influencing factors could play a role in a company's surrounding systems — especially ones that might not even be considered at first glance. In this respect, scenario planning consciously assumes that the the past is not necessarily an accurate predictor for the future. Scenario planning forms part of analyzing the macro environment, which means it is part of the first stage of the strategic process of designing the supply chain in Figure 2.1.0.1.

Several alternative scenarios explore the way that social, technical, (macro) economic, environmental or political trends may develop over time, and identify their drivers. There may be several drivers in each scenario, and each driver can also influence other drivers. For scientific discussion about scenario planning, see for example [Scho93]. Figure 10.4.3.1 shows the principle of scenario planning along a time axis.

S_i represents the scenario i, $i \geq 1$. The scenario drivers originate from a specific context in one or more of the company's surrounding systems. Individual drivers may also feature in several different scenarios. Scenarios may overlap for this reason, and also for other reasons. A funnel-shaped representation for each scenario shows its postulated development along the time axis. S_i', S_i", ... represent a revision of scenario S_i at time point t', t", ... This revision is based on a reassessment of the effective development of the scenario between the previous assessment and the assessment at the time of the revision. New scenarios can be added at any time (e.g. S_4 at time point t', S_5 at time point t"). At any stage, it may be decided that scenarios are no longer applicable (e.g. S_1 at time point t", indicated by Ω).

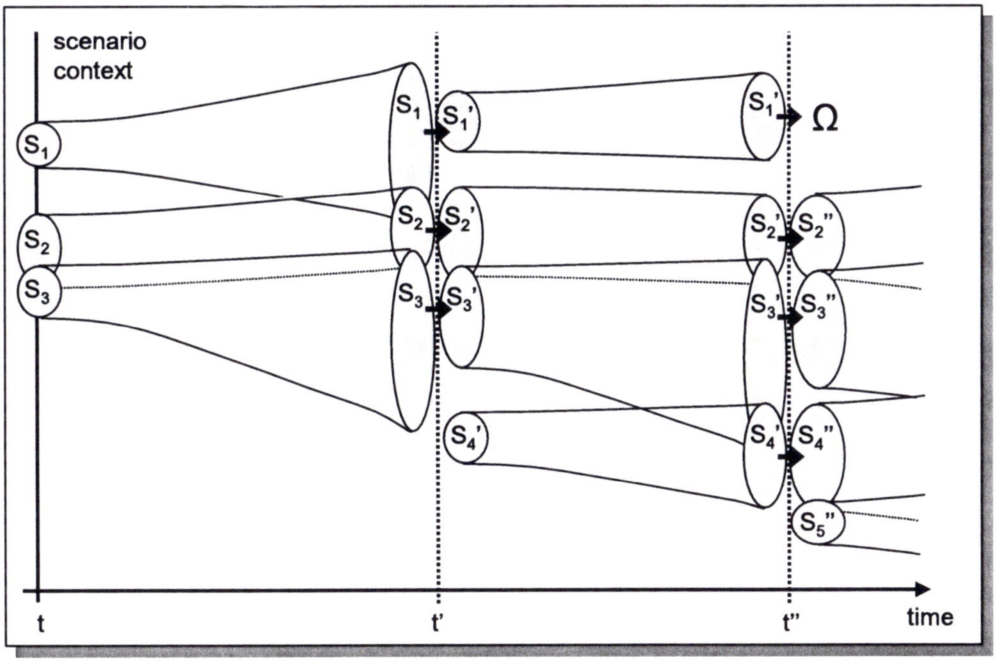

Fig. 10.4.3.1 Scenario planning along a time axis

One well-known example of a company that uses scenario planning is Royal Dutch/Shell. See also for example [Corn05]. In the late 1960s they developed various scenarios that could potentially affect their two most important planning variables, namely the demand for energy and the price of crude oil. These two initial and linearly independent planning variables, which also influence each other, largely determine the other planning variables used by

Shell. In a world that had until then been characterized by solid economic growth, one scenario considered a disruptive increase in the price of oil. In 1974, this became a reality when Arab oil-producing countries imposed an oil embargo following the Yom Kippur war.

The core components and processes used in scenario planning have a lot in common with systems engineering (see chapter 19.1). Figure 10.4.3.2 shows a possible approach.

Only after the completion or (possibly rolling) revision of the scenario planning should possible scenario forecasts for the microeconomic variables of interest be developed within the framework of the corporate strategy, e.g. long-term planning of demand for products or the long-term progression of procurement costs for raw materials. For scenario forecasts, intuitive procedures such as those described in 10.4.1 and 10.4.2 can be used.

The described procedure is complex. Scenario forecasting is therefore more suitable where there are fewer planning variables, and where there are also greater financial consequences if the forecasts are not accurate. This was the case, for example, with their use by Shell as described above. Developing the scenarios, on the other hand, provides extensive knowledge, which can be held "in stock". That in turn offers time savings if sudden and radically changed peripheral systems impact on the planning variables.

1. Determining the context and environment in which the scenario takes place, and identifying stakeholders with an interest in the scenario.
2. Identifying likely underlying trends and uncertainties, and their drivers. Suitable methods can be utilized for each trend, such as experiments, the Delphi method, other interviews and questionnaires or brainstorming.
3. Developing the scenario. One possibility would be to use workshops to illustrate the possible development whilst maintaining the current business strategy. All drivers and elements that have a positive effect can be used to develop a very extreme positive scenario, and all that have a negative effect can be used to develop to a very extreme negative scenario.
4. Plausibility and consistency checks: Are the assumed values for the variables that depict the drivers actually realistic? Are there any contradictions between drivers that influence each other and values for variables that describe other drivers? Could some of the trends or uncertainties be described more clearly?
5. Reviewing the completed scenario: Can the scenario be explained to the various stakeholders in a way that matches their cognitive perception? Is the scenario perceived as being inherently consistent and effective? Does it describe the different future that it portrays in comparison with other scenarios? Will it remain valid for a period of time?

Fig. 10.4.3.2 Procedure for scenario planning

10.5 Using Forecasts in Planning

Intended learning outcomes: Present the choice of suitable forecasting techniques. Describe consumption distribution and its limits, for continuous and discontinuous demand. Explain forecasting of variants of a product family. Present safety demand calculation for various planning periods. Disclose the translation of forecast into quasi-deterministic demand.

10.5.1 Comparison of Techniques and Choice of Suitable Forecasting Technique

Intended learning outcomes: Differentiate between various areas of applicability of forecasting techniques.

In Figure 10.5.1.1, the techniques discussed in Chapters 10.2, 10.3 and 10.4 are compared according to a number of criteria.[8]

When choosing a forecasting technique, it is crucial to find that technique (reasonable in use) that will provide the greatest accuracy of alignment to the demand structure. The following criteria also play a role:

- Adaptability to demand performance
- Possibility of forecast errors
- Aids required
- Expense for data collection and preparation for analysis
- Parameters available that describe the performance of the system to be forecast
- The purpose of the forecast and the importance of one material position
- Forecast time frame
- Transparency for the user

Feature / Technique	Demand Model					Weighting of data	Understandability of the technique	Storage required for necessary data	Processing time
	constant	with linear trend	with non-linear trend	with seasonal component	discontinuous, irregular				
Moving average	x					no	easy	large	short
1st order exp. smoothing	x					yes	easy	very little (2 values)	very short
2nd order exp. smoothing	(x)	x				yes	average	very little (2 values)	very short
Trigg & Leach adaptive smoothing		(x)	x			yes	average	very little (2 values)	very short
Exp. smoothing with seasonal influences				x		yes	difficult	little	short
Linear regression	(x)	x				no	easy	large	long to determine parameters, otherwise short
Extrapolation	x					no	easy	large	short
Delphi					x	—	easy	large	long

Fig. 10.5.1.1 Areas of applicability of forecasting techniques.

8 *Focus forecasting* is a system that allows the user to simulate the effectiveness of numerous forecasting techniques, enabling the selection of the most effective one ([APIC16]).

10.5.2 Consumption Distributions and Their Limits, Continuous and Discontinuous Demand

Intended learning outcomes: Present the consumption distribution. Explain its distribution function, expected value, and variance assuming a Poisson distribution for the frequency of events. Describe the demand filter to handle a discontinuous demand due to infrequent large issues. Disclose effects of length of statistical period on demand fluctuations.

The *distribution of forecast errors* is a tabulation of the forecast errors according to the frequency of occurrence of each error value (cf. [APIC16]).

The errors in forecasting are, in many cases, normally distributed, even when the observed data do not come from a normal distribution. Therefore, we now take a closer look into the origin of the observed values.

A *consumption distribution*, such as a statistic for order receipts allocated by time periods, can be understood as an aggregation of multiple individual events during each period. These individual events can be described by:

- The distribution of the frequency of the events themselves
- A distribution of characteristic values for an event, that is, order quantities

A combination of these two distributions results in consumption distribution.

Given the definitions in Figure 10.5.2.1 and a constant process (e.g., for constant demand), the formulas contained in Figure 10.5.2.2 are valid according to [Fers64]. Here, E stands for the *expected value*; VAR stands for the *variance*.[9]

$E(n)$, $VAR(n)$	Distribution parameters describing the frequency of events per statistical period
$E(z)$, $VAR(z)$	Distribution parameters of the characteristic values (here, the order quantity)
$E(x)$, $VAR(x)$	Parameters of the consumption distribution per period

Fig. 10.5.2.1 Definitions for a consumption distribution.

$$\begin{aligned} E(x) &= E(n) \cdot E(z) \\ VAR(x) &= VAR(n) \cdot E^2(z) + E(n) \cdot VAR(z) \end{aligned}$$

Fig. 10.5.2.2 Expected value and variance of the consumption distribution.

In a purely random process, the number of events per period has a Poisson distribution with distribution function P(n) and expected value = variance = λ. Knowing this, we can derive the formulas in Figure 10.5.2.3, where CV corresponds to the *coefficient of variation* for the distribution, that is, the quotient of standard deviation and expected value.

[9] In statistics, *variance* is a measure of dispersion, here the square of the standard deviation.

$$P(n) = e^{-\lambda} \cdot \frac{\lambda^n}{n!}$$
$$E(n) = VAR(n) = \lambda$$
$$E(x) = \lambda \cdot E(z)$$
$$VAR(x) = \lambda \cdot \left[E^2(z) + VAR(z)\right]$$
$$CV^2(x) = \frac{1}{\lambda}\left[1 + CV^2(z)\right]$$

Fig. 10.5.2.3 Distribution function, expected value, and variance of the consumption distribution under the assumption of a Poisson distribution for the frequency of events.

A few large issues can greatly influence the coefficient of variation for the *order quantity*. The square can very well take on a value of 3. If all issues are equally large, then the value is clearly at its minimum of 0 (e.g., the order quantity for service parts may always equal 1).

Even if the measured values of the consumption distribution allowed, based on the rules of statistics, the assumption of a normal distribution as such, a coefficient of variation of $CV \leq 0.4$ is a prerequisite for effective procedures in the stochastic materials management. The formula in Figure 10.5.2.3 says how many issues are necessary, so that such a small coefficient of variation results. Specifically, if 1 is assumed as the mean for the coefficient of variation of the distribution of the *order quantity*, then at least 12.5 orders or issues per period are needed, which can be high for a machine manufacturer ($\lambda \geq (1+1) / 0.16 = 12.5$).

The value for λ may vary very widely and may be quite small, particularly in the capital goods industry. This type of demand is referred to as *discontinuous* or *lumpy demand.* It is different from both *regular* demand (regularity as described in Section 10.3.1) and *continuous* (steady) demand. (See the definitions in Section 4.4.2).

From above observations, we can establish qualitatively that:

- The *discontinuous character* of a distribution is the result of a limited number of issues per time unit measured. With this, it is very difficult to calculate a forecast. Large coefficients of variation arise not least due to individual, perhaps rather infrequent, large issues. As Figure 10.5.2.4 indicates, large issues should be considered as outliers or as abnormal demand and should be taken out of a stochastic technique by a demand filter[10] and made available to deterministic materials management. This could be achieved by increasing delivery periods for large orders, for example.

- In the case of a stationary process, for example, constant demand, the relative forecast error depends heavily on the number of events, such as the number of orders. Generally, the actual forecast error is larger than that calculated by extrapolation. This is so because changes in the underlying regularities increase error, given that the number of events is small.

---

10 A *demand filter* in the forecast model is expressed by some factor times ± the standard deviation.

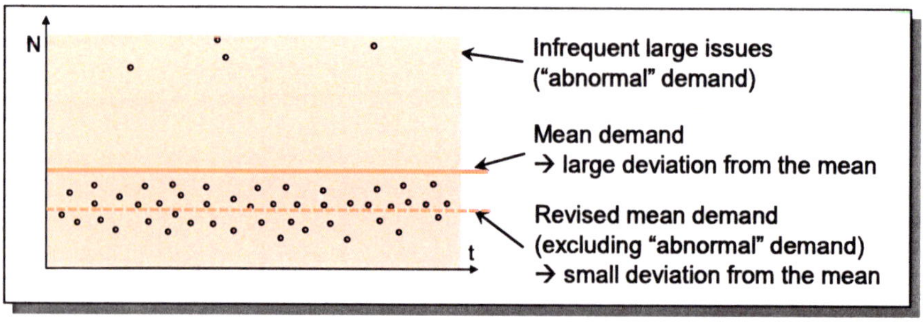

Fig. 10.5.2.4 Demand filter to handle a discontinuous demand due to infrequent large issues.

Whether demand will appear as continuous or discontinuous also depends upon the choice of the length of the statistical period. Figure 10.5.2.5 shows this effect.

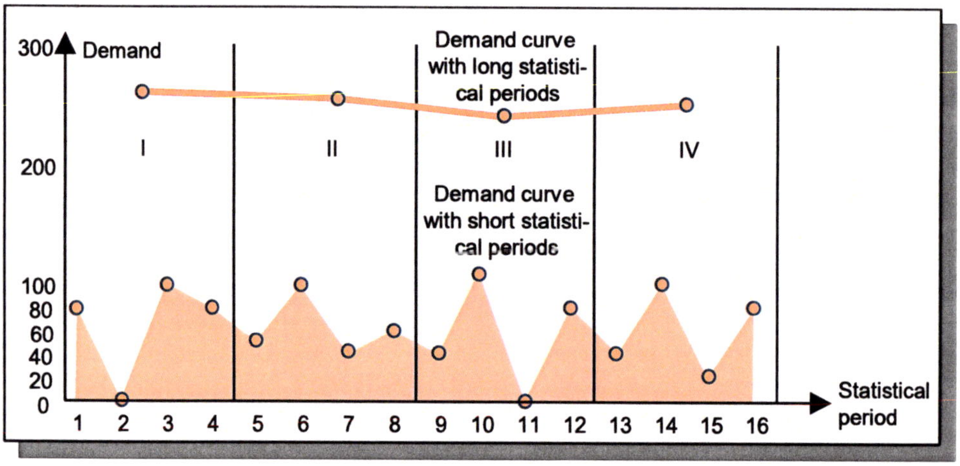

Fig. 10.5.2.5 Effects of length of statistical period on demand fluctuations.

If the statistical period chosen is too short, this quickly results in discontinuous demand values. These fluctuations are exaggerated and can be leveled by extending the statistical periods. However, the result in materials management may be an increase in levels of goods in stock or work in process, especially if the lead times are shorter than the statistical periods. For practical reasons, a unified length of the statistical period for the entire product range is required. Often, a period of one month is chosen.[11]

Even if there is rather discontinuous demand for an individual item, demand for the entire item family may be continuous. In this circumstance, the forecast would be accurate enough

[11] The leveling of demand fluctuations is necessary, for example, for simple control techniques such as Kanban, in which continuous demand is a prerequisite for their functioning. Enlarging the statistical period may sometimes be sufficient.

for rough-cut planning. If the need for more detailed information should arise, the allocation of the forecast to the various items of the family may be difficult. See also Section 13.2.

10.5.3 Demand Forecasting of Variants of a Product Family

Intended learning outcomes: Describe the option percentage. Explain the formulas for forecasting demand for variants.

Often, a standard product with options or a product family are derived gradually from one basic type . Often, a forecast can predict the total demand for a product family. Deriving the demand for components that are the same for all variants is no longer very difficult.

Demand forecasting for variants is more difficult. If the number of delivered items for each variant of a product family is large enough (this is the case for low-variety manufacturing — see Sections 7.1.1 and 7.2), the use of variant items — related to 100 units of the product family, for example — can be recorded in a statistic and used for management.

The *option percentage* OPC is the frequency with which a variant item is used within a product family.

This percentage varies from time period to time period and is therefore a stochastic variable that can be described with expected value and variance.

Often, in practice, the dispersion of the option percentage is not taken into consideration; that is, E(PF) is treated as a quasi-deterministic value. This increases the risk of stock failures. To calculate option percentages, sales are subdivided by statistical period. For each period, we determine the actual frequency of use and calculate mean and standard variation from the results of multiple periods. Linking the forecast for the product family with the option percentage for demand for variants is achieved by using the formulas in Figure 10.5.3.1. These formulas are used for every periodic demand.

$$E(OD) = E(PFD) \cdot E(OPC)$$

$$CV^2(OD) = CV^2(PFD) + CV^2(OPC) + CV^2(PFD) \cdot CV^2(OPC)$$

where OD = option demand

PFD = product family demand

OPC = option percentage

Fig. 10.5.3.1 Forecasting demand for variants.

Thus, the independent demand for a variant is equal to the option percentage multiplied by the demand for the product family. Because of safety demand calculation, the sum of variant demands is greater than the demand for components not dependent upon variants.

The following derivation (*) according to Prof. Alfred Büchel for multiplicative coupling x of two independent distributions y and z, x= y*z, provides a more in-depth consideration of the formulas in Figure 10.5.3.1. See also [Fers64]. Multiplication of a particular value Y of y by z results in a linear transformation of z with the following parameters:

$$E(Y^*z) = Y * E(z)\,,\ VAR(Y^*z) = Y^2 * VAR(z)\,.$$

The distribution obtained in this way is weighted by f(y) and summed (or, with continuous distributions, integrated) to create a mixed distribution. The zero moments are to be applied for this. The result of the individual linear transformations for the second zero moment — defined as $E(u^2) = E^2(u) + VAR(u)$ — are as follows:

$$E((Y^*z)^2) = E^2(Y^*z) + VAR(Y^*z) = Y2 * E^2(z) + Y^2 * VAR(z)$$
$$= Y^2 * (E^2(z) + VAR(z)) = Y^2 * E(z^2)\,.$$

The summation produces the following result:

$E(x) = E(y) * E(z)\,,\ E(x^2) = E(y^2) * E(z^2)$, and so the following hold:

$$VAR(x) = E(x^2) - E^2(x) = E(y^2) * E(z^2) - E^2(y) * E^2(z)$$
$$= [E^2(y) + VAR(y)] * [E^2(z) + VAR(z)] - E^2(y) * E^2(z)$$
$$= E^2(y) * VAR(z) + VAR(y) * E^2(z) + VAR(y) * VAR(z).$$

$$CV^2(x) = VAR(x) / E^2(x)$$
$$= [E^2(y) * VAR(z) + VAR(y) * E^2(z) + VAR(y) * VAR(z)] / [E^2(y) * E^2(z)]$$
$$= CV^2(z) + CV^2(y) + CV^2(y) * CV^2(z).$$

Note: The formulas in Figure 10.5.2.2 can be derived analogously. Linear transformations are replaced by the distributions for the sum of multiple issues per period (so-called convolutions), whose parameters are determined as follows:

$$E(n^*z) = n * E(z); \qquad VAR(n^*z) = n * VAR(z)$$

A general statement as to the form of the distribution cannot be made; a log-normal distribution (which becomes a normal distribution with small coefficients of variation) represents a useful approximation for practical application. With many periods with zero issues (low issue frequency), special consideration of the choice of "risk" may be required. This is also true for Section 10.5.2. However, the planning periods, not the statistical periods, are decisive.

10.5.4 Safety Demand Calculation for Various Planning Periods

Intended learning outcomes: Identify variables for safety demand calculations. Explain expected value and standard deviation with continuous demand and over n statistical periods.

A *planning period* represents the time span between "today" and the point in time of the last demand that was included in a specific planning consideration.

Figure 10.5.4.1 provides a few definitions required for the following discussion.

In a forecast calculation, we determine expected value and standard deviation for a particular statistical period, for example, the SP. In materials management, however, it is necessary to have values for various planning periods.

SP	=	Length of the statistical or forecast period (in some time unit)
PP	=	Length of the planning period
E(DSP)	=	Expected value of the demand in the statistical period
E(DPP)	=	Expected value of the demand in the planning period
σ (DSP)	=	Standard deviation of the demand in the statistical period
σ (DPP)	=	Standard deviation of the demand in the planning period
Z	=	Issue quantity (designated as in Section 9.5.3)
λ	=	Number of issues in the statistical period

Fig. 10.5.4.1 Definitions of variables for safety calculations.

If, for example, the planning period is the lead time, then we have to take the total forecast demand during the lead time into consideration. Usually this is up until the receipt of the production or procurement order.[12]

We can infer the formulas shown in Figure 10.5.4.2 on the basis of the models developed in Section 10.5.2; the formulas are also valid for the non-integral proportions of PP:SP.

$$\text{Number of issues during the planning period} = \lambda \cdot \frac{PP}{SP}$$

$$E(DPP) = \lambda \cdot \frac{PP}{SP} \cdot E(z) = \frac{PP}{SP} \cdot E(DSP)$$

$$\sigma(DPP) = \sqrt{\lambda \cdot \frac{PP}{SP}\left[E^2(z) + VAR(z)\right]} = \sqrt{\frac{PP}{SP}} \cdot \sigma(DSP)$$

$$\underbrace{VAR(DSP) = \lambda \cdot \left[E^2(z) + VAR(z)\right]}$$

Fig. 10.5.4.2 Expected value and standard deviation with continuous demand.

In a nonstationary process, different expected values or standard deviations arise for various time periods in the future. Assuming independent forecast values in individual periods, the expected values and variances of demand can be added during the planning period. For n statistical periods, this produces the formulas shown in Figure 10.5.4.3.

$$E(DPP) = \sum_{i=1}^{n} E(DSP(i))$$

$$\sigma(DPP) = \sqrt{\sum_{i=1}^{n} \sigma^2(DSP(i))}$$

Fig. 10.5.4.3 Expected value and standard deviation over n statistical periods.

---

12 The planning horizon is a further example of a planning period.

We can also use these formulas for certain periods, usually in the near future, where the demand has been established deterministically, that is, through customer orders, for example. The demand for these periods demonstrates a 0 variance. Similarly, a linear interpolation of the expected value and variance is used to determine intermediate values during a period.

10.5.5 Translation of Forecast into Quasi-Deterministic Demand and Administration of the Production or Purchase Schedule

Intended learning outcomes: Identify safety demand. Explain independent demand as total demand including safety demand, taken as a function of the planning period to be covered.

The (stochastic) independent demand to be considered for further planning steps results as the total demand from adding the expected value to the safety demand for the planning period to be covered.

The *safety demand* is the product of the safety factor and the standard deviation during the planning period to be covered.

Figure 10.5.5.1 shows the total demand to be considered as a function of the planning period to be covered. For products manufactured in-house, this total demand belongs to the *production schedule*. For purchased items, the independent demand belongs to the *purchase schedule* for salable products.

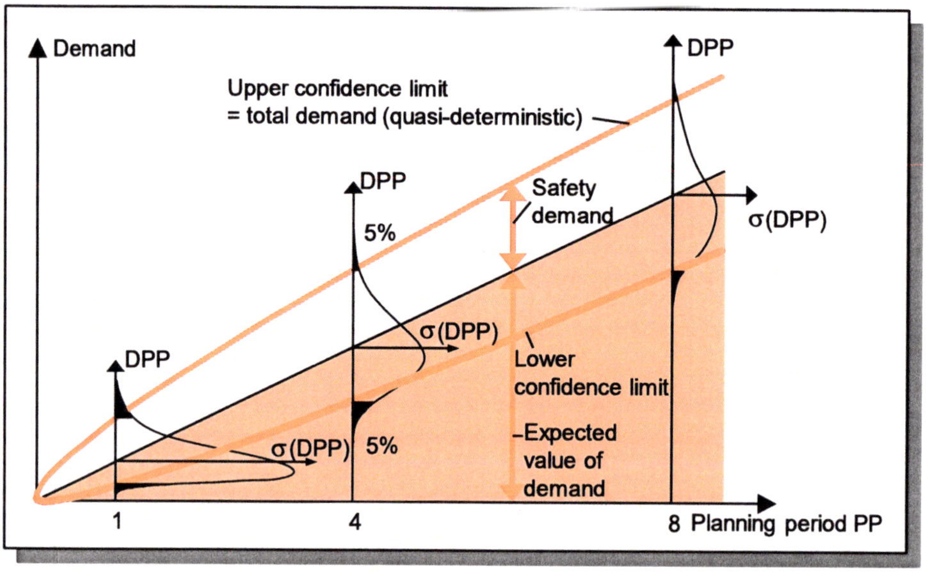

Fig. 10.5.5.1 Independent demand as total demand, taken as a function of the planning period to be covered.

If the total demand is subdivided into various partial demands later (for example, the annual demand into 12 monthly demands), a larger share of the safety demand needs to be included in the earlier partial demand. The order point technique discussed in Section 11.3 adds the safety demand de facto to the first partial demand.

Note: As presented in connection with Figure 5.3.2.2, the first step in determining high-cost dependent, but discontinuous or unique demand for an item is to stochastically determine the independent demand belonging to it. After this, the dependent demand is calculated using quasi-deterministic bill-of-materials explosion. In this way, the dependent demand contains the safety demand needed to produce the safety demand for the independent demand.

For *administrating independent demand,* an order-like *object class forecast demand* or *independent demand* is used, with at least the attributes

- Forecast or independent demand ID (similar to an order ID)
- Item ID or item family ID
- Planning date for the demand or its periodicity
- Forecast quantity (an item issue)
- Quantity of the forecast already "consumed" by orders (see Section 12.2.2)

A negative forecast demand is also conceivable. It would express receipt of an item, and serves, e.g., as a substitute for a purchase system that is lacking, or to eliminate an overlap effect on lower structure levels from higher structure levels (see, for example, Section 7.2.1).

There are a number of ways to change or delete a forecast demand:

- By manual administration.
- By periodic recalculation, e.g., according to the principle contained in Figure 10.1.1.1. This is particularly important for demand serving as input to subsequent stochastic materials management.
- With independent demand in the true sense: by successive reduction due to actual demand (e.g., customer orders). If the actual demand reaches the forecast, or if the forecast lapses into the past and is no longer to be considered, the corresponding forecast demand object is automatically deleted. See also Section 12.2.2.

10.6 Summary

A demand forecast is an expression of the probable course of demand along the time axis. A demand must be forecast if the cumulative lead time is longer than the customer tolerance time. Such a situation occurs, for example, in trade in consumer goods, in components for services, or in single parts of investment goods. Forecasts are transformed into demand for resources later and then compared with the organization's supply capacity. However, every forecast is associated with uncertainty. Therefore, forecasts must be compared to demand continually, e.g., in a rolling manner. A significant deviation in demand may require the selection of a different technique.

We distinguished two basic types of forecasting techniques: historically oriented and future-oriented. Both basic types are further subdivided into mathematical, graphical, or intuitive techniques. The selection of a technique is made according to a series of criteria intended to produce a reasonable alignment of the forecast to the demand, at reasonable expense.

Historically oriented techniques calculate demand based on consumption with the help of mathematical statistics (extrapolation of time series). There are simple techniques for continuous demand, such as moving average or first-order exponential smoothing. For linear trends, we may make use of linear regression or second-order exponential smoothing. In addition, the Trigg and Leach adaptive technique examines and adapts the parameters used in exponential smoothing. All the techniques may be expanded to account for the effect of seasonality. Trend extrapolation, the Delphi method, and scenario forecasts were discussed as future-oriented techniques, although these also contain historically oriented elements.

The more discontinuously consumption occurs, the more difficult it is to forecast reliably. The definition of consumption distributions as an overlay of the distribution of consumption events and the distribution of consumption quantities per event helps describe discontinuous conditions. A suitable length of the statistical period can lead to a smoothing of demands. Where there are few variants and repetitive production, forecast for variant demand of a product family may be calculated using option percentages. This is a stochastic variable with an expected value and standard deviation.

In all cases, larger fluctuations in demand lead to safety demand, which is calculated on the basis of standard deviation. The expected value and standard deviation are related to the statistical period, while independent demand is related to the planning period. The conversion of expected value is proportional to the ratio of the two time periods, whereas in the standard deviation the conversion is proportional to its square root. The expected value of the demand increased by safety demand is set as independent demand per planning period; the latter is then available as stochastic demand for further handling in the context of materials management. When dependent demand is calculated later, using a quasi-deterministic bill-of-material explosion, it will contain the corresponding safety demand.

For each independent demand, the item ID, the forecast quantity, and the quantity of the forecast already "consumed" by orders are recorded, as well as the planning date. The total of all independent demands belongs to the production schedule, or, when referring to trade items, the purchase schedule. Independent demand can be recalculated or canceled by rolling planning, either manually or with automated techniques. In general, actual demand successively replaces or reduces independent demand.

10.7 Keywords

adaptive smoothing, 443
consumption distribution, 454
curve fitting, 428
Delphi method, 449
demand model, 428
distribution of forecast errors, 454
expected value, 430
extrinsic forecasting techniques, 427
first-order exponential smoothing, 435
forecast error, 424
forecast horizon, 428
forecast interval, 428

forecasting, 424
forecasting model for constant demand, 431
forecasting technique, 424
graphic forecasting techniques, 427
historically oriented forecasting techniques, 426
intrinsic forecasting techniques, 427
intuitive forecasting techniques, 427
life-cycle analysis, 447
mean, 430
mean absolute deviation (MAD), 435
mean deviation, 444
moving average, 432
moving average (forecasting technique), 432
option percentage, 457
planning period, 458
probability, 433
qualitative forecasting techniques, 427
quantitative forecasting techniques, 427
regression analysis, 439
safety demand, 460
scenario planning, 450
seasonal fluctuations, 445
seasonal index, 445
seasonality, 445
second-order exponential smoothing, 441
single (exponential) smoothing, 435
smoothing, 431
smoothing constant γ, 444
smoothing constant α, 436
standard deviation, 432
statistical period, 428
time series, 428
tracking signal, 444
trend, 438
trend extrapolation, 447
trend forecasting model, 438
variance, 454
weighted moving average, 434

10.8 Scenarios and Exercises

Intended learning outcomes: Choose an appropriate forecasting technique. Calculate an example for moving average forecasting and for first-order exponential smoothing. Differentiate between moving average forecast and first-order exponential smoothing.

10.8.1 Choice of Appropriate Forecasting Techniques

Intended learning outcomes: Propose a forecasting technique for different products to apply to forecast future demand.

Figure 10.8.1.1 shows historical demand curves for four different products. What forecasting technique for each product do you propose to apply to forecast future demand?

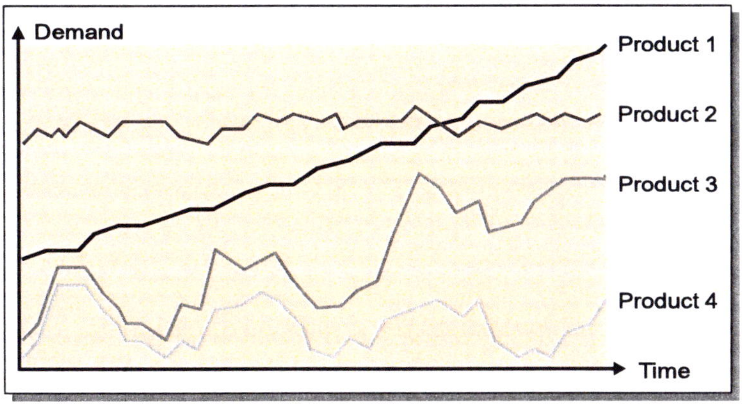

Fig. 10.8.1.1 Historical demand curves for four products.

Solution:

- Product 1: demand with linear trend → linear regression
- Product 2: constant demand without trend → moving average forecasting or first-order exponential smoothing
- Product 3: seasonal fluctuations with trend → linear regression or second-order exponential smoothing with seasonality
- Product 4: constant demand with seasonal fluctuation → moving average forecasting, or first-order exponential smoothing, with seasonality

10.8.2 Moving Average Forecasting Technique

Intended learning outcomes: Calculate forecasts with moving average forecasting.

The person in your firm responsible for forecasting has been absent for three months, so your supervisor asks you to forecast the demand of the most important product. The information you get is a table (see Figure 10.8.2.1) showing the historical data on the demand for the product (January to October) and the forecast for the period January to July based on the moving average forecasting technique.

	Jan.	Feb.	Mar.	Apr.	May.	Jun.	July.	Aug.	Sept.	Oct.
Demand	151	135	143	207	199	175	111	95	119	191
Forecast	183	195	177	155	159	171	181			

Fig. 10.8.2.1 Demand and forecast with moving average forecasting technique.

Moreover, your supervisor asks you to:

a. Forecast the demand just as your colleague does. Therefore, you have to calculate the parameter n from the historical forecast data.
b. Calculate the forecast for August, September, and October as well as for the following month, November.
c. Compute the standard deviation σ of the forecast from January to October and decide if the applied technique fits this product.

Solution:

a. $n = 4$
b. Forecast August = (207+199+175+111) / 4 = 173; forecast September: 145; forecast October: 125; forecast November: 129.
c. $\sigma = 53.87$ and variation coefficient = 53.87 / 152.6 ≈ 0.35. A variation coefficient of 0.35 stands for a relatively low quality of the forecast. Therefore, the applied technique is not appropriate for this product. Try a value other than $n = 4$, or with additional seasonal index.

10.8.3 First-Order Exponential Smoothing

Intended learning outcomes: Calculate forecasts using first-order exponential smoothing.

When you report to your supervisor that the moving average forecasting technique is not suitable for the product, he remembers that your colleague in charge of forecasting had been working on introducing the first-order exponential smoothing technique for this product. Therefore, your supervisor gives you the information in Figure 10.8.3.1.

	Jan.	Feb.	Mar.	Apr.	May.	Jun.	July.	Aug.	Sept.	Oct.
Demand	151	135	143	207	199	175	111	95	119	191
Forecast	187	176	164	158	172	180	179			

Fig. 10.8.3.1 Demand and forecast using first-order exponential smoothing technique.

The Figure shows the demand for the product (January to October) and the forecast using the first-order exponential smoothing technique with $\alpha = 0.3$ of the product (January to July). To evaluate your supervisor's suggestion, you execute the following steps:

a. Compute the forecast for August, September, and October and for the following month, November.

b. Calculate the mean absolute deviation (MAD) for November assuming MAD(Jan) = 18 and the smoothing parameter α.

c. In the preceding exercise, could you have obtained a result comparable to the one for the parameter α calculated above by changing n, that is, the number of observed values?

d. Decide whether the chosen first-order exponential smoothing technique with parameter α calculated above is appropriate for this product.

e. What can you say in general about the choice of α depending on the product life cycle?

Solution:

a. Forecast August = 0.3*111+0.7*179 ≈ 159; forecast September: 140; forecast October: 134; forecast November: 151.

b. MAD(Feb) = 0.3*(187 – 151)+0.7*18 ≈ 23 → MAD(Mar) = 29, MAD(Apr) = 26, MAD(May) = 33, MAD(Jun) = 31, MAD(Jul) = 23, MAD(Aug) = 37, MAD(Sept) = 45, MAD(Oct) = 37, MAD(Nov) = 43.

c. Yes, by choosing a value of n = (2 – 0.3)/0.3 = 5.67 (see the formula in Figure 10.2.3.1).

d. Since the demand fluctuates, it would be better to increase α. Moreover, the first-order exponential smoothing technique does not fit this demand curve well. Therefore, it is worth considering another forecasting technique, e.g., with short-term seasonality.

e. At the beginning and the end of the product (market) life cycle, α should be relatively high, e.g., $\alpha = 0.5$. For a well-established product, the α often chosen is around 0.1.

10.8.4 Moving Average Forecast versus First-Order Exponential Smoothing Forecast

Intended learning outcomes: Differentiate between moving average forecast and first-order exponential smoothing using various numbers of observed values or smoothing constants α.

Figure 10.2.2.6 showed the effect of different values of the smoothing constant α. Figure 10.2.3.1 shows the necessary relationship between the number of observed values and the smoothing constant α. You can view the animated comparison on the Internet at URL:

https://intlogman.ethz.ch/#/chapter10/demand-forecasting/show

In the red section at the top of the Web page, you can choose different values for the smoothing constant α. In the lower, green section you can choose either a different value for the smoothing constant α for comparison with the red curve or choose the number of values for the moving average forecast and compare the results of the technique with exponential smoothing (the red curve). Clicking on the "calculate" icon executes your input choice.

10.9 References

APIC16 Pittman, P. et al., APICS Dictionary, 15th Edition, APICS, Chicago, IL, 2016

BoJe15 Box, G.E.P., Jenkins, G.M., Reinsel, G.C., Ljung, G.M., "Time Series Analysis: Forecasting and Control," 5th Edition, John Wiley & Sons, 2015

Corn05 Cornelius, P. et al., "Three Decades of Scenario Planning in Shell," California Management Review, 48.1, pp. 92-109, 2005

Fers64 Ferschl, F., "Zufallsabhängige Wirtschaftsprozesse," *Physica*, 1964

Gahs71 Gahse, S., "Mathematische Vorhersageverfahren und ihre Anwendung," Verlag Moderne Industrie, München, 1971

GaKe89 Gardner, E.S, Jr., McKenzie, E., "Seasonal Exponential Smoothing with Damped Trends," *Management Science* (Note), 35(3) 1989, pp. 372–375

IBM73 IBM, COPICS, "Communications Oriented Production Information and Control System, Bedarfsvorhersage," IBM, Deutschland, 1973

Lewa80 Lewandowski, R., "Prognose- und Informationssysteme und ihre Anwendungen," de Gruyter, Berlin, 1980

Scho93 Schoemaker, P., "Multiple Scenario Development: Its Conceptual and Behavioral Foundation," Strategic Management Journal, 14, pp. 193-213, 1993

TrLe67 Trigg, D.W., Leach, A.G., "Exponential Smoothing with an Adaptive Response Rate," *Operations Research Quarterly*, 1967, pp. 53–59

WhMa97 Wheelwright, S.C., Makridakis, S., "Forecasting Methods for Management," 3rd Edition, John Wiley & Sons, New York, 1997

11 Inventory Management and Stochastic Materials Management

Intended learning outcomes: Describe usage statistics, analyses, and classifications. Explain in detail the order point technique. Disclose how to calculate safety stock. Differentiate various batch or lot sizing techniques.

This chapter deals with the translation of forecasted demand into production or procurement proposals through the function of materials management in the stochastic case. The relevant tasks and processes are shown on a dark background in Figure 11.0.0.1.

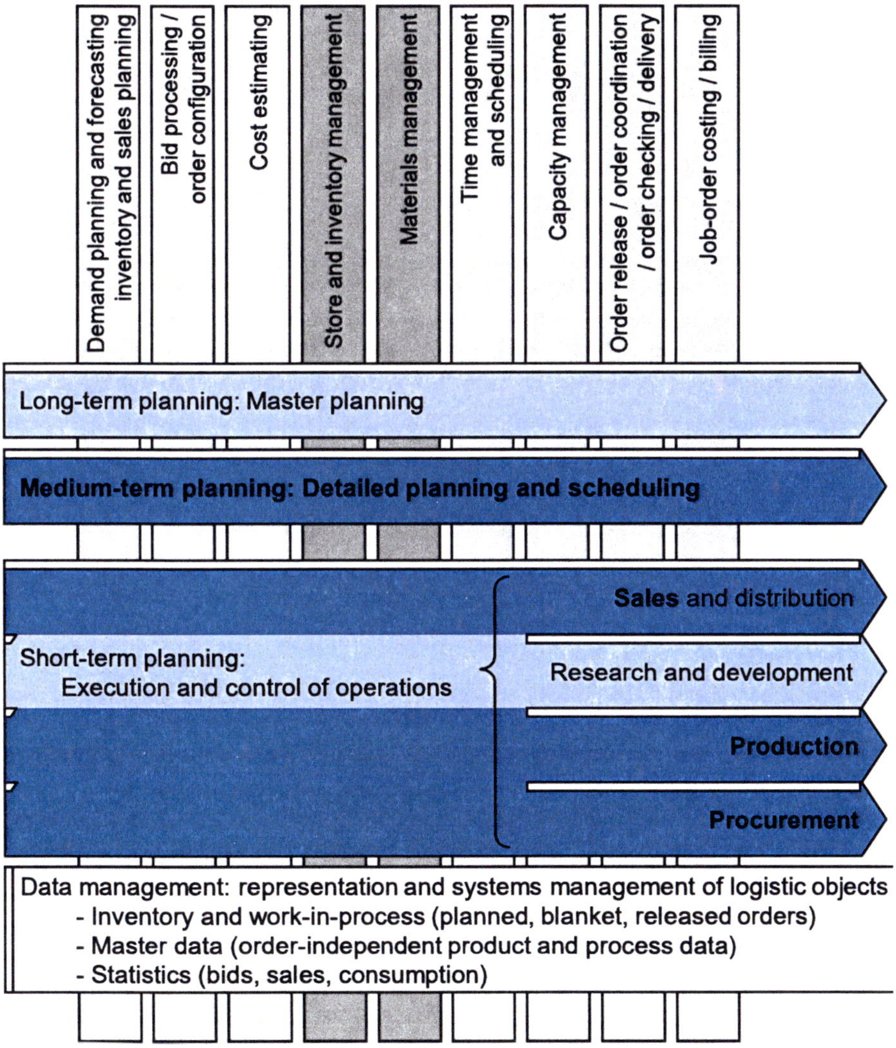

Fig. 11.0.0.1 The parts of the system discussed in this chapter (shown on darker background).

P. Schönsleben, *Handbook Integral Logistics Management*,
https://doi.org/10.1007/978-3-662-65625-9_11

The tasks and processes refer to the reference model for business processes and planning & control tasks in Figure 5.1.4.2. Sections 5.3.1, and 5.3.2 provide useful overviews.

Inventory has a buffer function, to achieve synchronization between use, on the one hand, and design and manufacturing, on the other. This makes inventory management another important task for planning & control. Inventory transactions are the basis for usage statistics. Together with ABC and XYZ analyses, and also other evaluative procedures, usage statistics build the principle of techniques of stochastic materials management —in particular demand forecasting.

For goods upstream or at the order penetration point, production or procurement orders must be released prior to customer demand. Inventory in stock or work in process must cover total demand up to the point when newly proposed orders will be filled. Here, because of its simplicity, the order point technique is widely used. The order point technique proposes orders with a quantity and a completion date. In medium-term planning, the proposals serve to reconcile inventory to blanket orders. In short-term planning, they trigger order releases. In the case of a production order, the proposal yields the requirements for components that, in turn, come under the direction of materials management.

Because of the inexact nature of demand forecasting and lead time, safety stock is carried to protect against the differences between forecast and actual usage and fluctuations in lead time. The level of safety stock thus affects stockout probability, carrying cost, and eventually the fill rate (also called customer service ratio).

In materials management, lot or batch size mainly affects costs. In scheduling and capacity management, additional considerations reveal the effect of batch size on lead time and flexibility. In the stochastic case, the composition of customer demand over time is unknown. This leads to imprecise proposals. The stochastic calculation technique presented in this chapter is robust at least in the face of forecast errors and incomplete parameters.

11.1 Stores and Inventory Management

Intended learning outcomes: Present characteristic features of stores management. Produce an overview on inventory transactions. Describe physical inventory and inventory valuation.

Inventory is one of the most important instruments of logistics planning & control. Although inventory of work-in-process items is sometimes linked to the production process, such physical inventory as well as stored inventory is — from the standpoint of value adding — often unnecessary (considered a non-value added or a waste) and costly in terms of time and money (tied-up capital). As discussed in Section 1.1.6, inventory is unavoidable if customer tolerance time is shorter than the cumulative lead time. A further reason for stockkeeping, however, lies in planning & control itself. Stocks provide for the storage of goods over time. They create degrees of freedom that allow for the matching of capacity (humans, machines, tools) to the demand for goods.

11.1.1 Characteristic Features of Stores Management

Intended learning outcomes: Present characteristic features such as storage location ID, storage type, valuation basis, stock organization, embedding in the flow of goods, storage management principle, inventory issuance principle, and inventory control principle.

Stores management, in particular, determines the values of characteristic features related to storage of goods. The choice of values is heavily dependent on the characteristic features of planning & control within supply chains listed in Section 4.4, but particularly on the order penetration point.

The *stockkeeping unit (SKU)* is an inventory item at a particular geographic location.

For example, a shirt in six colors and five sizes would represent 30 different SKUs. A product stocked at the plant and at six different distribution centers would represent seven SKUs. See [APIC16].

Figure 11.1.1.1 presents specific characteristic features of stores management. Definitions of some of the features and values follow:

Characteristic	→	**Expression**					
Identification (storage location)	→	Geographic identification of place of storage					
Type of storage	→	Floor	Rack	Shelving	Refrigerator	Tank	...
Valuation basis	→	Number	Value	Surface area	Volume	Weight	...
Stock organization	→	Single stock	Multiple stock	Variant stock, single	Variant stock, multiple		
Embedding of the store in the flow of goods	→	Centralized	Decentralized	Floor stock			
Storage management principle	→	Fixed location	Random location				
Issuance principle / Inventory valuation method	→	Unordered / average cost system	FIFO	LIFO	Order specific		
Inventory control principle	→	Centralized	Decentralized				

Fig. 11.1.1.1 Characteristic features for stores management.

The *identification* or storage location identifies the geographic place of storage. This will generally refer to the layout in a warehouse and include the warehouse ID, its different floors, and, for each floor, the coordinates row (x-axis), shelf (y-axis), and level (z-axis).

Storage type describes the infrastructure available for physical storage: floor storage, refrigerated storage, storage in special tanks, silos, and so on.

Cost accounting uses the feature *valuation basis* for distributing the costs of storage to their source, the stored goods, as accurately as possible.

Stock organization:

- *Single-stock organization* stores the entire stock of a particular item, or good to be stored, at one single stock location. Conversely, it is possible to store several different items at the same stock location, that is, under the same geographic identification.
- *Multiple-stock organization* keeps the inventory of a particular item at various stock sites. Each partial inventory corresponds to a different stockkeeping unit, according to the definition of this term.
- *Variant-stock organization* uses a concept that provides for storage of all variants of the same item family under one item ID. If, for example, the varying dimensions of a particular type of screw make up a family of screws, then every dimension of the screw is one variant of the same item family. The item family as a whole is then stored at one or several stock sites, while inventories for each variant are tracked separately.

Embedding (of the store) in the flow of goods:

- A *centralized store* is usually remote from the flow of goods. Between the centralized warehouse and the user operations, inventories are transferred using an inventory issue slip. Inventory receipts also generate a stock receipt slip. The responsibility rests with an organizational unit (usually centralized) created for that purpose.
- A *decentralized store* is located directly at the shop floor or production line. The (decentralized) responsibility for and management of this store lies with production.
- A *floor stock* is a stock of low-cost items held in the factory, from which production workers can draw without requisition.

Storage management principle:

- *Fixed-location*, or *"on site" storage*, is arranged according to a particular sequence. All items that logically belong together can be picked up one after the other.
- With *random-location storage*, any one storage location can hold the stock of one item or another. Warehouse personnel do not try to find logical locations for new stock to be stored, but simply place it in the next available location. While random storage requires a locator file to identify parts locations, it often requires less space than fixed-location.

Inventory issuance principle and inventory valuation method:

- With an *unordered issuance principle* it makes no difference what portion of stock should be issued. An appropriate valuation method is the average cost system: When a new order is received, a new weighted average unit cost value is computed as follows: (1) The value of the order is added to the value of the on-hand balance (valued at the current average unit cost value), and (2) the resulting value is divided by the sum of the units on hand plus those just received.
- The *FIFO issuance principle* / valuation method (first in, first out) or a LIFO issuance principle / valuation method (last in, first out) results in the removal from stock of that partial quantity that was received first — or last, respectively. For this we need proof of the time at which each quantity was placed into stock. Batch control, as described in Section 8.2.3, provides such data.

- The *order specific issuance principle* issues items produced or procured by specific order. The related order-specific valuation method assigns to these items a value that equals the actual costs of the respective order. For this, lot control has to be provided.

Inventory control principle:

- With *centralized inventory control*, one office or department is responsible for inventory decision making (for all SKUs) for the entire company.
- With *decentralized inventory control*, inventory decision making (for all SKUs) is exercised at each stocking location for SKUs at that location (see [APIC16]).

Optimal inventory organization is tuned to the characteristic features of planning & control currently valid in the supply chain. Just as the value of each of these features may change with the organization's policies, the value of each inventory management feature can also change. Inventory organization must therefore remain flexible. Rather than forming a constraint for logistics, it must ensue from the type of logistics chosen.

11.1.2 Inventory Transactions

Intended learning outcomes: Differentiate between perpetual inventory and book inventory. Produce an overview on the sources of planned and actual inventory transactions.

Inventory management includes the tasks involved in the handling of inventory transactions.

An *inventory transaction* alters the stored or in-process inventory. This can be a planned or executed inventory transaction.

Perpetual inventory is an inventory record-keeping system where each transaction in and out is recorded and a new balance is computed.

Book inventory is an accounting definition of inventory units or value obtained from perpetual inventory transaction records rather than by actual (physical) count (cf. [APIC16]).

Figure 11.1.2.1 shows an overview of the types and origins of important inventory transactions in an industrial organization, both planned (for example, an allocation) and executed.

An exact and well-documented book-inventory recording system is the basis of all inventory management. Appropriate organizational measures must make it possible to record accurate, up-to-date information on book inventory, even with thousands of transactions per week and numerous employees. Book inventory should either equal physical inventory or deviate from it in a controllable and traceable manner. Measures to this purpose include:

- Ensuring that there are no uncontrolled inventory issues or receipts. Generally, this means that there will be "closed," or separate, warehouses or accurately controlled buffers, such as in container units. Transactions are recorded at the moment the goods leave or enter the warehouse. It is important to keep the administrative costs low for putting into and issuing from stock for inexpensive parts (screws, nuts, springs, and the like). For this reason, decentralized small-parts stocks are often located directly at the production facilities.

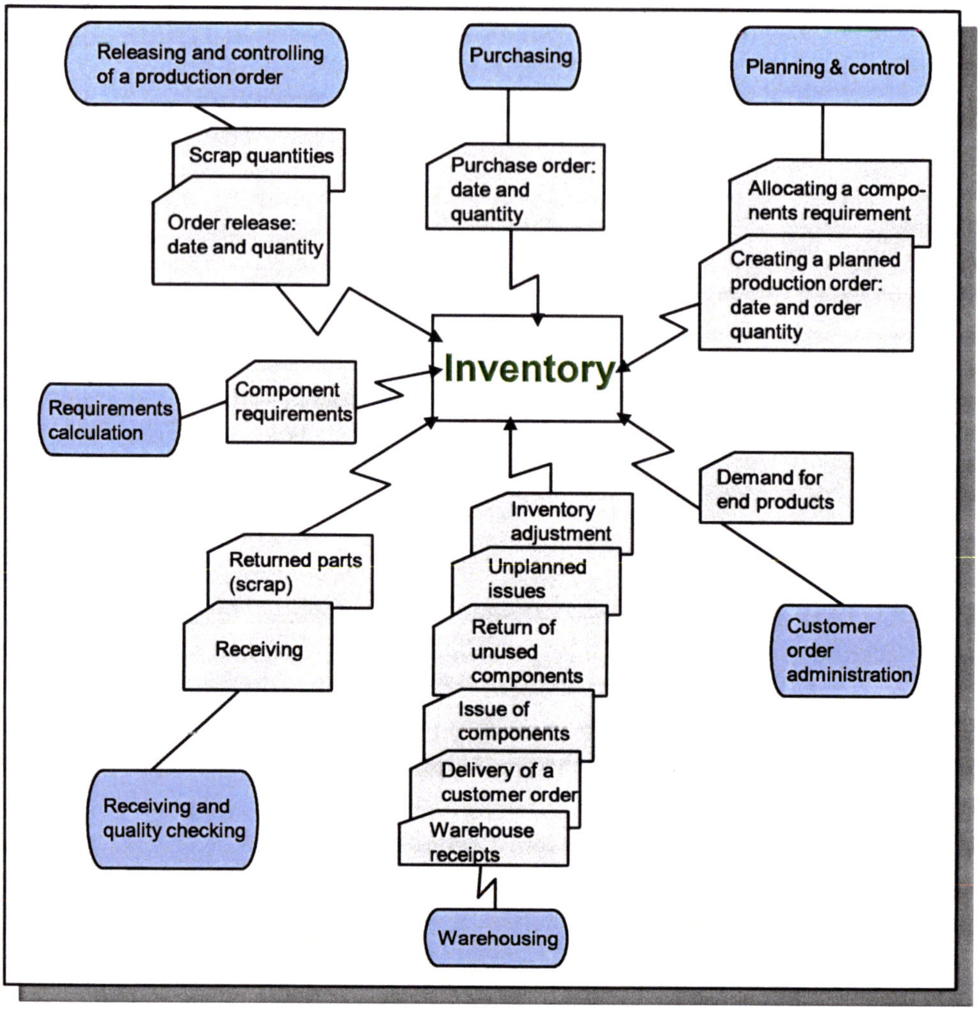

Fig. 11.1.2.1 Overview of the sources of planned and actual inventory transactions.

- Guaranteeing the identification of goods by accurate specification of item identification and storage location. This is one of the main purposes of automated inventory organization through, for example, IT-supported warehouse transport systems. Interactive verification guarantees accuracy without the use of paper records. Inventory management should perform plausibility tests, such as:

  Firstly, correct item identification. A number can contain control digits. This avoids recording errors such as reversed digits or data entry of, say, a 2 rather than a 3.

  Secondly, correct quantity. The transaction quantity (receipt or issue) should be below a particular amount. This limit quantity should either be defined manually or adjusted continually in dependency upon the average inventory movement (receipts or issues). In doubtful cases, an IT-supported system can request explicit double entry of quantities.

With tangible goods, bar codes can collect item identification. However, if the transaction quantity deviates from the planned quantity, it must be registered manually. This contrasts with product sales in the grocery or clothing industries, where each issue represents exactly one unit quantity, making quantity recording unnecessary.

To avoid recording lists of components for a production order, recording is required only for deviations from the picking list. The other positions are booked automatically by using the allocated quantity as the issued quantity as soon as the picking list is designated as issued.

11.1.3 Physical Inventory and Inventory Valuation

Intended learning outcomes: Identify physical inventory, inventory adjustment, and inventory valuation. Present an example of a stock inventory list. Differentiate between the periodic physical inventory procedure and the cycle counting procedure

Inventory accounting is dealing with valuing inventory (cf. [APIC16]).

Physical inventory is the process of determining inventory quantity by actual count. (Note: The term *physical inventory* can also mean the actual inventory itself. See Section 12.1.1.)

Inventory adjustment is a change made to an inventory record to correct the balance, to bring it in line with actual physical inventory balances.

Inventory valuation involves determining the value of the inventory at either its cost or its market value (cf. [APIC16]).

Physical inventory, inventory adjustment, and inventory valuation are needed to ensure goods on hand, for example. Furthermore, inventory is a company asset: One of the entries on the assets side of the balance sheet is the value of on-hand balance and in-process inventory. Tax authorities demand an exact physical inventory as well. Figure 11.1.3.1 presents an example of an inventory list that shows the value of the inventory stocks.

The lists generally class items according to group. Additional statistics at the end of the list, not shown here, group product range items according to certain other criteria.

Even with a very precise recording of book inventory, errors are possible — particularly in the case of unplanned, or unannounced, transactions:

- Errors in the data media recording inventory transactions
- Recording of erroneous quantity numbers
- Duplicate entry or failure to record a transaction
- Incorrect physical counts at the time of stock receipt
- Errors in the physical assignment of storage areas (stock sites are entered into the computer that in reality contain no stock)
- *Shrinkage*, or reduction of quantities in stock by pilferage, deterioration, or misuse

These errors are relatively difficult to detect. Physical inventory counts are necessary if users are to retain their trust in *record accuracy*, that is, the accuracy of the data in the computer.

Depending on the results of physical counts of inventory, new controls may be established, or controls that have proved to be unnecessary may be dropped.

Part ID	Description	Unit of measure	Stock	Entry	Issue	Avail-able	Or-dered	Allo-cated	Cost / unit	Inventory value	Stock range (months)
1348	Control-box	Pc	1499		850	649		600	1.45	941.05	
1349	Control-box	Pc	3314		1700	1614	560		0.59	952.26	
1425	Tank	Pc	2224		800	1424	2150	400	3.61	5140.64	1
1444	Horn	Pc	550	100	500	150			2.35	352.50	
2418	Hose 3 IN	Pc	7499		4200	3299	250		0.16	527.84	
2419	Hose 2 IN	Pc	7799	500	4400	3899		125	0.13	506.87	
2892	Closure	Pc	3058			3058	200	100	0.08	244.64	30
3010	Plate	Pc	918	315	525	708	175	110	0.15	106.20	1
3011	Gasket	Pc	5082	100	3185	1997	175		0.15	299.55	
3012	Spring	Pc	13500		7500	6000	100	500	0.07	420.00	
3021	Cartridge	Pc	1575		750	825	110		1.85	1526.25	1
3024	Cylinder	Pc	1978		1100	878		400	0.05	43.90	
3025	Pump	Pc	4			4			23.25	93.00	
3370	Motor	Pc	1350		750	600	3100	1200	7.25	4350.00	
3462	Pedal	Pc	100			100			1.53	153.00	999

Fig. 11.1.3.1 Example of a stock inventory list.

Particularly difficult is the inventory of items like coffee beans, leaves, seaweed, or gasoline. Such items change weight or volume significantly over time due to moisture or temperature.[1]

Periodic inventory is a physical inventory taken at a recurring, fixed interval, usually at the end of the organization's fiscal period (for example, the end of the calendar year).

Periodic inventory follows the procedure outlined in Figure 11.1.3.2.

- Shut down the warehouse.
- Physically count the stock quantities of randomly selected partial item quantities or all items. Check the results.
- Compare physically counted quantities to the quantities recorded in the inventory accounting system. Perform a deviation analysis.
- In case of significant deviations, first verify correct entry of inventory quantities. If this produces no results, re-perform the entire physical inventory, including the deviation analysis.

Fig. 11.1.3.2 Periodic physical inventory procedure.

The partial quantities of items to be inventoried are chosen in such a way that any deviations within these partial quantities will be representative for the entire quantity of the items.

1 Even roasted coffee loses weight over time. It gives off carbon dioxide, or "outgases," until it is stale.

For some firms, it is too costly to shut down the warehouse entirely, even for a few days. Sometimes the production rhythm does not permit it, or there is a lack of qualified employees for the physical inventory. Here, cycle counting, or even perpetual inventory, is important.

In *Cycle counting*, according to [APIC16], physical inventory is counted on a cyclic schedule, a regular, defined basis (often more frequently for high-value or fast-moving items and less frequently for low-value or slow-moving items).

The items determined by the cyclic schedule are mostly counted at the end of a workday, by a procedure similar to the one outlined in Figure 11.1.3.3. The method of comparison is the same as the one described above. A deviation analysis is performed for every counting cycle. It is also possible to count a random selection of all items for each cycle. After correction of any counting errors, the analysis is accepted, and the items can once again be released.

- Count every item periodically, in fixed cycles. The length of a period may vary, depending on the type and importance of the item. Count high-cost items more often than cheap ones.
- During the counting procedures, put a transaction freeze on only those items that are being inventoried at that particular moment. This will be a small percentage of all items. Furthermore, generally perform the physical count at the end of the working day, in other words, at a time at which the inventory transactions for the current day have already been executed.
- Select employees for the task who are trained and experienced. This reduces the probability of errors.

Fig. 11.1.3.3 Cycle counting procedure.

Some firms close the warehouse at the end of a working day for half an hour. They then inventory a random quantity of items and perform the deviation analysis. Usually, the same employees who have worked with receipts and issues during the day perform the counting.

11.2 Usage Statistics, Analyses, and Classifications

Intended learning outcomes: Present statistics on inventory transactions, sales, and bid activities. Explain the ABC Classification and the Pareto Chart. Describe the XYZ Classification and Other Analyses and Statistics.

11.2.1 Statistics on Inventory Transactions, Sales, and Bid Activities

Intended learning outcomes: Differentiate between usage and turnover statistics. Identify an outlier and abnormal demand. Differentiate between sales statistics and bid statistics.

Statistics on particular events can provide an important basis for various calculations in requirements planning and inventory management.

Usage statistics analyze the quantity of all inventory transactions.

For each transaction, the following attributes should be recorded:

- Date of transaction
- Identification of the item or the item family
- Moved quantity
- Employees responsible for the recording of the transaction
- The two customer, production, or purchase orders or inventory stock positions (target and actual, "before" and "after" position of the transaction)

As the number of recorded transactions is usually very large, in practice it is often impossible to make older transactions available for online queries. Moreover, too much time could be required to process many queries, particularly those pertaining to entire groups of items.

Turnover statistics condense the most important data on inventory transactions to gain rapid information about an item's movements.

Turnover statistics are updated, for example, daily, to include all transactions for that day. Managers maintain sales records for every item over the last statistical period, for example, the last 24 months and also over the three previous years. For all these periods, the following data are recorded:

- Total inventory issues, that is, items released from an inventory for use or for sale
- Partial inventory issues
- Inventory issues that were sold
- Total *inventory receipts*, that is, items released from an inventory for use or for sale
- Partial inventory receipts
- Inventory receipts that were purchased or produced

For each of these attributes, depending on need and the data storage capacity of the system, the following can be recorded:

- Number of transactions
- Turnover expressed in quantity
- Turnover expressed in value

The attribute *partial issues* allows to record outliers or abnormal demand. (The analogous considerations apply to the attribute *partial receipts*.)

An *outlier* is a data point that differs significantly from other data for a similar phenomenon. *Abnormal demand* — in any period — is demand that is outside the limits established by management policy. See [APIC16].

For example, if the average sales for a product were 10 units per month, and one month the product had sales of 500 units, this data point might be considered an outlier. Abnormal demand may come from a new customer or from customers whose demand is increasing or decreasing. In general, outliers and abnormal demand should not be taken as a basis for demand forecasting. Care must be taken to evaluate the nature of the abnormal demand: Is it a volume change, is it related to the timing of some orders, or a change in product mix?[2]

Usage and turnover statistics do not always suffice. This is the case when a relatively large time span lies between the demand to estimate and measured usage. A good example is capital goods having a considerable lead time of several months. In this case, we need statistics that are constructed in principle in the same way as the usage and turnover statistics above, but relate to more current events. A favorite measurement time point is the moment of sale or — even more up-to-date — the moment of bidding.

A *sales transaction* of an item records the sending of the order confirmation and thus the moment the customer order is accepted. *Sales statistics* analyze all sales transactions.

Sales statistics are more up-to-date than usage statistics — by the amount of the lead time for the order. However, the corresponding sales data files tend to be less precise, for customers may cancel or alter placed orders.

A *bid transaction* of an item records the sending of a bid to the customer. *Bid statistics* analyze all bid transactions.

Bid statistics are even more up-to-date than sales statistics due to the time that on average lapses between the formulation of the bid and the sale. But again, the corresponding data are less precise. In fact, the order success probability (see Section 5.2.1) shows the approximate percentage of bids that translated into sales. However, the latter cannot be ascertained reliably for every individual product, or even for every individual product family.

11.2.2 The ABC Classification and the Pareto Chart

Intended learning outcomes: Explain the principle of the ABC classification, shown as a Pareto chart. Describe the ABC classification for each ABC category.

Up to now we have stressed the "importance" of an item in relationship to all items as a whole. Turnover, which generally refers to past usage, can yield the importance of an item. However, forecasts rather than turnover may also yield this information.

In all types and sizes of organizations, it can be observed that a small number of products make up the largest portion of the turnover.

ABC classification divides a set of items into three classes, specifically A, B, and C.[3]

Figure 11.2.2.1 illustrates the principle of this classification and possible limits for a change of class (break points) in the form of a *Pareto chart*:

---

2 The *product mix* is the proportion of individual products that make up the total production or sales volume (cf. [APIC16]).

3 ABC classification is an application of Pareto's law to inventory management. *Pareto's law* is the concept that 20% of any entity represent the very important few and the remaining 80% are less important (cf. [APIC16]).

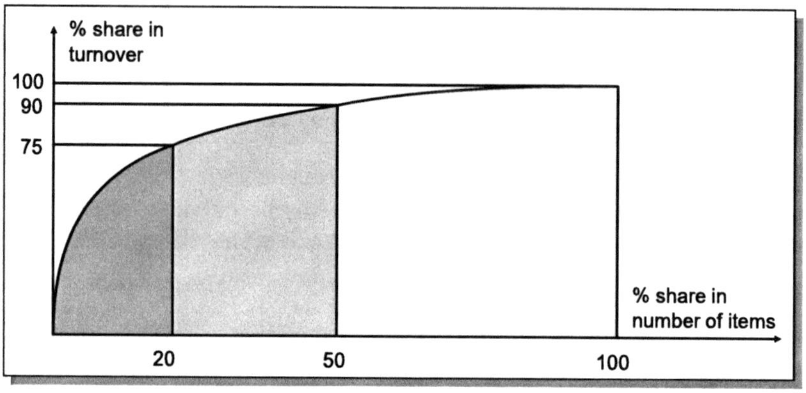

Fig. 11.2.2.1 The principle of the ABC classification, shown as a Pareto chart.

- In the example, A items, that is the A class, is composed of 20% of the items, which account for 75% of total turnover.
- B items, that is, the B class, is made up of 30 to 40% of the items, which comprise approximately 15% of total turnover.
- *C items*, that is, the remaining items, which make up a large part of the product range, here 40 to 50% of the items, only account for about 10% of total turnover.

The precise shape of the Pareto curve and the break points between classes will vary among firms, but the point that a small percentage of items make up most of the importance (or value) remains generally true. Prioritizing inventory items according to the ABC classification allows targeted implementation of appropriate materials management measures.

- It is much more important to reduce inventory for A items than for C items. In addition, since A items are more limited in number, close follow-up is much easier. They are ordered in frequent small batches. Purchase orders are placed only after intensive evaluation. Production orders are closely reviewed and expedited with high priority. All these measures increase ordering costs and administrative costs.
- It is important that C items are always available. Under no circumstances should an item that costs only a few cents be allowed to delay the delivery of a machine that may have a value of hundreds of thousands of dollars. Management releases procurement orders early, with ample margins as to quantity and time. This increases storage costs only slightly, since the items are inexpensive ones. Ordering costs for C items can be kept low, since large quantities are ordered at one time. It may sometimes even be possible to trigger orders automatically, without the intervention of a planner, by using an IT-supported system.
- Generally, management handles B items between the above two extremes.

The ABC classification thus provides the foundation for various parameters in materials management. Since goods have different importance, depending on their type, most organizations have separate ABC classifications for each item type, as outlined in Section 1.2.2 (final products, subassemblies, individual parts, raw materials, and so on). This is especially important when the value added is high. In that case, a sole ABC classification for the entirety of the item range would tend to classify all final products as A items and all purchased items as C items. However, this would defeat the objective of the ABC approach.

The *ABC category* is the identification of the set or group of items grouped together for an ABC classification.

Therefore, first all items are assigned to an ABC category. Then, the ABC classification is completed in two stages, as outlined in Figure 11.2.2.2.

In a first stage, sort all items of an ABC category to calculate 100% of the selected classification criterion (a measure of importance, or value), such as turnover.

In a second stage, handle all the items in descending order of the chosen criterion by adding up the criterion. Compare the partial sums of the items handled to the 100% figure.

- All items that are handled at the beginning in accordance with this descending order receive the classification A.
- If, for example, the partial sum exceeds, say, 75% (A break point) of the total quantity of 100% that was determined in the first stage, assign the items that follow to classification B.
- If the partial sum exceeds, say, 90% (B break point) of the total quantity of 100% that was determined in the first stage, assign the items that follow to classification C.

Fig. 11.2. 2.2 The ABC classification for each ABC category.

11.2.3 The XYZ Classification and Other Analyses and Statistics

Intended learning outcomes: Explain the XYZ classification. Present exception lists.

XYZ classification distinguishes items with regular or even continuous demand (X items) from those with completely irregular, lumpy, or unique demand (Z items). Y items lie between the two extremes.

The decision about the assignment of an item results from analysis of the demand quantities per statistical period. Thus, the dispersion of the demanded quantities is a measure for the classification. For example, for an item in the X class we could require that the deviation from average consumption should not be larger than 5% per week or 20% per month.

Materials management sets its policies according to the XYZ classification. It also determines whether important materials management parameters should be calculated automatically (for example, using forecast data) or set manually.

An *exception list* contains goods that do not "normally" pass through the company.

Exception lists can be based on inventory transactions, such as

- *Shelf warmers*, that is, items that have not moved during a certain number of months
- Items that do not show a sufficient turnover
- Items whose inventory value exceeds a particular total

Exception lists serve to sort out items that are in an exceptional state according to a particular criterion. Even in the case of computer supported planning & control systems, users can usually define such exception lists themselves.

We will discuss the entire category of exception messages that affect production and procurement orders in the course of this chapter as well as in Chapter 12.

11.3 Order Point Technique and Safety Stock Calculation

Intended learning outcomes: Explain the (re-)order point technique and variants thereof. Describe the safety stock calculation with continuous demand. Disclose the determination of the service level and the relation of service level to fill rate.

11.3.1 The Order Point Technique

Intended learning outcomes: Present in detail characteristic data for the (re-)order point technique. Explain the (re-)order point calculation. Identify the criterion for the release of a production or procurement order.

The *order point technique*, or *order point system*, is used for items with stochastic demand that is relatively continuous along the time axis. The characteristic inventory curve is the *saw-toothed curve* as shown in Figure 11.3.1.1.

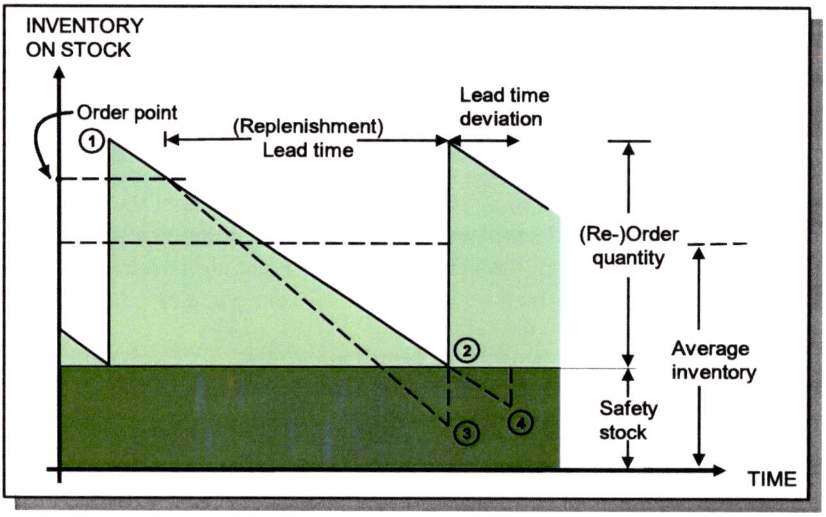

Fig. 11.3.1.1 Characteristic data for the order point technique.

* After stock entry (point 1), the stock falls gradually until it is below a quantity called the *order point*. At this time, a production or procurement order is generated.

- The inventory level sinks continually during the replenishment lead time, that is, the total period of time from the moment of reordering until point 2, where the replenishment order quantity is available for use (determining this batch size is the subject of Section 11.4). After the stock entry, the cycle begins anew at point 1. The decline between the points 1 and 2 represents the demand during the lead time. This demand is a stochastic value.
- If the actual demand is larger than the expected (forecast) demand, the inventory level curve corresponds to the dashed line that leads to point 3. If no safety stock was maintained, there will be a stockout.
- If the actual lead time is longer than the (expected) lead time, then the inventory stock curve corresponds to the dashed line that leads to point 4. If no safety stock was maintained, there will be a stockout.

The *order interval* or *order cycle* is the time period between the placements of orders.

Cycle stock is the component of inventory that depletes gradually as customer orders are received and is replenished cyclically when supplier orders are received (cf. [APIC16]).

Safety stock is the component of inventory that serves as a buffer to cover fluctuations in lead time and in the demand during the lead time. Statistically, we need to draw on safety stock in half of all procurement cycles. For definitions, see Section 11.3.3.

This system is more difficult to manage in the case of discontinuous but regular demand (the case, for example, with seasonal components). The saw-toothed curve then has a shape that reproduces the seasonality of the demand (see Section 10.3.4).

The area under the saw-toothed curve, multiplied by a cost rate, yields the carrying cost for this item per time unit. These are equal to the storage costs for the mean stock per time unit.

We can derive *average inventory* for the order point technique in Figure 11.3.1.1 by using the following formula (Figure 11.3.1.2):

$$\text{average inventory} = \text{safety stock} + \frac{\text{order quantity}}{2}$$

Fig. 11.3.1.2 Average inventory.

The *order point*, or *reorder point*, is calculated from safety stock and expected (forecast) demand during the procurement period according to the formula in Figure 11.3.1.3.

Calculation of the order point is executed after calculation of the demand forecast and always at the end of a statistical period. Order point calculation should be executed more frequently in cases of discontinuous demand, longer statistical periods, and shorter lead times, because the forecast may change significantly over the course of time.

$$\begin{aligned}&\text{order point}\\&\quad=\text{safety stock}\\&\quad+\text{demand forecast during the lead time}\\&\text{where}\\&\quad\text{demand forecast during the lead time}\\&\quad=\frac{\text{lead time}}{\text{length of the statistical period}}\cdot\text{demand forecast during the statistical period}\end{aligned}$$

Fig. 11.3.1.3 Order point calculation.

In addition to physical inventory, also *scheduled receipts* cover the demand during the lead time. These include firmly or released ordered quantities (see the definition in Section 12.1.1), since these will all arrive during the lead time. If the formula contained in Figure 11.3.1.4 holds, a new production or procurement order should be released.

$$\text{Physical inventory}+\sum\text{Scheduled receipts}<\text{Order point}$$

Fig. 11.3.1.4 Criterion for the release of a production or procurement order.

For management purposes, a periodically produced list contains each the items for which the criterion in Figure 11.3.1.4 is satisfied, together with an order proposal that contains all the required information, such as the predicted receipt to stock, the batch size, and information regarding earlier productions or procurements. Such an order proposal also serves to specify purchase blanket orders more precisely. Since the procurement decision must be made without delay, the proposal also contains bids from suppliers.

11.3.2 Variants of the Order Point Technique

Intended learning outcomes: Identify the criterion for the release of a production or procurement order, if the customer allows a minimum delivery lead time. Explain the criterion for an early issuance of a production or procurement order. Produce an overview on the min-max (reorder) system. Describe the double order point system.

If the customer allows a *minimum delivery lead time*, then we know all the *allocated quantities* or *reserved quantities* (i.e. the demand that is linked to released customer orders or assigned to production orders; see the definitions in Section 12.1.1) during this period of time. This is true for all customer or production orders that require the item. Thus, we can choose the time to release according to the formula in Figure 11.3.2.1.

$$\begin{aligned}&\left[\begin{aligned}&\text{Physical inventory}+\sum\text{Scheduled receipts}\\&\quad-\sum\text{Allocated quantities during minimum delivery lead time}\end{aligned}\right]<\text{Reduced order point}\\&\text{where}\\&\quad\text{Reduced order point}=f(\text{reduced lead time})\\&\quad\text{Reduced lead time}=\text{lead time}-\text{minimum delivery lead time}\end{aligned}$$

Fig. 11.3.2.1 Criterion for the release of a production or procurement order, if the customer allows a minimum delivery lead time.

Since the demand that is to be determined stochastically must now cover only a reduced lead time, the technique becomes more deterministic and precise — particularly in the case of trends that are not considered by the forecast model.

Production or procurement orders can be released earlier than necessary:

The *anticipation horizon* refers to the maximum anticipated time for consideration of early release of a production or procurement order.

Figure 11.3.2.2 shows a formula to identify the items that are candidates for an early release. For techniques with an early issuance of production orders, see Section 15.1.3.

$$\text{Physical inventory} + \sum \text{Scheduled receipts} - \sum \text{Allocated quantities during the anticipation time} < \text{Order point}$$

Fig. 11.3.2.2 Criterion for an early issuance of a production or procurement order.

The saw-toothed curve — which stands for the optimal functioning of the order point technique — is best attained if the issue quantities are small compared to the production or procurement batch size. If instead they are relatively large, a chopped-off saw-toothed curve results. For issue quantities on the order of the production or procurement batch size, the resulting curve looks more like the shape of human teeth with gaps between them. Then, the order point technique no longer yields satisfactory results. Here see Section 12.3.1.

A variant of the order point technique described above is the min-max (reorder) system.

With the *min-max (reorder) system*, the "min" (minimum) is the order point, and the "max" (maximum) is the *order-up-to level* or *target inventory level*. The order quantity is variable and is the result of the max minus physical inventory minus scheduled receipts. An order is recommended when the sum of the physical inventory plus scheduled receipts is below the minimum. The *periodic review system* is a variant of the min-max system in which an order is placed every fixed number of time units. The order quantity is variable and essentially replaces the items consumed during the current time period. Cf. [APIC16].

These techniques define maximal storage space requirements. This is particularly important for racks and shelves in supermarkets, for example. Another variant of the order point technique is a system that is used frequently for management of distribution inventory.

The *double order point system* has two order points. The smallest equals the traditional order point, which covers the demand forecast during the replenishment lead time. The second, higher order point is the sum of the first order point plus the demand forecast during the replenishment lead time *of the preceding structural level*, most usually the production lead time or the purchasing lead time. Cf. [APIC16].

Figure 11.3.2.3 shows the principle for applying the double order point system. RLT1 is the replenishment lead time of the traditional order point technique, and RLT2 is the replenishment lead time of the preceding structural level.

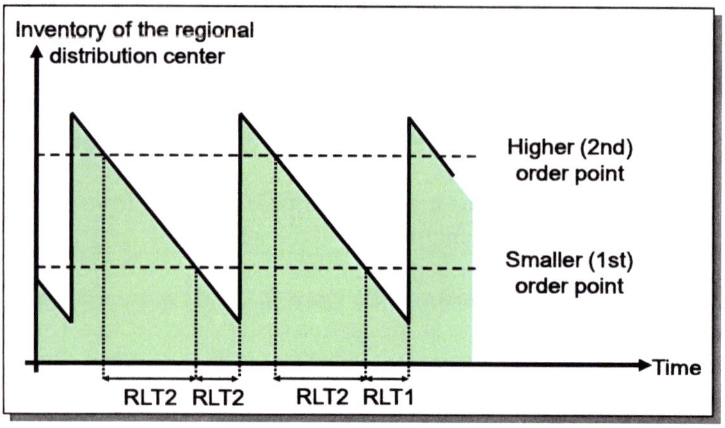

Fig. 11.3.2.3 The double order point system.

As soon as inventory at the regional distribution center drops and reaches order point 2, the information is sent to the central warehouse as an order proposal, which the regional distribution center would have to release at about this time if it were ordering directly from the manufacturer or supplier instead of from the central warehouse.

The central warehouse has now got advance warning that an order is pending. It enables the central warehouse to forewarn the manufacturer of future replenishment orders. The advantage is that in theory, no safety stock needs to be held at the central warehouse.

11.3.3 Safety Stock Calculation with Continuous Demand

Intended learning outcomes: Describe different techniques for determining safety stock. Identify different patterns of the deviation of demand from forecast. Explain safety stock in relation to service level. Disclose the normal integral distribution function (service function) and the Poisson distribution integral function. Present the formula for safety stock.

Figure 11.3.1.1 indicates that without safety stock, there will be a stockout in half of the cycles defined by the saw-toothed curve. This results in backorders.

Safety stock or *buffer stock* serves to cushion the impact of forecast errors or deviations in the lead time as well as in the demand during the lead time.

Anticipation inventories is a similar term, used in the management of distribution inventory. It means additional inventory above basic pipeline stock to cover projected trends of increasing sales, planned sales promotion programs, seasonal fluctuations, plant shutdowns, and vacations ([APIC16]).

Figure 11.3.3.1 shows different techniques for determining safety stock depending on the nature of the item.

The first two techniques determine safety stock in a largely intuitive manner. For the statistical derivation, however, there are formal techniques available, as described in the following:

Technique	Safety stock	Typical use
Fixed	Set (manually) quantity	New and old items, discontinuous or lumpy demand patterns, low-cost items
Time period	Determine by forecasts for future periods	Critical components, new and old items, discontinuous or lumpy demand patterns
Statistical	Calculate via statistical method based on history	Mature items, continuous or regular demand patterns, deviations in predictable range

Fig. 11.3.3.1 Different techniques for determining safety stock.

1. *Statistical Fluctuations in the Lead Time*

Fluctuations in the lead time due to unplanned delays in production or procurement, for example, are absorbed by a safety lead time.

The *safety lead time* is an element of time added to normal lead time to protect against fluctuations. Order release and order completion are planned for earlier dates (before real need dates), according to the time added.

Safety stock due to fluctuations in lead time is calculated simply as the demand forecast during this safety lead time. This technique is often used, because it is easily understood.

2. *Statistical Fluctuations in Demand*

For purposes of absorbing demand fluctuations, safety lead time is not a sufficient basis for calculation.

Fluctuation inventory, or *fluctuation stock*, is inventory that is carried as a cushion to protect against forecast error ([APIC16]).

Figure 11.3.3.2 shows the pattern of demand for two items with the same demand forecast, but different demand fluctuations.

The fluctuation inventory for the item in Situation B must be larger than that for the item in Situation A. A pattern of demand that has only a small dispersion around the demand forecast will result in a smaller quantity of safety stock; one with large variation will require a larger quantity of safety stock.

The *service level*, or *cycle service level*, or *level of service*, is the percentage of order cycles that the firm will go through without stockout, meaning that inventory is sufficient to cover demand.

The *probability of stockout* is the probability that a stockout will occur during each order cycle before a replenishment order arrives.

According to these definitions, the following relationship holds (see Figure 11.3.3.3).

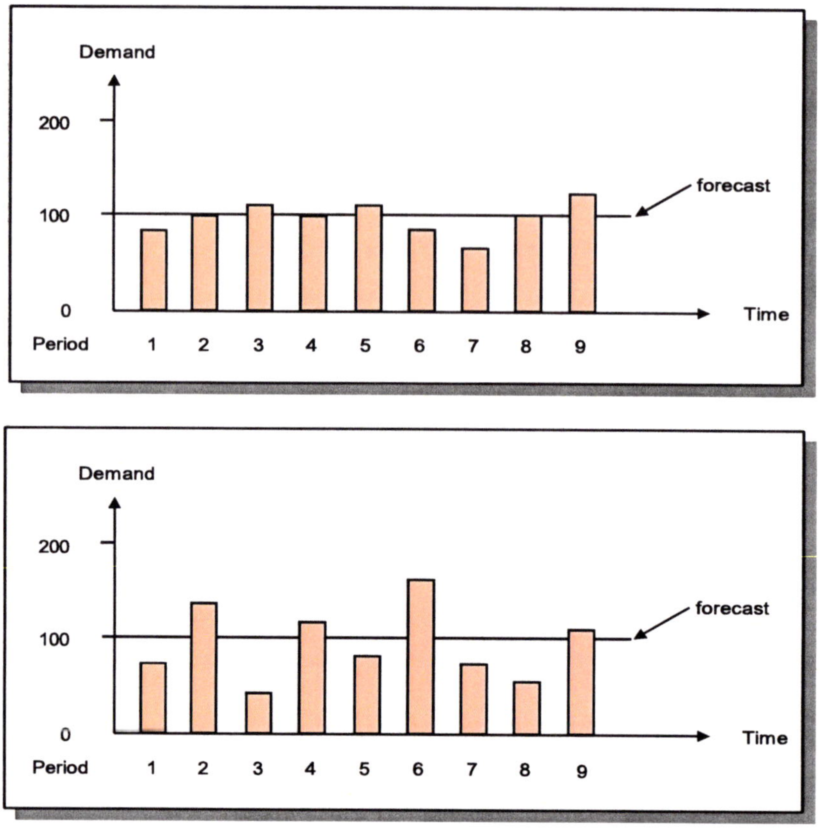

Fig. 11.3.3.2 Different patterns of the deviation of demand from forecast.

service level = 100% - probability of stockout per order cycle

Fig. 11.3.3.3 Service level expressed as the complement of probability of stockout.

With the order point technique, fluctuating demand can be satisfied from stock even without safety stock in about half of all cases. For this reason, the service level using this technique can be assumed to be at least 50%.

Safety stock — and with it carrying cost — grows quantitatively in dependency upon service level, as Figure 11.3.3.4 shows. Once the desired service level is set, safety stock can be estimated accurately through statistical derivation.

The *safety factor* is the numerical value, a particular multiplier, for the standard deviation of demand.

The *service function* is the integral distribution function, for which the integral under the distribution curve for demand up to a particular safety factor s corresponds to the service level.

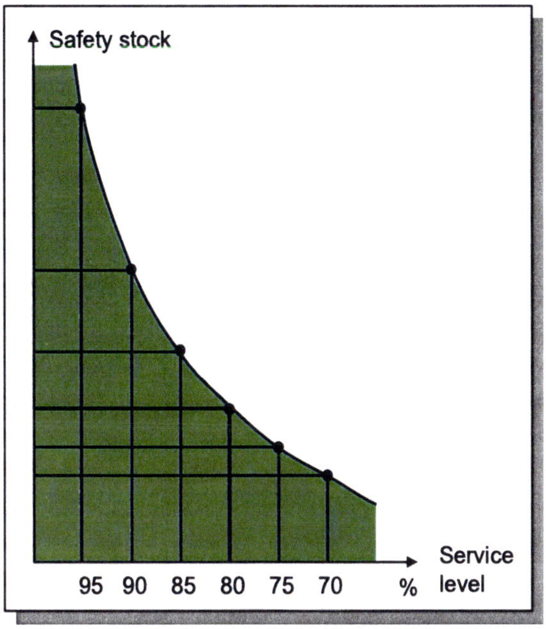

Fig. 11.3.3.4 Safety stock — and thus carrying cost — in relation to service level.

If demand follows a *normal distribution*, or a bell-shaped curve, the service level corresponding to the safety factor s is the area shown in gray in Figure 11.3.3.5.

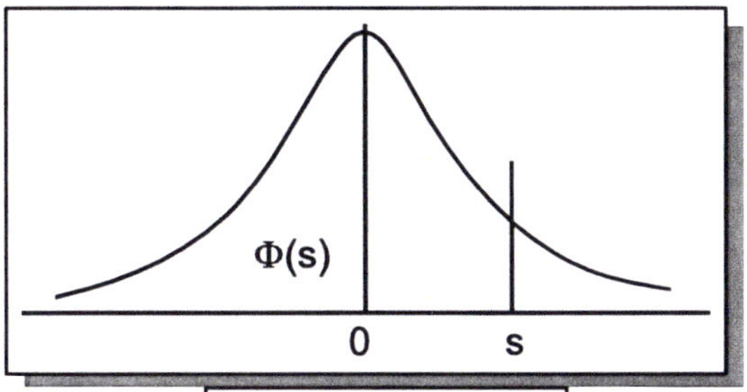

$$\Phi(s)=\frac{1}{\sqrt{2\pi}}\int_{-\infty}^{s} e^{-\frac{1}{2}x^2}\,dx$$

Fig. 11.3.3.5 Normal integral distribution function (service function).

Therefore, the safety factor is also the inverse function of the integral distribution function. It is the numerical value used in the service function (based on the standard deviation of the forecast) to provide a given level of service.

Figure 11.3.3.6 reproduces examples for corresponding values of the service level and the safety factor. They can be read from tables, such as the following table from [Eilo62], p. 26.

Figure 11.3.3.7 shows the resulting formula for safety stock. With a normal distribution, it is possible to use 1.25 * MAD (mean absolute deviation) instead of the standard deviation.

Safety factor	Service level %
0	50.00
0.5	69.15
1	84.13
1.5	93.32
2	97.73
2.5	99.38
3	99.86
4	99.997

Service level %	Safety factor
50	0
65	0.385
80	0.842
90	1.282
95	1.645
98	2.054
99	2.326
99.9	3.090

Fig. 11.3.3.6 Service level and safety factor when demand follows a normal distribution. (From [Eilo64], p. 26.)

safety stock
= safety factor · standard deviation of the demand during the lead time

where
safety factor = g(service level)
g = inverse function of the integral distribution function chosen

and
standard deviation of the demand during the lead time

$$= \sqrt{\frac{\text{lead time}}{\text{length of the statistic period}}} \cdot \text{standard deviation of the demand during the statistic period}$$

Fig. 11.3.3.7 Formula for safety stock.

In particular for small demand quantities, we cannot always assume that demand is normally distributed. Sometimes, we could assume a Poisson distribution instead. However, with a mean value (average demand quantity) of merely 9 units, the upper part of the Poisson distribution curve is very close to the curve of the normal distribution. This is particularly true for larger safety factors and high service levels. See also Figure 10.5.5.1.

Figure 11.3.3.8 shows an example of the *Poisson distribution* and its integral function. Depending upon the mean value λ, a different curve and a different inverse function will result.

Figures 11.3.3.9 and 11.3.3.10 show pairs of values of the service level and safety factor for means of $\lambda = 4$ and $\lambda = 9$, respectively.

For small consumption quantities, the cost of a stockout often does not depend so much on the quantity not delivered as upon the fact that there is a failure to meet the full quantity. Thus, with small usage quantities the tendency is to choose a high service level, which in turn results in a high safety factor. The calculated safety factor that is based on a Poisson distribution is then generally fairly equivalent to the one based on a normal distribution.

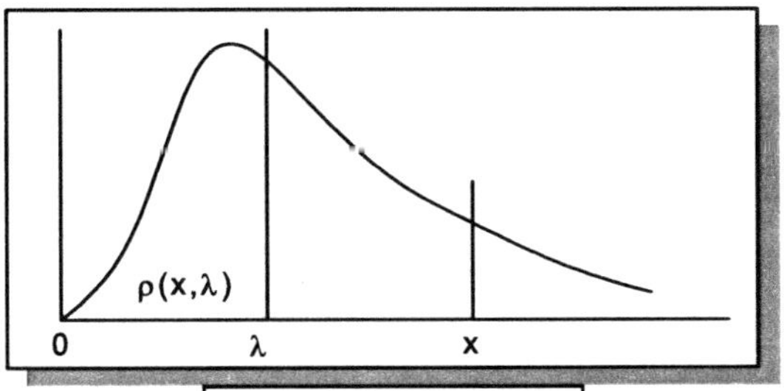

$$\rho(x,\lambda) = e^{-\lambda} \cdot \sum_{k=0}^{x} \frac{\lambda^k}{k!}$$

Fig. 11.3.3.8 Poisson distribution integral function.

Safety factor	Service level %
0	43.35
0.5	62.88
1	78.51
1.5	88.93
2	94.89
2.5	97.86
3	99.19
4	99.91

Service level %	Safety factor ≈
50	0
65	0.6
80	1.1
90	1.6
95	2.1
98	2.7
99	2.9
99.9	3.9

Fig. 11.3.3.9 Table of values for the Poisson cumulative distribution with a mean demand value of λ = 4 and standard deviation $\sqrt{\lambda}$ = 2 units per period. (From [Eilo62], p. 84 ff.)

Safety factor	Service level %
0	45.57
0.5	64.53
1	80.30
1.5	89.81
2	95.85
2.667	98.89
3	99.47
4	99.96

Service level %	Safety factor ≈
50	0
65	0.5
80	1.0
90	1.5
95	1.9
98	2.4
99	2.8
99.9	3.8

Fig. 11.3.3.10 Table of values for the Poisson cumulative distribution with a mean demand value of λ = 9 and standard deviation $\sqrt{\lambda} = 3$ units per period. (From [Eilo62], p. 84 ff.)

However, based on *probability* of stockout alone, we cannot say anything about the stockout *quantity*, the stockout *percentage*, or backorder *percentage*. Thus, service level is not the same as *fill rate*, which only measures what actually happens when demand occurs. See also [Bern99], [Chap06].

Like fill rate (see the definition in Section 5.3.1), service level is the quantitative application of the answer to the following question: What are the costs of not meeting customer demands from stock? Both measures, fill rate and service level, are thus estimates of opportunity cost. To achieve a specific fill rate, however, it is generally sufficient to set a smaller number as the service level, or desired probability that demand can be met from stock. The relationship between the two measures, fill rate and service level, as well as ways of determining the appropriate service level, are examined in Section 11.3.4.

11.3.4 Determining the Service Level and the Relation of Service Level to Fill Rate (*)

Intended learning outcomes: Describe the order point technique where the length of order cycle is a multiple of the lead time. Explain the stockout probability depending on stockout costs per unit. Present the service function of the stockout quantity coefficient in dependency upon the safety factor. Present examples of the relation between fill rate and service level.

Figure 11.3.4.1 shows a typical order cycle using the order point technique shown in Figure 11.3.1.1, in which the *length of order cycle*, that is, the length of time the batch size will provide stockout coverage, is a multiple of the lead time. The batch size itself is a multiple of the expected demand during the lead time.

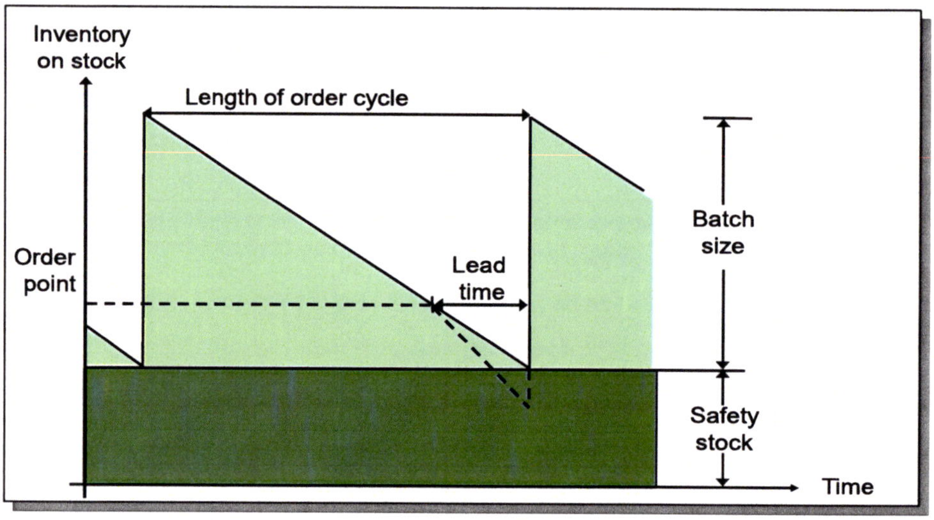

Fig. 11.3.4.1 Order point technique with an order cycle where the length of order cycle provided by the batch size is a multiple of the lead time.

If the length of order cycle divided by lead time equals 10, for example, and demand is not too discontinuous, then 90% of the batch size can be covered without stockout. Stockout will only occur for demand during the lead time, or for 10% of the batch size. If no safety stock were carried (safety factor is 0, that is, a service level of only 50%), the fill rate would be approximately 90% and higher. This shows that service level can usually be a percentage that is significantly smaller than the desired fill rate (which in most cases must be set at close to 100%; see the discussion in Section 5.3.1).

As mentioned above, determining the desired fill rate and service level has to be the quantitative application of the qualitative answer to the question of what stockouts will cost. Thus, fill rate and service level express an estimation of opportunity cost.

Stockout costs are the economic consequences of stockouts.

Stockout costs can include extra costs for express/emergency production or procurement or customer delivery, but also penalty costs, loss of sales, loss of contribution margin, loss of customer goodwill, and all kinds of associated costs. See the discussion in Section 1.3.1.

The following shows the derivation of two methods of determining the desired service level:

1. The first method is based on the assumption that opportunity costs can be assigned directly to each unit not filled.
2. The second method is based on the assumption that the total opportunity costs can be assigned to the fill rate during a particular time period (a year, for instance).

1. *Determine service level on the basis of stockout costs for each unit of an item not filled.*

Where stockout costs can be expressed as costs per unit (of measure) not delivered, [Cole00], [SiPy98], and [Ters93] offer the following direct calculation of the *optimum probability of stockout* (see Figure 11.3.4.2). Because a stockout can only happen at the end of an order cycle, the number of stockouts cannot be greater than the number of order cycles. Often the period chosen for the calculation is one year.

$$\text{Number of stockouts per year} = \text{Probability of stockout per order cycle} \cdot \text{Number of order cycles per year} \leq \text{Number of order cycles per year}$$

with

$$\text{Number of order cycles per year} = \frac{\text{Average annual consumption}}{\text{Batch size}}$$

$$\text{Optimum number of stockouts per year} = \frac{\text{Carrying cost per unit and year}}{\text{Stockout costs per unit}}$$

thus

$$\text{Optimum probability of stockout per order cycle} = \frac{\text{Carrying cost per unit and year}}{\text{Stockout costs per unit}} \cdot \frac{\text{Batch size}}{\text{Average annual consumption}}$$

Fig. 11.3.4.2 Probability of stockout in dependency on stockout costs per unit.

As a consequence, the *optimum service level* results directly from the relation in Figure 11.3.3.3. Section 11.4 discusses determination of batch size, which often precedes safety stock calculation.For example, if there are five order cycles per year (the average annual consumption is five times the batch size) and stockout costs per unit are four times greater than carrying cost, the resulting optimum probability of stockout is 0.05 and the optimum service level is 95%.[4]

2. *Determine service level on the basis of fill rate.*

If a certain stockout percentage or backorder percentage has been set on the basis of estimated annual stockout costs, then the service level can be derived from the fill rate by estimating the stockout quantity per order cycle. See also [Brow67] and [Stev14].

For a *particular safety factor*, from now on called s, the stockout quantity is the product of all possible not-filled quantities and their probability of occurrence. A specific not-filled quantity is the quantity m, which exceeds the expected quantity of demand plus s times the standard deviation of demand during the lead time. Proportional to the standard deviation, this quantity can be expressed as (t–s) times the standard deviation σ, for each $t \geq s$. p(t) is then, for example, the normal probability density function. Instead of the quantity, the factor of proportionality with its probability of occurrence yields the stockout quantity coefficient.[5]

The *stockout quantity coefficient* P(s) is the factor that, multiplied by the standard deviation of demand per lead time, yields the expected stockout quantity subject to the safety factor s.

The formula for the stockout quantity coefficient in Figure 11.3.4.3 is similar to the formula in Figure 11.3.3.5. P(s) is the integral, for all possible $t \geq s$, of the factor of proportionality (t – s) of the standard deviation of demand during lead time multiplied with p(t).

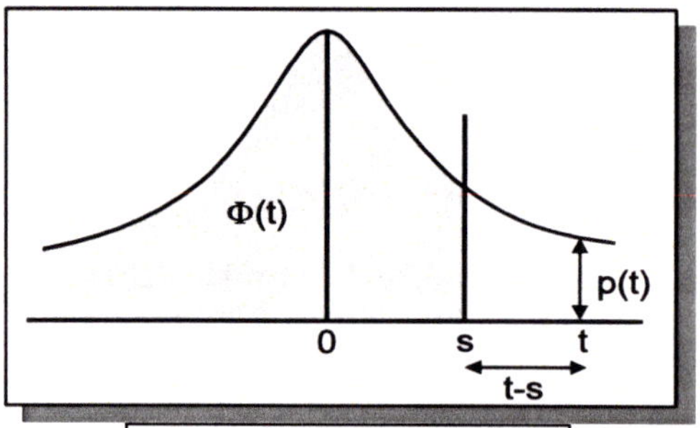

$$P(s) = \int_{t=s}^{\infty} (t-s) \cdot p(t)\,dt$$

Fig. 11.3.4.3 Service function (of the stockout quantity coefficient) P(s) in dependency upon the safety factor s.

---

4 In cases where the formal probability of stockout calculated in this way should be greater than 0.5, the lowest reasonable service level should be assumed (usually 50%).

5 This transformation of the quantity m can be confusing: m becomes $(t-s) \cdot \sigma$. The formula for calculating t, belonging to a specific m, is then $t(m) = (m + s \cdot \sigma) / \sigma$. This unusual method may be one of the reasons why, in the literature, the relation between service level and fill rate is often explained only superficially or not at all.

Figure 11.3.4.4 shows examples of corresponding values of safety factor s and stockout quantity coefficient P(s). The values can be determined by table look-up; see, for example, tables in [Brow67], p. 110, or [Stev14].

Stockout quantity coefficient P(s)	Safety factor s	Service level in %
0.8	–0.64	26.11
0.4	0	50
0.2	0.5	69.15
0.1	0.9	81.59
0.05	1.26	89.61
0.01	1.92	97.26
0.005	2.18	98.53
0.001	2.68	99.63
0.0001	3.24	99.95

Service level in %	Safety factor s	Stockout quantity coefficient P(s)
30	–0.52	0.712
50	0	0.399
65	0.385	0.233
80	0.842	0.112
90	1.282	0.048
95	1.645	0.021
98	2.054	0.008
99	2.326	0.003
99.9	3.090	0.0003

Fig. 11.3.4.4 Safety factor s and stockout quantity coefficient P(s) with normally distributed demand. (Following [Brow67] or [Stev14].)

Thus, the expected stockout quantity per order cycle can be calculated from safety factor s via the stockout quantity coefficient P(s).

According to the definition in Section 5.3.1, the stockout quantity per order cycle is also the product of batch size and stockout percentage (that is, the complement of fill rate). This yields formulas, shown in Figure 11.3.4.5, that relate *service level* to *fill rate.*

Let us look at an example that illustrates the relation between fill rate and service level.

Say the batch size is 100 units, and the standard deviation of demand during the lead time is 10 units. What safety stock should be carried to provide a desired fill rate of 99.9%? The stockout quantity coefficient P(s) is 0.01 (Figure 11.3.4.5), and the safety factor is thus 1.92 (Figure 11.3.4.4). Therefore, the resulting safety stock is 1.92 times 10 = 19.2 units (Figure 11.3.3.7).[6]

Figure 11.3.4.6 shows that the quotient resulting from the standard deviation of demand during lead time divided by batch size (following Figure 11.3.4.5) has a leverage between service level and fill rate. The smaller this quotient is, the higher — at a constant service level — the expected fill rate. That means that with a service level of 50% (that is, no safety stock) and a quotient of 1/5, a fill rate of more than 92% is achieved, while with a quotient of 1/10 (as in the example above), the fill rate achieved is about 96%. With a service level of 80%, a quotient of 1/10 results in a fill rate of over 98.8%.

---

6 It is interesting to note that setting a low service level results in a safety factor of less than 0 (as Figure 11.3.4.4 shows).

stockout quantity per order cycle = standard deviation of demand during lead time • P(s)
= batch size • (1 – fill rate)

On the basis of the fill rate, determine

$$P(s) = \frac{\text{batch size}}{\text{standard deviation of demand during lead time}} \bullet (1 - \text{fill rate})$$

And safety factor s and service level can be determined from P(s) through table look-up (such as Figure 10.3.3.4) or directly from safety stock according to the formula in Figure 10.3.3.7.

On the basis of the service level, safety factor s and P(s) can be determined through table look-up (such as Figure 10.3.3.4) and thus the expected fill rate can be calculated according to the formula:

$$\text{Fill rate} = 1 - \frac{\text{standard deviation of demand during lead time}}{\text{batch size}} \bullet P(s)$$

Fig. 11.3.4.5 Relation between fill rate and service level.

Service level in %	Standard deviation of demand during lead time / batch size	Fill rate in %
50	1/5	92.05
50	1/10	96.01
50	1/100	99.60
50	1/200	99.80
80	1/5	97.76
80	1/10	98.88
80	1/100	99.89
80	1/200	99.94

Fig. 11.3.4.6 Examples of the relation between service level and fill rate.

And finally, consider an example that links stockout costs per unit, via the optimal service level derived using method 1 above, with the fill rate calculated with method 2 above.

In this example, annual carrying cost per unit is 1, the batch size is 100, average annual demand is 500, and the standard deviation of demand during the lead time is 10. What is the expected fill rate based on the given carrying cost per unit of 4? The optimum probability of stockout in each order cycle is 0.05 (Figure 11.3.4.2), which results in an optimum service level of 95% following Figure 11.3.3. Following Figure 11.3.4.4, this corresponds to the stockout quantity coefficient P(s) = 0.021. Following Figure 11.3.4.5, this yields a fill rate of 99.79%.

According to the formulas in both method 1 and method 2 above for calculating the desired service level, the service level and safety stock both decrease with increasing batch size. For this reason, it would be desirable to set the batch size as large as possible. For production orders in particular, however, as Chapter 13 will show, the cumulative lead time often grows disproportionately as batch size increases, making it necessary to apply stochastic models of demand and to include the standard deviation. From this perspective, a small batch size is desirable. In practice, then, batch sizes and safety stock must be determined simultaneously (*de facto* in iteration).

11.4 Batch or Lot Sizing

Intended learning outcomes: Produce an overview on production or procurement costs, batch-size-dependent unit costs, setup and ordering costs, and carrying cost. Explain optimum batch size, optimum length of order cycle, the classic economic order quantity formally and in practical application. Disclose extensions of the batch size formula.

Batch sizing, or *lot sizing*, is the process of, or techniques used in, determining batch or lot size ([APIC16]).

11.4.1 Production or Procurement Costs: Batch-Size-Dependent Unit Costs, Setup and Ordering Costs, and Carrying Cost

Intended learning outcomes: Differentiate between batch-size-dependent and batch-size-independent production or procurement costs. Explain carrying cost and carrying cost rate. Present capital costs, storage infrastructure costs and the risk of depreciation.

Lot-size inventory is inventory that results whenever quantity price discounts, shipping costs, setup costs, or similar considerations make it more economical to purchase or produce in larger lots than are needed for immediate purposes ([APIC16]).

Batch sizes that are not specified by the user lead to longer lead times and procurement deadlines and should therefore be avoided, as discussed in the Lean/JIT concept. Even there, batch sizes have to be accepted because of setup costs. In this section, we will examine the arguments that tend to favor either smaller or larger batch sizes.

There are *batch-size-dependent production or procurement costs* for every produced or procured unit of measure of the order, that is, the *batch-size-dependent unit costs*.

Batch-size-dependent production or procurement costs are

- In the case of external procurement, acquisition cost per procured unit quantity plus eventual additional costs that are proportional to quantity (for example, customs, shipping, and so on).
- In the case of in-house production, the sum of the costs of the components and operations needed to produce a unit quantity. The unit cost for an operation are

calculated as "run load per unit · cost rate for internal labor costs," whereby the cost rate generally includes full costs (fixed and variable costs).

> *Batch-size-independent production or procurement costs* are incurred with the order, even with lot size one (batch size one).

Batch-size-independent procurement costs are mainly:

- *Ordering costs for procurement*, which are the administrative costs of purchasing divided by the number of purchases. Administrative costs of purchasing also include the costs of receiving stock and stock control. Batch-size-independent procurement costs also include all costs per order that are independent of quantity, such as shipping and handling costs. In the extreme case, these are dependent on the suppliers and the delivered items. To avoid large volumes of data, however, these costs are often added to purchasing costs.

  Procurement costs can also be tapped by item class, such as according to the ABC classification. This results in varying batch- size-independent procurement costs for each ABC category (for example, higher costs for A parts than for C parts). For a more precise determination, see Section 16.4 (activity-based costing).

Batch-size-independent production costs are mainly:

- *Ordering costs for production*, that is, the administrative costs of planning & control and other office functions.
- Possible overhead costs of production that are independent of quantity (transportation, control, putting into and issuing from stock). Usually, they also count as part of the ordering costs.
- *Setup costs* (= setup load · the cost unit rate for internal labor costs) for the various operations (machine adjustments, tool assembly, start-up process, loss of materials at start-up, and so on). For this, management must decide whether to include full costs or only variable costs (essentially wages) in the calculations; this may influence the batch sizes.

> *Carrying cost*, or *holding costs*, are all costs incurred in connection with holding inventory.
>
> *Carrying cost rate*, or *holding cost rate*, is the rate for the carrying cost, usually defined as a percentage of the dollar value of inventory per unit of time (generally one year).

See also Section 1.1.6. Carrying cost includes:

- The *costs of financing* or *capital costs:* Inventory ties up financial resources. Calculation using an interest rate yields the costs of immobilizing money in inventory. This rate corresponds to either the percentage of the mean return on investment if the inventories are financed using internal capital resources, or to the bank interest rate, if the inventories are financed by a third party. For calculation purposes, take interest rate values between 5 and 15% of the average value of the inventory.

- The storage infrastructure costs are incurred for the infrastructure necessary to store a particular product: buildings, installations, warehouse employees, insurance, and so on. The costs for inventory transactions, in contrast, are seen as ordering costs.

  The first cost driver for storage infrastructure costs is batch size, as enough surface area or volume for the whole batch size must be provided. In a first approach, it is possible to express storage infrastructure costs proportionally, as a percentage related to the average inventory, because the average inventory corresponds — apart from safety stock — to half of the batch size, according to the formula in Figure 11.3.1.2. More commonly used is a percentage related to the mean inventory value. In the machine tool industry, percentages between 1 and 3% are common.

  Further cost drivers are storage type and valuation basis (see Section 11.1.1). The storage infrastructure costs rate can be much higher for inexpensive and voluminous products (insulation materials and other construction materials) than for very expensive and possibly easy-to-store products. For more precise figures, then, the calculation should include at least some separate values, such as for information and documents, raw materials, purchased parts, semifinished goods, and end products. However, there are limits to diversifying storage infrastructure costs into as many different storage unit cost rates as possible, due to the expense involved in recording the incurred costs per separate category as well as for data maintenance, if, for example, a separate storage cost percentage were kept for each item.

  A large part of these costs is out of proportion to the value of the stored goods. Since warehouses involve specialized constructions, building a warehouse represents a long-term investment. A company will make the investment if it has exhausted existing warehouse volumes. This leads to a jump in costs. In contrast, reducing inventory value does not automatically lead to a reduction of personnel needed for warehouse management. Even so, in practice, a proportional relationship is common.

- The *risk of depreciation*: This is again expressed as a percentage of the inventory value. It includes, firstly, *technical obsolescence* that results from changes in standards or the emergence of improved products on the market. Secondly, it includes expiration due to *perishability:* Certain items can be stored only for a particular, limited period of time (shelf life). This is the case with "living" products such as groceries or biological pharmaceuticals, but also with "nonliving" products such as certain electronics items. Thirdly, it includes *damage*, *spoilage,* or *destruction* due to unsuitable handling or storage such as, for example, the rusting of sheet metals.

  The percentage of the risk of depreciation may be very large under certain circumstances. For short-lived items, it must be set at 10% or more. However, the percentage is generally dependent on the duration of storage.

It is not unusual for the carrying cost rate to be on the order of 20%. For goods with a high risk of depreciation, it may reach 30% and higher.

11.4.2 Optimum Batch Size and Optimum Length of Order Cycle: The Classic Economic Order Quantity

Intended learning outcomes: Explain economic order quantity (EOQ), variables for the EOQ formula and the EOQ formula. Describe the cost curves as a function of batch size. Present the optimum length of order cycle.

Most methods for determining batch sizes minimize the expected total costs. In dependency upon batch size, these are essentially composed of the costs mentioned in Section 11.4.1:

1. *Batch-size-dependent unit costs*. Mostly the price per produced or procured unit quantity does not change with increasing batch size. However, this is not true in case of allowance for discounts or changes in the production process from a certain batch size upward.
2. *Inventory costs*. These are all the costs incurred in connection with ordering *and* holding inventory. Thus, inventory costs are the following costs:
   a. Setup and ordering costs: These are incurred only once per production or procurement event. In the simplest and most common case, they are independent of the batch size. Thus, the larger the batch size, the smaller is the share in such costs that accrues to each unit. However, there may be an upward jump in costs if a certain batch size requires the choice of another production procurement structure (such as a different machine or means of transport).
   b. Carrying cost: With increasing batch size, the average physical inventory increases, together with carrying cost. For the sake of simplicity, these costs are often set as proportional to batch size, that is, proportional to the value of goods in storage. As was shown in Section 11.4.1, this is only valid provided that the following restrictions hold: Firstly, the carrying cost must be independent of the storage duration. Secondly, an entry in stock only occurs following the issue of the last piece. Issues occur regularly along the time axis. Thus, if X is the batch size, on average, X/2 pieces are in stock. Thirdly, there must be sufficient warehouse space. This means that the size of the batch does not necessitate new installations.

In the simplest case, application of these principles leads to the so-called economic order quantity.

The *economic order quantity (EOQ)*, or the *optimum batch size*, or the *economic lot size*, is the optimal amount of an item to be purchased or manufactured at one time.

The economic order quantity is calculated with respect to a particular planning period, such as one year. The variables for its calculation are listed in Figure 11.4.2.1.

The equation for calculating total costs is shown in Figure 11.4.2.2.

Since the objective is to minimize the total costs, the target function is as shown in Figure 11.4.2.3.

The economic order quantity X0 is the lot size with the minimum of total costs, and it results from deriving the target function and setting it to zero, as shown in Figure 11.4.2.4.

CU	= batch-size-dependent unit costs	\$ / unit
CS	= setup and ordering costs per production or procurement	\$
p	= inventory interest rate = i + s + r	1 / year
i	= interest rate used in calculating (capital costs)	1 / year
s	= storage infrastructure costs rate	1 / year
r	= depreciation risk rate	1 / year
X	= lot or batch size	unit
AC	= annual usage	unit / year
C1	= batch-size-dependent unit costs per year	\$ / year
C2	= carrying cost per year	\$ / year
C3	= setup and ordering costs per year	\$ / year
CT	= total costs of production or procurement per year	\$ / year

Fig. 11.4.2.1 Variables for the EOQ formula.

$$CT = C1 + C2 + C3,$$

where

$$C1 = AC \cdot CU$$

$$C2 = \frac{X}{2} \cdot \left(CU + \frac{CS}{X} \right) \cdot p = \frac{X}{2} \cdot CU \cdot p + \frac{CS}{2} \cdot p$$

$$C3 = \frac{AC}{X} \cdot CS$$

Fig. 11.4.2.2 EOQ formula: total costs equation.

$$CT = \min!$$

Fig. 11.4.2.3 EOQ formula: target function.

$$\frac{dCT}{dX} = \frac{CU}{2} \cdot p - \frac{AC}{X^2} \cdot CS$$

For the optimum batch size X_0, the following holds:

$$\frac{dCT}{dX} = 0$$

$$\Rightarrow X_0 = \sqrt{\frac{2 \cdot AC \cdot CS}{p \cdot CU}}$$

Fig. 11.4.2.4 EOQ formula: determining the optimum batch size.

EOQ (economic order quantity) formula is another name for the X_0 formula.

Figure 11.4.2.5 shows the cost curves that correspond to the values for C1, C2, C3, and CT as a function of batch sizes.

These cost curves are typical of the EOQ formula. The minimum point for total costs lies exactly at the intersection of the curves for setup and ordering costs and carrying cost.

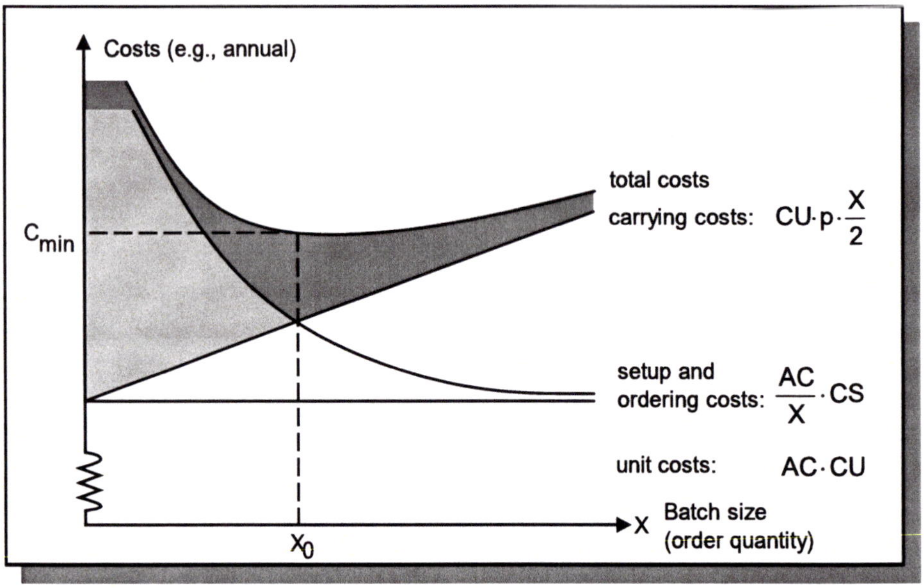

Fig. 11.4.2.5 Cost curves as a function of batch size.

Instead of an optimum batch size, we can also calculate an optimal time period for which an order or a batch covers demand.

The *optimum order interval* or *optimum length of order cycle* is an optimum period of time for which future demand should be covered.

This length is defined according to the formula in Figure 11.4.2.6. From this formula, it is immediately apparent that the optimum length of the order cycle — and the optimum batch size in Figure 11.4.2.4 — rises less than proportionally with increasing setup costs, and declines less than proportionally with increasing turnover. Thus, for example, if we set the value for the root of (2 · CS/p) at 40, the characteristic figures for optimum length of order cycle as a function of the value of turnover are those in Figure 11.4.2.7.

$$LOC_0 = \frac{X_0}{AC} = \sqrt{\frac{2 \cdot CS}{p \cdot CU \cdot AC}} = \sqrt{\frac{2 \cdot CS}{p}} \cdot \frac{1}{\sqrt{C1}}$$

Fig. 11.4.2.6 Optimum length of order cycle.

C1 ($)	LOC_0 (years)
400	2.0
1600	1.0
6400	0.5
25600	0.25

Fig. 11.4.2.7 Sample characteristic figures for length of order cycle as a function of the value of turnover.

Unless we can reduce setup costs decisively, a very large length of order cycle will result in low turnover. In practice, however, when the range of demand coverage is very long, the depreciation risk increases disproportionally. For this reason, upward limits are set for the length of the order cycle, and thus as well for the batch sizes, for items with a small turnover. This is, incidentally, the simplest and most common method in practice to control nonlinear patterns of carrying cost: for example, carrying cost that jumps steeply when inventory exceeds a particular volume. The consideration of the length of order cycle is also an important batch-sizing policy in deterministic materials management (see Section 12.4).

11.4.3 Optimum Batch Size and Optimum Length of Order Cycle in Practical Application

Intended learning outcomes: Present in detail the sensitivity analysis of the EOQ calculation. Produce an overview on the practical implementation of the EOQ formula. Identify several factors that influence a maximum or minimum order quantity.

Unfairly, the EOQ formula has recently been held responsible for large batches. However, a closer look at practice reveals that the formula was often used with carrying cost unit rates that were much too low, or it was applied to deterministic materials management, for which other techniques are better suited (see Section 12.4).

In any case, the EOQ formula basically provides "only" an order of magnitude, not a precise number. The total costs curve shown in Figure 11.4.2.5 is very flat in the region of the minimum, so that deviations from the optimum batch size have only a very small effect on costs. The following *sensitivity analysis* shows this "robust" effect. Beginning with a quantity deviation as given in Figure 11.4.3.1, and the fact that the formula in Figure 11.4.3.2 holds for the optimum batch size X_0, the cost deviation formula is shown in Figure 11.4.3.3.

$$v = \frac{X}{X_0} \quad \text{or} \quad v = \frac{X_0}{X}$$

Fig. 11.4.3.1 Sensitivity analysis: quantity deviation.

$$a \equiv \frac{X_0}{2} \cdot CU \cdot p \equiv \frac{AC}{X_0} \cdot CS$$

Fig. 11.4.3.2 Sensitivity analysis: carrying cost rates for optimum batch size.

$$b = \frac{C2 + C3}{C2_0 + C3_0} = \frac{\frac{v \cdot X_0}{2} \cdot CU \cdot p + \frac{AC}{v \cdot X_0} \cdot CS}{\frac{X_0}{2} \cdot CU \cdot p + \frac{AC}{X_0} \cdot CS} = \frac{v \cdot a + \frac{1}{v} a}{2a} = \frac{v + \frac{1}{v}}{2}$$

Fig. 11.4.3.3 Sensitivity analysis: cost deviation.

For example, a cost deviation of b = 10% results for v = 64% as well as for v = 156%, which means that the relationship shown in Figure 11.4.3.4 is valid:

$$\boxed{64\% \leq v \leq 156\% \Rightarrow b \leq 10\%}$$

Fig. 11.4.3.4 Sensitivity analysis: quantity deviation given a cost deviation of 10%.

This sensitivity analysis reveals the surprising robustness of the calculation technique, which indeed rests on very simplified assumptions. Extending batch size formulas to include additional influencing factors produces an improvement in results that is practically relevant only in special cases. In any event, we may round off the calculated batch size, adapt it to practical considerations, and, in particular, make it smaller if a shorter lead time is desirable.

This robustness increases even further if we include not only C2 and C3, but also the actual costs of production or procurement C1 in the division for b given in Figure 11.4.3.3. If C1 is much larger than C2 + C3 — which is usually the case — even bigger changes to batch size do not have a strong effect on the total production or procurement costs.

In a similar way, we can show that errors in determining setup and ordering costs, the carrying cost rates, or the annual consumption in the cost deviations make as little difference as a quantity deviation does. Among other things, the EOQ formula is thus not very sensitive to systematic forecast errors. This means that very simple forecasting techniques, such as moving average value calculation, will generally suffice when determining batch sizes.

In the case of produced items, the reduction in costs for in-process inventory achieved through smaller batches is thus negligible in most cases. More significant is the fact that smaller batches may lead to *shorter lead time*. In addition to this improvement in the target area of delivery, there are also positive effects in the target areas of flexibility and costs. The positive effects discussed in Chapter 6 are lacking in the classic EOQ formula. However, as we will show in Section 13.2, smaller batches only result in shorter lead time if — on the one hand — the run time is long in relation to the lead time, particularly in line production (in classic job shop production this proportion is likely to be of the order of magnitude of 1:10 and less), and — on the other hand — the saturation of a work center does not have the effect of creating longer queues for the entire collection of batches.

Thus, the longer the run times — often required when much value is added — the higher the costs for goods in process are. In such cases we should choose rather lower values for batch sizes than those recommended on the basis of the EOQ formula (see also lead-time-oriented batch sizing in Section 11.4.4). For work-intensive operations especially, shorter operation times can contribute to harmonizing the content of work, which in turn leads to a further reduction in wait times, and thus lead times, as explained in Section 13.2.2. As Figure 6.2.6.2 illustrated, at lower production structure levels a reduction in lead time is likely to result in lower safety stocks, and thus *cost savings*. If for some reason storage is not possible at all, shorter lead times can even achieve additional sales.

A practical implementation scheme, which takes both total costs *and* short lead time into account, is provided in Figure 11.4.3.5.

1. Determine the optimum batch size using the EOQ formula by using a sufficiently large carrying cost rate. For not fully utilized work centers, consider only variable costs (essentially wages) in the calculations.
2. If production is not fully utilized: Because of the low-cost sensitivity of the EOQ formula at the optimum, we can vary the batch sizes generously by x % where x is variable for every item category and can be chosen freely to be on an order of magnitude of 64 to 156%.
3. For manufactured articles, we should instead round off the batch sizes. In the case of large run times and larger value added, we may also choose a smaller percentage due to the effects of shorter lead times, even less than 50% under certain circumstances.
4. Include differentiated considerations concerning the minimum and maximum (see below).

Fig. 11.4.3.5 Practical implementation of the EOQ formula.

The *minimum order quantity* (or *maximum order quantity*) is an order quantity modifier, applied after the lot size has been calculated, that increases (or limits) the order quantity to a pre-established minimum (or maximum) ([APIC16]).

Differentiated considerations concerning the minimum and maximum order quantity can be found in Figure 11.4.3.6, as, for example, related to item groups or even individual items.

- Space requirements in warehouse (maximum)
- Length of order cycle (maximum)
- Product shelf life: obsolescence, perishability (maximum)
- Blocking of machine capacities (maximum)
- Limits of tool use (maximum)
- Liquidity problem (maximum)
- For purchased items: shortages or price increases to be expected (minimum)
- For purchased items: minimum order volumes (minimum)
- Coordination with transport and storage units (maximum or minimum)

Fig. 11.4.3.6 Several factors that influence a maximum or minimum order quantity.

In the literature, there are models that take more operating conditions into consideration. We will present several of these in Section 11.4.4. Because of its simplicity, however, the EOQ formula is used frequently in current practice. Even if the simplified model assumptions that underlie it are not given in the concrete case, the formula is very robust in the face of such deviations, as we have shown. Before applying a more complicated calculation method, materials management should clarify whether the more costly batch size determination truly offers crucial advantages over the simple implementation considerations outlined above.

11.4.4 Extensions of the Batch Size Formula (*)

Intended learning outcomes: Present lead-time-oriented batch sizing. Describe batch sizing considering discount levels. Produce an overview on joint replenishment: kit materials management and collective materials management.

1. *Lead-time-oriented batch sizing* is a generalization of the simplified approach using the EOQ formula for production, taking the cost of work in process into consideration.

As a complement to the variables in Figure 11.4.2.1, we add the variables shown in Figure 11.4.4.1. Most of these data come from the route sheet.

CU_M = materials costs per unit	\$ / unit
FD = flow degree = lead time / operation time	dimensionless
SUMRT = sum of run times per unit = $\sum_{1 \le i \le n} RT[i]$	working days / unit
NBRWKD = number of working days per year	working days / year

Fig. 11.4.4.1 Additional variables for lead-time-oriented batch sizing.

The EOQ has results according to the formula given in Figure 11.4.4.2. For details of the derivation, see [Nyhu91], p. 103. The denominator under the radical is significantly larger than the one in classic batch sizing only for a long manufacturing lead time.

$$\Rightarrow X_0 = \sqrt{\frac{2 \cdot AC \cdot CS}{p \cdot \left(CU + (CU + CU_M) \cdot AC \cdot FD \cdot \frac{SUMRT}{NBRWKD}\right)}}$$

Fig. 11.4.4.2 Lead-time-oriented batch sizing: determination of the minimum.

2. *Batch sizing considering discount levels* is a generalization of the simplified approach using the EOQ formula.

Figure 11.4.4.3 illustrates the decreasing batch-size-dependent unit costs as a function of the lot size, as well as the resulting total costs curves.

Batch-size-dependent unit costs CU are dependent on the purchased quantity. This is particularly valid for procured goods.

A *quantity discount* is a price reduction allowance on orders over a certain minimal order quantity or value.

For example, a supplier may offer a quantity discount for the whole order quantity with three discount levels; that is, reduced unit costs CU2 as soon as the quantity exceeds X_m2, CU3 as soon as the quantity exceeds X_m3, CU4, as soon as the quantity exceeds X_m4.

Every total costs curve for the various values of cost per piece demonstrates a minimum within the range of its validity. This is either the minimum of the corresponding total costs curve (X_02 in Figure 11.4.4.3), or it lies on the border of a discount level curve (X_m3 in Figure 11.4.4.3). If discounts are not large, we may also argue that the batch sizes for the different discount levels according to the EOQ formula will lie very close to each other. We may thus calculate the optimum batch size by selecting a particular mean cost per piece, and then rounding it up to the next discount level.

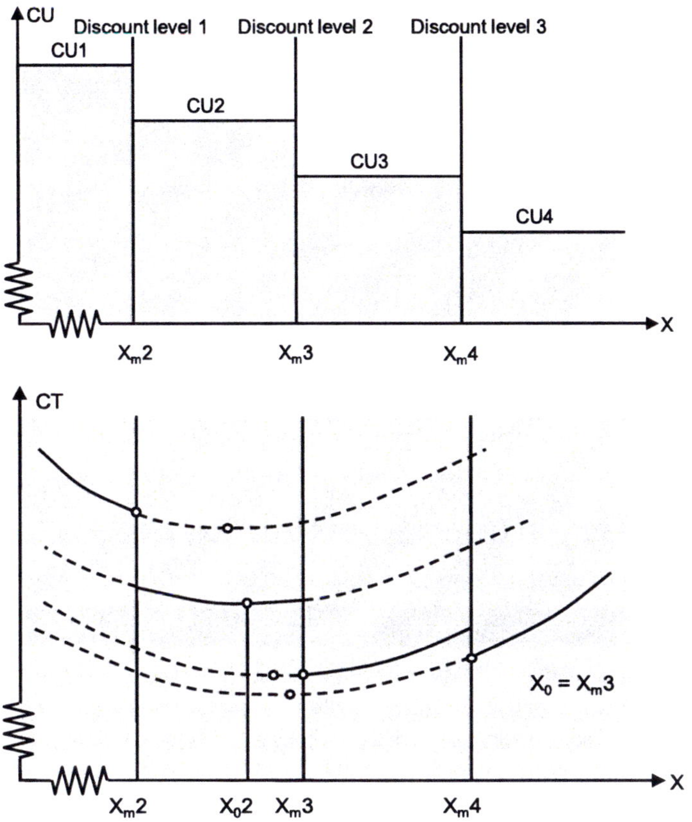

Fig. 11.4.4.3 Total costs curves, taking discount levels into consideration.

A similar line of thinking is followed when evaluating economic efficiency and batch sizing in the case of alternative (less expensive) production processes using larger batch sizes.

3. *Joint replenishment* is joint planning for a group of related items, treating them as an item family.

Two examples of management of sets of items follow.

3a. In *kit materials management*, various goods are combined into a so-called *(material) kit* (because of their joint use in particular assemblies or products) and managed as a group.

The individual optimum batch size for an element i from a kit S with annual consumption AC of S results from the formula in Figure 11.4.4.4.

Instead of these individual batch sizes, we may determine a kit batch size X_S using the compromise formula in Figure 11.4.4.5.

If the component kits are very heterogeneous with respect to the two factors in the batch size formulas above, we can form more homogeneous planning subgroups that are then used for separate batch sizings. Another possibility is to form an economic batch for the most value-intensive components. We then set the batch size of less value-intensive materials positions as whole-number multiples of this batch for correspondingly less frequent procurement.

$$M_i := \text{Number of parts per element } i \text{ in kit S}$$
$$X_i = \sqrt{\frac{2 \cdot M_i \cdot AC_S \cdot CS_i}{p \cdot CU_i}} = \sqrt{\frac{2 \cdot AC_S}{p}} \cdot \sqrt{\frac{M_i \cdot CS_i}{CU_i}}$$

Fig. 11.4.4.4 Individual optimum batch sizes for an element i of kit S with annual consumption AC_S.

$$X_S = \sqrt{\frac{2 \cdot AC_S}{p}} \cdot \sqrt{\frac{\sum (M_i \cdot CS_i)}{\sum CU_i}}$$

Fig. 11.4.4.5 Kit batch size X_S.

3b. In *collective materials management*, we form material groups, or planning groups, whose setup and ordering costs can be reduced, if the batches are ordered collectively.

Valid criteria for collective materials management include:

- The same supplier for purchased parts (taking advantage of simplified administration and/or a total invoice discount)
- The same production technique for in-house production (e.g., for a product family), whereby simplified machine setup achieves a reduction in the total setup costs

In the case of collective materials management, within a planning group materials managers must determine an average reduction in the setup and ordering costs as a percentage. As soon as an item is to be ordered, a check is made of all other items of the same planning group. If the order of a batch is due in the near future anyway, it can be ordered now through an *early order release*. This should be a reduced batch size, which is calculated by using the reduced setup and ordering costs.

11.5 Summary

Inventories form buffers for logistics within and among organizations. Inventory management is thus another important instrument for planning & control. Categorizing and typing storage and warehouses facilitates detailed inventory management. A physical inventory count of stored and in-process inventory verifies the accuracy of book inventories as a prerequisite of accurate inventory valuation.

An important basis for various calculations in demand forecasting and in materials management is provided by statistics that analyze particular events such as inventory transactions, sales, and bid activities. These statistics contain information on quantities and values as well as on the number of transactions.

The ABC classification according to various measures of value, such as turnover, determines the importance of items in a product line. For this, the item range is first divided into different ABC categories. The XYZ classification distinguishes items with regular or even continuous demand from those with lumpy/erratic or discontinuous or unique demand. Additional statistics sort out items that are exceptions according to some criterion.

Stochastic materials management aims to produce production or procurement proposals prior to actual demand resulting from customer orders. In most cases, a demand forecast is the sole basis for both the proposed quantity (the batch) and the proposed time of receipt.

The most familiar technique for stochastic materials management — particularly for continuous demand — is the order point technique. The order point is the expected value of demand during the lead time. Safety stock is carried to absorb deviations from the expected value, and safety lead time, which is also translated into a safety quantity, is used to absorb deviations from the lead time. If forecast parameters change, both order point and safety stock must be recalculated.

In the simplest case, materials management determines the batch size that will yield a minimum of setup and ordering costs and carrying cost. However, in the *stochastic case*, there is as yet no concrete customer demand, so that the optimum batch size can only be derived (the economic order quantity EOQ) from a long-term forecast of total demand. In the final reckoning, however, this calculated quantity merely indicates the order of magnitude, and thus it can be rounded up or down generously. The order of magnitude is robust in the face of errors in quantity or cost forecasts. However, the formula does not take into account the effects of shorter lead times with smaller batches. In practice, other constraints exert an important influence on the final selection of minimum or maximum batch size. These include storage space requirements, storability, minimum order volumes, speculation, and so forth. Extensions to the simple batch size formula arise when taking into account lead time, quantity discounts, and kit or collective management.

11.6 Keywords

ABC category, 479
ABC classification, 477
ABC-XYZ analysis, 509
abnormal demand, 476
anticipation horizon, 483
average inventory, 481
batch sizing, 495
bid statistics, 477
carrying cost, 496
carrying cost rate, 496
cycle counting, 475
double order point system, 483
economic order quantity (EOQ), 498
fixed-location storage, 470
inventory accounting, 473
inventory adjustment, 473
inventory costs, 498
inventory issuance principle, 470
inventory valuation, 473
issuance principle, 470
joint replenishment, 505
lot sizing, 495
lot-size inventory, 495
maximum order quantity, 503
minimum order quantity, 503
min-max (reorder) system, 483
multiple-stock organization, 470
optimum length of order cycle, 500
optimum probability of stockout, 491
optimum service level, 491
order point, 481
order point technique, 480
Pareto's law, 477
periodic inventory, 474

perpetual inventory, 471
physical inventory, 473
random-location storage, 470
replenishment lead time, 481
replenishment order quantity, 481
safety factor, 486
safety lead time, 485
safety stock, 484
sales statistics, 477
service function, 486
service level, 485
setup costs, 496
single-stock organization, 470
stock organization, 469
stockkeeping unit (SKU), 469
stockout costs, 491
turnover statistics, 476
unordered issuance principle, 470
usage statistics, 475
valuation method, 470
XYZ classification, 479

11.7 Scenarios and Exercises

Intended learning outcomes: Calculate examples for the ABC Classification. Disclose the ABC-XYZ analysis in materials management. Differentiate between safety stock variation and demand variation. Determine batch size depending on stockout costs. Assess the effectiveness of the order point technique.

11.7.1 The ABC Classification

Intended learning outcomes: Determine meaningful ABC categories of some items and perform the ABC classification.

This exercise refers to Section 11.2.2. Perform an ABC classification for the items shown in the table in Figure 11.7.1.1, separately for two ABC categories 1 and 2. Class A accounts for 75% of sales turnover, and items in the B class account for 90% of turnover. Why does it often make sense to perform separate classifications for two or more ABC categories? Is your classification of the items as A, B, or C the only possible solution?

Item ID	Sales ($)	ABC category
4310	10	1
4711	1	2
5250	0	2
6830	6	2
7215	30	1
7223	2	1
7231	84	1

Item ID	Sales ($)	ABC category
8612	70	1
8620	13	2
8639	1	2
8647	3	2
8902	4	1
8910	0	1
9050	1	2

Fig. 11.7.1.1 Sales and ABC categories of some items.

Solution:

ABC category	Item ID	Sales ($)	Sales cumulated	% Share on cum. sales	ABC classification
1	7231	84	84	42	A
	8612	70	154	77	A
	7215	30	184	92	B
	4310	10	194	97	C
	8902	4	198	99	C
	7223	2	200	100	C
	8910	0	200	100	C
2	8620	13	13	52	A
	6830	6	19	76	A
	8647	3	22	88	B
	4711	1	23	92	B
	8639	1	24	96	C
	9050	1	25	100	C
	5250	0	25	100	C

The division of the items into two categories for a meaningful ABC classification is necessary so that like items can be compared; the categories will reflect different types of items, such as individual parts and final products.

The classifications in the solution above do not represent the only possible solution. Certain classifications can be problematic around the break points. For example, why should item 4711 receive the classification B, while items 8639 and 9050 are assigned to classification C?

11.7.2 The ABC-XYZ Analysis and its Use in Materials Management

Intended learning outcomes: Explain the combined ABC-XYZ classification. Describe its use in materials management. Identify the appropriateness of Kanban control as the result of an ABC-XYZ analysis.

This scenario refers to Sections 11.2.2 and 11.2.3.

The *ABC-XYZ analysis* consists of a combined ABC-XYZ classification. It allows decision making as to the appropriate method of materials management for individual items.

Mark the areas (items) in the matrix in Figure 11.7.2.1 for which Kanban control would be appropriate. Explain the reasoning behind your answer.

Continuousness of demand \ Consumption value	A High	B Medium	C Low
X High	high value continuous demand	medium value continuous demand	low value continuous demand
Y Medium	high value regular, or fluctuating demand	medium value regular, or fluctuating demand	low value regular, or fluctuating demand
Z Low	high value discontinuous demand	medium value discontinuous demand	low value discontinuous demand

Fig. 11.7.2.1 The ABC-XYZ analysis — Combined ABC-XYZ classification.

Solution:

The prerequisite for the Kanban technique is continuous demand along the entire value chain. X items are particularly suitable for production in a Kanban system. For the Y group, A items should not be controlled by Kanban, for their consumption value is high, and fluctuating demand leads to lower stock-inventory turnover and thus longer storage time. For the same reason, Kanban control is as a rule not appropriate for Z items, whereby an exception can be made for C items, as carrying costs for C items may be lower than the costs of a more expensive control technique.

11.7.3 Safety Stock Variation versus Demand Variation

Intended learning outcomes: Find out whether the safety stock level increases with increasing demand.

True or false: The safety stock level increases with increasing demand.

Solution:

As the formula in Figure 11.3.3.7 shows, this statement is generally not correct. The safety stock depends on the *standard deviation* of the demand during the lead time. Increasing demand does not automatically increase either the standard deviation during the statistical period or the lead time.

11.7.4 Batch Sizing Depending on Stockout Costs (*)

Intended learning outcomes: Determine the batch size in dependence of carrying costs and stockout costs, average annual consumption, as well as the standard deviation of demand during lead time.

The carrying costs for a certain article are 2 per unit and year. Stockout costs are 5 per unit. The average annual consumption amounts to 1000, and the standard deviation of demand during lead time is 10. No safety stock is intended. Normal distribution is assumed.

a. How large should the batch size be, considering the optimum stockout probability? Can the fill rate target of 99% be met? What are the carrying costs per year?

b. Assume a batch size of only 250. What are the values for safety stock and fill rate corresponding to the optimum probability of stockout per order cycle?

c. Now assume a safety stock of 20 units. Again, the batch size is 250. What are the values for service level and fill rate?

Solution:

a. Zero safety stock entails a service level of 50% (see Figure 11.3.3.6, for example) and — by Figure 11.3.3.3 — a probability of stockout per order cycle of 50%. Because stockout can be expressed as cost per unit, the formulas in Figures 11.3.4.2, 11.3.4.4, and 11.3.4.5 apply. Therefore,
   - Batch size = 1000 * 50% * (5/2) = 1250.
   - Stockout quantity coefficient P(s) = 0.399.
   - → Fill rate = 1 – ((10/1250) * 0.399) = 99.68% > 99%.
   - Average inventory = 1250/2 = 625.
   - → Carrying costs per year = 625 * 2 = 1250.

b. Again, the formulas in Figures 11.3.4.2, 11.3.4.4, and 11.3.4.5 apply:
   - Optimum probability of stockout = (2/5) * (250/1000) = 10%.
   - → Optimum service level = 1 – 10 % = 90%.
   - → Safety stock = 1.282 * 10 {note: the standard deviation} ≈ 13.
   - → Stockout quantity coefficient P(s) = 0.048.
   - → Fill rate = 1 – ((10/250) * 0.048) = 99.81%.

c. Applying the formulas in Figures 11.3.4.4, 11.3.4.2, and 11.3.4.5:
   - Standard deviation = 10; => safety factor = 20/10 = 2.
   - → Service level ≈ 98%.
   - → Stockout quantity coefficient P(s) = 0.008.
   - → Fill rate = 1 – ((10/250) * 0.008) = 99.97%.

11.7.5 Effectiveness of the Order Point Technique

Intended learning outcomes: Explore the changing shape of the inventory curve for continuous and less continuous demand.

Figure 11.3.1.1 shows the famous saw-tooth-shaped curve that is characteristic of the order point technique. You can view the curve on the Internet at the following URL:

https://intlogman.ethz.ch/#/chapter11/order-point-technique/show

Explore the changing shape of the inventory curve for continuous and less continuous demand (moving your cursor over the gray icon executes your input choice). Try out different parameters to calculate lot size and service level. Try other consumption values. Observe the effect of the consumption values on the order of the production or procurement batch size. Again, touching the "calculate" icon executes your input choice. The initial demand values are automatically reentered by moving your cursor over the gray demand shape icon.

11.8 References

APIC16 Pittman, P. et al., APICS Dictionary, 15th Edition, APICS, Chicago, IL, 2016

Bern99 Bernard, P., "Integrated Inventory Management," John Wiley & Sons, New York, 1999

Brow67 Brown, R.G., "Decision Rules for Inventory Management," Holt, Rinehart and Winston, New York, 1967

Cole00 Coleman, B.J., "Determining the Correct Service Level Target," *Production and Inventory Management Journal*, 41(1) 2000, pp. 19–23

Chap06 Chapman, S., "The Fundamentals of Production Planning and Control," Pearson Education, Upper Saddle River, NJ, 2006

Eilo62 Eilon, S., "Industrial Engineering Tables," Van Nostrand, London, 1962

Eilo64 Eilon, S., "Tafeln und Tabellen für Wirtschaft und Industrie," R. Oldenbourg, München, 1964

Nyhu91 Nyhuis, P., "Durchlauforientierte Losgrössenbestimmung," *Fortschrittberichte VDI*, Reihe 2, No. 225, Düsseldorf, 1991

SiPy98 Silver, E.A., Pyke, D.F., Peterson, R., "Inventory Management and Production Planning and Scheduling," 3rd Edition, John Wiley & Sons, New York, 1998

Stev14 Stevenson, W.J., "Operations Management," 12th Edition, Irwin, Boston, 2014

Ters93 Tersine, R.J., "Principles of Inventory and Materials Management," 4th Edition, Prentice Hall, North-Holland, New York, 1993

12 Deterministic Materials Management

Intended learning outcomes: Produce an overview on demand and available inventory along the time axis. Describe deterministic determination of independent demand. Explain the deterministic determination of dependent demand (Material Requirements Planning, MRP). Differentiate various lot sizing techniques. Disclose how to analyze the results of the MRP.

Deterministic techniques are used in materials management whenever portions of the cumulative lead time remain within the customer tolerance time. This is the case, for example, with the assembly stage during the manufacture of capital goods. During this time, production, procurement, or services are dependent on customer demand. Figure 12.0.0.1 shows the relevant planning & control tasks and processes on a dark background.

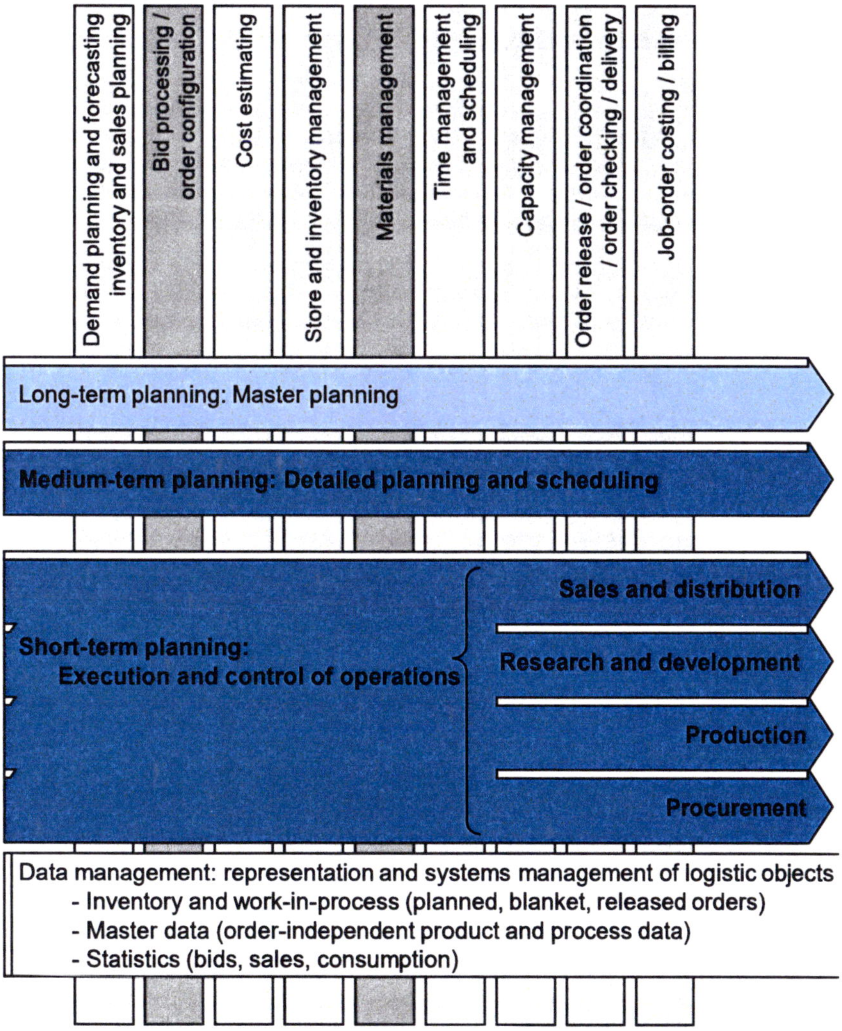

Fig. 12.0.0.1 The parts of the system discussed in this chapter are shown against a darker background.

P. Schönsleben, *Handbook Integral Logistics Management*,
https://doi.org/10.1007/978-3-662-65625-9_12

These tasks refer back to the reference model in Figure 5.1.4.2. Sections 5.3.1 and 5.3.2, in particular Figures 5.3.2.1 and 5.3.2.2, provide an introduction to the material in this chapter. We recommend that you read these Sections again before continuing to study this chapter.

So-called quasi-deterministic techniques are used in the stochastic case, that is, upstream from or on the order penetration point. Purely stochastic techniques outlined in Chapter 11 carry the risk that procured goods may not be used in time or that excessively large safety stock levels will have to be maintained. For high-cost discontinuous items both effects are not acceptable. The quasi-deterministic materials management is also used for long-term planning, particularly for budgeting personnel and other resources and for determining blanket contracts.

Section 5.2 presented deterministic techniques for *long-term* materials management. The present chapter sets out the techniques for medium and short-term planning. What distinguishes these techniques is that the demand for an item cannot simply be regarded as an average demand that is approximately constant over time, as is the case in long-term planning or stochastic materials management as described in Chapter 11. Instead, you know the exact point or limited period along the time axis at which each requirement will arise and then make use of this knowledge. This enables you to manage even lumpy demand efficiently.

Deterministic techniques are easy to understand. A long customer tolerance time favors their use. This is often given in producer dominated markets, but also holds for production or procurement orders that are customer specific, e.g., in special mechanical engineering and plant construction or in services. Also, deterministic techniques can be implemented more often if — through carefully thought-out methods — the lead time can be reduced.

12.1 Demand and Available Inventory along the Time Axis

Intended learning outcomes: Explain the projected available inventory and its calculation. Describe scheduling and cumulative projected available inventory calculation. Produce an overview on operating curves for stock on hand.

Both long-term management of resources, as outlined in Section 5.2.2, and stochastic materials management allow the demand for an item to be regarded as a scalar variable, that is, as a total, because the exact time at which the demand arises is either not relevant or was not the object of the estimate. What is estimated is the requirement quantity over a given time period. Thus, the shorter the selected period, the greater the scatter. At this level of inaccuracy, it is more sensible to assume that demand is uniformly distributed across the entire period.

However, when the exact point in time at which demand will occur within the customer tolerance time is known, it makes sense to utilize this information. Instead of relying on the order point technique (see Section 11.3), which only takes stock levels into account, you can now also consider future demand and deliveries.

Time phasing is a technique that divides the future time axis into time periods and considers stock levels for any desired point in the future ([APIC16]).

Time bucket is the chosen period for time phasing. It contains all relevant planning data summarized into a columnar display (for example, a weekly or monthly time bucket).

Time-phased order point (TPOP) is a concept that was used in the early version of the MRP (material requirements planning) technique as described in Section 12.3.2.

Considering time periods makes the technique easier to teach and learn. Also, calculation of the technique by hand is such a time-consuming procedure that it makes sense to produce a rough calculation according to time periods. This also held in the early days of ERP software, when access to the data media was very slow. Today, however, software packages produce calculations that are accurate at the event level.

The projected available inventory calculation described below forms the basis for deterministic materials management.

12.1.1 Allocated Quantities, and the Projected Available Inventory

Intended learning outcomes: Disclose the concept of allocated quantities. Explain projected available inventory and its components. Disclose planning-related events and their effect on available inventory.

Physical inventory is the actual inventory quantity determined by physical counting (cf. [APIC16]).[1]

Physical inventory is often also called *stock on hand* or *on-hand balance*.[2]

Precise physical inventories on their own are not enough to allow efficient inventory management, as the following example shows:

- "A customer orders a certain quantity of a product for delivery in one week's time. A check of the inventory shows that there is sufficient stock, and the order is confirmed. One week later, however, it emerges that the product cannot be delivered, because in the meantime the stock has been delivered to another customer."

Solving the problem requires taking future demand into consideration.

An *allocated quantity* is a quantity of items assigned to a specific customer or production order. It is also known as *reserved quantity*.

- A quantity ordered in a new customer order is thus not only compared against the physical inventory. It must also be compared against the physical inventory minus

---

1 This is one of two possible meanings. The other meaning is the process of determination of inventory quantity. See Section 11.3.1.

2 In practice, for the following calculations, physical inventory is often replaced by book inventory, assumed to be more or less accurate.

the sum of all reserved quantities. The customer requirements in question may only be confirmed if the result is sufficiently large.

On the other hand, it is also necessary to take quantities ordered through current *procurement orders* or *production orders* into account.

An *open order* is either a released order or an unfilled customer order.

An *open order quantity* is the quantity of an open order that has not yet been delivered or received.

A *scheduled receipt* is the open order quantity of an open production or procurement order with an assigned completion date.

- The customer demand in question can thus be confirmed on the date of the next scheduled receipt, provided that this date is sufficiently reliable and the expected quantity is sufficiently large.

This example gives us a definition for projected available inventory.

Projected available inventory or *projected available balance* is defined in Figure 12.1.1.1 for every future transaction or event that changes stock levels. The calculation also includes
- *planned demand*, i.e., the requirement for planned customer or production orders, and
- *planned receipts*, i.e., (anticipated) receipts associated with production or procurement orders that have not yet been released.

$$\text{Projected available inventory}(t) = \text{Physical inventory} + \left(\sum \text{scheduled receipts}\right)(t) - \left(\sum \text{allocated qty.}\right)(t) + \left(\sum \text{planned order receipts}\right)(t) - \left(\sum \text{planned gross requirements}\right)(t)$$

where

$\left(\sum \text{scheduled receipts}\right)(t)$	:=	the sum of all scheduled receipts where date of receipt ≤ transaction date.
$\left(\sum \text{allocated qty.}\right)(t)$	:=	the sum of all allocated quantities where issue date ≤ transaction date.
$\left(\sum \text{planned order receipts}\right)(t)$	:=	the sum of all planned order receipts where date of receipt ≤ transaction date.
$\left(\sum \text{planned gross requirements}\right)(t)$	:=	the sum of all planned gross requirements where issue date ≤ transaction date.

Fig. 12.1.1.1 Projected available inventory.

Projected available inventory is thus neither a scalar value nor an individually and directly manageable attribute. It changes with every planning-related event. Figure 12.1.1.2 shows the various planning processes or planning-related events or transactions that may change the values of the four totals and also the physical inventory (see also Figure 11.1.2.1):

1. *Increase in production plan:* Every forecast is a planned demand.
2. *Receipt of a customer order:* Every item ordered results in an allocated quantity.

Transaction	Physical inventory	Σ Scheduled receipts	Σ Allocated quantity	Σ Planned demand	Σ Planned receipts
1. Increase in production plan				+	
2. Receipt of customer order			+		
3. Delivery of customer order	–		–	(–)	
4. Creation of a planned order					+
5. Creation of dependent demand				+	
6. Release of an order		+			(–)
7. Allocation of a components requirement			+	(–)	
8. Issue of an allocated quantity from stock	–		–		
9. Unplanned return or issue	+/–				
10. Scrapping during production		–			
11. Checking of goods received	+	–			
12. Inventory adjustment	+/–				

Fig. 12.1.1.2 Planning-related events and their effect on available inventory.

3. *Delivery of a customer order:* Stock quantity is reduced. Reserved quantity and, if necessary, a forecast quantity are also reduced (see also Section 12.2.2).
4. *Creation of a planned production or procurement order:* The planned receipts total is increased.
5. *Creation of (dependent) demand for each component of a planned production order:* The total of planned demand is increased (see also Section 12.3.3).
6. *Release of a production or procurement order:* The scheduled receipts total is increased. If the order already exists as a planned order, then the planned receipts quantity is reduced.
7. *Allocation of a components requirement*: Planned demand in planned production orders is translated into allocated quantities.
8. *Issuance of an allocated quantity from stock:* The stock quantity and the allocated quantities total are reduced when an allocated quantity is issued from stock.
9. *Unplanned returns or issues*: Such transactions occur during distribution and procurement, as well as during production. They may relate to equipment overheads for offices and workshops or to items for R&D, or may be sent as samples, and so on.
10. *Scrapping during production:* Quality control determines the scrap quantity, which reduces scheduled receipts.
11. *Checking of goods received:* Physical receipts into stock raise the stock quantity and reduce the scheduled receipts total.
12. Physical inventory alters the stock quantity in both directions.

It is important that available inventory be changed by only one of the transactions listed above. For this reason, the physical inventory or the four summed quantities are never simply corrected. This conforms to the principles of financial accounting, which in turn adhere to the legal requirements.

12.1.2 Projected Available Inventory Calculation

Intended learning outcomes: Differentiate between the spreadsheet representation and the graphical representation (or inventory curve) of the projected available inventory calculation.

As described above, projected available inventory changes with every transaction, so there are as many projected available inventory figures as there are transactions for one item.

The *projected available inventory calculation* considers future changes in the projected available inventory, beyond a time horizon that incorporates at least the cumulative lead time.

The *inventory curve* is another term for the graphical representation of the projected available inventory calculation.

Figure 12.1.2.1 shows the conventional graphical representation, the spreadsheet, depicting the availability of an item along the time axis. It generally takes the following form:

Date	Entry	Issue	Balance	Text	Order ID
06.01.			1200	Physical inventory	
06.19.		500	700	Bernard	26170
07.31.	3000		3700	Stock replenishment	86400
08.02.		300	3400	Dow	27812
08.04.		2500	900	Sosa	26111
08.18.	3000		3900	Stock replenishment	87800
08.19.		2000	1900	Thomas	26666
09.24.		1000	900	Zoeller	25810

Fig. 12.1.2.1 Projected available inventory calculation (spreadsheet representation).

- The first row provides the current physical inventory.
- The other rows list the various transactions one after the other, in ascending order of transaction date. Quantities received and issued are recorded in the second and third columns. The fourth column shows the balance, that is, the quantity available after the transaction. The other columns describe the transactions.

Example problem: Using the spreadsheet in Figure 12.1.2.1 describing a possible actual situation for projected available inventory calculation, find an answer for the following important questions:

- What partial quantity is available on a particular date? The aim here is to determine the minimum available quantity — starting from the specified date.
- When will the entire quantity be available? Identify the earliest date after which the available quantity will no longer be smaller than the required quantity.

The contents of the graph shown in Figure 12.1.2.2 are exactly the same as in Figure 12.1.2.1. This qualitative view, however, allows fast, intuitive answers to the two questions addressed above. The necessary planning decisions can be made in a fraction of the time required when viewing the spreadsheet version.

The projected available inventory calculation presented in this section corresponds to the calculation of the ATP quantity (available-to-promise) presented in Section 5.3.5.

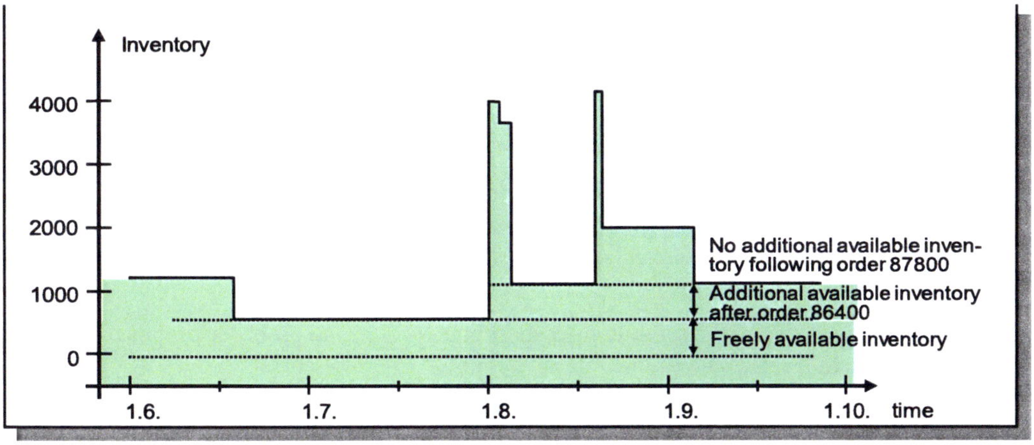

Fig. 12.1.2.2 Projected available inventory calculation (graph) or inventory curve.

12.1.3 Scheduling and Cumulative Projected Available Inventory Calculation

Intended learning outcomes: Explain scheduling projected available inventory calculation (spreadsheet and graphical representation). Describe the cumulative projected available inventory calculation (graph) or store throughput diagram.

The *scheduling projected available inventory calculation* attempts to assign the associated scheduled or planned receipt to every requirement.

Figure 12.1.3.1 shows the previous example using this type of calculation, where customer order 25810 has been moved forward to June 10.

*	Date	Entry	Issue	Balance	Text	Order ID
	06.01.			1200	Physical inventory	
	06.10.		1000	200	Zoeller	25810
*	07.31.	3000		3200	Stock replenishment	86400
	06.19.		500	2700	Bernard	26170
	08.02.		300	2400	Dow	27812
*	08.18.	3000		5400	Stock replenishment	87800
	08.04.		2500	2900	Sosa	26111
	08.19.		2000	900	Thomas	26666

Fig. 12.1.3.1 Scheduling projected available inventory calculation (spreadsheet).

Again, demands are listed in order by date. Receipts, on the other hand, are sorted by the date on which they will be needed in order to have projected available inventory. The following situations result in lists of exceptions (only the first one appears in Fig. 12.1.3.1):

- A demand can only be covered by bringing forward a corresponding receipt. Two receipts of this kind are indicated by an asterisk (*) in the first column in Figure 12.1.3.1.
- A receipt can be deferred, since the associated requirements have a later date than the date of the receipt.
- There are demands without corresponding receipts, so an order proposal should be generated.
- Planned or released orders without assigned demands may be canceled, if necessary.

Thus, the scheduling projected available inventory calculation also creates a link between materials management and scheduling by providing proposals to speed up or slow down production or procurement orders.

Conversely, if the production or procurement orders cannot be speeded up, the scheduling projected available inventory calculation indicates which requirements will have to be delayed. The orders associated with these demands should then be slowed down temporarily and then speeded up again as soon as the demands become available.

The scheduling projected available inventory calculation can also be shown in graph form. The graph in Figure 12.1.3.2 has the same contents as the spreadsheet in Figure 12.1.3.1. Negative projected available inventory corresponds to a backlog and is shaded accordingly, and the two extreme responses — delaying an allocated quantity or speeding up a production or procurement order — are shown as examples.

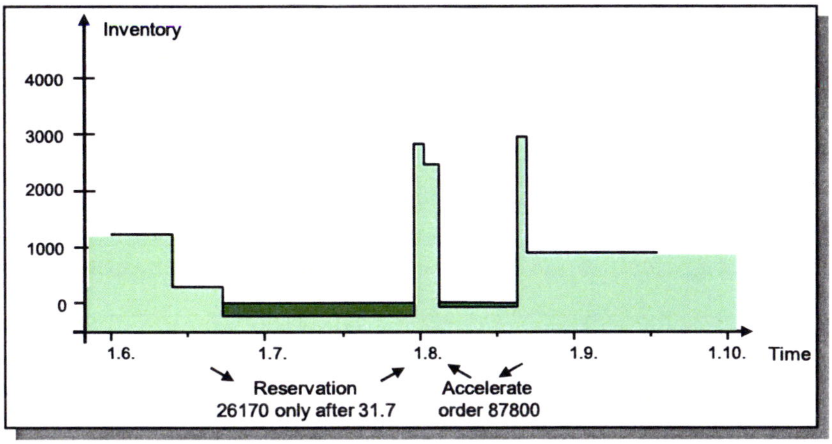

Fig. 12.1.3.2 The scheduling projected available inventory calculation (graph).

The *cumulative projected available inventory calculation* contains the same information as the noncumulative calculation, but it also provides the cumulative totals for entries and issues along the time axis.

Store throughput diagram is another name for the graphical representation resulting from the cumulative projected available inventory calculation.

This is illustrated in Figure 12.1.3.3. It is more difficult to represent, because the values along the vertical axis are sometimes very large.

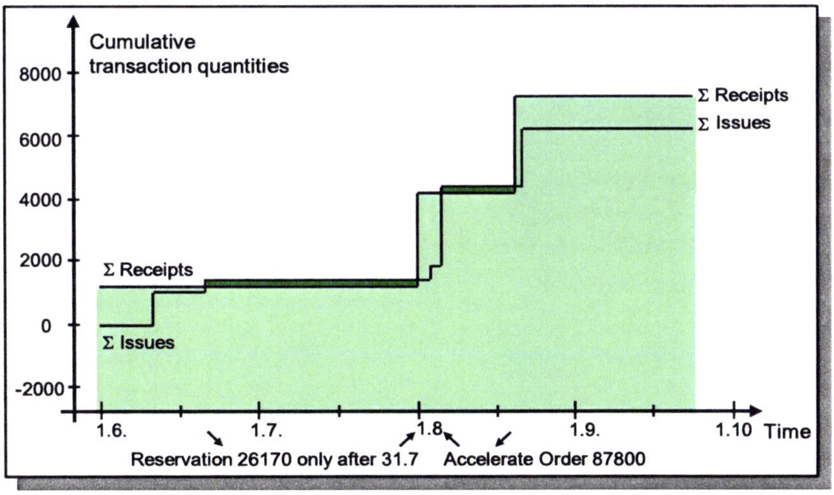

Fig. 12.1.3.3 The cumulative projected available inventory calculation (graph) or store throughput diagram.

The expected projected available inventory is shown as a vertical difference. If the cumulative issues curve is higher than the cumulative receipts curve, then we should expect a negative projected available inventory. This will correspond to the expected backlog and is again shaded accordingly.

12.1.4 Operating Curves for Stock on Hand

Intended learning outcomes: Explain the derivation of an operating curve for stock on hand from the store throughput diagram.

Operating curves for stock on hand describe delivery delays and time in storage in relation to the inventory.

Operating curves for stock on hand are created by representing different inventory statuses in condensed form as a curve. Figure 12.1.4.1 shows how the operating curves for the stock on hand of an item can be derived from the store throughput diagram (see Figure 12.1.3.3). See also [Wien97].

Inventory stock at a given point in time corresponds to the vertical distance between the stock receipts and stock issues curves. By considering the size of these areas, we can then calculate performance indicators such as mean inventory stock, mean time in storage, and mean delivery delay. See also [Gläs95].

Figure 12.1.4.1a shows the store throughput diagrams for three different inventory statuses. These statuses differ primarily with respect to mean inventory stock.

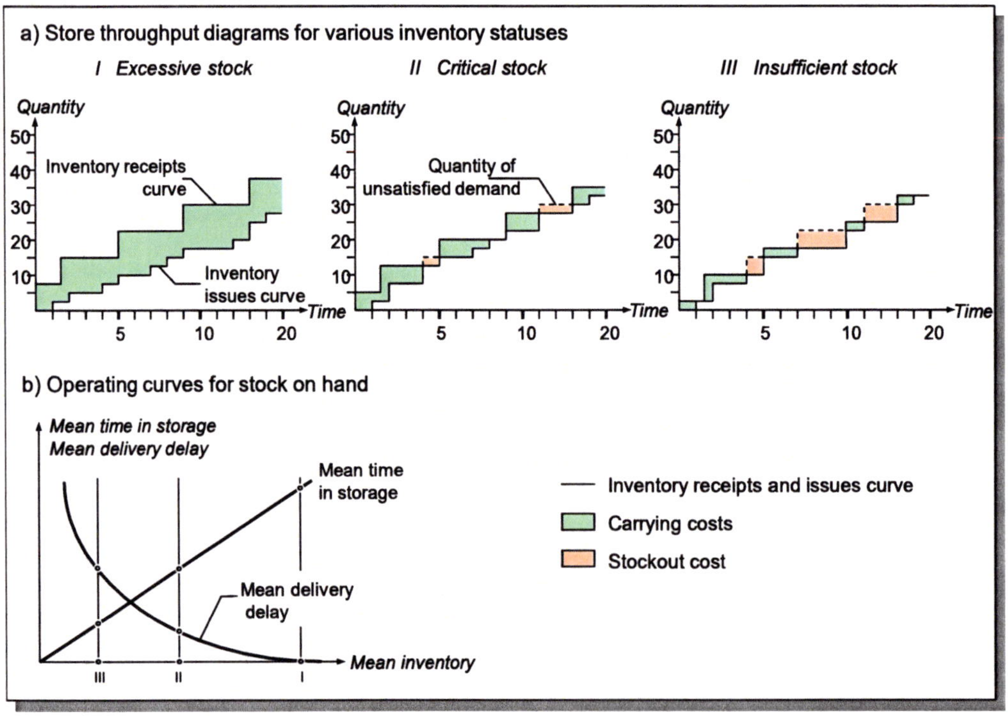

Fig. 12.1.4.1 Derivation of an operating curve for stock on hand from the store throughput diagram (see [Wien97], p. 173).

- Inventory status I has a high stock level. There are no delivery delays, because any demand can be fulfilled immediately. The mean time in storage is very long, however.
- For inventory status II, the mean time in storage is much shorter than for inventory status I. However, there are occasional supply bottlenecks; that is, periods in which demand cannot be satisfied.
- In inventory status III, no stocks are available over relatively long periods. Further demand cannot be satisfied, which leads to very long delivery delays.

Let us now consider Fig. 12.1.4.1b. Applying the three inventory statuses and their performance indicators — mean time in storage and mean delivery delay — to inventory stocks, we obtain the associated operating curves for stock on hand by joining up the points. This type of curve can be created in practice using analytical methods or by simulation. See [Gläs95].

The use of operating curves for stock on hand thus enables us to represent the interdependencies between quantitatively determinable logistic performance indicators in graph form. Operating curves for stock on hand enable us to derive target values for the important cost factor of inventory stock for the purposes of inventory control. This is analogous to the use of logistics operating curves for work stations (see Section 13.2.4). This form of graphical representation is useful for evaluating and improving procurement processes, analyzing capability when selecting suppliers, and comparing the power of different inventory control techniques. Typical examples include:

- The flatter the increase in the mean time in storage curve, the higher the stock-inventory turnover.
- The closer the mean delivery delay curve is to the two axes, the more closely inventory entries mirror inventory issues (and thus demand).

12.2 Deterministic Determination of Independent Demand

Intended learning outcomes: Present the customer order and distribution requirements planning (DRP). Disclose the consumption of the forecast by actual demand.

12.2.1 Customer Order and Distribution Requirements Planning (DRP)

Intended learning outcomes: Explain distribution requirements planning DRP, using an example.

Deterministic independent demand is independent demand where quantity, date, and physical characteristics are all known (see also Section 5.2.1).

For demand external to the organization, this means end products or service parts on order; that is, the individual positions on a customer order. The following are usually handled in a manner similar to the handling of customer demand:

- *Warehouse demand*; that is, demand for replacing inventory in a warehouse
- *Interplant demand*; that is, one plant's need for a part or product that is produced by another plant or division in the same organization

A specific position in the customer order exists at least until it is delivered and invoiced during distribution control. If the items or their components are not available from stock, then the "life span" of a position in the customer order incorporates the "life spans" of all the production and procurement orders needed to cover this deterministic independent demand. It should be possible to establish a connection between these orders and the underlying independent demand at any time, so that control of operations can respond to any deviations from the schedule. The consequences that production or procurement delays or changes in quantities will have on customer orders must be apparent.

Strictly speaking, deterministic independent demand arises only when the order is confirmed, since this is the first document to contain a legally binding description of the items ordered, their quantities, and delivery dates. Nevertheless, despite its legally binding nature, independent demand is still not deterministic as defined above at this point. Depending on supply and demand, the customer is still in a position to vary the quantity or defer the due date, despite divergent legally binding agreements. Here, the customer may be required to pay a previously agreed-upon penalty.

One important factor when scheduling customer demand is the organization's distribution network structure as determined by distribution planning. The due date for the independent customer demand is the date of shipment. The distribution network structure determines how far in advance of the delivery date to the customer this date is.

In-transit lead time is the time between the date of shipment (at the shipping point) and the date of receipt (at the receiver's dock) ([APIC16]).

In-transit lead time includes preparation for delivery from the plant, transportation to distribution warehouses, and distribution to the customer. These times are determined by distribution planning. For a distribution network structure with limited capacity, such as truck fleets, the date of an independent demand is often determined by the cycles used to cover certain routes. For very large or high-cost items in particular, route planning also determines the order in which parts are assembled (for customer production orders) or commissioned (for customer orders from stock), in addition to the delivery dates. See also Section 15.4.

The duration of the data flow accompanying the customer order is another important aspect of the distribution network structure. This applies to delivery documents and transportation documentation, such as for customs purposes. This aspect of the data flow should be planned carefully, as there are cases in which the data flow can last longer than the associated goods flow, particularly with respect to service parts. Solutions based on the latest communications technology help to speed up the process. Examples include fax, EDIFACT, and so on.

With a multi-echelon distribution network structure, customer demand at each intermediate level can be handled as independent. For management of distribution inventory, the order point technique can be used. However, if demand fluctuates widely, the distribution requirements planning technique has proved practical.

> *Distribution requirements planning (DRP)* translates planned orders of the various levels of warehouses in the distribution network directly into planned orders of the central distribution warehouse.

Figure 12.2.1.1 shows distribution requirements planning for an example item with the item identification 4711.

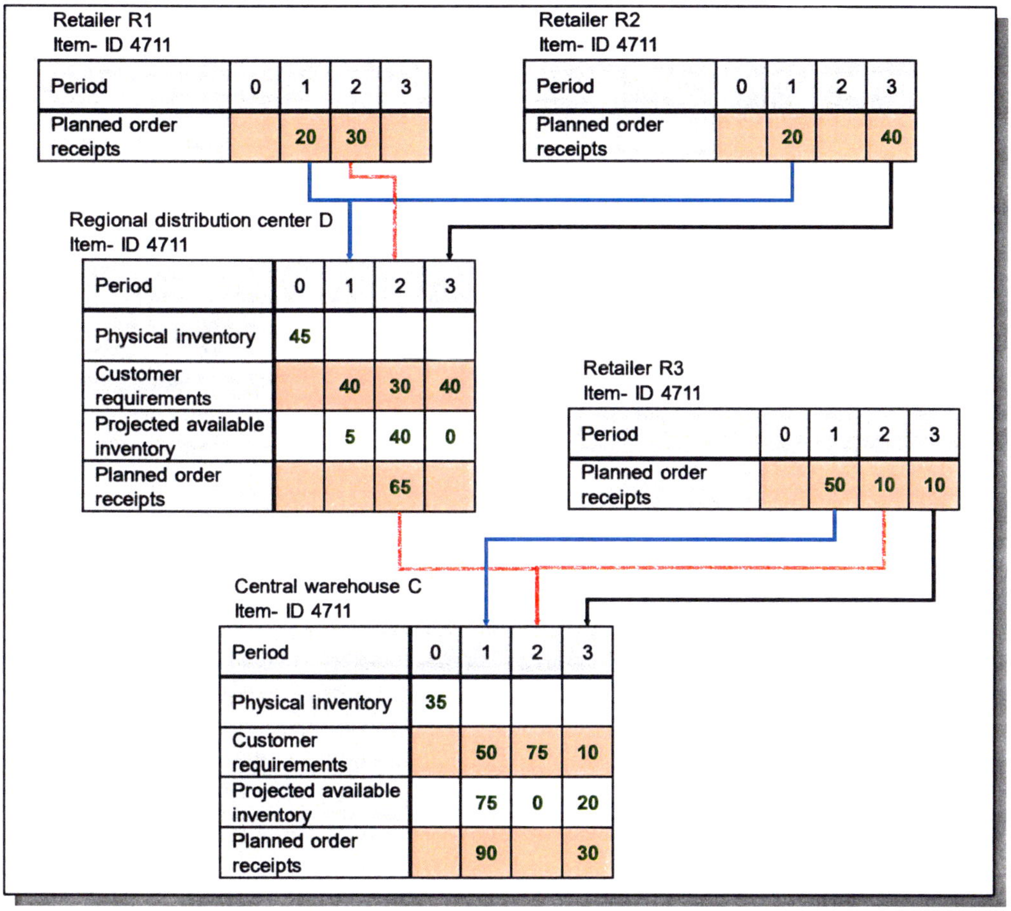

Fig. 12.2.1.1 Distribution requirements planning DRP (example).

Planned receipts of the central distribution warehouse — in the example in Figure 12.2.1.1 the 90 units in period 1 and the 30 units in period 3 — are the same as the gross requirements on the production source or sources that supply the central distribution warehouse.

The advantage of the DRP technique over multilevel application of the order point technique along the distribution chain is the elimination of safety stock at the individual levels. With this, however, all demand — on the distribution chain and on the supplying sources — in principle becomes dependent demand. The DRP technique thus corresponds to the MRP technique described in Section 12.3. For that reason, we will not go any further into the logic and the details of the DRP technique, such as determination of planned receipts from the projected available inventory.

12.2.2 Consuming the Forecast by Actual Demand (*)

Intended learning outcomes: Differentiate between the spreadsheet representation and the graphical representation (or inventory curve) of the projected available inventory calculation.

Consuming the forecast or *forecast consumption* is the process of reducing the forecast by customer orders or other types of actual demand as they are received ([APIC16]).

Independent demand determined by stochastic methods, that is, a forecast, can be used as an alternative to customer demand not yet received. Viewed quasi-deterministically, it enables deterministic dependent demand to be calculated at lower product structure levels by exploding the bills of material to trigger production or procurement of these items in good time and in sufficient quantity.

The forecast is gradually replaced or "consumed" by actual demand, that is, by customer orders. The actual (deterministic) demand thus "overlays" the stochastic independent demand, which either immediately precedes it along the time axis or is the earliest forecast along the time axis that has not yet been completely replaced by customer demand.

The resulting forecast consumption rules are as follows:

1. If a customer demand is canceled, the demand forecast remains unchanged.
2. If a customer demand is issued, it "overlays" the corresponding forecast and thus "consumes" its open quantity, which is then regarded as "issued." There are two variations of this. In variation 2.1, the demand forecast that immediately precedes it on the time axis is reduced. In variation 2.2, all the forecasts preceding the customer demand whose forecast quantities have not yet been reduced — in chronological order — are reduced.
3. *Option overplanning*: If the sum of the customer demands is too large, the quantity by which it exceeds the forecast quantity is regarded as net requirements.

The adjustments yield the value of the remaining forecast for each period. Figure 12.2.2.1 shows the principle of forecast consumption, both before and after the issue of two customer demands. This is variation 2.1.

The *demand time fence* (DTF) is that point in time inside of which the forecast is no longer included in total demand and projected available inventory calculations. Inside this point, only customer orders are considered ([APIC16]).

With option overplanning, an order may only be planned in the period where new customer orders are currently being accepted. This is typically just after the demand time fence.

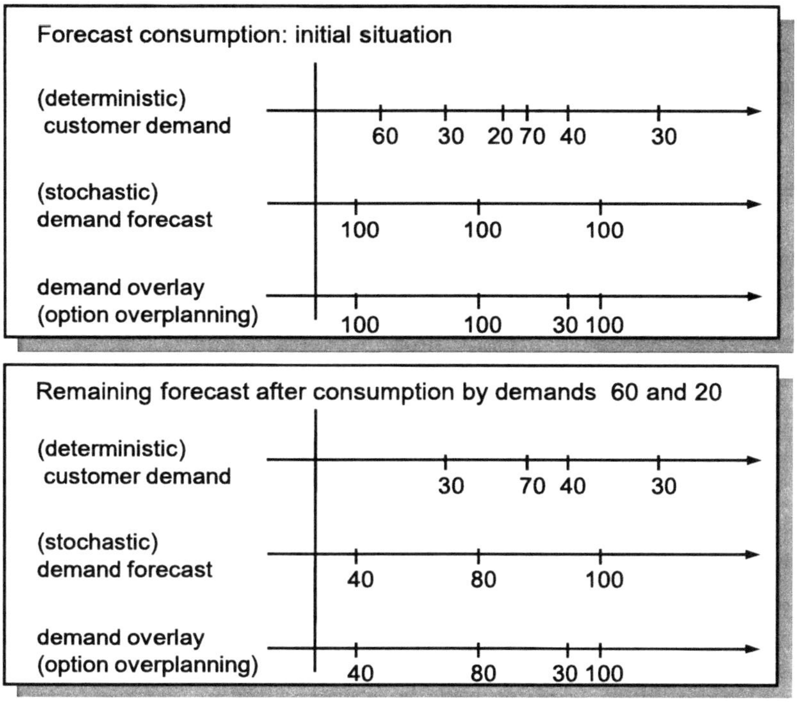

Fig. 12.2.2.1 The principle of forecast consumption.

12.3 Deterministic Determination of Dependent Demand

Intended learning outcomes: Describe characteristics of discontinuous dependent demand. Explain material requirements planning (MRP) and planned orders. Disclose the determination of the timing of dependent demand and the load of a planned order.

12.3.1 Characteristics of Discontinuous Dependent Demand

Intended learning outcomes: Describe the principle of lumpy dependent demand due to batch sizes at higher structure levels. Differentiate between two techniques for inventory management of components with lumpy demand.

If dependent demand is continuous or regular, analytical forecasting techniques may be used to determine demand, and, if necessary, the (stochastic) order point technique may be used for materials management. This applies to purchased parts, such as screws and nuts, or raw materials, such as sheet metal, which are of a very general nature and appear as components in various higher-level products. Demand for such commodities is very frequent, sometimes extremely high, and is distributed along the time axis such that a relatively continuous pattern of demand is obtained overall. The individual demands are also relatively small in relation to the batch size of the production or procurement order.

However, the need for components of manufactured products often arises discontinuously, rather than continuously. Under these circumstances, we will first see several periods with no demand, followed by a large demand resulting from a production or procurement batch for the product at a higher structure level, as Figure 12.3.1.1 shows. In this case, the quantities issued will typically be of the same order of magnitude as the production or procurement batch for the component.

Feature \ Period	0	1	2	3	4	5	6
Physical inventory	35						
Safety stock	5						
Allocated quantities		10	12	12	14	12	12
Planned order receipts				30		30	
Component requirements			30		30		

Fig. 12.3.1.1 Lumpy dependent demand due to batch sizes at higher structure levels.

Where the demand for components can be derived from the requirements for higher-level subassemblies, the order point technique is unsuitable for control purposes, because the carrying cost is too high. Figure 12.3.1.2 illustrates this point (the shaded areas represent the carrying cost).

- There is a demand for component C as soon as an order for assembly A is received. Thus, the demand for component C is not continuous. There is no point in maintaining a safety stock of 20 units of C, for example, if the lumpy demand is for 100 units.
- The order point technique results a large physical inventory of C, which must be kept until the next order is received for higher-level assembly A.
- The ideal situation is the one shown at the bottom part of Figure 12.3.1.2. The production or procurement order for C should occur immediately before the demand for component C arises. In this case, component C is stored in the warehouse either for a very short time or not at all. This type of planning is the explicit objective of the MRP (material requirements planning) technique.

The MRP technique calculates dependent demand on the basis of higher-level independent demands. In principle, this technique requires no safety stock to be kept in stock. On the other hand, a safety lead time must be incorporated into the lead time in order to absorb the effects of late deliveries.

If a small safety stock of components is kept to cover such fluctuations, its purpose is to enable any parts that have to be scrapped during production of higher product structure levels to be replaced as quickly as possible. Similarly, scrap and yield factor can also be considered for every batch that is released. For example:

Batch size (= expected yield):	100
Scrap factor:	5%
⇒ Yield factor:	95%

⇒ Order quantity to be released: 100/ 95% = 105.26 → 106

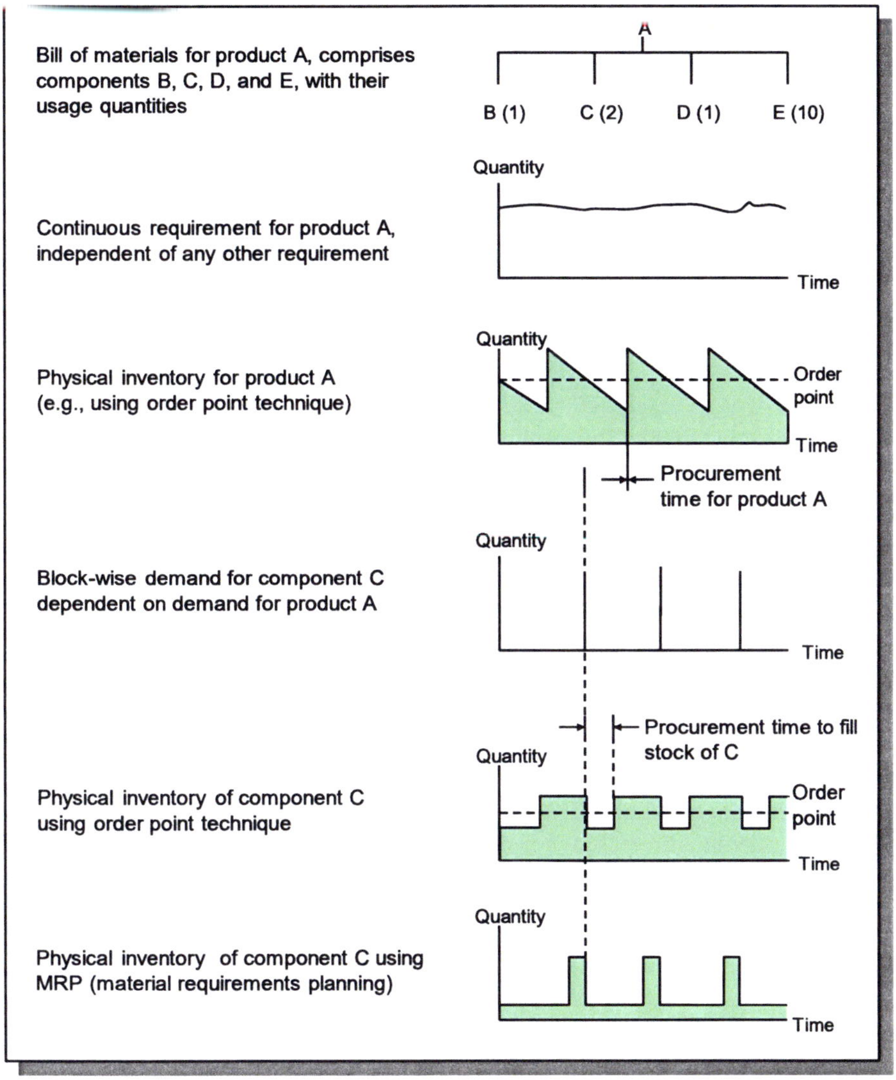

Fig. 12.3.1.2 Two techniques for inventory management of components with lumpy demand.

However, if the demand is a *stochastic independent demand*, that is, a forecast, then a safety demand will already have been included in the (quasi-deterministic) independent demand, as described in Section 10.5.5. In this case, the bills of material explosion transfers this safety demand to the lower structure levels.

12.3.2 Material Requirements Planning (MRP) and Planned Orders

Intended learning outcomes: Disclose a schematic representation of the MRP technique. Explain the determination of net requirements and batch sizes, using an example.

The *MRP technique (material requirements planning)* for calculating dependent demand is defined below. See also [PtSm11] and [Orli75]. *Net requirements planning* is another term for MRP (see also Section 5.1.2).

Four steps are carried out for each item, *in ascending order of their low-level code* (see Section 1.2.2). The four steps thus start with the end products and finish with the raw materials and purchased parts. Repeating the four steps for every item results in a multilevel procedure, as shown in Figure 12.3.2.1. Let us now consider the four steps in detail:

1. *Determine gross requirements:*

Gross requirement is the time-phased sum of independent and dependent demand of the respective period.

- At the highest level, that is, for end products, the gross requirement is independent demand. This main input for the MRP technique stems, in general, from the master production schedule (MPS) and is made up of, on the one side, customer orders (the "original" requirement, this is deterministic independent demand), and, on the other side, sales forecasts (the supplementary requirement, this is stochastic independent demand, which entails quasi-deterministic materials management).

- At the lower levels, that is, for assemblies and parts, the gross requirement often consists of just one of the two classifications of demand, namely, of independent demand or dependent demand. For service parts, for example, it will be made up of both classes. The so-called *service parts demand* is demand for service parts that are sold as such. Thus, it is forecasted independent demand. Demand for service parts that are integrated into higher-level products is calculated as dependent demand derived from the demand for the higher-level product in step 4. Thus, it is derived by a deterministic technique; in the case of stochastic independent demand, by a quasi-deterministic technique. If the gross requirement consists of both classes, a multilevel master schedule may have to be used.[3]

2. *Determine the net requirements:*

Net requirements are the time-phased negative projected available inventory.

- Figure 12.3.2.2 shows a common situation for any given item. The safety stock is subtracted from projected available inventory right at the start. As a result, production or procurement orders are then scheduled such that they enter into stock when the projected available inventory falls below zero.

---

[3] A *multilevel master schedule* allows management of components at any level of an end product's bill of material as master schedule items (cf. [APIC16]).

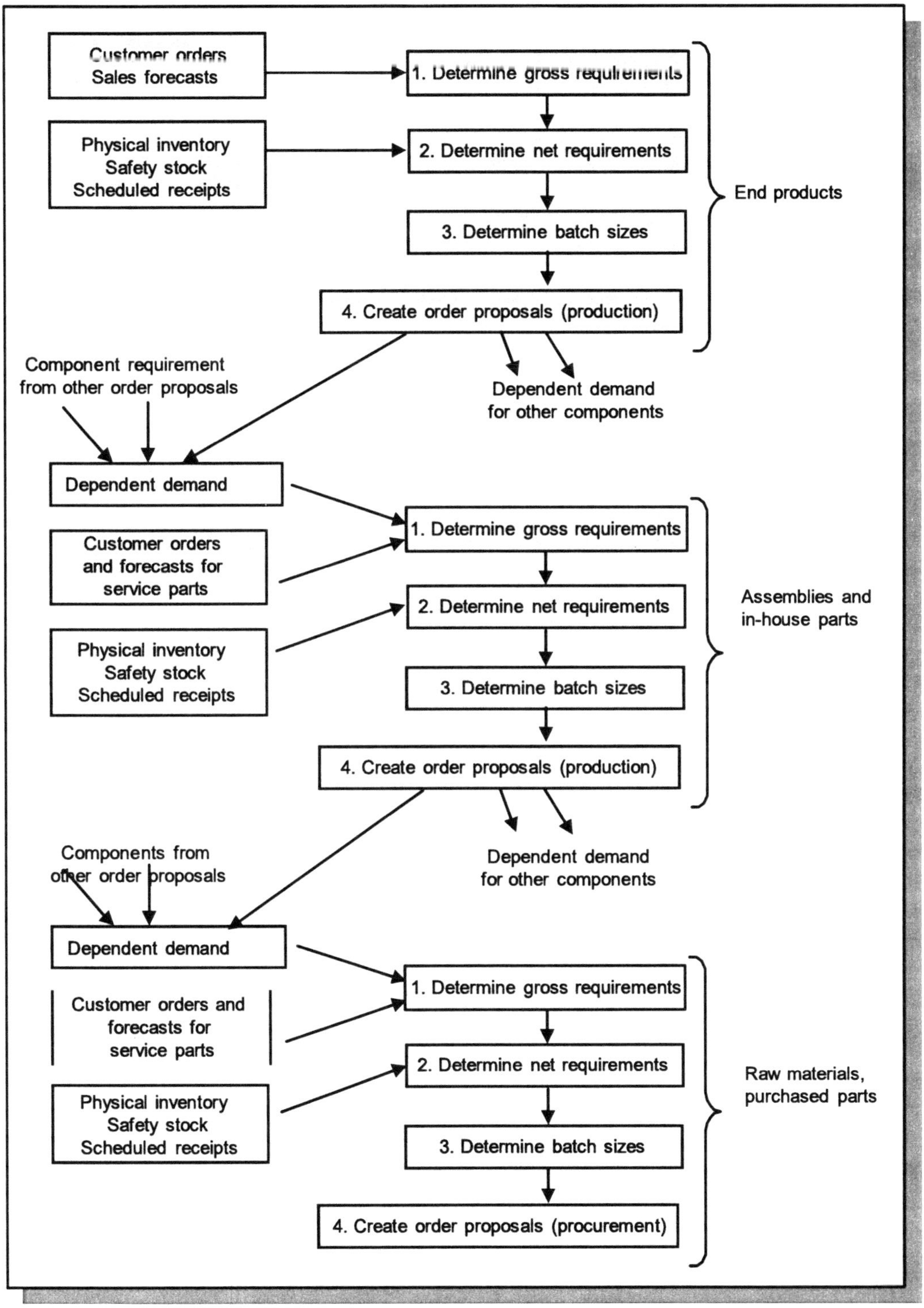

Fig. 12.3.2.1 Schematic representation of the MRP technique.

Feature / Period		0	1	2	3	4	5	6	7	8	9
Physical inventory	(+)	50									
Safety stock	(−)	20									
Scheduled receipts	(+)				65						
Allocated quantities	(−)		15	0	10	0	0	0	0	0	0
Planned gross requirements	(−)		5	0	40	25	0	20	15	0	10
Projected available inventory	(=)	30	10	10	25	0	0	0	0	0	0
Net requirements (negative projected available inventory)	(+)		0	0	0	0	0	20	15	0	10
Batch size / planned receipts								35			35
Planned releases					35			35			

Fig. 12.3.2.2 Determination of net requirements and batch sizes (example).

- It is assumed that receipts occur at the beginning of a time period and that issues occur during a period. Receipts and issues are now added or subtracted over time, and the available quantity is calculated along the time axis. This results in the net requirements: a series of negative available inventories after each period. The sum of all these negative available inventories along the time axis is known as net requirements.
- Step 3 of the MRP technique (see Figure 12.3.2.1), determination of batch sizes, has already been carried out by way of example. In step 4, planned orders are generated from the batches. *Planned release*, that is, the scheduled release of a planned order, is thus the planned receipt brought forward by the lead time (here, by three periods).
- Of course, it would also be possible to use the same graphical representation for listing every planning-related event individually, rather than in a *bucketed system*, that is, combining them in periods or time buckets. Such a *bucketless system* could result in a very large list, however (or a large number of columns in Figure 12.3.2.2).

3. *Determine the batch sizes:*
   - There are a number of batch-sizing policies for combining net requirements into batch sizes. These are described in Section 12.4.
4. *Create an order proposal, that is, a planned order for every batch:*
   - The first step is to calculate the lead time to determine the point in time at which the order should be released.
   - For a planned production order, the next step is to determine — from the routing sheet of the product to be manufactured — the planned operations and thus the planned load of the work centers (see also Section 12.3.3).

- For a planned production order, this also includes a requirements explosion to schedule the demand for components (see also Section 12.3.3). This (dependent) demand is the batch size multiplied by the usage quantity. It is also the gross requirement for the component and is one of the quantities to be determined in step 1 for the component in a subsequent MRP stage. This is the final MRP planning stage.

If the order proposals are not subsequently released, they are automatically adapted to take account of the current situation the next time that requirements are calculated. This generally means deleting all the planned orders and then recalculating them in a comprehensive rerun of the MRP algorithm.

If the independent demand changes only slightly, the *net change MRP technique* is usually faster. This technique attempts to consider only those net requirements that have changed. The four-step procedure is applied only to those articles whose projected available inventory has changed since the last MRP run. If planned production orders are changed, this will also affect the dependent demands for components, so that the MRP procedure must be repeated for each component. If a large number of items are affected, the entire order network will have to be recalculated — effectively a comprehensive rerun of the MRP algorithm.

12.3.3 Determining the Timing of Dependent Demand and the Load of a Planned Order

Intended learning outcomes: Differentiate between mean lead time calculation and detailed planning calculation of the timing of dependent demands.

Order proposals are compared against the net requirements, which are broken down into meaningful batch sizes. For a purchased item, generating an order proposal essentially means calculating the order point with due regard to the lead time (which is part of the master data for the item). For an item produced in-house, the start date can also be determined by subtracting the lead time from the completion date. The dependent demands for all the components will be needed on the start date. This is how the conventional MRP technique works.

A more detailed and comprehensive technique calculates the process plan (see Section 1.2.3) of the item's final production stage. At the same time, planning data are generated for materials management, time management, and scheduling and capacity management:

- The load that this order will generate at the various work centers: by multiplying the order quantity by the operation load for each operation (see also Chapter 14).
- The time at which a load arises: by means of a lead-time calculation starting with the order completion date (see also Chapter 13).
- The start date for the order (see also Chapter 13).
- The dependent gross requirement (or dependent demand): by multiplying the order quantity by the usage quantity for each position on the bill of material.
- The time at which a dependent demand arises, taking into account the start date for the operation that processes the demand.

Figure 12.3.3.1 compares the conventional MRP technique (variation 1), that is, the mean lead time, with the above-mentioned more comprehensive technique (variation 2). The example calculates the timing of the dependent demands for a product A, which is made up of components B and C.

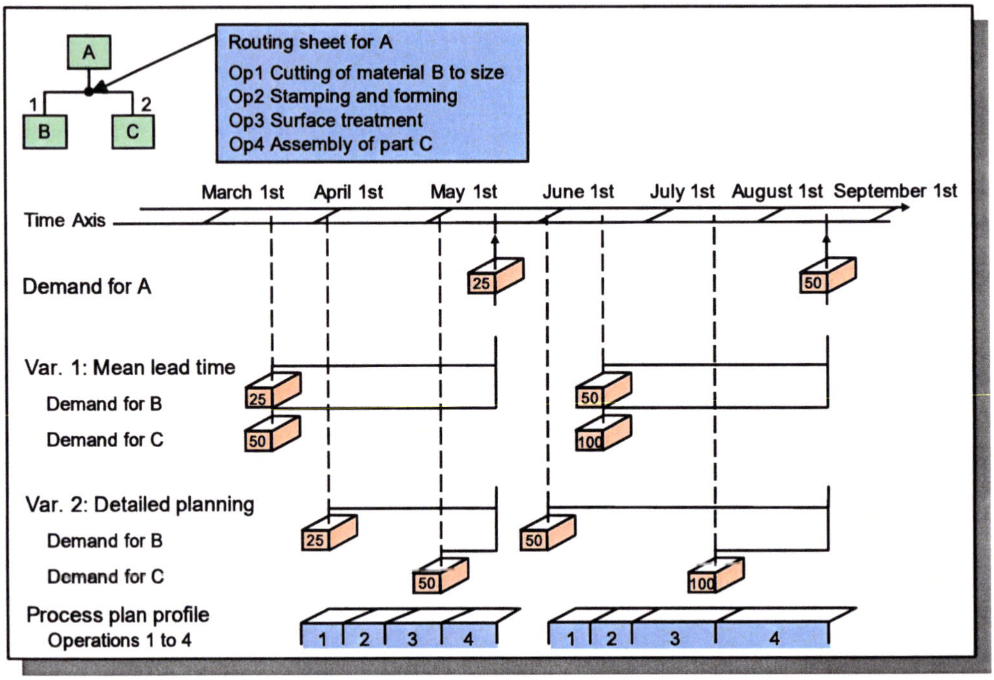

Fig. 12.3.3.1 Calculating the timing of dependent demands.

In variation 1, it is assumed that the average lead time for producing A is two months. The timing of the dependent demands for components B and C is thus the planned order completion date for A minus its average lead time.

Variation 2 shows the more comprehensive and detailed technique. The process plan for product A was included in the calculation. The first difference is that the lead time for batches of 25 is just 1.5 months, whereas it rises to 2.5 months for batches of 50. In addition, demand for C does not arise until the fourth operation, which should start half or one month before the order completion date, depending upon the batch size.

Figure 12.3.3.1 shows how this affects the way in which the timing of dependent demands is calculated in variation 2. If B and C are very high-cost items, the detailed procedure would help to allow the components to be channeled into production exactly when they are needed. This can reduce both the volume and the value of goods in process.

If we compare the two variations, we can see that the more general variation 1 is very suitable both for (long-term) master planning and for medium-term or short-term planning for inexpensive and low-volume components. In all other cases, variation 2 is more suitable, although calculation requires much more processing power and more complex algorithms, which may also be more prone to error.

By the way: The multilevel available-to-promise *(MLATP)* technique uses variation 1, the capable-to-promise (CTP) technique uses variation 2. For details refer to Section 5.3.5.

12.4 Batch or Lot Sizing

Intended learning outcomes: Explain combining net requirements into batches. Differentiate between different batch-sizing policies.

12.4.1 Combining Net Requirements into Batches

Intended learning outcomes: Disclose various batch sizing policies, namely lot-for-lot, optimum number of (merged) demands, optimum number of partial lots, optimum batch size (economic order quantity, EOQ), the period order quantity (or optimum length of order cycle), part period balancing (PPB), dynamic optimization. Explain the "blowthrough" technique linked with the lot-for-lot sizing policy.

A *batch-sizing policy* or *lot-sizing policy* is a set of techniques that create production or procurement batches from net requirements.

In practice, there are various possible batch sizing policies:

1. *Lot-for-lot*: every net requirement translates into just one planned order. Variation: if the component batch sizes fall below a certain quantity, a "blowthrough" of the component requirements right into the requirements given by its bill of material and its routing sheet may take place (see description below).
2. A dynamic lot size, made up of an *optimum number of merged demands* taken together. If this number is 1, then the situation is again one of make to order.
3. A dynamic lot size with an *optimum number of partial lots*. This policy suggests splitting the demand into several orders. Another attribute determines the minimum deferral time between two of these orders.
4. A *fixed order quantity*, known as the *optimum batch size*, either determined manually or calculated using the EOQ (economic order quantity) formula, for example (see Section 11.4.2). If two orders are closer together than the specified *minimum deferral time*, they are procured in a single batch (multiples of the EOQ).
5. A *dynamic lot-sizing* technique, known as *period order quantity*, which combines various demands into one batch over the course of an optimum number of time buckets. This corresponds to the optimum period of time for which future demand should be covered, that is, the *optimum order interval* or the *optimum length of order cycle* in Figure 11.4.2.6. It is calculated, in principle, by dividing the optimum batch size by the average annual consumption.

6. *Part period balancing*, another dynamic lot-sizing technique. For the first period's demand, an order is planned. For every further period's demand, the carrying cost that will be incurred from the time of the last planned order is calculated. If these costs are lower than the setup and ordering costs, then every further period's demand is added onto the last planned order. Otherwise, a new order is scheduled for every further period's demand.

   Variant: If the *cumulative* carrying costs of all the period's demand since the last planned order are greater than the setup and ordering costs, a new order is scheduled.

7. *Dynamic optimization* (as described by [WaWh58]). This relatively complicated technique calculates the various totals for setup and carrying costs resulting from different combinations of net requirements to form batches and determines the minimum costs from these totals. This technique for identifying minimum costs is illustrated in the example below.

All batch-sizing policies, except the fourth, result in so-called discrete order quantities.

A *discrete order quantity* is an order quantity that represents an integer number of periods of demand. That means that any inventory left over from one period is sufficient to cover the full demand of a future period.

The following additional aspects of the various batch-sizing policies should be considered:

- *The "blowthrough" technique linked with the lot-for-lot sizing policy:* Designers tend to define structural levels that correspond to the modules of a product. However, in the production flow, the modules are not always meaningful, since some products are manufactured in one go. This is often the case with single-item production, and requires — *de facto* — extended *phantom bills of material.* The blowthrough technique drives requirements straight through the phantom bill of material to its components and combines the operations in a meaningful order. Thus, the technique converts several design structure levels to a single production structure level (see the definitions of these terms in Sections 1.2.2 and 1.2.3). At the same time, the multilevel *design bill of material* is transferred to the associated single-level *production bill of material.* Figures 12.4.1.1 and 12.4.1.2 show as an example product X, which is made up of two longitudinal parts L and two transverse parts Q, each made from the same raw material. The information is shown before and after the "blowthrough" of requirements through L and Q. See also [Schö88a], p. 69 ff.

- For the 2nd to the 5th batch-sizing policies, you can also specify whether the optimum values should be calculated or set manually. Maximum and minimum values can be assigned to restrict these optimum values if the calculation returns unusual values.

- The 2nd and the 3rd batch-sizing policies are particularly important for harmonious or rhythmic production, in which a certain quantity leaves production during each unit of time. The components should be procured at a similar rate.

- The 3rd batch-sizing policy, or batch splitting, is used if the specified requirement in total is not needed all at the same time. For an assembly batch of 100 machines, for example, not all the components will be needed at once, since the machines are assembled one after the other. Thus, two partial batches could be created, if necessary, for producing or procuring components, and the second partial batch could be channeled into the assembly process some time after assembly starts.

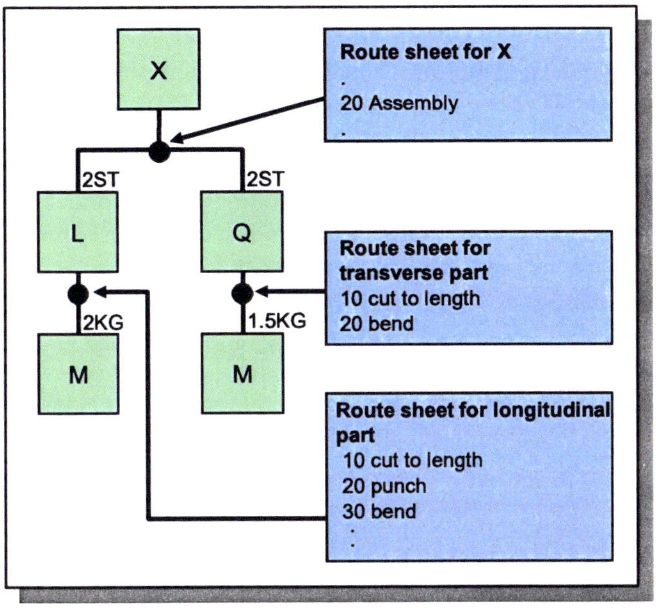

Fig. 12.4.1.1 Bills of material and route sheets for a product X from the viewpoint of design.

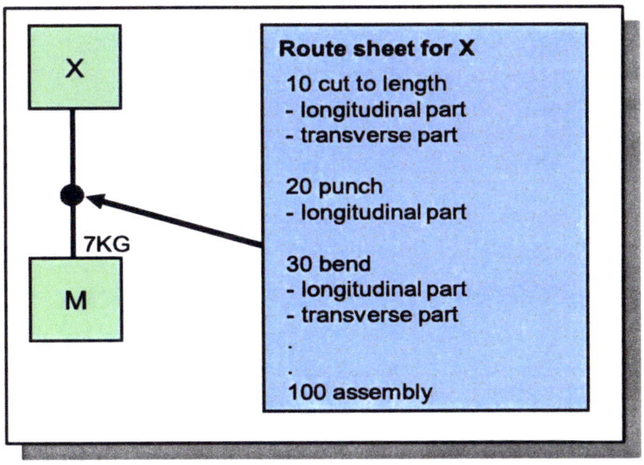

Fig. 12.4.1.2 Bills of material and route sheets for a product X: structure from the production viewpoint, after "blowthrough" of requirements through L and Q.

- *With the 4*[th] *batch-sizing policy,* or fixed order quantity, physical inventory is inevitable, since more items are generally procured than are needed to satisfy demand. This policy should therefore only be used if the inventory level will actually be reduced, that is, when it is safe to assume that demand will really occur in the future. This is the case if future demand can be determined on the basis of past consumption — at least where demand is regular. This batch-sizing policy is therefore not economically viable for lumpy demand.

- *5th, 6th, and 7th batch-sizing policies:* policies 5 and 6 are generally used in deterministic materials management. Policy 7 is the most complicated, and, although it produces a precise and optimum solution, it is unfortunately not very robust. The accuracy obtained and thus the economic viability of policies 5, 6, and 7 increase in ascending order. Unfortunately, the complexity and processing power required also increase accordingly, especially if the techniques are applied to precise events, rather than time periods. On the other hand, the robustness decreases in ascending order, which means that, if the quantity or date of a demand within the planning horizon changes, policy 7 will require complete re-calculation, while a change in demand will not necessarily have severe consequences for policy 5.
- *7th batch-sizing policy:* Figure 12.4.1.3 shows the steps of the dynamic optimization technique described by [WaWh58]. They should be studied in conjunction with the example in Figure 12.4.2.1.

1. The first batch should be determined at the start of the first period.
2. In every subsequent period, a new batch should be established as an alternative. The initial costs are determined from the minimum total costs for all variations to date (rows in the table) in the preceding period, plus the batch-size-independent production or procurement costs for establishing a new batch for the current period.
3. The minimum cost is the minimum value of the total costs in the previous period.
4. Starting from this minimum value, the lots are put together "backward," by seeking the way to achieve this minimum.
5. To reduce the processing required, the following simplification can be applied to each variation (row in the table): When the carrying cost for a demand in a given period exceeds the batch-size-independent production or procurement costs, it is no longer worth adding this demand to the batch. It will not be possible to establish a minimum value, even if this variation (row) is subjected to more extensive calculation of the total costs for a subsequent period.

Fig. 12.4.1.3 Dynamic optimization technique as described by [WaWh58].

12.4.2 Comparison of the Different Batch-Sizing Policies

Intended learning outcomes: Explain, using an example, the comparison of various batch-sizing policies, namely dynamic optimization, part period balancing (PPB), the period order quantity (or optimum length of order cycle), and the optimum batch size (economic order quantity, EOQ).

Batch-sizing policies 7, 6, 5, and 4 described in Section 12.4.1 are compared below. These policies are

- Dynamic optimization
- Part period balancing (PPB)
- The optimum length of order cycle or the optimum order interval
- The optimum batch size (economic order quantity, EOQ)

The following assumptions apply:

- Net requirement: 300 units of measure divided between six periods (for example, 2-month periods) giving 10, 20, 110, 50, 70, 40 units
- Batch-size-independent production or procurement costs: 100 cost units
- Carrying cost
  - per unit of measure and period: 0.5 cost units
  - per unit of measure over six periods: 3 cost units
- An order receipt is assumed at the start of a period. Carrying cost is always incurred at the start of the next period.

Based on these assumptions, you can thus calculate the following values:

- Optimum batch size using the economic order quantity (EOQ) (see Figure 11.4.2.4):

$$X_0 = \sqrt{2 \cdot 300 \cdot \frac{100}{3}} = \sqrt{20000} = 141.42 \approx 140$$

- Optimum length of order cycle or the optimum order interval (see Figure 11.4.2.6):

$$\begin{aligned} LOC_0 &= \frac{141.42}{300} \cdot 6 \text{ periods} \\ &= 0.47 \cdot 6 \text{ periods} \\ &= 2.83 \text{ periods} \\ &\approx 3 \text{ periods} \end{aligned}$$

In Figure 12.4.2.1, the total setup and ordering costs as well as the carrying cost are calculated for the various batch-sizing policies.

Every policy yields a different result in specific cases, although this is not necessarily so in the general case. The results obtained with these techniques tend to improve in the order given above. Indeed, the optimum batch-size technique can be used only if the quantity of the last batch does not exceed the net requirement. But, even under these circumstances, the technique produces unsatisfactory results when applied deterministically.

12.5 Analyzing the Results of Material Requirements Planning (MRP)

Intended learning outcomes: Present projected available inventory and pegging. Produce an overview on action messages.

12.5.1 Projected Available Inventory and Pegging

Intended learning outcomes: Differentiate between pegging (or requirements traceability) and demand coverage traceability.

	Period	1	2	3	4	5	6	Total costs per lot	Total costs cumulative
	Net requirements	10	20	110	50	70	40		
Dynamic optimization	Cumulative carrying and setup costs	100	110	220	295				
			200	255	305	410			
				210	235	305	365		
					310	345	385		
						335	355		
	Batch sizes	30		160		110			355
Part period balancing	Carrying cost per net requirement	0	10	(110)				110	
				0	25	70	60	255	
	Batch sizes	30		270					365
Optimum length of order cyle	Cumulative carrying and setup costs	100	110	220				220	
					100	135	175	175	
	Batch sizes	140			160				395
Economic order quantity	Cumulative carrying and setup costs	100	110	220				220	
					100	135	155	155	
							100	100	
	Batch sizes	140			140		140 (20)		475

Fig. 12.4.2.1 Comparison of various batch-sizing policies.

Pegging or *requirements traceability* determines the independent demands that give rise to a dependent demand or a production or procurement order.

Pegging is one of the most important analyses for delayed orders, for example. It can be regarded as active where-used information. It determines the source of demand requirements, determining whether the underlying independent demands are customer orders or whether they stem from uncertain forecasts in the master plan.

To carry out this type of investigation, objects are created in the course of MRP for order connection purposes, specifically between item issues (demand positions in an order) and item receipts (positions for demand coverage). These objects can then be used to derive the desired pegging.

Pegging is equivalent to an allocation algorithm that assigns demand (item issues) to orders (item receipts). It is sometimes possible to cover every demand with several positions from different production or procurement orders. Conversely, every position in a production or procurement order can be used for several demand positions in various orders.

Creating the *order connection* object during MRP results in four types of action messages, or exception messages:

- Order to be pushed forward (speeded up)
- New order proposal
- Order to be deferred (slowed down)
- Superfluous order

The *rescheduling assumption* assumes that it is more promising to speed up an order already in process than to create a new order, since the remaining lead time is shorter.

As a consequence of this assumption, MRP logic tends to push forward orders that have already been released before it proposes a new order:

For the purposes of pegging, the order identification concerned is entered. One of the algorithms corresponding to the multilevel where-used list (see Section 17.2.3) calculates all the independent demands that are affected by this order. This results in multilevel pegging, which identifies all the intermediate demands and orders. The "leaves" of the resulting tree structure are then independent demands: forecasts, genuine customer demands, or unplanned orders for end products or service parts. For example:

- For *bottom-up rescheduling*, the planner uses pegging to solve material availability or similar problems. This can entail compressing lead time, cutting order quantity, or making changes to the master schedule.

For quick decision making in procurement situations, it may be necessary to identify the types of independent demand that give rise to a dependent demand, without the help of a pegging algorithm. A possible technique to solve this problem can be found in [Schö88a], p. 117 ff.

The structure of the *order connection* object can also be used for the opposite purpose.

Demand coverage traceability specifies all the (dependent) demands or orders that are at least partly caused by a particular (independent) demand.

A demand coverage list may be needed if, for example, you have to change the date or quantity for an independent demand (such as a customer order) and want to assess the consequences of this change. The algorithm is thus equivalent to the algorithm that generates a multilevel bill of material (see Section 17.2.3).

12.5.2 Action Messages

Intended learning outcomes: Describe various possible action messages, or exception messages.

An *action message*, or *exception message*, is an output of a system that identifies the need for and the type of action to be taken to correct a current or potential problem ([APIC16]).

The MRP technique essentially yields planned orders with planned gross requirements for their components and loads at the work centers. The order completion date is calculated so that at least part of the batch will be used in a higher-level order or for a sales order as soon as it is produced or procured. For this reason, the start date of the production or procurement order should always be met. Exception messages should thus report the following problems associated with orders:

- Planned orders whose start date has passed
- Planned orders whose start date will pass in the immediate future, such as within a week
- Open orders that should be speeded up or slowed down due to changes in the projected available inventory or too-fast or too-slow progress of the production or procurement order

The main problem with exception messages is that there are so many of them. Sorting and selection of exception messages is important to ensure that the right people receive the right messages. The most urgent messages should arrive first. Sorting and selection can be performed at the least according to the classification of items into groups and subgroups that reflect the structural organization of the planners. The ABC classification is another possible sorting criterion.

Some dependent demand is not due at the start date of an order, but at the start date of a later operation. Therefore, to obtain accurate dates for dependent demands, a scheduling technique should be used that calculates the start date of each operation. This will also reveal the planned load at the work centers, which can then be compared against planned capacity. See also Chapters 13 and 15.

The planners check the number and order quantity of the proposed orders. If the proposals relate to purchased items, they also select the suppliers. Proposals for new orders must then be released — see Section 15.1.

12.6 Summary

This chapter describes the deterministic materials management technique for medium-term and short-term planning. The unique aspect of this technique is that the demand for an item is not simply regarded as a total that, *de facto*, can be evenly distributed along the time axis, as is the case with long-term planning or even stochastic materials management. In contrast, you take advantage of the fact that you know the precise time of every demand and thus the limited period it will take up along the time axis. Lumpy demand can be managed particularly efficiently in this way.

Purely deterministic materials management requires the independent demands to be precisely known. Dependent demands are then derived from them by exploding the bill of material. Since the cumulative lead time remains within the customer tolerance time, the exact demand for procured and produced goods is known.

An attempt should be made to use quasi-deterministic materials management techniques if components at lower levels have to be stored, but demand is only discontinuous. The independent demand is then calculated using stochastic techniques. On the other hand, dependent demand is again calculated by exploding the bill of material.

The starting point for deterministic materials management is the projected available inventory. This is not a scalar variable — it changes after every transaction or every future event that changes stock levels. At any given time, the projected available inventory is defined as the physical inventory plus all open and planned receipts minus all allocated quantities minus all planned demands up to this point.

The projected available inventory calculation thus shows the projected available inventory defined in this way along the time axis. This is useful, for it provides information on the possible demand coverage (quantity and timing, and partial demands, if necessary) for any new demand. The scheduling projected available inventory calculation attempts to bring forward or put back orders in process or allocated quantities so as to maintain a positive projected available inventory at all times. Operating curves for stock on hand describe delivery delays and time in storage in relation to inventory.

Lumpy dependent demand often arises as a result of batch size creation at higher levels, often regardless of whether the independent demand was determined stochastically or deterministically. If stochastic materials management techniques were to be used in this situation, they would result in excessively large inventory stocks and carrying cost. The deterministic MRP (material requirements planning) technique ensures minimum stocks for production or procurement orders that are received in good time.

The MRP technique consists of four steps that are applied to every item in ascending order of their low-level code — starting with the end products, followed by the assemblies and semifinished products, through to the purchased goods.

- The 1st step is to determine the gross requirement, which may be made up of independent and dependent demands. The gross requirement is a data set, rather than a scalar variable. If the calculation is applied to precise periods, there will be exactly one gross requirement per period. If the calculation is applied to precise events, every demand corresponds to a gross requirement.
- The 2nd step is to determine the net requirement by offsetting the physical inventory, safety stock, open orders, and allocated quantities. The net requirement can be made up of individual net requirements. If the calculation is applied to precise periods, there will be exactly one net requirement per period. If the calculation is applied to precise events, every demand may give rise to a net requirement.
- The 3rd step is to combine the individual net requirements to form batches. The conventional EOQ formula is not suitable here, because its batch sizes are fixed.

Techniques that use dynamic lot sizes are much more appropriate here, since the demands are known.

- The 4th step is to convert the batch sizes into order proposals. The start date is determined by scheduling. For in-house production, the work center load and the quantity and date of each component demand are determined from the routing sheet and bill of material. These are dependent demands and can thus be used to calculate the first of the four MRP steps for each component.

MRP generates exception lists containing orders to be released, speeded up, slowed down, or canceled, in addition to order proposals. Pegging and a demand coverage list help to identify orders that are interdependent within the order network.

12.7 Keywords

action message, 541
allocated quantity, 515
batch-sizing policy, 535
blowthrough, 536
consuming the forecast, 526
cumulative projected available inventory calculation, 521
demand coverage traceability, 541
demand time fence, 526
discrete order quantity, 536
distribution requirements planning, 525
DRP (distribution requirements planning), 525
exception message, 541
fixed order quantity, 535
forecast consumption, 526
interplant demand, 524
multilevel master schedule, 530
net change MRP technique, 533
net requirements, 530
open order, 516
optimum number of merged demands, 535
optimum number of partial lots, 535
part period balancing, 535
pegging, 540
period order quantity, 535
physical inventory, 515
planned receipt, 516
planned release, 532
projected available inventory, 516
quantity allocated, 515
quantity reserved, 515
requirement traceability, 540
requirements explosion, 533
reserved quantity, 515
scheduled receipt, 516
service parts demand, 530
time bucket, 515
time-phased order point (TPOP), 515
warehouse demand, 524

12.8 Scenarios and Exercises

Intended learning outcomes: Calculate projected available inventory. Determine net requirements and planned release using the MRP technique. Differentiate between order point technique and MRP technique.

12.8.1 Projected Available Inventory Calculation

Intended learning outcomes: Calculate projected available inventory by completing a grid.

Complete the grid in Figure 12.8.1.1.

Date	Entry	Issue	Balance	Text	Order ID
01 Jan			1000	stock on hand	
05 Jan	100		?	replenishment	101 2897
14 Jan		1050	?	customer Smith	102 8972
15 Jan	?	?	500	?	102 9538
16 Jan		150	?	customer Adams	103 2687

Fig. 12.8.1.1 Projected available inventory calculation

a. What is the available inventory without any restrictions along the time axis?

b. What is the additional available inventory after order 102 9538?

c. Which receipt could be deferred?

d. Furthermore, the following orders are planned:

- Customer order ID 104 2158 of 500 units on January 20
- Stock replenishment order ID 104 3231 of 500 units on January 22

Does this situation lead to a problem? If so, how can it be solved?

Solutions:

Date	Entry	Issue	Balance	Text	Order ID
01 Jan			1000	stock on hand	
05 Jan	100		1100	replenishment	101 2897
14 Jan		1050	50	customer Smith	102 8972
15 Jan	450		500	replenishment	102 9538
16 Jan		150	350	customer Adams	103 2687

a. 50

b. 300 (= 350 – 50)

c. Stock replenishment order ID 101 2897 could be deferred to Jan. 14.

d. Yes, there will not be enough available inventory on Jan. 20. Expediting order ID 104 3231 by at least two days could solve this problem.

12.8.2 MRP Technique: Determining Net Requirements and Planned Release

Intended learning outcomes: Determine net requirements and planned release using the MRP technique, whereupon the physical inventory and gross requirements are given.

Following the example in Figure 12.3.2.2, determine net requirements and planned releases for item ID 4711. Assume an optimum order interval (or optimum length of order cycle) of 3 periods. The production or procurement lead time for item ID 4711 is 2 periods.

Given data or assumptions: a physical inventory of 700 (no safety stock) and the planned gross requirements by period of time as in Figure 12.8.2.1.

Feature \ Period	0	1	2	3	4	5	6	7	8	9	10	11	12	13	14	15
Planned gross requirements		250	200	125	150	150	175	200	220	225	240	250	250	225	225	210

Fig. 12.8.2.1 Gross requirements.

As for the planned available inventory, please enter the result, including the planned receipts in each period.

Solution:

Feature \ Period	0	1	2	3	4	5	6	7	8	9	10	11	12	13	14	15
Physical inventory	700															
Planned gross requirements		250	200	125	150	150	175	200	220	225	240	250	250	225	225	210
Projected available inventory without order receipts	700	450	250	125	0	0	0	0	0	0	0	0	0	0	0	0
Net requirements (negative projected available inventory)		0	0	0	25	150	175	200	220	225	240	250	250	225	225	210
Batch size / planned receipts					350			645			740			660		
Planned releases			350			645			740			660				
Projected available inventory with order receipts	700	450	250	125	325	175	0	445	225	0	500	250	0	435	210	0

12.8.3 Order Point Technique versus MRP Technique

Intended learning outcomes: Explain how discontinuity or lumpiness of the demand influences the sum of carrying costs and setup and ordering costs, comparing the MRP technique with the order point technique.

Section 12.3 presents the MRP technique. It is clear why, in the comparison in Figure 6.5.2.1, the MRP is rated to be complicated with regard to the order point technique or the Kanban technique. Section 12.3.1 explained why discontinuous demand is a main reason for the need of the MRP technique for determining stochastic dependent demand (or quasideterministic demand). We created an example that will give you a sense of how discontinuity or lumpiness of the demand influences the sum of carrying costs and setup and ordering costs, comparing the MRP technique with the order point technique. You can view the animation at the following URL:

https://intlogman.ethz.ch/#/chapter12/order-point-vs-mrp/show

Note that to compare the two techniques a safety stock of the same size as for the order point technique has been introduced for the MRP technique. It is correct to do so, because in the quasi-deterministic case, a safety demand has to be introduced for the independent demand at the end product level (see Section 10.5.5). Through the MRP algorithm, this safety demand is — in fact — always present at some stage on the value chain, just as the safety stock is present in the order point technique for a specific component. Therefore, for comparison of the two techniques, we can assume the safety demand on the component — like a safety stock.

Now, find out how the shape of the of the inventory curve according to the two techniques changes for continuous and less continuous demand (running your cursor over the gray icon shape bar will execute your input choice).

Try out different parameters to calculate the lot size or choose a different initial inventory or service level. Running the cursor over the gray icon either leads you to a specific window where you can enter your input data or executes your input choice.

The "costs" icon opens a window with the carrying costs as well as the setup and ordering costs for the two techniques. Discuss whether for the given demand pattern with less continuous demand there is sufficient reason to prefer the MRP technique. Consider that the calculated costs do not take into account either the batch-size-dependent unit costs — which is the same for both techniques, but generally by far higher than the sum of carrying, setup, and ordering costs — or the administration costs for the implementation and use of the specific materials management technique.

Try out other demand values. Observe the effect of issue quantities on the order of the production or procurement batch size. Again, use the "calculate" icon to execute your input choice. The initial demand values are automatically re-entered by touching the gray demand icon. Note what happens with the curves as you continue to enter sequences of two or more periods with zero demand, interrupted by one or two periods with very high demand. You will see that the order point technique will not be able to handle this demand pattern. The projected available inventory level will sometimes fall below zero, engendering opportunity costs that we did even not consider in the costs comparison.

12.9 Literaturhinweise

APIC16 Pittman, P. et al., APICS Dictionary, 15th Edition, APICS, Chicago, IL, 2016

Gläs95 Glässner, J., "Modellgestütztes Controlling der beschaffungslogistischen Prozesskette," Fortschritt-Berichte VDI Reihe 2 No. 337, VDI, Düsseldorf, 1995

Orli75 Orlicky, J., "Material Requirements Planning," McGraw-Hill, New York, 1975

PtSm11 Ptak, C.A., Smith, C., "Orlicky's Material Requirements Planning," 3rd Edition, McGraw-Hill, New York, 2011

Schö88a Schönsleben, P., "Flexibilität in der computergestützten Produktionsplanung und -steuerung," 2nd Edition, AIT-Verlag, D-Hallbergmoos, 1988

WaWh58 Wagner, H.M., Whitin, T.M., "Dynamic Version of the Economic Lot Size Model," *Management Science*, 1958, pp. 89–96

Wien97 Wiendahl, H.-P., "Fertigungsregelung: Logistische Beherrschung von Fertigungsabläufen auf Basis des Trichtermodells," Hanser, München, 1997

13 Time Management and Scheduling

Intended learning outcomes: Present the elements of time management. Explain buffers and queues. Disclose algorithms for scheduling of orders. Describe splitting and overlapping.

Planning & control in organizational logistics aims to deliver products and orders reliably by the specified due date. Time management and scheduling are first and foremost a matter of medium-term and short-term planning (during order release), although there are some long-term elements. Figure 13.0.0.1 shows the reference model for business processes and the tasks of planning & control introduced in Figure 5.1.4.2, highlighting the tasks and processes in time management and scheduling on a darker background.

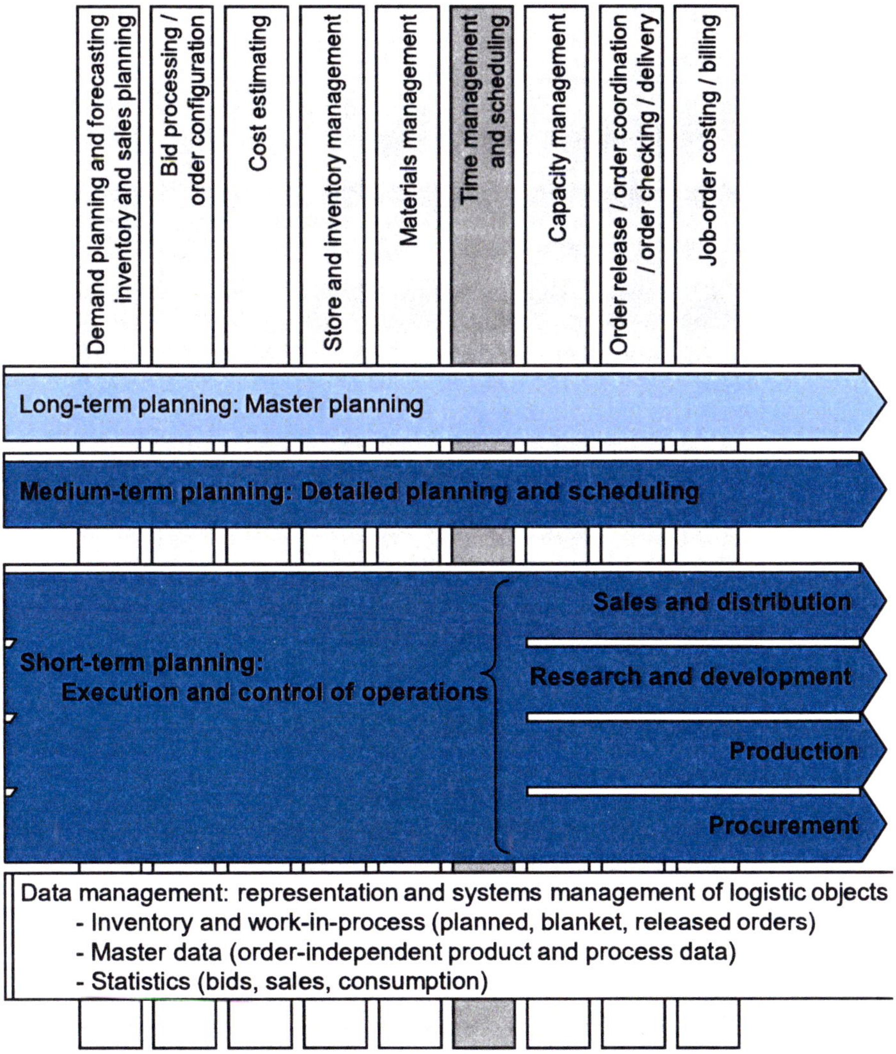

Fig. 13.0.0.1 The part systems examined in this chapter are shown on a darker background.

P. Schönsleben, *Handbook Integral Logistics Management*,
https://doi.org/10.1007/978-3-662-65625-9_13

For an overview of the material in this chapter, see also Sections 1.2.3, 5.3.3, and 5.3.4. We suggest that you reread these sections before studying Chapters 13 through 15.

The first step in time management and scheduling is to estimate the lead time for an order. This chapter views and analyzes lead time as a composite of time elements. We will pay particular attention to unproductive interoperation times and examine difficult-to-estimate wait times for work centers statistically. From the results, we will derive means to reduce wait times. This chapter also presents various scheduling techniques and their areas of application — specifically, forward, backward, central point, and probable scheduling — and discusses effects such as splitting and overlapping.

13.1 Elements of Time Management

Intended learning outcomes: Describe the order of the operations of a production order, operation time and load, as well as interoperation, administrative and transportation time.

Time management is the observation, control, and manipulation of time elements. *Time elements* are the duration of operations, interoperation times, and administration times.

In typical job shop production, the focus is on interoperation times, since they make up more than 80% of the total lead time. However, in line production, observation of the duration of the operations themselves is also of particular interest.

13.1.1 The Order of the Operations of a Production Order

Intended learning outcomes: Describe a sequence of operations. Differentiate between a directed and an undirected network of operations. Identify a synchronization point.

In materials management, *lead time* (see Sections 1.1.6 and 1.2.3) is a basic attribute of both manufactured and purchased products. With this data, the start date of a production or procurement order — starting from the due date — can be calculated, and rudimentary scheduling can be performed.

The value for lead time can be a value based on prior experience. However, for effective planning, particularly of production orders, such more or less arbitrary values are often not precise enough:

- Some components do not need be reserved for the start date of an order, as they are only needed for a later operation.
- For exact capacity planning, we need to know the point in time at which the work center will be loaded by work to be executed and thus a start date for each operation.

For a detailed calculation of *manufacturing lead time*, the essential elements are attributes of the bills of material and routing sheets. We can develop the process plan from these elements (see also Figure 1.2.3.3). Manufacturing lead time is the sum of the three different time elements that are defined in Section 1.2.3:

- Operation time (see Section 13.1.2)
- Interoperation time (see Section 13.3.1)
- Administrative time (see Section 13.1.4)

Lead time calculated on the basis of the lead times for individual operations is only an estimated value, since — especially for interoperation times — it is dependent on assumed average values. In this case, lead time calculation does not take into account the definite capacity utilization of work centers, which can dramatically affect wait time estimates (see also Section 13.2.). However, the "normal" lead time calculated in this way is accurate enough for several planning methods, and especially for rough-cut planning.

Lead time calculation is based on the *order of the operations* of the routing sheets.

A *sequence of operations* is the simplest order of operations. It is illustrated in Figure 13.1.1.1. In this simplest case, lead time is merely the sum of the time elements.

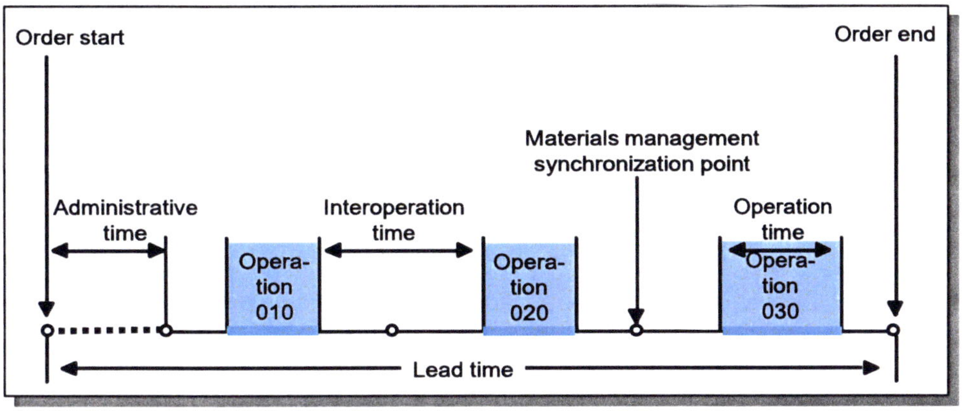

Fig. 13.1.1.1 A sequence of operations.

Besides the simple sequence of operations, there are also more complex structures, which can be portrayed as networks.

- In a directed network of operations, no operations are repeated. We can identify the operations in ascending order (in a semiorder). Lead time corresponds to the longest path through the network.
- In an *undirected network of operations*, sequences of operations within the network may be repeated. In this case, we can calculate lead time only if we know the number of repetitions or other constraints.

Figure 13.1.1.2 shows a typical example. In a *directed network of operations*, the lead time corresponds to the longest path through the network.

A process plan for multistage production, such as in Figure 1.2.3.3, corresponds to a directed network if a joint start event links together the open arborescent structure at the left.

A *synchronization point* is a link between the routing sheet and the bill of material, and thus between time management and materials management.

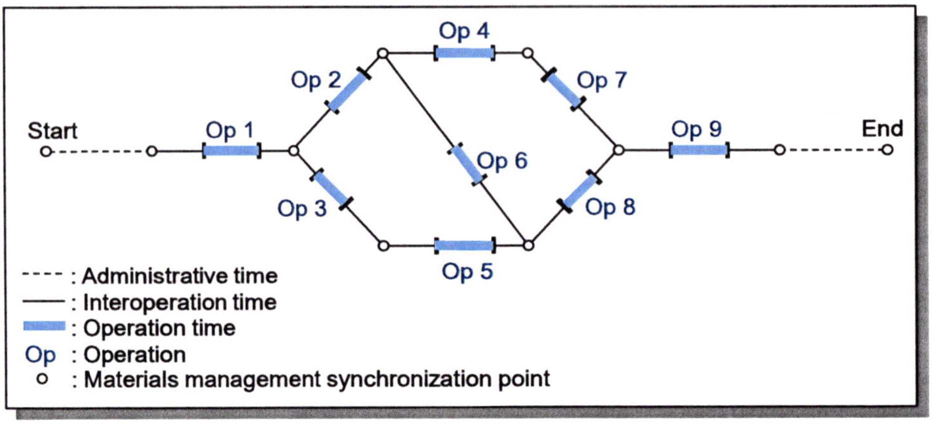

Fig. 13.1.1.2 A network of operations.

In Figures 13.1.1.1 and 13.1.1.2, circles designate the synchronization points at transitions between individual operations. At these points, we may channel in goods taken from a warehouse, directly procured, or taken from another, synchronous production order. At the same time, the circles represent an *intermediate stage* of the manufactured product. This can also be a partially completed product stage stocked as an in-house item. This means that these points in time on the time axis are also the planning dates for the necessary components.

13.1.2 Operation Time and Operation Load

Intended learning outcomes: Present the simplest formula for operation time using a graphic representation. Explain the simplest formula for operation load.

Operation time is the time required to carry out a particular operation. It is defined in Section 1.2.3 as the sum of *setup time* for machines and tools and the *run time* for the actual order lot.[1] The latter is the product of the number of units produced (the *lot* or *batch*) and the run time for a unit of the lot produced (the *run time per unit*).

The simplest formula for operation time occurs when run times are scheduled serially following the setup time, as in Figure 1.2.3.1. Figure 13.1.2.1 shows the formula for operation time as a graphic representation.

The formula for calculating operation time becomes more complicated when we include special effects such as splitting or overlapping. See also Section 13.4.

---

[1] Setdown time also belongs to operation time. In practice, however, it is generally short and therefore ignored.

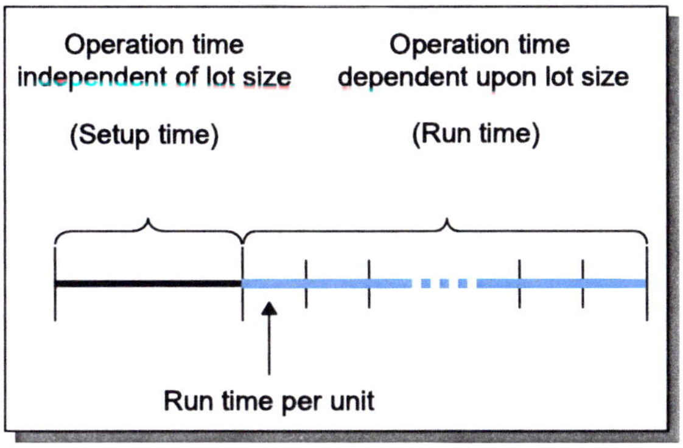

Fig. 13.1.2.1 The simplest formula for operation time (graphic representation).

Operation load is the work content of the operation, measured in the capacity unit of the work center used for the operation. In Section 1.2.4, we saw that operation load is the sum of the *setup load* — the work content that is *independent of batch size* — and the *run load* for the actual order lot.[2] The latter is the product of the number of units produced (the *lot* or *batch*) and the *run load per unit* for a unit of the lot produced.

Figure 13.1.2.2 shows the formula for the operation load shown in the simplest case. Compare here the formula given in Figure 1.2.3.1.

$$\text{Operation load} = (\text{setup load}) + \text{lot size} \ (\text{run load per unit})$$

Fig. 13.1.2.2 The simplest formula for operation load.

Often, the capacity unit for the work center used for the operation is a unit of time. In this case, setup time and run time are generally identical with setup load and run load. There are, however, instances in which the operation time bears no relationship to the operation load.

- For subcontracted operations, e.g., a cost unit may be chosen as the capacity unit.
- For operations with an extremely complicated execution or for purely fictitious "waiting operations," which have no influence upon the load of a work center or upon manufacturing costs, the chosen operation time must be different from the operation load.

If the interoperation times exert the dominant influence on total lead time, scheduling does not require exact knowledge of the operation time. For purposes of capacity management, however, planners need the exact value of the operation load to gain a meaningful load profile for a work center. If they are now able to derive the operation time from the operation load, they can calculate the precise operation time as well as the operation load.

---

2 Setdown load also belongs to standard load. In practice, however, it is generally short and therefore ignored.

13.1.3 The Elements of Interoperation Time

Intended learning outcomes: Identify the elements of interoperation time. Differentiate between technical wait time and nontechnical wait time. Describe transportation time.

Interoperation time occurs before or after an operation (see definition in Section 1.2.3). Figure 13.1.3.1 shows the *elements of interoperation time*:

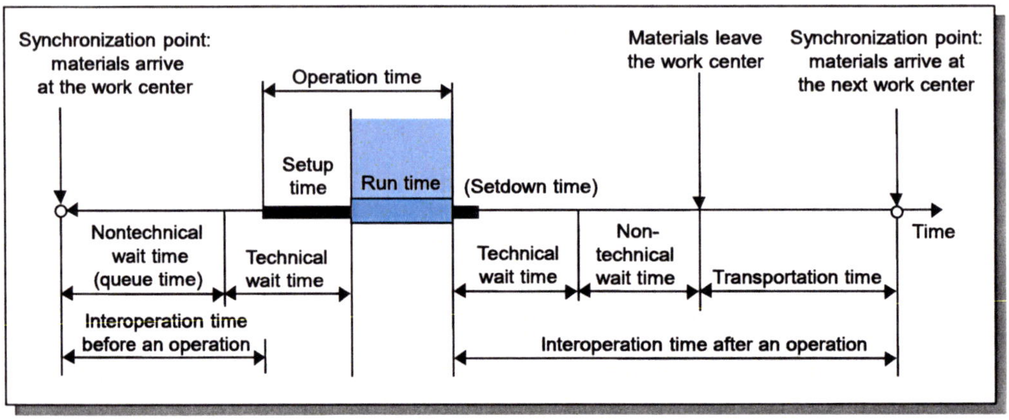

Fig. 13.1.3.1 The elements of interoperation time.

- *Technical wait time after an operation* describes the time required to complete testing, a chemical reaction, a cool-down period, or other things. It is an attribute of the operation. As is true of the operation itself, it is not generally possible to shorten this wait time, for example, to accelerate the order.
- *Nontechnical wait time after an operation* is the wait time incurred before the lot is collected for transport. It is dependent on the work center and can be an attribute of this object or be included in transportation time.
- *Transportation time*, also called *move time*, or *transit time*, is the time needed to transport the lot from the current work center to the work center that will carry out the subsequent operation. This time is dependent on both work centers. There are various techniques for determining this time (see Section 13.1.5).
- *Nontechnical wait time before an operation* is made up of the so-called queue time, that is, the amount of time a job waits at a work center before setup or work is performed on the job. This includes preparation time for the operation, as long as it is not counted as a part of the actual setup time. This time is dependent on the work center and is an attribute of that object (see Section 13.2).
- *Technical wait time before an operation* is made up of the operation-specific preparation time, such as a warm-up process, which does not yet load the work center. In practice, this time is of minor significance. It is an attribute of the operation.[3]

---

[3] Technical wait time is also called *technical idle time*.

All components of interoperation time, with the exception of technical wait times before and after the operation, are "elastic": We can lengthen or shorten them depending on the load at the work center and the order urgency (compare Section 13.3.6). Therefore, the values specified in the master data are only average values, and they can fluctuate widely.

13.1.4 Administrative Time

Intended learning outcomes: Describe the need and the various kinds of administrative time.

Administrative time is the time needed to release and complete an order (see definition in Section 1.2.3).

Administrative time at the beginning of an order is required for order release. This comprises availability control, decision making as to type of procurement, and the preparation time that the production or purchasing office needs for the order. It is also a lead time for the data or control flow (i.e., without flow of goods).

Buffer times added to this administrative time wherever possible will serve to control fluctuations in the effective loads of work centers. This will keep the capital-intensive lead time for goods as short as possible. Schedulers can use the play resulting from this buffer to move the entire order forward or backward on the time axis, according to the load of the work centers at the time of order release.

In addition, schedulers should plan administrative time for coordination purposes for each partial order. This time can also include a "normal" stock issue time for components, as long as it has not already been accounted for in the routing sheet as an independent operation, called "stock issue," for example.

Similarly, at the end of each partial order, there is administrative time that generally includes time to place the completed order in stock or to prepare it for shipping. This time may also include a "normal" control time, provided that schedulers do not want to account for this in the routing sheet as an independent operation, called "final control," for example.

13.1.5 Transportation Time

Intended learning outcomes: Explain the transportation-times matrix. Disclose a robust approximation of transportation time.

There are different techniques to determine transportation time between work centers (also called move time or transit time):

- Simple, but inexact: As a scheduling rule, planners use one single time that is not dependent on the work centers.
- Exact, but complex: A matrix of transportation times contains an entry for every combination: "preceding work center ⇔ following work center." This matrix should be maintained in the form of a table in a separate entity class. It is a square matrix containing zeros on the diagonal. If it is not dependent on the direction of the transport, the matrix will be symmetrical (see Figure 13.1.5.1). The difficulty with

this technique lies in maintaining the two-dimensional table, since the number of work centers and the transportation times are continually changing.

	A12	B18	A16	C5	C6	...
A12	0	10	1	4	4	
B18		0	9	4	4	
A16			0	4.5	4.5	
C5				0	0.5	
C6					0	
...						

Fig. 13.1.5.1 Transportation times matrix.

An efficient compromise between these two extremes is to use an approximation based on an analysis of transportation times, and that experience has been shown to be reliable, as in Figure 13.1.5.2.

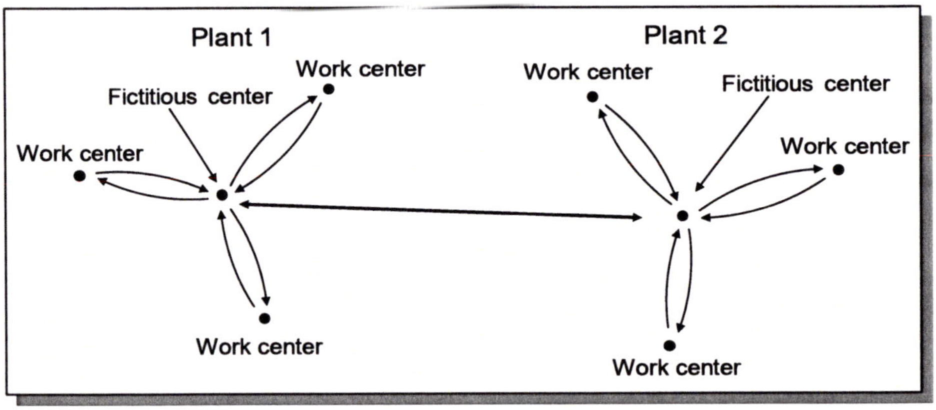

Fig. 13.1.5.2 Approximation of transportation time.

- *Within a plant*, planners define a fictitious center and assume that each shipment must pass through this center. With this, the transportation time from one work center to another becomes the sum of the transportation time from the first work center to the fictitious center and the transportation time from the fictitious center to the other work center. As a result, you only have to register two attributes for every work center, and their values are not dependent on the other work centers.

  This approximation is reliable, because the *loading and unloading* of the means of transportation comprise the greatest portion of transportation time. Actual transportation time from one work center to another varies little in relation to this.

- *Between* the fictitious centers of *two plants,* planners assume an additional transportation time. Again, for production facilities *in the same region,* this approximation is reliable, because loading and unloading of the means of transport make up most of the additional move time. In relation, the actual transportation time between the plants varies little.
- Characterizing plants by the attribute "region" will distinguish among plants in differing geographic areas. This allows differentiation among regional and interregional or even national and international shipments.

13.2 Buffers and Queues

Intended learning outcomes: Explain wait time, buffers, the funnel model, and queues as an effect of random load fluctuations. Present conclusions for job shop production. Produce an overview on logistic operating curves.

Nontechnical wait time before an operation is a difficult element of interoperation time to plan. It arises if the processing rhythm of the operations of a work center does not correspond to the rhythm of the receipt of the individual orders. This can happen in job shop production, for example, if the work center receives orders randomly from preceding operations. Queuing theory is a collection of models to deal with the resulting effects — buffers and queues.

A *buffer* or a *bank* is a quantity of materials awaiting further processing.

A buffer can refer to raw materials, semifinished stores or hold points, or a work backlog that is purposely maintained behind a work center (cf. [APIC16]).

A *queue* in manufacturing is a waiting line of jobs at a given work center waiting to be processed.

As queues increase, so do average queue time (and therefore lead time) and work-in-process inventory (cf. [APIC16]).

Queuing theory or *waiting line theory* is the collection of models dealing with waiting line problems, e.g., problems for which customers or units arrive at some service facility at which waiting lines or queues may build up ([APIC16]).

13.2.1 Wait Time, Buffers, and the Funnel Model

Intended learning outcomes: Describe inventory buffers to cushion disturbances in the production flow. Explain the buffer model, the reservoir model and the funnel model.

Scheduling may deliberately plan in buffers and wait times before a work center for organizational purposes.

Inventory buffer is inventory used to protect the throughput of an operation or the schedule against the negative effects caused by statistical fluctuations (cf. [APIC16]).

Such buffers should absorb potential disturbances in the production process, that occur, for example, in line production or Kanban chains. Figure 13.2.1.1 considers two adjacent workstations.

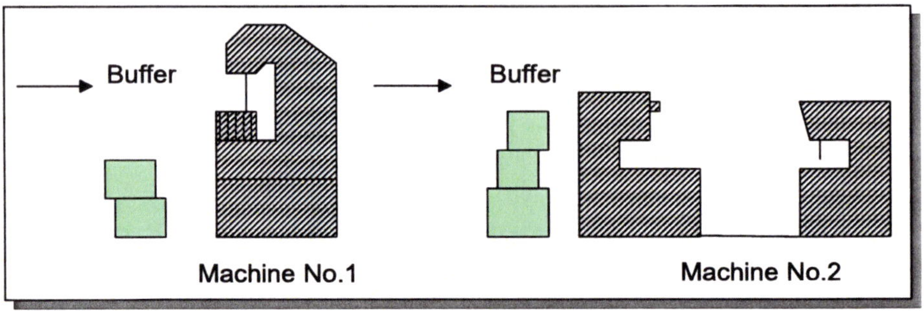

Fig. 13.2.1.1 Inventory buffers to cushion disturbances in the production flow.

If both workstations were perfectly synchronized, a waiting line would be unnecessary. However, a disturbance may occur at either of the two work systems as a result of, for example

- Overloading, scrap, or reworking
- Material shortage, breakdown, or absence of workers

The size of the inventory buffer in front of a work center depends on the degree of synchronization that can be maintained with the previous workstation in practice.

- If the work process on the first machine is disrupted, the queue waiting for the second machine is reduced. In this case, the second machine may become idle.4
- If the work process on the second machine is disrupted, the queue waiting for the second machine increases, as does the buffer before the second workstation. This may lead to a bottleneck at the second machine.

Scheduling may also plan buffers for *economic reasons*. By skillfully sequencing operations from the buffer inventory, you can save valuable setup times. Such setup time savings may occur, for example, in processing products from a single product family. Depending on the circumstances, it is possible to provide directly for such sequencing in detailed planning and scheduling. In practice, however, order lead times of unequal length or highly varied order structures limit the extent to which you can plan. As a result, you can often only optimize the sequence of operations at the workstation itself via finite forward scheduling.

4 *Idle time* is time when operators or resources (e.g., machines) are not producing product because of setup, maintenance, lack of material, lack of tooling, or lack of scheduling ([APIC16]).

Another *economic reason* for having a buffer in front of a work center is the psychological effect of the buffer on the efficiency of the workers:

- If the buffer is too small, the workers begin to slow down, fearing that their hours will be cut or even that they will not be needed at the work center. Small buffers make it look like there is not enough work. Therefore, efficiency decreases.
- Up to a certain point, long queues have a positive influence on efficiency. However, if the queue is too long, it can have a demoralizing effect on workers. The quantity of work to be performed seems insurmountable. Efficiency sinks.

In summary, a buffer in front of a work center is often tolerated or even planned deliberately. However, in evaluating buffers, and in particular their economic repercussions, it is important to take into account the double-negative effect of buffers, specifically 1.) an increase in lead time, and 2.) an increase in work in process and thus tied-up capital.

The buffer model and the *funnel model* below are concepts of the levels of work in process that are waiting at the workstations.

Figure 13.2.1.2 shows the buffer as a reservoir. This conceptualization is quite old (see [IBM75]).

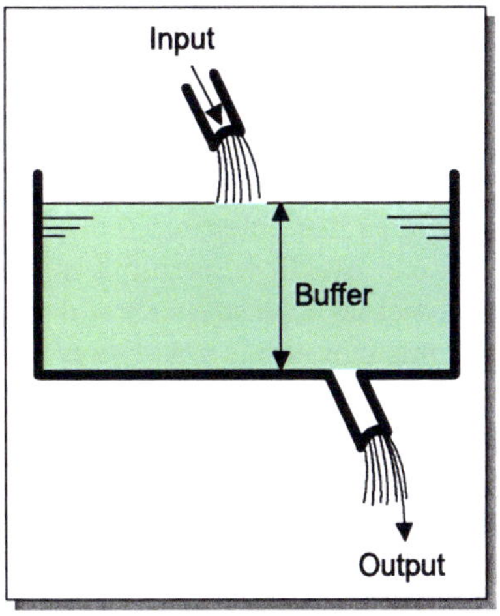

Fig. 13.2.1.2 Reservoir model.

A more recent conceptualization of the buffer is the funnel model (see [Wien95]). Each work center is viewed as a funnel, as illustrated in Figure 13.2.1.3.

The objective is to align the mean output of the work center with its mean load. The funnel volume is used to bring variations of the mean load under control. This means that there must be continual measurement of the mean load, its variation, and the mean output.

If we see total production as a system of work centers, or funnels, that are linked together by output flows, it becomes evident that there are basically two ways to adjust the system:

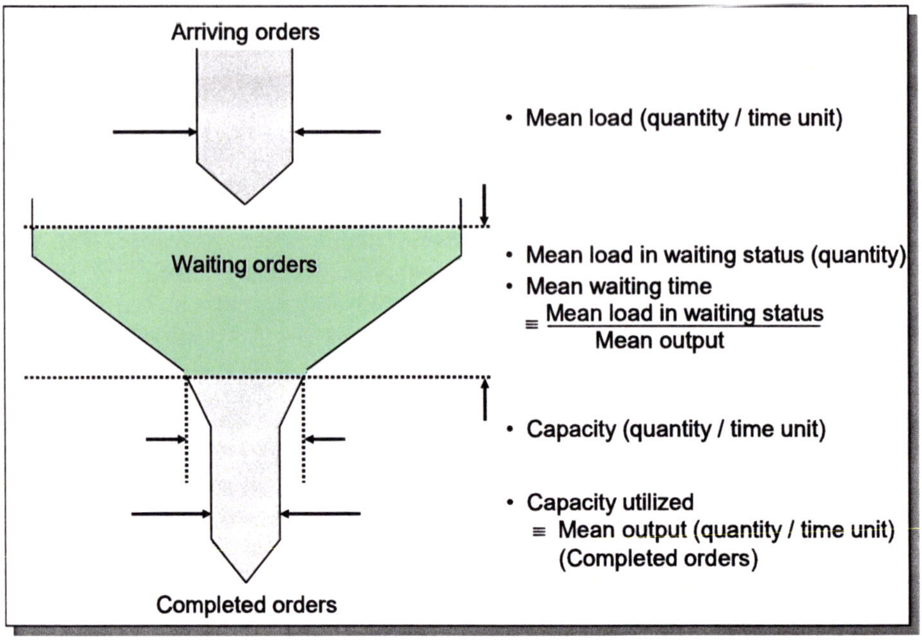

Fig. 13.2.1.3 Funnel model.

- Change capacity, or rather the capacity utilized for each individual funnel. However, it is not always possible to alter capacity short-term.
- Regulate the number of orders that enter into the system. If too many orders are on hand, individual funnels can overflow, resulting in blocked shop floors and poor delivery reliability. In this case, schedulers should decide what orders to withhold from production. Again, this measure is not always possible.

13.2.2 Queues as an Effect of Random Load Fluctuations

Intended learning outcomes: Describe job shop production as a network with work centers as nodes. Explain the average wait time as a function of capacity utilization. Produce a summary of relevant formulas in queuing theory.

With the exception of continuous production, there is no production type in which the capacities of machines and workstations following one another in the process are completely synchronized. As Figure 13.2.1.1 shows, even in other cases of line production, synchronization is not always possible. Thus, to a certain extent, buffers serve to balance the differing output rates of the work centers and to ensure continual load of the individual work centers over a certain period of time.

These buffers are queues formed in front of a workstation; the size of the queues changes over time. Particularly in job shop production, there is great variation in the behavior of the buffer, since a queue is fed from many locations. We can view job shop production as a network with work centers as nodes, as represented in Figure 13.2.2.1. In the figure, the nodes represent work centers, which are classified as homogeneous. The arrows represent the flow of goods or information between these work centers. In the discussion below, the focus is on "Node I" of this network.

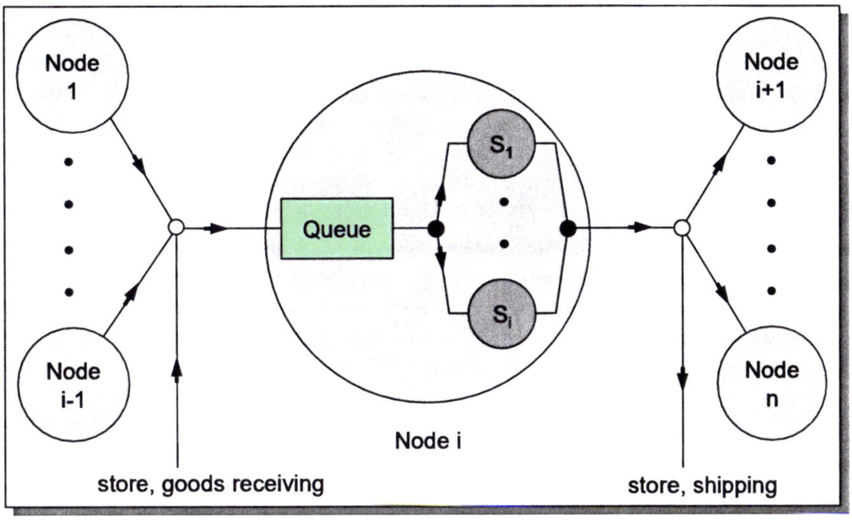

Fig. 13.2.2.1 Job shop production as a network with work centers as nodes.

Input enters from various nodes and sometimes also from the outside (from a store or a receiving department, for example). This input arrives at a joint queue in front of one of the various workstations (S_1, S_2, . . . , S_i) of work center i. After completion of the operation in Node i, the orders flow to other nodes or toward the outside, either in part or in their entirety (after a final operation), depending on the specification in the routing sheet. In line production, there is essentially a sequence of nodes rather than a network.

As mentioned above, determining the size of a buffer is an optimization problem. Queuing theory provides some fundamental insights into the way that job shop production functions and, to a certain extent, how line production functions as well. Here we limit our discussion to the stationary state of a queue, that is, the state after an infinite time period and with fixed constraints.

For the following discussion, Figure 13.2.2.2 sets out several definitions of variables from queuing theory.

s	= Number of parallel stations (e.g., workstations per work center)
ρ	= Capacity utilization of the work center ($0 \leq \rho \leq 1$) = $\frac{\text{load}}{\text{capacity}}$
CV	= Coefficient of variation (ratio of standard deviation to mean) of a distribution
OT	= Operation time
WT	= Waiting time per order in the queue

Fig. 13.2.2.2 Definitions of queuing theory variables.

To simplify the discussion, assume the following:

- Arrivals are random; that is, they follow a Poisson distribution with the parameter λ. λ is the average number of arrivals per period under observation.
- Arrivals and the operation process are independent of one another.
- Execution proceeds either in order of arrival or according to random selection from the queue.
- The duration of the operations is independent of the order of processing and is subject to a determinate distribution with mean M(OT) and coefficient of variation CV(OT).

Figure 13.2.2.3 shows the average wait time as a function of capacity utilization for a model with one station (s = 1, where a queue feeds only one operation station, i.e., one workstation or one machine). We assume the coefficient of variation CV(OT) for the distribution to be 1, which is the case with a negative exponential distribution, for example.

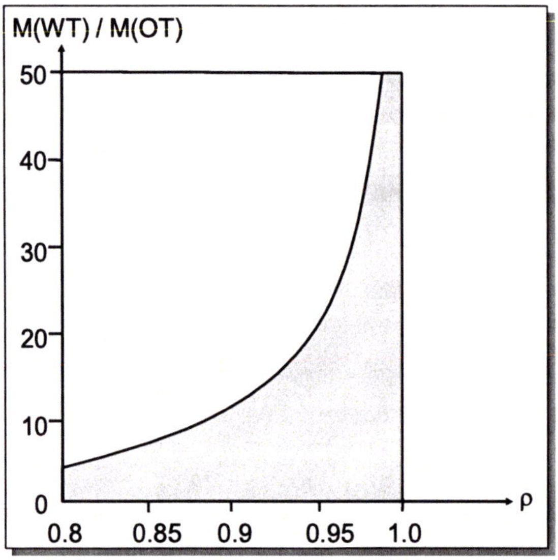

Fig. 13.2.2.3 Average wait time as a function of capacity utilization: special case s = 1, CV(OT) = 1.

Figure 13.2.2.4 presents the relevant formulas of queuing theory for the average case, with references to their original sources in the literature, specifically [GrHa18], [Coop90], and [LyMi94], including page and formula numbers. For further aspects of theoretical mathematics, the reader can consult [Fers64] and [Alba77]. It is important to note, however, that for multiple-station models (s = arbitrary), the relationships based on numerical calculation only approach validity under conditions of extensive capacity utilization.

	$s = 1$ $0 < \rho < 1$	s = arbitrary $\rho \to 1$
CV(OT) = 1	$M(WT) = \frac{\rho}{1-\rho} \cdot M(OT)$ [GrHa08], p. 77, formula (2.30)	$M(WT) \approx \frac{\rho}{1-\rho} \cdot \frac{M(OT)}{s}$ [Coop90], p. 487, formulas (5.22), (5.23), (5.36) and $\rho = a/s$
CV(OT) = arbitrary	$M(WT) = \frac{\rho}{1-\rho} \cdot \frac{1+CV^2(OT)}{2} \cdot M(OT)$ [GrHa08], p.256, formula (5.11) or [LyMi94], p. 191, formula 6	$M(WT) \approx \frac{\rho}{1-\rho} \cdot \frac{1+CV^2(OT)}{2} \cdot \frac{M(OT)}{s}$ [Coop90], p. 508, formula (9.3)

Fig. 13.2.2.4 Summary of relevant formulas in queuing theory.

Figure 13.2.2.5 shows wait time as a function of operation time for selected values of s and CV(OT).

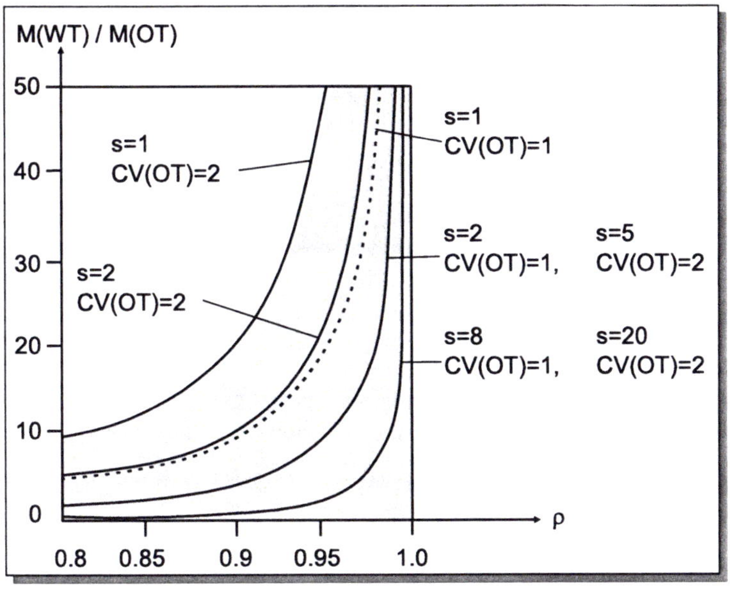

Fig. 13.2.2.5 Average relative wait time as a function of capacity utilization: selected values (following an unpublished slide of Prof. Büchel, ETH Zurich).

13.2.3 Conclusions for Job Shop Production

Intended learning outcomes: Present qualitative findings of queuing theory for job shop production and, in part, for line production. Describe the measures indicated by the qualitative findings of queuing theory.

It is not possible to apply quantitative results of queuing theory to job shop production directly, since some of the specified conditions are not satisfied. For example:

- The arrival process may be *short term*, a purely random process. However, scheduling can shield production from large capacity utilization peaks, and the delivery rates of supplying nodes will limit arrival rates at a work center (= network node). Therefore, medium-term fluctuations will be somewhat smaller than in the case of a purely random process.
- There is no independence between the execution and the arrival process. Since the negative consequences of large queues are undesirable, scheduling will spare no effort to avoid extreme situations. It manipulates the processes by:
  - Subcontracting individual orders
  - Subcontracting individual operations
  - Raising the capacity of operating facilities with overtime or shift work
  - Advancing or postponing individual operations

The result is not a stationary state, but rather a series of transitional states, which are characterized by varying values of the parameters and distributions that specify a queuing process. Nevertheless, queuing theory yields qualitative findings for job shop production and, in part, for line production:

1. *High capacity utilization ⇔ large queues:* In a rigid queuing system, particularly with a one-station model, it is not possible to achieve both good utilization of the capacities and short lead times simultaneously. The higher the capacity utilization desired (in the absence of capacity adjustments from planning interventions), the larger the average queue must be.
2. *High capacity utilization ⇔ (wait time >> operation time):* Wait time in the queue is significantly larger than operation time in the case of high capacity utilization.
3. *Shorter lead time ⇐ fewer operations:* Fewer operations mean fewer queues. In industrial production, this is achieved by a greater versatility of machine tools, such as numerically controlled machines or machining centers, and in services and administration by a reduction of extreme division of labor. However, it is important to ensure that the total operation time with a reduced number of operations is shorter than that with a larger number of operations. Otherwise, no positive effect will result, since wait time increases with prolonged operation time.
4. Large queues result from
   - Prolonged operation time
   - Extremely varied operation times
   - Few parallel workstations, or only one workstation

The qualitative findings of queuing theory indicate the following measures:

- *A reduction of setup time, which will reduce batch sizes and hence cut the average operation time.* However, direct reduction of batch size without reducing setup time increases manufacturing costs. It is only productive if the work center is not fully utilized, that is, if the larger setup time resulting from splitting the operations does not lead to overloading or nearly full utilization of the work center.

- *Equal contents for all operations, to avoid markedly different operation times.* Schedulers can reduce the coefficient of variation for operation times, that is, the difference in the duration of operations, by, for example, splitting up orders with long standard times. This results in a reduction in the mean operation time as well. However, in fully utilized production, increased setup can negate the positive effect.
- *A reduction in utilization*, which can be achieved by holding overcapacity. Schedulers may also transfer employees to those work centers where capacity utilization threatens to become too large.

All these measures are starting points or basic principles of the lean / just-in-time concept. The general, dominant tendency today is to move away from production as a system with fixed constraints. The more successful this move is, the shorter the wait times resulting from the queuing effect will be. As a result, organizational intent — rather than chance — increasingly determines lead times.

13.2.4 Logistic Operating Curves

Intended learning outcomes: Produce an overview on logistic operating curves. Explain an example of logistic operating curves.

Logistic operating curves are ways to summarize the facts of an operation, as shown in Figure 13.2.4.1 (see [Wien95]).

Logistic operating curves aid evaluation of production processes in the framework of production control. Logistic operating curves express a comparison of logistic performance indicators.

- In Figure 13.2.4.1, *performance* is the output, that is, the load processed by the work center (see also [Wien95]). Thus, the performance curve corresponds to the *capacity utilization* curve (see also Figure 1.4.3.4 or Figure 1.4.4.4). A particular output is achievable only if the waiting work in process is of a particular size. As output approaches its maximum, you can only increase it if you increase the inventory of work in the queue over proportionally. This logistic operating curve shows in its upper part roughly the same situation as in Figure 13.2.2.3, where the axes are reversed.
- The *range (of inventory)* is the length of time required to process the inventory at the workstation. Accordingly, the *mean range* is the mean of the wait time, as in Figure 13.2.2.4, plus the operation time. This mean has a minimum, which is influenced, among other things, by the operation times and their variances. For job shop production, the level of waiting work determines the inventory or work-in-process to a large degree. See also the performance indicator *work-in-process-inventory turnover* in Figure 1.4.3.2.

From capacity utilization, then, we arrive at work-in-process and from there to the mean wait time (which, in job shop production, makes up a large proportion of lead time). The three inventory levels I, II, and III represent, respectively, an underloaded work center, an appropriately loaded work center, and an overloaded work center. Thus, the logistic operating curves indicate how much play there is to reduce queues, and hence wait times, without endangering capacity utilization.

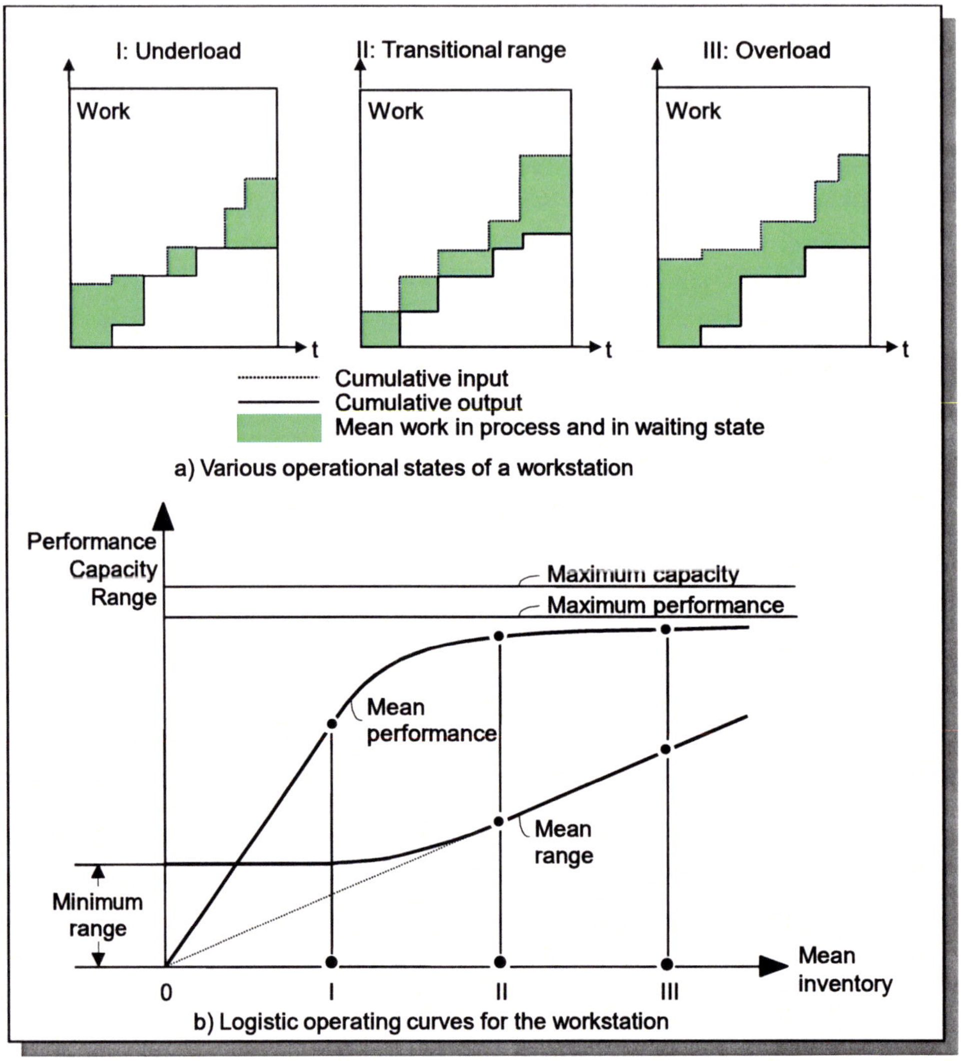

Fig. 13.2.4.1 An example of logistic operating curves (following [Wien95]).

In the following, we present suitable measures to alter the logistic operating curve so that the dangerous curve occurs as late as possible. In addition, the slope of the straight lines representing mean work on hand should be as small as possible. Lean/JIT concepts (see Chapter 6), for example, can create the potentials for achieving these aims. Through the use of these potentials, the logistic operating curves change, and new degrees of freedom arise that allow for a decrease in orders waiting to be processed.

13.3 Scheduling of Orders and Scheduling Algorithms

Intended learning outcomes: Describe the calculation of the manufacturing lead time. Differentiate between backward scheduling and forward scheduling. Explain network planning, central point scheduling and probable scheduling. Present scheduling of process trains.

Scheduling of orders starts from customer-set order deadlines and determines the other required deadlines for feasibility decisions, loading of capacity, and reservations of components.

A *scheduling algorithm* is a technique of calculation designed to support scheduling of orders.

Scheduling of orders is mainly the job of the personnel involved in the placing and execution of the order. For these purposes, they should have access to appropriate tools, such as to information technology in the form of ERP software.

Scheduling of orders is based on knowledge of and calculations of lead time. However, time management reveals that there are limits to the accurate estimation of lead times. Not all time elements can be estimated precisely, and perhaps most difficult to assess is queue time. Of additional concern are unanticipated factors that may arise during actual production. Rescheduling is often the necessary consequence.

Rescheduling is the process of changing order or operation due dates, usually as a result of their being out of phase with the time when they are needed ([APIC16]).

Even though it is important to build up potential for reactive rescheduling, we also need some approximation of cumulative lead time to set in relation to delivery lead time. We need this information proactively, that is, during scheduling of orders. In the short term, this allows decisions to be made to accept or refuse orders. In the medium term, we can get an idea of the probable utilization of the work centers along the time axis.

13.3.1 The Manufacturing Calendar

Intended learning outcomes: Present characteristics of the manufacturing calendar, or shop calendar. Explain an example of a manufacturing calendar.

Measures of the load and capacity of a work center are often in units of time. In other cases as well, time quantities are necessary rather than load, at least for calculating lead time. It is a problem, however, that according to the Gregorian calendar, a week does not always contain the same number of (e.g., five) working days.

The *manufacturing calendar* or *shop calendar* counts working days only and omits nonworking days, such as vacations, holidays, or weekends.

The *manufacturing date* of the manufacturing calendar begins on day "zero," which corresponds to a particular Gregorian date. For each working day, you add the value of one.

Figure 13.3.1.1 shows an excerpt from a manufacturing calendar.

Gregorian Date	Day	Type of Day	Manufacturing Date
2015.05.10	Sunday	Weekend	879
2015.05.11	Monday	Workday	880
2015.05.12	Tuesday	Workday	881
2015.05.13	Wednesday	Workday	882
2015.05.14	Thursday	Holiday	882
2015.05.15	Friday	Workday	883
2015.05.16	Saturday	Weekend	883
2015.05.17	Sunday	Weekend	883

Fig. 13.3.1.1 The manufacturing calendar.

A manufacturing calendar allows addition or subtraction of a certain number of working days to or from a given Gregorian date. Scheduling of orders often uses these calculations.

In addition, to gain the load profile of a work center when we want to compare the load over a particular time period with the capacity available, this calendar takes only working days into consideration.

13.3.2 Lead Time Scheduling: Calculating the Manufacturing Lead Time

Intended learning outcomes: Produce an overview on lead time scheduling. Identify definitions for the elements of operation time. Present the start date as a function of completion date. Differentiate between manufacturing lead time, cycle time and throughput time.

For every production order, the planner should know the load of each operation and the point in time at which the work center will be loaded. To determine these factors, planning uses lead-time scheduling techniques.

In *lead time scheduling*, a schedule is developed by calculating the lead time. This calculation includes the duration of all operations, interoperation times, and administrative times.

Let us assume that there is a production order with n operations. They are numbered throughout with the numerator i, where $1 \leq i \leq n$. The abbreviations mentioned in Figure 13.3.2.1 stand for the elements of *manufacturing lead time* introduced in Section 13.1 (generally measured in industrial units, i.e., hundredths of hours):

The *operation time* for operation i:

LOTSIZE	≡	lot size ordered
ST[i]	≡	setup time for operation i
RT[i]	≡	run time per unit produced for operation i
OT[i]	≡	operation time for an operation i = ST[i] + LOTSIZE * RT[i]

The interoperation times for operation i:

INTBEF[i]	≡	interoperation time before the beginning of operation i (zero, if two successive operations are performed at the same work center) = transportation time from the fictitious center to the work center + nontechnical wait time before the beginning of the operation (queue time)

INTTEC[i]	≡	technical interoperation time after the completion of operation i
INTAFT[i]	≡	nontechnical interoperation time after the completion of operation i = transportation time from the work center to the fictitious center + transportation time from the fictitious center to the subsequent work center

The administrative times:

ADMPORDBEG	≡	administrative time for the partial order at the beginning = administrative time for the order release + (possible) materials requisition
ADMPORDEND	≡	administrative time at the end of the partial order = administrative time at the end of the partial order + (possible) time for final control + (possible) time for stocking or preparing the order for shipment
ADMORD	≡	order administrative time for release of the entire order

Fig. 13.3.2.1 Definitions for the elements of operation time.

In practice, we distinguish between two different values for ADMPORDBEG and ADMPORDEND: that which takes the possibilities mentioned in Figure 13.3.2.1 into account and that which does not.

For a *sequence of operations* as the order of the operations, the lead time for an order (abbreviated by LTI) is equal to the sum of all operation times, interoperation times, and administrative times, as the formula given in Figure 13.3.2.2 expresses:

$$\text{LTI} = \sum_{1 \le i \le n} \{\text{INTBEF}[i] + \text{OT}[i] + \text{INTTEC}[i] + \text{INTAFT}[i]\} + \text{ADMORD} + \text{ADMPORDBEG} + \text{ADMPORDEND}$$

Fig. 13.3.2.2 Lead time formula (first version).

LTI corresponds to the lead time for a product with lot size LOTSIZE. The lead time will vary if the lot size is different. If we sum up the elements according to the formula in Figure 13.3.2.3, the result is LTI as a linear function of lot size, as shown in Figure 13.3.2.4.

You can save as data the partial sums from the lead time formula as attributes of the product. They can then be recalculated following each modification of the routing sheet by summing up all the values for the individual operations.

This procedure is the most efficient way to recalculate the lead time for a production order of any particular order quantity. Instead of having to read the operations, you need only refer to the product data. For a rapid calculation of secondary requirements, you can now calculate lead time simply according to the formula in Figure 13.3.2.4 and plan all reservations for components on the basis of the start date for the order as in Figure 13.3.2.5:

$$\begin{aligned}
\text{SUMINT} &= \text{ADMPORDBEG} + \text{ADMPORDEND} + \sum_{1 \le i \le n} \{\text{INTBEF}[i] + \text{INTEND}[i]\} \\
\text{SUMTEC} &= \sum_{1 \le i \le n} \text{INTTEC}[i] \\
\text{SUMST} &= \sum_{1 \le i \le n} \text{ST}[i] \\
\text{SUMRT} &= \sum_{1 \le i \le n} \text{RT}[i]
\end{aligned}$$

Fig. 13.3.2.3 Partial sums for the lead time formula.

$$\text{LTI} = \text{ADMORD} + \text{SUMINT} + \text{SUMTEC} + \text{SUMST} + \text{SUMRT} \cdot \text{LOTSIZE}$$

Fig. 13.3.2.4 Lead time formula (second version).

$$\text{Start date} = \text{completion date} - \text{LTI}$$

Fig. 13.3.2.5 Start date as a function of completion date.

In a *directed network of operations* as the order of the operations, the lead time for the order is the sum of the operations along the critical, that is, the longest, path. In some cases, this is dependent on lot size. Thus, the partial sums of the lead time formula are relevant for a particular lot size interval. This upper, or lower, limit of the lot size for a simplified calculation of lead time must be part of the product data.

Also, the meaning of the following terms is similar to manufacturing lead time, even though their formal definition differs:

- *Cycle time*: This is the time between completion of two discrete units of production. For example, the cycle time of motors assembled at a rate of 120 per hour would be 30 seconds (cf. [APIC16]). Cycle time is an important variable in connection with single-item-oriented line production, particularly with control via production rates.[5]
- *Throughput time* (sometimes also called "cycle time"): In materials management, throughput time refers to the length of time from when a material enters a production facility until it exits (cf. [APIC16]). Throughput time plays a role in connection with logistic operating curves and the expected value of wait time in the context of production controlling (see Section 13.2.4).

13.3.3 Backward Scheduling and Forward Scheduling

Intended learning outcomes: Explain forward scheduling and backward scheduling. Describe a simple algorithm for backward scheduling.

---

[5] *Takt time* is a *set* cycle time to match the rate of customer demand. *Flow rate* is the inverse of cycle time. In the example above, "120 units per hour," or "two units per minute" is the flow rate.

The two most important scheduling techniques are the following:

> *Backward scheduling*, or *back scheduling*, begins with the set (that is, the *latest* acceptable) *completion date* for the order (that is, the *order due date*), and calculates — for each operation — the latest (acceptable) completion date (that is, the *operation due date*) and the latest (possible) start date (that is, the *operation start date*), as well as the *latest (possible) start date* for the order.
>
> *Forward scheduling* begins with the set (that is, the *earliest* acceptable) *start date* for the order and calculates the earliest (acceptable) start date and the earliest (possible) completion date for each operation, as well as the *earliest (possible) completion date* for the order.

These two techniques assume the following definitions:

> The *latest date* is a date that we cannot exceed in execution and control of operations. Similarly, we cannot allow a date to fall before the *earliest date*.
>
> A *set date* is set "externally" and cannot be changed by means of the scheduling algorithm.

Figure 13.3.3.1 illustrates the two principles.

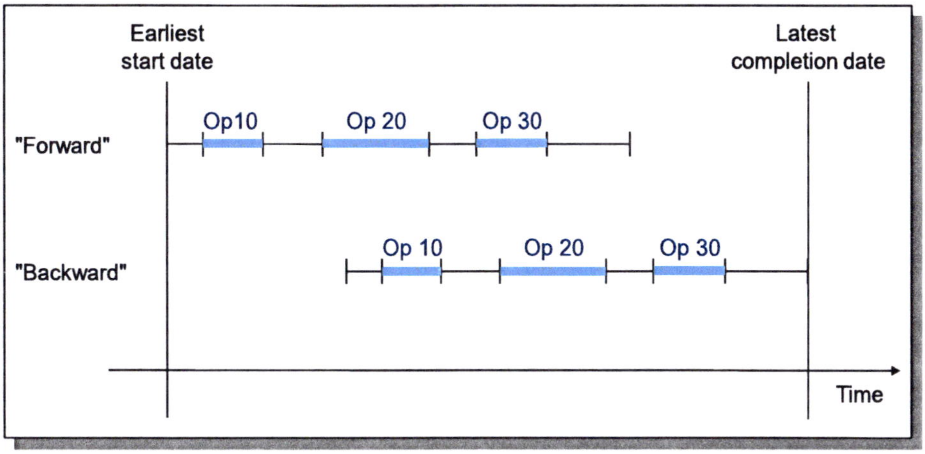

Fig. 13.3.3.1 Forward scheduling and backward scheduling.

Figure 13.3.3.2 shows the simplest algorithm for backward scheduling (the algorithm for forward scheduling has a similar structure):

1. The order of the operations is assumed to be a sequence of operations.
2. The production order consists of one single partial order.
3. All n operations are included in the lead time scheduling; that is, the order has not yet begun.
4. The interoperation times are weighted with a factor of 1; that is, they are assumed to be "normal."

0 Initialize the start date for the order:
- ESD[order] := max{ESD(set)[order], “today”

1. At the beginning of the partial order:

a. Calculate the completion date for the partial order:
   - LCD[partial order] := LCD(set)[partial order].
   - If LCD(set)[order] < LCD(set)[partial order], then
     - LCD[partial order] := LCD(set)[order].

b. Calculate the completion date of the last operation:
   - LCD[n] := LCD[partial order]-ADMPORDEND-INTAFT[n]-INTTEC[n]
   - If LCD(set)[n] < LCD[n], then LCD[n] := LCD(set)[n].

2. Loop: for operation i, n ≥ i ≥ 1, in descending order:

a. Calculate the start date for the operation:
   - LSD[i] := LCD[i] - OT[i].

b. If i > 1, then calculate the completion date for the preceding operation:
   - LCD[i-1] := LSD[i] - INTBEF[i] - INTAFT[i-1] - INTTEC[i-1]
   - If LCD(set)[i-1] < LCD[i-1], then LCD[i-1] := LCD(set)[i-1]

c. Otherwise (i = 1) calculate the start date for the partial order:
   - LSD[partial order] := LSD[i] - INTBEF[i] - ADMPORDBEG

3. At the end of the partial order:

a. Calculate the start date for the order:
   - LSD[order] := LSD[partial order] - ADMORD
   - If LSD[order] < ESD[order], then message: *start date too early*

b. For all components j, 1 ≤ j ≤ m, calculate the reservation date (the start date):
   - i := operation for which the components j will be needed
   - ESD[j] := LSD[i] - INTBEF[i] - ADMPORDBEG

End of algorithm

Fig. 13.3.3.2 Simple algorithm for backward scheduling.

The formal description of this scheduling task is as follows:

- Take a production order consisting of one partial order with n operations i, $1 \leq i \leq n$, and m components j, $1 \leq j \leq m$, as given. The operation numbers stand in a semiorder; if, for example, $i_1 < i_2$, then operation i_1 is performed before operation i_2.
- Beginning with the set (that is, the latest acceptable) order completion date, we calculate the following “latest” dates:
  - Start and completion dates for the individual partial order
  - Start and completion dates for the individual operations
  - Reservation dates (= start date) for the components
  - Start date for the order, with an exception message if it is earlier than a set (earliest) start date

As data specifications, the following notations are used:

- x ≡ order, partial order, or one position in the partial order (component or operation)
- LCD[x] ≡ latest completion date for x
- ECD[x] ≡ earliest completion date for x
- LSD[x] ≡ latest start date for x
- ESD[x] ≡ earliest start date for x
- OT[i] ≡ operation time for operation i
- INTBEF[i] ≡ interoperation time before operation i
- INTAFT[i] ≡ interoperation time after the end of operation i
- INTTEC[i] ≡ technical interoperation time after operation i
- ADMPORDBEG ≡ administrative time for the partial order at the beginning
- ADMPORDEND ≡ administration time for the partial order at the end

Remarks:

- For comparing the date attributes with one another, we will use the standardized "ISO" format, that is, YYYYMMDD.
- A date is calculated either by the scheduling algorithm or given as a set date. We distinguish the latter from the former by the addition of (set), for example, LCD(set)[x].

13.3.4 Network Planning

Intended learning outcomes: Explain network planning and the critical path method. Present an example of a scheduled network. Describe a network algorithm for backward scheduling.

Site production, or project manufacturing, uses mainly scheduling techniques proper to project management.

Network planning is a generic term for techniques that are used to plan complex projects (cf. [APIC16]).

Project routings, a project task, or a work package have directed networks of operations, such as in Figure 13.1.1.2, instead of simple operation sequences. For network planning, the simple algorithm in Figure 13.3.3.2 will not do.

The *critical path method (CPM)* is used for planning and controlling the activities in a project. It determines the *critical path* that is the path with the longest duration, which identifies those elements that actually constrain the cumulative lead time (or critical path lead time) for a project (cf. [APIC16]).

Scheduling is done forward *and* backward. Figure 13.3.4.1 shows the results of scheduling the network in Figure 13.1.1.2 with set values for ESD and LCD. The difference between

ESD and LSD is the lead-time margin.[6] On the critical path, it always has the same value (generally close to or equal to zero) and is also called *path float* or *slack time*.

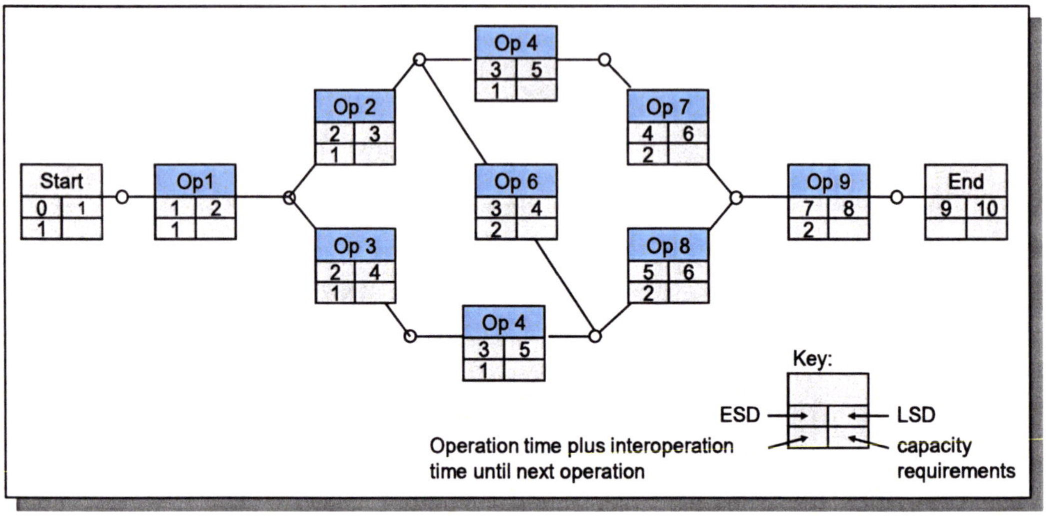

Fig. 13.3.4.1 Scheduled network.

Further network work techniques are:

- The program evaluation and review technique (PERT) is a network analysis technique in which each activity is assigned a pessimistic, most likely, and optimistic estimate of its duration. The critical path method is then applied using a weighted average of these times for each node. PERT computes a standard deviation of the estimate of project duration ([APIC16]).
- The critical chain method is an extension of the critical path method that was introduced in the theory of constraints, which considers not only technological precedence but also resource constraints (cf. [APIC16]).

Figure 13.3.4.2 shows an effective network algorithm for backward scheduling. It is formulated as a generalization of the algorithm in Figure 13.3.3.2. If BEGIN is the start and END the conclusion of the routing sheet, then:

- prec(i) designates the quantity of all operations, which precede operation i or END.
- succ(i) designates the quantity of all operations, which follow operation i or BEGIN.

An operation that precedes (or follows) a particular operation i bears a smaller (or larger) operation number than i. Thus, we can treat operations in an ascending (or descending) order. Usually this type of semiorder establishes itself naturally. Otherwise, it can be calculated easily by using the function prec(i) (or succ(i)).

---

6 In the critical path method, slightly different terms are used instead of ESD, ECD, LSD, and LCD: *early start date (ES)*, *early finsh date (EF)*, *late start date (LS)*, and *late finish date (LF)*.

0. Initialize the start date for the order:
- ESD[order] := max{ESD(set)[order],"today"}.

Initialize the completion date for the partial order and all operations:
- LCD[x] := min{"9999.99.99", LCD(set)[x]}.

1. At the beginning of the partial order:

a. Calculate the completion date for the partial order:
- If LCD(set)[order] < LCD(set)[partial order], then
  - LCD[partial order] := LCD(set)[order].

b. For each previous operation $i_1 \in$ {prec(END)}, calculate its completion date:
- LCD[i_1] := LCD[partial order]-ADMPORDEND-INTAFT[i_1]-INTTEC[i_1]
- If LCD(set)[i_1] < LCD[i_1], then LCD[i_1] := LCD(set)[i_1].

2. Loop: for operation i, $n \geq i \geq 1$, in descending order:

a. Calculate the start date for the operation:
- LSD[i] := LCD[i] - OT[i].

b. For each operation $i_1 \in$ {prec(i)}, $i_1 \neq$ BEGIN, calculate its completion date:
- LCD'[i_1] := LSD[i] - INTBEF[i] - INTAFT[i_1] - INTTEC[i_1].
- If LCD'[i_1] < LCD[i_1], then LCD[i_1] := LCD'[i_1].

c. For $i_1 \in$ {prec(i)}, i_1=BEGIN, calculate the start date for the partial order:
- LSD[partial order] := LSD[i] - INTBEF[i] – ADMPORDBEG.

3. At the end of the partial order:

a. Calculate the start date for the order:
- LSD[order] := LSD[partial order] - ADMORD
- If LSD[order] < ESD[order], then message: ***start date too early.***

b. For all components j, $1 \leq j \leq m$, calculate the reservation date (of the start date):
- i := operation for which the components j will be needed
- ESD[j] := LSD[i] - INTBEF[i] – ADMPORDBEG.

End of algorithm

Fig. 13.3.4.2 Network algorithm for backward scheduling.

Omitting all set dates, the above network algorithm is also able to calculate the critical path. For each operation i, the attribute CRIT[i] specifies the operation following i on the critical path. An analogous attribute specifies the first operation on the critical path in the item master data. In step 1b, all the last operations are assigned CRIT[i_1]= "END." Wherever the "<" condition appears in step 2b, CRIT[i_1] is replaced with "i."

13.3.5 Central Point Scheduling

Intended learning outcomes: Explain central point scheduling. Describe several possible solutions in a directed network of operations.

Central point scheduling is a combination of forward and backward scheduling. Figure 13.3.5.1 shows the underlying concept.

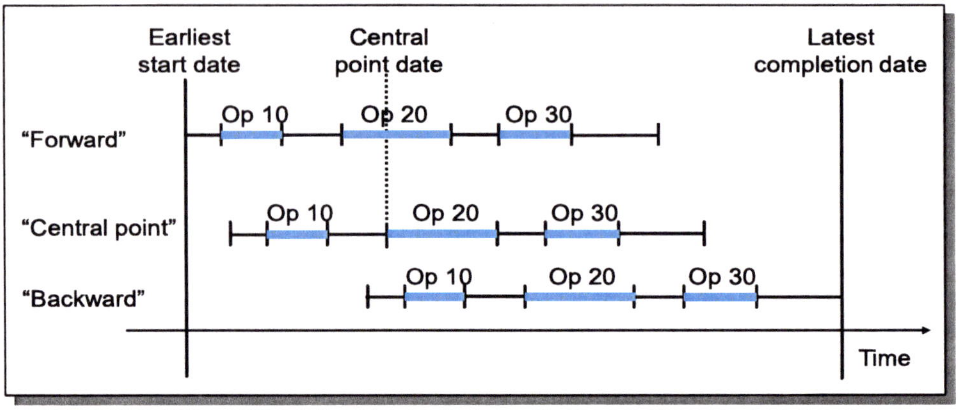

Fig. 13.3.5.1: Central point scheduling.

The central point date is the start date for a particular operation. This is usually a critical operation (i.e., an operation at a fully utilized work center — often a bottleneck capacity). The critical operation determines the order schedule and therefore both the start and the completion dates. The relationship of this technique to the two scheduling techniques introduced earlier is as follows:

- For the critical operation and all subsequent operations, we use forward scheduling; for the operations previous to the critical operation, we use backward scheduling.

In this way, central point scheduling provides the latest start date and the earliest completion date. This proves to be quite simple in the case of a sequence of operations with exactly one central point as shown in Figure 13.3.5.1.

Other cases are more complicated and lead us to several possible solutions. For example:

- If a sequence of operations has more than one central point, it is unclear whether planning should apply forward or backward scheduling between two central points.

In a directed network of operations, there are several possible solutions:

- If there is one central point and it lies on the critical path, the latest start date and the earliest completion date appear as they do in a sequence of operations. Planners schedule the network operations that are not time critical using either forward scheduling beginning with the latest start date or backward scheduling beginning with the earliest completion date.
- If there is a central point lying on a path that is not time critical, it will affect either the forward scheduling branch or the backward scheduling branch of the time-critical network path. Here, the simplest procedure is to choose between the following two basic options. Firstly, backward scheduling beginning with the central point — this will provide a latest start date, and the entire network is scheduled

forward from this date. Secondly, forward scheduling beginning with the central point — this will provide an earliest completion date, and the entire network is scheduled backward from this date.

- Where there are multiple central points located arbitrarily within the network, central point scheduling becomes more complex.

To eliminate ambiguities in central point scheduling within networks, it is useful to determine a so-called mid-level rather than a central point. The mid-level consists of a number of operations for which a start date is chosen in such a way that, without these operations, the beginning and end are no longer connected.

13.3.6 The Lead-Time-Stretching Factor and Probable Scheduling

Intended learning outcomes: Present order urgency and slack time. Differentiate between forward, backward, and probable scheduling. Explain the role and how to calculate the lead-time-stretching factor. Describe the equation for its recalculation.

In practice, the urgency of an order is often more important than an absolute date.

Order urgency is the urgency of the order's operations compared with those of other orders.

A possible measure for order urgency is the lead-time-stretching factor, which is introduced in the following.

For backward scheduling, *slack time* is the difference between the latest (possible) start date and the earliest (acceptable) start date; for forward scheduling, it is the difference between the earliest (possible) completion date and the latest (acceptable) completion date.

Therefore, slack time provides an element of flexibility in planning. Positive slack time allows an increase in lead time, while negative slack time requires that it be shortened.

In *probable scheduling*, we take slack time into account to increase or decrease lead time.

Figure 13.3.6.1 illustrates the principle of probable scheduling using an example with three operations ("op") and positive slack time. In contrast to forward or backward scheduling, the operations are distributed evenly between the earliest start date and the latest completion date. Then, the start or the completion date of each operation is its *probable start date* or *probable completion date*.

Since the technical process itself determines the duration of operations and the technical interoperation time, we can only modify slack time by increasing or reducing either the nontechnical interoperation times or the administrative times. All of these time elements are attributes of the product's master data, its routing sheet, and the work centers. Their values are averages, determined through measuring or estimating.

The *lead-time-stretching factor* is a numerical factor by which the non-technical interoperation times and the administrative times are multiplied.

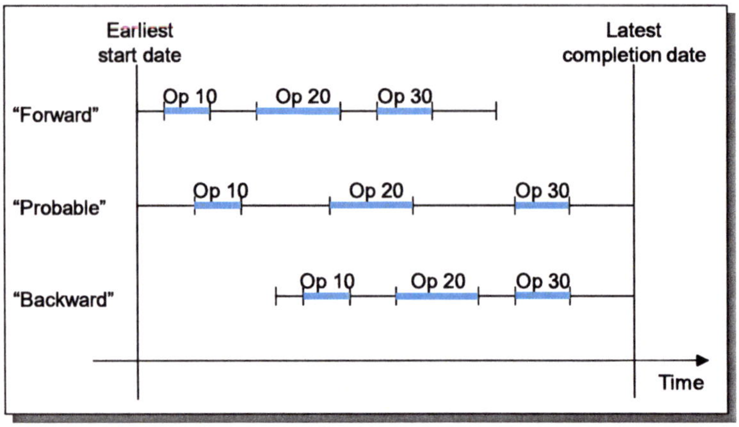

Fig. 13.3.6.1 Forward, backward, and probable scheduling.

The choice of the lead-time-stretching factor has the following effects on the scheduling algorithm:

- A factor greater than 1 results in increased lead time.
- A factor equal to 1 results in "normal," or average, lead time.
- A factor between 0 and 1 results in reduced lead time.
- A factor equal to 0 results in a minimal lead time, in that only the duration of the operations and technical interoperation times are strung together.
- With a factor of less than 0, the operations overlap.

Probable scheduling takes the latest completion date and the earliest start date as givens and calculates the lead-time-stretching factor. This is the starting point in the cases that follow.

- *Customer production orders with a set due date:* This due date is the latest acceptable completion date for scheduling. Because delivery dates are often very short term, the earliest start date becomes *de facto* "today." The scheduling algorithm calculates the lead-time-stretching factor (less than 1) needed to shorten the interoperation times so that the order can be completed between "today" and the delivery date. In this case, the lead-time-stretching factor indicates the feasibility of completion of the order cycle (where sufficient capacity is available, of course).
- *Orders in process:* The earliest start date for the first of all remaining operations is "today." The latest completion date is generally the date specified when the order is released. Rescheduling calculates the lead-time-stretching factor required for order completion on time. This is very useful if, for example, there are delays after the order is released. A lower lead-time-stretching factor gives this order immediate urgency.
- *Early released orders*: The earliest start date is provided by the date the order is released; the latest completion date is the date on which warehouse stocks will probably fall below the safety stock level. Again, probable scheduling will calculate the lead-time-stretching factor required for timely order completion. This factor can then serve as a priority rule for queues at the work centers (see also Section 15.3.1).

The lead-time-stretching factor is calculated using an iterative forward or backward scheduling process as follows:

0 Choose a lead-time-stretching factor, such as 1 (randomly) or the last valid factor used (in a previous scheduling process).

1 Schedule forward (or backward) using the chosen lead-time-stretching factor. At the same time, calculate the earliest completion date (or the latest start date) using the lead-time-stretching factor 0, and thus the lead time required for the duration of operations and technical interoperation times.

2 If the difference between the earliest completion date and the latest completion date in forward scheduling (or the earliest start date and the latest start date in backward scheduling) is approximately zero, then we have found the appropriate lead-time-stretching factor and the process is finished.

3 If the difference is not approximately zero, choose a new lead-time-stretching factor according to the formulas in Figure 13.3.6.3. Begin again with step 1.

Figure 13.3.6.2 shows the result of each iteration in Step 3, in *forward scheduling*.[7]

Iteration of the forward scheduling algorithm calculates the earliest completion date using the currently valid lead-time-stretching factor. The same iteration of the algorithm calculates the earliest completion date using the lead-time-stretching factor 0. The result yields the minimum load time without an overlapping of the operations. The objective of probable scheduling is, by recalculation of the lead-time-stretching factor, to eliminate the difference, that is, the slack time, between the earliest completion date and the latest completion date. This is shown in Figure 13.3.6.2.

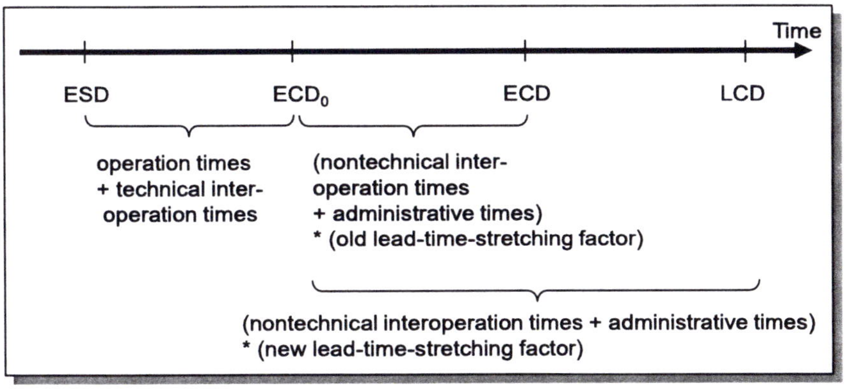

Fig. 13.3.6.2 The role of the lead-time-stretching factor in probable scheduling.

---

7 Key: ESD stands for the earliest start date, ECD for the earliest completion date, ECD_0 for the earliest completion date calculated with lead-time-stretching factor 0, and LCD for the latest completion date (see definitions in Section 13.3.3).

Since this involves a multiplication factor, the equation is a proportional relationship, as shown in Figure 13.3.6.3.[8]

Backward scheduling	$\frac{\text{STREFAC [new]}}{\text{STREFAC [old]}} = \frac{LSD_0\text{[order]} - ESD\text{[order]}}{LSD_0\text{[order]} - LSD\text{[order]}}$
Forward scheduling	$\frac{\text{STREFAC[new]}}{\text{STREFAC[old]}} = \frac{LCD\text{[order]} - ECD_0\text{[order]}}{ECD\text{[order]} - ECD_0\text{[order]}}$

Fig. 13.3.6.3 Equation for recalculation of lead-time-stretching factor.

For a production contract with a limited number of serially executed operations, probable scheduling using the formula in Figure 13.3.6.3 usually yields the exact solution after only one iteration subsequent to the initial step. In a network structure, however, there may be a different number of operations with varying interoperation times in each branch of the network. In any case, there are always situations where one iteration alone does not produce an immediate, exact solution with a slack time of approximately zero. The reasons for this and some suggestions for solving the problem are as follows:

- The lead-time-stretching factor was too inexact. Another iteration of the process will yield a more exact result, namely, a slack time close to zero.
- The calculations were inexact, which we can correct by, for example, calculating to finer units, such as to tenth-days instead of half-days.
- Because of the new lead-time-stretching factor, another path in the network of operations has become time critical; that is, it is now the longest path. A further iteration of the algorithm would yield precise results, provided that the critical path remains the same.
- There is a negative lead-time-stretching factor, and the scheduling algorithm cannot accommodate the operations between the earliest start date and the latest completion date. It is even possible that one of the operations itself is longer than the difference between these two set dates. In both cases, only lengthening the time span will resolve the situation.

13.3.7 Scheduling Process Trains

Intended learning outcomes: Differentiate between reverse flow scheduling, forward flow scheduling, and mixed flow scheduling.

---

[8] Key: STREFAC is the lead-time-stretching factor, LSD is the latest start date, LSD_0 is the latest start date (calculated with lead-time-stretching factor 0), and LCD is the latest completion date (see the definitions in Section 13.3.3).

Process trains were introduced in Chapter 8. A process train is a representation of the flow of materials through a process industry manufacturing system that shows equipment and inventories.

To schedule the process train, we need to know the order in which to schedule the stages of the process train. Take, for example, a process train with three consecutive stages 1, 2, 3. There are three possible scheduling techniques:

- *Reverse flow scheduling* (3, 2, 1) starts with the last stage and proceeds backward (countercurrent to the process flow) through the process structure. It supports demand-based planning.
- *Forward flow scheduling* (1, 2, 3) starts with the first stage and proceeds sequentially through the process structure until the last stage is scheduled. It supports supply-constrained planning, such as short harvest cycle in the food industry.
- *Mixed flow scheduling* (2, 1, 3 or 2, 3, 1) supports planning where stage 2 is the logical focus of attention for scheduling because of processing capacity or material supply constraints. In general, detailed scheduling starts at each bottleneck stage and works toward the terminal process stages or another bottleneck stage.

It is easy to see that these three scheduling techniques have much in common with backward, forward, and central point scheduling.

13.4 Splitting, Overlapping, and Extended Scheduling Algorithms

Intended learning outcomes: Explain order or lot splitting, and overlapping. Present an extended formula for manufacturing lead time and extended scheduling algorithms.

13.4.1 Order or Lot Splitting

Intended learning outcomes: Explain reducing lead time for operation i by using a splitting factor >1. Describe how the split offset factor offsets the split lots in time.

Order splitting or *lot splitting* means distributing the lot to be produced by an operation among two or more machines or employees at a work center for processing. This implies *split lots*.

Splitting reduces lead time, but it incurs additional setup costs, since employees must set up multiple machines. Figure 13.4.1.1 shows the situation.

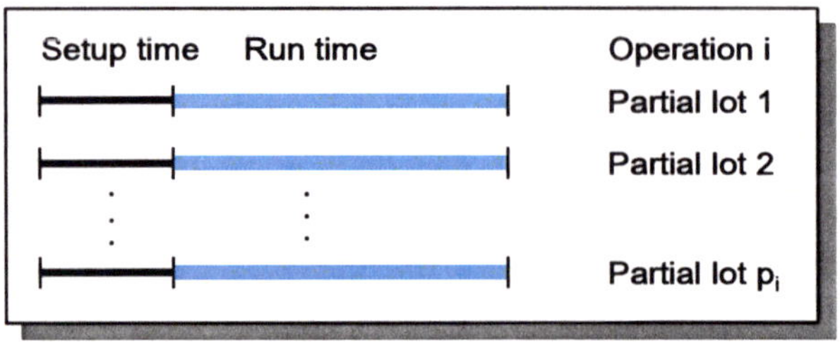

Fig. 13.4.1.1 Reducing lead time for operation i by using a splitting factor > 1.

The *splitting factor* for an operation expresses the degree of its potential splitting.

The initial value of the splitting factor is 1, that is, "no splitting." Where a splitting factor > 1 is given, run time is divided by this value. To calculate the costs of the operation, however, setup load must be multiplied by the splitting factor.

The split lots may be worked on in parallel or be finished at points that are offset in time.

A *split offset factor* expresses the possible temporal shift of the split lots, according to the principle illustrated in Figure 13.4.1.2.

The split offset factor is expressed as a percentage of the operation time after splitting. The initial value of this factor is zero, that is, "no split offset."

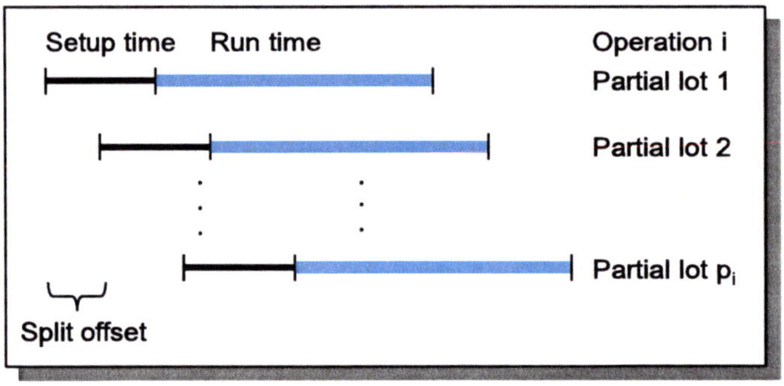

Fig. 13.4.1.2 The split offset factor offsets the split lots in time.

13.4.2 Overlapping

Intended learning outcomes: Explain the principle of overlapping within an operation. Describe the principle of operation overlapping.

We speak of *overlapping within an operation* when the individual units of a lot are not produced sequentially, or one after the other, but rather overlap one another.

Consider the example of an assembly operation for machines. The operation may comprise several partial operations. Figure 13.4.2.1 shows the situation for the lot as a whole.

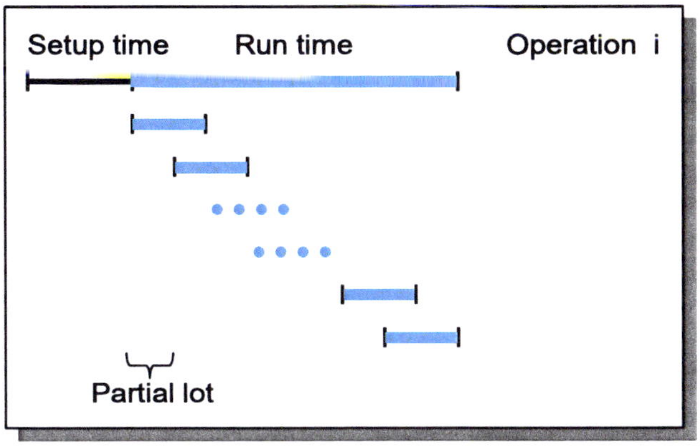

Fig. 13.4.2.1 The principle of overlapping within an operation.

A later partial operation on the first machine of the lot may be worked on parallel to the first partial operation on a subsequent machine of the lot.

The *run time offset*, or *offset of the next run time*, is a measure for the overlapping within an operation.

Run time offset is expressed as a percentage of run time. The standard value for run time offset is 100%, or "no overlapping."

For some production processes, you can overlap entire operations.

In an *operation overlapping* or an *overlapped schedule*, we begin the next operation on a portion of the lot before the entire lot is completed with the previous operation.

Figure 13.4.2.2 shows an example. Schedulers can use operation overlapping to accelerate a production order.

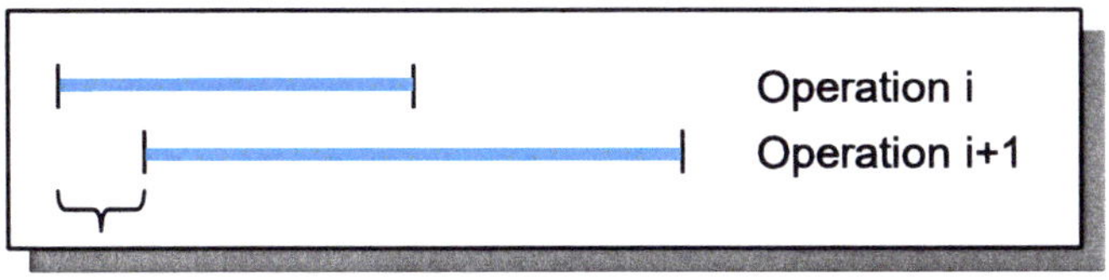

Fig. 13.4.2.2 The principle of operation overlapping.

The *maximum offset of the next operation* is a measure of operation overlapping. It is based on one operation and shows the maximum lapse of time before the next operation begins.

In practice, the next operation begins immediately after the setup time and run time for the first unit (or first units) of the order lot. (See, for example, near-to-line production in Figure 6.2.2.2).

The initial value of the maximum offset of the next operation is infinite, that is, "no overlapping." If the time we calculate (based on operation time and interoperation times) until beginning the next operation is smaller than the actual value, we take the smaller time as the new offset time.

13.4.3 An Extended Formula for Manufacturing Lead Time (*)

Intended learning outcomes: Explain the extended lead time formula in its first and second version. Disclose the influence of overlapping of operations upon lead time.

The following lists the definitions set out in Section 13.3.2 for the components of operation time. Here, we have added the following abbreviations for the elements defined above.

LOTSIZE	:=	lot size ordered
ST[i]	:=	setup time for operation i
RT[i]	:=	run time per unit produced for operation i
STREFAC	:=	lead-time-stretching factor
SPLFAC[i]	:=	splitting factor for operation i
SPLOFST[i]	:=	split offset factor expressed as a percentage
RTOFST[i]	:=	run time offset for operation i expressed as a percentage
MAXOFST[i]	:=	maximum offset of the operation immediately following operation i (a duration)

We can express the operation time for an operation i, OT[i], by the formula shown in Figure 13.4.3.1. This formula is much more complex than the one in Section 13.3.2.

$$\mathrm{OT}[i]=\left\langle \mathrm{ST}[i]+\mathrm{RT}[i]\cdot\left(1+\left(\frac{\mathrm{LOTSIZE}}{\mathrm{SPLFAC}[i]}-1\right)\cdot\frac{\mathrm{RTOFST}[i]}{100}\right)\right\rangle \cdot\left\langle 1+\left(\mathrm{SPLFAC}[i]-1\right)\cdot\frac{\mathrm{SPLOFST}[i]}{100}\right\rangle$$

Fig. 13.4.3.1 Extended operation lead time.

For a *sequence of operations* as the order of the operations, the formula in Figure 13.4.3.2 yields the lead time for the order.

$$\begin{aligned}\mathrm{LTI}=&\ \mathrm{STREFAC}\cdot(\mathrm{ADMORD}+\mathrm{ADMPORDBEG}+\mathrm{INTBEF}[1])\\&+\sum_{1\le i\le n-1}\min\left(\mathrm{MAXOFST}[i];\mathrm{OPD}[i]+\mathrm{INTTEC}[i]+\mathrm{STREFAC}\cdot(\mathrm{INTAFT}[i]+\mathrm{INTBEF}[i+1])\right)\\&+\mathrm{OT}[n]+\mathrm{INTTEC}[n]+\mathrm{STREFAC}\cdot(\mathrm{INTAFT}[n]+\mathrm{ADMPORDEND})\end{aligned}$$

Fig. 13.4.3.2 Extended lead time formula (first version).

LTI represents the lead time for LOTSIZE and will vary when lot sizes are different. In Figure 13.4.3.3, we attempt to define partial sums to express lead time as a linear function of lot size.

As in Figure 13.3.2.3, we can store the partial sums in the lead time formula as attributes of the product and recalculate them after each modification of the routing sheet. Correspondingly, the formula according to Figure 13.3.2.4 holds.

$$\begin{aligned}
\text{SUMINT} &= \text{ADMPORDBEG} + \text{ADMPORDEND} + \sum_{1\le i\le n} \left[\text{INTBEF}[i] + \text{INTEND}[i]\right] \\
\text{SUMTEC} &= \sum_{1\le i\le n} \text{INTTEC}[i] \\
\text{SUMST} &= \sum_{1\le i\le n} \left\langle \text{ST}[i] + \text{RT}[i]\cdot\left(1-\frac{\text{RTOFST}[i]}{100}\right)\cdot\left(1+\left(\text{SPLFAC}[i]-1\right)\cdot\frac{\text{SPLOFST}[i]}{100}\right)\right\rangle \\
\text{SUMRT} &= \sum_{1\le i\le n} \left\langle \text{RT}[i]\cdot\frac{1}{\text{SPLFAC}[i]}\cdot\frac{\text{RTOFST}[i]}{100}\cdot\left(1+\left(\text{SPLFAC}[i]-1\right)\cdot\frac{\text{SPLOFST}[i]}{100}\right)\right\rangle
\end{aligned}$$

Fig. 13.4.3.3 Extended partial sums for the lead time formula.

$$\text{LTI}' = \text{STREFAC}\cdot(\text{ADMORD} + \text{SUMINT}) + \text{SUMTEC} + \text{SUMST} + \text{SUMRT}\cdot\text{LOTSIZE}$$

Fig. 13.4.3.4 Extended lead time formula (second version).

Because of the overlapping of operations, which is expressed in the formula for LTI in Figure 13.4.3.2 as a minimization, LTI is not equivalent to LTI': For either one or the other operation, the *maximum offset of the next operation* is smaller than the sum of the other time elements (the "normal" time period until the beginning of the next operation).

Figure 13.4.3.5 shows a possible plotting of the two lead times as functions of lot size.

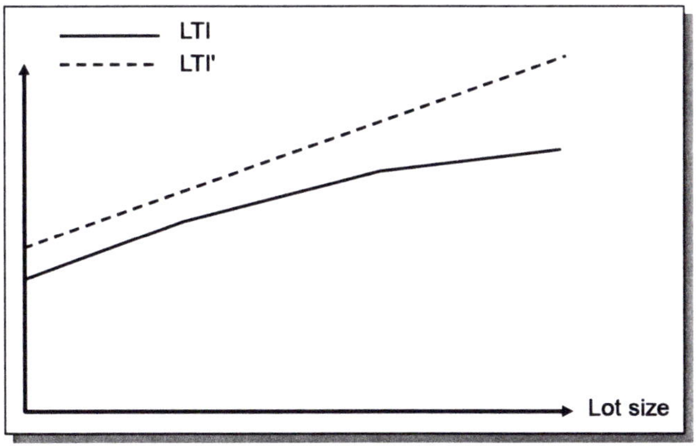

Fig. 13.4.3.5 Influence of overlapping of operations upon lead time.

In most circumstances LTI' is precise enough and certainly suffices for rough-cut planning. If necessary, we can set a *lot size limit for the lead time formula*. If the lot size is less than or equal to this quantity, we calculate lead time according to the "quick" lead time formula (the second version in Figure 13.4.3.4). Otherwise, we apply the more involved, "slow" formula in Figure 13.4.3.2.

In a *directed network of operations* as the order of the operations, similar considerations to those examined in Section 13.3.2 apply.

13.4.4 Extended Scheduling Algorithms (*)

Intended learning outcomes: Identify various possible extensions of the scheduling algorithms. Describe possible cases arising in process industries.

We can now extend the scheduling algorithms presented in Section 13.3.3 to include the definitions introduced in the subsections above. These include:

- The introduction of a lead-time-stretching factor that multiplies interoperation times
- The introduction of splitting and overlapping and an expanded formula for lead time
- The inclusion of multiple partial orders for each production order
- The inclusion of divergent product structures, as — for example — the case of temporary assembly
- Ongoing planning for released orders with work remaining to be done

We can derive a generalized algorithm from the algorithm presented in Section 13.3.3, for both a sequence of operations and for a *directed* network of operations. This would complicate the algorithm further, and we will not present it here in detail.

The extensions introduced thus far may not be sufficient for lead time scheduling in every potential scenario. A first case is the *undirected network of operations* with a *repetition of operations*. During a chemical process or in the production of electronic components, for example, production has to repeat certain operations. This may be because inspection has uncovered defects in quality. Here, the number of iterations and the individual operations to be repeated become evident only during the course of work and cannot be planned in advance. In this case, it is not possible to calculate lead time precisely. Instead, we have to use expected mean values for the number of iterations and accompanying deviation. However, we have to take into account that each calculation of lead time itself is based on estimations of the time elements, particularly wait time in front of the work center.

Another case arises in process industries. The processor-oriented concept implemented in these industries may require sequencing or, more precisely, the planning of optimum sequences of operations, as early as the phase of long- and medium-term planning. Because of the extremely high setup costs, planners should establish suitable lots even prior to order release to keep changeover costs at a minimum. To this category belongs, for example, cut optimizations for glass, sheet metals, or other materials. The scheduling of an individual order will depend on whether it may be combined with other orders and with what orders, to achieve optimal usage of the raw material, the reactors, or processing containers.

13.5 Summary

The ordering party sets the latest acceptable completion date and sometimes the earliest acceptable start date for a production order. The planner must establish start and/or due dates of the operations as well as the latest possible start date and the earliest possible completion date in advance, in order to obtain an initial estimate of feasibility and in preparation for work center loading and the setting of reservation dates for components.

For this, time management divides the lead time into meaningful time elements that can be measured or estimated relatively simply. Planners make use of the order of the operations (sequence or network of operations) of the product to be manufactured. Each operation has an operation time, and there are interoperation times before and after the operation. In addition, there are administrative times for each partial order and for the order in its entirety.

In job shop production, unproductive interoperation times make up the major proportion of total lead time. Simple models for estimating transportation times allow sufficiently precise estimates to be made without expending a lot of time and effort on data management. However, it is difficult to determine the adequate size of buffers or queues at the work centers. Statistical analysis of queues as the effect of random load fluctuations yields useful information with regard to reducing wait times: High loading as well as long or highly varied operation times lead to long wait times. This underlines the conflict between the entrepreneurial objectives of "low costs" and "short lead time" as set out in Section 1.3.1.

Scheduling management starts out from the dates set by the ordering party and calculates the other dates required for determining feasibility, loading capacity, and reserving components. The following list shows the scheduling techniques discussed in the chapter (for sequences as well as directed networks of operations), comparing data input with data output:

- Forward scheduling:
    - *Input*: earliest order start date, lead-time-stretching factor
    - *Output*: earliest order completion date, earliest start and completion dates for each operation, earliest reservation date for each component
- Backward scheduling:
    - *Input*: latest order completion date, lead-time-stretching factor
    - *Output*: latest order start date, latest start and completion date for each operation, latest reservation date for each component
- Central point scheduling:
    - *Input*: central point date, lead-time-stretching factor
    - *Output*: latest order start date and earliest order completion date; latest start and completion date for each operation as well as latest reservation date for each component *before* the central point, earliest start and completion date for each operation as well as earliest reservation date for each component *after* the central point

- Probable scheduling:
  - *Input*: earliest start date and latest completion date for the order
  - *Output*: lead-time-stretching factor, probable start and completion date for each operation, probable reservation date for each component.

Splitting and overlapping are techniques frequently used to reduce lead time. Their incorporation into the lead time formula, as well as the attempt to include other effects, reveals the limits to lead time estimation. Not all time elements can be estimated accurately, and only a modest degree of complexity can be expressed as a formula. Moreover, there are unforeseen factors that can always arise during actual production. On the other hand, planners must have a fair idea of cumulative lead time so that they can set it in relation to the customer tolerance time. With this, in the short term, the basic decision can be made to accept or decline an order. In the medium term, it allows planners to sketch out a possible load profile for the work centers along the time axis.

13.6 Keywords

backward scheduling, 571
buffer, 557
central point scheduling, 576
cycle time, 570
earliest date, 571
forward scheduling, 571
funnel model, 559
idle time, 558
inventory buffer, 558
inventory range, 565
latest date, 571
lead-time scheduling, 568
lead-time-stretching factor, 577
manufacturing calendar, 567
move time, 554
network planning, 573
nontechnical wait time
  after an operation, 554
  before an operation, 554
operation overlapping, 583
order splitting, 581
overlapping
  of operations, 583
  within an operation, 582
probable scheduling, 577
queue, 557
queue time, 554
queuing theory, 557
run time offset, 583
scheduling algorithm, 567
sequence of operations, 551
set date, 571
slack time, 577
splitting, 581
synchronization point, 551
takt time, 570
technical wait time after an operation, 554
technical wait time before an operation, 554
throughput time, 570
transit time, 554
transportation time, 554

13.7 Scenarios and Exercises

Intended learning outcomes: Assess queues as an effect of random load fluctuations. Calculate examples for network planning, backward scheduling, forward scheduling, the lead-time stretching factor, and probable scheduling.

13.7.1 Queues as an Effect of Random Load Fluctuations (1)

Intended learning outcomes: Answer questions using the relevant formula in queuing theory.

Answer the following questions using the relevant formulas in queuing theory (refer to Figure 13.2.2.4):

a. How many parallel workstations are needed to have an expected wait time of less than 10 hours, if capacity utilization is 0.95, the mean of the operation time is 2 hours, and the coefficient of variation of the operation time is 1?

b. The capacity is 10 hours. How much does the expected wait time increase if load rises from 4 to 8 hours?

c. How is the expected wait time affected when the coefficient of variation increases from 1 to 2?

Solutions:

a. $s = 0.95 / (1 - 0.95) * (1 + (1 * 1)) / 2 * 2 / 10) = 3.8$. Thus, with *four* workstations, the expected wait time will be 9.5 hours.

b. Capacity utilization increases from 4/10 to 8/10. Therefore, the respective factor in the formula for the expected wait time increases from $0.4 / (1 - 0.4) = 2/3$ to $0.8 / (1 - 0.8) = 4$. The new factor is $4 / (2/3) = 6$ times greater than the old factor. Thus, the expected wait time increases by a factor of 6.

c. The respective factor in the formula for the expected wait time increases from $(1 + (1 * 1)) / 2 = 1$ to $(1 + (2 * 2)) / 2 = 2.5$. Thus, the expected wait time increases by the factor 2.5.

13.7.2 Queues as an Effect of Random Load Fluctuations (2)

Intended learning outcomes: Experience wait time as a function of capacity utilization in a job shop environment with random arrivals, execution in order of arrival as well as operation times subject to a determinate distribution with mean and coefficient of variation.

Figure 13.2.2.3 shows the average wait time as a function of capacity utilization in a job shop environment with random arrivals, execution of operations in order of arrival (or according to random selection from the queue), as well as operation times (OT) subject to a determinate distribution with mean M(OT) and coefficient of variation CV(OT). We reproduced the effect shown in Figure 13.2.2.3 by means of a simulation, which you can view at this URL:

https://intlogman.ethz.ch/#/chapter13/queuing-theory/show

Start the simulation by clicking on the given arrival rate and execution (service) rate on the gray button to the far left at the bottom of the figure and watch the number of elements in the system. Stop the simulation by clicking on the middle of the three buttons (or empty the system by clicking the button to the far right). Now change the input rate to bring it closer and closer to the execution rate and observe the rising number of elements in the queue. You will see the exploding number of elements in the system as soon as, for an execution rate of 60 per unit of time, the arrival rate is 58 and higher.

13.7.3 Network Planning

Intended learning outcomes: Calculate a scheduled network with incomplete data for six operations and a start operation (administration time. Determine the critical path.

Figure 13.7.3.1 shows a scheduled network with incomplete data for 6 operations and a start operation (administration time).

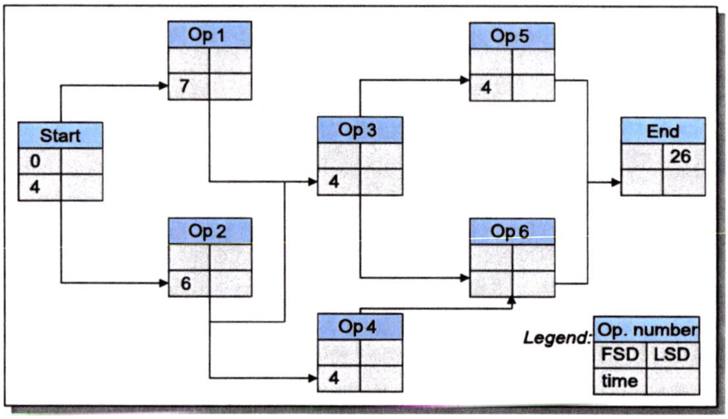

Fig. 13.7.3.1 Scheduled network (for you to complete).

a. For each process, please fill in the earliest start date (ESD) and the latest start date (LSD) in the scheduled network. What is the critical path, that is, the path with the longest duration? What is its lead-time margin, that is, the slack time?

b. The operation time for operation 6 has not yet been determined. What is the longest possible time for operation 6 (lead-time margin = zero)?

Solutions:

a. For the time being, as long as the time for operation 6 is still open, the longest path is (start – op1 – op3 – op5 – end). Lead-time margin = 7.

b. As soon as the time for operation 6 is greater than 4, the longest path is (start – op1 – op3 – op6 – end). The longest possible time for op6 is 11.

13.7.4 Backward Scheduling and Forward Scheduling

Intended learning outcomes: Explain and solve the forward and backward scheduling problems that is calculation of start and completion dates for the order and each operation, as well as the critical path and lead-time margin.

Here, you will practice some backward and forward scheduling. Figure 13.7.4.1 presents a simple network, including a legend showing the lead- time elements used.

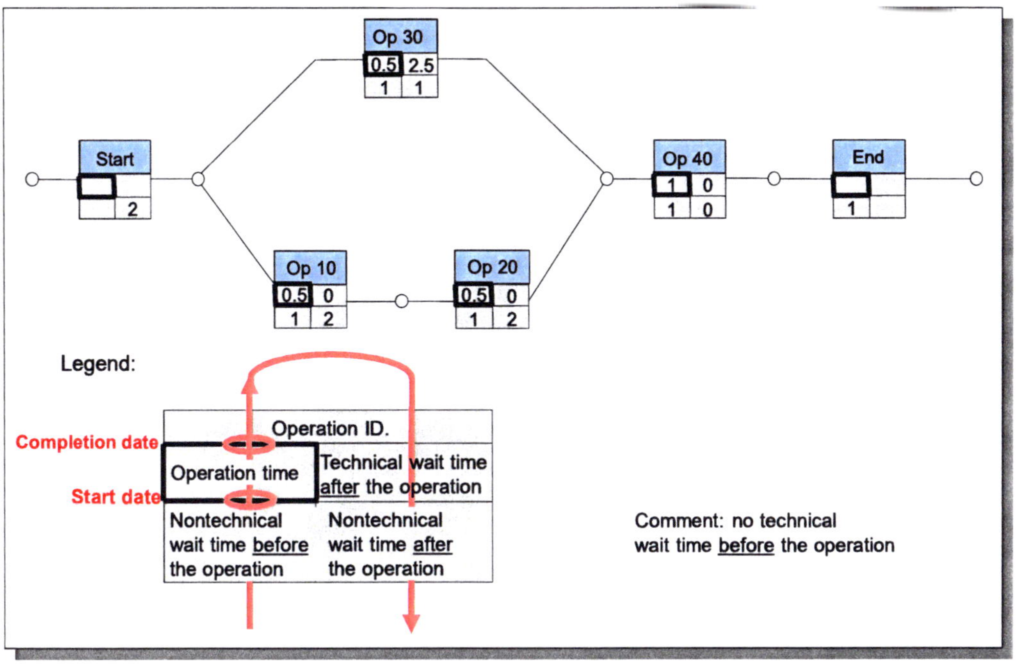

Fig. 13.7.4.1 Scheduled network.

Solve the forward and backward scheduling problems (calculation of start and completion dates for the order and each operation, as well as the critical path and lead-time margin) listed in Figure 13.7.4.2:

a. Common forward scheduling.

b. Common backward scheduling.

c. Forward scheduling with a different lead-time-stretching factor, that is, a different order urgency, to accelerate or slow down the order.

d. Forward scheduling with lead-time-stretching factor = 0, which results in the lead time as the sum of operation times plus the technical interoperation times.

Some common problems in the calculation process lead to the following potential errors:

- Calculating incorrect start date and due dates, not respecting interoperation times multiplied by the stretching factor
- Multiplying technical waiting time by the stretching factor
- Incorrectly calculating the longest path in a network
- Not understanding the principle of forward or backward scheduling

a) forward scheduling
earliest start date:0
stretching factor: 1

Operation	earliest start date	earliest completion date
10		
20		
30		
40		
Order		

b) backward scheduling
latest completion date:16
stretching factor: 1

Operation	latest start date	latest completion date
10		
20		
30		
40		
Order		

c) forward scheduling "express"
earliest start date: 0
stretching factor: 0.5

Operation	earliest start date	earliest completion date
10		
20		
30		
40		
Order		

d) forward scheduling without non-technical interoperation time
earliest start date: 0
stretching factor: 0

Operation	earliest start date	earliest completion date
10		
20		
30		
40		
Order		

Fig. 13.7.4.2 Various forward and backward scheduling problems.

Solutions:
(ESD stands for earliest start date, ECD for earliest completion date, LSD for latest start date, LCD for latest completion date.)

a. ESD(op10) = 3, ECD(op10) = 3.5; ESD(op20) = 6.5, ECD(op20) = 7; ESD(op30) = 3, ECD(op30) = 3.5; ESD(op40) = 10, ECD(op40) = 11; ESD(order) = 0, ECD(order) = 12. Note the critical path in determining the ESD(op40): The *lower* path is critical. The lead-time margin of the *upper* path is 2.

b. LCD(op40) = 15, LSD(op40) = 14; LCD(op30) = 9.5, LSD(op30) = 9; LCD(op20) = 11, LSD(op20) = 10.5; LCD(op10) = 7.5, LSD(op10)=7; LCD(order) = 16, LSD(order) = 4. Note that — again — the *lower* path is critical. The lead-time margin of the *upper* path is again 2.

c. ESD(op10) = 1.5, ECD(op10) = 2; ESD(op20) = 3.5, ECD(op20) = 4; ESD(op30) = 1.5, ECD(op30) = 2; ESD(op40) = 5.5, ECD(op40) =6.5; ESD(order) = 0, ECD(order) = 7. Note that both paths are critical.

d. ESD(op10) = 0, ECD(op10) = 0.5; ESD(op20) = 0.5, ECD(op20) = 1; ESD(op30) = 0, ECD(op30) = 0.5; ESD(op40) = 3, ECD(op40) = 4; ESD(order) = 0, ECD(order) = 4. Note that the critical path has changed. The *upper* path is now critical. The lead-time margin of the *lower* path is 2.

13.7.5 The Lead-Time-Stretching Factor and Probable Scheduling

Intended learning outcomes: Explain and practice the use of the lead-time-stretching factor as well as probable scheduling.

The following exercise will allow you to practice the use of the lead-time-stretching factor as well as probable scheduling. It uses the same network example as in Figure 13.7.4.1.

Solve the two probable scheduling problems shown in Figure 13.7.5.1. Hint: First, calculate a new lead-time-stretching factor using the formula in the lower part of Figure 13.3.6.3, based on an appropriate solution of one of the four problems in the previous exercise (13.7.4) as an initial solution.

a) probable scheduling
"customer order with priority to meet"
earliest start date: 0
latest completion date: 6

stretching factor = —·0.5 = — 0.5 =

Operation	earliest start date	earliest completion date
10		
20		
30		
40		
Order		

b) probable scheduling
"capacity filling" stock replenishment
order earliest start date: 0
latest completion date: 16

stretching factor = —·1= —1=

Operation	earliest start date	earliest completion date
10		
20		
30		
40		
Order		

stretching factor = —· = —· ≈

Fig. 13.7.5.1 Two probable scheduling problems.

Some common problems in the calculation process that can lead to errors are

- Not understanding the goal and principles of probable scheduling
- Not understanding the formula for recalculation of the lead-time-stretching factor in probable scheduling
- Not choosing the most appropriate last calculation as initial solution for recalculation of the lead-time-stretching factor

Solutions:
(Again, ESD stands for earliest start date, ECD for earliest completion date, LSD for latest start date, LCD for latest completion date, STREFAC for lead-time-stretching factor.)

a. Use problem (c) in the previous exercise (13.7.4) as an initial solution.
STREFAC(new) = (6 – 4) / (7 – 4) * 0.5 = 2/3 * 0.5 = 1/3. =>
ESD(op10) = 1, ECD(op10) = 1.5; ESD(op20) = 2.5, ECD(op20) = 3; ESD(op30) = 1, ECD(op30) = 1.5; ESD(op40) = 4.7, ECD(op40) =5.7; ESD(order) = 0, ECD(order) = 6.
Note that the *upper* path is critical. The lead-time margin of the *lower* path is 2/3 = 0.667.

b. Use problem (a) in the previous exercise (13.7.4) as an initial solution.
STREFAC(new) = (16 – 4) / (12 – 4) * 1 = 12/8 * 1 = 1.5. =>
ESD(op10) = 4.5, ECD(op10) =5; ESD(op20) = 9.5, ECD(op20) = 10; ESD(op30) = 4.5, ECD(op30) =5; ESD(op40)=14.5, ECD(op40)=15.5; ESD(order) = 0, ECD(order) = 17 (!).
Note that the *lower* path is critical. The lead-time margin of the *upper* path is 4.
Because the desired ECD(order) of 16 has not been met (can you say why this is the case?), an additional iteration is necessary: recalculation with
STREFAC(new) = (16 – 4) / (17 – 4) * 1.5 = 12/13 * 1.5 ≈ 1.4
will yield the desired solution.

13.8 References

APIC16	Pittman, P. et al., APICS Dictionary, 15th Edition, APICS, Chicago, IL, 2016
Alba77	Albach, H. (Hrsg.), "Quantitative Wirtschaftsforschung," in Ferschl, F., Ed., "Approximationsmethoden in der Theorie der Warteschlangen," p. 185 ff., Verlag Mohr, Tübingen, 1977
Coop90	Cooper, R.B., "Queueing Theory," Chap. 10, in Heyman, D.P., Sobel, M.J., Eds., "Stochastic Models," North-Holland, Amsterdam, 1990
Fers64	Ferschl, F., "Zufallsabhängige Wirtschaftsprozesse," *Physica*, 1964
GrHa18	Gross, D., Harris, C.M., Shortle, J.F., Thompson, J.M., "Fundamentals of Queueing Theory," 5th Edition, John Wiley & Sons, New York, 2018
IBM75	IBM, "Executive Perspective of Manufacturing Control Systems," IBM Brochure G360-0400-12, 1975
LyMi94	Lynes, K., Miltenburg, J., "The Application of an Open Queueing Network to the Analysis of Cycle Time, Variability, Throughput, Inventory and Cost in the Batch Production System of a Microelectronics Manufacturer," *International Journal of Production Economics*, 37, 1994
Wien95	Wiendahl, H.P., "Load-Oriented Manufacturing Control," Springer, Berlin, New York, 1995

14 Capacity Management

Intended learning outcomes: Present fundamentals of capacity management. Explain in detail load profile calculation and infinite loading. Disclose finite loading. Describe rough-cut capacity planning.

Work center capacity utilization, like levels of inventories in stock and work-in-process, increases the flexibility of logistics planning & control. Capacity resources have to be estimated for every planning term. Flexibility in medium- and short-term planning often requires long-term arrangements. In Figure 14.0.0.1, the tasks and processes covered in this chapter are shown against a darker background, referring back to the basic model introduced in Figure 5.1.4.2. Sections 1.2.4, 5.3.3, and 5.3.4 provide useful overviews for this chapter.

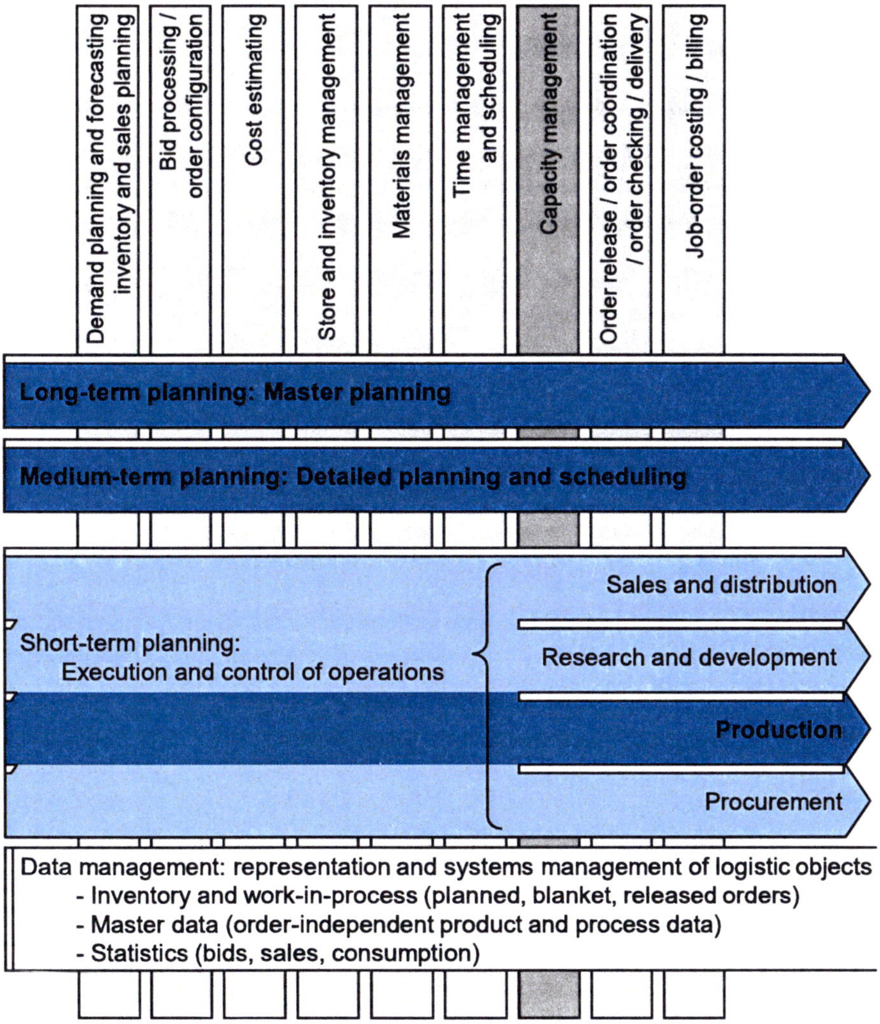

Fig. 14.0.0.1 The parts of the system discussed in this chapter (shown on darker background).

P. Schönsleben, *Handbook Integral Logistics Management*,
https://doi.org/10.1007/978-3-662-65625-9_14

Unlike delivery lead time and delivery reliability rate, the efficient use of capacity is not directly observable by the customer. Nonetheless, it is an extremely important factor, since it enables the company to cut costs, ensure prompt delivery, and increase flexibility.

We will begin by reexamining the basic concepts presented in Chapter 4 and in Chapter 5 on the nature of capacity and on well-known types of capacity management. A detailed look at well-known techniques follows, broken down by objectives, basic characteristics, methods and procedures, range of application, and other factors.

Rough-cut capacity planning deserves special attention. It can be used for both long- and short-term planning. For short-term planning, it supports rapid order promising. The techniques differ according to whether they give greater weight to due dates or capacity limits.

14.1 Fundamentals of Capacity Management

Intended learning outcomes: Produce an overview on capacity, work centers, capacity determination, and capacity management techniques.

14.1.1 Capacity, Work Centers, and Capacity Determination

Intended learning outcomes: Explain the determination of rated capacity as the product of theoretical capacity, capacity utilization, and work center efficiency. Produce an overview on other capacity definitions and their relationship to each other, such as demonstrated capacity, productive capacity, protective capacity, excess capacity, idle capacity, and activation.

Section 1.2.4 already presented basic definitions around *work center* and *capacity*. This chapter presents them in a more detailed way.

Depending on the type of work center, different capacities will be used *as the primary basis* for capacity management and for allocating costs:

- Machine capacity (referred to as machine hours, when using hours as the capacity unit), that is, the capacity of machines and equipment to produce output, is frequently used for parts manufacturing.
- *Labor capacity* (referred to as *labor hours*, when using hours as capacity unit), that is, the capacity of workers to produce output, is frequently used for assembly or stores.

These concepts are part of the *capacity determination*, as shown in Figure 14.1.1.1.

Theoretical capacity is the maximum output capacity, with no adjustments for unplanned downtime, determined by the number of shifts, the capacity theoretically available for each shift, and the number of machines and workers. The value thus determined applies up to a given boundary date, after which the calculation factors may change.

Shift no.	No. hours per shift	No. machines	No. workers	Daily capacity machines hours	Daily capacity workers hours	Correction factor
1	8	10	6	80	48	
2	8	10	6	80	48	
3	4	10	1	40	4	
Theoretical capacity				200	100	
(multiplied by) capacity utilization, subdivided in						
- availability (in capacity)						90%
- tactical underutilization (desired):						75%
Subtotal:				135	67.5	
(multiplied by) work center efficiency:						120%
Rated capacity:				162	81	

Fig. 14.1.1.1 Determination of capacity. Rated capacity is the product of theoretical capacity, capacity utilization, and work center efficiency.

Theoretical capacity can also vary from one week to the next in response to *foreseen*, overlapping changes that must be taken into account, such as

- *Scheduled downtime*, that is, downtime due to individual workers' vacations or for preventive maintenance, for example.
- *Scheduled overtime* due to additional shifts, for instance.

Planned capacity utilization is a measure of how intensively a resource should be used to produce a good or service. Traditionally, it is the ratio of actual load to theoretical capacity. There are two distinct factors in capacity utilization:

- *Availability (in capacity)*: Downtime due to breaks, cleaning tasks, clearing up, unplanned absences, breakdowns, etc., must be considered for each work center. These losses are considered by the availability factor (hours actually worked / hours available).
- *Tactical underload* or *tactical underutilization*: To avoid long queue times (see Section 13.2.3) or for non-bottleneck capacities or nonconstraint work centers, the desired capacity utilization should generally be less than 100%.

The measurement of *actual capacity utilization* cannot as a rule be broken down according to the two factors. This is the main reason for capturing availability and tactical underload in one factor, namely, capacity utilization.

The *work center efficiency* (or *efficiency rate*) is the ratio of "standard load to actual load," "standard hours produced to actual hours worked," or "actual units produced to standard units to produce" (see [APIC16]), averaged over all the operations performed at the center.[1]

[1] This factor can be greater than, equal to, or less than 1. The actual choice reflects different worker motivation concepts. Cost unit rates for load standards thus also differ accordingly.

Rated capacity is the expected output capability of a work center. It is defined as theoretical capacity times planned capacity utilization times work center efficiency.

We should therefore consider *standard load* to be scheduled (that is, load on the basis of standard setup and run loads) to *rated capacity*, and not to theoretical capacity.

The *overall equipment effectiveness (OEE)* includes the achieved quality. OEE can be defined as planned capacity utilization times work center efficiency times the yield factor.

In principle, rough-cut planning uses the same attributes, usually applied to fully utilized work centers at the level of the department or entire plant. The capacity of a rough-cut work center is thus not necessarily equal to the sum of all the individual capacities concerned.

There are other capacity-related terms that are useful for capacity management. Figure 14.1.1.2 shows possible relations among the terms. The definitions are based mainly on [APIC16]. Barry Firth, CPIM, Melbourne, contributed the figure and the explanations.

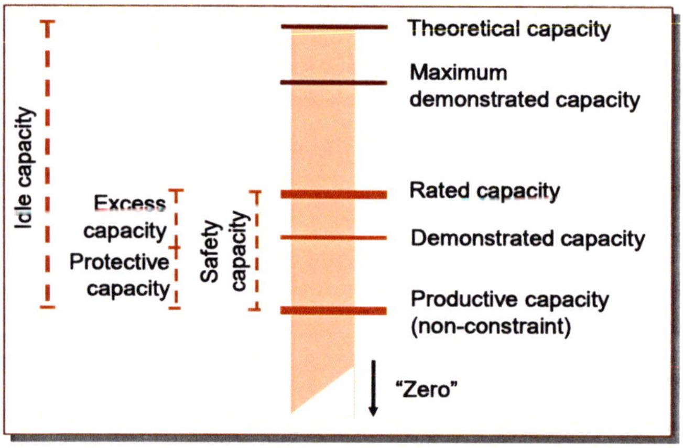

Fig. 14.1.1.2 Some capacity definitions and their relationship to each other.

Demonstrated capacity is proven capacity calculated from actual performance data, expressed in standard hours (for job shop) or production rate (for flow shop).

Maximum demonstrated capacity is the highest amount of actual output produced in the past, when all efforts have been made to optimize the resource.

Demonstrated capacity is a practical measure of capacity available in job shop manufacturing. The alternative of working with rated capacity (see below) is not as easy as it seems, because there are practical difficulties in measuring the utilization and efficiency factors.

Productive capacity is the maximum of the output capabilities of a resource (or series of resources) or the market demand for that output for a given time period.

Where the productive resource or system of linked resources is identified as the system constraint, its productive capacity is its maximum achievable output and should usually be based on 168 hours of available time per week (24*7; otherwise, TOC (theory of constraints) practitioners would say that this is not a true constraint. Where the system constraint is the market demand, productive capacity may be relative to a smaller number of hours per week.

Protective capacity is quantifiable capacity that is or can be made available at a nonbottleneck capacity to protect against fluctuation (idle time) of the bottleneck capacity. Technically, protective capacity provides contingency against unplanned events only, such as breakdowns and rework requirements.

Safety capacity is quantifiable capacity that is available over and above productive capacity that includes an allowance for planned events, such as on-shift plant maintenance and short-term *resource contention* (that is, simultaneous need from a common resource), and for unplanned events. It includes "protective capacity."

Excess capacity is defined as output capability at a non-constraint resource that exceeds the productive and protective capacity required.

Idle capacity is defined as capacity that is generally not used in a system of linked resources. It consists of protective capacity and excess capacity.

Activation is defined as the use of non-constraint resources to produce above the rate required by the system constraint, in this context a bottleneck capacity.

Budgeted capacity is the volume and mix of throughput on which financial budgets were set, for the purpose of establishing overhead absorption rates for calculating standard costs of products, expressed in standard hours. This really should be called *budgeted load*.

14.1.2 Overview of Capacity Management Techniques

Intended learning outcomes: Differentiate between infinite loading and finite loading. Identify the characteristics of algorithms for sufficient overall capacity planning flexibility and little overall capacity planning flexibility. Produce an overview on techniques for infinite loading and for finite loading.

If the (quantitatively) flexible capacity along the time axis is more important than the flexibility of the order due date (see Section 5.3.4), then *infinite loading* techniques should be used. In the reverse case, *finite loading* techniques are more appropriate.

If there is *sufficient overall capacity planning flexibility,* a computer algorithm can generally load all the orders in question with no regard to their sequence. The planner becomes involved only afterwards, to schedule capacities on a daily or weekly basis, for example. Exceptional situations will be brought to the planner's attention selectively in lists or graphs. If there is *little overall capacity planning flexibility,* planning takes place "order for order" (order-wise). Each new order is individually integrated into existing scheduled orders. The planning process is thus "interactive"; that is, in extreme cases the planner may intervene after each operation and change set planning values (completion date or capacity). Existing scheduled orders may have to be replanned.

Section 14.2 discusses (order-oriented) infinite loading. For special techniques see Sections 6.3 (Kanban) and 6.4 (cumulative production figures principle, CPFP). Section 14.3 discusses (operations-oriented, order-oriented, or constraint-oriented) finite loading. For special techniques see Section 15.1 (Loor, Corma). The techniques can all be used regardless of what organizational unit carries out planning & control. Thus, they can be found in all types of ERP and SCM software packages, electronic control boards *(Leitstand)*, simulation software etc.). Entirely different techniques are possible for short-term and long-term planning.

It is becoming increasingly important to plan machine tool capacities due to the increasing use of CNC and robot-controlled production. The methods are the same as those used to manage machine and labor capacities. On the other hand, tools to be produced or procured should be regarded as goods and represent a position on the order bill of material.

14.2 Infinite Loading

Intended learning outcomes: Present load profile calculation and problems associated with algorithms for load profile calculation. Explain methods of balancing capacity and load. Describe order-wise infinite loading.

The primary objective of *infinite loading* is to achieve a high delivery reliability rate, i.e., to meet the due date for production or procurement orders. Secondary objectives are low levels of goods in stock and work in process and short lead times in the goods flow. High capacity utilization is less important. Indeed, there can be good strategic reasons for maintaining overcapacity (meeting due dates).[2]

Overview: This section describes the generally applicable order-oriented method.[3] Load profiles are calculated for all the orders together after scheduling, and each scheduled operation represents a load at the specified work center and in the time period containing its start date. The sum of all these loads is compared to the available capacity for each time period. This yields load profiles showing the overcapacity or undercapacity for each work center and time period. Subsequent planning then attempts to balance capacity against load.

This most commonly used *order-oriented-infinite-loading* technique is also called *capacity requirements planning (CRP).*

Planning strategy: The objective is to manage fluctuating capacity requirements by having flexible capacities available. Both long-term and short-term actions are possible.

---

[2] An example is *surge capacity*, which is the capacity to meet sudden, unexpected increases in demand (cf. [APIC16]).

[3] Section 6.3 examined the popular *Kanban* technique, where due dates are fixed, that is, inflexible, and capacities are always modified to suit the load. Kanban can be used only for production or procurement with frequent order repetition.

14.2.1 Load Profile Calculation

Intended learning outcomes: Explain an example of a work-center-load profile. Present an example of a load profile known as an overload or underload curve along the time axis.

The *load profile* is a display of the work center load and capacity over a given span of time (Section 1.2.4). For *load profile calculation* we assume, as an approximation, that operations will be executed as scheduled (see Section 13.3). Thus, in the simplest case, the method places the operation load in the time period that contains the start date of the operation.

Figure 14.2.1.1 shows a load profile over two time periods for six production orders, P1, . . ., P6, each with operations at two different work centers, work center A and work center B.

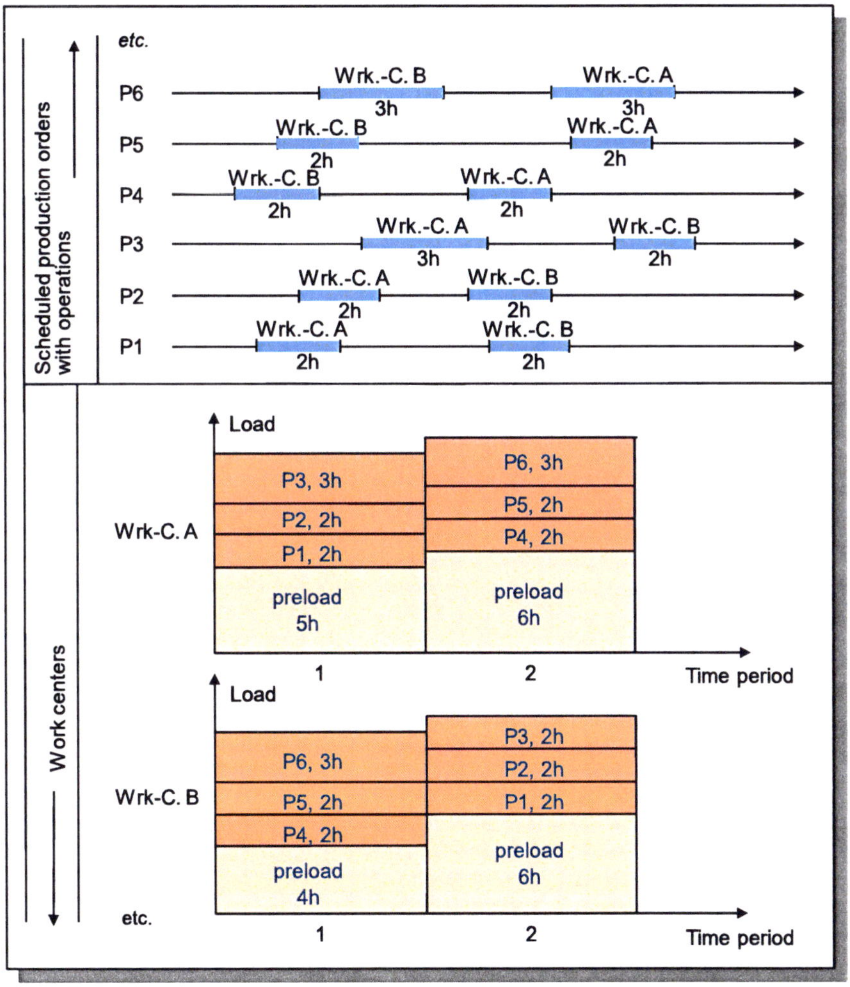

Fig. 14.2.1.1 Example of a work center load profile.

At the top of Figure 14.2.1.1, we can see the orders corresponding to the results of lead-time calculations. Each operation has a start date, which may be the earliest, latest, or probable date, depending on the scheduling technique used.[4]

The bottom of the figure shows the loads for these operations along the vertical axis. The "preload" represents operations for orders that were loaded *before* orders P1, . . . , P6. The method then adds together the operation loads in each time period on the planning horizon to create a load profile.[5]

Figure 14.2.1.2 provides an example of a load profile known as an *overload* or *underload* curve along the time axis.[6]

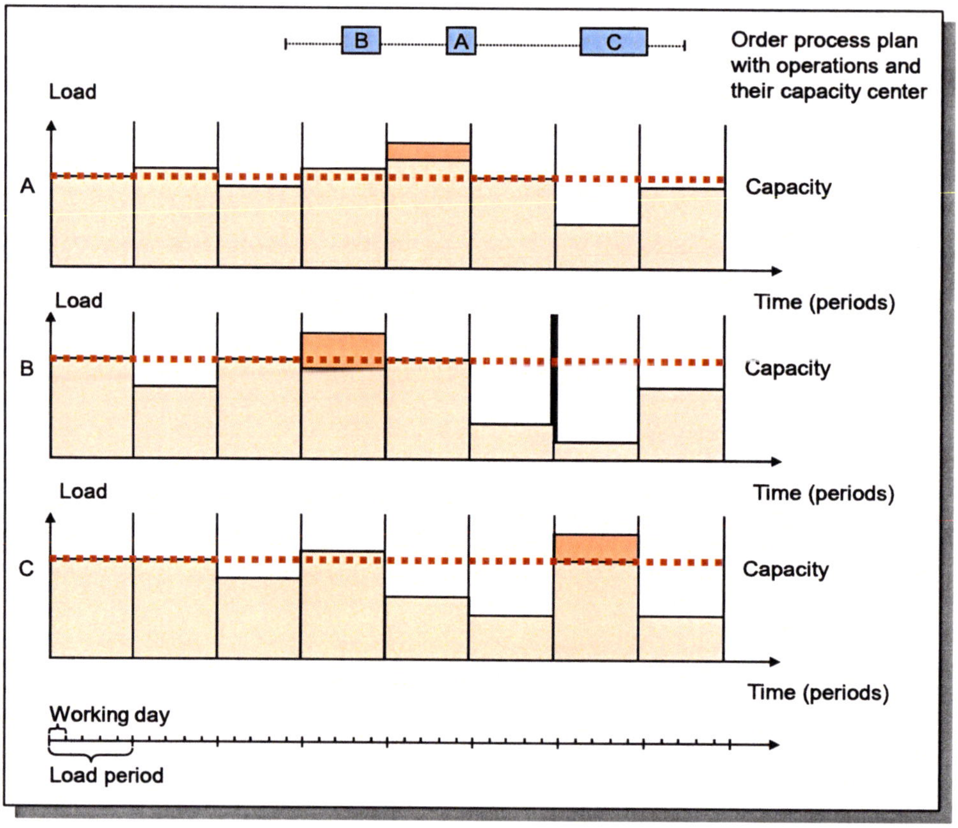

Fig. 14.2.1.2 An example of a load profile.

---

[4] We could also create load profiles for all the scheduling techniques discussed in Section 14.3, as long as the corresponding scheduling calculations have been executed.

[5] The time periods on the planning horizon are not necessarily of the same length. They may vary according to the type of work center. For example, some planners may use shorter time periods for the near future and longer periods for the more distant future.

[6] In this example, the capacities are of equal size for each period. The horizontal line is always obtained when capacity is viewed as equal to 100% in each period.

The use of different colors or hatching patterns make individual orders stand out in the profile. This can also highlight partial sums for particular order categories, such as:

- *Scheduled load*, caused by released orders (released orders with *provisional completion date* can be highlighted by an additional category)
- *Firm planned load*, caused by planned orders with fixed completion dates
- *Planned load*, caused by planned orders with provisional completion dates

The information may change according to the length of the chosen periods, that is,

- by selecting shorter load periods, the overload and underload curve is more precise
- longer time periods reveal a longer-term trend, with the short-term fluctuations evened out

14.2.2 Problems Associated with Algorithms for Load Profile Calculation

Intended learning outcomes: Describe the problem of calculating capacity per load period. Explain the problem of load assignment for one operation during the load periods. Present the issue of calculating the load in a given time period when various operations occur only partly within the time period.

A load profile calculation is not more than an approximation and must be interpreted as such. Thus, if interoperation times fluctuate widely (see Section 13.2), it will be difficult if not impossible to execute the operations as scheduled.

Further inaccuracies arise from the quality of the algorithms employed by ERP software or electronic control boards. Figure 14.2.2.1 shows a first problem associated with algorithms: assigning capacities to each time period on the planning horizon.

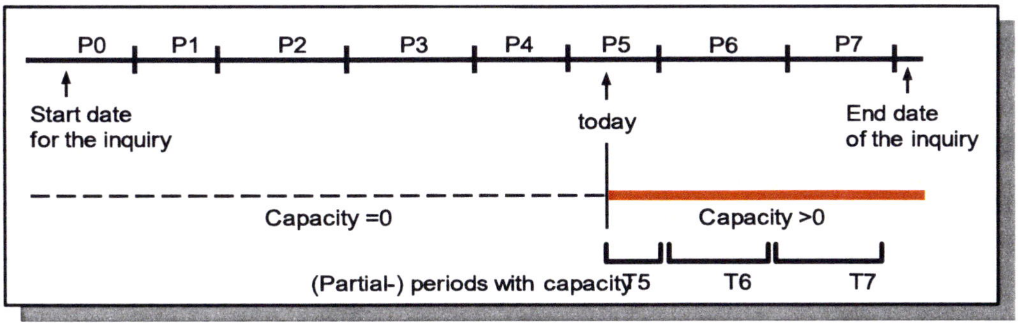

Fig. 14.2.2.1 Calculating capacity per load period.

The length of the load periods may vary. Also, we need flexible selection of the start date for the first and the end date for the last period under consideration. Then, loads that lie in the past cannot be compared against capacities, since the capacities are available only from "today" onward. "Today" may also fall within one of the time periods, in which case capacity is available only for the time remaining from "today" to the end of the time period.

Another problem is that a simple but imprecise method will assign the load to the time period containing the start date for the operation. An operation can extend across several load periods, however. Figure 14.2.2.2 shows this problem and offers an improved procedure.

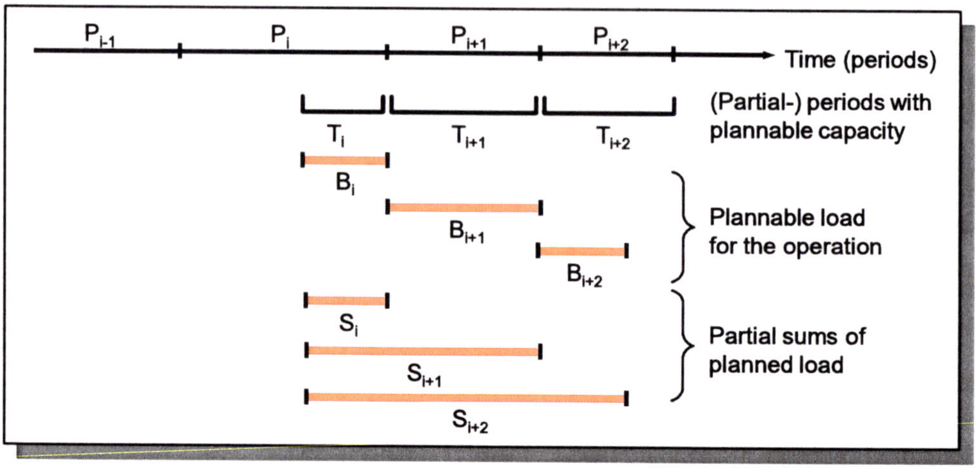

Fig. 14.2.2.2 Load assignment for one operation during the load periods.

The possible load per time period is obtained from the theoretical capacity (Fig. 14.1.1.1) and from the operation (splitting factor). The start date falls within a given period i. From this we can determine the time remaining until the end of period i and calculate the possible load B_i. The partial sum S_i is the sum of all the operation loads B_j, with $j \leq i$ of the operations that have already been assigned to the periods up to and including i. The load for the last period is the remaining load for the operation. At this point, S_i represents the entire load.

A third problem is how to determine all the operations that occur in a given time period [start, end]. Figure 14.2.2.3 shows that various operations occur only partly within the time period. In practice, we arrange the open (or open and planned) operations by start date. We consider any operation whose start date is earlier than the end date of the time period and whose completion date is later than the start date of the time period. The load of the operation occurring within the time period is assumed to be proportional to the lead time occurring within the time period.

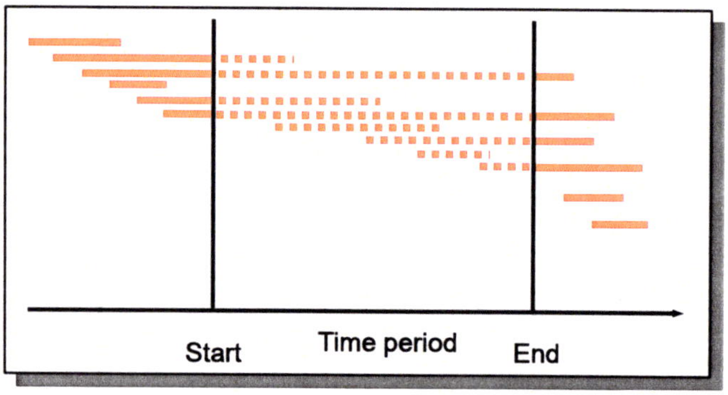

Fig. 14.2.2.3 Operations to be included in the load for a work center.

14.2.3 Methods of Balancing Capacity and Load

Intended learning outcomes: Describe the analysis of the load profile. Explain possible strategies for capacity planning, in case of a trend toward persistent overload or underload, or in case of frequent and infrequent self-compensating fluctuations. Differentiate between adapting capacity to load and load leveling. Present an evaluation of the technique. Identify its limitations and typical areas of application.

The cumulative illustration of loads and capacities along the time axis presented in Figure 14.2.3.1 is also suitable for analyzing the load profile. We can see the overload or underload along the vertical axis, between the curves for capacity and load. The maximum possible movement of the load in one or the other direction can also be seen along the horizontal axis.

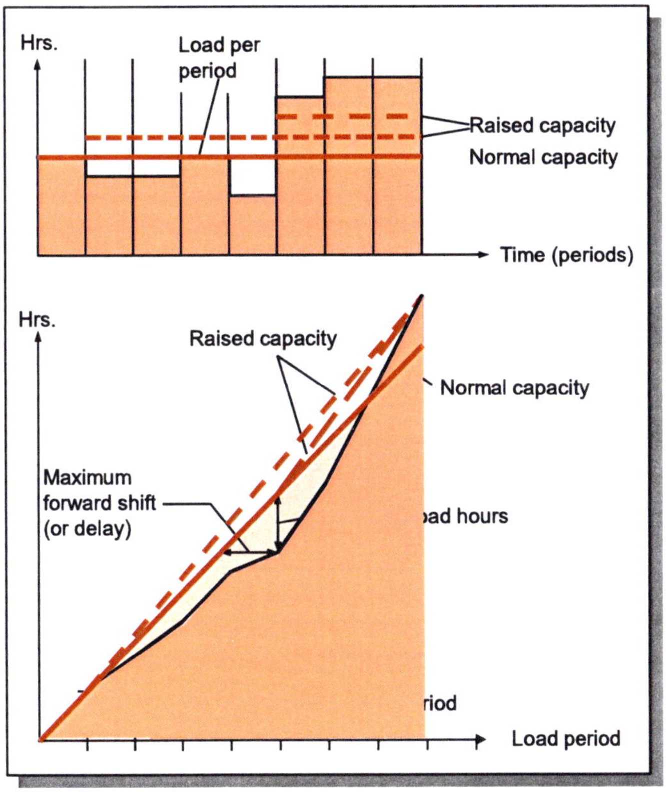

Fig. 14.2.3.1 Analysis of the load profile.

The load profile displays easily, directly, and accurately the overload and underload that would arise if our scheduling assumptions were totally accurate. Everything covered up to this point is not capacity planning, strictly speaking. However, the farther into the future we can identify the overload or underload, the less it need occur in reality, since the calculated operation start dates may be incorrect as a result of upstream bottleneck capacities, unplanned reworking, or unplanned operations due to rush orders, for example.

In the simplest case, one response would be to (manually) plan to increase or reduce capacity. Figure 14.2.3.2 shows possible methods.

1. Frequent self-compensating fluctuations, meaning that the interoperation times are longer than or roughly equal to the fluctuation frequency. No action is required. The time buffer can absorb these fluctuations without jeopardizing dates.

2. Trend toward *persistent overload*:
   2a. Long-term action (in the master planning): acquire additional production infrastructure (workers or machines) in good time. Other typical long-term responses are blanket orders to *subcontractors*, that is, sending production work outside the company ("extended workbench" or "outsourcing" principle) or arrangements with employment agencies (temporary workers).
   2b. Short-term action: arrange overtime or implement long-term blanket agreements as mentioned above.

3. Trend toward *persistent underload*: in principle, the action required is the opposite of that described under point 2.
   3a. Long-term action: cut back production infrastructure or reduce blanket agreements (insourcing).
   3b. Short-term action: cancel overtime, arrange short-time working, or cancel outsourced work.

4. Infrequent self-compensating fluctuations, meaning that the interoperation times are shorter than the fluctuation frequency:
   4a. Flexibly adapt capacity to load, alternating the steps described under points 2 and 3. For example, in the short-term case, this could mean arranging for and then cutting back on overtime.
   4b. *Load leveling*, that is, spreading orders over time so that the amount of work tends to be distributed evenly, resulting in a *level schedule*. This measure is associated with inflexible capacity, however, and thus actually belongs to finite loading. With a computerized system, we can move an operation forward or back and immediately see the consequences in a revised load profile. However, we must also take into account the work centers for the upstream and downstream operations for the order. Overload situations may now arise at other work centers precisely because the order was moved. Since the completion date is not flexible, this may require considerable manual replanning, order by order. See also Section 14.2.4.

Fig. 14.2.3.2 Infinite loading: methods for adapting capacity to load.

Finally, a list of available work supports analysis of the individual operations as well as *priority control*, that is, the communication of start and completion dates for execution in the shop floor.

Available work or *work on hand* or *load traceability* for a work center is a list of the operations to be carried out at that work center over a given time period.

This list is sorted according to a suitable strategy, which should also mirror the order in which the operations are carried out. Possible strategies include:

- Anticipated start date for the operation
- Operation time (SPT, shortest processing time rule)
- Order urgency (SLK, shortest slack time rule; see also Section 13.3.6)
- *Order priority*, that is, the preferred status of the customer

Evaluation of the technique: The following *prerequisites* must be satisfied before we can use planning methods for infinite loading:

- Capacities must be quantitatively flexible. Loads occur randomly according to the order situation. Replanning orders is time consuming and far too expensive given the often limited value-added.
- The technique only delivers good results if the collected shop floor data tracks work progress precisely. Also, no large load should lie in the past; otherwise, the backlog will be so great in the first period that the load profile no longer makes sense.

The following *limitations* also apply:

- The further we plan into the future, the lower the likelihood that the planning forecasts will be accurate; unforeseen breakdowns or variance in actual quantities will already affect accuracy. The technique merely predicts the probable capacity utilization so that sufficient capacity can be made available.
- The less that we know about the actual progress of the order, the more that actual control will have to be *ad hoc* on-site in response to the constantly changing dates and the mix of the orders.

The following are typical *areas of application*:

- For customer order production or where the mix of orders fluctuates, that is, in a *buyer's market*. Today, this is typically the case in the manufacture of capital goods or in discrete production and services in almost any industry.
- For all planning periods, particularly the long term. For execution and control of operations, this does not provide a precise program of work, but rather acts as a basis for situational planning of capacities and priorities at the work floor level.

14.2.4 Order-Wise Infinite Loading

Intended learning outcomes: Describe order-wise infinite loading. Identify the suitability of order-wise infinite loading.

Order-wise infinite loading loads the orders individually, order by order. The necessary planning measures are determined continuously, during, or after the loading of an order.

Order-wise infinite loading is necessary where there is little flexibility in terms of capacity and, at the same time, order due dates are inflexible. As Figure 14.2.1.2 showed, particular emphasis is placed on the new order. Planning takes place after loading of the entire order or after each operation. As soon as an overload occurs, it is important to check all the work centers concerned and take the steps outlined in Figure 14.2.3.2.

Order-wise infinite loading is extremely time consuming, especially if there are lots of operations or if starting with step 4b in Figure 14.2.3.2 (moving operations). It is possible that operations of other orders will also have to be moved. Capacity may even become saturated and thus inflexible, after a certain time. In this case, if the order due dates are inflexible, no further planning will be possible.

It thus follows that this type of planning is suitable only for firms with few and high value-added orders. One example is special machine construction in small or medium-sized firms.

14.3 Finite Loading

Intended learning outcomes: Explain operations-oriented, order-oriented, and constraint-oriented finite loading.

The primary objective of *finite loading* is high capacity utilization (see Section 1.3.1). The main target is *not* low levels of goods in stock and work-in-process, short lead times in the flow of goods, high fill rates, and delivery reliability rates. These are secondary objectives (see Section 1.3.1). With finite loading, the customer must be prepared to accept a longer delivery lead time and possible changes in the agreed-upon dates.

Essentially, there exist one *operations-oriented* and several *order-oriented* techniques. The operations-oriented technique is actually a simulation of the possible production processes that assumes — hypothetically — that all the planning data are correct. Some of the order-oriented techniques yield practically the same results as the operations-oriented technique. Others, however, tend to assume that capacities are not always fully utilized, which increases the delivery reliability rate and reduces levels of work in process.

14.3.1 Operations-Oriented Finite Loading

Intended learning outcomes: Produce an overview on the technique (the algorithm) for operations-oriented finite loading, also called operations sequencing. Explain an example of operations-oriented finite loading. Describe various priority rules. Present an evaluation of the technique. Identify its limitations and typical areas of application.

Operations-oriented finite loading aims to minimize possible delays to individual operations and thus the average potential delay of the entire production order.

Operations sequencing and operations-oriented finite loading are synonymous.

Overview: The individual operations are planned time period by time period on the basis of orders, starting from the start date determined by lead-time scheduling (Section 13.3.3).

Planning strategy: This means establishing meaningful rules of priority for the order in which operations are scheduled, with the aim of achieving maximum throughput. The queues waiting upstream of the work centers are monitored and adjusted.

Technique: The planning horizon is divided into time periods. The operations to be scheduled are then assigned to work centers, period by period, until the capacity limit is reached, regardless of the order to which they belong. Figure 14.3.1.1 demonstrates the principle of the resulting algorithm. This includes the following aspects:

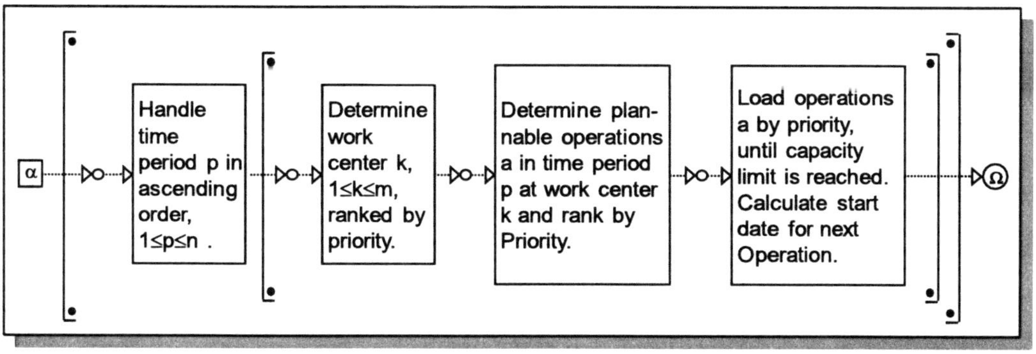

Fig. 14.3.1.1 Technique (algorithm) for operations-oriented finite loading.

- *Work center priority:* The order of the work centers becomes important as soon as there is more than one operation to be scheduled for an order in each time period. Possibly, the subsequent operation then relates to a work center whose planning has already been carried out for this period and now must be revised.

- *Determine the operations to be scheduled in the first time period;* typical operations are, firstly, every (subsequent) operation waiting for execution, for orders already started (the data on order progress identifies these operations), as well as, secondly, every first operation for orders not yet begun whose start date — calculated using a scheduling method (Section 13.3) — lies within the first time period.

- *Determine the operations to be scheduled in time period i, $2 \leq i \leq n$;* candidates are, firstly, all operations not scheduled in the previous time periods; then, secondly, those operations for which the previous operation was scheduled in an earlier time period and whose start dates lie within time period i, as well as, thirdly, every first operation for orders not yet begun whose start date — calculated using a scheduling method (Section 13.3) — lies within the time period i.

- *Arrange the plannable operations by priority.* The following secondary objectives may be applied to the selected order:

  A. Minimize the number of delayed orders
  B. Apply an equal delay to all orders
  C. Minimize the average wait time for operations
  D. Minimize the number of orders in process

- The following priority rules may be applied (see also [RuTa85]):

  1. The order in which the operations arrive (FIFO, "first in, first out")
  2. Shortest processing time rule (SPT)
  3. Proximity of the order due date (EDD, earliest due date)
  4. The ratio "remaining lead time for the order divided by the number of remaining operations"

5. The ratio "remaining lead time for the order divided by the time still available for the order" (SLK, shortest slack time rule, ≈ order urgency; see also Section 13.3.6)
6. The ratio "remaining lead time for the order divided by the remaining operation time for the order"
7. (External) order priority
8. Any combination of the above

Rules 1 and 2 are the easiest to apply in control of operations, because the information is immediately available: it is physically visible "locally." It is not necessary to consult a computer or a list. The other rules may require complicated calculations.

Every priority rule takes into account one or another secondary objective. Rule 1 is often used, since it minimizes the wait time upstream of the work center and thus the average order delay (objectives A and B). If capacity is utilized more fully, the strategy changes, and rule 2 is chosen. This accelerates the largest possible number of orders and thus reduces the value of goods in process (objectives C and D).

- *Load the operations in order until the capacity limit is reached:* If an operation exceeds the capacity limit, we transfer any as yet unscheduled operations to the next time period. The capacity used for the overlap load for the last operation is then no longer available in the next time period.

  One variation is not to schedule the operation that exceeds the capacity limit. However, this will use up remaining capacity only if an operation with a smaller load can be scheduled. This variation requires a more complicated algorithm.

- *Calculate the start date for the next operation:* After loading the operation, we calculate its completion date and the start date of the next operation on the basis of the interoperation time. To avoid problems with the algorithm (see "priority of the work centers" above), it may be useful to use the start of the next time period as the earliest start date.[7]

Figure 14.3.1.2 shows the result of operations-oriented finite loading using the orders in Figure 14.2.1.1, specifically P1, . . . , P6, and the same work centers, namely, work center A and work center B. Priorities were assigned in ascending order of order ID. Again, "preload" represents operations for orders that were loaded *before* orders P1, . . . , P6.

In contrast to the load profile in Figure 14.2.1.1, in finite loading we display the loads rotated 90° toward the time axis, whereby the height of the bar is equal for all work centers. The period length is then standardized at 100% capacity over the time period. This technique is possible because the load does not usually exceed capacity. We can then enter a number of work centers along the vertical axis. Utilization of the entire system is evident at a glance.

---

[7] If each operation is allowed a specific time period, such as a day or a week, we speak of *block scheduling*.

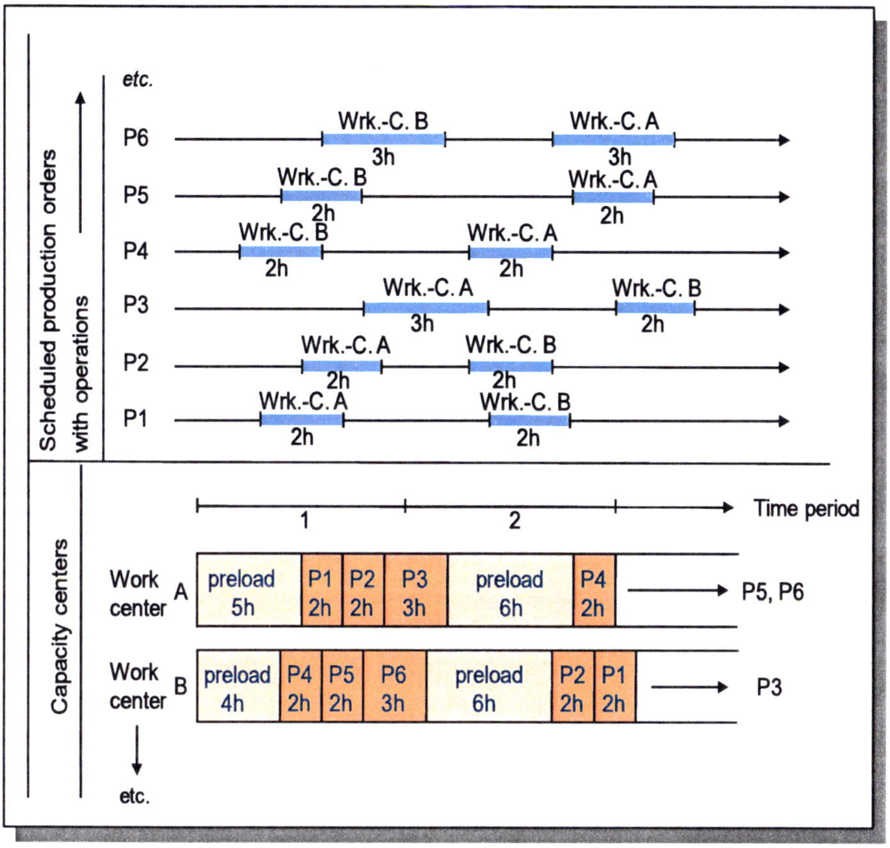

Fig. 14.3.1.2 Example of operations-oriented finite loading.

Evaluation of the technique: The following *prerequisites* must be met to use this technique:

- Capacities and loads must be sufficiently reliable, that is, the planning data and reported work progress must "tally." Otherwise, errors can accumulate very rapidly in the calculated dates.
- Due dates must be sufficiently flexible: We set the completion date for an order randomly on the basis of the existing utilization of production capacity. Lead times can be considerably longer than originally planned, however.
- It must be possible to limit the optimization of set-up times to the operations within a given period.

This creates the following *limitations:*

- The further we plan into the future, the smaller our chances that the planning forecasts will prove correct, if only due to unforeseen breakdowns or incorrect load specifications. For this reason, the technique is only sufficiently exact for short planning horizons, and it must be repeated at regular intervals.

- To be able to work to schedule in subsequent periods, any scheduled operations must be completed during this period. The technique does not allow reactive replanning locally.
- The level of goods in process is of secondary importance, both financially and with respect to volume. The planner monitors and adjusts the queues upstream of the work centers. Capacity is relatively inflexible, however, so orders must be held back, i.e., not released in good time. With long lead times in particular, however, order release can occur at the first identification of a bottleneck. This will physically hold up the production plant. Choosing a "neutral" priority rule will distribute the delay more or less evenly among all the orders.

The following are the typical *areas of application*:

- For batch production over a long period or in a *monopoly situation;* that is, in a *seller's market*. In such cases the date of delivery, for example, to the end products store or to the customer, is less important. Some typical industries that belong here today are the chemical and food processing industries and niche capital goods markets.
- The operations-oriented finite loading technique *simulates* a situation that may arise in job shop or even line production. The operations for an order are executed in a more or less random order, in competition with other such orders. For execution and control of operations, this type of planning provides a process simulation for the coming days and weeks; that is, an actual working program for the shop floor.

14.3.2 Order-Oriented Finite Loading

Intended learning outcomes: Produce an overview on the technique (the algorithm) for order-oriented finite loading. Describe various priority rules as well as dealing with exceptions. Explain an example of order-oriented finite loading. Present an evaluation of the technique. Identify its limitations and typical areas of application.

Depending on the technique that is used, *order-oriented finite loading* achieves maximum capacity utilization or ensures that as many orders as possible are executed on time with low levels of goods in process.

Overview: Orders are scheduled in their entirety, one after the other, in the time periods. If the period begins with an empty load, any orders that have already started are scheduled first, and only those operations that have not yet been carried out are considered.

Planning strategy: The objective is to find priority rules that will enable as many orders as possible to be completed. Special attention is given to those orders that cannot be scheduled, and whose start and completion dates must be modified as a result.

Technique: The planning horizon is once again divided into time periods. Individual orders (and all their operations) are scheduled in the order determined by the specified priority, without intervention by the planner. If the capacity limit for an operation is already exceeded, there are three possible responses: load the operation, defer it, or refuse the order. Once every order has been either planned or rejected, the planner handles the exceptions. The algorithm then attempts to plan rejected orders or those whose completion dates have been altered. Figure 14.3.2.1 illustrates the principle of the resulting algorithm.

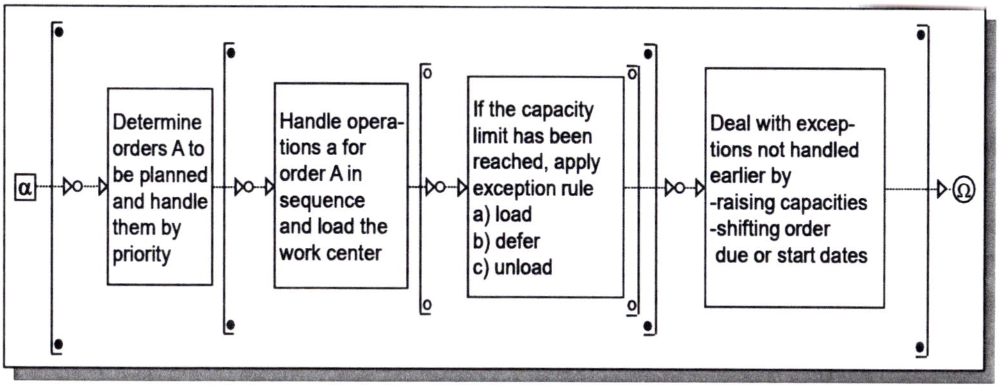

Fig. 14.3.2.1 Technique (algorithm) for order-oriented finite loading.

The details of the individual steps of the algorithm are as follows:

Determine the orders to be scheduled and treat them according to priority; typical orders are, firstly, all orders already begun (we know what operation is waiting to be carried out next from the order progress data;[8] all outstanding operations should be scheduled), and, secondly, all orders not yet begun whose start dates lie within an arbitrarily chosen time limit (this limit defines the anticipation horizon, which should ideally be smaller than or equal to the planning horizon; the start date should also be set or calculated using a scheduling method).

The possible *priority rules* are similar to those presented in Section 14.3.1, although here they apply to the entire order and not just to the individual operations:

- Proximity of the start date for the order (orders with fixed start dates can be loaded first)
- Proximity of the order due date (EDD, earliest due date)
- Ratio "remaining lead time for the order divided by the time still available for the order" (SLK, shortest slack time rule, ≈ order urgency; see Section 13.3.6)
- Ratio "remaining lead time for the order divided by the number of remaining operations"
- (External) *order priority*
- Any combination of the above

Handle and load operations in order: All operations are loaded at the corresponding work centers for the time period in question, working forward, beginning with the earliest start

[8] If the delivery date was promised on the basis of earlier planning and cannot be changed, no new scheduling can be performed. Probable scheduling (see Section 13.3.6) is one exception to this rule.

date, or backward, beginning with the latest completion date. Interoperation times are also considered, *but queue times are not.*

Deal with exceptions: If an operation falls within a time period during which the associated work center's capacity is already fully utilized, the following three possibilities can be applied:

a. Load without considering available capacity: This option is suitable for orders already begun or for relatively short operation times. Some general reserve capacity is thus kept free for the latter operations.
b. Defer the operation until the next period with available capacity (defer with forward scheduling, move forward with backward scheduling).
c. Unload the entire order, to give priority to other orders.

Deal with all exceptions that could not be handled earlier: If the steps described above have been carried out for all orders, the following contingencies requiring action may arise, depending upon which exception rule is applied:

a. For every capacity that is overloaded in a particular time period, either provide more capacity or unload orders accordingly.
b. (1) Backward scheduling: The resulting latest start date for an order lies before the earliest start date. Unload this order and then try again using forward scheduling, beginning with the earliest start date. (2) Forward or probable scheduling: The resulting earliest completion date for an order lies after its latest completion date. If the order due date is flexible, defer the order accordingly. Otherwise, it may be necessary to deliberately increase the fully utilized capacity to first unload the order.
c. For every unloaded order: It may be possible to bring forward the start date. If the order due date is flexible, defer the order. If the fully utilized capacities are at least a bit (quantitatively) flexible, they may be increased accordingly.

The unloaded orders are then scheduled in another iteration of these steps of the algorithm. This technique could quite conceivably be applied interactively, that is, "order by order": If an operation falls within a time period in which the capacity limit is already exceeded, the planner can immediately decide on the appropriate action.

Figure 14.3.2.2 shows the results of order-oriented finite loading after the first iteration, using exception rule (c). This example uses the same orders as in Figures 14.2.1.1 and 14.3.1.2, specifically P1, . . . , P6, and the same work centers, namely, work center A and work center B. Priority was assigned in ascending order by order ID. Again, "preload" represents operations for orders that were loaded *before* orders P1, . . . , P6.

Exception rule (b) would have produced results similar to those in Figure 14.3.1.2, that is, similar to operations-oriented finite loading. The more that exception rule (a) is applied or capacities are increased in the last step, the more infinite loading is obtained.

The following *prerequisites* must be met to use this planning technique:

- Capacities and loads must be sufficiently reliable, that is, the planning data and reported work progress must "tally." Errors can accumulate very rapidly in the calculated dates if this is not the case.

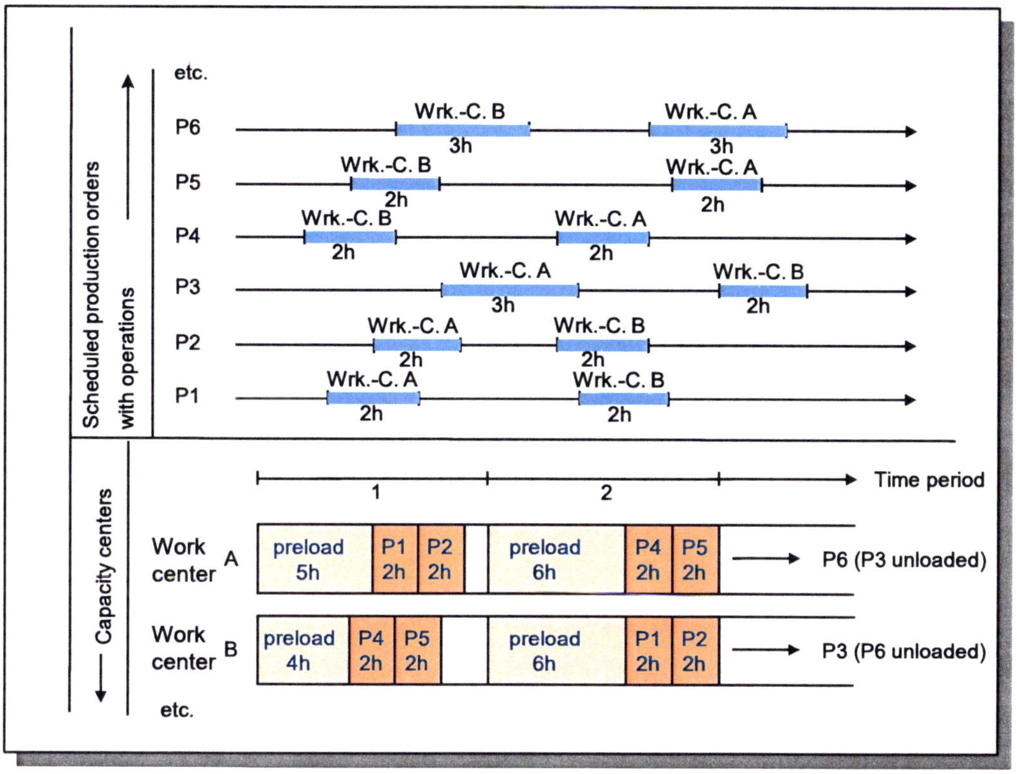

Fig. 14.3.2.2 Example of order-oriented finite loading, exception rule (c): unloading.

- Due dates must be sufficiently flexible — especially for exception rule (b). The order completion date results randomly on the basis of the existing utilization of production capacity. Lead times can sometimes be much longer than normal.
- Exception rules (a) and (c) are suitable for order due dates that are relatively inflexible. For these, however, the capacities must have some flexibility; otherwise, the administrative effort needed to regularly change dates would become unmanageable or so imprecise that capacities would be poorly utilized.

This creates the following *limitations:*

- The farther we plan into the future, the smaller our chances that the planning forecasts will prove correct. For this reason, the technique is only sufficiently exact for short planning horizons, and it must be repeated at regular intervals.
- In long-term planning, the technique calculates a *permissible* plan, in the full knowledge that it will change in the short term. Regular and efficient replanning is thus needed as the term becomes shorter.
- In short-term planning, for exception rule (b), any scheduled operations must once again be completed during this period. The technique does not allow local, reactive replanning. Exception rules (a) and (c) do, however, allow some potential degrees of freedom for reaction if capacity is not fully utilized.

- Exception rule (b) leads to the best possible utilization of capacity. As with operations-oriented finite loading, long queues may arise. Goods in process then tie up capital and even hold up the entire production plant. Choosing a “neutral” priority rule will distribute the delay more or less evenly among all the orders.
- Exception rule (c) loads production only with the orders that it is capable of processing. It thus results in lower levels of work-in-process and shorter lead times. Successfully planned orders are completed on time. Exception rule (c) essentially uses the model of the queue presented in Section 13.2.1, that is, the reservoir or open funnel model. If the funnel does not overflow, the production plant will not be held up. Thus, if further processing of an order is delayed excessively (e.g., over at least one time period), it should be rejected, rather than loaded.
- With inflexible capacities, on the other hand, exception rule (c) leads to lower utilization of capacity as soon as completion dates have to be deferred. This is because the load that would have been caused by operations earlier along the time axis is now missing. If there are no other orders, the capacity is wasted. Deferred orders will have long delays, and it may even become impossible to accept new orders.
- If the time between the earliest start date and the latest completion date is longer than the required lead time, then a start date and an end date that falls between these two extremes may be more suitable for the overall mix of orders. It is worth considering the load-oriented order release and capacity-oriented materials management (Corma) techniques outlined in Sections 15.1.2 and 15.1.3. Load-oriented order release, in particular, can actually be regarded as a generalization of order-oriented finite loading with exception rule (c).
- Interactive planning, that is, order by order, is only efficient if relatively little effort is needed to load an order compared to its value added. In addition, we need continuous knowledge of the total load on the work center resulting from previous orders, so that a very fast database is required. We also have to keep load *totals* for each time period. To create sufficiently simple and rapid algorithms, the length of the time periods for each work center and along the time axis must then be defined as fixed.

Typical *areas of application* are as follows:

- As with operations-oriented finite loading, exception rule (b) is suitable for *batch production* over a long period or in a *monopoly situation* or *seller's market*. Typical industries here are chemical and food processing industries and niche capital goods markets.
- Exception rules (a) and (c) are suitable for many discrete manufacturing industries, wherever there is the minimum required level of (quantitatively) flexible capacity. This is more often the case than we might at first suppose, even in short-term planning.
- For short-term planning and control. For this planning range, the technique provides, firstly, with exception rule (b), an actual work program for the next few days, and, secondly, with exception rules (a) and (c), an acceptable work program that also

allows a degree of situational planning. The horizontal bar chart provides a rapid overview of all work centers and all orders, as it requires little space. It corresponds to the familiar planning board in production control. Individual orders can often be replanned very efficiently — in the case of the electronic control board (*Leitstand*), through the click of the mouse.

- For long-term planning of few orders with high value-added and regular planning and replanning. For replanning individual orders, the advantages are again the clear display and ease of manipulation mentioned above.

14.3.3 Constraint-Oriented Finite Loading

Intended learning outcomes: Identify bottleneck capacities and the drum-buffer-rope technique. Describe the drum, the buffer, and the rope. Present an evaluation of the technique. Identify its limitations and typical areas of application.

In *constraint-oriented finite loading*, orders are planned around *bottlenecks*, or *bottleneck capacities*, which are work centers with a capacity utilization of 100% or more.

Bottlenecks depend on the given order volume and not upon the master data for the work center.

Drum-buffer-rope is a technique of production control that accords with the theory of constraints (TOC). See also Section 5.1.5 and [GoCo14].

The drum-buffer-rope technique includes the components shown in Figure 14.3.3.1.

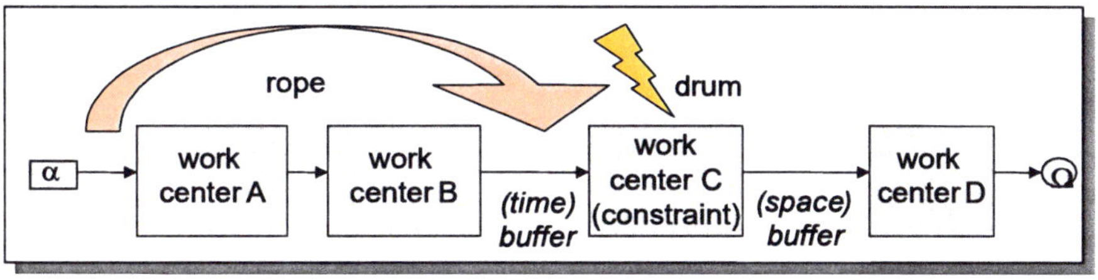

Fig. 14.3.3.1 The drum-buffer-rope technique.

The *drum* stands for the rate or pace of the system. The "drumbeat" results from the *drum schedule*, that is, the master production schedule for the system, set by the throughput of the constraint, which should be balanced with the customer demand. The constraint controls the throughput of all products that it processes. *Feeder workstations*, that is, work centers feeding bottlenecks, should be scheduled at a rate that the bottleneck can process.

A *buffer* in front of the constraint absorbs potential disturbances during a certain period of time. *Buffer management* expedites material in *time buffers* (that is, protection against uncertainty that takes the form of time) in front of constraints and helps to avoid idleness at the constraint. To avoid idle time due to disturbances of the succeeding operations, buffer management can also include the maintenance of a *space buffer* after the constraint. Compare here Figure 13.2.1.1.

The *rope* is an analogy for the communication process: the set of planning, release, and control instructions for bringing the necessary material for production to the constraint in due time. This can be achieved using any technique: pull (e.g., Kanban or reorder point type, or *generic Kanban*, i.e., a card that only serves to release the next order, for whatever item ID) or push (MRP type; for example, by releasing material at the right time into the system), or any other appropriate intuitive or heuristic technique for the specific case.

Evaluation of the drum-buffer-rope technique: The following *prerequisites* must be satisfied for use of this planning technique:

- Our picture of the capacities and loads must be accurate, that is, the planning data and reported work progress must "tally."
- Bottleneck capacities must be known and remain stable. Any change requires replanning of the elements drum, buffer, and rope.
- The order due dates must have a degree of flexibility, since the completion date for an order is determined by the way in which the orders come together at bottleneck capacities and by the subsequent forward scheduling.
- In fact, most capacities must be somewhat (quantitatively) flexible, or they would all become bottleneck capacities.

The following *limitations* apply:

- There must not be too many bottleneck capacities. In particular, the technique is unsuitable for situations where for a single order there are multiple bottleneck capacities, which may not follow in succession or may even be located at other production stages. Otherwise, it would become difficult or even impossible to determine the "rope" part of the technique in detail. This means that the techniques are applicable mainly for simple — for example, one-level — product structures.

Typical *areas of application* are the following:

- The technique is suited to mature line production running at a fixed rate, for example, simple chemical products, food processing, or production of simple parts.
- The technique is especially suited for *machine-limited capacity*, or a production environment where a specific machine limits throughput of the process (cf. [APIC16]).

14.4 Rough-Cut Capacity Planning

Intended learning outcomes: Describe rough-cut network plans and load profiles. Explain rough-cut infinite loading and rough-cut finite loading.

Rough-cut planning allows quick establishment of feasible variations of the master plan for many orders in *long-term planning* and quick determination of delivery dates for customer orders in *short-term planning*.

Efficient scheduling in the short term requires long-term overall coordination of load and capacity. If all the rough-cut structures are correct and sufficiently detailed and include all the goods to be procured through blanket contracts, then rough-cut planning of resources is entirely sufficient for long-term planning. At times it can even make shorter-term planning unnecessary or more straightforward.

Very simple rough-cut planning is possible wherever the total load for an order is sufficient for rough-cut planning.

With *capacity planning using overall factors (CPOF)*, the quantities of master schedule items are multiplied by the total load required by each item. This yields the total load of the master schedule. Historical percentages for each work center then provide an estimate of the required capacity of each work center to support the master schedule.

Figure 14.4.0.1 shows the (average) load of the master schedule with three items, I1, I2, and I3. Suppose that two work centers WC-1 and WC-2 are involved. Historical percentages allow quick assignment of the total load to the two work centers.

Assembly \ Week	1	2	3	4	Load per unit	Histori-cal %
I_1	60	60			0.75	
I_2			60	12	0.60	
I_3				48	0.50	
Total load (in h)	45	45	36	31.2		100
Required capacity on WC-1	29.25	29.25	23.4	20.28		65
Required capacity on WC-2	15.75	29.25	12.6	10.92		35

Fig. 14.4.0.1 Rough-cut capacity planning using overall factors: total load and estimation of the required capacity on work centers WC-1 and WC-2.

However, if knowledge of the load on each individual work center is necessary, rough-cut capacity planning gets more complicated. We look at this case in the following section.

14.4.1 Rough-Cut Network Plans and Load Profiles

Intended learning outcomes: Present an example of a rough-cut network plan with two rough-cut work centers. Explain the resource profile of both rough-cut work centers as well as the resource profile for the combination of the two rough-cut work centers.

The *rough-cut process plan* for a product is the rough-cut production structure along the time axis.

Section 1.2.5 introduced rough-cut bills of material and rough-cut routing sheets. These are either derived from the detailed structures of a product or determined and maintained "manually." These rough-cut structures allow us to derive a rough-cut process plan with lead-time setoff for components or operations. As Section 13.3.3 also shows, a rough-cut process plan can easily form a directed network of operations.

Figure 14.4.1.1 shows a production order in a form similar to the familiar network plan. Rough-cut order structures are often represented in this way. In our example, we have combined the work centers into two rough-cut work centers.

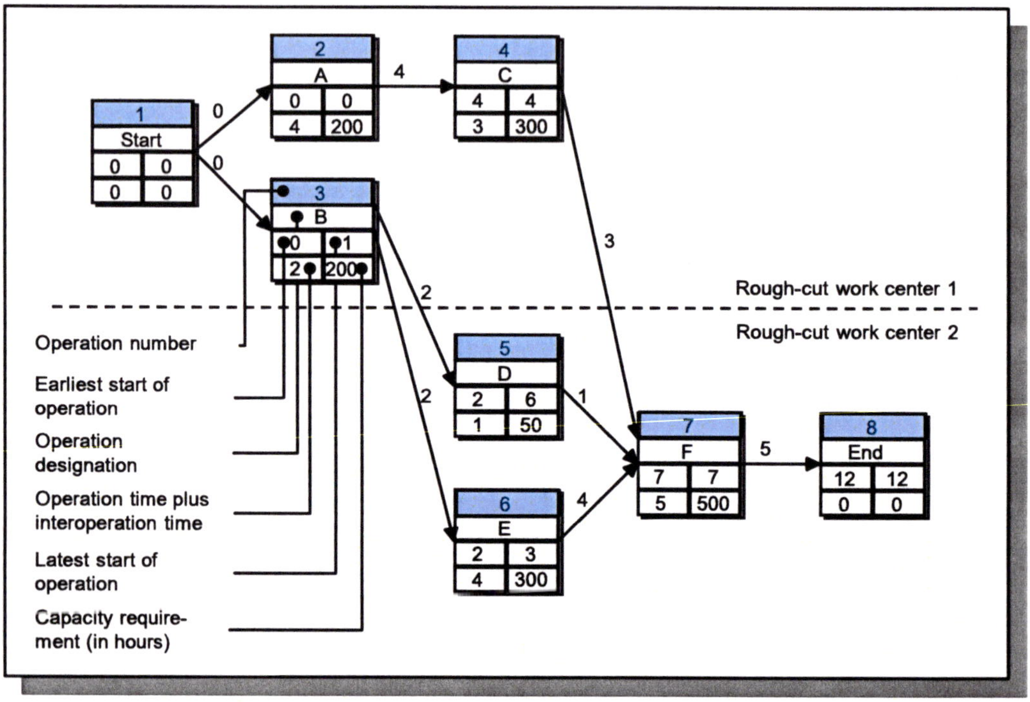

Fig. 14.4.1.1 Rough-cut network plan with two rough-cut work centers.

> A *resource profile* is essentially a load profile, that is, standard hours of load placed on a resource by time period, for rough-cut capacity planning.

Figures 14.4.1.2 and 14.4.1.3 show the resource profile derived from the rough-cut process plan or from the *rough-cut network plan*. Figure 14.4.1.4 shows how they are combined to form a single rough-cut work center.

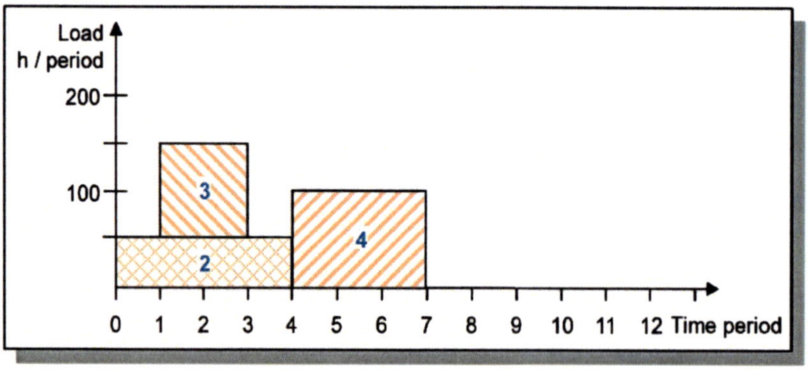

Fig. 14.4.1.2 Resource profile of rough-cut work center 1 as shown in Fig. 14.4.1.1.

For the sake of simplicity, in rough-cut planning we can regard the load as a rectangular distribution over the duration of the process. Indeed, this interpretation is also common in detailed planning.

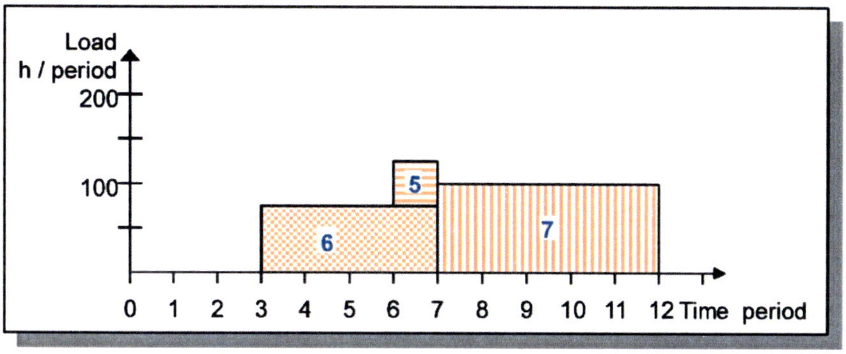

Fig. 14.4.1.3 Resource profile of rough-cut work center 2 as shown in Fig. 14.4.1.1.

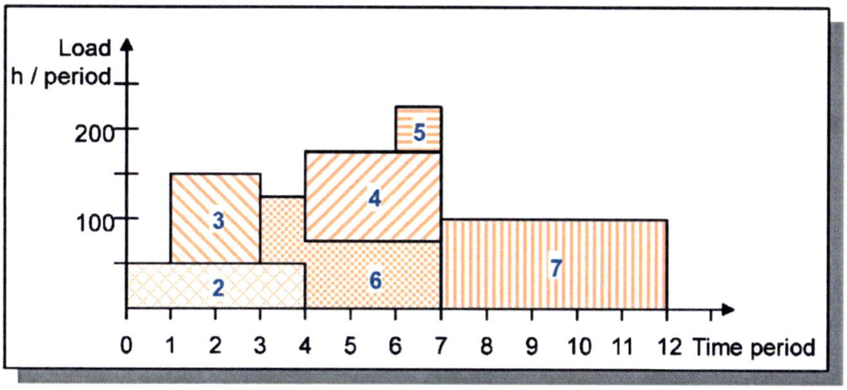

Fig. 14.4.1.4 Resource profile for the combination of rough-cut work centers.

If we chose the technique shown in Section 1.2.5, using lead-time setoff as the data structure behind the resource profile, we lose the typical information concerning operations before and after each operation in the network. Keeping the information in the data model is conceivable, however, and it would make the load and adjustment algorithms more flexible. However, the algorithms would also be more difficult to implement, which could result in longer response times.

Rough-cut planning is extremely interactive, that is, it requires the planner to intervene and make decisions. It is not surprising, therefore, that rough-cut planning often works using the simplest data models, that is, ignoring the interdependencies among operations.

The problem of taking account of demand derived from bids, described in Section 5.2.1, also arises in rough-cut capacity planning. Regardless of whether we are planning for limited or infinite loading, the procedure to deal with this problem entails the following steps:

- The simplest method multiplies the product load profile by the *probability of order success* ("devalues" it) and thus loads only the resulting reduced load. Validation of the order success probability is a key factor here.
- Bids must be confirmed at an early stage or must be unloaded in order to make room for orders requiring definitive planning. The bid should therefore be assigned an expiration date. From that date on, we designate the bid as inactive or defer the promised delivery date by a sufficient number of periods.
- If a very large number of bids have already been planned, it will be difficult to assign a reliable delivery date to new bids. The completion date determined in planning is only a possible completion date. Additional information is required, such as a "maximum" completion date, for example, which is calculated assuming that all bids (or a significant proportion thereof) will be accepted. We do this by adding together the unloaded portions of the bids after calculating the probability, and dividing this figure by the capacity available per period. This gives the number of periods that must be added to the probable date in order to arrive at the "maximum" date.

14.4.2 Rough-Cut Infinite Loading

Intended learning outcomes: Present an example of a resource profile for one rough-cut work center. Describe the result of rough-cut capacity planning with deferred completion date. Explain rough-cut capacity planning with two rough-cut work centers and the result after moving the completion date of one operation.

Rough-cut infinite loading corresponds to infinite loading, in this case based on the resource profile for rough-cut work centers presented in Section 14.4.1.

Here, we multiply the product load profiles related to a particular batch size (generally = 1) by the batch size and add a desired completion date. The orders thus defined are then considered in a particular planning order. The priority can be determined by

- The latest completion date
- The latest start date
- The external priority (importance) of the order

When all orders have been loaded in this way (with no intervention from the planner) onto an existing "preloading," we obtain a resource profile that is typical of infinite loading. Figure 14.4.2.1 shows, as an example, the resource profile presented in Section 14.4.1. with loading from the earliest start date.

The planner then either defers the capacities or brings them into line with one another:

- If rough-cut planning is performed in the master planning, that is, in the long term, capacities are flexible with respect to quantity. Indeed, determining such quantities is one of the objectives of long-term planning.
- If rough-cut planning is applied to the medium or short term, it provides decision support to accept, reject, or defer a waiting order. Capacity is then somewhat less

flexible, which means that it will not always be possible to meet the desired completion date. In this case, the loading proceeds individually, order by order, with possible intervention by the planner after every order. If there are just one or two rough-cut work centers, this does not entail a lot of work, even if there are a lot of orders.

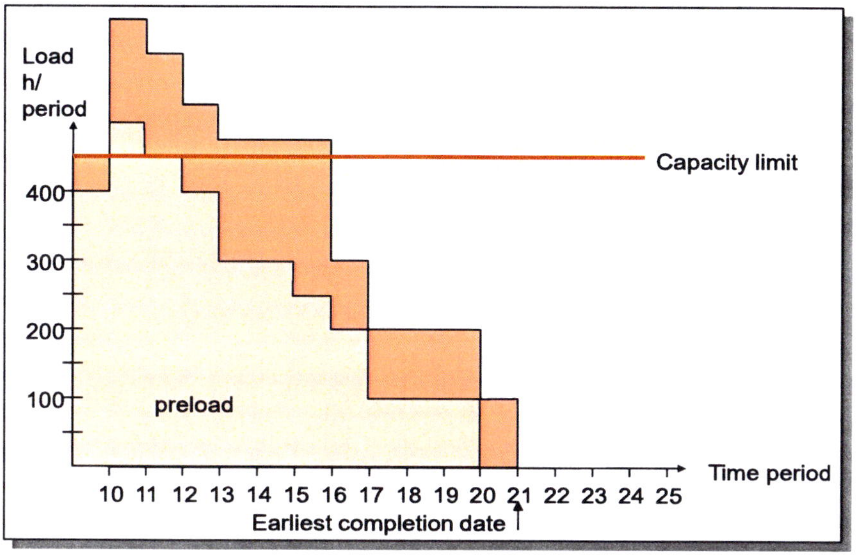

Fig. 14.4.2.1 Example of a resource profile for *one* rough-cut work center.

Figure 14.4.2.2 shows a possible response the planner can take to the overload shown in Figure 14.4.2.1. The planner can push the completion date back to the closest date that would produce an acceptable overload. Efficient algorithms highlight the order to be deferred by means of special graphical attributes (such as different colors) and automatically adjust the resource profile after rescheduling (generally using the mouse).

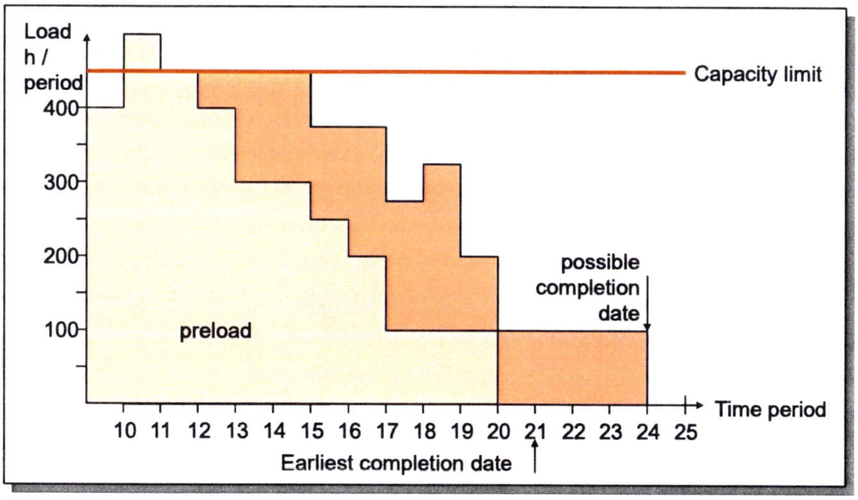

Fig. 14.4.2.2 Result of rough-cut capacity planning with deferred completion date.

Figure 14.4.2.3 shows the same resource profile, this time with two rough-cut work centers. The desired completion date causes an overload at rough-cut work center 2. In Figure 14.4.2.4, it has been brought into line by deferring the completion date by two periods.

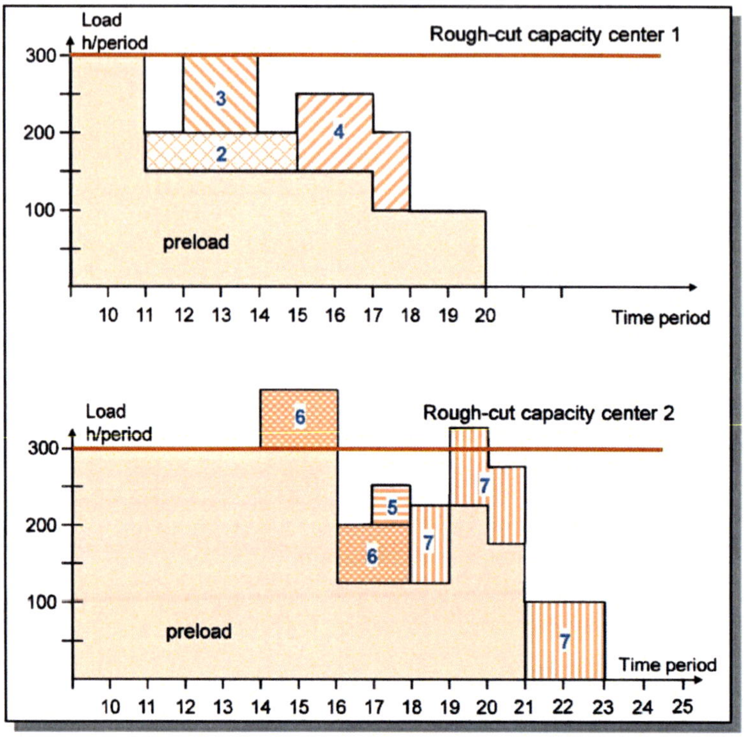

Fig. 14.4.2.3 Rough-cut capacity planning with two rough-cut work centers.

Detailed planning may nevertheless still be needed if we use short-term rough-cut planning. The individual rough-cut operations for the order under consideration can then be shown separately. It is only by recording interdependencies between operations in the network plan that we are able to move the fifth operation forward from the period [19–20] (overload) to the period [18–19] (no overload), as shown in Figure 14.4.2.4. To be able to work efficiently and interactively, it must be possible to view all the work centers at the same time. The entire resource profile for an order can then be moved at all the work centers simultaneously.

14.4.3 Rough-Cut Finite Loading

Intended learning outcomes: Present the cumulative resource profile of an order as well as cumulative load and capacity before loading the order. Explain rough-cut planning of the cumulative load and capacity after loading the order.

Rough-cut finite loading corresponds to finite loading, but it is based on the resource profiles for rough-cut work centers presented in Section 14.4.1.

If order due dates are flexible or if variations in capacity are undesirable or unfeasible, we can also use planning with rough-cut finite loading. In this case, order-oriented techniques are relatively simple, since rough-cut planning places the emphasis on approximate, rather than exact maintenance of capacity. The sum of any over- and underloads should then be smoothed out over a sufficiently short time horizon.

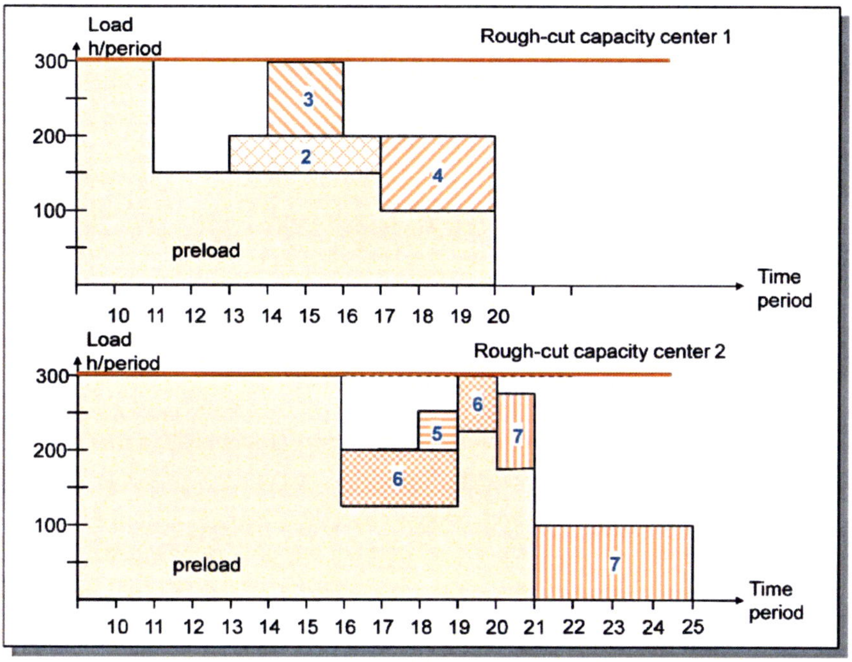

Fig. 14.4.2.4 Rough-cut planning: result after moving the completion date and operation 5.

It is always the cumulative capacity and cumulative "preload" that are considered. We also create a cumulative resource profile for the new order to be loaded. As an example, Figure 14.4.3.1 shows such a profile for one rough-cut work center. The important variable here is the cumulative load at the end of the profile.

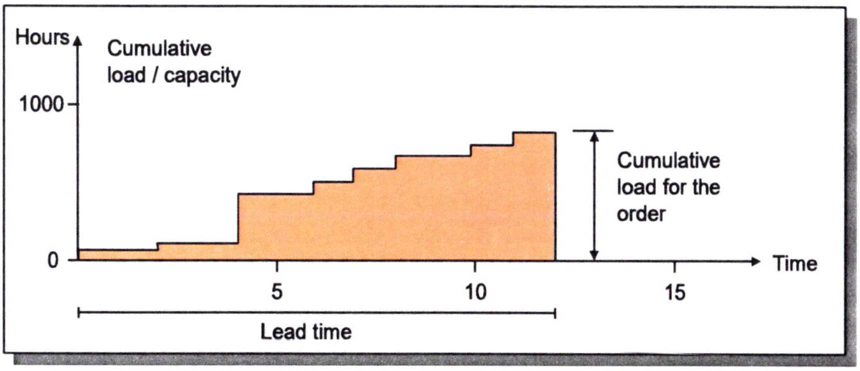

Fig. 14.4.3.1 Cumulative resource profile.

Figure 14.4.3.2 compares cumulative capacity and cumulative "preload" and yields the following result for the new order.

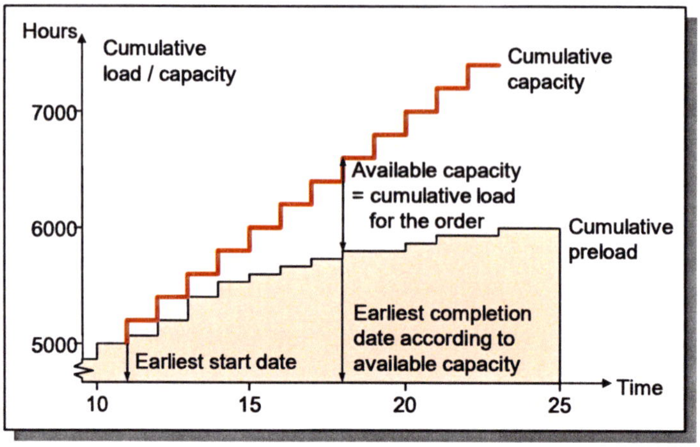

Fig. 14.4.3.2 Rough-cut planning: cumulative load and capacity before loading the order.

- The *earliest start date* is the period with the first available capacities that will not be used in subsequent periods.
- The *earliest completion date according to available capacity* is the end of the period in which, for the first time, the available capacity *permanently* exceeds the cumulative "preload" plus the cumulative load for the order, that is, in which the available capacity is not less than this total.

Figure 14.4.3.3 also shows the newly loaded order.

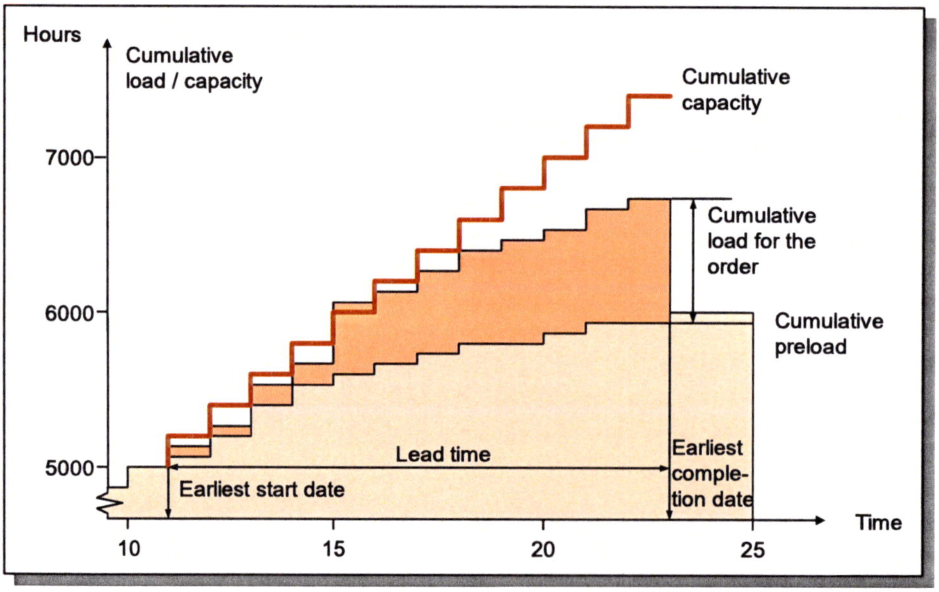

Fig. 14.4.3.3 Rough-cut planning: cumulative load and capacity after loading the order.

- The *earliest completion date* is the maximum total of the earliest completion date according to available capacity and the completion date obtained by adding the lead time to the earliest start date.

Capacities will be locally underused or exceeded. If the overload or underload frequency is relatively intense, that is, continues over just a few periods, it can be compensated by means of control of operations. This is possible because we are dealing here only with rough-cut structures. The same applies to long-term infinite loading as described in Section 14.4.2, in this case simply because the loading is long term.

14.5 Summary

Capacities are workers or machines that can carry out work in order to produce goods or services. To calculate the rated capacity in order to add value, we have to know the utilization and the work center efficiency (or efficiency rate). Planning in two dimensions (time and quantity), as described in Section 5.3.3, is a fundamental problem when planning capacity requirements. Depending on the situation, we must choose one of these dimensions to set the direction, which leads us to various classes of techniques.

Infinite loading is primarily a load profile calculation. The start date for an operation, which results from scheduling an order, determines the timing of an individual load. All the loads for each work center and time period are then added together to obtain the capacity requirement, which, in turn, is compared against rated capacity to obtain an overview. A closer look reveals a number of problems with the algorithms that have to be overcome. The load profile is then used for planning capacities, and the emphasis is placed on measures to alter capacity that are appropriate for the particular planning term. If there are only few orders to be planned, an additional measure — move the operations — is conceivable, although this can be difficult.

The chapter presented three finite loading techniques. The operations-oriented technique plans as many operations as possible for each work center from the perspective of the time axis. Priority rules are applied in order to decide among all the plannable operations in each time period. The result is high-capacity utilization, but some orders will be kept waiting. On the other hand, use of the FIFO priority rule will distribute the delay equally among all the orders.

The order-oriented technique plans entire orders according to a particular priority and all the operations for each order. If there is no more capacity available for an operation, we can defer the remaining operations. The consequences for the performance indicators are similar to those of the operations-oriented technique. Another response is to unload such an order in its entirety. In this case, the remaining orders will be executed on time (according to schedule), with a lower level of goods in process and less favorable capacity utilization than with the operations-oriented technique. However, we still need to find a later completion date for the unloaded orders, which may lead to long delays and possibly even to the loss of these orders. If there are only a few orders to be planned, we can also attempt to move individual operations forward or to defer other orders, although this can be a very time-consuming "manual" task.

A constraint-oriented technique is called drum-buffer-rope. The drum stands for the rate or pace of the bottleneck. This constraint controls the throughput of all products that it processes. A time buffer in front of the constraint absorbs potential disturbances during a certain period of time and helps to avoid idleness at the constraint. A space buffer after the constraint helps to avoid idle time due to disturbances of the succeeding operations. The rope is an analogy for the communication process: the set of planning, release, and control instructions for bringing the necessary material for production to the constraint in due time.

For rough-cut capacity planning, we first create a rough-cut network plan for each product family and derive the resource profile for each rough-cut work center needed to manufacture the product family. Infinite loading is first applied, as in the detailed technique. Order-by-order planning then enables us to defer the entire profile, so we can decide whether to accept the order in the short term, for example. With finite loading, we first determine the earliest completion date according to available capacity and then add the lead time to the first date for which any capacity is available. The later of the two dates thus calculated is the earliest completion date for the order.

14.6 Keywords

activation, 599
actual capacity utilization, 597
availability (in capacity), 597
available work, 606
buffer management, 617
capacity determination, 596
capacity requirements planning (CRP), 600
constraint-oriented finite loading, 617
drum-buffer-rope, 617
finite loading, 608
firm planned load, 603
idle capacity, 599
infinite loading, 600
labor capacity, 596
load leveling, 606
load profile calculation, 601
load traceability, 606
machine capacity, 596
operations sequencing (syn. operations-oriented finite loading), 608
order-oriented finite loading, 612
order-wise infinite loading, 607
overall equipment effectiveness (OEE), 598
priority rule, 609
rough-cut finite loading, 624
rough-cut infinite loading, 622
space buffer, 617
subcontracting, 606
time buffer, 617
work on hand, 606

14.7 Scenarios and Exercises

Intended learning outcomes: Determine capacity. Execute algorithms for load profile calculation. Elaborate an example for rough-cut capacity planning.

14.7.1 Capacity Determination

Intended learning outcomes: Determine theoretical capacity, availability, rated capacity, demonstrated capacity, actual utilization, work center efficiency, etc.

The following exercise was developed on the basis of a communication from Barry Firth, to whom we extend many thanks.

A plant runs 10 * 8 hour shifts per normal week. A work center in the plant has 5 identical machines, each requiring one operator to run it. This is a machine-paced work center (a machine capacity). Operators get a total of 1 hour for breaks, and they usually take their breaks at the same time. Each machine requires one episode of planned maintenance per week of 3 hours, scheduled by the planner. During the last 6 weeks, the performance data in Figure 14.7.1.1 were recorded:

Feature \ Week No.	1	2	3	4	5	6
Number of working days	5	4	5	5	5	5
Actual machine hours (setup+run)	260	200	280	320	260	280
Maintenance time in machine hours	15	12	18	15	15	15
Standard machine hours produced	220	160	240	280	220	220

Fig. 14.7.1.1 Capacity performance data.

Questions:

a. What is the *theoretical capacity* in machine hours *per normal week* (5 days)?

b. Taking into account scheduled nonproduction events, what is the *availability* (as a percentage) of machine time *per normal week*, without considering operator constraints?

c. What is the availability (as a percentage) of machine time *per normal shift*, taking into account the normal working conditions for operators?

d. If the tactical utilization is targeted to be 90%, what value should be used for the utilization factor of machine time for capacity rating purposes?

e. What is the *demonstrated capacity per normal week* of this work center? (Adjust the data for week 2 to correct for the short week.)

f. What was the *actual utilization* (as a percentage) through the 6 weeks in review?

g. What was the *work center efficiency* through the 6 weeks in review?

h. If *planned efficiency* is targeted to be 85%, and taking into account your answer to question (d), what was the *rated capacity per normal week*?

i. Compare your answers to questions (a), (e), and (h). What should we do now?

Solutions (see also the definitions in Section 14.1.1):

a. Theoretical capacity = 400 hours per normal week
= (5 machines) * (10 shifts) * (8 hours per shift and machine).

b. Downtime due to maintenance is 15 hours per week. Therefore, the availability factor is (400 – 15) / 400 = 96.25%.

c. Downtime due to operator breaks is 1 hour per shift of 8 hours. Therefore, the availability factor is 7 / 8 = 87.5%.

d. Assuming that maintenance cannot be effected during operator breaks, the utilization factor is 87.5% * 96.25% * 90% ≈ 75.80%.

e. Demonstrated capacity is expressed as standard hours produced (row 4 in the table above). The adjusted output for week 2 is 160 * 5 / 4 = 200 hours. Over 6 weeks, the mean is (1340 + 40) / 6 = 230 standard hours per week.

f. During the 6 weeks in review, production has run for 1600 machine hours (row 2 in the table above) out of a possible 2320 hours (= 5*400 + 320). Therefore, actual utilization = 1600 / 2320 ≈ 69.0%.

g. Actual efficiency = standard hours produced divided by actual hours worked
= 1340 / 1600 = 83.75%.

h. Rated capacity = 400 hours * 75.8% * 85% ≈ 258 (standard) hours.

i. Demonstrated capacity (230 hours) is too low compared to rated capacity (258 hours). However, in week 4, the output (280 hours) exceeded 258 hours. Check whether the measurements are still required. If so, check for exceptional events, calculating actual utilization and efficiency for each week. Decide whether to make adjustments to planned utilization or efficiency.

14.7.2 Algorithms for Load Profile Calculation

Intended learning outcomes: Explain load profile calculation extending across several load periods.

One of the problems associated with the use of simple algorithms is that an operation can extend across several load periods (see Figure 14.2.2.2). This exercise will examine how manual or computer algorithms establish capacity and load in a load profile.

Use Figure 14.7.2.1 to enter the capacity or load curve (continuous or rectangular distribution within a time period) for a work center, given the problem outlined below.

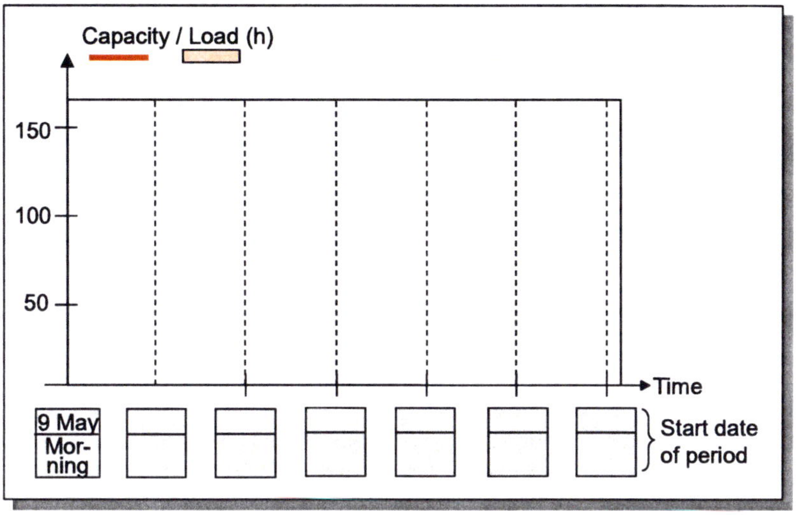

Fig. 14.7.2.1 Load profile calculation.

a. Determine the start date of each period and enter it into the figure above, given 2 weekly periods of 3.5 days each (½ calendar week): Sunday morning to Wednesday noon and Wednesday noon to Saturday evening. The load profile starts with Sunday morning, May 9 (as indicated in the figure). The load profile covers 6 periods (3 weeks).

b. Allocate theoretical capacity to each of the 6 time periods, respecting the following data: At the work center, the plant runs one 8-hour shift per normal workday (8 a.m. to 12 p.m., 1 p.m. to 5 p.m.). The work center has 5 identical machines. Saturdays and Sundays are off. Furthermore, May 13 and May 24 are public holidays (in practice, these dates would change each year). Note that "today," or the moment of the inquiry, is 7 a.m. on Wednesday, May 12.

c. Assume no existing load on the work center. For the following operation, allocate its standard load to the work center: Operation start date is Friday morning, May 14. Standard load (including setup) is 81 hours. The operation can be split on 2 machines, maximum.

Solutions:

a. The second period starts at Wednesday noon, May 12. The third period starts on Sunday morning, May 16. The fourth period starts at Wednesday noon, May 19. The fifth period starts on Sunday morning, May 23. The sixth period starts at Wednesday noon, May 26. The load profile ends before Sunday morning, May 30.

b. Note that there is either a Saturday or Sunday in each period of ½ calendar week. Thus, theoretical capacity per period with normal working days is

(5 machines) * (8 hours per day and machine) * (2.5 working days) = 100 hours.

Note that, in the first period, only 20 hours of capacity are left, because it is already Wednesday morning, May 12. Furthermore, in the second and the fifth periods there is one less working day due to public holidays, which results in only 60 hours of capacity for each of these periods.

c. The load has to be distributed to different periods. Only one working day is left during the period in which May 14 falls (the second period). Because only 2 machines can be used, a maximum of only 16 standard hours (note: not 40) can be loaded. During the third period, 2.5 working days on two machines allow the load of 40 hours. The same would be possible for the fourth period. However, only 25 hours are left to be loaded.

14.7.3 Rough-Cut Capacity Planning

Intended learning outcomes: Calculate, in a rough-cut network plan, the earliest start date and the latest start date for each operation. Determine the resource profiles as well as the load and the deferred earliest completion date for the operations without overloading the capacities.

Figure 14.7.3.1 shows the network plan for a production order.

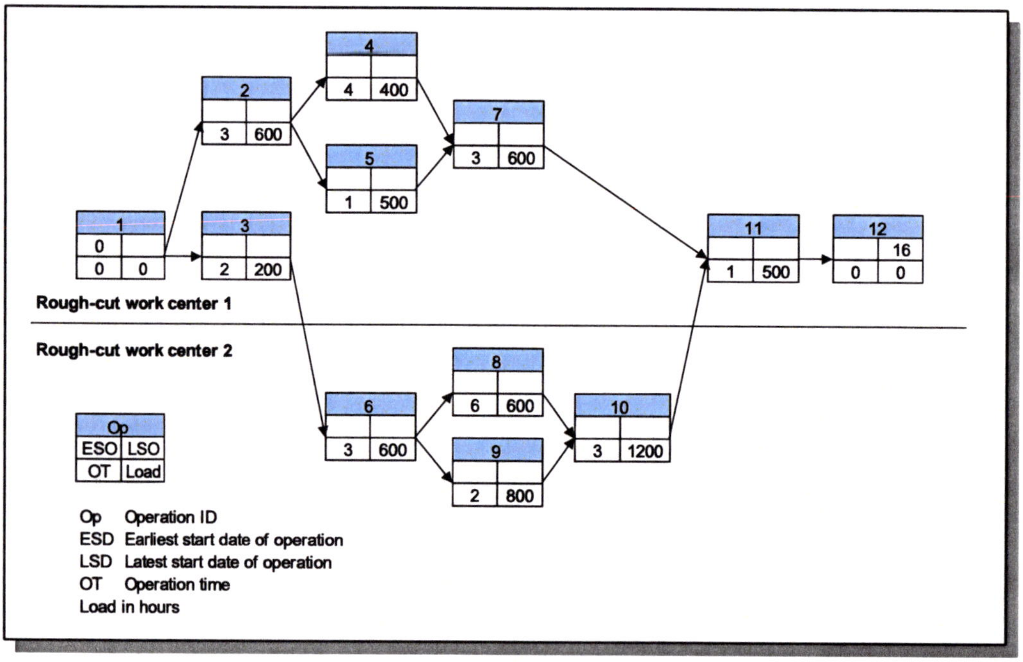

Fig. 14.7.3.1 Rough-cut network plan with two rough-cut work centers.

a. Complete the network plan: Calculate the earliest start date and the latest start date for each operation. What is the lead-time margin (the slack time), and what is the critical path? Determine the slack of all operations not on the critical path.

b. Following the technique introduced in Section 14.4.1, determine the resource profiles for rough-cut work centers 1 and 2, as well as the resource profile for the combination of rough-cut work centers 1 and 2.

c. Figure 14.7.3.2 shows the preload of rough-cut work center 2. Load the resource profile for rough-cut work center 2 with infinite loading. Determine the earliest completion date for the operations of rough-cut work center 2. Further, determine the load and the deferred earliest completion date for the operations of rough-cut work center 2 without overloading the capacities.

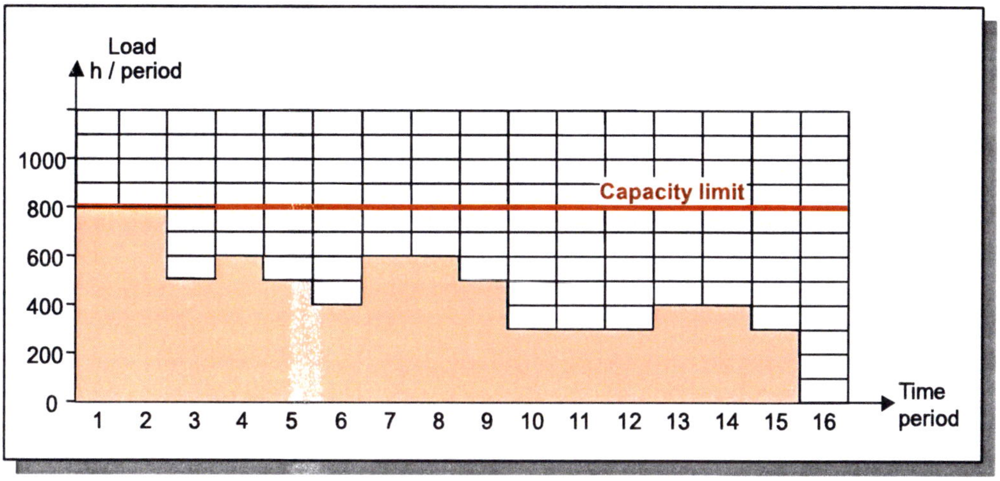

Fig. 14.7.3.2 Preload of rough-cut work center 2.

Solutions:

a. Lead-time margin is 1. Operations 1, 3, 6, 8, 10, 11, and 12 make up the critical path. Operations 2, 4, 7, and 9 could be deferred by 4 time periods, operation 5 by 7 periods.

b. The figure that follows shows the results for rough-cut work center 2 as well as for the combination of both rough-cut work centers 1 and 2. The length of the arrow indicates the number of time units for a possible deferring of the start date of operations not on the critical path.

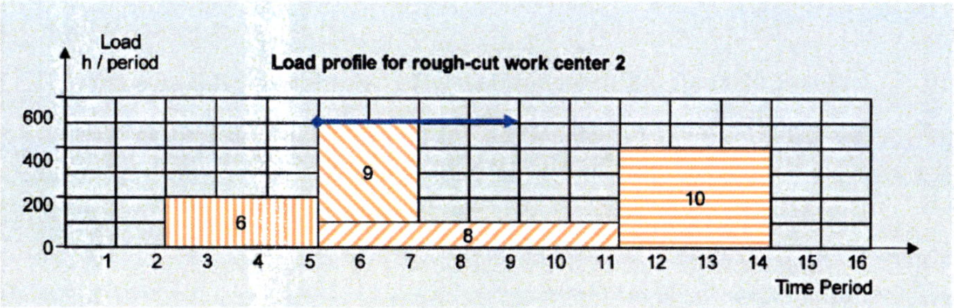

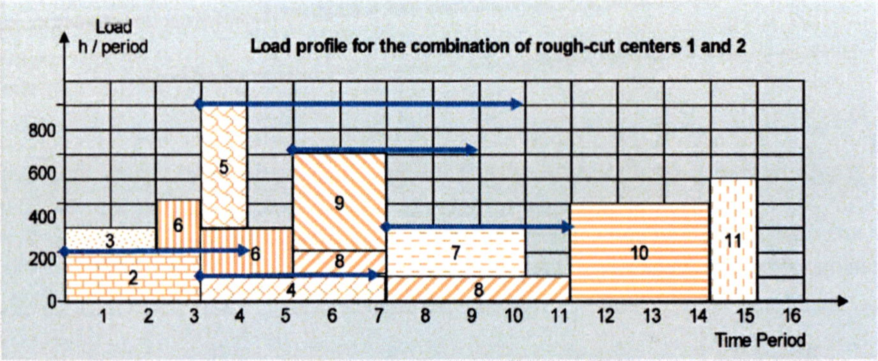

c. The earliest possible completion date for the operations of rough-cut work center 2 using infinite loading is — as the above figure shows — at the end of period 14. For finite loading, the next figure shows the result: an earliest possible completion date at the end of period 15. Note: Because operation 9 is not on the critical path, parts of its load can be deferred to later periods in order to prevent overload.

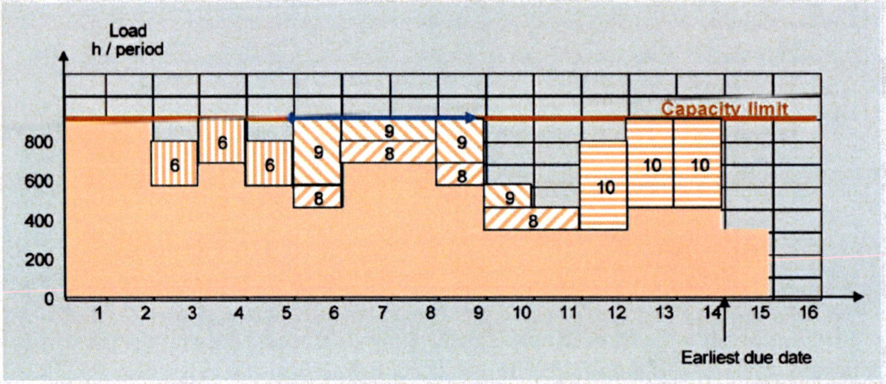

14.8 References

APIC16 Pittman, P. et al., APICS Dictionary, 15th Edition, APICS, Chicago, IL, 2016

GoCo14 Goldratt, E., Cox, J., "The Goal: A Process of Ongoing Improvement," 30th Anniversary Edition, North River Press, Norwich, CT, 2014

RuTa85 Russell, R.S., Taylor III, B.W., "An Evaluation of Sequencing Rules for an Assembly Shop," *Decision Sciences*, 16(2) 1985

15 Order Release and Control

Intended learning outcomes: Differentiate various techniques for order release. Explain in detail shop floor control. Present methods and techniques used for order monitoring and shop floor data collection. Describe distribution control.

Using the reference model for business processes and planning & control tasks from Figure 5.1.4.2, Figure 15.0.0.1 highlights the tasks and processes (darker background) that are the focus of this chapter. Sections 1.2.3, 5.3.3, and 5.3.4 also serve as introductions to this material. We suggest that you review these Sections before studying this chapter.

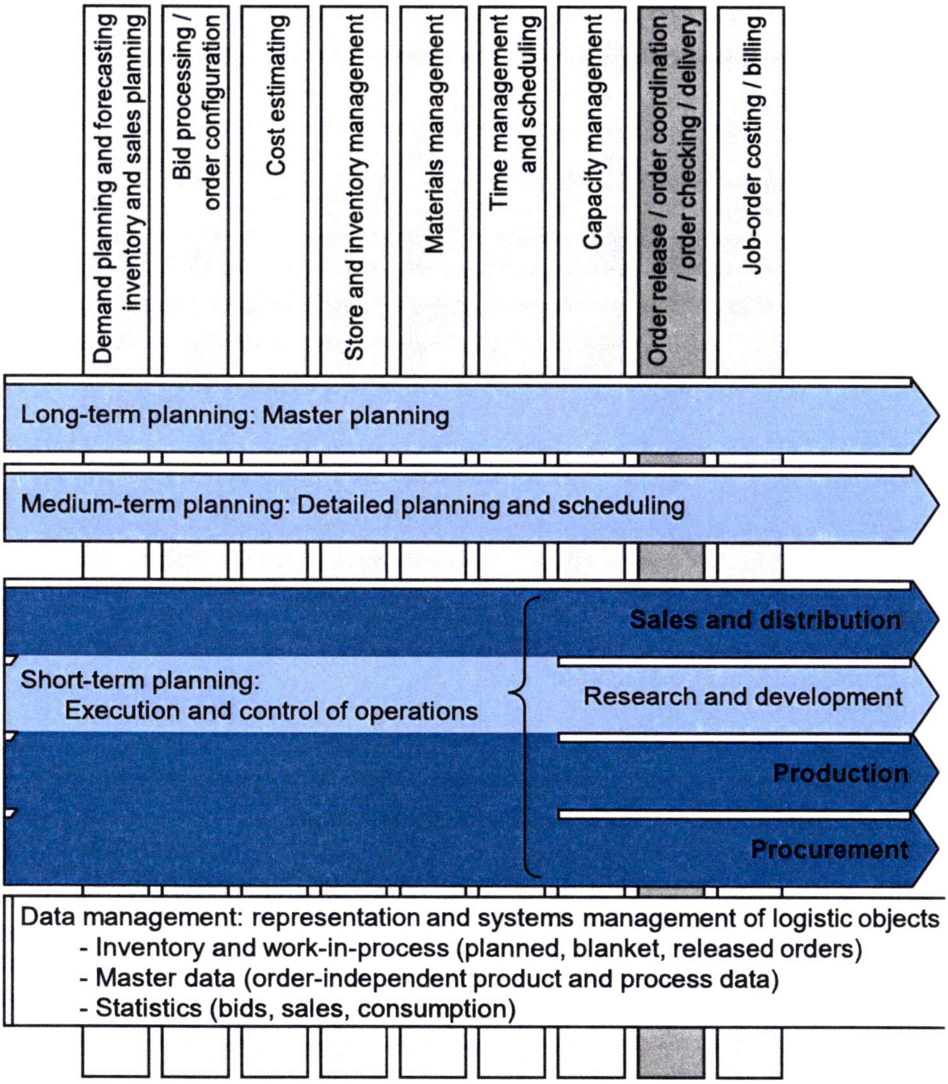

Fig. 15.0.0.1 The parts of the system treated in this chapter.

P. Schönsleben, *Handbook Integral Logistics Management*,
https://doi.org/10.1007/978-3-662-65625-9_15

Chapters 11 and 12 on materials management showed the derivation of resource requirements for long-term and medium-term planning from independent demand (customer orders and forecasts). This results in order proposals for production and procurement. These are proposals for blanket orders or proposals for specific orders for a product, depending on the temporal range. The present chapter now turns to planning & control tasks in the *short-term planning horizon*, i.e., order release and control, in the areas of distribution, production, and procurement. Section 5.4 discussed possible concepts and methods in the area of R&D.

Control, used in a traditional sense, means the regulation and coordination of orders to achieve successful order completion, following the flow of goods from controlled release of order proposals through value-adding activities to manufacture and distribution of salable products. (See here the discussion on the term *control* in Section 1.1.8.)

Each order release entails a *new* scheduling calculation and availability test of the needed resources using techniques of materials management, scheduling, and capacity management. If orders compete, there are techniques for choosing those that should be released.

Then, orders are controlled through the areas (job shops for parts production, assembly, and so on, or for procurement). Electronic control boards (*Leitstand*), or planning boards, are also used here. Accompanying documents are prepared. Order control also includes the loading of infrastructures for picking and distribution. A shop floor data collection system records progress reports and resources consumed. Finished products or received goods are inspected, supplied to further production, distribution, or stock and prepared for invoicing.

15.1 Order Release

Intended learning outcomes: Describe order proposals for production and procurement as well as order release. Explain load-oriented order release (Loor) and capacity-oriented materials management (Corma).

The *order release* changes the status of an order from "proposed" to "released" and triggers the flow of goods to a procurement process or production process.

Generally, for an order release, the availability of all resources needed to execute the order must be checked, in particular the availability of components and capacity. For a comprehensive discussion of the topic, see also the book of Lödding [Loed11].

15.1.1 Order Proposals, Order Release for Procurement and Production, as well as Accompanying Documents and Container Logistics

Intended learning outcomes: Describe the reasons for order proposals for production or procurement. Differentiate between the dealing of order proposals for C items and of other items. Explain purchase and production order release. Describe the availability test of resources. Disclose issues linked with allocation and staging. Identify accompanying documents such as the traveling card, and container logistics such as the two-bin inventory system.

The *order proposal*, or *planned order*, states the goods to be produced or procured, the order quantity, the latest completion date, and — often given implicitly — the earliest start date.

The reasons for order proposals for production or procurement vary:

- An unplanned demand is submitted, that is, demand for customer or production orders not covered by projected available inventory or scheduled or planned receipts. In certain cases, the proposal corresponds to the customer's demand in terms of quantity as well as delivery date. In other cases, order proposals stipulate production or procurement of a larger batch.
- A purchase requisition is submitted. This is an authorization to the purchasing department to purchase specified materials in specified quantities within a specified time, usually short term (cf. [APIC16]).
- Stock of an item falls below the order point. Here the planned order proposal stems from medium-term planning. See also Section 11.3.1.
- *Net requirements planning specifies production or procurement of a lot of an item.* This type of order proposal stems from *long- or medium-term planning*. See also Section 12.3.1.

There are two possible forms of presenting order proposals: either as simple lists of proposals or as planned orders in the order database. In the case of direct procurement for a customer order, the identification of the order proposal should refer directly to the identification of the order position in the customer order.

If there are only a few order proposals to release, or proposals based on unplanned customer demand, planners will release these individually. Keeping track of orders to be released among numerous order proposals is difficult. It is useful to sort the orders by planner and weekly time window:

- An order proposal for C items in an ABC classification, particularly for goods to be purchased, can be released directly — with the proposed order quantity, proposed latest completion date, and standard supplier. Spot verifications will suffice for these automatically released orders.
- For other items the rhythm of selection, and thus ordering, depends on the items' importance. This rhythm may be periodic, such as daily, weekly, biweekly, or monthly. However, an order may be released as soon as a demand event occurs.

In collective materials management, all items belonging to the same planning group are checked at the moment of order release. Joint ordering avoids procurement costs that are lot-size independent, but leads to additional carrying costs due to premature procurement.

Purchase order release does not necessarily have to be a formal procedure. According to specific agreements, certain suppliers can take on themselves the restocking of C items in the warehouse. This is common practice not only in the grocery retail trade but also with suppliers of consumable material in industrial production (compare the Kanban technique in Section 6.3).

A *purchase order*, for procurement both of goods and work (see Section 1.2.1), essentially corresponds in form and structure to a customer order.

As the current trend is to shorten administrative times between manufacturer and supplier, techniques supported by information technology are coming into increasing use. The detailed order data structures behind the "order" business object in Section 1.2.1 have undergone increasing standardization. This has led to the development of the EDI/EDIFACT interface. Thanks to Java programming and the CORBA standard (Common Object Request Broker Architecture), an increasing number of organizations are now making use of transmissions via the internet.

Production order release reasonably comprises, for every order, a check on availability, at least of critical resources. This also holds for order proposals stemming from long, or medium-term planning, even if availability checks have been carried out earlier.

The *availability test* consists of:

- Calculation of lead time, to determine the start dates for operations at critical work centers, as well as dates for the demand of critical components. We presented the techniques for calculating lead time in Sections 13.3 and 13.4.
- Availability test of the components on the start date of the operation for which they are needed. We outlined the techniques for this in Section 12.1. As an aside, in releasing contract work, the availability of any accompanying material that must be provided for the (external) operations must also be checked.
- *Availability test of required capacity* on the start date of operations using techniques that were outlined in Sections 14.2, 14.3, 15.1.2, or 15.1.3.

Even with computer support, checking resource availability is a complicated and lengthy process. It is often impossible to gain rapid and exact results. As a common compromise solution, planners will test at least component availability *at the start date* of the order or the relevant partial order.

Allocations and staging in production order release can entail problems:

An *allocation* is the classification of quantities of items that have been assigned to specific orders but have not yet been released from the stockroom to production.

Staging is pulling material for an order from inventory before the material is required (cf. [APIC16]).

If one of the resources is not available, the remaining components and production facilities nevertheless remain allocated for the order. Staging has the same effect: The order waits for missing resources and, moreover, blocks the plant.

Thus all the operations for an order should be released at the same time and only when all resources are fully available. However, some of such allocated resources may be used instantly in other production orders. Then, instantly available capacity may be used, which is very important for well-utilized capacities. But assigning resources to other orders would further increase the problems for the waiting order.

To achieve an acceptable lead time for production order release as well as to make use of available capacity, a compromise — though less than optimal, since it leads to waiting work-in-process — can be reached through the following measures:

- Release only a partial quantity of the order lot.
- Release the first operations only, if the missing components are not required until later operations.
- Designate the planned order as *firm planned order*: With this, the allocation of components and production facilities on hand is also designated as "firm allocated." The necessary organizational discipline then ensures that the "firm allocated" resources are not withdrawn for other orders. In addition, it is important that computer programs, such as the MRP technique (material requirements planning), do not change this type of orders automatically.

There are different kinds of *accompanying documents* and *container logistics* in production and procurement. The two following cases can be distinguished. *First case:* The contents of an order change remain the same each time, with the possible exception of the order quantity:

With a *traveling card*, a variable order quantity can be set. The due date automatically results, in relation to the date on which the traveling card is sent.

If the entire inventory of an item can be carried in two bins, a visual control system can be installed, namely, the following efficient "traveling" system with fixed order quantity:

With a *two-bin inventory system*, the replenishment quantity is ordered as soon as the first bin (the working bin) is empty. During replenishment lead time, material is used from the second (reserve) bin, which has to contain enough items to cover the demand during lead time, plus a safety demand. At receipt of the replenishment quantity, the reserve bin is filled up. The excess quantity is put into the working bin, from which stock is drawn until it is used up again.

The traveling card and the two-bin inventory system have come into renewed favor with the Kanban technique and its two-card Kanban system.

Second case: The contents of an order change each time. In this case, a formal order from the ordering party (the sales department or the production management) to the order recipient (production or the supplier) is necessary. For a production order, the people at the shop floor responsible for execution require precise instructions as to the nature of the work to be executed and the components to be built in. See Section 15.2.1.

15.1.2 Load-Oriented Order Release (Loor)

Intended learning outcomes: Produce an overview on the principle of the technique and the planning strategy. Describe the regulator analogy for load-oriented order release. Differentiate between time filter and load filter. Explain the steps of load-oriented order release. Present an evaluation of the technique. Identify its limitations and typical areas of application.

Load-oriented order release (Loor) [Wien95] has — for *planning of limited capacity* — high load as its *primary objective*. Further important objectives are low levels of work-in-process, short lead times in the flow of goods, and delivery reliability.

Principle of the technique: This heuristic technique is based on the funnel model (see also Section 13.2.1). Essentially, its aim is to adapt the load to the capacity actually available. It can be perceived as a generalization of the technique that we presented in Section 14.3.2, variation (c), because thanks to a clever heuristic the matching of load to capacity can be limited to one time period.

Planning strategy: Planning releases only those orders that can actually be handled by the work center without resulting in excessive queues. Processing of waiting work-in-process, and thus production control, proceeds according to the *first in, first out* (FIFO) principle.

Technique: Figure 15.1.2.1 illustrates the technique using the analogy of the funnel model. Starting from the uppermost funnel containing all known orders, two filtering techniques are used to determine the orders to be released.

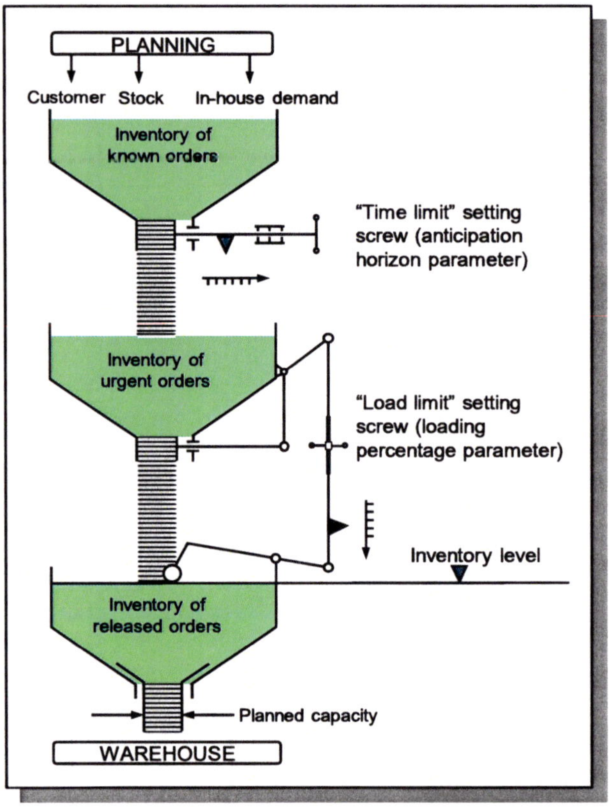

Fig. 15.1.2.1 Regulator analogy for load-oriented order release. (From: [Wien95]).

The *time filter* permits only those orders to flow into the urgent order book that fall within the *time limit*, that is, within the anticipation horizon.

The *load filter* releases only the amount of work that will maintain constant mean inventory, that is, the desired work on hand, for a work center. The *load limit* is equal to the product of capacity during the anticipation horizon and the *loading percentage*.

Although instructions are available for determining the anticipation horizon and the loading percentage, in practice the values chosen are often based on experience or arbitrary.

Load-oriented order release is performed perhaps weekly for a specific planning horizon. It comprises the following steps (basically the same as those outlined in Figure 14.3.2.1).

- Determine the orders to be included in planning and rank them by priority. The candidates are, on the one hand, all orders that have already been begun: The order progress report shows the next operation waiting to be performed; all remaining operations are to be planned. On the one hand, the candidates are all orders that have not been begun, and for which the start date of the first operation lies within the time limit; Backward scheduling with standard lead times (see Section 13.3.3.) will determine the start date. All of these candidates are classed "urgent" and are ordered by start dates, whereby already begun orders are loaded first.

- *Handle and load operations in series order*: The heuristic balances capacity for a single time period multiplied by the loading percentage against loads that will arise not only during this period, but also in later periods. This coarsening is the crucial idea. To this end, subsequent operations are not loaded with full work contents:

  The *conversion factor* progressively converts the loads of subsequent operations.

  For example, if the conversion factor is 0.5 (= 1 / 200%), then the cumulative conversion factor for the first operation of an order is 1, for the second 0.5, for the third 0.25 (= 0.5 · 0.5), for the fourth 0.125 (= 0.5 · 0.5 · 0.5), and so forth. If, on the other hand, the first operation of the order has already been completed, the second operation is next. Thus, the cumulative conversion factor for the second operation is 1, for the third 0.5, for the fourth 0.25, and so on.

- Use the exception rule: If one operation makes use of a work center whose load limit has already been exceeded (due to orders released earlier), unload the entire order, so that other orders are given priority.

- *Deal with all exceptions:* After having loaded all orders, list those orders that were unloaded or set aside. This list contains the identification of the order in question, the workload (e.g., in hours), and the work center that caused the order decline. Check whether the following possible measures can be applied: First, advance the start date of the order. Second, if there is flexibility in the timing of the order due date, postpone it. Third, if the fully utilized capacities are at least somewhat (quantitatively) flexible, then deliberately increase the capacities. By re-performing all the steps for the orders set aside, the orders can possibly now be released.

Figure 15.1.2.2 illustrates the steps using an example taken from [Wien95]. Assume that there are 5 orders to be added to an existing workload.

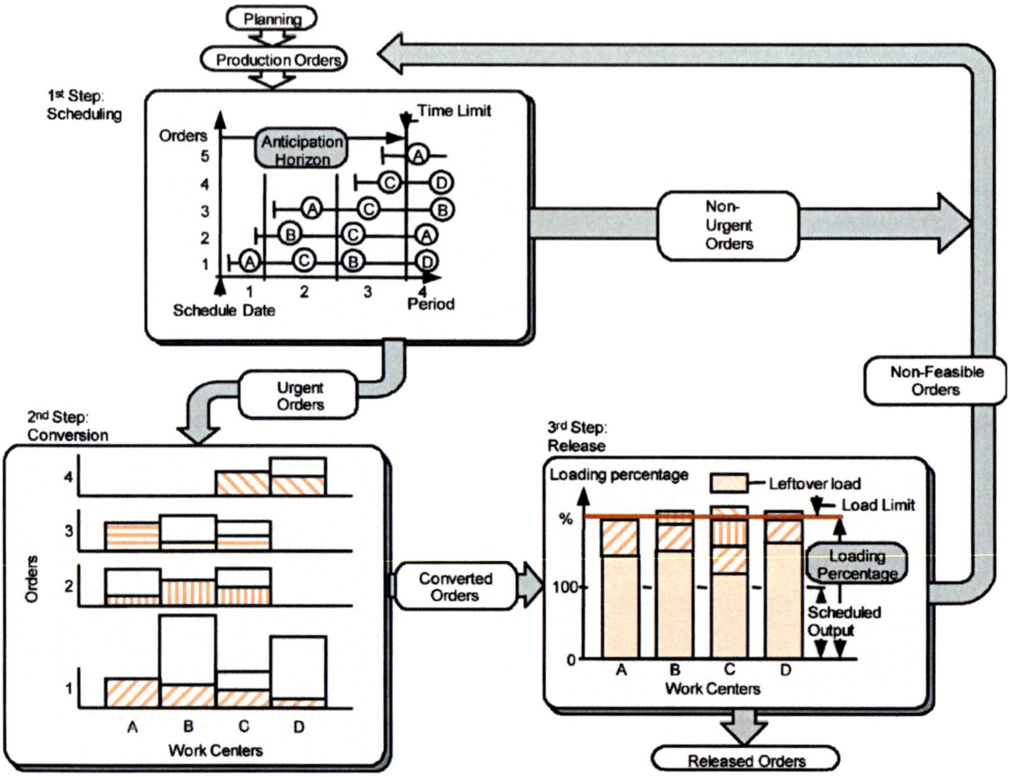

Fig. 15.1.2.2 Steps of load-oriented order release. (From: [Wien95]).

- In step 1 ("scheduling"), these 5 orders are shown together with their operations on the time axis. Each operation bears the work center (A, B, C, D, respectively) that is intended to execute the operation. Each order has its scheduled start date. The time filter of Loor (in fact a time limit, calculated by a given anticipation horizon) eliminates each order where the start date of the first operation is later than the time limit. In the example, the filter eliminates order 5, declared as not urgent. All other orders are declared as urgent and passed to step 2.
- In step 2 ("conversion"), the load of subsequent operations is converted progressively by the conversion factor, which in this example is 50%. That means that the load of the first operation is taken into consideration with 100%, the load of the second operation only with 50%, the load of the third operation only with 25%, and so on.1 In the graph, every order is now shown by its load profile (original and converted). The operations do not appear in the sequence of their execution, but in the sequence of the work center. This is done in preparation for the next step.

1 The further ahead you are trying to plan, the less certain you are that the planned load of an individual job will actually consume the planned capacity within the anticipation horizon. The conversion factor takes this into account. This heuristc looks reasonable — the more steps that are involved, the less certain we are that we can keep to the plan as expected. However, there is no methodical proof.

Take — as an example — order 2. In Figure 15.1.2.2, the load of this order is shown with vertical shading (not only in step 2, but also in step 3). From step 1 we know that the first operation will be executed at work center B. Therefore, the load is shown in step 2, converted by 100% (that is, the full load). Again, step 1 shows that the second operation will be executed at work center C. Therefore, the load is shown in step 2, converted by 50% (that is, half the load). The empty load in the pillar (that is without shadings) corresponds to the other 50% of the load, which will not be taken into account for step 3. Again, step 1 shows that the second operation will be executed at work center C. Therefore, the load is shown in step 2, converted by 50% (that is, half the load). The empty "load" (that is, without shadings) corresponds to the other 50% of the load, which will not have been taken into account for step 3.

- Step 3 ("release") shows first the existing (pre-)load of all workstations before loading the four new orders. This preload stems from different periods on the time axis. This is why it can be greater than the scheduled output capacity for one time period. Arbitrarily, then, a loading percentage of 200% is chosen.2 This yields the load limit for every work center. The orders are then loaded in the sequence of their start date.3 The load of every operation is added to the preload. As soon as an operation has to be loaded on an already overloaded work center, the whole order is unloaded. Thus, the load limit has the effect of a load filter.

  In the example, the load filter accepts first orders 1 and 2, with order 2 overloading work center B slightly (the algorithm accepts the first overload for each work center. But work center B is now declared to be unavailable for all subsequent orders). It then eliminates and unloads order 3 because of already fully loaded capacity at work center B by order 2. Finally, the load filter accepts order 4, for which work center B is not used.[4] Orders 1, 2, and 4 can thus be released, whereas Order 3 is non-feasible and becomes an item for a further step, when exceptions are dealt with.

There is no relationship between the conversion factor and the loading percentage, and the values should not be linked. However, in the original literature on Loor, these values often appear to be reciprocal. Furthermore, in practice, the values chosen for anticipation horizon, loading percentage, and the conversion factor are often based on experience or are arbitrary.

Case example: The Siemens Electronics Plant in Amberg, Germany, manufactures electronic components in customer-independent production to stock. The comprehensive range of

---

2 The loading percentage takes into account the aggregation of capacity of multiple periods (here about 3) over the anticipation horizon. This aggregation makes sense, as you can never be sure that a planned job will occur in precisely that time period that you expect. You then expect to get a more precise view of a forecast when it is aggregated over many periods, compared to considering it over one period. Simple statistics show this by looking at the reduction in the forecast error with a normal distribution — and this reduction in forecast error can be used as a basis for choosing the loading percentage (Mark Bennet, CPIM, Perth, personal communication, 2001).

3 Note that the height of the load might differ at each work center from that which is shown in the figure for step 2, because it has to be normalized with regard to the 100% of the capacity measure. However, the shading of the load of each order is the same as in step 2.

4 Note that order 4 overloads work centers C and D for the first time. Therefore, if there remained more orders to load, these two work centers would now also be unavailable.

components allows the customer to obtain the optimal configuration of programmable SIMATIC control and monitoring operator panels for automated systems. Approximately 500 components are manufactured and available for 24-hour delivery from stock. One production order consists of 10 to 20 operations. The number of machines in the area of load-oriented order release is 20. The main objective in implementing Loor was to limit work-in-process inventory, thus reducing lead times and releasing no orders for production for which capacity was not available. The Amberg Electronic Plant itself took over the task of programming the algorithm. Implementing Loor has brought the expected advantages.

Evaluation of the technique and organizational aspects:

- The debate over this technique is highly polarized. Perhaps, this arises because Loor is too readily presented as generally valid and scientific, as if it were a statistical technique. Thus, the conversion factor is often compared to a probability measure. Critics can easily take this to the point of absurdity. They construct an extreme case in which Loor loads operations that have an execution probability of 0 (zero), but does not release more urgent operations. However, Loor is not an analytical technique; it is a simple heuristic, limited to just a few control parameters. As with every heuristic technique, its applicability will depend on an organization's strategies.

For implementation of the Loor technique, the following *prerequisites* must be met:

- Order due dates must be at least somewhat flexible to provide the scope for dealing with exceptions.
- Capacities must be at least somewhat flexible. Otherwise, the administrative effort to make the numerous deadline alterations will be prohibitive, or the calculations so imprecise that the capacities are only poorly loaded.
- It must be possible to determine the parameters of anticipation horizon, load percentage, and conversion factor in every organization empirically — in some cases through the aid of simulations. The parameters are dependent on the desired work level and the size of the chosen planning period.

The following *limitations* result:

- Orders that fall outside the load limit are generally moved beyond the anticipation horizon, which may result in an unacceptable delay. Releases based on additional information (such as high external priority, rejections due to capacity overloads very far in the future, or similar information) are generally not provided for.
- In the medium term, the available capacity or the capacity that has been made available must be at least as large as the load. Otherwise, more and more orders will fall outside the load limit.
- Loor only loads production with orders that can be processed. It thus leads to low levels of work in process and to short lead times. Scheduled orders are finished on time. However, if the capacity is not flexible, the technique leads to low loading of capacity where completion dates must be pushed back in time. This is because the load that would have occurred far along on the time axis is now missing. If there are no other orders in line, the capacity is missed out (see also [Knol92]).

- In cases of underload, the parameter "anticipation horizon" must not be altered to make the best use of the available capacity. Otherwise, it will result in too early completion dates and possibly unneeded warehouse stocks.

The following *areas* lend themselves to *application* of the Loor technique:

- in branches of discrete manufacturing, particularly when there is a need for simplicity and robustness in the face of errors in planning dates or changes in order levels.
- in short-term planning and control, load-oriented order release provides a reliable work program that permits a considerable degree of situational planning on the spot.

15.1.3 Capacity-Oriented Materials Management (Corma)

Intended learning outcomes: Identify the calculation of anticipation time in the stochastic case. Produce an overview on the three parts of the generic principle of the technique. Explain the critical ratio of an order. Describe rescheduling of orders in process according to current materials management status. Present an evaluation of the technique. Identify its limitations and typical areas of application.

Mixed manufacturing or *mixed production* is concurrent make-to-stock production and make-to-order production, using a single set of plant and equipment.

Mixed-mode manufacturers are manufacturing companies with mixed production.

Mixed-mode manufacturers produce and sell standard products whereby stocks are carried at various levels of production, including the final product. Standard product manufacturing aims for *maximum possible utilization of capacity* (cost objective). At the same time, mixed-mode manufacturers also produce goods to customer order, often in one-of-a-kind production. Here, the manufacturer aims for the shortest possible lead times (delivery objective).

The main objective of mixed-mode manufacturers is *on-time delivery*. The on-time delivery of customer production orders takes high priority. Stock replenishment orders must be fulfilled on time — as soon as stocks have been depleted. The volume of orders of both types of orders is about the same. Simple logistics would call for separation and segmentation of the production resources. However, the very strength of some medium-sized organizations lies in their flexible planning & control, which allows them to make use of one and the same production infrastructure. They manufacture a relatively wide range of products based on specialized competence in a relatively small number of production processes.

Planning strategy: Manufacturing firms with mixed production require a flexible planning strategy. By observing the natural logics of production management as practiced in medium-sized mixed-mode manufacturers, the following generic principle could be derived. For convenience, it is called capacity-oriented materials management, or Corma.

Capacity-oriented materials management (Corma) is an operations management principle that enables organizations to play off work-in-process against limited capacity and short lead times for customer production orders. See [Schö95b].

Corma makes intelligent use of capacity that generally is fully utilized but available short term, which leads to *balanced loading*. This helps to reduce queuing and thus lead times. Essentially, stock replenishment orders are viewed as "filler" loadings, whenever such a capacity is looking for an order. Orders are released periodically, in "packages", providing for optimal order sequencing, which reduces setup times. Thus, shorter lead times in make-to-order production confront a higher level of work-in-process in make-to-stock production.

The *generic principle* consists of three parts:

1. A *criterion for order release* that releases stock replenishment orders earlier than needed. An early order release is considered as soon as there is available capacity in otherwise well-utilized work centers.
2. A *scheduling technique* that for such orders entails work-in-process rather than early stock replenishment. Still, orders will be completed on time. However, customer production orders can be delivered with a minimum lead time. The key is continual reassigning of order priorities by estimating order slack time by (re-)calculation of either the critical ratio or a suitable lead-time-stretching factor of all orders.
3. A *mechanism that couples shop floor scheduling with materials management.* This is done by continually rescheduling stock replenishment orders according to the actual usage. The current physical inventory is converted into an appropriate latest completion date for the open replenishment order.

Thus, the Corma principle not only serves to release orders but also supports overall short-term planning & control from the order release to the moment when the goods either enter stock or are shipped to the customer. Long-term planning for goods and capacity is carried out independently of this. It can be based on traditional forecasting techniques: based on historical data for production with frequent repetition, or based on future projections for one-of-a-kind manufacture, for example.

Technique: In general, the generic principle is implemented manually. To do this, the planner uses a set of known planning and control techniques. Each of these techniques can (but does not need to) be supported by functions of conventional PPC software, or simply by personal implementation using Microsoft Excel or similar software. The following describes the techniques of the three parts of Corma in greater detail.

Corma, Part 1: Criterion for early order release. The planner regularly checks the loading of *generally well-utilized capacity*. As soon as short-term unused capacity is discovered, he checks on the availability of the products manufactured using this capacity. A work center where-used list can provide essential information for this first step. It is as if capacity is on the lookout for an order (hence, the term *capacity-oriented* materials management). If an "agent" is assigned to each capacity, agent-based systems may also be applied here.

In practice, it often happens that a particular product family is manufactured in a group of just a few work-centers. If one of the work centers of this group — in particular the *gateway work center* that performs the first operation of a particular routing sequence — is not being utilized, quite often the others are not in use either. An early order release thus usually means that several operations can be performed in advance.

Which of the products thus identified are candidates for early order release? The planner finds the answer by calculating the anticipation time for each possible item.

The *anticipation time* for an item is the time that will probably elapse before a production or procurement order must be released.

Figure 11.3.2.2 provides a formula for determining the articles that are candidates for an early release in the deterministic case. It takes into consideration all known transactions in the near future. Figure 15.1.3.1 shows a graphical representation of anticipation time in the stochastic case. This is the time expected to elapse before inventory falls below its order point, assuming average usage for the near future.

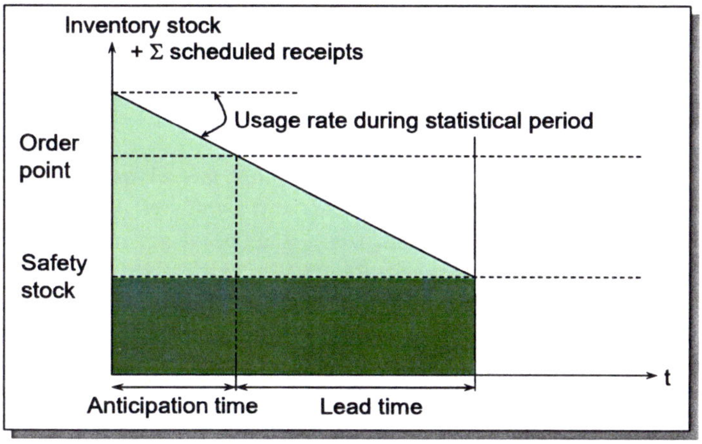

Fig. 15.1.3.1 Anticipation time in the stochastic case.

Figure 15.1.3.2 shows the formula for calculating the anticipation time.

$$\text{anticipation time} = \frac{\text{inventory stocks} + \sum \text{scheduled receipts} - \text{order point}}{\text{usage rate during statistical period}}$$

Fig. 15.1.3.2 Calculating anticipation time in the stochastic case.

If there is more than one candidate for early order release, the product having the shorter anticipation time gains priority. Clearly, software can aid the planner in efficient calculation and decision-making.

Corma, Part 2: Scheduling technique for control of operations. New customer orders continually alter the workload. They also “hinder” the progress of stock replenishment orders, and vice versa. In this situation, the planner continually reassigns the priority of all orders in process by estimating order slack times. A rough-cut estimation of order slack time is the following critical ratio.

The *critical ratio of an order* is obtained by dividing the time left until the order due date by the standard lead time of work left on the order.

A ratio < 1.0 indicates that the order is behind schedule; a ratio > 1.0 indicates that the job is ahead of schedule. The lower the result, the higher the order urgency in sequencing the operations of the order compared to those of other orders. Generally, the critical ratios of the orders can be obtained by an inquiry of the order database. The planner transfers a resulting priority to the production order as soon as he considers the difference compared to the actual order priority to be significant. As a result, this technique either accelerates or slows down the orders. It gives priority to early-released orders only when needed.

A more detailed and accurate measure of order urgency is obtained by implementing *probable scheduling* for shop floor control. Here, the key is the calculation of a suitable lead-time-stretching factor. See Section 13.3.6. This factor is a more accurate measure for the order slack time than the critical ratio of the order, as it is defined as a numerical factor by which only the nontechnical interoperation times and the administrative times are multiplied. Since the technical process itself determines the duration of operations and the technical interoperation time, we can only modify slack time by increasing or reducing either the nontechnical interoperation times or the administrative times.

Corma, Part 3: Coupling shop floor control with materials management. To do this, the planner checks the inventory on an ongoing basis and calculates the point in time at which inventory will fall at zero, assuming average use. This point in time becomes the probable date on which the replenishment order should arrive in stock. Clearly, software can provide for easy calculation here. The planners (or software) transfer this date to become the latest completion date for the replenishment order as soon as they consider the difference to the actual latest completion date to be significant. The following situations may arise:

- The latest completion date will be pushed forward, if inventory stock is being depleted at a rate faster than the statistical average for the period up to the point of order release. Rescheduling then calculates a smaller lead-time-stretching factor. This results in higher priority, and the order is accelerated.
- The latest completion date is postponed if inventory stock is being depleted at a rate slower than the statistical average for the period up to the point of order release. Rescheduling generates a higher lead-time-stretching factor. This results in lower priority, and the order is slowed down.

To show the effects of the Corma principle, let us look at a stock replenishment order with three production operations. Figure 15.1.3.3 shows four possible situations.

- First situation: Because of the early order release, all three work operations are evenly distributed between the earliest start date (i.e., the earliest possible start date of the order, which is originally the date of the early release and then moves — in fact — forward along the time axis with the "today" date) and the latest (acceptable) completion date for the order (that is, the *order due date*). They are all scheduled, but — in this situation — without priority. As a result, they are performed as soon as there are no more urgent operations waiting to be processed at the work station.

- Second situation: The mixed-mode manufacturer accepts an unplanned customer order with a high priority. Then the stock replenishment order in process will wait. Not even the first operation is performed. However, the ongoing rescheduling "discovers" any order that has waited for too long, and the latest start date, that is, "today," is being pushed closer to the latest completion date. Rescheduling then calculates a smaller lead-time-stretching factor. This gives the order higher priority.

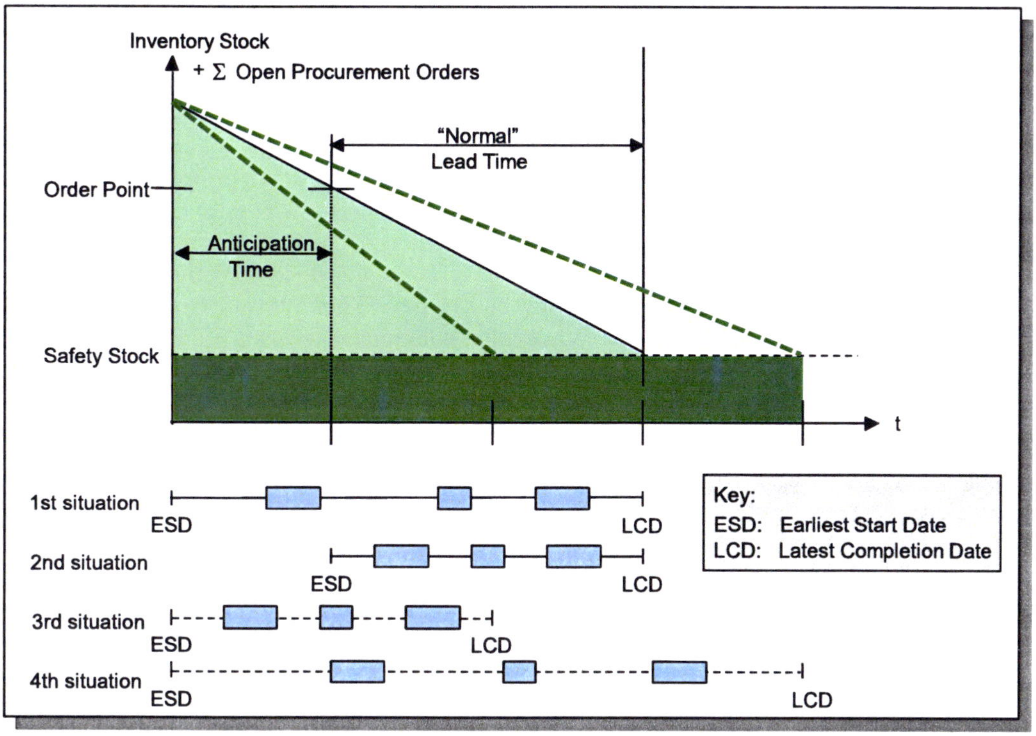

Fig. 15.1.3.3 Rescheduling of orders in process according to current materials management status.

- Third situation: The inventory stocks fall faster than expected. The latest completion date is therefore brought forward. Rescheduling calculates a smaller lead-time-stretching factor, and the order is accelerated by expediting.[5]

- Fourth situation: The inventory stocks fall slower than expected. Thus, the latest completion date is postponed. Rescheduling calculates a higher lead-time-stretching factor, and the order is delayed.

The third and fourth situations in Figure 15.1.3.3 illustrate the most important aspect of the third part of Corma. Stock replenishment orders will receive the same priority as customer production orders if stock falls below safety stock. If the demand is lower than expected, however, stock replenishment orders will not even start, or will be halted. Alterations in the due date of a customer production order may also lead to rescheduling, with consequences similar to those in situations 3 and 4 above.

---

[5] To *expedite* means to rush or chase production or purchase orders needed in less than the normal lead time, to take extraordinary action because of an increase in relative priority ([APIC16]).

An example: *Trox Hesco Corp. (Rüti, CH-8630, Switzerland).* Trox Hesco (200 employees) develops, produces, and distributes ventilation products, such as air diffusion lattices and fire dampers. Trox Hesco manufacturing is based on high competency in a relatively small number of production processes. 500 different stock line items make up approximately 60% of sales volume. The same items, but made-to-order according to customer requirements with respect to dimension, color, and so on, make up the other 40% of sales. Product structures and routings are of moderate complexity, with one to two production stages and about a dozen items in the bill of material and fewer than a dozen operations per stage.

As customer tolerance times are short, planning & control gives high priority to special customer orders. At the same time, however, stock replenishment orders must also be completed on time to prevent shortages. Stock replenishment orders can therefore compete with special customer orders. As demand for stock items is variable, the stock depletion date estimated at the moment of order release must now be verified. This allows determination of the priority of the replenishment order. While segmentation of the two production processes would make for simple logistics, this flexible planning & control of resources enables Trox Hesco to make use of the same production infrastructure for both modes of production.

Assessment of the technique and organizational considerations: Prerequisites for Corma:

- The increase in work-in-process, which results from the early release of stock replenishment orders, must be feasible economically and manageable in terms of volume. Corma does not result in premature inventory in stock, however.
- Early order release has to be possible to a sufficient degree. Orders that are released early are stock replenishment orders or customer production orders that start in advance of the latest start date.

There are some *limitations* involved in applying Corma:

- The focus has to be on a *more balanced* utilization of capacity, not maximal utilization. Load fluctuations will remain.
- Planners "on site" must be able to deal with constantly changing order inventories. They have to understand how to make the best use of the Corma recommendations, which may entail changing the sequence of operations that Corma proposes to accommodate additional, situation-specific information known to the planner.

Therefore, Corma is useful for the following *areas of application*:

- in addition to mixed production, in all cases where due dates must be met and, nonetheless, the system must be robust in the face of errors in planning dates or alterations in orders on hand;
- for self-regulating shop floor control (for mixed-mode manufacturers, for example), assuming that the data collected on order progress are precise enough. Because the basic premise of Corma is a constantly changing order backlog, it is robust enough to handle situational planning "on the spot," which in this case is desirable;
- as a self-regulating system for short-term materials management. Owing to its continuous coupling with materials management, an order may change its latest completion date multiple times. A stock replenishment order may change its completion

date up to the moment when inventories fall below the safety stock. From that moment onward, the replenishment order must be assigned to ongoing customer orders, since the replenishment order will serve to cover such customer orders. Since customer orders must have confirmed due dates that can no longer be changed, the replenishment order must also be given a *fixed, or definitive,* latest completion date.

15.2 Shop Floor Control — Production Activity Control and the Manufacturing Execution System

Intended learning outcomes: Describe the issuance of accompanying documents for production. Explain operations scheduling, dispatching, and finite forward scheduling. Present sequencing methods.

Shop floor control comprises the essential functions of production order processing, dispatching and sequencing, order coordination, and shop floor data collection (work-in-process, order progress checking, actual use of resources, performance indicators such as inventory turnover, work center efficiency, and capacity utilization). *Production activity control (PAC)* is a more generalized term, including also supplier control. See here also [BaBr94].

A *manufacturing execution system (MES)* is the corresponding *IT-supported information system* that manages production in factories. It supports all functions described in this section as well as in Section 15.3.

15.2.1 Issuance of Accompanying Documents for Production

Intended learning outcomes: For a production order, identify the shop packet. Describe shop order routing, operation card, parts requisitions and picking list.

For a *production order,* people at the shop floor responsible for execution require precise instructions as to the nature of the work to be executed and the components to be built in. They require a *shop packet*, which is a set of comprehensive technical descriptions and administrative documents. Among the latter are

- The *shop order routing* accompanies the order during the entire production process. It records the administrative course of the order in detail, and serves as a data collection document for the report of order termination and/or placement in stock. It is printed for each partial order and includes all operations and often all reservations.
- An *operation card* for each operation to be performed, and thus for each position on the shop order routing. Generally speaking, the operation card will contain the same information that the shop order routing itself contains. Their primary purpose is shop floor data collection. A template for time stamps is on the reverse, so that it can serve as *time ticket*. If there is automated shop floor data collection, the operation card is often no longer necessary.
- The *parts requisitions* relates to the reservation of an individual component of a production order and serves as the authorization for its issuance from stock. Parts

requisitions are most often produced for raw materials or for components that cannot be itemized reasonably on a picking list.

- The *picking list* includes all components that are to be issued. It provides for efficient (order) picking (see Section 15.4.1). It sorts reservations according to an issuance sequence that is optimal from a functional and technical perspective. Its identification also provides for an efficient shop floor data collection.

As to the point in time when accompanying documents should be printed:

- Individual execution deadlines for each operation as well as the assigned date of issuance for each reservation should not appear on the documents. This is because the execution dates are subject to change after the order release. Logically, it is not absolutely necessary to wait for the scheduling of individual operations and other time-intensive work in order to issue accompanying documents. Thus, they can be prepared immediately after the order release.
- Picking lists and parts requisitions are printed together with the shop order routing.
- Generally speaking, there are two possible ways to print operation cards. They may be printed either together with the shop order routing or at each work center within a particular window in time, in accordance with the scheduling currently in effect.

15.2.2 Operations Scheduling, Dispatching, Finite Forward Scheduling, and the (Electronic) Control Board (Leitstand)

Intended learning outcomes: Produce an overview on Operations scheduling and dispatching. Explain finite forward scheduling. Describe the loading of production resources in the form of a planning board before and after a new entry to orders on hand. Identify the (electronic) control board (Leitstand).

Operations scheduling is the actual assignment of starting or completion dates to operations or groups of operations (cf. [APIC16]).

The result of operations scheduling shows when these operations must be done if the production order is to be completed in time. These dates are then used in the dispatching function.

In *dispatching*, each operation is assigned to the individual workstations . Also, employees, production equipment, and other work aids are definitively assigned to the operation.

Dispatching is a part of production control. It is based on the inventory of work on hand or on the work program produced by detailed planning and scheduling (see Sections 14.2.3, 14.3.1, and 14.3.2). The latter is a time window, such as the coming week, for the inventory of work on hand at the work center.

Shop floor employees generally have the specific knowledge needed for dispatching. They know the secondary constraints in detail, for example

- The individual pieces of equipment at a work center: Not every machine in the work center can perform exactly the same jobs. Certain orders may require machine tools that can be mounted only on certain machines.

- The *qualification of employees*: Not all workers are qualified to perform exactly the same jobs. Certain orders may demand minimum qualifications.

Dispatching draws on large stores of fragmentary knowledge or knowledge by analogy to earlier cases. Such experience-based knowledge in the heads of supervisors or foremen is usually not structured or available in explicit form. Therefore, in most cases, the function of dispatching is a mental process — albeit supported by the algorithms of capacity planning (Sections 14.2 and 14.3). These algorithms show the probable consequences of prospective dispatching to individual machines in the context of the current situation.

Finite forward scheduling is a scheduling technique for production equipment and other aids, for the individual machines,[6] and possibly also for the workers and other resources, that builds a schedule by proceeding sequentially from the initial period to the final period while observing capacity limits. (cf. [APIC16]).

Production equipment includes machine tools, devices, NC programs, and equipment for measuring and testing. Aids include drawings.

Finite forward scheduling is based on the current inventory of work on hand at the work center, from medium-term planning within a particular time window. The technique further requires detailed information on the availability of each individual resource. For the needs of the technique, any operations too roughly defined in medium-term planning must be broken down into individual operations and further detailed to individual workstations.

Just as in the case of dispatching, employees who work at the work centers have important knowledge of the situation in their heads. These people tend to be able to make the best decisions about control of operations. For precisely this reason, excessively detailed planning for the medium and long term makes little sense.

For representing the results of operations scheduling and finite forward scheduling, a Gantt chart is appropriate. A suitable planning board permits the individual loads to be moved around among the workstations in a flexible way. Figures 15.2.2.1 and 15.2.2.3 show an example of finite forward scheduling with six work centers (WC). The second work center has three workstations (WS), the fourth two. A calendar showing available days for these work centers is shown across the top; the work centers are available only five days per week. Bold areas on the bars mark the related operations of a specific production order. In two cases, an interoperation time has to be respected.

In the scenario in Figure 15.2.2.2, there is an additional order to load. The due date is "as soon as possible." Existing scheduled jobs are not to be changed. The result of finite forward scheduling of this order is shown in Figure 15.2.2.3. Please note:

- The job is scheduled to start on August 11.

---

[6] Note: *Machine loading* is the accumulation by workstation, machine, or machine group of the hours generated from the scheduling of operations for released orders by time period (cf. [APIC16]). Machine loading does not use the planned orders, but operates solely from released orders.

- Both operations are scheduled to run on two workstations.
- Operation 320 is scheduled to begin on August 25.
- The scheduled completion date for order 4711 is Sept. 1 (or close of business day August 30).

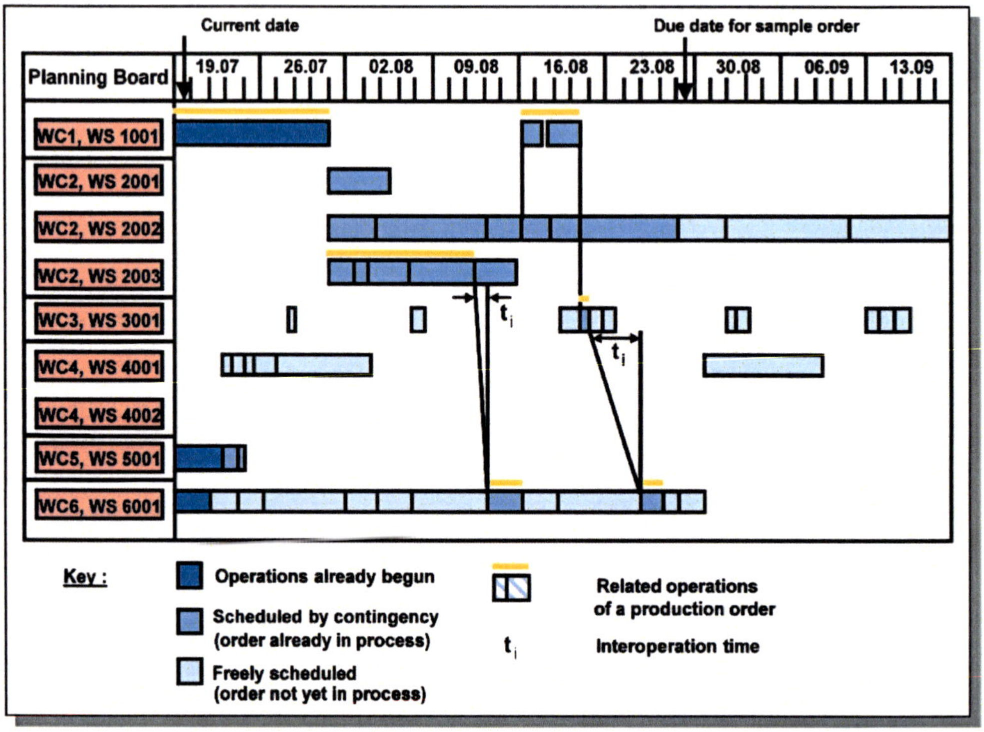

Fig. 15.2.2.1 Loading of production resources in the form of a planning board.

Orders on hand for planning as of 19.07:	
Order 4711, with a sequence of the following two operations:	
Operation 310	Work center 2, standard time 9 days, splittable on work stations of the same work center, can be interrupted by other jobs ESD = 11.08, LCD = 20.08
Interoperation time:	3 workdays
Operation 320	Work center 4, standard time 7 days, splittable on work stations of the same work center, can be interrupted by other jobs ESD = 25.08, LCD = 03.09
(Key: ESD = Earliest Start Date; LCD = Latest Completion Date)	

Fig. 15.2.2.2 New entry to orders on hand: order 4711.

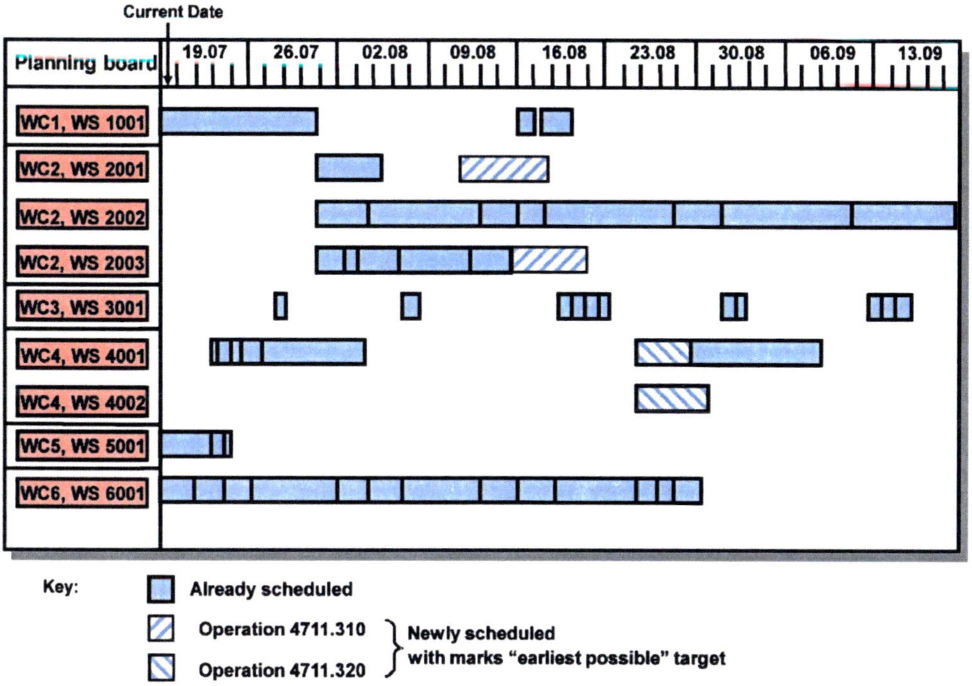

Fig. 15.2.2.3 Loading of production resources in the form of a planning board, situation following loading of the new order 4711.

For finite forward scheduling, an electronic control board offering graphic capabilities may come into use.

An *(electronic) control board (Leitstand)* essentially simulates a *planning board*. At the same time, good electronic control instruments provide an overview of the previous and subsequent operations and thus give information about the consequences of shifting the operations in various ways.

However, such software algorithms do not always lead directly to the objective, so that finite forward scheduling may involve some manual work or reworking. Thus, finite forward scheduling using a planning board is suitable only for production with operations of longer duration.

In summary, finite forward scheduling yields individually released operations together with their sequencing. It may cause aids to be made available and, in the case of disturbances in the process, provide suggestions for potential replanning, such as an altered assignment of personnel or orders to the individual workstation.

15.2.3 Sequencing Methods

Intended learning outcomes: Explain sequencing and its objectives.

Sequencing arranges the jobs in the inventory of work on hand in a particular series order.

A clever sequence of operations can reduce setup time. This is one of the essential objectives of sequencing. If the individual orders arrive randomly, in order to choose a sequence with minimal setup time, there has to be a queue. However, this increases lead times, which is not an option with certain critical products. Frequently, however, one of the first operations, if not the first itself, often uses fully utilized capacities, so that saving setup time due to good sequencing can contribute significantly to reducing lead time. Here, it is appropriate to sort and combine the orders waiting for release right at the point of release according to the sequencing criterion for the corresponding operations. In other cases, the situation is more complex. Sequencing is thus often a compromise among the various aspects and criteria of shop floor control.

If setup time reduction is not the chosen strategy, other objectives and rules of priority (see Section 14.3.1) will be selected. In shop floor control, these must be transparent and understandable for all persons involved. They can have the opposite effect, if they are applied incorrectly.

Detailed sequencing is indispensable for flexible manufacturing systems (FMS) under high load conditions, since the aim, due to costs and especially deadline reasons, is to avoid production interruptions due to order changes. Since information technology support in flexible manufacturing systems means that all data on the necessary time requirements are always available, a company may consider automating sequencing. This would still have to be an interactive process, however, since many decision-making rules arise on an *ad hoc* basis, grounded in the experience of the machine operators. These cannot be translated explicitly into rules that can be applied automatically.

Algorithms for sequencing constitute a field of study in operations research and in *artificial intelligence* and will not be presented here. Refer to [Sche98b].

15.3 Order Monitoring and Shop Floor Data Collection

Intended learning outcomes: Describe recording issues of goods from stock and completed operations. Produce an overview on progress checking, quality control, report of order termination, and automatic and rough-cut data collection.

Shop floor data collection provides for the reporting of all events relevant to planning and accounting during the value-added chain.

From this feedback, the exact state of orders can be derived, so that shop floor data collection additionally serves order monitoring and order checking as well as order coordination among orders that belong together in sales and distribution, R&D, production, and procurement.

15.3.1 Recording Issues of Goods from Stock

Intended learning outcomes: Differentiate between unplanned issuances and planned issuances. Produce an overview on backflush technique and critical point backflush technique.

From central warehouses, goods may be withdrawn only upon presentation of a parts requisition or a picking list. The data that appear on a parts requisition should include

- Order ID and order position
- Item ID
- Reserved quantity in stock units
- Reserved quantity in picking units. For example, an item may be carried in stock in kilos, but picked in meters (for example, materials in bars) or in number of sheets (sheet metal). The factor required for this conversion is maintained as an attribute of the bill-of-material position or, if it is the same for every possible issuance, as an attribute of the item master data.

For *unplanned issuances*, the parts requisition must be filled out in its entirety.[7] An availability check must precede every unplanned issuance, so that already confirmed reservations of physically available warehouse stocks for other orders can be taken into account (see Section 12.1).

For *planned issuances* from stock, the data that have to be collected are limited to the actually issued quantity, recorded either in converted units or in stock units. If the issued quantity corresponds to the reserved quantity, the only fact reported is that the material was "issued."

For a picking list, in a first step only those positions for which the issued quantity differs from the reserved quantity are recorded. Then the so-called backflush technique is used:

The *backflush technique* reports the picking list itself as "issued," whereby every (remaining) position on it is reported as issued in the reserved (or produced) quantity.[8]

A *critical point backflush technique* is a backflush technique performed at a specific point in the manufacturing process, at a critical operation, or at an operation where key components are consumed ([APIC16]).

15.3.2 Recording Completed Operations

Intended learning outcomes: Describe the data collected after the completion of an operation. Identify demonstrated capacity and downtime.

Among the data that are printed on an operation card are

- Order ID and order position

---

[7] If the issue concerns overhead costs, the cost center ID should be given instead of the order ID.

[8] *Post-deduct inventory transaction processing* is used as a synonym for backflush. In contrast, *pre-deduct inventory transaction processing* reduces the book inventory of the components at the moment of the order release for the product.

- ID of the assigned work center
- ID of the assigned machine or tool
- Quantity to be processed
- Standard setup load
- Standard run load
- If needed, the quantity to be produced in a unit that differs from the one on the order. For example, orders may be for pieces, but production is in meters (for sheet metal trimming, for example). The necessary conversion factor is an attribute of the *operation* object.

If the execution matches the standard, the only recorded fact is the execution of the operation. By collecting the number of finished items and the number of produced scrap items, rated capacity can be compared with demonstrated capacity.

Demonstrated capacity is proven capacity calculated from actual performance data, usually expressed as the average number of items produced by the standard load hours per item ([APIC16]).

Furthermore, actual operation load, measured in capacity units, can be collected, as well as effective times. Standard operation time can then be compared with actual operation time. In addition, downtime might be of interest:

Downtime is time when a resource is scheduled for operation but is not producing for reasons such as maintenance, repair, or setup ([APIC16]).

For statistical and accounting purposes, the ID of the worker goes on record. In multiperson servicing, various operation cards are recorded, all referring to the same operation. If the work center or other planning data change during the execution of the job, the altered data must be registered. The order ID is also recorded for every unplanned executed operation.

Also conceivable is a separate recording of the actual quantities and the fact that the operation was completed. This may be necessary because of the legal situation (labor unions). In this case, recording includes only the number of produced items (good items and scrap) on the operation card. Separate collection documents then keep note of the actual loads. These summarize the activity of the personnel along with their other activities (training, illness, vacation, and so on).

15.3.3 Progress Checking, Quality Control, and Report of Order Termination

Intended learning outcomes: Produce an overview on progress checking, quality control and the quality control sheet. Identify the anticipated delay report and the order termination report.

Progress checking monitors the execution of all work, in terms of quantity and delivery reliability, according to a plan.

Progress checking allows determination of the position of a production order in process at a specific moment. Every time a parts requisition or operation card is reported, the administrative status of the position changes into "issued" or "executed." A strictly maintained reporting system is the prerequisite for exact control. It is important to report every operation as "executed" immediately upon completion. This ultimately serves for order coordination. In turn, the meaningfulness of scheduling and capacity planning is maintained. The system is transparent and finds acceptance with the users.

The recorded actual load of an operation permits statistical evaluation and determination of the average work center efficiency overall. Modifications to the standard load for an operation may result.

Quality control checks every produced or purchased product according to a more or less explicit or detailed quality control sheet.

A *quality control sheet* is a routing sheet that holds the process for quality assurance.

With production orders, quality control can take place after each operation. Ideally, the person performing the operation should carry out quality control. However, quality control can also take place at the end of production. It may also serve to estimate process capability.[9] For purchase orders, the *receiving department* inventories incoming receipts as to identity and quantity before transferring them to the quality control unit.

The production resources used for quality control are called *quality control materials*. The produced lot is designated "finished" or "received," but also "in quality control". The availability date is, e.g., the received date plus the lead time for completion of the quality control sheet. During execution of the control operations, errors are recorded.

An *anticipated delay report* is a report to materials management, regarding production or purchase orders that will not be completed on time.

Besides the new date, the anticipated delay report has to explain why the order is delayed.

The *order termination report* is the message that an order was completed. It contains the results and states that all resources used were recorded.

For logistics purposes, the final stage of the quality check judges the portions of the procured order lot as accepted or as rejected as scrap. The *scrap* (that is, the material outside of specifications) goes back to production for *rework* (that is, reprocessing to salvage the defective items, if this seems practical), or back to the supplier for replacement (or reduction of the total of the receipts).[10] The *yield* (or the *"good" quantity*, that is, the acceptable material) moves to its destination: to stock, to a production process, or to sales.

---

9 *Process capability* refers to the ability of the process to produce parts that conform to (engineering) specifications. *Process control* is the function of maintaining a process within a given range of capability by feedback, correction, etc. (cf. [APIC16]).

10 The manufacturer may keep these items at his site as *inventory returns*.

Order termination is reported only when all resources used for a production order have been recorded, and when the accounting check for a purchase order has been performed. The latter is the comparison between the usable quantity of a shipment received and the corresponding purchase order position quantity.

15.3.4 Automatic and Rough-Cut Data Collection

Intended learning outcomes: Produce an overview on Automatic identification and data capture (AIDC), bar codes, RFID. *Describe the principle and issues of rough-cut data collection.*

Manual shop floor data collection, which uses operation cards, parts requisitions or picking lists, is slow, particularly for short operation times. Prompt recording of transactions requires additional administrative personnel in the job shops. In addition, there is a great danger of erroneous data entries. For these reasons, one tries to record shop floor data automatically.

*Automatic identification and data capture (AIDC) is a set of technologies that collect data about o*bjects and send these data to a computer without human intervention. Examples are:

- Bar codes: Information is coded in a combination of thick and thin lines. A light-sensitive pen reads and transfers this information to a computer.
- Radio Frequency Identification (RFID) is an automatic identification technique, relying on storing and remotely retrieving data using RFID tags as transponders. A transponder is an electronic transmitter. An RFID tag can be attached to or incorporated into an object product, animal, or person for the purpose of identification using radio waves. Electronic product codes (EPCs) are used with RFID tags to carry information on the product to support warranty programs.
- *Badges*: A badge is generally a card with a magnetic strip. The strip contains information that can be read with a device and sent to a computer.

The solutions developed thus far focus on the following techniques:

- The use of bar codes or RFID to identify the operation or the allocation directly on the shop order routing or picking list. The use of operation and parts requisition cards is reserved for unplanned issuances or operations. The human operator is identified by means of his or her badge. This is usually the same magnetic card used for measuring the employee's work hours.
- A clock in the data processing system runs together with the transaction and determines the actual time used through automatic recording of the start time and end time for the operation. The difference between start time and end time yields time used, or the actual load. However, an unplanned issued quantity must still be recorded by hand. With this, a small source of error remains. In contrast to the grocery trade, for example, issuances in industrial production are not in units; under certain circumstances a large set of units may be issued instead.
- Linking the data collection system to sensors that automatically count the goods produced or taken from stock. Such systems can be valuable for any kind of line production as well as for CNC or robot-supported production.

Rough-cut data collection takes into account the fact that the results of the entire operation are more important than the success of a single order.

The costs of data collection must stand in healthy relation to the benefits of data collection itself — namely, better control of the production and the procurement process. This condition is difficult to meet for all extremely short operations where the administrative time needed to record the operation is in the same range as the operation time itself:

- Collective data collection for entire groups of short operations is possible. However, this requires the recording of the operations represented by this group or by collective data collection, so that the time recorded can ultimately be distributed among the individual operations according to a key. Since we often cannot determine this grouping in advance, it must be recorded at some point during the process. This quickly results in a quantitative data collection problem.

For group work, the recording of the actual processing time is often possible only for rough-cut operations, that is, for a combination of individual operations. This can only deal with all participating persons together and includes interoperation times as well.

- This combination may correspond to a rough-cut operation, which is sufficient for long- or medium-term planning. It may, however, be even rougher and cover operations for multiple orders, as was shown above for short operations. In all these cases, accounting for individual orders is questionable. Instead, accounting for the entire group over one time period replaces this; the presence times of the group members and the actual times for the rough-cut jobs delivered are placed in relation to the corresponding standard times. This is also precise enough for payroll purposes (compensation); moreover, "success" is measured not only in terms of actual processing times, but also includes interoperation times.
- For the detailed operation, it is not possible in this way to compare the standard load to the actual load. In the case of well-tuned production — or procurement — with frequent order repetition this is actually not necessary, not even for cost estimation. The measure of success becomes the efficiency rate of the entire group (which is all the standard load divided by all the actual load; see Section 1.2.4), and not the costing of single jobs.

For machine-oriented work centers, especially for NC, CNC, and flexible manufacturing systems (FMS), as well as for automated stock transport systems, the solution for the future lies in inexpensive sensors and in the link to the computer that performs shop floor control.

For manual work centers, it is important that the workers do not need to leave their posts for data entry purposes and that they do not need to enter their identification anywhere. The company can introduce inexpensive data collection units that make use of bar code readers or transponders. These data collection units should be located right at the workstation and linked to an intranet. The employee badge identifies the individual employee.

There is an observation with all the techniques used for measuring job shop processes: Collection of excessively detailed data can influence processes to such an extent that without measurement the outcome as a whole would be different. This type of measurement falsifies the process (by slowing it down, for example) and should not be implemented.

15.4 Distribution Control

Intended learning outcomes: Explain order picking, packaging, load building, and transportation to receiver.

Distribution, or *distribution control*, comprises the tasks involved in distributing (moving) finished goods from the manufacturer to the customer.

The shipping department readies finished products for delivery according to sales orders transmitted in the form of delivery proposals by sales and distribution handling. Where appropriate, sales and distribution handling monitors production or procurement orders and transfers finished goods or received incoming goods directly to the shipping department.

The sales orders are readied for shipment according to delivery notes.[11] They are handled in sequence or grouped together for one-time picking, mainly depending on the confirmed delivery date. Determination of delivery dates depends to a significant extent on the available distribution system. Decisions on the type of distribution network are made in the context of facilities location planning.

Warehousing describes the activities related to receiving, storing, and shipping materials to and from production or distribution locations ([APIC16]).

The distribution network structure (see Figure 3.1.3.1) determines the shipping distances for delivery of orders and the likely means of transportation for delivery. Although operations planning of shipments will take place later on (see Section 15.4.3), it is important to take operations into account as early as delivery confirmation, for it will affect the delivery date. Depending on the means of transportation, delivery dates are not arbitrary, for deliveries are grouped together, or collated, in delivery "tours" that are usually served cyclically.

Flexible distribution control is capable of monitoring customer orders, or the confirmed delivery dates of individual positions on orders, by continuous checks on the progress of production and procurement orders. This is similar to the "freight train" of customer order processing described in Section 1.3.3, which halts at particular stations to monitor the supply of goods and information from other trains. Changes in production or procurement completion dates require adjustments of the planned transports.

The actual shipping process encompasses order picking, packaging, assembling the shipment, and transport to the receiver. This is accompanied by administrative activities,

---

[11] A *delivery note* more or less corresponds to the customer order form.

such as preparation of supporting documents and packing slips; maintaining transport statistics; complaints about handling of hauling (damage) claims, and much more.

15.4.1 Order Picking

Intended learning outcomes: Explain discrete order picking and batch picking. Describe sequential picking and zone picking. Differentiate between decentralized goods preparation and centralized goods preparation.

Order picking, or simply *picking*, is the issuance of items from stocking locations for delivery. Items are issued according to a particular picking strategy.

The *picking strategy* is the type of order picking chosen.

The order picking process typically includes making goods available in storage units, picking the required quantities of goods, consolidating the goods according to the picking sequence, transporting the result to shipping, and return of part-picked storage units to storage.

Order picking facilities find implementation mainly in the distribution of finished goods and in the shipping of spare parts, but they are also needed internally for supplying assembly or production. There are four picking strategies, depending on the type of stocking system and the replenishment techniques employed, as shown in Figure 15.4.1.1:

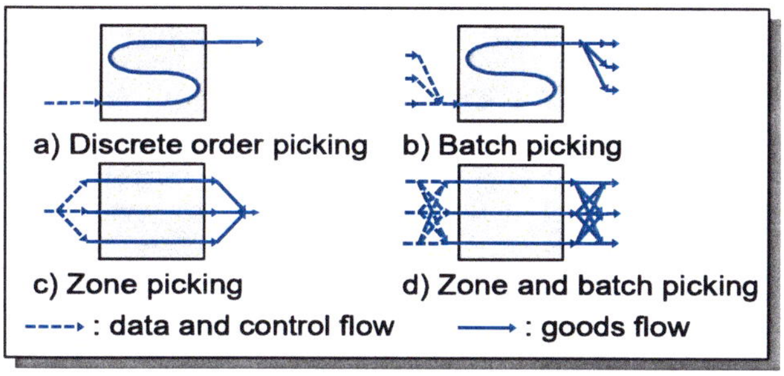

Fig. 15.4.1.1 Picking strategies. (Following [RKW-Ha]).

- In *discrete order picking*, orders are picked one after the other. The required accompanying document is a delivery note with the positions sorted in an optimal picking sequence. This optimal sequence is the shortest driving route through the warehouse. Accordingly, the specific sorting of the items of the delivery note is called a *picking list* (see also Section 15.2.1).
- *Batch picking* pools a number of orders and sorts the positions of all the corresponding items of the delivery notes together in an optimal sequence. The resulting picking list then permits all the products for delivery to be picked up in one trip through the warehouse. The individual shipments are then put together following the single, completed trip in a special secondary warehouse, in an *order picking store*, or in a *commission stock.* This procedure makes high-performance order picking possible, but it entails higher costs than discrete order picking, as both higher capital and operating costs result. Only where the range of products is very

large and there is a high volume of orders with few positions, such as in mail-order businesses, is batch picking a cost-effective solution.

There are further picking strategies dependent on both the size of the warehouse and the product structure:

- *Sequential picking*: A trip is made through the warehouse for each single order or batch of orders in its entirety.
- *Zone picking*, or *parallel picking*: The warehouse is segmented into a number of picking zones, and the single order or batch of orders is split into partial orders picked in parallel. In a further operation in a separate area of the warehouse, all partial orders are then placed together. This makes sense for very large warehouses to shorten the routes of individual pickers. There is also segmentation of the warehouse according to product, such as segmentation into different temperature zones for refrigerated and frozen goods or separate areas for flammable or hazardous chemicals or for products that should not be stored together in proximity.

Other distinguishing criteria are the type of storage and the movement of the picker:[12]

- In *decentralized goods preparation*, the goods are stored at constant locations, and the picker moves from one picking position to the next (called routing "people to the product"). The picker then moves the picked goods to an order consolidation area. Depending on the warehouse layout, the picker moves back and forth, or — in the case of multilevel storage with platforms that can be elevated and lowered or mezzanines with materials elevators — back and forth and up and down. This very common picking method is relatively simple to realize.
- In *centralized goods preparation*, goods are conveyed from storage to a permanent picking station, or kitting area (called routing "product to the people"). Conveyor belts, rack operation equipment, or stacking cranes move the goods. An important decision criterion for this type of goods storage is the issue of how to handle part-picked storage units from which items have been picked for an order. Part-picked pallets of product, for example, can remain at the pick station, but for space reasons, they often must be transported back to storage or to a special storage area. Control units of modern picking systems decide on the frequency of demand for the item (for example, using an ABC classification). Frequently demanded goods remain at the pick station, while other goods return to storage until needed.

Picking can range from manual to fully automated, depending on warehouse layout and type of picking system. Automation is possible for retrieving storage units, transporting them back to storage, dividing storage units (multiple-unit cartons or pallets) for single item picking, and finally, transporting the commissioned unit. Automation does not always require the use of robots. Separating single items can be accomplished by "automatic moving" rather than by gripping robots, whereby goods are lifted from a carton flow channel or caused to slide out of the channel.

---

[12] The "picker" can be the person removing the goods or various technical aids such as a stacker crane, grab, or picking robot.

Fully automated picking is a special case that is typically found in the pharmaceuticals industry and in mail-order businesses. For order consolidation, picking robots, automated conveyors, and other technical devices replace people entirely. This is only possible, however, if the items to be picked have similar dimensions (geometry) and are stable in shape (stiffness). In addition, the goods must be stored in precise arrangements, meaning that items must be stored in predetermined and dedicated storage locations and particular orientations to allow retrieval by automated equipment. Full automation is cost effective, however, only with high turnover and steady load of the facility.

For optimum processing of order picking, increasingly complex IT-supported control systems are being implemented. Warehouse management systems collate orders, create picking lists, calculate optimum picking routes, control and monitor traffic in the picking system (for example, the movements of picking robots), and, finally, document completion of the order. Modern systems also shorten retrieval times and retrieval routes by calculating the optimum design for the warehouse (that is, minimization of routes and replenishment efforts, with good utilization of floor space) and, when capacity is available, automatically trigger restocking of piles for faster goods retrieval.

The picking process ends when the item quantities have been consolidated to fulfill the order. The goods, usually not yet packaged, must now be prepared for shipment by the packaging department. Exceptions, however, are "pick and pack" operations, where goods are packaged during the picking process. If picking is incomplete, that is, not all the positions on a delivery note can actually be issued — whether foreseeable from the start or due to errors in inventory information — then the remaining positions of the order waiting for delivery can be split and put on a separate backorder.[13]

Order picking should choose an analogous procedure for putting accompanying materials together for *contract work*, that is, external operations. The delivery of accompanying materials is a legally binding event, just as true sales orders are. The only difference is that no invoicing results from this, since the accompanying materials ultimately remain the property of the company and are only temporarily "loaned out."

15.4.2 Packaging and Load Building

Intended learning outcomes: Present a conceptual framework to handle the diverse functions and requirements of packaging. Explain load building. Identify the levels of aggregation in load building.

> *Packaging* is the enclosure of goods for protection or other functions.
>
> The *packaged good* is the packaged product or product to be packaged.
>
> *The packing unit* is the quantity of packaged items per package, with reference to the item unit (for example, a case of 12 bottles).
>
> The *packaging function* is the reason for packaging.

[13] Split orders may have to be recombined for invoicing purposes.

Packaging plays a crucial role in logistics, as it is often only packaging that makes the goods produced distributable. Packaging has no function on its own; it is the goods packaged that determine the function. As soon as the product arrives at the place of consumption, packaging has fulfilled its purpose and becomes waste or material for re-use and recycle. The many possible functions of packaging can be grouped into five areas (see Figure 15.4.2.1).

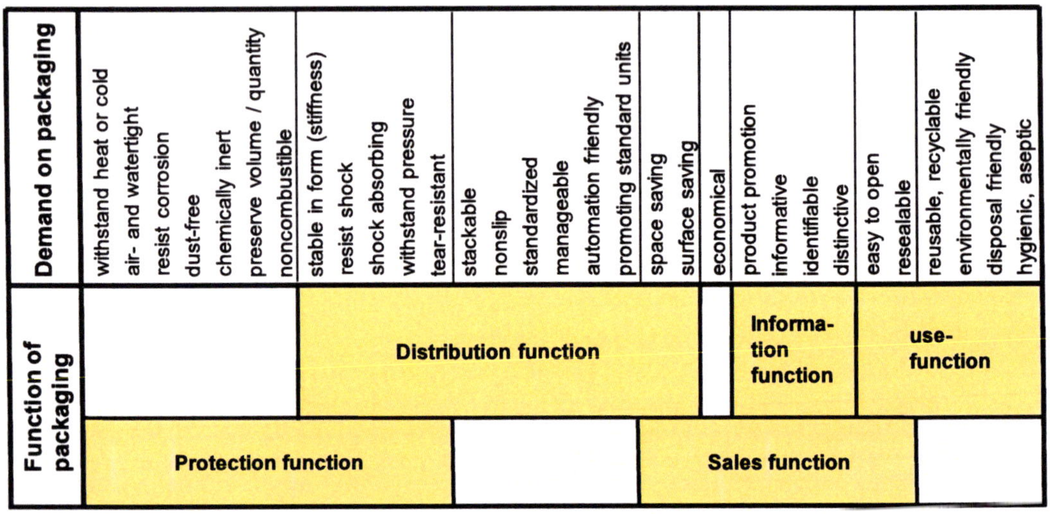

Fig. 15.4.2.1 Conceptual framework to handle the diverse functions and requirements of packaging. (Following [JuSc00]).

- The *protective function* is the classical function of packaging. *Active protective functions* ensure that the product reaches the end user undamaged. Packaging must protect packaged goods from inner and outer stresses: mechanical, chemical, physical, and biological. In addition, packaging can reduce pilfering. *Passive protective functions* protect the people, facilities, and other goods that are involved in distribution of the products.
- The *distribution function* of packaging supports storage, transport, and transshipment/reloading. The type of packaging has a significant effect on handling in the warehouse and utilization of storage and transport space. Well-thought-out packaging decisions can improve stackability, optimize space utilization, and simplify the implementation of technical devices. Reduction of the weight of packaging can reduce freight costs. The right packing — for example, the use of standardized load carriers like pallets and containers — can considerably improve cargo transshipment from one means of transportation to another at loading stations. For the most efficient packing, the dimensions of packaging will conform to standardized load carriers (for example, the 800 times 1200-mm European pool pallet).[14]

[14] The European pool pallet is a standardized block pallet introduced by the European railroads after World War II. This is the only pallet that should be referred to as a "EuroPallet." These pallets are produced by licensed manufactures and bear the "EUR" logo.

- Labeling and stamping produce the *information and sales promotion functions* of packaging. Legally required declarations of contents for foodstuffs or hazardous materials belong here, as do also printed instructions for transport, handling, or storage. Moreover, packaging can serve marketing purposes. This promotion function gains in significance the closer a product comes to being a consumer good. Self-service sales, for example, where there is no contact at all between producer and customer, rely on modern packaging, and it is the most significant component of the company's product-market positioning policy. Packaging attracts the attention of the customer and creates an association to the product. More and more, manufacturers mark products with EAN or UCC/UPC identification numbers[15] or suggest retail prices for easier handling on the part of retailers and customers.
- The *use function* refers to two things: the customer's handling of the packaging and the reusability and recyclability of packaging. Environmentally friendly packaging is becoming ever more important. Multiway, or return, packaging is gaining customer acceptance.
- The *sales function* overlaps with all the other functions listed above but adds the demand for economical design of packaging for cost-reduction purposes. There is increasing cooperation between industry and commercial enterprises in the area of sales and retail packaging. This is particularly important for self-service sales, because unpacking shipping cartons, stocking the shelves of the store, and labeling products and display shelves are extremely labor intensive. For this reason, packaging is becoming much more store-friendly (store-ready shipments for better flowthrough) and is even being designed to fit the dimensions of store shelving units (shelf-ready for better presentation).

The *packaging system* comprises the packaged good, the packaging or packaging materials, and the packaging process.

The three elements of the packaging system are closely intertwined. The choice of packaging is determined by the characteristics of the packaged goods and the functions that packaging must fulfill. The packaging materials in turn determine the packaging process. For example, they determine the type of machine required for forming, filling, and sealing. Conversely, to allow automated processing at all, packaging machines place much higher demands on the packaging materials than manual packing does.

Packaging is produced from *packaging materials*.

Various materials are used for packaging: paper, cardboard, corrugated board, plastics, metal (steel), aluminum, glass, wood, rubber, textiles, and multilayer materials (composites). Each packaging material has its own unique properties that can be utilized to fulfill the packaging function. The choice of packaging materials is also determined by recycling and return

---

[15] EAN is the European Association of Numbers. In the United States, retail items are identified with UPC codes (Uniform Product Code), which are created using a membership number provided by the UCC (Uniform Code Council). The 12-digit-long UPC-A bar codes were for a long time not the same as 13-digit EAN13 bar codes used in retail point of sale everywhere else in the world.

packaging considerations. Return and recycling can engender additional costs that may be prohibitive.

Packing and marking is made up of all necessary activities to package the good.

These activities include supplying empty packaging and the goods to the packing facilities, setting up and filling the packaging, marking and labeling, and preparation of the packaged units for transport. Support through packaging machines is common. Some examples are bagging machines, filling machines, form-fill-seal machines, can-filling machines, flat-bag packaging machines, palletizing machines, and overwrapping machines.

Load building is the grouping and consolidating of items for transport.

The *load unit* is the grouping of packing units for transport.

Packing units are placed on or in unit-load supports, such as pallets, trays, or containers, and secured with load stabilizers (bands, lashing belts, adhesives, stretch wrap, and so on), to facilitate handling, storage, and transport. The choice of the unit loading aid is highly dependent on the specific means of transportation (see Section 15.3). For truck trailer transport, for example, pallets are used, whereas containers are frequently used for air or sea freight.

The *transport unit* is the number of load units per unit of the means of transportation (container, truck, rail wagon, etc.).

This process of successive consolidating is shown in Figure 15.4.2.2.

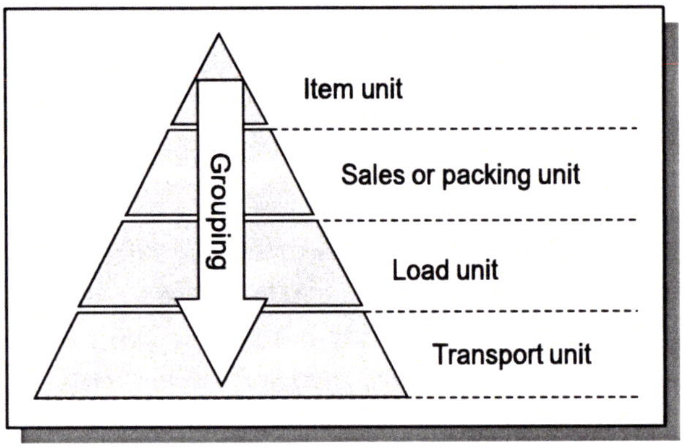

Fig. 15.4.2.2 Levels of aggregation in load building.

The necessary accompanying documents must be readied during the packaging process or, at the latest, during load building. These may include item-related instructions for use or transport-related documents such as bills of delivery, export authorization, export transfer notes, certificates of origin, international customs declarations, and the like.

Interactions between packing units, load units, transport units, and the logistics system will significantly influence economic efficiency. For example, optimal packaging allows improved and more cost-effective transport, savings in (intermediate) storage costs, and even improved sales of the packaged product. Understanding and taking into account all of these factors can result in satisfactory economic efficiency.

15.4.3 Transportation to Receiver

Intended learning outcomes: Describe the selection of the mode of transport. Explain the transport chain from supplier to receiver and various kinds of organizing transshipment points. Present routing and scheduling and the optimization of loading space.

Following picking of the goods to be shipped and packing, the next step is planning the transport of the goods to the receiver. Transport itself is often outsourced to a third-party logistics provider. The specific distribution network structure resulting from storage locations planning determines the distances of the routes and the choice of means of transport.

Transport planning and scheduling involve finding solutions in three problem areas: selection of the mode of transport, shipping route planning, and loading space optimization.

Fair and Williams (in [Ross15]) define a number of objectives that should be achieved through transport planning and scheduling. The most important objectives are most continuous flow of goods through the distribution network; optimal, load-specific selection of the mode of transport; minimization of number of vehicles; standardization of loading aids (pallets, containers); and maximization of capacity utilization (capital, equipment, personnel).

Figure 15.4.3.1 shows interactions and mutual influences among the three main transport planning and scheduling tasks.

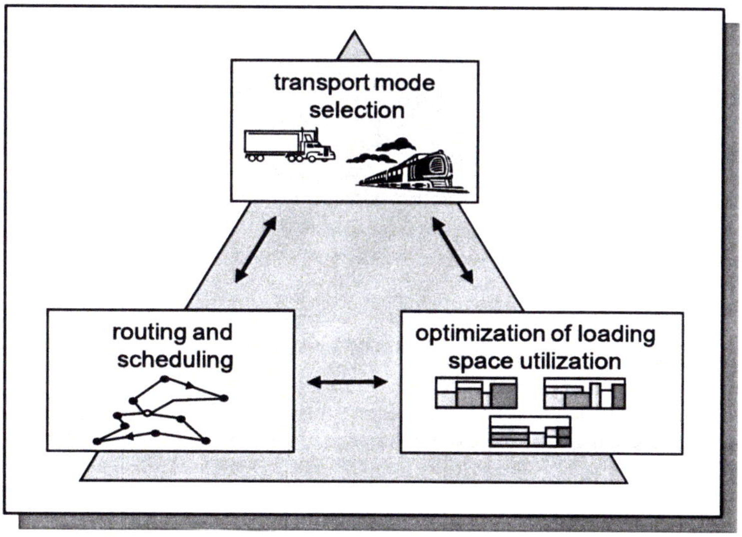

Fig. 15.4.3.1 Problems of transport planning and scheduling. (Following [Stic04])

The *selection of the mode of transport* is largely determined by the type of load. Bulk goods loads, that is, unpackaged substances in the form of solids, liquids, or gases, entail other requirements as to the mode of transport than loads that are made up of discrete standard loads like containers, packages, pallets, or sea containers. The specific nature of the goods to be shipped stipulates further requirements: The goods may be perishable, combustible, explosive, sensitive, or prone to shrinkage.

Possible modes of transport away from the company are road motor vehicles, railways, ships, and aircraft. The transport chain may integrate both company-owned vehicles (such as trucks) and public modes of transport. For bulk goods, pipeline systems are possible. Bowersox (in [Ross15]) outlines six criteria that influence the decision on mode of transport: speed, "completeness" (using the least possible number of different modes within one distribution channel), dependability, capability (not all goods can be transported via all modes of transport), transport frequency, and costs.

For the delivery of a transportation order, a combination of transport modes can form a transport chain. A distinction is made among *direct course* (no interruption from supplier to receiver), *"pre" course* (from supplier to transshipment point), *"post" course* (from transshipment point to receiver), and *main course* (from transshipment point to transshipment point). The advantages of individual modes of transport can be utilized for the various legs of the transport. Because of their flexibility, trucks are often used for *"pre"* and *"post" course*, whereas for the main course over great distances, the choice falls on air or water transport. Figure 15.4.3.2 shows some examples of transport chains.

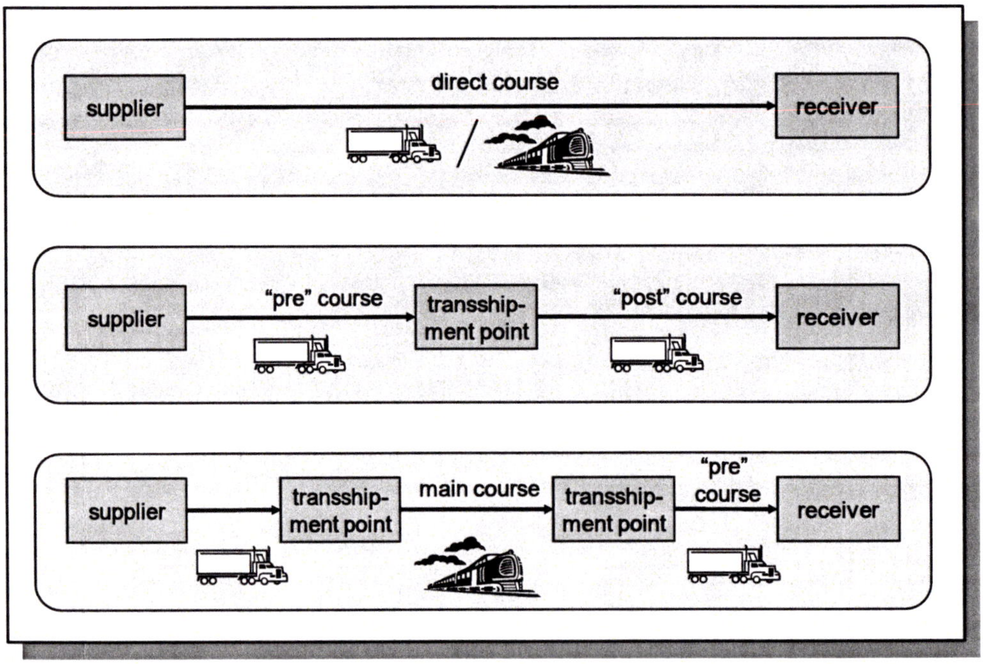

Fig. 15.4.3.2 Transport chain from supplier to receiver.

When changing the means of transportation, there is the problem of getting the goods from one modality to another. While the transfer of the goods can be simplified through the use of standardized loading aids like containers or pallets, transferring goods entails special handling equipment (gantry crane, winch, lifting platform, chute, and so on), time and personnel, and associated costs. The following concepts are gaining in importance:

- *Cross-docking*, or *direct loading*, is the concept of load building on the incoming vehicle so that the packaged goods can be easily carried at the transshipment point to the outgoing vehicle, without being stored in temporary inventory (cf. [APIC16]).
- The objective of *combined transport* is to transport goods using two modes of transport in combination, such as rail/road, in the best way possible so as to utilize the advantages of each. This is achieved through the use of intermodal transport units (container, swap body, or semi-trailer/goods road motor vehicle).
- *Trailer on flat car [TOFC] transport* is a synonym for road transport that is in part moved by rail. Semitrailers or entire road trains (drawbar-trailer combinations) are loaded onto trains. The major part of the journey is by rail; the final leg for delivery to the customer is carried out by road.

Routing and scheduling determine the order in which a means of transportation will reach the individual stations (customer, transshipment stations, warehouses, and so on).

The objective is to deliver to all customers in a delivery area at minimal cost. Strategic route optimization is required. The return movements of empty road vehicles must also be planned (direct return following delivery, loading of a return load, empty load minimization). Routing and scheduling are complex optimization problems that must be solved, taking account of numerous constraints and restrictions, such as weight and volume constraints, distances, delivery time windows, and more. Planners often implement algorithms from "operations research." Using so-called "opening heuristics," an initial tour is scheduled and then optimized through the use of "improving heuristics." For a detailed description of this procedure, see, for example, [DoSo10].

Optimization of loading space is closely connected to the two problems outlined above.

Selection of the means of transportation and scheduling of the tour results in the assignment of a definite number of load units to each means of transportation (for example, truck or rail wagon). The next task is to load the goods into the units with optimal utilization of the loading space, leaving the least possible space unused. Planners may again use heuristic methods from the field of "operations research" to achieve optimization of the loading space.

Transport planning and scheduling produces transportation orders, which trigger the physical transportation of goods.

The *transportation order* stipulates the time and place for pickup of a particular number of load units and the time and place they will be delivered.

This order can contain a single load unit, several load units referring to a unique delivery order, or the optimized, combined transportation of several delivery orders. The transportation order usually sets a delivery time window or a maximum delivery lead time.

Consolidation is a term for packages and lots that move from suppliers to a carrier terminal and are sorted and then combined with similar shipments from other suppliers for travel to their final destination (see [APIC16]).

For the supplier, the advantage of this transshipment is daily deliveries of various goods to various receivers. This is to the customer's advantage, as well. However, the advantage must be weighed against the costs of trans-shipment. Routing and scheduling with consolidation is usually a job for the transport company, freight forwarder, or third-party logistics provider. Simple cases of consolidation are called *milk runs*, or regular routes for pickup of mixed loads from several suppliers (see [APIC16]).

Transportation control consists of monitoring the route movements of the transport units, monitoring traffic conditions and delays, and registering and evaluating disturbances.

Transport controlling of external transport systems is usually managed via a control center that controls, monitors, and coordinates transports in dependency on actual conditions and contingencies. Today, drivers can communicate with central control via mobile data terminals, cellular phones, and personal digital assistants. The wireless Internet allows for the use of small handheld devices for interfacing with central control and accessing traffic and weather information. Some vehicles are also equipped with satellite navigation systems, such as global-positioning-system (GPS) receivers, that compute the location of vehicles in real-time and allow tracking by central control.

Whereas the delivery note used to suffice as the supporting document, today the entire goods flow is usually managed electronically. This makes standardization of communication means essential, so that uniform monitoring is possible in intermodal transport chains.

- Via scanning of the *bar code* on the goods, the legal "passing of the risk" that occurs at transshipment is documented.
- *EDIFACT* (electronic data exchange for administration, commerce, and transport) is one of the format standards that have been created for information technology support of transport control.
- *RFID* or another transponder technique for worldwide self-identification of goods and use electronic product codes (EPCs).
- *Tracking and tracing* of package deliveries is now offered by many transport service providers. Using the World Wide Web via the internet, customers can view information on their shipments (identified by transponders, for example).

Transport planning must also consider outsourcing logistics tasks to specialized distribution companies (self-owned or third-party logistics providers). With their focus on core competency and due to consolidation effects, the result can be a significant reduction in operating costs as well as improved efficiency, service, and flexibility. Companies can put these advantages to good use, for they stand under the growing pressure to lower costs and, at the same time, to meet higher demands from customers regarding service, price, and delivery capability. Courier services and express carriers are becoming more and more a part of logistics chains, in particular to fulfill just-in-time deliveries. For an extensive discussion of distribution tasks, see, for example, [Ross15], [Pfoh18], and [MarA95].

15.5 Summary

In the short-term time horizon, the company releases order proposals that stem from long- or medium-term planning. In the same time horizon, unplanned sales are also realized that have to be delivered as soon as possible. If available, the items can be delivered from stock. Otherwise, creation and release of production or procurement orders is necessary.

An order release generally includes a test of resource availability. In addition, the techniques of materials, scheduling, and capacity management are put to use, independently of the type of data processing support available or the person performing the function. Scheduling calculations provide the necessary start dates for the operations. The components and the capacity must be available at the start date.

Specific techniques were developed for the release of multiple orders. *De facto*, they are techniques of control of operations as well. Load-oriented order release (Loor) is a generalization of order-by-order planning with limited capacity. For each work center, the capacity of a planning period is first multiplied by a loading percentage and then balanced with the load for all future periods. The load of later operations is thereby subject to a conversion. Operations that cannot be scheduled due to lead time calculations result in rejection of the order. Capacity-oriented materials management (Corma) provides an early release of stock replenishment orders when generally fully utilized capacities are available. More urgent customer production orders may interrupt the processing of these orders. Continual rescheduling, which uses the technique of probable scheduling, is meant to speed up or slow down the orders in a timely fashion. Furthermore, completion dates for stock replenishment orders are adapted to the actual consumption on an ongoing basis.

Shop floor control includes the issuing of accompanying documents for procurement and production. At least, there is an order document, possibly in electronic form. In production, there are also shop floor routings, picking lists, parts requisitions, and operation cards. Scheduling of detailed operations, the allocation of work to the single persons and machines, the assignment of production equipment, and the sequencing of the orders for each workstation then follow. Ideally, the person executing the operation should perform these functions.

Shop floor data collection records the use of resources. It includes the issuance of goods and the executed (internal and external) operations. It also yields information about the progress of orders, if this is not tracked separately. Shop floor data collection is necessary to ensure updated planning of the availability of goods and capacities. It also serves as a preparation for order costing, readjusting load standards, and quality assurance. Completed or incoming orders undergo quality control, sometimes using a quality control sheet. Results determine whether parts of the lot are accepted or rejected and sent on to rework or replacement.

Automatic shop floor data collection offers speed, but it usually incurs higher costs as well. The benefits gained through precise data collection, which lie in better knowledge and control of the production processes, must justify the cost. For short operations or in group-work organization, the actual time can be measured at reasonable expense only for rough-cut operations. The measure of performance is then the efficiency rate for the group as a whole and not the costing of a single job.

Distribution control encompasses the tasks involved in distributing finished goods to the customer. Once the distribution network structure has been chosen, the tasks include order picking, packaging and load building, and transport to the receiver. Picking can follow various strategies and may be order or item oriented. Packaging serves a number of functions, such as protection, distribution, information and promotion, use, and sales. Load building is the successive grouping of production units in packing units, load units (on pallets, for example) depending on the mode of transport, and finally in transport units. For transport to the receiver, solutions must be found for choice of transportation mode, routing and scheduling, and optimization of the loading space.

15.6 Keywords

allocation, 638
anticipation time, 647
automatic identification and data capture (AIDC), 660
availability test, 638
backflush technique, 657
badge, 660
bar code, 660
batch picking, 663
control, 636
control board (Leitstand), 655
Corma (capacity-oriented materials management), 645
critical ratio, 647
demonstrated capacity, 658
discrete order picking, 663
dispatching, 652
distribution control, 662
downtime, 658
EAN (European Association of Numbers), 667
electronic product codes (EPCs), 660
finite forward scheduling, 653
firm planned order, 639
gateway work center, 646
load filter, 641
load-oriented order release (loor), 640
manufacturing execution system (MES), 651
mixed manufacturing, 645
mixed-mode manufacturer, 645
operations scheduling, 652
order picking, 663
order release, 636
order termination report, 659
packaging, 665
packing and marking, 668
parts requisition, 651
picking, 663
picking list, 652
process control, 659
progress checking, 658
purchase requisition, 637
quality control, 659
returns, 659
RFID, 660
rough-cut data collection, 660
routing and scheduling, 671
sequencing, 655
shop floor control, 651
tracking and tracing, 672
transponder, 660
transport planning and scheduling, 669
transport unit, 668
transportation order, 671
traveling card, 639
two-bin inventory system, 639
unplanned demand, 637
UPC (uniform product code), 667
warehousing, 662
zone picking, 664

15.7 Scenarios and Exercises

Intended learning outcomes: Calculate examples for load-oriented order release (Loor) and for finite forward scheduling. Assess characteristics of capacity-oriented materials management (Corma) and of order picking.

15.7.1 Load-Oriented Order Release (Loor)

Intended learning outcomes: Explain the Loor algorithm with given data for a scheduling problem, considering anticipation, loading percentage, and conversion factor.

The first table in Figure 15.7.1.1 shows five orders with their sequence of operations. The data for each operation include the work center, the standard load (e.g., setup plus run time), and a blank column for entering the converted load.

The second table in Figure 15.7.1.1 shows parameters for load-oriented order release, as introduced in Section 15.1.2, as well as their values given for this exercise. The third table holds data for each work center, namely, the weekly capacity, the existing (pre-)load before loading the five orders, a blank column for entering the capacity upgraded by the loading percentage, and blank columns for the summarized load after releasing orders 1 to 5 (that is in the sequence given by the Loor algorithm).

		1st operation			2nd operation			3rd operation			4th operation		
Order no.	Start-date	Work center	Standard load	Converted load	Work center	Standard load	Converted load	Work center	Standard load	Converted load	Work center	Standard load	Converted load
1	16.06.	A	100		B	60		C	480		D	240	
2	18.06.	B	40		C	120		A	120				
3	22.06.	A	40		C	30		B	20				
4	29.06.	C	40		D	60		A	20				
5	06.07.	A	30		B	40		D	100		C	120	

Today:	14.06.
Time period:	1 week
Anticipation horizon:	3 weeks
Loading percentage:	200%
Conversion factor:	50%

Work center	Weekly capacity	Cap. with loading %	Pre-load	Summarized load including orders 1	2	3	4	5
A	200		265					
B	100		150					
C	300		340					
D	100		160					

Fig. 15.7.1.1 Given data for a Loor problem.

a. Load the five orders according to the Loor algorithm.

b. What would have happened if for operation 3 of order 2 the standard load had been 200 units of time instead of 120?

c. Discuss whether in your solution the treatment of order 3 was efficient.

d. What would have happened if order 3 had been loaded before order 2?

Solutions:

a. The time filter eliminates order 5. This order is declared as not urgent. For the other orders, the conversion factor is applied to their operations. In the third table, the loading percentage multiplies the weekly capacity. Then, order 1 is loaded, followed by order 2. Order 2 is accepted, but it overloads work center B (220 units of time against 200 units resulting from the loading percentage). Hence, order 3 cannot be loaded, because its last operation is at work center B. However, order 4 can be loaded, since it has no operation at work center B.

b. Order 2 would have overloaded work center A. Hence, order 4 would not have been loaded.

c. The converted load of order 3 on work center B had only 5 units of time. This would have changed the total load only very slightly. As there was no overloading of other work centers by orders 1, 2, and 4, it might have been wise to release order 3 as well.

d. Order 3 would have overloaded work center A (405 units of time against 400 units resulting from the loading percentage). Therefore, the algorithm would formally reject both orders 2 and 4. This would result in a low utilization of the other work centers B, C, and D.

15.7.2 Capacity-Oriented Materials Management (Corma)

Intended learning outcomes: Describe results of applying the capacity-oriented materials management (Corma) principle in order release.

Applying the capacity-oriented materials management (Corma) principle has which of the following results?

I Evenly distributed extension of the manufacturing lead time for all the orders

II Minimum amount of work in process

III Maximum utilization of the generally well-utilized work centers

a. II only

b. III only

c. I and II only

d. II and III only

Solution:

The answer is (b), or "III only." In fact, the early release of an order implies an extension of its lead time, because it will wait as soon as there are (unplanned) customer orders. The latter will be performed with minimal lead time. Thus, I is not true. II is not true, either, because of the very presence of early released orders. However, III is true: A bottleneck capacity is loaded with nonurgent (i.e., early released) orders as soon as there is available capacity.

15.7.3 Finite Forward Scheduling

Intended learning outcomes: Perform finite forward scheduling for eight products manufactured on three machines by using a Gantt-type chart.

Your company owns one lathe (M1), one milling machine (M2), and one drilling machine (M3). A working day lasts eight hours. As Figure 15.7.3.1 shows, eight products (P1, P2, P3, . . . , P8) are manufactured on these machines. Each product loads these machines in a different sequence. For simplicity, assume that there is no interoperation time.

Product	1st operation		2nd operation		3rd operation	
	Machine	Load (h)	Machine	Load (h)	Machine	Load (h)
P1	M1	3	M2	4	M3	5
P2	M2	2	M1	3	M3	2
P3	M3	4	M1	3	M2	1
P4	M2	3	M3	2	M1	4
P5	M3	3	M2	3	—	—
P6	M2	4	M1	3	M3	3
P7	M3	1	M1	2	—	—
P8	M1	3	M3	4	M2	3

Fig. 15.7.3.1 Eight products manufactured on three machines.

Perform finite forward scheduling for the next three days. The normal working time of 8 hours per day has to be respected, as do the sequence of the operations for each order given by Figure 15.7.3.1 and the following three priority rules:

1. No idle time on the machine
2. Operation with the shortest processing time
3. Longest remaining lead time for the order

The Gantt-type chart planning board in Figure 15.7.3.2 will help you to perform the task. Note the first orders on each machine. The order for product P1 has been chosen for machine M1 because of the third priority rule.

Discuss whether other priority rules would result in a better solution with regard to work in process.

Solution:

The total load is 21 hours on machine 1, 20 hours on machine 2, and 24 hours on machine 3. Thus, machine 3 is fully loaded, and priority rule 1 makes full sense. There are solutions for this problem that schedule the other two machines without idle time, respecting the sequence of operations for all eight orders. One of these solutions can be found by simply following the priority rules.

Replacing the second and the third priority rule by the rule *shortest remaining lead time* would result in considerably less work in process. However, strict application of this rule not only results in idle time on machine 3, but also creates delays for order 3 and order 6: They cannot be finished at the end of the third day. Both effects cannot be tolerated because these orders are started too late. As a consequence, there must be some rule giving them priority at some time, thereby augmenting work in order.

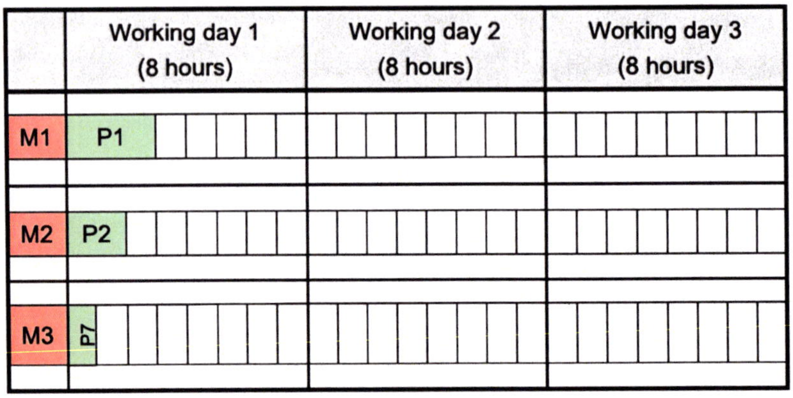

Fig. 15.7.3.2 Gantt-type chart for finite forward scheduling.

15.7.4 Order Picking

Intended learning outcomes: Differentiate between the main characteristics of several picking strategies, by listing the advantages and disadvantages of each, and deriving possible fields of application.

As depicted in Figure 15.4.1.1, discrete order picking, batch picking, sequential picking, and parallel, or zone, picking result in four common picking strategies. Point out the main characteristics of the following picking strategies. List the advantages and disadvantages of each. Derive possible fields of application:

a. Sequential, discrete order picking

b. Zone, or parallel, batch picking

Solution:

a. Sequential, discrete order picking

Characteristics:

- Most common method of picking
- Pickers fill all open positions of an order before work on picking the next order can begin
- Based on a picking list that contains an optimal routing

Advantages:

- Maintains order integrity
- Minimum of organizational efforts
- Simple to execute and easy to control
- Direct fill responsibility

Disadvantages:

- Required time for picking
- Decreasing efficiency with growing order size
- Large number of pickers needed

Possible fields of application:

- Small warehouses, low inventory turnover, low performance, small orders

b. Zone, or parallel, batch picking

Characteristics:

- Several orders are aggregated by product (as batch), the entire batch withdrawn, and the discrete orders reassembled in a consolidation area
- Batches are picked parallel in different zones of the warehouse and then merged in the consolidation area

Advantages:

- Reduced travel and fill times
- Low picking time due to parallel zones
- Improved supervision of order completion in consolidation area
- Increased picking accuracy and productivity due to zones
- Picker familiarity with zone products

Disadvantages:

- Double handling and sorting in the consolidation area
- Space and labor for consolidation area
- Difficult tracing and control of orders
- Requires high-volume picking

Possible fields of application:

- Large orders, high number of orders, large warehouses, products with different storage requirements (e.g., flammable goods, refrigerated goods)

15.8 References

APIC16 Pittman, P. et al., APICS Dictionary, 15th Edition, APICS, Chicago, IL, 2016

BaBr94 Bauer, S., Browne, S., Bowden, M., Duggan, A., Lyons, J., "Shop Floor Control Systems," 2nd Edition, Chapman & Hall, London, 1994

DoSo10 Domschke, W., Scholl, A., "Rundreisen und Touren," 2. Band der Reihe Logistik, 5. Auflage, Oldenbourg Verlag, München, 2010

JuSc00 Jünemann, R., Schmidt, T., "Materialflusssysteme: Systemtechnische Grundlagen," 3. Auflage, Springer, Berlin, 2000

Knol92 Knolmayer, G., "A Widely Acclaimed Method of Load-Oriented Job Release and Its Conceptual Deficiencies," Arbeitsbericht No. 29 des Instituts für Wirtschaftsinformatik, University of Bern, 1992

Loed11 Lödding, H., "Handbook of Manufacturing Control: Fundamentals, Description, Configuration," Springer, Berlin, 2011.

MarA95 Martin, A., "Distribution Resource Planning," 3rd Edition, John Wiley & Sons, New York, 1995

Pfoh18 Pfohl, H.-Ch., "Logistiksysteme: Betriebswirtschaftliche Grundlagen," 9. Auflage, Springer, Berlin, 2018

RKW-Ha RKW-Handbuch Logistik, "Integrierter Material- und Warenfluss in Beschaffung, Produktion und Absatz," ergänzbares Handbuch für Planung, Einrichtung und Anwendung logistischer Systeme in der Unternehmenspraxis, Erich Schmidt Verlag

Ross15 Ross, D.F., "Distribution: Planning and Control," 3rd Edition, Springer Netherlands, 2015

Sche98b Scherer, E., "Shop Floor Control — A Systems Perspective: From Deterministic Models towards Agile Operations Management," Springer, New York, 1998

Schö95b Schönsleben, P., "Corma: Capacity Oriented Materials Management," *Proceedings of the APICS World Symposium*, Auckland, Australasian Production and Inventory Control Society, 1995

Stic04 Stich, V., "Industrielle Logistik," 8th Edition, Wissenschaftsverlag Mainz, 2004

Wien95 Wiendahl, H.P., "Load-Oriented Manufacturing Control," Springer, Berlin, New York, 1995

16 Cost Estimation, Job-Order Costing, and Activity-Based Costing

Intended learning outcomes: Produce a review on costs, cost elements, and cost structures. Explain cost estimation in detail. Describe job-order costing. Disclose activity-based costing.

Figure 16.0.0.1 shows the reference model for business processes, and planning & control tasks, from Figure 5.1.4.2, and highlights the tasks and processes that we will examine in this chapter. Section 5.1.2 provided an overview of the topic discussed here.

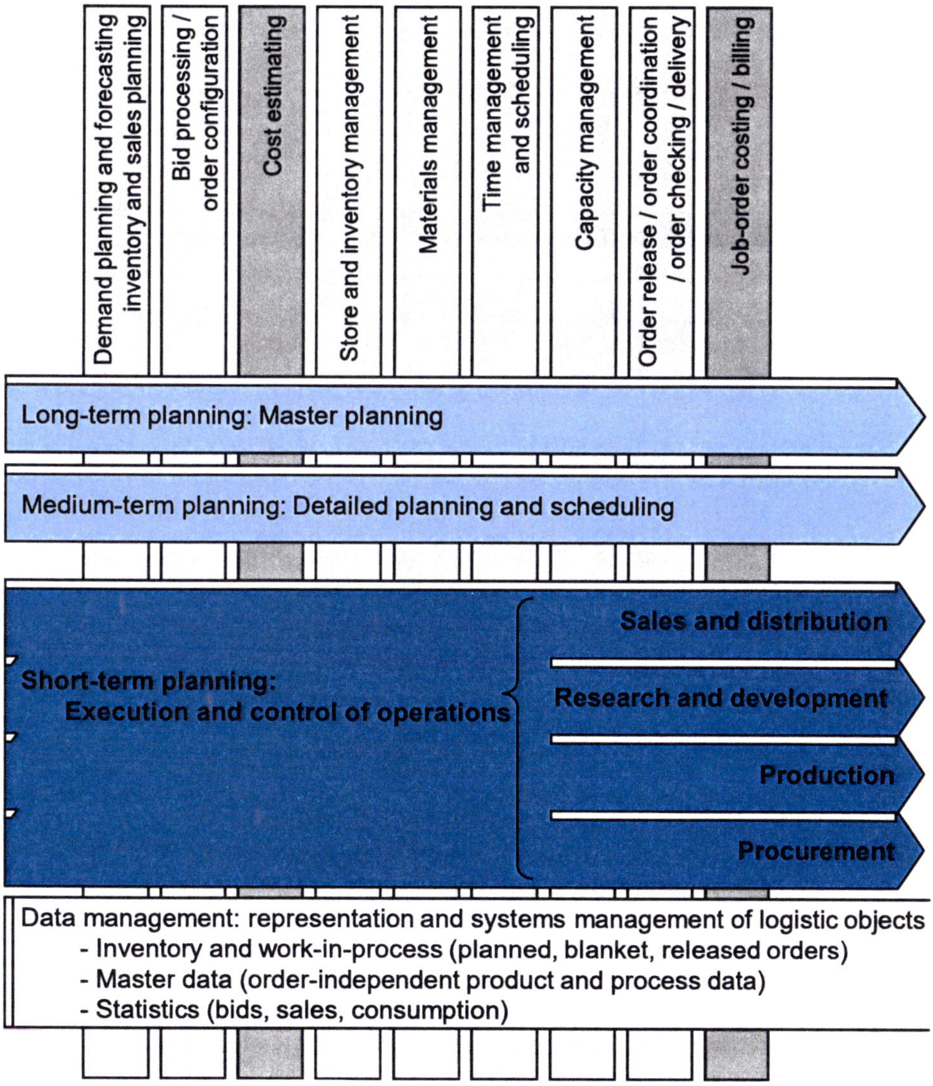

Fig. 16.0.0.1 The parts of the system discussed in this chapter (shown on darker background).

P. Schönsleben, *Handbook Integral Logistics Management*,
https://doi.org/10.1007/978-3-662-65625-9_16

Information on costs and pricing is vital to improve managerial decision-making in the area of sales and marketing: What is the cost of goods manufactured? How large is the profit resulting from an order, or, at the least, what fixed costs contribution margin does the order generate? How will varying the consumption of resources affect the costs of individual products or the total costs for the organization?

This chapter does not aim to provide an overview of financial and cost accounting, nor does it provide a detailed presentation of the various financing, costing, and cost accounting methods. See here, amongst others, [KeBu93], [Habe08]. However, since all cost object accounting, and thus also product and project costing, is based on the planning & control system — or more precisely on master data or order data — the chapter will address the issue of how administrative logistics manages and determines the various elements needed to calculate the cost of goods manufactured.

Job-order costing identifies and accumulates all the costs generated by an order.

Job-order costing on an ongoing basis allows comparing the costs incurred during production or procurement against target, or estimated costs. *Feedback*, or data flow from the shop floor data collection system, immediately signals any variances from these standards. Retrospective cost accounting systems generally have the disadvantage that they are applied too long after the actual events, when it is often impossible to identify the causes of the variances.

Cost estimation for a product or order identifies and accumulates all the costs likely to be incurred when manufacturing a batch.

As the most detailed master data are captured in the ERP system, it is possible to perform a simulation of the orders. So it is easy to perform preliminary calculations in advance for any variations in bills of material, routing sheets, or cost elements.

One of the major problems in identifying and accumulating costs is how to assign fixed costs, or overhead costs, to cost objects. Conventional cost systems assign these costs in relation to the number of product units manufactured, using, for example, direct-labor hours or direct material costs as a basis to assign production overhead. Activity-based costing, or activity-based cost accounting (ABC), is an instrument that focuses on the fixed costs (overhead) of repetitive processes. It is a more accurate costing method, for it traces expense categories to the particular cost object, making "indirect" costs "direct." ABC is based on management of the highly detailed master data in the planning & control system. The chapter will provide a detailed example to show what introducing ABC as a costing method entails.

16.1 Costs, Cost Elements, and Cost Structures

Intended learning outcomes: Produce an overview on actual, direct, and overhead costs. Differentiate between average costs and standard costs as well as between variable costs and fixed costs. Explain the cost accumulation breakdown that is the cost breakdown structure of a product.

16.1.1 Actual, Direct, and Overhead Costs

Intended learning outcomes: Describe actual costs. Differentiate between direct costs and overhead costs.

The *actual costs* of an item are the costs that were incurred when that item was last produced or procured. They refer to the item's unit of measure.

The revenue generated by a customer order can be compared against its costs. This allows determination of the profitability of the order. This is particularly useful if the sales price fluctuates greatly or the cost of procured materials used in production varies — such as when the purchasing department takes advantage of large quantity discounts or special offers.

In contrast to costs, the revenue generated by a sale is often easy to identify. Cost accounting subdivides expenditures into a number of alternative pairs, such as direct and overhead costs.

Direct costs are the costs that can be identified specifically with or traced to a given cost object (a product, service, or order, for example).

Direct costs are, for example, costs for *direct labor*, such as wages or external operations, or for *direct material*, such as purchased components needed to produce the order.

Overhead costs cannot be identified specifically with or traced to a given cost object. Such costs must be allocated across the various cost objects (e.g. products, services, or orders).

Typical examples include the costs of plant and operating equipment (machinery, devices, tools), depreciation, rent, lighting and heating, and management and administration costs.

In practice, actual costs may change frequently over the course of a year. Irregularities in procurement (breakdowns, scrap, discounts, special promotions) cause the actual costs to fluctuate considerably. There are also fundamental problems associated with calculating the cost of a sales order on the basis of the actual costs:

Firstly, many of the costs incurred within an organization are of an indirect nature, even overhead costs. To allocate overhead costs to the individual products or orders, we need some sort of "fair" distribution formula as a basis for apportionment. This is often a percentage of sales, measured against direct costs. An alternative is to base allocation on cost rates per labor or machine hours, based on forecasts.

Secondly, when items are issued for sale or for assembly in a higher-level end product, it is important to specify the associated production or procurement order to allow further calculation of the actual costs of the orders. To be able to do this, inventory management must keep accounting records according to production or procurement batch or charge. Issues are then always allocated to a particular batch. Lot control will provide the necessary documentation (indeed, this procedure is mandatory in the process industry).

Thirdly, job-order costing must be carried out as soon as possible after the order is completed. Invoices, the source of information on actual costs for external operations and for components purchased directly for the order, must be received within a reasonable period. Analysis of variances from budgeted costs becomes more difficult as more time elapses between the cost event and cost control. Various data regarding the event, particularly informal data, are often not registered and are thus no longer available at the time of the analysis.

16.1.2 Average Costs and Standard Costs

Intended learning outcomes: Differentiate between standard costs and average costs. Explain standard cost rates.

Many organizations have introduced standard cost accounting systems because of the difficulties associated with actual costing systems.

Standard costs are an estimate, a prediction, of actual costs.

Standard costs are used as the basis for budgeting and for analyzing variances (the differences that arise between targeted and actual results) in job order costing. Standards for costs, quantities, and times are also a useful means of cost estimation for a new product, particularly if it is comparable to previous products. In general, standard costs are determined on the basis of the average costs.

The *average costs* for an item are the average last-in costs of this item. They refer to the item's unit of measure.

Average costs can be determined using the same techniques that were described for historically oriented forecasting in Section 10.2.

At the end of the budget period, such as at the end of every year, the average costs are carried over as the new standard costs. Here it is important to consider factors similar to those outlined in Chapter 10 for forecasting techniques, in particular for trend forecasting. At this point, cost accounting also determines the new standard cost rates.

Standard cost rates for labor costs per work center include, firstly, as direct costs, the expected wage rate for the workers, and secondly, overhead costs, for which cost accounting establishes the depreciation requirements and divides them by the load forecast expressed in capacity units, for the new budget period.

For every operation, the same principle applies: calculation of the average values for the standard load of an operation, the setup load of an operation, the setup time, the run load of an operation, and the run time (see Section 13.1.2 for an explanation of these terms) on the basis of the actual load recorded during the processes. These values are then combined with other measurements to determine standard quantities and standard loads.

As far as possible, standard costs, cost rates, quantities, and times should not change over the course of a budget period. However, it may be necessary to modify standard values over a budget period if the average values vary widely from these standard values.

As a *prerequisite* for calculating standard costs and quantities, the processes must be easy to measure and occur sufficiently frequently to allow the calculation of a statistical mean. They must also exhibit a degree of continuity, so that the predetermined standard quantities, times, costs, and cost rates will still be meaningful in the future.

16.1.3 Variable Costs and Fixed Costs

Intended learning outcomes: Differentiate between variable costs and fixed costs. Describe step-function costs. Identify full costs.

Variable costs for a product or an order only incur if the company actually makes or buys a product. Variable costs include all costs generated directly by the order, e.g., the wages of production workers or salespeople, the cost of raw materials or purchased components, subcontracted operations, or electric power to run machines.

As a rule of thumb, the following statement applies:

"Variable costs are all those costs that would not be incurred if we did not produce or procure anything."

The *fixed costs* for a product or order are the costs that are not variable; they remain the same regardless of the level of production and sales.

Fixed costs remain constant even when activity levels change. Some typical examples include the production infrastructure (buildings, depreciation, property taxes, mortgage payments, insurance, salaries of foremen or departmental managers, heating), R&D, and so on.

Of course, fixed costs are "fixed" only for a certain period of time. Above this time threshold, they show step-wise jumps.

Step-function costs or *semifixed costs* have a habit of jumping in a step-wise fashion over time. For example, demand — and thus production — may increase and require a company to purchase new production equipment or to rent or purchase an additional building. Investments in infrastructure improvement or the hiring of new personnel will result in smaller step-wise jumps in the cost curve.

It is common practice to capitalize and depreciate investments that will be used for more than one year. Depreciation costs and ongoing fixed costs per year must be allocated to individual orders using a formula or measure as the basis of apportionment. See Section 16.1.4.

In most cases, direct costs are variable costs as defined above. Overhead costs are normally fixed costs.

However, costs are defined as fixed or variable with respect to specific cost objects. Therefore, some *overhead costs* can be *variable*, such as the cost of the energy used directly for the production process. A (rare) example of *direct fixed costs* is the capital costs that can be allocated directly to a production contract, such as fixed annual license fees.

Full costs for a product or order are the sum of the variable costs plus a reasonable portion of the fixed costs.

This reasonable allocation of fixed costs to products or orders entails the same problem as the "fair" distribution of fixed costs does. It is not possible in this chapter to go into the advantages and disadvantages of variable costing or full costing; a large body of literature is available on the topic. In general, though, it is important that the company can perform calculations using both of these costing principles. Furthermore, there are specific requirements for external financial reporting.

16.1.4 Cost Accumulation Breakdown: The Cost Breakdown Structure of a Product

Intended learning outcomes: Explain cost accumulation breakdown. Present in detail material costs, external and internal labor costs. Produce an overview on tooling costs and general fixed manufacturing costs. Differentiate between variable manufacturing costs and (full) manufacturing costs. Identify general and administrative expenses (G&A) and cost of sales. Describe value added.

The *cost accumulation breakdown*, or *cost breakdown structure of a product*, is the accumulation of manufacturing costs in various subdivisions of costs, or *cost types*, according to the product structure.

Figure 16.1.4.1 shows an example cost accumulation breakdown for calculating the product costs. It is a breakdown of costs used by a manufacturing company.

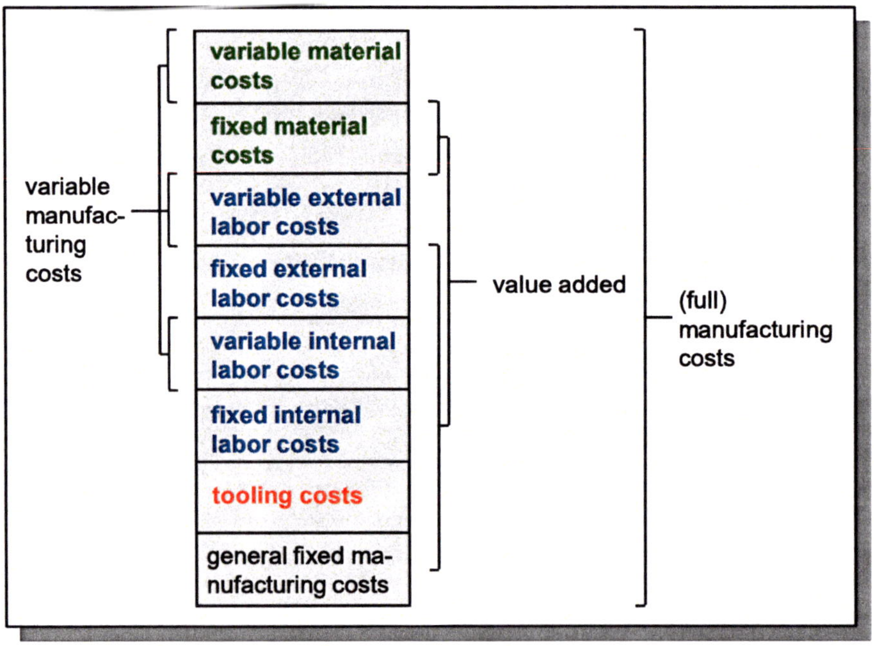

Fig. 16.1.4.1 Example cost accumulation structure, or cost breakdown structure of a product.

Material costs (green) are the costs associated with purchased components.

Material costs are subdivided into two cost subtypes:

- The *variable material costs* for a product are the sum of, firstly, the cost of purchased components (the true procurement costs), and, secondly, the variable material costs for all components produced in-house.

  Note: In many American costing systems, material costs also include the full cost of goods manufactured for components produced in-house (variable and fixed), rather than just the variable material costs (principle of the make-or-buy decision). In that case, the term *component costs* would be more appropriate than *material costs* (see the definition of the terms *material* and *components* in Section 1.1.1.

- The fixed material costs comprise, firstly, the costs of supplier and component qualification, secondly, the purchasing costs, thirdly, the carrying costs, and fourthly, the costs of receiving and inspecting purchased goods.

  The simplest way to express fixed material costs is as a percentage by dividing the total fixed costs by the total turnover of goods with variable costs. This calculation is carried out at the end of a budget period, using the data for the period just ended. It serves as the forecast for the next period. It is possible to use different percentages as a function of the storage type (special packaging or buildings, refrigerators, etc.) or as a function of the type or value of the goods (such as iron, gold, wood). However, the fixed costs must then be recorded separately for the various categories. The more exact and "fairly" costs should be allocated to products, the more expensive costing becomes. This also holds for activity-based costing (Section 16.4).

External labor costs **(blue)** are the costs associated with the subcontracting of work.

Work may be subcontracted because the necessary production techniques, infrastructure (special machines), or capacity are not available in-house. For financial accounting, special cost centers are defined for such operations. The ID used for the cost center may be the same as the supplier identification. External labor costs are further subdivided into two subtypes:

- *Variable external labor costs* are the sum of all invoices arising from work subcontracted to suppliers, and they contain the suppliers' fixed costs. For the subcontracting company, on the other hand, these costs are variable costs.

- *Fixed external labor costs* are the various costs generated by the subcontracting of work, particularly cost of shipping and transporting goods to and from the supplier, cost of receiving and inspecting the goods processed by subcontractors, as well as administrative expenses associated with subcontracting work (evaluation, writing the order, and so on).

  Usually, fixed external costs are expressed as a percentage in relation to the total invoiced amount for subcontracted work. We can apply different percentages to different categories of suppliers, in which case the fixed costs must be recorded separately for each category. As with material costs, the percentages are calculated at the end of a budget period and then serve as forecasts for the next budget period.

Internal labor costs **(blue)** are the sum of the costs for all in-house operations to manufacture the product.

Every internal operation is assigned to a work center,[1] for which two cost rates are established. A cost rate is related to a capacity unit, that is, the unit of measure of the capacity for a work center (mostly an hour).

- The *cost rate for variable internal labor costs*. This includes the costs for wages, plant utilities, plant supplies used, and so on that are needed to carry out the operation. The cost rate is essentially determined either directly or by measurement.
- The *cost rate for fixed internal labor costs*. This includes the depreciation costs for both machinery and infrastructure and tools and devices, provided that tools and devices are not depreciated independently of the machinery. It also includes ongoing costs, such as operations management. The cost rate is always calculated at the end of a budget period and is used as the forecast for the next period. The total fixed costs are then divided by the forecast load quantity for the next budget period.

The variable and fixed costs of an operation are calculated by multiplying the *standard load of an operation* (see Figure 13.1.2.2) by the cost rate for variable or fixed costs.

The *tooling costs* (red) for an operation are the costs incurred by the use of tools (a production equipment) during that operation.

In the past, tooling costs were regarded as part of the fixed costs for a capacity unit. Today, they represent such a large proportion of the costs and often differ so widely for each manufactured product that it is more sensible to set them out separately. The following technique, which accords with the activity-based costing approach (see Section 16.4), provides an illustration:

- The tooling costs per operation are calculated by multiplying the batch size by the cost rate per tool use, which is part of the master data for the tool (see also Section 17.2.7). We calculate the cost rate per tool use by dividing the amount to be depreciated by the expected number of uses of the tool.
- The actual number of uses of the tool (a cost driver) is recorded by the shop floor data collection system during the operation in question and is then stored in the master and inventory data for the tool. We can thus compare the actual number of uses against the budgeted number of uses for the tool. The cost rate can then be adjusted depending on the results of the comparison.

The *general fixed manufacturing costs* are the (fixed) costs for everything not associated directly with the manufacturing process or production infrastructure.

Typical general fixed manufacturing costs include licenses as well as general planning & control, manufacturing process design, and head of production.

These are usually calculated using one or more percentages that relate to the sum of these costs. The sum of all of the general fixed manufacturing costs is divided by the full cost of

---

[1] In some cases, the operation is assigned to two work centers: the machine and the person.

goods manufactured mentioned above. Again, this calculation takes place at the end of a budget period and serves as the basis for the forecast for the next period.

Variable manufacturing costs is the sum of all the variable costs (material and labor) of a specific product.

(Full) manufacturing costs, also called *cost of goods sold*, is the sum of all the variable and fixed costs (labor, material, and overhead) of a specific product or for a given period of time.

In addition to the fixed costs mentioned above, there are also:

General and administrative expenses (G&A) are the sum of (fixed) costs for R&D, administration, marketing and sales, and general management.

G&A are expressed as a percentage in relation to the (full) cost of goods sold. This percentage is calculated by dividing the accumulated G&A by the full cost of goods sold during the budget period, again at the end of the budget period. The result is used as the basis for the forecast for the next period.

The *cost of sales* is the sum of the costs of goods sold and the G&A for a specific product or for the products sold during a given period of time.

Finally, value-added is an organization's own output.

The *value added* of a product is defined as the full cost of goods manufactured minus variable material costs, minus variable external production costs, minus a part of the general fixed manufacturing costs (such as licenses). [2]

The complement of value-added are purchased products or services. This definition of added value also serves as the basis for some aspects of taxation.

The variable costs of goods manufactured serve as the short-term lower limit for the sales price (variable, costing) or partial costing, while the full cost of goods manufactured can be regarded as the medium-term lower limit for the sales price (full costing, absorption costing). The sales price then — ideally — includes a profit margin in addition to the cost of sales. For complete costing, costs must be broken down into all eight cost types for each item. The full manufacturing costs can then be derived simply by adding the cost types together.

---

2 This is the added value from the point of view of the manufacturer, in contrast to added value from the customer's viewpoint (see Section 4.1.2).

16.2 Cost Estimation

Intended learning outcomes: Present an algorithm for cost estimation of goods manufactured. Produce a representation of the cost accumulation and an overview of the comprehensive calculation for a product line.

16.2.1 An Algorithm for Cost Estimation of Goods Manufactured

Intended learning outcomes: Present in detail a ball bearing as an example product. Describe the algorithm for estimation the cost of a product (shown for three cost types). Explain the data flow of the cost-estimation algorithm.

Cost estimation for cost of goods manufactured is based on the master data. We can illustrate this using an example product, a *ball bearing*, according to Figure 16.2.1.1.

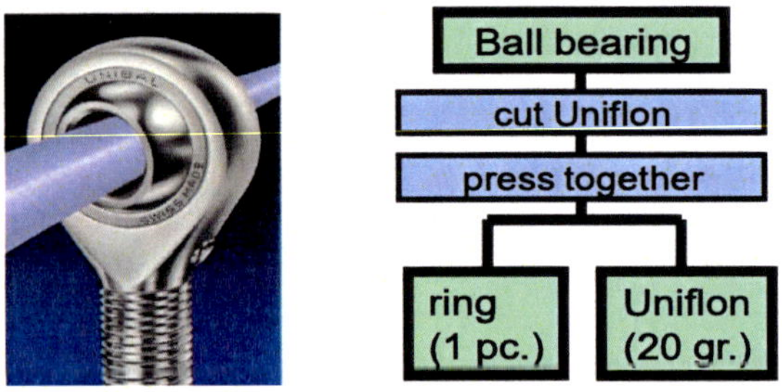

Fig. 16.2.1.1 A ball bearing as an example product.

The *ball bearing* (Item ID 83569) consists of two components, a *ring* (item ID 83593, a semifinished product manufactured in-house) and *Uniflon* (item ID 83607, a purchased raw material). The bill of material for the product thus has two positions. It is produced in two operations: *cut Uniflon* (position 250 at work center ID 907501, "manual production") and *press together* (position 270 at work center ID 908301, "special pressing"). The routing sheet for the product thus contains two operations.[3]

To obtain the *costs per unit produced*, we must either add together the costs for the entire batch and divide them by the batch size, or divide the setup load of each operation by the batch size.

To estimate the costs, we must then calculate the costs for each of the cost types in Section 16.1.4. For the sake of simplicity, the algorithm in Figure 16.2.1.2 uses only three cost types as illustrations.

---

[3] In the case under consideration, there are other components and operations. For the sake of simplicity, however, only these two components (respectively, these two operations) are listed here.

1 **Variable material costs**
- Treat each component of the bill of material for the product as follows:
  - Determine the *item* object associated with each component and determine the cost rate for the variable material costs for one unit of measure of that item.
  - Calculate the component costs by multiplying the quantity of the component incorporated into the product by this cost rate.
- Add together the component costs of all the components in the bill of material.

2 **Internal labor costs per unit of measure at this structure level**
- Treat each operation on the route sheet for the product as follows:
  - Determine the standard load for the operation.
  - Determine the *work center* object associated with each operation and determine the cost rate for the variable internal labor costs and the fixed internal labor costs for one capacity unit.
  - Calculate the variable and fixed operation costs by multiplying the standard load for the operation by the relevant cost rate.
- Add together all the variable and fixed operation costs for all operations on the routing sheet.

3 **Internal labor costs per unit of measure at all structure levels**
- Calculate the labor costs per unit produced at all lower production structure levels as follows:
  - Handle every component of the bill of material for the product as follows:
    - Determine the *item* object associated with each component and determine the cost rate for the variable internal labor costs and the fixed internal labor costs for one unit of measure of that item.
    - Calculate the variable and fixed labor costs for the component by multiplying the quantity of the component built into the product by the relevant cost rate.
  - Add together the variable and fixed labor costs of all components on the bill of material.
- Add to these the variable and fixed operation costs of all operations on the routing sheet for this level, as specified in step 2.

End of algorithm

Fig. 16.2.1.2 Algorithm for estimating the cost of a product (shown for three cost types).

Figure 16.2.1.3 shows the data flow of the cost-estimating algorithm described above.

The three steps outlined above are shown in the colored sections. The spreadsheet section shows the *item* (first table with three objects) and *work center* (fourth table with two objects) business objects. The *bill of material* business object (second table) is divided into detailed logistics objects, specifically into bill-of-material positions corresponding to the components. The operations are shown for the *routing sheet* business object (third table). See also the detailed description of the object and entity classes in Sections 17.2.1 to 17.2.8, in particular the Figures 17.2.1.1 and 17.2.8.1. The arrows in Figure 16.2.1.3 indicate the sources and usage of the data in the individual calculations.

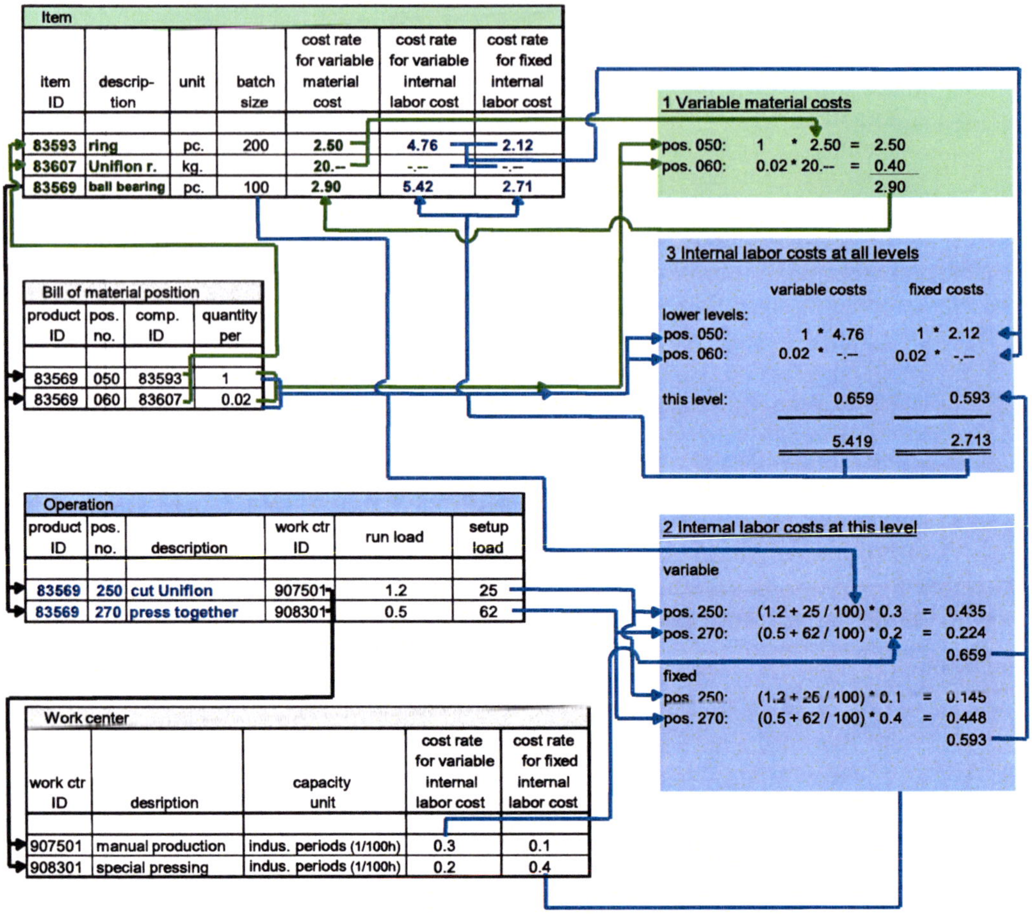

Fig. 16.2.1.3 Algorithm for estimating the cost of a product.

16.2.2 Representation of the Cost Accumulation and Comprehensive Calculation for a Product Line

Intended learning outcomes: Explain a graphical representation of the cost accumulation for a product. Disclose how to proceed if the entire line of products has to be recalculated.

Figure 16.2.2.1 shows one possible way of representing the results of the (single-stage) *cost accumulation* for an individual product. Here, again, the *ball bearing* product from Section 16.2.1 is used as an example. The cyan-colored lines contain blue *and* green types of cost.

In this graphical representation, you can see that this is an *estimated-cost accumulation*, as only the *target costs* column has been completed. For *ongoing job-order cost accumulation*, we would enter data collected from the shop floor into the *actual* column. Division by the batch size is performed only at the very end. However, first the run load per unit must be multiplied by the batch size. Compare the results of the calculation for batch size 5000 with the calculation in Figure 16.2.1.3 (where batch size is 100).

Cost Accumulation

Product ID: 83569 Description: Ball bearing Batch size (order quantity): 5000 Actual quantity: 5000

pos.	text	fix. var.	comp. ID Wrk.-C. ID	setup load	quantity per / run load per unit	total qty. / load		unit	cost per unit	cost	
						target	actual			target	actual
1	ring (materials)	var.	83593		1	5000	0	pc.	2.50	12500.--	0
2	ring (labor)	var.	83593		1	5000	0	pc.	4.76	23800.--	0
3	ring (labor)	fix.	83593		1	5000	0	pc.	2.12	10600.--	0
4	Uniflon (materials)	var.	83607		0.02	100	0	kg.	20.00	2000.--	0
5	cut uniflon	var.	907501	25	1.2	6025	0	Pe	0.3	1807.50	0
6	cut uniflon	fix.	907501	25	1.2	6025	0	Pe	0.1	602.50	0
7	press together	var.	908301	62	0.5	2562	0	Pe	0.2	512.40	0
8	press together	fix.	908301	62	0.5	2562	0	Pe	0.4	1024.80	0

costs per batch / order quantity		costs per batch / actual quantity			costs per batch	
target	actual	target	actual	cost type	target	actual
2.90	0	2.90	0	variable material costs	14500.--	0
5.22	0	5.22	0	variable internal labor costs	26119.90	0
8.12	0	8.12	0	variable cost of goods manufactured	40619.90	0
2.45	0	2.45	0	fixed internal labor costs	12227.30	0
10.57	0	10.57	0	(full) cost of goods manufactured	52847.20	0

Fig. 16.2.2.1 Graphical representation of the cost accumulation for a product.

If the bill of material for a product contains components produced in-house, the costs must be estimated for these items first. Only then should we calculate the costs for the product itself into which the components are built. This is best achieved by estimating the costs for all components, vertically along the tree structure, using a depth-first search. Once we have estimated the costs for all the components at one level, we can estimate the costs for the higher-level product when we return to the next highest level of the tree structure.

If the entire line of products has to be recalculated, it is more efficient to take the individual items in descending order of their low-level code. We start by calculating the costs for individual parts and subassemblies at the lowest possible level and end with the finished product. We can proceed in this order, because we have already calculated the level codes.

For components produced to order, which are produced on demand for the higher-level product rather than being stored, we can integrate the cost accumulation for each component directly into the cost accumulation for this product. Since the batch that is produced depends on the product batch, the result will be different every time.

If the end product is a product family with many variants, rather than a stock item, we can combine different parameter values in the cost estimation. In this way, we can calculate various points of support for product costs in the n-dimensional parameter space. These combinations of parameter values should then be stored in parameter value lists under the *item* object and introduced into the estimated-cost accumulation as shown in Figure 16.2.2.1.

16.3 Job-Order Costing

Intended learning outcomes: Describe actual quantities and actual costs. Explain cost analysis. Produce an overview on the interface from order management to cost accounting.

16.3.1 Actual Quantities and Actual Costs

Intended learning outcomes: Identify actual quantities and actual order costs. Produce an overview on backflush costing. Differentiate between standard costing, normal costing, and actual costing in determining actual order costs.

The *actual quantities* are the quantities of components and capacity used for an order.

The shop floor data collection system (see Section 15.3) provides the data on the actual quantities for a production, procurement, or R&D order. The actual quantities are generally used as a factor in calculating actual costs:

The *actual order costs* are the costs generated by an order.

In simple cases, we can determine the actual order costs as follows:

Backflush costing is the application of costs based on the output of a process. It works backward to flush out the costs for the units produced, applying costs using standard costs. Backflush costing is usually associated with repetitive manufacturing environments.

In all other cases, we determine actual order costs by an accumulation of job-order costs according to the following *cost identification techniques*:

- *Standard costing* or *standard cost (accounting) system*: actual (used or consumed) quantities times standard cost rates for variable and fixed costs.
- *Normal costing* or *normal cost system*: invoiced amounts or actual cost of wages for variable costs, actual (used or consumed) quantities times standard cost rates for fixed costs.
- *Actual costing* or *actual cost system*: invoiced amounts or actual cost of wages for variable and fixed costs.

We thus obtain a total for each of the individual cost types in correspondence to the cost accumulation breakdown shown in Figure 16.1.4.1. The algorithm for job-order costing corresponds to the procedure illustrated in Figure 16.2.2.1. Here, the data are taken from the business object *order*, rather than from the master data (see Section 17.1 for further details). In the costing method shown in Figure 16.2.2.1, the *actual values* are entered into the columns on an ongoing basis (ongoing job-order cost accumulation). The values listed correspond to usage by the reported operations and the parts issued. In this way, we can continuously track the costs of every production order and compare them against the *target values*. Continuous comparison is particularly important for production according to customer orders, since these are subject to a budget. This will identify the likely profit or loss at a relatively early stage, enabling us to take corrective action in good time.

For the comparison to be meaningful, the cost identification techniques used for cost estimating and job-order costing must be the same. However, for some types of costs, this may not be the case:

- For actual costing, the invoices for materials or external operations may arrive much too late for efficient control of internal operations. If this is the case, we can then fall back on the standard load or the actual quantities valued at standard cost rates.
- Global invoicing may sometimes make it difficult to assign costs fairly to individual resources obtained externally, which means that standard cost rates may prove to be just as accurate. These standard cost rates are again multiplied by the actual quantities.
- The valuation of material costs on the basis of standard cost rates may be inaccurate due to large fluctuations in the cost of purchased items. Under these circumstances it may be necessary to use the average costs as a basis or to value certain materials at the actual cost of the procurement batches.

If actual costing is chosen as the cost identification technique, then the estimated-cost accumulation essentially reflects the most recent order. We may, however, impose a budget on the individual cost types that does not necessarily correspond to the total standard costs for the underlying operations or individual items issued. If the budgets correspond to the expected revenue, then the ongoing comparison of estimated cost (budget) against job-order cost accumulation leads directly to the expected revenue from the order.

16.3.2 Cost Analysis

Intended learning outcomes: Differentiate between volume variances in an internal operation, volume variances for a component, and variances in the costs per unit produced. Disclose various cost variances.

Cost analysis seeks to reveal *significant variances* (i.e., variances that exceed established thresholds) of actual costs of an order (the actual order costs) from target costs.

Volume variances occur when the resources consumed deviate in quantity from planned quantities.

There are various causes for volume variances:

- *Volume variances in an internal operation.* Here, the actual load differs from the standard load because:
  - Unanticipated incidents occur during production.
  - The work center efficiency or efficiency rate (in a time period) is better or worse than expected.
  - The specified quantity of standard capacity requirement is wrong, or the quantity consumed is recorded incorrectly.
  - Additional operations are needed for reworking.

- *Volume variances for a component* or *an external operation*. The quantities consumed differ from the quantities specified on the bill of material or route sheet, because:
  - The wrong standards (estimates) were used.
  - Goods are lost or scrapped.
- *Variances in the costs per unit produced*. If scrap is produced, the quantity actually produced may be less than the quantity ordered, in which case the cost of goods manufactured *per unit produced* will be higher than expected, because most of the components and resources were used for the initial operations in accordance with the original quantity ordered.

Standard costing reveals all these variances through a simple comparison of the job-order cost against the estimated cost accumulation. Since the underlying cost rates remain the same, the job-order cost accumulation highlights any volume variances.

Cost variances are deviations between actual and standard costs.

Cost accounting analyzes the various cost variances, namely:

- Variances between the actual costs of the purchased components and the standard costs for the same items.
- *Variances of the actual costs of a capacity unit of a work center*. The costs per capacity unit are predicted for the future based on past values in the form of a forecast. At the end of the budget period, this reveals variances arising from undercapacity or overcapacity, meaning that fixed costs should actually have been divided by a different load.

When basing costing on the actual costs, comparison of job-order and estimated-cost accumulations yields variances that encompass both volume and cost variances. To show these variances separately, we must add a third column that captures "actual quantities at standard cost rates." However, we can only do this if we know the cost rates when we carry out the estimated-cost accumulation. But, if we specify only the total budget for each cost type, then we cannot show volume variances separately from cost variances.

16.3.3 The Interface from Order Management to Cost Accounting

Intended learning outcomes: Produce an overview on various outputs from cost accounting. Present the various transactions handled by costing software.

Carried out in the context of production order management, *cost object accounting*, e.g., *product costing* or *project costing*, is in essence job-order costing as described above.

Cost accounting also performs cost object accounting. Other outputs from cost accounting are cost center accounting and cost object group accounting. To be accurate, all costing systems, in particular, *costing software*, require a regular input of production order data and shop floor data. These data-capturing systems provide the interface to the cost accounting system and allow accumulation of the necessary cost data.

Costing software also manages the value of work in process. The cost accounting department requires a report of every transaction associated with a production order. These transactions include:

- Release or amendment of a production order.
- Every stock issue. Each stock issue increases the value of the work-in-process and reduces the value of inventory by the actual costs.
- Every execution of an operation. The actual cost of the operation is added to the value of the work-in-process. The load on the corresponding work center is reduced.
- Every invoice for delivery of goods or external subcontracting of work. We can also allocate the costs to a dummy inventory account or cost center, rather than the work-in-process, which will then be unloaded by a corresponding issue at standard cost rates.
- Completion of the order. The accumulated value for work-in-process for the order, together with the fixed costs, is charged either to the inventory account or directly to the expense account for customer production orders.

Transactions can be carried over every day. If cost accounting is carried out on a monthly basis, however, the data are transferred immediately before the accounting starts.

Note: At the end of every accounting period — at month's end, for example — all the actual values (such as the quantity consumed or actual costs) must be stored temporarily in a "quantity consumed to end of accounting period" attribute. This can be accomplished by a program that is run at the end of each accounting period. In this way, when the cost accounting department receives the data on the fifth of the month, for example, it receives the values stored in the temporary attributes. This is because the actual "quantity consumed" attribute now contains the usage that has accumulated in the new accounting period.

16.4 Activity-Based Costing

Intended learning outcomes: Disclose the limits of traditional product costing. Explain activity-based costing: aim, basic premise, requirements, and technique. Present typical processes (activities) and process variables as well as the activity-based product cost estimation.

16.4.1 Limits of Traditional Product Costing

Intended learning outcomes: Describe allocating fixed costs to products with conventional cost accounting using two cost types. Explain the potential for error in traditional product costing.

Often, job-order costing allocates fixed costs (overhead) by an extra charge, expressed as a percentage of the variable costs of labor and materials.

In the simplest case, this percentage is either a single percentage or multiplication factor of the variable manufacturing costs or two different percentages for material costs and labor costs, as shown in Figure 16.4.1.1. This traditional overhead-cost-allocation process thus allocates overheads to products using direct material and labor costs (for example, labor hours or machine hours) as the basis for allocation.

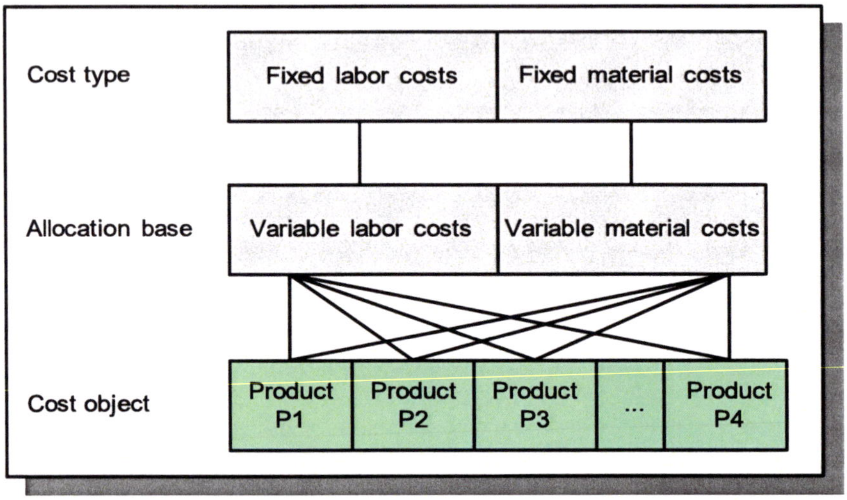

Fig. 16.4.1.1 Allocating fixed costs to products with conventional cost accounting using two cost types.

There has been a rapid explosion of the value of these simple job-order costing factors over the past two decades, mainly because internal labor costs have moved rapidly in the direction of fixed internal labor costs (machines, tools, etc.). Today, there are some companies with a ratio of fixed to variable costs of 10 to 1, meaning that variable costs represent just 10% of the manufacturing costs for the entire organization. The remainder is made up of fixed costs of various types (see also Figure 16.1.4.1), specifically:

- Material procurement and storage costs
- The cost of managing subcontracted operations
- Machinery, tool, production facility, and infrastructure costs
- The costs of R&D, licensing, product and process design, planning & control, etc.

Problems arise with conventional costing, for the focus remains on the variable costs. Often, the reduction in variable costs will merely increase the multiplication factor, since the same fixed costs are then simply distributed among fewer variable costs. If the organization has a broad product concept, such as mixed manufacturing with products ranging from products made to customer specification (which may change from one order to the next) to standard products with no variants, this results in the distortion shown in Figure 16.4.1.2.

As a result, too much overhead is attributed to products produced with high variable costs — often standard products — and too little overhead is attributed to products with low variable costs — often products according to (changing) customer specification. In the example, P1 — having high variable costs (the black portion) — is overcosted with fixed costs (the white portion), and P2 — having low variable costs, is undercosted with fixed costs. See here also [CoSt93].

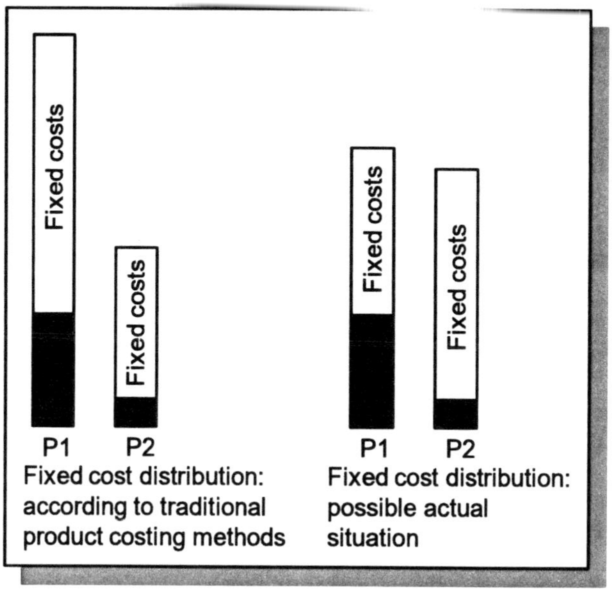

Fig. 16.4.1.2 Potential for error in traditional product costing.

Since the cost of goods manufactured is used as the basis for pricing, this misallocation of fixed costs would tend to result in less complex products (technically and logistically) being put on the market at too high a price, while too low a price would be charged for complex products. This would mean that a company could lose its competitive edge for series and mass-produced items — not because of the high cost of wages or other factors, but due to the costing system itself!

The problems with conventional costing came to light with respect to investment in qualification of employees and machinery. These investments raised the fixed costs and had a disproportionately large effect on the very product range for which the investments had targeted more efficient production. It is not surprising that the new production methods resulted in a demand for a new way of thinking about costs. The new type of costing system proposed and developed was activity-based costing (ABC).

16.4.2 Introducing Activity-Based Costing: Aim, Basic Premise, and Technique

Intended learning outcomes: Produce an overview on activity-based costing. Identify ABC process, process variable and process cost rate. Describe the ABC process plan and the process quantity. Explain allocating fixed costs using activity-based cost accounting.

Activity-based costing, or activity-based cost accounting (ABC), is a cost accounting system designed to allocate fixed costs (overhead) as fairly and realistically as possible to the business processes.

The *aim* of activity-based costing is thus essentially not new. However, achievement of the aim means better performance on the following tasks:

- Process management: Planned investments can be linked to specific processes right from the outset. The resulting investment costs can be converted into corresponding process costs and then compared with the previous process costs.
- Support for decision making in product design: Developers are informed of the consequences of their choice of purchased components or new design and manufacturing processes at a very early stage. This information usually provides a comparison of different technologies or shows the consequences of changing the product design. This type of information is very important, for cost of a product has essentially been determined by the end of the design phase. After that, very little can be done to influence cost.
- *Product cost estimation:* Activity-based costing, just like conventional costing, is a useful technique for estimating costs. Pricing will be much more accurate if the costs are estimated correctly.

With respect to the *basic premise* behind activity-based costing, the need for a "fair" distribution formula for overhead also means finding a suitable measurement, or allocation base. For this reason, we need to examine our fixed costs in greater detail, tracing them back to the underlying processes — or even subprocesses or individual activities. In Section 16.1.4, we demonstrated how fixed material costs can be calculated differently for different groups of materials or cost centers. This is one step toward the principle illustrated below.

An *ABC process* is a process or activity that incurs extensive fixed costs in the company and thus is allocated to business processes using ABC.

The *process variable* is a unit against which we can measure the costs for the ABC process or activity in a suitable manner.

This is called the *activity cost driver*. ABC uses activity cost drivers to allocate the costs to cost objects, such as products, in relation to the resources consumed.

In most cases, the activity cost driver is not associated with variable costs or an underlying time unit. Instead, activity cost drivers are, for example, the number of purchase orders, the number of items received, or the number of components for an assembly. If business processes or activities are identified and broken down with sufficient detail into subprocesses, the cost driver is usually easy to identify the fixed costs can now be related to the products by using the cost drivers. The methods used here are similar to those that were used in traditional *time studies* in process planning for establishing *time standards*: count, measure, and calculate average.

The *process cost rate*, or *planned cost rate*, for every ABC (sub-)process is the cost rate for an activity cost driver.

An ABC process or subprocess thus not only represents an actual process. Together with its process cost rate, it also represents a traditional work center or cost center and the associated cost rate. Processes can also be recorded in this way, especially in IT-supported systems.

The *ABC process plan* for each product is a list of all the ABC processes (activities) that a product requires while it is being produced or procured.

The *process quantity* is the quantity as measured by the activity cost drivers that is likely to be used in an ABC process for the product.

The structure of an ABC process plan is similar to that of a routing sheet (see Section 1.2.3).[4] One ABC process plan position is assigned to every ABC process required to produce or procure the product. These positions correspond to the operations. The process quantity corresponds to the standard load of an operation. This means that we can keep ABC process plans in exactly the same way as routing sheets, particularly if the system is IT supported.

Activity-based costing is thus based on the calculation of standard cost rates (see Section 16.1.2). The basic premise for such activity-based costing is that the ABC processes must be clearly measurable and repetitive. Such processes can be found in the operational management of an organization, in logistics, and in accounting. It is in these areas that activity-based costing can be implemented successfully. The technique is more difficult to implement at the strategic level, since few repetitive ABC processes can be identified at this level (or, if they do exist, they relate to an extremely long period of time). Even if we could identify an activity cost driver, it would still not be possible to accurately determine the process quantity per product, that is, the usage of process variables.

Figure 16.4.2.1 shows several examples of processes (activities) and process variables (activity cost drivers) in the areas of purchasing and production.

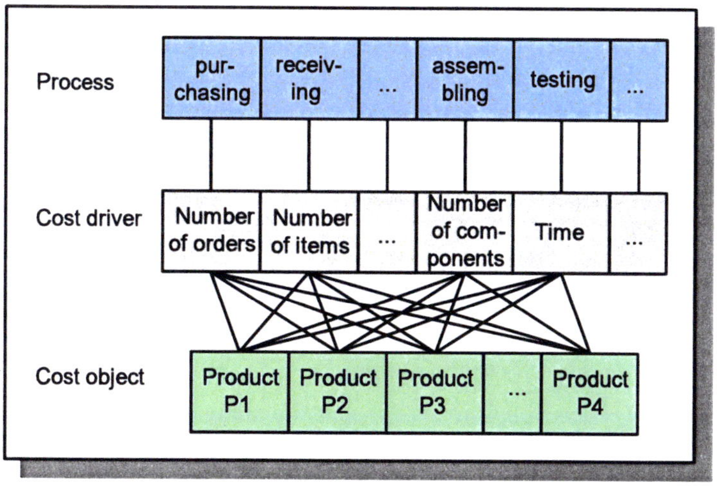

Fig. 16.4.2.1 Allocating fixed costs using activity-based cost accounting.

Examples of process cost rates associated with the process variables in Figure 16.4.2.1 are

4 The term *ABC process plan* is used here mainly to emphasize that this is not the same as the process plan introduced in Section 1.2.3. The latter process plan includes the product structure and the time axis in addition to the routing sheet.

- x dollars per order
- y dollars per item in receiving
- z dollars per component in assembly
- u dollars per run time unit in the testing process

The separation of tooling costs from fixed internal manufacturing costs described in Section 16.1.4 is one example of activity-based costing. There, the (ABC) process of tool utilization is considered separately. The activity cost driver may be the same as the tool utilization time or even, as suggested, simply the use of that tool to manufacture one unit of the batch.

To introduce activity-based costing into the company, the following steps are necessary:

1. *Determine the areas* in which activity-based costing is to be used.
2. *Determine the ABC processes, broken down into subprocesses (activities).* A meaningful ABC (sub-)process has at least the following characteristics:
   - The costs of the process are significant.
   - The process corresponds to a specific task within the process organization.
   - The various products (cost objects) should use the process to varying degrees (different process quantities).
3. *Determine the process variable (activity cost driver) for each process.* A good process variable has at least the following characteristics:
   - It is so closely related to the process costs that the process quantities can be based upon this unit variable.
   - It is self-explanatory to the people concerned within the organization, since it appears to be a natural variable within the operational process.
   - It should also appear to be a natural variable when options for different design variants or production methods are compared against one another.
   - The process quantities and cost rate per unit (process cost rate or process rate) can, as far as possible, be automatically calculated from the operational data.
4. *Determine the process cost rate for each ABC process.* This is done by dividing the fixed costs resulting from the process by the likely future process quantities.
5. *Specify the ABC process plan for each product and the process quantity for each ABC process in the ABC process plan.*
6. *Calculate the process costs for the product* by analyzing the ABC process plan (and the bill of material, of course) with the same algorithm used for production or procurement costs calculated using the traditional order costs or job order costing technique (see Section 16.2).
7. *Job-order costing and analyzing variances:* As in conventional costing, the volume variance can now be calculated for a particular order by recording the actual usage of process variables. Activity-based costing should thus identify any deviation from planned unit cost rates and compare actual process costs against the budgeted costs. This type of measurement is rather illusory, however, since small process quantities would take much too long to measure.

16.4.3 Typical Processes (Activities) and Process Variables

Intended learning outcomes: Explain determining main processes and subprocesses, using the example of circuit board assembly. Describe determining main processes and subprocesses using the example of procurement.

The following example, taken from [Schm92], shows how ABC is used in practice in the areas of production and purchasing. Figure 16.4.3.1 shows assembly of a printed circuit board, together with the main processes and subprocesses (activities), as well as the associated process variables.

Production: circuit board assembly		
Main process	**Subprocess, activity**	**Process variable**
Automatic assembly	DIP insertion AXIAL insertion ROBOTIC insertion SMT insertion	Insertions Insertions Insertions Insertions
Manual insertion	Setup Manual insertion IC programming	Components Insertions Seconds
Soldering	Wave soldering Infrared	Piece (circuit board) Piece (circuit board)
Testing	ATS operation ATS engineering	Tested components Test adapter
Reworking		Time

Fig. 16.4.3.1 Determining main processes and subprocesses; example: circuit board assembly.

Figure 16.4.3.2 shows activities in a conventional purchasing department.

16.4.4 Activity-Based Product Cost Estimation

Intended learning outcomes: Produce an overview on Determining the process cost rate and process quantity for supplier management. Explain determining the process cost rate as well as the process costs for external procurement of a single item: "standard component" versus "exotic component." Present in detail an ABC process plan and activity-based product cost estimation for a produced item as well as for a procured item.

Figure 16.4.4.1 shows another example, taken again from [Schm92]. Here, "supplier management" is a subprocess of the main process "materials purchasing."

The process costs are recorded for a period of time (in this case, six months). The process variable is the supplier. The process cost rate is determined by dividing the process costs by the number of suppliers. The number of different items procured from the supplier and the usage for each of these items over the time period determines process quantity. This provides the process quantity of each component that is incorporated into one product.

Purchasing			
Main process	Cost distribution %	Subprocess, activity	Process variable
Inventory management	50	Order management	Order
	50	Inventory management	Item
Materials purchasing	70	Supplier management	Supplier
	30	Order management	Order
Trading goods	70	Order management	Order
	30	Inventory management	Product
Specifying parts	100		Bill-of-material entry
Materials engineering	50	Supplier qualification	Supplier
	50	Component quality checking	Component
Planning	70	Assembly management	Assembly
	30	Order planning	Production order
Warehousing	50	Managing stockroom	Different item ID
	50	Issuing / receiving	transactions
Order Picking	100		Product
Packaging	100		Crate / box
Shipping		International/ local freight	Distance / weight

Fig. 16.4.3.2 Determining main processes and subprocesses; example: procurement.

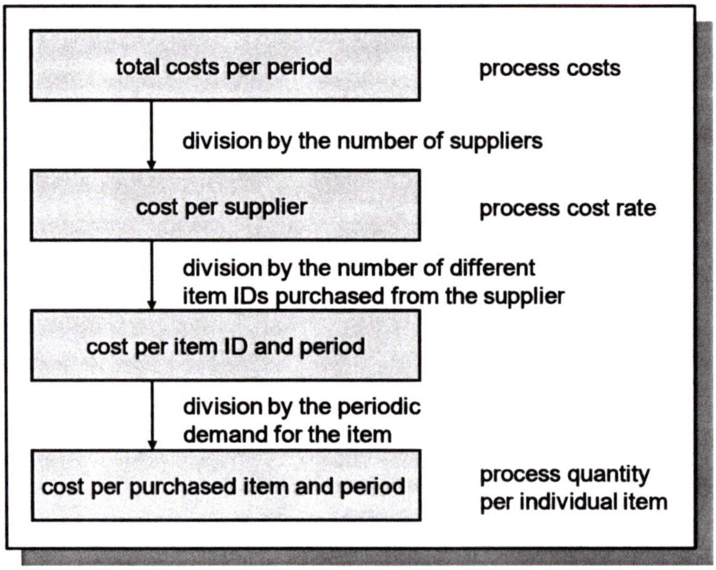

Fig. 16.4.4.1 Determining the process cost rate and process quantity for supplier management.

Figure 16.4.4.2 provides a quantitative example using simple data for illustrative purposes. It reveals the difference between the process costs for supplier management with just one supplier and a large number of purchased items with a high turnover and the process costs for one supplier with just a few purchased items and a correspondingly lower turnover.

Supplier management costs of a exemplary component		
Process costs	500'000. -	
Number of suppliers	100	
Process cost rate	5'000. -	
Feature	Standard component	"Exotic" component
Number of different purchased item IDs	200	5
Ø periodic requirements per article	1000	50
Process quantity per individual purchased item	$\frac{1}{200 \cdot 1000}$	$\frac{1}{5 \cdot 50}$
Process costs per individual purchased item	0.025	20.0

Fig. 16.4.4.2 Determining the process cost rate for a single item in the "supplier management" process: "standard component" versus "exotic component."

Figure 16.4.4.3 extends the supplier management example to include a costing for the entire purchasing process.

Again, the difference in process costs mentioned above stands out clearly. With conventional costing, in contrast, if the same additional percentage for material costs is applied to both the standard component and the exotic component, the loaded fixed material costs will be the same, even though the one is much more expensive to purchase than the other.

The next two examples show the *activity-based product cost estimation* for a manufactured item and a purchased item, each based on an ABC process plan. Figure 16.4.4.4 relates to the main processes and subprocesses of a product *manufactured in-house* that were shown in Figure 16.4.3.1. The individual positions are very similar to those that would be found on a normal routing sheet. In this case, however, "Process ID" replaces the work center. The administrative process plan positions for order management and stock issues/receipts, for example, would also be shown in addition to the operations. To calculate the cost of goods manufactured, we would also normally include the operations found on the normal routing sheet. They would be used only to calculate the variable costs, however.

Divisors for process quantities		Process for a standard component			Process for an "exotic" component		
- Number of different purchased item IDs per supplier		200			5		
- ∅ Periodic requirements for each item		1000			50		
- ∅ Number of items per inventory transaction		50			10		
- Number of orders per period		2			1		
Main process / subprocess	**Cost driver**	**Quantity**	**cost rate**	**costs**	**Quantity**	**cost rate**	**costs**
Materials purchasing:							
- Supplier management	Supplier	$\frac{1}{200 \cdot 1000}$	5000	0.025	$\frac{1}{5 \cdot 50}$	5000	20
- Purchase order management	Order	$\frac{2}{1000}$	30	0.06	$\frac{1}{50}$	30	0.6
Materials engineering:							
- Supplier qualification	Supplier	$\frac{1}{200 \cdot 1000}$	2000	0.01	$\frac{1}{5 \cdot 50}$	2000	8
- Component quality checking	Part ID	$\frac{1}{1000}$	300	0.30	$\frac{1}{50}$	300	6
Warehousing:							
- Managing stockroom	Item ID	$\frac{1}{1000}$	100	0.10	$\frac{1}{50}$	100	2
- Issuing / receiving	Trans-action	$\frac{1}{50}$	4	0.08	$\frac{1}{10}$	4	0.4
Total process costs for each individual item				0.575			37.0

Fig. 16.4.4.3 Determining the process costs for external procurement of a single item: "standard component" versus "exotic component."

Figure 16.4.4.5 represents the ABC process plan and the activity-based product cost estimation for a *purchased* item. Main process and subprocesses correspond to those shown in Figure 16.4.3.2, using the example in Figure 16.4.4.3. Therefore, we should allocate $37 to fixed material costs for each built-in "power supply" component. The similarity to a routing sheet is obvious. Standard ERP software can be used to store the ABC process plan.

Item ID: "PC Board"						
SEQ	Opera-tion	Description	Process ID	Process quantity	Process cost rate	Process costs
010	4411	Preforming	4311	48.0000	0.05	2.40
020	4401	DIP Insertion	4312	110.0000	0.15	16.50
030	4402	Axial Insertion	4313	163.0000	0.10	16.30
050	4400	Manual Insertion	4315	109.0000	0.20	21.80
060	4404	IC Programming	4316	0.1210	200.00	24.20
070	4405	Process Soldering	4317	1.0000	1.50	1.50
080	4407	ATS Engineering	4324	0.0050	5000.00	25.00
090	4408	Board Repairing	4322	0.0500	40.00	2.00
095	4409	ATS Operating	4318	459.0000	0.01	4.59
Total process costs						114.29

Fig. 16.4.4.4 An ABC process plan and activity-based product cost estimation for a produced item.

Item ID: "Power Supply"						
Divisors for process quantities:		Number of different purchased item IDs per supplier: 5 Average periodic requirements: 50 Number of orders per period: 1 Number of items per stock transaction: 1				
SEQ	Ope-ration	Description	Process ID	Process quantity	Process cost rate	Process costs
540	2400	Supplier management	4460	0.004	5000.00	20.00
545	2405	Purchase order management	4460	0.020	30.00	0.60
530	2300	Supplier qualification	4451	0.004	2000.00	8.00
535	2305	Component quality checking	4452	0.020	300.00	6.00
550	2500	Managing stockroom	4520	0.020	100.00	2.00
555	2505	Issuing / receiving	4520	0.100	4.00	0.40
Total process costs						37.00

Fig. 16.4.4.5 An ABC process plan and activity-based product cost estimation for a procured item.

16.5 Summary

In a sense, estimated cost and job-order cost accumulations are "by-products" of master data and production order management. The job-order cost accumulation is always current, which is not always the case where costing software is run on a monthly basis, for example. This is just one of the reasons why estimated and job-order cost accumulations are incorporated into computer-supported planning & control systems.

The actual costs cannot always be determined early on. We therefore have to use average costs and standard costs when estimating product costing. These also provide more robust estimates along the time axis. For short-term and long-term pricing purposes, we classify costs as variable and full, that is, variable and fixed costs.

A cost accumulation breakdown is made up of the cost types associated with a product, such as material costs, labor costs, and general costs, and differentiates between the fixed and variable parts of each type. These costs can also be used to calculate added value.

Cost estimation for a product is thus an algorithm. It calculates the cost of materials from the positions on the bill of material (and the associated component entities) as well as the labor costs from the operations (and the associated work centers) and bill-of-material positions (for components produced in-house). This means that every component that goes to make up a product is included in the estimated-cost accumulation.

For job-order cost accumulation, we must compare the actual quantities and actual costs collected from the shop floor against target quantities and costs. It is not always possible to determine the actual costs, however. For fixed internal manufacturing costs, "only" standard cost rates are available, and these have to be predetermined at the start of an accounting period. Generally, at least part of the standard cost rate has to be extrapolated from the past. Using standard costs instead of actual costs enables us to distinguish variances in cost and variances in quantity. Every transaction that affects value must be passed on to the costing function. If costing is carried out only on a periodic basis, it is absolutely essential to identify precisely those transactions that belong to a previous period.

Activity-based costing (ABC) is designed to assign fixed costs (overhead) to individual items in a targeted manner. Blocks of fixed costs are subdivided into main processes and subprocesses (or activities) to a level of detail that allows identification of a characteristic activity cost driver (or process variable) for each activity. The activity cost driver is the measure that allows us to relate costs to products. The block of fixed costs is broken down into costs per activity for each of these activities by the shop floor data collection system. This provides a process cost rate for each activity cost driver. The number of item IDs and ultimately the number of items affected by a cost driver unit also have to be determined. The reciprocal of the product of these two numbers is thus the process quantity per item.

An ABC process plan is then assigned to every item in the master data. The ABC process plan contains as many "operations" as there are ABC processes needed in order to produce or procure the item. The standard load is thus the process quantity per "operation." The actual ABC process itself plays the part of a "work center," since it has a unit, an activity cost driver, and a process cost rate. The algorithm for calculating product costs otherwise corresponds to the algorithm used for job-order costing.

ABC is less likely than conventional costing to undercost or overcost products and may lead to improvements in pricing. Experience shows that ABC is successful wherever there are repetitive fixed cost processes that are comparable over a long period (that is, at the operational level). If this is not the case, calculating process cost rates and process quantities on an ongoing basis as well as the amount of resources required to keep the ABC database current would be disproportionately expensive relative to the benefit gained from allocating the fixed costs to cost objects more correctly.

16.6 Keywords

ABC (activity-based costing), 700
activity cost driver, 700
actual costing, 694
actual quantity, 694
average costs, 684
backflush costing, 694
component costs, 687
cost accumulation, 692
cost accumulation breakdown, 686
cost analysis, 695
cost estimation, 682
cost identification technique, 694
cost object accounting, 697
cost of goods sold, 689
cost of sales, 689
direct costs, 683
general and administrative expenses (G&A), 689
general fixed manufacturing cost, 688
job-order costing, 682
labor costs, 687
manufacturing costs, 689
material costs, 686
normal costing, 694
overhead costs, 683
planned cost rate, 701
process cost rate, 701
process quantity, 701
process variable, 700
product costing, 697
project costing, 697
standard cost rate, 684
standard costing, 694
step-function costs, 685
tooling costs, 688
variable costs, 685
variable manufacturing costs, 689

16.7 Scenarios and Exercises

Intended learning outcomes: Elaborate job-order costing and activity-based costing using an example. Compare job-order costing and activity-based costing.

16.7.1 Job-Order Costing

Intended learning outcomes: Calculate the cost accumulation for two products, given the material consumption as well as the standard time for the operations and the fixed and variable costs of two work centers. Explain the graphical representation of the cost accumulation.

Two products A and B are produced from material Z with a batch size of 40. Consumption is the same for each product: 50 g per product A or B. The cost of 1 kg of material Z is $20.

For the sake of simplicity and comparison in our example, the *manufacturing process* is the same for products A and B: two operations (1 and 2) at two work centers (WC1 and WC2).

The standard time for each operation is 1 hour per 40 units. Assume that setup time is negligible.

To calculate the costs of the manufacturing process, it is important to take into account the costs of the two work centers WC1 and WC2 in addition to the standard times. As Figure 16.7.1.1 shows, WC1 is more machine intensive, while WC2 is more employee intensive. The investments will be depreciated in 5 years, assuming 1000 productive hours per year. Further, assume that these costs make up the full manufacturing costs.

	Work center 1	Work center 2
Variable costs	$20.- / hour (labor cost)	$40.- / hour (labor cost)
Fixed costs	$300,000.- (investitures in machines and tools)	$150,000.- (investitures in machines and tools)

Fig. 16.7.1.1 Work center costs data.

Following the principle of job-order costing, determine the cost accumulation values for products A and B marked "?" in the tables in Figures 16.7.1.2 and 16.7.1.3 (compare Figure 16.2.2.1).

Cost Accumulation

Product ID: 4711 Description: Product A Batch size (order quantity): 40 Actual quantity: 0

pos.	text	fix. var.	comp. ID Wrk.-C. ID	setup load	quantity per / run load per unit	total qty./ load target	total qty./ load actual	unit	cost per unit	cost target	cost actual
1	Material	var.	Z		?	?	0	?	?	?	0
2	Operation 1	var.	WC 1	0	?	?	0	?	?	?	0
3	Operation 1	fix.	WC 1	0	?	?	0	?	?	?	0
4	Operation 2	var.	WC 2	0	?	?	0	?	?	?	0
5	Operation 2	fix.	WC 2	0	?	?	0	?	?	?	0

costs per batch /order quantity target	costs per batch /order quantity actual	costs per batch/actual quantity target	costs per batch/actual quantity actual	cost type	costs per batch target	costs per batch actual
?	0	?	0	Variable material costs	?	0
?	0	?	0	Variable internal labor costs	?	0
?	0	?	0	Variable cost of goods manufactured	?	0
?	0	?	0	Fixed internal labor costs	?	0
?	0	?	0	(Full) cost of goods manufactured	?	0

Fig. 16.7.1.2 Graphical representation of the cost accumulation for product A.

Cost Accumulation											
Product ID: 4712		Description: Product B			Batch size (order quantity): 40				Actual quantity: 0		
pos.	text	fix. var.	comp. ID Wrk.-C. ID	setup load	quantity per / run load per unit	total qty. / load target	total qty. / load actual	unit	cost per unit	cost target	cost actual
1	Material	var.	Z		?	?	0	?	?	?	0
2	Operation 1	var.	WC 1	0	?	?	0	?	?	?	0
3	Operation 1	fix.	WC 1	0	?	?	0	?	?	?	0
4	Operation 2	var.	WC 2	0	?	?	0	?	?	?	0
5	Operation 2	fix.	WC 2	0	?	?	0	?	?	?	0

costs per batch/ order quantity		costs per batch / actual quantity			costs per batch	
target	actual	target	actual	cost type	target	actual
?	0	?	0	Variable material costs	?	0
?	0	?	0	Variable internal labor costs	?	0
?	0	?	0	Variable cost of goods manufactured	?	0
?	0	?	0	Fixed internal labor costs	?	0
?	0	?	0	(Full) cost of goods manufactured	?	0

Fig. 16.7.1.3 Graphical representation of the cost accumulation for product B.

Hint: The full cost of goods manufactured will be the same for both product A and B (why?): $4.75 per unit produced, or $190 for a batch size of 40.

16.7.2 Activity-Based Costing

Intended learning outcomes: Calculate the cost accumulation for two products, given the material consumption as well as the standard time for the operations and the fixed and variable costs of two work centers, differentiating the fixed costs between investitures in machines and in in tools and devices.

Think again about products A and B described above. After reading Section 16.1.4, you know that tooling costs make up a sizable proportion of the fixed costs. If the costs of the tools used for products A and B are different, this should be apparent in the cost accumulation. However, that can only be achieved if we view tool utilization as a process in its own right. Following the principle of ABC and the steps involved (see Section 16.4.2), the characteristic variables for this process are defined as follows:

- ABC process: tool utilization, or use.
- Process costs: the manufacturing or procurement costs of the tool.
- Activity cost driver: the number of units produced with the tool. Why? Usually, it is not the length of time that a tool is utilized that determines its wear, but rather

production of a certain number of units of the product. A good example would be pressing tools.

- *Process cost rate:* process costs divided by the total quantity of product units that are produced using the tool until the tool is used up or worn out.

Figure 16.7.2.1 shows a breakdown of the fixed costs in machine costs and costs for tools and devices.

	Work center 1	Work center 2
Variable costs	$20/hour (labor cost)	$40/hour (labor cost)
Fixed costs: investitures in machines	$200,000.-	$100,000.-
Fixed costs: investitures in tools and devices	Tool T1: $4000 (used to manufacture product A) Tool T2: $16000 (used to manufacture product B)	Tool T3: $2000 (used to manufacture product A) Tool T4: $8000 (used to manufacture product B)

Fig. 16.7.2.1 Work center costs data.

As in exercise 16.7.1 above, the investitures in machines will be depreciated in 5 years, whereby 1000 productive hours are assumed annually. It is further assumed that a tool can be used to manufacture 20,000 products A or B before it is used up or worn out, no matter whether it is an expensive or inexpensive tool.

Since one hour of capacity is utilized for 40 units of products A or B, 200,000 products can be manufactured in 5000 productive hours. This means that, in that period, 10 tools will be required.

In the following, assume also that the same number of units of products A and B is manufactured. In this case, work center 1 will use 10 tools (5 T1 and 5 T2 tools), which represents an investment of $100,000. Work center 2 uses 10 tools (5 T3 and 5 T4 tools), which represents an investment of $50,000. The sum of fixed costs is thus the same as in exercise 5 above.

Determine the values marked "?" in the cost accumulation tables in Figures 16.7.2.2 and 16.7.2.3 below (compare Figure 16.2.2.1).

To calculate the process cost of the tool, use the following:

- The process quantity or quantity per for the ABC process "tool use for operation 1 (or 2)" is 1 (one use per unit produced).
- The total (target) quantity is the number of units produced.
- The process variable, or activity cost driver, is the "use of the tool."
- The process cost rate (or cost per unit) is the cost of the tool divided by the number of units that can be produced until the tool is used up.
- The process costs (target) are the product of the total (target) quantity times the cost per unit.

Cost Accumulation

Product ID: 4711 Description: Product A Batch size (order quantity): 40 Actual quantity: 0

pos.	text	fix var.	comp. ID Wrk.-C. ID	setup load	quantity per / run load per unit	total qty./ load		unit	cost per unit	cost	
						target	actual			target	actual
1	Material	var.	Z		?	?	0	?	?	?	0
2	Operation 1	var.	WC 1	0	?	?	0	?	?	?	0
3	Operation 1	fix.	WC 1	0	?	?	0	?	?	?	0
4	Tool use for op. 1	fix.	T1	0	?	?	0	?	?	?	0
5	Operation 2	var.	WC 2	0	?	?	0	?	?	?	0
6	Operation 2	fix.	WC 2	0	?	?	0	?	?	?	0
7	Tool use for op. 2	fix.	T3	0	?	?	0	?	?	?	0

costs per batch / order quantity		costs per batch / actual quantity			costs per batch	
target	actual	target	actual	cost type	target	actual
?	0	?	0	Variable material costs	?	0
?	0	?	0	Variable internal labor costs	?	0
?	0	?	0	Variable cost of goods manufactured	?	0
?	0	?	0	Fixed internal labor costs	?	0
?	0	?	0	(Full) cost of goods manufactured	?	0

Fig. 16.7.2.2 Graphical representation of the cost accumulation for a product A.

Cost Accumulation

Product ID: 4712 Description: Product B Batch size (order quantity): 40 Actual quantity: 0

pos.	text	fix var.	comp. ID Wrk.-C. ID	setup load	quantity per / run load per unit	total qty./ load		unit	cost per unit	cost	
						target	actual			target	actual
1	Material	var.	Z		?	?	0	?	?	?	0
2	Operation 1	var.	WC 1	0	?	?	0	?	?	?	0
3	Operation 1	fix.	WC 1	0	?	?	0	?	?	?	0
4	Tool use for op. 1	fix.	T2	0	?	?	0	?	?	?	0
5	Operation 2	var.	WC 2	0	?	?	0	?	?	?	0
6	Operation 2	fix.	WC 2	0	?	?	0	?	?	?	0
7	Tool use for op. 2	fix.	T4	0	?	?	0	?	?	?	0

costs per batch / order quantity		costs per batch/actual quantity			costs per batch	
target	actual	target	actual	cost type	target	actual
?	0	?	0	Variable material costs	?	0
?	0	?	0	Variable internal labor costs	?	0
?	0	?	0	Variable cost of goods manufactured	?	0
?	0	?	0	Fixed internal labor costs	?	0
?	0	?	0	(Full) cost of goods manufactured	?	0

Fig. 16.7.2.3 Graphical representation of the cost accumulation for a product B.

Problem-solving hints:
The full cost of goods manufactured will *not* be the same for products A and B (why?): In fact, we calculate $4.30 per unit produced of product A (or $172 for a batch size of 40), and $5.20 per unit produced of product B (or $208 for a batch size of 40).

16.7.3 Comparing Job-Order Costing and Activity-Based Costing

Intended learning outcomes: Differentiate between job-order costing and activity-based costing, extending the results of the two preceding exercises.

a. Why is the cost per unit produced in the conventional job order costing exercise 16.7.1 ($4.75) exactly the mean of the costs per unit of the two products in the ABC exercise 16.7.2 ($4.30 and $5.20)?

b. What product pricing considerations would you take into account on the basis of the results when calculating manufacturing costs by ABC?

c. Would a change of the batch size (40 in both exercises) imply different results? Is this generally the case in the world of practice? What assumption made in the problem description for the sake of simplicity led to the special case of the two exercises?

16.8 References

CoSt93 Cokins, G., Stratton, A., Helbling, J., "An ABC Manager's Primer — Straight Talk on Activity-Based Costing," Institute of Management Accountants, 1993

Habe08 Haberstock, L., "Kostenrechnung I," 13. Auflage, Erich Schmidt Verlag, Berlin, 2008

KeBu93 Keller, D.E., Bulloch, C., Shultis, R.L., "Management Accountants' Handbook," 4th Edition, John Wiley & Sons, New York, 1993

Schm92 Schmid, R., "Activity-based Costing im praktischen Einsatz bei Hewlett Packard," Practitioners Meeting, Production and Information Management, BWI/ETH Zurich, Nov. 26, 1996

17 Representation and System Management of Logistic Objects

Intended learning outcomes: Describe order data. Explain master data for products and processes. Disclose extensions arising from the variant-oriented and the processor-oriented concepts. Produce an overview on the management of product and engineering data.

In Figure 17.0.0.1, the dark shaded box contains the logistical objects associated with the tasks and processes of the reference model for business processes and planning & control tasks in Figure 5.1.4.2.

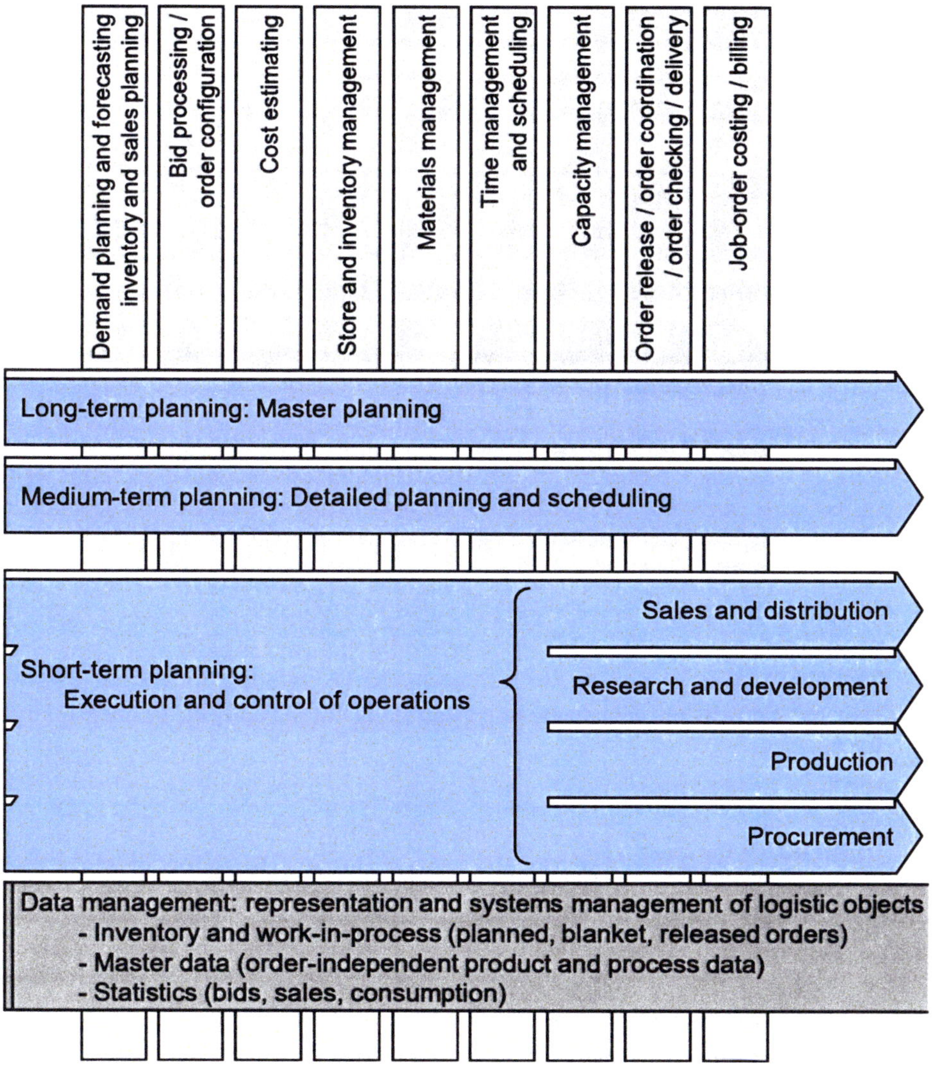

Fig. 17.0.0.1 The subject covered in this chapter is shown against a darker background.

P. Schönsleben, *Handbook Integral Logistics Management*,
https://doi.org/10.1007/978-3-662-65625-9_17

This chapter describes in detail the business objects introduced in Sections 1.2 (order and master data), 11.1 (inventory), and 11.2 (statistics). The discussion is also structured from the viewpoint of an information system with a view to IT support. In addition, the chapter covers tasks that can be summarized as obtaining information from an information system using suitable queries.

17.1 Order Data in Sales, Distribution, Production, and Procurement

Intended learning outcomes: Present the data structure of customers and suppliers. Describe the data structure of orders in sales and distribution, production, and procurement. Disclose the data structure of the order and partial order header as well as the order position.

The *order* business object was introduced in Section 1.2.1. It describes all types of order within the supply chain. This section describes in detail the order objects used in sales and distribution, production, and procurement. Section 17.5 (The Management of Product and Engineering Data) describes the R&D order in greater depth. Stock status and statistics are also discussed in this section since they are related to order objects.

17.1.1 Customer and Supplier Data, and the Corporate Group Structure

Intended learning outcomes: Present the business object "business partner": customer and supplier data. Identify the corporate group structure.

The *business partner* business object of a company was introduced in Section 1.2.1 as a general term to describe an external or internal customer or a supplier. In terms of their property as business objects, both *customer* and *supplier* may be defined as a specialization of *business partner*. The customer and supplier classes are thus both specializations of the business partner class. Most of the attributes of the customer object class correspond to those of the supplier object class. The most important common attributes are

- The business partner ID. This ID is generally "not meaningful" (see also Section 20.3.2). Changes to the identification should be avoided during the life of the business partnership. The business partner ID is unique and acts as the primary key for the class.
- Business partner name, address, and country, and optionally a delivery address: these attributes act as "secondary keys" enabling a particular customer to be easily traced within the class.
- Communication details (telephone, fax, e-mail, and Web site).
- Various codes used to classify the business partner.
- Credit limit, bank details.
- Codes for handling the business partner with respect to the tax authorities.
- Codes for order processing, shipping, and incoming goods.

Various types of sales statistics are kept for each business partner. These are generally administered in separate object classes.

Business partners may be incorporated into an overall company hierarchy.

A *corporate group structure* is the set of all business partners belonging to a corporate group business partner.

This business object enables general analyses (consolidations), for example, to be carried out for all the companies in exactly the same way as for the individual business partners.

Aspects of *computerized administration*: The business partner is normally identified by a dummy identification, which is allocated by the information system. A business partner entity, as a data set, may not be physically deleted while it continues to appear in an order or in statistics. A business partner ID is normally allocated for many years, even if the connection with that business partner no longer exists.

17.1.2 The General Structure of Orders in Sales and Distribution, Production, and Procurement

Intended learning outcomes: Explain the data structure of an order in sales and distribution, production, or procurement. Disclose the basic object classes in an order database.

The examples in Figures 1.2.1.1 and 1.2.1.2 show that the *order* is a relatively complex business object. The *order data* that combine to form the *order* business object include:

- The order header: This is the data that appear at the top or bottom of each order, including the principal, contractor, and the order validity date. Each order has precisely one order header.
- The order item or order position. An order may contain any number of this object. Each is assigned a suitable position number and appears in a specific order. Every line describes an object that must be scheduled or controlled within a company's logistics, or may be used for text only.

  In Figure 1.2.1.1, these objects are, without exception, *order positions (of the) item (type)* that pass from the supplier to the customer. From the supplier's viewpoint, they are *(item) issues*, whereas a customer regards them as *(item) receipts* or *entries*.

  Figure 1.2.1.2 also shows item issues, although in this case the contractor — the garage — also supplies *order positions (of the) work* or *order operation (type)*. This means individual pieces of work that the customer purchases as part of the service, but which never assume the character of a product. In this case, they are carried out directly on the object that characterizes the order, that is, the car. The other positions listed under the "Work" heading are an item issue (small items and cleaning materials) and an *order position (of the) use-of-production-equipment (type)*. A courtesy car was provided to fulfill the order. The courtesy car is an investment on the part of the garage, just like any other device, machine, or tool.

Figure 17.1.2.1 shows the general structure of an order in sales and distribution, production, or procurement that arises from these observations.

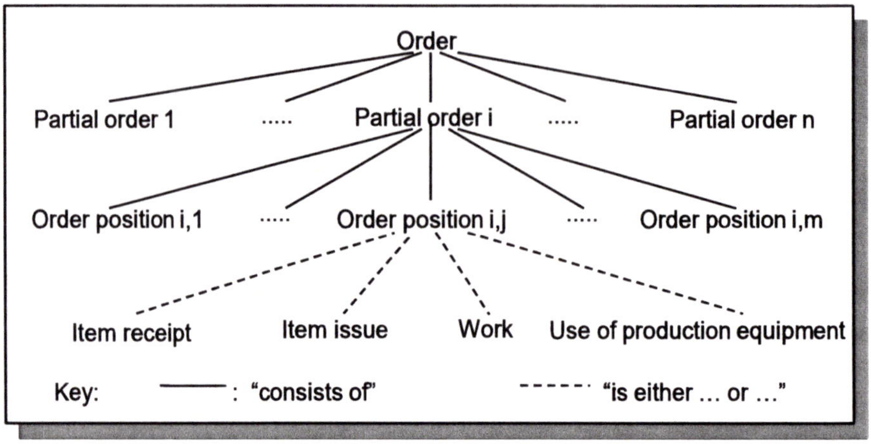

Fig. 17.1.2.1 The general structure of an order in sales and distribution, production, or procurement.

Here, the observations from the examples in Figures 1.2.1.1 and 1.2.1.2 are supplemented with a further level.

A *partial order* is an order object within an order, which is complete with respect to content but is not regarded as a separate business object.

Several partial orders may logically be combined under a single order.

- For example, the partial orders in a sales or procurement order may be sets of order positions that will be procured at different times, but together form a whole, for example, with respect to order billing.
- In addition, certain partial orders in a production order may result in semifinished goods, which, in turn, may appear as item issues in other partial orders. In this case, a first partial order is used to produce a lower structure level, for example. Its result is not stored temporarily, but rather is immediately used in the partial orders for the upper structure levels. This creates a network of partial orders.

In principle, all types of order position may appear in sales, production, and procurement orders.

- Sales orders generally relate to item issues, although in service companies they may also involve work and the production equipment used.
- Procurement orders generally contain *item receipts*, although purchased services can also involve the *work* and *equipment* types.
- Production orders are more complicated from the viewpoint of a firm's logistics: There is often only one item receipt, that is, the manufactured and salable product. This goes either into store or to shipping and thus is passed on to the sales department, which placed the order. In other situations, the item receipt is a semifinished good that is placed in stock. It is also possible for several different item receipts to arise from the same production process (see Chapter 8).

From the logistics viewpoint, the commodities used in the production process are also item issues; for example, issues from the raw materials or semifinished goods store. A production order is characterized by operations and the production equipment used, that is, tools, devices, and machines.

Figure 17.1.2.2 contains a formalized order structure with the same content as Figure 17.1.2.1, in this case as an entity or object model for an information system. For the definition of these terms, and all other technical terms of the domain of information management, see Chapter 20. The special graphical structures are defined as follows:

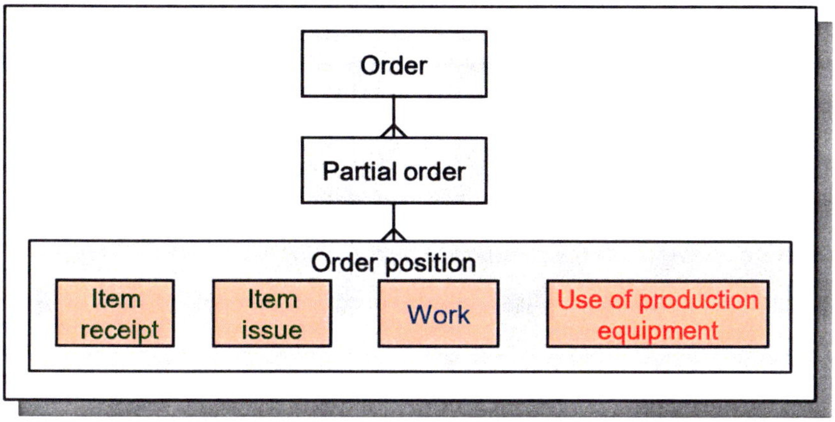

Fig. 17.1.2.2 The basic object classes in an order database.

- The "fan" symbol describes a *component hierarchy*, or *whole-part hierarchy*. An object of the *order* class consists of n different objects of the *partial order* class, and an object of the *partial order* class consists of n objects of the *order position* class.
- The way in which the *item receipt*, *item issue*, and *work* and *production equipment* class symbols are nested within the *order position* class describes a *specialization hierarchy*. An item receipt, item issue, or work or production equipment *is a* specialized order position.

The individual object classes that make up the *order* business object are discussed in depth later in this chapter.

17.1.3 The Order and Partial Order Header

Intended learning outcomes: Produce an overview on and present the attributes of the object classes "order header", and "partial order header".

The order header class combines all the data that represent the order as a whole. Its attributes can essentially be divided into the following subsets:

Firstly, attributes that describe the business partner. For a sales order, this is the customer. For a procurement order, it is the supplier; for a production order, it is the sales, R&D, or logistics department. These attributes include:

- Business partner ID and the address of the business partner
- Business partner's object for which the order is used

Secondly, attributes used to administer the order. These are attributes associated with the status of an order, typically:

- *Order ID*, that is, the order identification
- *Validity date* of the order (tender date, date on which order was issued, etc.)
- *Kind of order* (e.g., customer, procurement, production, or overhead order)
- *Costing unit* of the order (in order to combine orders to compare costs against profits) and other attributes to prepare for job-order costing
- *Billing address*
- *Order status*, that is, the administrative status of the order (e.g., in preparation, scheduled, released, started, cancelled, completed, inspected, deletable)
- *Order conditions* and other information that appears at the bottom of the order; allocation to the order header means that a separate *order footer* class may be omitted

Thirdly, the attributes that concern planning & control of the order include:

- A *flag* to indicate whether it is a *simulated* or *effective* order
- The *order priority*
- The *order urgency*
- The *order start date* and the *order end date* or *order completion date*
- A *flag* indicating whether the *dates* are *firm* or may be postponed

The *partial order header* object class essentially incorporates the same attributes as the third subset of attributes for the *order header* class, plus the order ID, partial order ID (generally a consecutive number that supplements the order ID), and a description of the partial order.

17.1.4 The Order Position

Intended learning outcomes: Present the object class "order position" and its attributes.

The *order position* class is comprised of all the attributes (information) that appear on each line of an order. One object is stored in each order position. The attributes may be divided into the following subsets:

Firstly, the identifying attributes, which include

- *Order ID*
- *Partial order ID*
- *Order position ID*, generally a number; for work, it may correspond to the sequence in the routing sheet, whereas for items or production equipment it is a relative position within a picking list determined using suitable logic (e.g., the order in which they are taken from stock)
- *Type of order position*: item receipt, item issue, work or production equipment

- *Position status*, that is, the administrative status of the position (e.g., scheduled, reserved, released, partly executed, fully executed, administration complete)
- *Flag* to indicate whether the *dates* are *firm* or may be postponed

Secondly, specific attributes that differ according to the type of order position. For the *item* order position, that is, for item receipts or item issues, these include:

- *Item ID*
- *Reserved* or *allocated quantity*
- *Quantity issued* or *effective quantity*
- *Billed quantity*
- *Reservation date* or the *earliest start date*
- *Item description*, a set of attributes that may be used for more detailed identification and classification (see Section 17.2.2)
- *Position-specific item description* within the current order; that is, the position of an electronic component
- Information for stockkeeping and accounting, which generally means a set of the attributes described in detail in Section 17.2.2.
- *Work order position ID*, that is, the operation ID for which an item issue is required or which results in an item receipt

The following attributes apply to the *work* order position (or *order operation*):

- *Work center ID* (or capacity ID), that is, the identification of the location or group of machines where or with which this operation is used for production
- *Work description*
- *Standard load* in capacity units (defined in the same way as the load of an operation in Sections 1.2.4 and 13.1.2)
- *Setup load* and *run load*
- *Actual load* in capacity units
- *Billed load* in capacity units
- *Lead time* and, if necessary, lead time components
- *Start date* (the *operation start date*), for example, the earliest, latest, or probable date
- *End date* (the *operation due date*), for example, the earliest, latest, or probable date
- *Work center description* and other data used to identify and classify the organizational unit carrying out the work; see also Section 17.2.4
- *Costs* and *availability data* for the work center; this is a set of the attributes described in detail in Section 17.2.4

For a *use-of-production-equipment* order position, it relates to the specific attributes of:

- *Production equipment ID*

- *Reserved* or *allocated quantity*
- *Quantity issued* or *effective quantity*
- *Billed quantity*
- *Production equipment description* and other attributes for identifying and classifying the production equipment
- *Work order position ID* (the operation ID that uses the production equipment)
- *Costs per issued quantity* and other attributes used for billing
- *Start date*, for example, the earliest, latest, or probable date
- *End date*, for example, the earliest, latest, or probable date
- *Quantity of available production equipment*, their *costs*, and other attributes used for billing; see also Section 17.2.6

Any amount of text may be assigned to each *order position* object.

17.1.5 Inventories and Inventory Transactions

Intended learning outcomes: Present the object classes for inventories and their attributes.

The following objects are grouped into logical units (object classes) for the purposes of administering inventories:

- Storage location for administering the various stock locations within the firm. The attributes of this object class are the location ID, description, various classifications, attributes for modeling the different features described in Section 11.1.1, and so on.
- *Physical inventory* for administering the various stocks of storable items for accounting purposes. The attributes of this object class are the identifications of the administered item and of the stock location, the quantity stocked expressed in the unit of measure for the item, date of the last receipt into and issue from stock, and so forth.

These two classes are not sufficient on their own to represent stocks of batches or variants, however. The extensions required for the processing industry and for production with a wide range of variants are discussed in Section 17.4.2. According to [Schö01], Ch. 8, a stock of batches or variants ultimately becomes a specialization of an order position.

All item movements, particularly the inventory transactions, are defined in a *transaction* class. See also Section 11.1. This class may be analyzed using any number of criteria, e.g., consumption, sales, or bid statistics (see Section 11.2). The attributes of this class include:

- Transaction date
- Item ID or item family ID
- Moved quantity
- Persons responsible for recording the transaction
- *Two* customers, production, or procurement order positions or stock level positions concerned ("from" and "to" positions associated with the transaction)

17.2 The Master Data for Products and Processes

Intended learning outcomes: Describe master data of products and operations. Explain the data structure of item master, bill of material, where-used list, work-center master, operation, routing sheet, production equipment, bill of production equipment, and bill of tools.

The term *master data* comprises the data of all the order-independent business objects discussed in Section 1.2 (see Section 5.1.4).

This section first introduces the master data for the conventional MRP II concept, which is intended for products with convergent product structures. Section 17.4 discusses the extensions arising from the processor-oriented concept (divergent product structures). The extensions arising from the variant-oriented concept are described in Section 17.5.

17.2.1 Product, Product Structure, Components, and Operations

Intended learning outcomes: Present the concept of master data. Explain the production order as a collection of master data. Describe a simple product structure. Identify the intermediate product used simultaneously as a component in higher-level products.

Master data are created as the result of product and process design that is not associated with a specific customer order. A suitable customer, production, or procurement order can then be repeatedly derived from these master data if an order quantity and date are added to the product and process description.

This can be compared to a recipe in a cookbook since the recipe is developed on its own, that is, independently of the subsequent cooking processes. Such a recipe may be used repeatedly for preparing meals, and different order quantities (= number of people) may be applied. The cook book contains the following information:

- The ingredients are shown in a list (recipe).
- The sequence of individual working tasks is also given in the form of a list that describes how to arrive at the result; that is, the finished meal, starting from the ingredients.
- The cooking utensils, such as knives, pans, and so forth, are mentioned in the description of the work. They are sometimes also summarized in a list.
- The cooking device required, for example, stove, oven, sink, is mentioned in the description of the work.

The same concept can be applied to the description of the product and production process within a company, using generalized or specific terminology:

- The result is a *product* or a *parent item*.
- The ingredients become components and the recipe becomes a bill of material.
- The work becomes operations and the sequence of operations becomes a routing sheet or process plan.

- The cooking device and other cooking utensils become machinery and other production equipment.
- The actual kitchen becomes a work center with one or more workstations at which the individual operations are carried out.

Figure 17.2.1.1 shows, by way of example, the composition of master data in the form of a production order of the ball bearing of Figure 16.2.1.1. It specifies an order quantity (a lot) of 100 units of measurement (in this case "piece"). No dates are specified, however. The only other information is certain characteristic data and positions.[1]

PRODUCT (POTENTIAL ITEM RECEIPT)

Product ID	Order quantity or lot	U/M	Description	Dimension
83569	100	Pce	Ball bearing	12 mm

BILL OF MATERIAL WITH POSITIONS (COMPONENTS OR POTENTIAL ITEM ISSUES)

Position	Component ID	Total usage quantity	U/M	Description	Dimension
050	83593	100	Pce	Ring	12 mm
060	83607	2	KG	Uniflon-R	67/3000 mm
⋮	⋮		⋮	⋮	⋮
	⋮				

ROUTING SHEET WITH POSITIONS (OPERATIONS OR POTENTIAL WORK TASKS)

Position	Work description	Standard time	U/M	Work center / Description
250	10.5 x 67 mm Cut Uniflon	1.45	h	907501/Manual production
270	Press together	1.12	h	983001/Special presses
⋮	⋮		⋮	⋮

Fig. 17.2.1.1 The production order as a collection of master data.

- The *ball bearing* product (item ID 83569) is a potential item receipt and consists of the two components *ring* (item ID 83593, an in-house made semifinished good) and *Uniflon* (item ID 83607, a purchased raw material). The bill of material for the product thus has at least the two specified positions. These are potential item issues.
- The ball bearing (item ID 83569) is produced by the two operations *Cut Uniflon* (position 250 at work center ID 907501, "Manual production") and *Press together* (position 270 at work center ID 908301, "Special presses"). The routing sheet for the product thus has at least the two specified operations. These are the potential *work* order positions (or order operations).

Figure 17.2.1.2 shows the simple, single-level *convergent product structure* that occurs in the initial stages. See also Figures 17.1.2.1 and 1.2.2.2. All the resources needed to manufacture the product are listed as positions in the product structure. Such a position may thus be a component, an operation, or production equipment.

[1] In the case under consideration, there are other components and operations. For the sake of simplicity, however, only these two components (respectively, these two operations) are listed here.

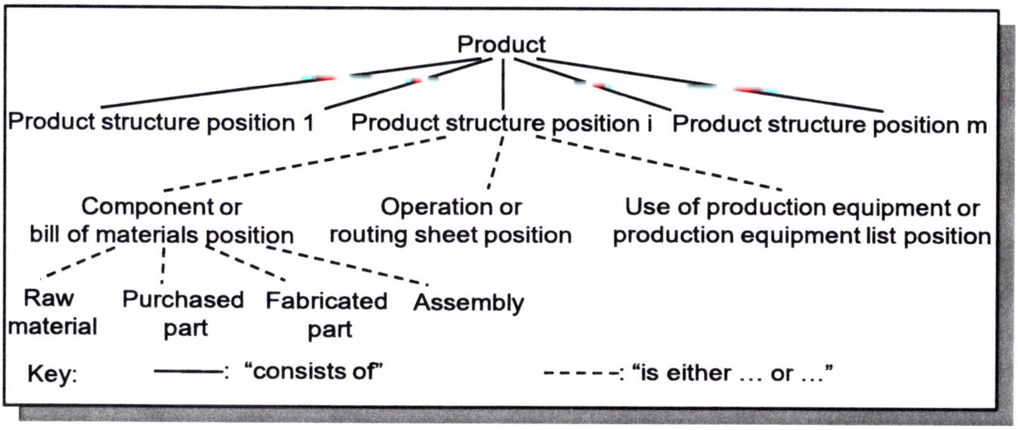

Fig. 17.2.1.2 A simple product structure.

According to Figure 1.2.2.1, a component may first be a raw material or a purchased part. In reality, a product often has hundreds, or even thousands, of such components. These are grouped into product modules or intermediate products (in-house parts [that is, parts produced in-house], semifinished goods, or assemblies). This takes place for various reasons:

- A module may be used in several different products. Under certain circumstances, it is sensible to produce or procure this intermediate product with a logistics characteristic different from that of the higher-level products.
- A module may be either produced in-house or purchased and thus acts as a point of differentiation for logistics purposes.
- A module corresponds to a design structure level or production structure level.

An intermediate product may itself be made up of different components and may also be used as a component of various higher-level products. Figure 17.2.1.3 formalizes this fact in two different hierarchies,[2] which refer to the upper and lower levels of the multilevel bill of material. See also the two intermediate products in Figure 1.2.2.2.

The creation of intermediate products may be repeated in several levels. Intermediate products lead from the single-level product structure to a multilevel product structure. To illustrate this, the cookbooks of a professional cook will contain multilevel recipes, that is, semifinished goods as components of the menu that are prepared in advance or purchased.

17.2.2 Item Master

Intended learning outcomes: Present the concept of the item master record. Describe the attributes of the technical information and the stockkeeping information of the item master record. Identify attributes of the item master record for information on costs and prices.

---

[2] Both cases refer to *"component is simultaneously an intermediate product"* as an association class. The "lower" case is also a composition ("whole-part" association).

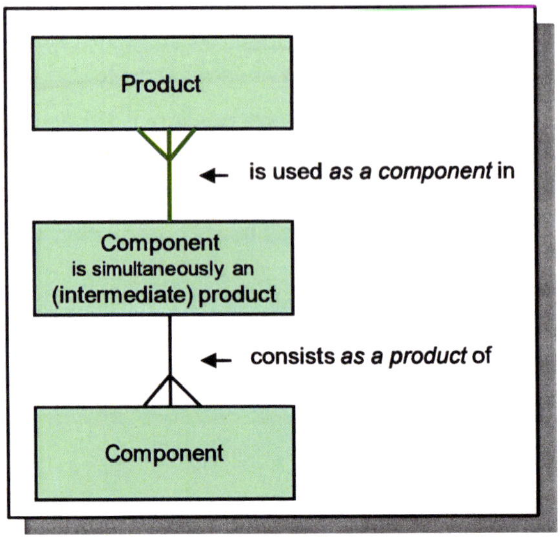

Fig. 17.2.1.3 The intermediate product used simultaneously as a component in higher-level products.

The various specializations of the *item* business object are summarized in Figure 1.2.2.1. This section provides a detailed description of the object, particularly its important attributes.

> An *item master record* contains the master data for an item.
>
> An *item master file* is a file containing all item master records.

Each record contains three different types of information: technical information, stock-keeping information, and information on costs and prices. The three types are often administered by different offices within the company. If this is the case, they must be coordinated by an organizational procedure (e.g., using workflow techniques).

The *technical information* has at least the following attributes:

- The item ID, that is, the item identification. If computerized, it should, if possible, be "not meaningful" and allocated by the information system. The item ID is a primary key and is thus unique. It must not be changed during the product life cycle. See also Section 20.3.2.
- The *EAN/UPC code*. This is a re-identification of the item ID for automatic shop floor data acquisition, and its structure is based on international standards.
- The *drawing number* or *technical reference number*. This also helps people within the company to identify the item. As a *secondary key*, however, it does not necessarily have to be unique. Its value can also change over the product life cycle, which may be necessary if the drawing numbers are reorganized, for example.
- The *item description*. This often has different attributes, which also act as secondary keys for quick and easy searching; for example, a verbal description that may be in different languages, the item abbreviation or acronym used to describe the item within the company, or the item's dimension or dimensions.

- The *item type*, that is, its specialization (end product, semifinished good, raw material, document, information, etc.).
- A flag to indicate whether the item is *purchased* or *produced in-house.*
- *Classification codes* that group items together for certain statistics.
- The *low-level code*; see Section 1.2.2.
- A flag to indicate a *by-product* or a *waste product.*
- The *units of measure*, for example, the storage unit, the unit to which costs and prices relate, the purchasing unit, or the weight unit.
- *Conversion factors* for converting from one unit of measure to another.

The *stockkeeping information* has at least the following attributes:

- The *reason for order release* (see Section 4.4.4), order release by demand (technique: MRP), order release by prediction (technique: MRP), order release by consumption (technique: order point or Kanban).
- The stock location or stockkeeping location. A separate class is needed for administering the storage locations of an item with *multiple stock organization* (see Section 11.1.1). See also Section 17.1.5.
- The *lead time.*
- The *production or procurement size.* This is a quantity (batch size), a time period, or a number of requests, depending on the *batch-sizing policy* (see Section 12.4.1).
- The *mean consumption* and the attributes used to update this value (see Section 10.2.1). Cumulative past consumption values are generally administered using separate classes (see Section 11.2.1).

Attributes for *information on costs and prices* are generally as follows (see also Chapter 16):

- The *manufacturing* or *procurement costs*: full or variable, standard, average, real or updated, simulated.
- The *cost types* taken from the cost breakdown structure of a product (the cost accumulation structure): cost of materials, direct labor costs and overheads, or fixed and variable labor costs.
- The various *selling prices*: Different prices for each market segment; previous, current, and future price (and optionally the date of validity).

Aspects of computerized administration:

- For certain large-scale modifications, however, it may be sensible to record the amendments in advance and then to run them in background mode using a batch procedure. A typical example is a change of selling prices: If the new prices are not derived from the old prices using a formula, the only possible solution is to record the new prices for each item online as separate attributes. At the key date, all the prices will then be changed in a few seconds by overwriting the "Price" attribute with the value of the "New price" attribute.

- It is necessary to record the latest modifications in the item master data if different users are able to modify the same data. It will thus be possible to identify who modified which data and when.
- When entering the data for a new item into the item master data, it is generally convenient first to copy all the attribute values of an existing item to the attributes of the new item and then to change the values.
- An item may not be physically deleted while it still occurs as a component, product, or reservation in an order or consumption statistics. An item ID is normally reserved for several years, even if the associated item is no longer physically present within the company.

17.2.3 Bill of Material, Bill-of-Material Position, and Where-Used List

Intended learning outcomes: Present the concepts of the bill of material, the bill-of-material position, and the where-used list. Differentiate between the single-level, the multilevel, and the summarized bill of material, or between the respective where-used lists. Describe the bill-of-material position and the where-used list position as logistical objects.

By way of example, Figure 1.2.2.2 shows a bill of material, that is, a convergent product structure with two structure levels. The conventional method used to model the *bill-of-material* business object does not represent the object as a whole. Instead, it defines a detailed logistical object for that business object.

A *bill-of-material position* is a product ↔ component connection within a bill of material.

Here is an example. Figure 17.2.3.1 contains five items — the three components x, y, and z, each of which occurs in products 1 and 2.

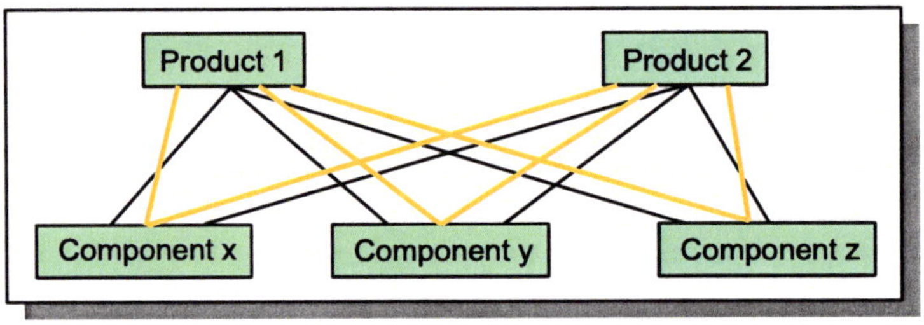

Fig. 17.2.3.1 Representation of two bills of material, each with three components.

The two bills of material lead to detailed objects, that is, *six* bill-of-material positions. These represent the six connections shown in Figure 17.2.3.2 from the product viewpoint (black) and from the component viewpoint (red).

Breaking down bills of material into their individual positions leads directly to further logistical objects. They are all derived from the bill-of-material positions by means of algorithms.

A *where-used list* indicates the way in which a component is used in different products, taking the structural levels into account (see Section 1.2.2).

Product viewpoint	Component viewpoint
Product 1 ↔ Component x	Component x ↔ Product 1
Product 1 ↔ Component y	Component x ↔ Product 2
Product 1 ↔ Component z	
	Component y ↔ Product 1
Product 2 ↔ Component x	Component y ↔ Product 2
Product 2 ↔ Component y	
Product 2 ↔ Component z	Component z ↔ Product 1
	Component z ↔ Product 2

Fig. 17.2.3.2 Detailed logistical objects: the six bill-of-material positions as connections in two bills of material, each with three components.

The component viewpoint in Figure 17.2.3.2, that is, the bottom to top viewpoint in Figure 17.2.3.1, leads to three where-used lists — for the components x, y, and z, each with two uses in products 1 and 2.

Different forms of these bills of material and where-used lists are needed, depending on the application. Each "product ↔ component" connection should only be defined or stored once, however. The only exception to this rule is for components that occur several times in the same product (but are different each time). These occurrences can be differentiated using a relative position number (see below).

A *single-level bill of material* shows all the components of a product.

Figure 17.2.3.3 shows the three single-level bills of material, each with two bill-of-material positions, as implicitly defined by the example in Figure 1.2.2.2.

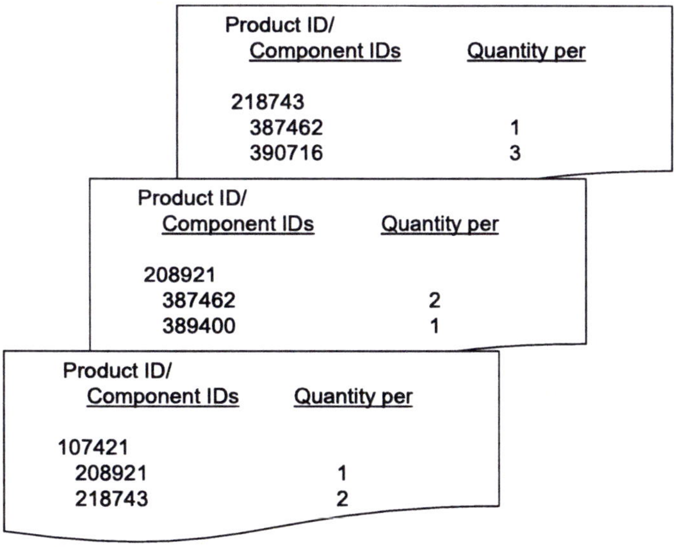

Fig. 17.2.3.3 Single-level bills of material.

The *multilevel bill of material* or *indented bill of material* shows the structured composition of a product, across all the levels.

Figure 17.2.3.4 shows the indented bill of material for the example in Figure 1.2.2.2.

Product ID/ Component IDs	(Cumulative) Quantity per
107421	
208921	1
387462	2
389400	1
218743	2
387462	2
390716	6

Fig. 17.2.3.4 Indented bill of material (multilevel bill of material).

In this form, the content corresponds exactly to the graphical representation of a product as a tree structure, again as shown in Figure 1.2.2.2.[3] The quantity per is always the cumulative quantity of the component used at this point in the product (by way of contrast to the graphical form in Figure 1.2.2.2).[4] An algorithm can also be used to generate a multilevel bill of material from the single-level bills of material.

The *summarized bill of material* is a condensed multilevel bill of material in which each component occurs only once, although the total quantity per is specified.

Figure 17.2.3.5 shows the summarized bill of material for the example in Figure 1.2.2.2.

Product ID/ Component IDs	(Total) Quantity per
107421	
208921	1
218743	2
387462	4
389400	1
390716	6

Fig. 17.2.3.5 Summarized bill of material (condensed multilevel bill of material).

---

3 This tree structure occurs naturally for a product with an assembly orientation of its product structure.

4 Of course, the cumulative quantity per can also be shown in the graphical form.

The quantity per is the cumulative quantity of components used in the product. A summarized bill of material is used for manual cost estimation or for quickly calculating the number of components to be bought for a lot of end products. A summarized bill of material can also be generated from the single-level bills of material using an algorithm.

Similar algorithms can also be used to create various types of where-used lists from the bill-of-material positions.

The *single-level where-used list* shows all the products that are integrated directly into a component.

Figure 17.2.3.6 shows the five single-level where-used lists implicitly defined by the example in Figure 1.2.2.2.[5]

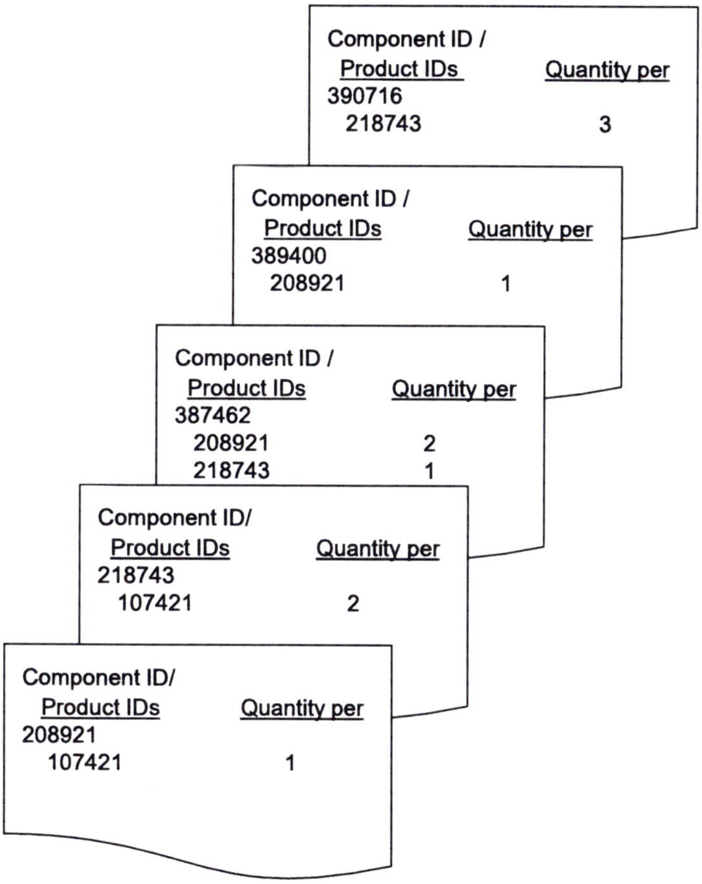

Fig. 17.2.3.6 Single-level where-used list.

5 The where-used list for an end product is empty; that is, there is no where-used list.

Figure 17.2.3.6 contains exactly the same number of connections as Figure 17.2.3.3, that is, six. Although these are the same connections, here they are taken from the component view in Figure 17.2.3.2. In this case, the quantity per is the quantity of components integrated directly into the product. The single-level where-used list is useful because it provides a picture of a certain component.

The *multilevel where-used list* or *indented where-used list* shows, in structured form, how a component is used across all the levels, right down to the end products.

Figure 17.2.3.7 shows the multilevel where-used list for the component with item ID 387462 from the example in Figure 1.2.2.2.

Component ID/ Product IDs	(Cumulative) Quantity per
387462	
208921	2
107421	2
218743	1
107421	2

Fig. 17.2.3.7 Indented where-used list (multilevel where-used list).

Here, the quantity per is the cumulative quantity of this component that is integrated into the product at this point. An indented where-used list is useful for assessing the possible consequences of a *substitution*, that is, the replacement of an unavailable primary product or component by a non-primary item.

The *summarized where-used list* is a condensed multilevel where-used list in which each product occurs only once, together with the cumulative quantity of that component incorporated into the product.

Figure 17.2.3.8 shows the summarized where-used list for the component with item ID 387462 from the example in Figure 1.2.2.2.

Component ID / Product IDs	(Total) Quantity per
387462	
208921	2
218743	1
107421	4

Fig. 17.2.3.8 Summarized where-used list (condensed multilevel where-used list).

In this case, the quantity per is the total quantity of components that are integrated into the product. A summarized where-used list is needed to draw up a procurement plan, for example, or to estimate which end products will be affected by replacing an item at a lower level.

The *bill-of-material position* logistical object appears in the type of formalized product structure shown in Figure 17.2.3.9.

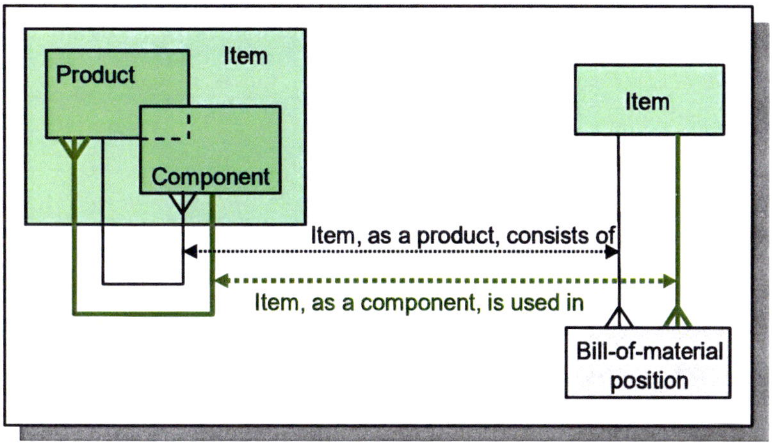

Fig. 17.2.3.9 The bill-of-material position logistical object.

The left-hand side of Figure 17.2.3.9 shows the content of Figure 17.2.1.3, as shown in Figure 1.2.2.1. The *item* class is thus in an n-to-n association with itself.

- A product may have different components. Expressed formally, this means that an object of the *item* class, in its specialization *as a product, consists of* n different objects from the item class, *component* specialization.
- A component may occur in different products. Expressed formally, this means that an object of the *item* class, in its specialization *as a component, is used in* n different objects from the item class, *product* specialization.

This n-to-n association is then shown on the right-hand side of Figure 17.2.3.9, broken down into the two corresponding 1-to-n associations. This results in an additional object class, namely, the bill-of-material position, which determines the "product ↔ component" connection or association between two items. This association may be either *"item, as a product, consists of"* or *"item, as a component, is used in,"* depending on which side we start with. A bill-of-material position is thus simultaneously a where-used-list position.

The *where-used-list position* is a different view of the bill-of-material position.

The view of the bill of material can be described as follows:

- All n bill-of-material positions can be reached from a product, and all these positions lead to a component that is built into the product. Taken together, all these positions with the information they contain on the component form the bill of material.

The view of the where-used list can be described as follows:

- All n where-used-list positions can be reached from a component, and all these positions lead to a product in which the component is used. Taken together, all the positions of the where-used list with the information they contain on the product form the where-used list.

The most important attributes that have to be administered for a bill-of-material position are:

- *Product ID* (the product identification); this is an item ID
- *Component ID* (the component identification); this is an item ID
- *Quantity per*, that is, the number or quantity of components that is needed to produce a single unit of the product
- *Sequential number of the position within the bill of material* (for sorting and identification purposes)
- *Operation ID* for which the component is needed (see Section 17.2.6)
- *Lead-time offset*, that is, the difference in time relative to the product completion date before which the components must be made available (see Section 1.2.3)
- *Effectivity (dates)* or *effective dates (start and stop)*, that are the dates on which a component is to be added or removed from the bill of material; effectivity control may also be by engineering change number or serial number rather than date

Again, these are only the most important attributes for the elementary functions associated with the bill of material and where-used list. Additional attributes and even additional logistical objects must be represented for more complex applications, for example, bills of material for a *product family with many variants*. See also Chapter 7 and Section 17.3.

In historic and generic terms, the *bill-of-material position ID* (bill-of-material position identification) combines the *product ID* and *component ID* attributes. Today, it is more often the union of the *product ID* and *sequential number of the position within the bill-of-material* attributes, however.

The advantage of the second definition is that the same component can occur more than once in the same bill of material. The components may also be sorted into a logical order that does not correspond to the component ID. This does have the disadvantage that the number of possible components of a product is limited by the number of possible relative position numbers. In addition, to keep a certain degree of order, any "holes" must be filled in the order of relative position numbers. This can be done by first allocating every tenth number and then periodically reorganizing the numbering.

Aspects of *computerized administration*:

- Certain transactions enable whole or partial bills of material for one assembly to be copied to another assembly. There are also transactions that allow large-scale modifications to be carried out, e.g., by replacing a certain component with a different component in every bill of material (batch procedure running in background mode).

- Another algorithm periodically calculates the *low-level code* of all items. It can also check whether a multilevel bill of material is actually a product structure without loops. This test is often rather time consuming and is difficult to carry out online while administering the bills of material. See also Section 8.3.3.

17.2.4 Work Center Master Data

Intended learning outcomes: Present the work-center business object. Describe the attributes of the work-center master record relating to capacity, costs, and lead time.

The *work center* business object is introduced in Figure 1.2.4 together with the other business objects. This section provides a detailed description of the object, particularly its most important attributes.

The *work center* object class generally comprises different types of information relating to capacity and costs, plus information used for scheduling, particularly for calculating lead times. These different types of information may, in turn, be administered by different people, depending on how the company is organized.

The *information relating to capacity* includes the following attributes:

- Work center ID
- Work center description
- Position within the hierarchy of workshops (see also Section 17.2.5)
- Work center type (store, parts production, assembly, external, etc.)
- Number of work centers or machines
- Number of working hours per shift and per day (often measured in 1/100 hour or industrial periods)
- Capacity unit (see Section 1.2.4)
- Number of capacity units per shift and per day (machine capacity or labor capacity, depending on the work center type
- Number of shifts per day
- Various factors: *capacity utilization, work center efficiency*, or the *efficiency rate*; see Section 1.2.4)

Capacity may change its value after a certain date. Capacities that change over the course of time are administered in a separate object class.

The *information concerning costs* includes at least the following attributes (see also Section 16.1.4):

- *Fixed labor costs* per capacity unit *for personnel*
- *Variable labor costs* per capacity unit *for personnel*
- *Fixed labor costs* per capacity unit *for machinery*
- *Variable labor costs* per capacity unit *for machinery*

This information is needed in order to analyze the standard or actual times for cost estimation or job-order costing. Conversion factors and different overhead rates are also needed when operating multiple machines or if the machines are operated by several people. It may also be necessary to specify different overhead rates for the setup time.

The following attributes are administered for time management (see Section 13.1), particularly for calculating the *lead time* (see Section 13.3.2):

- The move time from and to the work center. This time incorporates both the actual handling time (administration and transportation) needed to move a commodity from one work center and another and the time needed to move it between two successive operations. See also Section 13.1.5.
- The *nontechnical wait time before the operation* or *queue time*, that is, the average time a job remains in the queue upstream of the work center before being processed.

Other attributes concern alternative work centers, for example. As for the *item,* it is also possible to record the most recent modifications.

17.2.5 The Work Center Hierarchy

Intended learning outcomes: Present the concepts of workstation and cost center. Explain the work center hierarchy.

Figure 17.2.5.1 shows an example of a work center hierarchy within a company. It often corresponds to the company's structural organization. As already mentioned, a work center comprises several similar or identical workstations or machines.

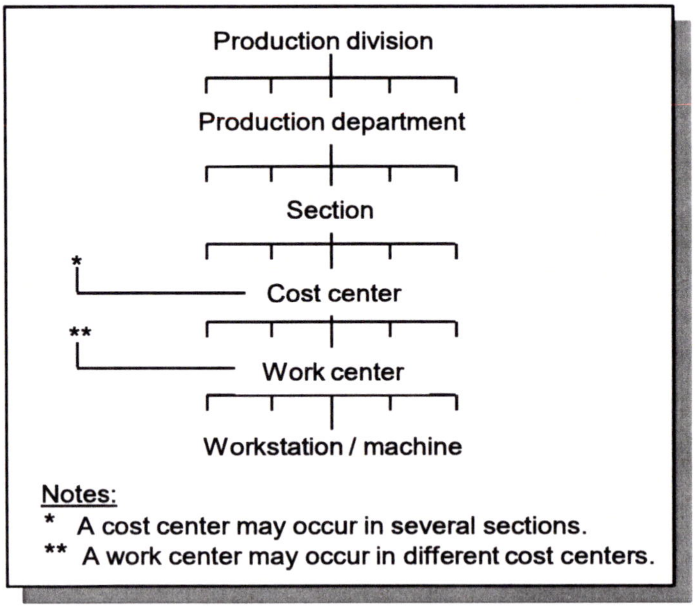

Fig. 17.2.5.1 The work center hierarchy.

A *cost center* is a unit made up of work centers with the same costs.

The work centers of a cost center are also often of the same type. Work centers are defined by the production carried out at them, while the cost center is an accounting term and is therefore defined for financial purposes.

A *section* is made up of several cost centers or work centers. It is managed by a foreman.

A *production department* is a factory, for example, which is managed by a production director. The *production division* comprises all the factories of a company.

The levels described above are needed for various analyses with different levels of compression (reserve capacity, comparison of capacity and load). The same analysis may be needed for a work center considered in isolation and for a group of work centers at any level in the work center hierarchy.

The simplest structure is that of a strict hierarchy (tree structure). In many cases, however, a network is created, as indicated in the notes to Figure 17.2.5.1. In fact, it may be necessary to define the same work center for several sections or cost centers. This applies, for example, if the same machine is used in different sections and the machine from one section can easily be used as an alternative machine in another section.[6]

17.2.6 Operation and Routing Sheet

Intended learning outcomes: Present the operation business object in association with the routing sheet. Describe its attributes. Identify the work center where-used list.

Section 1.2.3 introduces the *operation* business object in association with the routing sheet and resource requirement or process plan business objects. This section provides a detailed description of the operation object, particularly its most important attributes. An operation is described by at least the following attributes (see also Figure 17.2.1.1):

- *Product ID* (the product identification); this is an item ID
- *Sequential number* or *operations sequence number*; this defines the order in which the operations are carried out
- *Work center ID* of the *primary work center*, that is, where the operation is normally scheduled to be performed
- *Work center ID* of the *alternate work center*, that is, where the operation is not normally scheduled to be but can be performed
- *Operation description*, which may consist of several lines; this is ideally a typical concise description, followed by detailed information
- Standard *setup load* (see Section 13.1.2)
- Standard *run load per unit* (see Section 13.1.2)

---

[6] In this case, the identification or primary key for the *work center class* is made up from the identifications for the work center, cost center, and section classes. The load and capacity can then be compared for each combination of "section — cost center — work center," as well as for identical work centers in the various sections.

- *Setup time* and *run time per unit* or the *formulas for converting* from setup load and run load to setup time and run time
- *Technical wait time after the operation* (see Section 13.1.3)
- *Effective dates (start and stop)*. On these dates, the operation is to be added or removed from the routing sheet; effectivity control may also be by engineering change number or serial number rather than date

The *operation ID* is the union of the *product ID* and *operation number* attributes.

The *routing sheet* or *routing* can be derived from its operations, just as the bill of material can be derived from its bill-of-material positions. A product forms a 1-to-n association with its operations.

An *alternate routing* is a routing, usually less preferred than the primary routing, but resulting in an identical item.

An *alternate operation* is a replacement for a normal step in the manufacturing process.

Alternate routings and operations may be maintained in the computer or offline via manual methods, but the computer software must be able to accept alternate routings and operations for specific jobs (see [APIC16]).

A *work center where-used list* provides an indication of how a work center is used in products, or more precisely in the operations for products.

As with the where-used list for components, the work center where-used list addresses the operations from the work center viewpoint, as a supplement to the product viewpoint. See Figure 17.2.3.2. A work center also forms a 1-to-n association with the operations.

Aspects of *computerized administration*:

- There are transactions that allow the entire routing sheet for an assembly or a partial routing sheet to be assigned to another assembly. There are also transactions that allow large-scale modifications to be carried out, for example, by replacing a certain work center with a different work center in every operation (batch procedure running in background mode).
- A batch procedure can periodically calculate the sum of certain elements of the lead time and insert the results into the routing sheet in order to quickly recalculate the rough-cut planning (namely, the sum of the setup times, the sum of the run times (for an mean lot size), and the sum of the interoperation times; see Figure 13.3.2.4).

17.2.7 Production Equipment, Bill of Production Equipment, and Bill of Tools

Intended learning outcomes: Present the concepts of bill of production equipment and bill-of-production-equipment position as well as production equipment where-used list. Produce an overview on collective tool, bill of tools, bill-of-tools position, and tool where-used list.

The *production equipment* business object was introduced in Section 1.2.4 together with the work center and routing sheet objects. This section provides a detailed description of the production equipment object, together with some additional logistical objects and their most important attributes.

Production equipment means machines, devices (e.g., jigs, fixtures), and tools. These objects can no longer simply be mentioned in a work instruction. We are now interested in, f.ex.:

- How a certain tool will be used in the operations, for example, in order to plan an alternative for that tool or to determine the load on a tool
- Utilization of a tool, in order to calculate depreciation and to schedule maintenance

The *technical information* for production equipment is essentially the information that is administered as attributes for the item.

The *information concerning depreciation* of production equipment uses attributes similar to the cost attributes of the item. Additional specific attributes must also be administered, such as the depreciation rate and planned and effective utilization.

The *information concerning the capacity of a tool or device* uses attributes similar to those for the work center. Today, however, a tool is no longer necessarily associated with just one machine or work center. Flexible work cells often allow tools to be used flexibly.

The load and capacity of a machine are subsets of the load and capacity of the entire work center to which the machine belongs.

A *bill of production equipment* for a product is made up of various bill of production equipment positions. A *bill of production equipment position* is production equipment that is used in a specific operation.

A production equipment position has about the same attributes as a bill-of-material position.

A *production equipment where-used list* shows the usage of production equipment within products; more correctly within operations, for manufacturing products.

In analogy to the where-used list of components, the production equipment where-used list is a view of the production equipment on the operations, thus complementing the view of the product. A production equipment is in a 1-to-n association with the operations.

A *collective tool* or *toolkit* is the combination of a set of tools.

A *bill of tools* describes the individual tools that make up a toolkit.

Collective tools are particularly important in machining centers, for example. The structure of a bill of tools is similar to that of a bill of material with its bill-of-material positions (see Section 17.2.3).

A *tool where-used list* shows the usage of a tool within collective tools.

Bills of tools and tool where-used lists can be compared to bills of material and where-used lists of items. The possible variants of such bills and lists (single-level, multilevel, etc.) correspond to those in Section 17.2.3.

17.2.8 Composition of the Basic Master Data Objects

Intended learning outcomes: Explain the breakdown of the master data into individual classes and their associations. Describe the basic object classes for planning & control.

Figure 17.2.8.1 shows, by way of example, a breakdown of the master data for the ball bearing shown in Figure 17.2.1.1 into the four most important classes, namely, item, bill-of-material position, work center, and operation. The arrows point to the associations between the logistical objects discussed above, that is, to:

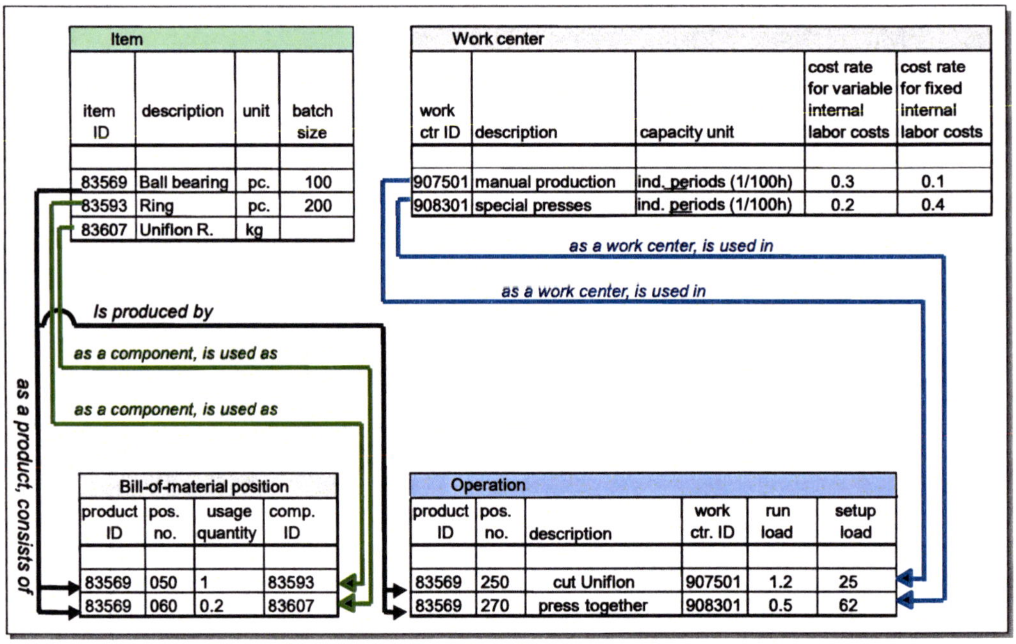

Fig. 17.2.8.1 Breakdown of the master data into individual classes and their associations using the example of the ball bearing (see Figure 17.2.1.1).

- the two 1-to-n associations shown in Figure 17.2.3.9 between the item and the bill-of-material position that determines the "product ↔ component" connection between two items. These connections are "as a product, consists of" (product viewpoint) or "as a component, is used in" (component viewpoint), depending on which side we start with. See Figure 17.2.3.2. They emerge from breaking up a reflexive n-to-n association according to Section 20.3.8.
- the two 1-to-n associations between the item and work center to the operation (see Section 17.2.6). These connections are *"is produced by"* (product viewpoint) or *"as*

a work center, is used in" (work center viewpoint), depending on which side we start with.

Figure 17.2.8.2, as a generalization of Figure 17.2.8.1, shows all the fundamental logistical object classes for the master data, together with their associations, for products with *a convergent product structure*. The representation corresponds to the type of data model used in ERP software today.

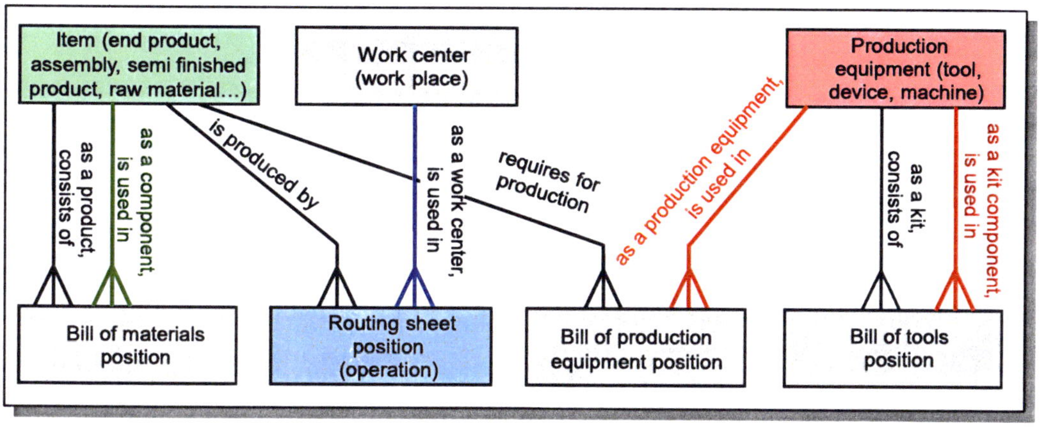

Fig. 17.2.8.2 The basic object classes for planning & control.

Depending on how it is organized, the master data are administered partly by a central standardizing committee and partly directly by the sections in which the data concerned arise, that is, design or production equipment.

It should be noted that the objects relating to production equipment are similar to the objects relating to the item (see Section 17.2.7). Collective tools or toolkits and their tool bills of material behave in the same way as products and their bills of material. Their use in operations, however, is similar to the use of a work center.

17.3 Extensions Arising from the Variant-Oriented Concept

Intended learning outcomes: Produce an overview on knowledge-based systems. Explain production rules. Present a data model for parameterized representation of a product family.

Chapter 7 covered the various techniques for planning & controlling product concepts such as product families and products produced to customer specification. Variants in bills of material and routing sheets were introduced in Section 7.3 as the production rules of an expert system for handling product families with many variants. This section explains the extensions arising from this approach, that is, the associated tools and objects.

17.3.1 Expert Systems and Knowledge-Based Systems

Intended learning outcomes: Explain the organization of an expert system (or knowledge-based information system). Describe a production rule.

It is hard to find a precise definition of the term *expert system* in the literature (see [Apel85]). One practical definition relates, in particular, to the way in which an expert system works:

Expert systems are *knowledge-based information systems*. Such systems firstly attempt to represent large amounts of knowledge concerning a limited application in a form that is suitable for the particular problem; secondly, help to acquire and modify this knowledge, and thirdly, at the user's request, draw conclusions from the knowledge and make the result available to the user.

Here, the term *knowledge* incorporates all the stored information that is needed to answer queries. Most expert systems differentiate among *facts*, *rules* (i.e., knowledge about the facts), and *metarules*, i.e., knowledge about the rules.

The term *fact base* is used to describe the rules as a whole.

The term *rule base* designates the rules as a whole.

The *inference engine* is a programming logic that applies rules to facts in order to derive new facts so as to answer questions.

Figure 17.3.1.1 illustrates the interaction between the various components of an expert system and its users for the purposes of design and operation.

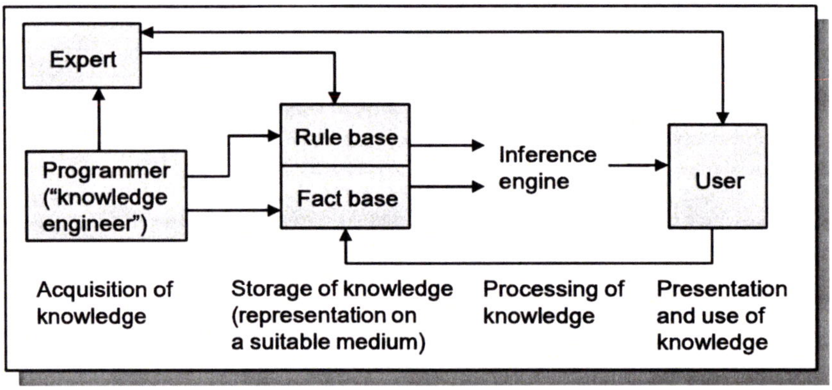

Fig. 17.3.1.1 Organization of an expert system (or knowledge-based information system).

- A programmer is responsible for designing the system.
- An expert drafts and maintains the rules and any metarules.
- The user records and maintains the facts.
- The user starts the inference engine in order to make a query.

It must be possible to operate an expert system without the help of programmers. It must be possible to make queries on the expert system without the help of experts. In practice, however, there is periodic contact between the users and experts in order to supplement or modify the rule base. The inference engine is independent of knowledge and facts. If the knowledge changes, the inference engine must not change in any way.

The *rules* of a knowledge base can be presented in different ways. The simplest and the most intuitive form is the production rule.

A *production rule* is a statement of the "if (condition), then (action)" type, i.e.,

- *If* a certain situation is true (a number of facts), *then* conclude (infer) various actions (a certain number of facts).

The structure of positions in a bill of material and routing sheet, which is made conditional though the use of IF clauses (see the example in Figure 7.3.2.1), precisely corresponds to the structure of production rules in an expert system expressed regressively (from effect to cause): Here, a *production rule in the true sense of the word*, that is, of a product to be manufactured, corresponds to a *production rule within the expert system* in the applied sense.

The *facts* of the expert system are formed by the item, production equipment, and work center logistical objects and by the values assigned to the query parameters (e.g., for an existing order). The *experts* are the designers and process planners within the company. The *users* are the people who issue, monitor, and produce the orders. See Section 7.3.2.

The *inference engine* works on the chaining principle: The inferred facts can in turn occur in rules (e.g., in the IF clause of a production rule). Further facts can be inferred if the engine is then applied iteratively, particularly to this type of rule. In this case, the inference engine is generally only needed for forward chaining. By analyzing production rules containing IF clauses with the relevant parameters, it is able to return the order bill of material and order routing sheet applicable to the specified parameter values.

A more complex expert system also contains a *declaration component* which makes the rules transparent to the user. In practice, however, most bill-of-material positions and operations are self-explanatory. Some expert systems also suggest methods for handling incomplete knowledge or knowledge arising from conclusions by analogy. They belong to the methods of artificial intelligence.

Artificial intelligence comprises computer programs that can learn and reason in a manner similar to humans (cf. [APIC16]).

17.3.2 Implementation of Production Rules

Intended learning outcomes: Differentiate between the representation of the bill of material or routing sheet for a product with options and the standard version without variants.

The following structure with three objects illustrates a production rule using object classes (see Sections 17.2.1, 17.2.3, 17.2.6, and 17.2.8):

a. The conventional item business object for items and item families, for products and components.

b. The *bill-of-material position variant* or *operation variant*. This is the conventional object bill-of-material position or operation, plus a variant number, which also belongs to the bill-of-material position ID or operation ID.

   The assembly has, for example, u positions, where $u \geq 1$. For each position x, $1 \leq x \leq u$, there are thus v_x variants, $v_x \geq 1$. If there is only one variant, then there is equality; this is the conventional situation with an unconditional bill of material.

c. The IF clause. This is a logical expression in parameters such as "type," "length," and so forth.

The three objects — product family, position variant, and IF clause — are linked together to form a production rule.

- "If product (a) and IF clause (c) are true, then the position variant (b) applies in the bill of material or routing sheet. In the case of the bill of material, it is thus true (or "is inferred") that the component in (b) is a (new) fact."

If analysis of the rule "infers" a component and thus adds to the fact bank, and this component is an intermediate product, then a further pass of the inference engine activates all the rules, and processes those that are assigned to the intermediate product (a). Such forward chaining corresponds to the processing of a multilevel bill of material (see Section 17.2.3).

The structure shown below is an extension of the traditional bill of material and routing sheet. The generalized structure and special case often encountered in the past are shown in graphical form in Figure 17.3.2.1 for ease of understanding. If we select $v_x = 1$ and no clause for all x, $1 \leq x \leq u$, then we obtain the conventional case of the "unconditional" bill of material or routing-sheet position.

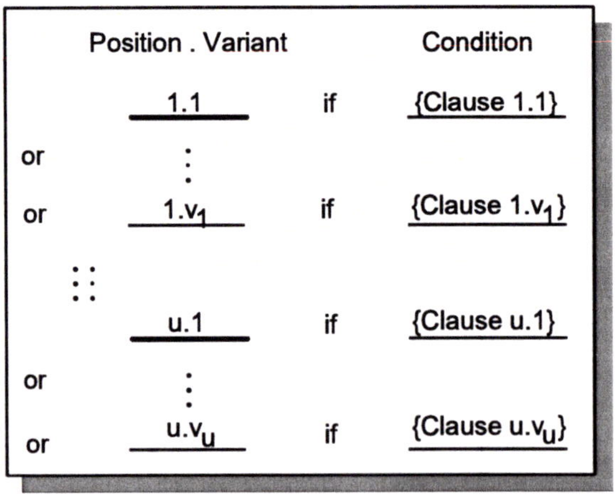

Fig. 17.3.2.1 Representation of the bill of material or routing sheet for a product with options (thick lines: standard version without variants).

Figure 17.3.2.2 shows this extension by supplementing the primary key of the conventional bill-of-material position object with the variant number. The simplest version of the IF clause is a succession of simple logical expressions; for example, connections such as type = 2 and order quantity > 100, linked with "and" or "or" in the manner of the disjunctive or conjunctive normal form. See also [Schö88a], p. 49 ff. For more complicated connections, it is better to use a formula scanner, which will create the logical expression in free form using the rules of Boolean algebra. In cases where the clauses are *not* self-explanatory, text linked to a production rule may be used to store a declaration component, in addition to its true function as a means for the operation description.

	Assembly	Position	Variant	Component	Quantity per	etc.
Before	69015	040		16285	2	
	69015	050		14216	15	
	Primary key (Assembly + Position)					
After	69015	040	01	16285	2	
	69015	040	02	16285	1	
	69015	050	01	14216	15	
	69015	050	02	14216	18	
	Extended primary key (Assembly + Position + Variant)					

Fig. 17.3.2.2 Extended primary key for a bill of material with options.

Reference is made to Figure 7.3.3.1, to demonstrate the *way the inference engine works*. It keeps the variants within a position in the best order for the query by counting the variants selected in previous queries and periodically rearranging these variants, sorting them by frequency of occurrence. For his part, the expert selects a criterion, for example, a lexicographical criterion, suitable for administering and arranging the variants.

17.3.3 A Data Model for Parameterized Representation of a Product Family (*)

Intended learning outcomes: Produce an overview on parameter and parameter class. Differentiate between primary and secondary parameters. Explain plausibility or compatibility test. Identify various types of formulas.

The production rules introduced in Section 7.3.2 as an extension of conventional bill of material and operation positions form the basic idea behind the generative technique for product families with many variants. Additional object classes are needed for a complete model. With respect to the information system, see also [Pels92], p. 93 ff., [Veen92], [Schö01], Section 13.3, or [Schi01]. See also [SöLe96] for a comprehensive application in the insurance industry. For an application in the banking industry and in the event of uncertainty, see [Schw96].

The master data model introduced in Section 17.2 must be supplemented with at least the following object classes:

- *Parameter*: This is used to define the distinctive characteristics of an item, for example, dimensions and options.
- *Parameter class*: A product family is described by an "item" entity. The specific products are also characterized by parameters or features. These are combined to

form parameter classes for structuring the set of all parameters. The item ID of the product family, together with a value for each parameter of the assigned parameter classes, then defines a product as a specific feature of the product family.

Parameters may be subdivided into:

- *Primary parameters,* which directly characterize the product family.
- *Secondary parameters,* which can be derived from the primary parameters using a formula whose range of values is thus totally dependent on the primary parameters. Secondary parameters are always needed if facts expressed by primary parameters can be better expressed for certain people using a different term.

The range of values that a parameter can assume may also be partly dependent on other parameters of the same class. In this case, we speak of:

- A *plausibility test*, or *compatibility test*. This can take the form "If ...," for example, "If width > 1000, then height < 500," or "If type = 2, then width ≤ 1500 and height ≤ 1500." The simple logical expressions in the IF and THEN clauses may, however, become very complex.

The components of product families may, in turn, belong to a product family, even one with different parameter classes. It must therefore be possible to transfer parameter values from one parameter class to another, so parameter classes are declared in the form of bills of material:

- *Bill of parameter class position*: This defines how a parameter from a (lower level) class is derived from the parameters of another (higher level) class. As with a secondary parameter, the parameter is derived using a rule or formula. The rule or formula may also be directly linked to the bill-of-material position that links the component to the product. If this is the case, the rule or formula can only be used to transfer the parameter values of this component from those of the higher-level product family.

It has been demonstrated in practice that, for complex connections, the quantities used, setup loads and loads per piece, and setup times and times per piece are dependent on the parameters, rather than being constant. Each of these master data attributes should therefore be linked to an arithmetical formula that expresses this dependency.

The *formula* is a logistical object for defining expressions that are dependent on parameters.

These *formulas* are maintained by the users and must therefore be easy to use. There are *formulas* for:

- *IF or THEN clause, a production rule, and a compatibility test.* If these contain only one parameter, then they can be represented by a table. Otherwise, they are logical expressions in disjunctive or conjunctive normal form or free form, which can be evaluated using a formula interpreter using the rules of Boolean algebra.
- *Range of values.* This may be a table or a general free-form logical expression.

- *Free-form numerical or alphanumerical expression, which nevertheless uses a standardized syntax.* Such an expression may be part of a logical expression or a formula for calculating attributes. A formula interpreter analyzes the algebraic expression using the basic operators, brackets, functions, and constants, with variable parameters, in accordance with the rules of arithmetic.

An object class that saves the parameter values of a specific product from a product family for an order or query is needed as an extension of the object classes for representing orders described in Section 17.1.

- *The parameter value* object is linked to an item receipt order position and defines the value of a parameter for a product family. The value is taken from a range of values. Sets of frequently recurring parameter values, for example, for a cost estimation of "interpolation points" for a product family, may also be defined as part of the master data.

While of relatively low degree of complexity, product configuration using knowledge-based techniques has become important within expert systems, particularly since the product with a wide range of variants has become a significant marketing strategy.

17.4 Extensions Arising from the Processor-Oriented Concept

Intended learning outcomes: Explain objects such as process, technology, and the processor-oriented production structure. Describe objects for lot control.

The various planning & control techniques for process industries were discussed in Chapter 8. This section covers processor-oriented production structures in detail. These can actually be regarded as an extension of the conventional production structure described in Sections 1.2.3 and 17.2.8. This extension is very important since it is likely that the processor-oriented production structure will become the most common model in the future. The conventional, convergent production structure, which is thus linked to a (single) product with its bill of material and routing sheet, will then become an important special case. In the future, lot control will become the general administration of on-hand balances. Proofs of origin are an increasingly common requirement in the field of logistics and in assembly-oriented systems.

17.4.1 Process, Technology, and the Processor-Oriented Production Structure

Intended learning outcomes: Disclose the object "technology and process". Explain the process train (processor-oriented production structure) and its master and order data.

As already mentioned in Section 8.2.1, product design requires a knowledge of the technologies that can be used in manufacturing processes. Such technologies and processes must be defined in a suitable manner. Figure 17.4.1.1 contains a simple structure.

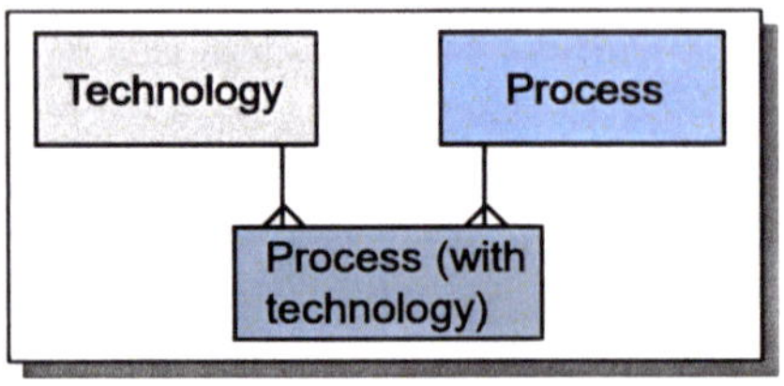

Fig. 17.4.1.1 Technology and process.

A *processor-oriented production structure* (or a *process train*) is a combination of objects described in Section 8.2.2, such as process stage, basic manufacturing step, and resource.

Figure 17.4.1.2 contains a data model for the processor-oriented production structure.

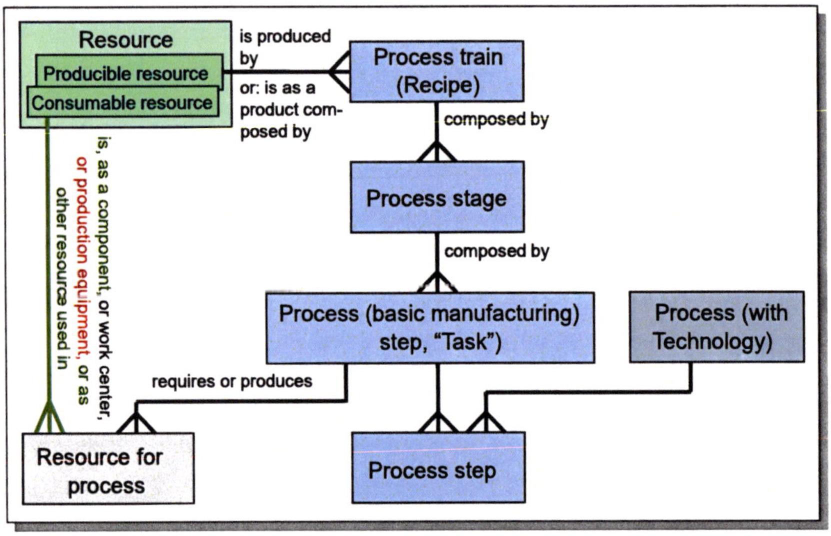

Fig. 17.4.1.2 Process train (processor-oriented production structure, recipe): objects for master data and order data.

The processor-oriented production structure defined in this way may be regarded as an extension of the model of a convergent product structure in Figure 17.2.8.2. Interestingly, the processor-oriented production structure also corresponds to the processor-oriented order structure. In this case, a stage corresponds to a partial order. An order position is now always work (an operation) to which the other order positions (resources) are assigned.[7]

17.4.2 Objects for Lot Control

Intended learning outcomes: Describe the objects for lot control in inventory management.

---

[7] However, the conventional production structure in Figure 17.2.8.2 (bills of material and routing sheet) does *not* correspond to the associated order structure in Figure 17.1.2.2.

Figure 17.4.2.1 shows the objects used for lot control (see Section 8.2.3).

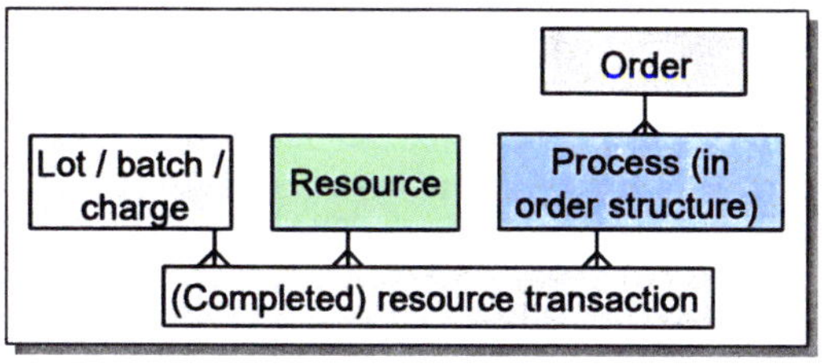

Fig. 17.4.2.1 Objects for lot control in inventory management.

The objects introduced in Section 17.4.1 must therefore be supplemented with the two objects *charge* and *completed resource transaction*. The latter object is still associated with traditional order administration. Transactions are not only used for legal reasons. They also ensure data integrity and are used in statistics concerning inventory transactions.

With this model, the structures of the two objects *on-hand balance* and *order* increase in similarity: in fact, the charge may also be regarded as the re-identification of an order ID. Placing a charge in stock simply means placing a production or procurement order in stock, where it remains identifiable as such.

17.5 The Management of Product Data and Product Life Cycle Data

Intended learning outcomes: Present product life cycle management and engineering data management. Explain the engineering database. Describe the data and functional model for EDM tasks. Present object classes and functions for release and engineering change control.

Section 5.4 discusses business methods for planning & control in the field of R&D. This essentially means project management for integrating the various tasks that take place during the business process. The interesting aspect here is the simultaneous engineering during both time to market and delivery lead time. Integration is more difficult because the various people involved have different views of the business objects. This section deals with the IT support of the efforts for integration.

17.5.1 Product Life Cycle Management and Engineering Data Management

Intended learning outcomes: Describe the concept and present the tasks of engineering data management (EDM). Identify the CAx technologies.

Product life cycle management (PLM)) is the process of facilitating the development, use, and support of products that customers want and need. PLM helps professionals envision the creation and preservation of product information, both to the customer and along the reverse-logistics portion of the supply chain ([APIC16]).

Engineering data management (EDM) is an earlier concept that focused more on a company's procedures to be integrated — all across the company. *Product data management (PDM)* is a synonymously used term.

A *product database*, or *engineering database*, is a database for commonly used information that can communicate with all information systems in the various areas.

CIM (computer-integrated manufacturing) is understood as a concept for information technology support of integrated business processes, based on the integration of the total manufacturing organization through information technology (IT).

Figure 17.5.1.1 shows the concept of engineering data management (EDM).

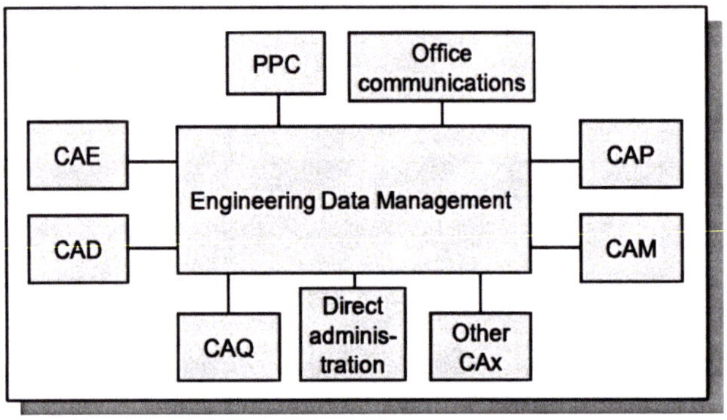

Fig. 17.5.1.1 The concept of engineering data management (EDM).

The following technologies are used in the CIM areas related to design and product:

- *CAE (computer-aided engineering)*: Computer tools to generate and test specifications, used in the product design phase.
- *CAD (computer-aided design)*: Computer tools to design and draw.
- *CAP (computer-aided process design)*: Computer assistance in defining production processes/routing sheets as well as in programming numerically controlled machines, facilities, and robots.
- *CAM (computer-aided manufacturing)*: The use of computers to program, direct, and control manufacturing through numerically controlled machines, robots, or entire flexible work cells.
- *CAQ (computer-aided quality assurance)*: Computer-aided quality assurance of the manufacturing process.

In production-related areas of CIM, there exist the following technologies:

- Computer-based planning & control systems, often called in shorthand ERP or SCM software, refer to Chapter 9.
- Computer-aided costing

To integrate the various IT-supported technologies the product database must contain all the information that is used by various IT-supported technologies or has to be transported from one to another, for example, the master data and technical product descriptions. EDM can also be associated with general office communications. This enables information and proposed action to be passed on to other areas, particularly to the company management and the planning and administration departments.

Figure 17.5.1.2 shows a possible structure of the tasks of EDM.

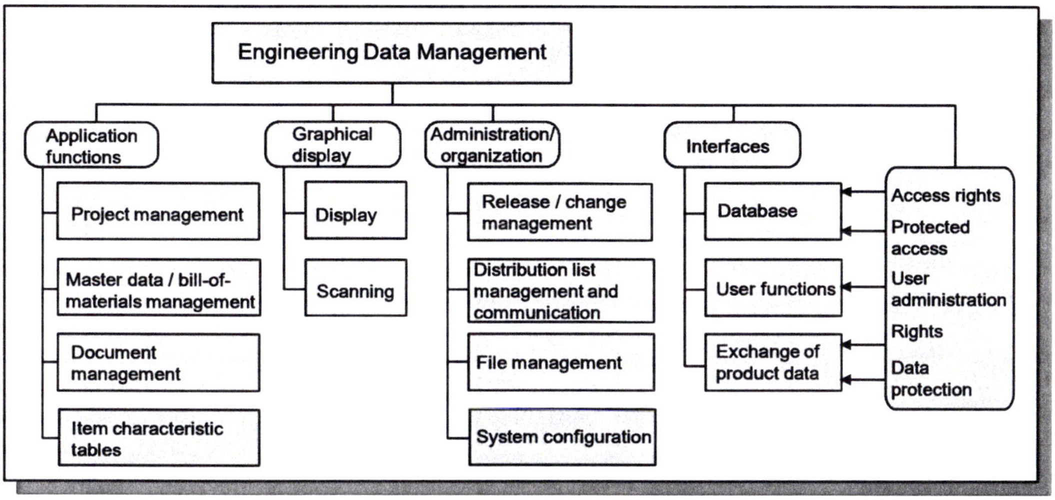

Fig. 17.5.1.2 Tasks of Engineering Data Management (EDM). (See [EiHi91]).

For a *detailed EDM model*, the basic idea behind the integration also means that the technical and commercial areas of the company must agree on a common functional and data model to represent the company's products. For example, if the design department requires a certain functionality, then it must be comprehensible to planning & control, and vice versa. Viewed pragmatically, EDM, computerized planning & control, and CAD must ultimately be adapted to one another (see also Figure 5.4.3.2). This will often already apply since, ultimately, the same products are represented and handled in each case.

17.5.2 The Engineering Database as Part of a Computerized System

Intended learning outcomes: Describe the integration of order processing by an EDMS (engineering data management system).

When implementing EDM, there are and always have been various options concerning the conceptual and technical aspects (see Section 5.4.1). The functionality of the individual links may differ greatly, depending on the direction of each link.

An *engineering data management system* (EMDS) is a database management system that links physically separate databases using the principle of a *data warehouse* as shown in Figure 17.5.2.1.

The principle works as follows: Data are stored in the databases provided by the local software. Whenever data are modified, the changes are transferred to the local database.

When a department requests data from the EDMS, it knows the location of all the data in the local databases, but is not aware of the values. The EDMS queries the local database to determine the value of the data and transfers the answers to the system. If there are m IT-supported technologies, then there will be up to m interfaces. Frequently requested data are also kept in a redundant central database that is connected to the EDMS. If there is no online interface, the data are transferred in batch mode by extraction programs and declared free-format files.

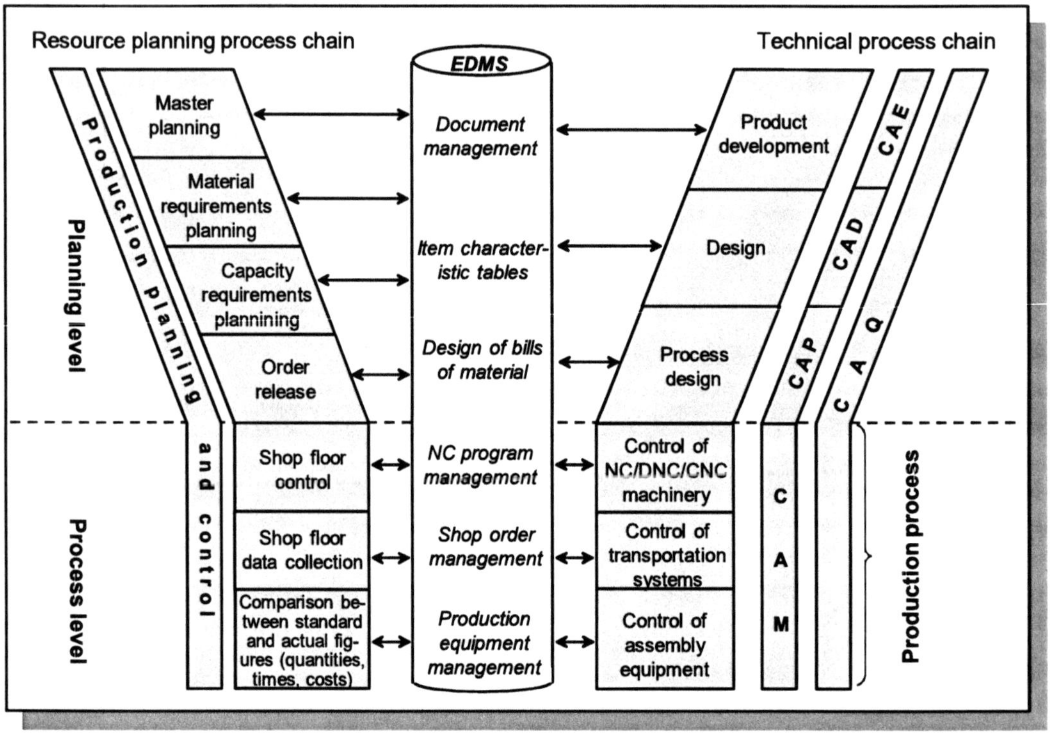

Fig. 17.5.2.1 Integration of order processing by an EDMS (engineering data management system). (From [EiHi91]).

17.5.3 Data and Functional Model for General EDM Tasks

Intended learning outcomes: Identify the object classes for engineering data management (EDM). Produce an overview on the classification system to aid the designer's work.

Engineering data management (EDM) is used to manage the technical data that describe a product, together with the relevant standards and classification. Many of these classes can be compared to the master data for planning & control described in Section 17.2:

- *Item master file*: All the technical data used to describe and classify items. This category includes data for defining the release and transfer of data to the corresponding IT-supported technologies. Search criteria are used to find items on the basis of different attributes. The item ID may first be assigned provisionally by the designer. However, the standardizing committee within the company must define an appropriate identification before the item may be definitively released. This ID is then used for planning & control.

- *Drawing directory*: This contains additional, item-related data, that is, data that are usually shown in the drawing header. The attributes are a description; the date on which the drawing was created, checked, or printed; and the people responsible for all these actions. A list of revisions is also provided.
- Special object classes for works standards, for example, DIN standards, may be kept in separate object classes.
- *Bill of material* (actually the *bill-of-material position*): This comprises the attributes described in Section 17.2.3. These include the "relative position in the drawing," which generally incorporates the relative position number. This forms the bill-of-material position ID together with the product ID. Other attributes include the date and person responsible for all this information.
- *Work center*, with the attributes shown in shown in Section 17.2.4.
- *Production equipment* and *bill of tools* (see Section 17.2.7).
- *Operation* (see Section 17.2.6).

There is also a classification system to aid the designer's work. This enables an item to be traced using a standardized, hierarchical classification. The classification system shown here should ideally be filled in using standardized information, for example, conforming to DIN 4000. The bottom level of DIN 4000 corresponds to an item family and is linked to the class list of characteristics.

An *(item) characteristic* is a parameter or criterion typically associated with this item family.

A *class list of characteristics* is a set of typical attributes for an item family, i.e., a description of a specific item from an item family using values for various item characteristics.

The item characteristics and class list of characteristics should ideally follow standards, for example, in Europe, in accordance with DIN 4000.

Multilevel bills of material or where-used lists would be needed to search the bill of material, as would tests for product structures with loops. Queries will also be needed for the standardized classification system and class-list-of-characteristics hierarchies.

17.5.4 Object Classes and Functions for Release and Engineering Change Control (*)

Intended learning outcomes: Produce an overview on the engineering change number. Identify the object classes for release and engineering change control (ECC).

The *EC number*, or *engineering change number*, is a standard concept in release and engineering change control (ECC). This is a unique and *ascending* number assigned to every modification or redesign project.

In principle, a new object is defined for every item belonging to a certain release. This new object has the *same item ID* but is suffixed with a new EC number.[8] A new item should be defined as soon as the function's forward compatibility can no longer be guaranteed. This means that the new item cannot replace the old item in every situation. On the other hand, backward compatibility is not required, that is, it does not have to be possible to install the old item in place of the new item.

The following object classes could be used for administrative checking by the project manager for release and engineering change control (ECC):

- *Project header*, with attributes such as description of the release, EC number, status and other data for staggered release, in each case indicating the person responsible.
- *Project operation*, defining one of the various stages and works required for release, with attributes such as the EC number, position, description, status, start date, and end date, in each case indicating the person responsible.
- *Project bill-of-material position*, specifying all the items belonging to the release, in each case with the status, date, and personnel responsible for release of the item; as well as its drawing, bill of material, and routing sheet. There are different pairs of "date / person responsible" attributes for different release stages.

The following functional model could be used for *release control*:

Firstly, definition of a new version, that is, of a new release or EC (engineering change):

- Enter in the project header the date and person responsible.
- Enter the items belonging to the release, each with date and person responsible for the various tasks, for example, creating or modifying drawings, bill of material, routing sheet, and item as a whole.
- Enter the various tasks involved in the release, each with start date, end date, and person responsible.

Secondly, progress and release:

- Enter the progress (with status changes) and the end of individual activities, plus correction of the status at a higher level.
- Allow for (staggered) release of bills of material, routing sheets, items or entire release (of the new version), with automatic correction of the higher-level activity list.

Thirdly, queries, for example, sort work in process by person responsible or various statuses, monitor deadlines, and indicate the content of a release (of the associated items and activities).

The data could be transferred from and to the IT-supported technologies; for example, for linking CAD and PPC software via the engineering database, using the following functions:

---

8 The EC number can thus be regarded as a mandatory parameter for a product. Depending on this parameter, different bill-of-material positions and operations can be defined.

- Transfer bills of material and any variants online, either from the CAD to the engineering database by a "drawing release" process or from the engineering database to the PPC software by a "production release" process (or, in both cases in the opposite direction by a revision process).
- Transfer all: Transfer any data that has not yet been transferred.
- Similar functions for the item master data, often in the opposite direction — from the PPC software via the engineering database to the CAD system. One example would be the transfer of all item descriptions modified after a certain date, but which have not yet been transferred to the engineering database or other CAx systems.
- Transfer order data from the PPC software to the CAD system: Transfer the item and order ID, optionally with lists of parameter values (see Section 17.3.3), as a request to create a drawing.

17.6 Summary

An order is a complex business object. It is made up of an object for data that is entered once only for each order (order header or footer), various partial orders for each order, and various order positions for each partial order. An order position is an item receipt, an item issue, work, or an order operation or production equipment.

A product or process design process creates order-independent data, known as master data. The most important object classes are the item, work center, and production equipment. The bill-of-material position, operation, and production equipment position object represent links between objects of the specified classes, thus enabling products and processes to be represented. Single-level or multilevel bills of material or where-used lists can be derived from the bill-of-material positions. Operations can be combined to form routing sheets or work center where-used lists.

Extensions arising from the variant-oriented concept concerns knowledge-based techniques for representing conditional positions in the bill of material and routing sheet. Product families can thus be suitably represented in a data model.

Extensions arising from the processor-oriented concept concern processor-oriented production structures and lot-control objects, in particular.

Product life cycle management (PLM), or engineering data management (EDM), bring together aspects of organization and processes as well as technical aspects of networking of IT systems. People have to agree on common data and functional models for general EDM tasks. These include class list of characteristics, object classes, and functions for release and change management.

17.7 Keywords

alternate routing, 738
bill of parameter class position, 746
bill of tools, 739
bill-of-material position, 728
collective tool, 739
corporate group structure, 717
cost center, 736
effective date, 734
engineering data management (EDM), 749
indented bill of material, 730
indented where-used list, 732
inference engine, 742
item ID, 726
item master record, 726
multilevel bill of material, 730
multilevel where-used list, 732
order header, 717
order item (syn. order position), 717
order operation, 717
order position, 717
parameter, 745
parameter class, 745
product life cycle management (PLM), 749
production size, 727
single-level bill of material, 729
single-level where-used list, 731
specialization hierarchy, 719
summarized bill of material, 730
summarized where-used list, 732
tool where used list, 739
toolkit, 739
unit of measure, 727
where-used list, 728
whole-part hierarchy, 719
work center where used list, 738

17.8 Scenarios and Exercises

Intended learning outcomes: Elaborate different forms of representing bills of material and where-used lists. Identify basic master data objects.

17.8.1 Different Forms of Representing Bills of Material

Intended learning outcomes: Identify various types of bills of material, given a graphical representation of the bill of material of two products.

Figure 17.8.1.1 shows the bill of material for products A and K represented in the form of the familiar arborescent structure.

In parentheses, you see the quantity per of a component, if it is not equal to one. For example, product K is assembled from two units of component D, one unit of component 5, and three units of component 1. From the two bills of material above, derive the following forms of representation, according to Section 17.2.3:

- All single-level bills of material
- Two multilevel bills of material for final products A and K
- Two summarized bills of material for final products A and K

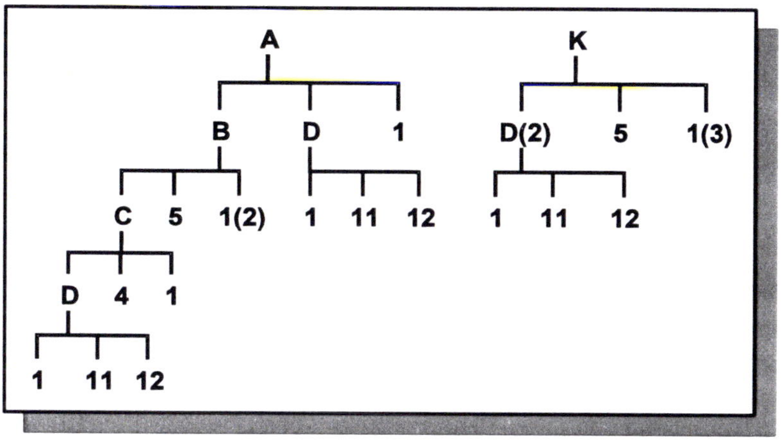

Fig. 17.8.1.1 Graphical representation of the bill of material of products A and K.

17.8.2 Where-Used Lists

Intended learning outcomes: Identify various types of where-used lists, given a graphical representation of the bill of material of two products.

On the basis of Figure 17.8.1.1, derive all types of where-used lists following the forms of representation in Section 17.2.3:

- All single-level where-used lists
- Multilevel where-used list for component 1
- Summarized where-used list for component 1
- Arborescent structure of the multilevel where-used list of component 1 (*hint*: it looks similar to Figure 17.8.1.1)

How can where-used lists be derived from bills of material?

17.8.3 Basic Master Data Objects

Intended learning outcomes: Describe how to transfer given data into the fundamental logistical object classes for the master data.

Take products A and B, as they were defined in the exercise in Section 16.7.2 (in other words, with the individual tools). Transfer the given data into the fundamental logistical object classes for the master data, as was shown in Figure 17.2.8.1 or 16.2.1.3, namely:

- Item
- Bill-of-material position
- Work center
- Operation

To enter all the data, you will need two additional classes that were mentioned in Figure 17.2.8.2, namely:

- Production equipment (tool, device, machine)
- Bill of production equipment position

Determine all the necessary attributes and their values for the individual objects (entities) in these six classes.

Hints: The number of objects per class is as follows:

- Item: 3
- Work center: 2
- Production equipment: 6 (2 machines and 4 tools / devices)
- Bill-of-material position: 2
- Operation: 4
- Bill of production equipment position: 8 (2 products, each with 4 equipments)

17.9 References

Apel85 Appelrath, H.J., "Von Datenbanken zu Expertensystemen," Informatik Fachberichte 102, Springer, Berlin, 1985

APIC16 Pittman, P. et al., APICS Dictionary, 15th Edition, APICS, Chicago, IL, 2016

EiHi91 Eigner, M., Hiller, C., Schindewolf, S., Schmich, M., "Engineering Database: Strategische Komponente in CIM-Konzepten," Hanser, München, 1991

Pels92 Pels, H.J., Wortmann, J.C., "Integration in Production Management Systems," North-Holland, Amsterdam, 1992

Schi01 Schierholt, K., "Process Configuration — Mastering Knowledge-Intensive Planning Tasks," vdf Hochschulverlag, Zurich, 2001

Schö01 Schönsleben, P., "Integrales Informationsmanagement: Informationssysteme für Geschäftsprozesse — Management, Modellierung, Lebenszyklus und Technologie," 2. Auflage, Springer, Berlin, 2001

Schö88a Schönsleben, P., "Flexibilität in der computergestützten Produktionsplanung und -steuerung," 2. Auflage, AIT-Verlag, D-Hallbergmoos, 1988

Schw96 Schwarze, S., "Configuration of Multiple Variant Products," BWI-Reihe Forschungsberichte für die Unternehmenspraxis, vdf Hochschulverlag, Zürich, 1996

SöLe96 Schönsleben, P., Leuzinger, R., "Innovative Gestaltung von Versicherungsprodukten: Flexible Industriekonzepte in der Assekuranz," Gabler, Wiesbaden, 1996

Veen92 Veen, E.A., "Modelling Product Structures by Generic Bills of Material," Elsevier Science, Amsterdam, 1992

18 Quality Management — TQM and Six Sigma

Intended learning outcomes: Produce an overview on concept and measurement of quality. Explain quality management tasks at the operations level. Describe quality management systems.

This chapter provides an overview of the management system called TQM (*total quality management*). In recent years, the Six Sigma program has come to the fore in the quality movement. Figure 1 in the introduction showed "quality" as a task that is oriented to company performance. To achieve goals in the area of quality, it is necessary to master the specific elements of the management concept for quality and to integrate these goals appropriately in all management systems along the value chain. Beyond that, Total quality management and Six Sigma are systems within comprehensive company management, which also focuses on the needs and expectations of internal and external stakeholders.

The first section in this chapter deals with the concept of quality and its measurability. Part two provides a summary of the tasks of quality management at the operations level. Part three looks at the more strategic tasks as related to Total quality management and Six Sigma.

18.1 Quality: Concept and Measurement

Intended learning outcomes: Produce an overview on the quality of processes, products and organizations as well as its measurability. Present the concept of quality measurement and Six Sigma.

The historical development of the topic of quality management resulted in very different ideas about what the term means. In common usage, quality means the good characteristics of an object. For instance, we speak of a "quality object" and mean that an object is well-made. In that usage, quality is equated with good quality, the term being used to indicate a positive value. However, particularly in the field of economics, it has become customary to use the term *quality* as a neutral term, following its original definition.

The term *quality* originated in the sixteenth century, derived from the Latin "qualis," meaning to be made in some way or being in a certain condition or state. According to dictionaries, quality always refers to an object and stands for its state, feature characteristic, or nature.

It is only according to the sense of this definition that degrees of quality (excellence) can be stated, that we can speak of "quality improvement." However, in many discussions on quality, it is apparent that people have certain angles of vision or standpoints. There are indeed different ways to view quality. For example, Joseph M. Juran defines quality as *"fitness for use"* [Defe16, see also www.juran.com]. Consumers and suppliers have different understandings of quality, as the different organizational units within a company usually do. "Quality" is hence a multilayered term. It is not by chance that the following four disciplines,

P. Schönsleben, *Handbook Integral Logistics Management*,
https://doi.org/10.1007/978-3-662-65625-9_18

among others, have dealt with the concept of quality: philosophy, business sciences, marketing, and operations management.

In the company environment, quality can relate to various objects. At the foreground stand the processes and the products of an enterprise or a service-providing public organization. In the sense of total quality, however, the company or service provider as a whole is such an object.

18.1.1 Quality of Processes

Intended learning outcomes: Produce an overview on process, service, and a service provided to dependents. Present the characteristics of the quality of processes. Identify process quality, process time, and process load.

A *process* comprises certain activities that lead from a beginning state to an end state and thus to certain functions. Examples of processes include:

- An assembly process, through which an assembly is built from various components
- A procurement process, through which various materials are purchased
- A quality control process, through which procured or manufactured parts are tested and verified for features and characteristics to specified requirements.

Processes come in different degrees of complexity. A process may be an individual, elementary activity, or a large-scale business process designed to produce a significant business outcome, called a product. Certain processes have a special structure.

> A *service* is a process that a customer views as the performance of some useful function.

Examples of services include

- Installing equipment and bringing it into service at the customer location
- Service and maintenance during use of a product
- Business consultancy in the broadest sense, particularly also sales advising and the sale itself

It makes a difference to the customers whether they buy a finished product and can judge only the quality of the outcome or whether they experience a process itself and thus can judge its quality. With a view to quality management, it is of interest that customers buying products increasingly want to observe exactly the processes that result in the products. Therefore, Figure 1 in the introduction postulates a supplier (a business partner) management system, which among other objectives aims toward knowledge of the supplier's processes.

> *Services provided to dependents* is a process in which the customer is not only the object on which the process occurs; the customer also ends up with a limited capacity to act, to which the service provider has contributed.

Examples of these processes include

- Processes in training and education
- Processes in connection with patients in health care
- Treatment of delinquents by the justice system

In these cases, the affected persons have restricted free will, which can lead to their treatment not as customers, but more as the objects of guardianship. However, the affected persons in this situation are particularly positioned to form judgments about the quality of a service.

Process quality is the quality of processes.

Process quality is judged according to certain subjective or objective characteristics of quality of processes. To these belong the features shown in Figure 18.1.1.1.

- Accuracy: precision in meeting expectations
- Reliability: consistency, for example, of same process when repeated
- Safety: for example, as related to undesirable side effects
- Competence: skill, expertise, professionalism in execution (sovereignty)
- Courtesy: friendliness and comfort (for example, of a service)
- Load: amount of work content required (often measured in time units)

Fig. 18.1.1.1 Characteristics of the quality of processes.

Process time is the period of time during which the process runs.

Process load, that is, the burden that the process places on the customer, is the work or effort content through which the characteristic effect of the process is achieved.

Process load should not be confused with process time. Process time can be shortened, for example, by putting more people to work (splitting) or by executing sequential work steps in an overlapping fashion. By doing so, the process load is higher, but for a shorter time. Process time also encompasses waiting times: When are people ready to begin the service? When is a means of transport available to take a person from A to B? Process time, which is of interest to the customer in addition to process load, is influenced by factors that lie outside of the nature of the process, namely, in the area of logistics management.

Increasing quality with regard to individual process characteristics can lead to greater work content. Increased work content as a rule results in longer process times. Here a conflict between the target areas quality and delivery becomes apparent.

18.1.2 Quality of Products

Intended learning outcomes: Identify product quality. Differentiate between a simple product and a product in a broad sense. Present the characteristics of the quality of products.

Product quality is the quality of products.

Products can be either of a material or a nonmaterial nature, for example:

- Raw materials, purchased parts, semiprocessed items, finished items in an industrial or commercial enterprise
- Insurance products, banking products, consulting products, travel arrangements in service industries.

Generally, a product represents the outcome of processes. Here, we are not interested in the quality of the processes, but rather only the quality of the product according to the characteristics listed below.

The second group of examples above also shows that services performed, that is, the outcome of the process with the customer, can be viewed as products. In this kind of process, products can also be used as components. Consider a trip by train or plane. Here various products can complement the primary service, such as meals or travel items. In some cases, especially when various service providers have the same process quality, these products — although secondary at first glance — can be deciding factors.

In a buyer's market, a product provider has to offer ever more services and advice along with the product. The product supplier in this way becomes a real systems supplier of a general contractor type. The outputs of the company are then the products as well as the processes that provide the products to the customer. The product concept shifts more and more to a product in a broad sense. In the insurance industry, for instance, the core product is a specifically assembled insurance policy. But it is complemented by services, so that in the end there is a package that is being offered as a product and perceived by the customer as such. Product and process thus stand ultimately in a dual relationship.

Product quality is judged according to certain subjective or objective *characteristics of quality of products*. Figure 18.1.2.1 shows several features of quality.

- Resource consumption
- Effect, function
- Consistency, durability, and reliability
- Conformance to pre-established or expected standards
- Features, workmanship
- Ease of use and aesthetics
- Recyclable, disposability

Fig. 18.1.2.1 Characteristics of the quality of products.

Quality targets are derived from these characteristics. Costs and delivery lead time, in contrast, do not belong to the characteristics of the product, as long as we are not viewing the product in its most comprehensive sense. Costs and delivery lead time can be influenced rather by logistics management; for example, by type of stockkeeping or resource use.

18.1.3 Quality of Organizations

Intended learning outcomes: Identify organizational quality. Describe the concept of quality toward the stakeholders of an organization.

Interested and affected parties have a perception of an enterprise that goes beyond the company's products or processes. This is the perceived quality of the company's work as a whole. This is true for any type of organization, including organizations in the public sector.

Organizational quality is the quality of organizations, meaning the quality of the organization as a whole.

Anyone who has an interest or stake in a business is a *stakeholder* (compare Figure 1 in the introduction). Stakeholders include employees, suppliers, creditors, customers, shareholders, local communities, and anyone else who is affected by the operations of the business. Each stakeholder will have a subjective — frequently also self-centered — perception of the quality of an organization. Generally, the stakeholders define their requirements of the organization independently of one another.

Organizational quality also is judged according to certain subjective or objective *characteristics of the quality of organizations*. Figure 18.1.3.1 assigns the various characteristics to meaningful groups.

- Quality in view of business partners
- Quality in view of employees of the organization
- Quality in view of shareholders
- Quality in view of society and environment / nature

Fig. 18.1.3.1 Quality toward the stakeholders of an organization.

- *Quality in view of business partners*: What is the customer's perception of the company's performance? Some criteria of the processes are already listed in Figure 18.1.1.1. Additional criteria apply to the organization as a whole, such as responsiveness, credibility, accessibility and communication, and understanding the customer. The characteristics of products are mentioned in Section 18.1.2. Moreover, customer satisfaction is more than satisfaction with the products and processes offered; there is a higher level of customer satisfaction that is the perception of receiving total care. In sellers' markets, the company must treat its suppliers similarly, attending to what is called "supplier satisfaction."

- *Quality in view of employees in the organization*: The summary criterion of "employee satisfaction" encompasses a whole host of characteristics, such as compensation, the type of leadership in the organization, executability of tasks, flexibility and options for creativity in plans-of-work and work hours, material safety, and so on.

- *Quality in view of shareholders*: Certainly owners and shareholders will judge the quality of their company mainly according to financial results. On closer inspection, however, money also stands for deeper needs, such as owners' individual financial security or independence.

- *Quality in view of society and the environment*: Society as a whole places requirements on an enterprise even if it is not the owner of the company in the literal sense. These requirements are often set down in laws or codes of conduct. The quality of a company is proportionate to how well its processes, products, and conduct fit into the given framework. The characteristics are, for example, the safety of society and the protection of the integrity and property of its citizens. In the general sense, the same holds for the environment, where laws are given as natural laws. In practice, the requirements of the environment become manifest only in the consciousness of the other stakeholders mentioned. The quality of an enterprise is then evaluated according to whether it adheres to these laws as society demands. Characteristics are, for example, protection of the environment and responsible use of resources.

18.1.4 Quality and Its Measurability

Intended learning outcomes: Explain the problems of the measurability of indicators and the step from measurement to corrective actions. Describe the issue using the example of the measurement of customer satisfaction.

The International Organization for Standardization (ISO) gives a formal definition of quality.

"*Quality* is the totality of characteristics of an entity that bear on its ability to satisfy stated and implied needs" [ISO8402].

In contrast to quantities (amounts), measurability is not contained in the term quality (nature, quality, character) from the start. But the nature, quality, and character of an object are nevertheless assessed. The measurability of quality could be advantageous for quality management: "You can only improve what you measure," say some executives. However, measurability requires a measurement system.

The *measurement system* contains a goal or target that is to be achieved through the measurement (*measurement objective*), from which an *indicator* must be derived.

The indicator must be scaled appropriately, that is, divided into units of measure, and sensors collecting data to obtain a measurement in these units of measure must be made available. Moreover, the indicator must be of a kind that can be translated into concrete corrective actions. Figure 18.1.4.1 shows well-known problems that arise with this endeavor.

- Easily measurable indicator can be disadvantageous in that it is not clear what caused the measurements obtained and therefore what actions and measures should now be taken.
- The other way round, indicators can be identified based on potential corrective actions for improvement. However, their measurement can exceed the budget and resources available, or the budget and resources required may be unforeseeable.

Fig. 18.1.4.1 Problems of the measurability of indicators and the step from measurement to corrective actions.

Characteristics that are relatively easy to measure are the physical characteristics of products and processes. This is the realm of traditional quality inspection and quality assurance. Both favorable outcomes and failures can be measured.

If the quantitative measures and the desired values are laid down in the product requirement specifications, the object can be measured accordingly. It is more difficult to determine whether the measurements obtained also satisfy stakeholders. It can happen that certain characteristics that are of crucial importance to stakeholders have not even been identified.

In connection with people, the measures are frequently combined and are general in content. Take, for instance, the characteristic *customer satisfaction*. Here it is not sufficient to measure some general value. The problem to be mastered is assessment of the customer's judgment of performance as to individual quality characteristics, while keeping cost and effort within reasonable bounds. On the time axis, the assessment should take place, where possible, in an events-related manner (for example, in reference to products or services sold). However, in the area of consumer goods particularly, many characteristics of customer satisfaction lie within the individual realm of the customer and may even be subconscious on the part of the customer. Measurement that provides objective and interpretable cause and effect analyses is thus often an illusion.

Similarly difficult is the measurement of *employee satisfaction*. People are not easily willing or even capable of openly explaining their conscious or unconscious needs. However, the effort that is required should not prevent us from measuring that which is feasible to measure.

It is interesting that frequently there will be some employees that have exact knowledge of their own needs and the needs of other stakeholders. It is therefore very important that these people be involved in the development and use of measurement systems in those areas.

18.1.5 Quality Measurement and Six Sigma (6σ)

Intended learning outcomes: Describe the metric of six sigma (6σ) and the sigma conversion table. Identify Six Sigma Quality. Differentiate between three Sigma and Six Sigma process reliability.

Six Sigma (6σ) had its origins in the 1970s in Japan, in shipbuilding and in the electronics and consumer goods industry. In the second half of the 1980s, Six Sigma — pioneered first by Motorola — was introduced as a program to reduce defects in the manufacturing of electronic components. It included a set of methods and techniques focused on quality improvement. Later, the Six Sigma philosophy came to be applied to other business processes for the same purpose, namely, to achieve reliable processes. The aim is to reduce variation and defects in all areas of company performance. Today, Six Sigma is important

- as a metric
- as a problem-solving methodology, or method for improving performance
- as a management system

This section focuses on the first definition.

The term "sigma" is often used as a scale for levels of "goodness" or quality.

Sigma, the eighteenth letter of the Greek alphabet used as a mathematical symbol, was employed for many years by statisticians, mathematicians, and engineers as a unit of measurement for the standard deviation.

Six Sigma (6σ) *as a metric* is a specific scale for measuring the number of successful products, events, processes, operations, or opportunities.

Figure 18.1.5.1 shows the conversion table for one sigma to six sigma.

Sigma	Rate of successful opportunities in %	Failure rate per million opportunities
1	30.9	691,462
2	69.1	308,538
3	93.3	66,807
4	99.4	6210
5	99.98	233
6	99.99966	3.4

Fig. 18.1.5.1 The sigma conversion table.

The conversion table shows an exponential scale, which does not, however, accord with the standard deviation of the normal distribution, as it is often assumed (a glance at the tables in Section 11.3.3 provides easy confirmation of this). Motorola defined the Six Sigma level as equal to 3.4 DPMO. In the world of practice, however the mathematical explanation of the conversion table does not stand at the center of attention.

Six Sigma (6σ) *Quality* is defined as a level of quality that represents no more than 3.4 DPMO (defect parts per million opportunities).

Figure 18.1.5.2 compares Three Sigma and Six Sigma process reliability, considering examples given by Motorola.

Reliability 99% (~ Three Sigma)	Reliability 99,9999% (~ Six Sigma)
20,000 pieces of mail lost every hour	7 pieces of mail lost every hour
Unsafe drinking water almost 15 min. every day	1 min. unsafe drinking water every 7 month
5000 incorrect surgical procedures every week	1.7 incorrect surgical procedures every week
2 critical landings at major airports each day	2 critical landings at major airports every 5 years
200,000 incorrect drug prescriptions filled / year	68 incorrect drug prescriptions filled every year
Almost 7 hours without electricity every month	Almost 1 hour without electricity every 34 years

Fig. 18.1.5.2 Three Sigma and Six Sigma process reliability.

18.2 Quality Management Tasks at the Operations Level

Intended learning outcomes: Produce an overview on the Deming Cycle (PDCA Cycle) and the Shewhart Cycle as well as the Six Sigma phases. Present the phases of quality planning, control, assurance, and activation of the Deming Cycle. Describe the Six-Sigma phases of define, measure, analyze, improve, and control. Differentiate between continual improvement and reengineering.

Quality management is a set of actions of the general management function that determines the quality policy and aims and responsibilities, and realizes them by means of quality planning, quality control, quality assurance, and quality improvement within the framework of the quality management system [ISO8402].

See here also [GrJu00], [PfSm14], and [PfSm15]. Six Sigma focuses on the *objects* of quality management.

Six Sigma (6σ) *as a problem-solving methodology, or method of improving performance*, is a method for improving products, procedures, processes, operations, or opportunities.

Within this context, the aim is to understand customer requirements and to improve the business processes that fulfill those requirements as rapidly and as sustainably as possible. Beyond the methods of quality management, the Six Sigma methodology attaches great importance to utilizing rigorous data analysis to minimize variation in business processes. The improvement processes are more precisely specified in measures of Six Sigma; this allows standard implementation.

18.2.1 The Deming Cycle (PDCA Cycle) and the Shewhart Cycle

Intended learning outcomes: Produce an overview on The Shewhart cycle developed in statistical quality control. Present the Deming cycle. Describe quality management tasks in the Deming cycle.

The Deming cycle, also called the Plan, Do, Check, Act cycle (PDCA cycle, also known as PDSA cycle, where S stands for Study), was an early means of representing the task areas of traditional quality management. The cycle is sometimes referred to as the Shewhart / Deming cycle since it originated with physicist Walter Shewhart at the Bell Telephone Laboratories in the 1920s. W. Edwards Deming modified the Shewhart cycle in the 1940s and subsequently applied it to management practices in Japan in the 1950s.

The Shewhart cycle ([Shew03], p. 45) is defined in Figure 18.2.1.1.

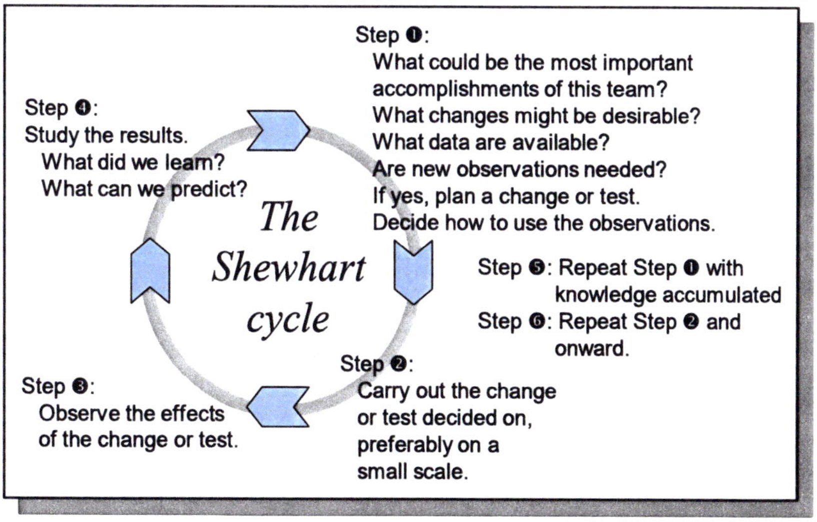

Fig. 18.2.1.1 The Shewhart cycle developed in statistical quality control.

The *Deming cycle* ([Demi18], p. 88), shown in Figure 18.2.1.2, is the application of the Shewhart cycle.

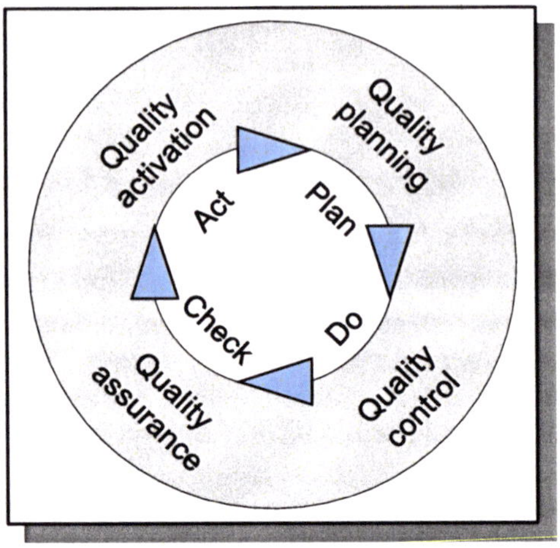

Fig. 18.2.1.2 Quality management tasks in the Deming cycle.

Figure 18.2.1.3 describes in greater detail the logical sequence of the four cyclical tasks in the spirit of continuous quality improvement.

- *Plan* task: Develop plan for quality improvement. In the value chain, this takes place mainly during product and process design.
- *Do* task: Execute the plan in order to control quality. Implement change measures that impact the flow of goods along the value-added chain, that is, during procurement, production, and distribution.
- *Check* task: Check and study the effects of implementation. This involves the tasks of measurement and testing in classical quality assurance.
- *Act* task: Take action to activate, that is to standardize the process improvements. The focus is on acting on what was learned from the changes. Thus, In general, the results lead to an improvement in quality. The result may lead to no change at all. Then, it is the confirmation of the results already found that equals quality improvement.

Fig. 18.2.1.3 Description of quality management tasks in the Deming cycle.

The fourth task in particular gives an indication of why, in total quality management, this same model is applied not only to systems in the value chain, but also to systems that impact stakeholders (see Figure 1 in the introduction). In those cases, the behavioral aspect — that is, making the changes a routine part of activity — has special significance (here see Section 18.3).

18.2.2 The Six Sigma Phases

Intended learning outcomes: Present DMAIC, the Six Sigma phases. Describe the tasks in the Six Sigma phases. Differentiate between DMAIC, RDMAIC, DMAICT, and DMADV.

The Six Sigma methodology is a sequence of phases, usually called DMAIC.

DMAIC is an acronym for process improvement consisting of the following phases: Define, Measure, Analyze, Improve, and Control.

These phases are usually depicted with a beginning and an end, as shown in Figure 18.2.2.1.

Fig. 18.2.2.1 DMAIC, the Six Sigma phases.

These phases consist of the tasks shown in Figure 18.2.2.2.

- *Define*: Define the project goals and the requirements of internal and external customers.
- *Measure*: Measure the performance of the process involved.
- *Analyze*: Analyze the data collected to determine root causes of defects.
- *Improve*: Improve the process by elimination of the defects.
- *Control*: Control the improvements to keep the process on the new course.

Fig. 18.2.2.2 Description of tasks in the Six Sigma phases.

As compared to the Shewhart or Deming cycle, it is noticeable that the Six Sigma phases are not arranged in a circular form. This is in accordance with the view that a Six Sigma project is run through once to achieve a result. A further rotation of the Deming cycle type forms a new Six Sigma project of its own. Overall, the effect achieved is similar to the continual improvement type of management system.

In the following, we will see that the five Six Sigma phases can be assigned quite well to the four tasks in the Deming cycle. However, the Six Sigma phases provide additional action catalogs and checklists that make operationalization generally easier. For each phase, there is a list of results and control questions that is intended to ensure the comprehensiveness of the approach.

There are variants of DMAIC. *RDMAIC* is an acronym that stands for a DMAIC process that adds Recognize as an initial phase. As a part of this phase, company management seeks to identify opportunities for improvement. In many cases, this phase is a part of the Define phase. *DMAICT* is an acronym for a DMAIC process with a subsequent Transfer phase. In this phase, best practices are transferred, or spread, to other areas of the organization.

Another variant focuses on product design. *DMADV* is an acronym for an improvement process that proceeds through the phases Define, Measure, Analyze, Design, and Verify. *DFSS (Design for Six Sigma)* comprises methods and instruments for ensuring that products and processes are designed at the outset to meet Six Sigma requirements. Through DFSS and DMADV, the aim is to make later DMAIC processes less frequently necessary. As the two initial Ds indicate, the methods and tools largely correspond to those in DMAIC.

18.2.3 Quality Planning: PDCA Plan Task — DMAIC Define Phase

Intended learning outcomes: Identify the cause of differences between stakeholders' expectations and actual product or process characteristics. Explain quality function deployment – the house of Quality and 10 steps of implementation. Describe the SIPOC diagram and the CTQ matrix.

> *Quality planning* is a term used today for all planning activities prior to the start of production; quality planning sets goals and works toward achieving the goals and preventing failures.

Analogously to this definition, in the PDCA Plan task as well as in the Six Sigma Define phase, the project team identifies what is important to the customer (captures the "voice of the customer"[1]), the goals, and the scope and boundary of the project. Inclusion of stakeholders in quality planning means that, for all of these tasks and activities, the quality of the outcome must be evaluated to determine whether it satisfies stakeholders' needs. Figure 18.2.3.1 shows potential discrepancies between stakeholder needs and product characteristics that can arise during execution of the whole task from subtask to subtask.

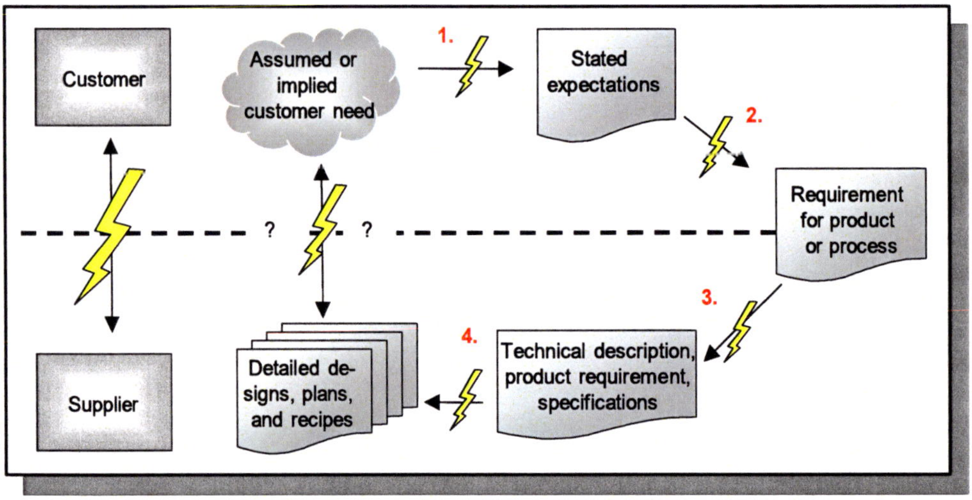

Fig. 18.2.3.1 Cause of differences between stakeholders' expectations and actual product or process characteristics.

Discrepancies can arise due to the following:

1. Assumed or implied needs have to be translated into words or symbols — that is, identified and established — in the language used by stakeholders. Here, there is the danger that the translation will fail to be accurate.

2. The stated expectations determined have to be broken down into ideas or expectations concerning requirements for the product and the process to be developed. This

---

[1] *Voice of the customer* (VOC) is the term for customer descriptions in words for the functions and features customers desire for goods and services See also [APIC16].

is often connected with a transition from relatively general quality characteristics to more specific ones. The result is a detailed, functional picture or functional model, which again is expressed in the language used by stakeholders.

3. The functional pictures or models determined for the product and the process are translated into specific quality requirements, but now in the language used by the provider/supplier. Finally, the requirements are described in specifications, called product requirement specifications, which are more technical descriptions.

4. The technical descriptions are transferred into designs, plans, and recipes. This is the actual development and design of product and process. The output then undergoes validation, which is the process of ensuring that the product conforms to the original stakeholder needs and requirements.

A typical method used in the quality planning phase is quality function deployment.

Quality function deployment (QFD) is *step-by-step development of quality functions.* The QFD process uses matrices to translate customer requirements into technical design parameters or characteristics.

To do this, a quality chart called the *"House of Quality"* is employed as a correlation matrix linking quality characteristics and target values and their tendency. See Figure 18.2.3.2.

The 10 steps of implementation in Figure 18.2.3.2 are taken from [Guin93]:

1 Gather customer requirements for the product or service.
2 Customers weight the importance of each of the requirements.
3 Ask customers to rate competitors' products or services.
4 Technical descriptors. Translate customer requirements into quality characteristics.
5 Determine relationship matrix between customers' needs and technical descriptors.
6 Estimation of the importance of the technical descriptors.
7 Conduct analysis of competitor's technical descriptors.
8 Estimate degree of difficulty, technical feasibility.
9 Determine target values for each technical descriptor.
10 Determine variation for each technical descriptor and examine mutual reaction.

The *first-pass yield* (FPY) is the percentage of results (i.e., units) that pass on first test — that is, without requiring rework.

An increased FPY entails reduced costs due to rework. Development is successful if the defect rate can be rapidly reduced once the product is introduced or if it is zero from the start (*zero-defect rate*). As the development process is essentially a creative one and can contain errors, defects can always be expected with an innovation. The need to reduce development time and development costs also speaks against a zero-defect rate. For these reasons, defects will be accepted at first, and importance will be placed on reducing this rate rapidly once the product is introduced. Then, it is important to have sufficient capacity for rapid revision and a comprehensive information system for capturing the responses of the first customers.

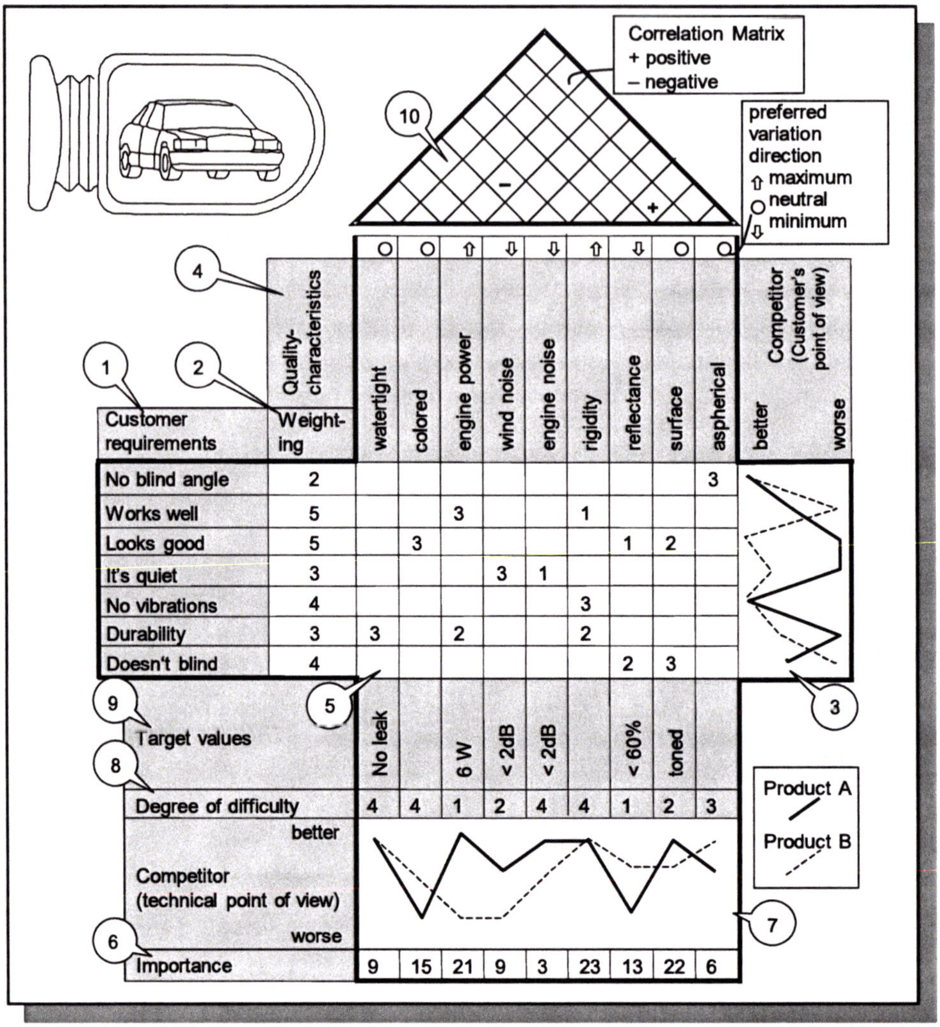

Fig. 18.2.3.2 Quality function deployment: House of Quality and 10 steps of implementation (Source: [Guin93]).

For example, during quality planning, quality requirements — together with the original ideas about the requirements — are translated into an offer to customers that describes the company's product or service. This description, which is often a component of a contractual agreement, can already deviate decisively from the customer's expectations, so that at this point in time at the latest, a decision must be made as to whether the individual steps of quality planning should be repeated (non-first-pass yield).

With this, the Define, or quality planning, phase entails capturing the relevant processes. In Six Sigma, this is represented in SIPOC diagrams.

The *SIPOC* diagram shows the system with Inputs, Process, and Outputs as well as the Suppliers and the Customers. See Figure 18.2.3.3.

In this phase, the diagrams show the actual state, before all work begins. On the basis of the diagramming of the process, the critical elements will then be worked out.

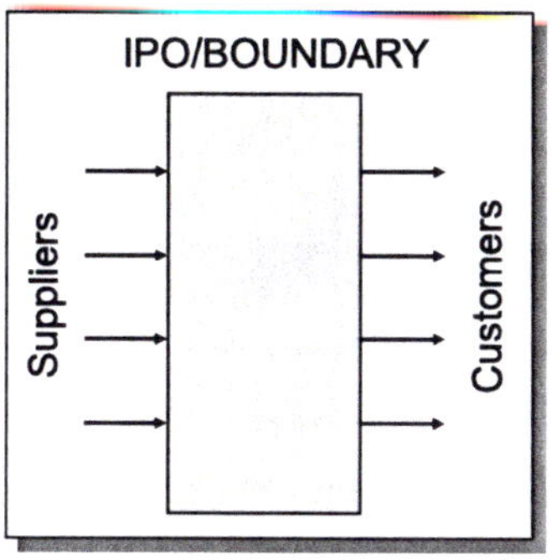

Fig. 18.2.3.3 SIPOC diagram.

CTQs (Critical to Quality) are key measurable characteristics (e.g. regarding quality, costs, or delivery), elements of the process, or practices that have a great and direct effect on the customer's perception of the quality of a product or service. See Figure 18.2.3.4.

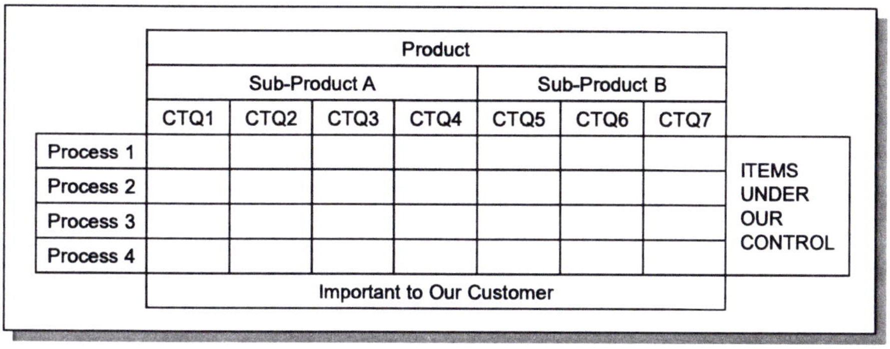

Fig. 18.2.3.4 CTQ matrix.

CTQs are usually represented in a matrix, or CTQ tree, that shows the (sub-)products and their critical attributes on the horizontal axis. The processes that can lead (or not lead) to these critical characteristics, process elements, or practices are shown on the vertical axis.

The Six Sigma method revisits the outcomes to be delivered (deliverables) again and again:

- Are the project teams well trained and motivated?
- Have the customers been identified and CTQs defined?
- Has the project management handbook been drawn up?
- Have the business processes been diagrammed appropriately (e.g. using SIPOC)?

Each of these questions is revisited repeatedly throughout the entire phase in greater detail, in order to ensure that they are handled comprehensively.

18.2.4 Quality Control: PDCA Check Task, Part 1 — DMAIC Measure Phase and Analyze Phase, Part 1

Intended learning outcomes: Produce an overview on tools used in quality control in their original usage stemming from production engineering. Identify deliverables of the Measure phase as well as the Analyze phase.

Quality control encompasses the operational techniques and the activities used to fulfill and verify requirements of quality [ISO8402].

Quality control aims to implement predefined targets from quality planning in reality; that is, it measures for conformance to quality requirements. The techniques of quality control can be used for both monitoring a process and correcting or eliminating defects or failures.

In the PDCA cycle, the Check task comprises the necessary tasks of quality control. In order to control the success of the Do task, the respective tools have to be implemented before the Do task and used for rechecking afterwards. In the Six Sigma method, quality control comprises two of the DMAIC phases, namely, Measure, and Analyze. Again, in order to control the success of the Improve phase, the respective tools have to be implemented before the Improve phase and used for rechecking afterwards.

In the Measure phase, the task is to determine how the spoken needs of the customers, the CTQs, will be specified in measurable terms using tools. The appropriate measurement system is then installed or an existing system improved. Here see Section 18.1.4. Further, actual current performance is quantified and the target goal determined (for example, increase process stability from three sigma to four sigma).

Some quality control tools and tests for this task are, for example, ABC classification or Pareto chart, sampling plans, and *statistical process control* to determine process capability and process performance.

The deliverables of the Measure phase can be reviewed and revisited as follows:

- Is there agreement on the critical characteristics, and is there a detailed description of their measurability?
- Has a plan been drawn up showing what data will be captured and what measurement system will be used? Have the data been gathered?
- Has the current variation of the process (current sigma level) been calculated and opportunities for improvement defined?

In the Analyze phase, the task is to identify root causes of variation and defects. Now it is important to provide statistical evidence of current deviations and to then formulate options for improvement (improvement goals). Thus, the Measure and Analyze phases both encompass activities of quality control.

Quality control in its original usage stems from production engineering. The tools used include risk analysis, such as failure mode and effects analysis (FMEA), design of experiment (DOE), and hypothesis testing, such as analysis of variance (ANOVA) and multivariate analysis of variance (MANOVA). Further tools for this step are cause and effect diagram (fishbone, or Ishikawa diagram), histograms, quality control charts, correlation diagrams, checklists, and general graphical representations, such as time series diagrams, pie charts, bar charts, Gantt charts, or network diagrams.

The deliverables of the Analyze phase can be reviewed and revisited as follows:

- Were data and process analysis conducted and the gaps between actual and target process performance determined?
- Have the root causes of variation and defects been found and prioritized according to importance?
- Were the performance deficits communicated and converted to financial quantities? (Here see the discussion in Section 1.3.1 on opportunity costs.)

18.2.5 PDCA Do Task — DMAIC Improve Phase

Intended learning outcomes: Produce an overview on tools of the jidoka concept, such as Andon (or a visual control system in a job shop) and Poke yoke (or fail-safe techniques).

Both the Do task of the PDCA cycle and the Improve phase of the DMAIC method can be regarded as the most creative part of quality management tasks; namely, solution finding. The task is to generate a number of possible solutions that counteract the root causes of variations and defects.

In the manufacture of physical products, the individual process steps are usually described in quite great detail. The process step instructions often include measures with tolerances that must be adhered to during machining. The same should hold for processes in the information flow of a company, for example, for order processing. Exact descriptions are just as necessary for service processes, even though here it can be considerably more difficult to determine targets and variations.

For this creative process, the following principles hold (these are realized, for example, in the jidoka concept; see definition in Section 6.1.1):

- Possible defects should be identified at the source of origin as early as possible in the process. The problems or defects can be identified by the human eye or by specialized sensors.
- All components and units should be checked 100% to ensure complete faultlessness.
- Direct intervention prevents further subsequent mistakes. With jidoka, any worker can stop the line by pulling a cord; this is then signaled on visual display devices called Andon boards that are visible to all.
- The processes must be made "foolproof" (poka yoke).

Andon is a visual signal, or visual control system. With jidoka, Andon are electronic display boards that show the status of the processes in a job shop or production line as well as information for coordination of the connected workstations. Commonly used colors to indicate status are green (OK), / yellow (needs attention), or / red (stop). See here [Toyo98].

Figure 18.2.5.1 shows an example of numbered Andon lights. The yellow light shows a problem at workstation 1. There the worker has pulled the cord to stop the line.

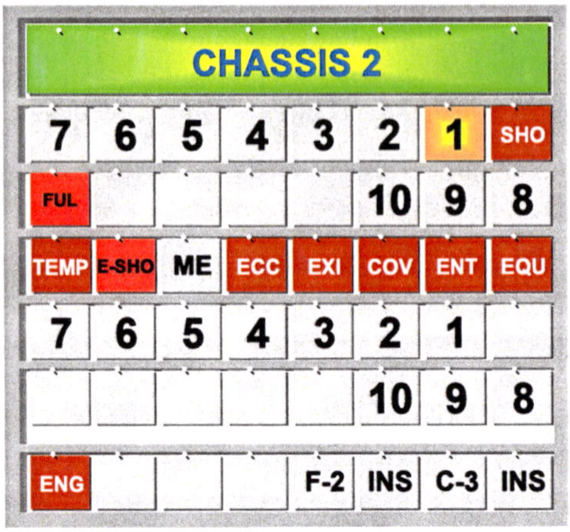

Fig. 18.2.5.1 Andon: visual control system in a job shop.

> *Poka yoke*, or *fail-safe techniques*, means to avoid (yokeru) inadvertent errors (poka). The basic principles of poka yoke advocate designing or developing tools, techniques, and processes to detect defects, thus relieving people of error-prone tasks such as repetitive monitoring of the same states or the checking of a great many details [Kogy90].

Some examples of such simple mechanisms and devices are:

- Sensors that detect missing or defective components and shut down the process automatically.
- Guide pins on parts that prevent assembly in the wrong orientation.
- Construction of parts and corresponding clamping and mounting apparatuses that prevent backward insertions.

Applying poka yoke to information flows represents a considerable challenge. However, for the area of order processing, computer-supported computer programs that can monitor the completeness of information and the correct sequence of functions are conceivable. Indeed, computer programs have always had the task of recognizing and preventing all possible defective constellations of data. Workflow techniques are now used to ensure correct implementation of additional sequences of ordered tasks, even where several people are involved in the process. Because of the many unforeseeable influences on information processes, however, these techniques are likely to be of help only for simple and highly repetitive processes.

Both the Do task of the PDCA cycle and the Improve phase of the DMAIC method require reviewing and revisiting the deliverables as follows:

- Have sufficient solutions been generated?
- Were the solutions tested? Has the best solution been selected based on these tests?

- For the selected solution, were target processes and cost-benefit analysis drawn up?
- Was a plan drawn up for introducing the selected solution?

18.2.6 Quality Assurance: PDCA Check Task, Part 2 — DMAIC Measure Phase and Analyze Phase, Part 2

Intended learning outcomes: Identify changing definitions of terms over time in quality assurance and quality management.

Quality assurance as it was used originally corresponds to today's term quality inspection (see Section 18.2.4).

Use of the term *quality assurance* — like use of the term *quality management* — has changed over time as shown in Figure 18.2.6.1.

	Aspect Defined	ISO Vocabulary (International Organization for Standardization)
Up to 1987	Wider term	quality assurance
Since 1987	Wider term Quality management system demonstration	quality management quality assurance

Fig. 18.2.6.1 Changing definitions of terms over time (Source: [Verb98]).

- Up until 1987, the term *quality assurance* was used as generic term for all activities with regard to quality.
- After 1987, the term *quality management* was introduced as the wider term. Quality assurance was now used for concrete quality management system demonstration.

Quality assurance can be understood today as *active* risk management for the purpose of reducing the probability of quality defects and of mitigating the consequences of defects (passive risk management would be insuring or covering against risk).[2]

Quality assurance as defined today first of all involves quality inspection to determine whether the quality targets for the individual quality characteristics are actually met. Such inspection measures include:

- Tests of incoming goods to ensure that procured goods are free of defects
- Supplier ratings, based on delivery quality
- Design reviews during the R&D process
- Early warning systems that detect defects in new products at an early stage

---

2 Where *risk* is defined as venture, danger, or possible losses in an insecure action, specifically as the product of the probably of occurrence of an event times the probable extent of the effect, that is, deviation from a goal.

- Testing of administrative processes, in particular of completeness of information and delivery reliability

In an analogy, the task in the Do task of the PCDA cycle or of the Improve phase in the Six Sigma method is implementation of one or more of the solutions found. After this, quality assurance means a check providing statistical evidence that the solutions are achieving the desired results.

For the quality assurance step, all of the tools used in quality control (see Section 18.2.4) are available. For quality assurance of organizations or of complex sequences of processes — especially whole business processes — evaluation methods in the form of assessments stand in the foreground. Assessments are used in the evaluation of the quality management system itself. They will be discussed in connection with TQM models in Section 18.3.

As for any type of organization, quality assurance must also not be simply a control mechanism; the point is to enable and motivate people appropriately to deliver defect-free products and processes. The focus today is usually on self-inspection. To avoid unnecessary slowing down of the value-adding processes, quality tasks, as far as possible, should be carried out by the same persons who are also responsible for the operational added value. For this, they require training in the relevant quality techniques.

To complement self-inspection, third parties (superiors, other internal parties, or external parties) conduct suitable inspections. The task of a company-wide post for quality is to advise the persons accountable for producing the goods and services as to selection of the quality assurance tools and, in difficult cases, to take on an advising and coordinating function in the quality assurance process.

18.2.7 PDCA Act Task — DMAIC Control Phase

Intended learning outcomes: Produce an overview on tools used in quality activation. Identify deliverables of the Control phase.

Quality activation means activation of quality improvement.

The Act task means active follow-up: The changes introduced have to be evaluated. The knowledge gained during the quality assurance phase must be compared with the targets set in the Plan task. Afterward, the decision can be made as to whether the change was good and should be continued, or whether it should even be applied to further activities, products, or processes, and what improvements must be made before doing so. This can possibly mean continuing on as before without implementing the change or making it standard work practice. Furthermore, the results have to be communicated, so that any subsequent iteration of the Deming cycle will be higher-level and thus achieve improvement a priori.

Analogous with this, the task in the Control phase of the Six Sigma method is to integrate the results into daily operations and document and communicate them within the organization. In addition, however, Six Sigma demands that measures be taken to maintain the gains of the processes in future. Here again, statistical evidence that the improvements are maintained is required.

The representation tools listed in Sections 18.2.3 to 18.2.6 can be applied. These include affinity diagrams (meaningful groupings of ideas to refine when brainstorming or moderating), relationships diagrams (such as mind maps), matrix diagrams, decision trees, network plans, decision tables, and flowcharts of some type, such as Process Decision Program Charts (PDPC). See here also [Mizu88]. These tools and methods are general in nature; that is, they are also utilizable in other management systems.

The deliverables of the Control phase can be reviewed and revisited as follows:

- Has a system for monitoring consistent use of new methods been documented and implemented?
- Have the new process steps, standards, and documentations become standard work practices?
- Has the knowledge gained regarding the new processes been documented and shared in the organization?
- Have responsibilities and accountability been identified, understood, and communicated in the organization?
- Has ownership and knowledge been handed over to the process owner and his or her team, and the project officially closed down?

18.2.8 Project Management, Continual Improvement, and Reengineering

Intended learning outcomes: Differentiate between continual process improvement and (business) process reengineering.

Responsibility for projects in quality management can be assigned to everyone involved in the process, the "process team." It is advantageous to define a "process owner" as a coordinator. Well-oiled teams of persons are preferable to individual specialists. The observation is generally valid that defects arise particularly if a process is beset with interfaces, where the process is handed over from one person to another person who acts independently of the first. Experience has shown that even if the interface is defined as specifically as possible, errors occur here, only because of the tendency of people to close themselves off from other organizational units.

The proponents of Six Sigma recognized early on that for successful project management for quality improvement, people must be trained and awarded special certification. Names that were chosen for the different levels of certification acknowledge Japanese origins: *Green Belts, Black Belts, Master Black Belts*. The first Black Belts acquired certification in the early 1990s, thus marking the beginnings of formalization that led to accredited certification programs in Six Sigma methods.

Because in Europe for a long time the Deming cycle did not catch on as an advantageous method, a new term was sought that would express an understanding of the Deming cycle as a permanent task. For this, the term CPI was coined.

Continual process improvement (CPI), or simply *continual improvement*, is a never-ending effort, a culture, in which improvement — usually in small steps — becomes the guiding principle: The journey is the objective!

With the introduction of CPI, the Deming cycle, once understood in a more static way, was made into a dynamic circle. Ultimately, the greatest potential can be set free only through influencing the behavior of the collaborating persons. Organizational measures can promote collaborative behavior, such as the collecting of proposals and suggestions in the firm, quality circles of employees, periodical goal and measures planning, campaigns, and so on. However, implementing the concept of CPI and the culture connected with it is difficult.

As customer needs change sooner or later, products and processes must change over time. Each change, however, entails the risk of errors. While quality control and quality assurance promote stability in the company, they are a priori hostile to change and therefore also hostile to improvement. The compromise solution may be to continuously improve performance through continuous incremental changes, without having to take too great a risk. For this, the Japanese use the term *Kaizen* [Imai94]. Here, the focus is not on achieving a specific level of quality, but rather a certain degree of continual improvement of quality.

Continual improvement as a whole is a continuous task and it does not have the character of a project. The individual incremental improvement measures as such, however, are usually carried out in the form of projects. For example, a project of this kind may attempt to:

- Increase customer benefit. The additional expenditure has to be able to be covered by either higher prices or lower costs. Higher prices can usually be realized only if customer satisfaction can also be improved in a durable way.
- Reduce the defect rate. The expenditure for the project and the connected investments must be covered by continuous cost savings created by fewer defects.

In contrast to CPI, what happens in new development is innovation on a grand scale.

Reengineering means fundamentally rethinking the company's options for designing products and processes.

Business process reengineering (BPR) is improvement of business processes in big steps by fundamentally redesigning the processes.

Improvement in big steps through radical changes is then the task of quality planning in the first iteration of the corresponding Deming cycle. In the course of the further product and process life cycle, continuous incremental changes serve to improve company performance.

18.3 Quality Management Systems

Intended learning outcomes: Present standards and norms of quality management: ISO 9000:2015. Describe the optimization paradigm: models and awards for Total Quality

Management. Produce an overview on audits and procedures for assessing the quality of organizations as well as on benchmarking.

Whereas for a long time the term *quality* was understood in America and in Europe as quality assurance in production, a management-oriented quality concept achieved dominance in the 1950s in Japan. Two Americans, W. E. Deming and J. M. Juran, developed this concept.

Total quality management is defined as a management approach of an organization centered on quality, based on the participation of all its members, and aiming at long-term success. This is achieved through customer satisfaction and benefits to all members of the organization and to society [ISO8402].

There is an analogous management-oriented understanding for Six Sigma. Motorola, for instance, learned early on that disciplined application of metrics and the improvement methodology alone are not sufficient to achieve big breakthroughs and sustainable improvements.

Six Sigma (6σ) *as a management system* is a framework for assigning resources with priority to projects that result in rapid and sustainable improvement of business results.

Here metrics and improvement methodology are implemented to tackle the problems in connection with company strategy in the correct sequence. In this way, the results should be evident at all levels of the company and ultimately in company results. In 1989, Motorola received the Malcolm Baldrige National Quality Award.

18.3.1 Standards and Norms of Quality Management: ISO 9000:2015

Intended learning outcomes: Disclose standards in the DIN ISO 9000:2015 series (without 10001 to 10004). Differentiate between fulfillment paradigm and optimization paradigm.

The *ISO 9000:2015 Standards* are shown in Figure 18.3.1.1.

ISO 9000:2015	QM Systems – Fundamentals and vocabulary
ISO 9001:2015	QM Systems – Requirements
ISO 9004:2009	QM Systems – Guidelines for performance improvements
ISO 19011:2018	Guidelines on Quality Management Systems Auditing
ISO 10005:2018	QM - Guidelines for quality plans
ISO 10006:2017	QM - Guidelines to quality in project management
ISO 10007:2017	QM - Guidelines for configuration management
ISO 10012:2003	Quality assurance requirements for measuring equipment
ISO 10013:2001	Guidelines for quality management system documentation
ISO 10014:2006	Guidelines for managing the economics of quality
ISO 10015:1999	Quality management - Guidelines for training
ISO 10017:2003	Guidelines for the use of statistical techniques
ISO 10019:2005	Guidelines for the selection of consultants in QM systems

Fig. 18.3.1.1 Standards in the DIN ISO 9000:2015 series (without ISO 10001 to 10004).

The standards pay attention to the ability of the organization to fulfill the requirements of various stakeholders, such as customers, employees, and investors, as well as for continual improvement. A major revision is scheduled for the end of the year 2015.

Today, there exist two strategies for deriving a quality management system. The two strategies can also be seen as paradigms:

- The *fulfillment paradigm* leads to systems that contain a set number of quality assurance standards, or rules and measures. Certification by an impartial third party aims to guarantee mutual trust among business partners as to the required quality of products or services. All organizations that achieve a specified level of quality receive certification. ISO 9000:2015 is a quality management system of this type.
- The *optimization paradigm* leads to comprehensive concepts that aim for outstanding quality performance. Here, independent associations evaluate the degree to which a company recognizes quality to be the crucial factor for all its activities and makes it the focus of attention of business activity. Only the best organizations receive a Quality Award (QA). Corresponding quality management systems are introduced in Section 18.3.2.

The advantage of awards over certification is that the demands to be met for awards actually increase over time, as the companies improve. This means that awards result in best practices instead of merely sufficient levels. Awards are the embodiment of a continual improvement philosophy (optimization relating to goals), whereas certification according to a standard results "only" in achievement of a specific level (fulfillment of requirements).

Durable improvement of company performance is a question of company culture, and thus of the behavior of the individual, the individual organizational units, but also the organization as a whole. The desired culture and the corresponding behavior are generally laid down in a strategy or policy. Strategic management then builds up the corresponding management systems ("structure follows strategy"), such as, for example, a quality management system. The aim of establishing such management systems is to influence the individual and thus to achieve the desired behavior ("culture follows structure").

18.3.2 Models and Awards for Total Quality Management

Intended learning outcomes: Present the Deming Prize and the Malcolm Baldrige National Quality Award (MBNQA). Produce an overview on Deming Prize examination criteria and the structure of the MBNQA. Explain the EFQM Excellence Model and the EFQM Excellence Award.

In the 1950s, together with the development of the management-oriented quality concept, the Union of Japanese Scientists and Engineers established a national prize to provide an incentive for the continued development of total quality management in Japan, the annually awarded (for the first time in 1951 in Osaka) Deming Prize. See here also www.deming.org.

The *Deming Prize* uses the examination criteria shown in Figure 18.3.2.1.

The reaction of the American government, science, and economy to the ever stronger Japanese competition occurred in the form of various initiatives and agreements under the lead of Malcolm Baldrige, who served as US Secretary of Commerce. This resulted in the Malcolm Baldrige National Quality Improvement Act of 1987. The Act established the *Malcolm Baldrige National Quality Award* (MBNQA). Compare also [Verb98] and [Zink94], and www.nist.gov/baldrige. The MBNQA evaluation model gives the company a good opportunity to determine its own standing with regard to quality management.

- Understanding and enthusiasm
- Policies
- Organization and operation
- Information
- Standardization
- Human resources development and utilization
- Quality assurance activities
- Maintenance and control activities
- Improvement activities
- Implementation and evaluation
- Social responsibilities
- Effects
- Future plans

Fig. 18.3.2.1 Deming Prize examination criteria.

Figure 18.3.2.2 shows the *Malcolm Baldrige National Quality Award* (MBNQA) criteria for performance excellence.

The European response to the challenge is the *European Foundation for Quality Management (EFQM.)*. The EFQM was founded in 1988 in Brussels. As one of its main activities, the EFQM presents the EFQM Excellence Award (formerly, European Quality Award [EQA]). The award, presented for the first time in 1992, recognizes excellence in the organization's ability to realize outstanding quality and comprehensive customer satisfaction. See more at www.efqm.org.

The *EFQM Excellence Model* and the *EFQM Excellence Award* are based on the evaluation elements shown in Figure 18.3.2.3.

Based on the EFQM Excellence Model, excellent leadership is seen as a prerequisite for customer satisfaction, employee satisfaction, and positive results for society. To this purpose, the organization must develop a quality-conscious policy and strategy, utilize resources efficiently, and choose an employee-oriented way of proceeding. Only in this way, under consideration of all processes, can sustainable performance results be achieved.

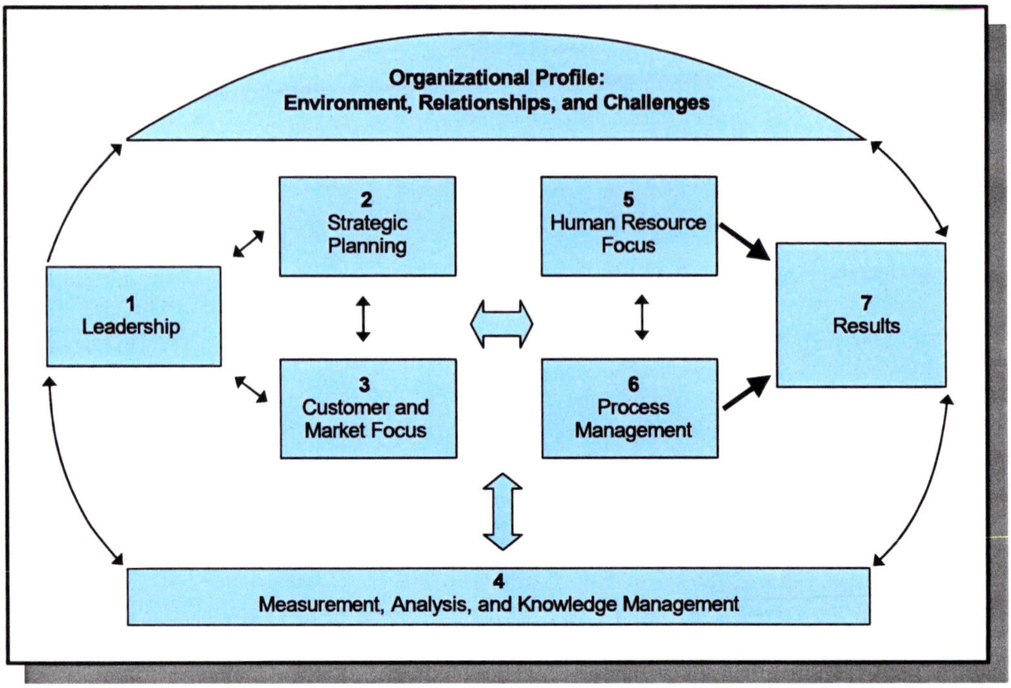

Fig. 18.3.2.2 Structure of the Malcolm Baldrige National Quality Award (based on [NIST06]).

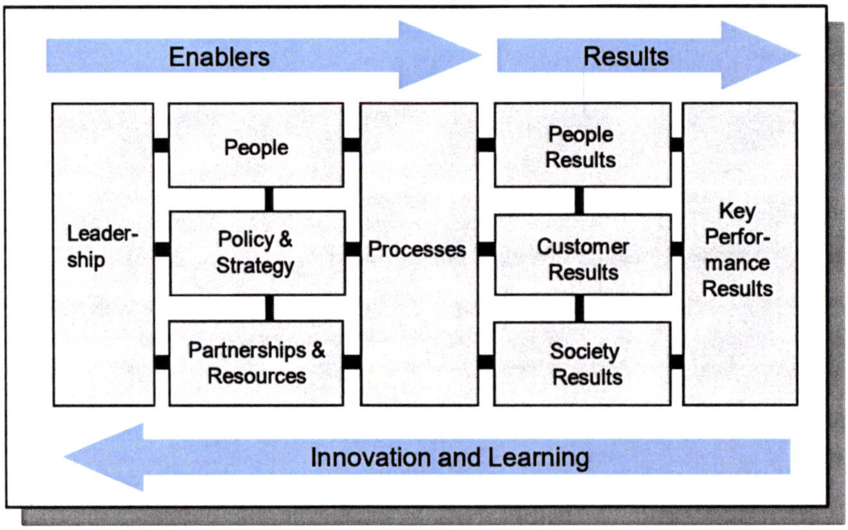

Fig. 18.3.2.3 The EFQM Excellence Model (based on www.efqm.org).

18.3.3 Audits and Procedures for Assessing the Quality of Organizations

Intended learning outcomes: Identify the quality audit. Differentiate between first-party or internal audit, second-party audit, and third-party audit.

An *audit* is a formal examination [MeWe18]. This is generally understood as an act of hearing that is conducted by third parties according to well-defined criteria and rules.

ISO 8402 [ISO8402] defines a *quality audit* as a "systematic and independent examination to determine whether quality activities and related results comply with planned arrangements and whether these arrangements are implemented effectively and are suitable to achieve objectives."

Probably for all areas of life, individuals and organizations are accustomed to periodically taking stock of where they stand, deriving opportunities and goals for improvement, and checking progress toward achieving these objectives.

Possibilities for conducting an assessment include external procedures of the audit type on the one hand and self-assessment on the other. Self-assessment results in strong identification, strong learning effects, and considerable self-motivation. The disadvantage can be that people, especially if they have little experience in assessing, tend toward misjudgments and can consciously distort the facts. But external audits also have the subjectivity of persons as a fundamental disadvantage, of those that establish the measurement criteria as well as of those that perform the evaluation according to these criteria. For this reason, the culture of the assessors appears to be the greatest factor influencing the assessment process.

Three types of audits provide an option for more formal assessments (see [Pira97] for further details):

- A *first-party*, or *internal audit* is performed by trained auditors who assess individual company areas. The auditors are selected from within the company but are independent of the area being audited.
- A *second-party audit* is performed by customers who evaluate their suppliers. Customers are typically interested in details of the product and process design. If the supplier has a quality management system in place, this type of audit can be limited to assessing that system.
- A *third-party audit* is performed by external agencies that are specialized in this task. This type of audit can be conducted in connection with quality awards, but it generally does not entail durable improvements. Instead, company-internal employees should be trained in assessment. External agencies can then, for example, participate in the company's internal audit in an advising role on methodology.

18.3.4 Benchmarking

Intended learning outcomes: Differentiate between benchmarking in very different industries and of different sizes at the level of the general management, and benchmarking with other firms in the same industry sector.

Benchmarking means identifying best practices that result in outstanding performance.

If companies compare themselves to firms in very different industries and of different sizes at the level of the general management, they will discover that they have strengths in different areas. A company can in this way obtain information about the improvement strategies set by very different firms. Occasionally, this can give rise to ideas on how practices in other industry sectors and companies might be applied to their own business activities. Nevertheless, benchmarking on too broad a base has only limited effectiveness.

Just as important, thus, are comparisons with other firms in the same industry sector. Once comparable processes, products, or organizational units are available, measurement criteria can be set up that should be included in the comparison procedure. Benchmarking extends to any aspect of the company that represents a best practice. Once the benchmarking partners and the objects to be compared have been established, it is possible to examine how the reference partner achieves outstanding performance. What key processes are involved? What is the company culture behind this? On the basis of the answers to these questions, companies can derive their own new goals. On implementation, see, for example, [Camp94].

Herein are the limits of benchmarking with competitors within the same branch of industry. Direct competitors will hardly be willing to reveal the secrets of their success. Information on competitors' best practices should probably be acquired from third parties. Direct collaboration in benchmarking among competitors only makes sense if it results in a win–win situation for both partners. This can be the case when otherwise competing suppliers in one geographic region decide to take on the suppliers in another geographic region.

For these reasons, the tendency is for groups of companies to form that are not competitors on the market but that have comparable processes, company structures, and stakeholders. *Functional benchmarking* compares similar processes or functions. Here a rather broad spectrum of companies can be examined. For example, companies might compare functions of operations and supply chain management. *Generic benchmarking* is comparing not only individual functions, but whole business processes, such as the R&D process. Here, the selection of comparable companies will, of course, be smaller.

18.4 Summary

Quality management encompasses concepts, methods, tools, procedures, and techniques that aim to improve the company performance. Quality in the company can refer to its processes or products. Organizational quality must be oriented to the various company stakeholders. Easily measurable quantities can have the disadvantage that it is not clear what, exactly, led to the test results (e.g. when it comes to people's perceptions) and thus not clear what actions must be taken. In reverse, the quantities to be measured can be determined on the basis of possible actions for improvement. But their measurement can entail excessive expense.

The Deming cycle, or Shewhart cycle, gathers together the traditional tasks of quality management, namely, quality planning, quality control, quality assurance, and quality activation (Plan, Do, Check, Act). Today, in addition, the Six Sigma method is also very well known. The Six Sigma method is divided into steps that, taken together, correspond with the tasks of the Deming cycle. For each task, there is a set of tools. Particularly well-known tools are the House of Quality in quality planning and *poka yoke* in quality control. For quality assurance, besides the statistical methods, a wealth of representation tools are available. These and further tools can also be utilized in quality activation. When the Deming cycle is conducted repeatedly, or a number of Six Sigma projects are carried out, this leads ultimately to continual process improvement. However, innovation on a grand scale, such as new development of products, breaks with that process, and a new process of continual improvement begins.

Total Quality Management (TQM) is a management-oriented quality concept. In the case of management systems that follow the fulfillment paradigm, all organizations that achieve a certain level of quality receive certification. The ISO 9000:2015 series belongs here. In the case of management systems that follow the optimization paradigm, the best organizations receive an award. The various prizes here include the Deming Prize, the Malcolm Baldrige National Quality Award (MBNQA), and the EFQM Excellence Award. In comparison, awards have the advantage that they promote best practices, which are found in very few organizations, whereas certification indicates a satisfactory level at many organizations. To determine the standing of a company in quality management, there are various assessment methods. The main method is self-assessment. Using benchmarking, companies compare their performance with others in their search for best practices.

18.5 Keywords

andon, 775
audit, 784
benchmarking, 785
continual process improvement (CPI), 779
Deming cycle, 768
Deming Prize, 782
DMADV, 769
DMAIC, 769
DMAICT, 769
E.F.Q.M.Excellence Model, 784
first-pass yield, 771
ISO 9000.2015 (series of standards), 781
kaizen, 780
Malcolm Baldrige National Quality Award, 783
organizational quality, 763
poka yoke, 776
process quality, 761
product quality, 761
quality, 759
  according to ISO8402, 764
quality activation, 778
quality assurance, 777
quality audit, 784
quality control, 774
quality function deployment, 771
quality management, 766
quality planning, 770
RDMAIC, 769
Shewhart cycle, 767
Six Sigma
  as a management system, 781
  as a metric, 765
  as method of improving performance, 767
stakeholder, 763
total quality management (TQM), 781

18.6 References

APIC16 Pittman, P. et al., APICS Dictionary, 15th Edition, APICS, Chicago, IL, 2016

Camp94 Camp, R.C., "Benchmarking," Carl Hanser Verlag, München, 1994

Defe16 Defeo, J.A., "Juran's Quality Handbook — The Complete Guide to Performance Excellence," 7th Edition, McGraw Hill, New York, 2016

Demi18 Deming, W.E., "Out of the Crisis," Center for Advanced Engineering Study, MIT, Cambridge, MA, Press Edition 2018

GrJu00 Gryna, F.M., Juran, J.M., "Quality Planning and Analysis: From Product design through Use," 4th Edition, McGraw Hill, New York, 2000

Guin93 Guinta, L.R., Praizler, N.C., "The QFD Book — The Team Approach to Solving Problems and Satisfying Customers through Quality Function Deployment," American Management Association, New York, 1993

Imai94 Imai, M., "Kaizen, The Key to Japan's Competitive Success," Random House, New York, 1994

ISO8402 International Organization for Standardization, "ISO 8402:1994: Quality Management and Quality Assurance; Vocabulary," 1994

Kogy90 Kogyo, N., "Poka-Yoke," Productivity Press, Cambridge MA, 1990

MeWe18 Merriam-Webster, "Merriam Webster's Collegiate Thesaurus," 2nd Revised Edition, Merriam-Webster, Inc., Springfield, MA, 2018

Mizu88 Mizuno, S., "Management for Quality Improvement: The 7 New QC Tools," Quality Press, Milwaukee, 1988

NIST06 National Institute of Standards and Technology, "Malcolm Baldrige National Quality Award: 2006 Criteria for Performance Excellence," ASQC Customer Service Department, Milwaukee, WI, 2006

PfSm14 Pfeifer, T., Schmitt, R. (Hrsg.), "Masing Handbuch Qualitätsmanagement," 6. Auflage, Carl Hanser Verlag, München Wien, 2014

PfSm15 Pfeifer, T., Schmitt, R., "Qualitätsmanagement: Strategien, Methoden, Techniken," 5. Auflage, Carl Hanser Verlag, München Wien, 2015

Pira97 Pira, A., "A Self-Assessment Approach for Implementing Total Quality Management (TQM) in Hospitals," Proceedings of the 9th Quest for Quality & Productivity in Health Services Conference, St. Louis, September 1997.

Shew03 Shewhart, W. A., "Statistical Method from the Viewpoint of Quality Control," Graduate School, Department of Agriculture, Washington, 2003

Toyo98 "The Toyota Production System," Toyota Motor Corporation, Public Affairs Divison, Toyota City, Japan, 1998

Verb98 Verbeck, A., "TQM versus QM — Wie Unternehmen sich richtig entscheiden," vdf Hochschulverlag, Zürich, 1998

Zink94 Zink, K.J., "Business Excellence durch TQM — Erfahrungen europäischer Unternehmen," Carl Hanser Verlag, München, 1994

19 Systems Engineering and Project Management

Intended learning outcomes: Explain systems engineering. Describe project management.

System is defined in [MeWe18] as a regularly interacting group of items (or set of elements) forming a unified whole.

Merriam-Webster lists further definitions, which all revolve around "arrangement," or "organization." In the narrower sense, a system describes complex phenomena in the real world, such as the solar system or the periodic table of the chemical elements. But abstract phenomena can be also described as systems, such as:

- Numerical systems or systems of equations in mathematics
- Theories and models
- Electrical, pneumatic, or hydraulic systems
- Social systems
- The organization of a company or the national economy

In the following, a *company* is understood *as a sociotechnical system*. The elements themselves — that is, the human beings (the focus of the *social* part system), production machines, materials, and so on (the focus of the *technical* part system) — and their relationships both within the system and with the surrounding systems (environment) are complex in nature. Parts of a company — for example, production — can also be viewed as systems.

Systems theory deals with general characteristics of systems. For example, a *dynamic system* refers to a system with interactions among the elements of the systems. *Open dynamic systems* refers to systems in which the elements also interact with other elements in the system environment (as opposed to closed systems). Systems thinking in general systems theory can be applied to the special company systems theory. Production in an industrial company, for example, is typically an open dynamic system. The interactions are formed through the flows of goods, data, and information. See here also [Züst04] and [HaWe18].

In analogy to the product life cycle, the system "company," or its part systems, also have a life cycle. What makes up the system life cycle? What problem-solving techniques are used? These questions will be covered in the section below on systems engineering.

Systems engineering (SE) is a method, based on some models and procedural principles, for enabling appropriate and efficient realization of complex systems, to which systems in a company also explicitly belong [HaWe18].

Systems engineering is thus a *systemic* method for the realization of systems.

A *project* is a scheme, plan, or planned undertaking (from the Latin *projectum*, or that which is thrown forward) [MeWe18]. For use of the term in practice, [PMBOKD] defines project as a temporary endeavor undertaken to create a unique product, service, or result. [APIC16] defines a project as an endeavor with a specific objective to be met within the prescribed time and dollar limitations and that has been assigned for definition or execution.

P. Schönsleben, *Handbook Integral Logistics Management*,
https://doi.org/10.1007/978-3-662-65625-9_19

In contrast to operations or processes in the company, which are recurrent and ensure "normal" business operation, such as, for example, the processing of sales orders, projects are undertakings that

- have a definite start and end
- create something new and in that sense are unique
- require resources (such as persons, equipment, money) that are most often limited in availability, absolutely and also often on the time axis.

Some examples of projects from business life and personal life are

- Introduction of a new business process
- A change in structural organization
- Development of a new product
- A research project to investigate a certain phenomenon
- Planning a trip around the world
- Redecoration of a room in your house

While a project may contain parts that were already parts of other projects, the result of a project as a whole is unique. For instance, any single bridge can be seen as a unique construction, although some components can be identical in many bridges.

Project management is the organizing, planning, scheduling, directing, controlling, monitoring, and evaluating of prescribed activities to ensure that the stated objectives of a project are achieved. See also [APIC16].

Project management is thus a systematic approach to ensure the effectiveness of a project and the efficient use of resources. On the basis of these definitions, we can derive the following connections:

- The realization of a system in a company virtually always has a unique character, creates in addition something new, and requires limited resources. The realization of a system can be seen as a set of projects, which can be handled — in particular where system complexity is high — by project management.
- Not every project has to be seen as realization of a system. The planning of a world trip or a research project investigating a specific phenomenon, for example, utilize methods and techniques of project management but not necessarily the methods and techniques of systems engineering.
- Simple projects do not necessarily have to be accompanied by project management. In private life especially, the project may affect only one single person. A person redecorating a room at home — this being the realization of a system, even — often takes a less systematic approach.

The two sections that follow below look at the methods of systems engineering for the realization of systems and the methods of project management for effective and efficient execution of projects.

19.1 Systems Engineering

Intended learning outcomes: Explain systems thinking and the top-down approach. Describe phases of life of a system and the system life cycle. Present in detail the problem solving cycle. Disclose the differences between software engineering and classical systems engineering.

In all realizations of systems, typical problems arise. These problems are dealt with by systems engineering, independently of the type of system. Figure 19.1.0.1 shows the characteristics of systems engineering following [HaWe18], [HaWe05], or [Züst04]. The most important principles are then summarized below.

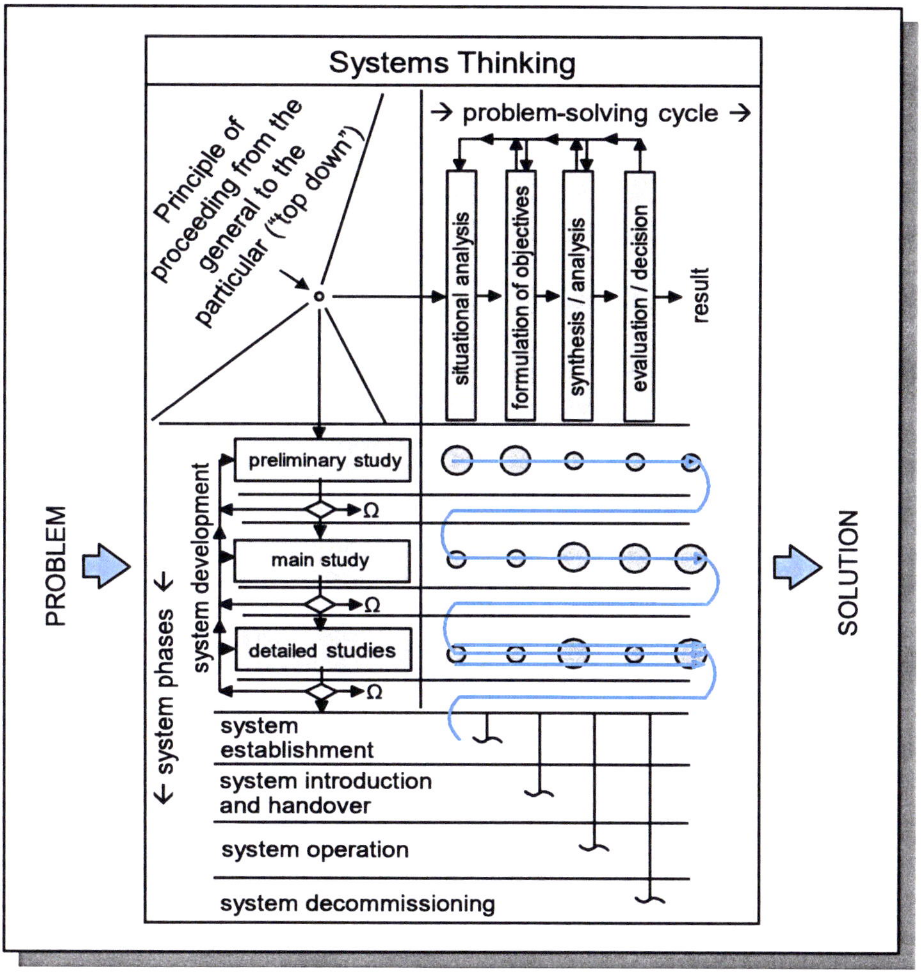

Fig. 19.1.0.1 Systems engineering: overview based on [HaWe18]. The symbol Ω stands for end, that is, termination of system development.

The essential, core ideas in systems engineering can also be applied without difficulty to the life cycle phases of all types of systems. In quite a lot of cases, however, there are specific differences. Section 19.1.4 provides an example: the development of IT-supported information systems, or software engineering. Software engineering differs from classical systems engineering in some important aspects.

19.1.1 Systems Thinking and the Top-Down Approach

Intended learning outcomes: Describe systems thinking and the proceeding from the general to the particular (top-down approach). Produce an overview on creating and evaluating variants at each level of the system.

In *systems thinking*, or *systems-related thinking*, the goal is to understand the issue to be solved as a system with its elements and interactions both within the system and with the *surrounding systems* — that is, the system environment.

The basic idea of *proceeding from the general to the particular* (top-down approach) demands that the system be observed at different levels. This can be at the highest level, meaning the whole system, or within *subsystems*, that is, at lower levels.

Also of interest are *part systems*, such as, for example, the flow of goods, data, or information. Figure 19.1.1.1 illustrates the terms.

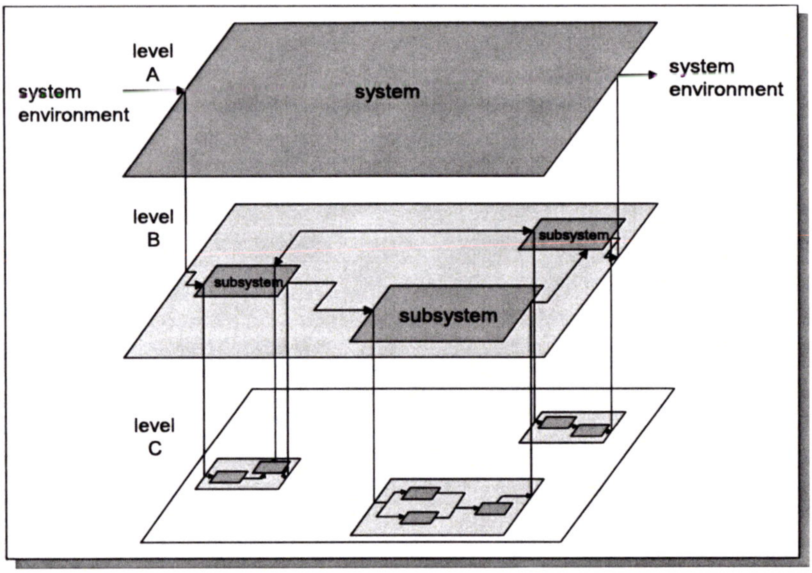

Fig. 19.1.1.1 Proceeding from the general to the particular (top-down approach).

First, the whole system at the highest level has to be formulated in its interaction with the systems in the system environment. At this point, the subsystems remain "black boxes," meaning that input, output, and the function of the black box are specified but not the mechanisms by which the function will be realized. In a subsequent phase, each subsystem, or black box, will be handled in the same way as the system. The highest level or levels generally describe the generalist's point of view on solving the problem, while the lower levels refer more to the structure of the problem and therefore resemble the specialist's point of view. For each aspect that is to be considered, it is important that the discussion is conducted at the correct system level. Usually, there are various possible ways to design the system at each level, especially regarding definition of the subsystems. These possibilities result in a range of variants at each level, as Figure 19.1.1.2 shows.

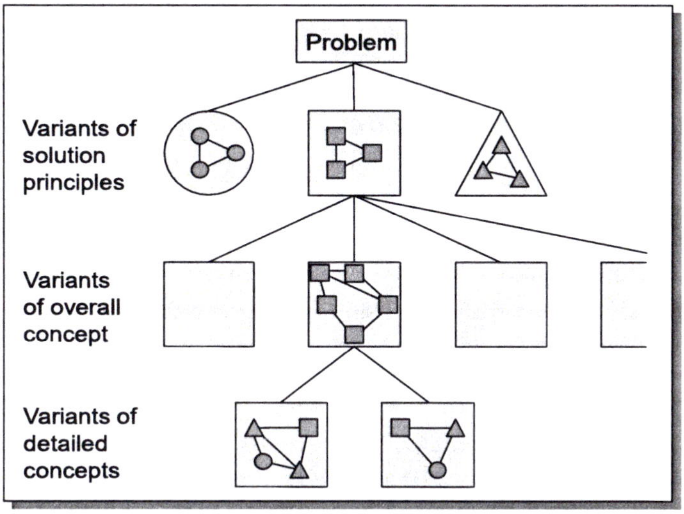

Fig. 19.1.1.2 Creating and evaluating variants at each level of the system.

Ideally, evaluation of variants and selection decision on which variants to retain should take place before designing a system or subsystem at a lower level:

- *Advantage:* Variants that solve the problem insufficiently can be determined and eliminated within a reasonable period of time.
- *Disadvantage:* Whether the postulated mechanism for a black box at a lower level can actually be realized can sometimes be determined only through detailed study at the lower system level.

Therefore, the effort required to work out a variant at a lower level stands opposite to the risk of making an erroneous decision at a higher level.

19.1.2 Phases of Life of a System and System Life Cycle

Intended learning outcomes: Differentiate between system development and system implementation. In system development, present in detail the content of the preliminary study, the main study and the detailed studies. In system implementation, produce an overview on system establishment, system introduction and handover, system operation, and system decommission.

The *system phases*, or *phases of life of a system*, encompass three concept phases and four implementation phases as shown in Figure 19.1.0.1. The *system life cycle* comprises all of the life-cycle phases of a system.

In *system development*, there are the three following concept phases.

Phase 1: The *preliminary study*, where the goal is, within a reasonable time frame, to acquire insights regarding:

- whether there exists a real need for a new system or modification of a system
- whether there exists a real need for a new system or modification of a system,
- whether the approach to the problem is correctly selected,
- what the boundaries of the given system are,
- what the most important functions of the system are, and
- what the principle solution variants for the problem are.

The client's decision regarding feasibility and thus continuation or termination of the project and decision regarding the variants to be selected form the end of the preliminary study.

Phase 2: The *main study*, which comprises study of the whole system. If a subsystem can be assessed only with great difficulty, it can be necessary to conduct a detailed study already at this point. The result of the main study is a comprehensive concept of the system, and — depending on the type of system — it is in the form of a plan of tasks or activities, of construction plans, of a verbal description, or of other suitable means.

- With these results, it is now possible to make decisions regarding the investments, to define the subsystems, and set priorities in the realization of the detailed studies. Setting priorities is particularly important, as the task is now to evaluate and plan project resources (financial and human resources). In addition, it is necessary to first develop the important subsystems to which the less important subsystems are oriented. While this may reduce the degrees of freedom in the realization, it does accelerate the realization of the other subsystems, through, for example, possibly copying relevant subsystems developed first.
- The result of the main study may require a return to the previous concept phase, for example, if the requirements are not precise or not feasible.

Phase 3: *Detailed studies*. The result of detailed studies encompass, for one thing, the detailed concept for the subsystems and the final decision on the variants to be selected, and, for another, specification of the description of the individual part systems. The description will be precise enough to allow the system to be built without interpretation problems.

- There are various detailed studies, namely, of the individual subsystems and at different levels. So the next task is to reintegrate the individual systems, proceeding from the particular to the general (bottom up). Thus, the entire function and the interplay of the subsystems in their supersystems can be tested. This is a process that can reveal ways in which the detailed studies or also the main study should be modified.

During each of the concept phases, decisions must be made regarding the selection of the variants, end of the study, or return of the project to the next higher-level phase. Factors in these decisions are — in addition to the factors that relate to the functional objectives of the project — the expected costs and benefits. Costs and benefits often determine the variant that is to be selected among variants that appear to be equally good functionally. Or they may provide impetus to search for other variants, to terminate system development (the symbol Ω in Figure 19.1.0.1), or to continue on to the next life cycle phase of the system.

In *system implementation*, there are the following four implementation phases.

Phase 1: *System establishment* describes elaboration of the system functions, for example:

- production of a machine and its documentation
- drawing up of a process or an organization
- coding and documentation of the programs in an IT-based information system
- elaboration of the organization to operate the system. In the example of an information system, what is needed are, among other things, documentation for the user, exact description of the procedures for data acquisition — especially from and to the surrounding systems — and the procedures for the use of information, actions to be taken in the case of system crisis, and training of users.

Phase 2: *System introduction and handover* is the transition to the production phase of the system. Often, especially with large projects, one subsystem is introduced after another, as there are always various unforeseeable factors to allow for. In this usually relatively short phase of introducing the system, systems corrections are often required that are very time critical. It is also possible to put some subsystems into operation while keeping others in the detailed concept phase, especially if the concept of the latter subsystems can be influenced by experience gained during operation of subsystems already introduced.

Phase 3: During *system operation*, there must be periodic and constructive review of the following points:

- Is the system really functioning as conceptualized? The answer to this question can serve as a source of experience for later, similar projects and provide the basis for a process of modification/correction.
- Are the commercial objectives being met as foreseen? Any deviations observed can aid improved assessment of costs and benefits in future projects.

Phase 4: Usually, the decision of *system disposal*, or of *system decommission,* is made concurrently with the introduction of a replacing system. For systems that are highly intertwined with daily operations, such as software systems at the operations level of a company, system replacement is often no light matter. Generally, the requirements are:

- that daily processes must not be interrupted for longer than a very short period;
- that the data should be transferred from the old system to the new system automatically, if possible.

System decommissioning for hardware systems (e.g., computers, terminals) involves physical disposal of the single components. This can be a big challenge, both technically and in terms of costs. Thus, system disposal should be a part of systems engineering from the start.

If the system will be in operation for a certain period of time only, all of the life cycle phases of the system can be seen as and conducted as one single project. Otherwise, the concept phases can be taken together with elaboration of the system and introducing the system as one project, and system decommissioning as a new project.

19.1.3 The Problem Solving Cycle

Intended learning outcomes: Identify the problem-solving cycle. Present in detail situational analysis, formulation of objectives, synthesis of solutions, analysis of solutions, evaluation of solutions, and decision in the problem-solving cycle.

As seen in Figure 19.1.0.1, the three concept phases (during system development) stand alongside the problem-solving cycle.

The *problem-solving cycle* is defined as the six steps described in the following, which are conducted during the three concept phases of system development.

The importance of each step is indicated in Figure 19.1.0.1 by small versus large circles, and the required expense and effort is indicated by single versus triple arrows.

Step 1: *Situational analysis* aims to understand the situation and identify the problem and its causes and consequences. Situational analysis distinguishes among at least four aspects:

- System-related: Determination of the system and subsystems, with their elements and interactions
- Diagnostic: Determination of the symptoms of the unsatisfactory solution, derivation of the causes
- Therapeutic: Finding of possible corrections and applying corrections to the relevant elements
- Time-related: Does the situation develop on the time axis with / without correction?

During situational analysis, the bounds and constraints for a possible solution have to be defined, such as:

- Bounds and constraints arising from the system environment (social, technological, regulatory, and so on)
- Bounds and constraints stemming from decisions made earlier, which cannot be modified at present
- Bounds and constraints coming from fixed parts of the situation, that is, parts that for some reason must remain as they are

It is advantageous to summarize the situational analysis in an analysis of strengths, weaknesses, opportunities, and threats, often abbreviated as *SWOT analysis*. The analysis of strengths and weaknesses refers to the system under review at the present time. The analysis of opportunities and threats refers to the system environment: How will changes in the environment expected in the future impact the system, if it is left *unchanged*?

Step 2: *Formulation of objectives* generally comprises functional, commercial, and time-related objectives. The objectives must be solution-neutral, complete, precise, understandable, and realistic. They must relate to the elements in the SWOT analysis; that is, they must be coherent with the system analysis. Generally, two classes of objectives are distinguished:

- Mandatory objectives, that is, objectives that must be achieved in any case to solve the problem ("need to have").
- Preferred objectives, that is, objectives that are to be achieved if possible ("nice to have"). These goals will eventually serve as a catalog of criteria for decision making among several acceptable variants.
- In the end, the client has to approve the formulation of objectives. This is because unanticipated factors may make it necessary to change the formulation of objectives.

Step 3: *Synthesis of solutions* is conception of the possible solutions. Synthesis has to be sufficiently precise to allow comparison of the various variants. All required functions and available resources have to be taken into account. Synthesis is the creative part of the work, and therefore it is usually also the most difficult part of the problem-solving cycle.

Step 4: *The analysis of solutions* is a kind of test of the synthesis. Is the solution concept comprehensive (that is, does it meet all objectives)? Is it realizable (that is, have all conditions and constraints been complied with)? It is sometimes difficult to differentiate between the two steps of synthesis and analysis in the problem-solving cycle. That is because analysis often already has begun at the birth of an idea for a solution concept.

Step 5: *Evaluation of solutions* select quantitative methods for measuring the efficiency or quality of a possible solution per se and as compared to other variants. The methods are usually similar to those otherwise used for a cost-benefit analysis, such as factor rating. The criteria come from the catalog of objectives, possibly enhanced by detailed technical criteria.

Step 6: The *decision in the problem-solving cycle* step refers to both selection of the variants and the decision to repeat this or a preceding concept phase. The decision is made jointly by specialists, the people responsible for the system, and the client. Reasons for repeating a concept phase are, among others:

- The situational analysis is not precise enough for derivation of a solution.
- The results of the analysis show that the concept does not in all parts meet the need and the constraints.
- New objectives are added.
- The objectives are changed, as no solution is possible.
- New variants should be developed for evaluation.
- New weightings are given to the criteria used for evaluation of the variants.

19.1.4 Differences between Software Engineering and Classical Systems Engineering

Intended learning outcomes: Describe system development by strictly following to sequence of life cycle phases and the waterfall model. Explain iterative system development, prototyping and the spiral model.

The majority of experiences in the past have shown that software engineering cannot be executed with exactly the same strictness of the *sequence* of life cycle phases that classical systems engineering likes to stipulate. There are two reasons for this:

* Firstly, the organizers and industrial engineers, with their often incomplete knowledge of what can be done on the information technology side, cannot formulate system objectives precisely and understandably enough to allow the software developers, who are not sufficiently familiar with the business processes, to design the correct information technology view of the system. As a result, the formulation of objectives is unstable through the course of the project. In the literature, this challenge is called the evolutionary character of software engineering.
* Secondly, a critical factor is the customer or sponsor contracting the project. Customers want "to see something." However, the system development for complex software systems is very long; it can take months or even a year or more. In the end, customers have little possibility to closely evaluate the quality of the work.

In both of these cases, therefore, the attempt is to solve the problem, for example, by what is called iterative system development.

Iterative system development means that software engineers break up the traditional systems engineering, so that not each life cycle phase of the system must be strictly completed before the next is initiated. For certain system functions in the preliminary, main, and detailed studies, *prototyping* is done, that is, a provisional and rough version of the system is built.

Iterative system development does not, however, shorten the system development as compared with the traditional sequential model. On the contrary, time is lost in the process. But this loss can be made up for by the following advantages:

* The user interface and process scenarios can be shown early on. This helps to avoid misunderstandings that can occur already in an early phase of the system, and this increases certainty. Often, business people and software developers only come to recognize that they have different understandings when they can "see the thing."
* The prototype strengthens users' and clients' trust in the project and in the project actors. This creates a boost in motivation.

In software engineering, system development following the *sequence* of life cycle phases is also called the *waterfall model*, as shown in Figure 19.1.4.1.

Opposed to that, the *spiral model* is a cyclic process, as shown in Figure 19.1.4.2. Here, the system is built, through prototyping, in every phase. In each cycle, the results of earlier phases are refined and extended.

In the iterative development process following the spiral model, the software is not specified, programmed, and taken on as a "complete" product but instead is developed in functional capability increments. The main advantage of the spiral model is that the application of successive increments provides future users with results that they can evaluate, so that the software developers can incorporate their feedback. The trend toward the spiral model also developed not least of all because of advances in programming techniques, in particular in connection with CASE (computer-aided software engineering) tools.

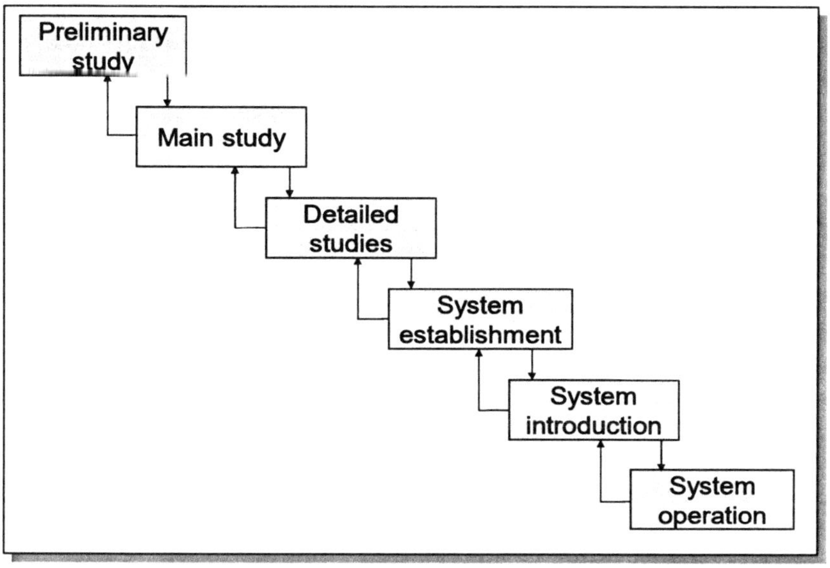

Fig. 19.1.4.1 The waterfall model.

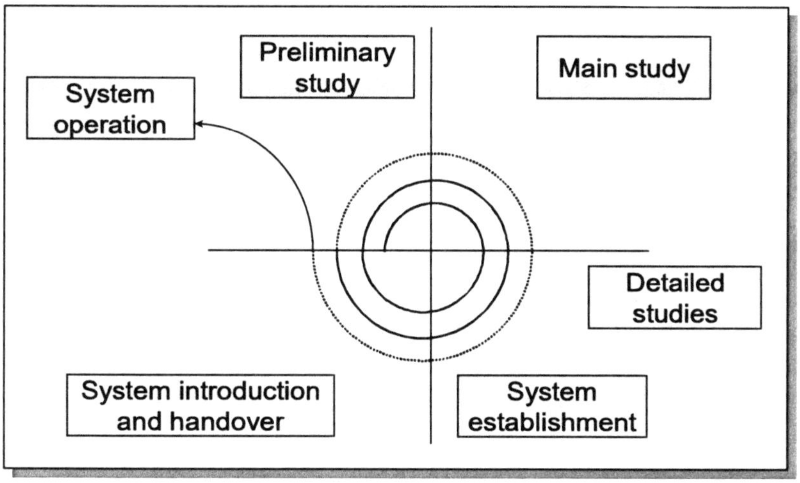

Fig. 19.1.4.2 The spiral model.

19.2 Project Management

Intended learning outcomes: Present goals and constraints of a project. Describe project phase, project life cycle, and work breakdown structure. Explain scheduling and effort planning as well as organization of a project. Differentiate between cost, benefits, profitability, and risk of a project.

The following overview of project management is limited to project goals and constraints, nature and content of a project, organization and process planning, and project costs, benefits, profitability, and risk. For details of the areas presented here in overview, the reader may want to refer to specific works on project management and on tasks such as project cost management, project human resources and project quality management, project information and communications management, and project procurement management (see, for example, [Kerz17], [KuHu15], [PMBOKD], and [PMBOK]).

19.2.1 Goals and Constraints of a Project

Intended learning outcomes: Produce an overview on project performance and project deliverables. Differentiate between external constraints and internal constraints in project management.

Every project has certain goals and objectives; the project performs these.

Project performance comprises achievement of the project objectives in the areas of quality, cost, delivery, and flexibility.[1]

The following describes target areas both generally and taking the example of developing and preparing a machine for marketing:

- In the target area of quality, the goals include functional objectives, such as the use of product and process technologies, or the way that the organization functions. Examples: exact specifications of the functions of the machine, decisions regarding the principles of construction and production.
- In the target area of delivery, the goals include, for example, realization by the planned date to meet the "time to market."
- In the target area of cost, the goals are commercial, such as meeting the cost budget, or realization of expected financial benefit. Example: determining cost per unit and the sales price of the machine, setting the development budget.

The goals themselves are expressed as what are called deliverables.

A *deliverable* is a tangible result created by the work of a project.

Deliverables are, for example, a product prototype, a software package introduced to the market, or a new organization of an area in the company. But a deliverable can also be a study, guidelines, or documents.

Generally, projects have to be performed and delivered under certain constraints.

- *External constraints* are constraints in the project environment over which one can have little or no influence. In the case of system development, these are mainly the surrounding systems. External constraints of business projects can be legal /

---

[1] Compare here the definition of company performance in Section 1.3.1.

regulatory but also political, sociological, environmental, or economic. External constraints are also the scarcity of goods or customers' quality requirements.

- *Internal constraints* are issues in the world of the project that can be influenced and changed — sometimes easily. Internal constraints can be aspects of business and business management but also qualitative abilities and quantitative availability of persons involved in the project. Internal constraints are also the complexity of the project and the technologies implemented, or the quality and delivery reliability of procured goods.

Once project goals and deliverables have been defined and approved, in principle they may not be changed, even if the constraints change. This is one of the main tasks of project management.

19.2.2 Project Phase, Project Life Cycle, and Work Breakdown Structure

Intended learning outcomes: Produce an overview on the project phases in a generic project life cycle. Describe the increasing degree of detail of tasks and work packages in a work breakdown structure. Present an excerpt from a work breakdown structure for the preliminary study for a building conversion.

Deliverables are produced at the end of a project but also as the result of individual phases within the project.

A *project phase* is a major part of a project. Collectively, the project phases are called the *project life cycle.*

Figure 19.2.2.1 shows the project phases in a sample generic project life cycle. See here [PMBOKD or PBMOK, Section 2.1].

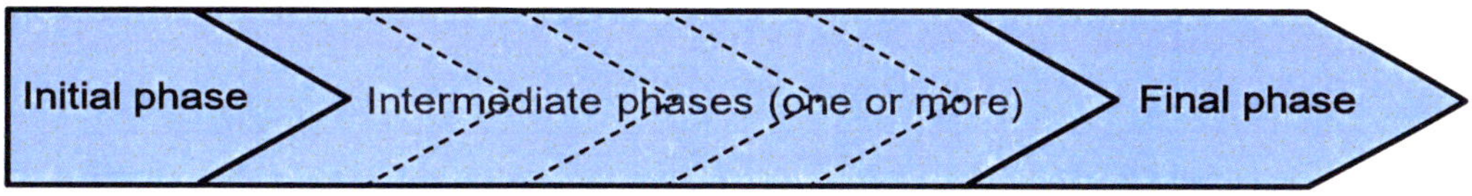

Fig. 19.2.2.1 Project phases in a generic project life cycle (Source: PMBOK).

The intermediate phases will differ depending on the type of project. For example, if the goal is realization of a system, the different life cycle phases shown in Figure 19.1.0.1, from preliminary study to establishment of the system, can be seen as system development. Figure 19.1.4.1 showed the life cycle phases of a project in software development. Possible life cycle phases of classical *product design* are concept development, product planning, process planning, building prototypes, pilot production, and ramp up.

The product life cycle in Figure 1.1.5.1 can be accompanied by several project life cycles. An initial project handles product design and a further project the development of services, that is, additional services connected with the product. Another project can aim at further development.

In project management, a *program* is a group of related projects. The term is then synonymous with a project, mostly a large project.

An example of a program is the NASA Space Shuttle program. The project itself is subdivided into smaller units.

A *(project) task* is a subset of a project, having a duration of a number of months, for example, and carried out by a certain group or organization. A task can also be subdivided into a number of subtasks.

A *work package* is a set of activities assigned to the manager of a component of the project and, if possible, also to an organizational unit. Work packages are deliverables, defined in as much detail as possible. A work package has a cost budget, scheduled start date, scheduled finish date, and *project milestones*, that is, the specific events in the project — usually completion of major deliverables.

Whenever possible, a project should begin with a statement of work.

A *statement of work* is the "first project planning document that should be prepared. It describes the purpose, history, deliverables, and measurable success indicators for a project. It captures the support required from the customer and identifies contingency plans for events that could throw the project off course" ([APIC16]).

The statement of work thus serves management as the basis for decision making. The logical relationships of a project, that is, the tasks and work packages, are called the work breakdown structure.

Work breakdown structure (WBS) is a hierarchical description of tasks and work packages of a project, whereby "each descending level represents an increasingly detailed definition of a project component" [PMBOK].

Figure 19.2.2.2 shows a formal representation of a WBS.[2]

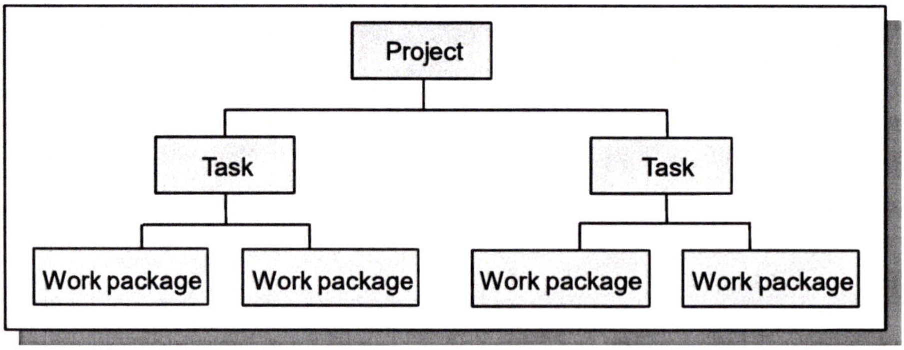

Fig. 19.2.2.2 Increasing degree of detail in a work breakdown structure.

[2] In other standards (e.g., in European Union [EU] projects), work packages are placed above tasks.

This representation corresponds to the product structure (bill of material) in Figure 1.2.2.2, or more precisely, a convergent product structure, or tree structure. In the place of manufacturing components, the work breakdown structure in Figure 19.2.2.2 has tasks or work packages, which are processes. Figure 19.2.2.3 shows a Work Breakdown Structure for a part of a sample project, here the preliminary study for the conversion of a building.

This representation corresponds to the multilevel bill of material in Fig. 17.2.3.4. Again, instead of components, there are tasks and work packages. Instead of item IDs, there are task and work package IDs; in the example in Figure 19.2.2.3, it is a lexicographical numbering.

With a view to project scheduling and rapid project completion, it is advantageous when tasks and work packages are defined such that as many as possible can run concurrently. In addition, they should be allocated the necessary resources, and there should be measurable indicators for success of the tasks and packages.

Building Conversion Project

Task Work package	Identification
Develop plans	1
Generate ideas	1.1
Describe work	1.2
Select suppliers	2
Conduct supplier research	2.1
Solicit bids	2.2
Secure financing	3
Estimate financial requirements	3.1
Obtain credit	3.2
Evaluate credit institutions	3.2.1
Apply for credit	3.2.2
Obtain building permit	4
Determine applicable regulations	4.1
Construction stakeout	4.2
Procure materials	4.2.1
Stakeout proposed construction	4.2.2
Prepare application for building permit	4.3
	...

Fig. 19.2.2.3 Excerpt from a work breakdown structure for the preliminary study for a building conversion.

19.2.3 Project Scheduling and Effort Planning

Intended learning outcomes: Present in detail the schematic display of project effort per organizational unit. Explain an excerpt of the Gantt chart for the project "preliminary study for building conversion".

Most representations used in project scheduling are graphic displays.

A *Gantt chart* is a planning board of schedule-related information, showing scheduling of tasks, work packages, or operations in the form of a bar chart.

Figure 19.2.3.1 shows a possible Gantt chart for the project "preliminary study for building conversion." This Gantt chart also shows the project milestones MS (start), M1, M2, and ME (end). A master schedule in the form of a Gantt chart that identifies milestones only is called a milestone chart.

A *milestone chart* shows the major deliverables on the time axis.

Under certain conditions, which do not have to apply in every project, a *network planning* technique can be used to aid project *scheduling* and control:

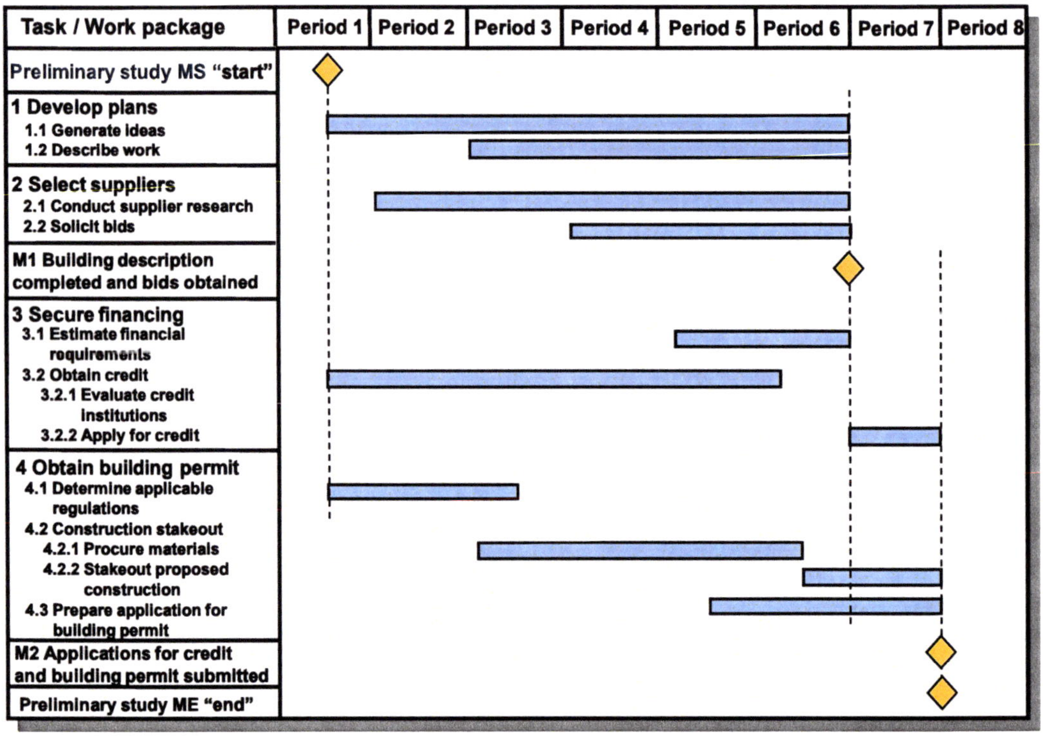

Fig. 19.2.3.1 Gantt chart for the project "preliminary study for building conversion" (excerpt).

- For every task or every work package, early start date and early finish date, as well as late start date and late finish date, can be calculated.
- For every activity within a task or work package, the durations can be determined precisely enough.
- The activities can be ordered in sequence, that is, for each activity, the activities that logically precede and succeed an activity can be displayed schematically. In Figure 19.2.3.1, starting out from the start, it must be determined for each activity what other activity triggers it and what activities it may trigger, or whether it leads to the end (the same is done for work packages).

Network planning techniques mainly determine the *critical path*. The critical path is the series of activities that determines the earliest completion of the project [PMBOK]. The critical path may change with time, especially when certain tasks, which lie on paths that were not yet critical in the first estimate calculated, are completed behind schedule.

Well-known network planning techniques are the *Critical Path Method* (CPM), the *Program Evaluation and Review Technique* (PERT), and the *Critical Chain Method*. These network planning techniques are described in more detail in Section 13.3.4.

Figure 19.2.3.2 shows a simple schematic display of *project effort*, which can be expressed in staff months.

Task / Work Package	Effort Group A	Effort Group B	Effort Group C	Effort Work Package
1 Develop plans				
1.1 Generate ideas	1	2	4	7
1.2 Describe work	3	1	1	5
L1.2 Building description	1	5		6
2 Select suppliers				
2.1 Conduct supplier research	1	2	3	6
2.2 Solicit bids	5	1	1	7
L2.2 List of suppliers and bids	2	1		3
3 Secure financing				
3.1 Estimate financial	1	1	1	3
requirements		2		2
L3.1 Budget				
3.2 Obtain credit				
3.2.1 Evaluate credit				
institutions		2		2
3.2.2 Apply for credit	1	1	1	3
L3.2.2 Credit application		1		1
4 Obtain building permit				
4.1 Determine applicable				
regulations	1		5	6
4.2 Stakeout proposed				
construction				
4.2.1 Procure materials		3	1	4
4.2.2 Stakeout proposed	1		4	5
construction	2			2
L4.2.2 Construction stake	1	4	1	6
4.3 Prepare building permit		2		2
application				
L4.3 Building permit application				
Total effort	20	28	22	70

Fig. 19.2.3.2 Effort per organizational unit.

This schematic representation also lists the deliverables L1.2, L2.2, and so on, as positions in project effort. This is in accordance with the fact that the completion of a deliverable can be connected with particular effort. In the example of Figure 19.2.3.2, this is almost always the effort required for the preparation of a document that requires the consent of all parties involved. Another example is the effort required for the stakeout of the proposed construction.

Effort is measured, for example, in staff days and can be added up for each task or work package. In the present example, the effort for subproject management for a task could be, for example, assigned to the group that has the greatest effort. In this case, that would be Group A for task 2, Group B for tasks 1 and 3, and Group C for task 4.

Based on the schematic representation, project resources, for example, can be released, in whole or part, for example, after completion of milestones, possibly in dependency on the quality of the deliverables. The consumption of resources can be measured accordingly.

19.2.4 Project Organization

Intended learning outcomes: Differentiate between project coordination in a functional, or line, organization and project management in a project-based organization. Describe project management in a strong matrix organization.

There are various possibilities for the organization of a project. Similar descriptions of the variants mentioned in the following are found in [PBMOK]. Figure 19.2.4.1 shows project organization in a functional, or line, organization.

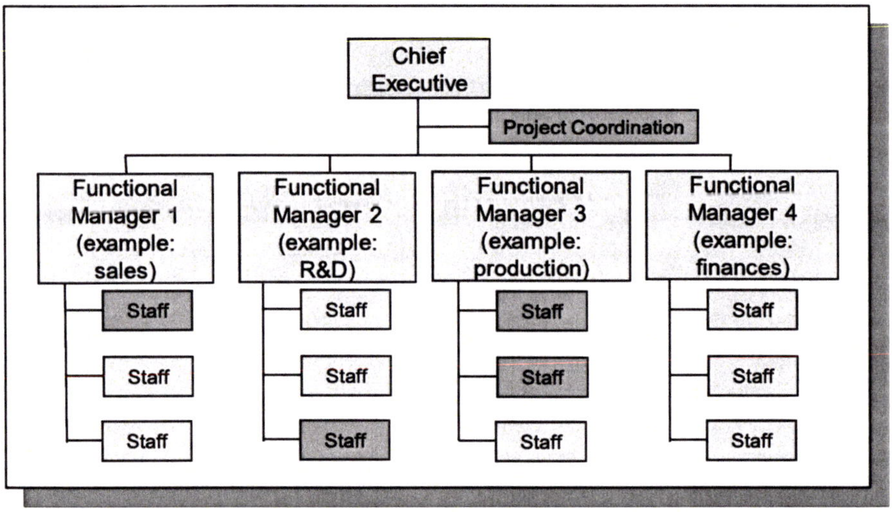

Fig. 19.2.4.1 Project coordination in a functional, or line, organization.

In this type of project organization, which is typical in small and medium-sized companies (SME), the project manager (see shaded box) has little authorization. The project manager usually works on the project part time, alongside his or her other tasks. The authority of the project manager is more or less limited. Various staff members from different functional areas (also shown in shaded boxes) are involved in the project. However, line managers always give priority to their original functional tasks. The connection among the people involved is given by the definition of the project. The project manager coordinates the project activities and the line managers.

As a variant of this type of project organization, there is the so called weak matrix organization, in which project organization is conducted by the persons involved in the project only. Here, the role of the project leader is mostly only coordination. Another variant is what is called the balanced matrix, in which the project manager, or project officer, answers to one of the line managers and from there acts with low to moderate authority across the organization. In Figure 19.2.4.1, one of the "staff" boxes would be labeled "project management."

A second possibility is the organization of the entire company based on projects, as shown in Figure 19.2.4.2.

This organizational structure is typical for a company that primarily sells and implements projects, such as, for example, in consulting businesses or in engineering. A *project manager* — usually assigned full-time to project work — manages all persons and the other resources that are required to complete the project within the project area (shaded boxes).

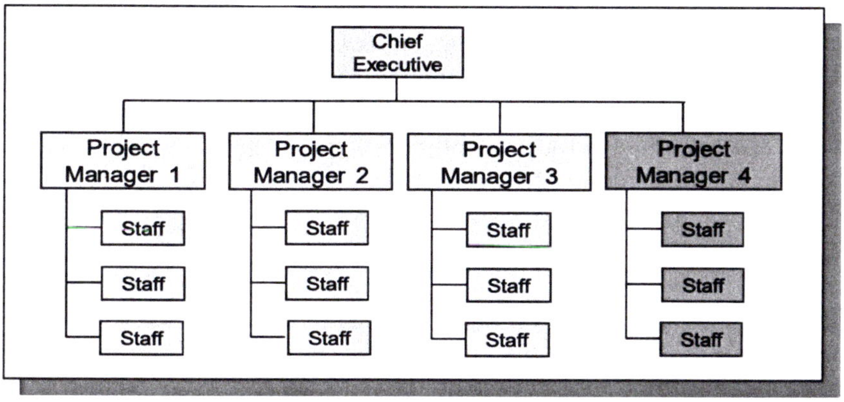

Fig. 19.2.4.2 Project management in a project-based organization.

A further possibility is the strong matrix organization shown in Figure 19.2.4.3.

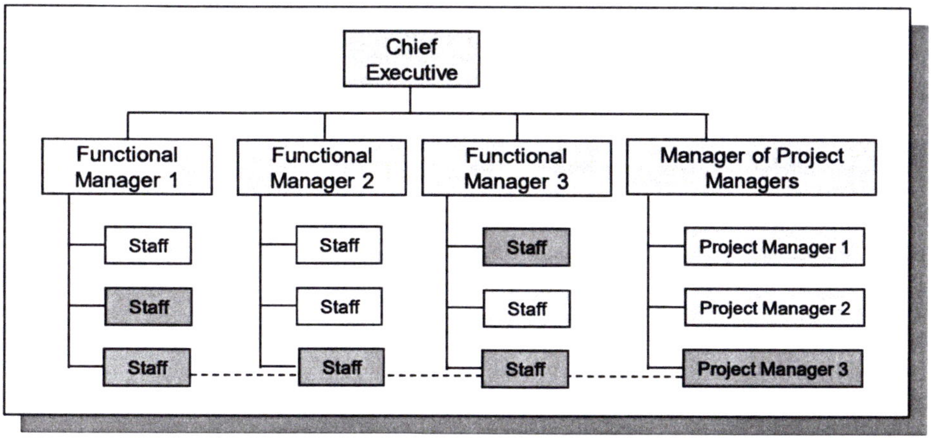

Fig. 19.2.4.3 Project management in a strong matrix organization.

This organization is found in larger companies. Again, the company areas that are involved in a particular project are seen in the shaded boxes in the figure. Important resources are allocated to the project manager as well as a moderate-to-high authority to issue instructions in the horizontal organization, which is indicated in the figure by the broken line. Other people may also be involved in the project, but as in Figure 19.2.4.1 their first work priority is assigned to their line tasks. Composite forms are feasible, especially if several projects are being conducted simultaneously, some of them possibly with a weak matrix organization.

Depending on the type of project organization, the project leader or manager has varying responsibilities and resources at his or her disposal. In any case, the project manager is responsible for motivating and inspiring people in relation to the project. The issues are after all unique, whereby a project team made up of people from different company areas usually must work together. For this, the project manager has to have a high degree of social competency, a wealth of ideas, and good communication abilities. For a high-performance project team, the same is required of all team members. For a detailed discussion of this aspect of project management, see, for example, [Kerz17] or [KuHu15].

19.2.5 Project Cost, Benefits, Profitability, and Risk

Intended learning outcomes: Produce an overview on the total cost of ownership of a project. Explain the Matrix for estimating the project benefit of an investment in a software system as well as the graphic representation in overlay of nine profitability calculations, for cumulative benefits with degrees of realization 1 to 9. Identify NPV, the net present value technique. Present the issue of project risk management.

For the decision to be made to conduct a project, generally the benefits have to be greater than the costs. A profitability calculation is also a basis for prioritizing several possible projects. As the projects of interest here are mostly the realization of systems, the decision techniques are illustrated, taking the example of the introduction of an ERP software system.

Project costs, also called *total cost of ownership of a project*, encompasses both for the initial investment and for running costs incurred in accomplishing work during a given time period.

The difficulty of estimating the different project costs varies. Relatively good estimates can be made of the *initial investment*. In the ERP example, the initial costs generally include:

- the required hardware, system software, and application software
- premises and installations for machines and people
- internal startup costs for the persons assigned to the project
- decommissioning of the existing, old system
- first training of users in mastering the selected business processes; that is, the organizational solution
- first training of users in mastering computer support of the business processes
- external startup costs; for example, for consultants

Running costs of maintaining the operation of the IT system should not be underestimated. In the case of introducing software, these include

- service and maintenance of hardware and software

- ongoing training of users

Estimation of expenses to avoid opportunity costs is not simple. These expenses arise through evaluation of customer requirements for the new system in the target areas of quality and delivery. They therefore concern *system risks* and are sometimes difficult to estimate. In the case of investment in information technology, the costs are also called "*total cost of computing.*" System risks include the costs of nonaccessibility for one, and faultiness of hardware and software, for another. The opportunity costs include lost profit contributions from customer business. Depending on the application, opportunity costs are very high as compared to other costs, or they are inconsequential. Examples:

- A bank that deals in online stock trading has to equip the system for extreme loads that can unexpectedly and rapidly occur if there are new issues or the stock market crashes. System overload or even system failure at this time inevitably leads to a loss of a large part of the customers. To reduce system risks, and thus avoid opportunity costs, the information technology system can at best be duplicated; that is, mirrored, as a backup, which leads to increased investment costs.
- Tax collection agencies do not have to deal with the problem of losing "customers." A system failure of short duration is unproblematic and does not result in opportunity costs. In other words, no additional costs arise in order to avoid opportunity costs.

Project benefit is the financial return that arises through the realization of the project.

As with project costs, the difficulty of determining project benefit can vary. A fundamental difficulty here is the following problem: many aspects of benefit — particularly in the target areas of quality and delivery — are not expressed primarily in monetary terms and therefore have to be converted in the end to financial quantities, that is, returns.

This is also the case with introduction of a new software system. The following description is taken from [IBM75a]. It can be applied easily to other types of projects. Here, three types of benefit are distinguished:

1. *Direct benefit through savings*: Example: Reduction of administrative personnel by one job, reduction of the expensive maintenance costs of previous machines requiring higher maintenance.
2. *Direct benefit through additionally achieved profit contributions:* For example, higher business volume through the processing of additional contracts with customers via EDI (Electronic Data Exchange) or via the World Wide Web, improved payment practices by customers through the charging of 0.5% interest on the invoiced amount for late payment.
3. *Indirect benefit:* For example: Reduction of physical inventory by 3% through more exact, complete, and detailed information, 2% increase in utilization achieved in the same way, faster lead times in the flow of goods.

Figure 19.2.5.1 shows these three types of benefits in a matrix, in comparison with high, medium, or low probability of realization.

	Probability of Realization		
	high	medium	low
1. Direct benefit through savings	1	2	4
2. Direct benefit through additionally achieved profit contributions	3	5	7
3. Indirect benefit	6	8	9

Fig. 19.2.5.1 Matrix for estimating the benefit of an investment in a software system.

The idea is to enter all expected benefits into the cells of the matrix, together with the year in which the respective benefit will occur (that is, the year of its realization), calculated from the time of introducing the software system. Sometimes, two or more cells must be used.

- For example, it is estimated that a 1% reduction in physical inventory will occur starting in year 2 with high probability, 1% starting in year 3 with medium probability, and 1% starting in year 4 with low probability. In this case, the value and the year are entered into cells 6, 8, and 9.

This technique takes the observation into account that benefits — in greater variation than costs — can be estimated pessimistically, realistically, or optimistically. The numbers from 1 to 9 indicate the sequence in which the expected benefits will be included in the calculation of the cumulative benefit.

Cumulative benefit with degree of realization d, $1 \leq d \leq 9$, is defined as the addition of the benefits in cells 1 to d.

The different degrees of realization allow estimation of risk in the form of a sensitivity analysis.

Project profitability is comparison of the costs and benefits of a project.

This calculation is also called capital budgeting. Following [IBM75a], the cumulative benefits with degree of realization 1 to 9 are entered on the time axis, as shown in Figure 19.2.5.2. This results in nine different benefit curves. The cost curve is entered into the same graph. At time point 0, this is the initial investment; in the following years, this is initial investment plus running costs. The result obtained is the *payback period,* or, in other words, the *breakeven point* of the investment: (and this is what makes the technique intuitively simple) de facto for nine profitability calculations shown in overlay.

According to the example graph in Figure 19.2.5.2, for degree of realization 6, the payback period is approximately two years. For cumulative benefits with degree of realization 7, the payback period is approximately one and a half years.

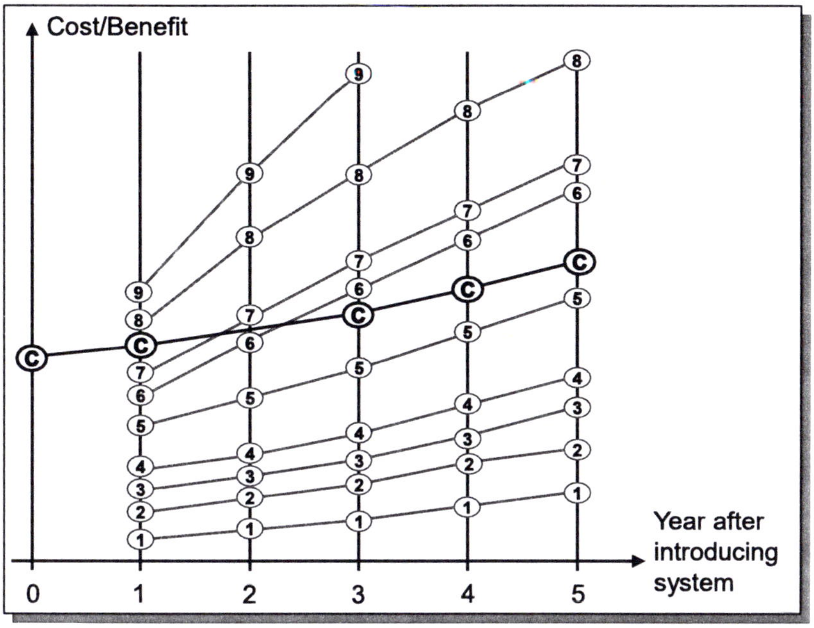

Fig. 19.2.5.2 Graphic representation in overlay of nine profitability calculations, for cumulative benefits with degrees of realization 1 to 9.

Up to now, costs and benefits were included in the comparison without adjustment. However, tallying up the costs and benefits can also be done with *discounting*, which means converting all costs and benefits to their present value (at the time of introducing the system). For this, standard methods for analyzing the profitability of investments can be used. The net present value (NPV) formula, shown in Figure 19.2.5.3 is one. See here also [Kerz17, Ch. 14].

$$\mathrm{NPV} = \sum_{t=0}^{T} \frac{(\mathrm{B_t} - \mathrm{C_t})}{(1+\mathrm{r})^{\mathrm{t}}}$$

NPV: Net present value

t: Period of realization (e.g., year after system introduction)

C_t: Sum of costs in period t (C_0: initial investment)

B_t: Sum of benefits in period t ($B_0 = 0$, in general)

r : Investment interest rate (e.g., 10% = 0.1)

Fig. 19.2.5.3 NPV, net present value technique.

The net present value formula takes into account the fact that one Euro is worth more today than one Euro a year from now. This is because of interest payments on capital. At an interest rate of 10%, one Euro today is equal in value to 1 Euro * (1+0.1) = 1.1 Euro a year from now. Vice versa, one Euro one year from now is only worth 1 Euro / (1+0.1) = 0.909 Euro today. Projects with a net present value greater than zero are profitable. As most projects have many of the investment costs up-front or in year one, the NPV formula favors de facto projects with a short payback period.

Project risk refers to events that impact the profitability of a project.

In accordance with Figure 19.2.5.3, the project risk can be that the cost objectives of the project or the expected benefits are not achieved. As the costs of the projects are generally far better known than the project benefits, the risk analysis is usually restricted to the amount of the benefit and the year in which it occurs (realization). One way to estimate the risks is sensitivity analysis, which becomes possible by performing several profitability calculations with inclusion of cumulative benefits with different degrees of realization, as is the case in Figure 19.2.5.2.

In practice, for most investments with strategic importance, the big benefits are often seen only with cumulative benefits with degree of realization 6 and higher — that is, they include indirect benefits. In these cases, project management must take caution. The realization of an indirect benefit depends, namely, not only on realization of the actual investment, but also and primarily on whether the selected organizational solution as such is appropriate for handling the business task and whether staff know how to use it. When an IT system to support business processes is introduced, for example,

- the physical inventory level also depends on the general situation concerning orders and the competitiveness of the company. The inventory that results from these influencing factors can surpass by far any possible reduction of inventory due to the software investment.
- the lead time for goods also depends on whether information made available by a software application can also be implemented on time. Rapid information flow — information concerning a late order, for example — is no use, if there is no one at the workplace to process the information.

If the choice is to be made among a number of possible projects, varying risk behavior is seen in the degree of realization of the cumulative benefits that one wants to include.

The reasons for project risk are many and diverse. The fundamental difficulty of estimating costs and benefits is made even more difficult, among other things, by inadequately defined project objectives, poor project organization, inadequate human and other resources, inadequate project management, and insufficient motivation on the part of the people. Appropriate *project risk management* includes the use of methods such as assessments or audits, which are discussed in Section 18.3.3.

19.3 Summary

Systems engineering is a systemic method of realizing systems. The basic idea of proceeding from the general to the particular (top-down approach) requires that the system be viewed at different levels, that is, with its subsystems and environment. The system life cycle encompasses three concept phases (system development) called preliminary study, main study, and detailed studies, and four implementation phases — establishment of the system, introducing the system and handing it over, operation of the system, and disposal, or decommissioning, of the system. The three concept phases run through the problem-solving cycle with the following six steps: situational analysis, formulation of objectives, synthesis, analysis, evaluation, and decision. In the preliminary study, this is predominantly system analysis and formulation of objectives; in the main study and in the detailed studies, this is synthesis, analysis, evaluation, and decision.

Individual phases in the system life cycle can be viewed as projects. Here, project management is a systematic approach to ensure project effectiveness and the efficient use of resources. A project follows certain objectives, the achievement of which often culminates in deliverables and is subject to external constraints and conditions from the environment, such as regulatory requirements, and internal constraints and conditions, such as deadlines and cost and capacity limitations. The project life cycle encompasses initiation, a number of intermediate phases, and an end phase. A work breakdown structure subdivides a project into levels of tasks and work packages, whereby the degree of detail increases with each descending level. For project scheduling, the durations for each task or work package in the work breakdown structure can be graphically shown as date-placed along horizontal bars. The resulting Gantt chart also contains the project milestones. Under certain conditions, network analysis techniques, such as the critical path method, can be used for scheduling. The role of the project manager can be that of coordination in a line organization, or direct management in a project-based organization, or one of limited authority in a matrix organization. Project costs can generally be determined relatively easily. Estimating benefits, however, is difficult in many cases. Often, project profitability can be established only when indirect benefit is included. Profitability itself can be determined using the payback period method or the net present value (NPV) formula. Multiple profitability calculations with inclusion of cumulative benefits with different degrees of realization can provide the required basis for decision making. This procedure also yields a sensitivity analysis that can be used in the estimation of project risk.

19.4 Keywords

deliverable, 800
detailed study, 794
Gantt chart, 803
main study, 794
preliminary study, 793
problem-solving cycle, 796
project, 789
project benefit, 809
project costs, 808
project life cycle, 801
project management, 790
project phase, 801
project profitability, 810
project risk, 811
system, 789
system disposal, 795
system establishment, 794
system introduction and handover, 795
system life cycle, 793
system operation, 795
system phase, 793
systems engineering, 789
work breakdown structure, 802

19.5 Literaturhinweise

APIC16	Pittman, P. et al., APICS Dictionary, 15. Auflage, APICS, Chicago, IL, 2016
DuFr15	Duden 05, „Das Fremdwörterbuch“, 11. Auflage, Bibliographisches Institut, Mannheim, 2015
DuHe14	Duden 07, „Das Herkunftswörterbuch“, 5. Auflage, Bibliographisches Institut, Mannheim, 2014
HaWe05	Haberfellner, R., De Weck, O., „Agile Systems Engineering versus Engineering Agile Systems“, INCOSE 2005 World Conference, Rochester, NY, 2005
HaWe18	Haberfellner, R., De Weck, O. et al., „Systems Engineering“, 14. Auflage, Verlag Industrielle Organisation, Zürich, 2018
IBM75a	IBM (Hrsg.), „Datenverarbeitung - Gewinnquelle des Unternehmens - Nutzenanalyse einer Wirtschaftlichkeitsrechnung für Datenverarbeitungsanlagen“, IBM Form GE 12-1307-01, 1975
Kerz17	Kerzner, H., „Project Management: A Systems Approach to Planning, Scheduling and Controlling“, 12th Edition, John Wiley & Sons, Hoboken NJ, USA, 2017
KuHu18	Kuster, J., Huber, E. et al., „Handbuch Projektmanagement — Agil – Klassisch – Hybrid“, 4. Auflage, Springer, Berlin, 2018
PMBOKD	„A Guide to the Project Management Body of Knowledge: PMBOK Guide“, Deutsche Übersetzung, 5. Auflage, Project Management Institute, PMI Publications, Newton Square, PA, USA, 2014
PMBOK	„A Guide to the Project Management Body of Knowledge: PMBOK Guide“, Project Management Institute, PMI Publications, Newton Square, PA, USA, 5th Edition, 2013
Züst04	Züst, R., „Einstieg ins Systems Engineering“, 3. Auflage, Verlag Industrielle Organisation, Zürich, 2004

20 Selected Sections of Information Management

Intended learning outcomes: Produce a review of important terms in information management. Disclose the modeling of operational information systems. Explain in detail the modeling of information systems in the data view and object view.

From information management, characteristics and ways of thinking can be taken over that are important for the development and implementation of a system for *planning & control in supply chains*. In this chapter, important definitions are presented, followed by basic considerations on the modeling of operational information systems. We then turn to modeling from the data and the object views, from which techniques and methods of correct representation of logistic business objects can be gained. Practical experience has shown that this type of modeling requires special attention to methodology, as a complement to process modeling, which is more directly accessible for company employees.

20.1 Important Terms in Information Management

Intended learning outcomes: Produce definitions for terms from colloquial language that are easily understood by employees, such as for information, data, information technology, information system, database.

For use in the company, terms from colloquial language that are easily understood by employees are needed.

Information is defined by [MeWe18] as intelligence, news, facts, data, and also as the communication or reception of knowledge or intelligence; information is also the attribute inherent in and communicated by one or more alternative arrangements of something (such as binary digits in a computer program) that produce specific effects. [APIC16] defines information as data that have been interpreted and that meet the needs of one or more managers.

These definitions correspond to the imprecise use of the term in practice.

Data (literally, *something given/facts*) is defined by [MeWe18] as factual information (as measurements or statistics); information output by a sensing device or organ that includes both useful and irrelevant or redundant information that must be processed to be meaningful; information in numerical form that can be digitally transmitted or processed.

Today, data *processing* (DP) is technology-supported.

Information technology, or simply *IT*, comprises *hardware*, i.e., computers, telecommunications, and other devices, and *software*, i.e., programs and documentation for using computers, for collecting, storing, encoding, protecting, processing, analyzing, transmitting, and retrieving data or information. Here see [APIC16].

These definitions lead to the term "information system."

P. Schönsleben, *Handbook Integral Logistics Management*,
https://doi.org/10.1007/978-3-662-65625-9_20

An *information system* is made up of interrelated computer hardware and software along with people and processes designed for the collection, processing, and dissemination of information for planning, decision making, and control ([APIC16]).

This definition refers to an *IT-supported information system*. In practice, an information system can be set up without the IT tools. Data can be kept on paper in suitably organized card indexes, for example. IT-supported information systems in particular have a more or less structured character depending on the organization of their information storage.

A *database* contains data in defined structures, independently of the programs that use or exploit the data.

A *database management system* (*DBMS*) organizes and protects data and allows data to be retrieved according to certain criteria. It regulates access to the data by various users or applications programs at the same time, using diverse routes of access and selection criteria.

A *data definition language* (*DDL*) is a set of functions that allow the data in the database to be described (static component of the DBMS, also called *data dictionary*).

A *data manipulation language* (*DML*) is a set of functions that allow the data in a database to be manipulated. These include inserting and retrieving (dynamic component of DBMS).

A *data warehouse* is a special database that is pulled together especially for query (according to criteria yet to be determined) and reporting purposes. A data warehouse facilitates *pattern recognition*, or the discovery of correlations between facts and events (*data mining*).

As a sociotechnical system, an information system must be carefully integrated into the company.

Information management is a broad term for all activities and functions in an organization concerned with managing the *resource* "information" in the company — that is, strategic and operational management of (1) the information itself, (2) the life cycle of information systems, and (3) the information technologies (IT management).[149]

The subtasks of (1) include identification, acquisition, storage, processing, transmission, presentation, and use of the information. See here, for example, [Schö01, Section 1.4]. These subtasks present a typical design problem. By means of suitable descriptive methods, the individual operational areas are converted into models. These also form a necessary interface to the technical part of an IT-supported information system.

Theoretically, information management should be well coordinated with overall management of the company. Of course, overall management and *business process reengineering* should respond rapidly to changing market situations. In contrast, changing from one IT to another is difficult and often takes too long. This is because it generally requires a large investment, particularly in new knowledge. This is not feasible within a very short time and not always financially feasible. In the world of practice, IT-supported information systems not infrequently survive for 20 to 40 years! Thus, today, an IT strategy

---

149 The term *information system architecture* addresses, in practice, similar tasks.

is de facto often longer-term than a rapidly changing company strategy. This is a problem that is well-known to become acute when companies merge. Information systems that are not compatible can become a knock-out criterion for mergers or company acquisitions. An information system design with as much flexibility as possible for changes in the company strategy is therefore important. This requires high quality in the data and object modeling.

20.2 Modeling Enterprise Information Systems

Intended learning outcomes: Present basic principles of modeling. Differentiate various dimensions in the modeling of information systems for business processes. Describe the dimension of hierarchy creation and the dimension of various views in modeling.

Particularly with IT-supported information systems, exact depiction of both objects and processes of the enterprise is indispensable for purposeful processing of information. Because of the complexity of the sociotechnical system *enterprise*, it is not surprising that modeling is used for this. The first section below therefore first of all presents the basic principles of modeling.

Because no single and generally valid model for an enterprise information system can be found, the sections further below present a common framework within which the different models can be arranged. The framework comprises three different dimensions.

20.2.1 Basic Principles of Modeling

Intended learning outcomes: Describe, as a basic problem in modeling, how the quality of a system image is affected by the person who designs the model.

The storing, processing, and presenting of information are associated with typical design tasks. To carry out these tasks effectively, it is necessary to have an exact idea of the items or facts to which the information refers.

A *model* is defined by [MeWe18] as a description or analogy used to help visualize something; a system of postulates data, and inferences presented as a mathematical description of an entity or states of affairs; also: a computer simulation based on such a system.

The term *modeling* denotes the procedure that leads to the result, namely, the model. A *modeling method* is a planned and systematic approach used for modeling.

Models for information systems are not material but are instead the mental models that a person has of objects or circumstances/issues. People develop their ideas of things based on personal experience and learning about the system itself (for example, about the company), its goals, structure, and processes, but also based on concrete visible objects and their functioning. This leads immediately to one of the main difficulties: A model tends to carry the personal stamp of the person who designed the model. Figure 20.2.1.1 shows this problem.

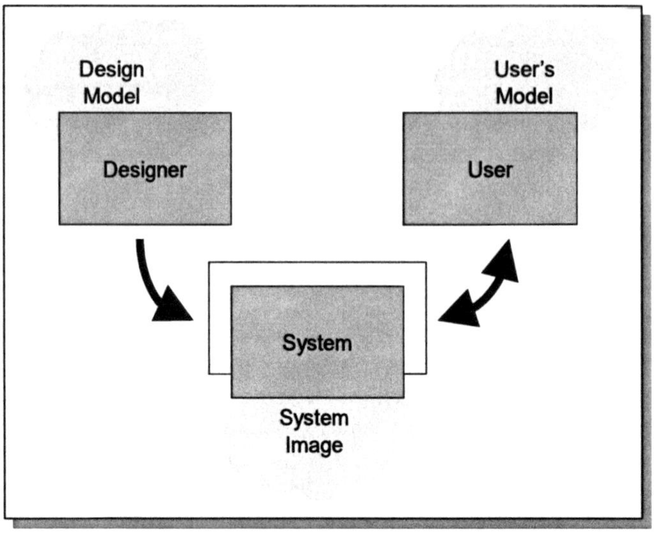

Fig. 20.2.1.1 A basic problem in modeling: The quality of a system image is affected by the person who designs the model. From [Norm13].

According to [Norm13, p.16], "the designer expects the user's model to be identical to the design model. But the designer doesn't talk directly with the user — all communication takes place through the system image. If the system image does not make the design model clear and consistent, then the user will end up with the wrong mental model." When modeling physical systems, such as goods flow in the company, it is already difficult to achieve models that everyone sees as being generally the same. This is all the more difficult when modeling information systems. There are few restrictions placed on imagination and creativity. But this is a disadvantage for communication among several staff members.

Reality shows that the desired single and generally valid model for an information system of a company cannot be found. The modeling of operational information systems is conducted from different views, which accord with the different mental models. This corresponds to the complexity of such sociotechnical systems. In the following, the objective is to find a common framework in which the various models can be arranged.

The different methods of modeling are themselves different, depending on the purpose. Each method has specific advantages. Thus, if due to personal background or experience individuals view a certain aspect as especially important, they will find the corresponding method to be especially suitable.

20.2.2 Various Dimensions in the Modeling of Information Systems for Business Processes

Intended learning outcomes: Present the various dimensions of the ARIS tool set for modeling information systems for business processes. Explain the dimension of four views and the dimension of three descriptive levels.

In past years, there have been several attempts to translate the complexity of an information system into a concept for development of information systems. A research consortium in an EU project developed the modeling framework called CIMOSA (Computer Integrated Manufacturing Open System Architecture). See here [Espr93]. The CIMOSA approach is the basis of numerous further models and tool sets. Probably the best known of the tool sets is the ARIS (Architecture of Integrated Information Systems) Tool Set [Sche00]. See here also [Sche98c]. Figure 20.2.2.1 shows the ARIS model as a house with three dimensions:

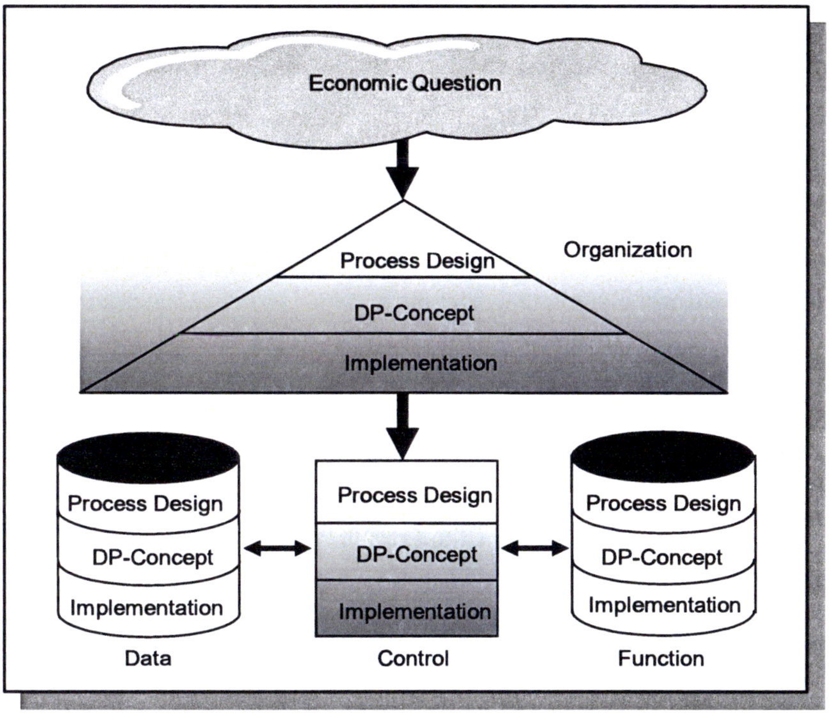

Fig. 20.2.2.1 ARIS Tool Set for modeling information systems for business processes: The dimension of four views and the dimension of three descriptive levels.

First, the *dimension of hierarchy creation*, or the *principle of proceeding from the general to the particular*, as mentioned also in Section 19.1.1. This dimension does not find direct expression in Figure 20.2.2.1. But it is realized in the ARIS Tool Set in that each process in the control view can be broken down further into its subprocesses in a new diagram. This corresponds to creating a component hierarchy, as shown in Section 20.2.3.

Second, the dimension of the *four views of the information system*, namely, the organization view, data view (object view is a more extensive term), function view, and control view (often called the process view). For each of the four views, there are different modeling methods. Certain methods handle only one view, and other methods attempt to combine several views. In addition, certain tools handle further views. The advantage of methods that deal with only one view is mostly their simplicity. But the combining of the four views must then be carried out by the observer of the model, using the support tools. In the ARIS Tool Set the combination takes place, for example, via the control view, or process view.

Third, the dimension of *system phases*, or *phases of life of a system*, is identified here with the three descriptive levels *process design, data processing (DP) concept*, and *(technical) implementation*. With reference to the phases in Section 19.1.2, the process design corresponds to the preliminary study and the main study. The DP concept corresponds to the main and detailed studies, and the technical implementation to system establishment.

20.2.3 Dimension: Hierarchy Creation

Intended learning outcomes: Differentiate between the component hierarchy, or whole-part hierarchy, and the specialization hierarchy. Explain the association hierarchy, or determination hierarchy.

> Through *hierarchy creation*, in a view of the information system, "lower" items belonging to a "higher" item are grouped/classified.

Hierarchy creation is a very important construct in modeling. It pervades and characterizes modeling, for which reason it can actually be called a dimension of modeling. Through hierarchy creation, elements belonging together are arranged, grouped, or classified in sets. The following three types of semantics in hierarchy creation are observed frequently in practice:

- Component hierarchy, or whole-part hierarchy
- Specialization hierarchy
- Association hierarchy, or determination hierarchy

> In the *component hierarchy*, or *whole–part hierarchy*, a hierarchically superior element G is composed of (consists of) elements T_1, T_2, T_3, …

In reverse order, each of the hierarchically subordinate elements T_1, T_2, T_3, … is a part or component of a superior element G. Figure 20.2.3.1 illustrates the semantics.

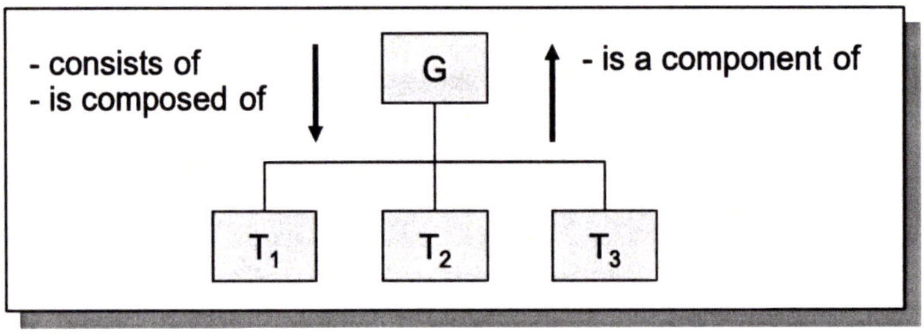

Fig. 20.2.3.1 Component hierarchy, or whole–part hierarchy.

Examples of component hierarchies in the material world are:

- A bicycle (as a whole) with its components: handlebars, frame, and wheels
- A transport (as a whole) with its phases: load, transport, and unload.

Examples of component hierarchies in information systems are:

- A system with its subsystems
- An object with its components
- A process with its subprocesses
- A function with its subfunctions
- A task with its subtasks.

The dimension of the principle of proceeding "from the general to the particular" in systems engineering (Section 19.1.1) and in different modeling tool sets corresponds to a component hierarchy.

In a *specialization hierarchy* the elements S_1, S_2, S_3, ... represent special manifestations of the superior element G. In other words, each of the elements S_1, S_2, S_3, ... is a (special) G.

In reverse order, the hierarchically superior element G is a generalization of all subordinate elements S_1, S_2, S_3, ... In other words, an element G can be an element S_1 or an element S_2 or an element S_3, and so on. Figure 20.2.3.2 illustrates the semantics.

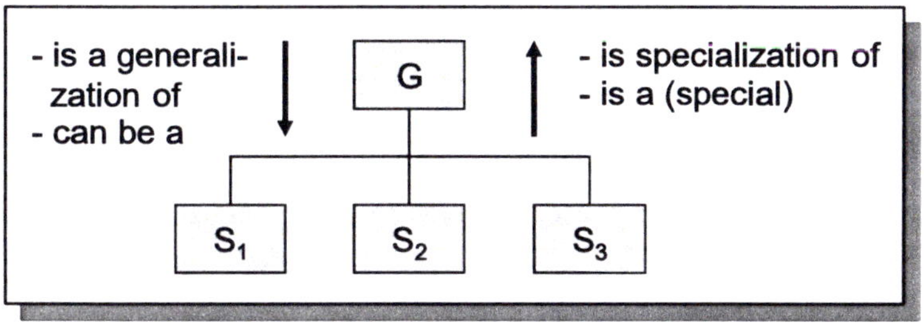

Fig. 20.2.3.2 Specialization hierarchy.

Examples of specialization hierarchies in the company environment are

- Employees of a company (generalization) with the various manifestations "salaried employee," "blue-collar worker," and "trainee."
- A shipping service (generalization) with its various options: "express," "normal," or "lowest-cost" processing.

Examples of specialization hierarchies in the modeling of information systems are

- An object with its specialized manifestations
- A process with various processing options

In an *association hierarchy*, or *determination hierarchy*, all elements G_1, G_2, G_3, ... that jointly define or determine a subordinate element A, or due to whose association an element A comes into existence or is generated, are placed superior to A.

In other words, to exist, element A needs all elements G_1, G_2, G_3,... jointly. We can also say that A belongs to the elements G_1, G_2, G_3, ... together, or that A comes into existence because of their association. In reverse order, the hierarchically superior elements G_1, G_2, G_3, ...

jointly "generate," "determine," or "possess" a subordinate element A. Figure 20.2.3.3 illustrates the semantics.

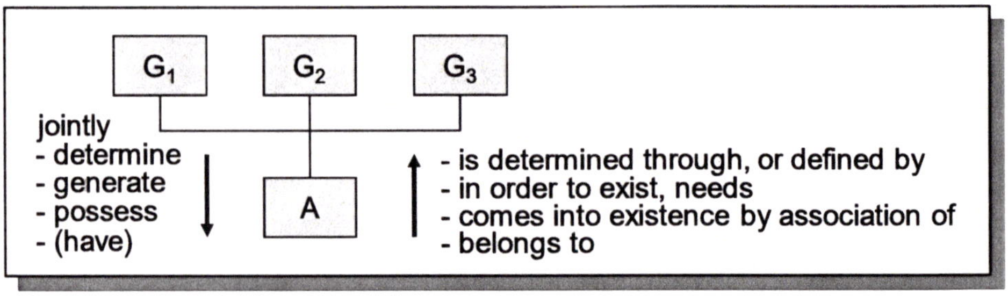

Fig. 20.2.3.3 Association hierarchy, or determination hierarchy.

Examples of association hierarchies, or determination hierarchies, are:

- To exist, a child needs a father and a mother. Or the child comes into existence by their association.
- A taxpayer is defined by a person and a political unit (municipality)
- A customer order is determined by a customer and a date.
- Sales and distribution are determined by marketing (which is strategically superior) and long-term corporate planning.

Examples of association hierarchies in the modeling of information systems are:

- To exist, an object needs various other objects
- A process is determined by several other processes

The different verbs used to describe the semantics of the association hierarchy, point up different possible degrees of intensity of determination: from generation to possession. However, what they all always share is that the subordinate element cannot exist without the elements that determine it. The most intensive form is a "parent/child" relationship.

The least binding form of the association hierarchy can be expressed by the verb "to have," or in reverse order "belongs to." However, the semantics are too general to characterize the association hierarchy, as it can also be used for the component hierarchy or for another hierarchical or nonhierarchical relationship.

The various semantic constructs of creating hierarchies are recursive. Thus, multilevel constructs can be built:

- Each component itself can in turn be composed of subordinate elements.
- Each specialized element itself can in turn be specialized into further elements, possibly according to another criterion.
- Each element defined through a superior element can in turn, possibly together with further elements, determine a subordinate element.

Familiar examples:

- A *bill of material* or *nomenclature* is the structured list of components making up a product. It is a multilevel component hierarchy. See here also Section 1.2.2.
- A *classification system* is a multilevel specialization hierarchy. An example of this is the German DIN 4000 standard, a standardized classification guide aiding the work of the designer, enabling the designer to systematically trace items — typically semiprocessed items and single parts. See here also Section 17.5.3.
- The classical *chain of command* in a company or another human or mechanical organization is a multilevel determination hierarchy in which in each instance only one determining element has an influence on one subordinate element.

20.2.4 Dimension: Various Views in Modeling

Intended learning outcomes: Explain the four views of information systems for business processes. Differentiate between process-oriented modeling, function-oriented modeling, object-oriented modeling, and task-oriented modeling.

[Spec05] presents a newer treatment of the *various views of an information system*. Figure 20.2.4.1 is taken from [Spec05].

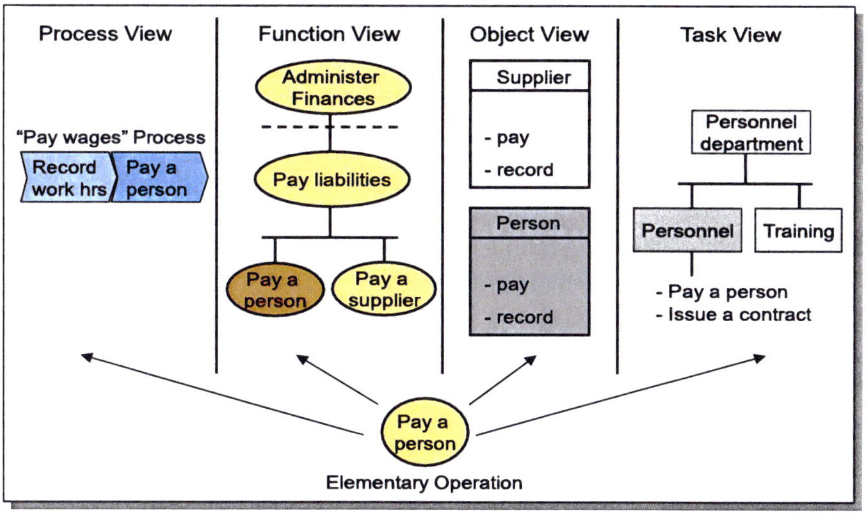

Fig. 20.2.4.1 Four views of information systems for business processes (example taken from [Spec05]).

- The *process view* is naturally suitable for an understanding of operations. A process groups several jobs, such as tasks, subtasks, functions, and subfunctions, and carries them out in a certain order. As an example, Figure 20.2.4.1 shows a possible representation of the process of paying wages to a person. This view corresponds to the control view in the ARIS Tool Set (Figure 20.2.2.1).

- The *function view*: Functions are grouped according to, basically, any characteristic under a corresponding superior function. The resulting hierarchy is possibly a multi-level hierarchy, a tree structure. In the example in Figure 20.2.4.1, the lower level shows all functions that "pay liabilities to others." On the higher level, all functions are grouped that in some way "administer finances." Often the grouping characteristic for functions is that the functions belong to a certain process. If we order the functions in the tree structure in the order in which they appear in the process, then in the function view we can essentially also read the process. Turned around, we can also say that the function view can be perceived like the process view.

- The *object view*: This view summarizes data and functions or processes that concern the same object. In the example in the figure, the objects are person and supplier. An important part of the object view is the *data view*, which is the description of each object by a set of attribute values. Objects form the basis for data processing, as they contain the data. Here it is not important whether the object descriptions are in a card index or stored in electromagnetic format. Appropriate tasks, functions, or processes can be assigned to an object. This makes sense as soon as they essentially concern only this object. As the processes, tasks, or functions have to be described in some way, this description can be seen as data and assigned to the object like the description of the object itself.

- The *task view*: This view groups subtasks, tasks, functions, or processes into tasks. A number of meaningfully connected tasks can consequently be assigned to an organizational unit in the company. In the example, the tasks "payroll accounting" and "training" that are grouped in the organization unit "personnel department" are defined. However, personnel administration and training themselves can also be seen as organizational units. In reverse order, personnel department can be seen as a more comprehensive task. This view is what links the process organization with the structural organization of the company. This link is actually the reason why one so often finds, instead of the task view, the term *organization view*, such as in the ARIS Tool Set in Figure 20.2.2.1.

In the following, these different views will be modeled.

Process-oriented modeling deals with the process view as the primary view. It produces *process model.*

Function-oriented modeling deals with the function view as the primary view. It produces *function model.*

Object-oriented modeling deals with the object view as the primary view. It produces an *object model.* The more restricted view of *data modeling* produces a *data model.*

Task-oriented modeling deals with the *task* or *organization view* as the primary view. It produces a *task model* or *organization model.*

20.3 Modeling Information Systems in the Data View and Object View

Intended learning outcomes: Present terms such as object, attribute, object class, view, primary and secondary keys. Explain basic concepts such as association, association role and type. Describe the breaking up of an n-to-n association – in particular a reflexive one – and the association class. Disclose the use of the hierarchical constructs for developing a company-wide generic object model.

With the introduction of IT support, the modeling of information systems in the data and object views became decisively more important. Storing data on operations objects comprises very complex structuring of a multitude of classes of data stocks. Users have to understand data and object models, if they are to be able to use it efficiently in practical operation. Then, the models have to be implementable on the computer; they must therefore also satisfy requirements for accuracy and unambiguity. Data and object models belong to the conceptual interface between users without specific IT-knowledge and producers of information systems. The following terms pertaining to the design of such models are therefore taken from colloquial language wherever possible and should be understandable.

20.3.1 Object, Attribute, and Object Class

Intended learning outcomes: Differentiate between entity, object, and data record. Differentiate between attribute and data field. Differentiate between object class, file, and table. Explain the class customer as a table in the relational database model.

An *entity* refers to a thing [MeWe18]; the existence of a thing. It can have objective reality (such as a person or object) or conceptual reality (such as a concept or ideas).

An *object* is something physical or mental toward which thought or activity is directed, or a goal or end of an effort or activity [MeWe18]. A related term is *thing*.

Entity thus describes the existence of a thing, whereas object describes more the item or thing as an object for study. With information systems for business processes, the one aspect does not exist without the other, so that in this case the two terms can be used synonymously.

An *attribute* is a quality, also a word ascribing a quality ([MeWe18]).

A *domain* is the set of all possible values of an attribute.

To each object or each entity, a number of attributes belong that contain all information about the object or entity that is of interest in the context studied. Each attribute describes the entity or object in a certain way; it shows one aspect of it. For each attribute, there is a corresponding attribute domain; different attributes can have the same attribute domain.

A *data record* contains several data fields, arranged in a certain format, to store the data belonging to an object.

A *data field* is a special place in a data record to store the value of an attribute of an object.

Object data are thus coded information on the object or entity, that is, the data record or the set of all data fields of an object or entity.

An *object class* or *entity class*, abbreviated as *class*, is a set of entities or objects that in their essential qualities are described by the same attributes.

The term *entity* or *object* thus describes the individual value, that is, the occurrence or instance of a class, such as a person within the class of all employees of a company.

- In the entity-oriented approach, in practice the term *entity* is unfortunately used for both data structure (that is, entity = set of attributes) and the concrete value (that is, entity = set of attribute values). Thus, the entity *person* can be used as both an umbrella term for the set of the attributes belonging to the entity and a certain person, that is, a certain value for each attribute.
- In the object-oriented approach the term *instance* means only the concrete value, whereas the attributes are assigned to the class. The class is thus a structural term that can also be defined without concrete instances.

A *file* is defined by [MeWe18] as a collection of papers or publications usually arranged or classified; a collection of related data records (as for a computer); a complete collection of data (as text or a program) treated by the computer as a unit especially for purposes of input and output. In EDP, a file contains all data records on objects or entities belonging to a class.

Table is another term for *class* in the *relational database* model, which is probably the most well-known model in the entity-oriented approach.

The columns of a table correspond to the different attributes; the rows are the various entities of a class. Figure 20.3.1.1 shows an example where the class is *customer* with the attributes *customer ID*, *name*, *agent*, *sales*, *line of business*, and *location*.

Customer ID	Name	Agent	Sales	Line of Business	Location
3001	Novartis	Meier	4,000,000	Chemistry	Basel
3002	La Roche	Mueller	3,000,000	Chemistry	Basel
3003	CS	Hofmann	6,000,000	Bank	Genf
3004	IO	Oldenkott	1,500,000	Publishing	Zurich
3005	Migros	Alberti	3,600,000	Food	Bern
3006	Int. Discount	Dossenbach	2,400,000	Electronics	Aarau
3007	UBS	Sauter	500,000	Bank	Zug
3008	Continental	Zuber	340,000	Insurance	Zug
3009	Hey	Guebeli	70,000	Textiles	Zurich

Fig. 20.3.1.1 The class *customer* as a table in the relational database model.

Representing a class as a table is an appropriate and widespread practice. Incidentally, both the objects of a class in the object-oriented approach and the data records in a file in a data-oriented approach can be represented as tables.

20.3.2 View and Primary and Secondary Keys

Intended learning outcomes: Produce an overview on the view on a class. Differentiate between primary key and secondary key. Describe characteristics for a "good" primary key. Explain the representation of a class in an object-oriented approach (with example) and the representation of an object.

View on a class is an excerpt of the table defined by entities and attributes, which includes only certain attributes and entities and for which the entities are presented in a certain order.

The term *view* can be applied to a class in an object-oriented approach, with objects in the place of entities. In Figure 20.3.1.1, the view is complete regarding the attributes as well as the entities. The different entities are listed sequentially in ascending order sorted by the attribute *customer ID*.

- Another view could present the customers in descending order sorted by the attribute *sales*, present the attributes *name* and *sales* (in that order) and only for customers in certain *locations*.
- A further view could present the entities in ascending order sorted by the attribute *agent*, showing only the attributes *agent*, *name*, and *customer ID*.

In practical application, the view has become probably the most important construct in electronic data processing (EDP). And in fact, most information needs can be formulated as simple or complex views of certain data in a database. To express it pointedly, the main task of information systems for business processes consists in sorting and selecting objects. This is especially true in the case of IT-supported systems. To create views there are tools:

Structured Query Language (SQL) is the name of a well-known data manipulation language (DML). The clauses of an SQL query correspond in principle to the definition of a view.

The SQL instructions for the two examples above are:

- SELECT name, sales FROM client BY sales WHERE city ∈ {"Zurich," "Basel"}
- SELECT agent, name, customer ID FROM customer BY agent

Another important design element is the primary key of a class, also called ID key, or — simply — ID.

A *primary key,* or *identification key* of a class (also called ID key or simply ID) is a minimal number of attributes that together as a set uniquely identify an object. More formally, a primary key is subset X of attributes of a class that always — that is, for each possible combination of entities — satisfies the following conditions:

- X determines all attributes of the class, that is, for each value of X there exists an unambiguous value for the remaining attributes, or — and equivalently — all entities or objects that possess the same value for X also have the same value for the remaining attributes (that is, it is the same entity or same object that we are dealing with here);
- There is no subset Y of X, $Y \neq \emptyset$, such that Y already determines the remaining attributes (minimality property, or irreducibility).

In principle, it is not necessary to choose a primary key purely for storing the data. Nevertheless, the primary key has become an important design element. The main reason for this is that when modeling the real world, the semantics are considerably simpler for the user to understand. A "good" primary key should have the following four characteristics:

1. *Stable:* A primary key must not change during the lifetime of an object.
2. *Immediate assignment:* A primary key must be assigned immediately, as soon as an object is created.
3. *Short and simple:* A primary key should be relatively short for quick data entry.
4. *Meaningful:* A primary key should consist of typical qualities of the object.

Unfortunately, the fourth requirement (meaningful) is often not consistent with the first, stability, and in part also the second and third. Examples of partly meaningful keys are the EAN number for product identification, the social insurance number ("AHV" number) previously used in Switzerland, and also in many countries the student ID number. In practice, stability has priority. This is because with a meaningful primary key, when the number of digits has been used up for meaningful description, the key has to be lengthened by one or more digits. Because of the existing stored data, this so costly and prone to errors that instead a new key is chosen and changed to "not meaningful." This was the case, for instance, with the introduction of the new social insurance number in Switzerland. The same issue concerns the choice of a meaningful item ID.

A *secondary key* of a class is a subset of attributes that sorts the objects for views.

A secondary key differs from a primary key in that it does not have to identify an object uniquely. But it defines a view of the class such that it determines the order of the objects.

In the object-oriented notation following UML (see [UML13]) as shown in Figure 20.3.2.1, the primary key attribute is underlined. Several attributes together can define a primary key. Several possible primary keys are distinguished by means of a list.

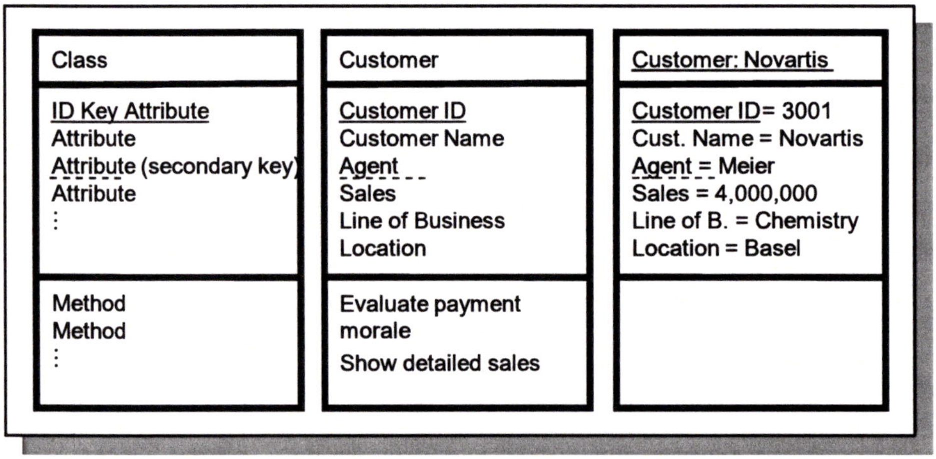

Fig. 20.3.2.1 Representation of a class in an object-oriented approach (with example) and representation of an object.

A view is marked by means of a dotted line underlining the attributes according to which the view is sorted. In Figure 20.3.2.1, for example, a view is defined by the attribute *agent*.

20.3.3 Association and (Association) Role

Intended learning outcomes: Differentiate between binary association and reflexive association. Explain Abrial's access function and the example of the association of the classes "book" and "customer" in library systems. Identify the (association) role.

An *association* is, according to [MeWe18], a relationship, link, connection, or uniting. In data-oriented and object-oriented modeling, an association is a relationship among classes, more precisely among the objects or entities of these classes.

A *binary association* or *two-sided association* is a relationship between exactly two classes, or more precisely between one object of each of two classes.

A *reflexive association* relates a class to itself; more precisely, it relates an object of a class to another object in the same class.

There are also associations among three or more classes. These are found rarely and moreover can always be traced to relevant binary associations. Reflexive associations are also found rarely, but some are typical. An important case is discussed in Section 20.3.8 below.

An association finds expression in the mutual assignment of the elements of classes. We perform data modeling and object modeling to identify these relationships and to depict them graphically to enhance understanding of the system. For this, formally, the Abrial's access function can be suitable.

Abrial's *access function* (see [Abri74]) is a binary relation (a mathematical relation!) that leads from the initial to the target class.

The formalism is shown in the case of binary association between classes X and Y in Figure 20.3.3.1. Here, f and g are binary relations, whereby g is the reverse relation to f. f leads from class X to class Y, and g leads in the reverse direction from Y to X.

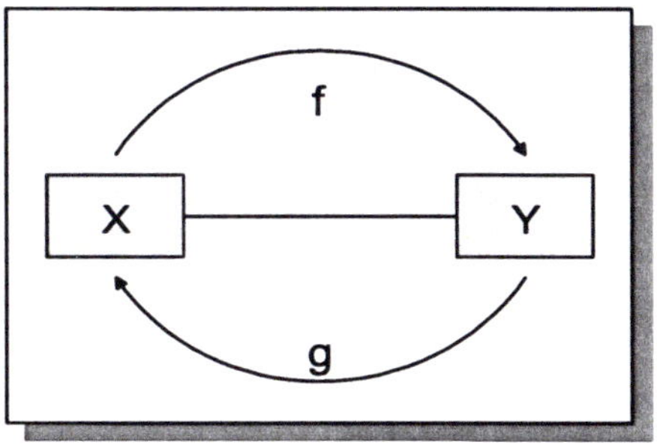

Fig. 20.3.3.1 Abrial's access function.

To exemplify the use of Abrial's access function, Figure 20.3.3.2 shows the association between the classes *book* and library *customer* in library systems.

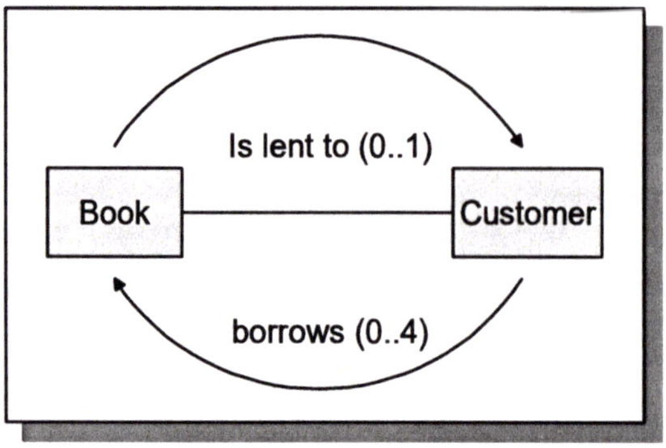

Fig. 20.3.3.2 The association of the classes *book* and *customer* in library systems.

The content depicted by the two Abrial's access functions is intuitively understandable: For one, a book can be lent to at most one library customer, but it can also be *not* lent out. For another, a library customer can borrow at most four books at a time. At a given point in time, however, a customer may have no books checked out. Therefore, the two Abrial's access functions stand for the two roles of a binary association.

An *(association) role* is a statement about the manner or reason for combining an object or entity of the initial class in an association with an object or entity of the target class.

The *role name* or *description of the access function* is the part of a role description that connects the subject of the sentence (namely, the object or entity of the initial class) with the object of the sentence (namely, the object or entity of the target class).

The role name, or description of the access function, thus contains at the minimum a verb and mostly contains the preposition used with that verb for the particular object. In the example of Figure 20.3.3.2, the role that corresponds to the access function f is called "*is lent to,*" and the role that corresponds to the access function g is called "*borrows.*"

20.3.4 Single-Valued and Multivalued Roles, and Total and Partial Roles

Intended learning outcomes: Differentiate between a single-valued role and a multivalued role. Differentiate between a total and a partial role, or access function.

The important properties of an association are made apparent with the following characteristics of its access functions:

A role, or access function, is *single-valued* if each object or entity in the initial class leads to at most one object or entity in the target class. It is *multivalued* if there are objects or entities in the initial class that lead to more than one object or entity in the target class.

The *maximum cardinality* of an access function is the maximum number of objects or entities in the target class to which an object or entity of the initial class can lead.

In practice, the difference between maximum cardinality 1 (one) and maximum cardinality >1 (many) is very important for data-oriented and object-oriented modeling. But the difference between larger maximum cardinalities is of no practical importance.

A role or access function is *total* or *strict*, if each object or entity in the initial class leads to at least one object or entity in the target class. It is *partial* or *nonstrict*, if there are objects or entities in the initial class that lead to no object or entity in the target class.

The *minimum cardinality* of an access function is the minimum number of objects or entities in the target class to which each object or entity of the initial class must lead.

In practice the difference between minimum cardinality 0 (partial role) and >0 (total role) is very important for data-oriented and object-oriented modeling. The difference between greater minimum cardinalities is less important. [150]

The example in Figure 20.3.3.2 shows the usual notation:

- Minimum and maximum cardinalities are in parentheses, and separated by two periods, next to the role names:

  "Role name (minimum cardinality.. maximum cardinality)"

- If there is no role name, it is assumed that the role will be understood. The cardinalities are not placed in parentheses.[151]

Additional conventions allow economical and clear notation:

1 ⇔ 1..1 : If the notation shows only the cardinality 1, then minimum cardinality is the same as maximum cardinality = 1. This very important case indicates hierarchies.

n ⇔ 0..n : If the notation shows only the cardinality n or *, this is the frequent case with minimum cardinality 0 and an in-principle undetermined maximum cardinality of >1. Incidentally, in this case the role is mostly partial.

20.3.5 Association Types

Intended learning outcomes: Differentiate between the 1-to-1 association, the 1-to-n association, the n-to-1 association, and the n-to-n association. Explain various examples of different binary associations.

The *association type* is an abbreviated characterization of the maximum cardinality of the two roles or access functions.

For a binary association, four association types were defined:

- *"One-to-one,"* or *1-to-1 association*: Both roles are single-valued.

---

[150] Incidentally, a mathematical function is a single-valued and total access function.

[151] Role names are generally not given in summaries of previously introduced detailed diagrams. There are also no role names in hierarchical relationships that are understandable based on the context.

- *"One-to-many,"* or *1-to-n association*: The access function f, or the one role, is multi-valued, and the reverse function, or reverse role, is single-valued.
- *"Many-to-one,"* or *n-to-1 association*: The access function f, or the one role, is single-valued, and the reverse function, or reverse role, is multi-valued.
- *"Many-to-many,"* or *n-to-n association*: Both roles, or access functions, are multi-valued.

Figure 20.3.5.1 shows examples of various binary associations and also introduces the notation.

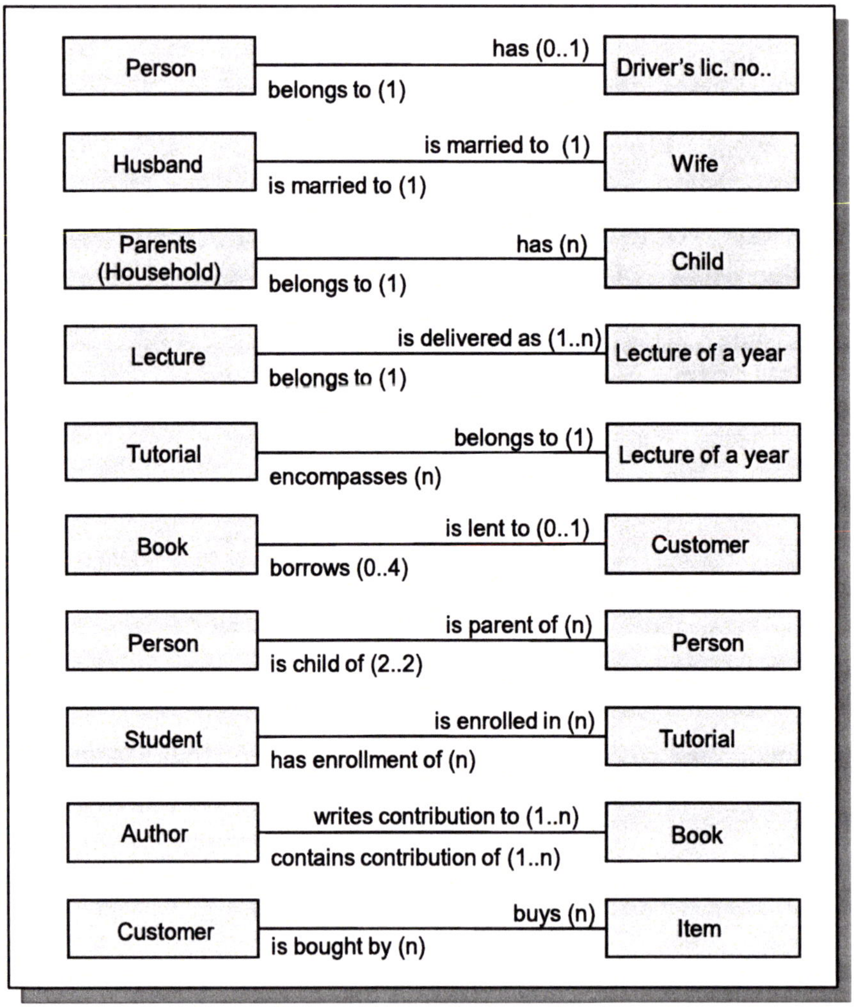

Fig. 20.3.5.1 Examples of different binary associations.

The line connecting the two classes stands for the two roles, or access functions. But each role name, together with its minimum and maximum cardinality, is next to the target class of the role. From top to bottom, the first two examples shown are 1-to-1 associations, the third and fourth are 1-to-n associations, the fifth and sixth are n-to-1 associations, and the last four examples are n-to-n associations.

20.3.6 Breaking up an n-to-n Association; Association Class

Intended learning outcomes: Explain the association class formed by breaking up an n-to-n association into a 1-to-n association and an n-to-1 association.

An *association class* is formed by breaking up an n-to-n association into two 1-to-n relationships.

Figure 20.3.6.1 shows the last three examples in Figure 20.3.5.1 with the new association class that is formed by splitting an n-to-n association.

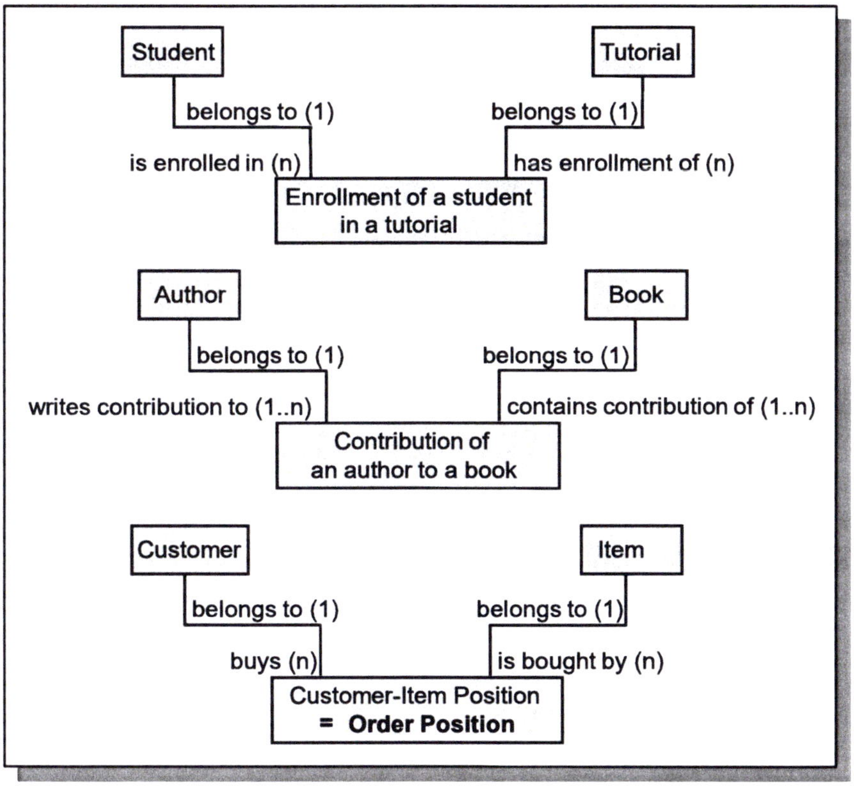

Fig. 20.3.6.1 Breaking up an n-to-n association into a 1-to-n association and an n-to-1 association.

Characteristic of this important phenomenon of breaking up is that a total and single-valued role always leads from the new class to the initial classes. This means that the role name chosen can always be "belongs to." The two initial roles lead from the two initial classes to the new class.

Each n-to-n association can be broken up with the help of an association class. In some cases, it is given naturally; in other cases, it must first be postulated artificially, but after doing so, it is frequently found that, although it is hidden, it really exists in reality or has meaning. This meaning becomes clear especially when attributes appear that cannot be assigned to either of the initial classes and for which the new class provides a location. For example, the attributes *order quantity* and *delivery due date* can be put into the association class *order position* in Figure 20.3.6.1. Order quantity and delivery due date are typical attributes of an order position. As each customer (it is hoped) places several orders, there are also several order quantities and delivery due dates. Consequently, neither of the two attributes can be entered into the class *customer*. The same holds analogously for the class *item*.

Incidentally, association class corresponds to the concept *entity relationship model* in Chen's entity-oriented modeling approach. See here [Chen76].

20.3.7 Different Notations and the Reidentification Key

Intended learning outcomes: Describe the commonly used crow's-foot notation for depicting associations in the entity-oriented approach. Identify the common notation for associations in the entity-oriented approach. Explain the Representation of associations in an object-oriented form, as well as the concept of the re-identification key.

Figure 20.3.7.1 shows a common notation for the entity-oriented approach.

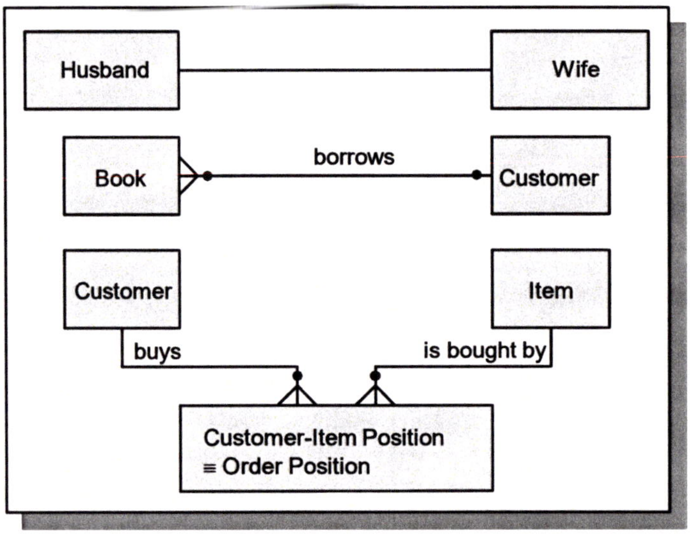

Fig. 20.3.7.1 Commonly used crow's-foot notation for depicting associations in the entity-oriented approach: Crow's-foot symbol for a multivalued role, and solid circle for a partial role.

In the characterization of the association, the multivalued role dominates here, in that role names are shown only for the multivalued role. The requirement is that the name of the single-valued role can be derived from the property of the associated classes. A multivalued role is indicated by a line ending (toward the target class) in a crow's-foot symbol, a foot with three toes (maximum cardinality is not shown). A partial role is represented by a solid circle before the crow's foot.

Figure 20.3.7.2 shows another notation style for the entity-oriented approach.

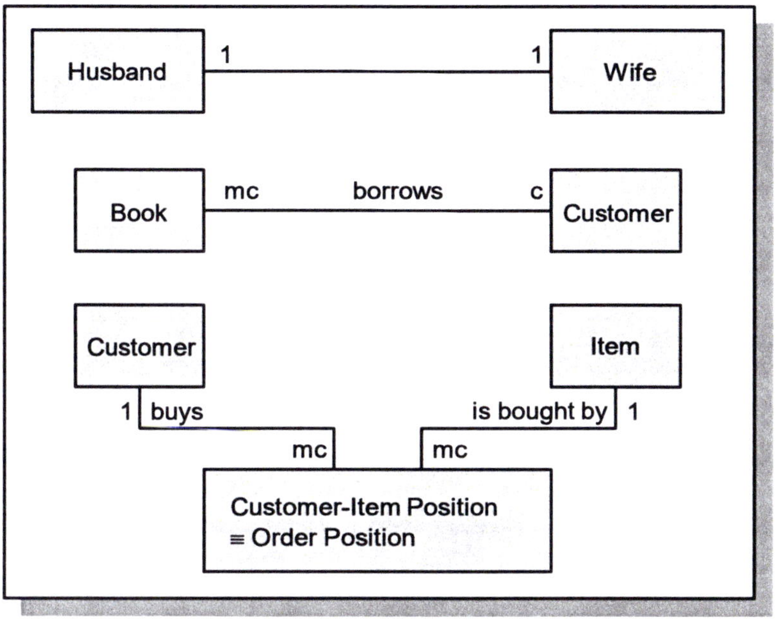

Fig. 20.3.7.2 Common notation for associations in the entity-oriented approach: m for a multivalued role, and c for a partial role.

The crow's-foot symbol is replaced by the letter m (for many). The number 1 stands for a total, single-valued role. If the role is partial, this is indicated by the letter c (for conditional). In contrast to the above definition of the single letter n in the 1-to-n association, the letter m without a c stands here for the total, multivalued role (which occurs rarely), and mc for partial, multivalued role (occurs frequently).

In addition, Figure 20.3.7.2, like Figure 20.3.7.1, shows the phenomenon of reidentification specifically for classes that were introduced to break up an n-to-n association.

A *reidentification key* is a primary key that consists of one single attribute and is equivalent in meaning to a primary key composed of more than one attribute.

The reidentification is usually already expressed in the name of the class, and it is indicated in the list of attributes by the symbol "≡." The original primary key attributes are placed in parentheses.

Figure 20.3.7.3 shows the breaking up of a many-to-many (n-to-n) association using object-oriented notation in accordance with UML (see [UML13]), and complemented by the notation for the primary and secondary keys and the re-identification. As mentioned above, because of the splitting, the single-valued roles can be named "belongs to."

It is typical of the association class that is formed by breaking up an n-to-n association that a *possible* primary key is always the set union of the primary keys of the classes that generated it, which in the example are *customer ID* and *item ID*. In practice, this primary key is often reidentified by a single attribute, in this case *order position ID*. On their own, both *customer ID* and *item ID* are secondary keys. Via *customer ID*, all items ordered by a customer can be obtained. Via *item ID*, all customers that have ordered that item can be identified. Furthermore, In Figure 20.3.7.3, the two attributes mentioned in Figure 20.3.6.1, *order quantity* and *delivery due date*, have also been added.

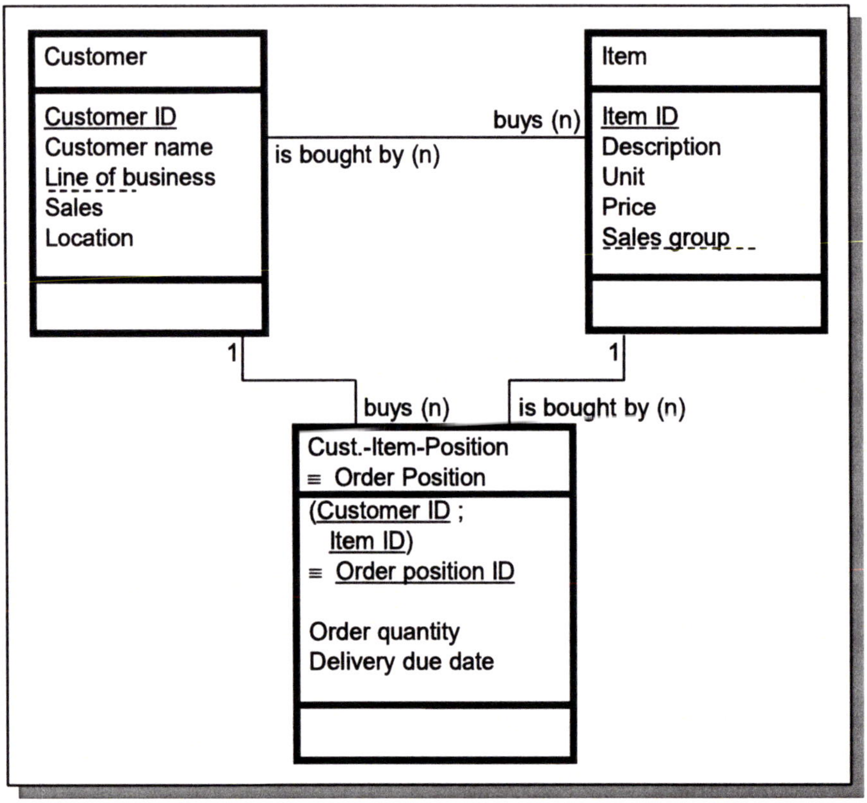

Fig. 20.3.7.3 Representation of associations: object-oriented form.

20.3.8 Breaking Up a Reflexive n-to-n Association

Intended learning outcomes: Explain Breaking up a reflexive n-to-n association, consisting of step 1: Duplicate the class to create its two subclasses, step 2: Classical breakdown and step 3: Merge the generating classes.

In a *reflexive n-to-n association*, a class is in relationship with itself, in that both access functions lead from a certain object in the class to many other objects in the same class.

Reflexive n-to-n associations are quite typically found in operational information systems of a company. Figure 20.3.8.1 shows an example.

Here the aim is to diagram the relationship of "superiors to subordinates" in a company. Both superiors and subordinates are objects in the class "person." The set of all superiors and the set of all subordinates are subsets of the whole set. These subsets are not disjunctive, as most superiors are themselves subordinate to persons of higher rank in the company hierarchy. This is an n-to-n association, as a superior can have many subordinates but a subordinate can also have more than one boss, such as when a secretary works for two bosses as two part-time jobs (two .50 FTE) at the same time in the same company. As Figure 20.3.8.1 shows, breaking up the association is accomplished in a roundabout way by first duplicating the original class, which results in two subclasses that correspond to the subsets mentioned.

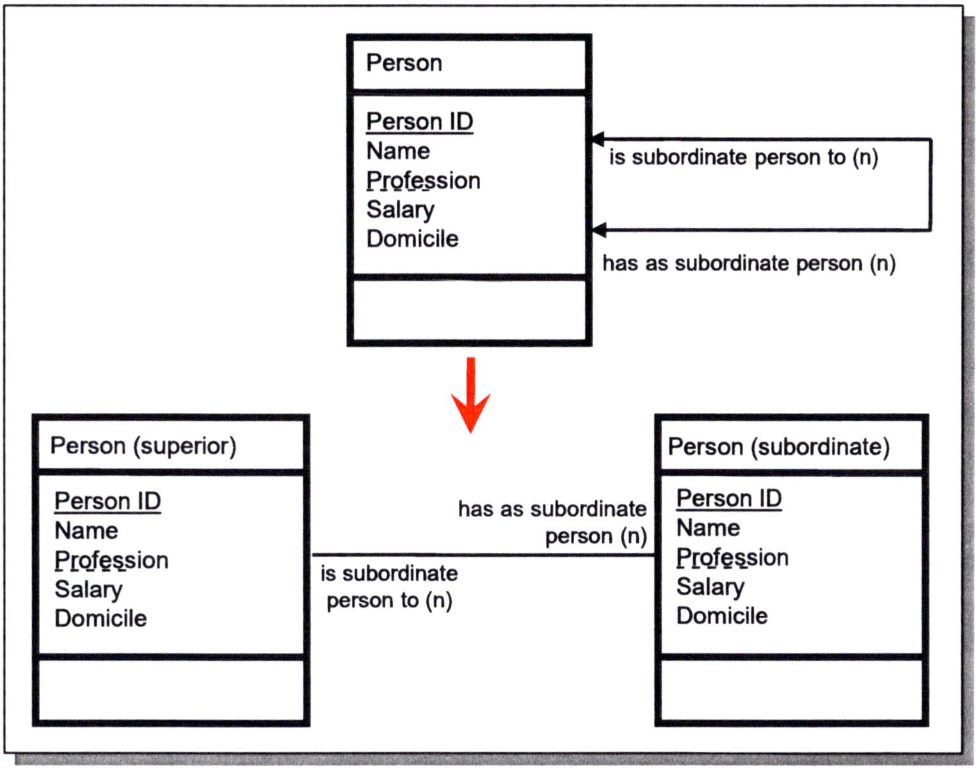

Fig. 20.3.8.1 Breaking up a reflexive n-to-n association. Step 1: Duplicate the class to create its two subclasses.

Figure 20.3.8.2 shows the second step. This is the classical breakdown of the n-to-n association as in Figure 20.3.7.3 above.

The result is the association class, here named personnel structure. As described in Figure 20.3.7.3, a possible primary key is always the set union of the primary keys of the classes that generated it, which in this example is *person ID*, twice, once from the point of view of the superior and once from the point of view of the subordinate. In this concrete case, it probably does not make sense here to consider reidentification of the primary key.

Attributes can be put into the association class that cannot be inserted into the class *person*. These are mainly *FTE in this subordination* (for example, as mentioned above, a secretary works for two bosses as two .50 FTE) and also the *type of subordination* (line, matrix).

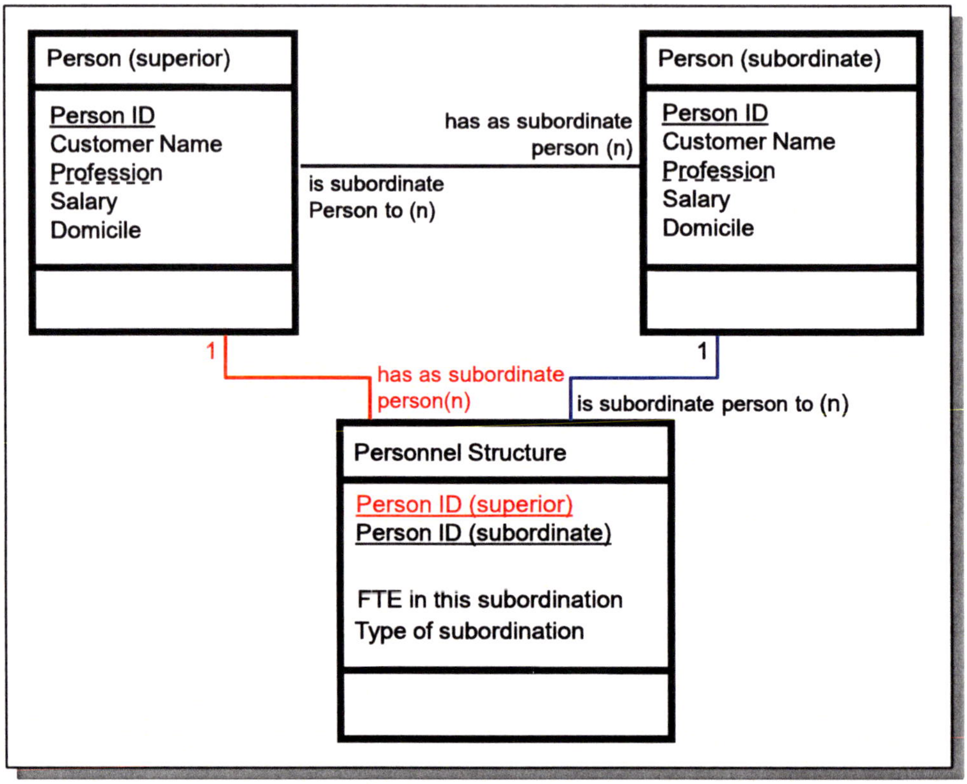

Fig. 20.3.8.2 Breaking up a reflexive n-to-n association. Step 2: Classical breakdown.

Figure 20.3.8.3 shows the final step, namely, merging the two generating classes to the original class. The result yields the typical form, where the original class is connected with the association class via two 1-to-n associations.

Starting from the association "has subordinate," via the first key *person ID* all objects are found in the association class that have the entry of the superior and in the second key *person ID* have the entry of the subordinate as well as the *FTE in this subordination*. But you are not taken to the other data for the subordinate person. Via the second 1-to-n association, or "is subordinate to," you then get to the object in the class "person" that corresponds to the subordinate person and it is there that you find the person's other attributes such as the person's last name, first name, and so on.

Analogously, you can start out from a subordinate to find all of the subordinate's superiors with the *FTE in this subordination*.

As mentioned, in operational information systems of a company the reflexive n-to-n association is typically found. Another example is the corporate group structure, when both a holding and the subsidiary companies appear as objects in the same class *customer*. Even more familiar is its use to model the structure of the bill of material or where-used list (see Figure 17.2.8.2).

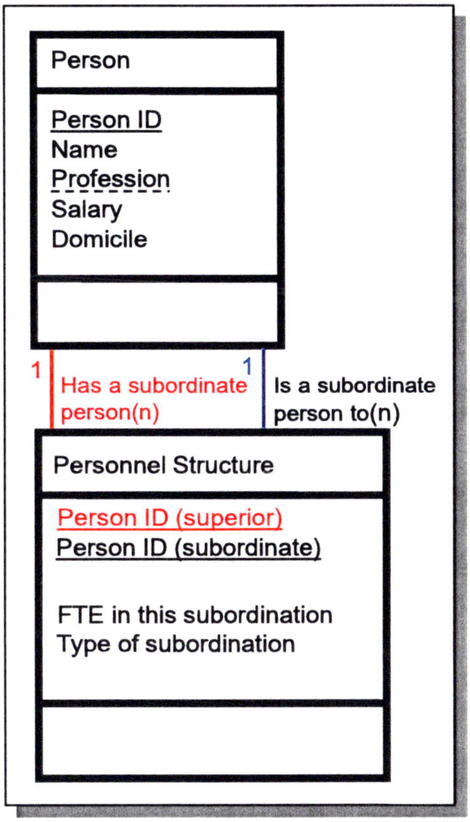

Fig. 20.3.8.3 Breaking up a reflexive n-to-n association. Step 3: Merge the generating classes.

20.3.9 Using the Hierarchical Constructs Based on Basic Object Classes: The Company-Wide Generic Object Model

Intended learning outcomes: Explain the basic object classes and possible subclasses (specializations) of an industrial company. Describe the concept of a company-wide data model, or generic model.

Figure 20.3.9.1 lists the most important basic data or object classes for an industrial company.

"Small classes" at the last line in Figure 20.3.9.1 can be classes for which, because of their few attributes, it would not be worthwhile to be in a class of their own (such as stock location). But often what we are dealing with is only the definition of the domain, or value range, of the attributes (such as time allowed for payment, type of costs). The ID key of the class "code" is made up of a name and numbers organized according to some criteria. The set of attributes comprises, for example, two or three attributes, whose domain is a simple sequence of numerical or alphanumerical characters.Together with the other kinds of hierarchy creation, creating association classes is a powerful tool for the development of operational information systems in a company. During the preliminary and main studies of the system, a company-wide *data model* or *object model* can be generated based on a few basic classes. One then speaks also of a *generic model* or *generic objects*. In the problem-solving cycle of systems engineering, this generic design process is a part of synthesis. For details, see here [Sche98a] or [Schö01, Sections 6.3 to 6.5].

Basic class	Possible subclasses (specializations)
Business partner	Customer, supplier
Item	Finished product, semifinished item, raw material, indirect material
Time	Date, calendar, clock time
Person	Salaried employee, blue-collar worker
Capacity	Internal (assembly, parts production, warehouse), external
Capital asset	Real estate, site, machine, device, tool
Location	Warehouse location, production location, workstation
Account	Financial account, operating account
Numbers/Counters	Bill-of-material position, routing-sheet number, routing-sheet position, parameter number, sequence number
Code	(This class generalizes many "small" classes)

Fig. 20.3.9.1 Basic classes and possible subclasses of an industrial company

20.4 Summary

This chapter discussed fundamental principles of modeling in information management. There exists no single and generally valid model for an operational information system. This is in accordance with the complexity of such sociotechnical systems. The goal is therefore to find a general framework in which the different conceptual models can be arranged and organized. To this purpose, three dimensions in the modeling of operational information systems were defined:

- Hierarchy creation, or the principle of proceeding from the general to the particular: This includes the component or whole-part hierarchy, the specialization hierarchy, and the association hierarchy or determination hierarchy.

- Four views in the modeling: Process view (sometimes also seen as control view), function view, object view (sometimes reduced to data view), and task or organization view.
- The phases of life of the information system: the concept phases preliminary study, main study, and detailed studies; system establishment (also called implementation); system introduction; system operation; and system disposal or decommission.

The basic design elements of data models and object models are object, attribute, and class, and in addition views of classes and the establishing of primary and secondary keys. From the user's point of view, a class is a two-dimensional table, showing the attributes in the rows and the objects of a class in the columns. The different generations of design methods have resulted in different notations for objects, attributes, classes, views, and keys.

Classes, more precisely objects of classes, can be connected with each other in associations. Mostly, exactly two classes are in an association. Such a binary association can be described by two roles, each in one direction. A role, also called access function, is a statement on the type or basis of the association. An access function can be single-valued or multivalued. The maximum cardinality of an access function is the maximum number of objects or entities in the target class to which an object or entity of the initial class can lead. An access function can be total or partial. The minimum cardinality of an access function is the minimum number of objects in the target class to which each object of the initial class must lead.

If both roles, or access functions, of an association are single-valued, the association is said to be a 1-to-1 association. If at least one role is multi-valued, an association is said to be a 1-to-n association, or an n-to-1 association, or an n-to-n association. Each n-to-n association can be broken down into two 1-to-n associations, and that from each of the original class to a new class, which is called the association class. In some cases, this is naturally given, but in other cases it has to be postulated artificially. A possible primary key for the new class is always a combination of the primary keys of the two original classes. That kind of composite primary key can be replaced by a re-identification key.

The chapter also focuses attention on the breaking up of a reflexive n-to-n association. The relatively frequent occurrence of this construct is characteristic of operational information systems. Together with the other kinds of hierarchy creation, creating association classes is a powerful tool for the development of a company-wide generic data model or object model.

20.5 Keywords

1-to-1 association, 831
1-to-n association, 832
association, 829
association class, 833
association hierarchy, 821
attribute, 825
data, 815
data view, 824
domain, 825
entity, 825
entity class, 826
function view, 823
identification key, 827
information, 815
information management, 816
information system, 816
information system architecture, 816

information technology, 815
model, 817
modeling, 817
n-to-1 association, 832
n-to-n association, 832
object, 825
object class, 826
object view, 824
object-oriented modeling, 824
organization view, 824
primary key, 827
process view, 823
reflexive n-to-n association, 836
reidentification key, 835
relational database, 826
role, 830
secondary key, 828
specialization hierarchy, 821
SQL (Structured Query Language), 827
table, 826
task view, 824
view on a class, 827
whole-part hierarchy, 820

20.6 References

Abri74 Abrial, J.R., "Data Semantics," in Klimbie, J. and Koffeman, K., Eds., "Data Base Management," North-Holland Pub. Co., Amsterdam, 1974, pp. 1-60

APIC16 Pittman, P. et al., APICS Dictionary, 15th Edition, APICS, Chicago, IL, 2016

Chen76 Chen, P., "The Entity-Relationship Model —Toward a Unified View of Data," ACM Transactions on DB-Systems, 1, 1976

Espr93 Esprit Consortium AMICE (Eds.), "CIMOSA, Open Systems Architecture for CIM," 2nd Edition, Springer, Berlin, 1993

MeWe18 Merriam-Webster, "Merriam Webster's Collegiate Thesaurus," 2nd Revised Edition, Merriam-Webster, Inc., Springfield, MA, 2018

Norm13 Norman, D., "The Design of Everyday Things," Revised and Expanded Edition, 2013, Basic Books, New York

Sche00 Scheer, A.-W., "ARIS - Business Process Modeling," 3rd Edition, Springer, Berlin, 2000

Sche98a Scheer, A.-W., "Business Process Engineering: Reference Models for Industrial Enterprises," Springer, Berlin, 1998

Sche98c Scheer, A.-W., "Benchmarking Business Processes," in Okino, N., Tamura, H., Fujii, S., Eds., "Advances in Production Management Systems," Chapman & Hall, London, 1998, pp. 133–136

Schö01 Schönsleben, P., "Integrales Informationsmanagement: Informationssysteme für Geschäftsprozesse — Management, Modellierung, Lebenszyklus und Technologie," 2nd Edition, Springer, Berlin, 2001

Spec05 Specker, A., "Modellierung von Informationssystemen: Ein methodischer Leitfaden zur Projektabwicklung," 2nd Edition, vdf Hochschulverlag, Zurich, 2005

UML13 OMG Object Management Group, "Unified Modeling Language Specification," Revised Version 2.5, www.omg.org, created 05.09.2013, accessed August 2019

Index

- **Page numbers in bold type** indicate the page on which a definition of the term can be found.
- Underlined page numbers refer to passages that contribute to an understanding of the term.
- Page numbers in normal type refer to important passages in which the term is used.
- "syn." refers to a synonym, used instead of the term.

1

1PL, first party logistics, **146**
1-to-1 association, **831**
1-to-n association, 733, **831**

2

2PL, second party logistics, **146**

3

3D printing, **51**
3M (muri, mura, muda), **295**
3PL, third party logistics, **146**

4

4PL, fourth party logistics, **146**

A

A item, 88, **478**
ABC (activity-based costing), **699**, 711
ABC category, **479**, 496, 508
ABC classification, 205, 318, **477**, 496, 508, 542, 774
ABC process, **700**
ABC process plan, **701**, 702
ABC-XYZ analysis, **509**
abnormal demand, 455, **476**
Abrial's access function, **829**
acceptance of ERP software, 415
access function
  Abrial's, **829**
  description of the (syn. role name), **830**
  multivalued, **830**
  nonstrict, **830**
  partial, **830**
  single-valued, **830**
  strict, **830**
  total, **830**
accompanying document, 636, 639, 663
  shop packet, **651**
accounting
  cost center, 696
  cost object, 682, **696**
  cost object group, 696
  inventory, **473**
act (tasks in the Deming cycle), **768**, 778
action message, 541
action state, **185**
activation, **599**
activity cost driver, **700**
activity-based cost accounting (ABC) (syn. activity-based costing), **699**
activity-based costing (ABC), 496, **699**, 711
activity-based product cost estimation, 705
actual capacity utilization, **597**
actual cost system, **694**
actual costing, **694**
actual costs, 471, **683**, 695
actual cumulative production figures diagram, **321**
actual demand, **17**, 526
actual load, **35**, 36, 597, 721
  of an operation, **35**
actual order costs, **694**, 695
actual quantity, **694**, 721
actual time, **31**, 658, 735
adaptive smoothing, **443**
adaptive technique, **339**
additive manufacturing, **51**
additive seasonality, **445**
administration cost rate, **56**
administration time, **32**, 298, 551, 555
administrative logistics, 682
advanced logistic partnership (ALP), **99**
advanced planning and scheduling (APS), 247, 405, 411
affinity diagram, 778
agency costs, **75**
agent-based system, 646
aggregate demand, **252**
aggregate forecast, **252**
aggregate plan, **252**
agile company, **48**, 111
agile competitor, **48**
agile manufacturing, **48**
agility, **48**, 190
allocated quantity, 276, 482, **515**
allocation, **638**
ALP model, **99**, 120
alpha factor (syn. smoothing constant α), **436**
alphabetical classification, 542
alternate operation, **738**
alternate routing, **738**
alternate work center, **737**
analysis
  ABC-XYZ, **509**
  basic process, **198**, 232
  cost, **695**
  cost-benefit, 159, 797
  life cycle, **447**

P. Schönsleben, *Handbook Integral Logistics Management*,
https://doi.org/10.1007/978-3-662-65625-9

of solutions, **797**
of variance, 774
multivariate, 775
regression (forecasting technique), **439**
risk, 812
sensitivity, 93, 160, 501, 810, 812
situational, **796**
SWOT, **796**
T, **205**, 341
time series, **429**
value stream mapping, **199**
VAT, **203**, 205, 341
what-if, **252**
analyze (Six Sigma phase), **769**
andon, 310, **775**
ANOVA (analysis of variance), 774
anticipated delay report, **659**
anticipation horizon, **483**, 613, 640
anticipation inventories, **484**
anticipation time, 483, **647**
APICS, V, VII
APO (software product), 406
approach
Japanese, 309
theory of constraints (TOC), **247**
transaction costs, **72**
APS (advanced planning and scheduling), 247, 405, 411
APS software, 247, 397, **405**
arborescent structure (syn. tree structure), 202, 551
upside-down, 203
Aris Tool Set, 819
artifact (syn. product), **10**
artificial intelligence, 656, **743**
ASCM, V, 22
assemble-to-order, 134, **207**, 217, 340
assemblies store, 207
assembling company, 208
assembly, **28**, 46, 305, 378, 700, 725
temporary, **374**, 586
assembly line, **210**
assembly orientation, **202**, 730, 747
assets
current, 44, 64, 137
fixed, 44, 60, 64
operating, **44**
association, **829**
1-to-1, **831**
1-to-n, 733, 740, **831**
binary, **829**
n-to-1, **831**
n-to-n, 733, **831**
reflexive, 740, **836**
reflexive, **829**
association class, 725, **832**
association hierarchy, **821**
association role, **830**
association type, **831**
attribute, 554, **825**
audit, **784**, 812
first-party, **785**
internal, **785**
automatic identification and data capture, **49**
automatic identification and data capture (AIDC), 660
automation, 12, 298
autonomation, **296**
availability (in capacity), 295, **597**
availability test, 241, **638**
available inventory
projected, **516**
available work, **606**
available-to-promise (ATP), 241, **275**, 286, 519
multilevel (MLATP), **277**
average
moving, **432**
weighted, 470, 574
weighted moving, **434**
average cost system, **470**
average costs, **684**
average inventory, **481**, 497

B

B item, **478**
Baan (software product), 403
back (syn. backward) scheduling, **571**
backflush costing, **694**
backflush technique, **657**
backlog, **275**, 520
backorder, 105, **263**
backorder percentage, **263**
backorder quantity, **263**
backshoring, 127
backward integration, **74**
backward scheduling, **571**, 590
badge, 49, **660**, 661
balance
on-hand (syn. physical inventory), **18**, 40, 138, 213, 470, 473, 515
projected available (syn. projected available inventory), **516**
balance sheet, 44, 66, 473
balanced scorecard, 52
balancing
line, **306**, 331
load with capacity, 258, 269
bank, **557**
banking industry, 353, 745
bar chart, 775
bar code, 49, 106, 316, 473, **660**, 672
base demand, **448**
base series, **445**
basic manufacturing step, **31**, 375, 380
basic process analysis, **198**, 232
basic producer, **369**
batch, **31**, 376, 381, 552, 683
batch identification, **381**
batch picking, **663**
batch procedure, 400
in background mode, 727, 734, 738
batch processing (syn. batch production), **224**
batch production, **224**, 320, 369, 393, 400, 612
batch size, 57, **214**, 329, 481, 688
dynamic, 535
large, 376
optimum, **498**, 535
batch size one, 51, **214**, 311, 355, 361, 496
batch size reduction, 299, 311
batch sizing, **495**
considering discount levels, **504**
depending on stockout costs, 510
lead time oriented, **504**
with collective management, **506**
with EOQ formula, 535
with kit materials management, **505**
batch-sizing policy, 268, 311, **535**, 727
BCRP (business continuity and resilience planning), 117

benchmarking, 53, **785**
functional, **786**
generic, **786**
bias, **443**
bid, **249**
bid processing, **238**, 403
bid proposal success rate, 48, **59**
bid statistics, **477**, 722
bid transaction, **477**
big data, **51**
bill of material, **29**, 46, 279, 347, 535, 690, 723, 733, 756, 802, 822
common parts, **343**, 346
design, 536
indented, **730**
maximum, 353
modular, **343**, 346
multilevel, 541, **730**, 802
parameterized, 352, 366
phantom, **343**, 353
extended, 536
planning, **345**, 346
plus/minus, **343**, 346
production, 536
pseudo, **343**
quasi-deterministic, 514
rough-cut, 619
single-level, **729**
summarized, **730**
super, **346**
variant, **342**, 351
bill of parameter-class position, **746**
bill of production equipment position, **739**
bill of resources, **38**, 252
bill of tools, **739**
billed load, 721
billed quantity, 721, 722
billing, **242**
billing address, 720
bill-of-material, 728
bill-of-material explosion, **239**, 266
quasi-deterministic, **267**, 461
bill-of-material position, 657, **728**, 739
as a logistical object, 733
variant, 744
bill-of-material position ID, **734**, 753
binary association, **829**
Black belt, 779
blanket order, 94, **213**, 260, 310, 606
customer, **238**, 250
for capacity, **213**
for goods, **213**
short range, 105, 250, **261**
supplier, 239
blanket order processing, **239**
blanket order quantity
minimum, **213**, 250, 260
blanket release, 105, **261**, 310
Blending (software product), 405
block scheduling, **610**
blowthrough, 343, **536**
book inventory, **471**, 515
bottleneck (capacity), **247**, 287, 576, 597, 599, 617
bottleneck management, 247
bottom line, **44**
double, **166**
triple, 92, 133, **163**
bottom-up rescheduling, **541**
Bpics (software product), 401
BPR (business process re-engineering), 303, **780**, 816
branch, **221**
banking, 353
insurance, 353
branch model, **221**, 227
breadth of qualification, **59**
break-bulk, **138**
breakeven point, **810**
bringing and picking up (customer involvement), **142**
brownfield planning, **152**
bucketed system, 532
bucketless system, 532
budget period, 684, 688, 696
budgeted capacity, **599**
budgeting
resource, **239**
buffer, 375, 384, **557**, 617
inventory, **558**
buffer management, **617**
buffer model, **559**
buffer stock (syn. safety stock), **484**
bullwhip effect, **105**, 295
business, **8**
environmentally responsible, **168**
business continuity and resilience planning, 117
business intelligence, 49
business interruption value, 116
business method, **10**
business model, 358
business object, **10**, 815
business partner, **25**, 716
business partner ID, 719
business plan, **41**
business process, **10**, 185
business process engineering, **193**
business process reengineering (BPR), 303, **780**, 816
business-to-business (B2B), 13, **49**, 139
business-to-consumer (B2C), **49**, 139
buy decision, 87, 208
factors for a, 73
buyer's market, 83, 86, 607
buyer-dominated relationship, 87, **89**
buy-side solution, 90, **95**
by-product, 202, **371**, 727
manufacture of, **371**, 394

C

C item, 88, **478**
CAD (computer-aided design), 353, **750**
CAE (computer-aided engineering), **750**
calculated capacity (syn. rated capacity), **36**
calculation
cumulative projected available inventory, **521**
lead time, 638
load profile, **601**, 603, 630
projected available inventory, **518**, 545
resource requirements, **241**, 337
scheduling projected available inventory, **519**
CAM (computer-aided manufacturing), 353, **750**
campaign, **383**
campaign cycle, **383**
campaign planning, **383**
CAP (computer-aided process design), **750**
capable-to-promise (CTP), 241, **277**, 535
capacity, **34**, 84, 337, 379, 468, 596
(quantitatively) flexible, **211**, 213, 251, 259, 271, 298, 304, 313

  bottleneck, **247**, 287, 576, 597, 599, 617
  budgeted, **599**
  calculated (syn. rated), **36**
  dedicated, **303**
  demonstrated, 598, **658**
  excess, **599**
  flexible capability of, **210**, 271, 304
  idle, 248, **599**
  labor, **596**, 735
  limited, 640
  machine, **596**, 735
  machine-limited, **618**
  maximum demonstrated, **598**
  nonbottleneck, 599
  productive, **598**
  protective, **599**
  rated, **36**, 598, 629, 658
  required, 247, **258**, 619
  safety, **599**
  surge, **600**
  theoretical, **34**, 596
  tool, 600
capacity determination, **596**
capacity management, **244**, 268, 402, 533, 596, 636
capacity of the means of transport, 139
capacity planning
  overall flexibility, **272**, 599
  using overall factors (CPOF), **619**
capacity profile, **34**
capacity requirements planning (CRP), 274, **600**
capacity unit, **34**, 596, 684
capacity utilization, 35, **55**, 57, 562, 597, 651, 735
  actual, **597**
  high, 40, 369, 413, 600
  maximum, 612
  planned, **597**
capital costs, 80, 251, 496
capital goods (syn. investment goods), **10**, 143
capital requirements, 74
CAQ (computer-aided quality assurance), **750**
card
  Kanban, **313**, 316, 333
  move, **314**
  operation, **651**
  production, **314**
  traveling, 313, **639**
  use (syn. move), **314**
carrier
  full-truckload, **148**
  less-than-truckload, 148
  package, **149**
carrying cost, 46, 135, 251, 267, **496**, 535, 536, 687
carrying cost rate, **496**, 501
cascade, **189**
CASE (computer-aided software engineeering), 799
cash and carry, 139
cash flow, 44, 52, 60
cash-to-cash cycle time (performance indicator), **60**
catalyst, **377**
catchment area, 139
cause and effect diagram, 775
cell, **302**
  work, **302**
cell driver, 304, 330
cellular manufacturing, 209, **303**, 330
central point scheduling, **576**
centralized distribution, **134**
centralized goods preparation, **664**
centralized inventory control, **471**
centralized organization, **75**
centralized production, **129**
  for the global market, **131**
  in part, for the local market, **132**
centralized service, **142**
centralized store, **470**
certificate of compliance, **168**
certification, 92, 781
certified supplier, **87**
chain of command, **823**
changeability, **47**, 130, 133
  supply chain, **41**, 48, 219
changeover (syn. setup), **31**
channel of distribution, **84**
characteristic, **753**
characteristic in planning & control, **200**
characteristics
  IHIP, **13**
  of quality of processes, **761**
  of quality of products, **762**
  of the quality of organizations, **763**
charge, **214**, 376, 381
  as an object, 749
chart
  bar, 775
  Gantt, 653, 775, **803**
  milestone, **803**
  organization-oriented process, **194**
chase production method, **259**
check (tasks in the Deming cycle), **768**, 774, 777
checking (order), **242**, 656
chemical production, 371
CIM (computer-integrated manufacturing), **750**
class, **825**
  entity, **825**
  object, **825**
class list of characteristics, **753**
Classical industry, **12**
classification
  ABC, 205, 318, **477**, 542
  alphabetical, 542
  of demand, according to accuracy, **262**
  of demand, according to its relationship, **263**, 530
  XYZ, 318, **479**, 509
classification system, 282, 753, 822
click and collect, 137
cloud computing, 400
CO_2 emissions, 133
code of conduct (CoC), **167**
  supplier (SCoC), **168**
CODP, customer order decoupling point, **46**
coefficient
  of variation, 454, 562
  stockout quantity, **492**
co-entrepreneur, **110**
cold chain, **52**
collaboration
  supply chain, **41**, 218, 422
collaborative planning, forecasting, and replenishment, **104**
collaborative process, **101**
collective materials management, **506**, 637
collective tool, **739**
combined transport, **671**
commission stock, **206**, 663
commodity, **84**, 90
common parts bill of material, **343**, 346
commonality, **30**, 340, 356
  high level, 46
company, **1**, 72
  agile, **48**, 111
  as a sociotechnical system, 1, 97, 237, **789**, 817

performance of a, 33, 185, 244
social competency of a, **57**
company performance, **40**
company strategy, **41**
company-specific software, **409**, 421
compatibility test, **746**
competitive advantage, **42**
competitiveness (of a company), **40**
competitor
agile, **48**
complaint rate, **54**
complete processing, **305**, 332
completion date
earliest, 571, 576, 626
latest, 240, 571, 613, 637
order, **25**, 720
component, **10**, 30, 378, 528, 728
as a business object, **28**
volume variance for a, 696
component (syn. whole-part) hierarchy, 719, **820**
component costs, **687**
component ID, **734**
computer-aided
design (CAD), **750**
engineering (CAE), **750**
manufacturing (CAM), **750**
process planning (CAP), **750**
quality assurance (CAQ), **750**
software engineering (CASE), 799
computer-integrated manufacturing (CIM), **750**
concept
ERP, 225, **235**
for planning & control, **225**, 231
jidoka, **296**, 775
just-in-time, 46, 177, 226, **296**, 311, 335, 384, 402, 565
MRP II, 225, **236**, 319, 402, 723
processor-oriented, 226, 247, **369**, 403, 586
product variety, **204**
variant-oriented, 46, 226, 247, **335**, 403
concurrent engineering, 102, **278**
confidence interval, 433
consigned stocks, **207**
consignment, **207**
consignment inventory, 89, **207**
consolidation, **671**
consortium, 111
consortium marketplace, **95**
constant
extrapolation, **448**
smoothing, α, **436**, 437, **444**
constant demand
forecasting model for, **431**
constraint, **247**, 405, 574
constraint-oriented finite loading, 274, **617**
consumable resource, **378**
consumer, **15**
consumer goods, **10**
consuming the forecast, **526**
consumption, **15**
mean smoothed, 435
consumption controlled materials management, **212**
consumption distribution, **454**
consumption statistics, 425, 722, 728
container logistics, 639
continual improvement (syn. continual process improvement), 104, 310, 779
continual process improvement (CPI), **779**
continuous demand, 84, 130, **203**, 455, 479
continuous flow production (syn. continuous production), **210**
continuous production, 202, **210**, 375, 393, 560
continuous replenishment, **104**
contract
item family, **88**
contract work, **15**, 638, 665
control, 22, **636**
distribution, 106, 524, **662**
inventory, 193, **245**, 326, 523
planning &, **22**, 397, 815
priority, 606
process, **659**
production activity (PAC), **651**
quality, 242, 280, 305, 375, 517, **659**
shop floor, **651**
transportation, **672**
control (Six Sigma phase), **769**
control block, **320**
control board (Leitstand), **666**
electronic, **655**
control flow, 16, **186**
control of operations, 610
execution and, **237**, 240
control view, 819, **823**
convergent product structure, 29, **202**, 225, 335, 400, 723, 741, 802
conversion factor, **641**
conveyor, 210
cooperative process, 117
coordination
order, 48, **242**, 337, 656
supply chain, **41**, 219, 252
Copics (software product), 400, 401
co-product, **299**
corba, 410, 638
core competency, 73, **185**
core process, **185**
Corma (capacity-oriented materials management), 271, 275, 616, **645**
corporate group, 111
corporate group structure, **717**, 839
Corporate Social Responsibility (CSR), **175**
cost
carrying, 46, 135, 251, 267, **496**, 535, 536, 687
of sales, **689**
of work in process, 504
unit, 59, 60, **205**, 265
batch-size-dependent, **495**, 498
cost accounting, 242, 683, 696
cost accumulation, **692**
estimated, 692
ongoing job-order, 692
cost accumulation breakdown, **686**, 727
cost analysis, **695**
cost breakdown structure, 74
of a product, **686**, 727
cost center, **75**, 697, 700, **736**
cost center accounting, 696
cost driver, 688, 702
activity, **700**
cost estimation, **682**, 690, 700, 731, 747
activity-based product, 705

cost identification technique, **694**
cost object accounting, 682, **696**
cost object group accounting, 696
cost of goods sold, **689**
cost of poor quality, 44
cost rate, 481, 683
  carrying, **496**
  for administration, 40, 413
  for fixed internal labor costs, **688**
  for variable internal labor costs, 496, **688**
  per tool use, 688
  planned, **700**
  process, **700**, 702
  standard, **684**, 694
cost type, 686, 727
cost variance, **696**
cost-benefit analysis, 159, 797
costing, 496, 702, 750
  activity-based, **699**, 711
  actual, **694**
  backflush, **694**
  full, 163, 175, 686, 689
  job-order, 31, 242, 661, 673, **682**, 702, 710, 735
  normal, **694**
  product, 682, **696**
  project, 682, **696**
  standard, **694**
  variable, 686, 689
costing software, 696
costing unit, 720
costs
  actual, 471, **683**, 695
  actual order, **694**, 695
  agency, **75**
  average, **684**
  component, **687**
  control, **73**, 80
  direct, **683**, 685
  facility, 143
  fixed, **685**
  full, **685**
  general fixed manufacturing, **688**
  holding (syn. carrying cost), **496**
  inventory, **498**
  labor, external, **687**
  labor, internal, **687**
  manufacturing, 553, **689**
  market transaction, **72**
  negotiation, **73**, 80, 91
  opportunity, **42**, 64, 491, 775, 808
  overhead, **683**, 685
  project, **808**
  running, **808**
  search and initiation, **73**, 80, 91
  semifixed (syn. step-function), **685**
  standard, **684**
  stand-by, 144
  step-function, **685**
  stockout, **491**, 510
  target, 692, 695
  tooling, **688**
  total, of ownership, **80**, 86
    of a project, **808**
  transaction, **72**, 80
  variable, **685**, 697
  variable manufacturing, **689**, 698
costs per unit produced, 690
  variances in the, 696
count point, 268, **320**, 321
counterdeals, 74
CPFR (collaborative planning, forecasting, and replenishment), **104**
CPI (continual process improvement), **779**
CPM, critical path method, 573, 804
critical chain method, 574, 804
critical path, 31, **573**, 575, 576, 580, 590, 804
critical path lead time (syn. cumulative lead time), **33**, 573
critical path method, 573, 804
critical point backflush technique, **657**
critical ratio, **647**
critical to quality (CTQ), **773**
CRM software, **407**
cross-docking, 149, **671**
cross-sectional task, 243
CRP (capacity requirements planning), 274, **600**
CRP (continuous replenishment planning), **104**, 407
CTQ (critical to quality), **773**
cumulative benefit, **810**
cumulative fill rate, **264**
cumulative lead time, **33**, 46, 239, 298, 468, 573
cumulative production figures, **320**
  curve, **320**
  diagram, **320**
  principle, 268, 274, **321**, 324, 600
cumulative projected available inventory calculation, **521**
cumulative quantity per, **29**, 730, 732
current assets, 44, 64, 137
current liabilities, 44
curve
  cumulative production figures, **320**
  inventory, **518**
  learning, 74
  logistics operating, 523, **565**, 570
  operating, for stock on hand, **522**
  saw-toothed, **480**
curve fitting, **428**
custom material, 86
custom semifinished goods, **84**, 90
customer, 10, **25**, 335, 719
  as business object, **716**
  external, 10, **25**, 241, 769
  internal, 10, **25**, 241, 769
customer bid, **249**
customer blanket order, **238**, 250
customer involvement in bringing and picking up, **142**
customer involvement in picking up, **134**
customer order, **25**, 249, 354, 515, 637, 683, 720
  open, **275**
customer order decoupling point, **46**
customer order production, 237, **335**, 607
customer order service, **249**
customer partnership (syn. customer-supplier partnership), **86**
customer procurement order, 207, **212**
customer production order, **212**, 578, 645
customer proximity, 130
customer relationship management (CRM)
  as an information system, **407**
customer satisfaction, 53, 765

customer service, **10**
customer service ratio, **56**, 413
high, 40
customer specification, 337
manufacturing according to, **336**
product according to (changing), **204**, 340, 354, 403, 698
customer support, **10**
customer tolerance time, **18**, 46, 130, 134, 207, 263, 312
customer-supplier partnership, **86**
customization, **46**
late, **46**, 132, 341
mass, **224**, 348, 354, 359
cut optimization, 586
cutting optimization, 51
cyber-physical system, **50**
cycle
campaign, **383**
delivery (syn. delivery lead time), **18**
product life, **15**, 128, **214**, 436, 726, 801
project life, **801**
system life, **793**
cycle counting, **475**
cycle order, **481**, 485, 490
cycle service level (syn. service level), **485**
cycle stock, 481
cycle time, **570**
cyclical component, 430

D

damage, 497
data, **815**, 825
data and control flow, 40, 265, 397
data collection, 50
shop floor, 242, **656**, 682
data definition language, **816**
data dictionary, **816**
data field, **825**
data flow, 16, **186**
data management, **238**, 245
data manipulation language, **816**, 827
data mining, **816**
data model, **824**, 840
data modeling, **824**
data protection, 51, 398
data record, **825**
data store, **187**
data view, 819, **824**
data warehouse, 751, **816**
database, 279, 751, **816**
engineering, **750**
order, 637, 719
product (syn. engineering database), **750**
relational, 400, **826**
database management system, 751, **816**
date, **25**
earliest, **571**
effective, **734**, 738
effectivity (syn. effective), **734**
end, **25**
latest, **571**
operation due, **571**, 721
operation start, **571**, 721
order completion, **25**
order due, **25**, 571, 609, 648
order end, **25**
order start, **25**
order validity, **25**, 717
start, **25**
date of receipt, 524
date of shipment, 524
date of validity, 727
DBMS (database management system), **816**
DDL (data definition language), **816**
decentralized distribution, **134**, 161
decentralized goods preparation, **664**
decentralized inventory control, **471**
decentralized organization, **75**
decentralized production, **129**
for the local market, **132**
in part, for the global market, **132**, 161
decentralized service, **142**, 161
decentralized store, **470**
decision in the problem-solving cycle, **797**
decision support system (DSS), 427
declaration component, 743
decomposition, **429**
decoupling, **19**
decoupling inventory, **19**, 46
decoupling point, **19**, 46
dedicated capacity, **303**
dedicated equipment, **303**
dedicated line, **210**
define (Six Sigma phase), **769**
degree of structuring of the process plan, **210**
deliverable, **800**, 805
delivery (syn. physical distribution), 242
last mile, 136, **148**
point of use, **306**
same day, 136
delivery cycle (syn. delivery lead time), **18**
delivery lead time, **18**, 189, 256
minimum, 482
required by the customer, 278
short, 40, 278, 296, 413
delivery note, 662, 672
delivery policy, **18**
delivery reliability, 313
delivery reliability rate, **56**, 413
high, 40, 267
delivery schedule, 105, **261**
delivery time (syn. delivery lead time), **18**
Delphi method (forecasting technique), **449**
demand, **17**, 262, 426, 448, 526, 637
abnormal, 455, **476**
actual, **17**, 526
aggregate, **252**
base, **448**
company-external, 263
continuous, 84, 130, **203**, 265, 313, 455
dependent, **263**, 527
deterministic, **262**
deterministic dependent, **266**
deterministic independent, **266**
discontinuous, 84, 130, **203**, 455, 528
for a product family, 457
for high-cost items, **265**
for low-cost items, **265**
forecasting model for constant, **431**
highly volatile (syn. discontinuous), 130, **203**
independent, **263**, 342, 457
interplant, **524**

lumpy (syn. discontinuous), **203**, 266, 348, 455, 479, 537
planned, **516**
regular, **203**, 265, 429, 455, 479
safety, 257, **460**, 529, 639
satisfied, 430
service parts, 263, **530**
steady (syn. continuous), **203**
stochastic, **263**
stochastic dependent, **267**
stochastic independent, **267**
unique, **203**, 266, 479
unplanned, 48, 105, **637**
warehouse, **524**
demand chain planning, 411
demand controlled materials management, **212**
demand coverage traceability, **541**
demand filter, **455**
demand fluctuation, 445, 485
demand forecasting, **18**, 238, 250, 424
demand lead time (syn. customer tolerance time), **18**
demand management, **248**, 425
demand model, **428**
demand pattern, 252, 253
seasonal (syn. seasonality), 254, **445**
demand planning, **238**, 249, 425
demand prognosis, **18**
demand time fence, **526**
demand variety, 117, **134**
demand volatility, **130**
Deming, 780
Deming cycle, **768**
Deming Prize, **782**
demonstrated capacity, 598, **658**
maximum, **598**
dependent demand, **263**, 527
dependent electronic marketplace, **95**
depreciation costs, 80
depth of added value depth optimum, 74, 101
depth of product structure, **202**
in the company, **207**
design, 15
process, 46, 98, 377, 723
product, 46, 98, 360, 723, 801
support for decision-making in, 700
re-, 21
supply chain, **71**
design and manufacturing, 15, 31
design bill of material, 536
design changeover flexibility, 47
design for Six Sigma, **769**
design for the supply chain, **279**
design rule, **350**
design structure level, **29**, 280, 536, 725
design/engineering participative, 102, **278**
destruction, 497
detailed objects planning, **237**
detailed planning resource requirements, **240**
detailed planning and scheduling, **236**, 262, 652
detailed production release, **280**
detailed study, **794**, 820
determination (syn. association) hierarchy, **821**
deterministic dependent demand, **266**
deterministic independent demand, 249, **266**, 523
deterministic materials management, **263**, 402
deviation, **430**
mean, **444**
mean absolute (MAD), 430, **435**
mean square, 430
standard, **432**, 444, 454
device, 34, 739
DFSS, **769**
diagram
affinity, 778
cause and effect, 775
cumulative production figures, **320**
actual, **321**
target, **321**
relationship, 778
store throughput, **521**
digitally transmittable item, **206**
direct costs, **683**, 685
direct fixed costs, 685
direct labor, **683**
direct loading (syn. cross docking), **671**
direct material, **84**, 90, 683
direct transport, **146**
disassembly, 21
discontinuous character (of a distribution), 455
discontinuous demand, 84, 130, **203**, 455, 528
discount, **504**
batch sizing considering d. levels, **504**
discounting, 811
discrete manufacturing, **202**, 335
discrete order picking, **663**
discrete order quantity, **536**
dispatching, **652**
dispersion, 428, **430**, 432, 435, 479
disposal, 15
distribution, 242, **662**
centralized, **134**
consumption, **454**
decentralized, **134**, 161
frequency, **428**
normal, 433, 487
of characteristic values for an event, 454
of forecast errors, **454**
of the frequency of the events, 454
Poisson, 454, 488
probability, 433
sample, 432
distribution center, **127**, 469
central, 138
regional, 138, 484
distribution channel, **84**
distribution control, 106, 524, **662**
distribution formula, 683
distribution function, 454, **666**
integral, 486
distribution inventory, **246**
distribution inventory management, 246, 483, 524
distribution network, **20**, 134, 138, 246, 405, 662
feature for designing, **134**
distribution network structure, 137, **138**, 249, 524, 662
distribution planning, **246**, 524
distribution requirements planning, **525**

distribution resource planning, **246**
distribution system (syn. distribution network), **20**, 246
distributor, 138
divergent product structure, 29, **202**, 226, 371, 586, 723
DMADV, **769**
DMAIC, **769**
DMAICT, **769**
DML (data manipulation language), **816**, 827
do (task in the Deming cycle), **768**, 775
domain, **825**
double bottom line, **166**
double order point system, **483**
downside supply chain adaptability (SCOR performance indicator), **60**
downtime, 247, 319, **658**
  scheduled, 597
drawing directory, 752
drawing number, **726**
drop shipping, **135**
DRP (distribution requirements planning), **525**
DRP II (distribution resource planning), **246**
drum, **617**
drum schedule, **617**
drum–buffer–rope, 274, **617**
DSS (decision support system), 427
dual sourcing, **85**, 88, 97
due date, **25**
  operation, **571**, 721
  order, **25**, 571, 609, 648
duration of one unit of harmonized content of work, **308**, 331
dynamic lot sizing, 535

E

E.F.Q.M. (European Foundation for Quality Management), **783**
E.F.Q.M. Excellence Award, **783**
E.F.Q.M. Excellence Model, **783**
EAN (European Association of Numbers), **667**, 726, 828
earliest completion date, 571
earliest date, **571**
earliest start date, 571, 721
early finish date (EF), 574, 804
early order release, 506, 578, 646
early start date (ES), 574, 804
early supplier involvement ESI, **102**
EC number, **753**
echelon, **138**
ecoefficiency, **174**
e-commerce, **49**
econometric model, 429
economic lot size, **498**
economic order quantity (EOQ), 268, **498**, 535
economic value added (EVA), **65**
economies of scale, **130**, 142
economies of scope, **130**, 142
EDD (earliest due date) priority rule, 609, 613
EDI (electronic data interchange), **49**, 94, 638
EDIFACT, **49**, 94, 638, 672
EDM (engineering data management), **749**, 752
EDMS (engineering data management system, **751**
EDP (electronic data processing), 827
effective date, **734**, 738
effectiveness, **19**, 40
  overall equipment (OEE), 304, **598**
effectivity (date) (syn. effective date), **734**
efficiency, **19**, 40
  eco-, **174**
  of the workers, 559
  work center, **36**, 597, 629, 695, 735
efficiency rate, **36**, 597, 661, 695, 735
elasticity of capacity, 271
electronic community, **49**
electronic control board (Leitstand), 603, 617, **655**
  software for, 275, 405, 600
electronic data interchange (EDI), **49**
electronic data processing (EDP), 398, 414, 417, 827
electronic marketplace, **95**
  consortium, **95**
  dependent, **95**
  horizontal, **96**
  independent, neutral, **95**
  private, **95**
  public, **95**
  vertical, **96**
electronic product codes (EPCs), **660**, 672
electronic signature, 49
e-mail, 49
embedding of the store in the flow of goods, **470**
employee, **33**
  empowerment of, 309
  motivation of, 309
  qualification of, 73, 192, 211, 309, 653
employee involvement (EI), 310
employee satisfaction, 765
empowerment of employees, 309
enabler objective, 40, 60, 96
enabling process for engineer-to-order, **358**
end date, **25**, 721, 754
  order, **25**, 720
end item (syn. end product), **28**
end product, 18, **28**, 83, 252, 383, 523
end products store, 206, 374, 612
end user (syn. consumer), **20**, 74, 666
energy (as a resource), **378**
energy management, **171**
energy-intensive industries, **168**
engineering
  business process, **193**
  concurrent, 102, **278**
  simultaneous, **278**
engineering change, 754
engineering change control (ECC), 241, **280**, 347, 753
engineering change number, 734, **753**
engineering data management (EDM), **749**
engineering data management system (EDMS), **751**
engineering database, **750**
engineer-to-order, 11, 134, **207**, 354
  basic, **356**
  complex, **355**
  non-competitive, **355**
  repeatable, **357**

engineer-to-order enabling process, **358**
enterprise, **1**
extended, **20**
lean, **296**
enterprise resources planning (ERP), 225, **235**
entity, 32, 745, **825**
entity class, **825**
entity relationship model (ERM), 833
entrepreneurial objective, **40**, 52, 64, 184, 412
primary, **44**, 52, 60, 64
environmentally responsible business, **168**
EOQ (economic order quantity) formula, **499**, 535
equal weighting, 430
equipment, 34, 379, 724, 739
dedicated, **303**
individual, of a work center, 652
production, **34**, 369
equity, 44
ERP (enterprise resources planning), 225, **235**, 411
ERP concept, 225, **235**
ERP software, 49, 397, **401**, 421, 706, 741, 808
acceptance of, 415, 420
range of implementation of, 415
estimated-cost accumulation, 692
ETA event tree analysis, 116
ethical standard, 92, **167**
Europe, 226, 309, 310, 403
European Foundation for Quality Management (E.F.Q.M.), **783**
European pool pallet, **667**
evaluation of solutions, **797**
event, **185**, 516
event tree analysis, 116
exception list, **479**
exception message, **541**, 572
excess capacity, **599**
excess inventory, 105
excessive stress (syn. overstraining), **295**
execution and control of operations, **237**, 240
expected value, **430**
expedite, **649**
expert system, 427, **742**
expiration date, 249, 622
explosion
bill of material, **239**, 266
quasi-deterministic, 461
requirements, **533**
exponential smoothing
first-order, **435**, 437, 465
second-order, **441**
exponential weighting, 430
extended enterprise, **20**
extended workbench, 606
external constraints, **800**
external customer, 10, **25**, 241, 716, 769
external setup (time), **301**
extranet, 49
extrapolation, 427, **430**, 438
trend (forecast technique), **447**
extrapolation constant, **448**
extrinsic forecasting techniques, **427**

F

facility, **34**, 127
multipurpose, 375, 389
production, **34**
single-purpose, 375, 389
facility costs, 143
facility layout, **208**, 375
facility location, **34**, 127
facility location planning, **127**
fact base, **742**
factor
conversion, **641**
lead-time-stretching, **577**, 648
location, **154**
potential, 271
safety, 460, **486**
particular, 486, 492
scrap, **54**, 528
split offset, **582**
yield, **54**, 304, 528, 598
factor rating, 92, **159**, 797
factor splitting, **582**, 584, 604
fail-safe technique, **776**
failure mode and effects analysis (FMEA), 115, 774
family
product, **30**, 204, 209, 346
fast moving item, 136
fault tree analysis (FTA), 115
FDA (food and drug administration), 375, 389
feature
for designing distribution networks, **134**
for designing production networks, **130**
for designing retail networks, **139**
for designing service networks, **142**
for designing transportation networks, **147**
of the characteristic in planning & control, **200**
feedback, 237, 656, 682
feeder workstation, **617**
FIFO (first in, first out), **470**, 609, 640
file, **826**
file transfer protocol (FTP), 49
fill rate, 41, **56**, 263, 489, 493
cumulative, **264**
high, 40, 135, 143
final assembly schedule (FAS), 256, 257, **340**, 344
fine planning, **237**
finish date
early (EF), 574, 804
late (LF), 574, 804
finished good (syn. end product), **28**
finishing lead time, **340**
finishing schedule, **340**
finish-to-order (syn. assemble-to-order), 46, **207**
finite forward scheduling, 277, 558, **653**, 677
finite loading, 272, 599, **608**
constraint-oriented, 274, **617**
operations-oriented, 274, **608**
order-oriented, 274, **612**
order-wise, **275**
rough-cut, **624**
firm, **1**, 72
firm planned load, **603**
firm planned order, **639**
first in, first out (FIFO), **470**, 609, 640
first-order exponential smoothing, **435**, 465
first-party audit, **785**
first-party logistics provider, **146**
first-pass yield, **771**
first-tier suppliers, **83**
fitness for use (Juran), **759**
five focusing steps, **248**
five Ss, **310**
fixed assets, 44, 60, 64

fixed costs, **685**, 688
direct, 685
fixed labor costs
for machinery, 735
for personnel, 735
fixed liabilities, 44
fixed material costs, **687**, 700
fixed order quantity, **535**
fixed-location storage, **470**
fixed-position layout, **208**
fixed-position manufacturing, **208**
flexibility, **47**, 269
in achieving customer benefit, 211, 348, 413
by product and process innovation, 40, 110
in the use of resources, 40, 211, 413
of the order due date, 213, 271
overall capacity planning, **272**, 599
to enter as a partner in supply chains, 40, 60, 97, 109, 413
to link others in a supply chain, 97
flexibility potential, 53
flexible capability of capacity, **210**, 271, 304
flexible capability of the production infrastructure, **211**, 375
flexible capacity (syn. quantitatively flexible capacity), **211**
flexible manufacturing systems (FMS), 299, **324**, 661
flexible workforce, **211**
floor stock, **470**
flow control, 224
flow manufacturing, **210**
flow of control, 16, **186**
flow of data, 16, **186**
flow of goods, 16, 40, **186**, 265
embedding of the store in the, **470**
flow rate, **570**
flow resource, **375**, 380, 388
flow shop, **210**, 386
flowchart, 778
fluctuation inventory, **485**
fluctuations (seasonal), **445**
FMEA (failure mode and effects analysis), 115, 774
focus forecasting, **453**
focused factory, **302**
food and drug administration (FDA), 375
food production, 376
forecast, **18**, 46, 687
aggregate, **252**
product familiy, **252**, 345
production, **345**
scenario -s, **449**
forecast bias, **443**
forecast consumption, **526**
forecast controlled materials management, **212**
forecast demand (object class), **461**
forecast distance, 448
forecast error, **424**, 430, 443
distribution of, **454**
forecast horizon, **428**
forecast interval, **428**
forecast order, **212**, 376
forecast value, 433, 448
forecasting, 238, 250, **424**
forecasting model
for constant demand, **431**
trend, **438**
forecasting technique, **424**, 443, 463
Delphi method, **449**
first-order exponential smoothing, **435**
linear regression, **439**
moving average, **432**, 437
pyramid, **345**
regression analysis, **439**
second-order exponential smoothing, **441**
trend extrapolation, **447**
forecasting techniques
extrinsic, **427**
future-oriented, 260, **427**
graphic, **427**
historically oriented, **426**, 684
intrinsic, **427**
intuitive, **427**
mathematical, future oriented, **427**
mathematical, historically oriented, **427**
qualitative, **427**
quantitative, **427**
subdivision of, 426
formula (as an object), **746**
formula (syn. recipe), **380**
formulation of objectives, **796**
unstable, 797
Forrester effect, **105**
forward chaining, 743, 744
forward flow scheduling, **581**
forward integration, **74**
forward scheduling, **571**, 573, 590
fourth-party logistics provider, **146**
free trade area (FTA), **76**
frequency
of customer demand, **203**, 213, 217, 265, 297
of order repetition, **213**, 339
of the events (distribution of the), 454
frequency distribution, **428**
FTA (fault tree analysis), 115
FTA (free trade agreement), **76**
FTA (free trade area), **76**
fulfillment paradigm, **781**
full costing, 163, 175, 686, 689
full costs, **685**
manufacturing, **689**
of goods manufactured, 687
of goods sold, **689**
full manufacturing costs, **689**
full truckload lot, **148**
full-truckload carrier, **148**
function, **8**
distribution, 454, **666**
information, **667**
integral distribution, 486
inverse, 487
packaging, **666**
protective, **666**
sales, **667**
sales promotion, **667**
service, **486**
use, **667**
function model, **824**
function view, 819, **823**
functional benchmarking, **786**
functional layout (syn. process layout), **209**
function-oriented modeling, **824**
funnel model, **559**, 640
future-oriented forecasting techniques, 260, **427**, 447

G

gamma factor (syn. smoothing constant γ), **444**

Gantt chart, 653, 775, **803**
gap method, 92
gateway work center, **646**
gemba, **310**
gembutsu, **310**
genchi genbutsu, **310**
general and administrative expenses (G&A), **689**
general fixed manufacturing cost, **688**
generative technique, **339**, 348, 353, 367
generic benchmarking, **786**
generic Kanban, **618**
generic model, 840
generic object, 840
geographic information system (GIS), 139
GIS (geographic informations system), 139
global market
 centralized production for the, **131**
 in part decentralized production for the, **132**, 161
global measure, **52**
global production plan, 161
global sourcing, **85**
GMP (good manufacturing practices), 131, 375
good, **10**
 intangible, **12**, 51, 206
 packaged, **665**, 667, 671
 tangible, **12**, 51
good quantity, 322, 659
goods, **10**
 capital (syn. investment), **10**, 143
 consumer, **10**
 custom semifinished, **84**
 investment, **10**, 20, 127
goods flow, 16, **186**
goods management (see also materials management), **244**
goods preparation
 centralized, **664**
 decentralized, **664**
goods store, **18**, 186
graphic forecasting techniques, **427**
graphical user interface (GUI), 400
Green belt, 779
greenfield planning, **152**
gross margin, 44
gross requirement, 252, **530**
gross requirements planning, **239**, 259
group production, 209
group technology (GT), **209**

H

hardware, 397, **815**
harmonizing
 the content of work, **306**, 331, 502
 the product range, **300**
hedge, **256**
heijunka, 295, **300**
hierarchy
 association, **821**
 component (syn. whole-part), 719, **820**
 determination (syn. association), **821**
 specialization, 719, **821**
 whole-part, 719, **820**
hierarchy creation, 819, **820**
high-cost item, 88, **205**
 demand for, **265**
highly volatile demand (syn. discontinuous), 130, **203**
high-variety manufacturing, **336**
high-volume line production, **209**, 375
histogram, **428**, 775
historically oriented forecasting techniques, **426**, 684
holding cost (syn. carrying cost), **496**
Hollerith, 398
horizon
 forecast, **428**
 planning, **239**, 256, 602
horizontal (electronic) marketplace, **96**
House of Quality, **771**
human being, 295, 326, 424, 789
hybrid production method, **259**

I

IBM (International Business Machines), 398, 401
ID (syn. identification key), **827**
 meaningful, **828**
identification (of the store), **469**, 722
identification key, **827**
idle capacity, 248, **599**
idle time, 301, 306, **558**, 617
IED (inside exchange of dies), **301**
IF clause, **350**, 743, 744
IGES, 49
IHIP characteristics, **13**
ILOG (software product), 406
improve (Six Sigma phase), **769**
inbound stockpoint, 306, 315
income statement, 44
incoterms, **18**
indented bill of material, **730**
indented where-used list, **732**
independent demand, **263**, 342, 426, 457, 526
 deterministic, 249, 523
 stochastic, 265, 530
independent demand (object class), **461**
indicator, 764
 leading, 427
 logistics performance, **52**, 565
 performance, **52**, 174
indirect material, **84**, 90
indirect transport, **146**
individual product, **204**
individualization of products, 11, 50
Industrial Internet Consortium, 50
industrial product-service system, **13**
industrial symbiosis, **172**
industrialization, **12**
industry, **12**
 banking, 745
 basic producer, **369**, 404
 classical (syn. industry), **12**
 energy-intensive, **168**
 insurance, 745
 process, 222, 237, **369**, 394, 404
 service, **12**
Industry 4.0, **50**
inference engine, **742**, 745
infinite loading, 272, 599, **600**
 order-oriented, 273, **600**
 order-wise, 274, **607**
 rough-cut, **622**
information, **815**
information function, **667**
information logistics, **22**, 397

information management, **816**
information system, 397, **816**
  geographic (GIS), 139
  IT-supported, 651, **816**
  knowledge-based, 350, **742**
information system architecture, **816**
information technology, 48, 50, 397, **815**
in-house part, **28**, 725
in-house parts store, 207
initial investment, **808**
in-memory computing, 400
innovative power, 40, **110**
in-process inventory, **18**, 40, 213, 473
  of work waiting, 565
inside exchange of dies (IED), **301**
insourcing, **72**, 606
insurance branch, 335, 353
insurance industry, 745
intangibles, **12**
integral distribution function, 486
integral logistics management, **21**
integrated resource management, **246**
integration
  backward, **74**
  forward, **74**
  of procedures, company-wide, 749
  vertical, **74**
interface, 188, 638, 825
intermediate product, **28**, 32, 378, 725
intermittent production, **209**
internal audit, **785**
internal constraints, **800**
internal customer, 10, **25**, 189, 241, 716, 769
internal labor costs, 691
internal order, **25**, 189
internal setup (time), **301**
internet, 49, 410, 638, 672
Internet of things, **51**
Internet protocol, **49**
interoperation time, **32**, 38, 298, 551, 554, 610
  elements of, 554
interplant demand, **524**
interplant order, **249**
intranet, 49
in-transit inventory, **18**, 40, 246
in-transit lead time, **524**
in-transit merge, **149**
intrinsic forecasting techniques, **427**
intuitive forecasting techniques, **427**
intuitive model, 429
inventory, **18**
  anticipation, **484**
  average, **481**, 497
  book, **471**, 515
  consignment, 89, **207**
  decoupling, **19**, 46
  distribution, **246**
  excess, 105
  fluctuation, **485**
  in-process, **18**, 40, 213, 473
    of work waiting, 565
  in-transit, **18**, 40, 246
  lot-size, **495**
  periodic, **474**
  perpetual, **471**
  physical, 468, **473**, **515**, 517, 722
  pipeline, **246**
  point of use, **306**
  projected available, **516**, 530, 637
  returns, **659**
  supplier-managed SMI (syn. VMI), **104**
  transportation, **18**, 138, 246
  two-bin system, **639**
  valuation method, **470**
  vendor-managed (VMI), **104**
  vendor-owned, 137, **207**
  zero, **296**
inventory accounting, **473**
inventory adjustment, **473**
inventory buffer, **558**
inventory control, 193, **245**, 326, 523
  centralized, **471**
  decentralized, **471**
inventory control principle, **471**
inventory costs, **498**
inventory curve, **518**
inventory investment, **19**
inventory issuance principle, **470**
inventory issue, 470, 471, **476**, 523
inventory management, **245**, 471
inventory plan, **252**, 287
inventory policy, **251**
inventory range, **565**
inventory receipt, 471, **476**
inventory transaction, **471**, 497, 722
inventory turnover, 651
  stock, **55**, 105, 523
  work-in-process, 55
inventory valuation, **473**
inventory value, 479, 497
investment
  initial, **808**
  inventory, **19**
  supply chain initiative (SCI), **44**, 46, 65, 220
investment goods, **10**, 20, 127
IPPS (software product), 403
Ishikawa diagram, 775
island production, **208**
ISO, 764
ISO 9000.2015 (series of standards), **781**
issuance
  planned, 657
  unplanned, 657
issuance principle, **470**
  FIFO, **470**
  LIFO, **470**
  order specific, **471**
  unordered, **470**
issue, 471, **476**
  cumulative production figure, 320
  item, **717**, 721
IT (syn. information technology), 50, 397, **815**
item, **10**, 28, 350, 378
  A, 88, **478**
  as a business object, **28**
  as an order position, **717**
  B, **478**
  C, 88, **478**
  digitally transmittable, **206**
  end (syn. end product), **28**
  fast moving, 136
  high-cost, 88, **205**
  low-cost, 88, **205**
  master schedule, **256**, 530, 619
  MRO, 18, **29**, 90
  nontransportable, **205**
  order, **25**, 42, 356, 717
  parent, **723**
  portable, **206**
  transportable, **206**
item characteristic, **753**
item description, 721, **726**
  position-specific, 721
item entry, **717**

item family, **29**, 37, 88, 426, 456, 743
item family contract, **88**
item family ID, 722
item ID, 721, **726**, 737
  meaningful, **726**, 828
item issue, **717**, 721
item master file, **726**, 752
item master record, **726**
item receipt, **717**, 721
item type, **28**, 478, 726
iterative system development, **798**
IT-supported information system, 397, 651, **816**

J

J.D. Edwards (software product), 401
Japan, 111, 297, 309
Japanese way of thinking, 309
Java, 410, 638
JDA solutions (software product), 405
jidoka concept, **296**, 775
JiT (Just-in-Time), 46, 177, 309
job enlargement, **211**
job enrichment, **309**
job order (syn. production order), **25**
job shop (syn. job shop production), **209**
job shop layout (syn. process layout), **209**
job shop production, 32, **209**, 301, 330, 386, 502, 550
job-order cost accumulation
  ongoing, 692, 694
job-order costing, 31, 242, 673, **682**, 702, 710, 735
joint replenishment, **505**
joint venture, 111, 152, 153
just-in-case mentality, 271
just-in-time concept, 46, 177, 226, **296**, 335, 384, 402, 565
just-in-time logistics, **309**

K

kaizen, 310, **780**
Kanban, 90, 267, 274, **313**
  generic, **618**
  one-card system, **315**
  two-card system, **314**, 639
Kanban card, **313**, 316, 333
Kanban feedback loop, **314**, 324, 326
Kanban rules, **315**
Kanban technique, **314**, 384, 456, 637
Keiretsu, 111
key
  identification, **827**
  primary, **827**, 828
  secondary, **828**
kind of order, **25**, 243, 720
kingdoms
  centralistic, 75
  within departments or foreman areas, 75
kit, **505**
kit materials management, **505**
knowledge, 48, **742**
knowledge-based information system, 350, **742**

L

labor capacity, **596**, 735
labor costs
  external, **687**
  external, fixed, **687**
  external, variable, **687**
  fixed, for machinery, 735
  fixed, for personnel, 735
  internal, **687**, 691
  internal, fixed, **688**
  internal, variable, **688**
  variable, for machinery, 735
  variable, for personnel, 735
labor hours, **596**
landed costs, **80**
large batch procurement, **214**
large batch production, **214**
last in, first out (LIFO), **470**
last mile delivery, 136, **148**
late customization, **46**, 132, 266, 341
late finish date (LF), 574, 804
late start date (LS), 574, 804
latest completion date, 571
latest date, **571**
latest start date, 571
layout, **196**, 231
  facility, **208**, 375
  fixed-position, **208**
  functional (syn. process), **209**
  job shop (syn. process), **209**
  process, **209**, 302
  product, **209**, 302
lead time, **18**, 31, 33, 322, 550, 569, 584, 721, 736
  critical path (syn. cumulative lead time), **33**, 573
  cumulative, **33**, 46, 239, 263, 298, 468, 495, 573
  delivery, **18**
    required by the customer, 278
    short, 278
  demand (syn. customer tolerance time), **18**
  finishing, **340**
  in-transit, **524**
  manufacturing, **31**, 550
  minimum, 646
  production, **31**
  purchasing, **32**, 310
  remaining, 609, 613, 677
  replenishment, **481**, 639
  safety, **485**
  short, 64, 329, 335, 397, 502
    in the data and control flow, 40, 397, 413
    in the flow of goods, 40, 265, 413
  supplier, **32**
leading indicator, 427
lead-time calculation, 602
lead-time offset, **33**, 277, 532, 621, 734
lead-time reduction, 312
lead-time scheduling, **568**, 608
lead-time-oriented batch sizing, **504**
lead-time-stretching factor, **577**, 648
lean enterprise, **296**
lean production, 190, 226, **296**
learning curve, 74
learning organization, **104**
Leegoo Builder (software product), 403
legacy system, **401**
Leitstand, **655**
Leitstand (electronic control board), 603, 617, 636
  software for, 275, 600
length of order cycle, 490
  optimum, **500**, 535

less-than-truckload carrier, 148
level (syn. structure level), **29**
level of service (syn. service level), **485**
level production method, **259**
level schedule, **259**, 606
leveling of the production, **300**
liabilities
current, 44
fixed, 44
life cycle
product, **15**, 128, **214**, 436, 726, 801
project, **801**
system, **793**
life-cycle analysis, **447**
LIFO (last in, first out), **470**
LIFO issuance principle, **470**
line, **210**
assembly, **210**
dedicated, **210**
L-shaped, 210, **303**
product (syn. product family), **30**
production, **210**
U-shaped, 210, **303**
line balancing, **306**, 331
line manufacturing (syn. line production), **306**
line production, **210**, 302, 386, 502, 550, 558, 660
continuous, **210**, 375, 393
high-volume, **209**, 375
single-item-oriented, **209**, 570
linear programming (LP), 153, **161**
linear regression (forecasting technique), **439**
linearity, **259**
linking of information systems, 279
load, **34**
actual, **35**, 324, 658, 721
of an operation, **35**
billed, 721
firm planned, **603**
operation, **34**, 533, 552, 553, 601, 658
planned, **603**
process, **761**
run, **34**, 552, 684, 721
scheduled, **603**
setup, **34**, 496, 552, 582, 684, 721, 737
standard, **35**, 658, 684
of an operation, **35**, 684, 721
work center, **35**, 601
load building, 149, **668**, 671
load center, **34**
load filter, **641**
load forecast, 684
load leveling, **606**
load limit, **641**
load periods, 603
load profile, **35**, 62, 553, 568, 601, 620
product, **38**, 252, 621
load profile calculation, **601**, 603, 630
load projection (syn. load profile), **35**
load traceability, **606**
load unit, 146, **668**
loading
finite, 272, 599, **608**
operations-oriented, 274, **608**
order-oriented, 274, **612**
order-wise, **275**
rough-cut, **624**
infinite, 272, 599, **600**
order-oriented, 273, **600**
order-wise, 274, **607**
rough-cut, **622**
machine, **653**
loading percentage, **641**
load-oriented order release (loor), 274, 616, **640**, 675
local market
decentralized production for the, **132**
in part centralized production for the, **132**
local measure, **52**
local sourcing, **85**, 98, 101
local supplier, 310
world-class, 101
location (syn. facility location), **34**, 127
picking, 142
storage, 52, **469**, 722
location configuration, **152**
location factor, **154**
location planning, **127**
location selection, **152**
logistics, **16**
administrative, 682
container, 639
effective, 190
information, **22**, 397
just-in-time, **309**
pull, **189**, 212
push, **190**, 212
reverse, **21**, 134
logistics management, **16**, 21
logistics network, **20**
logistics operating curve, 523, **565**, 570
logistics performance indicator, **52**, 565
logistics system, **185**, 187
long-term planning, 105, **236**, 238, 248, 319, 388, 661
loop
Kanban feedback, **314**, 324
product structure with, **215**, 376, 387, 753
product structure without, **215**, 735
loops in the order structure, **215**
loor (load-oriented order release), 274, 616, **640**
lot, **31**, 376, 381, 552
(full) truckload, **148**
split, **581**
lot control, 376, **381**, 470, 683, 748
lot number (syn. batch identification), **381**
lot size (syn. batch size), 57, **214**, 299, 311
dynamic, 535
economic, **498**
large, 376
lot size one (syn. batch size one), 51, 209, **214**, 311, 355, 361, 496
lot sizing policy (syn. batch-sizing policy), **535**
lot splitting, **581**
lot traceability, **214**, 376, 381
lot-for-lot, 311, **535**
lotless procurement, **214**
lotless production, **214**
lot-size inventory, **495**
lot-sizing, **495**
low-cost item, 88, **205**
demand for, **265**
low-level code, **30**, 387, 530, 693, 727
low-variety manufacturing, **338**, 457
L-shaped line, 210, **303**
lumpy demand (syn. discontinuous), **203**, 266, 348, 455, 479

M

machine, 34, 724, 739
machine capacity, **596**, 735
machine center, **34**
machine hours, **596**, 629
machine loading, **653**
machine-limited capacity, **618**
machinery costs (syn. labor costs for machinery), 735
MAD (mean absolute deviation), 430, **435**, 487
main study, **794**, 820, 840
maintenance, 144
  preventive, 34, **319**, 597
maintenance, repair, and operating (MRO) supplies, **29**
make decision, 86, 127, 208
  factors for a, 73
make-or-buy decision, **72**, 127
make-to-order, 134, **207**, 348
make-to-stock, **206**, 374
Malcolm Baldrige National Quality Award, **783**
management, **10**
  bottleneck, 247
  buffer, **617**
  capacity, **244**, 402, 533, 596, 636
  customer relationship (CRM), **407**
  data, **238**, 245
  demand, **248**, 425
  demand controlled materials, **212**
  energy, **171**
  information, **816**
  integral logistics, **21**
  inventory, **245**, 471
  logistics, **16**
  material group, **91**
  materials, **243**, 533, 550, 636, 646
  of distribution inventory, 246, 483, 524
  operations, **16**
  process, 187, 700
  procurement, 16, 83
  product data (PDM), **749**
  project, 280, 337, 355, 418, **790**
  quality, according to ISO8402, **766**
  resource, **246**, 400
  risk
    active, 777
    project, 812
    supply chain, **112**
  stores, **245**, 469
  supplier relationship (SRM), **94**
  supply (syn. procurement), 83
  supply chain, **21**
  supply chain event (SCEM), **49**, 66, 117
  time, **550**, 736
  total quality (TQM), 310, 759, 768, **781**
  value-added, **17**
management principle (storage), **470**
mandatory objective, **796**
MANOVA (multivariate analysis of variance), 775
manufacture of by-products, **371**, 394
manufacturer
  mixed-mode, **645**
  original equipment, 13, **83**
manufacturing, 15, 22
  according to customer specification, **336**
  cellular, 209, **303**, 330
  discrete, **202**, 335, 401
  fixed-position, **208**
  flow shop, **210**
  high-variety, **336**
  line (syn. line production), **306**
  low-variety, **338**, 457
  mixed, **645**
  process, **369**
  project, **208**, 573
  re-, 21
  repetitive, **224**, 293, 306, 340, 694
manufacturing calendar, **567**
manufacturing costs, 553, **689**, 727
  variable, **689**, 698
manufacturing date, **567**
manufacturing environment (syn. production environment), **206**
manufacturing execution system (MES), 172, **651**
manufacturing lead time, **31**, 550
manufacturing order (syn. production order), **25**
manufacturing planning and control (MPC), **22**
manufacturing process, **196**
manufacturing resource planning (MRP II), 225, **236**
manufacturing segmentation, **301**
manufacturing step (basic), **31**, 375, 380
manufacturing strategy, **251**, 253
manufacturing systems
  flexible (FMS), 299, **324**, 661
MAPICS (software product), 401
Marcam (software product), 405
market, 41, 699
  buyers', 607
  sellers', 612, 616
market driven, 335, 360
market share, 42, 44, 64, 73
market specificity of products, 130
market transaction, **72**
market-oriented relationship, **85**, 89
marketplace (electronic), **95**
MAS90 (software product), 403
mass customization, **224**, 266, 335, 348, 354, 359
mass production, **224**
master black belt, 779
master data, **245**, 412, 682, 723, 726, 741, 757
master planning, **236**, 248, 260, 388, 534, 606, 622
master production schedule (MPS), 239, **255**, 285, 340, 390, 530
  overstated, **259**
master schedule, 239, **255**
  mixed-model (syn. variant), **343**
  multilevel, **530**
  two-level, **345**
  variant, **343**, 347, 366
master schedule item, **256**, 530, 619
master scheduler, **256**
master scheduling, **239**, 255
material, **10**
  bill of, **29**, 46
  custom, 86
  direct, **84**, 90, 683
  indirect, **84**, 90
material costs, **686**
  fixed, **687**, 700
  variable, **687**, 691

material flexibility, 47
material group, 88, 506
material group management, **91**
material kit, **505**
material requirements planning (MRP), 268, 515, **530**, 639
material-dominated scheduling (MDS), 246, **385**
materials, 638
  quality control, 659
materials budget, 239
materials management, **243**, 533, 550, 636, 646
  capacity-oriented (Corma), 275, 616, **645**
  collective, **506**
  consumption controlled, **212**
  continuous coupling with, 650
  demand controlled, **212**
  deterministic, **263**, 312, 402
  forecast controlled, **212**
  kit, **505**
  quasi-deterministic, **263**, 382, 514, 530
  stochastic, **263**, 402
mathematical forecasting techniques
  future-oriented, **427**
  historically oriented, **427**
mathematical programming, **153**
maximum bill of material, 353
maximum cardinality (of an access function), **830**
maximum offset of the next operation, **583**, 585
maximum order quantity, **503**
maximum routing sheet, 353
MBNQA (Malcolm Baldrige National Quality Award), **783**
mean, **430**, 432, 435, 441, 727
  of the first-order means, 441
  sliding (forecasting technique), 437
mean absolute deviation (MAD), **435**
mean deviation, **444**
mean range (of inventory), **565**
mean smoothed consumption, 435
mean square deviation, 430
meaningful identification key, **828**
meaningful item ID, **726**, 828
means of transport, **146**, 498
  primary, **146**
measure
  global, **52**
  local, **52**
  performance, **52**
    logistics, **53**
measure (Six Sigma phase), **769**
measurement objective, 764
measurement system, **764**
  performance, **52**
MEDILS, **186**
medium-term planning, 105, **236**, 239, 260, 319, 661
merged transport (need for), **147**
MES (manufacturing execution system), 172, **651**
metarule, 742
method, **8**, 187
  chase production, **259**
  critical chain, 574, 804
  critical path (CPM), **573**, 804
  Delphi (forecasting technique), **449**
  deterministic, **263**
  gap, 92
  hybrid production, **259**
  inventory valuation, **470**
  level production, **259**
  scoring, 92
  statistical, 430
  stochastic, **263**
milestone
  project, **801**
milestone chart, **803**
milk run, 672
mind mapping, 778
minimum blanket order quantity, **213**, 250, 260
minimum blanket order quantity zero (= forecast), 213, 239, 250
minimum cardinality (of an access function), **831**
minimum deferral time, 535
minimum delivery lead time, 482
minimum order quantity, **503**
min-max (reorder) system, **483**
mix flexibility, 47
mixed flow scheduling, **581**
mixed manufacturing, **645**, 698
mixed production, 295, **645**
mixed-mode manufacturer, **645**
mixed-model master-schedule (syn. variant MS), **343**
mixed-model production, **209**, 295
mobility cost ratio of the service, **142**
mode of transport, 68, **146**
  selection of, 670
model, **817**
  ALP (advanced logistic partnership), **99**, 120
  data, **824**, 840
  econometric, 429
  forecasting, for constant demand, **431**
  function, **824**
  generic, 840
  intuitive, 429
  object, **824**, 840
  operator, **15**
  organization, **824**
  process, **824**
  SCOR, **22**, 58, 59, 66, 227
  spiral, **798**
  supply chain operations reference, **22**
  task, **824**
  trend forecasting, **438**
  waterfall, **798**
modeling, **817**
  data, **824**
  function-oriented, **824**
  object-oriented, **824**
  process-oriented, **824**
  task-oriented, **824**
modeling method, **817**
modification flexibility, 47
modular bill of material, **343**, 346
modular product concept, **46**, 300, 356
modular sourcing, **85**, 98
modular supplier, **83**, 97
module
  product, **32**, 46, 343, 725
monitoring (order), **242**, 656
monopoly situation, 612, 616
morphological scheme, 153, 200

motivation of employees, 309
move card, **314**
move time, **554**, 555, 736
moving average, **432**
(forecasting technique), **432**, 437
weighted, **434**
MPS (master production schedule), 239, **255**, 285, 390, 530
MRO item, 18, **29**, 90
MRO supplier, **84**
MRP (material requirements planning), 268, 311, 515, **530**, 546, 639
net change technique, **533**
MRP II (manufacturing resource planning), 225, **236**, 319, 385, 402
MRP II software, **401**
MS (master schedule), **255**
muda (syn. waste), **295**
multi echelon, **138**, 246, 524
multi-echelon, 145
multilevel available-to-promise (MLATP), **277**, 535
multilevel bill of material, 541, **730**, 802
multilevel master schedule, **530**
multilevel product structure, 725
multilevel where-used list, 541, **732**
multiple sourcing, **85**, 86
multiple stock organization, 727
multiple-stock organization, **470**
multiplicative seasonality, **445**
multipurpose facility, 375, 389
multisourcing (syn. multiple sourcing), **85**
multivalued role or access function, **830**
multivariate analysis of variance, 775
mura (syn. variation), **295**
muri (syn. overstraining), **295**

N

near-to-line production, **303**
net change MRP technique, **533**
net income, **44**, 60
net operating profit after tax (NOPAT), **44**, 65
net present value technique NPV, 160, **811**
net requirements, 526, **530**, 535, 546
net requirements planning, **240**, 259, 530
net working capital, 40, **44**, 66
network, 737
distribution, **20**, 134, 246, 405, 662
logistics, **20**
neural, 427
procurement (syn. logistics network), **20**
production (syn. logistics network), **20**
retail, 139
service, **20**, 142
transportation, 146
network algorithm, 575
network diagrams, 775
network of operations, 32
directed, **215**, 551, 570, 586, 619
undirected, **215**, 377, 551, 586
network plan, 337, 778
rough-cut, 620
network planning, 403, **573**, 590, 803
neural network, 427
neutral, or independent electronic marketplace, **95**
noise, 430
nomenclature (syn. bill of material), **29**, 822
nonbottleneck capacity, 599
nonconstraint work center, 597
nonlinear programming, 153
nonlinear usage quantity, **386**
nonrepetitive production, 108, 311, 335, 354, 403
nonstrict role or access function, **830**
nontechnical wait time after an operation, **554**
nontechnical wait time before an operation, **554**, 557, 736
nontransportable item, **205**
non-value-added, 310, 468
normal cost system, **694**
normal costing, **694**
normal distribution, 433, 487
North America, 225, 226, 309
n-to-1 association, **831**
n-to-n association, 733, **831**
reflexive, 740, **836**

O

object, **8**, 187, 825
business, 280
generic, 840
service, **10**
service, 141
object class, **825**
object model, **824**, 840
object view, 819, **824**
objective
enabler, 40, 60, 96
entrepreneurial, **40**, 52, 64, 184, 412
formulation of -s, **796**
unstable, 797
mandatory, **796**
preferred, **796**
primary entrepreneurial, **44**, 52, 60, 64
project, **800**
results, 40
object-oriented modeling, **824**
observation interval (syn. statistical period), **428**
obsolescence, 18
technical, 267, 497
OED (outside exchange of dies), **301**
OEE (overall equipment effectiveness), 304, **598**
OEM (original equipment manufacturer), 13, **83**
offset
lead time, **33**, 277
run time, **583**
offshoring, 127
Ohno's seven wastes, **295**
one less at a time, **299**
one-card Kanban system, **315**
one-dimensional variant structure, 346
one-of-a-kind production, **224**, 226, 278, 335, 340, 354, 403, 645
one-piece flow, **303**
on-hand balance (syn. physical inventory), **18**, 40, 138, 213, 470, 473, 515
open order, **516**
customer, **275**
open order quantity, **516**
operating assets, **44**

operating curve
for stock on hand, **522**
logistics, 523, **565**, 570
operating supplies, 18
operation, **31**, 34, 375, 533, 574, 651, 688, 723, 737
alternate, **738**
critical, 576
directed network of -s, **215**, 551, 570, 586, 619
order, **717**, 721, 724
rough-cut, **37**, 209, 661
undirected network of -s, **215**, 377, 551, 586
operation card, **651**, 657
operation description, 737
operation due date, **571**, 721
operation ID, 734, **738**
operation load, **34**, 533, 552, 553, 601, 658
operation overlapping, **583**
operation sheet (syn. routing sheet), **31**
operation start date, **571**, 721
operation time, **31**, 298, 329, 552, 584, 606, 658
short, 614, 660
operation variant, 744
operations, **16**
control of, 610
order of the, 32, 551, 569
operations management, **16**
operations research, 247, 427
operations scheduling, **652**
operations sequence number, 737
operations sequencing (syn. operations-oriented finite loading), **608**
operations-oriented finite loading, 274, **608**
operator model, **15**
OPP (order penetration point), **45**, 206, 250
opportuneness, **42**
opportunity cost, **42**, 54, 491, 775, 808
optimization
dynamic (Wagner–Whitin), **536**, 538
of loading space, **671**
optimization paradigm, **782**
optimum batch size, **498**, 535
optimum length of order cycle, **500**, 535
optimum number of merged demands, **535**
optimum number of partial lots, **535**
optimum order interval, **500**, 535
optimum probability of stockout, **491**
optimum service level, **491**
option, **30**
option overplanning, **526**
option percentage, 257, 285, 342, 366, **457**
Oracle (software product), 401
order, **8**, 25, 612, 716, 717
as business object, **25**, 716
blanket, 105, **213**, 260, 606
critical ratio of an, **647**
customer, **25**, 249, 354, 515, 637, 683, 720
customer procurement, 207, **212**
customer production, **212**, 645
firm planned, **639**
forecast, **212**
in process, 578, 647, 649
internal, **25**, 189
interplant, **249**
kind of, **25**, 243, 720
of the operations, 32, 551
open, **516**
customer, **275**
overhead, **25**, 720
partial, **718**
planned, **240**, 532, 637
production, **25**, 516, 639, 651, 720
purchase, **25**, 638
released, **242**, 516
replenishment, **212**, 316
scheduling of, **567**
stock replenishment, **212**, 267, 316, 645
transportation, **671**
type of, 192, **212**, 645
work, **25**
order bill of material, 743
order checking, **242**, 656
order completion date, **25**, 720
order condition, 720
order configuration, **241**, 337
order confirmation, **26**, 249, 524
order confirmation time, **58**
order coordination, 48, **242**, 337, 656
order costs
actual, **694**, 695
order cycle, **481**, 485, 490
length of, 490
optimum length of, **500**, 535
order data, 682, **717**
order database, 637, 719
order due date, **25**, 571, 648
flexibility of the, 213
order end date, **25**, 720
order entry, **249**
order fulfillment cycle time (SCOR performance indicator), **58**
order header, **717**, 719
order ID, 720
order interval, **481**
optimum, **500**, 535
order item (syn. order position), **25**, 42, 356, 717
order losers, **42**
order management, 187
parallel, in multiple levels, 190
order mix, 616
order monitoring, **242**, 656
order network, 533
order operation, **717**, 721, 724
order penetration point (OPP), **45**, 206, 250, 340
order picking, 242, **663**, 678
discrete, **663**
order picking store, **206**, 663
order point, 268, **481**, 646, 727
time-phased (TPOP), **515**
order point system, **480**
order point technique, 268, 325, **480**, 511, 527
time phased, 228
order position, **25**, 717, 720
item, **717**
production equipment, **717**, 721
type of, **720**
work, 28, **717**, 721, 724
order priority, **606**, 610, 720
order processing, 9, 104, 186
blanket, **239**
integrated, 278
order promising, **26**
order proposal, **240**, 482, 532, 637
order qualifiers, **42**
order quantity, **214**, 533
discrete, **536**
economic (EOQ), 268, **498**, 535

fixed, **535**
maximum, **503**
minimum, **503**
minimum blanket, **213**, 250, 260
open, **516**
period, **535**
replenishment, **481**, 639
order release, 241, **636**, 652
according to consumption, **212**, 298, 316, 325, 727
according to demand, **212**, 727
according to prediction, **212**, 727
early, 506, 578, 646
load-oriented (loor), 274, 616, **640**, 675
reason for, **212**, 376, 727
order routing sheet, 743
order service, **249**
order splitting, **581**, 665
order start date, **25**, 720
order status, **26**, 720
order structure
loops in the, **215**
processor-oriented, **380**, 748
order success probability, **249**, 408, 477, 621
order success rate, 48, **59**, 249
order termination report, **659**
order urgency, **577**, 592, 609, 648, 720
order validity date, **25**, 717
order winners, **42**, 357
ordering costs, 268, 498, 536
for procurement, **496**
for production, **496**
order-oriented finite loading, 274, **612**
order-oriented infinite loading, 273, **600**
order-specific issuance principle, **471**
order-up-to level, **483**
order-wise finite loading, **275**
order-wise infinite loading, 274, **607**
organization
centralized, **75**
decentralized, **75**
learning, **104**
of the production infrastructure, 34
process-focused, **75**
product-focused, **75**
stock, **469**
organization model, **824**
organization view, 819, **824**
organizational quality, 12, 40, 413, **763**
organizational units
semiautonomous, 75
organization-oriented process chart, **194**
orientation
assembly, **202**
of product structure, **202**
product family, **335**
variant, 204, **335**
orientation of product structure, 730
original equipment manufacturer, 13, **83**
OTIF (on-time and in-full), **56**
outbound stockpoint, 315
outlier, 455, **476**
outside exchange of dies (OED), **301**
outsourcing, **72**, 606
overall capacity planning flexibility, **272**, 599
overall equipment effectiveness (OEE), 304, **598**
overburdening, **295**
overcapacity, 36, 48, 304, 313, 696
overhead (syn. overhead costs), 300, **683**
overhead budget, 239, 252
overhead costs, **683**, 685
variable, 685
overhead order, **25**, 720
overlapped schedule (syn. operation overlapping), **583**
overlapping, 552
of operations, **583**, 585
within an operation, **582**
overload, 269, 602, 623
persistent, 606
overstated master production schedule, **259**
overstraining, **295**
overtime, 271, 564, 606

P

package carrier, **149**
packaged good, **665**, 667, 671
package-to-order, 46, **207**, 340
packaging, 46, 207, **665**
packaging function, **666**
packaging materials, **668**
packaging system, **667**
packing and marking, **668**
packing unit, **666**
parallel picking, **664**
parameter, 348, **745**, 753
parameter class, **745**
bill of, position, **746**
parameter value, 352, **747**
parameterized bill of material, 352, 366
parent item, **723**
parent version, 350, 354
Pareto chart, 477, 774
Pareto's law, **477**
part, **10**
as a business object, **28**
in-house, **28**, 725
purchased, **28**, 725
service, **28**, 228, 455
single, **28**
spare (syn. service part), 18, 28
part period balancing, **535**, 538
part system, **792**
partial costing, 689
partial order, **718**
partial order header, 720
partial order ID, 720
partial role or access function, **830**
participative design/ engineering, 102, **278**
particular safety factor, 492
partnership
customer-supplier, **86**
partnership relationship, **89**, 96, 99
parts production, 210, 307
parts requisition, **651**, 657
parts store
in-house, 207
purchased, 207
single, 207
path float, 574
pattern recognition, **816**
payback period, **810**
PDM (product data management), **749**
PDPC (process decision program chart), 779
pegging, **540**
people, 33, 184, 192
PeopleSoft (software product), 406
percentage
backorder, **263**

loading, **641**
option, 257, 342, 366, **457**
stockout, **263**
perfect order fulfillment (SCOR performance indicator), **58**
performance
of a company, **40**
supply chain, **40**
performance criterion (syn. performance indicator), **52**
performance indicator, **52**
administration cost rate, **56**
batch size, **57**
bid proposal success rate, **59**
breadth of qualification, **59**
capacity utilization, **55**, 57, 651
cash-to-cash cycle time, **60**
complaint rate, **54**
customer service ratio, **56**
delivery reliability rate, **56**
downside supply chain adaptability (SCOR), **60**
fill rate, **56**
lot size (syn. batch size), **57**
order confirmation time, **58**
order fulfillment cycle time (SCOR), **58**
order success rate, **59**, 249
OTIF (on-time and in-full), **56**
perfect order fulfillment (SCOR), **58**
response time, **58**
return on net assets (RONA), **60**
scrap factor, **54**
supply chain value at risk (SCOR), **60**
temporal flexibility, **59**
turnover
stock inventory, **55**
work-in-process inventory, **55**, 565
upside supply chain adaptability (SCOR), **60**
upside supply chain flexibility (SCOR), **59**
utilization of capacity, **55**, 57
value-added rate of lead time, **57**, 199
variance in work content, **57**
work center efficiency, **55**, 651
yield factor, **54**
performance indicators, 174
performance measure, **52**
performance measurement, **52**
performance measurement system, **52**
performance of a company, 33, 185, 244
performance of a supplier, 91
period order quantity, **535**
periodic inventory, **474**
periodic review system, **483**
perishability, 267, 497
perpetual inventory, **471**
personalized medication, **51**
personalized production, 11, 48
PERT, 804
phantom bill of material, **343**, 353
extended, 536
phase
of life of a system (syn. system phase), **793**, 820
project, **801**
system, **793**, 820
physical inventory, 413, 468, **473**, **515**, 517, 722
picking, 652, **663**
batch, **663**
order, **663**, 678
parallel, **664**
sequential, **664**
zone, **664**
picking list, 473, **652**, 657, 663, 720
picking location, 142
picking station, 664
picking strategy, **663**, 678
picking up (customer involvement in), **134**
pilot test, **209**
pipeline (production), 389
pipeline inventory, **246**
pipeline stock, **246**, 484
plan
aggregate, **252**
business, **41**
inventory, **252**, 287
network, 337, 778
rough-cut, 620
process, **33**, 62
ABC, **701**, 702
rough-cut, **37**, 62, 619
procurement, 426, 733
production, **251**, 426, 516
global, 161
sales, **251**, 287
strategic, **41**
tactical, **251**
plan (task in the Deming cycle), **768**, 770
plan process, 339, 533, 723
planned capacity utilization, **597**
planned cost rate, **700**
planned demand, **516**
planned issuances, 657
planned load, **603**
planned order, **240**, 532, 637
firm, **639**
planned receipt, **516**, 532, 637
planned release, **532**
planning, **21**
brownfield, **152**
business continuity and resilience, 117
cyclic, **299**
demand, **238**, 249, 425
detailed objects, **237**
distribution, **246**, 524
distribution requirements, **525**
distribution resource, **246**
facility location, **127**
fine, **237**
greenfield, **152**
location, **127**
long-term, 105, **236**, 238, 248, 319, 388, 661
master, **236**, 388
medium-term, 105, **236**, 239, 260, 319, 653, 661
network, 403, **573**, 803
production, **251**
release, 241, 280
resource requirements, **239**, 252, 388
rolling, 105, 253
rough-cut, **237**, 585, 620
route, 148
sales and operations, **239**
scenario, **450**
short-term, 105, **237**, 240, 616, 645
situational, 607, 616, 645
strategic, **41**
supply chain, **21**
tactical, **251**
three-level, according to temporal range, **236**
planning & control, **22**, 397, 815

short-term, 105, **237**, 240, 616, 645
planning and scheduling
detailed, **236**, 262, 652
transport, **669**
planning bill of material, **345**, 346
planning board, 385, 617, 653, **655**, 803
planning group, 88, 506, 637
planning horizon, **239**, 602
planning period, **458**, 460
planning time fence, **256**
plant construction, 153, 155
plausibility test, **746**
plus/minus bill of material, **343**, 346
plus/minus routing sheet, **343**
point of sale (POS), **106**, 139
point-of-use delivery, **306**
point-of-use inventory, **306**
Poisson distribution, 454, 488
poka yoke, 310, **776**
policy
batch sizing, 268, 311, **535**, 727
delivery, **18**
inventory, **251**
lot sizing (syn. batch sizing), **535**
portable item, **206**
portal, **408**
portfolio, 362
supplier, **88**, 100
supply chain, **113**
supply chain risk, **116**
position
bill-of-material, **728**
order, **25**, 717, 720
routing sheet (syn. operation), **31**
where-used-list, **733**
position in lot, **214**, 376
position status, **721**
post-deduct inventory transaction processing, **657**
postponement, **46**, 117, 132
potential factor, 271
PPC, **22**, 411
PPC software, 22, 353, **397**
pre-deduct inventory transaction processing, **657**
preferred objective, **796**
preliminary study, **793**, 820
preventive maintenance, 34, **319**, 597
primary entrepreneurial objective, **44**, 52, 60, 64
primary key, 726, **827**, 828
primary means of transport, **146**
primary parameter, **746**
primary product, **371**
primary work center, **737**
principle of proceeding from the general to the particular, 819
priority
order, **606**, 610, 613, 720
priority control, 606
priority planning, **242**, 246
priority rule, **609**, 613
Prism (software product), 405
private electronic marketplace, **95**
private trading exchange (PTX), **95**
proactive service, 48
probability, 264, **433**, 644
of stockout, **485**
optimum, **491**
order success, **249**, 408, 477, 621
probability distribution, 433
probable scheduling, **577**, 592, 648
problem-solving cycle, **796**
procedure, **8**
proceeding from the general to the particular, **792**
process, **8**, 187, 760
ABC, **700**
as an object, **378**
business, **10**, 185
collaborative, **101**
cooperative, 117
manufacturing, **196**
service, **196**
transaction, **72**
with direct contact with the object, **141**
with indirect contact with the object, **141**
with no object contact, **142**
process capability, **659**, 774
process configurator, 337
process control, **659**
process cost rate, **700**, 702
process costs, 702, 703
process decision program chart (PDPC), 779
process design, 98, 377, 723
process flow production, **209**
process flow scheduling, 226
process industry, 202, 222, 237, **369**, 394, 404, 586
process innovation
flexibility in achieving customer benefit by, 40, 110
process layout, **209**, 302
process load, **761**
process management, 187, 700
process manufacturing, **369**
process map, **194**
process model, **824**
process organization, 278
process owner, 779
process plan, **33**, 62, 339, 533
ABC, **701**, 702
degree of structuring of, **210**
rough-cut, **37**, 62, 619
process quality, 12, 40, 413, **761**
consistent, 130
process quantity, **701**, 702
process reengineering (business, BPR), 303, **780**, 816
process rule, **350**
process stage, 375, **380**, 581
process store, **187**, 320
process team, 779
process time, **761**
process time line, 199
process train, **380**, 581, 748
process unit, **380**
process variable, **700**, 712
process view, 819, **823**
process with technology, **378**
process-focused organization, **75**
processor (in the process industries), **369**
processor-dominated scheduling (PDS), **385**
process-oriented modeling, **824**
processor-oriented concept, 226, 247, **369**, 403
software for, **404**
processor-oriented production structure, **380**, 748
procurement, 16
customer order, 207
large batch, **214**
lotless, **214**
single-item, **214**
small batch, **214**
with frequent order repetition, **213**, 297, 313, 318, 661

with infrequent order repetition, **213**
without order repetition, **213**
procurement budget, 239, 252
procurement costs, 260, 727
batch-size-dependent, **495**
batch-size-independent, 324, **496**
lot-size-independent, 637
procurement management, 83
procurement market research, **91**
procurement network (syn. logistics network), **20**
procurement order (syn. purchase order), 516, 720
customer, 207, **212**
procurement plan, 287, 426
for components and materials, **239**, 252
for salable product, 239, **251**
procurement size, **727**
procurement without lots, **214**
producible resource, **378**
product, **10**, 378, 723, 728
according to (changing) customer specification, **204**, 340, 354, 403, 698
by-, **371**, 727
co-, **299**
end, 18, **28**, 30, 83, 252, 383
in a broad sense, **11**, 762
in the most comprehensive sense, **12**
individual, **204**
individualization of -s, 11, 50
intermediate, **28**, 378, 725
primary, **371**
standard, 131, **204**, 698
standard, with options, **204**
waste, **371**, 394, 727
product concept, 213, 297
modular, **46**, 300, 356
product configuration, 350
product configuration catalog, **346**
product configurator, 337, 361
product cost breakdown structure, **686**, 727
product cost estimation, 700
activity-based, 705
product costing, 682, **696**
product data management (PDM), **749**
product database (syn. engineering database), **750**
product design, 46, 98, 110, 377, 723, 801
customer-driven, 360
market-driven, 360
support for decision-making in, 700
product differentiation, **42**
product family, **30**, 37, 46, 204, 209, 226, 239, 340, 348, 457, 477, 646
with many variants, **204**, 311, 335, 340, 348, 364, 403, 734, 741
product family forecast, **252**, 345
product family orientation, **335**
product ID, **734**, 737
product innovation, 97, **277**
flexibility in achieving customer benefit by, 40, 110
product layout, **209**, 302
product life cycle, **15**, 128, **214**, 436, 726, 801
management of the (PLM), **749**
product line (syn. product family), **30**
product load profile, **38**, 252, 621
product mix, 414, **477**
product module, **32**, 46, 343, 725
product positioning, **42**
product quality, 12, 40, 413, **761**
product solution space, **348**, 359
product structure, **29**, 45, 83, 266
convergent, 29, **202**, 225, 335, 400, 723, 802
depth of, **202**
divergent, 29, **202**, 226, 586, 723
multilevel, 725
orientation of, **202**
rough-cut, **37**
single-level, 724
unfinished, **356**
with loops, **215**, 376, 753
without loops, **215**, 735
product variant (syn. variant), **30**, 346
product variety concept, **204**, 226, 227
product viewpoint, 740
product-focused organization, **75**
production, 16, 237
batch, **224**, 369, 400, 612
centralized, **129**
chemical, 371
continuous, 202, **210**, 393, 560
customer order, 237, **335**, 607
decentralized, **129**
food, 376
group, 209
high-volume line, **209**, 375
intermittent, **209**
island, **208**
job shop, 32, **209**, 301, 330, 386, 550
large-batch, **214**
lean, 190, 226, **296**
leveling of the, **300**
line, **210**, 302, 386, 558
lotless, **214**
mass, **224**
mixed, 295, **645**
mixed-model, **209**, 295
near-to-line, **303**
nonrepetitive, 354, 403
of prototypes, **209**
one-of-a-kind, **224**, 278, 340, 354, 403, 645
personalized, 11, 48
process flow, **209**
service, 13
single-item, **214**, 278, 536
single-item-oriented line, **209**, 570
site, **208**, 573
small-batch, **214**
stockless, **296**
synchronized, **248**, 315
variant, **335**
with frequent order repetition, **213**, 297, 300, 313, 318, 402, 661
with infrequent order repetition, **213**
with many variants, 348
without order repetition, **213**, 226
production activity control (PAC), **651**

production bill of material, 536
production card, **314**
production control, 565, 570
production costs
  batch-size-independent, 324
production department, **737**
production division, **737**
production environment, **206**, 227, 339, 348, 354, 374
production equipment, **34**, 369, 379, 653, 722, 739
production equipment ID, 721
production equipment order position, **717**, 721
production facility, **34**
production forecast, **345**
production infrastructure, **34**, 73, 304, 369, 645
  (quantitatively) flexible, **211**
  flexible capability of the, **211**, 375
  organization of the, 34
production lead time, **31**
production line, **210**
production model, **380**
production network (syn. logistics network), **20**
  feature for designing, **130**
production of a prototype, 104
production order, **25**, 516, 639, 651, 720
  customer, **212**, 578, 645
production order release, 638
production pipeline, 389
production plan, **251**, 287, 426, 516
  global, 161
  optimal, 253
production planning, **251**
production planning and control, 22
production rate, **209**, 224, 251, 570, 598
production release, **280**
  detailed, **280**
  rough-cut, **280**
production rhythm, 253
production rule, 351, **743**, 744
  within an expert system, 743
production schedule, **239**, 460
production segmentation, **301**
production size, **727**
production structure, **32**, 747
  processor-oriented, **380**, 748
  rough-cut, **37**
production structure level, **32**, 536, 725
production system (syn. production network), **20**
production type, **223**, 231, 296, 335, 339
production without lots, **214**
production-equipment where-used list, **739**
productive capacity, **598**
productivity (of a company), **40**, 85
product-service system, 11, **13**, 14
profit, 42, 44
profit after tax, **44**
profit center, **75**
profit margin, 44
prognosis, **18**
program, **801**
  quick response (QRP), **106**
program evaluation and review technique (PERT), 574, 804
progress checking, 242, **658**
project, **789**
project benefit, **809**
project bill-of-material position, 754
project costing, 682, **696**
project costs, **808**
project effort, **804**
project life cycle, **801**
project management, 280, 355, 418, **790**
project manager, **806**
  social competency of a, 807
project manufacturing, **208**, 573
project milestone, **801**
project objective, **800**
project operation, 754
project performance, **800**
project phase, **801**
project profitability, **810**
project risk, **812**
project risk management, 812
project task, 153, 573, **801**
projected available inventory, **516**, 530, 542, 637
  calculation of, **518**, 526
    cumulative, **521**
    scheduling, **519**
Protean (software product), 405
protectionism, 74
protective capacity, **599**
protective function, **666**
protocol, **49**
  file transfer (FTP), 49
  VoIP (Voice over Internet), 49
prototype (production of), **209**
prototyping, 798
pseudo bill of material, **343**
PSIpenta (software product), 403
public electronic marketplace, **95**
pull logistics, 48, **189**, 212
pull signal, **316**
punched card, 398
purchase order, **25**, 638
purchase order release, 637
purchase price, 80
purchase requisition, **637**
purchase schedule, **239**, 460
purchased part, **28**, 725
purchased parts store, 207
purchasing lead time, **32**, 310
push logistics, 48, **190**, 212
put-away time, 32
pyramid forecasting, **345**

Q

qualification of employees, 73, 211, 309, 653
qualitative forecasting techniques, **427**
quality, **759**, 777
  according to ISO8402, **764**
  consistent process, 130
  in view of business partners, **763**
  in view of employees, **763**
  in view of shareholders, **763**
  in view of society and the environment, **763**
  organizational, 12, 40, 413, **763**
  process, 12, 40, 413, **761**
  product, 12, 40, 413, **761**
quality activation, **778**
quality assurance, 309, 659, **777**
  as used originally, **777**
  computer-aided (CAQ), **750**

quality at the source, **309**
quality audit, **784**
quality circle, **310**, 780
quality control, 242, 280, 305, 375, 517, **659**
  according to ISO8402, **774**
quality control charts, 775
quality control materials, 659
quality control sheet, **659**
quality function deployment, **771**
quality management
  according to ISO8402, **766**
  total (ISO8402), 768, **781**
quality planning, **770**
quantitative forecasting techniques, **427**
quantitatively flexible capacity, **211**, 251, 259, 271, 298, 304, 313
quantitatively flexible production infrastructure, **211**
quantitatively flexible workforce, **211**, 309
quantity
  actual, **694**, 721, 722
  backorder, **263**
  billed, 721, 722
  moved, 722
  stockout, **263**, 492
quantity allocated, 276, **515**
quantity deviation, 501
quantity discount, **504**
quantity per, **29**, 345, 386, 734
  cumulative, **29**, 730, 732
quantity required (syn. quantity per), **29**
quantity reserved, 482, **515**, 657, 721
quasi-deterministic bill of material explosion, **267**, 461
quasi-deterministic materials management, **263**, 382, 530
queue, **557**
queue time, **554**, 597, 736
queuing theory, **557**, 564
quick response program (QRP), **106**
quotation, **249**
  request for, 59, 91, 94, 239, 249

R

R/3 (software product), 401
random component, 430
random variation, 430
random-location storage, **470**
range
  of implementation of ERP software, 415
  of inventory, **565**
    mean, **565**
rate
  bid proposal success, **59**
  delivery reliability, **56**, 413
  efficiency, **36**, 597, 661, 695, 735
  fill, 41, **56**, 263, 493
    cumulative, **264**
  flow, **570**
  order success, 48, **59**, 249
  production, **209**, 224, 251, 570, 598
rate-based scheduling, **209**
rated capacity, **36**, 598, 658
ratio
  critical, of an order, **647**
  customer service, **56**, 413
raw material, **28**, 36, 725
raw materials store, 207
raw materials supplier, **84**
RDMAIC, **769**
reason for order release, **212**, 376, 727
receipt, 471, **476**
  date of, 524
  item, **717**, 721
  planned, **516**, 532, 637
  scheduled, 482, **516**, 637
receiving department, 659
recipe, 29, **380**, 723
record accuracy, **473**
recyclability of packaging, 667
recycling, 15, 21
redesign, 21
reduction
  batch-size, 299, 311
  lead time, 312
  of variants, 300
  setup time, 299, 564
reengineering, **780**
  business process- (BPR), 303, **780**, 816
reference object, 53
reflexive association, **829**
reflexive n-to-n association, 740, **836**
regression analysis (forecasting technique), **439**
regular demand, **203**, 429, 455, 479
reidentification key, **834**
relational database, 400, **826**
relationship
  buyer-dominated, 87, **89**
  classification of demand according to its, **263**, 530
  market-oriented, **85**, 89
  partnership, **89**, 96, 99
  supplier-dominated, 87, **89**, 93
relationship diagram, 778
release, 754
  blanket, 105, **261**
  final, 280
  of contract work, 638
  order, 241, **636**, 652
  planned, **532**
  production, **280**
    detailed, **280**
    rough-cut, **280**
  provisional, 280
  staggered, 754
  stepwise, 280
release control, **280**, 753
released order, **242**, 516
remanufacturing, 21
reorder point (syn. order point), **481**
repair, 144
repetition of operations, 586
repetitive manufacturing, **224**, 293, 297, 306, 340, 694
replenishment, **192**, 298
  continuous, **104**
replenishment lead time, **481**, 639
replenishment order, **212**, 316
replenishment order quantity, **481**, 639
report
  anticipated delay, **659**
  order termination, **659**
request for quotations, 59, 91, 239, 249
required capacity, 247, **258**, 619
requirement
  gross, 252, **530**
  net, **530**
requirement traceability, **540**
requirements calculation
  resource, **241**, 337
requirements explosion, **533**
requirements planning
  gross, **239**, 259
  net, **240**, 259
  resource, **239**, 252, 388

requisition, 326, 470
part, **651**
purchase, **637**
rerouting flexibility, 47
rescheduling, **567**, 623
bottom-up, **541**
rescheduling assumption, **541**
reservation date, 721
reserved quantity, 482, **515**, 721
resilience, 47
business continuity and r. planning, 117
resource, **378**
bill of, **38**, 252
capacity, 595
consumable, **378**
flow, **375**, 380, 388
producible, **378**
resource budgeting, **239**
resource contention, **599**
resource management, **246**, 400
integrated, **246**, 253
resource profile, **620**
cumulative, 625
resource requirements calculation, **241**, 337
resource requirements planning, **239**, 243, 252, 260, 388
detailed, **240**
resource transaction
as an object, 749
completed, 381
response time, **58**
results objective, 40
retail network, 139
feature for designing, **139**
retrofitting, 142
return on investment (ROI), **64**, 496
return on net assets (RONA), **44**, 60
returns, 21, **659**
need for efficient, **134**
reuse, 21
reverse flow scheduling, **581**
reverse logistics, **21**, 134
rework, 659
RFID, 49, 97, **660**, 672
RFID tag, **660**
risk, **777**
project, **812**
supply-chain, 46, 86, **112**
system, 808
risk analysis, 774, 812
risk costs, 80
risk management
active, 777
project, 812
supply chain, **112**
risk of depreciation (by stocking), 251, 267, **497**
risk pooling, **117**
role, **830**
multivalued, **830**
nonstrict, 830
partial, 830
single-valued, **830**
strict, **830**
total, **830**
role name, **830**
rolling motorway, **146**
RONA return on net assets, **44**, 60
RoO (rules of origin), **77**
rope, **617**
rough-cut bill of material, **37**, 619
rough-cut capacity planning (RCCP), 239, **258**, 632
rough-cut data collection, **660**
rough-cut finite loading, **624**
rough-cut infinite loading, **622**
rough-cut network plan, 620
rough-cut operation, **37**, 209, 661
rough-cut planning, **237**, 250, 424, 585, 598, 620, 738
rough-cut process plan, **37**, 62, 619
rough-cut product structure, **37**
rough-cut production release, **280**
rough-cut production structure, **37**
rough-cut routing sheet, **37**, 619
rough-cut structure, 618
rough-cut work center, **37**, 238, 619
route planning, 148
routing (syn. routing sheet), **31**, 738
alternate, **738**
routing and scheduling, 136, 144, 148, **671**
routing sheet, **31**, 347, 535, 659, 738
maximum, 353
plus/minus, **343**
rough-cut, **37**, 619
variant, **342**, 351
routing-sheet position (syn. operation), **31**
rule, 742
design, **350**
EDD (earliest due date) priority, 609, 613
meta-, 742
priority, **609**, 613
process, **350**
production, 351, **743**, 744
scheduling, 555
shortest processing time (SPT), 606, 609
shortest slack time (SLK), 606, 609, 613
rule base, **742**
rule-based product and/or process configurator, 337
rules of origin (RoO), **77**
run load, **34**, 552, 684, 721
per unit, **34**, 496, 553, 737
run time, **31**, 552, 582, 684
offset, **583**
per unit, **31**, 329, 552, 738
running costs, **808**

S

safety capacity, **599**
safety demand, 257, **460**, 529, 639
safety factor, 460, **486**
particular, 486, 492
safety lead time, **485**, 528
safety stock, 257, 268, 316, 481, **484**, 510, 528
sales, 16, 312
cost of, **689**
sales and distribution, 237
sales and operations planning, **239**, 250
sales function, **667**
sales order (syn. customer order), **25**, 542
sales plan, **251**, 287
sales promotion, 408, 484
sales promotion functions, **667**
sales statistics, **477**, 722
sales transaction, **477**
same day delivery, 136
sample, 432
sample distribution, 432
SAP (software product), 401
satisfied demand, 430
saw-toothed curve, **480**

SCEM (supply chain event management), **49**, 117
scenario driver, **450**
scenario forecasts, **449**
scenario planning, **450**
schedule
  delivery, 105, **261**
  drum, **617**
  final assembly (FAS), **340**
  finishing, **340**
  level, **259**, 606
  multilevel master, **530**
  overlapped (syn. operation overlapping), **583**
  production, **239**, 460
  purchase, **239**, 460
  two-level master, **345**
scheduled downtime, 597
scheduled load, **603**
scheduled receipt, 482, **516**, 637
scheduling, **255**, 553, 636, 803
  backward, **571**, 573, 590
  block, **610**
  central point, **576**
  finite forward, 277, 558, **653**, 677
  forward, **571**, 590
  forward flow, **581**
  lead-time, **568**, 608
  material-dominated, 246, **385**
  mixed flow, **581**
  of orders, **567**
  operations, **652**
  probable, **577**, 592, 648
  processor-dominated, **385**
  rate-based, **209**
  re-, **567**, 623
    bottom-up, **541**
  reverse flow, **581**
  routing and, 136, 144, **671**
  supplier, 105, 108, **260**
scheduling algorithm, **567**, 586
scheduling projected available inventory calculation, **519**
scheduling rule, 555
scheduling technique, 568
scheme, 33, 35
  morphological, 153, 200
SCM, 411
  software, 49, 97, 397, **405**
SCOR model, **22**, 58, 66, 227
scoring method, 92
scrap, 322, 517, 659
scrap factor, **54**, 528
SCRM (supply chain risk management), **112**
seasonal component, 430
seasonal demand pattern (syn. seasonality), 254, **445**
seasonal fluctuations, **445**
seasonal index, **445**
seasonality, 425, **445**
  additive, **445**
  multiplicative, **445**
secondary key, 716, 726, **828**
secondary parameter, **746**
second-order exponential smoothing (forecasting technique), **441**
second-party audit, **785**
second-party logistics provider, **146**
second-tier supplier, **84**
section, **737**
seiketsu, **310**
seiri, **310**
seiso, **310**
seiton, **310**
self-assessment, 784
sellers' market, 83, 612, 616
sell-side solution, **95**
semiautonomous organizational units, 75
semifinished good, **28**, 36, 718, 725
semifinished good supplier, **84**
semifixed, 685
semiprocessed item, 823
sensitivity analysis, 93, 160, 501, 810
sequence of operations, 32, 304, **551**, 569, 584
sequencing, 299, 383, 586, **655**
  operations (syn. operations-oriented finite loading), 274, **608**
sequential number
  of the operation, 737
  of the position within the bill of material, 734
sequential picking, **664**
serial number, 734, 738
service, **9**, 25, 108, 205, 760, 801
  (customer) order, **249**
  centralized, **142**
  customer, **10**
  decentralized, **142**, 161
  in the originary sense, **10**, 141
  mobility cost ratio, **142**
  proactive, 48
  provided to dependents, **760**
service blueprinting, **197**
service center, 142
service function, **486**
service industry, **12**, 27
service level, 257, **485**, 493
  high, 265, 269
  low, 267
  optimum, **491**
service network, **20**, 142
  feature for designing, **142**
service object, **10**, 141
service part, **28**, 228, 455
service parts demand, 263, 523, **530**
service positioning, **42**
service process, **196**
service production, 13
service-oriented architecture (SOA), **400**
servitization, **11**
set date, **571**
setup, **31**
  external, **301**
  internal, **301**
setup costs, 268, **496**, 498, 506, 536, 586
setup lead time (syn. setup time), **31**
setup load, **34**, 496, 552, 582, 684, 721, 737
  of an operation, 690
setup time, **31**, 329, 552, 656, 684, 738
  reduction, 299
  short, 373
shareholder, **763**
shelf life, **18**, 376, 387, 497
shelf warmer, 46, **479**
Shewhart cycle, **767**
shipment, 662
  date of, 524
shitsuke, **310**
shop calendar (syn. manufacturing calendar), **567**
shop floor control, 237, **651**
shop floor data collection, 242, **656**, 682, 694
shop floor scheduling, 646
shop order (syn. production order), **25**
shop order routing, **651**, 660
shop-packet, **651**
shortest processing time rule (SPT), 606, 609

shortest slack time rule (SLK), 606, 609, 613
short-range blanket order, 105, 250, **261**
short-term planning and control, 105, **237**, 240, 645
short-time working, 559, 606
shrinkage, **473**
Siebel (software product), 407
sigma (as a measure of dispersion), **430**
sigma (as a scale), **765**
significant variances, **695**
simplex algorithm, 162
simulation, **252**
  process, 274, 612
simulation software, 275, 600
simultaneous engineering, **278**, 307
single (exponential) smoothing, **435**
single part, **28**, 823
single parts store, 207
single sourcing, **85**, 87
single-item procurement, **214**
single-item production, **214**, 278, 536
single-item-oriented line production, **209**, 570
single-level bill of material, **729**
single-level product structure, 724
single-level where-used list, **731**
single-minute exchange of dies (SMED), **301**
single-purpose facility, 375
single-source supplier, **85**
single-stock organization, **470**
single-valued role or access function, **830**
SIPOC, **772**
site production, **208**, 573
situational analysis, **796**
situational planning, 607, 645
Six Sigma
  as a management system, **781**
  as a metric, **765**
  as method of improving performance, **767**
Six Sigma Quality, **766**
SKU (stockkeeping unit), **469**
slack, 190, 192
slack time, 31, 574, **577**, 647
slack time rule, 606, 609, 613
small batch procurement, **214**
small batch production, **214**
small parts stock, 471
smart sensor, **50**
SME, 403, 417
SMED (single-minute exchange of dies), **301**
SMI (supplier-managed inventory, syn. VMI), **104**
smoothing, **431**
  adaptive, **443**
  exponential, first-order (forecasting technique), **435**, 437
  exponential, second-order (forecasting technique), **441**
smoothing constant γ, **444**
smoothing constant α, **436**, 437, 444
social competency
  of a company, **97**, 100
  of a project manager, 807
software, 397, **815**
  APS, 247, 397, **405**
  company-specific, **409**, 421
  costing, 696
  CRM, **407**
  ERP, 49, 397, **401**, 808
  for customer order production, **403**
  for electronic control board (Leitstand), 405
  for the processor-oriented concept, **404**
  for the variant-oriented concept, **403**
  MRP II, **401**
  PPC, 22, 353, **397**
  SCM, 49, 97, 397, **405**
  SRM, **94**
  standard, 400, **409**, 421
  standard PC, 410
software product
  APO, 406
  Baan, 403
  Blending, 405
  Bpics, 401
  Copics, 400, 401
  ILOG, 406
  IPPS, 403
  J.D. Edwards, 401
  JDA solutions, 405
  Leegoo Builder, 403
  Mapics, 401
  Marcam, 405
  MAS90, 403
  Oracle, 401
  PeopleSoft, 406
  Prism, 405
  Protean, 405
  PSIpenta, 403
  R/3, 401
  SAP, 401, 406
  Siebel, 407
  Solvaxis, 403
sole sourcing, **85**, 87, 93
solution space, 247
  product, **348**, 359
Solvaxis (software product), 403
space buffer, **617**
spare part (syn. service part), 18, 28
specialization hierarchy, 719, **821**
specification, **103**
  customer, 337, 741
    manufacturing according to, **336**
    product according to, **204**, 354, 403, 698
specificity
  market, of products, 130
  of location, 73
  of product and processes, 73
spiral model, **798**
split lot, **581**
split offset factor, **582**
splitting, 552, **581**, 665
splitting factor, **582**, 584, 604
spoilage, 497
SQL (Structured Query Language), **827**
SRM software, **94**
stage (process), 375, **380**, 581
staging, **638**
stakeholder, 39, 211, 452, **763**
standard containers, 313
standard cost (accounting) system, **694**
standard cost rate, **684**, 694
standard costing, **694**
standard costs, **684**
standard deviation, **432**, 454
standard hours (syn. standard time), **31**, 198, 251, 598
standard load, **35**, 36, 597, 658, 684, 695
  of an operation, **35**, 684, 688, 701, 721
standard parts supplier, **84**
standard PC software, 410

standard product, 131, **204**, 320, 698
  with options, **204**, 320, 340, 457
standard software, **409**, 421
standard time, **31**, 198, 565, 658, 735
standardization, 12, 22, 46, 228, 304, 356, 369, 669, 672
stand-by costs, 144
start date, **25**, 533, 542, 550, 569, 576, 602, 721, 754
  calculated, 613
  earliest, 240, 571, 613, 626, 637, 721
  early (ES), 574, 804
  late (LS), 574, 804
  latest, 571, 573, 576
  operation, **571**, 721
  order, **25**, 720
  set, 613
state, **185**
  action, **185**
  waiting, **185**, 186, 187
statement of work, **802**
station
  picking, 664
statistical decomposition, **429**
statistical period, **428**, 431, 456, 479
statistical process control, 774
statistics, **245**, 427
  bid, **477**, 722
  consumption, 425, 722
  sales, **477**, 722
  turnover, **476**
  usage, **475**
status
  of an order, **26**
steady demand (syn. continuous), **203**
STEP, 49
step-function costs, **685**
stochastic dependent demand, **267**
stochastic independent demand, 265, **267**
stochastic materials management, **263**, 402
stock
  commission, **206**, 663
  consigned, **207**
  cycle, 481
  floor, **470**
  fluctuation (syn. fluctuation inventory), **485**
  pipeline, **246**, 484
  safety, 257, 268, 316, 481, **484**, 510, 528
  small parts, 471
stock location, 727
stock on hand (syn. physical inventory), **515**
stock organization, **469**
  multiple, **470**, 727
  single, **470**
  variant, **470**
stock replenishment, **192**, 298, 376
stock replenishment order, **212**, 267, 316, 645
stocking level, **46**, 364
stock-inventory turnover, **55**, 105, 523
stockkeeping, 721, 727
stockkeeping unit (SKU), **469**, 470
stockless production, **296**
stockout, 104, **263**, 484, 485
  probability of, **485**
    optimum, **491**
stockout costs, **491**
  batch sizing depending on, 510
stockout percentage, **263**
stockout quantity, **263**, 492
stockout quantity coefficient, **492**
storage, **18**, 192, 469
  fixed-location, **470**
  on sight, **470**
  random-location, **470**
storage costs, 478
storage infrastructure costs, 251, **497**
storage location, 52, **469**, 722
storage management principle, **470**
storage type, **469**, 497
store, **18**
  assemblies, 207
  centralized, **470**
  data, **187**
  decentralized, **470**
  end products, 206, 612
  goods, **18**, 186
  in-house parts, 207
  order picking, **206**, 663
  process, **187**
  purchased parts, 207
  raw materials, 207
  single parts, 207
store throughput diagram, **521**
stores management, **245**, 469
strategic alliance, 111
strategic plan, **41**
strategic planning, **41**
strategy
  manufacturing, **251**, 233
  of a company, **41**
  picking, **663**, 678
  supply chain, **41**
  toward success, 277, 293
strengths in a company's negotiating position, 101
strict role or access function, **830**
structural organization, 278, 736
structure
  corporate group, **717**, 839
  cost breakdown, of a product, **686**, 727
  distribution network, 137, **138**, 249, 524, 662
  product, **29**, 45, 46, 83, 266
  production, **32**, 747
  rough-cut, 618
  supplier, **83**
  supply chain, 66, 78, 97
  work breakdown, **802**
structure level, **29**, 32, 137, 138, 202, 210, 340, 526, 528, 691, 718, 728
Structured Query Language (SQL), **827**
subcontracting, 271, **606**, 687
substitution, **732**
subsystem, **792**
success probability (order), **249**, 408, 477, 621
summarized bill of material, **730**
summarized where-used list, **732**
super bill of material, **346**
supermarket principle, 316
supplier, **25**, 208, 260, 504, 638, 703
  as a business object, **716**
  certified, **87**
  local, 310
  modular, **83**, 97
  MRO, **84**
  raw materials, **84**
  semifinished good, **84**
  single-source, **85**
  standard parts, **84**
  system, **83**, 90, 97, 208
  world-class, 101
  world-class local, 101
supplier blanket order, 239

supplier code of conduct (SCoC), 92, **168**
supplier evaluation, **91**
supplier lead time, **32**
supplier partnership, 111
supplier portfolio, **88**, 100
supplier relationship management (SRM), **94**
supplier scheduling, 105, 108, **260**
supplier selection, **92**, 239
supplier structure, **83**
supplier-dominated relationship, **89**
supplier-managed inventory SMI, **104**
supply, **17**
supply (syn. procurement) management, 83
supply and demand, 18, 42, 83, 524
  temporal synchronization between, **17**, 45, 192
supply chain, **20**, 99, 201, 218, 252
  design for the, **279**
  tariff oriented, 79
supply chain changeability, **41**, 48, 219
supply chain collaboration, **41**, 218, 422
supply chain community, **20**, 60, 130
supply chain coordination, **41**, 219, 252
supply chain council, 22
supply chain design, **71**
  cost optimal, 78
supply chain event management (SCEM), **49**, 66, 117
supply chain expertise, 146
supply chain initiative (SCI), **44**, 65, 220
supply chain management, **21**
supply chain operations reference model, **22**
supply chain performance, **40**
supply chain planning, **21**
supply chain portfolio, **113**
supply chain risk, 46, 48, 86, 88, 97, 110, **112**
supply chain risk catalog, **114**
supply chain risk management, **112**
supply chain risk portfolio, **116**
supply chain strategy, **41**
supply chain structure, 78, 97
supply chain value at risk (SCOR performance indicator), **60**
supply chain value contribution (SCVC), **66**
supply chain vulnerability, 113, **130**
surge capacity, **600**
surrounding system, 427, 450, **792**
sustainability, **163**
SWOT analysis, **796**
synchronization
  temporal, between supply and demand, **17**, 45, 192
synchronization point, **551**
synchronized production, **248**, 315
synthesis of solutions, **797**
system, **789**
  actual cost, **694**
  agent-based, 646
  cyber-physical, **50**
  database management, 751, **816**
  decision support (DSS), 427
  distribution (syn. distribution network), **20**, 246
  dynamic, **789**
    open, **789**
  engineering data management (EDMS), **751**
  expert, 427, **742**
  industrial product-service, **13**
  information, 397, **816**, 840
  logistics, **185**, 187
  min-max (reorder), **483**
  normal cost, **694**
  one-card Kanban, **315**
  order point, **480**
  packaging, **667**
  part, **792**
  performance measurement, **52**
  phase of life of a (syn. system phase), **793**, 820
  production (syn. production network), **20**
  self-regulating, 650
  standard cost (accounting), **694**
  sub, **792**
  surrounding, 427, 450, **792**
  two-bin inventory, **639**
  two-card Kanban, **314**, 639
  visual control, **326**, 775
system decommission, **795**
system development, **793**, 815, 840
  iterative, **798**
system disposal, **795**
system environment, 792, 796
system establishment, **794**
system implementation, **794**
system introduction and handover, **795**
system life cycle, **793**
system operation, **795**
system phase, **793**, 820
system risks, 808
system sourcing, **85**, 98
system supplier, **83**, 97, 208
systems engineering, 280, **789**, 791
systems thinking, **792**

T

T analysis, **205**, 341
table, **826**
tactical plan, **251**
tactical planning, **251**
tactical underload, **597**
tactical underutilization, **597**
takt time, **570**
tangibles (syn. tangible goods, **12**
target area
  costs, **40**, 86, 92, 98, 110
  delivery, **40**, 56, 86, 87, 92, 98, 110, 213, 226
  flexibility, **40**, 86, 87, 92, 96, 98, 110, 211, 296
  quality, **40**, 86, 92, 98, 110
  supply chain changeability, **41**
  supply chain collaboration, **41**
  supply chain coordination, **41**
target costs, 692, 695
target cumulative production figures diagram, **321**
target inventory level, **483**
tariff, **76**
task, **8**, 187
  project, 153, 573, **801**
task model, **824**
task view, **824**
task-oriented modeling, **824**
TCO (total cost of ownership), **80**

- technical reference number, 726
- technical wait time after an operation, **554**, 738
- technical wait time before an operation, **554**
- technique
  - adaptive, **339**
  - backflush, **657**
    - critical point, **657**
  - blowthrough, 343, **536**
  - cost identification, **694**
  - deterministic, **263**
  - double order point, **483**
  - drum–buffer–rope, 248, 274, **617**
  - exponential smoothing forecast
    - first-order, **435**, 437
    - second-order, **441**
  - fail-safe (poka yoke), **776**
  - forecasting, **424**, 443, 463
    - linear regression, **439**
    - pyramid, **345**
    - regression analysis, **439**
  - generative, **339**, 348, 353
  - Kanban, **314**, 384, 456, 600, 637
  - moving average forecast, **432**, 437
  - MRP (material requirements planning), 268, 311, 515, **530**, 546
    - net change, **533**
  - net present value NPV, 160, **811**
  - network planning, **573**, 803
  - order point, 268, 324, **480**, 511, 527
    - double, **483**
  - PERT, 574, 804
  - RFID, 49, 97, **660**, 672
  - scheduling, 568
  - service blueprint, **197**
  - stochastic, **263**, 265
  - variant-oriented, **339**
  - workflow, 726, 776
- technology, 747
  - as an object, **378**
  - information, 50, 397, **815**
- template, 356
- temporal flexibility, **59**
- temporal range, 236
- theoretical capacity, **34**, 596
- theory
  - queuing, **557**, 561, 589
  - waiting line (syn. queuing), **557**
- theory of constraints (TOC), **247**, 274, 286, 574, 617
- **third-party audit**, **785**
- third-party logistics provider, **146**, 672
- three-level planning according to temporal range, **236**
- throughput, 52, **247**, 274, 302, 558, 608
- throughput time, **570**
- time
  - actual, **31**, 658, 735
  - administration, **32**, 298, 551, 555
  - customer tolerance, **18**, 46, 130, 134, 207, 263, 312
  - cycle, **570**
  - delivery (syn. delivery lead time), **18**
  - down, 247, 319, **658**
  - idle, 301, 306, **558**, 617
  - interoperation, **32**, 38, 298, 551, 554, 610
  - lead, **18**, 31, 33, 322, 550, 569, 584, 721, 736
  - move, **554**, 555, 736
  - operation, **31**, 298, 329, 552, 584, 606, 658
  - process, **761**
  - queue, **554**, 567, 597, 736
  - run, **31**, 502, 552, 582, 684
    - per unit, **31**, 552, 738
  - setup, **31**, 329, 552, 656, 684, 738
  - slack, 574, **577**, 647
  - standard, **31**, 198, 565, 658, 735
  - takt, **570**
  - throughput, **570**
  - transit, **554**, 555
  - transportation, **554**, 555
  - wait, **32**, 589
- time bucket, **515**, 532, 535
- time buffer, 274, 606, **617**
- time element, **550**
- time fence, **256**
  - demand, **526**
  - planning, **256**
- time filter, **640**
- time limit, 613, 640, 641
- time management, **550**, 736
  - and scheduling, **244**, 268, 269, 402, 550
- time period, **25**, 321, 604
- time phasing, **515**
- time series, **428**, 432
- time series analysis, **429**
- time series diagrams, 775
- time standard, 700
- time study, **198**, 700
- time ticket, **651**
- time to delivery (syn. delivery lead time), **18**
- time to market, 73, 97, **277**
- time to product (syn. delivery lead time), 73, **278**
- time-phased order point (TPOP), **515**
- TOC (theory of constraints), **247**, 274, 617
- tool, 34
  - bill of, **739**
  - CASE (computer-aided software engineeering), 799
- tool capacity, 600
- tool where-used list, **739**
- tooling costs, **688**
- toolkit, **739**
- total care, **12**, 48
- total cost of computing, **808**
- total cost of ownership, **80**, 86
  - of a project, **808**
- total monetary costs, 80
- total quality management (TQM), 310, 759, 768, **781**
- total role or access function, **830**
- toyota production system, 198, **295**
- TQM (total quality management), 768, **781**
- traceability
  - demand coverage, **541**
  - load, **606**
  - lot, 214, 376, 381, 382
  - requirements, **540**
- tracking and tracing, **49**, 52, 672
- tracking signal, **444**
- trading company, 208
- trailer on flat car [TOFC] transport, **671**
- train (process), **380**, 581, 748
- transaction, 516
  - as a busineess object, **722**
  - bid, **477**
  - inventory, **471**, 722
  - sales, **477**
- transaction costs, **72**, 80
  - internal, **74**
- transaction date, 722
- transaction process, **72**
- transfer of the service object (need for repeated), **143**
- transformability, **47**

transit time, **554**, 555
transparency of product, process, and organization, 40, 413
transponder, 49, 97, **660**, 672
transport
  direct, **146**
  indirect, **146**
  means of, **146**
  mode of, 68, **146**, 670
  need for merged, **147**
transport and logistics costs, 80
transport planning and scheduling, **669**
transport unit, 134, **668**
transportability, **205**
transportable item, **206**
transportation control, **672**
transportation inventory, **18**, 138, 246
transportation network, 146
  feature for designing, **147**
transportation order, **671**
transportation time, **554**, 555
transshipment center, **147**
traveling card, 313, **639**
tree structure (syn. arborescent structure), 202, 541, 730, 802
  reversed, 371
trend, **438**
trend component, 429
trend extrapolation (forecasting technique), **447**
trend forecasting model, **438**
Trigg and Leach, 444
triple bottom line, 92, 133, **163**
truckload lot, **148**
trust-building measures, 100
turnover statistics, **476**
two-bin inventory system, **639**
two-card Kanban system, **314**, 639
two-level master schedule, **345**
two-sided association, **829**
type
  item, **28**, 478, 726
  of long-term order, **213**, 338
  of order, 192, **212**, 645
  of order position, **720**
  production, **223**, 296, 339
  work center, 596, **735**

U

UCC (uniform code council), **667**
undercapacity, 696
underload, 269, 602, 645
  persistent, 606
  tactical, **597**
underutilization
  tactical, **597**
unevenness, **295**
unfinished product structure, **356**
unique demand, **203**, 479
unit
  capacity, **34**, 596, 684
  load, 146, **668**
  packing, **666**
  process, **380**
  stockkeeping (SKU), **469**
  transport, 134, **668**
unit cost, 59, 60, **205**, 265
  batch-size-dependent, **495**
unit of measure, 29, 31, 34, 205, 495, 683, **727**
unordered issuance principle, **470**
unplanned demand, 48, 105, **637**
unplanned issuances, 657
UPC (uniform product code), **667**, 726
upside supply chain adaptability (SCOR performance indicator), **60**
upside supply chain flexibility (SCOR performance indicator), **59**
urgency
  order, **577**, 609, 648, 720
usage quantity (syn. quantity per), **29**, 256, 386, 533
  nonlinear, **386**
usage statistics, **475**
use, 15
  re-, 21
use card (syn. move card), **314**
use function, **667**
useless effort (syn. waste), **295**
U-shaped line, 210, **303**
utilization, **35**, 55, 269
  capacity, 35, **55**, 57, 562, 597, 735
  tactical under-, **597**

V

validity date, **25**, 717, 720
valuation
  inventory, **473**
valuation basis (of the stock), **469**, 497
valuation method, **470**
  average cost system, **470**
  FIFO, **470**
  LIFO, **470**
  order-specific, **471**
value chain, 16
value co-creation, 11, **13**
value content, **77**
value creation, **10**
value density, **131**, 134, 206
value stream, **199**
value stream mapping, **199**
value-added, **10**, 318, 478, 616, 689
  from the customer's view, 237, 295
value-added management, **17**
value-added rate of lead time, **57**, 199
variable bill of material, 346
variable costing, 686, 689
variable costs, **685**, 697
  manufacturing, **689**
variable labor costs
  for machinery, 735
  for personnel, 735
variable manufacturing costs, **689**, 698
variable material costs, **687**
variable overhead costs, 685
variable routing sheet, 346
variance, **454**
  cost, **696**
  in work content, **57**
  significant, **695**
  volume, **695**, 702
variant (syn. product variant), **30**, 204, 339, 457
variant bill of material, **342**, 351
variant master schedule, **343**, 347, 366
variant number, 744
variant orientation, 204, **335**
variant production, **335**
variant routing sheet, **342**, 351
variant structure
  one-dimensional, 346
variant-oriented concept, 46, 226, 247, **335**, 403

software for, **403**
variant-oriented technique, **339**
variant-stock organization, **470**
variation, **295**
variety of demand, 117, **134**
VAT analysis, **203**, 205, 341
vendor, **72**, 104, 191, 354
vendor-managed inventory (VMI), **104**
vendor-owned inventory, 137, **207**
vertical (electronic) marketplace, **96**
vertical integration, **74**
vicious circle, 269
view
control, 819, **823**
data, 819, **824**
function, 819, **823**
object, 819, **824**
of an information system, **819**, 823
of the production equipment, 739
on a class, **827**
organization, 819, **824**
process, 819, **823**
task, **824**
virtual (adjective), **109**
virtual enterprise, **109**
virtual service organization, 111
visual control system, **326**, 639, 775
VMI (vendor-managed inventory), 90, **104**, 406
VOI (vendor-owned inventory), 137, **207**
voice of the customer, **770**
VoIP (voice over internet protocol), 49
volatility of demand, **130**
volume flexibility, 47
volume variance, **695**, 702
for a component, 696
for an external operation, 696
in an internal operation, 695
vulnerability
supply chain, 113, **130**

…hitin, 536, 538

wait time, **32**, 589
average, 609
nontechnical, after an operation, **554**
nontechnical, before an operation, **554**, 736
technical, after an operation, **554**, 738
technical, before an operation, **554**
waiting line (syn. queuing) theory, **557**
waiting state, **185**, 186, 187
warehouse, **18**, 192, 469
central, 138
warehouse demand, **524**
warehousing, **662**
waste, 199, **295**, 310, 468
Ohno's seven, **295**
waste product, **371**, 394, 727
waterfall model, **798**
weighted average, 470, 574
weighted average costs of capital (WACC), 65
weighted moving average, **434**
weighting
equal, 430
exponential, 430
what-if analysis, **252**
where-used list, **728**, 734, 757
indented, **732**
multilevel, 541, **732**
production equipment, **739**
single-level, **731**
summarized, **732**
tool, **739**
work center, 646, **738**
where-used-list position, **733**
whole-part, 725
whole-part hierarchy, 719, **820**
wholesaler, 138
win–win situation, 99, 101
work, **8**, 25
as an order position, 28, **717**, 721, 724
contract, **15**, 638, 665
work breakdown structure, **802**
work cell, **302**
work center, **34**, 37, 298, 350, 554, 560, 596, 652, 688, 700, 724, 736
alternate, **737**
description, 721, 735
efficiency, **36**, 55, 597, 629, 651, 695, 735
gateway, **646**
ID, 721, **735**, 737
individual equipment of a, 652
load, **35**, 601
nonconstraint, 597
primary, **737**
rough-cut, **37**, 238, 619
type, 596, **735**
viewpoint, 738, 741
work description, 721
work on hand, **606**, 641, 652, 653
work order, **25**
work package, 573, **801**
work program, 652
work-center-where-used list, 646, **738**
worker (syn. employee), **33**
workflow technique, 726, 776
workforce
flexible, **211**
quantitatively flexible, **211**, 309
working day, 567
work-in-process (WIP), **18**, 413, 468, 565, 646
cost of, 504
work-in-process-inventory turnover, 55, 565
workstation, **34**, 561, 652, 724, 736
feeder, **617**
World Wide Web, 49, 672
world-class local supplier, 101
world-class supplier, 101

X

XML, 49, **95**
XYZ classification, 318, **479**, 509

Y

yield, 322, 659
operation/process, **386**
yield factor, **54**, 304, 528, 598

Z

zero inventories, **296**
zero-defect rate, 771
zone picking, **664**